Multiple Muscle Systems

Jack M. Winters   Savio L-Y. Woo
Editors

Multiple Muscle Systems

Biomechanics and
Movement Organization

Springer-Verlag
New York   Berlin   Heidelberg
London   Paris   Tokyo   Hong Kong

Jack M. Winters
College of Engineering and Applied Sciences
Department of Chemical, Bio., and Materials
   Engineering
Arizona State University
Tempe, Arizona 85287-6006, USA

Savio L-Y. Woo
Department of Surgery
Division of Orthopedics and Rehabilitation
University of California at San Diego
   School of Medicine
La Jolla, California 92093
USA

Library of Congress Cataloging-in-Publication Data
Multiple muscle systems / biomechanics and movement organization /
   Jack Winters, Savio Woo, editors.
        p. cm.
      Includes bibliographical references.
      Includes index.
      ISBN 0-387-97307-9 (alk. paper). — ISBN 3-540-97307-9 (alk.
   paper)
      1. Biomechanics. I. Winters, Jack, 1957–   . II. Woo, Savio L
   -Y.
      [DNLM: 1. Biomechanics. 2. Models, Biological. 3. Movement–
   –physiology. 4. Muscles—physiology. WE 103 M961]
   QP303.M85   1990
   612.7'6—dc20
   DNLM/DLC
   for Library of Congress   90-10128
                                                        CIP

$Pb\ 2906$

Printed on acid-free paper

Camera-ready copy provided by the editors.
Printed and bound by: Edwards Brothers, Inc., Ann Arbor, Michigan.
Printed in the United States of America.

9 8 7 6 5 4 3 2 1

ISBN 0-387-97307-9 Springer-Verlag New York Berlin Heidelberg
ISBN 3-540-97307-9 Springer-Verlag Berlin Heidelberg New York

Preface

The picture on the front cover of this book depicts a young man pulling a fishnet, a task of practical relevance for many centuries. It is a complex task, involving load transmission throughout the body, intricate balance, and eye-head-hand coordination. The quest toward understanding how we perform such tasks with skill and grace, often in the presence of unpredictable perturbations, has a long history. However, despite a history of magnificent sculptures and drawings of the human body which vividly depict muscle activity and interaction, until more recent times our state of knowledge of human movement was rather primitive. During the past century this has changed; we now have developed a considerable database regarding the composition and basic properties of muscle and nerve tissue and the basic causal relations between neural function and biomechanical movement. Over the last few decades we have also seen an increased appreciation of the importance of musculoskeletal biomechanics: the neuromotor system must control movement within a world governed by mechanical laws. We have now collected quantitative data for a wealth of human movements. Our capacity to understand the data we collect has been enhanced by our continually evolving modeling capabilities and by the availability of computational power. What have we learned? This book is designed to help synthesize our current knowledge regarding the role of muscles in human movement.

The study of human movement is not a mature discipline. For instance, within this book, respected leaders in the field find that understanding even simple, stereotyped movements is quite challenging. Similarly, researchers in robotics find that it is surprisingly difficult to get machines to move gracefully or to interact intelligently with a variable external environment. Yet a young child can perform a countless variety of tasks quite effortlessly. It is a field that consistently humbles the researcher.

Fundamental to our difficulties in identifying neuromotor control strategies is the inherent biomechanical complexity of the musculoskeletal system. Because of mechanical coupling, motion of one segment affects many others. The standing human can be considered as an interlinked inverted pendulum that must operate within a gravitational field. Such systems tend to be unstable. Thus, the neuromotor control system must not only plan neurocontrol inputs that will cause appropriate volitional movements, but must also assure that the overall system remains stable. It must contend with (and perhaps take advantage of) the inherent complexity of muscle actuators; dozens of muscles often contribute to even the simplest of movements. Furthermore, each of these actuators has exquisitely nonlinear properties. This book addresses this integration between biomechanics and neuromotor organizational strategies in-depth, and as such provides a unique perspective regarding our current state of knowledge.

In some ways the task of creating this book is analogous to planning and executing a movement. We first had to define our task and its goals. We then had to work out a strategy that could help us meet these criteria. Our goals were: *i)* to provide contributions of high quality that span the entire field of biomechanics/movement organization; *ii)* to serve both as a resource book for students and as a source for presentation of state-of-the-art research;

iii) to synthesize, as much as possible, the interrelationships between contributions; and *iv)* to suggest directions for future research that are likely to be fruitful.

Our strategy was manifold. First, we actively pursued leaders in the field. In this regard, we were remarkably successful. This has been clearly recognized by our contributors, and our impression has been that this situation leads to each group "putting their best foot forward". As a state-of-the-art resource book, each author was asked to include a selected review within their chapter (more than usually possible within refereed journal publications), and to keep the "Methods" section short whenever possible by referring to other publications. We also provide a number of chapters that are of a more basic nature, suitable for instruction. For instance, one of us (J.W.) will use this as the primary resource book for a graduate course entitled "Neuromuscular Control Systems"; Chapters 1, 5, 8–11, 23 and 35 will serve as a "core." To synthesize the many contributions presented here, we used two approaches. First, each author had access to outlines for all other chapters and was asked to reference other chapters as much as possible. Second, each part of the book starts with an overview chapter which attempts to synthesize information within that respective part of the book. Finally, we specifically asked each contributor to suggest future research directions, within either a separate section or a discussion section; these insights and recommendations will perhaps extend the useful lifetime of this book.

The book is organized into five parts. The first two parts emphasize modeling. Such models help document the state of current knowledge, provide predictions that can often be tested experimentally, and allow estimation of information that is difficult or impossible to measure experimentally (e.g., muscle forces). *Part I* addresses the properties of muscle, the biological actuator that allows voluntary movements to unfold. It turns out that movement strategies are quite sensitive to the properties of this unique actuator; thus a detailed consideration of muscle dynamic properties is a fitting way to start this book. Our emphasis here is on concrete models of muscle that are capable of representing salient muscle properties. We will constantly face the inherent tension that exists between model complexity and simplicity, and we will see that different contributors come to different solutions, based largely on their research goals. There is a natural progression within these chapters from complex models (Chapters 1–4) toward somewhat simpler models (Chapters 5–7). Chapters 1–4 challenge the backbone of simpler muscle modeling approaches, such as the Hill-based and variable spring models used throughout the rest of this book. Chapter 5 addresses these concerns, suggesting that appropriately used simpler models are sufficient for most of the questions of interest within *Parts III–V* of this book. Chapter 6 compares the predictions for various model formulations, while Chapter 7 contrasts technological actuators with biological muscle. *Part II* considers neuromusculoskeletal models. Chapter 8 overviews the foundations underlying musculoskeletal model development, Chapter 9 develops important theoretical foundations for musculoskeletal systems analysis, and Chapter 10 develops an approach for utilizing models in conjunction with experiments to gain insight into neuromechanical control systems.

The last three parts of this book emphasize the interplay between biomechanics and movement organizational strategies, specifically addressing the role of multiple muscles in this process. A wide variety of tasks are considered, with the emphasis on movements in humans. *Part III* concentrates on organizational strategies for upper limb movements. Such movement tasks typically involve tracking or manipulation, with the upper limb considered to be mounted on a base at the shoulder. Because of the variety of upper limb movements that occur throughout life, upper limb movements tends to serve as a testing ground for assessing principles of movement organization. In contrast, *Part IV* considers what happens at the other side of the shoulder, where issues such as postural stability and tissue loading become dominant. Control of posture often involves the whole body, from eye-head orientation to the maintenance of balance by effective action of limb and torso musculature. Chapters 24-28 focus on tasks involving the spinal musculature, while Chapters 29-33 emphasize whole body balance and the relationship between intentional movement and posture. These are both areas of tremendous complexity. They also represent areas of great importance with regards to understanding organizational strategies for the wealth of practical tasks of daily living. For tasks such as walking, it is often difficult to separate movement and postural aspects (Chapters 32-33, 43-44). However, for the repetitive, skilled cyclic and propulsive movements considered in *Part V*, postural concerns tend not to be the issue. Here the focus is on movement, and especially on the roles of lower limb muscles in such movements. Unlike upper limb movements, in which an isolated limb can be considered, coupling to the moving torso is typically quite important for lower limb movements. Joint and muscle loads tend to be high, and biomechanical issues related to inertial dynamic coupling, energy transfer, the role of spring-like muscle properties, and the "stretch-shortening cycle" tend to dominate. Most of the tasks under investigation are stereotyped and often cyclical in nature.

The book ends with an appendix that provides tables summarizing musculotendon parameters utilized by various groups. This represents the most complete resource for such information.

In conclusion, this book provides the fruits of a team effort by leaders in this fascinating (and humbling) field of movement biomechanics and motor control. Put together, it provides the reader with a wealth of insights and a unique global perspective on multiple muscle systems. At the very least, the book should give each of its readers a great appreciation for a task so simple as grasping a fishnet.

Acknowledgments

This book grew out of a conversation in April of 1989 between the two of us regarding the possibility of J.W. organizing a symposium for the *First World Congress of Biomechanics*, held in La Jolla, CA, August 30-September 4, 1990. The response within the community was greater than anticipated, with our preliminary plan evolving into two symposia with over 80 presentations total, one on *Multiple Muscle Systems* and one on *Multiple Muscle Movement Organization*, plus this book. Part of this enthusiastic response was due to good timing — a resource book in this area was sorely needed. It was also due to the combination of a major international meeting and our plan for making the book available in time for the meeting. We gratefully acknowledge the assistance of the World Congress organizers in this endeavor.

Our most valuable resource has been the contributors. It was emphasized early on that this project would involve a team effort. As an analogy, we used the muscles of the body. In order to complete a task, many muscles (contributors) must play a role. To be successful, muscles must be strong, somewhat attentive to the goals of the overall task, and fatigue-resistant. Fortunately, these attributes held true for our contributors. We especially thank them for their willingness to integrate their material so that it fit within the context of the whole book.

Muscles cannot function without a support system. We thank the staff of Springer-Verlag for their patience and enthusiasm. We also thank Jean George for her help with editing, Lesley Rathburn for her help with layout of the book, and Kathleen Winters for creating the index. Without their professional and cheerful help during various stages of this project, completion in time for the meeting would have been impossible.

Finally, we acknowledge Mother Nature for providing us with such a fascinating system to study.

Jack M. Winters
Tempe, AZ

Savio L-Y. Woo
La Jolla, CA

Contents

Part V: Principles Underlying Movement Organization: Propulsive and Cyclic Movements with Lower-Limb Emphasis

Contributors

S.V. Adamovich
Institute of Problems of Information
Transmission
Academy of Sciences
Ermolova 19, 103051
Moscow, USSR

Gyan C. Agarwal
Depts. of Electrical Engineering and
Computer Science, and Bioengineering
University of Illinois at Chicago
Chicago, IL 60680, USA

R. McNeil Alexander
Dept. of Pure and Applied Biology
University of Leeds
Baines Wing
Leeds LS2 9JT
ENGLAND

John H.J. Allum
Dept. of Otolaryngology
University Hospital
CH-4031 Basel
SWITZERLAND

Gunnar B.J. Andersson
Dept. of Orthopedic Surgery
Rush-Presbyterian-St.Luke's Medical Center
1653 West Congress Parkway
Chicago, IL 60612-3864, USA

Maarten F. Bobbert
Dept. of Functional Anatomy
Free Universiteit
van der Boechorststraat 9
1081 BT Amsterdam
NETHERLANDS

Simon Bouisset
Laboratoire de Physiologie du Movement
UA CNRS no. 631
Universite de Paris-Sud
F-91405 Orsay
FRANCE

John D. Brooke
Human Biology/Physics
University of Guelph
Guelph, Ontario
CANADA N1G 2W1

Arthur E. Chapman
School of Kinesiology
Simon Fraser University
Burnaby, British Columbia
CANADA V5A 1S6

Jill Conrad
Dept. of Physical Education
University of Wisconsin
Madison, WI 53706, USA

A.C.C. Coolen
Dept. of Medical and Physiological Physics
University of Utrecht
Princetonplein 5
NL-3584-CC Utrecht
NETHERLANDS

Daniel Corcos
Dept. of Physical Education
University of Illinois at Chicago
Chicago, IL 60612, USA

Pat E. Crago
Dept. of Biomedical Engineering
Case Western Reserve University
Cleveland, OH 44106, USA

P. Crenna
Politecnico di Milano
Dipartmento di Elettonica
Piazza Leonarda da Vinci 32
20133 Milano
ITALY

Joseph J. Crisco III
Dept. of Orthopedics and Rehabilitation
Yale Medical School
333 Cedar Street
New Haven, CN 06510, USA

Ronita L. Cromwell
Dept. of Physical Education
University of Illinois at Chicago
Chicago, IL 60680, USA

J.J. Denier van der Gon
Dept. of Medical and Physiological Physics
University of Utrecht
Princetonplein 5
NL-3584-CC Utrecht
NETHERLANDS

Marek Dietrich
Warsaw University of Technology
Institute for Aircraft Engineering
and Applied Mechanics
ul. Nowowiejska 22/24
00-665 Warsaw
POLAND

C.J. Erlkelens
Dept. of Medical and Physiological Physics
University of Utrecht
Princetonplein 5
NL-3584-CC Utrecht
NETHERLANDS

G.J.C. Ettema
Group in Functional Anatomy
Free University
van der Boechorststraat 9
1081 BT Amsterdam
NETHERLANDS

A.G. Feldman
Institute of Problems of Information
Transmission
Academy of Sciences
Ermolova 19, 103051
Moscow, USSR

Mark J. Fessler
Arizona State University
College of Engineering & Applied Sciences
Dept. of Mechanical and Aerospace Engineering
Tempe, AZ 85287-6006, USA

J.R. Flanagan
Dept. of Psychology
McGill University
Montreal, Quebec
CANADA H3A 1B1

Tamar Flash
Dept. of Applied Math and Computer Scien
Weizmann Institute of Science
Rehovot 76100
ISRAEL

James R. Gage
Kinesiology Laboratory
Dept. of Surgery
Newington Childrens' Hospital
Newington, CN 06111, USA

Stan Gielen
Dept. of Medical Physics and Biophysics
University of Nijmegen
Geert Grooteeplein Noord 21
NL 6525 EZ Nijmegen
NETHERLANDS

Gerald L. Gottlieb
Dept. of Physiology
Rush Medical College
1753 W. Congress Pkwy.
Chicago, IL 60612, USA

Serge Gracovetsky
Dept. of Electrical Engineering
Concordia University
1455 De Maisonneuve Blvd. West
Montreal, Quebec
CANADA H3G 1M8

Blake Hannaford
Dept. of Electrical Engineering
University of Washington
Seattle, WA 98195, USA

Zia Hasan
Dept. of Physiology
University of Arizona
Health Sciences Center
Tucson, AZ 85724, USA

Herbert Hatze
Abt. Bewegungslehre/Biomechanik
Institut fur Sportwissenschaften
der Universitat Wien
Auf der Scheltz 6
A-1150 Wien
AUSTRIA

David A. Hawkins
Dept. of Mechanical Engineering
University of California at Davis
Davis, CA 95616, USA

Hooshang Hemami
Dept. of Electrical Engineering
Ohio State University
Columbus, OH 43210, USA

Richard N. Hinrichs
Exercise & Sport Research Institute
PEBE 107
Arizona State University
Tempe, AZ 85287-0404, USA

At L. Hof
Laboratory of Medical Physics
University of Groningen
Bloemsingel 10
9712 KZ Groningen
NETHERLANDS

Neville Hogan
Dept. of Mechanical Engineering
Massachusetts Institute of Technology
Cambridge, MA 02139, USA

James C. Houk
Dept. of Physiology
Northwestern University Medical School
303 E. Chicago Avenue
Chicago, IL 60611, USA

Peter A. Huijing
Group in Functional Anatomy
Free University
van der Boechorststraat 9
1081 BT Amsterdam
NETHERLANDS

Maury C. Hull
Dept. of Mechanical Engineering
University of California at Davis
Davis, CA 95616, USA

Gerrit Jan van Ingen Schenau
Dept. of Functional Anatomy
Free Universiteit
van der Boechorststraat 9
1081 BT Amsterdam
NETHERLANDS

Steve C. Jacobsen
Dept. of Mechanical and Industrial Engineering
University of Utah
Salt Lake City, UT 84112, USA

Slobodan Jaric
Dept. of Biomechanics
Faculty of Physical Education
Blagoja, Parovica 156
11030 Belgrade
YUGOSLAVIA

H.J.J. Jonker
Dept. of Medical and Physiological Physics
University of Utrecht
Princetonplein 5
NL-3584-CC Utrecht
NETHERLANDS

Greg M. Karst
Dept. of Physiology
University of Arizona
Health Sciences Center
Tucson, AZ 85724, USA

Krzysztof Kedzior
Warsaw University of Technology
Institute for Aircraft Engineering
and Applied Mechanics
ul. Nowowiejska 22/24
00-665 Warsaw
POLAND

R.F. Ker
Dept. of Pure and Applied Biology
University of Leeds
Baines Wing
Leeds LS2 9JT
ENGLAND

Emily A. Keshner
Dept. of Physical Therapy (M/C 898)
College of Associated Health Professions
The University of Illinois at Chicago
1919 West Taylor Street
Chicago, IL 60612, USA

Zvi Ladin
Neuromuscular Research Center
Biomedical Engineering Department
Boston University
Boston, MA 02215, USA

Mark L. Latash
Dept. of Neurosurgery
Rush Medical College
1753 W. Congress Pkwy.
Chicago, IL 60612, USA

Steve L. Lehman
Dept. of Physical Education
103 Harmon Gym
University of California at Berkeley
Berkeley, CA 94720, USA

Michel A. Lemay
Dept. of Biomedical Engineering
Case Western Reserve University
Cleveland, Ohio 44106, USA

William S. Levine
Dept. of Electrical Engineering
University of Maryland
College Park, MD 20742, USA

Like Liu
Dept. of Biomedical Engineering
Case Western Reserve University
Cleveland, Ohio 44106, USA

Gerald E. Loeb
Bio-Medical Engineering
Queen's University
Kingston, Ontario
CANADA K7L 3N6

Colum D. MacKinnon
Dept. of Kinesiology
University of Waterloo
Waterloo, Ontario
CANADA N2L 3GI

W.E. McIlroy
Human Biology/Physics
University of Guelph
Guelph, Ontario
CANADA N1G 2W1

Thomas McMahon
Division of Applied Mechanics
Harvard University
Pierce Hall
Cambridge, MA 02138, USA

Sanford G. Meek
Dept. of Mechanical and Industrial Engineering
University of Utah
Salt Lake City, UT 84112, USA

Lee E. Miller
Dept. of Physiology
Northwestern University Medical School
303 E. Chicago Avenue
Chicago, IL 60611, USA

D.W. Moran
Arizona State University
College of Engineering & Applied Science
Dept. of Chem., Bio., and Materials Engine
Tempe, AZ 85287-6006, USA

David Morgan
Dept. of Electrical and
Computer Systems Engineering
Monash University
Clayton, Victoria
AUSTRALIA

Michael Mungiole
Exercise & Sport Research Institute
PEBE 107
Arizona State University
Tempe, AZ 85287-0404, USA

Sandra J. Olney
School of Rehabilitation Therapy
Queens University
Kingston, Ontario
CANADA K7L 3N6

Hoo Dennis Ong
Dept. of Electrical Engineering
Ohio State University
Columbus, OH 43210, USA

D.J. Ostry
Dept. of Psychology
McGill University
Montreal, Quebec
CANADA H3A 1B1

Sylvia Ounpuu
Kinesiology Laboratory
Dept. of Surgery
Newington Childrens' Hospital
Newington, CN 06111, USA

Marcus G. Pandy
Dept. of Kinesiology and Health Education
University of Texas at Austin
Austin, Texas 72712, USA

Manohar Panjabi
Engineering Laboratory for
Musculoskeletal Diseases
Section of Orthopedic Surgery
Yale Medical School
333 Cedar Street
New Haven, CN 06510, USA

Antonio Pedotti
Politecnico di Milano
Dipartmento di Elettonica
Piazza Leonarda da Vinci 32
20133 Milano
ITALY

Joe Peles
College of Engineering and Applied Sciences
Dept. of Chemical, Bio., and Materials
Engineering
Arizona State University
Tempe, AZ 85287-6006, USA

C.F. Ramos
The Neuroscience Institute
1230 York Avenue
New York, NY 10021, USA

Gordon K. Ruder
Dept. of Kinesiology
University of Waterloo
Waterloo, Ontario
CANADA N2L 3GI

David J. Sanderson
School of Physical Education & Recreation
University of British Columbia
Vancouver, British Columbia
CANADA V6T 1W5

A.G.-U. Sawa
Arizona State University
College of Engineering & Applied Sciences
Dept. of Chem., Bio., and Materials Engineering
Tempe, AZ 85287-6006, USA

Amir Seif-Naraghi
Samaritan Rehabilitations Institute
Good Samaritan Medical Center
1111 E. McDowell Road
Phoenix, AZ 85006, USA

Sheldon Simon
Dept. of Orthopedic Surgery
Ohio State University Hospitals
Columbus, OH 43210, USA

Arthur J. van Soest
Dept. of Functional Anatomy
Free Universiteit
van der Boechorststraat 9
1081 BT Amsterdam
NETHERLANDS

Lawrence W. Stark
Neurology Unit
483 Minor Hall
University of California at Berkeley
Berkeley, CA 94720, USA

Toine Tax
Dept. of Medical Physics and Biophysics
University of Nijmegen
Geert Grooteeplein Noord 21
NL 6525 EZ Nijmegen
NETHERLANDS

Marc Theeuwen
Dept. of Medical Physics and Biophysics
University of Nijmegen
Geert Grooteeplein Noord 21
NL 6525 EZ Nijmegen
NETHERLANDS

Scott C. White
School of Health-Related Professions
SUNY at Buffalo
Buffalo, NY 14260, USA

David A. Winter
Dept. of Kinesiology
University of Waterloo
Waterloo, Ontario
CANADA N2L 3GI

Jack M. Winters
College of Engineering and Applied Sciences
Dept. of Chemical, Bio., and Materials
Engineering
Arizona State University
Tempe, AZ 85287-6006, USA

John E. Wood
Center for Engineering Design
Dept. of Bioengineering
University of Utah
Salt Lake City, UT 84112, USA

Chi-haur Wu
Dept. of Electrical Engineering
and Computer Science
Northwestern University
Evanston, IL 60208, USA

Gary T. Yamaguchi
Arizona State University
College of Engineering & Applied Sciences
Dept. of Chem., Bio., and Materials Engineering
Tempe, AZ 85287-6006, USA

Kuu-Young Young
Dept. of Electrical Engineering
and Computer Science
Northwestern University
Evanston, IL 60208, USA

Tomasz Zagrajek
Warsaw University of Technology
Institute for Aircraft Engineering
and Applied Mechanics
ul. Nowowiejska 22/24
00-665 Warsaw
POLAND

George I. Zahalak
Dept. of Mechanical Engineering
Washington University
Campus Box 1185
One Brookings Drive
St. Louis, MO 63130, USA

Feliz Zajac
Dept. of Mechanical Engineering
Design Division
Stanford University
Stanford, CA 94305-4021, USA

Maurice Zattara
Laboratoire de Physiologie du Mo◂
UA CNRS no. 631
Universite de Paris-Sud
F-91405 Orsay
FRANCE

CHAPTER 1

Modeling Muscle Mechanics (and Energetics)

George I. Zahalak

1.1 Introduction

This book deals with an engineering perspective of the mechanics and control of movement in animals, particularly humans. The final effectors that actually produce movement are the actuators of the neuro-musculo-skeletal control system: the skeletal muscles. In the analysis of any control system, especially one as complex as that governing movement, it is essential to have a clear understanding of the physical nature of the actuators and also tractable mathematical representations of their dynamics. A satisfactory comprehension of movement is difficult to achieve without sophisticated model simulations, which require a very heavy computational effort for even severely reduced models, and which involve complex histories of muscle activation, force, and motion beyond the range of experience in simple laboratory experiments. Of the many predictions in a simulation of movement only a few are accessible to direct experimental verification; for example, limb positions, surface electromyograms (*EMG*s), and joint torques. Many other variables of interest, such as individual muscle forces, patterns of neural activation, and energy storage and liberation are accepted more or less on faith. Model redundancy is a serious impediment to understanding: many different internal models may be compatible with the limited measurements available for a few global external variables. Thus it is desirable that analytical representations of the elements of the motor control system, particularly representations of muscles, possess two attributes: *1)* they should be sufficiently simple that studies

involving large ensembles of such elements are analytically and computationally tractable, and *2)* their structure should embody as closely as possible the known physical, chemical and physiological character of the modelled elements, so that theoretical predictions are credible. Obviously there is a natural tension between these two requirements, and the proper balance of simplicity versus fidelity is a subject of continuing debate.

In this first chapter of **Part I** (muscle modeling) I will focus on mathematical representations of the dynamics of an isolated skeletal muscle. Subsequent parts of this book will deal with how such muscle models are integrated into multiple muscle systems. I will begin with a very brief general overview of the nature of muscle contraction which reflects the current consensus among muscle biologists. The purpose of this review is to provide a physiological framework for the subsequent discussion of muscle models, and it will avoid both excessive detail and controversial topics. This qualitative description will be followed by a brief exposition of biophysical cross-bridge models, first introduced by A.F. Huxley (1957) and elaborated by Huxley, T.L. Hill (1977), and many contemporary workers. After this introduction to current concepts of muscle structure and function I will trace the history of a more traditional view of muscle: the A.V. Hill model. The evolution of the concepts of *contractile element*, *series elasticity*, and *parallel elasticity* from the classic experiments will be developed, and generalizations to include behavior in stretch as well as shortening, length-dependent properties,

Multiple Muscle Systems: Biomechanics and Movement Organization
J.M. Winters and S.L-Y. Woo (eds.), © 1990 Springer-Verlag, New York

and time-dependent activation will be discussed briefly [see Chapter 5 (Winters) for a more detailed treatment of Hill–based muscle models].

At present the Huxley–type cross-bridge models are relied upon almost exclusively by muscle biophysicists and biochemists to understand the mechanisms of contraction at the molecular level and to interpret the results of mechanical, thermodynamic, and biochemical experiments on muscle. Elegant as they are, these models are much too complicated to serve directly as mathematical representations of the muscle actuators in motor control studies, and indeed they will be encountered rarely in the remainder of this book. On the other hand bioengineers and movement scientists rely almost exclusively on phenomenological A.V. Hill–type models for an understanding of the workings of whole muscles in multiple muscle systems. I will describe an approach with the potential to partially bridge these two disparate perspectives, called the Distribution-Moment (DM) Approximation, which extracts Hill-like state-variable models of whole muscle as mathematical approximations to Huxley-type cross-bridge theories. I close the chapter with a summary, and some personal thoughts on the future development of muscle models and their implementation.

Although many theories and models of muscular contraction have appeared, for purposes of this chapter I have deliberately chosen to concentrate on the two which seem to me most prominent: A. V. Hill–type macroscopic models and A. F. Huxley-type microscopic models. (The DM model is a mathematical approximation of the latter with some characteristics of the former.) In particular, I have avoided discussion of theories of contraction (Bornhorst and Minardi, 1970; Tirosh et al. 1978; Iwazumi, 1978) which, although they may be interesting and provocative, have no significant current support in the biophysics community. [But see Chapter 2 (Hatze) for a new unconventional theory of cross-bridge dynamics.] The Hill model is sanctified by long tradition going back to the 1930's, and is the basis of an enormous amount of experimental effort and data reported in the literature. Huxley-type models are used by almost all active workers in muscle biophysics and biochemistry, and therefore these models must be considered if one wishes to

develop mathematical representations of muscle that have some discernible connection with the currently perceived biological realities of contraction. It is far too much to expect that any such model be simultaneously simple, tractable, and accurately mimic all the known experimental facts about muscle including structure, mechanics, thermodynamics, biochemistry, and myoelectricity, under conditions of time-varying activation and loading. For example, both the traditional Hill and Huxley models when applied to whole muscle consider that contractile tissue is homogeneous, but it has been proposed that several features of a muscle's mechanical behavior are due to inhomogeneities in this tissue; these features and their importance for modeling muscle behavior in the living animal are discussed by Morgan in Chapter 3.

Modeling always involves a compromise between mathematical tractability and physical veracity, and I hope in this chapter to touch on some of the considerations bearing on this compromise in the case of skeletal muscle.

1.2 The Physiology of Muscle Activation and Contraction

Since the determination of the sub-microscopic structure of striated muscle by x-ray diffraction in the 1950's, an intense and continuing experimental effort has been mounted to discover the ultimate molecular mechanisms of contraction. Several good reviews of this rapidly evolving field are available (Huxley, 1980; Squire, 1981; Pollack and Sugi, 1984; Woledge et al., 1985; Cooke, 1986; Ruegg, 1986; Squire, 1986; Paul et al., 1989). A general consensus seems to exist among muscle physiologists concerning the broad outlines of the contraction process, although the fine details are the subject of active investigation and debate; in this section I will qualitatively sketch this consensus view.

Figure 1.1 is a cartoon illustrating the cascade of steps leading from neural excitation to contraction. A neural action potential is generated in a motoneuron by the spatial and temporal summation of incoming action potentials from other neurons impinging on the motoneuron. This electrical disturbance travels down the motoneuron's axon at a constant characteristic velocity of the order of 100 ms^{-1}, and invades the several branches of the neuron, each with a terminal that is closely apposed to a skeletal muscle fiber. The arrival of the neural action potential at the neuromuscular junction triggers a sequence of biochemical events that result in muscle contraction. First the action potential causes a secretion of the neurotransmitter acetylcholine into the narrow gap between the nerve terminal and a specialized region of the muscle fiber called the motor end-plate. The acetylcholine is secreted by exocytosis via small synaptic vesicles that package the neurotransmitter; this process is triggered by an influx of calcium ions made possible by a transient change of the nerve-membrane permeability to calcium which is associated with an action potential. The acetylcholine diffuses rapidly across the synaptic cleft separating the nerve ter-

minal and the motor end-plate and binds to channel-linked receptors in the muscle-fiber membrane. This binding results in a rapid influx of sodium into the fiber, locally depolarizes the post-synaptic membrane, and generates a propagated muscle action potential that travels down the fiber at a characteristic speed of the order of 5 ms^{-1}.

As the muscle action potential propagates down the length of the muscle fiber, a transmembrane electrical disturbance periodically invades the interior of the fiber via the transverse tubules which invaginate the fiber at the z-disks spaced approximately 2 μm apart. This internal electrical disturbance is communicated to the membranes of

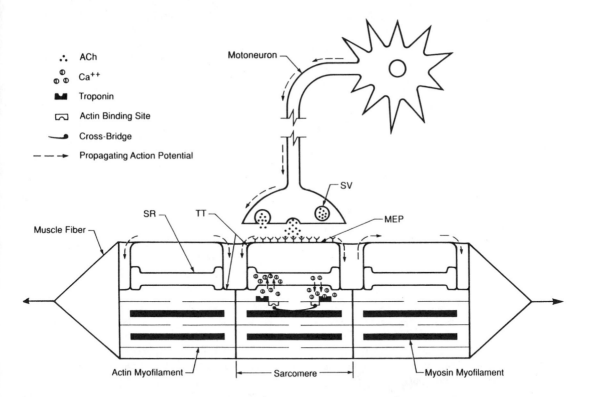

Figure 1.1: Cartoon illustrating the activation-contraction coupling sequence in a skeletal muscle fiber. See text for details. The following abbreviations are used: *ACh* (acetylcholine), *Ca* (calcium), *SV* (synaptic vesicle), *MEP* (motor end-plate), *TT* (transverse tubule), and *SR* (sarcoplasmic reticulum).

the sarcoplamsmic reticulum (*SR*), which increases the permeability of these membranes to calcium. In the absence of electrical stimulation the *SR*, a system of closed internal tubules, accumulates calcium ions by pumping them out of the surrounding cytoplasm (sarcoplasm); the protein pump which accomplishes this is located in the *SR* membrane and is *ATP*-driven. Thus under resting conditions there is a large concentration difference of calcium between the interior and exterior of the *SR*, but the permeability of the *SR* membrane to calcium is low. The transient increase of this permeability results in a rapid efflux of calcium ions into the sarcoplasm, where they bind reversibly to the regulatory protein troponin which is located on the thin actin myofilaments.

There is approximately one troponin molecule for each myosin head (cross-bridge) projecting from the thick myosin filaments which occupy the central region of a sarcomere and interdigitate with the thin actin filaments. In the absence of bound calcium troponin is believed to exert an inhibition against the binding of myosin cross-bridges to specific receptor sites on the actin filaments. But when the muscle action potential supplies bound calcium as described above, it is believed that this troponin-mediated inhibition is removed, perhaps by a conformational change, and myosin is free to bind to actin: the muscle tissue is "activated" and ready to contract. The subsequent binding of myosin cross-bridges to actin sites causes, through an incompletely understood series of steps discussed in the next section, muscle movement by relative sliding of the two interdigitating myofilament arrays; or, if movement is constrained, the actin–myosin interaction generates contractile force.

If electrical stimulation ceases then relaxation occurs by a reversal of steps listed above. First, there is a net flux of calcium out of the sarcoplasm and into the *SR*, because the calcium permeability of the *SR* membrane is now low but the *ATP*-driven calcium pump continues to operate. The reduced sarcoplasmic free calcium concentration promotes dissociation of calcium from troponin, which in turn reestablishes the inhibition against actin–myosin binding. Thus when the cross-bridges detach from actin they cannot re-attach, which causes motion to stop and muscle force to relax.

Chemical energy is consumed in both the activation and contraction processes. The immediate source of this energy is the hydrolysis of adenosine triphosphate (*ATP*), dissolved in the sarcoplasm, to adenosine diphosphate (*ADP*). In the activation process this energy is used by specialized proteins in the *SR* membrane to pump calcium ions from the sarcoplasm into the *SR* and thus establish the calcium concentration gradient. In the actin–myosin contractile interactions one *ATP* molecule is hydrolyzed for each cycle of cross-bridge attachment and detachment, and this is the energy source for the work performed by the muscle. As neither the conversion of *ATP* free energy into an electrochemical calcium potential difference across the *SR* membrane, nor its conversion into muscle work, is perfectly efficient, some of the available chemical energy is degraded into heat which is transferred across the muscle boundary or increases the temperature of the muscle. In spite of its rapid utilization, however, no decrease in the *ATP* concentration is usually detected in contractions of moderate duration. This is because the *ADP* generated by muscular activity is immediately re-phosphorylated back to *ATP* by the Lohmann reaction, at the expense of the hydrolysis of another chemical energy source available in the sarcoplasm: phosphocreatine (*PCr*).

This rather crude summary of molecular activation–contraction dynamics has glossed over an abundance of complex, incompletely understood detail. For example, the single step of translocating a calcium ion across the *SR* membrane may involve as many as twelve separate biochemical states in the pumping protein (Inesi, 1985). Or again, the assumption that all cross-bridges are detached in the absence of electrical stimulation may be an oversimplification: some experiments have suggested that a considerable number of cross-bridges may be instantaneously attached in relaxed muscle (at least at low ionic strength) but these attached cross-bridges are in a fast equilibrium with those detached and therefore produce negligible force unless a high stretch velocity is imposed (Schoenberg et al., 1984). Clearly, even without these considerations, the basic schema presented in this section is quite complicated and poses a severe challenge to anyone wishing to develop realistic models of muscle.

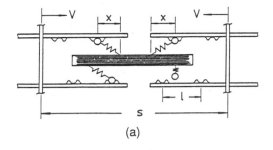

(a)

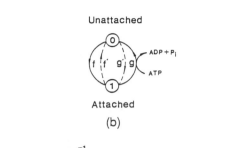

(b)

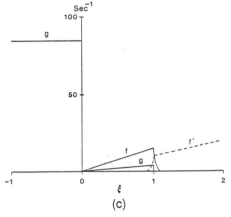

(c)

Figure 1.2: *a)* Schematic illustration of the interdigitating array of thick (myosin) and thin (actin) myofilaments. The cross-bridges are portrayed as balls attached by springs to the myosin backbone. *b)* Kinetic diagram for a two-state cross-bridge model showing two separate reaction paths between the detached (0) state and the attached (1) state. *ATP* splitting is assumed to occur on detachment via the *g*-path. *c)* Assumed dependences of the rate functions *f*, *f'*, *g*, and *g'* on the dimensionless bond length $\xi = x/h$. For reasonable values of ΔA, $\hat{\varepsilon}$ and $\hat{k}$, the rate function $g'(\xi)$ is negligible on the scale of the diagram. (From Zahalak and Ma, 1990; reprinted with permission of ASME.)

1.3 Biophysical Models of Contraction

I now turn to biophysical models of how force and motion are produced in activated muscle; that is in muscle where actin sites have been made accessible to the myosin cross-bridges by the prior binding of calcium to troponin. First I will review the original simplest model proposed by Huxley (1957), as amended to include the "self-consistency" requirement introduced by T.L. Hill (Hill et al., 1975). This model assumes that a myosin cross-bridge can exist only in one of two distinct biochemical states: either *(1)* attached to an actin site, or *(2)* detached from actin. This is illustrated in the cartoon of Figure 1.2(a). The cross-bridges are assumed to be attached to the thick myosin filaments by an elastic linkage of some sort (the precise nature of which has yet to be determined), and to be in random thermal motion about a neutral equilibrium position. From time to time each cross-bridge, which in the simplest theory is assumed to function independently of all other cross-bridges, will bind to a nearby actin site and the hypothetical elastic linkage will be stretched or compressed; this, in turn, will produce an elementary force of interaction between the actin and myosin filaments, which may act in either direction. The sum of all these elementary forces of over all the cross-bridges of the contractile tissue is the force exerted by the muscle.

T.L. Hill (1977) pointed out that in order for cross-bridges to produce net work in a sequence of attachment and detachment there must exist at least *two* distinct reaction paths between the attached and detached states. This is illustrated in Figure 1.2(b), where the rate constants for attachment via the two paths are denoted by *f* and *g'*, whereas the corresponding reverse rate constants for detachment are denoted by *f'* and *g*; I will refer to these two reaction paths as the "*f*-path" and the "*g*-path". Such a situation could arise if, for example, either the attached or detached state were actually two distinct states (related by a conformational change) which were in a fast equilibrium with each other, or if one of these two sub-states were a transient intermediate (Hill, 1977). The critical ingredient which makes the Huxley theory work is that the rate constants *f*, *f'*, *g*, and *g'* are not, in fact, constants but rather functions of *x*, the displacement of a cross-bridge from its neutral equilibrium position. These functions are assumed to be such that a cross-bridge has a higher probability of binding with $x > 0$, and thus producing a tension in the contractile tissue, than with $x < 0$, which would produce compression. A typical set of rate functions is illustrated in Figure 1.2(c). It is believed that one molecule of ATP from the solution bathing the myofilaments is hydrolyzed to ADP each time a cross-bridge passes through a complete cycle of states. This is indicated in Fig. 1.2(b) by coupling the hydrolysis of one ATP molecule for each cross-bridge detachment via the g-path; attachment via this path would therefore require the synthesis of ATP from ADP in solution, which is deemed very unlikely for the normal operation of muscle.

Under special restrictive assumptions one can actually deduce the dependence of the rate functions on x from macroscopic muscle experiments (Lacker and Peskin, 1986), but in general this is not possible; the rate functions are assumed a priori in this theory. But only two of the four rate functions of a two-state model are independent due to a thermodynamic restriction which T. L. Hill has dubbed "self-consistency". This restriction may be shown to take the form (Hill et al., 1975)

$$f/f' = e^{-(A_1 - A_0)/KT} \quad \text{and} \quad g/g' = e^{-(A_0 - A_1 - \hat{\epsilon})/KT} \quad (1.1)$$

where A_0 and A_1 are, respectively the free energies of the detached and attached states, T is the absolute temperature, K is Boltzmann's constant, and at $\hat{\epsilon}$ is the free energy of hydrolysis of one ATP molecule. At constant temperature and ligand concentrations A_0 is assumed to be constant. But the free energy of the attached state, A_1, is assumed to include the strain energy of the elastic link of the cross-bridge; indeed, if the cross-bridge elasticity is assumed linear with a spring constant $\hat{k}$, then

$$A_1(x) = A_1^0 + \frac{1}{2}\hat{k}x^2 \quad (1.2)$$

where A_1^0 is a function of temperature only. From Eq. 1.1 and Eq. 1.2 it is clear that when a linked cross-bridge is severely deformed ($|x| >> \sqrt{KT/\hat{k}}$ then detachment is favored over attachment in both the ATP-independent f-path and the ATP-dependent g-path. Which path will predominate depends on the detailed shape of the rate functions and the ex-

ternal constraints imposed on the muscle.

The Huxley theory turns on the determination of a *bond-distribution function*, $n(x,t)$, which gives the fraction of cross-bridges at time t that are bonded to actin with bond lengths (i.e. cross-bridge displacements) lying in the range $(x, x + dx)$. Assuming (**1**) independent cross-bridges, (**2**) rigid myofilaments, (**3**) a fixed number of active cross-bridges, and (**4**) first-order chemical kinetics, it can be shown that n satisfies the equation

$$\left(\frac{\partial n}{\partial t}\right)_x - v(t)\left(\frac{\partial n}{\partial x}\right)_t = \quad (1.3)$$

$$[f(x) + g'(x)][1 - n(x,t)] - [f'(x) + g(x)]\, n(x)$$

where $v(t)$ is the relative velocity of sliding between the myofilaments (shortening positive) (Hill et al., 1975). *Eq. 1.3* is the same as Huxley's original celebrated rate equation (Huxley, 1957) if $f + g'$ and $g + f'$ are replaced respectively by f and g. Once $n(x,t)$ has been found by solving *Eq. 1.3*, all macroscopic quantities of interest can be found via appropriate integrals involving n. Thus, for example, the force, P, per unit current cross-sectional area, A, (the Cauchy or "true" stress) is given by

$$(P/A) = C_1 \int_{-\infty}^{\infty} x\, n(x,t)\, dx \quad (1.4)$$

whereas the rate of energy release by ATP hydrolysis during cross-bridge cycling is given by

$$\dot{E} = C_2 \int_{-\infty}^{\infty} \{\, g(x)\, n(x,t) - g'(x)\, [1 - n(x,t)]\, \}\, dx \quad (1.5)$$

where C_1 and C_2 are constants depending on the myofilament structure, the cross-bridge stiffness, and the ATP-hydrolysis energy (Zahalak, 1981). A typical solution of *Eq. 1.3* is illustrated in Figure 1.3 by the solid curves, which show the evolution with time of the bond-distribution function when a cylinder of contractile tissue obeying the Huxley model is either shortened or stretched at a constant speed.

In his classic 1957 paper Huxley showed that his model was consistent with the broad features of muscle behavior known at that time, including the force-velocity relation and the rate of heat production during shortening. Further detailed investigation of muscle behavior, however, revealed several phenomena which appeared to be incompatible with a simple two-state cross-bridge model. Thus, for example, Huxley and Simmons (1971) found that in order to explain the transient tension changes following step shortening or lengthening of muscle fibers (the $T_1 - T_2$ phenomenon) it appeared necessary to postulate at least two distinct attached states of a cross-bridge, which resulted in a three-state model. Another deficiency of the basic Huxley model was its failure to predict a maximum in the steady-state heat-plus-work rate of shortening muscle, measured by A.V. Hill (1964). In order to account for these and other features the number of states attributed to a cross-bridge has steadily increased. Further, accumulating knowledge of the biochemical kinetics of actin–myosin in solution has motivated an expansion of the cross-bridge models: as more biochemical states were revealed, more corresponding cross-bridge states were postulated. Indeed the central theme of contemporary muscle biophysics is a correlation of the solution biochemistry of muscle with cross-bridge mechanics (Eisenberg, 1986). The result is that published models of contraction dynamics have grown from the original two states (Huxley, 1957), to three (Huxley and Simmons, 1971), four

(Eisenberg et al., 1980), five (Wood and Mann, 1981), seven (Pate and Cooke, 1986), eight (Eisenberg and Greene, 1980), and even to eighteen (!) states (Propp, 1986). While these multiple states may be necessary or convenient to understand the many complex fine details of the mechanical, biochemical, and thermodynamic behavior of muscle in terms of fundamental cross-bridge mechanisms, clearly not all the states are required to describe the broad features of macroscopic contraction dynamics under normal conditions. In fact, considering its extreme (relative) simplicity, a two-state model does a remarkably good job, which may be attributed to the fact that it captures the most essential feature of muscle contraction: actin–myosin binding.

In addition to a proliferation of states, several other refinements of the basic cross-bridge model have been proposed. The original Huxley model and its direct descendants assumed that each cross-bridge acted independently of all others, and that at each time instant each cross-bridge had a significant probability of binding to only one actin site. Some later elaborations have included multiple actin sites (Wood and Mann, 1981) and

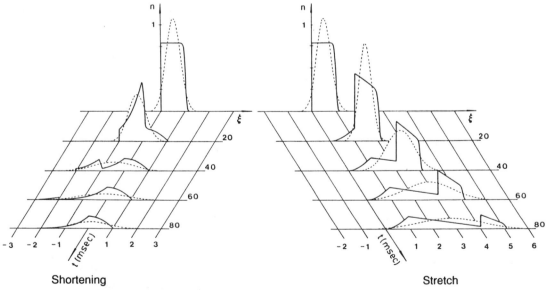

Shortening   Stretch

Figure 1.3: Time evolution of the bond-distribution function $n(\xi,t)$ when a block of maximally activated ($r = 1$) Huxley-type contractile tissue, characterized by the rate of functions of Figure 1.2(c), is either shortened or stretched at the maximum speed of unloaded shortening. The heavy solid lines represent exact solutions of the Huxley rate equation (*Eq. 1.3*; see also *Eqs.*

1.12–1.13), and the thin dashed lines represent the Gaussian approximations associated with the *DM* equations (*Eqs. 1.25*). Although the Gaussian curves are not good pointwise approximations to n their first three moments are excellent approximations to the corresponding moments of n.

competition between myosin heads for actin sites (Tozeren and Schoenberg, 1986). Further, almost all cross-bridge models assume a constant stiffness for the putative elastic link, but a recent paper has explored the effect of nonlinear cross-bridge elasticity (Harry et al., 1990). These perturbations on Huxley's basic schema do not appear to radically alter the predicted behavior.

While they present an elegant unifying view of muscle function, published biophysical cross-bridge models have not yet adequately addressed several issues of importance in whole muscle behavior. These issues include: *1)* the interaction of electrical stimulation and calcium activation with contraction (although some attempts have been made), *2)* possible dependence of activation dynamics on muscle length, and *3)* the effects of sarcomere inhomogeneity. Nevertheless these models are so well established among muscle researchers that they should serve as at least a conceptual foundation for any approximate engineering model of the muscle actuator.

1.4 Classic Macroscopic Muscle Mechanics: The A.V. Hill Model

The current views of how muscle works, as presented in the preceding sections, were almost all developed since the 1950's, and depended on modern experimental equipment and techniques. But a considerable amount had been known about the macroscopic mechanical behavior of whole muscle since the early 1920's, due primarily to the work of A.V. Hill and his associates. In this section I will briefly trace the development of the phenomenological muscle model to which the name of Hill is usually attached. This model is so influential and important that a separate chapter [Chapter 5 (Winters)] is devoted to a discussion of its fine details and modern ramifications.

The origins of the Hill model can be traced to careful measurements of the mechanical behavior of isolated maximally stimulated muscle which were made by several investigators between 1920 and 1930. Gasser and Hill (1924) imposed very rapid force and length changes on frog sartorius and concluded that the transient responses of active muscle resembled those of a passive viscoelastic system. (They went so far as to build a model consisting of a rubber tube filled with a viscous fluid which could mimic most of their observations on muscle). Shortly thereafter Levin and Wyman (1927) published studies employing

their new ergometer which could apply precisely controlled shortening and lengthening velocities to muscle, and concluded in agreement with Gasser and Hill that stimulated muscles did indeed behave mechanically like viscoelastic bodies. These authors seem to have been the first to suggest explicitly that the structure of the viscoelastic model should consist of two elements in series: an undamped elastic element, and a damped elastic element [Figure 1.4(a)]. Electrical stimulation was assumed to somehow rapidly transform the spring of the damped element from a compliant state into a stretched tension-generating state, which would then cause the muscle to contract. Subsequent experiments by other workers produced results that seemed to support this viscoelastic theory of muscle (e.g. Bouckaert, Capellen, and de Blende, 1930; see Hill, 1938 for a bibliography).

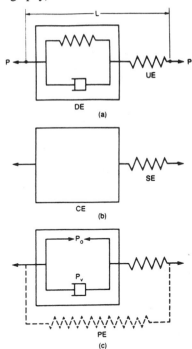

Figure 1.4: Macroscopic muscle models. *a)* The Levin-Wyman 1927 model consisting of an undamped elastic element (*UE*) and a damped elastic element (*DE*). *b)* The A.V. Hill 1938 model consisting of a series elastic element (*SE*) and a contractile element (*CE*). *c)* Internal structure of the *CE* of the Hill model showing the active state force, P_0, and the quasi-viscous internal resisting force P_v. Shown in dashes is the parallel elastic element (*PE*) which may be added to model the passive elastic properties of unstimulated muscle.

Continued experimentation in the next decade (Fenn and Marsh, 1935; Hill, 1938) led to a rejection of the viscoelastic theory in favor of one that held that shortening and work production were governed by rates of energy-yielding chemical reactions rather than release of elastic energy. This change of view was based particularly on measurements of the heat produced by muscle during contraction, and culminated in A.V. Hill's famous 1938 paper which first explicitly proposed a prototype of the muscle model bearing his name. Although this paper was primarily concerned with heat measurements, its most important result was a model for the *mechanical* behavior of muscle.

The major experiment reported in Hill's 1938 paper was the isotonic quick-release, and it illustrates succinctly the rationale for his model. In this experiment a muscle is held at a fixed length (close to the mean length of the muscle in the body) and stimulated maximally until it attains a steady isometric force, P_0. Then the applied force is suddenly decreased to a lower value, P (the "after-load") and held constant [Figure 1.5(a)]. As shown in Figure 1.5(b) the muscle length changes in this experiment can be idealized as occurring in two distinct phases. The first is an almost instan-

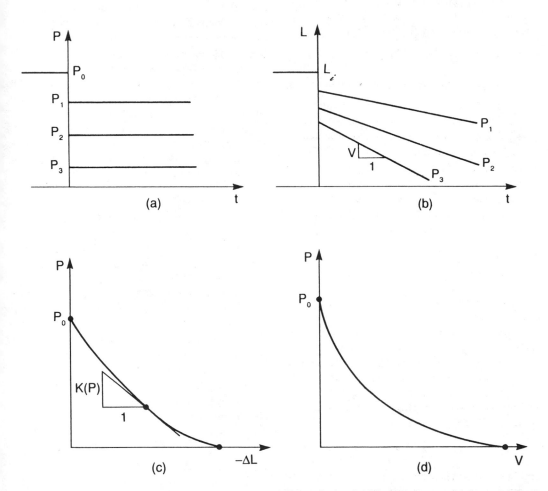

Figure 1.5:   The isotonic quick-release test. *a)* Applied force as a function of time, for three after-loads: P_1, P_2 and P_3. *b)* Muscle length as a function of time for the three after-loads. *c)* After-load as a function of the initial "intantaneous" shortening step magnitude. $K(P)$ is the tangent stiffness. *d)* After-load as a function of the steady velocity of shortening in the second phase.

taneous shortening of the muscle, of magnitude $-\Delta L$, which coincides with the almost instantaneous change in force level; $-\Delta L$ increases with $(P_0 - P)$ [Figure 1.5(c)]. Following this initial response the muscle proceeds to shorten for some distance at almost constant speed. The speed of shortening in this second phase is an increasing function of $(P_0 - P)$ [Figure 1.5(d)].

Hill applied a new thermopile, which he had designed, to measure the heat produced by muscle during the second phase of steady shortening and made two important observations:

a) The heat liberated during shortening in excess of the isometric ("maintenance") heat was proportional to the distance shortened, independent of force and distance. Thus

$$\Delta H = a\,(-\Delta \hat{L}) = a\,(L_i - L + \Delta L) \qquad (1.6)$$

where ΔH is the increment of additional heat liberated, $-\Delta \hat{L}$ is the amount of shortening in the second phase, and a is a constant characteristic of a given muscle which Hill dubbed the "heat of shortening."

b) The rate at which extra energy (that is, heat plus work in excess of the maintenance heat) was released during shortening, $\Delta \dot{E}$, was a linear function of the force applied to the muscle. In mathematical form

$$\Delta \dot{E} = b\,(P - P_0) \qquad (1.7)$$

where b represents another constant characteristic of the muscle.

These observations on the energetics of muscle can be immediately converted to a statement about its mechanics. Differentiating Eq. 1.6 with respect to time, recognizing that $-(d\hat{L}/dt) = V$, the velocity of shortening, and further that the rate of external work production by the muscle is PV, one obtains from Eq. 1.7 that

$$(a + P)\,V = b\,(P_0 - P) \qquad (1.8)$$

But Eq. 1.8 establishes a relation between the purely mechanical variables P and V [Figure 1.5(d)], which can be verified directly by mechanical measurements without recourse to thermal measurements, and indeed this equation was found to provide an excellent fit to the measured data.

Thus as a simple model that exhibits the behavior of muscle in an isotonic quick-release test Hill proposed the one shown in Figure 1.4(b). As

in the Levin–Wyman model muscle is assumed to be a series combination of two elements. The first is called the "series elastic element" (SE) and is assumed to be a (possibly nonlinear) spring, with a length and stiffness determined by the instantaneous muscle force in accordance with Figure 1.5(c). The second element is called the "contractile element" (CE) and is assumed to be characterized by a unique relation (the "force–velocity relation") between its instantaneous speed of shortening and the instantaneous muscle force. In an isotonic experiment where the muscle force is constant, the Hill model implies that the muscle velocity is the same as the CE velocity. This in turn implies that the force–velocity relation of the CE is nothing but Eq. 1.8. In the context of the Hill model the CE embodies the processes of chemo-mechanical energy conversion. This separation of the elasticity and contractility of muscle into two different phenomenological entities connected in series is a fundamental characteristic of the Hill model. (Note, however, that from the viewpoint of conventional cross-bridge theory the separation is artificial and unwarranted because the cross-bridges are simultaneously contractile and elastic structures distributed uniformly throughout the contractile tissue).

When it was first introduced Eq. 1.8 appeared to establish an appealing, albeit not well understood, connection between muscle mechanics and energetics via the heat of shortening constant a. Unfortunately Hill's later more refined heat measurements (Hill, 1964) severed this connection by showing that the heat of shortening is not constant and is, in general, not equal to a in the force-velocity relation, Eq. 1.8; the latter equation, however, still remains a good empirical fit to measured force-velocity data.

It is worth pointing out that Eq. 1.8 can be rewritten in a way which assigns internal structure to the CE, as follows

$$P = P_0 - P_V = P_0 - C(V; P_0)\,V \qquad (1.9)$$

where

$$C(V; P_0) = \frac{P_0 + a}{V + b} \qquad (1.10)$$

This algebraic manipulation can be interpreted as follows: the force, P, generated by the CE is the sum of an internal contractile force, P_0, and an internal viscous resisting force, P_V, which depends nonlinearly on the velocity. Figure 1.4(c) is a diagram of the Hill model incorporating this internal decomposition of the *CE*.

Combining the dynamics of the *CE*, as expressed by the force-velocity relation *Eq. 1.8*, with the force-extension characteristics of the *SE* one may write a differential equation governing the basic Hill model of muscle

$$\dot{L} = \dot{P}/K(P) - V(P; P_0) \qquad (1.11)$$

where $K(P)$ is the force-dependent tangent stiffness of the *SE*, *L* is the overall muscle length, and the superposed dots denote time differentiation.

In addition to the force-velocity relation Hill introduced another important concept - that of the *active state*. At first this idea was used in a qualitative sense to mean the internal state of a stimulated muscle which is capable of producing force and motion. But gradually the term was assigned quantitative meaning by defining it to be the force that a muscle exerted when the *CE* was neither shortening nor lengthening – that is, when the *CE* velocity is zero (Hill, 1949). Therefore, according to *Eq. 1.8*, the active state is to be identified with the internal force, P_0. This force was assumed to depend on the history of stimulation, and considerable effort was expended to measure its time course, particularly in the experimentally important cases of isometric tetanus and twitch. A discussion of these experiments may be found in Aidley (1971) and McMahon (1984) and they revealed several quantitative difficulties with the notion of active state, which suggested that this variable should best be regarded as only a qualitative measure of muscular activity. Indeed Aidley (1971) has gone so far as to state "we must conclude that the concept of the 'active state' leads more to confusion than enlightenment." Nevertheless, this concept remains an integral part of the Hill muscle model, and is generally accepted by bioengineers as providing at least an approximate measure of muscular activation [see also Chapter 5 (Winters)].

The experiments that led to the Hill model were quite restricted, involving muscle that shortened (not lengthened) at maximal (not partial) activation over a limited range of muscle lengths near the "optimal" length at which the maximal isometric force is generated. Successive investigators have attempted to generalize the model so that it is useful under circumstances more repre-

sentative of those encountered in-vivo. A simple enhancement is to add a third "parallel elastic" (*PE*) element [Figure 1.4(c)] to model the fact that passive, unactivated muscle can resist stretch; it has been found, however, that this element usually generates substantial forces only at long muscle lengths, and is often ignored in simulations. It has long been known that isometric muscle under constant stimulation produces an active force that is maximal at a length L_0 close to the mean length of the muscle in the body – the optimal length mentioned above – and decreases at shorter and longer lengths (Ramsey and Street, 1940); this behavior illustrated schematically in Figure 1.6(a), was elegantly explained in terms of the cross-bridge theory by Gordon, Huxley, and Julian (1966). In order to incorporate this tension–length characteristic into the Hill model it is assumed that the active state, P_0, depends on *CE* length as well as stimulation history [Ritchie and Wilkie, 1958; see also Chapter 5 (Winters)].

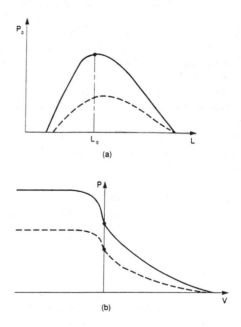

Figure 1.6: *a)* Solid curve: isometric force, P_0, versus muscle length, *L*, at maximal stimulation. The force is maximum at the "optimal" muscle length, L_0. Dashed curve: relation between P_0 and *L* at constant submaximal stimulation. *b)* Force-velocity relations of the *CE* for both positive and negative shortening velocities, *V*. The upper solid curve is for constant maximal stimulation, and the dashed curve is for constant submaximal stimulation.

Even with the elaborations listed above the Hill model for shortening could not simulate the normal function of muscle *in vivo* (or even the complete force history in an isometric twitch), because muscle often lengthens as well as shortens under load. Thus it was necessary to generalize the force–velocity relation to include lengthening as well as shortening. This was done on the basis of early isotonic quick-release experiments by Katz (1939) on maximally stimulated frog and tortoise muscles, who found a relation between the applied force and the *steady* velocity of shortening or lengthening roughly resembling that shown in Figure 1.6(b). It is worth noting that Katz did not observe a unique relation between force and lengthening velocity in this isotonic experiment (Katz, 1939, Figure 4). Rather a dynamic transient of variable duration interposed itself between the initial elastic response and an eventually attained steady lengthening velocity. If the afterload was too large (greater than 1.8 P_0 for frog sartorius) then no steady lengthening velocity was attained — the muscle simply "gave" by extending rapidly. Recently Harry et al. (1990) have performed the converse experiment on tetanized frog sartorius, wherein constant shortening or lengthening velocities were imposed and the corresponding forces were measured; they found a force–velocity relation like that shown in Fig. 1.6(b), but with a maximum force in lengthening of *1.5 P_0* rather than *1.8 P_0*. Based on a limited number of similar experiments, it is usually assumed in Hill models that at maximal stimulation the force–velocity relation of the *CE* has the shape of the solid curve in Figure 1.6(b), and further that this relation scales down with submaximal stimulation as suggested by the dashed curve in this figure.

In this introductory chapter I have limited myself to a broad overview of the Hill model. Chapter 5 (Winters) is devoted entirely to this topic, and gives a comprehensive discussion of its fine details, extensive references, and modern developments. In closing this section, however, it should be emphasized that in spite of its wide acceptance and use by biomechanists there are substantial questions about how well the Hill model represents the dynamics of muscle. Indeed, there is direct experimental evidence (Joyce and Rack, 1969; van Ingen Schenau et al., 1988) that the fundamental assumptions of the Hill model are not true in general — in particular, that for some skeletal muscles at constant activation and length there is no unique relation between muscle force and velocity under isotonic conditions.

1.5 The Distribution-Moment (DM) Model: A Bridge Between Molecular and Macroscopic Muscle Mechanics

In the preceding sections I have reviewed the two main currents in muscle modeling: Huxley-type molecular models and Hill-type phenomenological models. I will now describe an approach to muscle modeling which, in a sense, provides a bridge between these two viewpoints, and yields low-order state-variable models for whole muscle via a mathematical approximation of the cross-bridge theories. This approach has been called the Distribution-Moment (*DM*) Approximation and has been presented in several previous publications (Zahalak, 1981, 1986; Ma and Zahalak, 1987; Zahalak and Ma, 1990); the objective of this theory is to generate macroscopic models of whole muscle approaching in simplicity the Hill model, while retaining much of the presumed biological veracity of the Huxley model. The most recent and complete development of the *DM* model is contained in the paper by Zahalak and Ma (1990). While this paper is based on a two-state Huxley model, it extends the classic Huxley theory by integrating a model for electrical stimulation and calcium-activation dynamics into the cross-bridge model. I will sketch the main results of the theory, but must refer the reader to the references for detailed supporting arguments and literature citations.

To lay the foundations of the model it is assumed, as in the original Huxley theory, that cross-bridges act independently and that at any given time each cross-bridge interacts with only one actin site — the "closest". But it is assumed further that for each instantaneous (myosin cross-bridge)–(actin binding site) pair there is one molecule of troponin that regulates the availability of that site for binding. Thus attention is focussed on the dynamics of the Myosin–Actin–Troponin (*MAT*) complex. While it is known that each troponin molecule binds four Ca^{++} ions, the model assumes that only the two so-called low affinity binding sites are important for the control of contraction; the kinetics of the two calcium binding steps are assumed to be identical. Figure 1.7 shows a kinetic diagram for two possible activation schemes that were considered in the model. Both schemes assume that myosin cannot bind to actin until two calcium ions are bound to troponin. But the "loose coupling" scheme assumes that the

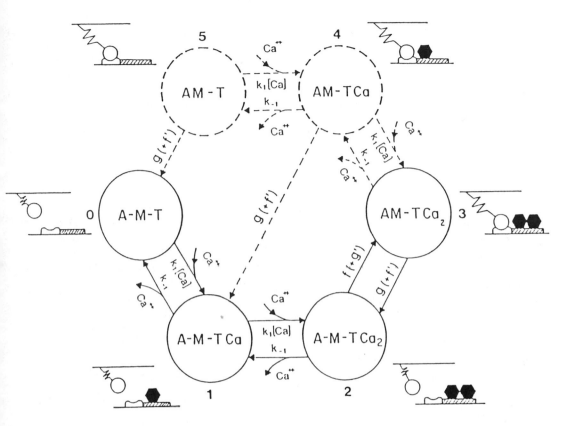

Figure 1.7: Kinetic diagram for two calcium-activation schemes. In both schemes a cross-bridge (ball and spring) can bind to an actin site only if the associated regulating troponin molecule (hatched rectangle) has bound two calcium ions (filled hexagous). In the "tight coupling" scheme (solid large circles) a calcium ion can unbind from troponin only if myosin is detached from actin, but in the "loose coupling" scheme (solid and dashed large circles) detachment of calcium from troponin is unaffected by the state of actin myosin binding. [From Zahalak and Ma (1990); reprinted with permission of ASME.]

binding and unbinding of calcium and troponin proceeds completely independently of the state of binding between actin and myosin, whereas in the "tight coupling" version calcium can unbind from troponin only if myosin is detached from actin. The binding of calcium to troponin is assumed to be fast on the time scale of cross-bridge reactions, so that these two species can be considered to be in chemical equilibrium at all times. The model assumes also that only a fraction α $(0 < \alpha < 1)$ of the cross-bridges in a mass of contractile tissue are available to interact with actin at any given time. This fraction is assumed to be a function of muscle (more properly, contractile tissue) length; it may be less than one at long muscle lengths because of reduced myofilament overlap, and also at short muscle lengths because of steric interference.

If the assumed calcium activation dynamics described above are combined with the basic postulates of the Huxley theory it can be shown (Zahalak and Ma, 1990) that the Huxley rate equation (*Eq. 1.3*) is modified to

$$\frac{\partial n}{\partial t} - v(t)\frac{\partial n}{\partial x} = (f + g')(r\,\alpha - n) - (f' + g)\,n \quad (1.12)$$

for loose coupling, and

$$\frac{\partial n}{\partial t} - v(t)\frac{\partial n}{\partial x} = r\,(f + g')(\alpha - n) - (f' + g)\,n \quad (1.13)$$

for tight coupling. The function r appearing in the above equations is called the *activation factor*, and is a pure function of $[Ca]$, the free calcium ion concentration in the sarcoplasm (outside the *SR*). Specifically this function is defined by

$$r([Ca]) = \frac{k_1^2 [Ca]^2}{k_1^2 [Ca]^2 + k_1 k_{-1}[Ca] + k_{-1}^2} \quad (1.14)$$

where k_1 and k_{-1} are, respectively, the rate constants for binding and unbinding of calcium and troponin. It is the activation factor r that couples cross-bridge contraction dynamics to the action-potential mediated calcium-activation dynamics. A comparison of *Eq. 1.3* with *Eq. 1.13* shows that the effect of activation in the tight coupling version of the model is simply to change the effective binding rate constant in the Huxley model from (f + g') to r (f + g'), while leaving the detachment rate constant (g + f') unchanged.

At this point the *DM* approximation can be applied to the Huxley rate equation modified for calcium activation, either *Eq. 1.12* or *Eq. 1.13*. To explain the approximation we first consider the normalized moments of the actin-myosin bond-distribution function, $n(x,t)$. These are defined by

$$Q_\lambda(t) = \int_{-\infty}^{\infty} \xi^\lambda n(\xi,t)\, d\xi \quad (1.15)$$

where $\xi = x/h$, and h is a constant scaling parameter that measures the range of x over which there is a significant probability of cross-bridge attachment [see, for example, Figure 1.2(c)]. In addition to their microscopic significance as moments of the molecular distribution function the Q_λ's have important macroscopic interpretations. For a two-state cross-bridge model with constant cross-bridge stiffness, $\hat{k}$, it is easy to show (Zahalak and Ma, 1990) that the first three moments, Q_0, Q_1, and Q_2, are proportional, respectively, to: *1)* the instantaneous *stiffness* of the contractile tissue[1]; *2)* the muscle *force*; and *3)* the *elastic energy* instantaneously stored in the cross-bridges.

---

[1] Q_0 corresponds to the operationally defined "series elastic stiffness" of the Hill model, or at least that part of it which resides in the contractile tissue rather than the tendon. This stiffness is the ratio of force to displacement which would be measured in an instantaneous stretch or shortening of the contractile tissue. Q_0 is not, however, the *same* as Hill's series elasticity because the latter is determined uniquely by the instantaneous muscle force, whereas the former is not.

By integrating the rate *Eq. 1.12* or *1.13* with respect to the bond-length variable, x, one can obtain an infinite system of coupled equations on the moments of the form

$$\dot{Q}_\lambda = \alpha r \beta_\lambda - \phi_\lambda - \lambda u(t) Q_{\lambda-1} \quad (1.16)$$

for loose coupling, or

$$\dot{Q}_\lambda = \alpha r \beta_\lambda - r \phi_{1\lambda} - \phi_{2\lambda} - \lambda u(t) Q_{\lambda-1} \quad (1.17)$$

for tight coupling, with $\lambda = 0, 1, 2, \ldots$ in each case. In the above equations the superposed dot denotes time differentiation, $u(t)$ represents the scaled myofilament sliding velocity, $v(t)/h$, and $Q_{\lambda-1}$ is defined to be zero for convenience. The β's and ϕ's are functionals of n and the rate functions defined as follows:

$$\beta_\lambda(t) = \int_{-\infty}^{\infty} \xi^\lambda [f(\xi) + g'(\xi)]\, d\xi$$

and

$$\phi_{1\lambda}(t) = \int_{-\infty}^{\infty} \xi^\lambda [f(\xi) + g'(\xi)]\, n(\xi,t)\, d\xi$$

$$\phi_{2\lambda}(t) = \int_{-\infty}^{\infty} \xi^\lambda [g(\xi) + f'(\xi)]\, n(\xi,t)\, d\xi \quad (1.18)$$

and

$$\phi_\lambda = \phi_{1\lambda} + \phi_{2\lambda}$$

Up to this point there has been no approximation of the cross-bridge theory. The DM approximation is introduced by assuming *a-priori* a *functional form* for n which depends on the moments. Zahalak (1981) postulated as reasonable and mathematically convenient the Gaussian form

$$n(\xi,t) \doteq \frac{Q_0}{\sqrt{2\pi}\, q} e^{-(\xi-p)^2/2q^2} \quad (1.19)$$

where $p \doteq Q_1/Q_0$ and $q = \sqrt{(Q_2/Q_0) - Q_1/Q_0)^2}$ are respectively the mean and standard deviation of the approximate bond distribution. If now the Gaussian approximation, *Eq. 1.19*, is inserted into *Eqs. 1.16* or *1.17* via *Eqs. 1.18*, and the first three equations of either set retained, there results a set of three coupled nonlinear ordinary differential equations for the approximate determination of Q_0, Q_1, and Q_2. These are the distribution-moment

(DM) equations corresponding to the chosen cross-bridge model:

$$\dot{Q}_\lambda = r\,\alpha\,\beta_\lambda - \phi_\lambda(Q_0,Q_1,Q_2) - \lambda\,u(t)\,Q_{\lambda-1} \quad (1.20)$$

for loose coupling, and

$$\dot{Q}_\lambda = r\,\alpha\,\beta_\lambda - r\,\phi_{1\lambda}(Q_0,Q_1,Q_2) -$$

$$\phi_{2\lambda}(Q_0,Q_1,Q_2) - \lambda\,u(t)\,Q_{\lambda-1} \quad (1.21)$$

for tight coupling, with $\lambda = 0,1,2$ in each case. The β's and ϕ's can be constructed as explicit functions of the moments (and the parameters of the bonding and unbonding rate functions) once the cross-bridge model has been specified; several specific cases have been calculated and published (Zahalak, 1981; Ma and Zahalak, 1987; Zahalak and Ma, 1990).

A representative comparison between exact solutions of the rate equation Eq. 1.13 and the Gaussian approximation of Eq. 1.19 with moments given by the DM Eqs. 1.21 is shown in Figure 1.3: the solid curves are exact solutions and the dashed curves are the approximations. Even in this case of non-smooth rate functions where the Gaussian curves are not particularly good pointwise approximations to the exact solution, it may be shown nevertheless, that the DM approximations of the exact moments are good (Ma and Zahalak, 1987; Zahalak and Ma, 1990). Extensive computational experience indicates that this is generally the case. Thus the DM equations define a quantitative model of contractile tissue with essentially the same macroscopic behavior as the parent cross-bridge model, but the former is much more tractable computationally than the latter. Further the DM equations yield directly the macroscopic variables of interest (stiffness, force, elastic energy) without requiring first the determination of essentially unmeasurable bond-distribution functions.

Eqs 1.20 or 1.21 embody the contraction dynamics portion of the DM model. As shown in Zahalak and Ma, 1990, the DM approximation can also be applied to the calcium activation dynamics, and this results in the calcium-rate equation

$$\frac{d}{dt}[Ca]_t = R_0\left(1 - \frac{[Ca]}{[Ca]^*}\right)\chi(t) - V_m\frac{[Ca]}{[Ca]+K_m} \quad (1.22)$$

with

$$[Ca]_t = [Ca] + m\,r\,\{\,2 + k_{-1}/k_1[Ca]\,\} \quad (1.23)$$

for loose coupling, and

$$[Ca]_t = [Ca] + (mh/l)\,\{\,2Q_0 +$$

$$r\,(2 + k_{-1}/k_1[Ca])\,(l/h - Q)\,\} \quad (1.24)$$

for tight coupling, where $[Ca]_t$ is the total sarcoplasmic calcium concentration (free plus bound to troponin), m is the concentration of cross-bridges, and l is the distance between successive actin binding rites. The second term on the right side of Eq. 1.22 is the calcium uptake rate into the SR, which is assumed to obey simple Michaelis-Menten kinetics with parameters V_m and K_m. The first term on the right side of Eq. 1.22 represents the rate of efflux of calcium out of the SR into the sarcoplasm; R_0 is the concentration change of calcium injected into the sarcoplasm by a single action potential in a rested muscle, and $\chi(t)$ is a pulse train representing a sequence of action potentials. The term $(1 - [Ca]/[Ca]^*)$ is an attenuation factor introduced ad-hoc to account for the fact that the size of the calcium pulse injected by successive action potentials decreases with repeated stimulation, presumably because the concentration gradient driving calcium release diminishes. The parameter $[Ca]^*$ has a value of the order of magnitude of the mean concentration of calcium in the muscle. (This calcium-activation model tries to avoid detailed representations of both the action-potential dynamics and the calcium transport dynamics, as these details do not appear to be of primary importance for contraction dynamics). Note that the contraction dynamics are coupled to those of calcium activation in the tight-coupling version of the model through the presence of Q_0 in Eq. 1.24; no such coupling exists in the loose-coupling version, Eq. 1.23. Although both loose- and tight-coupling theories have been developed, both experimental evidence and model simulations (Zahalak and Ma, 1990) suggest that the tight-coupling model is a better representation of reality, and only this model will be discussed further.

If a passive elastic tendon is assumed to be connected in series with a cylinder of contractile tissue then one more state variable and one more rate equation must be added to the model

(Zahalak, 1986); this variable is L, the muscle length. If this is done, and all variables except time are rendered dimensionless by appropriate scaling parameters one obtains the following normalized equations of the *DM* model (Zahalak and Ma, 1990)

$$\dot{\Lambda} = \kappa(Q_1)\,\dot{Q}_1 - \gamma\,u(t)$$

$$\ddot{C} = \rho\,(1 - c/c^*)\,\chi(t) - \tau_0^{-1}\left[\frac{c}{c + k_m}\right] \qquad (1.25)$$

$$\dot{Q}_\lambda = \alpha\,r\,\beta_\lambda - r\,\phi_{1\lambda}(Q_\lambda) - \phi_{2\lambda}(Q_\lambda) - \lambda\,u(t)\,Q_{\lambda-1}\,,$$

where $\lambda = 0,1,2$, with

$$C = c + 2\,b\,Q_0 + r\,(2 + \mu/c)\,(1 - b\,Q_0) \qquad (1.26)$$

and

$$r(c) = c^2/(c^2 + \mu\,c + \mu^2) \qquad (1.27)$$

These are a set of five coupled nonlinear first-order ordinary differential equations, derived directly from cross-bridge theory, on five state variables of a muscle: Λ, the dimensionless muscle length; c, the dimensionless sarcoplasmic free calcium concentration; and Q_0, Q_1, and Q_2, the dimensionless stiffness, force, and elastic energy. It is not possible within the confines of this chapter to give a full discussion of the various parameters entering these equations, and the reader must be directed to the references. Once the normalized equations (1.25) have been solved the corresponding physical quantities can be recovered as:

$$[Ca] = (m/N)c \qquad L = L_0\Lambda$$

$$K_c = (\Gamma/\gamma L_0)\,Q_0 \qquad (1.28)$$

$$P = \Gamma\,Q_1 \qquad U_c = (\gamma\,\Gamma\,L_0/2)\,Q_2$$

where N is Avogadro's Number, L_0 is the length of the muscle in a standard reference state, and γ and Γ are scaling parameters involving muscle geometry and cross-bridge stiffness (Ma and Zahalak, 1990).

Numerous simulations carried out with this *DM* model, some of which have been published in the cited references, have shown that it reproduces many of the complex experimental behaviors of muscle rather well, including some that are impos-

sible with the traditional Hill model. In Figures 1.8 through 1.11 are shown a few comparisons of *DM* model predictions with the corresponding experiments. Figure 1.8 shows steady-state force–velocity relations in isovelocity shortening and stretch at three constant values of activation. The inset shows experimental data (Joyce, Rack, and Westbury, 1969) for the cat soleus muscle. The model correctly predicts the non-uniqueness of the relation between force and velocity at submaximal activation; this is associated with the "yielding" phenomenon wherein there is a sharp decrease in the force when a muscle is stretched at constant velocity (Joyce, Rack, and Westbury, 1969, Figure 3).

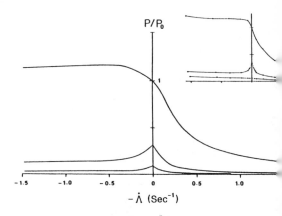

Figure 1.8: Steady-state force–velocity relations for a *DM* model subjected to constant velocity stretch or shortening, at three constant activation levels ($r = 1.0$, 0.2, 0.05). The inset shows corresponding experimental data (of Joyce et al., 1969) for cat soleus at 37° (From Ma, 1988).

Figure 1.9 illustrates an interesting experiment by A.V. Hill (Hill, 1970). A frog sartorius muscle was held isometric and subjected to a single supramaximal stimulus pulse. At various times after the pulse a fast exponential step stretch was imposed on the muscle. The inset of Figure 1.9 shows how the measured muscle force varied in this experiment. The interesting observation is that if the stretch was imposed soon after the pulse then the force remained above the isometric twitch force, but for later stretches the force dropped below the isometric twitch force after an initial rise. Figure 1.9 shows that the *DM* model response shares these characteristics.

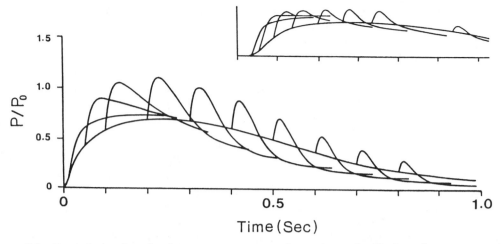

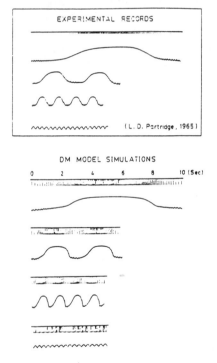

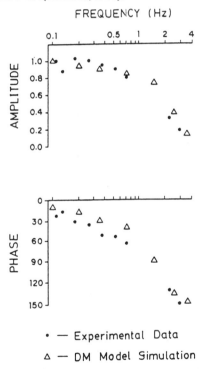

Figure 1.9: Force-time trajectories for a *DM* model subjected to a fast exponential stretch at various times following the application of a single stimulating pulse to an isometric muscle. The inset shows corresponding experimental data (of A. V. Hill, 1970) for frog sartorius at 0°C. (From Ma, 1988).

Figure 1.10: *DM* model simulation of the force generated by an isometric muscle which is subjected to sinusoidally modulated trains of stimulating pulses. The main figure shows pulse trains at four different modulation frequencies together with the resulting force traces. The box shows corresponding experimental data (of Partridge, 1965) for cat triceps surae at 37°C. (From Zahalak and Ma, 1990; reprinted with permission of ASME.)

Figure 1.11: Frequency-response characteristics of the *DM* model for the conditions of Figure 1.10. Shown versus frequency are the normalized amplitude ratio of force to the magnitude of the stimulation frequency modulation (upper graph), and the phase lead of stimulation frequency modulation on force (lower graph). Filled circles represent experimental data, and open triangles represent *DM* model simulations. (From Zahalak and Ma, 1990; reprinted with permission of ASME.)

Finally, Figures 1.10 and 1.11 exhibit some frequency-response data of the *DM* model, where sinusoidally modulated pulse trains are applied to an isometric muscle. The simulations are compared to data of Partridge (Partridge, 1965) for cat triceps surae. It can be seen that both the waveforms and the frequency response characteristics show reasonable agreement.

The preceding summary of the *DM* model has been confined to activation and contraction dynamics. But one of the most attractive features of Huxley-type models is that they inherently contain much of the biochemical energetics of contraction. Recently (Ma and Zahalak, 1990) equations have been developed describing the rate of chemical energy release, $\dot{E}$, and the rate of heat production, $\dot{H}$, for the *DM* model of *Eq. 1.25*. These energy equations follow naturally from the cross-bridge model under the *DM* approximation and take the dimensionless form

$$\left(\frac{\dot{E}}{P_0\dot{L}_m}\right) = \left(Q_1^{(0)}\dot{\Lambda}_m\right)^{-1}\left[\hat{\nu}_0\tau_0^{-1}\left(c/(c+k_m)\right)+\hat{\nu}_1\Psi(\Lambda,c,Q_\lambda)\right.$$

$$+\hat{\nu}_2\varsigma(c)\left\{\tau_0^{-1}\left(c/(c+k_m)\right)-\rho(1-c/c^*)\chi(t)\right\}$$

$$+\hat{\nu}_3\left\{\rho(1-c/c^*)\chi(t)-\tau_0^{-1}\left(c/(c+k_m)\right)-\dot{c}\right\}$$

$$\left.+\hat{\nu}_4\dot{Q}_0-\gamma\dot{Q}_2/2\right] \qquad (1.29)$$

and

$$\left(\frac{\dot{H}}{P_0\dot{L}_m}\right) = \left(\frac{\dot{E}}{P_0\dot{L}_m}\right)+\left(Q_1^{(0)}\dot{\Lambda}_m\right)^{-1}\left[Q_1\dot{\Lambda}-\kappa Q_1\dot{Q}_1\right]$$

where

$$\Psi = \int_{-\infty}^{\infty}\left[\{g(\xi)+r(c)g'(\xi)\}(Q_0/\sqrt{2\pi}q)\exp\left\{-(\xi-p)^2/2q^2\right\}\right.$$

$$\left.-\alpha(\Lambda)r(c)g'(\xi)\right]d\xi$$

$$(1.30)$$

Again, space limitations preclude a full discussion of *Eqs. 1.29*, but we note that the terms involving the parameters $\hat{\nu}_0$ through $\hat{\nu}_4$ (these parameters can be estimated from published biochemical and thermodynamic experiments) are associated with the following contributions to the energy: *ATP* splitting in *SR* calcium pumping, *ATP*

splitting in cross-bridge detachment, flow of calcium ions across the electrochemical potential gradient at the *SR* membrane (this term appears to be negligible), calcium binding to troponin, and cross-bridge binding to actin at zero strain; the term $\gamma\ddot{Q}_2/2$ represents the rate of change of linked cross-bridge strain energy. The last two terms on the right side of the second of *Eqs. 1.29* correspond, respectively, to the external work done by the muscle and the elastic energy stored in the passive series tendon. It must be emphasized, however, that these equations deal with short-term energy transfers, and do not account for slow oxidative recovery processes that recharge depleted phosphocreatine stores; a consistent model for oxidative recovery remains to be developed.

1.6 Conclusion

At this point I will attempt to summarize the advantages and disadvantages of the three approaches to muscle modeling that have been considered, both from a general point of view and, more specifically, with regard to their utility as representations of the muscle actuators in studies of multiple muscle systems.

Advantages of Biophysical Cross-Bridge Models

The primary virtue of these models is that they are concise mathematical descriptions of the way muscle is believed to actually work by the great majority of biologists. Thus users of such models have immediate contact with and access to the large body of current experimental work on the important relations between mechanics, energetics, and chemical kinetics in muscle. It is, after all, in terms of these detailed mechanisms that the normal and abnormal function of muscle as a prime mover must be ultimately understood. These models link the mechanics, chemistry, and structure of muscle into an integrated whole. All the variables appearing in these theories posess an independent physical existence apart from the model itself. At present these cross-bridge models can be ignored only at the risk of biological irrelevance.

Disadvantages of Biophysical Cross-Bridge Models

From the viewpoint of macroscopic neuro-musculo-skeletal mechanics the main drawback of the cross-bridge models is their extreme complexity. Even a two-state model, such as that discussed in Section 1.3, leads to a nonlinear partial integro-differential equation if it is coupled to a visco-elastic-inertial external load. If several muscles act about a joint then we have a kinematically coupled system of such equations, one for each muscle. If more than two cross-bridge states are deemed necessary to represent a muscle, then an additional Huxley-type rate equation must be added to the system for each cross-bridge state above two. Clearly the computational burden can become extremely heavy. But the interpretive burden can also become onerous, and physical insights about movement may be lost in a morass of bond-distribution functions that, at present, no one knows how to measure directly. (Indeed, a desire to avoid computing bond-distribution functions is one of the main motivations for the *DM* model).

There is at this time no consensus among biophysicists about the fine details of the cross-bridge cycle. Thus it has not yet been conclusively decided whether or not cross-bridges rotate in order to generate force (Cooke, 1986). The exact number of distinct states in the cycle has not been decided, but this number is probably a function of the amount of experimental detail that one wishes to reproduce with the model. Further, the biophysical modeling efforts have been focussed on simple experimental situations — mainly isometric or isotonic shortening conditions. Time varying activation, length-dependent activation, variable external loading, and electrical stimulation have not yet been addressed seriously in the published cross-bridge models, but all these phenomena are of immediate importance in macroscopic muscle modeling. One may expect, however, that these problems will be dealt with in due course.

Advantages of the Hill-Type Models

The greatest virtues of these models are simplicity, familiarity, and a direct connection with macroscopic muscle experiments. The traditional Hill model represents the contractile dynamics of muscle with one simple nonlinear first-order ordinary differential equation; inclusion of electrical stimulation/calcium activation usually calls for one more such equation. The model has a long history going back fifty years, and became firmly established during the decades when A. V. Hill dominated muscle physiology, so that at present it is almost universally accepted by engineers (and even some biologists) as an appropriate mathematical representation of muscle mechanics. The parameters of the model have simple relations to standard experiments, as explained in Section 1.4, and lend themselves to easy identification - a great practical advantage. Over the years a large body of experimental data based specifically on this model has accumulated; thus values of P_0, a, and b are known for many muscles [see also Chapter 5 (Winters)]. The Hill model is certainly sufficiently simple to occasion no undue difficulties even if large numbers of such models are used in studies of multiple muscle systems [e.g. see Chapter 8 (Zajac and Winters)].

Disadvantages of the Hill-Type Models

The most obvious deficiency of the Hill–type models is that they are basically viscoelastic analogies that have little connection with the underlying physiological mechanisms of muscle contraction. Experiments have shown that the fundamental assumption of the Hill model — namely, that the instantaneous muscle force and the *CE* velocity are uniquely related at a given muscle length and level of activation — is not generally true (Katz, 1939; Joyce and Rack, 1969; van Ingen Schenau et al., 1988). Further, it should be recognized that the "series elastic element" and the "active state" are purely conceptual constructs that have no independent physical existence: they cannot be defined and measured in general without invoking the Hill model. Thus, for example, a force change per unit length change can be measured in a quick stretch or release of an initially isometric muscle, and is found to be an increasing function of muscle force. This function of muscle force is then assigned as the stiffness of the *SE*, and one *assumes* in the Hill model that it will be equal to the measured force change per unit length change in a rapid perturbation of the muscle about *any* operating condition. According to the cross-bridge theory this view is fallacious, because a part of the stiffness in a quick stretch or release depends on the instantaneous number of linked cross-bridges, which is not determined uniquely by the muscle force but rather by the entire preceding history of muscle motion and activation. Indeed, experiments in which the stiffness of shortening muscle is measured by high-frequency-vibration (Julian and Sollins, 1975) indicate that a muscle still has considerable stiffness (about one third of that in an isometric tetanus) when it is shortening so fast that it produces no external force; this is consistent with cross-bridge theory, but seems at odds with the Hill model which would have this stiffness equal to the very low value of unactivated muscle.

It is difficult to see what, if any, physical variable in the cross-bridge theory could correspond quantitatively to the Hill "active state," although several variables bear a qualitative resemblance, including sarcoplasmic free calcium, the activation factor and the fraction of troponin molecules saturated with calcium. Further, there is an insidious temptation for engineers to take the Hill model literally and use it in situations when the viscoelastic analogy does not apply. For example if, as is usual, the CE is conceived of as a parallel combination of force generator producing the "active state" and a nonlinear viscous damper [see Figure 1.4(c)] then one might attempt to calculate the heat production (in excess of isometric) during shortening as the "energy dissipated" in the damper. On this basis one would conclude that the heat production rate during unloaded shortening of a tetanized muscle is $P_0 V_m$; the measured value in frog sartorius at 0°C is about $0.2 P_0 V_m$ — a 400% discrepancy. Indeed one of the significant deficiencies of the Hill-type models is that they give no information about biochemical energetics; the latter are usually appended ad-hoc by postulating a "heat of shortening" or "heat of lengthening," which is measured under isotonic steady-state conditions, and assumed to apply under all conditions.

Advantages of DM Models

DM models, as strictly mathematical approximations to Huxley-type cross-bridge models, retain much of the presumed physical veracity of the latter while remaining mathematically tractable. But the DM theory focusses directly on the quantities of interest in whole muscle behavior — force, stiffness, etc. — without first solving for unmeasured bond distribution functions. With appropriate chocies for the parameters DM models offer quite reasonable predictions of most of the measured behaviors of muscle, both in shortening and stretch, both of mechanics and energetics. The state variables of the DM model are all measurable physical quantities in principle and do not include internal variables, like "active state" which depend on the model for their definition. The moments have simultaneously a macroscopic and microscopic significance and thus provide a bridge to the understanding of muscle function of both levels. Mathematically, the equations of the DM model, Eqs. 1.25, can be written in normal state-variable form and their numerical integration by standard techniques is easy even when the muscles are coupled to arbitrary external impedances (a formidable computational problem for cross-bridge models). The simplification achieved over cross-bridge models in the representation of contraction dynamics permits the addition of rational models for both calcium activation and energetics without making the enhanced model intractable. Further, as these models include, at least in rudimentary form, the basic facts of actin–myosin binding and calcium activation, there is some basis for accepting their predictions beyond the narrow range of experimental experience.

Disadvantages of DM Models

Just as the DM models share in the virtues of the biophysical cross-bridge models, so also they share some of the shortcomings of the latter. Even when based on a minimal two-state cross-bridge model the resulting DM model is significantly more complex than the Hill representation. From the practical viewpoint of application to studies of multiple muscle systems a serious deficiency is the large number of parameters and the fact that they enter the model in a complicated nonlinear manner. Parameter identification is a tedious iterative process that requires a large body of structural, chemical, mechanical, and thermodynamic data for a single muscle; indeed, reasonably firm parameter estimates have been obtained for only a few muscles (Ma and Zahalak, 1990) which are well documented in the experimental literature. Further, the DM models which have been worked out explicitly so far have been based on the simplest assumptions about muscle architecture: uniform sarcomeres, fibers parallel to the muscle axis running the whole length of the contractile tissue, and a purely elastic series tendon. These restrictions can be removed, but of course at the cost of complicating the model structure.

In summary, it is probably safe to assume that no tractable model will predict with quantitative accuracy *all* the known features of muscle behavior. Imperfect models must be used and it is therefore important to understand and make allowances for their imperfections. In particular, specific predictions derived from the employment of such models in studies of multiple muscle systems must be taken with the appropriate number of grains of salt; the further the model from biological reality and the further the simulation from direct experimental experience, the less confidence one can place in conclusions reached. Due to their simplicity and familiarity Hill-type models are likely to continue to be the workhorses of motor-control studies into the foreseeable future, in spite of their manifest deficiencies. Due to their formidable complexity and incomplete development biophysical crossbridge models are very unlikely to find employment directly as mathematical representations of muscle actuators in studies of animal movement (although their prospects for direct applications to cardiac muscle, where the history of activation and contraction is much more restricted and stereotyped than in limb motion, seem somewhat better). The role of *DM* models in multiple-muscle studies is unclear at present, and will probably depend on the extent to which they can be adapted to human muscles, and the associated parameters identified. It would seem possible, however, to apply these models to restricted systems involving only a few muscles - with the aim of checking analogous calculations based on Hill-type models, and also of interpreting to some (admittedly imperfect) extent the actions of the participating muscles in terms of underlying physiological mechanisms.

1.7 Future Directions

As a coda to this chapter I will offer some personal speculations on future developments which I think ought to occur in muscle modeling and its applications. This assessment will focus on Hill-type and *DM* models because, as noted above, cross-bridge models are unlikely to see direct application to multiple muscle systems. A rather immediate problem which should be resolved is the importance of sarcomere inhomogeneity to the overall function of muscle in vivo. Sarcomere inhomogeneity, discussed in Chapter 3 (Morgan), is a challenge to all the models presented in this chapter, as they do not take this factor into account. Although most of the known behaviors of muscle seem to be capable of explanation on the basis of a uniform sarcomere distribution, certain features appear to be best explained as arising from a nonuniform distribution. It is important to define clearly when sarcomere inhomogeneity should be incorporated into muscle models and how this should be done and, further, to estimate the magnitudes of the errors incurred relative to other errors inherent in the models if this nonuniformity is ignored.

It appears to me that after fifty years Hill-type models, especially in their most recent incarnations [see Chapter 5 (Winters)] have reached their highest appropriate level of development, and achieved the best balance of simplicity and fidelity of which they are capable; further elaboration would lead to a complexity comparable to that of the *DM* models, without the benefit of the biophysical foundation of the latter. Perhaps one exception is that the Hill models could be adapted to more complex fiber architectures, which have become the subject of recent interest (Richmond et al., 1985), and this is certainly a possible avenue of development for the *DM* models also.

There are several other developments possible for the *DM* models which may be worth pursuing on the grounds these models are actually good approximations to the cross-bridge models. One improvement could be to incorporate the "doublet effect" [Burke et al., 1976; see also Chapter 7 (Hannaford and Winters)] produced in some muscles by a single extra pulse following the onset of stimulation; this phenomenon may have functional significance, and although it can certainly be simulated by *DM* models in an ad-hoc manner, to the best of my knowledge there is no model that incorporates this effect rationally. Another obvious improvement would be the inclusion of long-term oxidative recovery in the energy equations (Woledge et al., 1985), an effect that is so far absent from the *DM* models. It may also be worthwhile to explore the application of the *DM* approximation to cross-bridge models with more than two states, or to use approximating functions different than the Gaussian. But probably the most important direction for future improvement would be to combine the existing DM fiber model with volume-conductor theory to produce an integrated model, founded on basic principles and mechanisms, which rationally links muscle mechanics to the ubiquitous electromyogram. This is certainly a difficult, but probably feasible, project.

In order to ensure that future muscle modeling efforts do not stray too far from reality, there will be a need for many more experimental data than are currently available on the behavior of various muscles under histories of activation, force, and motion that are more complex than in the standard isometric and isotonic experiments. These considerations, however, fall outside the scope of this chapter devoted to muscle modeling.

References

Aidley, D.J. (1971) *The Physiology of Excitable Cells*, Cambridge Univ. Press, London, UK.

Bornhorst, W.J. and Minadri, J.E. (1970) A phenomenological theory of muscular contraction: Parts I and II. *Biophys. J.* **10**:137-171.

Bouckaert, J.P., Capellen, L., and de Blende, J. (1930) The visco-elastic properties of frog's muscles. *J. Physiol.* **69**:473-492.

Burke, R.E., Rudomin, P., and Zajac, F.E. (1976) The effect of activation history on tension production by individual muscle units. *Brain Res.* **109**:515-529.

Cooke, R. (1986) The mechanism of muscle contraction. *CRC Crit. Rev. Biochem.* **21**: 53-118.

Eisenberg, E. (1986) How ATP hydrolysis drives muscle contraction. *Lec. Math. Life Sc.* (Amer. Math. Soc., Providence, RI) **16**:19-55.

Eisenberg, E. and Greene, L.E. (1980) The relation of muscle biochemistry to muscle physiology. *Ann. Rev. Physiol* **42**:293-309.

Eisenberg, E., Hill, T.L., and Chen, Y.D. (1980) Crossbridge model of muscle contraction: quantitative analysis. *Biophys. J.* **29**:195-227.

Gasser, H.S. and Hill, A.V. (1924) The dynamics of muscular contraction. *Proc. Roy. Soc.* B **96**:398-437.

Fenn, W.O. and Marsh, B.S. (1935) Muscular force at different speeds of shortening. *J. Physiol.* **85**:277-297.

Gordon, A.M., Huxley, A.F., and Julian, F.J. (1966) The variation in isometric tension with sarcomere length in vertebrate muscle fibers. *J. Physiol.* **184**:170-192.

Harry, J.D., Ward, A.W., Heglund, N.C., Morgan, D.L., and McMahon, T.A. (1990) Crossbridge cycling theories cannot explain high-speed lengthening behavior in frog muscle. *Biophys. J.* **57**:201-208.

Hill, A.V. (1938) The heat of shortening and the dynamic constants of muscle. *Proc. Roy. Soc., B* **126**:136-195.

Hill, A.V. (1964) The effect of load on the heat of shortening of muscle. *Proc. Roy. Soc., B* **159**:297-318.

Hill, A.V. (1970) *First and Last Experiments in Muscle Mechanics*, Cambridge University Press, London, 120.

Hill, T.L. (1977) *Free Energy Transduction in Biology*, Academic Press, New York, NY.

Hill, T.L., Eisenberg, E., Chen, Y., and Podolsky, R.J. (1975) Some self-consistent two-state sliding filament models of muscle contraction. *Biophys. J.* **15**:335-372.

Huxley, A.F. (1957) Muscle structure and theories of contraction. *Prog. Biophys. and Biophys. Chem.* **7**:257-318.

Huxley, A.F. (1980) *Reflections on Muscle*, Princeton University Press, Princeton, NJ.

Huxley, A.F. and Simmons, R.M. (1971) Proposed mechanism of force generation in striated muscle. *Nature* **233**:533-538.

Inesi, G. (1985) Mechanism of calcium transport. *Ann. Rev. Physiol.* **47**:573-601.

Iwazumi, T. (1978) A new field theory of muscle contraction, In: *Cross-Bridge Mechanisms in Muscle Contraction* (H. Sugi, and G.H. Pollack, Eds.), Univ. Tokyo Press, 611-632.

Joyce, G.C. and Rack, P.M.H. (1969) Isotonic lengthening and shortening movements of cat soleus muscle. *J. Physiol.* **204**:475-491.

Joyce, G.C., Rack, P.M.H., and Westbury, D.R. (1969) The mechanical properties of cat soleus muscle during controlled lengthening and shortening movements. *J. Physiol.* **204**:461-467.

Julian, F.J. and Sollins, M.R. (1975) Variation of muscle stiffness with force at increasing speeds of shortening. *J. Gen. Physiol.* **66**:287-302.

Katz, B. (1939) The relation between force and speed in muscular contraction. *J. Physiol.* **96**: 45-64.

Lacker, H.M. and Peskin, C.S. (1986) A mathematical method for unique determination of cross-bridge properties from steady-state mechanical and energetic experiments on macroscopic muscle. *Lec. Math. Life Sc.* (Amer. Math. Soc., Providence, RI), **16**:121-153.

Levin, A. and Wyman, J. (1927) The viscous elastic properties of muscle. *Proc. Roy. Soc., B* **101**:218-243.

Ma, S. and Zahalak, G.I. (1987) A simple self-consistent distribution-moment model for muscle: chemical energy and heat rates. *Math. Biosc* **84**:211-230.

Ma, S. (1988) Activation dynamics for the distribution-moment model of muscle, D. Sc. Dissert. Dep. Mech. Eng., Washington University, St. Louis, MO., 73-115.

Ma, S. and Zahalak, G.I. (1990) A distribution-moment model of energetics in skeletal muscle. *J. Biomech.*, (in press).

McMahon, T.A. (1984) *Muscles, Reflexes, and Locomotion*, Princeton Univ. Press, Princenton, NJ.

Partridge, L.D. (1965) Modification of neural output signals by muscles: a frequency response study. *J. Appl. Physiol.* **20**:150-156.

Pate, E. and Cooke, R. (1986) A model for the interaction of muscle cross-bridges with ligands which compete with ATP. *J. Theor. Biol.* **118**:215-230.

Paul, R., Elzinga, G., and Yamada, K., Eds. (1989) *Muscle Energetics*, Alan R. Liss, New York, NY.

Pollack, G. and Sugi, H. Eds. (1984) *Contractile Mechanisms in Muscle*, Plenum Press, New York, NY.

Propp, M.B. (1981) A model of muscle contraction based upon component studies. *Lec. Math. Life. Sc.* (Amer. Math. Soc., Providence, RI), **16**:61-119.

Ramsey, R.W. and Street, S.F. (1940) The isometric length-tension diagram of isolated muscle fibers of the frog. *J. Cell. Comp. Physiol.* **15**:11-34.

Richmond, F.J.R., MacGillis, D.R.R., and Scott, D.A. (1985) Muscle fiber compartmentalization in cat splenius muscles. *J. Neurophys.* **53**:868-885.

Ritchie, J.M. and Wilkie, D.R. (1958) The dynamics of muscular contraction. *J. Physiol.* **143**:104-113.

Ruegg, J.C. (1986) *Calcium in Muscle Activation*, Springer-Verlag, New York, NY.

Schoenberg, M., Brenner, B., Chalovich, J.M., Greene, L.E. and Eisenberg, E. (1984) Cross-bridge attachment in relaxed muscle. in *Contractile Mechanisms in Muscle* (G. Pollack and H. Sugi, Eds.), Plenum Press, New York, NY, 269-284.

Squire, J.M. (1986) *Muscle: Design, Diversity, and Disease*, Benjamin/Cummings, Menlo Park, California.

Squire, J.M. (1981) *The Structural Basis of Muscular Contraction*, Plenum, New York, NY.

Tirosh, R., Liron, N., and Oplatka, A. (1978) A hydrodynamic mechanism for muscular contraction. In: *Cross-Bridge Mechanisms in Muscle Contraction* (H. Sugi and G.H. Pollack, Eds.), Univ. Tokyo Press, 593-609.

Tozeren, A. and Schoenberg, M. (1986) The effect of cross-bridge clustering and head-head competition on the mechanical response of skeletal muscle fibers under equilibrium conditions. *Biophys. J.* **50**:873-884.

van Ingen Schenau, G.J., Bobbert, M.F., Ettema, G.J., de Graaf, J.B., and Huijing, P.A. (1988) A simulation of rat EDL force output based on intrinsic muscle properties. *J. Biomech.* **21**:815-824.

Woledge, R.C., Curtin, N.A., and Homsher, E. (1985) *Energetic Aspects of Muscle Contraction*, Academic Press, New York.

Wood, J.E. and Mann, R.W. (1981) A sliding-filament cross-bridge ensenble model of muscle contraction for mechanical transients. *Math.Biosc.* **57**:211-263.

Zahalak, G.I. (1981) A distribution-moment approximation for kinetic theories of muscular contraction. *Math. Biosc.* **55**:89-114.

Zahalak, G.I. (1986) A comparison of the mechanical behavior of the cat soleus muscle with a distribution-moment model. *J. Biomech. Eng.* **108**:131-140.

Zahalak, G.I. and Ma, S.-P. (1990) Muscle activation and contraction: constitutive relations based directly on cross-bridge kinetics. *J. Biomech. Eng.* **112**:52-62.

CHAPTER 2

The Charge-Transfer Model of Myofilamentary Interaction: Prediction of Force Enhancement and Related Myodynamic Phenomena

2.1 Introduction

The exact mechanism of myofilamentary energy conversion and force production in skeletal muscle remains shrouded in mystery. Although a large amount of information is available on the structure and function of the tension-generating subunits of the actin and myosin filaments, the intricate processes of inter-molecular force production are still not understood. On the contrary, established theories and models of muscular contraction are now being seriously challenged on the grounds of new experimental evidence (Pollack, 1983).

This applies in particular to the model of A.F. Huxley (1957) and its numerous versions (Hill et al., 1975; Hill, 1977; Huxley and Simmons, 1971; Zahalak, 1981, etc.). Recently, some of the basic assumptions underlying these models have been directly contradicted by experimental findings. We briefly mention two of these assumptions and the corresponding experimental evidence:

1) The number of energy-providing *ATP* molecules that are split during contraction is directly related to the number of completed cross-bridge cycles.

2) The major part of the sarcomere compliance resides in the cross-bridges (Huxley and Simmons, 1971).

Studies carried out by Gilbert et al. (1971) as well as many subsequent investigations have shown that during contraction the energy output

(heat plus work) of a muscle is much higher than the energy input derived from the splitting of high-energy phosphate, thus invalidating assumption *1*. As regards assumption *2*, Tawada and Kimura (1984) performed cross-linking experiments on rabbit psoas fibers in rigor by fixing the *S2*-moiety of the myosin cross-bridge onto the surface of the myosin backbone. The increase in rigor stiffness observed was 20 to 30%, regardless of the amount of filamentary overlap, and was therefore not due to the cross-linking of the *S2*-portion. From this result the authors conclude that the fixation onto the thick filament of the *S2*-portion did not make the cross-bridges stiffer, i.e., the *S2*-moiety of the cross-bridges is not compliant as had been assumed in the Huxley–Simmons model. However, since an approximately linear dependence of the rigor stiffness on the sarcomere length had been found (Figure 3 of Tawada and Kimura, 1984), Tawada and Kimura assumed that the compliance observed must necessarily reside in the cross-bridges and hence in the myosin head (subfragment *S1*). This conclusion may not be correct. As will be shown in another section of this chapter, an exponential stress-strain relation of a sarcomere series elasticity which is *not* located within the cross-bridges implies the observed linear dependence of the stiffness on the degree of filamentary overlap via the declining tension resulting from the decreasing number of attached bridges.

Therefore, only a small amount of the total sarcomere compliance may, in fact, reside in the

Multiple Muscle Systems: Biomechanics and Movement Organization
J.M. Winters and S.L-Y. Woo (eds.), © 1990 Springer-Verlag, New York

cross-bridges. A similar conclusion has been reached by Sugi (1979). We shall elaborate on this point later.

The two contradictions mentioned between experimental results and the Huxley model are not the only ones. Reports have appeared in the literature on other phenomena that do not support the theory (for details see Pollack, 1983).

The criticism levelled against the Huxley model does, of course, apply to most other contraction theories as well. The model of A.F. Huxley has been selected for the discussion because of its popularity and widespread use. It should be mentioned that Huxley (1988, p. 7) himself has recently emphasized that his theory is "a skeleton theory" and "incomplete." Furthermore, he points out a number of shortcomings of his theory (Huxley, 1988, p. 7) and that it "contains no specific statements about the nature of the actin–myosin bond, the nature or location of the elastic element, or the mechanism for breaking the bond at the right time."

Similar reservations about some of the most widely held features of the theory were expressed by Pollack in 1983. He suggested to begin afresh and listed a number of items he would consider worthy of incorporation into any new theory (Pollack, 1983, p. 1104). From this list and the comments made by A.F. Huxley, it becomes clear that today any thrust directed towards the creation of a successful new theory of muscular contraction must begin at the level of the molecular substructures that constitute the contractile proteins.

The intended purposes of this chapter are: *i)* the presentation of a model of myofilamentary force generation based on intra-molecular charge transfer; *ii)* the description of the model responses in a variety of contractive modes; and *iii)* an outline of the development of a global myocybernetic model of skeletal muscle based on the charge-transfer model.

No specific section will be included on the assessment of pertinent literature as such an assessment of existing theories and models of muscular contraction can be found in Chapter 1 (Zahalak) of this volume. The references relevant to the development of the present model will be quoted as the need arises.

2.2 A Hypothesis and Model of Muscular Contraction by Intra-Molecular Charge Transfer

This section is devoted to the description of the structural and functional details of the proposed model. Except for a few assumptions on the intramolecular field distribution, only features of molecular structures and functional interactions that have been verified experimentally will be utilized in the formulation of the contraction hypothesis and the construction of the model.

2.2.1 Substructure of the Actomyosin Complex

Today it is generally accepted that the crosslinking of the actin and myosin filaments by so-called cross-bridges is responsible for the production of contractive force in skeletal muscle fibers. Experimental evidence in support of this assumption is overwhelming indeed: If an activated muscle fiber is stretched rapidly, the transient tension increase is many times larger than that in the resting fiber (e.g., see Hill, 1968). Concurring results were obtained by oscillation and quick release experiments (Kawai and Brandt, 1980; Ford et al., 1977). Further though indirect support for the notion of tension generation by cross links comes from time-resolved X-ray diffraction studies on frog sartorius muscle (H.E. Huxley, 1984). It therefore appears to be fairly well established that the active force produced in a muscle fiber upon stimulation originates from the interaction, via cross links, of the actin and myosin filaments. Since it is also known that the energy-providing adenosine triphosphate (ATP) binds to a specific site on the subfragment 1 (S1) of the myosin molecule, we are led to the conclusion that the actomyosin complex (S1 bound to two actin monomers) is the most likely location of energy transduction and force generation. For this reason we shall concentrate on the fine structure of these molecular entities.

The *thick (myosin) filaments* are composed of myosin molecules, each of which consists of *two* globular heads (two S1) connected to a long tail. The latter consists of heavy meromyosin (HMM) and light meromyosin (LMM) polypeptide chains which aggregate to form the thick filament, the latter having a diameter of approximately 20 nm. From the surface of the thick filament protrude the S1–moieties (globular heads) of the myosin molecules at regular, helically arranged axial intervals of about 42.9 nm, allowing for an actin-bound cross-bridge periodicity of 14.3 nm. The globular heads are approximately pear-shaped and have a dimension of roughly 18 nm (length) by 9 nm (width). Their connection to the tail is highly flexible. The distal part of the tail region (subfragment 2) has a second flexible link, about 40 nm distant from the *S1*-link. This permits a translocation of the globular heads away from the surface of the thick filaments towards binding sites at the actin filaments. The translocation distance is on average 13 nm, varying from about 8 nm at long sarcomere lengths to about 15 nm at short sarcomere lengths.

The *thin (actin) filaments* consist of monomers arranged in double-stranded helix form with an axial repeat of about 38 nm. Notice that this repeat is comparable with that of the myosin projections on the thick filaments. The diameter of the thin filament varies between 6.5 and 9.5 nm. Attached to the actin strands are the regulatory proteins tropomyosin and troponin. Tropomyosin is a long, rod-like molecule that binds to one strand of the actin helix. Hence there are two tropomyosin molecules on opposite sides of the helix. Troponin, on the other hand, is a globular protein that binds to tropomyosin (but not to actin) at regular axial intervals of about 38 nm (7 actin monomers to 1 troponin molecule). It consists of three components: *TN–T* that binds to tropomyosin, *TN–C* that binds the calcium ions, and *TN–I* which constitutes the inhibitory component. More detailed information on the fine structure of the contractile proteins can be found in Pepe (1983).

Most crucial to the development of the present theory of contraction and the associated muscle model is the substructure of the subfragment 1 (*S1*) of the myosin molecule. Yamamoto and Sekine (1979), Mornet et al. (1981a), and others have performed tryptic digestion experiments to decompose the *S1* into its various domains. As the result of this decomposition emerged a domain structure consisting of a moiety connected to the *HMM* tail (*S2*) and having a molecular weight of 23 kiloDaltons, a portion of molecular weight 50 kiloDaltons, and connected to the *23k*-moiety, and a fragment of weight 27 kiloDaltons linked to the *50k*-portion. The three domains will be labelled *S1–23k*, *S1–50k*, and *S1–27k* respectively.

The *S1–27k* domain contains the *NH2*-terminal

and, as has been revealed by photoaffinity-labelling studies performed by Szilagyi et al. (1979), most likely also the nucleotide (*ATP*) binding site. It does, however, not bind to actin. On the other hand, both the *S1–23k* and the *S1–50k* domains have been shown to bind to actin (Mornet et al., 1981b). Moreover, the *S1–23k* contains the two reactive *SH* groups *SH1* and *SH2*, which are probably involved in free electron donation (Ladik, 1982).

The binding of the myosin head (i.e, of the *S1–23k* and *S1–50k* domains) to two actin monomers on opposite strands of the actin helix has been confirmed by Wakabayashi et al. (1984) in *3–D* electron microscopic image reconstructions of the actotropomyosin complex.

Finally, it is well known that negative charges are situated on the surfaces of polypeptide chains (Kodama 1985, p. 514) and, in particular, on the actin filaments (Pollack 1983, p. 1103).

The above-mentioned details of the submolecular structures of the myosin subfragment 1 and the actin helix permit the construction of a fairly detailed morphological model of the actomyosin complex. This model is shown in Figure 2.1 below. It should be noted that our present knowledge of the probable shape and dimension of *S1*, the connective arrangement of its subfragments and the configurational appearance of the actomyosin complex does not allow for much freedom in the design of such a model.

2.2.2 Formulation of the Contraction Hypothesis

On the grounds of the morphological details depicted in Figure 2.1 and by utilizing further experimental findings relating to the contractile events taking place within and around the actomyosin complex, a hypothesis can be advanced of how energy conversion and force production may be effected.

There are several lines of evidence which strongly suggest that intra-molecular transfer of potential energy in the form of charge transfer could be the mechanism responsible for the development of contractive force. First, when certain negatively charged ions (Br^-, I^-, etc.) were used in muscle fibers to replace *ATP*, contractions could be induced (Bowen, 1957; Laki and Bowen, 1955; Mandelkern and Villarico, 1969). This indicates that negative charges could be involved in the contraction process. Secondly, as Kodama (1985, p. 540) points out, *ATP* is bound to myosin with an extraordinarily high binding constant, so that in this phase the major free-energy change occurs. Kodama refers to this phenomenon as "energy trapping." The subsequent hydrolysis of *ATP* into *ADP* + P_i is almost isoenergetic, so that the free-energy change for the hydrolysis of myosin-bound *ATP* is negligible. It therefore appears that upon hydrolysis the trapped energy is converted into some transit form which by repetition of the process could possibly lead to energy storage within the *S1–27k* subunit of *S1*. Indeed, the existence of such a mechanism has been proposed by Pollack (1983, p.1104) in order to provide an explanation for the so-called **unexplained energy**. He suggests that "potential energy may be stored in the contractile proteins before contraction."

A plausible hypothesis about energy storage mechanisms in hydrogen-bonded peptide groups such as α-*helix* proteins has recently been formulated by Lomdahl (1984). This author proposes that *amide-I*, a characteristic intra-molecular vibrational mode of all peptide groups, could serve as a "basket" for energy storage. Resonant energy transfer between identical peptide groups would constitute a dispersion mechanism which can be approximated by electrostatic dipole-dipole interaction. This mechanism, in turn, would be counteracted by the coupling between the *amide-I* vibration and other intra- and intermolecular excitations, and would correspond to a focusing of *amide-I* energy. The competition between dispersion and focusing of *amide-I* energy will lead in α-*helix* proteins (of which large amounts are contained in *S1*) to the formation of a soliton-like object travelling along the peptide chains without altering their shape (Lomdahl, 1984, p.144).

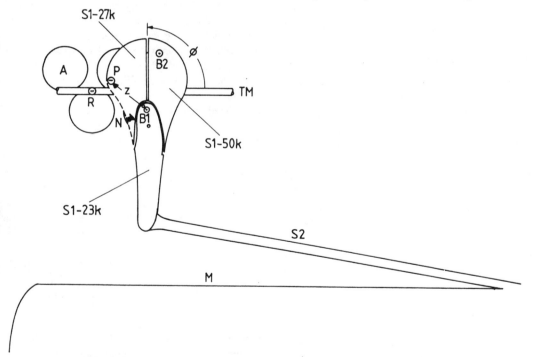

Figure 2.1: Morphological model of the actomyosin complex drawn to scale. The symbols have the following meaning: *M*: myosin filament; *S2*: rod-like part of the cross-bridge; *S1–23k*, *S1–50k*, *S1–27k*: the three interconnected moieties of the globular head *S1*; *A*: actin monomer; *TM*: tropomyosin molecule; *N*: nucleotide (*ATP*) binding site; *R*: negative reference charge situated on the surface of the tropomyosin molecule; *P*: principal electron charge; *B1*, *B2*: actin binding sites; *z*: distance of electron charge *P* from *B1*; ϕ: angle between *S1* major axis and the axis of the actin helix. The two small circles above and below *B1* indicate the location of the *SH1* and *SH2* groups respectively.

At this point it is of importance to emphasize the fact that *solitons may carry negative charges* (Bishop, 1984, p. 158). Davydov (1982) proposed that biological solitons may be triggered by the hydrolysis of ATP bonds and propagate along linear molecules (α-*helix* portions of proteins). Disturbances of electronegative charge fields or mechanical oscillations of peptide chains can also account for soliton propagation (Hameroff et al., 1984).

A fairly detailed model for the energy transduction process from *ATP* splitting into a protein has recently been suggested by Scott (1984). He proposes the introduction of a *proton soliton* as an intermediate state in the process of resonant energy transfer. His model involves the following steps (Scott 1984, p. 140).

$$ATP^{-4} + H_2O \rightarrow ADP^{-3} + HPO_4^{-2} + H^+$$

$$H^+ + OH^- \rightarrow H_2O^*$$

$$H_2O^* \rightarrow (1 \text{ or } 2) S_p + H_2O$$

$$S_p \rightarrow S_D + \text{phonons.}$$

In this scheme, the *ATP* hydrolysis first produces an extra hydrogen ion (H^+) which carries the free energy of 0.49 ev = 7.84 x 10^{-20} J derived from the *ATP* splitting. In the second reaction, H^+ combines with OH^- to generate an excited water molecule (denoted by an asterix) that carries two vibrational quanta (0.41 ev) of free energy in *H-O-H* binding. Next, the free energy of one or both of these *H-O-H* vibrational quanta is transferred to one or two proton solitons S_p. Finally, the free energy of a proton soliton (0.205 ev = 3.28 x 10^{-20} J) is transferred directly to a Davydov soliton S_D. If in the third reaction only one proton soliton is produced, the process has an efficiency of 42%, otherwise one of 84%.

We shall assume that the latter is the case, i.e., an amount of 6.56 x 10^{-20} J of free energy is available from the hydrolysis of one *ATP* molecule under physiological conditions.

Following now the arguments of Lomdahl (1984, p. 144), whose work is based on Davydov's theories, it appears likely that the energy released during *ATP* hydrolysis is stored in the form of an intra-molecular *amide-I* vibrational mode. As Scott (1984, p.135) points out, localized *amide-I*

bond energy acts to generate longitudinal sound waves along α-*helix* portions which, in turn, react to trap the bond energy and prevent its dispersion. In all likelihood, this process of intra-molecular storage of vibrational energy is repeatable, leading to an accumulation of considerable amounts of *ATP*-derived energy within the substructures of *S1*. This is the transit form of the molecular energy referred to earlier.

Kodama (1985, p. 492) has emphasized the importance of the functional connection between the nucleotide and the actin-binding sites on the myosin head. Binding at one site influences the other antagonistically. In the presence of *ATP*, for instance, the strong association between actin and myosin is weakend by 3 orders of magnitude. Conversely, actin weakens the interaction between *ATP* or its hydrolysis products and the nucleotide binding site.

Furthermore, Kodama (1985, p. 517) stresses the fact that there is strong interaction between the two reactive *SH* groups in the *S1–23k* and the nucleotide binding site on the *S1–27k*. This implies that there is close spatial disposition between these domains (Kodama 1985, p. 517), a fact which has been taken into account in constructing the model displayed in Figure 2.1. As already mentioned in Section 2.2.1, the two *SH* groups are probably donors of free electrons (negative charges).

Also of importance are the intra-molecular changes that occur in *S1* on binding to actin. The *S1–23k* to *S1–50k* joint is strengthened while the *S1–50k* to *S1–27k* connection is attenuated (Yamamoto and Sekine 1979). In other words, on binding to actin, the joint connecting the two domains of *S1* associating with actin becomes stiffer, while the *S1–27k* subunit becomes more loosely connected to the rest of *S1* (see Figure 2.1).

After having presented the experimental evidence and the biophysical notions relevant to the present *contraction hypothesis*, we may proceed to the formulation of the latter. The following events are postulated to take place.

1. Alternating action of the two globular myosin heads

Upon activation, one of the two globular heads linked to the tail (*S2*) of the myosin molecule attaches to actin, dissociates P_i, and executes the power stroke. Finally, *ADP* is dissociated from the head. Simultaneously, the other head is in a detached state and performs random motions while hydrolyzing *ATP*. The random oscillations occur within a cone of about 40° full angle (Cooke et al., 1984). Each *S1* contains an intra-molecular energy depot in which *amide-I* vibrational energy derived from the hydrolysis of several *ATP* molecules is stored. Activation occurs by influx of Ca^{2+} ions into the interfilamentary space. These calcium ions bind to troponin and presumably remove the steric block on actin by initiating the movement of the tropomyosin molecules into the helix groove. In addition, Ca^{2+} and the energy-releasing dissociation of P_i are assumed to trigger the transduction of stored energy from the intra-molecular depot to electrical potential energy after the binding of *S1* to actin has taken place. Such a mechanism concurs with the finding of Schoenberg et al. (1984, p. 278) that relaxation in muscle seems to be brought about by blocking an energy supplying step which is subsequent to attachment but prior to force generation.

2. The charge-transfer step

As mentioned above, the step subsequent to actin binding of the force-generating *S1* is the conversion of intra-molecularly stored energy to a form of potential energy which enables the actomyosin complex to perform mechanical work. There are many different modes in which such a transduction could occur on a molecular scale. However, as outlined at the beginning of this section, experimental evidence strongly supports the concept of an intra-molecular transfer of negative charges from the interior to the boundary of the *S1-27k* domain. This process is indicated in Figure 2.1. Upon initialization by actin binding and the subsequent action of Ca^{2+} and the energy-supplying dissociation of P_i, an amount of about 6.56×10^{-20} J of the stored energy is set free and assumed available in the form of electric potential energy. The corresponding field is centered at or near the region where the *S1-23k* actin binding site *B1*, the two *SH*-groups, and the nucleotide binding site on *S1-27k* are situated. That there is indeed a close spatial disposition between these sites on the different domains has been clearly demonstrated by Wells and Yount (1979).

The field is now assumed to transfer electron charges, donated by the *SH2* group, from a position near *B1* to the position *P* on the boundary of the *S1-27k* domain (see Figure 2.1). The transfer is effected over a distance of about 5 nm, possibly by means of a charge-carrying soliton. Work has to be done in this process against the repulsive forces created by the surface charge *R* that is situated laterally on the tropomyosin molecule as depicted in Figure 2.1. As will be shown in the next section, the amount of work done (neglecting small intra-molecular resistive losses) is approximately equal to 5.90×10^{-20} Nm, i.e., equal to the amount of energy set free in the form of eletric potential energy. Note that this amount is comparable to the energy derived from the hydrolysis of one *ATP* molecule.

3. Contraction in the isometric mode

If no length change occurs in the sarcomere, the contraction is called isometric. The true isometric state is attained after the contractile structures have extended the series elastic components of the sarcomere to an equilibrium position.

In this state, the average angle of attachment of *S1* relative to the axis of the actin filament is postulated to be 90° (see Figure 2.1). The transfer of one electron charge and thus force production commence simultaneously with the completion of the P_i-dissociation step. Subsequently, *ADP* is dissociated.

Due to intra-molecular losses (Scott, 1984, p. 134), the potential that transferred the principal charge *P* to its terminal position declines as a function of time. Consequently, the charge *P* retracts in a hopping motion (Bishop, 1984, p. 158) to its original position near *B1* which results in a concomitant decrease of the contractile force. The energy lost in this process is identified as the *maintenance heat*.

Because of the strong interaction between the field created by the potential near *B1* and the bond orbitals at the actin binding site *B1* (Hofacker, 1982, p. 235), the declining potential also results in a decrease of the bond strength. This process is supposed to trigger *ATP* binding at the nucleotide site and thus leads to the dissociation from actin of both the *B1* and *B2* binding sites. The detached *S1* goes into a state of Brownian motion while hydrolyzing *ATP*. The second *S1* of the same myosin molecule attaches and commences the cycle.

It should be remarked that according to the present hypothesis the myosin head does not change its attitude during the attached state in isometric contraction, a fact which has been confirmed experimentally by measurement of polarized fluorescence from fluorescent *ATP* bound to *S1* (Yanagida, 1984) and by electron paramagnetic resonance spectroscopy (Cooke et al., 1984).

4. Contraction in the shortening mode

If the sarcomere is not kept isometrically during tension development but allowed to shorten, the present hypothesis implies that rotation of the attached *S1* must occur about an axis through *B1*. Such cross-bridge rotations have indeed been observed in muscle (Borejdo et al., 1979). The angular movement is assumed to extend over a range of about 48°, from 90° average attachment angle to approximately 42° detachment angle (see Figure 2.1). This corresponds to a linear filament excursion of about 9.7 nm per cross-bridge stroke.

Attachment of *S1* to actin and charge transfer are assumed to take place as in the isometric case once a suitable actin binding site has become available to the searching *S1* on the moving actin filament. Obviously, some search period will elapse before attachment can occur. Likewise, once a suitable site has become available, the process of binding and charge transfer will take some time during which the filaments continue to slide past each other. Thus, after completion of the binding and charge transfer process, the cross-bridge must rotate rapidly through a certain slack angle before producing contractive force and dissociating *ADP*. At maximum shortening velocity, this slack angle is equal to the total angular range of 48° so that no force is produced at all. However, as is apparent from Figure 2.1, during its angular excursion the *S1*, which rotates about *B1*, must rotate or (and) deform the actin monomer at binding site *B2*. That this rotation actually occurs within the muscle fiber has recently been demonstrated by Yanagida (1984) by observing the change in orientation of phalloidin-FITC-labelled actin monomers. The work done in rotating (or deforming) the monomers is degraded into heat and identified with the *shortening heat*.

By definition of the detachment angle of 42°, the contractile force becomes zero at this position and dissociation of *S1* from actin occurs. The second head becomes active and begins a new cycle by searching for an attachment site on actin.

Energetically, the situation is somewhat different from that in the isometric case. During one cross-bridge excursion, the decline in the potential due to internal losses is comparatively small, except at very low shortening velocities. However, because of the strong interaction between the field and the actin bond at *B1*, zero contractive force at the end position of the power stroke implies a reduction of the field to minimal values upon detachment. The rest energy not converted into work and heat during the power stroke is assumed to be returned to the intra-molecular energy depot by the collapsing field to be used anew in the next charge transfer.

5. Contraction in the lengthening mode

If the length of the sarcomere in the activated muscle fiber is increased, the attached *S1* is forcibly rotated by a small amount in the direction opposite to its normal contraction mode (see Figure 2.1). The increase in the angle ϕ prompted by the lengthening of the sarcomere results in a decrease of the distance separating the principal charge *P* on *S1* and the surface charge *R* located on the actin helix. This, in turn, produces a dramatic increase in the electrostatic repulsive force acting between the negative charges. As will be shown in the next section, this force increases in a highly nonlinear fashion according to the square of the inverse of the separation distance.

The large increase in the contraction force produces a high strain on the binding sites, predominantly at *B1*. As already mentioned at the beginning of this section, disturbances in the mechanical environment at the binding sites cause changes in the electric field. It will be assumed that the strain occurring at *B1* leads to bond deformations which, in turn, increase the intra–molecular potential. This is equivalent to saying that mechanical deformation energy produced by the stretch is, at least partly, converted into intra–molecular field energy. Since, eventually, the bonds break and then reform at different sites along the actin helix, the process is iterative.

The postulated increase in the potential is assumed to elicit the transfer of additional charges from the initial towards an intermediate position. In this way, contractive force is enhanced, even after the lengthening action has ended, since the increased potential only declines slowly.

The effect just described should be potentiated at greater sarcomere lengths where filamentary overlap decreases. This is because as the number of active cross-bridges decreases during lengthening, the instantaneous tension across the series elastic component of the sarcomere is suddenly distributed among fewer cross links. This implies that each *S1* instantaneously experiences an impulsive stretching action, in addition to that already present and corresponding to the given stretching velocity. Obviously, the stretch response of each of the active cross-bridges should be potentiated and the effect should increase, up to a certain limit, with decreasing filamentary overlap. That this is indeed the case has been amply confirmed in many studies investigating muscle responses to lengthening (Edman et al., 1978, 1984; Déléze, 1961; Hill, 1977, etc.).

Finally, the present hypothesis also predicts that very little angular change should occur in the position of *S1* during stretch and that the amount of *ATP* splitting should be greatly reduced in the lengthening mode. Both predictions are in accordance with the corresponding experimental findings (Cooke et al., 1984 and Curtin and Woledge, 1978, respectively).

2.2.3 The Charge-Transfer Model

After having completed the verbal description of the present contraction hypothesis, the corresponding biophysical model will be presented. It permits the quantitative testing of the hypothesis by comparing the model responses with the respective experimental results.

Model of Contractive Force Generation

Of primary interest is, of course, the contractive force produced by one *S1* during one complete cycle comprising the search for a suitable binding site on actin, attachment, the power stroke, detachment, and the inactive period. In certain contractive modes (such as lengthening), one or more of these phases may be absent. The inactive period, during which the first *S1* is dissociated and hydrolyzes *ATP*, is of no significance in the present context since the second myosin head is active during this period of time. Indeed, Greene (1981) has convincingly demonstrated that in the presence of nucleotide (*ATP*) only one of the two *S1* can bind to actin and that the binding mode is the same for either of the two myosin heads. This

feature of the present hypothesis is therefore strongly supported by direct experimental evidence.

The actual contractive cycle of one *S1* therefore comprises the following periods: τ_1, the velocity-dependent search time for an actin binding site; τ_2, the time required for binding, charge transfer and slack range translocation; and τ_3, the time of the power stroke. Since it is assumed that all cross-bridges (actin-bound myosin heads) work asynchronously, the **average contractive force** $\tilde{f}$ generated by one bridge over one complete cycle will be representative of the "overall" force production during that cycle period. Thus

$$\tilde{f} = \frac{1}{\tau_1 + \tau_2 + \tau_3} \int_0^{\tau_3} f(\phi(z), z(t))\, dt \qquad (2.1)$$

where the contractive force $f(\phi(t), z(t))$ is non-zero only during the power stroke period τ_3 and ϕ and z denote respectively the *S1* angle and the distance *B1-P* of the principal charge *P* from the binding location *B1* as shown in Figure 2.1.

An expression for the contractive force can be derived from the geometry of the actin-bound *S1* displayed in Figure 2.1, and from the known repulsive forces acting between the negative charges *P* and *R*. Referring to Figure 2.1, let $a_l = 13$ nm and $a_s = 6.5$ nm denote the distances from the *S1–23k–S2* link to *B1*, and from *B1* to *B2* respectively. Furthermore, let f_a be the repulsive force between the charges *R* and *P*; f_s the intermolecular resistive force opposing rotation and deformation of the actin monomer at bond *B2*; *m* the moment of the force f_a relative to the pivot point *B1*, ϕ the angle between the *S1* major axis and the actin axis, and θ the angle between the *S2* moiety and the myosin filament axis.

As can be verified from Figure 2.1, the contractive force *f* is given by

$$f = \left(\frac{m}{a_l} f_a - \frac{a_s}{a_l} f_s \right) \sin(\phi + \Theta) \cos \Theta \qquad (2.2)$$

where the constraint

$$d + 5 = 40 \sin \Theta + 13 \sin \phi \qquad (2.3)$$

defines the angle Θ in terms of ϕ and the surface distance *d* (in nm) between actin and myosin filaments. The distance *d* varies between 17.1 nm at short sarcomere lengths (60% of rest length), 13 nm at rest length (full filamentary overlap), and 10.1 nm at the terminal length of 1.8 times the rest length (no overlap). The variation in d is due to the constant volume relation (Dragomir 1970).

It can be shown that f does not change by more than 5% on average if the filament distance is varied over the above range. The influence on the force production of the varying interfilamentary lattice spacing may therefore be neglected and an average value of $\Theta = 5°$ assumed corresponding to the sarcomere rest length.

Referring again to Figure 2.1, let $\underline{r} = (r_x, r_y)$ and $\underline{z} = (z \cos \alpha, z \sin \alpha)$ denote the position vectors of R and P respectively relative to a space-fixed Cartesian coordinate system with origin at $B1$. The angle α is defined by

$$\alpha = \phi + \bar{\alpha} \tag{2.4}$$

$\bar{\alpha}$ having a value of $49°$. Then the moment produced by the force f_a about $B1$ is given by the vector cross product of $\underline{z}$ and $\underline{f_a}$, i.e.

$$-mf_a = (\underline{z} \times \underline{f_a}) \underline{k} \tag{2.5}$$

where $\underline{k}$ is a unit vector normal to the xy-plane and the vector of repulsive forces is given by

$$\underline{f_a} = (\underline{z} - \underline{r}) f_a / |\underline{z} - \underline{r}| \tag{2.6}$$

The moment arm function $m(\phi, z)$ may now be computed from $Eq.\ 2.5$ and $Eq.2.6$ and is found to be

$$m(\phi, z) = z [(r_y \cos \alpha - r_x \sin \alpha] / y \tag{2.7}$$

where

$$y = [(z \cos \alpha - r_x)^2 + (z \sin \alpha - r_y)^2]^{\frac{1}{2}} \tag{2.8}$$

For the present case the numerical values are $r_x = -5.7$ nm and $r_y = 1.9$ nm for the location of the reference surface charge R on the actin helix.

Hameroff et al. (1984, p. 577) describe the non-linear electrodynamics in cytoskeletal protein lattices including soliton propagation in structures containing contractile proteins. They use the Coulomb forces acting between neighboring electrons in their automaton model. A similar situation is anticipated for the interaction between the charges R and P in the present model. The magnitude of f_a is therefore given by

$$f_a = \frac{1}{4\pi\varepsilon_o} \frac{e^2}{y^2} \tag{2.9}$$

where the electron charge e has a value of $1.6 \times 10^{-19} C$, $1/(4\pi\varepsilon_o) = 9 \times 10^{-9}$, and y is the distance between the charges as given by $Eq.\ 2.8$. It follows that

$$f_a = 230.4 \times 10^{-30} / y^2 \tag{2.10}$$

Finally, an expression will be required for the force f_s that resists the rotation and deformation of the actin monomers which occurs during contraction. It has already been mentioned that such rotations have indeed been observed experimentally (Yanagida, 1984). Hofacker (1982, p. 233) gives an empirical relation for the potential describing the repulsive interaction between molecules. This potential function is exponential so that its first derivative with respect to the distance between the interacting groups must also be an exponential function. The most likely model for the force function $f_s(\phi)$ is therefore

$$f_s = a_1 (e^{a_2(\phi_o - \phi)} - 1) \tag{2.11}$$

where the attachment angle ϕ_o has a value of $\pi/2$ (90°), and the constants a_1 and a_2 have values of 1.1×10^{-13} and 4.24 respectively.

The work done by the cross-bridge in overcoming this resistive force in every power stroke is given by

$$s_b = \int_\phi^{\phi_o} f_s(\phi)\, a_s\, d\phi =$$
$$= a_s a_1 [e^{a_2(\phi_o - \phi)} - 1 + \phi_o - \phi] / a_2 \tag{2.12}$$

Using the present values for the constants a_s, a_1, a_2, and ϕ_o, and integrating over the full range of the power stroke from $\phi_1 = 0.733$ rad (42°) to $\phi_o = \pi/2$, a value of 5.86×10^{-21} Nm is found. This compares well with 5.51×10^{-21}J, the amount of shortening per heat per cross-bridge power stroke as calculated from the data of Homsher et al. (1983).

Since all functions and parameters appearing in $Eq.\ 2.2$ have now been defined, the contractive force $f(\phi, z)$ can be computed for given values of the cross-bridge angle ϕ and the position z of the mobile principal charge P. However, in order to evaluate the integral in $Eq.\ 2.1$, expressions for the periods τ_1, τ_2, and τ_3 need to be derived.

In the absence of detailed information on the velocity-dependent search time, τ_1, of a myosin head, it will be assumed that τ_1 is a linear function of the *normalized contraction velocity*

$$\eta = \dot{x} / |\dot{x}_{max}| \qquad (2.13)$$

where $\dot{x}$ is the filamentary sliding velocity of a half-sarcomere and $\dot{x}_{max}$ the corresponding maximum value. Note that in the present model shortening velocities are *negative* (as they ought to be) and lengthening velocities *positive*. Thus, for shortening, we have that $-1 \leq \dot{\eta} \leq 0$. The model for τ_1 is therefore

$$\tau_1 = c(-\dot{\eta}) \qquad (2.14)$$

where c denotes a constant.

During the period τ_2 actin binding, the charge transfer and the slack range translocation takes place, while the actual contractive work is performed during the subsequent power stroke period τ_3.

By definition of the maximum shortening velocity $\dot{x}_{max}$, the cross-bridges are unable to produce contractive force at that velocity. In terms of the present model this implies that during τ_2 the distance x_{max} covered by the filaments sliding past each other must be equal to the full range excursion of the $S1$. In this case, detachment would occur after the slack length had been taken up, without generation of any contractive force. Thus $\tau_2 = x_{max} / |\dot{x}_{max}|$. The expression for the linear cross-bridge excursion x which corresponds to an angular excursion from ϕ_o to $\phi \geq \phi_1$ is obtained from (see Figure 2.1)

$$x = a_l \int_{\phi_o}^{\phi} \sin \phi \; d\phi = a_l (\cos \phi_o - \cos \phi) \qquad (2.15)$$

Hence for the full range of angular cross-bridge excursion from $\phi_o = 90°$ (attachment angle) to $\phi_1 = 42°$ (detachment angle) and with $a_l = 1 \times 10^{-9}$ m, we have that

$$x_{max} = 13 (\cos \phi_o - \cos \phi_1) = -9.66 \text{ nm} \qquad (2.16)$$

which is the length of the cross-bridge stroke in the present model. Therefore,

$$\tau_2 = a_l (\cos \phi_o - \cos \phi_1) / |\dot{x}_{max}| \qquad (2.17)$$

At a contraction velocity of $|\dot{x}| < |\dot{x}_{max}|$ the moving myosin filament produces during the period τ_2 a slack length of

$$x_s = \dot{x} \tau_2 = a_l (\cos \phi_o - \cos \phi_s) \qquad (2.18)$$

which is taken up rapidly by the rotating $S1$ before force production begins at the angular position ϕ_s. Thus, at a certain, non-maximal shortening velocity $\dot{x}$, the total period of actomyosin association is $\tau_2 + \tau_3$, given by

$$\tau_2 + \tau_3 = a_l (\cos \phi_o - \cos \phi_1) / (\dot{\eta} |\dot{x}_{max}|) \qquad (2.19)$$

where *Eq. 2.13* has been used.

It should be pointed out that the resistive force f_s is still active during slack range translocation so that shortening heat is always produced, even at maximum shortening velocity. The speed of slack range uptake can be estimated from the superfast tension transients displayed in the records of Figure 7 of Van den Hooff and Blangé (1984) to be about 50 μm/s per half sarcomere, which is much larger than the maximum shortening speeds of muscle.

In the previous section it has been mentioned that in isometric contraction the charge P retracts after transfer to its terminal position, possibly in a hopping motion involving solitons (Bishop, 1984, p. 158). The phase period τ_3^o is the reciprocal of the *ATPase* rate in isometric contraction and detachment of $S1$ from actin must occur after a period τ_3^o. From *Eqs. 2.17* and *2.19* it follows that

$$\tau_3 = a_l (\cos \phi_1 - \cos \phi_o) (-1 - 1/\dot{\eta}) / |\dot{x}_{max}| \qquad (2.20)$$

In *Eq. 2.1* we therefore have for isometric contractions ($\dot{\eta} = 0$)

$$\tau_1 + \tau_2 + \tau_3 = a_l (\cos \phi_1 - \cos \phi_o) / |\dot{x}_{max}| + \tau_3^o \qquad (2.21)$$

while otherwise

$$\tau_1 + \tau_2 + \tau_3 = \qquad (2.22)$$
$$= c(-\dot{\eta}) + a_l (\cos \phi_1 - \cos \phi_o) / (-\dot{\eta} |\dot{x}_{max}|)$$

The relationship between $\phi(t)$ and $x(t)$ is obviously given by *Eq. 2.15* with ϕ_o replaced by the initial power stroke angle ϕ_s, i.e.

$$\phi(t) = \arccos(\cos\phi_s - x(t)/a_l) \qquad (2.23)$$

where for constant contraction velocities $x(t) = \overset{*}{x}\,t$, and ϕ_s follows from Eq. 2.18.

Thus, if the charge transfer function $z(t)$ is known, the average contractive force $\bar{f}$ of a cross-bridge in isometric and shortening contractions can be computed from Eq. 2.1 by using Eq. 2.2 and Eq. 2.21 or 2.22.

Microdynamics of Intra-Molecular Charge Transfer

A crucial part of the present contraction hypothesis and the associated model is the microdynamics of the charge transfer. As has been outlined in Section 2.2.2, the transfer of the negative charges (one principal charge in isometric and shortening contractions, two charges in the lengthening mode) from their initial to their terminal positions is effected by a combination of intra-molecular forces and potentials. In a heuristic manner we shall assume that driving potentials, long-range and short-range repulsive and attractive forces, and internal resistive forces determine the charge transfer. Hence a possible model for the charge transfer dynamics is given by

$$c_3\,\varepsilon(t) - c_1\overset{*}{z} - h(\phi,z) = 0 \qquad (2.24)$$

where $c_3\varepsilon(t)$ is the driving force function representing the field action, $c_1\overset{*}{z}$ is the internal resistive force, and $h(\phi,z)$ is the sum of the z-components of all repulsive and attractive electrostatic and dipole forces that act on the charges to be transferred. The value of the constant c_1 is about 10^{-6}.

The driving force function, in turn, is determined by the postulated field dynamics

$$\overset{*}{\varepsilon} = p(v)\,[u(t,\xi,v,\overset{*}{\eta}) - \varepsilon]; \quad \varepsilon(0) = \varepsilon^O \qquad (2.25)$$

where the transfer potential $u(.)$ is a function of the time t, the normalized sarcomere length ξ, the normalized contraction force $v = f/\bar{f}$, and the normalized half-sarcomere sliding velocity $\overset{*}{\eta}$. The symbol $p(v)$ denotes a function of the normalized contraction force (see Section 2.3.2).

In *isometric and shortening contractions*, only one electron charge is transferred to the terminal position at $\bar{z} = 5$ nm against the internal resistance $c_1\overset{*}{z}$ and the repulsive force in the z-direction. The expression for the latter has the form

$$\bar{h}(\phi,z) = f_a\,(1 - m^2(\phi,z)\,/\,z^2\,)^{\frac{1}{2}} \qquad (2.26)$$

where f_a and $m(\phi,z)$ are given by Eq. 2.10 and Eq. 2.7, respectively. By hypothesis, the potential $u(t)$ collapses at $t = \tau_g$, which leads to the retraction of the principal charge and to the dissociation of $S1$ from actin. During $0 \leq t \leq \tau_g$, however, $\overset{*}{\varepsilon} \approx 0$, i.e., $z(t) \approx \bar{z}$ for this interval.

For a given normalized contraction velocity $\overset{*}{\eta}$, $-1 \leq \overset{*}{\eta} \leq 0$, the corresponding contraction force $\bar{f}(\overset{*}{\eta})$ of a cross-bridge can therefore be computed from Eq. 2.1 with $z(t) = \bar{z}$. If this is done for the whole range of $\overset{*}{\eta}$, the shortening velocity-dependence function of the cross-bridge contractive force is obtained. The graph of this function will be shown later.

In the *lengthening mode*, the situation is profoundly different. Owing to the counterclockwise roation of $S1$, the repulsive force between R and P increases drastically, leading to a field potentiation which, by hypothesis, elicits the release of additional charges from their initial location near $B1$. This model bears some resemblance to the molecular contraction model of Hatze (1973). For the purpose of the present model we shall assume that only one additional charge is released. Since the principal charge P is already located at the terminal position $z_p = \bar{z}$ (stretch proceeds from the isometric condition), the differential equation for the motion $z(t)$ of the second charge follows from Eq. 2.24 as

$$\overset{*}{z} = [\,c_3\,\varepsilon - h(\phi,z)\,]\,/\,c_1; \quad z(0) = z_o \qquad (2.27)$$

where

$$h(\phi,z) = \bar{h}(\phi,z) + 230 \times 10^{-30}\,/\,(\bar{z} - z)^2 \qquad (2.28)$$

and c_3 is determined by the condition that $\varepsilon = 1$ under isometric conditions, i.e. $c_3 = h(\bar{\phi},0)$. The second term in Eq. 2.28 accounts for the fact that the principal charge P at the terminal position also produces a repulsive force repelling the second charge. The variable ε is determined by Eq. 2.25 as in the case of shortening contractions.

In the stretching mode, the myofilament series elastic component plays an important role as outlined in the next section. There it is shown that the major part of the series elasticity does most probably not reside within the $S1$ but in the bare zone of the myosin filament and, possibly, also in the actin strands and the z-discs.

Assuming the series elasticity to be located mainly in the bare zone of the myosin filament, the following normalized force relation must hold for each cross-bridge:

$$ f/\bar{f} = f^{SE}/\bar{f} \tag{2.29} $$

and also

$$ \dot{f}/\bar{f} = \dot{f}^{SE}/\bar{f} \tag{2.30} $$

where f denotes the instantaneous force produced by the "average" cross-bridge, $\bar{f}$ is the isometric force, and f^{SE} is the respective half-sarcomere part of the total filament elasticity. In the next section it will be shown that the load-extension curve of the series elastic componenet is exponential (see also Chapter 5 (Winters)).

Using Eqs. 2.2, 2.7, 2.8, 2.10, 2.11 and 2.37, and carrying out the differentiation of the terms in Eq. 2.30, a fairly complicated differential equation for ϕ of the form

$$ \dot{\phi} = [\,(\dot{\lambda}_h - \dot{x}_a)\,g_1(\phi,z,\xi) - \dot{z}\,g_2(\phi,z)\,]\,/\,g_3(\phi,z) \tag{2.31} $$

$$ \phi(0) = \phi^0 $$

is obtained. The symbols $g_i(.)$ denote functions of ϕ and z, while $\dot{\lambda}_h$ is the half-sarcomere stretching velocity and $\dot{x}_a$ the "slipping" velocity of myosin heads on actin. The relationship between the stretching velocity $\dot{\lambda}_h$, the extension velocity $\dot{\lambda}_s$ of the half-filament series elastic component, the velocity $\dot{x} = a_l\,\dot{\phi}\,\sin\phi$ of cross-bridge excursion, and the half-sarcomere slipping velocity $\dot{x}_a$ is given by

$$ \dot{\lambda}_h = \dot{\lambda}_s + a_l\dot{\phi}\,\sin\phi + \dot{x}_a \tag{2.32} $$

where λ_s and δ (see also Eq. 2.37) are related by $\delta = (\lambda_s - \lambda_{so})/\bar{\lambda}_h$, $\bar{\lambda}_h$ being the half-sarcomere rest length. The velocity $\dot{\lambda}_s$ can be expressed in terms of the remaining velocities and the resulting expression can then be used in Eq. 2.31.

It must be stressed that the contractile force f in Eq. 2.29 is composed of both the force due to the principal charge P and that due to the second charge.

The microdynamics of the stretching mode is therefore governed by the three differential equations described by Eqs. 2.27, 2.25 and 2.31, given an appropriate potential function $u(t,\xi,v,\eta)$ as input. Such a function will have to account for part of the force enhancement during stretch and, most importantly, for the potentiation of this effect at sarcomere lengths greater than the rest length.

To this end consider the force F^{SE} across the total filament series elasticity. This force must be equal to the sum $n\lambda_u f$ of all cross-bridge forces produced by the $n\,\lambda_u$ active bridges in the overlap region λ_u. Thus

$$ f = F^{SE}/n\,\lambda_u \tag{2.33} $$

The instantaneous force increment Δf on a cross-bridge that results from a given overlap decrease $-\Delta\lambda_u$ as a result of the stretch follows from above as

$$ \Delta f = \Delta F^{SE}/n\,\lambda_u + F^{SE}(-\Delta\lambda_u)/n\,\lambda_u^2 \tag{2.34} $$

Since at least a very small damping element must be present in the series elastic component, F^{SE} may be considered instantaneously constant and thus ΔF^{SE} nearly zero, i.e. the instantaneous force increment Δf is proportional to $1/\lambda_u^2$. As is easily shown, the normalized overlap length $\xi_u = \lambda_u/\bar{\lambda}_u$ is approximately given by

$$ \xi_u = 1 - a_u\,(\xi - 1) \tag{2.35} $$

where $a_u = 1.49$ and ξ is the normalized half-sarcomere length.

By hypothesis, the potential u is proportional to the instantaneous stretching stress imposed on the cross-bridge and to the slipping distance $s_1\dot{\eta}\,t$, through which the $S1$ is moved. In addition, a sizeable contribution $s_0(v,t)$ is due to the first breaking of the actomyosin bond. Thus, the potential function u can be represented as

$$ u(t,\xi,v,\dot{\eta}) = \varepsilon^o + s_0(v,t) + s_1\,\dot{\eta}\,t/\xi_u^2 \tag{2.36} $$

where ξ_u is given by Eq. 2.35, ε^o is the isometric initial condition of Eq. 2.25, and s_1 is a constant. The function s_0 will be specified later.

In the next section we discuss the probable location of the sarcomere series elastic component and its load-extension properties.

Location of the Sarcomere Series Elasticity

The precise location of the series elastic components in a sarcomere is of cardinal importance for any molecular model of muscular force production. Indeed, as has been pointed out in Section 2.1, for some theories (e.g. Huxley and Simmons, 1971) the assumption is fundamental that the major part of the sarcomere compliance resides within the cross-bridges.

However, experimental evidence suggests the contrary. Tawada and Kimura (1984) have demonstrated that the *S2* portion of the myosin molecule is not significantly compliant. They assume that the elasticity observed must therefore be situated in the *S1-23k* moiety of the globular head. Because an actin binding site is located on *S1-23k*, the free end from *B1* to the *S2-link* (see Figure 2.1) has a length of about 13 nm. The distance of quick half-sarcomere shortening required to drop the isometric tension to zero is about 1% of the half-sarcomere length, i.e., about 12 nm in frog muscle and 90 nm in crayfish muscle (Sugi, 1979). It is highly unlikely that the entire *S1-23k* portion should bend by an angle of 90° to account for the observed elastic extension of 12 nm in frog muscle.

The conclusions of Tawada and Kimura are based on the observation that sarcomere stiffness decreases approximately linearly with decreasing filamentary overlap. Such a decrease in rigor stiffness with sarcomere length can indeed be seen in their Figure 3 (Tawada and Kimura 1984). Since increasing sarcomere length above rest length implies a decreasing number of cross links, it is certainly tempting to attribute the reduced sarcomere stiffness to the reduced number of small series elastic components residing within the cross-bridges. Such a conclusion may, however, not represent the complete picture. Van den Hooff and Blangé (1984) have performed superfast release experiments on frog muscle fibers at sarcomere length l_o=2.1 μm and have plotted the relationship between tension drop and amount of release (see their Figure 5, black dots). The curve is exponential and well fitted by the equation

$$F^{SE}/\overline{F} = 0.158\,(e^{232\,\delta} - 1) \qquad (2.37)$$

where $F^{SE}/\overline{F}$ denotes the relative tension and $\delta = \Delta x / \overline{x}$ is the relative extension, $\overline{x}$ denoting the half-sarcomere rest length.

The experiment was repeated at a sarcomere length of 3.1 μm where the isometric force dropped to half its maximal value and the results were replotted (black triangles in Figure 5 of Van den Hooff and Blangé, 1984). After scaling the new values by the appropriate factor corresponding to the reduced tension, it turns out that the load-extension curve of the sarcomere series elastic component at l = 3.1 μm is identical with that determined at l = 2.1 μm, despite the fact that at 3.1 μm only about 50% of the total number of cross-bridges are attached. If there had been appreciable compliance in the cross-bridges, the load-extension curve should have changed as a result of decreased overlap. However, no such change could be detected.

The second implication of the exponential load-extension relationship of the sarcomere series elasticity found by Van den Hooff and Blangé is the linearity between stiffness and tension (see also Chapter 5 (Winters)). Assume the general form of the load-extension relationship to be

$$F = a\,(e^{b\delta} - 1) \qquad (2.38)$$

The stiffness is then given by

$$dF/d\delta = a\,b\,e^{b\delta} = ab + bF \qquad (2.39)$$

i.e., the relationship between stiffness and force is linear. Since there is also an approximately linear relationship between tension and amount of filamentary overlap above rest length, it follows that the stiffness is a linearly decreasing function of the sarcomere length with a certain residue value at zero overlap. This is exactly what is seen in Figure 3 of Tawada and Kimura (1984), i.e., the observed stiffness decrease may be attributable to the nonlinearity of the sarcomere series elastic component residing outside the cross-bridges and not necessarily to a reduced number of small series elasticities within the cross-bridges.

Such a conclusion is in line with the observations of Sugi (1979), who found that be major part of the sarcomere series elastic component resides most likely in the bare zone in the middle of the myosin filament and, possibly, also in the Z-discs and the actin filaments (see also discussion in Chapters 4 and 5).

To summarize, the major part of the sarcomere

series elastic component is most likely *not* situated within the cross-bridges but in series connecting structures such as the bare zone of the myosin filaments and the Z-discs. Futhermore, the load-extension curve of this component is well approximated by an exponential.

2.3 Verification and Validation of the Charge-Transfer Model

The *credibility* of a model is determined by its verifiable predictions. It is therefore imperative to test under as wide a variety of conditions as possible the charge-transfer cross-bridge model presented in the previous sections. Unfortunately, there is a severe restriction imposed on this attempt: Practically all experimental observations on muscular contractive behaviour are global and not made at the microstructural level.

For some contractive modes not involving the effects of intrafilamentary calcium dynamics, this problem can be partly circumvented by postulating the existence of an "average cross-bridge" whose behaviour can be considered representative of the asynchronously working bulk of the cross-bridges. This is the approach adopted here.

In order to simulate on the computer the responses of the charge-transfer model in a variety of contractive modes, the *ANSI-Fortran 77* computer program *MYOACT* implementing the conceptual model was developed. The computerized model has been verified, i.e., it has been established that it represents the conceptual model adequately, within specified limits of accuracy.

In this section we shall be concerned with the *validation* of the model, i.e., with the process of establishing its ability to predict system responses that agree reasonably well with the corresponding responses as observed on the real biosystem.

2.3. Velocity-Dependence of Cross Bridge Contraction Force in Shortening

As has been described in Section 2.2.3, the *Eq. 2.1* can be evaluated for various values of the normalized contraction velocity $\dot{\eta}$. The resulting force function $\tilde{f}(\dot{\eta})$ is then the velocity-dependence function of the contractile cross-bridge force in the isometric and the shortening mode. The function is sometimes termed somewhat loosely the force-velocity relation. Since it presents an overall property of the whole cross-bridge assemblage, it is representative of a whole-fiber property.

Equation 2.1 has been evaluated for 100 values of shortening velocity ($-1 \leq \dot{\eta} \leq 0$), in steps of 0.01. The result is shown in Figure 2.2a, while the corresponding normalized curve is shown in Figure 2.2b.

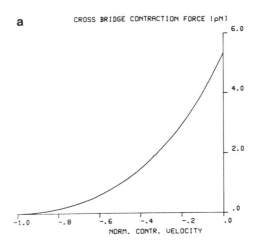

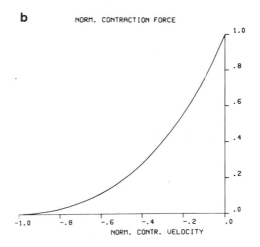

Figure 2.2: *a)* Velocity-dependence function of the contractile cross-bridge force in isometric and shortening contractions. The value of the isometric cross-bridge force is 5.33 *pN*. *b)* Normalized velocity-dependence function of the cross-bridge contraction force.

It is seen from Figure 2.2 that the predicted curve closely resembles the so-called force-velocity relation usually observed in contracting skeletal muscle. This is remarkable since in the present model this function is predominantly the result of the changing repulsive forces, during the angular cross-bridge excursion, between the reference charge and the transferred principal charge, i.e., an intrinsic property of the model. Also, the isometric cross-bridge force of 5.33 pN corresponds to about 33.6 N/cm^2 isometric fiber force (for 6.3 x 10^{12} myosin heads/cm^2 half-sarcomere), a value usually found in skeletal muscle fibers.

In addition, the work necessary to transfer the principal charge P from its initial to its terminal position against the repulsive forces $\bar{h}(.)$ given by *Eq. 2.26* can be computed by integrating over the whole path z. It is found to have a value of 5.90 x 10^{-20} Nm. This is slightly less than the energy value of 6.65 x 10^{-20} *J* derived from the splitting of one *ATP* molecule. In isometric contractions, where no mechanical work is performed, this amount of energy is degraded into heat which constitutes part of the maintenance heat. The rate at which this degradation occurs is a nonlinear function of the fiber's maximum shortening velocity (Elzinga et al., 1987, Figure 5)

2.3.2 Force Enhancement Occurring During and After Sarcomere Lengthening

Possibly the most intriguing phenomenon of muscular contraction is the behaviour of the contractile structures during and after sarcomere lengthening (see also Chapters 3-5). Depending on the stretching velocity, the force during sarcomere lengthening rises steeply above the isometric level in stimulated fibers. Upon cessation of the stretch, the force first declines rapidly to a level above isometric, to be followed by a very slow decrease of the residual force towards the tetanic level. The amount of force enhancement after stretch increases, while the rate of its decline decreases, with increasing sarcomere length above the rest length (Edman et al., 1978). Virtually no force enhancement after stretch is observable at sarcomere lengths below the rest length.

Edman et al. (1978, p. 153) have performed a thorough analysis of the above mentioned phenomena. They reached the conclusion that stretch during activity might alter "the function of the cross-bridges to enable them to generate more tension without altering the kinetics which determine v_{max}." This conclusion is in complete agreement with the predictions of the present charge-transfer model: Not an increased number of attached cross-bridges is responsible for force enhancement after stretch but the inherent functional properties of the actomyosin complex.

In order to simulate on the computer the responses of the charge-transfer model to stretches, we use the differential equations (*Eq. 2.25, 2.27, 2.31*), with the following specifications. By hypothesis, both the inverse $p(v)$ of the damping function and the initial-burst function $s_o(v,t)$ appearing in *Eq. 2.25* and *2.36*, respectively, are supposed to be influenced by the stress disturbances occurring at $B1$ during stretch (see Figure 2.1). They are therefore expected to be exponential functions of the normalized force $v = f/\bar{f}$, i.e.,

$$p(v) = p_1 (e^{P_2 v} - 1) \tag{2.40}$$

and

$$s_o(v,t) = e^{vlnk_o} = \begin{cases} k_o^v & 0 \leq t \leq \tau_b \\ \\ 0 & t \leq 0 \text{ or } t \geq \tau_b \end{cases} \tag{2.41}$$

where for the present model $p_1 = 8.12$ x 10^{-4}, $p_2 = 4.6$, $k_o = 20$, and τ_b denotes the time interval from the beginning of the stretch until the breakage of the first bond. For a given normalized stretching velocity $\dot{\eta}$, the value of the constant s_1 appearing in *Eq. 2.36* is computed from $s_1 = 0.8/(\tau_s \dot{\eta})$, τ_s denoting the stretching time.

An important point to be discussed is the modelling of the forcible rupture (the sarcomere 'give') of the actomyosin bond in stretches. That this phenomenon actually occurs has been clearly demonstrated by Flintney and Hirst (1978, Figure 6). Referring to Figure 2.1, sarcomere 'give' implies that in the first phase of the stretch the "average"-$S1$ rotates, in a counterclockwise direction, from its initial "average" angular position $\bar{\phi}$ towards the limiting position ϕ_l at which bond rupture occurs. It seems reasonable to expect the rate of change $\dot{\phi}_l$ of angular limit positions to be proportional to the stretching acceleration $\dot{v} = \ddot{x} = \ddot{\eta} /|\ddot{x}_{max}|$ and to the difference $\phi_{max} - \phi_l$ between maximal and the current limit angle. Thus

$$\dot{\phi}_l = k_1 \dot{v} \, (\phi_{max} - \phi_l) \qquad (2.42)$$

k_1 being a constant. Using *Eq. 2.13*, we may rewrite this as:

$$d\phi_l/d\dot{\eta} = k \, (\phi_{max} - \phi_l) \qquad (2.43)$$

which, for given constant values of $\dot{\eta}$, and initial value $\phi_l = \bar{\phi}$, may be solved to yield

$$\phi_l(\dot{\eta}) = \bar{\phi} + \overline{\phi_{max} - \bar{\phi}} \, (1 - e^{-k\dot{\eta}}) \qquad (2.44)$$

For the present model $\bar{\phi} = 1.253$ rad, $\phi_{max} = 1.423$ rad, and $k = 6.90$. Using *Eq. 2.40, 2.41, 2.44*, and the initial conditions $\varepsilon^o = 1$ in *Eq. 2.25*, $z_o = 0$ in *Eq. 2.27*, and $\phi^o = \bar{\phi}$ in *Eq. 2.31*, the respective differential equations can now be integrated over a specified time interval, subject to the following provisions.

During the **first stretching phase** at constant $\dot{\eta}$ from $t = 0$ to $t = \tau_b$, the angular movement of *S1* takes place from $\bar{\phi}$ to $\phi_l(\dot{\eta})$, where bond breakage occurs. During that period, *Eq. 2.31* determines the dynamics of $\phi(t)$, and $u(\cdot)$ appearing in *Eq. 2.25* is given by *Eq. 2.36* with $s_o(\cdot)$ expressed by *Eq. 2.41*.

During the **second stretching phase** lasting from $t = \tau_b$ to $t = \tau_s$ (end of stretch), "slipping" of myosin heads on actin occurs and $\phi(t) = \phi_l(\dot{\eta}) = $ constant, i.e., $\dot{\phi} = 0$, which replaces *Eq. 2.31* for this period of time. The function $u(.)$ is as given by *Eq. 2.36* with $s_o(.) = 0$.

In the **post-stretching phase** from $t = \tau_s$ to the final time of the simulation period, all cross-bridges in the half sarcomere rearrange, so that the dynamics of the variable ϕ of the "average" *S1* is determined by

$$\dot{\phi} = -k_d (\phi - \bar{\phi}) \qquad (2.45)$$

where $k_d = 6.2$. During this period, *Eq. 2.45* replaces *Eq. 2.31* and the function $u(.)$ controlling the dynamics in *Eq. 2.25* is now given by $u = \varepsilon^o$, due to the rapid collapse of the potential to its initial value at the end of the stretching phase.

Finally, it should be mentioned that, strictly speaking, the normalized stretching velocity $\dot{\eta}$ appearing on the right hand side of *Eq. 2.36* should be replaced by the normalized "slipping" velocity $\dot{x}_a / |\dot{x}_{max}|$, since it is the latter and not $\dot{\eta}$ which is

directly related to the postulated build-up of the potential $u(\cdot)$. However, the two velocities are identical after the extremely short initial phase τ_b, so that the error committed would be small. In the simulation program *MYOACT*, the correct "slipping" velocity was used.

Furthermore, a very precise treatment requires that the change in the properties of the common filament series elastic component due to the changing filamentary overlap be considered. As fewer cross-bridges are attached at increased sarcomere length, the apparent series elasticity per cross-bridge becomes stiffer. This implies a small alteration in the function $g_l(\cdot)$ appearing in *Eq. 2.31*. The complete form of the latter equation was, of course, implemented in the simulation program, but will not be elaborated on further in the present context.

In order to test the responses of the charge-transfer model to a variety of stretching modes, a value of $\dot{x}_{max} = 2 \times 1050$ nm/s, equivalent to two half-sarcomere lengths per second was chosen for the maximum shortening velocity. Stretching distances were 105 nm (10% of half-sarcomere length) and 210 nm (20%), while the values of the normalized stretching velocities were $\dot{\eta} = 0.1$ (10% of $\dot{x}_{max}$) and $\dot{\eta} = 0.2$. To demonstrate the effect of increased sarcomere length, stretching simulations were performed at the following values of the normalized half-sarcomere length: $\xi = 1$ (corresponding to the optimal length), $\xi = 1.2$, $\xi = 1.4$, and $\xi = 0.8$.

Figure 2.3 presents model responses to stretches performed over a distance of 105 nm, at a normalized velocity of $\dot{\eta} = 0.2$, and starting from normalized lengths of ξ both above and below the nominal length. All stretches begin in the isometric state where the normalized cross-bridge force equals unity.

From Figure 2.3 we see that the force produced by the present cross-bridge model first rises steeply from the isometric to a new level, and then either remains at that new level ($\xi = 0.8$, Figure 2.3d) or increases further during the second stretching phase ($\xi \geq 1$, Figures 2.3a-c). After cessation of the stretch at $t = 0.25$ s, the force first declines rapidly, which phase is followed by a period of very slow decline towards the isometric level. Thus force is enhanced after active stretch and this force enhancement increases dramatically with sarcomere length (compare Figures 2.3c and

2.3d). Furthermore, force enhancement also occurs during the second stretching phase and is also dependent on the sarcomere length at which the stretch was initiated (see Chapter 3 (Morgan) for an alternative (and perhaps compatible) approach for simulating force enhancement).

The present model predictions are well supported by experimental findings. The initial rapid force increase, its further rise during the second phase of the stretch at greater sarcomere lengths and the potentiation of the force enhancement after stretch at increased sarcomere lengths have all been observed by Déléze (1961), Edman et al. (1978, 1984), L. Hill (1977), and others. In fact,

Edman et al. (1984, Figure 7) have shown that there exists an approximately linear relationship between the force enhancement during the second phase of the stretch and the residual force enhancement after stretch. According to the present model this should be so because the charging process taking place during the second stretching phase implies the simultaneous increase in the force, an increase which persists in the post-stretching phase. Thus, the amount of force enhancement present after stretch should be directly related to the distance of the stretch but should be independent of the stretching velocity.

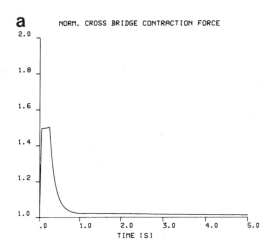

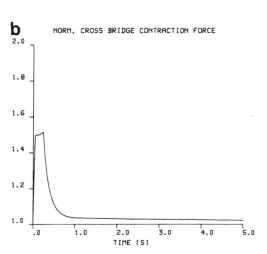

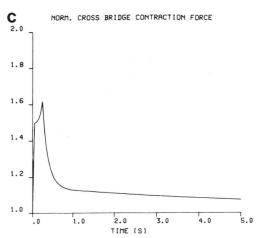

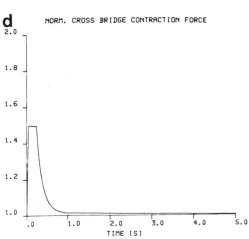

Figure 2.3: Cross-bridge model stretch response. Normalized cross-bridge contraction force as model response to a stretch performed over a distance of 105 nm, at a normalized stretching velocity of $\dot{\eta} = 0.2$, and at normalized starting lengths of: *a)* $\xi = 1$; *b)* $\xi = 1.2$; *c)* $\xi = 1.4$; *d)* $\xi = 0.8$. Simulation time is 5 *s* in all records.

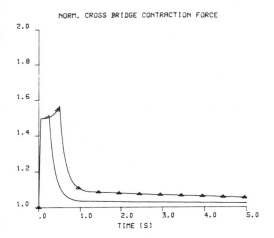

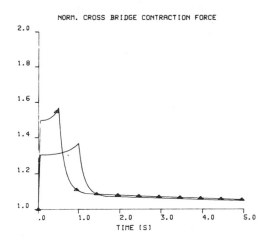

Figure 2.4: Cross-bridge model stretch responses for $\dot{\eta}$ = 0.2, ξ = 1.2 starting length, and stretching distances of 105 nm (unmarked curve) and 210 nm (curve marked by triangles) respectively.

Figure 2.5: Cross-bridge model stretch responses for 210 nm — stretches starting from ξ = 1.2 with stretching velocities of $\dot{\eta}$ = 0.2 (unmarked curve) and $\dot{\eta}$ = 0.1 (curve marked by triangles) respectively.

In order to test the corresponding model responses, two stretches were performed with $\dot{\eta}$ = 0.2 and from starting length ξ = 1.2, but with stretching distances of 105 nm and 210 nm respectively. The results are shown in Figure 2.4, from which it is clearly seen that the longer stretch (curve marked by triangles) also produced the greater force enhancement after stretch. This prediction is also seen experimentally by Edman et al. (1978, Figure 2C).

On the other hand, stretches performed at different velocities but over the same distance have been shown by Edman et al. (1978, Figure 2B) to produce practically the same force enhancement after stretch. Again, this is in agreement with the predictions of the present model. Figure 2.5 shows the model responses to two 210 nm stretches proceeding from a starting length of ξ = 1.2, but performed at normalized stretching velocities of $\dot{\eta}$ = 0.2 and $\dot{\eta}$ = 0.1. As can be seen,

a

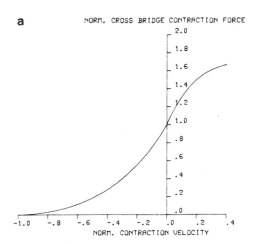

b

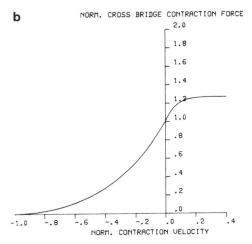

Figure 2.6: Extended velocity-dependence function of the normalized cross-bridge force including positive (stretching) velocities. *a)* Normalized maximum (asymptotic) stretching

force value is 1.722. *b)* Normalized maximum (asymptotic) stretching force value is 1.271. Values of constants used in *Eq. 2.44* were ϕ_{max} = 1.33 and k = 16.1.

there is virtually no difference between the after-stretch force enhancement at these two different velocities. There is, however, a marked difference observable in Figure 2.5 between the forces attained in the two curves after the initial force rise.

This difference is obviously due to the different stretching velocities and reflects the well-known dependence of the initial stretching force on the velocity (e.g., see Flitney and Hirst, 1978, Figure 4). Indeed, when simulations are carried out at various stretching velocities $\dot{\eta}$ and the resulting values of the initial stretching force are plotted, the graph so obtained closely resembles the experimental curves found by Flitney and Hirst (1978, Figure 4A-B). The graphs of the simulation results for positive (stretching) velocities may be combined with those for negative (shortening) velocities. This has been done in Figure 2.6 for two different values of normalized maximum stretching forces.

Finally, the observation made by Edman et al. (1978, Figure 7) that the contractile force is enhanced also in filament shortening following a stretch is confirmed by the predictions of the present model (Figure 2.7). As was found in the experiment, the value of the maximum shortening velocity is not affected by the mechanisms responsible for force enhancement.

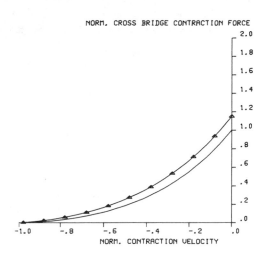

NORM. CROSS BRIDGE CONTRACTION FORCE

NORM. CONTRACTION VELOCITY

Figure 2.7: Normalized velocity-dependence function of the cross-bridge force without (unmarked curve) and with (marked curve) preceding stretch. Stretches started at $\xi = 1.4$ and extended over 105 nm at $\dot{\eta} = 0.2$. Force value taken 1 s after beginning of stretch.

In summary, it has been demonstrated that the present charge-transfer cross-bridge model exhibits all important features of the force enhancement occurring during and after stretch of the contractile structures.

2.4 Conclusions and Future Perspectives

This chapter has been devoted to the presentation of a model for myofilamentary force generation by means of negative charges transferred within the 27 kiloDaltons moieties of the globular myosin heads. The model can neither be classified as being unique nor is it likely to be representative of the actual events taking place within the subfragment 1 upon binding to actin and during contraction. However, since every attempt has been made to incorporate into the model as much as possible of the current knowledge available on the molecular fine structure and the functional properties of the actomyosin complex, there is some reason to believe that the present contraction hypothesis may indeed reflect some of the essential features of the real molecular contraction process. An encouraging sign in this direction is the considerable predictive power of the model that results from its intrinsic properties.

As regards the *transition from the charge-transfer model to a global myocybernetic model of skeletal muscle*, it is obvious that a model of the cross-bridge microdynamics alone is incapable of predicting global responses of skeletal muscle. The total contractive force produced by a muscle is the result of the neurally controlled action of myriads of single actomyosin complexes. Therefore, not only the *contraction dynamics* of the cross-bridges but also other phenomena must be considered in the construction of a global muscle model, including the *excitation dynamics* of the Ca^{2+} regulatory system, the nonlinear *recruitment dynamics* of whole motor units, and the influence of the series elastic elements residing outside the filaments (e.g. see also Chapters 1, 3-5). But not even the contraction dynamics is completely determined by the cross-bridge microdynamics since the latter does not take into account the effects of reduced filamentary overlap.

A detailed description of the transition from the charge-transfer model to a global muscle model would go far beyond the limits of this chapter. Suffice it to say that such a development is carried out similar to that described by Hatze (1981). The first step to be taken is the transition from the microdynamics of a single cross-bridge to the dynamics of a cross-bridge assemblage contained within a half sarcomere, taking into account the effects of changing filamentary overlap. Next, the nonlinear *excitation dynamics* of a single fiber has to be established and combined with its contraction dynamics, giving due consideration to all phenomena relating to force enhancement during and after stretch as well as to the additional series elastic structures that are present in muscle fibers, including tendinous tissue. Then the model must be extended to include single motor units of varying sizes and no longer single fibers as the smallest neurally controllable units in whole skeletal muscle. This step is one of the most difficult ones to perform since not only the sizes of the sequentially recruited motor units change in a highly nonlinear fashion but so do all their other properties such as maximum shortening velocity, twitch contraction time, etc. In other words, as motor units are recruited (or derecruited) during contraction in a whole muscle, the *contractive properties of that muscle change continuously and nonlinearly*. It is suggested that any model of skeletal muscle that claims to represent physiological reality should include this highly nonlinear *dynamics of motor unit recruitment*.

Finally, all parallel elastic structures present in whole muscle and the effects of oblique fiber arrangements must, of course, also be included in a model of whole muscle.

If one were to speculate about possible future developments regarding experiments on and the creation of contraction hypothesis and models of skeletal muscle, one would anticipate a high priority to be placed on the elucidation of the actual molecular events responsible for the contractile process. In all likelihood, the corresponding mathematical treatment will have to involve the methods of quantum mechanics. I am convinced that no really satisfactory global model of skeletal muscle can be developed until this basic problem has been solved.

Acknowledgment

I wish to thank Dr. A. Baca and H. Kain for their assistance with the computer simulations.

References

Bishop, A.R. (1984) Solitons in synthetic and biological polymers. In: *Nonlinear Electrodynamics in Biogical Systems* (Edited by Adey, W.R. and Lawrence, A.F.), pp. 155-175, Plenum, New York.

Borejdo, J., Putnam, S. and Morales, M.F. (1979) Fluctuations in polarized fluorescence: evidence that muscle cross-bridges rotate repetitively during contraction. *Proc. Natl. Acad. Sci. USA* **76**: 6346-6350.

Bowen, W.J. (1957) Adenosinetriphos hate and the shortening of muscular models. *J. Cell. Comp. Physiol.* **49**: Suppl. 267-290.

Cook, R., Crowder, M.S., Wendt, C.H., Barnett, V.A. and Thomas, D.D. (1984) Muscle cross-bridges: Do they rotate? In: *Contractile Mechanisms in Muscle* (Edited by Pollak, G.H. and Sugi, H.), pp. 413-423, Plenum, New York.

Curtin, N.A. and Woledge, R.C. (1978) Energy changes and muscular contraction. *Physiol. Rev.* **58**: 690-761.

Davydov, A.S. (1982) The migration of energy and electrons in biological systems. In: *Biology and Quantum Mechanics* (Edited by Holden, A.V. and Winlow, B.), Springer, Berlin.

Déléze, J.B. (1961) The mechanical properties of the semitendinosus muscle at lengths greater than its length in the body. *J. Physiol.* **158**: 154-164.

Dragomir, C.T. (1970) On the nature of forces acting between myofilaments in resting state and under contraction. *J. Theor. Biol.* **27**: 343-356.

Edman, K.A.P., Elzinga, G. and Noble, M.I.M. (1978) Enhancement of mechanical performance by stretch during tetanic contractions of vertebrate skeletal muscle fibers. *J. Physiol.* **281**: 139-155.

Edman, K.A.P., Elzinga, G. and Noble, M.I.M. (1984) Stretch of contracting muscle fibers: evidence for regularly spaced active sites along the filaments and enhanced mechanical performance. In: *Contractile Mechanisms in Muscle* (Edited by Pollack, G.H. and Sugi, H.), pp.739-749, Plenum, New York.

Elzinga, G., Lännergren, J. and Stienen, G.J.M. (1987) Stable maintenance heat rate and contractile properties of different single muscle fibres from *Xenopus Leavis* at 20°C. *J. Physiol.* **393**: 399-412.

Flitney, F.W. and Hirst, D.G. (1978) Cross-bridge detachment and sarcomere 'give' during stretch of active frog's muscle. *J. Physiol.* **276**: 449-465.

Ford, L.E., Huxley, A.F. and Simmons, R.M. (1977) Tension responses to sudden length change in stimulated frog muscle fibers near slack length. *J. Physiol. London* **269**: 441-515.

Gilbert,C., Kretzschmar, K.M., Wilkie, D.R. and Woledge, R.C. (1971) Chemical change and energy output during muscular contraction. *J. Physiol.*

London **218**: 163-193.

Green, L.E. (1981) Comparison of the binding of heavy meromyosin and myosin subfragment 1 to F-actin. *Biochemistry* **20**: 2120-2126.

Hammeroff, R.S., Smith, S.A. and Watt, R.C. (1984) Nonlinear Electrodynamics in Cytoskeletal Protein Lattices. In: *Nonlinear Electrodynamics in Biological Systems* (Edited by Adey, W.R. and Lawrence, A.F.), pp. 567-583, Plenum, New York.

Hatze.H. (1973) A theory of contraction and a mathematical model of striated muscle. *J.theor. Biol.* **40**: 219-246.

Hatze.H. (1981) *Myocybernetic Control Models of Skeletal Muscle.* University of South Africa Press, Pretoria.

Hill, D.K. (1968) Tension due to interaction between the sliding filaments in resting striated muscle. The effect of stimulation. *J. Physiol. London* **199**: 637-684.

Hill, L. (1977) A-band length, striation spacing and tension change on stretch of active muscle. *J. Physiol.* **266**: 677-685.

Hill, T.L., Eisenberg, E., Chen, Y.D. and Podolsky, R.J. (1975) Some self-consistent two-state sliding filament models of muscle contraction. *Biophys. J.* **15**: 335-372.

Hill, T.L. (1977) Free Energy Transduction in Biology. Academic Press, New York.

Hofacker, G.L. (1982) Wechselwirkungen zwischen Strukturbausteinen. In: *Biophysik* (Edited by Hoppe, W., Lohmann, W., Markl, H. and Ziegler, H.), pp. 232-239, Springer, New York.

Homsher, E., Irving, M. and Lebacq, J. (1983) The variation in shortening heat with sarcomere length in frog muscle. *J. Physiol.* **345**: 107-121.

Huxley, A.F. (1957) Muscle structure and theories of contraction. *Progr. Biophys. biophys. Chem.* **7**: 255-318.

Huxley, A.F. (1988) Muscular contraction. *Ann. Rev. Physiol.* **50**: 1-16.

Huxley, A.F. and Simmons, R.M. (1971) Proposed mechanism of force generation in striated muscle. *Nature* **233**: 533-538.

Huxley, H.E. (1979) Time resolved X-ray diffraction studies in muscle. In: *Cross-Bridge Mechanism in Muscle Contraction* (Edited by Pollack, G.H. and Sugi, H.) pp. 391-405, University Park Press, Baltimore.

Huxley, H.E. (1984) Time-resolved X-Ray diffraction studies of cross-bridge movement and their interpretation. In: *Contractile Mechanisms in Muscle* (Edited by Pollack, G.H. and Sugi, H.) pp. 161-168, Plenum, New York.

Kawai, M. and Brandt, P.W. (1980) Sinusoidal analysis: a high resolution method for correlating biochemical reactions with physiological processes in activated skeletal muscles of rabbit, frog and crayfish. *J. Muscle Res. Cell Motil.* **1**: 279-303.

Kodama, T. (1985) Thermodynamic analysis of muscle ATPase mechanisms. *Physiol. Rev.* **65**: 467-551.

Ladik, J.J. (1982) Charge-Transfer-Reaktionen in Biomolekülen. In: *Biophysik* (Edited by Hoppe, W., Lohmann, W., Markl, H. and Ziegler, H.), p.239, Springer, New York.

Laki, K. and Bowen, W.J. (1955) The contraction of muscle fiber and myosin B thread in KI and KSCN solutions. *Biochem. Biophys. Acta* **16**: 301-302.

Lomdahl, P.S. (1984) Nonlinear dynamics of globular proteins. In: *Nonlinear Electrodynamics in Biological Systems* (Edited by Adey, W.R. and Lawrence, A.F.) pp. 143-154, Plenum, New York.

Mandelkern, L. and Villarico, E.A. (1969) The effect of salts and adenosine 5'-triphosphate on the shortening of glycerinated muscle fibers. *Macromolecules* **3**: 394-401.

Mornet, D., Betrand, R., Pantel, P., Audemard, E. and Kassab, R. (1981a) Proteolytic approach to structure and function of actin recognition site in myosin heads. *Biochemistry* **20**: 2110-2120.

Mornet, D., Betrand, R., Pantel, P., Audemard, E. and Kassab, R. (1981b) Structure of the actin-myosin interface. *Nature* **292**: 301-306.

Pepe, F.A. (1983) Immunological techniques in fluorescence and electron microscopy applied to skeletal muscle fibers. In: *Handbook of Physiology, Section 10: Skeletal Muscle* (Edited by Peachy, L.D.) pp. 113-142, Am. Physiol. Soc., Bethesda.

Podolsky, R.J., Onge, R.St., Yu, L. and Lymn, R.W. (1976) X-ray diffraction of actively shortening muscle. *Proc. Natl. Acad. Sci USA* **73**: 813-817.

Pollack, G.H. (1983) The cross-bridge theory. *Physiol. Reviews* **63**: 1049-1113.

Schoenberg, M., Brenner, B., Chalovich, J.M., Greene, L.E. and Eisenberg, E. (1984). Cross-bridge attachment in relaxed muscle. In: *Contractile Mechanisms in Muscle* (Edited by Pollack, G.H. and Sugi, H.) pp. 269-279, Plenum, New York.

Scott, A.C. (1984) Solitons and Bioenergetics. In: *Nonlinear Electrodynamics in Biological Systems* (Edited by Adey, W.R. and Lawrence, A.F.), pp. 133-142, Plenum, New York.

Sugi, H. (1979) The origin of the series elasticity in striated muscle fibers. In: *Cross-Bridge Mechanism in Muscle Contraction* (Edited by Sugi, H. and Pollack, G.H.), pp. 85-102, Univerity of Tokyo Press, Tokyo.

Szilagyi, L., Bálint, M., Sréter, F.A. and Gergely, J. (1979) Photoaffinity labelling with an ATP analog of the N-terminal peptide of myosin. *Biochem. Biophys. Res. Commun.* **87**: 936-945.

Tawada, K., Kimura, M. (1984) Cross-linking studies related to the location of the rigor compliance in glycerinated rabbit psoas fibers: Is the SII portion of the crossbridge compliant? In: *Contractile Mechanisms in Muscle* (Edited by Pollack, G.H. and Sugi, H.). pp. 385-393, Plenum, New York.

Van den Hooff, H. and Blangé, T. (1984) Superfast tension transients from intact muscle fibers. *Pflügers Arch.* **400**: 280-285.

Wakabayashi, T., Toyoshima, C. and Katayama, E. (1984) Image analysis of the complex of actin-tropomyosin and myosin subfragment 1. In: *Contractile Mechanisms in Muscle* (Edited by Pollack, G.H. and Sugi, H.), pp. 21-26. Plenum, New York.

Wells, J.A. and Yount, R.G. (1979) Active site trapping of nucleotides by crosslinking two sulfhydrils in myosin subfragment 1. *Proc. Natl. Acad. Sci. USA* 76:4966-4970.

Yamamoto, K. and Sekine, T. (1979) Interaction of myosin subfragment-1 with actin. *J. Biochem.* **86**: 1855-1881.

Yanagida, T. (1984) Angles of fluorescently labelled myosin heads and actin monomers in contracting and rigor stained muscle fiber. In: *Contractile Mechanisms in Muscle* (Edited by Pollack, G.H. and Sugi, H.), pp. 397-408, Plenum, New York.

Zahalak, G.I. (1981) A distribution-moment approximation for kinetic theories of muscular contraction. *Math. Biosci.* **55**: 89-114.

CHAPTER 3

Modeling of Lengthening Muscle: The Role of Inter-Sarcomere Dynamics

David Morgan

3.1 Inter-Sarcomere Dynamics

Every muscle fiber is made up of a large number of sarcomeres connected in series, so the tension must be the same in all of them. If the sarcomeres are identical, then velocity will also be the same in all, so any externally applied movements will be equally distributed among all the sarcomeres of the fiber. This leads to the common procedure of modeling a muscle fiber, or even a whole muscle, as a scaled sarcomere. Most cross-bridge modeling of muscle makes this assumption implicitly, bearing in mind that many experimenters have gone to great lengths, by use of segment length clamps or diffraction measurements, to make it at least approximately true in their experiments. Note that throughout this chapter the sarcomere is taken to be the fundamental unit of contraction. In many ways, the half sarcomere would be a more appropriate unit, and all the comments and calculations referring to sarcomeres could equally be taken to refer to half sarcomeres.

However, it is most unreasonable to expect exact uniformity in the isometric tension-generating capability (herein referred to as strength) of the sarcomeres. There will always be some variation in cross-sectional area and in myo-filament overlap. The question is whether and when the non-uniformity is significant. Huxley and Peachey (1961) showed that fibers tetanized at long lengths exhibited internal motion, with small regions at the ends shortening and the rest of the fiber slowly lengthening. They postulated that this internal motion, together with the discontinuity of the

slope of the force-velocity curve, gave rise to the slowly rising phase (tension creep) of "fiber isometric" tension records at long lengths. A long length here refers to lengths such that at least some sarcomeres are on the descending limb of the sarcomere length-tension diagram (Gordon et al., 1966b; Julian and Morgan, 1979a, Figure 1).

The explanation runs as follows. Series connected elements with a length-tension diagram that shows tension decreasing with increasing length must be unstable. That is, inequalities in length will be accompanied by inequalities in tension generating capability that will lead to greater inequalities in length. The strong sarcomeres will shorten and get stronger, while the weak will be stretched and get even weaker. The force–velocity relation, however, shows an increasing force for increasing lengthening velocity. This will provide dynamic stability to the sarcomere length distribution. In fact for any achievable duration of tetanus, most of the sarcomeres will not have changed in length by more than a few tenths of a micron. The internal motion would not affect the tension (while all sarcomeres were on the linear descending limb) were it not for the change in slope of the force–velocity curve about zero velocity. This can be seen by considering two sarcomeres of unequal isometric tensions but a zero sum of velocities – that is, their total length is kept constant (Morgan, 1985, Figure 1). Their tensions must be equal and the velocities of equal magnitude and opposite direction. If the force–velocity curve were of constant slope, the tension that satisfied this condition would be the average of the two isometric tensions, and therefore independent of the amount of difference between the two sarcomere strengths, provided that both are on the descending limb. However, with a force–velocity curve that is steeper for lengthening than for shortening, the tension must be nearer to the isometric value of the shortening (stronger) sarcomere than to that of the lengthening (weaker) sarcomere in order to make the velocities of equal magnitude. This means that increasing the difference between the lengths and hence strengths of the two sarcomeres will increase the tension. This same argument can be generalized for many sarcomeres to show that the tension in an isometric fiber is not the average of the isometric tensions of the sarcomeres, but is closer to the isometric capability of the stronger sarcomeres.

The problem can be formalized as one of distributions and transformations. The first distribution is in sarcomere lengths. This is transformed into the distribution of sarcomere strengths by the length–tension diagram. If all sarcomeres are on the descending limb, then this is a linear transformation and the mean of the transformed distribution is equal to the transform of the mean of the original distribution. That is, the mean strength is the strength of the mean sarcomere length. The second transformation is from strength to velocity at a particular tension. This is essentially the force–velocity curve. Therefore the mean velocity at a given tension will only be the velocity appropriate to the mean sarcomere strength at that tension if the force–velocity transformation is linear. In particular the mean velocity will not in general be zero when the tension is equal to the mean strength if the force–velocity transformation is non–linear, as is found experimentally.

Gordon et al. (1966a, 1966b) showed that keeping the length of a more uniform central segment of a fiber constant, rather than the length of the whole fiber reduced the rate of rise of the "creep" tension, in agreement with the explanation offered. A smaller initial distribution of lengths and strengths is expected to lead to a lower tension early in the tetanus and less initial internal motion. This in turn means that the degree of non-uniformity and the amount of "extra" tension increase more slowly. Reducing the spread of sarcomere lengths has not been shown to reduce the final peak tension, nor does modeling predict that

it should (Morgan et al., 1982). If the tetanus is continued long enough, the non-uniformities will become large, and so will the tension.

Inter-sarcomere dynamics are also thought to be significant in slow shortening of muscle fibers (Julian and Morgan, 1979b, Figures 1 and 2) and whole muscles (Abbott and Aubert, 1952, Figure 5). Experimentally, the tension in a fiber held isometric after a slow active shortening from a long length is less than the tension generated when stimulation is commenced at the shorter final length. Rapid shortening, however, produces no tension deficit. In other words, a fiber is not able to shorten slowly up its length–tension curve, as would a crossbridge model of a sarcomere. Interrupting stimulation of the shortened fiber or whole muscle removes the deficit. Julian and Morgan showed that the slow shortening was largely absorbed by the end regions seen by Huxley and Peachey (1961). This means that most of the sarcomeres have not shortened, and so have not increased their tension generating capabilities. The initially shorter and stronger end sarcomeres have shortened out of the descending limb, and have become weaker again as they moved onto the ascending limb of the length–tension relation. However, there is always a transition zone of sarcomeres with lengths between long and short, and hence strengths greater than the tension. These are the ones that take up most of the imposed shortening. When the shortening is more rapid, the tension is less and the range of velocities for a given range of strengths is smaller. This leads to a more uniform shortening and so a final isometric tension nearer to that seen if stimulation is commenced at the final length. The model described here is able to simulate this experiment (Morgan, 1990). The depression of stiffness, or more accurately the lack of rise of stiffness, which is reported to accompany the lack of rise of tension (Tsuchiya and Sugi, 1988), is also compatible with this explanation. Most of the sarcomeres have not changed length, and therefore not stiffness, so that the fiber stiffness remains near to that at the original long length.

3.2 Sarcomere Behavior During Stretch

Despite the extensive attention given to the behavior of muscles during stretch in recent years (e.g., Cavagna et al., 1968; Huxley, 1971; Edman et al., 1978; Flitney and Hirst, 1978; Morgan et al.,

1978; Julian and Morgan, 1979a; Edman et al., 1982; Cavagna et al., 1986; Umazume et al., 1986; Colomo et al., 1988; Sugi and Tsuchiya, 1988; Tsuchiya and Sugi, 1988; Bottinelli et al., 1989; Harry et al., 1990 [see also Chapter 2 (Hatze)], relatively little consideration has been given to inter-sarcomere dynamics under these conditions. Lengthening muscle produces a number of unexpected results (see Sections 3.4 and 3.5) and also a remarkable inconsistency of results. This difficulty in reproducing results in even successive stretches on the same fiber could be seen to suggest that inter-sarcomere dynamics may be important, with the initial distribution of sarcomeres being the fundamental inconsistent factor. Woledge et al. (1985, p.71) highlighted this variability in respect to the yield ratio (the yield tension expressed in terms of the isometric tension) and change in slope of the force-velocity curve, but it is apparent in many other features of tension records during lengthening (see also Chapter 6 (Winters)).

The force–velocity curve of a fiber or whole muscle for lengthening is usually found to have a yield tension, meaning that increasing the velocity beyond a certain point does not increase the tension any further (Katz, 1939). This of course corresponds to zero incremental damping or viscosity. The statement above, that the force–velocity curve will stabilize the inherent instability of the sarcomere length distribution, is no longer true if the sarcomeres are stretching in this yielding condition. The instabilities will be catastrophic in the sense of producing very rapid changes in the sarcomere length distribution.

This becomes clear in a simple thought experiment. If a fiber is stretched at other than a very small velocity, the tension will rise as elastic structures are stretched, until it reaches the yield point of the weakest sarcomere. There will always be a weakest sarcomere, no matter how small the differences between the sarcomeres may be. At this point the weakest sarcomere will yield, that is, begin to stretch more rapidly than the others, without increasing tension. Elastic elements will also cease lengthening as the tension levels out. If the weakest sarcomere is on the descending limb of the length–tension relation, this increased lengthening will reduce its tension-generating capability. It will then be unable to support the existing tension at any velocity and so will lengthen

very rapidly indeed, limited only by inertial and passive viscous forces. Of course this lengthening will allow shortening of the rest of the fiber which will cause some reduction in tension. If the number of sarcomeres in the fiber is large, however, this reduction of the tension is likely to be less than the reduction in tension-generating capability of the weakest sarcomere, and so will not stop the rapid elongation. Eventually the passive structures within the sarcomere will produce a passive tension equal to the tension in the fiber, and extension of the sarcomere will stop. If the imposed stretch is continued, the tension will again rise until it reaches the slightly greater yield point of the next weakest sarcomere, which will then extend rapidly. This process will be repeated until the motion stops.

Our thought experiment suggests then that lengthening of a fiber will not be at all uniform, but will take place essentially by "popping" of sarcomeres, one at a time, in order from the weakest towards the strongest. This has far-reaching consequences, which will be discussed in Section 3.4. Where are the weakest sarcomeres likely to be? The shortest and strongest have been shown to be concentrated near the ends. It has been reported (Colomo et al., 1988), in accord with my own experience, that some fibers do have weak patches — that is, the weakest sarcomeres all in one part of the fiber. In these fibers the yield ratio is low, the continued rise during stretch is large, and nonuniform lengthening can be seen. Colomo et al. (1988, Figure 5) also reported a smaller change in slope of the force–velocity curve at zero velocity. In a more uniform fiber the weakest are likely to be randomly distributed throughout most of the fiber. Of course a fiber consists of many parallel myofibrils, able to move somewhat independently of each other. The popping of the randomly distributed weakest sarcomeres in a relatively uniform fiber is more likely to be a myofibrillar phenomenon than a fiber one. The weakest sarcomere in one myofibril may be at a different point along the fiber to the weakest sarcomere in the neighboring myofibril. This widespread distribution of elongated sarcomeres in myofibrils will make them difficult to detect by direct observation. No sarcomere will be extended all the way across the fiber, but scattered long sarcomeres extending only across one or a few myofibrils will be scattered in three dimensions throughout the fiber. This should be visible as increased disorder and skewing of sarcomeres. To the best of my knowledge, no quantitative measurements have been made of this. The mechanical consequences of the myofibrillar distribution of elongated sarcomeres will be small, unless significant forces are generated between myofibrils by the elongation of a sarcomere in one myofibril. Consequently the model assumes that each sarcomere has a unique sarcomere length applicable all across the fiber. In this sense it could be considered as a model of a myofibril.

3.3 The Computer Model

The model was closely based on that of Morgan et al. (1982), ran on a Macintosh computer (Apple Computer Inc., Cupertino, CA), and was written in Lightspeed Pascal (Think Technologies, Bedford, MA, now a division of Symmantec) using the Programmers Extender (Invention Software Corp, Ann Arbor, MI). One half of the muscle fiber was modeled as either 100 or 500 sarcomeres (or contractile units) connected in series. In order to accommodate different muscles and temperatures, the unit of time was defined as the time for all unloaded sarcomeres to shorten 1 μm. Thus the unloaded shortening velocity was 1 micron per sarcomere per time unit. For the usual frog single fibers, this means that one time unit corresponds to about 500 ms at 0°C, and about 50 ms at 20°C. Each sarcomere was represented by a Hill type model, consisting of a contractile component characterized by a force–velocity curve, a linear series elastic component (default stiffness required 0.024 μm per half sarcomere to drop the tension from isometric to zero), and an exponential parallel elastic component. The force–velocity curve was taken as the classic Hill–Katz curve as quantified by Morgan et al. (1982). Constants could be entered for a/F_o {where a is the Hill parameter and F_o is the Hill contractile element force intercept [see also Chapter 5 (Winters)]; default here: 0.25}, the change of slope between slow lengthening and slow shortening (default: 6 times higher for lengthening), the asymptote for lengthening (default: $1.8 F_o$), and a curvature coefficient for the lengthening region. The unloaded shortening velocities of all sarcomeres were the same. For each sarcomere the isometric tension (strength) was taken as the product of the length-tension

curve of Gordon et al. (1966b) and the specified isometric capability at optimum length F_{Oopt} of that sarcomere. The distribution of F_{Oopt} was specified as an exponential distribution with a random variation added. Values for the end sarcomere, the central sarcomere, the length constant of the exponential distribution, the random component amplitude, and the random number generator seed were all specified by the user. The random component was generated by smoothing a series of pseudo-random numbers generated by the computer, giving an approximately Gaussian distribution. The initial length distribution was very similarly specified. The basic passive tension curve was an offset exponential, specified by the slack length, a length constant of the exponential and the tension at some specified sarcomere length, and applied to the sarcomere at the center of the fiber end. The passive tension curves for the other sarcomeres were scaled from this so that the specified sarcomere length distribution produced the same passive tension in all the sarcomeres. No provision has been made for sarcomere lengths less than the slack length, so no simulations involving slack fibers can yet be run.

The length changes to be applied were specified by the times of beginning and ending the ramp, and the final average sarcomere length. The initial average sarcomere length was calculated from the initial sarcomere length distribution. The program included facilities to display, save, and print the length–tension relation, the force–velocity relation, the passive curve for any sarcomere, the movement being applied, and the distributions of F_{Oopt}, strength and of sarcomere length, as well as the tension–time record being produced. Sarcomere length, F_{Oopt}, and strength could also be displayed as histograms. In addition a "segment length" record was obtained by adding the sarcomere lengths of the "central" half of the fiber. The simulation could be stopped at any time to examine these curves and then resumed.

The solution proceeded iteratively as before (Morgan et al., 1982). The time intervals for the calculation were not equal, but varied automatically to accommodate the rate of change of tension. No interval greater than 0.0001 time units (less during very rapid ramps) was accepted if tension changed more than 2% of isometric tension or if the length of any sarcomere changed more than 0.2 μm within that time interval. An option allowed the fiber to be replaced by a single sarcomere, with parameters equal to the average of those for the fiber. Tension was plotted on the screen as simulation progressed.

Experience using the model justified several of the assumptions made in the thought experiment above. It was found that for realistic passive tension curves and initial sarcomere lengths, a popped sarcomere was extended well beyond zero overlap. Of particular importance was the confirmation that the sarcomere lengths are instantaneously unstable for any reasonable assignment of parameters and number of sarcomeres, even as low as 100. The only way to actually find a solution during the rapid elongation phase was to add a small amount of damping to the series elastic component of the sarcomere model. This meant that the tension fell to the minimum of the curve of total sarcomere tension against sarcomere length as each sarcomere popped. In practice this procedure led to a large number of very short time intervals and excessive calculation time. Consequently a "quick pop" option was provided, whereby no attempt was made to track a sarcomere through popping.

If the "quick pop" option was enabled, then in each time interval, after solving for the tension but before updating the sarcomere lengths, a check of the sarcomere nearest to its yield point was made to see if its updated yield point would be less than the existing tension, that is, whether the tension would need to be reduced in the next time interval. If so, the present time step was repeated with half the duration. When a tension decrease would be required and the time interval was at the minimum allowed (0.0001 time unit), then the sarcomere would be popped by setting its isometric capability (F_{Oopt}) to zero. This ensured that only its passive tension would be used from then on, and that during the next time interval, it would be stretched appropriately. This led to an instantaneous fall in tension, as the other sarcomeres shortened. This option reduced the calculations considerably, but was shown not to affect the results perceptibly other than during the actual "popping" of the sarcomere.

3.4 Modeling Results and Discussion

The model with 100 sarcomeres showed a large abrupt drop in tension each time a sarcomere popped. This enabled the pattern of popping to be easily observed, but produced unrealistic tension traces. Using 500 sarcomeres produced much more realistic records, though individual pops are still discernible. Most real muscle fibers have many more than 500 sarcomeres, so that an even smoother trace would be expected. Figure 3.1a shows the simulated tension during a stretch, and while isometric at the initial and final lengths. This figure shows that the model simulates a large number of the peculiar features of muscle being stretched.

3.4.1 Continued Tension Rise During Stretch

Most experimenters agree that stretching a muscle at constant velocity produces a tension that continues to rise throughout the stretch, whether the sarcomere lengths are on the plateau or descending limb of the length–tension relation (see in particular Harry et al., 1990). Shortening at constant velocity within the plateau of the length–tension diagram, on the other hand, produces a tension that is much more nearly constant. The lengthening behavior is inconsistent with "normal" cross-bridge models, as the distribution of cross-bridge extensions, and hence the tension, should reach a steady state after stretching more than a few cross-bridge strokes. On the descending limb, of course, tension should fall as the overlap of thick and thin filaments is reduced. This was seen when the model was set to single sarcomere, as shown in Figure 3.2.

If inter-sarcomere dynamics dominates, then the continued rise represents popping ever stronger sarcomeres. The continued rise is inherent in the model. The tension trace is just a series of yield points, each one greater than the last. On the descending limb of course there will be a countervailing effect of the slow reduction of strength of all the sarcomeres that are slowly lengthening below their yield point. This may account for occasional observations of tension falling during stretch, such as in Edman et al. (1978, Figure 2b). Variations in the pattern of sarcomere non-uniformity will account for the variable amount of rise seen between experimental records. A large spread of sarcomere lengths in the model can lead to a rounding of the yield corner as sarcomeres are popped from early in the stretch, as shown in Figure 3.2. During shortening the slope of the force–velocity curve ensures that the sarcomeres shorten much more uniformly, and the single sarcomere cross-bridge models are more nearly correct.

3.4.2 Permanent Extra Tension

The model records also show the phenomenon of permanent extra tension [Abbott and Aubert, 1952; Julian and Morgan, 1979b; Edman et al., 1982; Woledge et al., 1985; Sugi and Tsuchiya, 1988; Tsuchiya and Sugi, 1988; see also Chapter 2 (Hatze)]. After a stretch at long length, the tension does not fall to the level appropriate to the final length, regardless of how long the tetanus is

continued. The final tension is near that for an isometric contraction at the original length, and almost independent of the amplitude or speed of the stretch. These are exactly the results found in muscle fibers. Looking at the final sarcomere length distribution (Figure 3.1b) shows what has happened. Most of the lengthening has been taken up by the few popped sarcomeres, and the rest have only stretched a little. Furthermore the weakest sarcomeres have been "removed" from

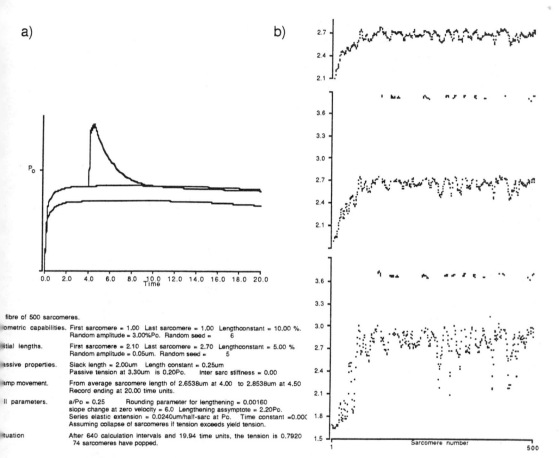

fibre of 500 sarcomeres.

ometric capabilities.	First sarcomere = 1.00 Last sarcomere = 1.00 Lengthconstant = 10.00 %. Random amplitude = 3.00%Po. Random seed = 6	
itial lengths.	First sarcomere = 2.10 Last sarcomere = 2.70 Lengthconstant = 5.00 % Random amplitude = 0.05um. Random seed = 5	
assive properties.	Slack length = 2.00um Length constant = 0.25um Passive tension at 3.30um is 0.20Po. Inter sarc stiffness = 0.00	
amp movement.	From average sarcomere length of 2.6538um at 4.00 to 2.8538um at 4.50 Record ending at 20.00 time units.	
ll parameters.	a/Po = 0.25 Rounding parameter for lengthening = 0.00160 slope change at zero velocity = 6.0 Lengthening assymptote = 2.20Po. Series elastic extension = 0.0240um/half-sarc at Po. Time constant =0.000 Assuming collapse of sarcomeres if tension exceeds yield tension.	
ituation	After 640 calculation intervals and 19.94 time units, the tension is 0.7920 74 sarcomeres have popped.	

Figure 3.1: The model behavior during stretch. *a)* Tension traces during fixed end activation at initial and final lengths and during an active stretch of 0.2 μm between these lengths. Note the following points. The longer length produces less isometric tension, owing to the length–tension curve. The isometric traces show a "tension creep" or slow rise phase followed by a slow decline. The tension continues to rise throughout the stretch. The popping of individual sarcomeres produces the ripple during stretch. The tension after the stretch does not fall to the tension generated while isometric at the same length. The details in the figure relate to the stretch record. *b)* The distribution of sarcomere lengths at the beginning of contraction, at the end of the stretch, and at the end of the simulation. Note the "popped" sarcomeres, scattered throughout the fiber, and extended beyond the length of filament overlap. After such a long tetanus, the sarcomeres are quite non-uniform.

the distribution of active sarcomeres, giving a slight tendency for the tension to rise above that at the original length. This is counteracted by the slight elongation of the other sarcomeres. Combined with the older evidence suggesting sarcomere non-uniformity (interrupting the stimulation long enough for the tension to fall, and then resuming produces the tension appropriate to the final length (Julian and Morgan, 1979b, Figure 6)), these records make the inter-sarcomere dynamics explanation of permanent extra tension

very attractive. Note that a decrease in stiffness during the stretch is also predicted by the model, as observed by Tsuchiya et al. (1988) and Sugi et al. (1988). The decrease in this model, however, is not due to a decreased overlap and hence number of crossbridges as they postulated, but to the increasing number of popped sarcomeres. The stiffness of the passive tension curve is less than that of an active sarcomere, so popping more sarcomeres decreases the stiffness, even though most of the sarcomeres have not lengthened.

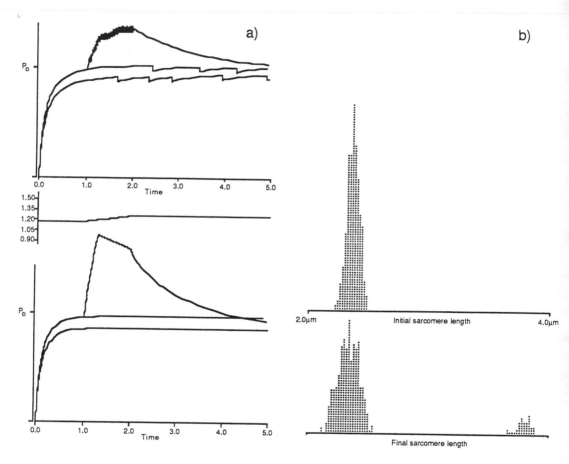

Figure 3.2: A fiber with a wider distribution of F_{Oopt} a) The upper traces are for a fiber of 500 sarcomeres with a 10% random variation of isometric capabilities and a 0.05 μm random variation in initial sarcomere lengths. The lower traces are for a single sarcomere with the average properties of those in the upper traces. The middle trace shows the movement, calibrated in mm for the fiber. Note the following points. The fiber has a smaller yield ratio than the sarcomere, and the yield point is much less distinct. The tension continues to rise throughout the fiber stretch but falls for the single sarcomere, owing to the sarcomere length–tension curve. Isometric contractions can also involve popping of sarcomeres. For the single sarcomere the tension decays towards that appropriate to the final length, but stays above for the fiber. b) The initial and final sarcomere length distributions shown as histograms. Initial sarcomere lengths 2.35 μm plus 0.05 m random variation. Slack length is 2.1 μm. Passive tension at 3.1 μm is 0.2 F_o. Stretch amplitude is 0.15 μm. Default force–velocity curve is utilized.

3.4.3 Force-Velocity Curve at High Speeds

Stretching the simulated fiber produces a tension that is almost independent of the stretch velocity except for very small velocities. This is to be expected, since the tension during stretch is determined mainly by the distribution of yield points for the various sarcomeres. The velocity has a minor effect through the lengthening of the other sarcomeres below their yield points. Experiments also show that the tension of a muscle or fiber during constant velocity stretch is essentially independent of velocity beyond the yield point. A cross-bridge model, however, can only show this under specific, rather unlikely assumptions. As a sarcomere is stretched at higher and higher speeds, the opportunity for cross-bridges to form becomes less and less, and in most models the number of attached sarcomeres decreases. The only way to avoid a fall in tension is to increase the average tension per cross-bridge, and hence the average extension of the cross-bridges. It has been shown (Harry et al., 1990) that imposing a limit on cross-bridge extension even as large as six times the maximum extension in an isometric muscle produces predictions that depart significantly from experiment.

Inter-sarcomere dynamics can easily accommodate a sarcomere force–velocity curve that falls at large velocities. Once a sarcomere yields and becomes unable to support the existing tension at any speed, the tension that it can support becomes essentially irrelevant. This is shown by the fact that using the "quick-pop" option in the model, equivalent to a sarcomere force–velocity curve that falls to zero immediately past the yield point, does not affect the overall tension trace. In this way the sarcomere force–velocity curve can fall at high speed, but the fiber force–velocity curve does not.

3.5 Explanations of Other Phenomena

The ideas behind the model can also be seen to provide explanations for other puzzles about lengthening muscle that have not yet been modeled.

3.5.1 Isotonic Experiments

Simulations of isotonic experiments are not appropriate to the present model, as they rely heavily on the moderately fast transient behavior of sarcomeres, which is not realistically modeled by a Hill model. The behavior to be expected from a consideration of inter-sarcomere dynamics can however be predicted, and compared to experiment. If a model fiber is subjected to a load greater than the yield point of some of its sarcomeres, all those sarcomeres unable to support the imposed tension will quickly pop, giving a rapid lengthening of the fiber. When that has happened, the rate of stretch of the fiber will drastically slow to that due to the sub-yield lengthening of the sarcomeres with yield points greater than the imposed tension. Such slowing of the rate of lengthening is the behavior that has been seen in real fibers, and has proved so difficult to explain by cross-bridge or any other models (Huxley, 1971, Figure 6; Huxley, 1980, p. 84). (How can enough cross-bridges form in a rapidly extending fiber to resist the imposed load when they were unable to do so in the isometric fiber?) This would also explain the difficulty experienced by Pollack's group (e.g., Granzier et al., 1989) in plotting the force–velocity curve using isotonic stretches. They found that increasing the isotonic load increased the amplitude of the immediate lengthening, but not the steady lengthening velocity that followed (Granzier et al., 1989). Sustained rapid isotonic lengthening does not occur experimentally.

3.5.2 Damage from Eccentric Contractions

Another peculiarity of muscle being lengthened while active is its propensity to damage. Step tests (Friden et al., 1983; Newham et al., 1983) and arm curls (Clarkson and Tremblay, 1988) in humans, and downhill running in rats (Armstrong et al., 1983), all produce damage with the following characteristics. Immediately after the exercise, the only changes seen are small areas of elongated sarcomeres, sometimes as small as one half sarcomere in one myofibril. In other cases a group of elongated sarcomeres, extending part or all of the way across the fiber, are seen. The next day, the muscles involved are painful, and histology shows damaged muscle fibers being replaced. The degree of damage is not related to the general fitness of the subject. Everyday experience also shows that sports involving eccentric contractions such as horse riding and mountain climbing often produce such "delayed-onset muscle soreness," while concentric exercise sports such as swimming and bicycle riding usually do not. The proposal of non-uniform lengthening provides the mechanism for the initial local damage, which can then lead to destruction of the fiber (Armstrong, 1984, esp. p. 35).

When a sarcomere "pops", it is extended, probably to the point where there is no overlap of thick and thin filament arrays. When the muscle relaxes, it is likely that the inter-digitating pattern of the filaments is not fully resumed immediately. (The question of the extent and time course of the recovery of a popped sarcomere is an area that needs more experimental investigation.) This provides a weak point during the next stretch, and increases the stress on the neighboring myofibrils at that sarcomere. In this way, repeated stretches can be envisaged to produce a microscopic tear in the fiber. At some point this tear damages sarcoplasmic reticulum or sarcolemma, allowing uncontrolled release of calcium, and "clot" formation.

In single fibers, stretching will sometimes kill fibers that have withstood many isometric and/or shortening contractions (personal observations). My own recent observations suggest that these fibers often have a low fiber yield ratio, consistent with a wide spread of sarcomere strengths. (See below and Colomo et al., 1988.)

3.6 Modeling Conclusions/Predictions

This modeling has led to several conclusions beyond the general principle of the non-uniform lengthening of muscle. If the yield point of the fiber is the yield point of the weakest sarcomere, but the isometric tension of the fiber is biased towards the isometric capability of the stronger sarcomeres, then the yield ratio of the fiber (yield tension divided by isometric tension) must be less than for its sarcomeres. This means that the yield ratio for a sarcomere must be greater than the highest value ever seen in a fiber of that type. That is, the true value for frog single-fiber sarcomeres should not be taken as the mean of the fiber observations (approx. 1.8), but as the largest, at least 2.1. In addition, it is concluded that a low yield ratio can be taken as indicative of a wide dispersion of sarcomere strengths.

In the model, a steep rate of rise of tension during a stretch is also indicative of a wide sarcomere strength distribution. Although the two parameters are measuring slightly different aspects of the distribution, a general correlation is still

predicted by the model. No specific experiments have yet been undertaken to test this point, but a brief examination of the literature provides some support. Certainly the rise during stretch tends to be greater at longer lengths, where non-uniformities may be expected to be greater.

Similarly, inter-sarcomere dynamics suggests that the change in slope of the force–velocity curve about zero for a fiber will always be less than the change for its sarcomeres, since the tension for a slow stretch will be measured later in the contraction, when the non-uniformity will be greater, and hence the yield point of the weakest sarcomere less. This similarly means that the sarcomere value for the slope change must be at least as high as the highest ever seen for a fiber. My measurements of Figure 5 of Colomo et al. (1988) suggest a value nearer 9 than the classically assumed 6. The isotonic measurements of Granzier et al. (1989) produced values even higher than that. If the difficulties found in measuring the force–velocity curve (continued rise of tension) by stretching at low velocities are due to non-uniformities, as seems likely, then the isotonic experiments, which quickly pop all the very weak sarcomeres before non-uniformities become worse, may be the best method of measurement. Note that with more sarcomeres in the fiber, the deviation of strengths required to ensure that at least some are below a given threshold will be less.

3.7 Future Directions

3.7.1 Experimental Confirmation

Perhaps the greatest objection to the suggestion that inter-sarcomere dynamics dominate the response of muscle to stretch is the absence of direct evidence of such large sarcomere non-uniformities. Clearly if the popped sarcomeres are few and widely scattered in individual myofibrils, then seeing them will not be easy. In particular their effect on a diffraction pattern is likely to be complex, and measurements of segment length are not appropriate.

Other indirect evidence can be sought. If the myofibrils of popped sarcomeres do not fully return to their inter-digitating pattern on relaxation, particularly after long and/or fast stretches, then cumulative effects should be observable. Preliminary experiments with Drs. Julian and Claflin have shown increased apparent series compliance and shifts in the fiber length for optimum tension. Both of these were permanent and cumulative, in that repeating the stretches increased the changes. Further experiments to explore the various parameters are under way.

Other experiments that could provide useful information include quantitative measurements of the disorder of sarcomeres, and looking for a correlation between the yield ratio and the rate of continued tension rise during stretch.

3.7.2 Relevance to Whole Mammalian Muscles

Nearly all the experiments described here have been for single frog muscle fibers fully tetanized at near-freezing temperatures (the exception being damage from eccentric contraction). How directly relevant are these ideas to whole mammalian muscle sub-maximally activated at normal body temperature?

Whole muscles have much more passive tension than single fibers, often to the point of not having a descending limb in a plot of total tension against muscle length. However, much of the additional elasticity is probably not effectively in parallel with the individual (half) sarcomeres, so that the sarcomeres probably do have a region of decreasing tension with increasing length. Permanent extra tension has certainly been shown in whole toad sartorius (Abbott and Aubert, 1952). Whole muscle does show a continuing rise during stretch, although the absence of a clear plateau of isometric tension without significant passive tension complicates the experiment.

The yield ratio of tetanized mammalian muscle is commonly less than for frog fibers. One possible interpretation is that mammalian muscles have a greater range of sarcomere strengths. However, differences in species and temperature make this suggestion rather speculative. The effects of submaximal activation are even more difficult to evaluate, partly because of the difficulty in doing experiments on sub-maximally activated single fibers. If the effect of motion on a submaximally activated sarcomere is similar to the tension traces observed in whole muscle, that is, a collapse of tension as the stretch is continued, then instability is very probable and inter-sarcomere dynamics are likely to be very important. However, inferring the effect of lengthening on sarcomere tension from observations of the effect of lengthening on fiber tension in this situation is

far from trivial.

3.7.3 Refinement of the Model

Future work in this area is currently concentrated on improving the sarcomere model along the lines discussed by Zahalak (Chapter 1) so that isotonic experiments can be simulated. Consideration of sub-maximal activation and a study of the energetics of eccentric contractions should also be facilitated by this development.

References

Abbott, B.C. and Aubert, X.M. (1952) The force exerted by active striated muscle during and after change of length1. *J. Physiol.* **117**:78-86.

Armstrong, R.B., Ogilvie, R.W. and Schwane, J.A. (1983) Eccentric exercise induced injury to rat skeletal muscle. *J. Appl. Physiol. Respirat. Environ. Exercise Physiol.*, **54**:80-93.

Armstrong, R.B. (1984) Mechanism of exercise-induced delayed onset muscular soreness: A brief review. *Med. Sci. Sports Exerc.* **16**:529-538.

Bottinelli, R., Eastwood, J. C. and Flitney, F. W. (1989) Sarcomere 'give' during stretch of frog single fibers with added series compliance. *Q. J. of Exp. Physiol.* **74**:215-217.

Cavagna, G.A., Dusman, B. and Margaria, R. (1968) Positive work done by a previously stretched muscle. *J. Appl. Physiol.* **24**:21-32.

Cavagna, G.A., Mazzanti, M., Heglund, N.C. and Citterio, G. (1986) Mechanical transients initiated by ramp stretch and release to P_o in frog muscle fibers. *Am. J. Physiol.* **251**:C571-C579.

Clarkson, P.M. and Tremblay, I. (1988) Exercise-induced muscle damage, repair and adaptation in humans. *J. Appl. Physiol.* **65**:1-6.

Colomo, F., Lombardi, V. and Piazzesi, G. (1988) The mechanisms of force enhancement during constant velocity lengthening in tetanized single fibers of frog muscle. *Adv. Exp. Med. Biol.* **226**:489-502.

Edman, K.A.P., Elzinga, G. and Noble, M.I.M. (1978) Enhancement of mechanical performance by stretch during tetanic contractions of vertebrate skeletal muscle fibers. *J. Physiol.* **281**:139-155.

Edman, K.A.P., Elzinga, G. and Noble, M.I.M. (1982) Residual force enhancement after stretch of contracting frog single muscle fibers. *J. Gen. Physiol.* **80**:769-784.

Flitney, F.W. and Hirst, D.G. (1978) Cross-bridge detachment and sarcomere give during stretch of active frog's muscle. *J. Physiol.* **276**:449-465.

Friden, J., Sjöström, M. and Ekblom, B. (1983) Myofibrillar damage following intense eccentric exercise in man. *Int. J. Sports Med.* **4**:170-176.

Gordon, A.M., Huxley, A.F. and Julian, F.J. (1966a)

Tension development in highly stretched vertebrate muscle fibers. *J. Physiol.* **184**:143-169.

Gordon, A.M., Huxley, A.F. and Julian, F.J. (1966b) The variation in isometric tension with sarcomere length in vertebrate muscle fibers. *J. Physiol.* **184**:143-169.

Granzier, H.L.M., Burns, D.H. and Pollack, G.H. (1989) Sarcomere length dependence of the force-velocity relation in frog single muscle fibers. *Biophy. J.* **55**:499-507.

Harry, J.D., Ward, A.W., Heglund, N.C., Morgan, D.L. and McMahon, T.A. (1990) Crossbridge cycling theories cannot explain high-speed lengthening behavior in frog muscle. *Biophys J.* **57**:201-208.

Huxley, A.F. (1971) The activation of striated muscle and its mechanical response. *Proc. Roy. Soc. Lond. B.* **178**:1-27.

Huxley, A.F. (1980) *Reflections on Muscle.* Liverpool University Press, Liverpool.

Huxley, A.F. and Peachey, L.D. (1961) The maximum length for contraction in vertebrate striated muscle. *J. Physiol.* **156**:150-165.

Julian, F.J. and Morgan, D.L. (1979a) Intersarcomere dynamics during fixed-end tetanic contractions of frog muscle fibers. *J. Physiol.* **293**:365-378.

Julian, F.J. and Morgan, D.L. (1979b) The effect on tension of non-uniform distribution of length changes applied to frog muscle fibers. *J. Physiol.* **293**:379-392.

Katz, B. (1939) The relationship between force and speed in muscular contraction. *J. Physiol.* **96**:45-64.

Morgan, D.L. (1985) From sarcomeres to whole muscles. *J. Exp. Biol.* **115**:69-78.

Morgan, D.L. (1990) New insights into the behavior of muscle during active lengthening. *Biophys. J.* **57**:209-221.

Morgan, D.L., Mochon, S. and Julian F.J. (1982) A quantitative model of inter-sarcomere dynamics during fixed-end contractions of single frog muscle fibers. *Biophys. J.* **39**:189-196.

Morgan, D.L. Proske, U. and Warren, D. (1978) Measurements of muscle stiffness and the mechanism of elastic storage in hopping kangaroos. *J. Physiol.* **282**:253-261.

Newham, D.J., McPhail, G., Mills, K.R. and Edwards, R.H.T. (1983) Ultra-structural changes after concentric and eccentric contractions in human muscle. *J. of the Neurol. Sci.* **61**:109-122.

Sugi, H. and Tsuchiya, T. (1988) Stiffness changes during enhancement and deficit of isometric force by slow length changes in frog skeletal muscle fibers. *J. Physiol.* **407**:215-229.

Tsuchiya, T. and Sugi, H. (1988) Muscle stiffness changes during enhancement and deficit of isometric force in response to slow length changes. *Adv. in Exp. Med. and Biol.* **226**:503-511.

Umazume, Y., Onodera, S. and Higuchi, H. (1986) Width and lattice spacing in radially compressed frog skinned muscle fibers at various pH values, magnesium ion concentrations and ionic strengths. *J. Muscle Res. Cell Motil.* **7**:251-258.

Woledge, R.C., N.A. Curtin, N.A. and Homsher, E. (1985) *Energetic Aspects of Muscle Contraction.* Academic Press, London.

CHAPTER 4

Architecture and Elastic Properties of the
Series Elastic Element of Muscle-Tendon Complex

Gertjan J.C. Ettema and Peter A. Huijing

4.1 Introduction

4.1.1 Functions of Series Elastic Element

Series elasticity in skeletal muscle is considered to be of great importance for muscle
functioning in several ways. For example, in
movement control studies, the musculo-skeletal
system is often modelled as a mass–spring complex, in which the stiffness characteristics of the
springs determine a joint equilibrium position
which will be obtained at certain activation levels
of the muscles (Schmidt, 1982). This type of
modelling is also applied for studying mammalian
running gaits with respect to movement speed,
type of gait and energy expenditure [McMahon,
1985; Chapter 37 (McMahon); Taylor, 1985].
Furthermore, the series elastic element (*SE*) takes
up part of length changes of the muscle–tendon
complex, which means that the contractile element
(*CE*) does not "see" all of the muscle–tendon complex movement [see also Chapter 38 (Hof)]. An
approach in principle similar to these behavioral
models can be applied to series elastic tendinous
structures (i.e., part of *SE*) and muscle fibers. This
is important, especially in reflex movements in
which information from muscle spindles play a
crucial role (e.g. Rack et al., 1983; Rack and Ross,
1984). In biomechanics and physiological studies
information about *SE* compliance is important for
estimating elastic energy uptake and release in
stretch-shortening cycles [e.g. Komi , 1984;
Cavagna 1977; Haan et al., 1989; Chapter 38
(Hof)].

4.1.2 Localizing and Modelling *SE*

SE is located in several compartments of the
muscle–tendon complex, acting differently in relation with muscle force. Within the scope of the
cross-bridge theory, compliance (short range) of
the muscle fibers was thought to be located mainly
in the cross-bridge attachments (e.g. Blangé et al.,
1972; Bressler and Clinch, 1974 and 1975; Ford et
al., 1981). However, in recent years some
evidence has been provided in favor of the
hypothesis that the myofilaments contain a considerable amount of fiber compliance (Sugi and
Tameyasu, 1979; Blangé et al., 1985). In the
literature considering whole muscle–tendon complexes, *SE* is usually assumed to be localized
mainly in tendinous structures and attached cross-
bridge linkages (e.g., Morgan, 1977; Proske and
Morgan, 1987; Ettema) and Huijing , 1990).
Tendinous structures can be considered undamped
elastic springs (Rack and Ross, 1984), while *SE* located in the fibers strongly depends on contraction
dynamics. Furthermore, within fiber compliance,
the influence of contraction dynamics is different
for myofilament and cross-bridge compliance
(Bressler and Clinch, 1974).

For pennate muscles, the angle of aponeurosis
(intra-muscular tendinous sheet) and muscle fibers
with the line of pull introduces an extra compliance on the level of the muscle–tendon
complex, which will be referred to as angular compliance in this chapter. Series elastic components
in pennate muscles have been incorporated in
modelling muscle–tendon complexes by Otten
(1985) and Zajac et al. (1986). However, Otten

Multiple Muscle Systems: Biomechanics and Movement Organization
J.M. Winters and S.L-Y. Woo (eds.), © 1990 Springer-Verlag, New York

(1985) only considered the elastic aponeurosis to study its effects on the muscle length–force relation. Zajac et al. (1986) used a model in which the entire tendinous structures run parallel with the line off pull, and consequently only short-range stiffness of the fibers was placed under an angle of pennation. In the present chapter emphasis will be put on the effects of pennation on the contribution of angular compliance. The model we used for this purpose is presented in Section 4.3.

It may be clear that for an accurate description, *SE* cannot simply be modelled as a single spring (either linear or non-linear, damped or undamped): force–elongation characteristics of *SE* located within cross-bridges differ, depending on the number of attached cross-bridges (type of active force exertion). The question remains which simplifications of modelling *SE* are justified for which purposes [see also Chapter 5 (Winters)]. We will distinguish the major parts of *SE* of a muscle–tendon complex and give a description of their behavior. We already made some simplifications of SE characteristics by merely discussing elastic properties, and neglecting viscous and other properties of *SE* [see Chapter 5 (Winters) for discussion of the effects of this assumption]. Furthermore, we assume fiber compliance to be entirely located in the cross-bridge linkages (i.e., that myofilaments are very stiff): the main purpose of this chapter is to study influence of muscle–tendon complex geometry on compliance, for which it is not necessary to model elastic fiber characteristics by means of two components.

4.2 Review of Pertinent Literature: SE in Tendinous Structures and Muscle Fibers

4.2.1 Methods

Several methods have been developed to measure SE characteristics. Three important methods are described briefly by Bahler (1967) and Close (1972) and discussed further in Chapter 5 (Winters): *i)* the (isotonic) quick release method; *ii)* a method using fast constant velocity releases; and *iii)* a method calculating compliance from the force–time curve of an isometric tetanic contraction. Application of such methods only yields results concerning compliance of the entire SE. In the past some estimates were made to determine the relative contribution of the tendinous structures to SE compliance (e.g. Jewell and Wilkie,

1958; Joyce and Rack, 1969).

More recently, some elegant methods were developed to distinguish the contribution of tendinous structures to *SE* from that of the fibers. Morgan (1977) modelled the muscle–tendon complex as consisting of two springs connected in series: one spring with a constant stiffness was assigned to the tendinous structures, while the other, with a stiffness proportional to force, was assigned to the intracellular structures (i.e., cross-bridges). By means of this model, stiffness measured during quick length changes can be separated into tendon stiffness and fiber stiffness. This method, the *alpha* method, is valid only for force levels above 20% of maximal isometric force (F_o), at which tendon stiffness was assumed to be more or less constant (Morgan, 1977; Proske and Morgan, 1987). However, for rat *GM* and *EDL* Ettema and Huijing (1990) showed that compliance of the tendinous structures decreases with increasing force even at force levels near F_o. They developed the so-called *extension* method, based on a similar model as used in the *alpha* method (Morgan, 1977), but they did not assume tendon compliance to be constant. In the *extension* method the elastic extension of all tendinous structures at F_o is experimentally determined (by means of photographic techniques). In this way estimates of tendon compliance at high as well as low force levels were obtained. Furthermore, use of photographic techniques allows consideration of differences between free tendon and aponeurosis. Especially for low force ranges, another method, the *spindle-null* method, was developed by Rack and Westbury (1984) for estimating compliance of intact tendinous structures. They applied small sinusoidal stretches to the muscle–tendon complex, and simultaneously modulated muscle activation sinusoidally. One particular modulation resulted in a silence of the discharge of the muscle spindles (the null–point). This indicates that fiber length changes were zero and that all stretch applied to the muscle–tendon complex should be assigned to the tendious structures.

4.2.2 Results

Jewell and Wilkie (1958) reported that about 50% of length changes of *SE* of frog sartorius muscle, during a fully isometric contraction, resided in the tendinous structures. For movements of interphalangeal joint of the human thumb, the tendinous structures of the flexor pollicis longus take up more than 60% of total muscle-tendon complex movement (Rack and Ross, 1984). Morgan et al. (1978) found about eight times more movement in the tendinous structures than in the fibers (within its short-range stiffness) during stretching of maximally contracting *GM* of the kangaroo. For cat soleus this amount is about 50% (Morgan, 1977). For rat *EDL* and *GM* Ettema and Huijing (1990) found that about 85% of *SE* extension at F_o resides in the tendinous structures. These results indicate that differences of elastic characteristics may exist between species and muscles. On the other hand, some of the differences may be explained methodologically. For example Ettema and Huijing, (1990) (*extension method*) included compliance of tendinous structures at low force levels, while Morgan (1977) and Morgan et al. (1978) (*alpha* method) did not. Table 4.1 gives an overview of the literature on this issue.

The three methods discussed here estimate compliance of the entire tendinous structures, but do not discriminate characteristics of aponeurosis from tendon characteristics. In their review, Proske and Morgan (1987) emphasized the possibility of differences between tendon and aponeurosis stiffness. The extension method gives the opportunity to distinguish tendon and (superficially located) aponeurosis characteristics.

Ettema and Huijing (1990) found significant differences between normalized tendon and aponeurosis extension for rat *GM* (7.97 ± 1.13% and 3.41 ± 0.48% respectively, *n* = 7). For rat *EDL* qualitatively similar results were found (2.68% and 1.89% extension for tendon and aponeurosis respectively), but these values could not be shown to be significantly different.

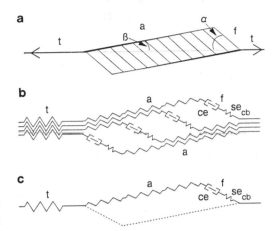

Figure 4.1: *a)* Representation of the planimetric model as used by Huijing and Woittiez (1984) and Otten (1988). *b)* Representation the geometrical organisation of the series elastic and contractile elements, within the planimetric model. *c)* Simplified geometry of the model, containing a single fiber.

Table 4.1: Literature on tendon and fiber compliance of *in situ* muscle-tendon complexes.

Reference	Species/Muscle	Method	Temp. (°C)	F (N)	F (%)	Compliance (mm/N) tendon	fiber
Morgan, 1977	Cat/Soleus	Alpha	-	22	100	0.03 - 0.08	0.04 - 0.05
Morgan et al., 1978	Kangaroo/GM	Alpha	30 - 33	135	100	0.04	0.007
Walmsley and Proske, 1981	Cat/Soleus	Alpha	36 - 38	15	-	0.05 ± 0.02	0.043 ± 0.009
				12	-	0.05 ± 0.02	0.053 ± 0.011
	Cat/GM	Alpha	36 - 38	15	-	0.06 ± 0.02	0.026 ± 0.002
				5	-	0.06 ± 0.02	0.079 ± 0.007
Proske and Morgan, 1984	Cat/Soleus	Alpha	35 - 37	22	100	0.074 - 0.096	0.065 ± 0.07
Rack and Westbury, 1984	Cat/Soleus	Spindle-null	35 - 37	5	25	0.05 - 0.17	-
				10	50	0.03 - 0.08	-
Ettema and Huijing, 1990b	Rat/EDL	Alpha Extension	25 25	3 3	100 100	0.092 ± 0.034 0.138 ± 0.063	0.128 ± 0.011 0.057 ± 0.047
	Rat/GM	Alpha Extension	25 25	11 11	100 100	0.035 ± 0.008 0.048 ± 0.003	0.042 ± 0.008 0.020 ± 0.005

4.3 Ongoing Efforts: Angular Compliance

4.3.1 Model Description and Methods

The fact that, in pennate muscle, muscle fibers are positioned under an angle with the line of pull of the muscle results in a transformation of the active fiber length-force curve to muscle length-force characteristics (Huijing and Woittiez, 1984 and 1985; Otten, 1985 and 1988; Woittiez et al., 1984). Huijing and Woittiez (1984) developed a planimetric model of a uniform unipennate muscle, incorporating a stiff aponeurosis (Figure 4.1a). Otten (1988) showed the transfer function of this model to be

$$F_m = F_f * \cos(\alpha+\beta)/\cos(\beta) \qquad (4.1)$$

$$dl_m = dl_f * \cos(\beta)/\cos(\alpha+\beta) \qquad (4.2)$$

where F_m and F_f are force exerted by the muscle and fiber respectively, dl_m and dl_f are infinitely small length changes of muscle and fiber respectively, and α and β are angles of fiber and aponeurosis with line of pull. This transformation implies a reduction of force, in exchange for an enhanced length change of the muscle compared to the fiber, and conforms to the requirement of constant muscle volume (i.e., area of the parallelogram in the planimetric model) as well as to the requirement of a work balance between the muscle and fiber. Another implication is that the slope of the length-force curve (dl/dF) of the muscle is increased with respect to that of the fiber (i.e.,elastic linkages of cross-bridges). This means that, with increasing pennation, muscle stiffness is reduced (i.e.,compliance is increased) compared to fiber stiffness. For the planimetric model, incorporating an infinitely stiff aponeurosis, the transfer function of fiber compliance (C_f) to muscle compliance (C_m) can be obtained mathematically. Mathematical analysis is much more complicated if elastic tendinous structures (especially regarding the aponeurosis) are considered. We tackled this problem by modelling a uniform unipennate muscle, according to the planimetric model, introducing an elastic aponeurosis and tendon connected in series with the muscle. Setting up a free-body diagram of the aponeurosis *Eq. 4.1, 4.3* and *4.4* are obtained, representing muscle force and force exerted on the aponeurosis and tendon as a function of fiber force:

$$F_a = F_f * \cos(\alpha+\beta) \qquad (4.3)$$

$$F_t = F_m \qquad (4.4)$$

F_a and F_t are force exerted on the aponeurosis and tendon respectively. Keeping the area of the muscle constant (equivalent to constant muscle volume in a three-dimensional model) in the model, for each condition of the muscle–tendon complex, lengths of the elements and forces exerted on them, as well as muscle geometry, can be found iteratively.

The rule of balance of work between the muscle–tendon complex and its elements appeared to be fulfilled if elastic work performed perpendicular to the aponeurosis is taken into account: in order to keep muscle volume constant when the aponeuroses is elastically stretched, an elastic deformation must also occur in the surface connecting the aponeuroses (elastic deformation of a constant volume body can only occur if more than one side is elastic, i.e.,if a body becomes longer it also becomes thinner). When elastic energy is stored in the aponeurosis, at the same time some energy will be stored in this surface connecting the aponeuroses. The amount of this energy is small but essential for obtaining the balance of work. It equals the product of force directed perpendicular to the aponeurosis (obstructing the aponeuroses to move towards each other) and the actual change of distance between the aponeuroses. The muscle model contains a simplification, since aponeurosis and fiber are kept straight and do not curve during muscle shortening. It is more likely that changes of the shape of the muscle induced by lengthening of the aponeurosis consists of curving of fibers (Otten, 1988).

We used this model to simulate quick release experiments performed on *GM* and *EDL* muscle–tendon complexes of the rat (Ettema and Huijing, 1990). Using estimated elastic characteristics for each element and for the entire muscle–tendon complex, shortening and force decrement during the release were calculated. In this way compliance of the entire complex and its elements could be compared. The difference is considered angular compliance (C_{ang}) induced by pennation:

$$C_{ang} = C_{cmplx} - (C_t + C_a + C_f) \qquad (4.5)$$

where C_{cmplx} represents compliance of the entire muscle-tendon complex and C_t, C_a, and C_f are compliance of tendon, aponeurosis, and fiber, respectively.

Table 4.2: Parameter constants of rat *GM* and *EDL* used in simulation models.

	GM	EDL
lt.eq (mm)	8.0	16.0
la.eq (mm)	19.2	21.2
lf.eq (mm)	14.5	12.4
lm.eq (mm)	32.7	33.2
α.eq (°)	16.1	11.6
β.eq (°)	12.1	6.8
lfo (mm)	14.5	12.4
lfs (mm)	8.7	7.44
Ffo (mm)	12.0	3.2
Compliance constants		
Δla = k√Fa: k	0.166	0.237
Δlt = r√Fm: r	0.115	0.268
Δlf = t: t	0.174	0.15
Δlf = t'Ff: t'	0.0145	0.0469

Parameters of rat *GM* and *EDL* muscle–tendon complex, derived from Ettema and Huijing (1990) and used in the model, are shown in Table 4.2. Elongation of the tendinous structures was assumed to be linearly related to the square root of force (Ingen Schenau et al., 1988). A value of 1.2% of fiber optimum length was assumed for extension of the elastic cross-bridge linkages at isometric force level. This is somewhat lower than the value found by Ettema and Huijing (1990) (i.e., 1.5%), because they included angular compliance in estimates of fiber compliance. The force–elongation relationship of a single cross-bridge was assumed to be linear (Blangé et al., 1972). We constructed force–compliance relationships by simulating quick release experiments for two conditions. *a)* Muscle force was changed by performing simulations for different muscle-tendon complex lengths, using maximal levels of stimulation. Muscle level was varied from just above optimum length to active slack length (i.e., the shortest length at which active force is generated). *b)* Muscle force was changed by varying stimulation level(s) between 0 to 1, where

$$F_f(s) = F_f(1) * s \tag{4.6}$$

Literature is not unequivocal about the ascending limb of the sarcomere length–force curve in terms of number of attached cross-bridges. Consequently, it is not clear in which way short–range stiffness (as determined by the number of attached cross–bridges) is related to fiber force at lengths below fiber optimum length. Therefore, we used two extremes for characteristics of fiber short-range stiffness for *condition A*. *A):* Fiber stiffness (i.e. number of cross-bridges) is linearly related to fiber force, resulting in the same extension (*t*) of the elastic linkage at isometric force level for all fiber lengths. *A'):* Fiber stiffness is constant for all fiber lengths and force levels, i.e., the number of attached cross-bridges does not decrease with diminishing fiber length (Stephenson et al., 1989). This yields

$$\Delta l_f = t' * F_{fi} \tag{4.7}$$

where Δl_f is elastic elongation of the cross-bridge linkage and F_{fi} is isometric fiber force. For *construction A* fiber compliance equals $t*F_{fi}$ and for *construction A'* it is equal to t'. Values t and t' were chosen such that at F_o fiber compliance was the same for both conditions. It should be noted that, by definition, the number of attached cross-bridges remains constant during the release for both conditions. Therefore, during the release, fiber compliance is determined by the number of attached cross-bridges prior to the onset of release, and by the elastic characteristics of a single cross-bridge (which has a constant compliance; see above). Consequently, short-range fiber stiffness does not change during the quick release, but is completely determined by the initial condition before the release.

The model was also applied to experimental results of quick-release experiments performed on three *GM* muscle–tendon complexes of the rat. In this case C_{cmplx} was determined experimentally rather than calculated by running the model. As a consequence fiber compliance could be calculated as model outcome rather than that a C_f value had to be assumed. Force–length relations of tendon and proximal aponeurosis were determined during an isometric contraction at muscle optimum length, using high speed film (200 *Hz*). *Equation 4.8* was fitted to these force–length data of tendon and aponeurosis and compliance was calculated for each quick release experiment by means of differentiation of *Eq. 4.8* with respect to force.

$$l_j = u * F_j^v + k \tag{4.8}$$

where l_j is length of element j, F_j is force exerted on element j (j = tendon or aponeurosis). The small proximal tendon of *GM* could not be detected on the photographs. Therefore, to be sure that all tendinous compliance "seen" by the muscle belly was included in the model calculations, elongation of the free tendon was measured by means of shortening of the muscle belly (which are the exact equal but opposite by sign in an isometric contraction). This way all length changes occurring within the equipment were included within tendon extension. Afterwards, tendon compliance was corrected for compliance of the equipment (0.014 mm/N). Using experimentally determined muscle–tendon complex compliance (measured by means of quick length decreases of 0.2 mm within 3 ms) and geometry measurements obtained from the high–speed film (i.e., length of muscle, distal fiber, proximal aponeurosis, and free tendon; accuracy of 0.05 mm) a model run resulted in an estimation of C_f and C_{ang} for each quick release experiment, which were performed at different muscle lengths ranging from a few millimeters above to about 8 mm below optimum length (*condition A* for the simulations).

4.3.2 Results

Simulation results of *GM* and *EDL* are shown in Figure 4.2. The choice of *construction A* or *A'* for fiber compliance does influence the relationship between force and compliance for the entire muscle–tendon complex (C_{oi}) extensively. The type of method used for regulation of muscle force (by means of *A:* muscle length or *B:* stimulation level) does influence force–compliance characteristics only to a small extent (Figure 4.2a). The small differences between *conditions A* and *B* for *GM* rat are mainly due to a higher angular compliance in the *A* condition (Figure 4.2d): in *condition B* the muscle–tendon complex remains at optimum length for all force levels, with relatively small angles of pennation. For rat *EDL* the differences are negligible, because of the small pennation angle. This was also found experimentally by Ettema and Huijing (1989) and indicates that *construction A* is more likely than *A'*, the latter one showing considerable differences with *condition b* (Figure 4.2a).

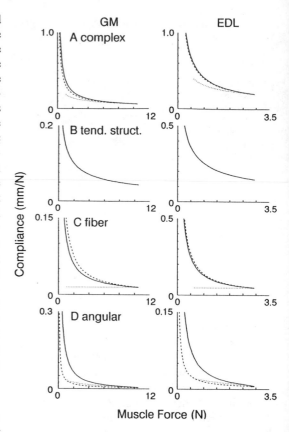

Figure 4.2: Compliance results of models of *GM* and *EDL* muscle–tendon complex and their elements, performing quick length changes of -0.05 mm. Solid line: model *construction A*, simulations performed at several muscle–tendon complex length at and below optimum length under supramaximal stimulation conditions. Dashed line: model *construction B*, simulations performed at optimum length with different submaximal activation levels. Dotted line: model *construction A'*, same as *construction A*, with altered elastic characteristics of the fiber. Model parameters are shown in Table 4.2.

The degree of pennation (i.e., muscle geometry), however, is not the only factor determining the relative contribution of angular compliance to total compliance. This is shown in Figure 4.3, where C_{ang}, expressed as percentage of muscle compliance (i.e., total compliance minus

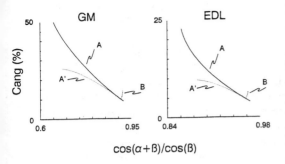

$$\cos(\alpha+\beta)/\cos(\beta)$$

Figure 4.3: Relative contribution of angular compliance to muscle belly compliance (compliance of free tendon excluded) as a function of pennation index. Model *constructions* (*A* and *A'*) and simulation *conditions* (*A* and *B*) are indicated.

C_t), is plotted as a function of the transfer function between fiber force and muscle force (*Eq. 4.1*), used as indication for the degree of pennation. The contribution of angular compliance is much lower for *construction A'* than for *A*. The compliance of the fiber is very small for all force

levels in *construction A'* (Figure 4.2c), whereas aponeurosis compliance is the same for both constructions (Figure 4.2b). Coincidentally, the pennation angle of the fiber is larger than that of the aponeurosis. This causes a smaller contribution of angular compliance in *construction A'*. Furthermore, *condition B* (all simulations at optimum length, i.e., relatively small angles of pennation) results in a quite different curve, even though force–compliance relations of the elements (i.e., fiber and aponeuorsis) are the same as in *condition A*. Such interaction effects of muscle geometry and elastic characteristics of the muscle elements on angular compliance can be proved mathematically for muscle models incorporating a stiff aponeurosis (Zajac et al., 1986).

Figure 4.4 shows experimental results of three *GM* muscle-tendon complexes, of which morphological and physiological data are presented in Table 4.3.

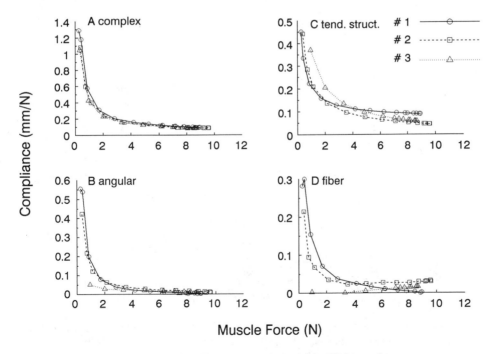

Figure 4.4: Compliance of three rat *GM* muscle–tendon complexes and their elements. Marker and line types indicate the different muscles.

Table 4.3: Morphological and physiological data of the experimental rat *GM* muscles.

	experiment		
	# 1	# 2	# 3
lt.eq (mm)	10.1	9.9	8.3
la.eq (mm)	16.6	19.9	20.1
lfo (mm)	11.1	11.7	11.1
lmo (mm)	28.0	30.6	30.9
αo (°)	13.9	18.0	19.6
βo (°)	8.9	10.5	10.3
Fo (N)	9.94	10.67	9.66
Δlt [Fo] (mm)	0.8	0.9	1.3
Δla [Fo] (mm)	0.6	0.5	1.0
Coi [Fo] (mm/N)	0.094	0.084	0.082

Compliances of the muscle–tendon complexes are quite similar. However, substantial differences occur for the compliances of the separate elements. Fiber compliance of *muscle 3* is very small for the entire force range (Figure 4.4d). This is caused by the rather high compliance values of the tendinous structures compared to those of *muscle 1* and *muscle 2* (Figure 4.4c and Table

4.3). These values are so high that they take up almost all of the muscle-tendon complex compliance. In similarity with the simulation results, this low fiber compliance results in a low angular compliance (Figure 4.4b). Considering the shape of the F-C_f curve (Figure 4.4d) results for *muscle 1* are more or less in accordance with results of *construction A*, while *muscle 2* and *muscle 3* agree with *construction A'*. For all muscles, however, a sizable contribution of angular compliance is seen. The relative contribution of tendinous structures, fiber, and angular compliance is plotted in Figure 4.5, for the experimental muscles and the *GM* model under *condition A* (both constructions). It is concluded that at optimum length (F_o) about 8% of muscle–tendon complex compliance is due to pennation, while at short muscle lengths (force levels of about 0.2 F_o) this is about 25%. For submaximally stimulated *GM*, pennation contributes approximately 12% at all force levels (model calculations).

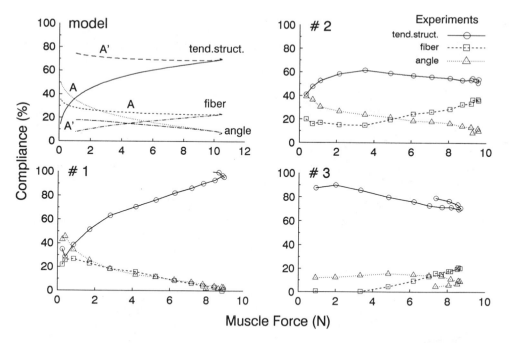

Figure 4.5: Relative contribution of the tendinous structures, fibers, and angular compliance to compliance of the entire muscle-tendon complex, for the *GM* model (*construction A and A'*) and three ex- perimental rat *GM* muscle–tendon complexes. Range of muscle force was induced by measuring and simulat- ing at different muscle–tendon complex lengths.

4.4 Discussion and Future Directions

4.4.1 Compliance Characteristics of Elements of *SE*

Within the scope of this chapter elastic characteristics of the tendinous structures can be described in a rather simple way, i.e. the force–elongation curve can be represented by a single non-linear function. Viscous behavior of these structures seems to be of minor importance compared to elastic behavior (Rack and Ross, 1984).

However, the fact that the aponeurosis forms a part of the surface of a constant volume body influences its effective compliance. Model calculations on work balance (see Section 4.3.1) point out that more elastic energy is stored in the muscle than in the elastic surfaces (i.e., aponeurosis and fiber). Generally speaking, the elastic characteristics of a body containing a constant volume differ from the elastic characteristics of the elastic surface of such a body, irrespective of architecture. For example, in order to lengthen an elastic hollow tube a certain amount, a certain force is necessary which is determined by the elastic characteristics of the tube. If the same tube is sealed at the ends, resulting in a constant volume body, a somewhat higher force is needed for the same elongation: the constant volume induces in an extra resistance against elongation of the tube. The elongation enforces a narrowing of the tube perpendicular to the direction of elongation. An extra amount of elastic energy will be stored in this narrowing deformation of the tube. This means that the longitudinal compliance of the constant-volume system will be lower than compliance of the tube itself. A similar situation occurs for the muscle and its aponeuroses: some elastic energy is stored in the muscle surface, by means of deformations directed perpendicular to the aponeurosis. In a similar way, as stated above, the effective compliance of a muscle with constant volume is somewhat lower than the total compliance of isolated elastic elements. Note that in the model the functional compliance is lowered because of this mechanism. The elastic surface connecting the aponeuroses is not "forgotten"; in the model the elastic characteristics of this surface is implicitly included by the assessment of the aponeurosis characteristics and conforming the model to the constraint of a constant muscle surface area. As a consequence of the lowered effective aponeurosis compliance, angular compliance is somewhat underestimated, since this lowering was not considered in *Eq. 4.5*.

Angular compliance of pennate muscles is not only determined by the geometry of the muscle (i.e., the degree of pennation expressed in fiber and aponeurosis angle with respect to the line of pull), but also by the compliance of fiber and aponeurosis. This is also true for the relative contribution of angular compliance to compliance of the muscle. A twofold increase of the compliance of the muscle elements does not simply result in a twofold increase of angular compliance. This is caused by interactions between elastic length changes of the elements and geometrical changes of the muscle. Furthermore, the ratio of fiber and aponeurosis compliance as well as angles will influence angular compliance. However, generally speaking, for unipennate muscles large pennation angles result in high angular compliance.

The dynamics of cross-bridge cycling and consequently its influence on the elastic characteristics of the fibers are beyond the scope of this chapter. Therefore we modelled two extreme circumstances for the fiber force–compliance relationship (*constructions A and A'*). For the experimental results fiber characteristics were obtained as calculated output of a model run, in contradiction with the original model for which fiber compliance was a prerequisite parameter. These experimental results also do not give a decisive answer on the issue concerning fiber compliance characteristics. This is not surprising, given the uncertainties in experiments on whole muscle–tendon complexes: in the first place measurement errors of characteristics of the elastic tendinous structures may obscure the exact force–compliance relationship of the rather stiff fibers. Furthermore, the fiber force–compliance curve is a characteristic of the entire fiber population of the muscle. Any non-uniformity within this population results in a transformation of the curve of the representative fiber. Moreover, the elastic characteristics found for the fibers might partially represent compliance characteristics of the myofilaments rather than cross-bridge linkages (Sugi and Tameyasu, 1979; Blangé et al., 1985), which complicates interpretation of the results. However, the results presented in this chapter do give clear information about the relative contribution of the fibers to total compliance of the muscle–tendon complex. For *GM* this contribution lies in the same order as that of angular compliance (about 10% to 25%).

4.4.2 Other Properties of *SE*: Shift of Aponeurosis Length

A peculiar behavior of the aponeurosis has not been mentioned yet: recent studies show that the length of the aponeurosis not only depends on force but also seems to depend strongly on muscle length (Huijing and Ettema, 1988/89; Ettema and Huijing, 1989). At short muscle lengths in active condition, the aponeurosis appeared to be shorter than at optimum muscle length in passive condition, even though total muscle force was highest in the former condition. Furthermore, we (Ettema and Huijing, 1989) showed that length differences between the aponeurosis in active short muscle (low force) and active muscle at optimum length exceeded extension of entire *SE* calculated on basis of compliance measurements. We hypothesized that muscle length itself determines the equilibrium length of the aponeurosis, which is thus shifted along with muscle length (the mechanism was not explained).

To explain shifts in muscle equilibrium length, Alexander and Johnson (1965) proposed a plastic deformation of frog muscle, induced by active force. Such a deformation of the aponeurosis, induced by muscle length changes, may be a mechanism explaining the shift of aponeurosis length. One can imagine this as follows: in passive muscle the aponeurosis length can easily be adapted by straightening the tendinous fibers within the meshwork of the aponeurosis. During muscle activity, however, high forces perpendicular to the aponeurosis (i.e., pressure forces) occur. This results in internal frictional forces within the meshwork of the aponeurosis, obstructing straightening the tendon fibers, which is imposed by tensile forces. In other words, unstraightened tendinous fibers might still be present in the aponeurosis, even though high tensile fibers are exerted. This is not possible in free tendon because of absence of the muscle pressure force. This hypothesis predicts a plastic-like deformation of aponeurosis, induced by lack of force, in contradiction to the plastic deformation as proposed by Alexander and Johnson (1965), for which a certain force level is needed to evoke the deformation. At this moment this hypothesis must be considered rather speculative, and clearly much more information about the morphological structure and behavior of the aponeurosis is needed to test such hypotheses. This hypothesis is in contradiction to results of a single *GM* muscle–tendon complex, indicating that during a slow concentric contraction similar length changes of the aponeurosis occur as between separate isometric contractions at different muscle lengths (Huijing and Ettema, 1988/1989).

Another explanation for the aponeurosis shift coinciding with muscle length changes is that muscle force does not reflect force exerted on the aponeurosis in any way. One way to explain the rather wide length-force curve of a rat *EDL* muscle concerns a distribution of lengths of the muscle–tendon complex, at which several groups of fibers act at their own optimum length. If such a distribution occurs, active fiber force is not exerted uniformly along the aponeurosis. It is even possible that at short muscle lengths the most distal fibers do not exert any force at all, while the whole muscle is still producing force. In this situation one can only speculate what will happen with the proximal aponeurosis, the aponeurosis of which the shift was measured (Bobbert et al., 1990). However, for *EDL* muscle–tendon complex, our present calculations indicate that such non-uniformities of the fiber population cannot explain the aponeurosis shift entirely (unpublished results). To be able to test any hypothesis along this line of thought, one should measure morphometrical changes of the distal as well as proximal fiber and aponeurosis. This way, information is obtained concerning non-uniformities within the fiber population of a muscle.

4.4.3 Consequences for *In Vivo* Movements and Motor Control

It should be noted that any change of compliance of a muscle–tendon complex due to the geometrical configuration of its elements does not influence the amount of elastic energy stored in the series elastic element. The rule of balance of work between the elements within the muscle and the muscle as one unity also holds in terms of compliance and elastic energy storage: the increased elastic elongation of the muscle, owing to pennation, is quantitatively related to reduced muscle force. As a result elastic energy calculated using muscle–tendon complex force–compliance characteristics is equal to elastic energy stored in the elements (fiber and aponeurosis). However, angular compliance may affect energetics of active muscle indirectly. For example, during eccentric contractions, angular compliance enhances the amount of lengthening of a muscle–tendon complex, which can be taken up by means of elastic elongation without stretching of the contractile element. In other words, angular compliance increases the possibility of a muscle–tendon complex to act within its short-range stiffness (i.e., no cross-bridge detachment during lengthening), avoiding loss of energy due to cross-bridge cycling. For similar reasons activity of muscle spindles of pennate muscle will be influenced by angular compliance during changes of muscle–tendon complex length. Therefore this certainly will have its consequences for movement control (Maier et al., 1972) in a similar way as tendon compliance has, particularly in reflex movements [Chapter 35 (McMahon)].

Within the scope of muscle compliance and motor control it is therefore interesting to draw more attention to the relation between architecture and specific function of a muscle–tendon complex.

References

Alexander, R.S. and Johnson, P.D. (1965) Muscle stretch and theories of contraction. *Am. J. Physiol.,* **208:** 412-416.

Bahler, A.S. (1967) Series elastic component of mammalian skeletal muscle. *Am. J. Physiol.,* **213:** 1560-1564.

Blangé, T., Karemaker, J.M. and Kramer, A.E.J.L. (1972) Elasticity as an expression of cross-bridge activity in rat muscle. *Pflugers Archiv.,* **336:** 277-288.

Blangé, T., Stienen, G.J.M. and Treijtel, B.W. (1985) Active stiffness in frog skinned muscle fibres at different Ca concentrations. *J. Physiol.,* **366:** 65P.

Bobbert, M.F., Ettema, G.J.C. and Huijing, P.A. (1990) The force-length relationship of a muscle-tendon complex: experimental results and model calculations. *Eur. J. appl. Physiol.,* accepted.

Bressler, B.H. and Clinch, N.F. (1974) The compliance of contracting skeletal muscle. *J. Physiol.,* **237:** 477-493.

Bressler, B.H. and Clinch, N.F. (1975) Cross bridges as the major source of compliance in contracting skeletal muscle. *Nature,* **256:** 221-222.

Cavagna, G.A. (1977) Storage and utilization of elastic energy in skeletal muscle. *Exercise Sport Sci. Rev.,* **5:** 89-129.

Close, R.I. (1972) Dynamic properties of mammalian skeletal muscles. *Physiol. Rev.,* **52:** 129-197.

Ettema, G.J.C. and Huijing, P.A. (1989) Properties of the tendinous structures and series elastic component of EDL muscle-tendon complex of the rat. *J. Biomech.,* **22:** 1209-1215.

Ettema, G.J.C. and Huijing, P.A. (1990) Contributions to compliance of series elastic component by tendinous structures and cross-bridges in rat muscle-tendon complexes. Submitted to *J. Biomech.*

Ford, L.E., Huxley, A.F. and Simmons, R.M. (1981) The relation between stiffness and filament overlap in stimulated frog muscle fibres. *J. Physiol.,* **311:** 219-249.

Haan, A. de, Ingen Schenau, G.J. van, Ettema, G.J., Huijing, P.A. and Lodder, M.A.N. (1989) Efficiency of rat medial gastrocnemius muscle in contractions with and without an active prestretch. *J. Exp. Biol.,* **141:** 327-341.

Huijing, P.A. and Ettema, G.J.C. (1988/89) Length-force characteristics of aponeurosis in passive muscle and during isometric and slow dynamic contractions of rat gastrocnemius muscle. *Acta Morphol. Neerl.-Scand.,* **26:** 51-62.

Huijing, P.A. and Woittiez, R.D. (1984) The effect of architecture on skeletal muscle performance: A simple planimetric model. *Neth. J. Zool.,* **34:** 21-32.

Huijing, P.A. and Woittiez, R.D. (1985) Notes on planimetric and three-dimensional muscle models. *Neth. J. Zool.,* **35:** 521-525.

Ingen Schenau, G.J. van, Bobbert, M.F., Ettema, G.J., de Graaf, J.B. and Huijing, P.A. (1988) A simulation of rat EDL force output based on intrinsic muscle properties. *J. Biomech.,* **21:** 815-824.

Jewell, B.R. and Wilkie, D.R. (1958) An analysis of the mechanical components in frog's striated muscle. *J. Physiol.,* **143:** 515-540.

Joyce, G.C. and Rack, P.M.H. (1969) Isotonic lengthening and shortening movements of cat soleus muscle. *J. Physiol.,* **204:** 475-491.

Komi, P.V. (1984) Physiological and biomechanical correlates of muscle function: effects of muscle structure and stretch-shortening cycle on force and speed. *Exercise Sport Sci. Rev.,* **12:** 81-121.

Maier, A., Eldred, E. and Edgerton, V.R. (1972) The effects on spindles of muscle atrophy and hypertrophy. *Exp. Neurol.,* **37:** 100-123.

McMahon, T.A. (1985) The role of compliance in mammalian running gaits. *J. Exp. Biol.,* **115:** 263-282.

Morgan, D.L. (1977) Separation of active and passive components of short-range stiffness of muscle. *Am.*

J. Physiol., **232**: C45-C49.

Morgan, D.L., Proske, U. and Warren, D. (1978) Measurements of muscle stiffness and the mechanism of elastic storage of energy in hopping kangaroos. *J. Physiol.*, **282**: 253-261.

Otten, E. (1985) Morphometrics and force-length relations of skeletal muscle. In *Biomechanics IX-AA* (ed. D.A. Winter et al.). Champaign, Illinois: Human Kinetic Publishers, pp. 27-32.

Otten, E. (1988) Concepts and models of functional architecture in skeletal muscle. *Exercise Sport Sci. Rev.*, **16**: 89-137.

Proske, U. and Morgan, D.L. (1984) Stiffness of cat soleus muscle and tendon during activation of part of muscle. *J. Neurophysiol.*, **52**: 459-468.

Proske, U. and Morgan, D.L. (1987) Tendon stiffness: methods of measurement and significance for the control of movement. a review. *J. Biomech.*, **20**: 75-82.

Rack, P.M.H., Ross, H.F., Thilmann, A.F. and Walters, D.K.W. (1983) Reflex responses at the human ankle: the importance of tendon compliance. *J. Physiol.*, **344**: 503-524.

Rack, P.M.H. and Ross, H.F. (1984) The tendon of flexor pollicis longus: its effects on the muscular control of force and position at the human thumb. *J. Physiol.*, **351**: 99-110.

Rack, P.M.H. and Westbury, D.R. (1984) Elastic properties of the cat soleus tendon and their functional importance. *J. Physiol.*, **347**: 479-495.

Schmidt, R.A. (1982) *Motor Control and Learning. A Behavioral Emphasis.* Human Kinetic Publishers, Champaign, Illinois. pp. 267-270.

Stephenson, D.G., Stewart, A.W. and Wilson, G.J. (1989) Dissociation of force from myofibrillar MgATPase and stiffness at short sarcomere lengths in rat and toad skeletal muscle. *J. Physiol.*, **410**: 351-366.

Sugi, H. and Tameyasu, T. (1979) The origin of the instanteneous elasticity in single frog muscle fibres. *Experientia*, **35**: 227-228.

Taylor, C.R. (1985) Force development during sustained locomotion: A determinant of gait, speed and metabolic power. *J. Exp. Biol.*, **115**: 253-262.

Walmsley, B. and Proske, U. (1981) Comparison of stiffness of soleus and medial gastrocnemius muscles in cats. *J. Neurophysiol.*, **46**: 250-259.

Woittiez, R.D., Huijing, P.A., Boom, H.B.K. and Rozendal, R.H. (1984) A three dimensional muscle model: A quantified relation between form and function of skeletal muscles. *J. Morphol.*, **182**: 95-113.

Zajac, F.E., Topp, E.L. and Stevenson, P.J. (1986) A dimensionless musculotendon model. *Proc. 8th ann. conf. of IEEE Engng. in Med. and Biology Soc.* IEEE, Piscataway NJ, pp. 601-604.

CHAPTER 5

Hill-Based Muscle Models:
A Systems Engineering Perspective

Jack M. Winters

5.1 Introduction

Chapter 1 (Zahalak) provided a brief historical treatment of the early findings that led to the muscle model structure first proposed by A.V. Hill (1938). From a "systems engineering" perspective, this is a phenomenologically based, lumped-parameter model that is based on interpretations of input-output data obtained from controlled experiments. Simply stated, this model consists of a *contractile element* (*CE*) that is surrounded, both in series and in parallel, by "passive" connective tissue (Figure 5.1). *CE* is furthermore characterized by two fundamental relationships: *CE* tension–length and *CE* force–velocity. Each of these is modulated by an activation input that is structurally distinct from the location for mechanical coupling between the muscle and the environment (Figure 5.1).

This model has been (and remains) the model of choice for most modeling studies of multiple muscle movement systems. For example, it is explicitly utilized in Chapters 6–8, 10, 19, 21, 31, 38–39, and 41–43.

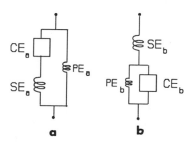

Figure 5.1: Classical structures for the Hill muscle model, with contractile element (*CE*), series element (*SE*, larger spring), parallel element (*PE*).

Part of the reason for the wide use of this model structure is that it describes most of the salient features of muscle mechanics quite well. A related reason is due to the ease with which model parameter values can be estimated; in fact, by my count over two dozen papers from over one dozen distinct groups have provided raw data explicitly in a form compatible with the Hill model structure.

However, as outlined in Chapter 1 (Zahalak), this model is far from perfect. Criticisms come from two extremes:

i) The model is *too simple*, and fails to capture certain fundamental features of real muscle. Indeed, there have been suggestions, based on results from certain classes of controlled experiments on isolated fibers or a single muscle, that this model and its fundamental structure is inherently flawed and inadequate (Chapters 1–2). Consequently, it is suggested that it should be discarded in favor of more realistic (and complex) models.

ii) The model is *overly complex*, with this complexity excessive for studies addressing underlying principles regarding multiple muscle movement organization. As such, it may limit one's vision. Rather, from this perspective muscle is viewed as either a force generator (e.g., based on interpretation of rectified and smoothed *EMGs*) or a spring (often with variable stiffness or rest length). This approach is used, either implicitly or explicitly, in Chapters 9, 12, 16–17, 24–27, 29–30, 32–33, 37, 40, 44–46).

Our primary goal in this chapter is to address these criticisms, with an emphasis on the first, which is a more serious charge. Such an assess-

Multiple Muscle Systems: Biomechanics and Movement Organization
J.M. Winters and S.L-Y. Woo (eds), © 1990 Springer-Verlag

ment of the strengths, weaknesses, and limitations of Hill-based models appears timely and especially appropriate within this book. Our secondary goal is to provide a reasonably deep foundation that will enable the novice to understand and effectively utilize (or reject!) this modeling foundation.

Our perspective will be that of a *"systems engineer"* interested in using a muscle model to help gain *insight into dynamic movement systems that involve multiple muscles*. We furthermore specify that adequate simulation only over a moderate operating range is unacceptable. More explicitly, we want to utilize a muscle model that we can *trust*: while not perfect, each muscle modeled will *never be far off* for *any* musculoskeletal movement task of potential interest. Such a model would have a high capacity for providing insight and a negligible capacity for providing misinformation of significance.

In Section 5.2 we start with a brief historical introduction to the Hill-based model from a "systems" perspective, complimenting the review in Chapter 1 (Zahalak). Since the series and parallel elements within this model represent connective tissue, we then review the basic mechanical properties of such tissues. We then briefly distinguish between the structural forms of current Hill-based models. With the foundation established, in Section 5.3 we consider muscle properties, especially as related to constitutive relations for the basic components of the model.

In Section 5.4 we address experimental results that challenge the Hill model foundation. Our goal here is to: *i)* outline the nature of any discrepancies between experiment and Hill-model predictions; *ii)* see whether techniques exist for enhancing Hill-based so that these discrepancies are insignificant; and *iii)* assuming discrepancies still exist (or elimination comes at too high a computational or conceptual cost), discern whether or not these discrepancies are physiologically relevant for *in-situ* musculoskeletal systems. In Section 5.5 we outline some of the insights that are possible when utilizing Hill-based models that might be "missed" if muscle is simply assumed to be an idealized force generator or a simple spring, or if one is interpreting *EMG* data without the aid of a mathematical model. Finally, in Section 5.6 a few possible future directions are suggested.

5.2 Hill-Based "Systems" Muscle Model

5.2.1 Modeling Foundations

"Input–Output" Experiments

The conceptual foundation behind the early experimental investigations of Hill, Fenn, Wilkie, and colleagues have their root in what would now be called "systems physiology" or "systems engineering": controlled "black box" systems identification with fairly well-defined inputs and outputs, leading to models with a predictive capacity. This "systems" approach differs somewhat from the "reductionist" approach on which biophysical models of muscle are based (e.g., Chapters 1–2). (Although these two approaches often evolve somewhat independently, both are necessary.)

As seen in Figure 5.2, there are three potential "inputs" to a muscle: muscle load, muscle length, and muscle stimulation (secondary effects such as temperature or extracellular ionic balances are ignored here). The traditional experimental approach has been to specify the input sequence for neural stimulation and then *either* force or length/velocity; the other of the latter two becomes the measured output trajectory. Typically, one input is held constant while the other is varied as an impulse, step, ramp, sinusoid, or a "white noise" (random signal). Notice that many combinations are possible, and, as seen in Figure 5.2, many of the classic combinations have been given specific names.

The above input–output description is a simplification. As shown in Figure 5.2f, in practice the controlling/measuring apparatus that is coupled to the system may possess significant dynamics, which here could represent the mass of the measuring apparatus. Mechanically, there is inherently bidirectional energetic coupling between the muscle and its environment [see Chapter 7 (Hannaford and Winters) and Chapter 9 (Hogan)], and thus the "output" measure may be influenced by the confounding effects such as apparatus mass/inertia. Of note is that the early pioneers were very much aware of this influence and went to lengths to minimize such effects [e.g. the widely used two-lever scheme developed by Bouckaert et al. (1930)]. For human studies, efforts were made by Wilkie (1950) to mathematically subtract out the effects of inertia.

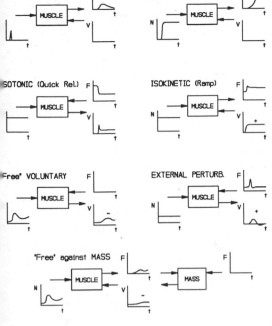

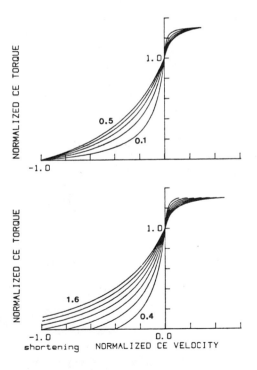

Figure 5.2: Input–output combinations for skeletal muscle. In each case the neural input (N) is on the left and the mechanical variable pair, force (F) and velocity (v), is on the right. Notice that in each case there are two input sequences and one output trajectory to be measured. An "impulse" is as an input of short time duration relative to the system response — the area under the curve determines the intensity. A "step" is a change in level that, relative to the overall system response, is rapid. A "ramp" is the derivative of a step.

5.2.2 Identification of Model Structure

Based on early experiments and observations on animal and human movements, primarily for contractions against inertial loads, Hill (1922) and Gasser and Hill (1924) viewed muscle as a spring-like structure working in a linearly viscous medium [see also Chapter 1 (Zahalak)]. Based on isokinetic experiments, Levin and Wyman (1927) proposed the developed a model consisting of a damped spring-like property in series with an undamped spring. Bouckaert et al. (1930) suggested that after the initial undamped change in length, the remaining shortening followed a logarithmic relation. Fenn and Marsh (1935), introducing the concept of afterloaded isotonic contraction as a technique for isolating the damped component from the undamped component, determined that the force–velocity property was nonlinear rather

than linear as Hill had originally suggested.

It was within this context that Hill published his famous 1938 paper which formally identified a lightly damped elastic element in series with a nonlinear CE. Furthermore, for maximal activation, the generated contractile force was found to decrease nonlinearly as the shortening velocity increased. This CE force–velocity behavior could be described, for shortening muscle, by a hyperbolic equation (other forms given in Section 5.3.3):

$$(F + a)(v + b) = (F_{max} + a)b \qquad (5.1)$$

where F is muscle tensile force, v is muscle shortening velocity, F_{max} is the maximum (tetanic) isometric force, and a and b are constants. The dimensionless "shape" parameter $a_f = a/F_{max} = b/v_{max}$, which specifies the hyperbolic concavity (Figure 5.2a), was found to have a value of about 0.25, approximately independent of length (however, see also Section 5.2.3). It was anticipated early on that the b parameter would have higher values for fast muscle fibers (Katz, 1939)). Hill also showed how CE could shorten during isometric contraction because of SE extension.

Figure 5.3: Effects of parameters on hyperbolic Hill equation (shortening to left). *a)* Variation in a_f, *b)* Variation in b (or v_{max} if a_f is assumed constant).

Katz (1939) showed that shortening contractions followed the Hill equation while lengthening contractions showed a greatly increased slope for lengthening muscle as opposed to shortening muscle and then exhibited significant "yielding," or "giving" (see also Chapters 1–3 and Section 5.4.1). This latter effect had been previously observed by others [e.g. Gasser and Hill (1924)] but not as succinctly described. Without the current insight into the molecular nature of the contractile machinery, as described in Chapters 1–4, explanations of this effect were clearly difficult.

In summary, the concept of a *CE* in **series** and in **parallel** (Figures 5–1a,b) with passive, lightly-damped elastic tissue was understood by 1939 [Katz (1939)]. However, since the parallel elasticity was typically small, it could effectively be ignored for most applications. This model structure, along with simplified versions of the activation process, can be considered as the foundation for the phenomenologically-based, lumped-parameter model of muscle. The concept behind the "active state" was in common use [e.g. "internal 'fundamental' mechanical change" in Gasser and Hill (1924), "active state" in Hill (1938)], but it was not until 1949 that Hill provided an explicit (albeit controversial) definition and a technique for estimating excitation–activation dynamics (see Section 5.3.4).

If the elements are linearized, Figures 5.1a and 5.1b can be shown to be equivalent. Defining the series and parallel springs as K_s and K_p, respectively, B as the internal Hill "viscosity" (see Figure 5.8), and F_o as the input to *CE*, the following relations hold between the elements of the two versions [e.g. McMahon (1984)]:

$$K_{s_b} = K_{s_a} + K_{p_a}; \qquad K_{p_a} = (K_{s_b} K_{p_b})/K_{sp_b}$$

$$\text{(5.2)}$$

$$B_a/B_b = K_{s_a}/K_{sp_b}; \quad F_o_a/F_o_b = K_{s_b}/K_{sp_b}$$

where $K_{sp_b} = K_{s_b} + K_{p_b}$

5.2.3 Models for Soft Connective Tissues

Since *CE* is connected in series and parallel with viscoelastic tissues, it is appropriate to develop an understanding of the basic properties for such tissues. Passive soft connective tissues,

ranging from tendon to skin to blood vessel, tend to have quasi-static mechanical properties such as shown in Figure 5.4a, in which the ⋅⋅⋅ness (slope of the curve) increase fairly line⋅ with force over the primary operating range [e.g. Fung (1967); Figure 5.4b]. This can be described mathematically:

$$\frac{dF}{dx} = K_1 F + K_2 \qquad \text{(5.3)}$$

where F is force, x is extension, and K_i are constants (often $K_2 \approx 0$). Solving this equation and applying boundary conditions, we arrive at the classic exponential relationship for soft connective tissues (e.g., Fung, 1969; Hatze, 1974):

$$F = K_3 (e^{K_1 \Delta x} - 1) \qquad \text{(5.4)}$$

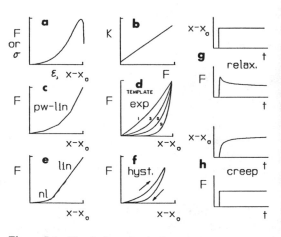

Figure 5.4: Classical mechanical properties of biological soft connective tissue. *a)* Typical quasi-static stress–strain or force–extension curves ($\sigma = F/(\text{cross-sectional area } A)$; $\varepsilon = \Delta x/(\text{rest length } L_o)$). *b)* Classic stiffness–force relation, which results in the exponential force–extension behavior of *d*). *c)* Concept of linear collagen fibers, of different initial orientation, starting to stretch at different extensions ("piecewise linear"). *d)* Exponential fit via "intuitive" parameters: a dimensionless "shape" parameter K_{sh} and a point on the curve (conveniently F_{max} and Δx_{max}). *e)* "Hybrid" curve: exponential "toe" segment followed by linear region. *f)* Hysteresis due to viscoelasticity for tissues under cyclic loading (constant velocity stretch followed by release). *g)* Force "relaxation" output trajectory due to a length step input. *h)* Length "creep" output trajectory due to a step change in force.

where Δx is the extension relative to the rest (i.e. zero force) length. For most tissues this behavior is due to wavy collagen fibers (which vary in number and orientation between tissues) gradually straightening out and bearing load (e.g., Fung, 1981). Thus, the overall curve could be exponentially shaped even if linear collagen properties are assumed (Figure 5.4c–d). This relation can be reformulated to possess a more convenient set of parameters (Hatze, 1981; Winters and Stark, 1985):

$$F_{se} = \left(\frac{F_1}{e^{K_{sh}} - 1} \right) (e^{(K_{sh}/\Delta x_1)\Delta x} - 1) \quad (5.5)$$

where the describing parameters are now *very intuitive*: a point on the curve (here F_1 and ε_1) and a dimensionless shape parameter (Figure 5.4d). In tendon and most ligaments most collagen fibers are initially orientated primarily along the long axis; this can result in flattening of the curve after the initial toe region (Woo et al., 1981). If desired, an exponentially-shaped toe region can easily be connected to a linear region, with the shape parameter set by the constraint of no discontinuity in slope (i.e. stiffness), as shown in Figure 5.4e.

Biological connective tissues are also inherently *viscoelastic*, i.e., they exhibit hysteresis during cyclic loading (Figure 5.4f), force relaxation when held at a constant length (Figure 5.4g), and length creep when held at a constant force (Figure 5.4h). Of note is that viscoelastic effects are seen within both short (milliseconds) and long (minute or hours) time periods. Thus, the amount of hysteresis (measure of energy loss) varies with both the speed of ongoing extension and the history of recent extensions; hence a biomechanical basis for "stretching" before exercise.

There are three classic models that have been used since the 1800's for describing basic viscoelastic behavior, with that of Figure 5.5c known to be more appropriate. (The "series" model of Figure 5.5a is inadequate during quasi-static conditions due to a tendency toward "drifting," while the "parallel" model of Figure 5.5b often poorly approximates initial transient behavior.) The series and parallel elements of the Hill model would be better represented by a Kelvin model; however, the added model complexity is usually considered not to be worth the burden. However, one must keep these *passive properties in mind* when *interpreting* muscle testing data, especially for experiments that unfold over *very fast* (order of milliseconds) or *very slow* (many seconds) time intervals. Interestingly, these are the time frames of most current muscle testing; hence the reason for considering passive properties in more depth than often seen in a "muscle mechanics" review. Over the moderate time range of most past musculotendon testing, long-term passive viscoelastic effects would appear as a slight drift.

5.2.4 Hill-Model Structural Extensions

Model structures for musculotendinous systems that have been commonly used are shown in Figure 5.1 and Figure 5.6. Essentially, the difference between these models is the arrangement of passive spring and dashpot elements. In both of these figures, springs with lower stiffness (higher compliance) are shown smaller. *Model d is a better approximation of physical reality. However, is it worth the cost of the added complexity? If dashpots are assumed negligible (as is common), it turns out that some of the springs are *not mathematically independent* of each other. This can be seen by using methods such as *bond graphs* [see Chapter 7 (Hannaford and Winters)] to identify independent energy storage elements. When this is the case, models *can be reduced* without any loss to dynamic performance, and in fact, extra springs may confound interpretation and make modeling less computationally efficient. However, if an *internal* node location, such as *node d* in Figure 5.6, happens to be of special interest (e.g. for sensory feedback of muscle length), then the added structure is worthwhile. My own experience with various formulations is that the parallel elasticity is best lumped with joint properties [see Chapter 8 (Zajac and Winters)]. We will also see in Section 5.3.3 that the *CE* tension–length relation, although

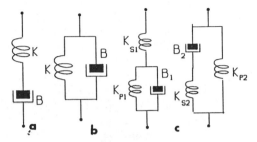

Figure 5.5: Classic models for describing viscoelastic properties. In each case the constitutive relation for the spring is $F = f(\Delta x)$, the dashpot is $F = f(\Delta v) = f(\Delta \dot{x})$. *a)* "Maxwell" series model. *b)* "Voigt" parallel model. *c)* "Kelvin," or "standard" models.

"spring-like", is best not treated as a spring in parallel with a dashpot representing the *CE* force–velocity relation – the spring and dashpot forces would then be additive, which is *not* compatible with the *CE* force– velocity–length surface that is traditionally assumed to exist for each activation level. Upon reduction, we are back to an "equivalent" spring in series with a large *CE* "force-velocity" dashpot – the basic model structure which will serve as our base as we now develop each of the classic muscle properties.

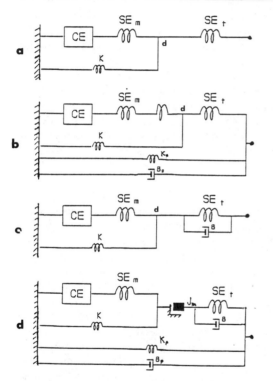

Figure 5.6: Common musculoskeletal model structures that go beyond those of Figure 5.1. *a)* Tendon in series with classic muscle structure. *b)* Combined muscle-tendon models with additional springs [see Hatze (1981) for details]. *c)* Simple model structure with a "lumped" *SE* viscoelastic element. *d)* "Complex" model which includes an internal node for muscle mass (the usual convention is to lump mass with the limb segment).

5.3 Model Parameter Identification

Since, mechanically, one can "see" the contractile machinery (for Hill-based models *CE*) only after connective tissue influence (even true for isolated muscle fibers), we start with *PE* and *SE* properties before considering *CE* properties.

5.3.1 Passive Parallel Element (*PE*)

This relation can be measured simply by pulling the passive muscle tissue at various specific (usually fairly slow) rates and measuring force. It is due primarily to the passive tissue within and surrounding muscle, and as with soft connective tissues in general, can be approximated by a viscoelastic model of one of the forms of Figure 5.5. The dashpot is typically neglected or assumed linear (Winters and Stark, 1985). For skeletal muscle the force developed by this element is insignificant except for extreme lengths (e.g over 1.2 times the rest length) which are often beyond the physiological range [e.g., Katz (1939, Bahler et al. (1967)]; hence its *lack of inclusion* within many models. For cardiac muscle, the parallel elasticity is significant [e.g., Parmley et al. (1970)]. For musculoskeletal systems the musculotendinous passive elements can be considered *separately* from the active muscle component, and lumped with other passive tissue in parallel if the passive element is considered to be a spring or a parallel arrangement of spring and dashpot and the model structure of Figure 5.1a is used [see also Section 8.3 of Chapter 8 (Zajac and Winters)]; this is the model structure that is recommended here.

5.3.2 Lightly Damped Series Element (*SE*)

Fundamental to the early work and all work since has been the concept of the lightly-damped elastic element is series with the contractile machinery. In fact it should be termed the series *compliance* element since the distinguishing feature is significant compliance, but in deference to history we will use the traditional terminology. Controversy surrounds the subtle details of this element [e.g. McMahon (1984)]; however, the fundamental existence of this type of property cannot be denied [Hill (1970)], and its use permeates this book (see especially Chapters 4, 34–37).

Three classic experimental techniques have been used to estimate this element [reviewed by Hill (1970)]: *i)* "controlled" stretch/release (Hill, 1950); *ii)* "quick release" (usually from tetanus [e.g., Wilkie, 1956)]; and *iii)* the isometric "calculation" method [Wilkie (1950)]. The second method has been most widely used. Some groups employing fast "controlled" stretch/releases have preferred terms such as "short-range" stiffness (Rack and Westbury, 1974) or "high frequency" stiffness (e.g., Cecchi et al., 1984) to distinguish the observed behavior from that of an ideally passive spring (the distinction is discussed later). In general, the results with the third method have predicted a more compliant relationship; reasons for this difference will become evident later. However, all techniques have produced concave upward load-extension curves, with peak element extensions between 2% and 8% [e.g., review by Close (1972)].

A fourth class of techniques involve consideration of muscle oscillation; authors' utilizing these methods have been hesitant to directly relate their "stiffness" findings to *SE*, but the estimated values is so compatible with those obtained from other techniques that it is hard to deny a correlation (e.g., see data of Zahalak and Heyman, 1979). When put together, there is remarkable consistency across the data, as will be seen below.

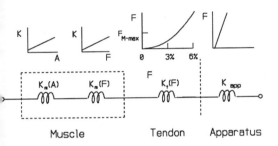

Figure 5.7: Conceptual model of four springs in series. *a)* The two left-most are within muscle tissue, with the stiffness (slope times initial length) potentially a function of both activation (far left, cross-bridges) and force (middle left, Z-discs and filaments). The stiffness for the middle-right curve, representing tendon, is a function of force, except perhaps at higher values. The right curve, representing the apparatus, is likely linear with potentially a slight offset due to stiction effects. *b)* Quasi-statically, the total curve is obtained by summing the individual extensions for each force level:

$$\delta L = (\delta L_{m_a} + \delta L_{m_p}) + \delta L_t + \delta L_a \qquad (5.6)$$

or:

$$= (L_{m_{o_a}} \varepsilon_m + L_{m_{o_p}} \varepsilon_m) + L_{t_o} \varepsilon_t + \delta L_a \qquad (5.7)$$

Thus, the relative extension due to each component is a function of the relative lengths. Also, because of the series arrangement, for a given force the individual compliances (inverse of stiffness) add; thus, the overall stiffness (compliance) is less than (more than) that of the individual components.

Overview of SE Structural Components. Figure 5.7 conceptually separates the "series element" into three components and an additional component for the apparatus. The complex aponeurosis structure (not shown) could perhaps be placed within the tendon, although as shown in Chapter 4 (Ettema and Huijing) this tissue has unique features, including apparent dependence on both force and muscle length. Also not included is extension (but not elastic energy storage) due to geometric factors [see Chapter 4 (Ettema and Huijing) for a quantification of these effects]. Of note is that if the curve shapes are not the same, the relative contribution between the subcomponents is a function of force (e.g., Stein and Gordon, 1986).

Isolated Tendon. Tendon properties are well documented, with the "toe" region (where the stiffness is increases approximately linearly with force, and thus is well represented by the classic exponential relation of *Eq. 5.3*) for about 3–4% extension, followed by a more linear region up to failure at about 7% (e.g. Butler et al., 1979). Although tendons possess viscoelastic properties, a high percentage (about 80%) of the stored elastic energy is recoverable for the typical physiological range of rates of loading (Alexander, 1988).

Tendon Extension with Muscle Contraction. Peak muscle-induced tendon extension is typically within 2–5% for most tendons [see Chapter 4 (Ettema and Huijing) and Chapter 36 (Alexander and Ker)]. The range of *in situ* applicability of the toe region, where the force–extension relation is nonlinear, is controversial. Based on assuming a factor of safety of at least 2 between the tendon failure load and the peak load generated by maximally contracting muscle (e.g., see Eliott, 1965), one would predict that typically a good portion of the tendon operating range during muscle contraction is within the "toe region" (see Figure 5.7). Indeed, some groups have found stiffness to increase (compliance to decrease) with increasing force for most of the range of loading [e.g., see Chapter 4 (Ettema and Huijing) for review of their work and that of Rack and Ross (1984)]. Winters and Stark (1985) assumed that peak extension was about 3-4%, and that for practical purposes this was within the toe region; thus they assumed an exponential relation. However, Morgan (1977) and Proske and Morgan (1987) suggest that tendon properties are linear above about 20% of maximum isometric tension. Zajac's group (Zajac, 1989), emphasizing this data, utilize a linear relation from 20% force onward, and many groups assume a linear force-extension relation.

Simulation experience from thousands of runs utilizing various nonlinear and linearized models suggest that this assumption matters. Thus, further investigation is needed, primarily along the lines of Chapter 4. However, it appears that the findings presented in Chapter 36 (Alexander and Ker) may help resolve this issue. Here they divide musculotendinous systems into three categories: *i)* long muscle fibers with short tendons; *ii)* short muscle fibers with long, thin, extensible tendons (e.g. 4–5% extension with F_{max} applied); and *iii)* short muscle fibers with long tendons and larger cross sections than would appear necessary for a reasonable mechanical factor of safety (thus extending only about 2% with F_{max}). They postulate different uses for these tendons – in particular, *ii)* is best for elastic energy storage (e.g. many lower limb actuators) while *iii)* is better for control transmission and impedance modulation (upper limb). This helps explain some of my own concerns regarding predicted *SE* properties that I and my colleagues have estimated for about 80 muscles (e.g., Winters and Stark, 1988). For instance, assuming a similar peak tendon extension for all maximally contracting muscles (i.e. ratio of muscle to tendon cross-sectional area is uniform), we find that when converted to (more intuitive) joint units, the muscles surrounding the ankle joint in particular were quite stiff (e.g. soleus peak extension of only 19°) while those surrounding the wrist were quite compliant (e.g. half over 50° peak extension). This is due primarily to differences in moment arm. In retrospect, Chapter 36 and common sense suggests that the former, which are more involved in elastic energy storage and utilization, have relatively thinner tendons (see Chapters 33–37), while the latter, more involved in fine control and impedance modulation, likely have relatively thicker tendons (as suggested by the data of Rack et al., 1984).

Physical Sources Within Muscle. Jewell and Wilkie (1958) showed via direct observation that significant series extension existed within muscle tissue as well as within tendon. They also suggested that *SE*, and originally identified the Z-discs (and perhaps the actin filaments) as the most probable site. However, these are relatively short and would have to extend considerably for overall

muscle extension to reach 4% (see form of *Eq. 5.7*). As suggested by Hanson and Huxley (1955) and Joyce and Rack (1969), it is now well accepted that part of the elasticity lies within the cross-bridge structure, likely including the *S1* bridge of myosin (e.g. Huxley and Simmons, 1971; Flitney and Hirst, 1978; Morgan, 1977). However, the range of strain proposed for the cross bridges [e.g. under 2% in Huxley and Simmons (1971) and up to 3% in Flitney and Hirst (1978), both relative to overall sarcomere length] seem small in comparison to data from whole muscle experiments (reviewed in Close, 1972). Evidence is also mounting that the myofilaments are also somewhat extensible [e.g., reviewed in Chapter 2 (Hatze); see also Chapter 4 (Ettema and Huijing); see Ford et al. (1977) for an alternate view]. Thus, *all components of muscle appear to show some extension.* In retrospect it would be surprising if such was not the case, especially since experimental investigations of whole muscle (often with little tendon included) have consistently found peak extensions of about 5% (reviewed in Close, 1972).

Relative Contribution of Each Component. What is the relative extension of the various contributions to series extension? Traditional input–output methods only measure a "lumped" value, but more recent techniques [see Chapter 4 (Ettema and Huijing)] allow more careful estimates. As reviewed in Chapter 4, for some muscles on the order of half of the *SE* extension is within muscle tissue (e.g. frog sartorius, cat soleus, human thumb). However, for many others, the majority of the *SE* extension of musculotendinous unit is due to the tendon. In a recent review article, Zajac (1989) suggested that, for most musculotendinous units, muscle series extension is virtually insignificant relative to tendon extension. When all put together, however, it seems evident that peak strains for both muscle and tendon are on the order of 3–5%, with only some of that within the muscle due to the cross-bridges.

Cross-Bridge Stiffness Vs. Passive SE. Based on intuition from molecular models (Section 5.3), the overall cross-bridge induced stiffness would be expected to increase as a function of the *number of attached bonds*. This is because cross-bridge stiffness can only be realized when the actomyosin complex is attached. This argument assumes an idealized mechanical view of cross bridges in series and in parallel with each other, which is probably reasonable (Rack and Westbury, 1974). Each attached bond then contributes a "short-range" stiffness for about 2% of the sarcomere length (Huxley and Simmons, 1971, Rack and Westbury, 1974). Since the number of attached bonds is a function of calcium activation, the "short range stiffness" seen during ramp stretches differs from a passive series *elastic* element (Rack and Westbury, 1974). This is because a cross-bridge, after being broken, can *reattach* at a *different* (e.g. lengthened) actin site — it "forgot" that it was storing elastic energy relative to a certain rest length. This in turn also raise questions regarding the adequacy of the Hill-based model structure which will be addressed in Section 5.4.3.

Estimating SE/Stiffness in Humans. The classical "quick release" and "controlled release" approaches for isolated muscle are difficult to apply on intact human systems, due primarily to the confounding effects of large limb inertias (Goubel et al., 1971). Two very different approaches emerge:

1. Calculation Methods. This is based the isolated muscle data described above: peak muscle and tendon extensions are both about 3–6% strain, and a dimensionless "shape" parameter (Figure 5.7d) on the order of 3–5 provides a reasonable prediction of curve concavity. This *material* information, combined with estimated musculotendinous *geometry* [e.g., as cataloged in the Appendix (Yamaguchi et al.)], can be utilized to estimate *structural* values (e.g., Alexander, 1983; Winters and Stark, 1985).

2. Experimental Oscillation Methods. One set of approaches is based on applying small, fairly high-frequency oscillations about an otherwise steady isometric contraction against a bias torque. Either *position oscillations* [e.g., Joyce and Rack (1974); Zahalak and Heyman (1979), Ma and Zahalak (1987)] or *force oscillations* (e.g., Agarwal and Gottlieb, 1977) can be applied; the other of these becomes the measured output. The estimated stiffness, as defined by the peak-to-peak force change divided by the peak-to-peak position, changes as a function of the bias force in a way that appears to provide a stiffness measure that is essentially a good prediction of the *SE* over the entire *SE* range. Angular stiffness is almost linearly related to the bias torque, just as suggested previously for muscle and tendon over, at minimum, the lower force operating range.

Another class of approaches, which tend not to estimate the whole *SE* curve but rather a certain operating range, estimate stiffness from a measured *resonance* frequency; this may be determined from an "elastic bounce" record (e.g. Cavagna et al., 1971) or a "natural" voluntary rhythm (Greene and McMahon, 1979; Hof and Van den Berg, 1981, Bach et al., 1983).

It turns out that overall structural properties, as determined by these two methods, are in good agreement with each other for systems such as elbow flexion-extension, wrist flexion-extension, and ankle dorsi-plantar flexion (Alexander, 1983,1988; Winters and Stark, 1985,1988). Why do these methods tend to work? Because, as shown in Figure 5.8, for minor perturbations the *CE* works as a viscous, "soft" ground. Also, the changes in activation (as measured by the *EMG* fluctuations relative to the bias *EMG* level) are small and of sufficiently high frequency to be fairly filtered by muscle dynamic properties. Finally, activation and force are at about the same relative level, and thus if anything the curve concavity should be greater than that seen for quick release studies, as is seen.

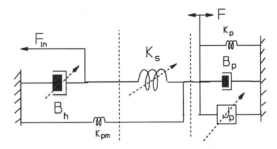

Figure 5.8: Muscle model isolating the series component but retaining gross properties to either side of the model. The lines through elements indicate sources of nonlinear influences.

Synthesis. The phenomena of variable series stiffness is admittedly a complex property that is not fully understood. In particular, the history dependence of series stiffness during dynamic movements, during which activation and force differ, requires further investigation, as do the effects of geometry. Nonetheless, there is a *remarkable consistency* between the information obtained by various methods, more so than generally recognized.

This important property is explicitly used in Chapters 1–6, 8–9, 17, 19–21, 30, 33–37, 41–43. These chapters, which span limb and torso systems, document a wide variety of consequences of series compliance, including: *i)* elastic energy storage; *ii)* modulation of operating ranges for the *CE*; *iii)* smoothing of movements; and *iv)* facilitation of dynamic stiffness/impedance modulation.

How should it be modeled? Possibilities include: *i)* a single linear spring element (used in many chapters); *ii)* a spring which an initial concavity followed over a large part of the operating range by a linear region (Zajac, 1989); *iii)* a traditional concave upward exponential curve, with a "shape" parameter of 3-5 (i.e., a linear relationship between stiffness and force); *iv)* a viscoelastic model including one of the aforementioned springs; *v)* a series connection of two (or more) nonlinear series springs (Stein and Gordon, 1986), perhaps with one spring modulated by force, the other via activation; and *vi)* a single stiffness value which is a weighted function of force and activation, with element extension determined via integration.

For those wanting a reasonable yet simple model, I suggest *iii* as opposed to *i)* or *ii)*. In particular, linearization of this element, while often sufficient for studies of lower-limb elastic energy storage or unloaded upper-limb movements, should be avoided whenever one desires to use the model to gain insight into *control* aspects of tasks involving dynamic interaction with the environment. Recent (as yet unpublished) simulations by myself also suggest that adding a viscous component (model *iv)*) has some interesting (yet subtle) consequences: *a)* a closer fit to quick-stretch and quick-release data (e.g. of Bagley, 1987); *b)* some capacity for showing history-dependent "force enhancement" tendencies (discussed later; see also Sprigings, 1986); and *c)* enhanced numerical stability of the muscle algorithm (which allows larger integration step sizes). The last reason alone is sufficient for employing a small damping element in optimization studies. Whether *SE* should be activation- and well as force-dependent requires further investigation by studies along the lines of Chapter 6 (Lehman).

5.3.3 The Contractile Element:
Tension–Length and Force–Velocity

Two properties define this element: *i)* *CE* tension–length; and *ii)* *CE* force–velocity. After developing the basic relationships, we will look at how these relationships combine and how the *CE* element is influenced by activation. Special issues such as "yielding" are postponed until Section 5.4.1.

Active CE Length–Tension Relation

It has long been known that isometric muscle force development is a function of length (e.g. discussed in Gasser and Hill (1924); Ramsey and Street (1940), Ralston et al. (1949)), with the muscle force maximum at an intermediary ("optimum") length (about 1.05 times the "rest" length) and lower for shorter or longer lengths, finally reaching near zero at about 0.4 and 1.5 times the resting length, respectively (e.g., Figure 5.9). Thus the slope, a measure of static stiffness, includes both positive and negative (potentially unstable) ranges. The lack of an abrupt end of force near zero force locations is likely due to cross-bridge non-uniformity [see also Chapter 3 (Morgan)].

Since the 1950s, isolated muscle *CE* tension–length and human joint moment–angle curves have been available within the literature for the special case of maximal activation and maximal effort, respectively [e.g. review by Kulig et al. (1985)]. Using techniques that allowed measurement at the microscopic level, Gordon et al. (1966) showed conclusively that this fundamental property was due to the amount of overlap between thick and thin filaments, as predicted by the sliding filament theory [Chapter 1 (Zahalak)]. A wide variety of empirical fits have been utilized to characterize the basic behavior.

The physiological operating range is always much less than the full range; that shown in Figure 5.9, with a positive slope over most of the *in situ* muscle range, is typical. However, there are notable exceptions (e.g. knee and elbow extensors).

Of note is that the curve shape for the *CE* is *not* the same as that for the whole muscle (e.g., see Figure 12 in Zajac, 1989 and Figure 5.10). This is because the extension of *SE* is a function of force, and the force is not constant . For this and perhaps other reasons (see Hatze, 1981), the musculotendinous "tension–length" curves for isometric twitches (about 15–30% of tetanic force) and those for isometric tetanus should also not be expected to be simple multiplicative scales of each other.

Two important practical questions remain:

1. Determination of in situ rest length. This is hard to estimate because most joints are crossed by multiple muscles. Figure 5.9 provides an example of simulation predictions of how *CE* tension–length and subsequently joint torque–angle relations change as the assumption of rest length changes. By making such

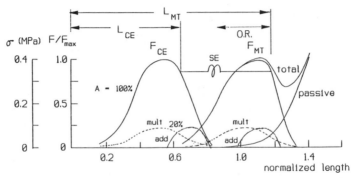
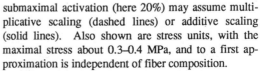

Figure 5.9: Idealized tension–length for tetanic (maximal) contraction (solid lines), as seen across the *CE* element (to left) and whole musculotendon unit (to right). This difference between the two is the extension of the *SE*. Family of tension–length curves with submaximal activation (here 20%) may assume multiplicative scaling (dashed lines) or additive scaling (solid lines). Also shown are stress units, with the maximal stress about 0.3–0.4 MPa, and to a first approximation is independent of fiber composition.

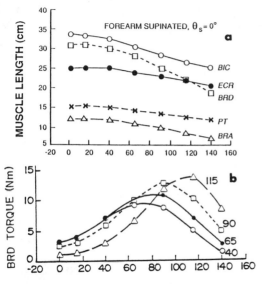

Figure 5.10: *a)* Sensitivity of muscle length to joint angles, here for elbow flexors as a function of elbow flexion angle. *b)* Changes in the simulated maximal isometric torque-angle relation as a function of the assumed muscle rest length, shown here for the brachioradialis (*BRA*). For the extended positions a curved (spherical arc) approximation was employed [method reviewed in Section 8.2.4 of Chapter 8 (Zajac and Winters)]. (Adapted from thesis by Kleweno, 1987).

predictions for all muscles and comparing to experimental data, rest lengths can be estimated (Kleweno and Winters, 1988).

2. CE tension-length curve for submaximal activation. This is less well documented and more controversial. Based on purely mechanical considerations, since *SE* extension increases with force and thus (usually) activation, the curve peak for whole muscle might be expected to shift toward longer lengths as the

steady activation level increases. Modeling assumptions may assume that the *CE* (or whole muscle) curve scales downwardly as a multiplicative factor, essentially modulating stiffness (e.g. Winters and Stark, 1985; dashed line in Figure 5.8) or that it shifts downward and to the right or left (dotted line in Figure 5.8). Notice that for lower activation levels the passive (parallel elastic) component has a relatively larger influence. These poorly understood subtle differences are especially important in relation to the theories of motor control discussed in Chapters 11–17.

CE Force–Velocity Relation.

Four classical methods have been utilized to experimentally estimate this behavior: *i)* Isometric-to-isotonic contraction (afterloaded); *ii)* isometric-to-isotonic contraction (not afterloaded); *iii)* isometric contraction dynamics (requires knowledge of *SE*); and *iv)* isokinetic contraction. Since first used by Fenn and Marsh (1935), isotonic methods (fortunately) dominated the early work. Ideally, measurement should be made at a specified position while the muscle force and velocity are **constant**, i.e. their derivatives (slopes) are zero. The classic isotonic studies of the human flexors by Dern et al. (1947) and Wilkie (1950) provided strong evidence in support of the basic Hill model structure, and further documented shapes of the force–velocity relation that were consistent with the Hill relation, especially once an estimate of the effects of forearm inertia is subtracted out.

Although related to muscle energetics by Hill, the *CE* force–velocity is essentially just a fit to experimental data. This relationship has been estimated by many groups, with the vast majority finding data compatible to the Hill hyperbola.

Other proposed force–velocity relations include:

$$F_{ce} = F_{max}\, e^{-a v_{ce}/v_m} - b\, v_{ce} \qquad (5.8)$$
$$\text{[Fenn and Marsh (1935)]}$$

$$F_{ce} = F_{max}\, e^{-(F_{ce} v_{ce}/b F_{in})} - V\, a/b \qquad (5.9)$$
$$\text{[Abbott and Wilkie (1953)]}$$

$$F_{ce} = F_{max}\, e^{-a v_{ce}/v_m} - b\, l_{ce} \qquad (5.10)$$
$$\text{[Aubert (1956)]}$$

where a and b are different constants in each case and F_{max} is the peak isometric force; all have three describing parameters.

Hatze (1981) has proposed an additional relationship in which the concavity of the curve changes for small shortening velocities (which is not true for the Hill hyperbola); it is too complex to present here. His rationale was the data of Edman et al. (1976), which suggested that a concavity occurred around 80% of the isometric force; Cecchi et al. (1984) presents similar results. Furthermore, a number of similar shapes have been obtained for human systems, primarily for knee rotation experiments utilizing isokinetic machines (e.g. Perrine and Edgerton (1978)) and wrist rotation (Baildon and Chapman, 1983). Reasons for these differences are unknown; however, many curve shapes can be estimated from the same human testing data set, especially for "isokinetic" testing (Yates and Kamon, 1983; Winters and Bagley, 1987). In fact, I have found that students presented with isokinetic "dynanometer" input–output data from a *model* which *includes* a Hill-based *CE* will produce a variety of curve estimates! Also, Zahalak et al. (1976) have shown that *EMG* activity tends to grow for slow shortening velocities; this may also help explain part of the human data. The fundamental problem is that *CE* tension–length, *SE*, *PE*, system inertia and activation all represent confounding influences. Results not following a Hill-like shape must be contrasted with the dozens of results, ranging from isolated fibers through whole muscle to muscle joint systems, where the Hill hyperbola has been a good fit for shortening muscle. Lengthening *CE* force–velocity behavior will be addressed further in Section 5.4.1–2.

The Hill hyperbolic relation has remained the standard not just because it adequately approximates a wealth of experimental data but also because the *describing parameters are intuitive*.

The dimensionless parameter a_f defines the hyperbolic shape and the parameter CE_{vmax}, the maximum unloaded *CE* velocity, defines velocity intercept (see also Figure 5.2). The last parameter, which describes the force intercept, is the hypothetical isometric force, which we will see comes from the *CE* tension-length relation. This relation can be rewritten into many forms, including (Wilkie, 1956):

$$\left(a_f + \frac{F_{ce}}{F_{max}}\right)\left(a_f + \frac{v_{ce}}{v_{max}}\right) = a_f + 1 \qquad (5.11)$$

One could also divide by a_f. Other normalized forms are (e.g., Fung, 1981; note symmetry):

$$\frac{v_{ce}}{v_{max}} = \frac{1 - (F_{ce}/F_{max})}{1 + (F_{ce}/(F_{max} a_f))} \qquad (5.12)$$

$$\frac{F_{ce}}{F_{max}} = \frac{1 - (v_{ce}/v_{max})}{1 + (v_{ce}/(v_{max} a_f))} \qquad (5.13)$$

To show explicitly that this equation describes the behavior of a force generator in parallel with a dashpot, as in Figure 5.11, the following form can be utilized [e.g. Cook and Stark (1968)]:

$$F_{ce} = F_{in} - B_h\, v_{ce} = F_{in} - \frac{F_{in}(1+a_f)}{a_f v_{max} + v_c}\, v_{ce} \qquad (5.14)$$

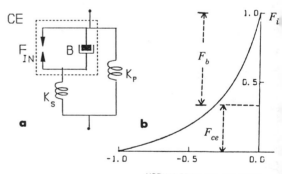

Figure 5.11: *a)* Hill model structure showing *CE* divided in a parallel dashpot representing force "lost" via the force–velocity relationship and a *CE* "sculpturing" product relationship for the *CE* tension–length relation. *b)* Explicit view of the force "lost" and the force "passed".

It is worthwhile to provide *material* property estimates for the two intercept parameters. In this way one can easily extend the results to *structural* properties by combining the materials properties with certain geometric features such as physiological cross-sectional area, pennation angle, and length (Winters and Stark, 1985; Zajac, 1989). The experimentally verified range for a_f is approximately 0.1 to 1.0, with slow muscle fibers between 0.1 and 0.25 and fast fibers greater than 0.25 (e.g., reviews by Close, 1972; Winters and Stark, 1985). CE_{Vmax} ranges from about 2 L_o/sec (muscle fiber lengths per second) for slow fibers to about 8 L_o/sec for fast. The composition of most muscles is mixed, and thus most muscles are between these extremes, and can be estimated by a linear functions of muscle fiber composition (Winters and Stark, 1985). Thus, using the Hill approach, the only new parameter needed to adequately describe muscle **material** properties is **muscle fiber composition**. Structural properties for a given musculotendinous unit of course scale with muscle geometry.

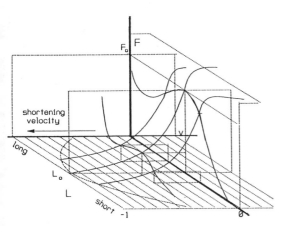

Figure 5.12: Three-dimensional view of the *CE* behavior for a certain activation level, showing a few cases for different lengths and velocities.

Interaction Between CE Tension–Length and CE Force–Velocity. The studies of Abbott and Wilkie (1953) for isolated frog muscle provided the conceptual foundation for allowing $F_{fv\text{-}in} = F_{tl\text{-}out}$, i.e. the *CE* force–velocity input force (y–intercept) equals the *CE* tension-length output (see also Figure 5.12). They also suggested that the parameters a_f and b (and thus v_{max} in Hill's equation) remained constant. These observations were supported by Bahler (1968) and are now a standard assumption. A three-dimensional force-velocity-length curve can be utilized to describe this relationship [Bahler et al. (1967), Partridge

(1979), Figure 5.12]. Mathematically, this takes the form (e.g., from *Eq. 5.11* and *Eq. 5.14*, respectively):

$$F_{ce} = \frac{a_f\,(1 - (v_{ce}/v_{max}))}{a_f + (v_{ce}/v_{max})}\,F_{fvin}$$

$$= \left[1 - \frac{1 + a_f}{a_f v_{max} + v_h} \right] F_{fvin} \qquad (5.15)$$

Notice that these forms display a **product** relationship between the two phenomena (assuming maximal activation for present):

$$F_{ce} = F_{fv}(v_{ce}, F_{fvin}) = F_{fv_1}(v_{ce})*F_{tl_1}(L_{ce})$$

$$\qquad\qquad\qquad (5.16)$$

$$= F_{fv}(v_{ce})*F_{tl}(L_{ce})*F_{max}$$

where $F_{fv}(v_{ce})$ and $F_{tl}(L_{ce})$ are dimensionless functions, each defined by two parameters. This shows that *CE* properties are best **not** represented by a parallel arrangement of a spring and dashpot – in such a case the relation would be additive [see also Figure 21.5 in Chapter 21 (Crago et al.)].

Scaling of Force-Velocity Parameters with Length. A_f is traditionally considered to be independent of length. Bahler et al. (1967) found that with maximal activation CE_{Vmax} peaks within the mid-operating range and drops off a shorter lengths where cross-bridge overlap is less. Data by Edman et al. (1968) on frog muscle fibers suggests relatively constant value over a medium range of lengths, a low value at small lengths, and a higher value at long lengths. However, over the typical physiological range of length changes (e.g., sarcomere lengths of about 5.8 to 2.8 μm), the value is relatively constant and probably consistent with the results of Bahler et al. Thus, variation with length over the physiological range appears marginal.

Scaling of the CE Force–Velocity–Length Relation with Activation. A question that the early pioneers had to address was how to modify *CE* force–velocity behavior at sub-maximal activation and muscle length. Wilkie (1956), summarizing frog data, and Bigland and Lippold (1954), using human calf muscle *EMG* measurement as an estimation of activation, suggested that the F_{fvin} parameter in the Hill equation could be scaled with activation, with the curve shape parameter (a_f) and the maximum velocity parameter (CE_{vmax}) remaining about the same. (However, the data of Bigland and Lippold (1954) for human calf muscles lacked data above about 40% of CE_{vmax}). The data of Bahler et al. (1967) further refined these concepts, which can be viewed as follows:

$$\frac{F_{ce}}{F_{max}} = F(A, l_{ce}, v_{ce}) = A * F_{max} * F_{tl}(l_{ce}) * F_{ce}(v_{ce})$$

where A is the normalized activation. Notice that the normalized CE length–tension and CE force–velocity effects are effectively decoupled (shown orthogonal in Figure 5.12), and that for a given activation, the relation is assumed *instantaneous*, i.e. statically satisfied. Furthermore it is still a *product* relationship. Notice that there are four interrelated variables: activation A, CE length L_{ce}, CE velocity v_{ce}, and muscle force F_m. If *any three* of these variables are known, the other can be determined. For example, if muscle force is unknown: *i)* A "sets" the CE tension–length relation, *ii)* given this relation, the "hypothetical isometric force" $(F_{tlout}$, which equals $F_{fvin})$ is read off from the appropriate length; *iii)* F_{fvin} the CE force-velocity curve; and *iv)* finding the appropriate CE velocity, the muscle force is read off. If activation is unknown: *i)* given F_m and v_{ce}, F_{fvin} is determined; *ii)* given an estimate of l_{ce} and F_{tlout} $(= F_{fvin})$, A is estimated.

Of note is that care must be taken if one chooses instead to represent the CE tension–length property by a spring in parallel with the force–velocity dashpot: the resulting CE then has dynamics, and thus the CE force–velocity–length relation is no longer instantaneous. Additionally, because the dashpot is traditionally assumed to be a function of F_{fvin}, force–velocity determination becomes awkward.

Is the parameter CE_{vmax} constant with activation? Some groups (e.g. Hatze, 1981; Audu and Davy, 1985) have followed the traditional line that says that it doesn't change with activation. However, Julian (1971) shows that this parameter is a function of the calcium concentration. On theoretical grounds, Julian and Sollins (1973) suggest that the level of activation affects the Huxley rate constant for breaking bonds which are opposed to shortening; this, in turn, affects CE_{vmax}. Both Zahalak et al. (1976, human elbow flexion) and Petrofsky and Phillips (1981; cat medial gastrocnemius) found that this parameter does scale significantly with activation. Of note is that some scaling would be

expected for a muscle with mixed fiber composition due to the assumption of an orderly recruitment of motor units (Henneman, 1965): *slower* muscle fibers are more likely to be active during *low* activation, with faster fibers *recruited* as activation increases (Winters and Stark, 1985). This suggests that for mixed-composition muscle (the majority) this parameter *should vary*, but perhaps not as significantly as F_{fvin} (e.g. as in Figure 5.13c). For instance, Winters and Stark (1985) assume as a default that CE_{vmax} drops at half the rate of F_{fvin}. Notice that if CE_{vmax} varies linearly with F_{fvin}, the relative shape of the force–velocity relation is independent of activation.

Zajac (1989) suggested that since the high-velocity region is of little practical importance for *in situ* functioning systems, the assumption for CE_{vmax} may not be of practical importance. Consider, however, the vertical lines in Figure 5.13, where for a given CE velocity (representing the low-velocity range of great practical importance), the assumption for CE_{vmax} affects not only the absolute force, but also the slope (and thus the instantaneous dashpot value). Also drawn in are the slopes connecting the intercepts, which have been utilized often as a mode of linearization (e.g. Baildon and Chapman, 1983; Zajac, 1989). Note that the intercept slopes and the curves slopes are *not* one-to-one, even though the same dimensionless shape parameter is assumed. This shows how difficult (and dangerous!) linearization of the force–velocity relation can be; if truly necessary, it is best accomplished uniquely for *each task* of interest [Seif-Naraghi and Winters, 1989)].

Is the Force–Velocity–Activation Relation "Instantaneous"? This question considers whether or not the force–velocity relation is satisfied at every instant, given an appropriate activation input. If so, the relation is static, and thus not history-dependent. (Of course, the overall model still produces history-dependent behavior due to the interaction between elements.) Jewell and Wilkie (1958) concluded that in frog sartorii a change in velocity follows a change in force very quickly ("probably in less than 1 msec, certainly in less than 6 msec"). Hill investigated this in more detail using various strategies, including both isotonic and isokinetic, and came to the conclusion that, for all practical purposes, the relation is uniquely satisfied for shortening muscle at each instant. The human elbow flexion data of Pertuzon and Bouisset (1973) provide further support for

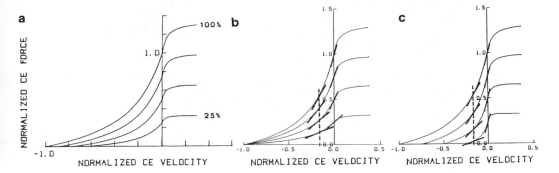

Figure 5.13: Maximal and submaximal *CE* force–velocity curves for assumptions of a variable versus an activation-independent $CE_{v_{max}}$ parameter. *a)* Default approach used by the author. *b)* and *c)* Two extreme approaches used in the literature. Notice that this assumption does make a difference, both for moderate velocities (mostly magnitude effect) and near zero velocity (mostly slope effect).

this assumption. It clearly is not satisfied for lengthening muscle (e.g. Joyce and Rack, 1969). (However, if "attachment" replaces "activation" as the input to the *CE*, as recommended in Section 5.3.1, then perhaps it might still be satisfied [Winters (1989)].

5.3.4 Excitation–Activation Dynamics from the Hill Model Perspective

In this section we cover material also presented in Chapter 1 (Zahalak), only from the "input–output" systems perspective characteristic of the Hill model approach. Activation dynamics is first covered; this is followed by the treatment of how *CE* is usually scaled by activation.

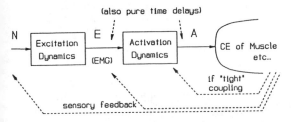

Figure 5.14: Block diagram of relationship between excitation, activation and muscle mechanics.

"Active State" and "Calcium Activation"

Hill (1949) defined the "active state" as the tension that the *CE* would generate, without lengthening or shortening, after the beginning of excitation. He considered activation to be vir-

tually instantaneous, with deactivation a much slower process. An activation "plateau" was often seen. The time course of the falling phase of the "active state" could be estimated experimentally [see Close (1972) for review]. As seen in Figure 5.14, activation is thus considered as the *output* of the excitation–contraction coupling dynamics and the input to *CE*. In 1960 Jewell and Wilkie questioned the concept of the active state, in particular as related to its assumed length-independence. Others, such as McCrorey et al. (1966), wondered whether some of the differences between series elastic and force–velocity relation obtained by isotonic versus isometric methods might not be due to active state assumptions.

In 1968, based on studies of the rate-limiting role of calcium in the muscle activation process, Ebashi and Endo defined the active state as the *relative amount of calcium bound to troponin*. This concept, since modified to represent evolving information on bonding site physiology [see Chapter 1 (Zahalak)], found immediate acceptance [e.g., Julian and Sollins (1972), Hatze (1974)], and is now a common conceptual (though immeasurable) assumption for Hill-based models.

For physiological temperatures, the data of Bahler et al. (1967) suggested an activation time on the order of 10 msec and deactivation on the order of 50 msec, and that active state dynamics could be adequately represented by a first-order system. Thus the simplest approach for modeling this dynamic process has been to assume unidirectional flow and first-order dynamics in which either the activation time constant is constant (linear model) or there are two time constant values, one for lengthening and one for shortening [e.g., Cook and Stark, 1958; Lehman and Stark, 1979; Hof and Van den Berg, 1981; Winters and Stark, 1985; Zajac, 1989]. Winters and Stark (1985) suggest that these time constants are a function of the muscle fiber composition (higher time constants with slower muscle fibers) and muscle size (higher time constants in larger muscle volumes). Ma and Zahalak (1987) and Zajac (1989) have suggested first-order nonlinear dynamics in which the time constant is variable (Section 5.7). Hatze (1981) assumes second-order dynamics for this excitation–activation relation.

Is muscle activation truly uni-directional, independent of muscle dynamics? Indirect evidence from experiments in muscle mechanics (e.g., Bahler et al., 1967; Rack and Westbury, 1969) suggests that it may be a function of length. Experiments by Fuchs (1977) suggested that calcium–troponin interactions may be coupled with actin–myosin interactions. Such observations prompted Ma and Zahalak (1987) to formally distinguish between the two possibilities by referring uni-directional coupling as *"loose" coupling* and bi-directional coupling as *"tight" coupling* [Chapter 1 (Zahalak)].

Of note is that the "loose" coupling assumption is compatible with *any* muscle model, and thus when employing this assumption *one can "mix and mesh" models of activation with models of muscle mechanics*. An example in the *PEXA* model of Hannaford that is described in Chapter 7 (Hannaford and Winters). However, combining an activation model with Hill-based models is not as theoretically elegant as it is when using the molecular models.

Excitation, Effort, and Neural Dynamics

As seen in Figure 5.14, excitation is defined here as the input to activation dynamics. Conceptually, the output of excitation dynamics can be thought of as similar to a lightly filtered, rectified *EMG* (Winters and Stark, 1985; Gottlieb et al., 1989; Chapters 14-15, 19). If muscle is electrically stimulated or the model input is an *EMG*, excitation dynamics are of no interest. However, in most simulation studies an idealized input representing the *CNS* output, such as a pulse, is assumed that differs in shape from an *EMG*. Excitation dynamics heuristically represents the very real fact that neural dynamics may be part of the rate-limiting process. Such dynamic effects

may include recruitment, the finite time nature of firing rate, any "smoothing" effects in the lower circuitry, and the simple fact that higher brain signals are not idealized. It may be modeled as a first-order filter (Hatze, 1981; Winters and Stark, 1985). Often excitation has been assumed to be negligible, especially when considering eye movements (e.g. Cook and Stark, 1968; Lehman and Stark, 1979).

Chapman and Calvert (1979) have suggested that perceived "*effort*," similar in concept to the "excitation" input, is fairly consistent between subjects. Our data with submaximal exercise (Silver-Thorn and Winters, 1988) suggests that subjects correlate perceived effort with force rather than *EMG*. This interesting concept needs to be further explored.

Hatze (1977, 1981) has separated out "recruitment" and "firing rate" effects. Although theoretically elegant, the latter approach is practically difficult since there are now two inputs per muscle, and it is difficult to separate out the two effects when using surface *EMG* electrodes. Nonetheless, his approach deserves further study, both for theoretical investigations and for studies related to muscle fatigue (e.g., due to muscle stimulation).

5.4 Challenges for Hill-Based Models (and Suggested Strategies)

In this section we deal with three classes of experimentally measured history-dependent phenomena that are difficult to describe within the context of the traditional Hill-based model structure: *i)* "yielding" during lengthening; *ii)* "force enhancement" after stretch; *iii)* subtle stiffness effects.

The concept of "*attachment*" (rather than "activation") as an input to the Hill-based contractile machinery is introduced as an approach for partially dealing with such phenomena. Additionally, it is suggested that passive mechanical sources for certain phenomena have not always been adequately addressed. Finally, it is shown that some of these phenomena are seen only in experiments of little relevance during normal human movement tasks.

5.4.1 Lengthening Muscle

It is well documented, at least for ramp stretches within certain muscles (e.g. cat soleus), that lengthening muscle exhibits history-dependent behavior that is incompatible with the classical *CE* activation–force–velocity–length approach of Figure 5.12 (Joyce et al., 1969; reviewed in Chapters 1–3). However, other muscles display a force saturation with little yielding (e.g. cat medical gastrocnemius); these muscles can be adequately approximated by a lengthening *CE* force-velocity fit.

The fact that the *CE* element in Hill-based muscle models is assumed to be instantaneous and single valued is one of the primary criticisms of such models (e.g., Zahalak, 1981 and Chapter 1). This is indeed a weakness. Most current *CE* force–velocity relations assume a saturation with progressively lengthening velocities at about 1.3 F_{max} (e.g., Hatze, 1981; Winters and Stark, 1985; see also Figures 5.2 and 5.12); however, values as low as 1.1 F_{max} [Chapter 36 (Hof and Van den Berg), experimentally estimated for ankle muscles] and as high as 1.8 F_{max} [Chapter 6 (Lehman)] have been used. Additionally, most models assume that the slope at low lengthening velocities is higher than that are low shortening velocities.

Of note is that most of the evidence for yielding unfortunately comes from experiments where the muscle is originally under *steady* activation for a significant period of time. The data of Cordo and Rymer (1982) indicate that yielding does not occur in *newly recruited* muscle fibers. This is of great importance since it suggests that the muscles causing changes in movement direction (involved in a "stretch-shortening cycle") are likely not to yield. On teleologic grounds alone this might be expected (Winters and Stark, 1987), especially given the proficiency of tasks in life (e.g. walking, running, kicking) which involve stretch-shortening muscle behavior.

Yielding is also is a function of *how* the muscle is transiently lengthened. For example, yielding is not seen below a certain velocity threshold, and furthermore sudden isovelocity ramps appear to cause more yielding than isotonic loads which induce lengthening [compare Joyce et al. (1969) to Joyce and Rack (1969)]. Of note is that because of the inherent inertia of musculoskeletal systems, the idealized isokinetic "hold-to-ramp" transition that induces yielding is not a good approximation of *in situ* human movements, except for tasks involving unexpected "impacts". Here is where yielding might be most expected. Anticipated impacts may also show yielding (e.g. striking a ball with a bat; kicking or hitting an object; being struck by an object).

A number of modeling approaches have been used to simulate yielding which take advantage of insights from Huxley-based molecular models, but without their complexity. Van Dijk (1978), representing the Huxley model by 3 differential equations, used a "compromise" lengthening force–velocity curve that had a high force-velocity slope for velocities under about 0.2 times V_{max}, then yielded dramatically. Crowe et al. (1980) provided a length-dependent feature by dividing attachment locations into 4 length "bins". They found reasonably good correlation to their experimental data. The most theoretically elegant approach, the Distribution Moment (*DM*) model of Zahalak (1981), is covered in Chapter 1. Here we consider a conceptually different approach that allows the Hill model to remain intact.

"Attachment" as the Input to the CE

Based initially on a "challenge" from George Zahalak, I recently developed a class of approaches for approximating "yielding" behavior from within the confines of Hill-based models (1988, 1989). Central to the conceptual foundation is a **distinction between "activation" and "attachment."** It is assumed that the "problem" with simulating yielding is not an inadequate *CE* model but rather an inadequate estimation of attached bonds; the instantaneous, single-valued *CE* relation is assumed to still hold for **those bonds that remain attached.** Attachment, the input to the *CE* element, is thus a function not only of the ongoing activation, but also the rate-of-change of activation and *CE* dynamics. Somewhat surprisingly, a number of heuristic relations were formulated that provide reasonably good simulations of this behavior. All involve increasing the order of the model by one due to the addition of a nonlinear first-order process relating attachment to activation. All start by assuming that at each instant the "attachment" (A_{att}) differs from the "activation" (A_{act}) by a history-dependent yielding function (F_y):

$$F_{att}(A,v_{ce}) = F_{act}(A) - F_y(A,v_{ce}) \qquad (5.18)$$

The form of F_y used in Winters (1989) is:

$$\tau \dot{F}_y + F_y =$$

$$\begin{cases} \left(\dfrac{v_{ce}}{v_{max}}\right)^{0.5} \left(\dfrac{kF_a + F_{max}}{k+1}\right)\{1-(A/A_{max})\} & v_e > v_{cr} \\ \\ 0 & v_e \le v_{cr} \end{cases}$$

$$(5.19)$$

where:

$$\tau_y(A, v_{ce}) = \frac{\tau_{yo}}{1 + (v_{ce}/v_{max})} \qquad (5.20)$$

Eq. 5.19 **empirically** combines into a product relationship three experimental observations (from left to right in *Eq. 5.29*): *i)* yielding increases as lengthening increases, with there being a "threshold" below which yielding does not occur (Joyce et al, 1969)); *ii)* the amount of yielding is inversely proportional to the current activation level; and *iii)* the amount of yielding is greatest with steady or decreasing activation and is nearly non-existent when muscle is increasing in activation (Cordo and Rymer, 1982).

Although the *CE* relation is **remains instantaneous** and single valued, a schematic of what the history-dependent range would have been if the F_{fvin} due to attachment was replaced by that which would have occurred due to activation is illuminating. Such a plot is shown in Figure 5.15.

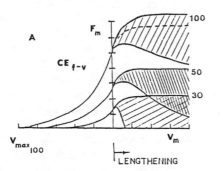

Figure 5.15: Schematic identifying of range of possible lengthening behavior *if* the y-intercept F_{fvin} due to activation were to replace that due to attachment (which is actually used). Three hypothetical activation levels are shown. The location within these ranges at a given time is history-dependent.

This distinction allows classic nonlinear "yielding" phenomena, as described by Joyce et al. (1969), to be closely simulated within the context of a Hill-based model structure that retains the instantaneous *CE* force–velocity–length element (Winters, 1989) and the *CE–SE* model structure. Figure 5.16a provides a dynamic example of how attachment differs from activation and consequently affects model performance; Figure 5.16b provides a summary of the observed model "ramp-and-hold" behavior, while Figure 5.16c shows that appropriate simulation of the sinusoidal oscillation data of Joyce et al. (1969). The latter case addresses another criticism of Hill-based models,

showing that, with our "attachment" approach, the average force (and for low activation even the peak force) can in fact fall below the isometric force (Figure 5.16b).

Our own investigations of lengthening muscle behavior in humans, concentrating on the "strategic" low-to-moderate velocity range, have produced mixed results. In a very careful study of alternate isokinetic-isometric (repetitive ramp-hold-ramp) sequences during "steady maximal" (perhaps 90% effort) elbow flexion, in which the effects of *CE* tension–length-related "drifts" were carefully isolated out, Bagley (1987) saw little yielding in most subjects but a definite indication of a change in *CE* force-velocity slope with increasing velocity (Figure 5.17a). Yet Silver-Thorn (1987), employing **submaximal** effort levels

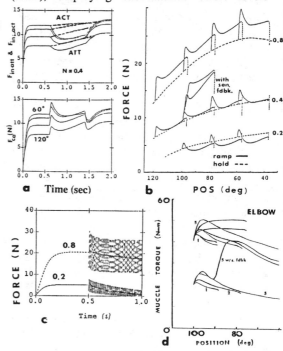

Figure 5.16: Summary of simulation results of classical cat soleus data displaying "yielding" and for elbow flexion. *a)* Muscle activation (dashed), attachment (solid) and force trajectories during a typical ramp-and-hold simulation. *b)* Summary of cat soleus force-position data for an assortment of "ramp-and-hold" experiments (compare to Figure 3 of Joyce et al., 1969). *c)* Cat soleus model response to 100 Hz vibration for two different initial activation levels. *d)* Elbow flexion ramp-and-hold simulations for two initial bias torques and in each case for speeds between 1 and 5 rad/sec. Top traces includes two parameter sets, one causing yielding and one not. Notice for both *b)* and *d)* that activation increases due to sensory feedback eliminated most of the yielding effect.

during cyclic isokinetic exercises of knee musculature, saw consistent, repetitive indications of yielding in *some* subjects (Figure 5.17b) that were clearly not due to changes in effort levels or reflex activity (as seen via *EMG*s). Others showed no such signs whatsoever. These preliminary observations suggest that there is *variability* in the amount of yielding seen in different subjects performing the same tasks. This makes modeling more difficult. It also suggests to me that one of the primary differences between "great" athletes and the rest of us may be related to the capacity to control yielding: yielding is undesirable during activation of a lengthening agonist undergoing "stretch-shortening", yet it is desirable to have quick yielding in an antagonist undergoing deactivation (e.g., see Figure 5.18).

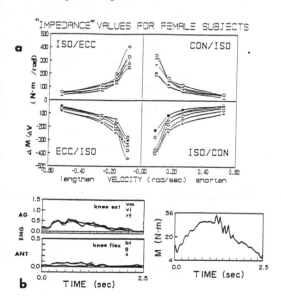

Figure 5.17: Sample results from experiments searching for yielding in humans. *a)* Summary of change in torque versus change in velocity (impedance units but not a true measure of impedance) versus Kincom (Chattecx) isokinetic velocity setting during the initial 80 ms intervals for eccentric⟷isometric and concentric⟷isometric transitions. Results for female subjects and for the "female" muscle-joint model (**o**) are shown here; results for males were very similar, only with "impedance" values nearly twice as high for all velocities. Adapted from Bagley (1987). *b)* Typical low-velocity experimental data which showed "yielding" during isokinetic knee extension movements (adapted from Silver-Thorn, 1987). The bell-shaped base in the torque curve is an artifact of start-up effects and the *CE* torque-angle property. Notice also the damped torque oscillation; this was typical.

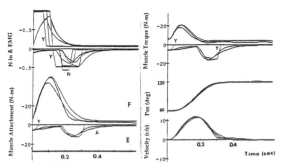

Figure 5.18: Model results for tasks with similar position trajectories but with and without the "attachment-yielding" addition to the model. With this addition, the antagonist "turns off" more quickly than with the default model, and consequently the agonist pulse width can be less. However, a larger antagonist "clamping" pulse was required for the yielding case.

Does this new approach "solve" the problem of simulating muscle yielding within the context of the Hill-based model? Yes and no. On the positive side, it provides a conceptual foundation for simulating yielding phenomena while retaining the Hill-based model structure and properties; "attachment" simply becomes the input to *CE*. On the negative side, it is a totally heuristic approach. Additionally, unlike parameters for shortening muscle or the other relations that have been discussed, we have not uncovered a systematic, rational basis for determination of task-independent parameter values for a range of muscles. Of course, our experimental data suggests due to muscle yielding variability this might be futile anyway.

5.3.2 "Force Enhancement"

Another phenomenon of muscle that has traditionally been difficult to model via Hill-based (and Huxley-based) models is force enhancement (Abbott and Aubert, 1951; Sugi, 1979; Edman et al., 1978; Julian and Morgan, 1979). When isometric conditions are reestablished after stretching a muscle (the "ramp-and-hold" protocol), the isometric force generated tends to be higher than that which would be seen for a normal isometric contraction under the same activation level and initial length. This phenomena is most likely when the muscle is at a length longer than the optimum length [Edman et al., 1978; see also Chapter 3 (Morgan)]. It is an open (and controversial) question regarding how "permanent" this "extra" force is, with some groups claiming that it is (Hill, 1977), while others suggest that it continues to decay for at least 1.5 seconds and eventually disappears (Van Atteveldt and Crowe, 1980). Although Sugi (1979) attributes this excess tension to forces exerted by cross-links that are "locked out" at small displacements, this phenomenon is difficult explain via traditional cross-bridge models (Edman et al., 1978; Atteveldt and Crowe, 1980).

Can this phenomenon be seen in human data? The data of Gielen and Houk (1987; wrist movements) show that it exists for at least a few seconds. In our investigations of maximal isometric force levels following slow lengthening of maximally contracting elbow flexors, we have seen clear indications of such phenomena in *some* subjects but not in others (Bagley, 1987); however, some of this effect is probably related to "drifts" in subject attention.

Can this phenomenon be modeled? Edman et al (1978) suggested a few possible sources, including a viscoelastic effect and changes in the force–velocity relation or activation level. Crowe et al. (1980), using four length "bins" to represent length-dependency, found that the basic behavior of enhancement could be simulated, but that the predicted delay was too fast (within 1 sec). Reasonable force enhancement can be also be obtained if *Eq. 5.18* is modified by having an empirical "enhancement" factor added to the right side (see also curve *E* in Figure 5.16a):

$$F_{att}(A, v_{ce}) = F_{act}(A) - F_y(A, v_{ce}) + F_{enh}(A, v_{ce}) \quad (5.21)$$

Sprigings (1986), assuming a linearized Hill model structure, computed the time constant B_h/K_{se} and showed that if this time constant could somehow increase after stretch, enhancement could be modelled. He suggested that K_{se} is attachment-dependent and thus, with fewer bridges attached, K_{se} would decrease (i.e. compliance would increase). However, my own evaluation suggests that the magnitude of the compliance change doesn't seen to be enough to cause responses on the order of seconds. It should be noted, however, that the slope of the lengthening CE force–velocity curve is very large during small lengthening velocities (e.g. Katz, 1939). With both B_h increasing and K_{se} decreasing, as would be predicted by modern Hill models, the time constant may indeed be on the order of seconds. Additionally, passive viscoelastic tissue exhibits inherent force relaxation that can be on the order of seconds and even minutes (e.g. Fung, 1970). Thus, the concept of this having, in part, a mechanical origin describable from within the Hill model framework cannot be eliminated. However, in my opinion other sources, and particularly the issue of series

sarcomere inhomogeneity addressed in Chapter 3 (Morgan), seems more appealing. There it is proposed that this phenomenon is related to cross-bridge uniformity, and there being an inherent tendency toward instability for muscle fibers longer than the rest length. Another approach towards modeling this effect, based on the "charge-transfer" model for muscle, is presented in Chapter 2 (Hatze).

5.4.3 Subtle Stiffness Effects

1. Sensitivity of SE/Stiffness Identification to Testing Method. A discrepancy between results from "quick release" experiments and the isometric "calculation" method (e.g., Parmley et al., 1970) that simple Hill models cannot explain. We saw in Section 5.2.2 that if the stiffness for each individual cross-bridge is assumed linear and the population in homogeneous), then for isometric contraction the stiffness would be a linear function of force, and thus the force–extension curve would then have an *exponential shape*, similar to that traditionally used for biological soft tissues (*Figure 5.7c-d, Eq. 5.8-5.10*), only here for a different reason.

In all other cases, however, force is of course *not* statically related to activation. When activation is higher than force (as during shortening of the *CE*), the force–extension curve would be expected to be more flat; when lower (e.g. lengthening *CE*), a more concave relationship would be expected. Since different methods for estimating *SE* utilize different activation operating ranges, this may help explain different results for different tests. As we saw in Section 5.2.2, the Hill *SE* relation can be made directly activation-dependent. Finally, Hill (1970) has suggested that part of the discrepancy may be due to passive viscoelastic effects; this possibility is best explored from within the framework of the Hill model structure.

2. Stiffness Variation During Isometric Contraction.

Gassar and Hill (1924) found muscle to be especially "rigid" right after the initiation of stimulation (leading muscle force). Cecchi et al. (1984) and Stein and Gordon (1986) have carefully quantified changes in stiffness during isometric contraction, showing that stiffness *leads* force during the rising phase of tetanic isometric contraction, as if it was in part a function of activation. However, stiffness *lags* behind force (and even more so activation) during relaxation; such results are not consistent with the concept of stiffness being a static function of force and activation, and thus this subtle effect is very difficult to incorporate within the Hill model structure [or certain forms of Huxley-based models (Bobet et al., 1990)].

3. Transient Force "Dip" During Fast Releases.
Another "unusual" effect is the transient force response

to very, very fast releases of low magnitude (< 0.5%), where an excessive force "dip" occurs over the time course of a few milliseconds and then recovers to a plateau before drifting back (Huxley and Simmons, 1971). This behavior cannot be represented by a single series stiffness element, and has been used as evidence of a mechanically induced change in attachment states [see McMahon (1984) for review]. McMahon also suggests that it cannot be explained by a Kelvin model of Figure 5.5c (however, at such high speeds both viscous and muscle mass effects will be more prominent and may play a role). Of note is that this phenomenon has little relevance to the study of multiple muscle movement systems because of the smoothing effects of system inertia and tendon compliance (Alexander, 1988).

5.5 Comparison to Simpler Models.

Simpler muscle models include: *i*) an idealized force generator; *ii*) a force generator that produces a smoothed (filtered) version of the rectified *EMG*; *iii*) a spring; or *iv*) parallel arrangement of spring and dashpot. Each of these approaches is fundamentally flawed, yet each is used successfully within this book. These models tend to be flawed because muscle force depends not only on the neural input, as represented by an *EMG*, but also on muscle length and velocity. Muscle dynamic properties clearly cannot be captured by a spring, nor in most cases by a parallel arrangement of spring and dashpot. Winters and Stark (1987) have shown, however, that higher-order Hill-based models can, over limited operating ranges, take on the input-output characteristics of lower-order models. Predictably, however, the required describing parameters change dramatically with the type of task, and even during a task.

Nonetheless, simpler models can be quite useful (and even preferable) as long as the range of applicability is firmly established. Of note is that whenever an *EMG* trajectory is causally related to a motor task in implicit model has been assumed. Table 5.1 provides a few insights, based on modeling experiences of myself and simulation results presented by other contributors, that could be used as "rules of thumb" to aid interpretation of experimental data.

Table 5.1: "Rules of Thumb" regarding Muscle Dynamics for Various Tasks

TASK	Parameter Sensitivity CE_{Vmax}	SE_{Xmax}	Comments
Shortening muscles in general	+ +	+	Lower force relative to *EMG* (reaching about 25% at higher velocities
Moderate- and high-speed "point-to-point" movements.	+ +	+	Assuming bell-shaped velocity patterns: *Agonist:* shape of force trajectory differs from *EMG* shape, with force especially low during higher velocity region and often showing a second "hump" even if *EMG* doesn't (e.g. Figures 19.6-19.11). *Antagonist:* force potentially relatively high if not turned "off" in time — a reason for the common observation of the antagonist turning "off" just before the agonist turns "on". Very effective at generating "clamping" force (assuming yielding doesn't occur).
Voluntary "stretch-shortening" movements	+ + +	+ +	*Antagonist-becoming-agonist:* CE length transition will lead that for the overall muscle as the muscle force goes into a favorable CE operating range and to some extent elastic energy is released; thus the muscle force will be relatively higher than the *EMG* trace might indicate (see Chapters 38-39). *Agonist-becoming-antagonist:* normally *EMG* will have subsided early enough such that the low level of *EMG* that is seen does in fact represent low muscle force levels.
Elastic Bounce (prepared impact)	+ +	+ + +	Pre-impact *EMG* helps set up elastic recoil potential and helps assure that CE and SE are within the effective operating range; the force and work done may be higher than might be predicted from the *EMG*s. Unlike the case for voluntary stretch-shorting, antagonist co-contraction, which increases impedance, may be utilized effectively.
Aggressive Manual Tracking	+ +	+ +	Nonlinear models are much more effective at tracking, especially when quick direction changes are desired. Expect pulse-like *EMG*s and variable cocontraction, features which can make the mechanical system "look different" at different times.
Movements under high bias loads (e.g. isotonic, torso during tasks)	+	+ +	Due to lower velocities, the forces during shortening movements can be much higher than during "free" movements with the same *EMG*s.
Postural/equilibrium	+	+ +	Often a static *EMG*-force relation is appropriate, but use caution — the CE force–velocity relation is quite steep near zero velocity, and hence even slow lengthening muscle may generate 20-30% more force than predicted from an isometric *EMG*-force calibration; the converse, though not as dramatic, holds for shortening muscle.
Manipulation Tasks/Interaction with Environment	+ +	+ +	Both sensory feedback and cocontraction will be used aggressively as necessary. For dynamic tasks, the SE plays a major role in this regard. Hard to generalize since manipulation tasks differ significantly and since the hand and upper limb (and torso) muscles will be used quite differently. *EMG* and muscle force may bear little resemblence to each other when the task is complex — modeling becomes necessary.

5.6 Future Directions

There is a great need for experiments in the *submaximal* activation region, which is of greatest importance for "normal" human movement. Most individuals go through their entire day without maximally contracting muscles. Questions remain regarding sub-maximal relations for *CE* tension-length, *CE* force–velocity, *SE*, yielding, force enhancement, etc. Perhaps an "unwritten law" could be "passed" making it illegal to study muscle mechanics for maximally contracting muscles? For example, for activation-dependent *SE* characterization, it is suggested that future experiments should consider applying quick releases to steady *sub-maximal* isometric contraction, following the early lead of Joyce and Rack (1969).

Our emphasis here has been on utilizing carefully controlled experiments to illuminate muscle structure and properties. In many cases experiments are purposely designed to isolate certain elements within the model or to dramatically show certain experimental "quirks." For the "systems biomechanist" studying multiple muscle systems, however, the bottom line is the use of the model for the realistic tasks of special interest. Hopefully, the model provides insights into internal behavior and makes predictions that can be tested experimentally. Of note is that a model is more than just the sum of its components. The real "beauty" lies in the fact that the various elements *interact dynamically* with each other. Often "intuition" will fail unless actual simulations are performed – hence part of their value. How do we get maximum use out of a model? By following up "core" simulations with aggressive use of advanced tools such as the systematic study of the *sensitivity of model behavior to model parameters*. Of note is that the relative sensitivity of model behavior to various parameters within a model is a function of the task. Another very helpful approach is maximizing insight is to employ task-specific linearization of a nonlinear model, followed by re-simulation. This process can be easily automated (Seif-Naraghi and Winters, 1989).

One of the best ways to simplify a model for a given class of tasks is to *start* with more complex nonlinear Hill-based models, as recommended here, and then identify insensitive model parameters # elements described by these parameters can then be linearized or eliminated with more confidence.

Ironically, significantly more effort has gone into *creating* models than in *utilizing* models. In this regard, this book is an exception, and perhaps a statement of what the future holds.

References

Abbott, B.C. and Aubert, X.M. (1951) The force exerted by active striated muscle during and after change in length. *J. Physiol. Lond.* **117**:77-86.

Abbott, B.C. and Wilkie, D.R. (1973) The relationship between velocity of shortening and the tension-length curve of skeletal muscle, *J. Physiol.* **120**:214-222.

Agarwal, G.C. and Golltieb, G.L. (1977) Oscillation of the human ankle joint in response to applied sinusoidal torque on the foot. *J. Physiol.* **268**:151-176.

Aubert, X. (1956) *Le Couplage Energetique de la Ccontraction Musculaire.* Brussels, Editions Arscia.

Alexander, R. McN. (1983) *Animal Mechanics* (Second Edition). Blackwell Sci. Publ., Oxford.

Alexander, R. McN. and Bennet-Clark, H.C. (1977) Storage of elastic strain energy in muscle and other tissues. *Nature* **265**:114-117.

Audu, M.L. and Davy, D.T. (1985) The influence of muscle model complexity in musculoskeletal motion modeling. *J. Biomech. Engng.* **107**:147-157.

Bagley, A.M. (1987) Analysis of human response to slow isokinetic movement. M.S. Thesis, Arizona State University.

Bahler, A.S. (1967) The series elastic element of mammalian skeletal muscle. *Am. J. Physiol.* **213**:1560-1564.

Bach, T.M., Chapman, A.E. and Calvert, T.W. (1983) Mechanical resonance of the human body during voluntary oscillations about the ankle joint. *J. Biomech.* **16**: 85-90.

Bahler, A.S. (1968) Modeling of mammalian skeletal muscle. *IEEE Trans. Biomed. Engng.* **BME-13**:248-257.

Baildon, R.W.A. and Chapman, A.E. (1983) A new approach to the human muscle model. *J. Biomech.* **16**:803-809.

Bennett, M.B., Ker, R.F., Dimery, N.J. and Alexander, R. McN. (1986) Mechanical properties of various mammalian tendons. *J. Zool.* **209**:537.

Bigland, B. and Lippold, O.C.J. (1954) The relation between force, velcoity and integrated electrical activity in human muscles. *J. Physiol.* **1253**:214-224.

Bobet, Stein, R.B. and Oguztoreli, M.N. (1990) Mechanisms relating force and high-frequency stiffness in skeletal muscle. *J. Biomech.*, accepted.

Bouchaert, J.P., Capellen, L. and de Blende, J. (1930) *J. Physiol.* **69**:473.

Butler, D.L., Grood, E.S., Noyes, F.R. and Zernicke,

R.F. (1979) Biomechanics of ligaments and tendon. *Exer. Sport Sci. Rev.* **6:**125-185.

Cavagna, G.A., Komarek, L., Citterio, G. and Margaria, R. (1971) Power output of the previously stretch muscle. *Med. Sport (Biomech. II)* **6:**159-167.

Cecchi, G., Griffiths, P.J. and Taylor, S.R. (1984a) The kinetics of crossbridge attachment studies by high frequency stiffness measurements. In *Contractile Mechanisms in Muscle* (Pollack, G.H. and Sugi, H., eds.), Plenum Press, pp. 641-655, New York.

Cecchi, G., Lombardi, V. and Menchetti, G. (1984b) Development of force-velcoity relation and rise of isometric tetanic tension measure the time course of different processes. *Pflugers Arch.* **401:**396-405.

Chapman, A.E. (1985) The mechanical properties of human muscle. *Exer. Sci. Sport Rev.* **13:** 443-501.

Chapman, A.E. and Calvert, T.W. (1979) Estimations of active-state from EMG recordings of human muscular contraction. *Electromyogr. Clin. Neurophys.* **19:**199-222.

Close, R.I. (1972) Dynamic properties of mammalian skeletal muscles. *Physiol. Rev.* **52:**129-197.

Cook, G. and Stark, L. (1968) The human eye movement mechanism: experiments, modelling and model testing. *Arch. Ophthal.* **79:**428-436.

Cordo, P.J. and Rymer, W.Zev (1982) Contributions of motor-unit recruitment and rate modulation to compensation for muscle yielding, *J. Neurophys.* **47:**797-809.

Crowe, A., Van Atteveldt, H. an Groothedde, H. (1980) Simulation studies of contracting skeletal muscles during mechanical strech. *J. Biomech.* **13:**333-340.

Dern, R.J., Levine, J.M. and Blair, H.A. (1947) Forces exerted at different velocities in human arm movement. *Am. J. Physiol.* **151:**415-437.

Ebashi, S.M. and Endo, M. (1968) Calcium ion and muscle contraction. *Prog. Biophys. Mol. Biol.* **18:**123.

Edman, K.A.P., Mulieri, L.A. and Scubon-Mulieri, B. (1976) Non-hyperbolic force-velocity relationship in single muscle fibres, *Acta Physiol. Scand.* **98:**143-156.

Edman, K.A.P., Elzinga, G. and Noble, M.I.M. (1978) Enhancement of mechanical performance by stretch during tetanic contractions of vertebrate skeletal muscle fibres. *J. Physiol.* **280:**139-155.

Elliott, D.H. (1965) Structure and function of mammalian tendon. *Biol. Rev.* **40:**392-425.

Fenn, W.O. and Marsh, B.S. (1935) Muscular force at different wpeeds of shortening. *J. Physiol. (Lond.)* **85:**277-297.

Flitney, F.W. and Hirst, D.G. (1978) Crossbridge detachment and sarcomere "give" during stretch in active frog's muscle. *J. Physiol.* **276:**449-465.

Ford, L.E., Huxley, A.F., and Simmons, R.M. (1977) Tension responses to sudden length change in stimulated frog muscle fibers near slack length. *J. Physiol.* **269:**441-515.

Fuchs, F. (1977) Cooperative interactions between calcium-binding sites on glycerinated muscle fibers - the influence of corss-bridge attachment. *Biochim. Biophys. Acta* **462:** 314-322.

Fung, Y.C. (1967) Elasticity of soft tissues in simple elongation. *Am. J. Physiol.* **213:**1532-1544.

Fung, Y.C. (1970) Mathematical representation of the mechanical properties of the heart muscle. *J. Biomech.* **3:**381-404.

Fung, Y.C. (1981) *Biomechanics.* Springer-Verlag, New York.

Gasser, H.S. and Hill, A.V. (1924) The dynamics of muscle contraction. *Proc. Roy Soc.* **96:**398-437.

Gielen, C.C.A.M. and Houk, J.C. (1987) A model of the motor servo: incorporating nonlinear spindle receptor and muscle mechanical properties. *Biol. Cybern.* **57:**217-235.

Gordon, A.M., Huxley, A.F. and Julian, F.J. (1966) The variation in isometric tension with sarcomere length in vertebrate muscles. *J. Physiol.* **184:**170-192.

Goubel, F., Bouisset, S. and Lestinne, F. (1971) Determination of muscular compliance in the course of movement. *Med. Sport (Biomech. II)* **6:**154-158.

Greene, P.R. and McMahon, T.A. (1979) Reflex stiffness of man's antigravity muscles during kneebends while carrying extra weights. *J. Biomech.* **12:**881-895.

Hanson, J. and Huxley, H.E. (1955) The structural basis of contraction in stiated muscle. *Symp. Soc. Exp. Biol.* **9:** 228-264.

Hatze, H. (1974) A model of skeletal muscle suitable for optimal motion problems. In: Biomech. IV, pp. 417-422, S. Karger, Basel.

Hatze, H. (1977) A myocybernetic control model of skeletal muscle. *Biol. Cybern.* **25:**103-119.

Hatze, H. (1981) *Myocybernetic Control Models of Skeletal Muscles.* Univ. of South Africa.

Henneman, E., Somjen, G. and Carpenter, D. (1965) Excitability and inhibitability of motoneurons of different sizes. *J. Neurobiol.* **28:** 599-620.

Hill, A.V. (1922) The maximum work and mechanical efficiency of human muscles, and their most economical speed. *J. Physiol.* **56:**19-45.

Hill, A.V. (1938) The heat of shortening and the dynamic constants of muscle. *Proc. Roy. Soc.* **B126:**136-195.

Hill, A.V. (1949) The abrupt transition from rest to activity in muscle. *Proc. Roy. Soc.* **B126:**399-420.

Hill, A.V. (1950) The series elastic component of muscle. *Proc. Roy. Soc.* **B141:**104-117.

Hill, A.V. (1970) *First and last experiments in muscle mechanics.,* Cambridge Univ. Press, Cambridge.

Hof, A.L. and Van den Berg, J. (1981) EMG to force processing. I. An electrical analogue of the Hill muscle model. *J. Biomech.* **14:**747-758.

Huxley, A.F. and Simmons, R.M. (1971) Proposed mechanism of force generation in striated muscle. *Nature* **233:**533-538.

Jewell, B.R. and Wilkie, D.R. (1958) An analysis of the mechanical components in frog striated muscle. *J. Physiol.* **143:**515-540.

Joyce, G.C and Rack, P.M.H. (1969) Isotonic lengthening and shortening movements of cat soleus muscle. *J. Physiol.* **204:**475-495.

Joyce, G.C., Rack, P.M.H. and Ross, H.F. (1974) The forces generated at the human elbow joint in response to imposed sinusoidal movements of the forearm. *J. Physiol.* **240:**375-396.

Joyce, G.C., Rack, P.M.H. and Westbury, D.R. (1969) The mechanical properties of cat soleus muscle during controlled lengthening and shortening movements, *J. Physiol.* **214:**461-474.

Julian, F.J. and Sollins, M.R. (1973) Regulation of force and speed of shortening in muscle contraction. *Cold Spring Harbor Symp. Quanrt. Biol.* **37:**635-646.

Katz, B. (1939) The relation between force and speed in muscular contraction. *J. Physiol.* **96:**45-64.

Kleweno, D.K. (1987) Physiological and theoretical analysis of isometric strength curves of the upper limb. M.S. Thesis, Arizona State University.

Kleweno, D.K. and Winters, J.M. (1988) Sensitivity of upper-extremity strength curves to 3-D geometry: model results, *Adv. in Bioengng.*, ASME-WAM, **BED-8:** 53-56.

Komi, P.V. (1973) Relationship between muscle tension, EMG and velocity of contraction under concentric and eccentric work. In *New Developments in Electromyography and Clinical Neurophysiology*, (Desmedt, J.E., ed.), S. Karger, Basel, pp. 596.

Kulig, K., Andrews, J.G. and Hay, J.G. (1984) Human strength curves. *Exerc. Sport Sci. Rev.* **12:** 81-121.

Levin, A. and Wyman, J. (1927) The viscous elastic porperties of muscle. *Proc. Roy. Soc.* **B101:**218-243.

Lehman, S.L. and Stark., L. (1979) Simulation of linear and nonlinear eye movement models: sensitivity analysis and enumeration studies of time optimal control. *J. Cybern. Inform. Sci.* **2:**21-43.

Lehman, S.L. and Stark, L. (1982) Three algorithms for interpreting models consisting of ordinary differential equations: sensitivity coefficients, sensitivity functions, global optimization. *Math. Biosci.* **62:**107-122.

Ma, S. and Zahalak, G.I. (1987) Activation dynamics for a distribution-momnet model of skeletal muscle. *Proc. 6th Int. Conf. Math. Model.* **11:**778-782., St. Louis.

McCrorey, H.L., Gale, .H. and Alpert, N.R. (1966) Mechanical properties of the cat tenuissimus muscle. *Am. J. Physiol.* **210:**114-120.

McMahon, T.A. (1984) *Muscles, Reflexes and Locomotion.* Princeton Univ. Press, Princeton.

Morgan, D.L. (1977) Separation of active and passive components of short-range stiffness of muscle. *Amer. J. Physiol.* **232:**C45-C49.

Otten, E. (1988) Concepts and models of functional architecture in skeletal muscle. *Exer. Sport Sci. Rev.* 89-137.

Parmley, W.W., Yeatman, L.A. and Sonnenblick, E.H. (1970) Differences between isotonic and isometric force-velocity relations in cardiac and skeletal muscle. *Am. J. Physiol.* **219:**546-550.

Partridge, L.D. (1979) Muscle properties: a problem for the motor controller physiologist. In *Posture and Movement* (Talbott, R.E. and Humphery, D.R., eds.), pp. 189-229, Raven Press, New York.

Perrine, J.J. and Edgerton, V.R. (1978) Muscle force-velocity and power-velocity relationships under isokinetic loading. *Med. Sci. Sports* **10:** 159-166.

Petrofsky, J.S. and Phillips, C.A. (1981) The influence of temperature, initital length and electrical ativity on the force-velocity relationship of the medidal gastrocnemius muscle of the cat. *J. Biomech.* **14:**297-306.

Pertuzon, E. and Bouisset, S. Instantaneous force-velocity relationship in human muscle. *Med. Sport, Biomech. III*, **8:** 230-234.

Proske, U. and Morgan, D.L. (1987) Tendon stiffness: methods of measurement and significance for the control of movement. A review. *J. Biomech.* **20:**75-82.

Rack, P.M.H. and Ross, H.F. (1984) The tendon of flexor pollicis longus: its effects on the muscular control of force and position at the thumb. *J. Physiol.* **351:** 99-110.

Rack, P.M.H. and Westbury, D.R. (1969) The effects of length and stimulus rate on tension in isometric cat soleus muscle. *J. Physiol.* **204:**443-460.

Rack, P.M.H. and Westbury, D.R. (1974) The short range stiffness of active mammalian muscle and its effect on mechanical properties. *J. Physiol.* **240:**331-350.

Rack, P.M. and Westbury, D.R. (1984) Elastic properties of the cat soleus tendon and their functional importance. *J. Physiol.* **347:**479.

Ralston, H.J., Polissart, M.J., Inman, V.T., Close, J.R. and Feinstein, B. (1949) Dynamic features of human isolated voluntary muscle in isometric and free contractions. *J. Appl. Physiol.* **1:**526-533.

Ramsey, R.W. and Street, S.F. (1940) The isometric length tension diagram of isolated skeletal muscle fibers of the frog. *J. Cell. Comp. Physiol.* **15:**11-34.

Seif-Naraghi, A.H. and Winters, J.M. (1989) Effect of task-specific linearization on musculoskeletal system control strategies. *ASME Biomech. Symp.*, **AMD-98:** 347-350.

Silver-Thorn, M.B. (1987) Muscle imbalance in osteoarthritis of the human knee. M.S. thesis, Arizona State University.

Silver-Thorn, M.B. and Winters, J.M. (1988) Muscle imbalance and osteoarthritis of the knee. *Adv. in Bioengng.* ASME Wint. Ann. Mtng., **BED-8:** 95-98.

Sprigings, E.J. (1986) Simulation of the force enhancement phenomenon in muscle. *Comput. Biol. Med.* **16:**423-430.

Stein, R.B. and Gordon, T. (1986) Nonlinear stiffness-force relationships in whole mammalian skeletal muscles. *Can. J. Physiol. Pharmacol.* **64:**1236-1244.

Sugi, H. (1979) The origin of the series elasticity in striated muscle fibers. In *Cross-Bridge Mechanism in Muscle Contraction* (Sugi, H. and Pollack, G.H., eds.), pp. 85-102, Univ. of Tokyo Press, Tokyo.

Van Atteveldt, H. and Crowe, A. (1980) Active tension changes in frog skeletal muscle during and after mechancial extension. *J. Biomech.* **13:**323-335.

Van Dijk, J.H.M. (1978) Simulation of human arm movements controlled by peripheral feedback. *Biol. Cybern.* **29:**175-186.

Wells, J.B. (1964) Comparison of mechanical properties between slow and fast mammalian muscles. *J. Physiol.* **178:**252-269.

Wilkie, D.R. (1950) The relation between force and velocity in human muscle. *J. Physiol.* **K110:**248-280.

Wilkie, D.R. (1956) The mechanical properties of muscle. *Br. Med. Bull.* **12:**177-182.

Winters, J.M. (1985) Generalized analysis and design of antagonistic muscle models: effect of nonlinear muscle-joint properties on the control of fundamental movements. Ph.D. Dissertation, Univ. of Calif., Berkeley.

Winters, J.M. (1988) Improvements within the A.V. Hill model structure: strengths and limitations. *Proc. IEEE Engng. Med. Biol.*, pp. 559-560, New Orleans.

Winters, J.M. (1989) A novel approach for modeling transient lengthening with a Hill-based muscle model. *XII Int. Congr. Biomech.*, Abstract 128., Los Angeles.

Winters, J.M. and Bagley, A.M. (1987) Biomechanical modelling of muscle-joint systems: why it is useful. *IEEE Engng. Med. Biol.* **6:** 17-21.

Winters, J.M. and Stark, L. (1985) nalysis of fundamental movement patterns through the use of in-depth antagonistic muscle models. *IEEE Trans. Biomed. Engng.* **BME-32:** 826-839.

Winters, J.M. and Stark, L. (1987) Muscle models: what is gained and what is lost by varying model complexity. *Biol. Cybern.* **55:** 403-420.

Winters, J.M. and Stark, L. (1988) Estimated mechanical properties of synergistic muscles involved in movements of a variety of human joints. *J. Biomech.* **21:** 1027-1042.

Winters, J.M., Stark, L. and Seif-Naraghi, A.H. (1988) An analysis of the sources of muscle-joint system impedance. *J. Biomech.* **21:** 1011-1026.

Woittiez, R.D., Huijing, P.A., Boom, H.B.K. and Rozendal, R.H. (1984) A three-dimensional muscle model: a quantified relation between form and function of skeletal muscles, *J. Morphol.* **182:**95-113.

Yates, J.W. and Kamon, E. (1983) A comparison of peak and constant angle torque-velcoity curves in fast and slow-twitch populations. *Eur. J. Appl. Physiol.* **51:**67-74.

Zahalak, G.I. (1981) A distribution-moment approximation for kinetic theories of muscular contraction. *Math. Biosci.* **55:** 89-114.

Zahalak, G.I., Duffy, J., Stewart, P.A., Litchman, H.M., Hawley, R.H. and Pasley, P.R. (1976) Partially activated human skeletal muscle: an experimental investigation of force, velocity and EMG. *J. Appl. Mech.* **98:** 81-86.

Zahalak, G.I. and Heyman, S.J. (1979) A quantitative evaluation of the frequency-response characteristics of active human skeletal muscle in vivo. *J. Biomech. Engng.* **28:** 28-37.

Zajac, F. (1989) Muscle and tendon: properties, models, scaling and application to biomechanics and motor control. *CRC Crit. Rev. Biomed. Engng.* **17:**359-415.

CHAPTER 6

Input Identification Depends on Model Complexity

Steven L. Lehman

6.1 Introduction

Biomechanical data are often the best evidence of motor control patterns available. However, the inverse problem in motor control, i.e., to determine the activation of muscles from torques or from kinematics, is made difficult by noise in the data and by imprecision in specifying the operator to be inverted. This chapter addresses the dependence of the inversion process on the operator.

Consider a model M that takes an input $u(t)$ and produces a trajectory $x(t)$: $Mu = x$. The inverse problem is to determine u from some experimentally determined trajectory x^*. How does the inferred input u depend on the model M? There appear to be no mathematical tools for finding the dependence of u on changes in the *structure* of M, for models of any generality.

This chapter takes a practical approach, based on simulation. I solved the inverse problem for three identified models of wrist mechanics, using the same data x^*. The results demonstrate that the inferred u depends strongly on the structure of M: there are *qualitative* differences between the inferred inputs. To show that these differences have some generality, I investigated the sensitivity of each model's output to small changes in its inferred input. Small changes in most input parameters yield large changes in each model's output: the different inferred inputs are not close to each other.

The three models are variations on an old standard in the motor control literature, chosen for its physiological sense. Each has a simple mechanical plant (an inertia) driven by two muscle models.

The muscle models are: a Hill model, with a nonlinear force–velocity curve and constant (tendon-like) series elasticity; an augmented Hill model (with the series stiffness a function of active state); and a linearized Hill model, with constant series stiffness and a constant force–velocity viscosity. In each case, the assumed input is active state, in the Hill sense [discussed in Chapter 5 (Winters)].

As model complexity increases, so does the similarity between the inferred active state and electromyograms. How big a model is needed to get a good estimate of active state? Inferred inputs to the two nonlinear versions of the Hill model differ most in their antagonist active states. It is just this situation − force production by lengthening partially activated muscles − that is likely to be poorly represented by a Hill model. I therefore simulated a fourth model, with Huxley two-state representations of muscle, modified for simulating lengthening as by Zahalak (1981). This model is not much harder to identify than the Hill versions, and it has the advantage of simulating force transients during lengthening [see also Chapter 1 (Zahalak)].

Parameters of the Huxley model are identified from the characteristics of a shortening muscle, so the force produced during shortening is the same as for the Hill model. The biggest differences should occur during rapid stretches. I found actual movements in which active muscles are stretched rapidly, and simulated these stretches using the Huxley model.

The movement data show that forces increase with velocity of stretch, even at high velocities. The force increases might be caused by cor-

Multiple Muscle Systems: Biomechanics and Movement Organization
J.M. Winters and S.L-Y. Woo (eds), © 1990 Springer-Verlag

responding increases in active state, but the *EMG* does not increase. The force increases might rather be caused by the increasing stretch velocities. Here, the transient force production, available to the Huxley model, differs from the steady-state, inherent to the Hill representation. The transient maximum force increases with lengthening velocity even for fast stretches, for which steady state force output levels off.

6.2 Methods

Flexion and extension of the wrist are good representatives of single–joint limb movements. Plant mechanics are simple in a sizable range around anatomical position — to a good approximation, the hand is an inertial load rotating about a single axis. Easily measured accelerations are therefore good measures of net joint torque. Flexion and extension torques are produced by relatively few muscles — three flexors and two extensors, compared, for example, with the five flexors of the forearm about the elbow. The muscles on each side are often turned on and off together (Litvintsev and Seropyan, 1977), so their forces can be lumped into one equivalent on each side. The wrist movers produce relatively small torques. For example, wrist flexors produce about 1/20 the torque of elbow flexors. Yet wrist movements may even have some generality: the ratio of the inertia of the hand about the wrist to that of the forearm about the elbow is about the same as the ratio of maximum torques.

6.6.1 Wrist Geometry

Estimates of muscle velocities from angular velocities of the hand, and of muscle forces from torque, became important in estimating tendon compliance, and in applying a Huxley model to wrist movements. The center of wrist rotation was determined by grasping a pencil in the fist and tracing an arc with the forearm fixed. Perpendicular bisectors of chords to this arc intersected at nearly the same point for rotations over the middle of the range of motion. I approximated the radius from the center of rotation to the carpal tunnels as approximately $w = 2$ cm, and the distance from the center to muscle insertions on the hand as about $r = 7$ cm. For small movements about resting position, relative muscle length change is approximately proportional to angular change: $\Delta L/L = w\Delta\theta/r$. The forces scale from

torque as $F = T * (L/rw \sin \theta)$ where F is force (N), T is torque (Nm), and L is the distance from wrist to tendon insertion. The geometry is detailed in Lehman and Calhoun (1990, in press). For the purpose of estimating sarcomere velocities from muscle velocities, the wrist flexors and extensors were assumed to be fusiform and approximately 10 cm long.

Table 6.1: Wrist Movement Model Parameters

Param	Value	Units	Remarks
m	0.0039	kg-m^2	Inertia of hand
M	0.0071	kg-m^2	Inertia of apparatus
F_{of}	21.0	N-m	Maximum torque, flexors
F_{oe}	12.0	N-m	Maximum torque, extensors
V_{maxf}	20.0	rad/s	Max. shortening vel., flex
V_{maxe}	20.0	rad/s	Max. shortening vel., ext
a/F_o	0.25		Hill a
k_{ce}	1.5	rad^{-1}	*CE* series stiffness per unit active state
k_t	10.0	Nm/rad	Tendon stiffness
D_o	0.035	s	*EMG* to movement delay
f_1	105	1/s	Huxley mod. attach. rt. const.
g_1	456	1/s	Huxley mod. detach. rt. const.
g_2	2198	1/s	Huxley mod. detach. rt. const.
g_3	280	1/s	Huxley mod. detach. rt. const.

6.2.2 Identification of Model Parameters

The hand acts as a nearly pure inertial load over a range of about 40° of either flexion or extension. The inertia can be measured by pushing the relaxed hand with a constant torque, and measuring acceleration. Estimates of the inertia by modeling the hand as a rigid body agree with such measurements. Passive elastic torques are small (less than 0.1 Nm) over the middle 80° of the hand's range of motion, but increase sharply at extreme wrist angles. Viscous torques for the relaxed wrist are very small. Full details of the identification of the plant and the Hill model parameters are in press (Lehman and Calhoun, 1990). Identified parameters are listed in Table 6.1.

Maximum voluntary torque is nearly independent of angle over the central 80° of the range of motion, but declines sharply (to 0.2 to 0.25 of the maximum) near either end of the range.

Maximum velocity varies with load in a hyperbolic curve. The relationship is well fit by Hill curves [described in detail in Chapter 5 (Winters)], with a/F_o approximately 0.25. For the wrist, unlike the forearm about the elbow, maximum velocities are reached and maintained during even the fastest isotonic contractions.

The lengthening force–velocity relationship was not measured. Rather, torque output of a stretched muscle was assumed to increase linearly for slow lengthening velocities, with a slope six times that for slow shortening, then to saturate at $1.8 F_o$ (Katz, 1939).

Using a quicki–release protocol, it is possible to measure the series elasticity of the wrist, given that the inertia is already known. Stiffness is a function of initial load (before the quick release). It is approximately proportional to torque for small initial torques (0.9 to 1.8 Nm/rad per Nm for small active state), but appears to level off at a stiffness comparable to tendon stiffness.

Derivation of a two–state Huxley model from the identified Hill model required identification of three parameters. These three rate constants were derived from the identified torque/angular velocity curves (specifically from V_{max}) using the assumptions regarding ratios of the constants originated by Huxley (1957).

I modified the Huxley model by increasing the release rate for long stretches, exactly as in Zahalak (1981). The additional rate constant was determined by simulating stretches at high velocities, and adjusting the rate constant to achieve a steady–state force of $1.8 F_o$.

6.2.3 Model Simulation and Inversion

Hill models integrated with time steps of 0.001 s, using fourth–order Runge Kutta integration. I used G.I. Zahalak's Distribution Moment Approximation [Zahalak, 1981; see also Chapter 1 (Zahalak)] to compute force for the modified Huxley model. Active state was assumed to modify only attachment rates, with no effect of attachment on activation.

Model inversion was accomplished by use of an optimization algorithm, which minimized the root mean square (RMS) difference between model

position and experimentally recorded position by varying the parameters of an input pattern consisting of two pulses per muscle. Pulse heights were allowed to vary between 0 and F_o, and pulse widths and delays varied between 0 and 500 ms. I performed the optimization using the same position data for each model.

6.2.4 Collection and Analysis of Movement Data

Wrist flexions were measured for five healthy adults. Subjects were seated at a table with the right forearm secured between wooden blocks. The forearm was supinated 90°, and the hand was placed in a manipulandum designed to allow only flexion and extension in the horizontal plane. The handle squeezed the hand between two plates, disallowing grasp.

An oscilloscope placed in front of the subject displayed a target and a cursor that moved proportional to wrist rotation. The target was a pair of lines separated by the equivalent of 0.2° wrist rotation.

Subjects began the task with the wrist close to anatomical position. When the subject entered the center target, it would jump to a position requiring flexion of 4 to 40°. Subjects were instructed to move as fast as possible to the target, and were informed of the maximum velocity after each movement. Movements at each amplitude were practiced until the subject felt proficient (about 20 trials). Then 15 movements were recorded at each amplitude.

Position and electromyograms were sampled 500 times per second. Position records were digitally differentiated to obtain velocities and accelerations.

6.3 Results

6.3.1 Identified Active State Depends on Model Structure

It was possible to fit each of the three models to position data with some allowable combination of pulse heights and widths (Figure 6.1). The error between the actual position and best-fit model trajectory was small in each case (less than 0.001 rad, about 0.8% of the RMS movement amplitude).

The inferred active states were quite different from each other. The linear model approximated the overshoot and undershoot of the actual trajectory by its mechanical resonance, a damped sinusoid approaching the final position. The active state includes only one antagonist pulse, and only a very small second agonist pulse. The nonlinear Hill model has two sizable pulses of agonist active state, but still only one long antagonist pulse. The augmented Hill model has two full pulses in each muscle. The agonist pulses are larger than those for the Hill model, but have about the same timing. The larger first agonist activity is offset by a strong antagonist pulse, as is the second agonist pulse.

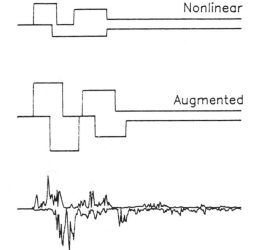

Table 6.2: Sensitivities of error to active state parameters.

Agonist				
Pulse height	A	2.1	1.3	3.1
	C	0.1	1.4	2.8
Pulse width	A	2.1	2.6	4.9
	C	0.1	3.2	4.6
Delay to	C	0.07	3.3	2.4
Antagonist				
Pulse height	B	0.5	2.3	4.0
	D	none	none	4.0
Pulse width	B	0.5	5.8	7.0
	D	none	none	4.1
Delay to	B	0.2	3.8	3.2
	D	none	none	0.8
Tonic level		none	0.4	0.4

6.3.2 Sensitivity Analysis

The relative sensitivity of *RMS* error to changes in each of the input parameters is indicated in Table 2.2. Relative sensitivity is the fractional change in *RMS* error ($\Delta e/E$), divided by the fractional change in the parameter value ($\Delta p/P$). The nominal value of error, E, is *0.119* rad, the *RMS* average position over the whole movement. Relative error is therefore also relative average position. The nominal parameter value, P, is its optimum value. The fractional change in parameter value ($\Delta p/P$) was always *0.1* — I varied each parameter by 10%, leaving all the others constant, and measured the relative change in error.

The output is generally quite sensitive to small changes in the nominal input, for each model. Most of the relative sensitivities are between 1 and 10. For example, the relative sensitivity of position error to the first agonist pulse height is 3.1 for the augmented Hill model: a 10% change in the pulse height would cause a 31% increase in relative error. The absolute change in error would be large — 31% of the *RMS* movement amplitude. Considering that the absolute error between the model position trajectory driven by the optimum input and the actual position was less than 0.001 rad for each of the models, the increase in error is

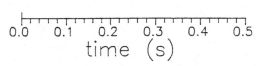

Figure 6.1: *Active state functions inferred from a wrist flexion.* Position (top) and electromyograms (bottom) for a fast-as-possible flexion of 12°, and inferred two-pulse active states for linear, nonlinear, and augmented Hill models. Antagonist active state and *EMG* are inverted for display.

enormous — some 37 times the original error, in this case.

Relative errors are quite different between models. The sensitivities are generally higher for the nonlinear Hill model than for the linear version, and still higher for the augmented Hill model. In all cases, the largest sensitivities are to pulse widths. For the two larger models, error is very sensitive to antagonist pulse widths. The two antagonist pulses of the augmented model could not be merged into a single pulse, like the input to the Hill model. The qualitatively different antagonist patterns are by no means interchangeable.

Error is insensitive to only a few parameters - the second pulse of agonist active state for the linear model, and the tonic level for the other two.

The optimum inputs are, according to the sensitivity analysis, well identified. Small changes in almost any parameter of any input would lead to large changes in the error.

6.3.3 Torque as a Function of Stretch Velocity in Fast Wrist Movements

The unrealistic simplifications of the Hill model are the separation of series elasticity from the contractile element, and the assumption of steady state in the lengthening force–velocity curve. Movements that push these limitations are those in which activated muscles are stretched rapidly.

In investigating the control of a range of fast movements, we (Lehman and Lucidi, 1990) found that peak velocity scales with amplitude for fast-as-possible wrist flexions, and the peak velocities exceed $0.5\ V_{max}$ for large (40°) movements.

Electromyograms show that these rapidly stretched muscles are activated. Of course, they do not produce their full torque at peak velocity (maximum velocity is reached at the zero crossing of acceleration, and therefore of torque, the hand acting as an inertia). However, they do produce their peak torques at high stretch velocities. Figure 6.2a shows maximum decelerating torques produced during movements of amplitudes from 12° to 40°, and the stretch velocity at which the peak torque is reached. The maximum shortening velocity of these muscle (V_{max}) is about 20 rad/s.

During part of the largest movements, the (active) antagonist muscle is therefore being stretched at about $0.25\ V_{max}$.

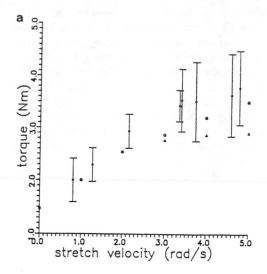

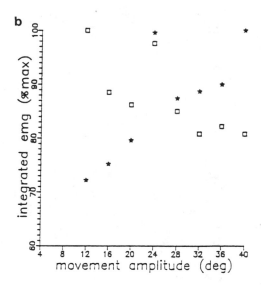

Figure 6.2: *a)* Maximum decelerating torque as a function of stretch velocity for fast-as-possible wrist flexions of 12 and 40°deg. Mean and standard deviations for ten movements at each amplitude. Triangles show torque developed at steady state by Huxley model. Circles show maximum torque developed by Huxley model. *b)* Integrated electromyograms for the same movements. Agonist *EMG* (stars) was integrated over a period including the first agonist pulse. Antagonist *EMG* (squares) was integrated over the whole of antagonist activity.

6.3.4 Torque as a Function of Stretch Velocity for the Modified Huxley Model

The torques produced by the Huxley model, stretched at the same rates and partially activated, are indicated on the same graph (Figure 6.2a: circles show maximum torque; triangles show steady–state torque). The steady–state torque has leveled off at 1.8 F_o before stretch velocity reaches its maximum. The transient maximum torque is still increasing.

Over the 4° to 40° range of fast-as-possible wrist flexions, peak accelerating torque is scaled to movement amplitude. Electromyographic evidence (Figure 6.2b) supports the hypothesis that this scaling is by activation. Peak decelerating torque is also scaled, over the same range. However, electromyograms decrease in amplitude, if anything, as torques increase.

6.4 Discussion

6.4.1 Why u Depends on M

In the parameter space based on pulse heights, pulse widths and inter–pulse delays, where altitude is the error between model and actual position, there seems to be a single deep valley for each model. The linearized Hill model resides in relative isolation, close to the other two in only two of 12 parameters. The other two are reasonably close to each other along the agonist coordinates, but far separated in antagonist space. The sensitivity analysis shows that the valley walls are steep. A large escarpment separates even the two nonlinear models.

It is likely that the linear model sits in its isolated pocket owing to differences in coupling. The linear model matches the oscillatory motion of the hand by mechanical resonance. In the nonlinear models, the force generator is not as coupled to the hand mass. At low active state, the Hill "viscosity" is small, so the series springs and mass are uncoupled from the contractile element. Active state forcing, and not mechanical resonance, must therefore determine the last trajectory. A second agonist pulse is necessary to drive the hand to its final position.

The differences between the two nonlinear Hill models are mostly in the antagonist active state. The augmented version, with inertia further uncoupled from the active state by an activation-dependent stiffness, can more freely modulate the activation of the lengthening muscle, which produces much higher forces than its shortening counterpart, even at low stretch velocities. The model assumes a 6:1 ratio of slopes of the force velocity curve lengthening:shortening (Katz, 1939).

With such a large initial slope, the lengthening force–velocity curve must level off quickly: the largest forces in steady state are at 1.8 F_o. Steady-state force from the stretched Huxley model also saturates at slow-lengthening velocities (Figure 6.2). In fast wrist movements larger than 20°, force continues to increase at these high velocities. Decelerating force calibrates precisely to accelerating force (Lehman and Lucidi, 1990). The precision of the correspondence argues against a neural mechanism, as does the last of scaling in antagonist *EMG*. The mechanism may therefore be mechanical.

6.4.2 Conclusion and Future Directions

A cross bridge representation in muscle mechanics may then be necessary, even for movements of intact humans. There are advantages to such a model: an automatic dependence of series stiffness on active state, separation of tendon and contractile element elasticities, a reasonable representation of the force–velocity relationship for lengthening, and the shortening force–velocity curve as an emergent property.

A clear experimental problem is set by the wrist movement results: to measure torques produced by stretching muscles of an intact wrist, at different levels of active state. An allied anatomical problem is to specify more accurately the wrist geometry and muscle architecture, in order to determine the correspondence between wrist angle and sarcomere length. The estimates made in the model presented here are conservative – changes in architecture or geometry are likely to increase the estimate of muscle velocities.

References

Huxley, A.F. (1957). Muscle structure and theories of contraction. *Progr. Biophys. Biophys. Chem.* 7:257-318.

Joyce, G., Rack, P.M.H, Westbury, D.R. (1969). The mechanical properties of cat soleus muscle during controlled lengthening and shortening movements.

J. Physiol. **204**:461-474.

Katz, B. (1939). The relation between force and speed in muscular contraction. *J. Physiol.* **96**:45-64.

Lehman, S.L., Calhoun, B.M. (1990). An identified model for human wrist movements. *Exp. Brain Res.* (in press).

Lehman, S.L., Lucidi, C.A. (1990). Control of a range of fast movements mediated neurally and mechani-

cally. *Exp. Brain Res.* (submitted).

Litvintsev, A.I., Seropyan, N.S. (1977). Muscular control of movements with one degree of freedom. *Avtomat. Telemakhan.* **5**:88-102.

Zahalak, G.I. (1981). A distribution-moment approximation for kinetic theories of muscular contraction. *Math. Biosci.* **55**:89-114.

CHAPTER 7

Actuator Properties and Movement Control: Biological and Technological Models

Blake Hannaford and Jack Winters

7.1 Introduction

Actuation is the process of conversion of energy to mechanical form. A device that accomplishes this conversion is an actuator. There are many types of actuators, with most including energy transformation through multiple forms. Of course an equally vital part of the definition of an actuator is controllability: the actuator's conversion of energy must be modulated by a control input.

Galvani demonstrated that muscle activity could be electrically modulated. We now know that this involves an electrical signal that, through a series of steps that are rate-limited by the influx–efflux of calcium, modulates acto-myosin interaction [reviewed in Chapter 1 (Zahalak)]. In the 1920s Hill, Fenn, and their colleagues broke new ground by subjecting muscle to the thermodynamic analysis developed for the rational design of energy conversion devices such as steam engines (Hill, 1922; Fenn, 1924). We now have a fairly good idea of how chemical energy stored in the form of *ATP* is converted to mechanical work (Chapter 1). Thus, we can view muscle as an actuator.

This chapter will examine the mechanical properties of muscle actuators in the context of technological actuators such as those used in robot manipulators. The goals of the chapter are: *i)* to elucidate and contrast the dynamic properties of various technological actuators, with concentration on how other actuators differ from biological muscle; and *ii)* to elucidate how actuator properties influence system control strategies.

7.2 Energy Conversion Devices

Just as muscle has developed into families of specialization to meet specific needs, no single technological actuator can be said to dominate in technological energy conversion. In fact, the variety of approaches to technological actuation exceeds that of muscle types which, despite their specialization, share the same cross-bridge architecture.

Table 7.1: *Effort–flow* definitions for various media. For solid mechanical systems, the mirror analogy, shown in parentheses, is also often used (e.g., Takahashi et al., 1972).

	Effort	*Flow*
Electrical	Voltage	Current
Hydraulic	Pressure	Flow
Pneumatic	Pressure	Mass Flow
Translational Mech.	Force	Velocity
	(Velocity)	(Force)
Angular Mech.	Torque	Ang. Vel.
	(Ang. Vel.)	(Torque)

We will describe seven forms of actuators in a unified format. The terminology of generalized dynamical systems from Paynter (1961) will be utilized, in which (for the mechanical end of the actuator), as shown in Table 7.1, we will use *effort* (*e*; also referred to as *"potential"* or *"across"*) to represent force or torque, and *flow* (*f*; also sometimes called *"through"*) to represent velocity or angular velocity (see Table 7.1). The product of *effort* and *flow* is **power**. The beauty of such a formulation is

Multiple Muscle Systems: Biomechanics and Movement Organization
J.M. Winters and S.L-Y. Woo (eds.), © 1990 Springer-Verlag

that transformation among different forms of dynamical systems is straightforward, with each mode including three types of passive elements (Figure 7.1): C (compliance or capacitance), I (inertance or inductance), and R (resistance). Energy can be stored in C and I elements and is dissipated in R elements (Figure 7.1). For each actuator, we will consider here two of the three defining or *"constitutive"* relations: *effort* vs. *flow* (*R–element*), and *effort* vs. *displacement* (*C–element*), where *displacement* is the integral of *flow* (see Figure 7.1). The dynamic relationship between effort and flow is termed impedance [$Z(t)$] or admittance [$Y(t)$], depending on causality

$$e(t) = Z(f(t)); \quad f(t) = Y(e(t)) \qquad (7.1)$$

or, for the linearized case

$$Z = \frac{e}{f} = \frac{1}{Y} \qquad (7.2)$$

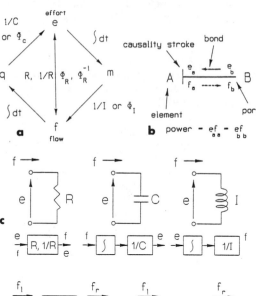

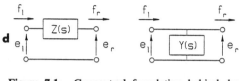

Figure 7.1: Conceptual foundation behind dynamic systems modeling approach. *a)* Schematic showing relationships between passive elements and variables. *b)* An idealized bond connects elements via ports. Information flows bicausally: at each port one of the *e–f* pair is designated an input, the other the output, as defined by the causality stroke convention. Energy is ideally passed across the bond without loss. *c)* Passive one-port elements, described by constitutive equations. Energy is stored by C and I elements, lost across R elements. *d)* 2-port representation of linear impedance Z and admittance Y.

By mis-matching impedances/admittances between parts of a system, inherently bi-directional power flow takes on the nature of uni-directional information flow.

In comparing these diverse machines, we need some measures with which to quantify their properties. One approach to this comparison would be to make a standard dynamic model and specify the parameters of each actuator. Because the appropriate dynamic model structure for these actuators differ, however, direct comparisons would be difficult and in many cases parameter vectors would be orthogonal; for instance, the *I-elements* may be in different structural locations. What we can do is note that all actuators lose heat (i.e. none are 100% efficient), and all have one primary power-dissipating *R-element*. This element is represented by a static *effort–flow* constitutive relationship. We will also find that it is instructive to provide the steady-state *effort–displacement* relationship (*C-element* property).

Finally, when appropriate we will supplement the above information with some basic practical performance-related information for each device: *i)* power output per actuator mass; *ii)* force per cross-sectional area (translational actuators) or torque per volume (rotary actuators); *iii)* efficiency of actuation; and *iv)* power source. The actuators we will consider are:

1. Internal Combustion Engine
2. *AC* Induction motor
3. *DC* Motor
4. Microstepping Motor
5. Hydraulic Cylinder (with/without spring return)
6. Pneumatic Cylinder (with/without spring return)
7. Braided Pneumatic (McKibbon-like) Artificial Muscles
8. Human Skeletal Muscle

Within each type of actuator there are many variations. For each case we will define a representative actuator and then present the simplified *effort–flow* and *effort–displacement* relations. Most treatments of a particular type of actuator (including muscle) strive for generality and thus use normalized units to describe constitutive relations. For example, the *force–velocity* constitutive relationship of muscle contractile tissue (which describes an *R-element*) is usually plotted with normalized axes, with the force normalized to the maximal isometric force and the velocity to the maximal unloaded peak velocity [e.g. Chapter 5 (Winters)]. For our purposes, normalization is harmful because it obscures the large differences between properties of different actuators. Consequently, we will plot the properties of a specific example of each actuator type in absolute coordinates so that they may be appropriately contrasted.

Based primarily on fundamental differences in the shape of *effort–flow* relationships, Holmes (1977) identified four classes of energy conversion devices:

1.   Self Induction Machines
2.   Slip Driven Machines
3.   Linear Effort Controlled Machines
4.   Linear Flow Controlled Machines

In Holmes' classification, *types 3*, and *4* are linear characteristics that differ only in slope, and there is no type which describes the concave *effort/flow* property of muscle. Thus, the following modified version of Holmes' scheme is proposed (Figure 7.2):

1.   Self Induction Machines
2.   Slip Driven Machines
3a. Linear Effort Controlled Machines
3b. Linear Flow Controlled Machines
4.   Concave Effort–Flow Machines
       (Muscle-like machines)

7.2.1 Internal Combustion Engine *(Type 1)*

Although the internal combustion engine (Figure 7.3a) has a power density considerably greater than that of the other actuators considered here, it is included in the comparison as a representative of the *type 1* actuator, the self induction machine. This property holds because the charge

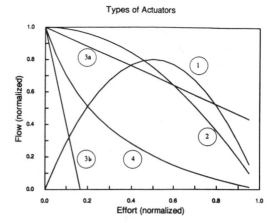

Types of Actuators

Figure 7.2:   Classification of energy conversion devices according to the shape of their *effort-flow* relationship (modified from Holmes (1977)). Types are: *1)* Self Induction Machines; *2)* Slip Driven Machines; *3a)* Linear Effort Controlled Machines; *3b)* Linear Flow Controlled Machines; *4)* Muscle-like Machines.

of air–fuel mixture is drawn into the cylinder by the engine's vacuum at a rate proportional to engine speed. Thus, the *i.c. engine*, like all *type 1* machines, cannot start under it's own power. This is evident from the *effort–flow (R-element)* curves and in applications, a constant velocity starter is used (Figure 7.3b). The internal combustion engine's torque is developed by smoothing a series of impulsive shocks (Figure 7.3c) through an inertial system (the flywheel).

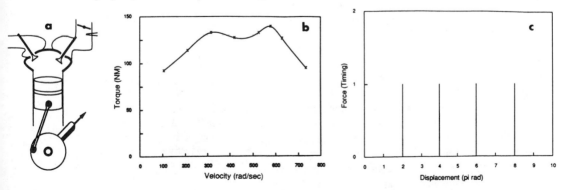

Figure 7.3:   Internal Combustion Engine: *a)* Schematic diagram of the internal combustion engine.   *b)* *Effort-Flow* data (1988 1.6 L Accura Integra).   *c)* Schematic depiction of the *Effort-Displacement* relation of an internal combustion engine without the filtering effect of it's flywheel.

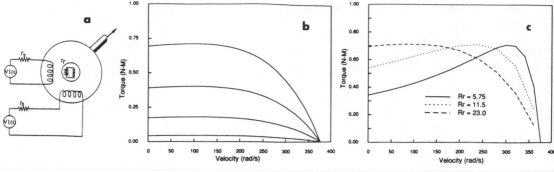

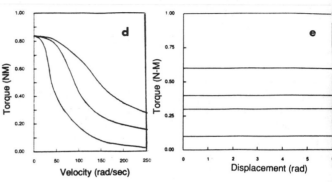

Figure 7.4: *AC* Induction Motor: *a)* Schematic diagram of a 2-pole *AC* induction motor. The rotor windings are not connected to a commutator; instead, winding current is induced by the changing magnetic fields created by the stator. *b)* Equilibrium *effort-flow* data for a 2 pole *AC* servomotor. *c) Effort-flow* relation for the *AC* induction motor for three values of armature resistance R_a at $V_c = 115$ *V*. *d) Effort-flow* relation for certain classes of micro-stepper motors (see text). *e) effort-displacement* data (idealized).

7.2.2 AC Induction Motor *(Types 1, 2)*

In the *AC* induction motor orthogonal field coils are excited with sinusoidal currents that are 90° out of phase in order to create a rotating magnetic field vector inside the motor (Figure 7.4a). The magnetic field vector rotates at the frequency of the field coil currents. The armature coils are short circuited. The rotating magnetic field vector induces a current in the armature coil proportional to the difference in velocity between the armature and magnetic field vector. The induced current causes a torque proportional to the flux, ϕ. The induced current is inversely proportional to the armature resistance R_a. Since the flux both induces the current and causes the resulting torque, the system is highly non-linear. The common *AC* induction motor (Figure 7.4) can be either a *type 1* or 2 machine depending on the value of its armature resistance. For low values of R_a, the *AC* induction motor behaves as a self induction (*type 1*) machine (Figure 7.4c). This mode of operation is used in constant velocity applications with a starting capacitor for high efficiency. In servo control applications, the *type 1* characteristic limits the controllability of the motor as well as its low velocity application, and thus a higher value of R_a is used to bring the torque–velocity curve into the

more gently convex shape of a *type 2* machine (Figure 7.4b).

The torque–velocity relation for the *AC* induction motor at constant velocity can be derived as (Krause and Wasynczuk, 1989):

$$T_e = P\left(\frac{x_{ms}^2}{\omega_e}\right)\frac{r_r'^2\,\sigma\,v_c^2}{r_s\,r_r' + \sigma(x_{ms}^2 - x_{ss}x_{rr}')^2} \quad (7.2)$$

where $\quad \sigma = \dfrac{\omega_e - \omega_r}{\omega_e}$

and

P = number of poles
$r_r' = (N_s/N_r)^2\,r_r$
lineup = rotor resistance reflected into stator circuit
r_s = stator resistance
ω_e = excitation frequency
ω_r = rotor angular velocity
X_{ws} = stator magnetizing inductive reactance = $\omega_e L_{ms}$
X_{ss} = stator self inductive reactance = $\omega_e (L_{ls} + L_{ms})$
X_{rr} = rotor self inductive reactance = $\omega_e (L_{ls} + L_{ms})$

For a selected example [Krause and Wasynczuk (1989)], the parameter values are:

$r_s = 24.5\ \Omega;\ r_r' = 23.0\ \Omega$
$L_{ls} = 27.0\ \text{mH};\ L_{ms} = 273.0\ \text{mH}\ ;\ L_{mr}' = 27.0\ \text{mH}$
$\omega_e = 377.0\ \text{rad/sec}$
$V_a = 0\text{--}115.0\ \text{V}$ $\quad (7.3)$

The torque–velocity curve for an *AC* servomotor with the above parameter values (Figure 7.4b) has a concave shape characteristic of the slip driven machine (*type 2*). To illustrate the strong dependence of this characteristic on the armature resistance, R_a, the same torque velocity curve at maximal driving voltage is plotted for three values of R_a (Figure 7.4c).

Until recently *AC* motors suffered from difficulties in appropriate control excitation hardware, quite low torques, and the control stability problem identified earlier. The power to mass ratio is typically under 0.1 W/g. However, the efficiency is quite high (e.g. 80%). Although these problems are now less severe, a recent assessment suggested that these motors will be used where cost and durability, as opposed to precision performance, are the primary considerations (Andeen, 1988; see also below). Of note is that *AC* and *DC* motors both represent relatively mature/stable technologies. The optimum selection therefore is likely to depend on available control technology, which is rapidly changing.

7.2.3 AC Micro-Stepper Motors (Type 4)

AC microstepper motors translate electrical pulses into mechanical movements in fixed increments. Conventional step motors in general are a poor choice as an actuator for servo control due to low resolution, roughness at slow speeds, ringing between steps, and resonance over a certain speed range. However, recent progress in control technology make microstepping motor systems a viable (though extremely expensive) option. These devices take a voltage and current input and

index in displacement increments of very small size. The *effort–flow* relations of Figure 7.5d, for a 1.1 kg motor (Compumotor C/CX 57-102) show that the force–velocity relation is concave, with the concavity a function of the *DC* input voltage and the *I-element*. The maximum power per unit mass is about 0.06 W/g. Of note is that these motors can be modeled as uni-causal devices, with position set by the electrical input.

7.2.4 DC Motor (*Type 3a*)

DC motors have been the most widely utilized actuator for robotic (Andeen, 1988) and artificial limb (Parker and Scott, 1986) design. In general they produce moderate torque, high speed, and continuous power output.

The separately excited *DC* motor (Figure 7.5), when driven by a voltage source, V_a, connected to the armature winding, develops a torque proportional to the armature current, i. This current is in turn proportional to V_a minus a voltage proportional to angular velocity. This "back-emf" reduces torque in direct proportion to angular velocity, ω, just as would a parallel viscous element. Notice that there is inherently bilateral coupling. Such a motor is commonly described by

$$L\frac{d\tau}{dt} + R_a\,\tau = K\,\phi\,(V_a - K\,\phi\,\omega) \qquad (7.4)$$

where ϕ is the magnetic flux density due to permanent magnets or constant current drive to the field windings, τ is the motor torque, R_a is the resistance, L the inductance of the armature windings, and K is the "motor constant", which is a function of the number of windings, their

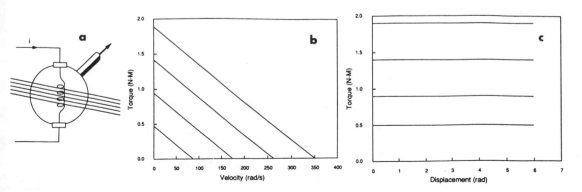

Figure 7.5: *DC* Servomotor: *a)* Schematic diagram: the *DC* servomotor consists of a stator, rotor, commutator, brushes (most cases), bearings, and a housing. The stator creates a magnetic field, with the rotor coils energized via the commutator and brushes. *b)* *Effort–flow* Curve. *c)* *Effort–displacement* data (idealized).

geometry, etc. To find the steady-state *effort–flow* relation, we neglect dynamic effects due to L and solve for the torque:

$$\tau = K \phi \frac{V_a}{R_a} - K^2 \phi^2 \frac{\omega}{R_a} \qquad (7.5)$$

Often *DC* motor specifications are given as:

$$\tau = K_1 \frac{c}{R_a} V_a - K_2 K_1 \frac{\omega}{R_a} \qquad (7.6)$$

where $0 < c < 1.0$.

For the selected *DC* servomotor, *Electrocraft model M-1140*, $K_1 = 9.1$ oz in/Amp, $V_{a_{max}} = 28.0\,V$, $K_2 = 67.4$ V/krpm, $R_a = 0.76\,\Omega$. Note that when converted to *MKS* units,

$$K_1 (\text{NM/Amp}) = K_2 (\text{V/r sec}^{-1}) = 0.064 \qquad (7.7)$$

i.e., the two commercial motor constants are the same as expected from *Eq. 7.5*. We can thus model the torque–velocity relation of this motor as (see Figure 7.5b)

$$\tau = 0.084\, V_c - 0.0054\, \omega \qquad (7.8)$$

Additional specifications: $\tau_{max} = 2.32$ Nm, $\omega_{max} = 430$ r/sec, weight = 2.25 kg.

For the representative motor described above the torque is approximately a linear function of the voltage input and the speed. In practise, a fairly wide variety of *DC* motor curves are available, and the curve may be somewhat concave in either direction. The subtle details of the curve magnitudes and shape depending on the structural arrangement of the windings [i.e. series (present case), shunt or compound]. Also, high-torque direct drive actuators are now in use in robotic systems. Of general importance is that reversal of the rotor current causes reversal of the motor torque, which is beneficial. These motors tend to be well-behaved near zero velocity. Transient behavior is limited primarily by the rotor inertia. Notice that the torque is approximately independent of angular displacement, i.e. there is a flat *effort-displacement* relationship (Figure 7.5c).

Maximum continuous power per unit mass is about 0.1 W/g, torque (stall torque) per mass is only on the order 0.04–0.1 MPa, and efficiency is typically about 75% (Shoemaker, 1988).

7.2.5 Hydraulic Actuator (*Type 3b*)

The hydraulic actuator is just one component of the full hydraulic system, which because of the need for a closed fluid system becomes rather involved (and expensive). Components include a pump, reservoirs, valves, tubing, the actuator, and circuitry. As with electric motors, hydraulic actuators are produced in many varieties. The variety considered here (Figure 7.6a) is the cylindrical piston driven by a constant displacement fluid pump. Counter to the name of the pump, due to the use of shunts and accumulators, the source of fluid to the control valve is commonly modeled as a source of constant pressure (*effort*). The control valve acts as a variable resistance to dissipatively control the rate of flow into and out of the cylinder, and thus can be thought of as controlling the piston velocity. The product of the pressure drop across the valve and the flow rate is the power dissipated across the *R-element*, which is seen as heat in the fluid and apparatus. This dissipative form of control causes relatively low energy efficiency (under 50%).

The working fluid (e.g. oil) is nearly incompressible, and thus the compliance is very low. Consequently, hydraulic systems are inherently set up for position regulation, with impedance modulation quite difficult (for high bandwidth compressibility must be considered). Operating pressures for hydraulic systems may range from 1–10 MPa (145–1450 psi). Systems designed for the higher-pressure range must naturally have thicker transmission lines and cylinders. The static relationship between shaft force and hydraulic pressure is:

$$F = PA \qquad (7.9)$$

where A is the cross-sectional area. The maximum continuous power output per mass ranges around 0.1-1 W/g, with the higher range for high-pressure systems. With high pressures, the force per area can be as high as 10 MPa, which is considerable.

Theoretical curves for a selected hydraulic actuator are shown in Figure 7.6b and d. Practical difficulties arising in the sharp edges of the spindle valves usually dictate that the spindles be somewhat smaller than the fluid ports. The "underlap", *UL*, can be as much as 20% and has the effect of introducing a shunt path for fluid flow at low valve openings. Two models are thus required (Rothbart, 1985), one for the flow in the underflap range (i.e. where x, the valve position, is less than the underlap, or $x < UL$) and another for the flow beyond that ($x > UL$):

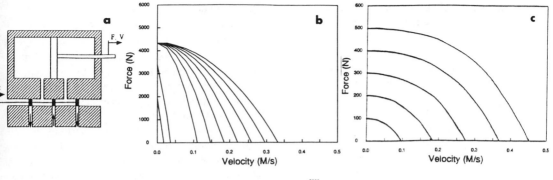

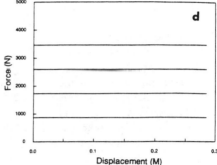

Figure 7.6: Hydraulic and pneumatic cylinders. *a)* Schematic diagram. *b)* Hydraulic *effort–flow* curves, modulated by changes in flow. *c)* Pneumatic *effort–flow* curves (35 mm dia cylinder, 1/8" ports), modulated by changes in pressures (up to 6 bar). *d)* Typical *effort– displacement* data (idealized; addition of spring-loading would cause a positive slope which would depend only on displacement).

$$F = AP_{max} \frac{1 - 2V^2}{V_{max}^2 \ x^2} \qquad (x > UL) \tag{7.10}$$

otherwise,

$$V = V_{max} \ \sqrt{(1 + x^2 - 2x\frac{F}{A} - (1 - x^2)\sqrt{1 - \frac{F^2}{A^2}}}$$

Thus, while the actuator is capable of considerable maximum force output for $x > UL$, the maximum force at $V = 0$ declines in the underlap region. This lowers the effective gain of the valve for stable operation in the region near $x = 0$.

The *effort–displacement* characteristic is constant over the length of the cylinder if the cylinder is not spring-loaded; otherwise there is a flat slope over the primary operating range. The series compliance to applied loads is quite low due to the low working compressibility of the fluid and for most applications quite rigid tubing walls.

7.2.6 Pneumatic Cylinders (Type 3)

Pneumatic cylinders are structurally similar to hydraulic cylinders, only usually lighter since the internal pressure range is typically lower (typically up to about 0.5–1 MPa). Also, the surrounding appararatus differs greatly, with the fluid flow exhausted directly to the environment. Pneumatic cylinders have been utilized in a number of artificial limb applications (e.g. Simpson and Lamb, 1965). Fluid inertance and viscosity are much lower than for hydraulic (oil or water) systems. Consequently, the peak flow rate velocity can be higher than for hydraulic fluid systems. Because of the compressibility of air, the compliance is high, and, furthermore, is a strong function of the

apparatus coupled to the actuator (e.g. length of lines, properties of control valves). Pressure–flow relations across control valve circuitry, and especially flow-restrictor valves, tend to be nonlinear (McCloy and Martin, 1973; Lord and Chitty, 1974). The force–pressure relationship is the same as described in Eq. 7.9, and since the internal pressure is less than for most hydraulic applications, the effective force is lower, on the interval 0.6 MPa for muscle-sized actuators. The power to mass ratio is typically about 0.2 W/g (Shoemaker, 1988).

For pneumatic systems supply lines, valves, etc. carry dynamic significance. To aid analysis, empirical flow coefficients (F_c) can be determined for all system components, including supply and exhaust ports (Yeaple, 1984). For example (Lansky and Schrader, 1986):

$$F_c = \frac{f}{K} \ \sqrt{\frac{G \ T}{(P_1 - P_2) \ P_2}} \tag{7.11}$$

where P_1 is the absolute inlet pressure, P_2 is the absolute outlet pressure, G is specific gravity (e.g. 1.0 for air), T is absolute temperature, and the range of applicability is for $P_2 > 0.53 \ P_1$. The *effort–flow* curve may then be estimated from empirical curves that relate this coefficient to cylinder diameter and/or port sizes. The shape of the curve for pneumatic systems (Figure 7.6c) is similar to that typical for hydraulic systems (Figure 7.6b). However, note that the magnitudes differ and that the curves scale differently.

Because exhaust time is often the limiting factor (e.g. on the order of 0.1 msec), "quick exhaust valves" are often used. Electro-pneumatic control may be "on-off" or proportional, and may control either pressure or flow. Of note is that on-off valves tend to bring out the nonlinear properties of air. When coupled to an appropriate mass or inertia, due to the inherent compliance of the fluid the overall system may have an inherent tendency to oscillate; thus a stability analysis is often necessary. A number of techniques are available to stabilize such systems (e.g., Burrows et al., 1974; Lord and Chitty, 1974). Experience with the *Utah-MIT* dextrous hand suggested in particular that pressure regulation, as opposed to flow regulation, was desirable (Jacobsen et al., 1986); we have also found this to be true.

A typical example is the double-acting, *Festo-DSNK* plastic cylinder (slightly larger that a biceps muscle), which weighs 0.37 *kg* and has a 25 *mm* (1 *in*) diameter and a 2 *cm* stroke (shown in Figure 7.7). The estimated curves are shown in Figure 7.6c. Notice that the *effort–displacement* curve is flat because the cross sectional area is independent of displacement (i.e. not spring-loaded).

7.2.7 Braided Pneumatic "Artificial Muscle" Actuators (Type 3-4)

Braided pneumatic actuators consist of an internal bladder surrounded by a braided mesh that is attached at either end to fittings or to some type of "tendon-like" structure (Figure 7.7). These actuators, first patented by Morin (1947 in France, 1953 in U.S.) and utilized extensively in artificial limb research in the 1950s and 1960s under the name "McKibben muscles", shorten and bulge out when pressurized. The static force–pressure relationship, can be approximated as (Schulte, 1961):

$$F = P A_d [3 \cos^2 \theta - 1] \qquad (7.12)$$

where θ is the angle between the elemental length of the helical fiber and the longitudinal axis of tubing and A_d is the cross-sectional area of the cylinder when $\theta = 90°$. This is the largest possible cross section, and consequently it turns out that the maximum tensile force is approximately 3 times that of a piston-type actuator having a cross-sectional area equal to the maximum stable area of the braid actuator; personal experience suggests that this is about twice the resting cross sectional area, and thus the predicted force levels can be quite high. The actuator is also much lighter. Setting $F = 0$ and solving for θ in *Eq. 7.12*, we see that, when pressurized, the angle of equilibrium with no applied force is on the order of 55°; a load on the actuator prevents this from being reached, and the actuator generates force. By changing the extensibility of the weave fibers (usually nylon has been used) and the resting weave angle (20°–30° is typical), a continuum of iso-pressure *effort–displacement* curves can be seen. Athough fabrication techniques have differed, we have found in our testing of various actuators, in compatibility with available literature (e.g., Schulte, 1961; Gavrilovic and Maric, 1969; *Bridgestone Acfas* Robot system manual), that the stiffness (inverse of compliance) increases approximately linearly with pressure. Also, the offset ("slack length") usually shifts to the left, with the amount being fabrication dependent (see Figure 7.8 for a typical example).

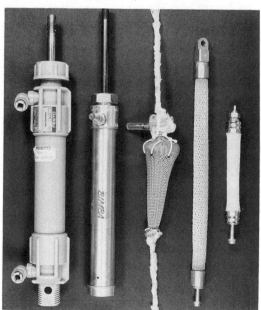

Figure 7.7: Photograph of representative pneumatic actuators. From left to right: double-acting plastic cylinder (*Festo-DSNK*), stainless steel single-acting cylinder (*Bimba*), fabricated braided musculotendon actuator, *Bridgestone* braided actuator, and an original McKibbon muscle (courtesy Rancho Los Amigos Rehabilitation Engineering Center, Ca).

Braided "rubbertuator" muscles (*Bridgestone Co.*), which we have found to be essentially identical in concept to McKibben muscles and in mechanical properties to the McKibben line of actuators with stiffer fibers (Schulte, 1961), are currently used primarily for robotic applications in which proportional electropneumatic valves are utilized. As an example, rubbertuator #5, which is the smallest that we have tested, weighs only 28 g (compare to electric motors and cylinders) and can generate a force (at rest length) of 800 N with an internal pressure of 0.6 MPa, which is quite impressive. The resting cross-section is approximately 0.5 cm^2, the maximal cross-section at the rest length is 2.0 cm^2, and the maximal cross-sectional area when fully pressurized and fully shortened is 3.0 cm^2. These are representative of a small-to-moderate sized muscle. Taking the second value, the force per area is 4 MPa. As with pneumatic cylinders, we have found it quite difficult to estimate the *effort–flow* relation for this actuator due to the limited excursion and the sensitivity of results to air line length and valve port effects. Based on crude "quick-release" tests from a pendulum apparatus, a curve shape similar to that shown in Figure 7.7b has been estimated; because of the sensitivity the actuator speed to the length and width of the air line, the *effort–flow* relationship depends on the application. The peak power per mass, for a short pneumatic line, is estimated as 10 W/g, which is certainly superior to conventional pneumatic cylinders and electric motors and somewhat greater than that seen for hydraulic cylinders.

Braided actuators with stiff fibers tend to be very stiff in passive extension. In part for this reason, one of the authors (JW) has been involved in research aimed at producing inexpensive autuators that more closely mimic muscle-like properties, including the passive compliance of resulting muscle and a separate series elasticity (Liang, 1989; Winters, 1990). These actuators are then placed in head-neck and upper-limb anthropomorphic replicas. The resulting units can be built for under $5 in readily-available parts plus a few hours of labor. Unlike previous designs, we do not use fittings at either end, rather choosing to directly attach artificial tendons in series to the braided structure (middle actuator in Figure 7.7). Additionally, we do not have the diaphragm extend the length of the unit since diaphragm expansion normally creates an axial force component that opposes shortening. The resulting actuators can weight as little as 10 g, yet can produce forces in excess of 300 N.

Of note is that the cylindrical McKibbon muscle design is not the only possible braided actuator, nor even the best design. It is, however, the easiest to fabricate and at present is the most reliable. Baldwin (1969) utilized a different design in which glass fibers were arranged along the long axis, with the overall actuator having an eliptical shape. The force produced by these actuators could be quite high, but in order to produce high forces, the amount of shortening was compromised.

More recently, Immega (1986) has introduced a new, ultra-strong actuator termed the *ROMAC* (*RO*bot *M*uscle *AC*tuator). Here an articulating polylobe bladder is surrounded by a flexible, inelastic sheath (e.g. Kevlar) and an inelastic harness (e.g. steel cable). When under pressure, the bladders force both lateral and outward deflection of the harness, causing the harness to exert large forces. This design results in axial shortening of up to 50% and very high loads [e.g. 2,000 kg for an actuator with a mass of only 0.3 kg)]. However, these high forces come at the cost of a high-volume actuator, which means greater filling time. Those actuators have been controlled by *EMG* signals (Grodski and Immega, 1988).

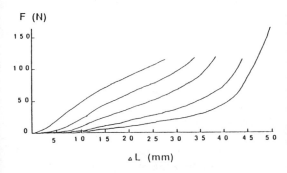

F (N)

Figure 7.8: *Effort–displacement* data for a typical fabricated braided musculotendon actuator [for splenius muscle, which is part of the head-neck anthrorobotic model shown in Figure 28.5 of Chapter 28 (Winters and Peles)].

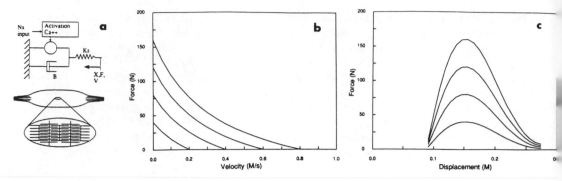

Figure 7.9: Representative skeletal muscle: long head biceps brachi: *a)* Schematic diagram. *b) Effort–flow* curve. *c) Effort–displacement* data (idealized).

7.2.8 Human Skeletal Muscle (Type 4b)

Human skeletal muscle mechanics have been discussed extensively in Chapter 5 (Winters). Shortening muscle has a concave force–velocity relation (Figure 7.9b) which declines sharply for low velocities and with a much shallower slope for higher velocities. This behavior can be approximated by Hill's equation (1938):

$$(F + a)(V + b) = b(cF_o + a) \qquad (7.12)$$

Where a and b are constants, F_o is the isometric force, and c is a dimensionless control input ($0 \leq c \leq 1$). This hyperbolic form can be expressed as (see Chapter 5):

$$F = cF_o - b_h(V)\,V = F_c - b_h(V)\,V \qquad (7.13)$$

$$\text{where} \quad b_h = \frac{F_c + a}{V + b} \qquad (7.14)$$

This form expresses the loss of force generated by the contractile element (*CE*) in terms of a nonlinear viscosity b_h (Figure 7.9b). In Figure 7.9b it is assumed that the maximum unloaded velocity scales proportionally with activation (i.e. b varies with activation), rather than staying constant or scaling less that proportionally – see Chapter 5 for further discussion. Locally, the damping of the muscle can be approximated by the slope of the force–velocity relation.

Parameter values have been estimated for a wide variety of skeletal muscles (e.g. Winters and Stark, 1988). Representative is the long head of the biceps brachi, where the parameters are:

$$a = 70\text{N}; \quad b = 1.0 \text{ m/s}; \quad F_o = 200 \text{ N}; \quad \text{weight} = 60\text{g} \qquad (7.15)$$

fiber length = 0.15; musculotendon length = 0.36

These parameters were used in *Eq. 7.13* to plot the *effort–flow* curves for muscle (Figure 7.11b).

The *effort–displacement* (*CE* tension–length) relation can be described by many empirical fits. In Figure 7.9c we use a fit described by Hatze (1981):

$$\tau = c \int 0.32 + 0.71\, e^{-1.11(X_s - 1)} \sin[3.77(X_s - 0.66)] \qquad (7.16)$$

$$\text{where} \quad 0.58 < X_s < 1.8 \text{ } \mu\text{m}$$

where X_s is the sarcomere length, and c is the dimensionless control input ($0 \leq c \leq 1$). Assuming uniform contraction, a rest sarcomere length of 1.0 µm, and the rest length of the biceps long head, then $X_s = X_b/(0.36 \times 10^6)$, where X_b is the muscle length.

As was discussed in detail in Chapter 5, skeletal muscle is unique in that the actuator includes viscoelastic tissue both in series and in parallel with the contractile machinery. The series compliance surfaces during the application of rapid forces or extensions – all actuators have some series compliance; in most technological actuators it is usually negligible. We will see later that this significant series compliance, which is an integral part of contractile tissue, is quite important. Here we note that the *SE* force–extension curve is non-linear, with the stiffness increasing with force for low and moderate forces (see Chapter 5). The muscle tissue is connected to tendon in series, and thus the overall series element extension is the sum of that due to both muscle and tendon. As described in Chapter 8 (Zajac and Winters), the actuator is the **musculotendinous unit**. This stiffness, seen especially during transient loading, is different from that of the static *effort–displacement* curve shown in Figure 7.9c, which is due to *CE* tension–length properties.

7.3 Coupling of Actuator to System

Actuators are bilaterally coupled to the mechanical linkage system, which in turn may be bilaterally coupled to the environment[(see also Chapter 6 (Zajac and Winters)]. Consequently, power (*effort–flow* product) can flow to and from the actuator. In our review of actuators, we have identified both angular (torque–angle) and translational (force–length) types of actuators. These structural differences influence the mode of transfer. Here we discuss these differences from the point of view of how these devices interface to linkage systems. We then consider how structural coupling (e.g. multiple actuator per kinematic degree of freedom, multi-link actuators) affects the dynamic process of actuation and consequently system control.

7.3.1 Transmission for Rotational Actuators

The three electric motors discussed here are inherently torque–angle devices (although translational versions exist). With the exception of recently introduced high-torque direct-drive motors, however, for servo-control applications of interest here their torques are too low. This is clearly the case for the examples of Sections 7.2.2–7.2.4. Consequently gear or cable/pulley transmissions are utilized to increase the torque at the sacrifice of speed:

$$M_1 \omega_1 = \{ n M_2 \} \{ \omega_2 / n \} = M_2 w_2 \qquad (7.17)$$

It is not uncommon for the gear ratio, n, to be over 100. For instance, for the three electric motor examples, gear ratios of 22, 71, and 60 would be required for the maximum joint torque due to a single actuator to reach 50 Nm, which is approximately the torque generated by elbow flexors. Such gearing adds inertia and potentially backlash, and thus in many cases the idealized instantaneous relationship is not quite realized.

Of note is that gearing can dramatically increase the effective actuator impedance Z by the square of the gear ratio, which can be considerable. The effects of such as increase are considered in Chapter 9 (Hogan). Finally, it should be noted that if desired the output of the gearing process can be a tension cable, and thus the torque motor may provide force–length transmission.

7.3.2 Transmission for Translational Actuators

The hydraulic and pneumatic cylinders discussed here supply translational power, typically via rigid rods connected to one or both ends of the cylinder. Transmission "gearing" is normally not necessary since actuator cross-section and length can be chosen independently. Coupling to linkage systems is typically either by a straight-line path between attachment sites on separate links or via a frictionless pully crossing the joint and connecting links. In the former case the moment arm $[R(\theta)]$, which is discussed in greater detail in Chapter 8 (Zajac and Winters), changes as a function of relative link angle θ. However, the idealized static *effort–flow* transformer relationship is generally preserved to a good approximation:

$$F V = \{ M R(\theta)^{-1} \} \{ \omega R(\theta) \} = M \omega \qquad (7.18)$$

where $R(\theta)$ is the moment arm. Although power is conserved, the equilvalent I and C elements scale with the square of the moment arm. As an example of this approach, the Utah-MIT Dextrous Hand couples the rod to a "tendon"-like cable, with these cables routed via pulleys (Jacobsen et al., 1986).

Eq. 7.18 also applies to braided actuators and muscles. These units have an added practical advantage of being inherently flexible. Furthermore, skeletal muscle and certain types of the braided actuators (e.g. Winters, 1990) utilize flexible tendons.

7.3.3 Local vs. Remote Transmission

As described in Chapter 36 (Alexander and Ker) for musculoskeletal systems, it is often desirable to place the mass of actuators for more distal joints within the body of more proximal links. This helps lower the effective inertia of the system, and is accomplished via some type of cable routing network or geared shafts (e.g. *Microbot*, *Unimation PUMA*, and *MIT-Utah Dextrous Hand* designs). Both transational and rotational actuators can be utilized for this purpose. If the shaft or cabling contains significant compliance, then a "series elastic" term exists.

7.3.4 Flexibility Gained by Actuators Crossing Multiple Links

Considerable flexibility (but also complexity) is gained when actuators cross multiple links. As seen in Chapter 9 (Hogan) and Chapter 41 (van Ingen Schenau et al.), the force vector of the end effector can be modulated more fully. Additionally, the shapes and priciple directions of the stiffness (*C-element*) and viscosity (*R-element*) fields for a given position of the end effector can be selectively modulated [Hogan, 1982; see also Chapter 17 (Flash)]. Multi-link actuators may also be able to work within more optimal regions of the *effort–flow* or *effort–displacement* curves during movements, especially when the links being crossed are rotating in opposite directions [e.g. during cycling (Chapter 40, Hull and Hawkins)]. Such actuators also turn out to be more effective at stabilizing multi-link inverted pendulum system structures [Chapter 26 (Crisco and Panjabi)]. Finally, such an arrangment can couple moments and power to other joints and furthermore create more energy efficient movements [Chapter 18 (Gielen et al.), Chapter 41 (van Ingen Schenau et al.)].

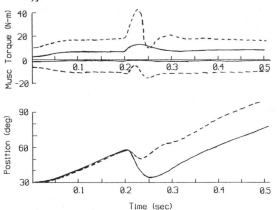

Figure 7.10: Simulation results for two overplotted runs of a nonlinear antagonistic muscle–joint model, as described in the text. For simplicity, a "base" elbow model with nonlinear *CE* and *SE* and no sensory feedback (or "yielding") is used [see Chapter 5 (Winters)]. Moment vs. time (top, with flexor upward and extensor downward) and position vs. time are plotted for the two isotonic flexion tasks described in the text, one with high ongoing cocontraction (dashed) and one with low concontraction (solid). The responses to the same sudden impact (30 N-m pulse at 0.2 sec, lasting 0.02 sec) differ dramatically despite the same initial position and torque differential.

7.3.5 Lumped Actuators: "The Whole is Greater Than the Parts"

Consider the simplest possible case for a single degree-of-freedom human movement: an antagonistic pair of muscles surrounding a joint. Make the classical assumptions that the series element stiffness increases with muscle force and that the *CE* force–velocity dashpot increases with activation and decreases with the absolute *CE* velocity. There is no unique torque-angular velocity (*effort–flow*) relation, **even under steady velocity and torque conditions**. This is because muscles can co-contract. Under these steady (isotonic) conditions, the three position "nodes" all move at the same velocity, and thus for this special condition the dashpot viscosities approximately add; this in general is not true, but does help show that typically the overall "viscosity" is higher than that of the individual actuator viscosities. Notice also that the transient "stiffness" due to sudden perturbation is the sum of the muscle stiffnesses. The stiffness in each muscle is predominantly due to the ongoing active muscle force level since the series elastic stiffness is much larger that either the parallel elastic stiffness or the *CE* tension–length relation for most of the joint operating range.

To help illuminate basic combined "equivalent" actuator properties, let's briefly consider two similar tasks (see Figure 7.10). For the first task, a voluntary movement occurs against a moderate (2%) isotonic load at a moderate effort level (10%), with no concontraction. A moderate shortening occurs which eventually approximates a constant velocity until a sudden load perturbation is applied. During the steady velocity phase the equivalent *effort–flow* viscosity is essentially that of the agonist because the low antagonist activation means low antagonist viscosity (a **nonlinear** slope change which makes the antagonist temporarily "disappear"). At the perturbation the transient muscle stiffness is proportional to 12% muscle force level. Now let's repeat this task, at the same torque and joint rotational speed in steady-state, only now with an additional 20% cocontraction between antagonists. Because of asymmetry between the shortening and lengthing *CE* force–velocity, the torque activation signals have to be adjusted slightly for the velocities to be the same as before. The new steady-state *effort–flow* viscosity (i.e. slope) turns out to be about 4 times higher than before, even though the movements proceed similarly until the impact! Furthermore, the lumped joint stiffness, a function of muscle forces, is 3 times higher! The response to the transient loading naturally differs dramatically, first because of the change in stiffness, and subsequently because the viscosity (a "soft ground" for the series element) differs.

This simple, almost trivial example hopefully brings home a key point: when you employ multiple actuators around a joint, *each with nonlinear properties*, the overall system properties (e.g. the *effort-flow* properties) can provide richer and more varied behavior than can be described by traditional *effort-flow* curves. The *CNS* clearly has the capacity to take advantage of such properties. An important observation is that this impedance modulation capacity disappears if the actuators are linear — superposition holds.

7.3.6 Comparing Technological Actuators with Skeletal Muscle

Effort-flow and effort-displacment relations do not tell the whole picture. Different actuators have fundamentally different structural arrangements and receive different types of inputs. Here we directly compare skeletal muscle to other actuators.

The internal combustion engine and *AC* motors have clearly different *effort-flow* relationships than skeletal muscle and need not be considered further.

On the surface, since microstepping motors and skeletal muscle have fairly similar *effort-flow* curves, one might be misled into thinking that they yield similar performance. Stepper motors can be viewed as stiff, quantized, displacement sources with a maximum torque. Torque is thus a function of the external load. Skeletal muscle is driven by a unicausal signal, but does not "set" a position or velocity. Rather, those are a byproduct of an energetic interaction between the actuator and the "load" (i.e. the rest of the system). Furthermore, stepper motors have no capacity for dynamic impedance modulation with the environment. Conversely, as is clear from Chapters 1–5, skeletal muscle has unique and interesting properties, especially when lengthened. Thus, while these actuators may have similar-shaped *effort-flow* curves, from a compliant control standpoint, they differ dramatically.

Traditional hydraulic and pneumatic cylinders have little in common with skeletal muscle. The *effort-flow* and *effort-displacement* relations are typically quite different. Hydraulic systems have little series compliance, and thus are very stiff in comparison to muscle. The stiffness of pneumatic cylinders, however, increases with pressure. An important structural difference is that the biological system has unicausal coupling from the motoneuron to the muscle (due to synapse properties), while for hydraulic and pneumatic systems the pressure "input" is itself part of a bilaterally coupled dynamic system. Thus, to help mimic sleletal muscle, the pressure (not flow) must be well regulated. However, for pneumatic cylinders to mimic a given skeletal muscle, the size would have to be disproportionate.

Two types of actuators have at least some potential to realistically "mimic" skeletal muscle properties: *DC* motors and braided pneumatic actuators. Some *DC* torque motors have *effort-flow* shapes that are at least in the ballpark. With appropriate gearing, the addition of a parallel torsional spring (to offset the normally flat *effort-displacement* relation), and a series cable with nonlinear compliance properties, there is potential for nearly muscle-like open loop behavior. Alternatively, in theory torque sensing coupled with a high bandwidth nonlinear controller could implement effectively equivalent properties. However, approximating the full capacity for impedance modulation would be difficult, and furthermore care would need to be taken to be sure that the system remains inherently passive [see Chapter 9 (Hogan)].

As documented by Winters (1990), braided acutators with series "tendons" can adequately mimic basic tension–length, series elastic and parallel elastic properties, and with additional research likely force–velocity properties. By adding internal pressure regulation circuitry (available from a number of manufacturers) to prevent the bicausal coupling effects from dominating, we find that there is great potential here, which is one reason why we chose to use these actuators on our existing anthro-robotic head-neck and limb models (Liang, 1989). Of interest is that the actuators have a physical shape during contraction and a general flexibility that are similar to biological muscle. More importantly, it turns out that for the same physical size and for the practically reasonable pressure input range, these artificial "muscles" have the same range of forces, stiffnesses, and velocities as skeletal muscle. Although the amount of active shortening and passive extensibility is in each case less than for skeletal muscle, it is clear that these should be amenable to logical extensions of the past braided actuator technology (Winters, 1990).

7.4 Unique Phasic Dynamics of Muscle Contraction

To this point we have not addressed the higher-order dynamics that influence fast transient events such as movement inititation, rather concentrating on steady-state *effort–flow* and *effort–displacement* behavior. Each actuator has its own unique start-up dynamics. For example, in the *DC* servomotor, armature inductance is often neglected in the dynamical analysis. For the majority of trajectories, this is justified by the very short (~4 *msec*) time constants due to motor armatures. However, for the fastest possible trajectories, armature inductance must be considered to have significant effects. Similar higher-order phenomena (e.g. due to fluid inertance, distributed coupling between properties) exist for all actuators. As suggested earlier, the unicausal neuromotor dynamics, followed by bicausal energetic muscle dynamics, are unique. Part of the temporal behavior that is seen for fast movement start-up is due to excitation-activation dynamics, which represent one part of the actuator. Here we develop a "systems-based" model formulation for helping illuminate these effects.

In this section, a model will be presented which illustrates the effect of short time scale dynamics and non-linearities of motoneuron response and muscle activation on force development in muscle. A fully detailed derivation of the model is available (Hannaford, 1990).

7.4.1 Experimental Foundation

Excellent experimental data describes the short time dynamics of these processes. These basic experiments measured responses of motoneurons when driven intracellularly by steps of transmembrane depolarizing current (Granit et al., 1963); Kernell, 1965), the force output of motor units when their axons were stimulated with different patterns of pulses (Burke et al., 1976), and the concatenation of these two paradigms, the force response of motor units in vivo to a step or ramp of depolarizing current injected into the controlling motoneuron (Baldissera et al., 1975, 1977, 1982, 1987).

7.4.2 Phasic Excitation/Activation (PEXA) Model

In building a model from the experimental studies we can anatomically divide the motor unit system into three sub-systems, the motoneuron, muscle activation, and muscle unit mechanics (Figure 7.11). In this case then we have data which give us the behavior of the individual components (Kernell, 1965; Burke et al., 1970) as well as of their concatenation (Baldissera and Parmiggiani; 1975).

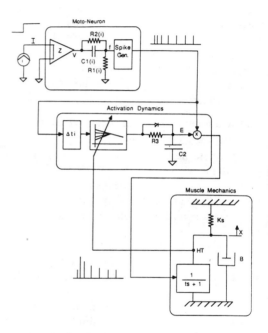

Figure 7.11: *PEXA* model block diagram: Model consists of three main blocks corresponding to motoneuron excitation and adaptation dynamics, activation dynamics, and muscle mechanics. Typical signals are plotted at the interfaces: current step input to motoneuron; impulse train with adapting (increasing) inter-pulse interval at moto-neuron output; and "enhanced" impulse train at muscle input.

The first model block, *Motoneuronal Dynamics*, consists of a trans-resistance amplifier (an amplifier with current input and voltage output), high pass filter, and spike generation implemented as a voltage to pulse rate converter. Its output is a series of unit value impulses which exhibit a decline in firing rate (adaptation) with constant current input.

The next block, labeled *Activation Dynamics*, consists of a pulse rate to voltage converter, piecewise linear relation, non-linear *RC* circuit, and multiplier. This block produces a multiplication of the effects of the action potential impulses to simulate the non-linear enhancement ("catch" property) found by Burke et. al. (1970); other techniques for describing activation dynamics for whole muscle are presented in Chapter 1 (Zahalak) and Chapter 5 (Winters).

Finally, the muscle is described by a second-order Hill-based non-linear model [e.g. see Cook and Stark; 1968; Chapter 5 (Winters)].

The first two blocks will be developed here; the third block is similar to that described by within a number of other chapters in this book which utilize Hill-based muscle models.

7.4.2 Motoneuronal Dynamics

The motoneuron model is designed to approximately reproduce current-step to pulse-rate responses. The basic form of the response to a step of current is a quick rise to an initial firing rate ι, followed by an exponential decline in firing rate, having time constant τ, to a steady state rate θ. Because of the motoneuron's non-linear response, the dynamic parameters of the step response; ι, τ, and θ, are functions of the current step amplitude, I. The firing rate step response [expressed in pulses per second (*pps*)] is modeled by the exponential form:

$$f_r(t,I) = \theta(I) + [\, \iota(I) - \theta(I)\,]\, e^{-t/\tau(I)} \qquad (7.19)$$

The computer simulation of this behavior was performed by numerical integration of the differential equations describing the equivalent circuit model in Figure 7.11.

7.4.3 Activation Dynamics

This section of the model represents the interval-dependent increase in the active state generated by each action potential in a sequence. Burke et. al. (1976) quantified this effect by plotting the enhancement of integral area (*g-sec*) as a function of inter-stimulus interval (*isi*).

A similar non-linear enhancement effect was simulated using a piecewise linear, *Area Enhancement Map*. In the experimental data, the enhancement function has a relatively fixed maximal value for intervals around the "contraction time" (the time to peak twitch tension), *tc*, of the

motor unit. For intervals greater than *tc*, the enhancement declines to unity as the effect is "forgotten." The enhancement due to intervals less than *tc*, is reduced according to the sequence in the pulse train, declining rapidly after the first few intervals.

To model this effect, we assume that for intervals shorter than the "contraction time" the phenomenon depends on the degree of muscle activation. In the piecewise linear approximation, the decline in enhancement due to repeated stimulation is controlled by the amount of muscle activation: a state variable in the non-linear muscle model.

The enhancement effect of small *isi*'s occurs instantaneously but persists for a time on the order of a second. The non-linear circuit used in the *PEXA* model has this asymmetrical dynamical property. Since the persistence time of the catch like enhancement, τ_{ca}, is quite long compared to the scale of fast movements, no effort has been made to precisely identify it and it is set at 500 msec.

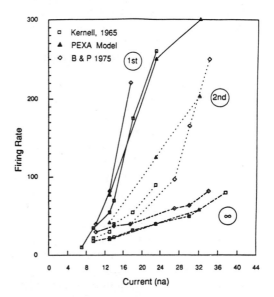

Figure 7.12: Current-to-Firing-Rate for simulation and experimental data. Motoneuron firing rate plotted against input depolarizing current. Rates are computed from the first (solid line), second (dotted line), and steady state (dashed line) pulse intervals. Experimental data replotted with permission from Kernell et al. (1965) (open squares) and Baldissera and Parmiggiani, (1975) (open diamonds) for comparison with simulation (filled triangles).

7.4.5 Basic Simulation Results:
Isometric Contractions

In the real motoneuron, the firing rate in response to current steps is characterized by an initially high rate (short inter-spike intervals) followed by adaptation in which the firing rate decreases to a steady state value. The effective firing rate corresponding to an interspike interval is the inverse of the time spanned by the interval. The adaptation process is illustrated by plotting the interval firing rate for sequential intervals as a function of depolarizing current (Figure 7.12). In a given experiment a fixed current step is applied to the neuron and output spike arrivals are recorded. The firing rates given by the first, second, and steady state intervals are plotted along a vertical line specified by the current step amplitude. The points can be connected in groups to illustrate the dependence of interval firing rate on interval number and step amplitude.

In the cat motoneurons studied by Kernell (1965) and Baldissera and Parmiggiani (1975) the first interval firing rate ("1st" group, solid lines) increased dramatically with depolarizing current step size, ranging from 10 pps to over 250 pps as current increased from 7 to 24 na. In the second interspike interval following the current step ("2nd" group, dotted lines) the firing rate rises more slowly with current, and the steady state firing rate ("infinity" group, dashed lines) rises gently to a maximum value of about 80 pps. The dynamics of adaptation are such that steady state firing rate is essentially reached after only two to four intervals: a time of less than 100 msec. When steps of current were applied to the simulated motoneuron, the firing rate of the model for the first, second, and steady state intervals closely matched the experimental data (Figure 7.13).

The next section of the model, firing rate to muscle force, was tested in a simulation experiment analogous to those performed in vivo by Burke, et.al. (1970) in which force output was re-

corded in response to pulse train inputs of varying frequencies. For each frequency, two pulse train stimuli were generated. One consisted of a simple train of pulses at that frequency. The second was the same but for the addition of an extra pulse 10 msec after the first of the train. In each force record (Figure 7.13, inserts), the higher amplitude signal came from the pulse train input containing the extra pulse, and the "catch like enhancement" persisted over approximately a full second.

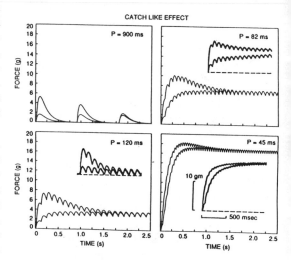

Figure 7.13: Model responses similar to "catch-like" effect. Each graph is the superposition of two responses: Force output in response to a step of constant rate pulses; and force output to the same pulse train with an additional pulse 10 msec after the first one. Insets show experimental data reproduced with permission of Burke, et. al. (1976). Note expanded force and time scales (lower right panel) for the four inserts.

The simulation was repeated using the *PEXA* model for basic pulse train periods of 900, 120, 82, and 45 msec. Force output was recorded in response to pulse train input to the "muscle activation" and muscle mechanics models (Figure 7.13, main traces). These compare well with the experimental data (1970) which are reproduced for comparison as inserts to Figure 7.13.

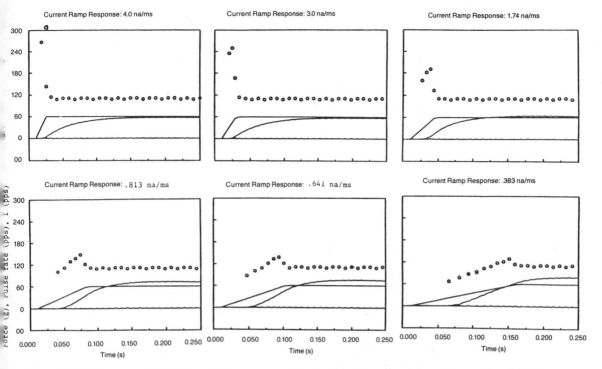

Figure 7.14: Response of *PEXA* model to current ramps. Plotted are simulations of the complete model driven by current ramps of six decreasing rates of change of current. Shown are current ramps and resulting muscle force (straight and curved solid lines respectively), and instantaneous firing rate of the motoneuron (circles).

Finally, we can simulate the complete experiment in which we inject current into the model motoneuron and observe force development at the muscle tendon. Baldissera et. al. (1982) injected ramps of depolarizing current into the motoneuron and observed instantaneous firing rate and tension development. The *PEXA* model was driven by current ramps whose slopes (di/dt) varied from 4.0 na/ms to 0.4 na/ms. The ramps began at $t = 0.01$ sec, and terminated at a maximum depolarizing current of 60 na. These values match those used Baldissera et al. (1982). The resulting current, pulse rate, and force outputs (Figure 7.14) show initial phasic responses in motoneuron firing rate (circles) whose peak output rate depends strongly on the current slope and ranges from 308 pps at 4.0 na/ms to 135 pps at 0.4 na/ms with the occurrence of the maximum firing rate ranging from the second interval at $t = 0.21$ ($di/dt = 4.0$ na/ms) to the ninth interval at $t = .152$ ($di/dt = 0.4$ na/ms). The force output slope was estimated by fitting a straight line to the force record up to the time of the decline in firing rate to its tonic level (there is no nerve conduction delay in the model). The slopes ranged from 1.11 gf/ms ($di/dt = 4.0$ na/ms) to 0.76 gf/ms ($di/dt = 0.4$ na/ms). The slopes

however saturated at about 1.1 gf/ms for the current slopes above 1.0 na/ms. Fitting a line to the force slopes below saturation gives a "dynamic gain" of 0.61 gf/na for the simulated motor unit, a value quite typical of the experimentally measured units (Baldissera and Campadelli, 1977).

Muscles contain a population of motor units having a distribution of twitch-speed / fatigue properties which is commonly approximated with a bimodal one having "fast" and "slow" peaks. Although fatigue properties are not modeled, parameter values relating to the muscle activation block and the muscle mechanics block change the twitch behavior of the *PEXA* model. It is initially assumed that motoneuron dynamical properties are invariant between fast and slow units. Thus the only changes were to the values of activation time constant, the "twitch time" parameter of the catch effect, tc, and the muscle equivalent viscosity, b. Two parameter vectors (Table 7.2) were chosen to simulate "fast twitch" and "slow twitch" fibers. Responses were plotted when the motoneuron model was driven by a depolarizing current of 20 na. The resulting adapted discharge rate was 33.33 pps (period = 30 ms) (Figure 7.15).

Table 7.2: Parameter values used to simulate fast (left column) and slow (right column) twitch muscles. Lines beginning in ' ' are comments.

# parameter file # FAST twitch unit # catch parameters		# parameter file # SLOW twitch unit # catch parameters		Units
tc	0.057	tc	0.057	sec
te	0.100	te	0.100	sec
Amin	1.0	Amin	1.0	-
Amax	3.0	Amax	3.0	-
htmax	20.0	htmax	20.0	Newtons
tca	0.50	tca	0.50	sec
#muscle		#muscle		
ta	0.02	ta	0.05	sec
b	100.0	b	340	$M\,sec^{-1}$
Ks	2458.0	Ks	2458.0	NM^{-1}
#misc		#misc		
dt	0.0005	dt	0.0005	sec
scale	4.0e-1	scale	4.0e-1	Newtons

The complete response of the two models consists of a train of action potentials (weighted with the catch-like enhancement), muscle activation ("active state"), and iso-metric muscle force. The longer activation time constant in the "slow" unit model is evident in the smaller excursions of activation (20–40 gf, steady state vs. 10–45 gf). Force output rises beyond 12.5 gf (50% of it's final value of 25 gf) in 21 msec for the fast unit vs. 87 msec for the slow unit.

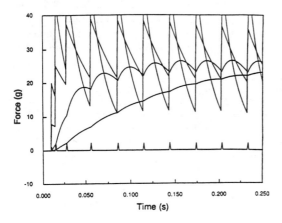

Figure 7.15: Fast and slow motor unit responses: Simulations using alternate parameter values for "fast twitch" and "slow twitch" motor units. Force outputs for "fast" and "slow" motor units driven by 20 na and 40 na depolarizing current steps.   Plotted are motoneuron output (small spikes), muscle hypothetical active state (jagged traces), and isometric force output (smooth lines).

The *PEXA* model has illustrated how the nonlinear dynamics of the activation process and the adaptation behavior of the moto-neuron work together to enable very fast rates of force development in muscle.   In the motoneuron, the firing rate adaptation is significant only in the first two or three firings after step excitation, but the "catch like effect" in the muscle activation process "remembers" the initial brief intervals for times on the order of 1 sec. The understanding of muscle excitation at these times scales is essential for accurate modeling of time optimal human movements. This study has been aided by consideration of analogous problems in the analysis of mechanical actuators.

7.5 Discussion: Future Directions of Actuators and Control

Technological actuators have "evolved" primarily to meet specific needs in single purpose applications. As technology competes for a larger share of precious resources such as energy and space (for example space in the home), the trend will be towards multi-purpose (and more "intelligent") technology. Skeletal muscle has evolved under constraints that require many types of muscles to perform a wide variety of functions. It also has evolved in a way that allows the mechanical system to be controlled to vary its mechanical properties to fit the task. A given joint may need, for example, to exhibit very high stiffness and positioning accuracy or very low stiffness force control. The concave *effort-flow* property of muscle, coupled with the unique (and nonlinear) steady-state and transient spring-like characteristics of the musculotendinous actuator, is fundamentally different in design that current technological actuators. Furthermore, this type of actuator is coupled to the skeletal frame in highly optimized ways, providing unique features such as reciprocal actuation and strategically placed multi-link actuators. The former allows joint impedance to be effectively modulated at will through co-contraction,   while   the   latter   allows great flexibility regarding the coupling of the end effector to the environment (primary upper limb benefit), the enhancement of efficient ways to stabilize multi-link "inverted pendulum" systems (primary torso benefit), and the distribution of moments, energy and power between segments (primary lower-limb benefit). Clearly linkage and actuator redundancy (and, perhaps, complexity) is not only tolerated but appears beneficial; we've barely scratched the surface in understanding how or why this is so, or how neural networks [e.g. Chapter 20 (van den Gon et al.)] enter into this picture.

As machines evolve towards multi-purpose designs, system properties that provide control strategy flexibility, such as those described here for musculotendinous actuators, will be increasingly needed, as is already the case in today's post-industrial robots. In conclusion, the inherent intertwining of neuromotor and robotic system research is mutually beneficial: the study of robotics and machinery has lead to an increased understanding of the role of muscle properties in the control of the movement of the skeleton, especially in regards to "asking the fundamental questions" [e.g. Chapter 9 (Hogan)], while the study of human movement provides us with glimpses of potentially better ways for designing and controlling robots and artificial limbs.

References

Andeen, G.B. (1988) *Robot Design Handbook, SRI Intern.*, McGraw-Hill, New York.

Baldwin, H.A. (1969) Realizable models of muscle function. In *Biomechanics, Proc. of First Rock Is. Arsen. Biomech. Symp.*, (Bootzin, D. and Muffley, H.C., eds.), pp. 139-148, Plenum Press, New York.

Baldissera, F. and Campadelli, P. (1977) "How Motoneurons Control Development of Muscle Tension," *Nature* **268**:146-147.

Baldissera, F., Campadelli, P., and Piccinelli, L. (1982) "Neural Encoding of Input Transients Investigated by Intracellular Injection of Ramp Currents in Cat Alpha Motoneurons," *J. Physiol.* **328**:73-86.

Baldissera, F., Campadelli, P., and Piccinelli, L. (1987) "The Dynamic Response of Cat Gastrocnemius Motor Units Investigated by Ramp-Current Injection into their Motorneurons," *J. Physiol.* **387**:317-330.

Baldissera, F. and Parmiggiani, F. (1975) "Relevance of Motoneural Firing Adaptation to Tension Development in the Motor Unit," *Brain Research* **91**:315-320.

Burke, D., Rudomin, P., and Zajac, F.E. (1970) "Catch Property in Single Mammalian Motor Units," *Science* **168**:122-124.

Burke, R.E., Rudomin, P., and Zajac, F.E. (1976) "The Effect of Activation History on Tension Production by Individual Muscle Units," *Brain Res.* **109**:515-529.

Burrows, C.R., Martin, D.J. and Ring, N.D. (1976) Responses of a pneumatically powered elbow-joint. In: *Human Locomotor Engng.*, Inst. Mech. Eng., pp. 136-144, London.

Cook, G. and Stark, L. (1968) "The Human Eye Movement Mechanism: Experiments, Modeling, and Model Testing," *Arch. Opthalmol.* **79**: 428-436.

Fenn, W.O. (1924) The relationship between the work performed and the energy liberated in muscular contraction. *J. Physiol.* **58**: 371-395.

Gavrilovic, M.M. and Maric, M.R. (1969) Positional servo-mechanism activated by artificial muscles, *Med. & Biol. Engng.* **7**: 77-82.

Granit, R., Kernell, D., and Shortess, G.K. (1963) "Quantitative Aspects of Repetitive Firing of Mammalian Motoneurons Caused by Injected Currents," *J. Physiol.* **168**:911-931.

Grodski, J.J. and Immega, G.B. (1988) Myoelectric control of compliance on a ROMAC protoarm. *Proc. Int. Symp. Teleop. and Control*, pp. 297-308.

Hannaford, B. (1985) *Control of Fast Movement: Human Head Rotation*, Ph.D. Thesis, Department of Electrical Engineering and Computer Science, University of California, Berkeley.

Hannaford, B. (1990) "A Non-linear Model of the Phasic Dynamics of Muscle Activation," *Accepted: IEEE Trans. Biomed. Engng.*

Hatze, H. (1977) "A Myocybernetic Control Model of Skeletal Muscle," *Biol. Cybern.* **25**:103-119.

Hill, A.V. (1922) The maximum work and mechanical efficiency of human muscles, and their most economical speed. *J. Physiol.* **56**: 19-45.

Hill, A.V. (1938) "The Heat of Shortening and Dynamic Constraints of Muscle," *Proc. Royal Soc.* **126**:136-195, London.

Holmes, R. (1977) *The Characteristics of Mechanical Engineering Systems*, Pergamon, Oxford, 1977.

Immega, G.B. (1986) Romax muscle powered robots. *Proc. Robotics Res. Manf. Eng.*, **MS86-777**: 1-7.

Jacobsen, S.C., Iverson, E.K., Knutyti, D.F., Johnson, R.T. and Biggers, K.B. (1986) Design of the Utah-MIT dextrous hand, *Proc. IEEE Robotics and Autom.*, pp. 1520-1532.

Kernell, D. (1965) "High Frequency Repetitive Firing of Cat Lumbosacral Motoneurons Stimulated by Long-Lasting Injected Currents," *Acta Physiol. Scand.* **65**:74-86.

Lansky, Z.J. and Schrader, L.F. (1986) *Industrial Pneumatic Control.* Marcel Dekker, New York. (

Lehman, S. and Stark, L. (1979) "Simulation of Linear and Nonlinear Eye Movement Models: Sensitivity Analysis and Enumeration Studies of Optimal Control," *J. Cyber. & Inf. Sci.* **4**:21-43.

Liang, D. (1989) Mechanical response of an anthopomorphic head-neck system to external loading and muscle contraction, M.S. Thesis, Arizona State

University.

Lord, M. and Chitty, A. (1974) Stabilization of pneumatic prosthetic systems.   In: *Human Locomotor Control*, Int. Mech. Engng., pp. 175-183, London.

McCloy, D. and Martin, H.R. (1973) *The Control of Fluid Power*. John Wiley and Sons, New York.

Morin, A.H. (1953) Elastic diaphragm.   U.S. patent 2,642,091.

Paynter, H.M. (1961) *Analysis and Design of Engineering Systems*, MIT Press, Cambridge.

Rothbart, H.A. (1985) *Mechanical Design and Systems Handbook*. McGraw-Hill, New York.

Schulte, R.A. (1961) The characteristics of the McKibbon artificial muscle. In *The Application of External Power in Prosthetics and Orthotics*, Lake Arrowhead, Publ. 874, NAS-NRC, pp. 94-115.

Shoemaker, P. (1989) Personal communication and unpublished manuscript.

Simpson, D.C. and Lamb, D.W. (1965) A system for powered prostheses for severe bilateral upper limb deficiency. *J. Bone & Joint Surg.*, **47**: 442.

Winters, J. (1987) "Biomechanical Modelling of the Human Head and Neck," in *Control of Head Movements*, ed. Peterson, B., Richmond, J.R.

Winters, J.M. (1985) *Generalized Analysis and Design of Antagonistic Muscle Models: Effect of Nonlinear Properties on the Control of Human Movement*, Ph.D. Dissertation, University of California, Berkeley, July, 1985.

Winters, J.M. (1990) Braided artificial muscles: mechanical properties and future uses in prosthetics/orthotics. *RENSA 13th Ann. Conf.*, Washington, D.C., pp. 173-174.

Winters, J.M. and Stark, L. (1988) Simulated mechanical properties of synergistic muscles involved in movements of a variety of human joints, *J. Biomech.* **12**: 1027-1042.

Yeaple, F. (1984) *Fluid Power Design Handbook*. Marcel Dekker, New York.

Zangemeister, W.H., Lehman, S. and Stark, L. (1981) "Sensitivity Analysis and Optimization for a Head Movement Model", *Biol. Cybern.*, **41**:33-45.

CHAPTER 8

Modeling Musculoskeletal Movement Systems: Joint and Body Segmental Dynamics, Musculoskeletal Actuation, and Neuromuscular Control

Felix E. Zajac and Jack M. Winters

8.1 Introduction

8.1.1 The Modeling Challenge: Keeping the Goal in Mind

It is doubtful that anyone would argue that biological motor control systems are less complex than robots. Given that modeling and designing robotic control systems that can walk or manipulate objects is quite challenging to engineers [e.g. (Lee, 1989)], is there any hope for those of us who wish to develop "adequate" models of biological motor control systems? The answer depends on the definition of "adequate".

Whether a model is adequate or not depends on whether it helps, hopefully significantly, in fulfilling the scientific or engineering *goal*. For study of movement the underlying goal is usually to maximize one's insight into the system and its behavior for a given movement task or class of tasks. For instance, suppose the goal is to understand how muscles work together [i.e., act in synergy, see Zajac and Gordon (1989) for discussion of synergistic and agonist/antagonist muscle-group definitions] to control elbow flexion and extension where, say, the shoulder and trunk are to be kept motionless. One paradigm is to design a shoulder and trunk harness to keep the shoulder and trunk stationary, in which case a model with just one body-segment and elbow flexor and extensor muscles would probably be adequate. Another paradigm is to allow the shoulder and trunk to be free to move. The subject must then coordinate the elbow muscles with the shoulder and trunk muscles to perform two sub-goals to accomplish the overall task. The subject must *a)* move the elbow as before, and *b)* maintain the shoulder and trunk stationary. Thus a model is needed to understand how the shoulder and trunk muscles act with the elbow muscles to perform these two sub-goals and a complex multi-jointed segmental model controlled by shoulder, elbow, and trunk muscles must be formulated. This model cannot be subdivided into two models, one for elbow control and one for shoulder and trunk control, because of the dynamical interactions occurring in multijoint motor tasks. For example, muscles crossing one joint act to rotate the other joints and these multijoint effects must be considered (Gordon and Zajac, 1989; see below).

8.1.2 Evolution of Models

It is generally accepted that the simplest model which fulfils the needs of the research and development project should be employed (cf. Hatze 1980a). So how does one determine the simplest model? In making a judgement call, the modeler makes a decision weighed by the current level of understanding of the *task* to be studied and the *system* to be involved in producing task behavior. These are interrelated. If understanding of the system is low, a simple model is appropriate; if understanding of both the task and the system is high, a more complex model is appropriate. If the task is not well understood, but a potentially relevant, though complex, model exists for the biomechanical system (perhaps because the model was used to study other tasks), one may wish to use techniques, such a task-specific sen-

wish to use techniques, such a task-specific sensitivity analysis, to systematically simplify the model (described below).

Traditionally, models have usually evolved towards complexity as we understand the task better (e.g., Zajac, 1985; Pandy et al., 1990; Mochom and McMahon, 1980; Chapters 38–43). Eventually, models become complex enough to provide insight into motor control issues that are unobtainable from experimental data alone. Often this is due to the model providing estimations of internal behavior that cannot be measured. It may also be due to sensitivity analysis results or optimization predictions. The next step in the modeling study is to assess how sensitive these *insights* are to fundamental body structural and functional parameters (e.g., muscle strength, type and assumed properties; body weight and height; mass distribution among the body segments). Experiments must then be designed to have sufficient resolution, which is definable from the sensitivity studies, so that the experimental data to be collected can be meaningfully analyzed to confirm or negate the hypotheses. If hypotheses are confirmed, the model can be used to gain an even deeper understanding of the motor task, and other experiments can be designed based on the parameters to which the insights are next most sensitive. On the other hand, if hypotheses are negated, then the next generation model is proposed, based on how the experimental and modeling data differ. The level of model complexity ultimately reached is thus limited by the level of confidence in the experimental data, the variables that can be recorded in the experimental laboratory, and computational limitations in modeling. Based on this rationale, models of animal motor tasks can evolve to greater complexity than models of human tasks because invasive techniques permit variables to be recorded from animals that are unobtainable from humans. See Chapter 10 (Loeb and Levine) for a foundation behind this approach, as applied to cat limb movements; here we emphasize human studies.

The above expose may convey the impression that modeling techniques and capabilities do not evolve. Actually, new modeling techniques evolve along with new experimental ones, and their evolution can go hand-on-hand. Furthermore, as techniques improve to study a specific task, better approaches to study other tasks may become apparent. Consider, say, that one desired to study the role of the leg musculature during dynamic lifting tasks. Such studies could probably take advantage of existing biomechanical and control models of jumping, as well as of the approach used to gain insight about muscular coordination in jumping. Thus, because of the existence of models, other motor tasks can also be studied.

8.1.3 Modeling Objectives

Since model structure depends on the specific research and development objective, let us briefly describe the goals of scientists and engineers studying motor control. Some wish to identify the neural networks essential to central nervous system (*CNS*) coordination of muscles, and then to discern how these networks function during task execution [e.g. Chapter 20 [Denier van der Gon et al.)]. In contrast to emphasizing the role of neural circuitry in the control of the movement, others desire to discover how the *CNS* output signals act on muscles to coordinate single- or multi-joint movements (e.g., Chapters 14–19, 27–28, 37–43). However, the overall neuromusculoskeletal control system is not open loop, as might be inferred from these two statements, but rather is closed-loop (Figure 8.1). The closed-loop property becomes especially apparent in practical applications, such as in functional neuromuscular stimulation [*FNS*; e.g. Chapter 21 (Crago et al.)]. So others try to decipher how these short- and long-loop feedback pathways participate in motor control (e.g., Chapters 10, 29–30, 46). Others focus on understanding the integrative properties of the "whole" neuromusculoskeletal control system (e.g. Chapters 28–34, 45).

One Approach: To Study Musculoskeletal Control

More effort has been expended to model mathematically the musculoskeletal system and how it participates in multi-muscle motor control than to model how neural networks control movement (Chapters 9, 17–19, 21–27, 30–31, 36–43; cf. Chapter 20 and parts of Chapter 10 − however, being a biomechanics book, this is perhaps a "biased" sample). One major reason is that quantitative data needed to model *CNS* neural network control has been quite limited in comparison to

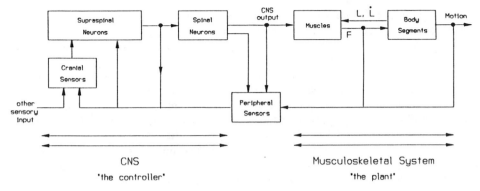

Figure 8.1: Conceptual diagram of the neuromusculoskeletal control system. The central nervous system (*CNS*), consisting of supraspinal neurons (supraspinal neural networks) and spinal neurons, including motoneurons (spinal neural networks), can be considered "the neural network controller," or just "the controller." The *Musculoskeletal System*, consisting of muscles, tendons, and the body segments, can be considered "the plant." The body dynamics account for the inertial properties of all the massy elements of the musculoskeletal system (e.g., muscles, skeleton, inter-nal organs of the torso) and their interaction with external forces (e.g. gravity). Motoneuronal output (*CNS* Output) excites the muscles. Receptors (*Cranial Sensors, Peripheral Sensors*) feed back proprioceptive, kinesthetic, visual, auditory information , etc., to the *CNS*, which "the controller" uses to modify the *CNS output*. Notice that the "muscles" and "body segments" have bilateral energetic mechanical coupling — i.e. information flows in both directions. In contrast, (short term) neural information usually travels unidirectionally due to the properties of the synapse.

musculoskeletal data [see also Chapter 10 (Loeb and Levine)]. *CNS* data is limited because the experimental techniques available to study single neurons, groups of similar functioning neurons, and the interactions among the different neural groups are at best few, and even then very difficult. The techniques available to study the function of single muscles, muscle groups, and the interactions among muscle groups are relatively many and more straightforward. Second, it can be argued that the properties of the musculoskeletal system (i.e., "the plant," to use engineering jargon) must be "adequately" understood before neural control properties (i.e., "the controller") can be elucidated (Figure 8.1). Engineers (cf. neuroscientists) have no difficulty in accepting this argument. The more an engineer knows about "the plant," the better the controller he can design.

Another Approach: To Study Neural Control

Neuroscientists have traditionally studied neuromotor function by documenting pathways between neurons or neural structures and by establishing time-locked, causal relationships between neural activity and muscle force or motion parameters. The result has been a vast neuroanatomical mapping of central and peripheral neural structures and between neural structures and classes of movements.

Engineers and biomathematicians now have renewed interest in neural network modeling. The reason is because of the ability of computers to simulate neural networks complex enough, it is believed, to be useful to the development of neural network theories [e.g. Chapter 20 (Denier van der Gon et al.)]. These neural network models are, however, quite "artificial" and must be distinguished from "biological neural networks." Because biological data is so scarce, few attempts have been made to actually develop "neural networks" based on biological phenomena (Pellionisz, 1988).

Nevertheless, neuroscientists could argue that continued engineering efforts to model "the musculoskeletal plant" are "overkill." Don't we already know *enough* about how the musculoskeletal system acts on neural signals to interpret how the system will respond to a given set of signals? After all, a flexor muscle trys to flex the joint it spans; an extensor muscle to extend the joint it spans. Aren't motor units that innervate a given muscle more-or-less recruited in a fixed order? Thus, if we know how neural networks pattern the *CNS* outflow to the musculoskeletal system, won't we be able to predict how the body segments will move?

Of course both the neuroscientist and the neural network engineer have valid points. The better each element of the "system" (Figure 8.1) can be understood, the better we can comprehend how the components interact to generate system behavior. Because the system has feedback and because muscle force generation is itself a function of limb movement, it is difficult to identify cause and effect (see Figure 8.1). The *CNS* "input" signals modify the "output", but the feedback signals modify the "input" signals. Thus perhaps the neural network is "the plant" and the musculoskeletal system is "the controller," and which is which becomes just semantics. So it comes down to one's bias.

Our Bias: To Study Musculoskeletal Control

We believe that the whole system needs to be studied and, as a first step, the characteristics of "the plant" must be understood, with "the plant" being the musculoskeletal system. Thus, to us, studies of neuromuscular control of movement are justified from which it is hoped that functional neural networks can be suggested [e.g. Chapter 10 [Loeb et al.]]. At times, neuromuscular control studies identify less well understood properties of the musculoskeletal system that are important to limb control but not well appreciated by medical scientists, including neuroscientists. For example, a biarticular muscle that develops an extensor torque at one joint and a flexor torque at the other spanned joint may not always act to accelerate the former joint towards extension and the latter joint towards flexion (Zajac and Gordon 1989; see below).

Below we review the engineering approach to modeling the musculoskeletal system in motor control studies when this system is viewed as "the plant." Using this approach, sensorimotor and *CNS* control are considered secondary issues to neuromuscular control, i.e., to understanding how "the plant" impacts on neural control. Especially as applied to studies of motor control of human extremities, emphasis on neuromuscular (multi-muscle) control is meaningful because experimentally measurable variables are indeed peripherally or externally located (e.g. *EMG* activity, joint rotation, body-segment position and orientation, ground or other external reaction forces). Only rarely can other variables be recorded from humans [e.g. tendon forces (Komi et al., 1987)].

In animals, sensory, motor, and *CNS* neural activity can be recorded [Chapter 10, (Loeb et al.)] and more complete models of the "whole" motor control system become justified, though implementation is still extraordinarily challenging (Chapter 10).

We should note that models of the musculoskeletal system are not always used to study motor control; that is, neural control of movement. In orthopaedics, musculoskeletal models are essential for estimating muscle and other soft tissue forces, which are needed to estimate joint contact forces. Such information is needed, for example, to design implant prostheses (e.g. Paul, 1974; Andriacchi et al., 1985) and to understand conditions such as low back pain (e.g. Chapters 24-25) or osteoarthritis (Chao, 1986; Fuller and Winters, 1987). To some extent, we review these as well.

In Section 8.2 we discuss how the constituents of the musculoskeletal plant are modeled in studies of motor control. We also discuss how neuromotor *CNS* circuitry is modeled and, in more detail than above, why there have been few attempts to add models of neuromotor circuitry to musculoskeletal models to study motor control. In Section 8.3 we discuss why multiarticular movements are complex dynamically. In Section 8.4 we show how models can be used to estimate muscle forces, while in Section 8.5 we discuss how such models can be used to illuminate fundamental principles of movement strategy, in particular, estimating *CNS* output signals that drive the musculoskeletal system to perform a specified motor task.

8.2 Modeling the Musculoskeletal System

Seven major steps are needed to synthesize a model of the musculoskeletal system to account for multi-muscle control of the motion of the body segments during a motor task: *(a)* The *body segments and joint kinematics* involved in the motor task must be specified, including the kinematic degrees of freedom of the joints interconnecting the segments and, if deemed significant, joint frictional losses. (For example, the thigh, shank, and foot of each leg may be a sufficient number of segments to model a person cycling a stationary bicycle. In this case the hips are assumed stationary, the feet as rigid bodies, and the patella as an extension of the shank. If the pelvis moves "significantly," then this segment would have to be included, and perhaps even other more rostral segments. If motion outside the sagittal plane is significant or of special interest, *3-D* segmental dynamics would need to be modeled.) *(b)* The **dynamical equations of motion** of the body segments must be derived. These equations depend not only on the assumed properties of the joints, but also on how the body segments are assumed to interact with the environment (e.g. in walking, the ground may be considered a rigid structure relative to the compliance of the foot and the body as a whole or, if walking on carpet, a damped compliant structure). *(c)* **Passive-tissue joint mechanics** must be modelled or assumed insignificant (e.g. limits on the joint range of motion via passive connective tissues). *(d)* The **geometric joint transformation**, which relates muscle force and length to body segmental torque and rotation, must be specified. This depends on assumed joint and musculoskeletal geometry (e.g., musculoskeletal moment arm relative to axis of rotation). *(e)* The **musculotendon** force generation process must be modeled [see also Chapter 5 (Winters)]. *(f)* **The neuromotor CNS circuitry** controlling muscle excitation must be modeled [cf. vestibulo-ocular reflex (Robinson, 1982)]. *(g)* The **complete musculoskeletal dynamical** model must be synthesized from these constituent parts.

We now examine each of these aspects in detail.

8.2.1 Modeling Body Segments and Joint Kinematics

Once the structures participating in the motor task have been identified, both those internal and external to the body, it is then necessary to specify how the body segments can move relative to one another; that is, what joint motion is permissible.

Degrees of Freedom (DOF) of "Joint"

The highest number of kinematic degrees-of-freedom (n_k) a joint can have is six (i.e., three to account for rotation and three for translation). However, for computational reasons, models of multi-muscle control of movement almost always assume less than six kinematic *DOF*. For example, the hip is assumed to have three, or less, rotational *DOF*. The knee is often assumed to have one rotational *DOF* (flexion/extension) in muscle coordination studies (Yamaguchi and Zajac, 1989), even though the other kinematic

DOF of the knee are well known and must be modeled when emphasis is on the stress and strain properties of the knee, such as in orthopaedics (Wismans et al., 1980). Of note is that the *DOF* that are kinematically constrained $(n_s = 6 - n_k)$ must be in equilibrium, with appropriate equal and opposite forces/moments at the joint (e.g. forces at the hip prevent translation of the femur relative to the pelvis).

Body Segments

Body segments are almost always assumed rigid in coordination studies (e.g. in this book). The implicit assumption made is that the effect of motion of structures internal to a segment (e.g., the motion of muscles, when activated, or the bending of bones) on intersegmental motion is insignificant. The "rigid body" assumption is reasonable, given our current knowledge of intermuscular coordination, except perhaps during impact of the body with very rigid objects (e.g., concrete floor). For modeling of the torso, a "rigid body" segment typically is either an individual vertebra [e.g. Chapter 27 (Dietrich et al.) or lumped segments of vertebrae [e.g. Chapter 28 (Winters et al.)], depending on the nature and goals of the study [see Chapter 23 (Andersson and Winters) for further discussion].

Joint Properties and Dynamical Equation Development

Virtually all studies of multi-muscle control of limb movement assume segments can rotate frictionlessly relative to each other. By and large, this is an excellent assumption for healthy young individuals (Fung, 1981). Consequently, if joint contact forces are not of interest then joint reaction forces do not have to be included in the equations of motion (Kane and Levinson, 1985), though they often are needlessly found in the derivation of these equations (for review, see Zajac and Gordon, 1989). Also, from the point of view of motion, the net effect of force in a muscle can be accounted for by the joint-torques it develops at the spanned joints.

If joint friction is high (e.g. in pathological conditions such as osteoarthritis), joint reaction forces have to appear explicitly in the equations of motion. Then, both the direction and magnitude of muscle force have to appear explicitly in these equations, and not just the moment of muscle force about the joint instantaneous axis of rotation. This is because frictional force is likely a function of the joint contact load, which depends, in part, on the muscle force vectors. It is important to distinguish the joint frictional forces, which are traditionally considered to be a function of contact loads as well as joint kinematics, from the viscoelastic properties of passive tissues crossing the joint (e.g. ligaments, passive muscle properties), which are usually assumed to be only a function of joint kinematics (see also Section 8.2.3).

8.2.2 Finding the Dynamical Equations of Body Segmental Motion

Many methods have been used to derive the equations of motion of the body segments (for review, see Zajac and Gordon, 1989), with the three most common classes of methods being Newton–Euler methods, Lagrange methods, and Kane's method. Assuming that at least one part of the body is stationary in an inertial reference frame, and all joints are rotary and frictionless, the dynamical equations of body–segmental motion can be written as:

$$\frac{d\omega(t)}{dt} = [J]^{-1} (\theta(t) \{ M(\theta(t) A \ T +$$

$$B(\theta(t)) + C(\theta(t), \omega(t) \} \tag{8.1}$$

where $\omega(t)$ is the angular velocity vector, $\theta(t)$ is the segment orientation vector, J is the mass-inertia matrix, A transforms joint torques into segmental torques, M is the moment arm matrix, B is a matrix containing gravitational contributions, $C(\theta(t), \omega(t))$ is a vector describing centrifugal and coriolis effects, and T is a vector containing externally applied joint-torques. Of importance is that as the order of the system increases, these equations become very complex. Commercially-available programs exist for creating complex multi-link models using any of the three methods; however, most of these were created for biodynamic impact investigations on mainframe computer systems and come with considerable "baggage".

The Newton–Euler method, though commonly used (e.g. Chapters 19, 27, 34), tends to generate more equations than are usually necessary. One advantage to this method is that it is quite intuitive for simpler systems and provides a natural extension from quasi-static analysis (e.g. Suh and

Radcliffe, 1978). Another advantage of this method is that computationally efficient recursive formulations of the equations often exist [e.g. for configurations such as the *3-D* open kinematic chain, of great interest for real-time robotic or prosthetic control (Walker et al., 1981)]. The primary disadvantage is that considerable effort is needed to find a reduced (minimal) set of equations for complex configurations.

The Lagrangian method (used in Chapters 21, 30) and Kane's method (used in Chapters 42–43) both generate a minimal set of equations once an appropriate minimal set of generalized coordinates is chosen. A set of generalized coordinates is one from which the positions and orientations of all the body segments and interacting body objects (e.g. bicycle crank) participating in the motor task can be found.

The number of *DOF* of a motor task is equal to the number of equations of motion corresponding to the minimal set, and can be less than, equal to, or greater than the total number of kinematic *DOF* of the joints involved in the task (Zajac and Gordon, 1989). Kane's method is particularly advantageous to the generation of a minimal set of equations of motion of the motor task because inexpensive, commercially available software (Autolev, ®, OnLine Dynamics, Inc., Sunnyvale, CA) exists to develop the equations, regardless of the number of *DOF* of the motor task (Schaehter and Levinson, 1988). The software can also generate computer code for use directly in computer simulations and analyses of movement. Kane's method is especially suited to those situations where non-contributory forces are of no interest (e.g., equal and opposite "resultant" forces acting at the joints).

The equations of motion describe how the segments will move (subject to a set of contributory muscle, passive tissue, and external force trajectories), given that the inertial parameters of the body segments are known. Studies have been conducted on adults and children (and on cadavers), from which techniques or models have become available to estimate the parameters of a subject from specific measurements of body physique (e.g., Hatze 1980b; Jensen 1989). How sensitive the conclusion of a study is to inertial parameter estimation is obviously important and should be addressed whenever possible (Jenson, 1989).

8.2.3 Modeling the Contributions of Passive Tissues Crossing the Joints

For a pin-joint, all passive connective tissues crossing the joint can be assumed to function in parallel. At times, the contributions of passive muscle tissues are accounted for by a musculotendon dynamical model, which then accounts singly for both the passive and the active force generated by a muscle (Zajac, 1989). At other times, as we do here, the passive muscle force contributions are accounted for separately and included, along with the non-muscle passive structures, as contributors to the "total passive joint moment." Quasi-statically, with these assumptions, the total passive joint moment ($M_{k\text{-}tot}$) can be written as the sum of the individual moments (M_i), which in turn are assumed to be a function of joint angle (θ):

$$M_{k\text{-}tot} = \sum_i (M_{i_m}) + \sum_j (M_{j_t})$$

$$(8.2)$$

$$= \sum_i [r_{i_m}(\theta) \times F_{i_m}(\Delta x_m(\theta))] + \sum_j [r_{j_t}(\theta) \times F_{j_t}(\Delta x_t(\theta))]$$

$$= \sum_k (f(\theta)) = \Phi_k(\theta)$$

where the subcripts m and t refer to muscle and other tissues, respectively; r is the vector from the joint center to the tissue at the free-body-diagram "cut"; and Δx is the difference between tissue length and rest length (typically zero if negative). Many tissues are commonly assumed to provide insignificant moments (e.g. skin, blood vessels, nerves).

Practically, it is possible to measure the *lumped* static contribution of all of the passive tissues crossing the joint by slowly rotating the joint of a relaxed individual through the range of motion in a plane orthogonal to gravity. There is thus little reason to distribute passive properties unless individual passive tissue loading, or joint contact loading, is of interest. [In fact, estimating individual contributions can be quite difficult (e.g., knee joint).] The effect of other joints on the passive joint moment, due to the coupling caused by multiarticular muscles, may not be significant and, if it is, can also be measured experimentally. Thus, the number of independent variables needed to specify all the individual passive spring (static) forces and their moments at a joint will be one (Figure 8.2). Joint angle can serve as the variable. The lumped static effect of all the springs can be given by a single, nonlinear torsional spring ($\Phi_k(\theta)$, see *Eq. 8.2*).

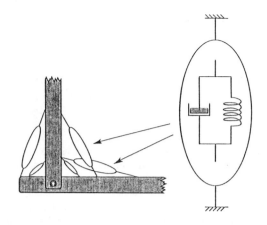

Figure 8.2: Technique for lumping passive viscoelastic tissues. Each tissue is represented by a simple parallel viscoelastic element (expanded to right). If the length of one uniarticular spring is known, the length of all other uniarticular springs can be determined. Thus, only one independent state variable (e.g. joint angle) is necessary for each kinematic *DOF*.

Typical passive joint stiffness values are low until the extremes of the range of motion are approached (e.g., 2–5 Nm/rad for most major joints over the mid-range; reviewed in Winters and Stark, 1985). Consequently, for tasks where joints do not approach the range of motion extremes, this static component is often ignored.

An approach similar to that outlined above can be utilized to model the lumped dynamic contribution, or passive joint viscosity, if a parallel (Voigt) model [as opposed to Kelvin model — see Chapter 5 (Winters)] is assumed. The actual relation is in theory a function of both angular position (θ) and velocity (ω), Usually position-dependence is assumed negligible and a linearized approximation is utilized for velocity-dependence. In summary, the passive viscoelastic component, which is rarely of great importance except at movement extremes and high joint angular velocities, can be lumped and represented as:

$$M_{PE\text{-}tot} = M_{K\text{-}tot} + M_{B\text{-}tot} = \Phi_k(\theta) + \Gamma(\theta,\omega)$$

$$(8.3)$$

$$\approx \Phi_k(\theta) + B\,\omega$$

A more detailed analysis, when desired, requires geometric transformation techniques, as will now be described for individual muscles (but is equally appropriate for passive tissues crossing joints).

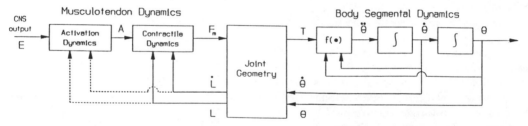

Figure 8.3: The musculoskeletal control system. *CNS Output E* (output of the motoneuronal pools as manifested in the *EMG* signals) are considered here as "the inputs" to the musculoskeletal system. These *CNS* output signals excite muscles. Typically musculotendon dynamics is decomposed into activation dynamics and contraction dynamics, as shown. Notice that uni-directional information flow is assumed from *CNS* output to activation *A*, i.e. "loose coupling" [cf. "tight coupling" of Chapter 1 (Zahalak)]. According to the classical Hill model structure (e.g., see Chapter 1 (Zahalak)], musculotendon dynamics are described by a second-order differential equation; state variables can be activation *A* and active muscle force F_m. The muscle forces cause angular acceleration of the segments, which may describe the complete motion of the segments by developing torques on the segments, as assumed here (see text). The torques from all the active muscle forces (plus all the passive components, not shown but presented in Figure 8.2) sum to produce the net angular acceleration of the segments ($\ddot{\theta}$). The torques developed by a muscle depend on how the musculotendinous structure is juxtaposed to the joints (see *joint geometry*, Section 8.2.4)]. The latter represents relationships that pass energy bicausally without loss; i.e. power is conserved (e.g. for uniarticular case for the *i-th* muscle and *j-th* joint: $F_i \dot{L}_i = [M_i/r_i(\theta_j)] [r_i(\theta_j) \dot{\theta}_j] = M_i \dot{\theta}_j$; see also Chapter 9 (Hogan)]. Past accelerations of the segments determine the current angular velocity (θ) and position (θ) of the segments, which are usually considered to be "the output" of the musculoskeletal system.

8.2.4 Modeling The Joint Geometric Transformation

How Muscle Force Actuates Body Segmental Motion

Because multi-muscle coordination studies usually assume frictionless pin and ball-and-socket joints, the moment of muscle force about the instantaneous axis of rotation is the transformation needed to relate how muscle force actuates segmental motion (for review, see Zajac and Gordon, 1989). Since rotation of the body joints are often chosen (included into) the set of generalized coordinates, muscle joint-torques are linearly related to muscle moments and can also be used to transform muscle forces into body segmental actuation (Figures 8.3 and 8.4; Zajac and Gordon, 1989). Because motors drive robots, joint-torques of motors are often used to represent the actuation of robots. This transformation, which is a function of joint geometry and also needed to describe the relationship between musculotendinous length (velocity) and joint angular length (velocity), can be represented as an instantaneous (no memory) relation with no bicausal energy loss (see Figure 8.3).

The approach of relating the moment of muscle force to joint-torque to prescribe muscle actuation may be incorrect under two conditions (Zajac and Gordon, 1989): *i)* if friction at the joints is significant; or *ii)* translation in addition to rotation of one body segment relative to another is believed significant (or, equivalently, the instantaneous axis of rotation varies, perhaps dramatically [see Chapter 28 (Winters and Peles)]).

These problems can be handled in a straightforward (albeit perhaps tedious) manner when utilizing Newton–Euler methods. When using Lagrangian or Kane's method, it may well be more convenient to choose a set of generalized coordinates (Kane and Levinson, 1985) and calculate the "generalized active force" of a muscle, which would appear in the equations of motion rather than the moment of muscle force (*T* in Figure 8.3 would represent the "generalized active force of muscle" and *joint geometry* would specify the transformation between actual muscle force and "generalized active force of muscle").

Muscle Path

Regardless of the approach used to transform muscle force to actuation of the body segments, the part of the musculotendon path crossing the joint is critical. Sensitivity analysis studies show that musculoskeletal model behavior tends to be very sensitive to the assumptions used to define the path (and consequently moment arm) near the joint (e.g. Winters and Stark, 1988).

Commonly, the *effective* path crossing the joint is defined by assuming *effective* origin and insertion points and drawing a straight line between them (e.g., Brand et al. 1982; Hoy et al. 1990; Chapters 42–43). As seen in Figure 8.4, the effective path (*path b*) may differ from the line connecting the actual origin and insertion (*path a*). Implicit in this approach are the assumptions that the "internal" forces required to curve the path (e.g. due to boney prominences, intermuscle contact) have a resultant direction that bisects the muscle path (i.e. the path is frictionless) and that the source of this resultant "internal" force is independent of the length or force of contributing nearby structures, including muscles, contributing to actuation of the segments. With respect to how muscle force actuates the body segments, therefore, it makes no difference whether the *non-*contributory portion of the musculotendon path is curved or not (Zajac and Gordon, 1989).

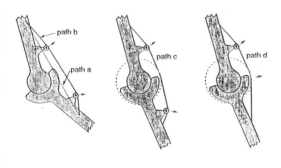

Figure 8.4: Schematic showing convenient assumptions for describing muscle paths. Left: difference between "straight-line" (*path a, dashed*) and "effective" (*path b, solid*) assumptions. Notice that the latter requires the identification of additional parameters (see text). Middle and Right: "Hybrid" possibilities that may be necessary as the joint angle increases. Notice the direction of the internal force required to curve the path; relative to movement dynamics, these forces usually needn't be considered.

Another useful approximation, complimentary to that just described, is to assume that, in the neighborhood of the joint, the muscle path arcs around a spherical shell once a certain "critical" moment arm is reached (e.g., *path c, path d* in Figure 8.4). Notice that the muscle length is a function of the assumptions made. By assuming frictionless contact, the resultant internal force passes through the joint center and thus produces no joint moment and thus is non-contributory to rotation. It turns out that the mathematical formulation for the *3-D* spherical (or also, for the *2-D* planar problem, cylindrical) shell is computationally efficient for both moment and length calculation since the critical path length (in *3-D*) or angle (in *2D*) where the arc is first contacted can be pre-calculated; the appropriate arc length is then simply added (Kleweno and Winters, 1987).

Decomposition of Certain Muscles Into Multiple Paths

The fibers or aponeurosis of a muscle may have such a wide insertion on the skeleton (e.g., gluteus maximus and medius) that a more complex representation may be warranted than one that assumes the muscle has one path (e.g. one point of origin). The common way to deal with this is to decompose the muscle into a few lumped compartments [e.g., decomposition of gluteus maximus into anterior, middle, and posterior compartments [Brand et al., 1982; used in Chapter 43 (Yamaguchi)]; similar decomposition strategy for deltoid [used in Chapter 22 (Meek et al.)]; and decompositon of muscles with origins within the lower spine (Chapters 25, 27) and the upper spine (Chapters 27–28)]. The other extreme is to compartmentalize the muscle into so many compartments that a true distributed representation is made. Such a representation does not yet seem justified to studies of intermuscular multi-joint coordination [cf. Chapter 3 (Morgan) for reasons to consider such models, and Chapter 27 (Dietrich et al.) for an effort in this direction].

8.2.5 Modeling Musculotendon Force Generation

Once the methodology of how muscle force actuates the body segments is delineated, a model of muscle force generation must be assumed. Since compliance of tendon and muscle aponeurosis impact on muscle force generation, the integrative properties of muscle and tendon must be modeled, and at times muscle architecture (pennation) as well [see Chapter 4 (Ettema and Huijing) for further discussion)]. The overall structural model combines models of musculotendinous tissue with architecture.

Models of Musculotendinous Tissue

Modeling muscles, tendons and their integrative properties is reviewed in detail elsewhere [e.g. Chapman, 1985; Winters and Stark, 1987; Zajac, 1989; Chapter 5 (Winters)] and only a few salient features are described here. Models describing the dynamical properties of the musculotendon actuator range from damped spring-like structures to those having dynamics representing both the excitation-contraction coupling of the muscle tissue and musculotendon contraction dynamics. Spring-like models of muscle, or combined muscle-reflex behavior, have been most widely utilized either *i)* to lay down a theoretical framework for musculoskeletal system behavior [e.g. Chapter 9 (Hogan), Chapter 12 (Feldman); see also review in Chapter 11 (Winters and Hogan)]; or *ii)* to explore the basic function of muscles (including musculotendinous architecture) during tasks involving energy storage [e.g. Chapter 36 (Alexander and Ker) and Chapter 37 (McMahon); see also review in Chapter 35 (Mungiole and Winters)]. These simpler models will not be discussed further.

As outlined in Chapter 1 (Zahalak), there are two classes of more complex models: *i)* models based on biophysical events at the molecular level (e.g. Huxley and Simmons, 1971); and *ii)* models based on an input-output descriptions of whole muscle behavior, as first formulated by Hill (1938). The latter are utilized heavily in this book and will be outlined here so as to help place into context how the musculotendon actuator structurally fits into the framework of the musculoskeletal system model [see Chapter 5 (Winters) for more details)].

As summarized in Figure 8.3, excitation-contraction coupling, termed *Activation Dynamics*, is typically modeled with a uni-causal (one way) linear or nonlinear lower-order (usually first-order) smoothing filter. The output of this process is typically termed **activation** (A), although the term **attachment** has been suggested for Hill-based models [Chapter 5 (Winters)]. The time course for increasing activation is on the order of 10 ms, while that for decreasing activation (deactivation) is on the order of 50 ms. The mechanical model structure which has evolved from the work of Hill (1938, 1970) consists of a contractile element (CE) in series and in parallel with viscoelastic tissue. Many subtle variations exist; these are outlined in Chapter 5. The following assumptions have been commonly employed: *i)* the passive parallel element (PE) can be assumed negligible, linear or nonlinear, and can be accounted for by the passive joint mechanics (not considered further here — see Section 8.2.3); *ii)* the series element (SE) is modeled as a nonlinear spring, often with an exponential force–extension curve (however, at times the force–extension curve is assumed linear for moderate and high forces, and sometimes a viscous element is added in parallel); and *iii)* the contractile element (CE) encompases both "force–length" and force–velocity" contractile properties that instantaneously satisfy a force–length–velocity relationship, which is scaled by the activation A [see Chapter 5 (Winters) for 3-D graphical representations]. The form of the CE can be thought of as a product relationship:

$$F_m = F_{fv}(A, \dot{L}_{ce}) * F_{tl}(L_{ce}) * F_{max} * A \qquad (8.4)$$

where F_m is active muscle force, F_{tl} is the dimensionless isometric force–length relation, F_{fv} is the dimensionless force–velocity relation, F_{max} is the maximum isometric force, and A is activation on the interval <0,1>). As described here the muscle contraction model is second-order (state variables associated with activation dynamics and the energy storing SE spring):

$$\frac{dA}{dt} = F_A(E) ; \quad \frac{dF_m}{dt} = F_m(A, F, L, \dot{L}) \qquad (8.5)$$

where L and $\dot{L}$ are state variables associated with joint passive elasticity and body segmental equations of motion (see Section 8.2.7), and E represents the net motoneuronal EMG output (Figure 8.3). The SE extension, or the length of the CE element, could also be utilized as the second state variable (e.g. one of us (FZ) uses force as the state variable, the other (JW) uses SE extension for numerical stability reasons). In any case, relevant internal variables, most especially L_{ce} and $\dot{L}_{ce}$, both of which appear in *Eq. 8.4*, can be directly (algebraically) obtained from these state variables. In cases where neural pulses are used in place of lightly filtered and rectified *EMG*'s, an additional uni-causal first-order equation is utilized, with its output, excitation *E*, the input to the activation process (Hatze, 1981; Winters and Stark, 1985); the active muscle model is then third-order. Finally, notice that musculotendon dynamics is history-dependent. That is, force cannot be determined by just knowledge of current muscle length, velocity, etc., but rate of change of muscle force can be (*Eq. 8.5*).

Musculotendon Structural Properties

To estimate musculoskeletal structural properties, muscle tissue properties must be combined with architecture. Maximal muscle force is estimated from an estimate of maximal muscle stress (e.g. 0.2–0.5 MPa) multiplied by a measure of the physiological cross sectional area, accounting for pennation [e.g. Hatze, 1981; Alexander and Vernon, 1975; Chapter 4 (Ettema and Huijing)]. Musculotendon length is determined from the path of the muscle and tendon from real origin to real insertion points (cf. effective origin and insertion points; see Figure 8.4) since the musculotendon path not crossing the joint, as well as that which does, affects the dynamical properties of the actuator because of musculotendon compliance. The ratio of muscle length to tendon length must be estimated since *CE* element length depends only on muscle length, while *SE* length depends on both muscle and tendon length.

The inputs to an isolated *musculotendinous* system are the excitation signals *E* to the muscles, and typically the muscle length *L* and velocity $\dot{L}$ the output is then muscle force F_m. When functioning as part of the *musculoskeletal* system, the excitation signals become a control input vector and muscle forces become an internal state vector; muscle length and velocity are instantaneously related to body segmental state variables, which in turn are a function of muscle force and (usually) external loading from the environment (e.g. picking up object) or kinematic constraints set by the environment [e.g. bicycle crank; Chapter 40 (Hull and Hawkins)]. Figure 8.3 shows how the body segmental dynamics, as given by the equations of motion of the body segments (see *Eq. 8.1*), and the musculotendon dynamics, as given by activation and contractile mechanics, are coupled. Here we

explicitly show all bilateral energetic coupling [see the dynamic systems framework established in Chapter 9 (Hogan)] via unicausal information flow lines. From this perspective, body segmental kinematics "feed back" and directly affect force generation. The dashed "feedback" lines to activation dynamics represent the "tight coupling" concept [Chapter 1 (Zahalak)], which has traditionally been assumed in musculoskeletal models to be negligible.

8.2.5 Modeling Neuromotor *CNS* Circuitry

The level of effort afforded to the modeling of the neuromotor *CNS* circuitry is small in comparison to the level afforded to the modeling of the musculoskeletal system. The basic reason is that the *CNS* structures (e.g., neurons, neuronal pools, neural nets) are not nearly as *identifiable* as the peripheral musculoskeletal structures. Even if the structures can be identified (e.g., *Ia* Inhibitory interneurons), their function is so uncertain in a specific task that investigators focus, at the moment, on their qualitative rather than quantitative properties. One reason for this uncertainty in the sheer volume of cells and the wealth of interconnections. A more tangible reason for the uncertainty is that the input-output "gain" of these interneurons, which are only one layer (synapse) removed from the final motoneuronal pathway, is modulated by higher *CNS* structures (Baldissera et al. 1981). The result is that, unlike the musculoskeletal system, the neuromotor parameters cannot be "set" with any confidence; in fact adaptive parameter modulation is an important aspect of neural control.

Modeling of Peripheral Sensor Dynamics

Muscle spindle sensors are embedded within the intrafusal (or fusimotor) system [see Baldissera et al. (1981), Brooks (1986), Loeb (1985) or Hasan et al. (1986) for recent reviews of anatomy and function]. Thus spindles do not measure absolute muscle length directly. Practical approaches involve breaking down the modeling process into components (Hasan, 1983): intrafusal dynamics, spindle transduction, and transmission time delay/filtering. [For tasks at steady activation (e.g. classic "ramp-and-hold" protocol), the first of these may not be important.] We now briefly consider each of these.

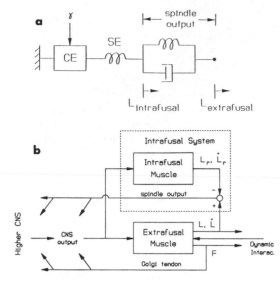

Figure 8.5: The muscle spindle receptor. *a)* The contractile portion of the intrafusal fibers (*CE*, contractile element) are excited by γ–motoneurons. The *SE* (series elastic element) is the elasticity of the intrafusal fibers that resides between the ends of the fibers and the central regions of the fibers, which is modeled by viscoelastic elements. Muscle spindle feedback (neural receptor discharge) is a function of the stretch of this central region. *b)* A model reference adaptive control system can be postulated by assuming that the intrafusal fiber system mimics the extrafusal system, with spindle output being the "error" signal. Notice that both the extrafusal and intrafusal systems have a common drive (*CNS* output) that links the γ motor-drive system to the α motor-drive system (called an "α–γ linkage").

Intrafusal Dynamics. One approach to modeling intrafusal muscle-fiber dynamics, based on known physiology, is to assume a model structure similar to the extrafusal muscle-fiber structure (Figure 8.5a). The *CE* is excited by the *CNS* γ–drive (cf. α–drive to extrafusal fibers). In addition to a series elastic element (similar to the *SE* used in models of the extrafusal fibers), a viscoelastic element in series models the central portion (e.g., the "bag") of the intrafusal fibers (cf. no such element for the extrafusal system). Spindle output (especially *1a* afferents) would ideally be related to stretch of this viscoelastic bag, which in turn is related to the difference between the extrafusal muscle length and the intrafusal length (once scaled for the "slack"

length of the bag). Interestingly, with this approach, nonlinear spindle output (e.g. the "product" transduction process described below) might result from a linear spindle transducer since "bag" extension would be sensitive to muscle mechanical phenomena. For instance, sudden stretch would cause an initially high force (due to high lengthening *CE* force–velocity slope) followed by "yielding" phenomena, which corresponds to initially high spindle sensitivity followed by the low-sensitivity "product" relationship.

The spindle dynamical model can be incorporated into a systems model of musculoskeletal control by recognizing the relation between the α–motor system and the γ–motor system (Figure 8.5b). Specifically, a low-order adaptive reference model can be proposed (Inbar, 1972; Winters and Stark, 1985b). A "tight" α–γ linkage is assumed, and the "spindle" output occurs when the difference between the length of the "intrafusal" (reference) fibers and extrafusal fibers exceeds a threshold.

Transduction. Mathematical representation of the spindle transduction process, given a length change, can be crudely classified into two approaches [Chapter 13 (Wu et al.)]: *i)* "additive" descriptions; and *ii)* "product" descriptions. A number of linear *additive* approximations have been formulated [e.g. reviewed in Gottlieb and Agarwal (1987)]. All possess in common a tendency for phase lead (strong derivative components) over the main operating range, with the following form typical:

$$\tau \, \frac{dN_{sf}}{dt} + N_{sf} = \{K_a \frac{d^2 \Delta L}{dt^2} + K_v \frac{d\Delta L}{dt} + K_p \Delta L + K_f \Delta F\}$$

$$(8.6)$$

where N_{sf} is the (as yet non-delayed) neural sensory feedback from the given muscle, on the interval <0,1> and the time constant τ is low in value (e.g. 10 ms) and often neglected. Typical "ballpark" values providing a moderate yet noticeable (and generally stable) feedback contribution for a variety of tasks, are, for the position (K_p), velocity (K_v), acceleration (K_a) and force (K_f) feedback gains, respectively, 0.1 /rad, 0.01 sec/rad,

0.0002 sec^2/rad and 0.2 /Fmax (Winters, 1985). In general, experimental data suggest higher values for lengthening muscle, lower for shortening, and changes in sensitivity with the length of stretch. Based on *EMG* data during small forced elbow oscillations, Cannon and Zahalak (1982) suggested a similar form, only with the addition of a square root operator on right side (to simulate a soft saturation) and a zero force gain. Once compensating for the square root effect, their gains are surprisingly similar. Why surprising? Because, unlike for musculoskeletal model parameters, these parameters can change at the "whim" of more central control commands. Also, one can regulate stiffness-like muscle properties by modulating the spindle gains versus the Golgi tendon gains [e.g. see Houk and Rymer (1981) for review].

The *"product"* approximation comes from the observation that experimental "ramp-and-hold" data, after the initital transient, can be well fit by the product of length multiplied by the velocity raised to a low power such as 0.3 (Houk, 1979). It is possible to employ a linear combination of both approaches (Winters, 1985). Hasan (1983) extended the "product" approach by utilizing a nonlinear first-order differential equation to dynamically relate the nerve ending stretch to muscle length [see *Eq. 13.1* of Chapter 13 (Wu et al.)]. He also developed a relation between stretch and reflex-induced muscle activation. In Chapter 13 both "additive" and "product" muscle–reflex models are considered for both ramp stretches and force step disturbances.

Transmission. The transmission of sensory information back to the spinal cord can be adequately modeled by a pure time delay and a simple low-pass filter with fairly fast (and thus perhaps insignificant) temporal dynamics (e.g. Hasan, 1983; Winters, 1985). This information must then be utilized.

Neuromotor   Utilization   of   Sensor Information. An engineering approach for utilizing sensory feedback is to assume that feedback from a given muscle not only goes back to the same muscle but also diverges to neighboring muscles and potentially other segmental levels [see also Figure 8.6 and Chapter 10 (Loeb and Levine)]. Each muscle could then receive con-

verging information from multiple muscles, as well as from higher structures. Whether this information converges additively or multiplicatively is an open (and controversial) question. For purposes of formulating a structure along classical engineering lines, we assume additive convergence. Based on many studies suggesting multiple "loops", with the "longer" loop gain more dependent on instruction (e.g. articles in book by Desmedt, 1978), we may assume two loops:

$$N_i = N_{cns_i} + \sum_j^n \{[(K_{s_{ij}} \ \delta(\ t - T_{s_{ij}}) \ + \ (K_{l_{ij}} \ \delta(t - T_{l_{ij}})] N_{sf_j}\}$$

(8.7)

for all $i \leq n$, where T_s and T_l represent matrices of short- and long-loop transmission time delays, respectively, K_s and K_l are feedback gain matrices, respectively, and $(\delta(\cdot)$ is the unit impulse function, i.e. $\delta = 1$ if $t = T_{ij}$, 0 otherwise). As shown, the relationship is static, i.e. it holds for each time step; one could easily add a "smoothing" effect [e.g. with a first-order filter with a suggested time constant of under 10 ms (Winters and Stark, 1985b)]. Unfortunately (for the modeller), the overall short- and (most especially) long-loop gains (K_s and K_l, respectively) are known to be quite variable even during simple tasks. They also tend to be low most of the time, as would be expected for a system with delayed feedback loops [see also Chapter 9 (Hogan)]. Furthermore, the traditional conceptual foundation of assuming that all joint "synergists" have similar gains and "antagonists" have different gains has been somewhat destroyed by the evolving experimental facts, which show that the concept of joint-based "synergists" and "antagonists" is itself fuzzy (e.g. Zajac and Gordon, 1989). *EMG* activity is often better correlated to global task measures than to individual joint motion measures (e.g., Soechting et al., 1981). Such observations suggest that feedback is organized by *task* rather than anatomical structure, prompting Loeb (1984) to propose the concept of "task groups". From a modeling perspective, this is somewhat of a "nightmare", albeit a fascinating one. Unlike musculoskeletal parameters, which in a reasonably sophisticated model should not need to be changed either during a task or (usually) for other tasks, feedback gain parameters are totally subject to the "whims" (needs) of the brain. Not only may they change as a function of task, but also may change during a task. It is for this reason, combined with the fact that feedback gains are often low anyway (presumably for stability reasons related to time delays), that feedback has rarely been added for investigations employing advanced neuromusculoskeletal modeling. A notable exception is provided by the work outlined in Chapter 10 (Loeb and Levine), where a concerted effort is underway to utilize optimal control techniques to estimate gain matrices as a function of task. It is worth noting that the motoneuron, and the *EMG*, represent a "final common pathway", and thus the signal *after* the integration of inputs from both the periphery and higher *CNS* structures.

8.2.6 Synthesizing a Dynamical Model of the Musculoskeletal System

The dynamical properties of the complete musculoskeletal system ("the plant") is specified by the interactive properties of the constituent components, which consist of the dynamical equations of motion of the body segments, the transformation describing muscle force actuation of the body segments, the dynamical equations of muscle force generation, and (potentially) sensory feedback (Figure 8.1). From these dynamical equations and transformations, the input vector $u(t)$, which consists of all the muscle excitation signals and the external forces acting on the body, can be defined. The state vector $x(t)$, which is a minimal (usually non-unique) set of state variables characterizing the musculoskeletal system, can also be defined. That is, all system variables are known because they are either the states themselves or functions of the states. We saw earlier that convenient state variables for a second-order muscle are muscle activation and muscle force. The state variable needed to define joint passive elasticity is joint angular displacement and, to define the body segmental motion, joint angular velocity is needed as well. The state equations can then be formed:

$$\frac{dx(t)}{dt} = F(x(t),u(t)) \tag{8.8}$$

Given the state equations, the significance of the state vector is that its value at time t_1 [i.e., $x(t_1)$] summarizes the past. In this way, future state

trajectories [i.e., $x(t)$, $t > t_1$] can be found, given knowledge of the state at time t_1, $x(t_1)$, and the input vector from then on [i.e., $u(t)$, $t > t_1$].

The number of state variables can very easily approach a high number (e.g., using the muscle model of Section 8.2.5, we would have $2m + 2n$ states, assuming m is the number of muscles and n is the number of body segments being modeled). Additionally, if the musculoskeletal system is coupled to a dynamic object (e.g. an electromechanical motor exercising a body part), state equations describing object dynamics (e.g. inertia, elasticity, motor inductance) may be required [see also Chapter 9 (Hogan)].

The outputs of the system [$y(t)$], which can be whatever happens to be of interest, are at each time step a function of state variables and inputs:

$$y(t) = G(x,u) \tag{8.9}$$

In summary, given a trajectory of controls (the time history of the muscular excitation signals, "the inputs") and the initial values of the states, the dynamical equations describing the system can be integrated to find, for example, all the muscle forces and activations, and the motion of the body segments (i.e., all the internal state and output trajectories). When the number of states is high, finding appropriate muscle excitation signals (the inputs) and interpreting the computer-derived trajectories of body segmental motion (the outputs) is extremely challenging (see below).

8.3 Complexity of Multiarticular Motor Tasks

The musculoskeletal system involved in a multiarticular motor task is a highly nonlinear system. System nonlinearities arise even in the absence of gravity. It is thus impossible to assume *a priori* that superposition will occur. Thus, the net motion of the body segments (i.e., typical output of the system) acted on by muscle excitation (inputs) will not be the sum of the motions that would occur should each muscle act in isolation.

From inspection of Figure 8.3, there are three fundamental sources of nonlinear properties: *i)* musculotendon dynamics; *ii)* the static transformation (mapping) between musculotendon force and velocity to joint torque and angular velocity, respectively; and *iii)* intersegmental dynamics and gravitational influences. The second of these will not be addressed since it is simply a straightforward static nonlinearity. The first and third, however, are nontrivial and indeed "intuition" can fail — simulations are necessary to fully grasp the significance of the interaction between the various components of the system. Our goal here is to identify a few salient features of the nonlinear musculoskeletal system that are perhaps nonintuitive.

8.3.1 Musculotendinous Actuators are More Than Simple Filters

For certain simple tasks, muscle dynamics can indeed take on the form of a uni-directional "smoothing" *EMG-to-force* filter. This, however, is rare, and simply assuming that muscle force is a smooth version of an *EMG* trajectory can be dangerous. As noted previously and discussed in detail in Chapter 5 (Winters) and Chapter 9 (Hogan), musculotendinous actuators *bi*causally interact with the skeletal system, with muscle force also a function of the muscle length history. To help visualize some of the roles and subtle effects of musculotendinous properties during single- and multiarticular movement, we start with the conceptual foundation of a Hill model with a nonlinear *CE* and *SE*. Here we will not present simulations but rather provide a brief overview of basic findings, most of which are addressed in other chapters.

Simple Voluntary Movements

The simplest possible situation is for an agonist *EMG* to change level (e.g. a step increase). It turns out that even for this case muscle "filter" properties are a strong function of the load on the system [e.g. see Chapter 5 (Winters)]. In particular, when the limb segment is unloaded ("free"), the muscle will shorten more quickly. Under these conditions the force trajectory that is generated will be lower than when loaded, such as (in the extreme) during an isometric contraction. This is simply a statement of the *CE* force–velocity property. Additionally, the **shape** of the curves are fundamentally different, with the force for the unloaded case likely showing a double peak [Chapter 5 (Winters)]. This is a function of combined properties. Furthermore, the time from *EMG* rise to force production above a threshold [the so-called "electromechanical delay" (Norman and Komi, 1979)] is clearly a function of the load. All

of these are in direct contradiction to the concept of a unicausal force generator. The next best model, a bicausal linear filter (e.g. linear *SE* spring and *CE* dashpot), is shown in Chapter 5 to be helpful but not satisfactory, especially for different voluntary movements, where it is seen that the parameters for the "best" linearized model fit change significantly as a function of movement amplitude, etc. Differences are especially evident during movement inititation, where the slopes for both the *CE* viscous element and *SE* are low.

What about antagonist activity? Nonlinear models allow for much lower antagonist force levels than linearized models. This is due to *CE* viscous-like properties being lower when activation is low. However, when it comes time to clamp (brake) a movement, nonlinear models predict a much greater capacity for attaining a significant braking force quickly (e.g. see Chapter 5).

Stretch-Shortening Tasks

Many tasks of life involve the excitation of muscles which lengthen first before contracting. Lengthening occurs because other, stronger forces act to pull on the muscle simultaneously. As reviewed in Chapter 35 (Mungiole and Winters) and 39 (Chapman and Sanderson), such a contraction allows a muscle to start the shortening process with a higher muscle force and with elastic energy stored in *SE*, and perhaps with a force enhanced over and above that prediced by traditional Hill-based models. The relative contribution of the various effects is debatable [e.g. compare Chapter 38 (Hof) to Chapter 39 (Chapman and Sanderson)]; what is not debatable is that muscle force generation depends on the recent history of both *EMG* and muscle length (i.e., muscle has dynamical properties).

Effects of Slow Postural Adjustments

Here the *SE* element is less important than *CE* behavior. Effects due to the *CE* tension–length relation are certainly straightforward, albeit perhaps troubling when this relation has a negative slope. Here concepts raised in Chapter 12 (Feldman et.al.) become important: musculoskeletal and neuromotor dynamics are best considered as a unit. However, perhaps less evident are the predictions of the *CE* force–velocity property during slow shortening and lengthening movements. Most experimental data for isolated muscle suggest that the slope near zero velocity is quite high, especially whenever there is co-contraction around joints. The end result is that even for slow movements the *EMG*–force relation is not that of a unicasual filter. Force changes are disproportionally higher for slow lengthening than for slow shortening. Certainly this nonlinear feature is important for limb stability, especially as related to working in conjunction with neural feedback to help prevent position "drifting".

Effects of Rapid External Perturbations

Many experiments have been performed during which muscles are: *i)* stretched or released at constant velocities, with force trajectories measured; *ii)* step changes in force are applied; *iii)* small force or velocity sinusoids are applied; or *iv)* impulses ("impacts") are applied. For such tasks the concept of an *EMG*–force relation is absurd – the *EMG* is constant (indefinitely for isolated animal experiments and for at least 30 ms for human experiments) and yet the muscle force shows immediate and often dramatic changes. It is these tasks which most completely show the need for bicausal muscle models and bring out most fully muscle nonlinear properties. Importantly, for single tasks over small operating ranges the input-output behavior sometimes appears linear. However, once the operating range changes, the describing parameters of lower-order linearized models change dramatically [Winters and Stark, 1987; Chapter 5 (Winters); Chapter 13 (Wu et al.)]. Explanations emerge from consideration of the Hill nonlinear model. For instance, when muscle force increases, muscle impedance increases because both the *SE* and *CE* force–velocity slopes increase, and in some cases the *CE* tension–length slope as well. This can be dramatic. Consequently, the muscle can be made to "look different" depending on the initital conditions and the type of perturbation.

Generalization of the "Perturbation" Concept

Of great importance is the following observation: As far as a given muscle is concerned, whenever a joint crossed by the given muscle undergoes rotation, the muscle experiences the ***equivalent of an external perturbation***. Interestingly, this perturbation may be a function of what this muscle has done in the past – it is one part of the bicausal, dynamic energetic mechanical system. This means that as far as, say, a shoulder or back muscle is concerned, it makes little difference whether a kinematic perturbation is due to contraction of an uni-articular muscle crossing the elbow, an impact to the forearm, or some combination of multiple sources (the usual case). [Of course, as part of the bicausal system, the dynamic interaction between these situations differ (Chapter 9, Hogan)]. This dynamic coupling is the primarily mode of interaction for most proximal limb muscles and for back muscles (cf. eye muscles and some finger and hand muscles). Since perturbations (and especially complex sequences of *EMG* and length histories) tend to most dramatically bring out the nonlinear properties of muscle, clearly nonlinear, bicausal muscle models are often necessary. The bottom line is that since, as seen in the next section, a muscle accelerates all segments and joints, it also provides significant perturbations to many other muscles, which in turn results in complex nonlinear responses, which in turn ... and so on. Certainly initution can fail and modeling becomes a necessity.

8.3.2 A Muscle Accelerates All Segments and Joints

Because the musculoskeletal system is multi-input, multi-output (*MIMO*), the position, velocity, and even the acceleration of any segment, and the angular position, velocity, and acceleration of any joint, depends on the excitation of all the muscles. For example, the joint-torque produced by a uniarticular muscle not only accelerates into rotation the joint it spans, but all other unspanned joints as well (Zajac and Gordon, 1989). And the muscle-induced angular acceleration of the unspanned joints can be much greater than that of the spanned joint. Thus, the force developed by a uniarticular muscle acts instantaneously to rotate both spanned and unspanned joints, and the effect at the unspanned joint can be significant.

To illustrate how potent a uniarticular muscle can accelerate an unspanned joint into rotation, consider soleus in the standing human. Though soleus is assumed to generate only an ankle extensor torque, it accelerates instantaneously into extension the unspanned knee as well as the spanned ankle (Figure 8.6). The induced angular acceleration at the knee can be two times as strong as the acceleration induced at the ankle, when the body is near the upright vertical posture (Figure 8.6, 180°). In comparison, biceps femoris (short head), assumed to develop only a knee flexor torque, accelerates the unspanned ankle into flexion, but the effect is only about half the flexor angular acceleration induced at the spanned knee (Figure 8.6). Though the model used to compute these data has only two segments to represent the body (Zajac and Gordon, 1989), the results are robust to the number of segments and to body physique, assuming the feet are flat on the ground during standing (Gordon et al., 1988).

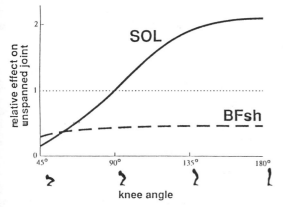

assumed to always develop a knee flexor torque, and at other times the ankle into flexion, even though gastrocnemius is assumed to develop always an ankle extensor torque (Figure 8.7). The

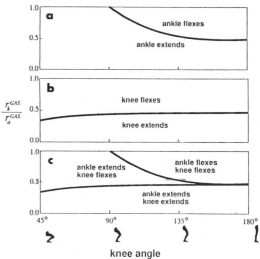

Figure 8.6: Angular acceleration of an unspanned joint relative to the spanned joint of two uniarticular muscles during flat-footed standing. A two segment (shank and thigh) model of the body is assumed. The mass of the head, arms, and torso are assumed to reside at the hip. Soleus (*SOL*) is assumed to span the ankle and generate only an ankle extensor (plantar flexor) torque; biceps femoris-short head (*BF$_{sh}$*) to span the knee and generate only a knee flexor torque. *SOL* (*BF$_{sh}$*) accelerates both the ankle and knee towards extension (flexion). The ratio of angular acceleration of the knee to the ankle produced by *SOL* per unit torque is higher at more erect postures (solid line). Notice that *SOL* extends (accelerates towards extension) the unspanned knee joint about 2x more powerfully than it does the ankle, the joint it spans, when the body is in the vertical upright posture (knee angle = 180°). In contrast, *BF$_{sh}$* flexes the unspanned ankle joint less than the joint it spans, the knee, regardless of the posture of the body (dashed line). (Modified from Zajac and Gordon, 1989).

Because of multijoint interactions, the force in a biarticular muscle can instantaneously accelerate the spanned joints in directions opposite to the directions of the developed joint-torques (Zajac and Gordon, 1989). For example, gastrocnemius in the standing human may at times accelerate the knee into extension, even though gastrocnemius is

Figure 8.7: Regions where the biarticular gastrocnemius (*GAS*) muscle flexes/extends (i.e., accelerates towards flexion/extension) the spanned ankle and knee joints. The regions are a function of body position (the knee angle for this two segment model) and how much flexor torque it develops at the knee relative to the amount of extensor torque it develops at the ankle, which is equivalent to the ratio of knee to ankle moment arm (r_k^{GAS}/r_a^{GAS}). *a*) Effect on the ankle. The bold line separates the regions where *GAS* flexes and extends the ankle. For high knee to ankle moment arm ratios, *GAS* will flex the ankle (region above the bold line) because the knee flexor torque, which flexes both joints, dominates the ankle extensor acceleration caused by the ankle extensor torque. *b*) Effect on the knee. For low knee to ankle moment arm ratios, *GAS* will extend the knee (region below the bold line) because the ankle extensor torque, which extends both joints, dominates the knee flexor acceleration caused by the knee flexor torque. *c*) Combined Effect. Notice that *GAS* can have three possible effects, depending on the posture of the body (the knee angle in this case) and the knee to ankle moment arm ratio. Notice that *GAS* can never simultaneously flex the ankle and extend the knee. When the knee is flexed no more than 45° (knee angle between 135–180°) the action of *GAS* to flex and extend the joints is very sensitive to moment arm ratios near 0.5, which are the ratios to be expected in humans (Hoy et al., 1990; modified from Zajac and Gordon, 1989).

action (angular acceleration) caused by gastrocnemius depends on *i)* the ratio of knee to ankle
moment arm, which may be affected by body position, including knee joint angle, and **ii)** knee joint
angle explicitly (i.e., even if moment arms do not
depend on knee angles). Notice that for moment
arm ratios around one-half, near the upright vertical posture, the action caused by gastrocnemius to
flex/extend the ankle and knee is very sensitive to
its moment arm ratio. If the ratio is above about
0.5, the knee flexor torque dominates, and gastrocnemius flexes both the knee and ankle (i.e.,
accelerates towards flexion the knee and ankle).
For ratios below about 0.5, the ankle extensor
torque dominates, and gastrocnemius extends both
the knee and ankle. And for a narrow range of
ratios near 0.5, gstrocnemius acts to rotate each of
the two spanned joints into the same direction as
the torque developed by gastrocnemius (i.e., into
knee flexion and ankle extension). Since moment
arm ratios near 0.5 are expected in humans (Hoy
et al., 1990), the action of gastrocnemius may differ among individuals while standing upright.
However, gastrocnemius would not be expected to
produce large accelerations, whatever direction it
is trying to rotate the joints, because the magnitude of the acceleration it induces will be small.
Thus, gastrocnemius and other multiarticular
muscles may be excited for reasons other than accelerating the body segments (e.g. see also
Chapters 9, 17–18, 24–28, 39–44).

Just as a muscle can accelerate all joints, and
thus all body segments, even those to which it
neither attaches nor spans, it delivers power to, or
absorbs power from, all the body segments [see
also Chapter 9 (Hogan)]. The power transferred
from a musculotendon actuator to a segment is
proportional to the segmental acceleration it induces and to the mass of the segment (Pandy and
Zajac, 1990). In fact, as one might expect, since
much of the body mass is in the trunk, leg muscles
deliver much of their power proximally to the
trunk in motor tasks where the trunk is free to
move [e.g., jumping, Pandy and Zajac, 1990;
Chapter 42 (Pandy)].

Simply because the action of one muscle to
move the body segments through a trajectory cannot be studied in isolation from the action of
another muscle, we believe that an understanding
of multi-muscle control of a multijoint motor task
requires an understanding of the complex dynami-

cal interactions acting among the muscles,
segments, and joints. For example, one may be
wrong to conclude from *EMG* and movement
records that a uniarticular extensor muscle (i.e., a
muscle that develops an extensor torque) is excited because it is needed to accelerate the
spanned joint into extension. That is, though experimental records may show that an extensor
muscle is active while the spanned joint is undergoing acceleration towards extension, it may be
more important to the motor task that this extensor
muscle be excited to accelerate joints unspanned
by this muscle [e.g., soleus acting to extend the
knee in standing, see above, or in walking,
Yamaguchi and Zajac 1990, Chapter 43
(Yamaguchi)].

8.4 Estimating Musculotendon Forces

Based on the previous section, it is clear that estimating muscle forces is nontrivial. Yet
knowledge of muscle and tendon force is desirable
because these forces are under the control of the
CNS and thus of interest to motor control investigators. In addition, muscle forces interest
orthopaedic biomechanicians because joint contact
forces, as well as muscle forces, must be estimated
to understand joint loading and pathology (Chao,
1986; Chapters 23–27). However, tendon force
has only rarely been recorded directly in humans
(Komi et al., 1987). Four classes of methods have
been developed to estimate muscle and tendon
forces during human movement: *(a)* heuristic
methods based on statics or inverse dyanamics
which are based on simple assumptions for load-
sharing (e.g. an "equivalent" knee extensor); *(b)* an
inverse dynamical approach involving processing
of experimental motion data, modeling and static
optimization to solve the muscle redundancy
problem; *(c)* an *EMG-to-force* processing approach;
and *(d)* a direct dynamical approach involving
model-driven simulations of the movement task.

8.4.1 Inverse Dynamical Reductionist ("Heuristic") Approach

This is certainly the oldest and most utilized approach, especially in textbooks. The inverse
dynamical approach uses body motion data alone,
or more desirably the motion data together with
external (e.g., ground) reaction force data, to compute the net torques developed at joints by muscles
(see review by Zajac and Gordon, 1989). A sig-

nificant subset of the inverse dynamics approach are tasks where a quasi-static analysis is assumed to be sufficent, often with the initital configuration assumed (e.g. classic textbook by Williams and Lissner, 1960). Muscle forces cannot usually be computed without additional assumptions because the number of muscles participating in the movement exceed the number of joints, which is often referred to as "muscle redundancy." Here a set of limited heuristic "rules" are employed which can range from a simple assumption of "equivalent" muscles (e.g., Williams and Lissner, 1960) to heuristic "rules" for load-sharing between assumed "synergists" (e.g. Paul, 1965, Morrison, 1970; Fuller and Winters, 1987). Such strategies may be quite appropriate when *3-D*, multi-link movements are performed that are essentially quasi-static and estimating muscle forces is not an end goal but an intermediate step that is necessary for estimating joint contact loads [e.g. in aged subjects with osteoarthritis performing various exercises slowly (Fuller and Winters, 1987)]. Of note is that heuristic "rules" which work for one task will not necessarily work for another.

8.4.2 Inverse Dynamical (Static) Optimization Approach

Static optimization procedures have been employed to find the muscle forces, given the net muscle torques obtained from an inverse dynamics analysis (Crowninshield, 1978; Herzog, 1987a,b). As with the above method, kinematic and joint torque data are assumed to be known (and with insignificant error). Solution of the force distribution problem at many instants in the motor task requires multiple solutions, one per each instant, to the static optimization algorithm. The solution solves the muscle redundancy by minimizing (or maximizing) a performance criterion (or "cost function") that is usually a subset of the form:

$$J_c = \sum_m (K_m F_m^{n_m} + K_s(F_m/A)^{n_s}) + \sum_j (K_j F_j^{n_j} + K_t M_j^{n_t})$$

$$(8.10)$$

where the measures presented here (muscle force (F_m), muscle stress (F_m/A), joint loading (F_j), joint constraint moments (M_j) have been used most often. Typically for a given "run" only one criterion is utilized (i.e. only one set of additive terms is al-

lowed to be non-zero), with muscle stress clearly the most popular choice (Pedotti et al., 1978; Crowinshield and Brand, 1981; Dul et al., 1985). A slightly different approach is to minimize the upper bound on muscle stresses (An et al., 1984). However, both Crowinshield and Brand (1981) and Dul et al. (1985) have suggested that muscles with a higher percentage of (fatigue resistant) slow muscle fibers are best represented by muscle stress raised to a higher power (e.g. $n_s = 3$). Multiple involved criterion are also possible (and usually preferable). This typically involves minimization of some combination of muscle, ligament and joint loading (e.g., Seireg and Arvikar, 1975, 1989; see also Chapter 23 (Andersson and Winters) for a review and Chapter 25 (Gracovetsky) for an example).

A general observation is that none of the criteria contain terms related to neural excitation or internal muscle activation. This is because, with few exceptions (e.g., Pedotti et al., 1978), these approaches do not consider muscle dynamical properties; i.e. muscle force estimation is assumed to depend only on the joint torque, the musculotendinous moment arm, and the cross-sectional area, and not on muscle dynamics.

Defining a suitable performance criterion is critical to this approach, and can be elusive. Interestingly, some investigators have shown that often the muscle force predictions are relatively insensitive, within limits, to the form of the cost function (Chao and An, 1978; Hardt, 1981). However, many criteria may yield the same shortcomings. For example, it is rare for any of the above cost functions to predict muscle coactivation of antagonists, a strategy that might indeed be preferred in some motor tasks (Chapters 9, 19, 23, 27–32 40–43), and may be especially important for postural stability of inverted pendulum-type systems [see Chapter 23 (Andersson and Winters), Chapter 26 (Crisco and Panjabi)].

Notice also that there is no integral sign within *Eq. 8.10*: Static optimization methods find the inputs at some single instant of time, given knowledge of the segmental kinematics (e.g., recordings of joint angles and velocities), and perhaps kinetic quantities (e.g., ground forces). These methods thus focus *only on the present*, solving an algebraic problem, and optimize the inputs according to some criterion without regard to where the segments will go as a consequence of the optimized current inputs. The algorithm must be run repeatedly to find the inputs at other instants. Interpolation can be used to find the input trajectories associated with the whole motor task. Of note is that theory behind static optimization is advanced and algorithms are readily available (for review, see Zajac and Gordon, 1989). However, many algorithms constrain the form of the performance criterion; for instance, the popular (and computationally efficient) method of linear programming (e.g. Chao and An, 1978; Bean et al., 1988; Seireg and Arvikar, 1989) requires that the exponent $n = 1$ for all terms, while others require that $n = 2$. The recent book by Serieg and Arvikar (1989) in particular exemplifies the wide range of problems, ranging from lower limb to lower back to upper limb to jaw movements, that can benefit from the utilization of this approach.

Of note is that "success" of a certain performance criterion has usually been based on qualitative compatibility between the predicted force patterns and the *EMG*. As noted in Section 8.3.1, this must be done with care. Estimation of *EMG* from muscle force, using a simple Hill model, is possible, but is rarely done. This would involve finding the series element extension by knowing the muscle force, then finding the muscle activation by taking the force and *CE* velocity (estimated from the *CE* position change), and then using this value and muscle length to estimate activation, given the assumed instantaneous *CE* force–length–velcocity–activation relation.

8.4.3 EMG Processing Approach

A method to estimate force in a single muscle is to process the *EMG* signal (Hof and van den Berg, 1981; for review, see Zajac and Gordon, 1989). A dynamical model of muscle activation and musculotendon contraction, such as Hill's model, must be used. The inputs to this model are (typically) the recorded *EMG* signal, after rectification and filtering, and the recorded length and velocity of the musculotendon actuator. Therefore, the model has the *EMG* signal, the (measured) joint angles and velocities as the inputs, and muscle (or tendon) force as the output. This *EMG* processing method has been appled to estimate muscle force and energy storage in tendon during tasks such as walking [Hof et al., 1983; Chapter 38 (Hof)] and jumping (Bobbert et al., 1986). Advantages of this approach are that it util-

izes a dynamic muscle model, takes advantage of all available information, doesn't depend on (imperfect) joint torque calculation via inverse dynamics, and is computationally simple enough to potentially be applied in real time. Disadvantages are that it places great trust in imperfect signals such as the *EMG* and in the muscle parameter values of the model.

A related approach that is under current investigation (Winters' group) is to use a hybrid optimization scheme which assumes that *EMG* and kinematic information are inherently imperfect and noisy, yet provide information that is useful as reference signals within the performance criterion. This provides some "slack" for the controller signal (i.e. the signal is allowed to be adjusted by the optimization scheme), and potentially will shorten the algorithm convergence time since the model starts close to the experimentally-predicted signal. By keeping the relative "weights" high for measured data that is trusted (which may vary from joint to joint), one can keep the final solution from deviating significantly from "trusted" experimental data and get reasonable muscle force predictions. If no subcriteria are active other than the *EMG* and kinematic error signals, one essentially is finding a "best match"; if in addition other subcriteria (e.g. muscle stress) are active, one is also potentially investigating issues related to neuromotor strategy.

8.4.4 Direct Dynamical Approach

The direct dynamical approach computes the force trajectory of each muscle and tendon during motor task execution as a byproduct of the musculoskeletal system response to neuromotor control inputs and any external loading; this is the approach emphasized in Section 8.2 and illustrated in Figure 8.3. This method can provide much more information than just the forces developed during the task, including the contribution of muscle force to accelerating and powering the body segments. This approach can be combined with sensitivity analysis and dynamic optimization methods. When used with dynamic optimization (discussed in detail in Section 8.5.4), muscle force is also a function of the goal of the task. We thus believe that it is the method of choice when a full understanding of motor control, including identification of neuromotor strategies, is sought. This issue will now be our focus.

8.5 Identification of Neuromotor Strategies

8.5.1 Purely Experimental Approaches

The experimental approach to finding how the *CNS* controls the musculoskeletal system is to record the *EMG* signals from muscles believed to be significantly involved in the execution of the motor task. Interpretation of these *EMG* records, together with kinematic and kinetic movement data also recorded during task execution, are often used in the absence of a mathematical theoretical basis to suggest *CNS* control mechanisms. This approach certainly has a place, and is utilized in a number of chapters in this book (Chapters 29–30, 46). Useful and thought-provoking data is obtained in these chapters. Of special importance are experiments in which perturbations (usually mechanical) are applied to an ongoing task and the neuromotor response is measured. However, interpretations of such data are difficult, and may not consider (or at least fully appreciate) the complex dynamical interactions existing among the segments and muscles (e.g., as discussed in Section 8.3). Thus we suggest that meaningful hypotheses will evolve more frequently when theoretically-based simulation data is collected and analyzed together with experimental data.

8.5.2 Static Optimization Method: A Limited Option

In comparison to dynamic optimization studies (Section 8.5.4), many more studies use static optimization methods to find the muscle excitation signals (inputs) (for review, see Zajac & Gordon 1989, Herzog 1987a,b, Serieg and Arvikar, 1989). Though static optimization methods can be appropriate for estimating muscle forces or joint loading; we suggest that they provide little insight into neuromotor strategy. The reason is that the performance criterion loses sight of the *task* because it is only solved at each instant. Also, with these methods, kinematics are assumed "set", and thus the structure of the approach does not permit a direct causal investigation of the effects of muscle activation on movement kinematics, which after all is how the human neuromotor system must learn movement organizational strategies. Dynamic optimization methods, based on direct dynamics simulations (described in detail below), do not assume knowledge of the segment trajectories [cf. Chapter 43 (Yamaguchi)], do consider the *future consequences* of the inputs, and do find both the input and output trajectories from the beginning to the end of the motor task so as to maximize the accummulated performance. From the context of illuminating neuromotor *strategy* for temporal movement *tasks*, dynamic optimization methods are clearly much more apropos to multi-muscle movement control than static optimization methods. Interestingly, the phases "cost function" and "performance criterion" have traditionally been used interchangeably; we suggest that for static optimization the former is a more appropriate term while for dynamic optimization the latter provides a more representative description.

8.5.3 Direct Dynamics Approach Without Dynamic Optimization: The Problem

The above observations suggest that direct (forward) dynamical simulations provide the preferred avenue for understanding human neuromotor control strategy. This theoretical approach is based, for most cases of practical interest, on the use of a *dynamical model* of the musculoskeletal system (e.g., Figure 8.3) to predict the muscle excitation signals (the inputs) that produce the movement (i.e. the "typical" outputs, as defined by experiments). However, finding appropriate muscle excitation signals is not easy, even for multi-muscle, single-joint systems, and especially for multijoint systems. A heuristic approach, even when the number of *DOF* being controlled is just a few (e.g., four), is difficult to impossible. The reason is because of the complex dynamical interactions. For example, if a specific joint needs to flex more during the simulation of a motor task but all other joints in the simulation follow desirable trajectories, it might be expected that increasing the excitation of the flexor muscles crossing the badly-simulated joint would lead to a better overall simulation. However, a worse simulation might result because, even if the previously badly-simulated joint now more closely resembles observations, the other previously well-simulated joints will likely be significantly disturbed (personal observations). Figuring out appropriate strategies by trial and error, while recommended as an occasional intellectual excercise, is difficult. Consequently, there is a great need for dynamic optimization.

8.5.4 Dynamic Optimization Method

The only viable theoretical approach to finding the muscle excitation signals and relating it to neuromotor strategy is to apply dynamic optimization (optimal control) theory (Athans and Falb, 1966). This approach demands, however, not only a mathematical (and typically deterministic) description of the system dynamics (the musculoskeletal system in this case) but a mathematical description of the *goal* (performance criterion) associated with the motor task. Any constraints on the ranges for control inputs (e.g. inputs on the order <0,1>) and state variables (e.g. finite range of muscle tensile force) must be specified, as must any mechanical coupling to the environment. The idea of motor tasks being goal-directed, as required when using dynamic optimization techniques, is compatible with the concept of a biocybernetic system. As such dynamic optimization provides a convenient (and as yet virtually untapped) link between biomechanists and psychophysicists. For dynamic optimization, this "*goal*" must be described in mathematical terms via a performance ("cost") criterion. This *scalar* criterion (J_c) in general includes multiple sub-criteria, some related to kinematic aspects (e.g. output) of the task and some to how much the controls or states are used (e.g., how much energy is expended; the magnitude and duration of forces developed by muscles, or in joints). The relative weights between (often competing) subcriteria help define the subtleties of what is really important within the task. The goal of the optimization algorithm is to determine a solution [e.g., set of control (input) vector trajectories] that minimizes the performance criterion. Importantly, a dynamic optimization algorithm is run "once" and the complete history of the input, state, and output signals associated with performance of the motor task is found (cf. the many "runs" of static optimization algorithms). The solution (control input) depends fundamentally on the entire time history of the specified task, including both past and future times. Thus, the solution represents not just "goal-directed" but also "skilled" behavior [cf. Chapter 10 (Loeb and Levine)]. Notice that the traditional solution to the "regulator" problem of determining optimal feedback gains depends only on past and present time.

Performance Criterion

A mathematical description of the goal of a motor task is somewhat ambiguous, which further complicates the use of dynamic optimization methods to study motor control. At times, an explicit goal can be rather easily identified and mathematically specified [e.g., mimumum time movements (Chapter 17), maximum-height jumping (Chapter 42), and pedaling as fast as possible (Zajac 1985, Sim 1988, Chapter 42 (Pandy)]. More often the goal is elusive, such as in arm movements or even walking (Stein et al., 1986). Usually a guess is made based on one's insight and bias [e.g., minimum jerk, Chapter 17 (Flash)]. Successful comparisons between experimental and modeling data then work to support, reject, or refine the goal that was chosen and defined mathematically. However, even if supported, one cannot be sure that this is a criteria actually used by the *CNS* in controlling the movement since multiple goals can have solutions with more-or-less the same movement trajectories (Stein et al., 1986). This is sometimes taken as a criticism of optimization methods. Another criticism is that optimal performance is difficult to prove or disprove; if the solution for one criterion doesn't match data, perhaps some other would. These criticisms are justified. Perhaps, however, there has been too much emphasis on searching for one (or a few) mysterious, illusive criteria.

It might be argued that as humans we are not optimized for any one task; for example, even after many thousands of practice attempts, a typical professional basketball player still misses 20% of his free throws. This suggests that movement is suboptimal (i.e., hard to "box" into our simplistic optimization criteria), but with practice can become close to "optimal". Also, a person has different goals throughout life, ranging from lifting objects (Chapters 23–26) to orienting to targets (Chapter 28) to sitting down (Chapter 31) to walking (Chapters 33–34, 36–37, 43–45) to running (Chapter 35–36) to throwing a ball (Chapter 40) to jumping (Chapters 40-42) to cycling (Chapters 31, 39–40). More importantly, a person can *change performance within a specific type of task at will*, whether it be the speed of a hand movement, the posture while sitting, the speed of running, or the height of a jump. Clearly the athlete involved in a long-distance event, whether running, biking or swimming, changes strategies as the finish line approaches. Perhaps, therefore, neuromotor systems have evolved to be fairly good at many tasks as opposed to being exceptional at a few. Another observation emerges: there is significant scatter even within well-controlled, practiced movements. Furthermore, there may be a few distinct ways to perform a given task [Chapter 34 (Pedotti and Crenna)].

To help develop these concepts, consider the following generalized performance criterion:

$$J_c = K_k J_{k\text{-}task} + K_{nm} J_{nm\text{-}p} + K_{jb} J_{jb\text{-}ld} \qquad (8.11)$$

where K_k is a scalar weighting kinematically-based "task" subcriteria, K_{nm} weights neuromuscular "penalty" subcriteria, and K_{jb} weights the loading or stresses in joints or along bones.

Kinematic–based task criteria ($J_{k\text{-}task}$) may include subcriteria such as movement time (MT, defined as staying within some specified interval), relative position error relative to some reference trajectory (Δx_{ref}, $\Delta \theta_{ref}$, etc.), minimum jerk (JK, third derivative of position [see also Chapter 17 (Flash)], or some extremum for a kinematic parameter (X_{extr}, e.g. representing center of mass change of a jump or peak velocity at release of a throw). Thus,

$$J_{k\text{-}task} = K_{mt}MT + K_{xtr}X_{extr} + \sum_j \{ \int K_{jk}JK_j dt$$

$$+ \int K_{xr}(\Delta x_{ref_k}) \, dt \} \qquad (8.12)$$

where the index j equals 1 if only end-effector (e.g. hand) movement is of interest or otherwise equals the number of joints. Such kinematic-based criteria are especially prevalent in investigations of upper limb movement organization.

Neuromuscular penalty ($J_{nm\text{-}p}$) may include subcriteria such as neural effort (NE, e.g. neural excitation), muscle stress (MS, force divided by cross-sectional area, which is essentially proportional to the dimensionless muscle force [Pedotti et al., 1978]), and has also been related to muscle "fatigue (e.g. MS^3 [Crowinshield and Brand, 1981]), and some type of measure of muscle energy loss (MD, e.g. "dissipation" across the contractile element):

$$J_{nm\text{-}p} = \sum_i \{ \int [K_{ne_i}NE_i + K_{mf_i}MF_i + K_{ms_i}MS_i +$$

$$K_{md_i}MD_i + K_{en_i}EN \,] \, dt \} \qquad (8.13)$$

where the last term represents general energy storage/loss, which has been formulated in a number of ways (e.g., Hatze, 1981; Oguztorelli and Stein, 1983). Notice that we have assumed that these "penalty" criteria are integrated over the entire time period and summed over all relevant muscles.

$J_{jb\text{-}ld}$ is a measure related to joint (e.g. cartilage stress) or bone loading (e.g. bending moment or bone surface stresses), which has been hypothesized by some to mitigate muscle origin-insertion sites, and are potentially of importance for movement strategies for individuals with osteoarthritis (Jensen and Winters, 1988) or lower back pain [Chapter 25 (Gracovetsky); Chapter 24 (Ladin)]. Chapter 25, utilizing static optimization, penalizes not only compressive and shear stresses but also seeks to maintain uniform stresses along a length of the lumbar spine.

Algorithm and Computational Limitations.

Though the concept of dynamic optimization is seemingly simple, implementation is another matter. Except for very simple, lower-order models (e.g. Nelson, 1983), analytical solutions are difficult [see Chapter 19 (Seif-Naraghi and Winters) for the range of currently possible analytical solutions]. Thus, numerical methods become essential, and simple algebraic methods such as linear programming are not a viable option. Will the algorithm converge to a solution? To a local or global maximum in performance (e.g., to the hole in a golf green, or to the valley in the fairway of a golf course)? How close to the local maximum in performance is the solution? Even for seemingly simple optimal control problems (e.g., maximum-height jumping), these questions are hard to answer and require much effort from optimization experts because the system dynamics are nonlinear, multidimensional, and constrained (Sim 1988) — one simply cannot systematically check every possibility within finite time (although this has essentially been done for minimum time, 10° eye movements [Lehman and Stark, 1979)].

The numerical algorithm is set up as follows: Once the dynamics, constraints, goal of a motor task, and assumed structure for the control inputs are defined, some algorithm must determine the solution within the specified control space n_c that minimizes (maximizes) the performance criterion, i.e. the lowest valley (highest hill) within n_c+1 space. Current methods can be roughly separated into two general categories: *i)* gradient-based methods; and *ii)* "random-based" search methods. Briefly, in the first method information regarding the "slopes" (partial derivatives) are utilized to enable the performance criterion to converge on a valley, while the second searches the n_c+1 criterion–control space, usually utilizing some method for systematically eliminating part (hopefully most) of the space from consideration. Many variations of each method exist, and some methods (including many current algorithms) are hybrids. Ideally, all algorithms should converge to the same (global) solution, and thus the form of the algorithm should not matter. Finding this "perfect" solution can be quite challenging. However, it has been suggested above that identification of the class (range) of "nearly optimal" solutions may be sufficient (and perhaps more interesting). In practice, computational limitations are a major factor that helps define what can and cannot be done utilizing dynamic optimization; hence the need to consider algorithms and computer hardware when formulating the problem. Of note is that there is no one "best" method for optimization. For instance, within this book gradient-based methods (e.g. Sim, 1989) are utilized in Chapter 10 (Loeb and Levine) and in Chapter 31 (Oh et al.), while versions of a "hybrid" method consisting of "educated" random searching with a certain gradient-like feature [extensions of the "Bremermann Optimizer" (Bremermann, 1970)] are utilized in Chapter 6 (Lehman) and Chapter 19 (Seif-Naraghi and Winters), and "dynamic programming" (Belman, 1957; essentially an intelligent random search) is utilized in Chapter 43 (Yamaguchi). Different problems suggest different methods, and furthermore the types of assumptions and simplifications within the modelling process are a function of the method utilized. As an example, dynamic programming was ideal for the *FNS*/gait study of Chapter 43 in part because a "coarse" control space was an adequate representation of current *FNS* control capabilities.

Some algorithms constrain the form of the performance criterion (e.g. to all terms being squared ($n=2$); however, most do not. For those that don't, it is important to realize that there is no *a priori*

reason to believe that a more elaborate performance criterion increases computational time or complexity. The performance criterion simply provides a scalar value that helps define the "landscape" in the n_c+1 cost-control space. In fact, experience shows that adding additional sub-criteria, especially for neuromuscular penalty, often dramatically **shortens** the convergence time, apparently by "smoothing out" the landscape and/or sculpting "flat" areas of the surface.

Interpreting Results and Comparing to Experimental Data

Once the control inputs (or range of inputs) are identified, all the internal state variable trajectories (muscle activations, muscle forces, muscle-fiber and tendon lengths and velocities), and all the output trajectories (e.g. segmental positions and velocities, and segmental orientations and angular velocities) are now available. The solution(s) to the dynamic optimization algorithm thus contain(s) a plethora of theoretical data, some of which can be compared to experimental data. In human motor tasks, for example, the body segmental trajectories and *EMG* signals can be compared. At times, such as in simulations of jumping and walking, other measurable quantities, such as ground reaction forces, can also be compared [e.g., Pandy et al. 1990; Chapter 43 (Yamaguchi)].

If the comparisons among the theoretical and experimental data are good, then all aspects of the motor task have been modeled, including the neural-output control strategy coordinating the muscles and the body segments. The complex dynamical interactions among the muscles and the body segments during coordination can then be analyzed to find, for example, how muscles accelerate the segments and deliver power to the segments [Pandy and Zajac, 1990; Chapter 42 (Pandy)]. However, since simulations are never perfect (though we engineers would like to believe otherwise), the real challenge is to identify significant imperfections in the model, perhaps through additional experiments. Computer-generated parameter sensitivity data can be used to assist in the design of the experiments.

If the comparisons between experimental and theoretical data are bad (i.e., qualitative differences exist), then the assumptions inherent to the model structure, including the assumed goal of the motor task, must be scrutinized. For example, most neuromusculoskeletal control models assume a muscle can be independently excited from all others. But that may not be the case, for example, due to *CNS* networks that constrain independent excitation of motoneuronal pools (e.g., *CNS* pattern generators, Grillner, 1981). A comparison of recorded to computed *EMG* signals, especially the timing of activity, might suggest that this assumption of independent control is invalid. In this case, performance of the task suggested by the model should be less than the experimental observation since performance will always be higher in the absence of constraints, including "*CNS* constraints."

8.6 Future Directions

8.6.1 Remarks by Felix Zajac

Understanding how the *CNS* controls the musculoskeletal system, so that complex motor tasks can be performed, challenges neuroscientists and engineers and is essential to the systematic development of rehabilitation strategies for persons disabled from neurological disorders. Similarly, understanding how the musculoskeletal system transforms neural signals into effective motor task execution equally challenges biomechanicians who desire to develop effective rehabilitation strategies for persons with motor disabilities resulting from musculoskeletal disorders. Clearly, the understanding of how both the musculoskeletal and neural systems influence motor control is needed, regardless of whether the motor disability has a neurologically or a musculoskeletal etiology, because of the complex feedback among these two systems (Figure 8.1).

The musculoskeletal system, in comparison to the neural system, is much more amenable to experiment and modeling. The computational tools available to model the complex properties of the musculoskeletal system are becoming quite powerful, with no end in sight as computer capabilities increase per dollar invested. I imagine, therefore, that we will soon make rapid advances to unraveling how the musculoskeletal system ("the plant"), particularly the extremities, transforms neural output signals to produce specific, complex *3-D* movements, especially those tasks where upper and lower extremities dominate the movement (e.g., Chapters 11–23, 34–46).

As the role of the musculoskeletal system in specific motor tasks becomes recognized, especially through computer modeling studies, I expect that the dynamical interactions among the body segments, joints, and muscles will be shown to be complex and important, and at times to be contradictory to intuition or bias. I also expect hithertofore unknown properties to be elucidated. As a result, I believe computer modeling will emerge as a necessary adjunct to experiment in the study of multijoint movement tasks.

As the musculoskeletal system becomes well understood, we will be ready to perform detailed, systematic modeling studies to decipher neural control mechanisms operational to specific motor tasks. Modeling studies should be emphasized because experimental techniques to elucidate human *CNS* neural networks are close to a null set. This is not to say that experiments will be unimportant, in fact, to the contrary. The message that I am trying to convey is that I believe theoretical and experimental studies will have to proceed hand-in-hand if systematic advances are to made in motor control research and its applications to rehabilitation.

Finally, the biggest challenge will be in synthesizing all the theoretical and experimental facts into an understanding of how *CNS* coordinates muscles to control, so eloquently and seemingly effortlessly, the myriad of complex movements that we perform daily. Perhaps even more perplexing is how the *CNS* is able to generate, seemingly at will, an emergent neural output pattern that functions well in getting the body to perform tasks that had presumably been novel to one's movement repertoire.

8.6.2 Remarks by Jack Winters

As a student of human movement, I remain both fascinated by this wonderful field and yet frustrated by how little we really understand about movement organization. As a community, we have collected a considerable volume of experimental data, encompasing many of the tasks of life plus a wide assortment of specific experimental protocols. Given general improvements in experimental measurement and reduction techniques within the biomechanical community, this trend is bound to continue. Yet myself and many others feel a lack of satisfaction. Progress in **understanding** (as opposed to **documenting**) human movement — and especially human movement organization — seems quite slow. In my opinion much (most?) of the *existing* data could benefit from additional interpretation. In this chapter we have emphasized the necessary role of biomechanical models in this process, and especially the need for direct dynamic simulations and dynamic optimization. However, dynamic models can potentially generate more curves than the full-time graduate student (much less the faculty member) can comprehend.

I suggest that that the future lies in dynamic optimization, at least as related to understanding human movement *organizational strategies*. There are six reasons for this belief: *i)* as computational power steadily increases, dynamic optimization will become more viable [see also discussion in Chapter 43 (Yamaguchi)]; *ii)* the intuitive form of a generalized performance criterion (Section 8.5.3), provides a (virtually untapped) natural tie between biomedical scientists (e.g., psychophysicists/neuroscientists studying of goal-directed human movement behavior or orthopedists studying patient compensation for deformity or pain) and biomechanists/bioengineers developing and utilizing complex musculoskeletal models — medical scientists can understand causal relationships between changes in relative weights between competing subcriteria and the changes that occur in basic control strategy (as visualized graphically); *iii)* although not all human movement behavior is tangibly goal-directed, many practically important tasks (e.g. in rehabilitation and sports) are; *iv)* virtually all tasks can be performed a few different ways, depending on subtle changes in performance goals — with dynamic optimization one can search for correlations between the variety of ways that a certain class of tasks can be performed and how various performance criterion predict that changes will occur; *v)* insight from optimization studies can occur both by obtaining "good matches" and "poor predictions" — for example, we are so used to expecting "base" *EMG* signals to be low that we may forget that without some form of neuromuscular penalty, low-level signals with occasional pulses are typically not the norm [see Chapter 19 (Seif-Naraghi and Winters)]; and *vi)* computers can be good teachers, and "the best course being offered" is dynamic optimization (e.g., I provide "goals" that I can understand, and then let the computer figure out

the best way to meet these goals — I am often initially surprised at the results until I think about them for a while; then comes the moment of illumination).

Two final predictions: First, it will turn out (to the possible frustration of neuroscientists) that neuromotor movement organizational strategies, especially for everyday movements, are a stronger function of joint and bone stresses than we usually currently assume. Perhaps muscle redundancy has evolved in (large?) part to help evenly distribute bone stresses. Second, stochastic optimization approaches, capable of addressing system stability issues, will take on importance in the future.

Acknowledgements

This work was supported by NIH grant NS17662 and the Rehabilitation R&D Service, Department of Veterans Affairs (FEZ).

References

Alexander, R.M. and Vernon (1975) The dimensions of the knee and ankle muscles and the forces they exert. *J. Human Movem. Stud.*, 1: 115-123.

An, K.N., Kwak, B.M., Chao, E.Y. and Morrey, B.F. (1984) Determination of muscle and joint forces: a new technique to solve the indeterminate problem. *Trans. of the ASME*, 106: 364.

Athans M. and Falb, P.L. (1966) *Optimal control: an introduction to the theory and its application.* McGraw-Hill, New York.

Baldissera, F., Hultborn, H. and Illert, M. (1981) Integration in Spinal Neuronal Systems. The Nervous System, *Vol. II, Sect. I: Handbook of Physiology.* (Edited by Brooks, V.B.), pp. 509-595, American Physiological Society, Bethesda, MD.

Bobbert, M.F., Huijing, P.A. and van Ingen Schenau, G.J. (1986) A model of the human triceps surae muscle-tendon complex applied to jumping. *J. Biomech.* 19: 887-898.

Brand, R.A., Crowninshield, R.D., Wittstock, C.E., Pederson, D.R., Clark, C.R. and van Krieken, F.M. (1982) A model of lower extremity muscular anatomy. *J. Biomech. Engng.* 104: 304-310.

Cannon, S.C. and Zahalak, G.I. (1982) The mechanical behavior of active human skeletal muscle in small oscillations. *J. Biomech.* 15: 111.

Chao, E.Y.S. (1986) Biomechanics of the human gait. In *Frontiers in Biomechanics,* edited by Schmid-Schonbein, G.W., Woo, S.L.Y., and Zweifach, B.W., pp. 225-244, Springer-Verlag, New York.

Chao, E.Y.S. and An, K.N. (1978) Determination of internal forces in human hand. *J. Eng. Mech. Div.*, 104: 255.

Chapman, A. E. (1985) The mechanical properties of human muscle. *Exercise and Sport Sciences*

Reviews (Edited by Terjung, R. L.), Vol. 13, pp. 443-501. Macmillan Publishing Company, New York.

Crowninshield, D. (1978) Use of optimization techniques to predict muscle forces. *J. Biomech. Engng.*, **100**: 88-92.

Crowninshield, D. and Brand, R. (1981) A physiologically based criterion of muscle force prediction in locomotion. *J. Biomech.*, **14**: 793.

Dul, J., Johnson, G.E., Shiavi, R. and Townsend, M.A. (1984) Muscular synergism II: A minimum-fatigue criterion for load-sharing between synergistic muscles. *J. Biomech.*, **17**: 675-684.

Fuller, J.J. and Winters, J.M. (1987) Estimated joint loading during exercises recommended by the arthritis foundation. *Advances in Bioengineering, ASME Winter Ann. Meeting*, **BED-8:** 159-162.

Gordon, M.E., Zajac, F.E., Khang, G. and Loan, J.P. (1988) Intersegmental and mass center accelerations induced by lower extremity muscles: Theory and methodology with emphasis on quasi-vertical standing postures. In *Computat. Meth. in Bioengng*, edited by Spilker, R.L., Simon, B.R.), **BED-9:** 481-492, Amer. Soc. Mech. Eng., New York.

Grillner, S. (1981) Control of locomotion in bipeds, tetrapods, and fish. The Nervous System, Vol. II, Sect. I: Handbook of Physiology. (Edited by Brooks, V.B.), pp. 1179- 1236, American Physiological Society, Bethesda, MD.

Hasan, Z. (1983) A model of spindle afferent response to muscle stretch. *J. Neurophys.*, **49**: 989-1006.

Hasan, Z., Enoka, R.M. and Stewart, D.G. (1983) The interaction between biomechanics and neurophysiology in the study of movement: some recent approaches. *Exer. & Sport Sci. Rev.*, **13**: .

Hatze H. (1980a) Neuromusculoskeletal control systems modeling - A critical survey of recent developments. *IEEE Trans. Auto. Control*, **AC-25**: 375-385.

Hatze, H. (1980b) A mathematical model for the computational determination of parameter values of anthropometric segments. *J. Biomech.* **13**: 833-843.

Hatze, H. (1981) Myocybernetic control models of skeletal muscle. *Univ. S. Africa.*

Herzog, W. (1987a) Considerations for predicting individual muscle forces in athletic movements. *Int. J. Sport Biomech.* **3**: 128-141.

Herzog, W. (1987b) Individual muscle force estimations using a non-linear optimal design. J. *Neurosci. Meth.* **21**: 167-179.

Hill, A.V. (1938) The heat of shortening and the dynamic constants of muscle. *Proc. Roy. Soc.,* **126B:** 136-195.

Hill, A.V. (1970) *First and last experiments in muscle mechanics.* Cambridge Univ. Press, Cambridge.

Hof, A.L. and van den Berg, J.W. (1981) EMG to force processing I: an electrical analogue of the Hill muscle model. *J. Biomech.* **14:** 747-758.

Hof, A.L., Geelen, B.A. and van den Berg, J.W. (1983) Calf muscle moment, work and efficiency in level walking; role of series elasticity. *J. Biomech.* **16:** 523-537.

Hollerbach, J.M. and Flash, T. (1982) Dynamic interactions between limb segments during planar arm movement. *Biol. Cybern.* **44:** 67-77.

Hoy, M.G., Zajac, F.E. and Gordon, M.E. (1990) A musculoskeletal model of the human lower extremity: the effect of muscle, tendon, and moment arm on the moment-angle relationship of musculotendon actuators at the hip, knee, and ankle. J. Biomechanics 23:157-169.

Jensen, R.K. (1989) Changes in segment inertia proportions between 4 and 20 years. *J. Biomech.* **22:** 529-536.

Kane, T.R. and Levinson, D. A. (1985) Dynamics: Theory and Applications. McGraw-Hill, New York.

Komi, P. V. and Norman, R. W. (1987) Preloading of the thrust phase in cross- country skiing. *Int. J. Sports Med.,* **8:** (Suppl. 1), 48-54.

Komi, P.V., Salonen, M., J_rvinen, M. and Kokko, O. (1987) In vivo registration of achilles tendon forces in man I. Methodological development. *Int. J. Sports Med.* **8:** 3-8.

Lee, C.S.G.(1989) Special Issue on Robot Manipulators: Algorithms and Architectures. IEEE Trans. Robotics and Automation, **RA-5:** 541-710.

Loeb, G. (1984) The control and responses of mammalian muscle spindles during normally executed motor tasks. *Exer. & Sprot Sci. Rev.*, **12:** 157-204.

Morrison (1970) The mechanics of the knee joint in relation to walking. *J. Biomech.*, **3:** 51-61.

Norman, R.W. and Komi, R.V. (1979) Electromechanical delay in skeletal muscle under normal movement conditions. *Acta. Physiol. Scand.*, **106:** 241-248.

Oguztorelli, M.N. and Stein, R.B. (1983) Analysis of a model for antagonistic muscles. *Biol. Cybern.*, **45:** 177-186.

Pandy, M.G., Zajac, F.E., Sim, E., and Levine, W.S. (1990) An optimal control model for maximum-height human jumping. J. Biomechanics (accepted).

Pandy, M.G. and Zajac, F.E. (1990) Optimal muscular coordination strategies for jumping. J. Biomechanics (accepted).

Paul, J.P. (1965) Bio-Engineering studies of the forces transmitted by joints: I. Engineering analysis. In: *Bimechanics and Related Bio-Engineering Topics.*, ed: Kenedi, R.M., Pergamon Press, Oxford.

Pedotti, A., Krishnan, V.V. and Stark, L. (1978) Optimization of muscle-force sequencing in human locomotion. *Math. Biosci.*, **38:** 57-76.

Pellionisz, A. (1988) Tensorial aspects of the multidimensional massively parallel sensorimotor function of neuronal networks. Progress in Brain Research. (Edited by Pompeiano, O. and Allum, J.H.J.), vol. 76, 341-354, Elsevier Science Publ., Amsterdam.

Robinson, D.A. (1982) The use of matrices in analyzing the three-dimensional behavior of the vestibulo-ocular reflex. Biol. Cybern. 46, 53-66.

Schaehter and Levinson (1988) Interactive computerized symbolic dynamics for the dynamicist. J. Astronautical Sci. 36, 365-388.

Seireg and Arvikar (1975) The prediction of muscular load sharing and joint forces in the lower extremities during walking. *J. Biomech.*, **8:** 89-102.

Seireg, A. and Arvikar, R. (1989) *Biomechanical analysis of musculoskeletal structure for medicine and sports.* Hemisphere Publ. Co., New York.

Sim, E. (1988) The application of optimal control theory for analysis of human jumping and pedaling. Ph.D. Thesis, University of Maryland, College Park, MD.

Stein, R.B., Oguztoreli, M.N. and Capaday, C. (1986) What is optimized in muscular movements? *Human Muscle Power* (Edited by Jones, N.L., McCartney, N. and McComas, A.J.), pp. 131-150, Human Kinetics, Champaign, IL.

Walker, M.W. and Orin, D.E. (1982) Efficient dynamic computer simulation of robotic mechanisms. *ASME J. Dynam. Sys., Meas. & Control,* **104:** 205-211.

Winters, J.M. (1985) Generalized analysis and design of antagonistic muscle models: effect of nonlinear properties on the control of human movement. Ph.D. Dissertation, Univ. of Calif., Berkeley.

Winters, J. M., and Stark, L. (1985) Analysis of fundamental human movement patterns through the use of in-depth antagonistic muscle models. *IEEE Trans. Biomed. Engng.*, **BME-32:** 826-840.

Winters, J.M. and Stark, L. (1987) Muscle models: what is gained and what is lost by varying model complexity. *Biol. Cybern.* **55:** 403-420.

Wismans, J. Veldpaus, F., Janssen, J., Huson, A. and Struben, P. (1980) A three- dimensional mathematical model of the knee joint. *J. Biomech.* **13:** 677-685.

Yamaguchi, G.T. and Zajac, F.E. (1989) A Planar model of the knee joint to characterize the knee extensor mechanism. *J. Biomech.*, **22:** 1-10.

Yamaguchi, G.T. and Zajac, F.E. (1990) Restoring unassisted natural gait to paraplegics via functional neuromuscular stimulation: a computer simulation study. *IEEE Trans. Biomed. Engng.* Sept. 1990.

Zajac, F.E. (1985) Thigh muscle activity in cats during maximal height jumps. *J. Neurophysiol.*, **53:** 979-993.

Zajac, F.E. and Gordon, M.E. (1989) Determining muscle's force and action in multi- articular movement. *Exerc. Sport Sci. Rev.* **17:** 187-230.

Zajac, F.E. (1989) Muscle and tendon: properties, models, scaling, and application to biomechanics and motor control. *CRC Crit. Rev. in Biomed. Engng.*, **17:** 359-411.

CHAPTER 9

Mechanical Impedance of Single- and Multi-Articular Systems

Neville Hogan

9.1. Introduction

The goal of this chapter is to consider some of the consequences of the dynamic interactions due to the transmission of power within the musculo-skeletal system and between it and its environment. Power transmission embodies a two-way or bi-causal interaction which may be characterized by mechanical impedance. Some theoretical considerations in modeling and analyzing mechanical impedances will be reviewed. Some properties of the mechanical impedance of the neuro-muscular system and their implications for organizing and executing motor behavior will be surveyed.

9.2. Single Muscle Behavior

One of the reasons why understanding muscle behavior is so challenging is because muscles are the primary organs by which the brain may influence the material world. They operate at the interface between the neural environment of the central nervous system and the mechanical environment of force and motion. Usually we think of these two environments in quite different ways. Major aspects of central nervous system function may be described as information processing: operating on input signals to produce output signals. Operations on signals imply a one-way interaction; inputs produce outputs *and not vice-versa*. This is consistent with the usual description of neural function: events at synapses induce changes in cell potential which in turn may[1] induce action potentials which travel down the axon and in turn influence other synaptic events. Under normal physiological conditions this is a decidedly one-way effect.

In contrast, the mechanical environment of forces and motions cannot be described adequately in terms of operations on signals. Muscles transfer power. In order to move a limb segment or any other object kinetic energy must be transferred to it (and/or removed from it). Unlike information transmission, *energy exchange fundamentally requires a two-way interaction* — an observation first made by Newton in his familiar third law: "For every action there is an equal and opposite reaction."

Muscles are often treated (either implicitly or explicitly) as force generators, organs that exert a force in response to inputs from the associated alpha-motoneurons. This description may suffice under isometric conditions, but aside from that special case, it is grossly inadequate. The force exerted by a muscle is a complicated function of many variables in addition to its neural input.

For any actuator, biological or artificial, it is important to distinguish between two aspects of its behavior. In engineering parlance they are termed the *"forward-path response function"* (or transfer function) and the *"driving point impedance."* In the case of muscle, those aspects of muscle behavior which depend on "external" mechanical variables such as muscle stretch should be distinguished from those which depend on "internal" physiological variables such as neural activation. The two are inter-related, being different facets of the same object, but the differences are significant. Neural activation changes the mechanical state of muscle, but the alpha-motoneuron does not respond to mechanical events in the muscle; this is a one-way effect. Muscle motion changes muscle force, but in general the converse is also true; this is a two-way interaction.

---

[1] Non-spiking events also occur and are the focus of intensive current research.

Multiple Muscle Systems: Biomechanics and Movement Organization
J.M. Winters and S.L-Y. Woo (ed.), © 1990 Springer-Verlag, New York

This observation may seem innocuous, even trivial; in fact it has several important consequences, some of which will be reviewed below.

9.2.1. Statics

Probably the simplest useful model of muscle describes its static behavior, a relation between force, length and neural input[2]. The dependence of muscle force on length is often referred to as "spring-like" behavior. Care is required in using this term: an important way in which muscle differs from a spring is that it may supply power indefinitely (at least over the time-scales relevant to normal motor behavior); it is an *active* object. In contrast, a mechanical spring is a *passive* object: only a finite amount of energy may be stored in the spring and the net energy it delivers to its environment may not exceed that amount[3]. The force–length relation for a muscle may vary as a function of neural input (or any of a large set of other variables). If those variables are held constant, the resulting behavior may be analogous to that of a spring, but the distinction between the passive behavior of a spring and the active behavior of muscle should be kept clearly in mind, for reasons that will be expanded below.

The defining property of a spring is its ability to store elastic energy, which is determined mathematically by integrating force with respect to length, so any integrable force–length relation may be spring-like. Thus if the nonlinear steady-state relation is

$$f = f(l,u) \tag{9.1}$$

where f is the force exerted, l is muscle length, and u is neural input, then the behavior is spring-like if the neural input is held constant.

$$f = f(l)\big|_{u = constant} \tag{9.2}$$

That is, the only requirement is that an unique force must be associated with every length. The converse need not be true: the same force may be associated with many different lengths; the force–length relation need not be linear nor monotonic. There is ample evidence that muscles may exhibit any or all of these features[4].

The spring-like behavior of a muscle is sometimes confused with its stiffness. The latter term usually refers to an incremental change of force due to an incremental change of length. For example, the nonlinear steady-state relation may be expanded about an operating length, l_0, and neural input, u_0, in a Taylor series.

$$f = f(l_0,u_0) + \frac{\partial f}{\partial l}(l_0,u_0)\, dl + \frac{\partial f}{\partial u}(l_0,u_0)\, du$$

$$+ \dots \text{(higher-order terms)} \tag{9.3}$$

The coefficients of the linear terms are a stiffness, k, and a gain, c, relating incremental force changes to incremental input changes.

$$k = \frac{\partial f}{\partial l}(l_0,u_0) \tag{9.4}$$

$$c = \frac{\partial f}{\partial u}(l_0,u_0) \tag{9.5}$$

Thus stiffness, which is often associated with a linearized model (e.g. obtained by neglecting higher-order terms) is merely one aspect of a more general spring-like behavior.

Another point of confusion is that muscle force is considered positive in tension and work done by the muscle is also considered positive; but if length increases while muscle force is non-zero, work is done on the muscle, which is considered to be negative (or eccentric) work. Thus it is sometimes useful to define a variable (e.g. $x = -l$) which increases as the muscle shortens so that positive dx corresponds to shortening.

9.2.2. Dynamics

Describing the static behavior of muscle as spring-like is not invalidated by the demonstrable dependence of muscle force on a large set of other variables. A prominent example is the decline in muscle force in relation to the rate of shortening. Subject to the caveats outlined above (i.e. all other variables influencing muscle force assumed constant) this behavior may be described as analogous to a mechanical damper, a viscous element [see Chapter 13 (Wu et al.) and Chapter 5 (Winters)].

---

[2] For simplicity, a single neural input variable per muscle is assumed throughout this chapter, but a multiple independent neural inputs could be treated similarly.

[3] A rigorous definition and detailed discussion of passivity may be found in Wyatt et al. (1981).
[4] The relation need not even be continuous, but the author knows of no evidence that muscles exhibit this behavior.

There is no reason to believe that the relation between muscle force and motion is confined to a dependence on length and its first time derivative. All of the motion-dependent effects may be conveniently summarized in the *mechanical impedance* of the muscle. A mechanical impedance is a dynamic operator which specifies the forces an object generates in response to imposed motions. It may be thought of as a dynamic generalization of the familiar notion of an elastic spring.

Linear Case

From this point of view, a visco-elastic effect may be thought of as a rate-dependent spring. Of course, motion-dependent effects need not be that simple — arbitrarily complex dynamics are theoretically possible. This is most easily understood when the relations are linear. In that case, impedance may be represented as a function of frequency similar to the familiar transfer–function, but relating output force to input velocity. The reason for using these two variables is that they are sufficient to determine the instantaneous mechanical power transferred to or from the muscle. The point to note is that a mechanical impedance need not simply be a combination of inertia, stiffness and viscosity. The impedance may have an arbitrary number of peaks and valleys and for even a simple mechanical structure it typically does. There is little *a priori* reason to believe that muscle should be any simpler.

An alternative representation of the same information is as a mechanical admittance relating output velocity to input force. Admittance is simply the inverse of impedance and in the linear case the two contain the same information. For this reason the term "impedance" is often used loosely as a general term referring to impedance and/or admittance.

Nonlinear Case

Muscle behavior is, of course, highly nonlinear. The concept of impedance may readily be applied to nonlinear systems[5] (Hogan, 1985a). For example, a state-determined representation suitable for a muscle model is:

$$dz/dt = z_r(z,v,u,t) \qquad (9.6)$$

$$f = z_o(z,v,u,t) \qquad (9.7)$$

where z is a finite-dimensional vector of state variables, v is velocity of shortening and $z_r(\cdot)$ and $z_o(\cdot)$ are algebraic (memoryless) functions. Note that the muscle models discussed elsewhere in this book may be cast in this general form.

9.3 Effects and Limitations of Feedback Control

Muscle is richly endowed with sensory organs such as muscle spindles and Golgi tendon organs which respond to mechanical events in the muscle. Among other functions, this afferent activity influences the efferent α–motoneuron activity through feedback loops operating at several levels in the central nervous system, from the spinal cord to the cortex. This feedback can play a significant role in shaping movement behavior, though its precise function is a topic of continuing research and debate. Comprehensive recent reviews exist (e.g. Loeb, 1984) which will not be duplicated here.

When feedback is used to modify a system's behavior there is an ever-present possibility of instability. In general, the greater the influence of feedback, the greater the likelihood of instability. This is a fundamental limitation of feedback control, the reason why it cannot be used to bring about arbitrarily large changes in behavior.

9.3.1. Destabilizing Effect of Transmission Delays

Neural feedback control of dynamic behavior is severely curtailed by the inevitable time delays associated with neural transmission. Too large a feedback gain in the presence of a transmission delay will give rise to instability. The limited response speed of muscle tends to offset this effect to some degree; in effect, the muscle responds in a somewhat sluggish manner to rapidly varying neural inputs, attenuating the feedback loop gain at the higher frequencies at which instabilities could occur and thereby permitting larger feedback gains at lower frequencies. Biological sensors also exhibit significant dynamic behavior which influences feedback stability, but the basic fact is that neural feedback can only be effective below a frequency determined by the magnitude of the transmission delay.

The greater the transmission delay, the lower the frequency. Drawing on engineering experience we would expect it to be about one

---

[5] Though, as usual, fewer general-purpose analytical tools are available for the nonlinear case.

twentieth of the inverse of the transmission delay. Using 30 ms as an approximate value of the transmission delay for spinal cord feedback loops in the human upper extremity yields an estimated frequency of 1.7 Hz; using 100 ms as an appropriate value for the trans-cortical loop delay yields a frequency of 0.5 Hz. Above these frequencies, neural feedback control of neuromuscular dynamics can be expected to be severely limited.

9.3.2. Intrinsic Muscle Impedance Governs Rapid Interactions

These estimates are crude, but the rapidity of the dynamic events involved in voluntary human motor actions can far exceed these frequencies. For example, Antonsson and Mann (1985) have reported that foot–floor interactions in normal, unhurried gait contain significant frequency components up to 15 Hz. In rapid manual tasks such as throwing, catching, or wielding tools, the frequency content may be significantly greater. If these rapid dynamic interactions are to be controlled, some alternative to neural feedback control is needed.

That alternative is found in the muscle itself. Intrinsic muscle impedance generates force extremely rapidly in response to imposed displacements, essentially without any time delay (see Chapters 1–5). Grillner (1972) made this argument and demonstrated that the intrinsic stiffness of muscle is responsible for the earliest response of the lower limb muscles to the disturbances encountered in walking; neurally mediated response also occur, but somewhat later.

9.3.3. Contact Instability

Robotic experience has revealed an important distinction between *isolated stability* and *coupled stability*. Robots that perform unconstrained maneuvers stably, smoothly, and rapidly may break into a sustained, pathological chattering on contact with an object, a phenomenon known as *contact instability*. This has been called one of the outstanding challenges of robotic research (Paul, 1987).

To understand the problem, consider a simple system comprised of a mass (which could be a crude model of a skeletal segment) subject to forces from the environment and from a controlled force source, spring and damper (representing the associated musculature). Using the Laplace variable, the dynamic response of the system may be described as follows.

$$(m s^2 + b s + k) x = c u - f \tag{9.8}$$

where m is mass, b is damping factor, k is stiffness, x is position, c is a gain, u is a controllable input (e.g. neural) and f is environmental force. The forward-path transfer function relating input, u, to position, x may be obtained by assuming environmental forces are zero.

$$\frac{x}{u} = \frac{c}{ms^2 + bs + k} \tag{9.9}$$

Now consider the following hypothetical control system which would act to regulate position by generating a control signal proportional to the time integral of any deviation between the position, x, and a reference value, r.

$$u = \frac{g}{s}(r - x) \tag{9.10}$$

where g is a feedback gain. With this controller the dynamic behavior is:

$$(m s^3 + b s^2 + k s + c g) x = c g r - s f \tag{9.11}$$

The closed loop transfer function relating the reference value, r, to position, x is:

$$\frac{x}{r} = \frac{cg}{ms^3 + bs^2 + ks + cg} \tag{9.12}$$

In steady state the position is identical to the reference value, which might be desirable for a regulator. Even this highly simplified control system is capable of unstable behavior if the gain, g, is too large. Assuming $m > 0$, $b > 0$, $k > 0$ it is straightforward to show that for the system to remain stable, the feedback gain must be restricted as follows:

$$g < \frac{bk}{cm} \tag{9.13}$$

However, this condition is not sufficient to ensure that the system will remain stable when it interacts with objects in the environment. Consider the effect of coupling an external mass, m_e, to the system (this might represent the effect of grasping an object). The dynamic behavior of the external mass couples the force and motion at the point of contact (the driving point) as follows:

$$f = m_e s^2 x \tag{9.14}$$

Substituting into *Eq. 9.11* the resulting behavior is:

$$[(m + m_e)s^3 + b s^2 + ks + cg] x = cgr \quad (9.15)$$

The closed loop transfer function relating the reference value, r, to position, x is now:

$$\frac{x}{r} = \frac{cg}{(m + m_e)s^3 + b s^2 + ks + cg} \quad (9.16)$$

The steady-state behavior of the system is as before, but the condition for stability has changed. If the system is to remain stable when coupled to a mass, m_e, the feedback gain must be restricted as follows:

$$bk > cg(m + m_e) \quad (9.17)$$

This means that for any value of the feedback gain (other than zero) which is chosen to ensure stability of the control system in isolation, there exists an entire range of masses that will destabilize the system on contact. Coupled instability will occur if m_e is sufficiently large:

$$m_e > \frac{bk}{cg} - m \quad (9.18)$$

9.3.4. Passivity and Coupled Stability

The distinction between the one-way and two-way effects in an actuator (the forward-path transfer function and the driving point impedance) is the key to understanding coupled stability. Recent work has established the conditions under which a system that is stable in isolation will remain stable when coupled to any passive, stable object; the condition is a constraint on the driving point impedance.

Most of the objects a limb contacts are passive; they may store energy (as when a spring is compressed) and that energy may subsequently be recovered (as when a spring is released) but the amount of energy recovered cannot exceed the amount stored (and because of dissipation is typically less). In contrast, an actuator can (at least theoretically) supply energy indefinitely. In the linear case, if an object is passive, e.g., a real spring, damper, etc., its driving point impedance is restricted. For continuous sinusoidal inputs, net power must be absorbed over each cycle. The phase angle between velocity and the force acting on the object and must lie between $\pm 90°$.

The necessary and sufficient condition for a system such as a robot or a human limb to avoid instability when coupled to any stable, passive object is simple: its driving point impedance must appear to be that of a passive system (Colgate and Hogan, 1987, 1989).

Returning to the simple example above, with no controller the driving point impedance relating velocity, v, to environmental force, f, may be obtained from *Eq. 9.8* by assuming the control input is zero.

$$\frac{f}{v} = -\frac{ms^2 + bs + k}{s} \quad (9.19)$$

Under these conditions, the system is clearly passive and the phase angle between velocity and the force acting on the system (the negative of the force, f) never exceeds $\pm 90°$. However, under the action of the controller defined in *Eq. 9.10* the driving point impedance is as follows.

$$\frac{f}{v} = -ms^2 + bs + k \quad (9.20)$$

In this case, at sufficiently low frequencies, the phase lag between force on the system and velocity exceeds $-90°$ and may be as large as $-180°$. With this controller, the driving point impedance is not that of a passive object, and as illustrated above, there exist objects which will induce instability on contact.

In general, active control systems do not result in passive driving point impedances, but they may be constrained to do so. Consider the following alternative control system in which the input is directly proportional to the error

$$u = g(r - x) \quad (9.21)$$

With this controller the dynamic behavior is:

$$[ms^2 + bs + (k + cg)] x = cgr - f \quad (9.22)$$

The closed loop transfer function relating the reference value, r, to position, x is:

$$\frac{x}{r} = \frac{cg}{ms^2 + bs + (k + cg)} \quad (9.23)$$

In steady state the position is proportional to the reference value, and as the gain increases the proportionality constant approaches unity, so this controller also acts to regulate position. However, unlike the previous case, this controller results in the following driving point impedance:

$$\frac{f}{v} = - \frac{ms^2 + bs + k}{s} \qquad (9.24)$$

In this case the the phase angle between force on the system and velocity always remains within ±90°. This *actively* controlled system has a *passive* driving point impedance. Consequently it will be stable on contact with any stable, passive object.

Neither of these controllers is offered as a plausible model of neural feedback; they are intended only to illustrate the point that dynamic interaction profoundly affects stability. This is one of the reasons why controlling the arms (or the torso or legs) is more complicated than controlling the eyes. Under normal circumstances, the eyes do not interact physically with the environment, and contact stability is not an issue. In contrast, the arms, hands and legs frequently interact with the environment; that is one of their primary functions. Similarly, the torso is loaded whenever the arms pick up an object; here too stability requires impedance modulation [Chapter 26 (Crisco and Panjabi), Chapter 23 (Andersson and Winters)]. Ensuring stability when coupled to a wide range of environmental objects is the *sine qua non* of movement control systems and this is a property of the neuro-muscular system's output impedance.

9.3.5. Feedback Regulation of Muscle Behavior

These considerations suggest a perspective on the action of feedback complementary to the usual control systems viewpoint: not only does feedback regulate quantities such as position, it acts to modify apparent behavior, e.g. mechanical impedance. Indeed, one of the greatest benefits of feedback is that as the gain of a feedback loop increases, the overall dynamic behavior of the controlled system depends less on the dynamics of the actuator inside the feedback loop and more on the feedback system.

Muscle receives a combination of negative position feedback from the muscle spindles and negative force feedback from the Golgi tendon or-gans. These two feedback loops might seem to counteract one another, but as the feedback gains increase, the overall stiffness of the system becomes less sensitive to variations in muscle behavior and is determined by the ratio of the feedback loop gains. Nichols and Houk (1973) originally proposed the idea that one of the major functions of the spinal cord feedback loops is to regulate or maintain muscle stiffness in the face of disturbing influences such as the yield in muscle force that accompanies rapid muscle stretch; a substantial body of evidence has since been amassed in support of this postulate (e.g. Nichols and Houk, 1976; Crago et al., 1976; Houk, 1979; Hoffer and Andreassen, 1981).

One important point to note is that although feedback may modify muscle impedance, the steady state effect of neural feedback to a muscle from sensors in the same muscle will always produce a spring-like behavior. Whatever the role of the muscle spindles and tendon organs, if all other variables (e.g. the descending commands to the spinal motorneuron pool, etc) are held constant, a given muscle length will correspond to a unique muscle force, and that is the only requirement for spring-like behavior.

9.4. Multi-Muscle, Single-Joint Systems

As outlined above, mechanical impedance is the quantity which determines how a limb interacts with its environment. The apparent dynamic behavior observable at any point on the musculoskeletal system (e.g. the hand) is determined by three major factors: *1)* the intrinsic mechanics of the muscles and skeleton; *2)* neural feedback; *3)* the geometry or kinematics of the musculoskeletal system.

The simplest example of the influence of kinematic factors is the antagonism of muscles acting about a joint. Muscles pull, hence in order to generate both flexive and extensive moments about a joint, they are usually deployed in groups such that at least two of the muscles oppose one another. Consider a hypothetical system of two muscles connected in opposition about a single joint. Each muscle force generates a moment about the joint proportional to the moment arm from its line of action (assumed to be unique) to the joint axis (assumed to be in a constant location). Assume (for simplicity) that the moment arms of the muscles are independent of the joint angle. Using subscripts *1* and *2* to denote agonist and antagonist respectively, the relations between muscle lengths and joint angles are:

$$l_1 = l_{10} - r_1 q \qquad (9.25)$$

$$l_2 = l_{20} + r_2 q \qquad (9.26)$$

where q is the joint angle, r_1 and r_2 are the moment arms, and l_{10} and l_{20} and the muscle lengths at some configuration designated as $q = 0$. The moment, μ, about the joint is the difference of the muscle tensions weighted by the moment arms.

$$\mu = r_1 f_1 - r_2 f_3 \qquad (9.27)$$

Assume (also for simplicity) a linear model for the steady-state behavior of the muscles.

$$f_1 = c_1 u_1 + k_1 l_1 \qquad (9.28)$$

$$f_2 = c_2 u_2 + k_2 l_2 \qquad (9.29)$$

The corresponding steady-state relation between moment and joint angle is:

$$\mu = [r_1 c_1 u_1 - r_2 c_3 u_2] - [r_{12} k_1 + r_{22} k_2] q \qquad (9.30)$$

This example illustrates several simple but important points. Whereas the net moment about the joint is a weighted *difference* of muscle tensions, the net stiffness about the joint is a weighted *sum* of the contributions each of the muscles. Secondly, the moment contributed by the one-way relation between neural input and force is proportional to the moment arm, while the contribution due to the two-way interaction between force and length is proportional to the *square* of the moment arm. In a more realistic model the relations will be more complex, (see below) but kinematic effects will always have a more pronounced effect on the output impedance than on the forward-path transfer function.

9.4.1. Controlling Mechanical Impedance

Not only is the human arm–torso system stable when coupled to passive objects, it can stabilize unstable objects, and routinely does so. A simple example is holding an object in an upright posture against gravity. If we include a gravitational load in our hypothetical two-muscle single-joint system the moment due to gravity, μ_g, is

$$\mu_g = W h \sin q \qquad (9.31)$$

where W is the weight of the limb plus load, and h the distance from the center of mass to the joint axis. The "gravitational stiffness", k_g, is

$$k_g = \frac{\partial \mu_g}{\partial q} = W h \cos q \qquad (9.32)$$

This tends to destabilize the system. Suppose the system is at equilibrium in, say, the upright position ($q = 0$). At equilibrium the net moment is zero and a necessary condition for stability is that the net stiffness be negative (i.e. displacements result in restoring moments).

$$k_g - r_{12} k_1 + r_{22} k_2 < 0 \qquad (9.33)$$

The larger the load held, the less stable the system. To maintain stability, it may be necessary to modulate mechanical impedance. This becomes especially important for inverted pendulum systems [a reasonable first approximation for the head-neck system [see Chapter 28 (Winters and Peles)] and low-back–torso system [Chapter 26 (Crisco and Panjabi)]; see Chapter 23 (Andersson and Winters) for further discussion.

In fact, modulating mechanical impedance is a general strategy for controlling interaction between both a limb and its environment (Hogan, 1985; Kazerooni, 1987) and a flexible base of support and the limb [Chapter 23 (Andersson and Winters)]. Robotic experience has shown that controlling mechanical impedance is effective in a wide variety of situations including stable control of contact tasks (Hogan, 1987). Controlling mechanical impedance also appears to be one of the strategies available to the biological system. Several researchers have reported task-dependent changes in the mechanical impedance of the limbs or the reflex feedback system which affects the mechanical impedance (e.g. Hammond, 1956; Evarts and Tanji, 1974; Dufresne et al., 1978, 1980; Akazawa and Milner, 1981; Lacquaniti et al., 1982; Lacquaniti and Soechting, 1983). A closely related strategy of controlling stiffness has been successfully applied in external control of limb movements by functional neural stimulation [Chapter 20 (Crago et al.)].

Synergistic Activation of Antagonist Muscles

Because the impedance of a muscle increases with activation level, one means of modulating mechanical impedance is to synergistically activate both muscles. Because the stiffness contributions of the muscles add while their moment contributions subtract, by carefully coordinating muscle activities, the net joint moment may be held constant while the net stiffness increases and this may be one of the ways the central nervous system controls its dynamic behavior. For example, Akazawa and Milner (1981) reported coactivation of antagonist muscles when the finger was coupled to an unstable load, and the level of muscle coactivation and limb impedance increased with the magnitude of the instability.

One drawback of this strategy is that it is energetically inefficient. By opposing one another, the muscles perform no net mechanical work, yet they consume metabolic energy. If we make the reasonable assumption that the biological system does not squander energy needlessly, one would expect synergistic activation of antagonists to be kept to a minimum except when the task demands it. That is, a compromise must be reached between minimizing energy expended and maintaining stability. This compromise can be analyzed using dynamic optimization theory (Hogan 1984). Optimization studies are a powerful tool for investigating motor control strategies and are discussed in Chapters 8, 10, 17, 19, 24, 25, 27, 28, 32, 42 and 43. If synergistic activation of opposing muscle groups is used to stabilize the limbs, including this consideration in the analysis may improve the ability of these approaches to predict the patterns of muscle activation.

Neural Feedback Complements Intrinsic Mechanics

Impedance modulation through synergistic activation of antagonist muscles does not preclude the action of neural feedback. Hoffer and Andreassen (1981) showed that feedback acts to enhance the apparent stiffness of muscle and maintain a constant relation between stiffness and force level. Other studies have shown that the stiffness about a joint increases with the moment about the joint (Joyce et al., 1974; Gottlieb and Agarwal, 1978; Zahalak and Heyman, 1979; Akazawa and Milner, 1981; Hunter and Kearney, 1982; Billian and Zahalak, 1983).

Given that the stiffness of a single muscle increases with force (whatever the mechanism), synergistic activation of antagonist muscles will modulate net joint stiffness. For the faster dynamic events, feedback will be ineffective because of the inevitable neural transmission delays, but intrinsic muscle mechanics are available for impedance modulation, though at a cost in energy consumption. At lower frequencies, feedback comes into play and significantly enhances the stiffness corresponding to a given muscle force; again, antagonist muscle coactivation may be used to modulate impedance and control dynamic interaction. In this way neural feedback may complement intrinsic muscle behavior, and synergistic activation of antagonist muscles may be a single unified strategy for controlling dynamic interaction at all frequencies.

By this argument, synergistic activation of antagonist muscles would be anticipated when the frequency content of a task exceeded that which could be dealt with via feedback. Humphrey and Reed (1983) applied periodic torques to a monkey's forearm and observed that at low frequencies (around 0.2 Hz) they were resisted by reciprocal activation of antagonists, but as the frequency of perturbation increased a progressive increase in coactivation was observed, becoming significant at frequencies as low as 1.0 Hz.

9.4.2. Minimum Model Requirements for Impedance Modulation

Modulation of mechanical impedance is a nonlinear effect that cannot be captured in a linear model. Consider a Taylor series expansion of a steady-state relation between force length and neural input (Eq. 9.3). If terms above first order in the differentials are neglected the resulting linear model produces a family of parallel force–length characteristics, each with the same stiffness.

To reproduce the observed modulation of stiffness with neural input, terms at least up to second order in the differentials must be included. In fact it is sufficient to include just one of those terms: the crossed partial derivative of force with respect to length and neural input.

$$f = f(l_0, u_0) + \frac{\partial f}{\partial l}(l_0, u_0)\, dl + \frac{\partial f}{\partial u}(l_0, u_0)\, du$$

$$+ \frac{\partial^2 f}{\partial l\, \partial u}(l_0, u_0)\, dl\, du + \ldots \qquad (9.34)$$

The result of neglecting all other terms in the expansion is a *bi-linear* model.

$$f_{bi\text{-}linear} \simeq (\alpha + \beta\, l)(\gamma + \delta\, u) \qquad (9.35)$$

where α, β, γ, and δ are constants derived from the first four terms in the Taylor series expansion. This produces a fan-shaped family of force-length curves, a model that has been used extensively in exploring the consequences of the spring-like properties of muscles. Though clearly a crude representation of neuro-muscular mechanics, it can be quite effective. For example, the stiffness of a bilinear model is proportional to neural input.

$$k_{bi\text{-}linear} = \frac{\partial f}{\partial l}_{bi\text{-}linear} = \beta\,(\,\gamma + \delta\,u\,) \qquad (9.36)$$

The force generated at any given length is also proportional to neural input; therefore the bi-linear model reproduces the relation between force and stiffness which has been reported by several investigators.

Another point to keep in mind is that a good mathematical model does not merely reproduce known behavior as accurately as possible. Enhancement of insight and understanding is equally important (perhaps more so) and is usually at variance with accurate prediction. More complex nonlinear models of muscle are clearly justifiable, but they are difficult to analyze. In contrast, though it is not widely recognized, a wide range of general-purpose mathematical tools have been developed for bi-linear systems.

This crude bi-linear model of impedance modulation has some counter-intuitive consequences. Just as stiffness is proportional to neural input, the effective gain relating force to neural input is proportional to muscle length.

$$c_{bi\text{-}linear} = \frac{\partial f}{\partial l}_{bi\text{-}linear} = (\,\alpha + \beta\,l\,)\,\delta \qquad (9.36)$$

That is, a longer muscle is more effective at generating force than a shorter muscle. Now suppose this muscle model is used in the hypothetical two-muscle single-joint system of section 4.1 and we analyze the requirements for maintaining an off-upright posture of the limb. To counterbalance the effect of gravity at a given posture a moment must be generated, hence the agonist muscle must be active. Suppose further that the resulting stiffness due to the active agonist muscle is sufficient to provide the necessary stability to maintain posture so that the antagonist may be inactive.

Consider the effect of deliberately co-activating the antagonists (i.e. tensing the arm) while maintaining posture. Because the agonist is nearer the longer end of its physiological range, a small change of its neural input produces a large change in its contribution to the joint moment. In contrast, the antagonist is at the shorter end of its physiological range, therefore in order to generate an equal contribution to the joint moment (which is necessary if posture is not to change) a much larger change of its neural input is required.

Therefore this model predicts a highly counter-intuitive result: at sufficient levels of co-activation, *the activity of the antagonist will exceed the activity of the agonist*, despite the fact that the agonist generates a greater moment. Recent experiments reported by Murray and Hogan (1988) using myoelectric activity as a measure of neural activity have unequivocally confirmed this prediction.

9.5. Multi-Joint Mechanics

The kinematics of the musculo-skeletal system is, of course, far more complex than the simple hypothetical model of Section 9.4. In the multi-joint case new factors must be taken into consideration. One of the most important is the possibility of interaction between different degrees of freedom. In a multiple degree of freedom system, force and motion are vector quantities with direction as well as magnitude. The impedance also has a directional property. In particular, the force and displacement need not be co-linear; both the magnitude and the direction of the output force vector resulting from an input displacement vector of a given magnitude will depend on the orientation of the displacement vector.

Modulating the directional character of the impedance may be important for performing interactive tasks. For example, the limb might be made compliant in one direction to accommodate an external kinematic constraint and stiff in another direction to minimize the effects of disturbing forces. This provides a different perspective on the problems of motor control. Coordination is usually regarded as a movement problem but modulating multi-joint impedance implies that muscle coordination is required to control postural behavior.

9.5.1. The Unique Contribution of Poly-Articular Muscles

For the purposes of motion control, muscle groups spanning a single joint would be sufficient to generate the necessary moments. Poly-articular muscles would appear to be an excess, a redundancy in the system. However, if only single-joint muscle groups were available, the ability to modulate impedance by synergistic muscle activation would be quite limited; there would be no way to modulate the interaction between joints. Neural feedback connections between the joints might play this role, but as argued above, modulation of impedance via neural feedback would be ineffective for rapid or high-frequency interactions.

Including the poly-articular muscles changes the situation considerably. Poly-articular muscles provide precisely the coupling between joints needed to modulate inter–limb interactions through muscle synergies (Hogan 1979, 1985b). From this point of view, the neuro-muscular system may be considerably less redundant than it would otherwise appear and poly-articular muscles may have an unique functional role to play. This contribution of poly-articular muscles to limb impedance would complement other functional roles they may play [see Chapter 18 (Gielen et al.), Chapter 41 (van Ingen Schenau et al.)].

9.5.2. Transformation Theory

The most effective way to understand kinematic effects such as muscle connectivity is to consider the transformation of the apparent mechanical behavior of the muscles into an equivalent behavior of the joints and into an equivalent behavior of the hand (or other point of contact with the environment). All of the transformation relations may be determined from the geometry of musculo-skeletal connections. The lengths of all the muscles, uni-articular as well as poly-articular, may be determined from the configuration of the limbs.

$$l = l(q) \tag{9.38}$$

where $l = [l_1, l_2, ... l_m]^t$ is an array (a "vector") containing a set of muscle lengths and $q = [q_1, q_2, ... q_r]^t$ is an array containing a set of numbers which define the limb configuration. Those numbers will usually be joint angles (e.g. defining relative rotation of adjacent limb segments) but that is not essential; there is an infinity of equivalent sets of configuration variables. As mentioned above, to keep a consistent sign convention we define a set of variables, x, which increase as the muscles shorten.

$$x(q) = -l(q) \tag{9.39}$$

The relation between muscle displacements and joint displacements is obtained by differentiating.

$$dx = [\partial x(q)/\partial q] \, dq = [-\partial l(q)/\partial q] \, dq$$

$$= j(q) \, dq \tag{9.40}$$

The Jacobian matrix $j(q)$ is a matrix of partial derivatives of muscle lengths with respect to joint angles. As there are typically far more muscles than degrees of freedom in the skeleton, the Jacobian matrix is usually not square (and hence usually cannot be uniquely inverted). The Jacobian also relates muscle and joint velocities.

$$v = j(q) \, \omega \tag{9.41}$$

where $v = dx/dt$ and $\omega = dq/dt$.

The connection of the muscles to the skeleton cannot of itself generate power, so the incremental work done by shortening the muscles is identical to the incremental work done by rotating joints.

$$dE = \mu^t \, dq = f^t \, dx \tag{9.42}$$

where $f = [f_1, f_2, ... f_m]^t$ is an array of muscle forces and $\mu = [\mu_1, \mu_2, ... \mu_r]^t$ is an array of moments about the joints. The relation between muscle force and joint moments is obtained by substitution.

$$f^t \, dx = f^t \, j(q) \, dq = \mu^t \, dq \tag{9.43}$$

$$\mu = j^t(q) f \tag{9.44}$$

This demonstrates that the elements of the Jacobian matrix are the moment arms of the muscles [see Chapter 8 (Zajac and Winters) for elaboration of this transformation].

Transforming Apparent Viscosity

These relations may be used to transform the mechanical impedance of the muscles into an equivalent mechanical impedance about the joints. For example, consider a relation between muscle force and velocity of shortening. If the individual muscle behaviors are described by relations $f_1 = f_{v1}(v_1), f_2 = f_{v2}(v_2), \ldots f_m = f_{vm}(v_m)$ the aggregate behavior of all of the muscles may be written concisely as follows.

$$f = f_v(v) \qquad (9.45)$$

Transforming to joint coordinates is a simple matter of substitution.

$$\mu = j^t(q) \, f_v\{j(q) \, \omega\} \qquad (9.46)$$

Viscosity is usually defined as the incremental change of force due to an incremental change of velocity. The apparent muscle viscosity is:

$$b = \frac{\partial f_v}{\partial v} \qquad (9.47)$$

The apparent joint viscosity due to muscle behavior is:

$$b_j = \frac{\partial \mu}{\partial \omega} = j^t(q) \, \frac{\partial f_v}{\partial v} \, j(q) \, \omega \qquad (9.48)$$

Thus the viscosity transforms as follows.

$$b_j = j^t(q) \, b \, j(q) \qquad (9.49)$$

As in the simple example above, the transformation of this component of the impedance is different from the transformation of forces or motions. The viscosity transforms as a second-rank tensor (in proportion to the "square" of the Jacobian) whereas the forces and velocities transform as a first-rank tensors (i.e. vectors).

Transforming Apparent Stiffness

One important point is that unlike viscosity, stiffness does **not** transform as a second-rank tensor. If the individual muscle behaviors are described by relations $f_1 = f_{e1}(x_1), f_2 = f_{e2}(x_2), \ldots f_m = f_{2m}(x_m)$ the aggregate behavior of all of the muscles may be written as follows.

$$f = f_e(x) \qquad (9.50)$$

The corresponding joint behavior is

$$\mu = j^t(q) \, f_e\{x(q)\} \qquad (9.51)$$

The apparent joint stiffness due to muscle behavior is

$$k_j = \frac{\partial \mu}{\partial q} \, \frac{\partial j^t(q)}{\partial q} \, f + j^t(q) \frac{\partial f_e}{\partial q} \qquad (9.52)$$

The second term transforms the muscle stiffness, k, the same way as the viscosity.

$$j^t(q) \frac{\partial f_e}{\partial q} = j^t(q) \, k \, j(q) \qquad (9.53)$$

The first term, however, is quite independent of the muscle stiffness. It is a "fictitious" joint stiffness due to the variation of the moment arms (the Jacobian matrix) with configuration. We may denote the array of second partial derivatives of muscle lengths with respect to joint angles by the symbol $\Gamma(q)$.

$$\Gamma(q) = \frac{\partial j^t(q)}{\partial q} \qquad (9.54)$$

Thus the stiffness transforms as follows.

$$k_j = \Gamma(q) f + j^t(q) \, k \, j(q) \qquad (9.55)$$

If the stiffness is computed at an operating condition corresponding to zero muscle force, the "fictitious" joint stiffness due to the variation of the moment arms vanishes. However, it becomes progressively more significant as the muscle forces at the operating condition increase, and this is precisely what happens if opposing muscles are synergistically activated.

9.5.3. Multi-Joint Spring-Like Behavior

In the case of a single degree of freedom, no matter how many muscles are involved, and no matter what neural feedback relation modulates the neural input to the muscles, the steady-state behavior of the system is always spring-like. That is because all that is required is that every angle to correspond to a unique moment. The converse is not necessary and indeed, because of nonlinear moment arms, is highly unlikely. For example, moment arm of biceps varies very strongly with angle (approaching low values at the limits of joint travel). Therefore even if the force–length behavior of biceps were monotonic, the moment–angle relation is not.

However, when the system has multiple degrees of freedom, the steady-state behavior is a vector relation between moment and configuration. The conditions for multi-variable spring-like behavior are more restrictive; a particular symmetry of interactions between degrees of freedom is required (Hogan 1985b).

The fundamental requirement for spring-like behavior is that a potential function analogous to elastic energy must be definable. It is a scalar function $E_p(q)$ of the configuration variables (e.g. the joint angles). The moment-angle relation is the gradient of that scalar function.

$$\mu = \text{grad } E_p(q) = \frac{\partial E_p(q)}{\partial q} \qquad (9.56)$$

But a basic fact from vector algebra is that the curl of a gradient is zero.

$$\text{curl } \mu = 0 \qquad (9.57)$$

This requires the crossed partial derivatives of moment with respect to angle to be equal.

$$\frac{\partial \mu_i}{\partial q_j} = \frac{\partial \mu_i}{\partial q_j}; \quad i, j = 1, \dots r; \quad i \neq j \qquad (9.58)$$

These quantities are the off-diagonal terms in the stiffness matrix k_j. Therefore to test rigorously for spring-like behavior in the multi-joint case it is sufficient to measure the linearized (or local) stiffness and check for symmetry. Furthermore, if the curl is zero in joint coordinates it will be zero in all coordinates (Hogan 1986), so it is sufficient to measure the linearized stiffness at some convenient point such as the hand.

The steady-state relation between force and displacement for ten normal human subjects was measured in experiments reported by Mussa-Ivaldi et al. (1985). Briefly, subjects were seated and held the handle of a planar two-link manipulandum. The elbow was supported so that the forearm and upper arm moved freely and comfortably in the horizontal plane. While subjects maintained an equilibrium posture of the limb and handle, controlled displacements in eight directions in the plane were applied using computer-controlled servomotors mounted on the manipulandum. When the limb reached steady state following this perturbation, but before the onset of any voluntary response, the applied displacement vector and the restoring force vector generated by the subject were measured.

Although the force–displacement relation is nonlinear for large displacements, for sufficiently small deviations about an equilibrium point it is approximately linear.

$$F = -K \, dX \qquad (9.59)$$

where F is the force at the hand, dX is the displacement of the hand and K is the stiffness at the hand, a square matrix of coefficients. The experimental observations were fit to a linear force–displacement relation using multivariable regression techniques. They were then partitioned into two components, one symmetric and one anti-symmetric.

$$K_s = \frac{K + K^t}{2} \qquad (9.60)$$

$$K_a = \frac{K - K^t}{2} \qquad (9.61)$$

The symmetric component, K_s, can be derived from a potential function; this is the spring-like part of the behavior. In contrast, the anti-symmetric component, K_a, cannot be derived from a potential function. The force due to this component is always directed at right angles to the displacement with a magnitude proportional to the displacement. It gives the corresponding force–displacement relation a rotational character (hence the term curl). If the hand were displaced along a closed path a net non-zero quantity of mechanical work would be done by (or on) the hand. Thus by repeatedly traversing this closed path the hand could be made to supply (or absorb) an arbitrarily large amount of work; this is a *non-passive* behavior.

Comparison of the relative magnitudes of the symmetric and anti-symmetric components of the stiffness provides a numerical quantification of the extent to which the neuromuscular system is spring-like. Experimental observations showed that the conservative component dwarfed the rotational component. If fact, in about half of the measurements, the curl was not statistically different from zero; in all cases, it was small. At least under the conditions of these experiments, the steady-state behavior of the neuro-muscular system was essentially indistinguishable from that of a multi-variable mechanical spring.

How surprising is this result? If the steady-state multi-variable behavior were simply the aggregate of the individual muscle behaviors, the symmetry requirement would always be met. For each individual muscle (including poly-articular muscles) a potential function could be defined by integrating its steady-state force–length relation.

$$E_{pi}(x_i) = \int f_i \, dx_i \qquad (9.62)$$

An aggregate potential function may be defined for the multi-variable system by summing the contributions for each muscle.

$$E_p(x) = \sum_{i=1}^{m} E_{pi}(x_i) \qquad (9.63)$$

Of course, the force–displacement relation for the entire limb depends also on feedback-generated interaction between degrees of freedom. If this feedback-mediated interaction were not bilaterally symmetric it would introduce a curl. For example, if rotation of the elbow produced a moment about the shoulder which was stronger (or weaker) than the moment about the elbow resulting from a corresponding rotation of the shoulder) the vector force–displacement relation for the neuromuscular system would not be integrable; a potential function analogous to elastic energy could not be defined and the neuromuscular system would not be spring-like.

Thus the experimental observation of zero curl means that either feedback interaction between the degrees of freedom was absent under the conditions of this experiment or it was exquisitely balanced. The latter seems more plausible. Neural feedback pathways which relate motions at one joint to the activity of muscles at another are known to exist (e.g. Eccles et al., 1957a,b; Eccles and Lundberg, 1958; Hongo et al., 1969). Given that the neural pathways exist, it would be puzzling if feedback was insignificant. If, instead, feedback interactions were present, but balanced and symmetrical, this would indicate that the neuromuscular system goes to some lengths to exhibit an apparently passive behavior. As mentioned above, an apparently passive impedance is sufficient to ensure that contact with passive objects of arbitrary complexity and nonlinearity will not result in unstable behavior (Hogan 1988a).

9.5.4. Posture Modulates Dynamics

To characterize interactions between the musculo-skeletal system and objects in its environment we need to determine the impedance at the point(s) of contact (e.g. the hand). The kinematic relation between joint configuration and muscle lengths is the critical information needed to transform muscle impedance into a corresponding dynamic behavior of the skeleton. In an exactly similar manner, the critical information needed to transform the joint impedance into an equivalent behavior at the hand is the kinematic relation

$$X = X(q) \qquad (9.64)$$

between joint configuration, q, and an array of variables, X, defining hand position and orientation. With this information, transformation relations exactly analogous to *Eqs. 9.40, 9.41, 9.44, 9.46, 9.49, 9.51* and *9.55* may be derived. An important point to note is that joint configuration variables, q, figure prominently in all of these transformation. Consequently, *posture profoundly modulates all aspects of dynamic behavior.*

Transforming Inertia

Perhaps the most striking example of the influence of kinematic effects on the apparent dynamic behavior of the musculo-skeletal system is in the modulation of apparent inertia. Unlike the mechanical behavior of muscle, the distribution of mass in the bone is not subject to voluntary control. In principle, neural feedback might be invoked to modulate apparent inertia, but inertial behavior is most important during the fastest transient dynamic events, such as the foot striking the ground or the hand catching a ball and for these actions neural feedback would be expected to be at least effective because of the inevitable transmission delays. However, the apparent inertia of a multi-joint system may be modulated over a wide range by selecting an appropriate posture of the limb segments.

The inertial behavior of the skeleton is described by a relation between velocity and (generalized) momentum.

$$\omega = I^{-1} \eta \qquad (9.65)$$

where I^{-1} is the inverse of the inertia tensor for the skeleton and η is the generalized momentum associated with joint motion. Differentiating *Eq. 9.65* yields a relation between velocities.

$$V = J(q) \, \omega \qquad (9.66)$$

where $V = dX/dt$ and $J(q)$ is a Jacobian matrix. An

argument analogous to that used the derive *Eq. 9.44* yields a relation between generalized momenta (Hogan 1985a).

$$\eta = J^t(q) P \qquad (9.67)$$

where P is the generalized momentum associated with hand motion. Substituting:

$$V = J(q) I^{-1} J^t(q) P = M^{-1}(q) P \qquad (9.68)$$

Eq. 9.68 describes the apparent inertial behavior at the hand; $M^{-1}(q)$ is the inverse of the apparent inertia tensor. Because it is a function of configuration[6] the biological system may modulate its inertial behavior by repositioning the limb.

In effect, repositioning the limb changes the lever arms through which forces at the hand generate joint moments. Because of this, changing the configuration of the limb modulates the apparent stiffness and viscosity at the hand — indeed, all aspects of the mechanical impedance — as well as its inertia. Consequently, appropriate choice of *posture* may be one of the most important strategies for controlling dynamic interactions between the limbs and its environment. Time will tell whether this concept, outlined here for the upper limb, is also of importance for trunk and whole-body posture.

9.6. Implications for Multi-Joint Control

The mechanical impedance of the neuro-muscular system coupled with the inertia of the skeleton define the dynamics of an *attractor*. If the descending commands to the motoneuron pools are held constant, the spring-like behavior of the neuromuscular system defines an equilibrium configuration for the skeleton. Particular properties of the mechanical impedance such as its apparent passivity ensure the stability of that configuration. Whenever the limb is transiently disturbed, it will return to that equilibrium posture.

The attractor dynamics may be *used to produce movement as well as to sustain posture*. If neural activities were changed so that the neurally defined equilibrium point coincided with a new position, the limb would move under the action of

the spring-like effect to acquire the new equilibrium posture. This idea has been investigated in depth by numerous researchers and is discussed in Chapters 11–14 and 17. Early thinking was that movement might be generated by an abrupt shift of the neurally-defined equilibrium point, but that turned out not to be the case. Experimental observations (Bizzi et al. 1984) showed that the *path* between two point exhibited a measurable degree of stability, not just the end-point [see also Chapter 17 (Flash)]. Thus, the production of movement appears to be accomplished by a progressive movement of the neurally–defined equilibrium posture, which has been termed a *virtual trajectory* (Hogan 1984).

The principal significance of this idea is that it implies a dramatic reduction in the computational complexity of movement generation (Hogan 1988b). If muscles merely produced forces on command, then it would appear to be necessary for the brain to perform some equivalent of the notoriously complex *inverse dynamics* computations (Asade and Slotine, 1986) so as to determine the appropriate forces needed to generate a particular movement. In contrast, if movement is executed through a neurally-defined virtual trajectory, the spring-like neuromuscular properties mean that the limb tends to follow the virtual trajectory without any explicit inverse dynamic computations. A series of simulations based on this theory have proven remarkably successful at reproducing observed features of upper-limb multi-articular motor behavior. A comprehensive review is presented in Chapter 16 (Flash).

9.9. Future Directions

Dynamic interaction between the limbs and the environment may profoundly influence motor behavior, possibly causing instability. The conditions for ensuring stability on contact with a stable, passive object have been established: the impedance of the limb must appear to be passive. Observations under restricted conditions (i.e. steady state) have shown that the impedance of the human upper limb is consistent with this passivity constraint. It would be informative to determine if this result generalizes to other conditions and other limb segments. Is the steady-state behavior of other multi-joint systems (e.g. the head and neck, the spine, the lower limb, etc.) spring-like in the sense defined above? Over what range of

---

[6] In general, the inverse inertia tensor of the skeleton may also be configuration-dependent, though that is strongly influenced by the variables chosen to describe the configuration (Hogan, 1985a).

amplitudes and frequencies of perturbation does the impedance of the limb appear to be passive? Do some neural disorders disrupt apparent passivity?

Apparently passive behavior is sufficient to ensure stability on contact with stable, passive objects. However, we humans are capable of much more robust behavior: we routinely grasp and wield active objects (such as power tools) and unstable objects (such as unsupported weights against gravity) and do so with unconcerned skill. Experimentally, what are the limitations of human performance in manipulating unstable, active objects? On the theoretical side, what are the conditions for preserving stability and performance with these classes of objects?

In the area of motor control and movement organization, much has been achieved but more needs to be done. What strategies are used to perform interactive tasks? Under what circumstances is impedance modulation the strategy of choice? Can the organizing principles which have been proposed for unrestrained motions (reviewed in Section III of this book) be applied (or suitably generalized) to describe interactive tasks? Preliminary investigations of how humans perform kinematically constrained motions have recently been reported, (Abul-Haj 1989, Russell and Hogan 1989) but to the author's knowledge, this area is largely unexplored.

9.8. Acknowledgements

Neville Hogan is supported in part by National Science Foundation Research Grants No. *EET-8613104* and *8914032-BCS*, National Institute of Neurological and Communicative Disorders and Stroke Research Grants *No. AR40029* and *NS 09343*, Office of Naval Research Grant No. *N00014-88-K-0372* and National Institute on Disability and Rehabilitation Research Grant No. *H133E80024-89*.

References

Abul-Haj, C. J. (1989) Experimental Evaluation of Control Systems for Cybernetic Elbow Prostheses. In: *Issues in the Modeling and Control of Biomechanical Systems*, J..L. Stein, J. A. Ashton-Miller and M. G. Pandy, eds. ASME, pp. 89-99.

Akazawa, K., and T. E. Milner (1981) Modulation of the Stretch Reflex in Human Finger Muscle. *Proc. Conf. Vocal Fold Physiology*, Wisconsin.

Antonsson, E. K. and R. W. Mann (1985) The Frequency Content of Gait. *J. Biomech.* **18**: 39-49.

Asada, H. and J-J. E. Slotine (1986) *Robot Analysis and Control*, Wiley, New York

Billian, C. and Zahalak, G. I. (1983) A Programmable Limb Testing System (and some Measurements of Intrinsic Muscular and Reflex-Mediated Stiffnesses. *J. Biomech. Eng.* **105**: ??

Bizzi, E., N. Accomero, W. Chapple and N. Hogan (1984) Posture Control and Trajectory Formation During Arm Movement. *J. Neurosci.* **4**: 2738-2744.

Colgate, J. E. and Hogan, N. (1987) Robust Control of Manipulator Interactive Behavior. In: *Modeling and Control of Robotic Manipulators and Manufacturing Processes*, eds. R. Shoureshi, K. Youcef-Toumi and H. Kazerooni, ASME, pp. 149-159.

Colgate, J.E., and Hogan, N., (1988) Robust Control of Dynamically Interacting Systems. *Int. J. Contr.* **48**: 65-88.

Crago, P, E., J. C. Houk and Z. Hasan (1976) Regulatory Actions of the Human Stretch Reflex. *J. Neurophysiol.* **39**: 925-935.

Dufresne, J. R., J. F. Soechting and C. A. Terzuolo (1978) Electro-myographic Response to Pseudo-random Torque Disturbances of Human Forearm Position. *Neurosci.* **3**: 1213-1226.

Dufresne, J. R., J. F. Soechting and C. A. Terzuolo (1980) Modulation of the Myotatic Reflex Gain in Man During Intentional Movements. *Brain Res.* **193**: 67-84.

Eccles, J.C., Eccles, R.M. and Lundberg, A. (1957a) The convergence of monosynaptic excitatory afferents on to many different species of alpha motoneurons. *J. Physiol. (Lond.)* **137**: 22-50.

Eccles, J.C., Eccles, R.M. and Lundberg, A. (1957b) Synaptic actions on motoneurones caused by impulses in golgi tendon organ afferents. *J. Physiol. (Lond.)* **138**: 227-252.

Eccles, R.M. and Lundberg, A. (1958) Integrative pattern of Ia synaptic actions on motoneurones of hip and knee muscles. *J. Physiol. (Lond.)* **144**: 271-298.

Evarts, E. V. and J. Tanji (1974) Gating of Motor Cortex Reflexes by Prior Instruction. *Brain Res.* **71**: 479-494.

Gottlieb, G. L. and G. C. Agarwal (1978) Dependence of Human Ankle Compliance on Joint Angle. *J. Biomech.* **11**: 177-181.

Grillner, S. (1972) The Role of Muscle Stiffness in Meeting the Changing Postural and Locomotor Requirements for Force Development by the Ankle Extensors. *Acta Physiol. Scand.* **86**: 92-108.

Hammond, P. H. (1956) The Influence of Prior Instruction to the Subject on an Apparently Involuntary Neuromuscular Response. *J. Physiol. (Lond.)* **132**: 17-18.

Hoffer, J. A. and S. Andreassen (1981) Regulation of

Soleus Muscle Stiffness in Premammillary Cats: Intrinsic and Reflex Components. *J. Neurophysiol.* **45**: 267-285.

Hogan, N. (1979) Adaptive Stiffness Control in Human Movement. In: *1979 Advances in Bioengineering*, M. K. Wells, ed., pp. 53-54, American Society of Mechanical Engineers, New York.

Hogan, N. (1984) Adaptive Control of Mechanical Impedance by Coactivation of Antagonist Muscles. *IEEE Trans. Auto. Cont.* **AC-29**: 681-690.

Hogan, N. (1985a) Impedance Control: An Approach to Manipulation: Part I - Theory, Part II - Implementation, Part III - Applications. *J. Dyn. Syst. Meas. Contr.* **107**: 1-24.

Hogan, N. (1985b) The Mechanics of Multi-Joint Posture and Movement Control. *Biol. Cybern.* **52**: 315-331.

Hogan, N. (1986) Multivariable Mechanics of the Neuromuscular System. *Proceedings of the 8th Annual Conference of the IEEE Engineering in Medicine and Biology Society*, pp. 594-598.

Hogan, N. (1987) Stable Execution of Contact Tasks Using Impedance Control. *Proceedings of the IEEE Conference on Robotics and Automation*, pp. 1047-1054.

Hogan, N. (1988a) On the Stability of Manipulators Performing Contact Tasks. *IEEE Journal of Robotics and Automation* **4**: 677-686.

Hogan, N. (1988b) Planning and Execution of Multi-Joint Movements. *Canadian Journal of Physiology and Pharmacology* **66**: 508-519.

Hongo, T., Jankowska, E. and Lundberg, A. (1969) The rubrospinal trace. II. Facilitation of interneuronal transmission in reflex paths to motoneurones. *Exp. Brain Res.* **7**: 365-391.

Houk, J. C. (1979) Regulation of Stiffness by Skeletomotor Reflexes. *Ann. Rev. Physiol.* **41**: 99-114.

Humphrey, D. R. and D. J. Reed (1983) Separate Cortical Systems for the Control of Joint Movement and Joint Stiffness: Reciprocal Activation and Coactivation of Antagonist Muscles. *Adv. Neurol.* **39**: 347-372.

Hunter, I. W. and R. E. Kearney (1982) Dynamics of Human Ankle Stiffness: Variation with Mean Ankle Torque. *J. Biomech.* **15**: 747-752.

Joyce, G., P. M. H. Rack and H. F. Ross (1974) The Force Generated at the Human Elbow Joint in Response to Imposed Sinusoidal Movements of the Forearm. *J. Physiol.* **240**: 351-374.

Kazerooni, H. (1987) Robust, Non-Linear Impedance Control for Robot Manipulators. *Proceedings of the IEEE Conference on Robotics and Automation*, pp. 741-750.

Lacquaniti, F. and J. F. Soechting (1983) Changes in Mechanical Impedance and Gain of the Myotatic Response During Transitions Between Two Motor Tasks. *Exp. Brain Res. Suppl.* **7**: 135-139.

Lacquaniti, F., F. Licata and J. F. Soechting (1982) The Mechanical Behavior of the Human Forearm in Response to Transient Perturbations. *Biol. Cybern.* **44**: 35-46.

Loeb, G. E. (1984) The Control and Responses of Mammalian Muscle Spindles During Normally Executed Motor Tasks. *Exer. Sp. Sci. Rev.* **12**: 157-204.

Murray, W. R. and Hogan, N. (1988) Co-contraction of antagonist muscles: predictions and observations. *Proceedings of the Annual International Conference of the IEEE Engineering in Medicine and Biology Society*, 10, 1926-1929.

Mussa-Ivaldi, F. A., N. Hogan and E. Bizzi (1985) Neural and Geometric Factors Subserving Arm Posture. *J. Neurosci.* **5**: 2732-2743.

Nichols, T. R. and J. C. Houk (1973) Reflex Compensation for Variations in the Mechanical Properties of Muscle. *Science* **181**: 182-184.

Nichols, T. R. and J. C. Houk (1976) Improvement in Linearity and Regulation of Stiffness that Results from Actions of Stretch Reflex. *J. Neurophysiol.* **39**: 119-142.

Paul, R. P. (1987) Problems and Research Issues Associated with the Hybrid Control of Force and Displacement. *Proceedings of the IEEE Conference on Robotics and Automation*, pp. 1966-1971.

Russell, D. and N. Hogan (1989) How Humans Perform Constrained Motions. In: *Issues in the Modeling and Control of Biomechanical Systems*, J.L. Stein, J. A. Ashton-Miller and M. G. Pandy, eds. ASME, pp. 13-19.

Wyatt, J. L., Chua, L. O., Gannett, J. W., G_knar, I. C. and Green, D. N. (1981) Energy Concepts in the State-Space Theory of Nonlinear n-Ports: Part I — Passivity. *IEEE Transactions on Circuits and Systems* **CAS-28**: 48-61.

Zahalak, G. I. and S. J. Heyman (1979) A Quantitative Evaluation of the Frequency Response Characteristics of Active Human Skeletal Muscle in vivo. *J. Biomech. Eng.* **101**: 28-39.

CHAPTER 10

Linking Musculoskeletal Mechanics to Sensorimotor Neurophysiology

Gerald E. Loeb and William S. Levine

10.1 Introduction

The general topic of motor systems and their control can be considered at three hierarchical levels: *i)* mechanical components and their properties; *ii)* control algorithms and their implementation in neural circuits; *iii)* general control strategies and emergent behavior of systems. Most of the preceding chapters are concerned with the first level; most of the subsequent chapters deal with the third level. This chapter considers the intermediate level, which to date has been largely the province of neurophysiologists.

There are two traditional approaches to understanding a control circuit by experimental probing of its connections and activity. The first involves collecting such data phenomenologically in the hope that the design principles are self-evident from the structure and function. The second involves intuiting (or more often borrowing) a control principle, devising a hypothesis regarding the expected response to a specific perturbation, and then confirming or rejecting the hypothesis experimentally. For sensorimotor systems as complex as whole limbs, both approaches encounter serious problems. The neural circuits now appear to be sufficiently complex that, by judicious selection, components can be found that appear to suggest, support, or reject almost any reasonable hypothesis of sensorimotor control. Furthermore, the diversity of architectural features and control problems to be found among the muscles of even a single limb during a single behavioral task suggests that control requires several different, parallel algorithms which must be coordinated among themselves (for review, see Loeb, 1987).

Our work is directed toward developing more objective and less intuitive processes for determining how the neural circuitry mediates between the mechanical properties of the musculoskeletal system and the performance goals of the whole organism. At the simplest level (Section 10.2), this involves facilitating the combination and presentation of diverse experimental data through model-based analyses and simulations and graphical displays. While such analyses are beginning to be common in human kinesiology, the very different technical demands of neurophysiological experiments on chronically instrumented, naturally behaving animals have posed a logistical barrier to combined studies. More ambitiously (Section 10.4), we are using engineering tools for optimal control to make formal predictions regarding the details of the control circuitry and exploring how changes in the mechanical properties or performance criteria of model systems are reflected in this predicted circuitry. In between (Section 10.3), we have discovered that the process of formalizing the properties of a model system is an exacting master that highlights deficiencies and inconsistencies in existing data, thus providing a demand and a fertile source of direction for new experiments on individual components of the system.

This report surveys our ongoing attempts to assemble such a modeling environment. Most of the component model relationships have been selected from prior efforts elsewhere on the basis of their supposed ability to capture those aspects of biomechanics and behavior that are most salient for neural control (for review, see Loeb et al., 1989, and other chapters herein). Our experimental work has concentrated on testing the validity

Multiple Muscle Systems: Biomechanics and Movement Organization
J.M. Winters and S.L-Y. Woo (eds.), © 1990 Springer-Verlag

and saliency of these models for the physiological range of natural, voluntary behaviors, often with surprising results requiring modification of the models and/or development of new experimental methodologies.

10.1.1 Heterogeneity of the Data

Sensorimotor neurophysiology traditionally has been organized around motor pools, the individual muscles or compartments of muscles that can be selectively activated during the performance of a behavioral task. Any hypothesis of control at this level must include a separate representation for each such entity. Thus, in addition to the traditional kinesiological data (e.g. joint angles, ground reaction forces) and limb parameters (e.g. mass, center of mass, rotational inertia), the models must include the parameters governing the dynamics of individual muscles and tendons, the sensitivity of their proprioceptors, and neurokinesiological data regarding their normal level of activation. For the hindlimb model described below, there are about 45 separate motor pools; it is not possible to obtain complete kinesiological, neurophysiological and morphometric data from any single animal. Thus, one of the main requirements of the modeling environment is that it facilitate the orderly collection, scaling, and combination of different types of data from different structures in different specimens and different trials of a range of behavioral activities.

10.1.2 Current State of the Cat Hindlimb Model

The first job of a systems modeler is to validate each of the modeled relationships by determining that its critical parameters can be measured with adequate accuracy vis-a-vis the sensitivity of the proposed relationships to uncertainties in those parameters. We have chosen locomotion in the cat hindlimb as our first model system because it has sufficient complexity and diversity to test these relationships thoroughly while providing a wealth of preexisting morphometric, neurophysiological, biomechanical and kinesiological data and well-developed methodologies for obtaining more such data. We believe that experience with this system will lead eventually to simpler and adequate procedures for modeling less accessible systems such as human patients with neuromuscular disorders.

It is possible to create a series of models of the cat hindlimb system with gradually increasing complexity. Each successive model requires additional data and computation while revealing how its additional terms contribute to the realism of the model behavior. Our simplest model (Figure 10.1) consists of four skeletal segments connected by three hinge joints moving only in the parasagittal plane under the control of 10 muscle groups. The model can be complexified by including the motion of the phalanges (obtainable by cinefluoroscopy) and adding and separating out the muscles that control the toe joints, most of which are now grouped together on the basis of their actions at the ankle joint. The model can also be complexified by adding a second degree of freedom at the hip to permit abduction/adduction. This would require separation of the hip extensors and flexors into subgroups based on the component of their moment arms normal to the sagittal plane; it would greatly improve the ability to evaluate the realism of different control strategies (see Section 10.4.5) for responding to displacements of the foot applied during standing, a well-studied paradigm (Macpherson, 1988a & b). A more incremental form of complexity can be introduced by dividing some muscle groups into their component motor pools (Section 10.3.2), thereby providing an environment in which to study specializations of muscle architecture and compartmentalized recruitment.

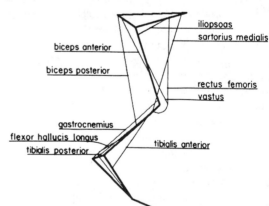

Figure 10.1: Computer-generated representation of a simplified model of the cat hindlimb consisting of four segments moving in the parasagittal plane (foot, shank and thigh articulated at hinge–joints; pelvis trajectory driven as a boundary condition) plus 10 muscle groups, labeled by name of one typical member of that group.

10.2 A Framework for Neuromusculoskeletal Modeling

10.2.1 Interactive Relational Database

The design goal is not a specific model but a work-space environment in which the scientist can choose freely the particular model relationships to be explored. Such an environment is similar to a computerized spread-sheet, in which input data are gradually combined and transformed into higher-order entities. However, the data heterogeneity problem described above (Section 10.1.1) creates unusually severe problems for assuring compatibility and traceability of data pooled from different experimental sources. Furthermore, this system must be accessible to neurophysiologists with limited expertise in mechanics and computer programming, adding a large requirement for intuitively clear, user-friendly interactive interfaces.

We have elected to develop a custom relational database on the Macintosh II personal workstation. This platform offers an acceptable compromise among computing power, cost, general availability, programmability and capabilities for real-time kinesiological data collection. Our database is designed like a spreadsheet, each of whose "cells" occupy a third dimension in which each time- varying state variable is represented by a standard 200 datapoint vector. Individual parameters (e.g. morphometric data, mechanical constants) are collected into sub-sheets that can be pointed to and nested so that the entire database of such data to date are available to the user and traceable to the original experimental source. The names of all parameters and state-variables are proscribed by a syntax generator and parser according to a standard nomenclature. This makes it possible to read the structure of the model directly from the equations that define each cell and to include high-level functions that automatically generate calls for the cells that are required to compute such functions (e.g. the intersegmental angle at joint n requires the x and y coordinates of joints n, $n-1$ and $n+1$).

10.2.2 Icon-Based Graphics

Because of the heterogeneity and complexity of kinematic data, it can be very difficult to appreciate important relationships by visual inspection of the usual multitrace graphs of various state-variables, particularly for users not well-versed in biomechanical terms and units. This is particularly true when dealing with different groups of uni- and multiarticular muscles and their direct and indirect effects on accelerating, coupled skeletal segments. For this reason, we use quantitative icons in which state-variables are depicted as mnemonic symbols arranged on sequential stick-figure depictions of the limb position (see Figure 10.2). For example, each joint torque is shown as a sector of a circle centered on the joint, with a

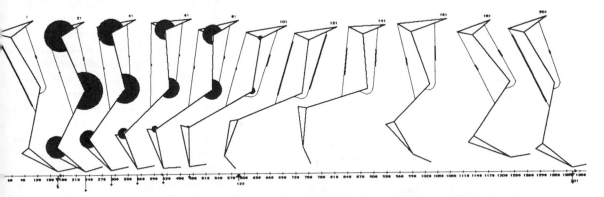

Figure 10.2: Computer-generated iconographic presentation of one cycle of slow walking by the cat hindlimb (modified from usual multicolor format). Stick figures taken at intervals of 20 phase units out of 200 for the cycle; ground force vectors (small arrows) and phase of footfall and footlift (large down and up arrows) shown along bottom scale. Typical *EMG* activity for two muscles, posterior biceps and anterior sartorius, is shown as the variable length of the thickened rectangle on the icons depicting their path from origin to insertion. Net torque at ankle, knee and hip joints attributable to the work of muscles (active and passive) is shown as sectors of a circle at those joints, with diameter and orientation corresponding to magnitude and sign (torques during swing are too small to be visible at this scale).

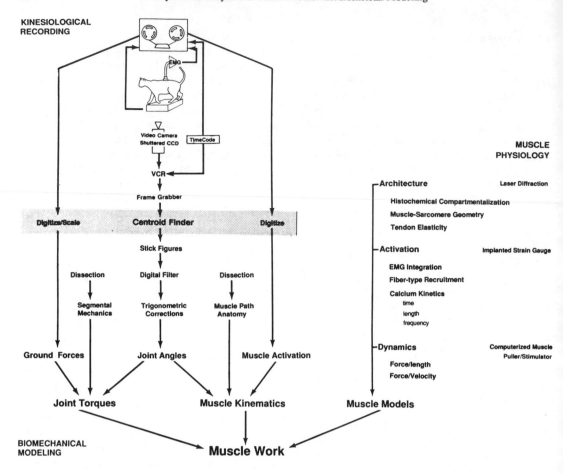

Figure 10.3: The data management scheme for the modeling project has been divided into three separate tasks: *a)* kinesiological recording to obtain state variables for normal behavior; *b)* biomechanical modeling whereby kinesiological data are transformed into the muscle kinematics and joint torques; and *c)* muscle physiology models whereby the *EMG* recordings are transformed into estimates of the time course of force generation. (Modified from Loeb et al., 1989).

diameter and direction indicating magnitude and sign of the torque. Variables related to a single muscle are depicted by a rectangle lying on a line showing the path of the muscle from origin to insertion; its length and width can be selected to represent kinematic terms such as muscle length or velocity and/or dynamic terms such as *EMG* activation or force output. Ground reaction forces are shown by appropriately oriented and sized arrows; translational accelerations can be shown similarly. The syntactically defined name for each state-variable is linked to its appropriate icon-type; its display can be selected with one of a range of distinguishing parameters such as color.

10.2.3 Automated Equation-of-Motion Writing

So far, we have generated Newtonian equations-of-motion by hand. In addition to being tedious and prone to error, this fixes the structure of the model in the database. The relational database approach offers the opportunity for the same software system to depict any mechanical system depending simply on the parameters describing the limb segments and linkages. At present, a two-dimensional model such as the cat hindlimb can be converted into any planar movement system with the same number of joints simply by changing the morphometric and mechanical parameters describing the skeletal segments and muscles, but it cannot be converted into a topologically different linkage. Chapters 8 (Zajac and Winters) and 43 (Yamaguchi) describes current approaches to the formidable algebra for general problems in dynamics, including emerging, promising equation–writing software. These will need to be coupled with a general notation for representing the nature of mechanical linkages, perhaps by building them up iconographically.

10.3 Biological Issues

A model is an approximation. Any model can be shown to contain errors simply by increasing the resolution or range of the experimental data. One can then respond by increasing the complexity of the model, but there must be some objective end–point to this infinitely recursive process. Our component models are being used in a model system with a definite range of behavior and resolution of measurement, namely locomotion in the cat hindlimb. Thus, our strategy is to pick models that capture adequately the behavior of these components for the range of conditions under which they operate during such behaviors. Tactically, this has meant starting with models from the literature and performing many experiments in acute and chronic animals to compare biological behavior with model predictions for the particular range of parameters normally experienced by each component. We have discovered that most literature models are actually summaries of experimental data obtained under very limited and often quite non-physiological conditions. This is because the models were often constructed to make inferences about underlying structure and function which were better revealed under such unusual conditions. For example, the cross-bridge mechanism in muscles is often studied with very short-range stretches and isotonic single twitches or tetanic activation, whereas muscle fibers usually operate over large movement ranges at intermediate frequencies of activation. Such models often exhibit three forms of inadequacy that must be corrected before using them for our purposes: *i)* they are undefined, unstable or incorrect when extrapolated to the pertinent range of behavior; *ii)* they include parameters that are difficult to measure with sufficient accuracy; and *iii)* they include unnecessary terms that complicate the mathematics (e.g. noninvertible relationships) but do not produce concomitant improvements in accuracy vis-a-vis the experimental data.

10.3.1 Musculoskeletal Scaling

In order to combine data from many subjects and trials, it is useful to convert these data into dimensionless units. The identification of such units is also a form of modeling in which testable assumptions are embedded about the range and significance of differences between subjects and trials, some of which are described below.

Morphometric

We have adopted the general strategy of assuming some degree of congruency in the form of our subjects, which are selected to be normal, young–adult domestic cats (*Felis domesticus*). Data that are related to individual length measurements are readily described as a fraction of those lengths; e.g. the origin of a muscle on a skeletal segment can be measured as the fraction of the length of that segment from the joint across which it operates. Lineal dimensions that relate to the limb as a whole (e.g. joint positions in space and stride length) can be normalized to an arbitrary standard; we use the sum of the lengths of the principle segments, namely foot plus shank plus thigh. Cubic terms such as segment mass and muscle volume are normalized directly by the total mass of the animal. However, this leaves an ambiguity regarding intermediate terms such as cross-sectional areas, which could be normalized to the square of the normalizing limb–length or to the resultant of total mass divided by limb–length. At the moment, we prefer the latter, but validation awaits compilation of sufficient morphometric data from our experimental animals to test the nature of their presumed congruency.

Dynamic

The fixed, 200-bin time-dimension of our relational database establishes a scaling of time in dimensionless phase units for a cyclical behavioral activity such as walking. The database contains one complete cycle of the behavior, which can be wrapped around during read-out to visualize better the phases near the beginning and end of the cycle; for inverse dynamical calculations, these ends must wrap smoothly to permit the calculation of the required time derivatives.

This approach requires a somewhat arbitrary division of the range of locomotor gaits into individual behaviors, each of which constitutes a separate kinesiological database into which data from many samples of such behavior are pooled. Fortunately, there is strong evidence that gait in most quadrupeds is discretely multimodal, with similar preferred-locomotor-speeds for each gait across similarly sized animals (Hildebrand, 1985). Some non-cyclical behaviors such as jumping also appear to be stereotypable (Zajac et al., 1981) and can be accommodated in this scheme by arbitrarily defining their phases to start and stop with similar, stable preparatory and terminal postures such as the sitting crouch.

The identification of the phase boundaries used for separating and combining individual cycles requires some attention. The classical phase boundaries for locomotion defined by Phillipson (1905) include two transitions that are abrupt and easily identified mechanically and kinesiologically (footlift and footfall), plus two that are smooth and continuous (mid-stance and mid-swing). The abrupt transitions are easily identified in the experimental data, so they make the best markers for normalizing small fluctuations in the relative duration of each phase of the behavioral cycle. However, their very abruptness makes them unsuitable for defining the beginning and end of the cycle because small uncertainties in timing translate into large discontinuities in many of the state-variables that must wrap smoothly around the cycle boundary. Therefore, we have picked the mid-stance transition (denoted E2-3 by Phillipson) as the cycle boundary, a phase when the limb is a stable support with the mass of the pelvic girdle centered above the support point, the ground-reaction force has a zero longitudinal component, and muscle recruitment is relatively constant; these conditions also characterize pre– and post–conditions for noncyclical behaviors such as jumping (Zajac et al., 1981).

In addition to the length, mass and time (phase) units described above, there are terms related to dynamic properties of musculotendinous components such as activation, elasticity, and viscosity that appear to be considered usefully in terms of frequency response of these components. See Zajac (1989) for a discussion of this approach.

10.3.2 Inverse Dynamics

The kinesiological and computational approaches to inverse dynamic analyses are well developed in human studies (e.g. Bernstein, 1935, 1940) and described in some detail elsewhere in this volume [e.g. Chapter 34 (Hinrichs)]. The application of such analyses to animals, particularly in conjunction with neurophysiological studies, is somewhat more recent (e.g. Hoy and Zernicke, 1985) and poses some special methodological problems that merit brief description here.

The collection of kinesiological data on limb position in animals differs from humans in several ways:

i) *Speed.* The standard 50 or 60 field/sec scan rate of video information provides sufficient temporal resolution for most forms of human locomotion, but it is only marginally adequate for the slower gaits of small animals. Optimal digital filtering tends to be limited by the amount of spatial noise in the original signal, which is also considerable unless centroids are determined mathematically rather than placed by eye at the closest pixel. Video systems with 200 fields/sec capable of on–line multicentroid extraction in *3-D* are just starting to become available and seem likely eventually to replace cumbersome high-speed cinephotography, but only when high contrast markers are feasible.

ii) *Skin slippage.* The amount of skin slippage relative to skeletal movement is higher in many animals, particularly at proximal joints. The severe problem at the knee in cats is typically resolved by trigonometric calculation based on ankle and hip position and thigh and shank lengths, but this puts a larger burden on ankle and hip location which may break down when studying movements that include components out of the sagittal plane. Cinefluoroscopy may be required, at least to validate procedures for estimating skeletal position from skin position; it is certainly required to resolve the position of small bones such as phalanges and vertebrae.

iii) *Gadget tolerance.* Most animals have little tolerance for anything attached to their skin, particularly if it produces noticeable drag and provides good footholds for claws and teeth. This virtually eliminates external sensors and emitters such as goniometers and light-emitting-diodes (*LEDs*), leaving only passively reflective markers which usually must be reapplied daily.

iv) *Concurrent physiological studies.* Animal kinesiology often includes other measurements at the same time, such as *EMG* and nerve recording and electrical stimulation. Mutual interference can take many forms, including electrical cross-talk (e.g. high-current pulses in strobed lights or *LEDs*) and optical confusion from randomly reflective enclosures and dangling cables.

10.3.3 Forward Dynamics

Several methods have been developed for resolving the net joint-torques calculated through inverse dynamic analysis into the work of individual muscles and muscle groups. These divide generally into optimization schemes and forward dynamic models [see also Chapter 8 (Zajac and Winters)]. Optimization has the advantage that it requires no additional kinesiological data. Instead, assumptions are made regarding the "least expensive" ways in which activation would be distributed to the available, overcomplete musculature to achieve the required net torques (e.g. Pierrynowski and Morrison, 1985). These assumptions should be driven by large amounts of data regarding the morphometrics and energetics of the different muscles and muscle-fiber distributions to be found in the limb; in practice, the models are often quite simple and based upon fragmentary data from other studies. Furthermore, the notion that the musculature is overcomplete is based on the presumption that the only task of the muscles is to produce the nominal torques necessary for normal, unperturbed locomotion. As analyzed in Chapter 9 (Hogan), the action of muscles produces a trajectory of limb impedance as well as position, and that impedance may be structured to produce an initial stability in the face of a range of perturbations that have been commonly experienced during a task such as locomotion. Thus, in the use of optimization to determine muscle work, there is embedded a large assumption about the nature of control, which is incompatible with our goal of using models to understand such control.

Forward dynamic models use a direct neurokinesiological measurement of muscle activation such as *EMG* to estimate the output of force by each muscle or muscle group (see Figure 10.4). In humans, this technique is limited by the poor quality of *EMG* data that can be gathered clinically. Skin surface electrodes are subject to crosstalk and cannot sample small or deep muscles; intramuscular needles are difficult to place reliably and tend to sample small, perhaps unrepresentative samples with unpredictable sensitivity. However, laboratory animals can be implanted with large arrays of carefully designed *EMG* electrodes that provide a reproducible picture of the relative activation of virtually all muscles (Loeb and Gans, 1986). Furthermore, despite the many complex factors that enter into the generation of the interference electromyogram, it appears that such recordings are readily processed into a remarkably accurate representation of the overall command signal to the pool of motoneurons (Hoffer et al., 1987), an important consideration when such models are subjected to perturbations to evaluate the "reflexes" produced by their model controllers. There are three general issues in implementing forward-dynamic models to predict muscle work.

Muscle Architecture

Perhaps the biggest surprise in the process of implementing our models of the cat hindlimb has been the inadequacy of the voluminous literature over the past century on functional anatomy. Not only are the necessary data often missing or incorrect, but the whole scope of muscle design variations appears to have been grossly underestimated in ways that have important and largely unexplored mechanical implications.

The largest problem involves the parallel-fibered muscles that seem, misleadingly, to have the simplest mechanics. Most of the architectural and theoretical studies have concentrated on the interesting but largely straight forward properties of pinnate-fibered muscles (Gans, 1982). Recently, there have been several important advances in our understanding of the details of contractile movements and forces within pinnate muscles, particularly regarding hetereogeneities in fiber lengths produced by curvature and elasticity of the aponeuroses into which they insert (Otten, 1988). However, from the perspective of modeling the work of such muscles as a whole, these improvements probably constitute refinements too small to warrant their computational complexity.

The problem in parallel-fibered muscles is the belated rediscovery of the prevalence of architectures in which the force-transmitting chain from boney origin to boney insertion is composed of more than one independently controllable entity. It is well known that many axial muscles are composed of a series of anatomically distinct compartments separated by tendinous inscriptions and innervated by separate nerves, often arising from motoneurons in separate spinal segments. However, until recently, it was generally assumed that in the cat hindlimb, only the semitendinosus muscle presented such an in-series muscle architecture and both of its sets of motoneurons were known to be intermingled in a single,

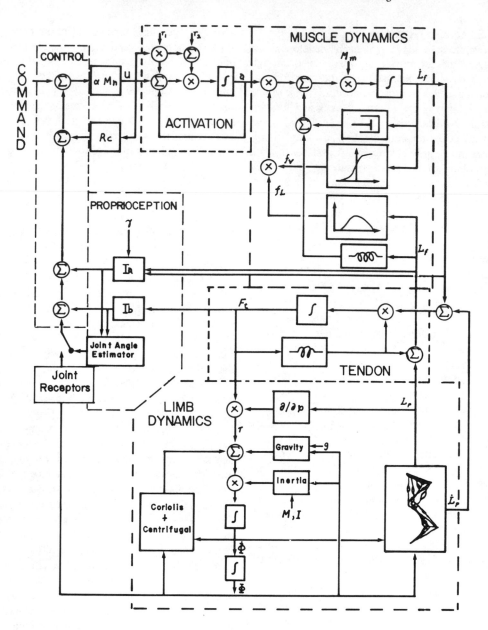

Figure 10.4: A block diagram of the complete cat hindlimb model arranged for forward (as opposed to inverse) dynamics, driven by alpha motoneurons (Mn) arranged into 10-45 recruitment groups, depending on the complexity of the model being used. The output of each motor pool is converted into muscle activation (which uses separate time constants for rise and fall time). Because the tendon is elastic and the force output of the muscle fibers is velocity dependent, the relationship between the muscle dynamics and tendon properties is recursive, with the force of the muscle calculated as an output of the tendon on the basis of two competing input velocities at its ends: L_f from the motion of the muscle fibers, and L_p from the motion of

limb segments defining the overall path length from origin to insertion. The limb dynamics accept muscle forces as inputs and produce limb motion as an output (ground reaction forces and pelvic girdle trajectory are boundary conditions acting as constraints on the motion). Proprioception uses muscle spindle primary afferents (*Ia*) driven by muscle fiber length and velocity and Golgi tendon organs (*Ib*) driven by tendon force; joint angle may be obtained from explicit receptors or estimated from *Ia* and *Ib* signals. The proprioceptive signals plus Renshaw feedback from motoneuron collaterals (RC) are distributed as feedback through the control matrix.

homogeneous motor nucleus in the lumbosacral spinal cord (Letbetter and English, 1981). It now appears that the fascicles of most, if not all, of the long, parallel-fibered muscles of the hindlimb are composed of relatively short muscle-fibers in an overlapped, series distribution (Loeb et al., 1987). The ends of these fibers are long tapers, devoid of myotendinous attachments, which appear to transfer tension by viscous shear into a surrounding matrix of randomly oriented filaments of collagen (Trotter, in press) muscles with long fascicles, such an arrangement may be necessitated by mechanical instabilities that would arise in individual muscle fibers attempting to span this distance because of the relatively slow conduction of action potentials along the sarcolemma relative to the rise-time of mechanical force. However, it is also seen within the relatively short single compartments of some inscripted axial muscles (Richmond et al., 1985).

At least in some muscles, there is evidence that many of the motoneurons branch to innervate long, narrow muscle territories that extend over the length of the fascicles (Loeb et al., 1987). This arrangement would assure a similar tension-generating capability for each motor unit over the length of the muscle, although the mechanism of coupling the tension over a sparse distribution of asynchronously active fibers has not been worked out and may involve considerable elastic and/or viscous compliance that would affect force–length–velocity relationships. Unfortunately, it now appears that many motor units in some such muscles may have motor territories that are highly asymmetrically distributed over the length of the muscle (Thomson et al., 1990). This poses physiological control problems for motor unit recruitment and reflex organization (see below) as well as methodological problems in the design of experiments to study the mechanics of such muscles or to stimulate them for clinical neural prosthetics.

Motor Pools and Groups

The recent interest in compartmentalization of muscles and motor pools raises difficult questions regarding the number of separate, component entities to be included in a model of the limb musculature (for review, see Loeb, 1989). It is not enough to take a list of muscles from an atlas of gross anatomy. Some of these muscles have obvious anatomical compartments with separate nerve branches and fascial boundaries that may be recruited quite homogeneously during some behaviors and heterogeneously during others (Chanaud et al., in press). There may be two or more anatomically continuous regions of the muscle that are separately recruited or a single region in which different populations of motor units are intertwined but recruited in different phases of the behavior or under different kinematic conditions (Hoffer et al., 1987b). The EMG in one part of a muscle may appear to be quite different from that in another part because of a heterogeneous distribution of motor unit types, but the whole muscle may actually consist of a single, homogeneously recruited motor pool (Chanaud et al., in press). Decisions on how to represent the musculature require careful and thorough anatomical analyses interpreted in the context of EMG data that may need to include comparison of multiple, simultaneous records from different regions of the muscle during a variety of voluntary and reflexively evoked behaviors. These decisions are often further complicated by the desire for simplified models in which muscles with some degree of differences are pooled into groups of actuators (e.g. Figure 10.1).

EMG-to-Force Prediction

The development of models of the force-generating process in muscle is one of the most active areas of biomechanical research, with various aspects covered in some detail in the preceeding chapters. As noted earlier, our interest in this stems from the need to resolve the net joint torques derived from inverse dynamic analyses into the work of individual muscles. Our approach is to use the gross EMG signal, recorded by bipolar, epimysial "patch" electrodes implanted in the fascial planes between muscles (Loeb and Gans, 1986) as a measure of relative sarcomere activation and to combine this with length and velocity data to predict the time-course of force output by the muscle as a whole. The absolute force output can be determined by calculating a weighting coefficient for each muscle using an error-minimizing procedure; the different time-courses of force predicted for each muscle then must add up to the net joint-torque over the complete behavioral task. In fact, it is possible to make fairly strong inferences about the absolute forces on the basis of the muscles' physiological cross-section areas and the absolute amplitudes of the *EMG*s, which empirically tend toward similar maxima of 10 mV when recorded with these highly standardized patch electrodes (5 mm contact spacing) and integrated into 2-ms bins (Bak and Loeb, 1979). This is not surprising, because the voltage amplitude depends essentially on the mean current density in the tissue, which in turn depends primarily on the percentage of the cross-sectional area of the muscle that is active, given similar ranges of muscle fiber diameters and asynchronous firing frequencies.

Identifying the form and parameter values for models of muscle dynamics is less straightforward. Chapters 1 and 5-8 deal with the competing demands of simplicity and accuracy in defining the minimum necessary complexity for a model, which depends on both the particular muscle and the particular behavior. Where possible, we have adopted an empirical approach to the specification and validation of models, subjecting individual muscles to electrical stimulation of single units and whole nerves at physiological firing rates while directly measuring force output during behaviorally relevant length and velocity excursions, particularly including active-lengthening (eccentric work; see Chapters 1 and 3) which has been quite neglected by muscle physiologists. For a number of poorly understood reasons, it seems that force output is highly dependent on recent history of movement and activation (Heckman et al., 1988); therefore, we have eschewed "general" techniques for system identification such as white-noise analysis in favor of more focused parametric studies.

Again where possible, these parameters are tested empirically by comparing the predicted force-output of the models to the directly measured forces in naturally behaving animals using strain-gauges chronically implanted on tendons (Weytjens and Loeb, 1987). These data are invaluable in assessing the sensitivity of models to changes in their form and parameters. For pinnate muscles such as the triceps surae, theory suggests (Zajac, 1989) and experiments confirm (Weytjens and Loeb, 1987) the importance of dynamic terms such as the slope of the force–velocity curve and the elasticity of the tendons and aponeurotic sheets, which together may comprise 80% or more of the length of the path from boney origin to boney insertion (see also Chapter 4 (Ettema and Huijing); Chapter 38 (Alexander and Ker)). Figure 10.4 shows schematically the general structure and computational form of a musculotendon model that seems generally adequate.

One troublesome problem with some formulations of the dynamic models of muscle has been stability. When the muscle goes from passive to active, the contractile element begins to shorten as it takes up slack in the series elastic element, which reduces contractile force output according to the force–velocity relationship. However, in the next iteration of the model, the reduced force is suddenly exposed to the larger elastic force stored in the series elasticity on the previous pass, which begins to lengthen the contractile element. The now-eccentric velocity produces an abrupt increase in force output of the contractile element, which pulls even more strongly on the slackened elastic element in the next pass. Regardless of the temporal resolution of the simulation, the situation is fundamentally unstable and produces large, unphysiological oscillations in the force output. This has led some investigators to restrict the minimal activation of the muscle force to some small, non-zero value to maintain tension on the series elastic element (Zajac, 1989); however, this is unacceptable for many cat hindlimb muscles which are used intermittently at low levels of recruitment during slow walking. We have found that the models become better behaved when they are made more realistic, by adding a viscous term in the tendon compliance that is known to exist (Woo, 1982; Hubbard and Chun, 1988; see also Chapter 5 (Winters)) and/or by adding in the mass of the muscle belly, which provides an inertial damping term.

Another issue in stability concerns muscles with multiple contractile elements in series, as described above. Even within a single muscle fiber, the series arrangement of sarcomeres can give rise to transient heterogeneity in sarcomere lengths as the wave of activation propagates from the neuromuscular end-plate near the center to the ends of the muscle fiber [see Chapter 3 (Morgan)]. The problem of distributing precisely equal amounts of tension along the lengths of muscles with separate or partially separate motor unit territories would seem to be insurmountable. In fact, the very steep (actually discontinuous) slope of the force–velocity relationship of active muscle may solve this problem quite neatly (albeit at the expense of introducing other control problems for voluntary transitions between concentric and eccentric work). If an over-excited part of the muscle begins to contract at the expense of an under-excited region in series, the shortening elements immediately decrease their force output while the lengthening elements become locked isometrically until the external force exceeds the breaking force for their cross-bridges. The backwards breaking force appears to be about 1.8 times their isometric force output (Joyce et al., 1969). The two effects produce about a 50% margin of error in allowable imbalance of activation, at least for the relatively brief contractions of locomotor and many other behaviors.

10.4 Control Theory

Even at a subjective level, an appreciation of the intrinsic mechanical properties of muscles and tendons and of the mechanical interactions of the limb segments during natural movements has provided a great deal of insight into the control problems that are presumably being addressed by sensory feedback and segmental reflexes. However, it now appears that it may be possible to formalize these relationships to test specific hypotheses about the organization of sensorimotor control.

As detailed elsewhere (Loeb et al., 1989; He et al., 1989), we have begun to apply a class of general engineering tools that have been used in the design of "optimal controllers" for complex systems. In the typical industrial application, a complex "plant" (e.g. roboticized assembly line, oil refinery, etc.) is equipped with a large number of sensors and actuators related to various aspects of the process. For particular performance criteria, an optimal configuration of feedback can be calculated, whereby signals from the sensors are weighted and combined into commands to the actuators (see Figures 8.4 and 8.5). Thus, three things are required as inputs:

i) a complete mathematical model of the plant, including the response of the actuators to command and feedback signals, as discussed above;

ii) models of sensors for all "state-variables" of the mathematical functions describing the plant or estimators for those state-variables built up from available sensors;

iii) performance criteria, which are often a weighted mixture of the "cost" of using the actuators to correct perturbations and the "penalty" for allowing each selected state-variable to stray from its nominal, desired value.

10.4.1 Models of Proprioceptors

We are starting with a simplified set of those mechanoreceptors that are generally believed to provide the most important feedback for segmental control: Golgi tendon organs (muscle force sensors), muscle spindle primaries (muscle length and velocity sensors), and joint receptors (joint angle and velocity sensors). Of these, only the first (*GTO*s) have a well-accepted mathematical representation (Crago et al., 1982). For spindles, we are using a highly simplified model of fusimotor control based on our previous recordings (Loeb and Duysens, 1979; Loeb et al., 1985; Loeb and Hoffer, 1985) and a signal–theory analysis (Loeb and Marks, 1985) of single primary afferents from various muscles during natural behavior. This model assumes that the static and dynamic components of the fusimotor system are adjusted to modulate the spindle output over its full dynamic range during one cycle of the behavior. The velocity and length terms are additive and each is scaled so that the range of its state–variable encountered during the locomotor cycle accounts for 50% of the dynamic range. More sophisticated models based on a more complete database are being developed (Prochazka et al., in press); these will be tested against ours to determine if they are significantly different in terms of their effects on overall control of the limb musculature.

Regarding joint afferents, there is a long and contentious literature on the nature and accuracy of their proprioceptive function (for review, see Ferrell et al., 1987). Optimal control theory mandates either a sensor or an estimator for joint angles and velocities, which are state-variables of the equations for limb dynamics. Therefore, we are comparing models in which this function is performed explicitly by joint sensors versus models in which this information is derived from a weighted admixture of muscle spindle and GTO signals (the latter are required to correct for the not-insignificant effects of tendon and aponeurotic sheet elasticity; Rack and Westbury, 1984).

One particularly interesting side-issue has arisen from the formalized definition of state–variables and their use in optimal controllers. Each mathematical function used in the model of the plant has inputs that depend on the time-varying state of the system; hence, these are state-variables. As noted above, there must be a signal available from a sensor or an estimator for each state-variable for use in the feedback matrix of the controller. The force output of muscle is related to an intramuscular "activation" that is a time-dependent function of motoneuronal commands involving factors such as the rise and fall of sarcoplasmic calcium. Thus, muscle activation is a state-variable that must be sensed or estimated. Curiously, this provides a formal requirement for an efference copy of the motor pool output that seems to be entirely compatible with the known properties of the Renshaw cells, which receive input from recurrent collaterals of alpha motoneurons. The available data on Renshaw cells suggests that the relative weighting of their input from various motor units is consistent with an estimation of overall muscle activation (Hultborn et al., 1988); even their temporal integration properties are not inconsistent with estimating the rise and fall of activation (Windhorst and Koehler, 1983; Windhorst et al., 1988), but this requires more specific experimentation.

10.4.2 Optimal Control

We are using optimal control as a tool to identify salient features of a complex system, much as one would use linear regression analysis, to which it is related. We use the linear quadratic method (LQ; Stein and Athans, 1987). It is linear because for small perturbations, the system to be controlled can be assumed to be linear; it is quadratic because the performance measure minimizes the sum of the squares of the perturbations in all of the state variables and inputs. The combination of a linear plant plus quadratic criteria implies linear state feedback control (Athans and Falb, 1969).

Like linear regression analysis, LQ design makes no assumptions about the sign or magnitude of the optimal relationship that is "predicted" between each sensor and each actuator. For example, the spindle afferents could inhibit the homonymous motoneurons if that contributed to stability of the overall system. In fact, the models consistently predict the experimentally observed homonymous pattern of positive feedback from spindles and negative feedback from GTOs. Like linear regression analysis, there are no presumptions that the neural circuits that embody the feedback network are individually linear or that they have a one-to-one correspondence with the terms of the feedback matrix. The calculated

feedback matrix is merely a summary of the net trends that should emerge from the spinal circuitry if it is attempting to maintain stability optimally in the face of small perturbations. Departures from this feedback that are observed experimentally can be evaluated in simulations to determine their effects on regulation.

LQ controllers are useful only for perturbations small enough that linear combinations of sensor signals can deal usefully with the underlying nonlinearities of the mechanical system. Larger perturbations (e.g. Forssberg's, 1979, stumbling corrective reaction) must be handled through special circuits capable of triggered, sequenced programs.

The LQ controller does not obviate the need for a central pattern generator (CPG), which still must generate the nominal sequence of states that the system is trying to achieve. In fact, the changing mechanical conditions of the limb (e.g. from stance to swing phases) suggest that the CPG must switch between two or more LQ control schemes (see Figure 10.5). The LQ feedback matrix also operates independently of sensory pathways that reset or otherwise modulate the CPG (for review, see Rossignol et al., 1988). Adopting the convention of Bryson and Ho (1975), the feedback matrix constitutes a "regulator" for maintaining the sequence of desired states, whose trajectory is determined by a separate "controller". The controller is the CPG, whose open-loop output has been shaped by a lifetime of motor learning and adaptation plus millions of years of evolutionary testing. Given appropriate performance criteria (no simple matter), its pattern also could be predicted using optimal control [e.g. Chapter 42 (Pandy); Chapter 19 (Winters and Seif-Naraghi)], but it seems prudent for now to accept the mean observed behavior of unperturbed locomotion as the empirical description of the controller and to concentrate on the narrower design problem of the regulator.

10.4.3 Performance Criteria and Feedback Matrices

In a model with 10 muscle groups and 3 joints, the multiple-input-to-multiple-output feedback matrix K has 36 rows (inputs from spindle primaries, Golgi tendon organs and Renshaw cells for each of 10 muscle groups plus joint angle and velocity sensors or estimators for each of 3 joints)

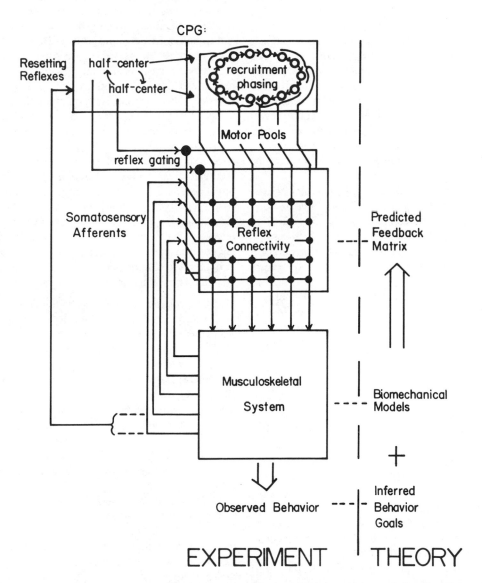

Figure 10.5: Because the mechanical conditions of the limb change drastically during the different phases of locomotion, the feedback provided through the control matrix must be changed by gating in different patterns of connectivity. This is presumably under the control of a central pattern generator (*CPG*), here shown as a system that has a few discrete states (e.g. two half-centers) plus a network for producing the more complexly phased patterns of recruitment in individual motor pools. The output of the motor pools thus reflects the sum of their *CPG* commands and their reflexive drive from somatosensory input through the control matrix, producing the observed movement after being transformed by the mechanics of the musculo-skeletal system. The resulting movements and forces are sensed by the afferents, some of which seem capable of resetting the *CPG* in the event that certain critical criteria indicate that it is not safe to proceed. The blocks at left represent anatomical structures and physiological processes that can be observed experimentally only through their outputs; the goal of the theoretician is to make inferences about the contents of those blocks, as shown at right.

by 10 columns (outputs to motoneuron pools for each of 10 muscle groups). All of the inputs represent state variables whose deviation from nominal trajectory the animal might wish to minimize. We have grouped them into two matrices representing performance criteria. The proprioceptive signals (muscle length and force sensors plus joint signals) are grouped into the Q matrix, where the weighting coefficients represent the relative importance of stabilizing that particular measure of limb trajectory. The muscle activation levels (Renshaw cells) are grouped into the R matrix, where the weighting coefficients represent the relative cost of using a muscle to restore the desired limb trajectory.

In order to calculate K, we must select particular values for R and Q. However, it is more instructive (albeit computationally expensive) to examine how K varies with changes in R and Q. Two basic types of change have been examined. The first is to force various elements in Q to zero, thereby simulating hypotheses such as "stiffness regulation," in which it is supposed that only a subset of the state variables are being used in control (e.g. muscle length and force without explicit consideration of joint angles or velocities). Normalizing assumptions about the relative sensitivity and dynamic range of the various sensors (see above, particularly regarding fusimotor control of spindle afferents) are important for informing the selection of the nonzero elements in Q. The second type of change involves a weighting coefficient for determining the ratio between R and Q. This reflects a cost-benefit consideration; a small ratio indicates relatively little cost in the recruitment of muscles to maintain the target trajectory whereas a large ratio indicates relatively high tolerance for deviations from nominal trajectory. As might be expected, feedback matrices K calculated for low R/Q values have large coefficients, particularly in the diagonal terms reflecting homonymous feedback, which produce large changes in muscle recruitment and rapid movements in response to perturbations. K-matrices calculated for high R/Q ratios have lower coefficients overall, which produce smaller and more widely distributed changes in muscle activation that result in less "stiff" response.

The various K matrices that can be produced by varying Q and R/Q ratio can be evaluated by comparing the simulated responses of the model limb

and its putative regulator to the actual responses of an intact, naturally behaving cat recorded neurokinesiologically (e.g. Macpherson 1988a,b). In general, we have found that realistic behavior can be obtained only when Q has no zero elements and for intermediate values of R/Q (He et al. 1989, Loeb et al. 1989). However, we have just begun to explore systematically the effects of various restrictions and weightings of performance criteria. One obvious issue is the selection of muscle activation (as estimated by the Renshaw system) as the measure of cost; it would be more realistic to estimate the thermodynamic cost of muscle activation based on the relative cross-sectional areas and the different economics of force generation under concentric and eccentric conditions (active shortening vs. lengthening, respectively). Another issue involves the difference between eliminating a state variable from the performance criteria (setting Q terms to zero) as opposed to eliminating its input to the feedback system (setting K terms to zero).

10.4.4 Hypothesis Development and Testing

Linear-quadratic methods for designing optimal regulators have been applied to a simplified model of the cat hindlimb during quiet standing. The resulting predictions regarding the distribution of sensory feedback among the motor pools can be changed by varying the performance criteria and the model representations of the muscles and sensors. This provides a subjective appreciation of the nature of the control problems posed by the cat hindlimb and the possible roles for segmental reflexes to stabilize its posture during small perturbations.

Once we have had sufficient experience with this approach to control and the robustness of its predictions in the face of uncertainties about the models of the actuators and sensors, we intend to look experimentally for the more interesting and robust of its predictions. Three classes of experiments have been proposed:

i) *Biological circuit tracing.* Classically, the connectivity of the spinal cord has been determined by painstaking, acute electrophysiological probing of surgically reduced and/or anesthetized preparations. Normally, it is difficult to get a sense of perspective from this very fine-grained approach, but it may be feasible to probe for the more robust and unusual circuits predicted by control theory. Importantly, this technique has been applied to neuronal systems that are cycling in an identifiable way through the locomotor pattern, using "fictive locomotion" in decerebrate, paralyzed preparations (Pratt and Jordan, 1987).

ii) Locomotor reflexes. Using implanted nerve cuff electrodes, electrical stimuli can be applied and accurately calibrated on individual nerves to activate particular classes of nerve fibers and to examine their reflex effects at different phases of natural behaviors such as locomotion (Pratt et al., in press). Similar stimulation of muscle nerves is possible, but unfortunately the spindle primary afferents, *GTOs* and alpha motoneurons (projecting to Renshaw cells) have similarly excitable axons, which will complicate interpretation of such data.

iii) Reflex plasticity. It is possible to change the mechanics of the musculoskeletal system by tendon transfers in animals at various stages of development. It seems likely that at least some of the specification of the sensory feedback matrix depends on motor experience and remains malleable to account for growth and development. The changes that should occur for a given tendon transfer can be explored through the model and then looked-for experimentally using the above paradigms. While there is evidence that the pattern of muscle recruitment during locomotion in rats and cats is not relearned after such transfers (Sperry, 1945; Forssberg and Svartengren, 1983), our conceptualization of motor control (Figure 10.4) suggests that the feedback pathways responsible for spinal reflexes should be separate from the *CPG* recruitment; they have yet to be examined in such a paradigm. Interestingly, Yumiya et al. (1979) identified learned changes in paw-placing reflexes following forelimb tendon transfers that were not associated with changes in motor cortical maps. Also, McMahon and Wall (1989) recently noted that some spinal cord reflexes associated with cutaneous and muscle afferents appeared to change in accordance with changes in the sources of their receptor modalities following distal nerve cross-untons.

10.5 Conclusions and Future Directions

Although it has been said before, one lesson bears reiterating in a volume such as this: the value of modeling comes largely from the process, not the results. The benefits may be public, i.e. the discovery or reappreciation of processes overlooked, or private, i.e. elevation of one's perspective from the parochial vantage point of one's training. Both accrue to the investigator during the iterative process of developing and testing the model, not merely reading about its "final" attributes. This philosophy underlies our attempt to construct a "modeling environment" rather than a specific model. Our goal is to produce a general tool that is accessible and meaningful to physiological experimentalists, both to facilitate our own ongoing modeling work and to build bridges among control theory, biomechanics, and sensorimotor neurophysiology.

Recently, this effort has gained momentum from two areas of practical application, robotics and neural prosthetics, in which it appears increasingly advisable to attempt to discover existing control principles in nature rather than to invent them *de novo*. Undoubtedly, this will continue to spur on many related projects such as those described elsewhere in this volume.

References

Athans, M. and Falb, P.L. (1969) *Optimal Control.* NY: McGraw Hill.

Bak, M.J. and Loeb, G.E. (1979) A pulsed integrator for EMG analysis. *Electroenceph. Clin. Neurophysiol.* 47:738:741.

Bernstein, N.A. (1935, 1940) Translation (1967), reprinted in *Human Motor Actions: Bernstein Reassessed* (Edited by Whiting, H.T.A.) pp. 77-120 and 171-222, Amsterdam: Elsevier.

Bryson, A.E. and Ho, Y. (1975) *Applied Optimal Control.* Waltham, MA: Blaisdell Publishing Company.

Chanaud, C.M., Pratt, C.A. and Loeb, G.E. (1990) Functionally complex muscles of the cat hindlimb. V. The roles of histochemical fiber-type regionalization and mechanical hetero-geneity in differential muscle activation. *Exp. Brain Res.* (in press)

Crago, P.E., Houk, J.C. and Rymer, W.Z. (1982) Sampling of total muscle force by tendon organs. *J. Neurophysiol.* 47(6):1069-1083.

Ferrell, W.R., Gandevia, S.C. and McCloskey, D.I. (1987) The role of joint receptors in human kinaesthesia when intramuscular receptors cannot contribute. *J. Physiol. (London)* 386:63-71.

Forssberg, H. (1979) Stumbling corrective reaction: A phase-dependent compensatory reaction during locomotion. *J. Neurophysiol.* **42**:936-953.

Forssberg, H. and Svartengren, G. (1983) Hardwired locomotor network in cat revealed by a retained motor pattern to grastrocnemius after muscle transposition. *Neurosci. Lett.* **41**:283-288.

Gans, C. (1982) Fiber architecture and muscle function. *Exercise and Sport Sciences Reviews.* **11**:160-207.

He, Jiping, Levine, W.S. and Loeb, G.E. (1989) Feedback gains for correcting small perturbations to standing posture. *Proc. 28th IEEE Conf. on Decision and Control.* pp. 518-526.

Heckman, C.J., Weytjens, J.L.F. and Loeb, G.E. (1988) The force-velocity and force-length behavior of single motor units of the medial gastrocnemius muscle of the cat. *Soc. Neurosci. Abst.* **14**:998.

Hildebrand, M. (1985) Walking and running. In: *Functional Vertebrate Morphology.* (Edited by Hildebrand, M., Bramble, D.M., Liem, K.F. and Wake, D.B.) pp. 38-57, John Wiley & Sons, New York.

Hoffer, J.A., Sugano, N., Loeb, G.E., Marks, W.B., O'Donovan, M.J. and Pratt, C.A. (1987a) Cat hindlimb motoneurons during locomotion: II. Normal activity patterns. *J. Neurophysiol.* **57**:530-553.

Hoffer, J.A., Loeb, G.E., Sunano, N., Marks, W.B., O'Donovan, M.J. and Pratt, C.A. (1987b) Cat hindlimb motoneurons during locomotion: III. Functional segregation in sartorius. *J. Neurophysiol.* **57**:554-562.

Hoy, M.G. and Zernicke, R.F. (1985) Modulation of limb dynamics in the swing phase of locomotion. *J. Biomech.* **18**:49-60.

Hubbard, R.P. and Chun, K.J. (1988) Mechanical responses of tendons to repeated extension and wait periods. *J. Biomech. Eng.* **110**:11-19.

Hultborn, H., Lipski, J., Mackel, R. and Wigstrom, H. (1988) Distribution of recurrent inhibition within a motor nucleus. I. Contribution from slow and fast motor units to the excitation of Renshaw cells. *Acta Physiol. Scand.* **134**:347-361.

Joyce, G.C., Rack, P.M.H. and Westbury, D.R. (1969) Mechanical properties of cat soleus muscle during controlled lengthening and shortening movements. *J. Physiol. (London)* **204**:461-474.

Letbetter, W.D. and English, A.W. (1981) The relationship between peripheral intramuscular "compartments" and spatial arrangement of bicepts femoris and semi-tendinosus motor nuclei in the cat lumbar spinal cord. *Soc. Neurosci. Abst.* **7**: 567.

Loeb, G.E. (1987) Hard lessons in motor control from the mammalian spinal cord. *Trends in Neurosciences* **10**:108-113.

Loeb, G.E. (1989) The functional organization of muscles, motor units, and tasks. In: *The Segmental Motor System* (Eds. M.D. Binder and L.M. Mendell) Oxford Univ. Press, New York, pp. 23-35.

Loeb, G.E. and Duysens, J. (1979) Activity patterns in individual hindlimb primary and secondary muscle spindle afferents during normal movements in unrestrained cats. *J. Neurophysiol.* **42**:420-440.

Loeb, G.E. and Hoffer, J.A. (1985) The activity of spindle afferents from cat anterior thigh muscles. II. Effects of fusimotor blockade. *J. Neurophysiol.* **54**:565-577.

Loeb, G.E., Hoffer, J.A. and Pratt, C.A. (1985) The activity of spindle afferents from cat anterior thigh muscles. I. Identification and patterns during normal locomotion. *J. Neurophysiol.* **54**:549-564.

Loeb, G.E. and Marks, W.B. (1985) Optimal control principles for sensory trans-ducers. In: *Proc. Internat. Sympos.: The Muscle Spindle.* (Edited by Boyd, I.A. and Gladden, M.H.) pp. 409-415, Macmillan Ltd., London.

Loeb, G.E. and Gans, C. (1986) *Electromyography for Experimentalists.* 373 pp., 140 figs., Univ. Chicago Press.

Loeb, G.E., Pratt, C.A., Chanaud, C.M. and Richmond, F.J.R. (1987) Distribution and innervation of short, interdigitated muscle fibers in parallel-fibered muscles of the cat hindlimb. *J. Morph.* **191**:1-15.

Loeb, G.E., He, J., Levine, W.S. (1989) Spinal cord circuits: Are they mirrors of musculoskeletal mechanics? *J. Motor Behavior.* **21**: 473-491.

Macpherson, J. (1988a) Strategies that simplify the control of quadrupedal stance. I. Forces at the ground. *J. Neurophysiol.* **60**:204-217.

Macpherson, J. (1988b) Strategies that simplify the control of quadrupedal stance. II. Electromyographicactivity.*J.Neurophysiol.***60**:218-231.

McMahon, S.B. and Wall. P.D. (1989) Changes in spinal cord reflexes after cross-anastomesis of cutaneous and muscle nerves in the adult rat. *Nature* **342**:272-274.

Otten, E. (1988) Concerts and models of functional architecture in skeletal muscle. *Exercise & Sport Sciences Reviews* **16**:89-138.

Philippson, M. (1905) L'autonomie et la centralisation dans le systeme nerveux des animaux. *Trav. Lab. Physiol. Inst. Solvay (Bruxelles)* **7**:1-208.

Pierrynowski, M.R. and Morrison, J.B. (1985) Estimating the muscle forces generated in the human lower extremity when walking: A physiological solution. *Mathematics & Bioscience* **75**:43-68.

Pratt, C.A., Chanaud, C.M. and Loeb, G.E. (in press) Functionally complex muscles of the cat hindlimb. I. Patterns of activation across sartorius.

Experimental Brain Research.

Pratt, C.A. and Jordon, L.M. (1987) Ia inhibitory inter- neurons and Renshaw cells as contributors to the spinal mechanisms of fictive locomotion. *J. Neurophysiol.* **57**:56-71.

Prochazka, A., Trend, P., Hulliger, M. and Vincent, S. (in press) Ensemble proprioceptive activity in the cat step cycle: Towards a representative look-up chart. *Prog. Brain Res.*

Rack, P.M.H. and Westbury, D.R. (1984) Elastic properties of the cat soleus tendon and their func- tional importance. *J. Physiol. (London)*, **347**:479- 495.

Richmond, F.J.R., MacGillis, D.R.R. and Scott, D.A. (1985) Muscle-fiber compart- mentalization in cat splenius muscles. *J. Neurophysiol.* **53**:868-885.

Rossignol, S., Lund, J.P. and Drew, T. (1988) The role of sensory inputs in regulating patterns of rhythmi- cal movements in higher vertebrates. *Neural Control of rhythmic movements in vertebrates.* (Edited by Cohen, A., Rossignol, S. and Grillner, S.) John Wiley and Sons, Inc.

Sperry, R.W. (1945) The problem of central nervous reorganization after nerve regeneration and muscle transposition. *Quart. Rev. Biol.* **20**:311-369.

Stein, G. and Athans, M. (1987) The LQG/LTR proce- dure for multivariable feed- back control design. *IEEE Trans. on Automatic Control*, **AC-32**:105- 114.

Thomson, D.B., Scott, S.H., Richmond, F.J.R. and Loeb, G.E. (1990) Complex motor unit architecture of anterior sartorius muscle in the cat. *Abstract to First World Congress of Biomechanics*, San Diego, California.

Trotter, J.A. (in press) Interfiber tension transmission in series-fibered muscles of the cat hindlimb. *J. Morph.* (in press).

Weytjens, J.L.F. and Loeb, G.E. (1987) An electromyogram-to-force processor and its testing in cat hindlimb muscles. *Soc. Neurosci. Abst.* **13**:1178.

Windhorst, U. and Koehler, W. (1983) Dynamic be- havior of alpha motoneurone sub- pools subjected to inhomogeneous Renshaw cell inhibition. *Biol. Cybern.* **46**: 217-228.

Windhorst, U., Rissing, R., Meyer-Lohmann, J., Laouris, Y. and Kuipers, U. (1988) Facilitation and depression in the responses of spinal Renshaw cells to random stimulation of motor axons. *J. Neurophysiol.* **60**:1638-1652.

Woo, S. L-Y. (1982) Mechanical properties of tendons and ligaments. I. Quasi-static and nonlinear vis- coelastic properties. *Biorheology* **19**:385-396.

Yumiya, H., Larsen, K.D. and Asanuma, H. (1979) Motor readjustment and input- output relationship of motor cortex following cross-connection of forearm muscles in cats. *Brain Res.* **177**:566-570.

Zajac, F.E. (1989) Muscle and tendon: Properties, models scaling and application to biomechanics and motor control. *CRC Crit. Rev. Biomed. Eng.* **17**:359- 411.

Zajac, F.E., Zomlefer, M.R. and Levine, W.S. (1981) Hindlimb muscular activity, kinetics and kinematics of cats jumping to their maximum achievable heights. *J. Exp. Biol.* **91**: 73-86.

CHAPTER 11

Principles Underlying Movement Organization: Upper Limb

Neville Hogan and Jack M. Winters

11.1 Introduction

It seems natural to assume that certain fundamental principles underlie the organization and performance of motor behavior. The search for governing principles has spanned many motor systems, ranging from isolated muscle contraction through locomotion in invertebrates and vertebrates, quadrupeds and bipeds, to whole-body posture in humans. More recent attempts to study the fundamental principles of upper-limb function are the topic of **PART III** of this book. These chapters provide a representative sample of the issues which have been raised and the progress which has been made towards gaining an understanding of this complex topic and applying that knowledge to the development of assistive devices for rehabilitation of upper-limb motor dysfunction.

Several features distinguish upper limb function from that of the lower limb or whole-body posture and balance:

1. The arm is the primary organ humans use to manipulate objects in the environment.

2. The function of the hands is as much sensory as it is motor.

3. The upper limb routinely interacts with a wide variety of physical objects. It must retain its stability and performance while doing so; yet robotic experience has shown that this requirement is anything but trivial.

4. The human upper limb typically does not need to support the weight of the body; hence inherent muscle patterns designed to hold and stabilize an inverted pendulum system are rarely necessary,

and concerns such as energy storage, energy transfer and joint loading take on less relative importance.

5. Rhythmic motion patterns typical of gait are less appropriate and less frequently observed.

This overview chapter will attempt to provide a perspective on the current state of knowledge. Our goal is not to attempt to survey the field; a number of fairly recent reviews exist from a variety of perspectives (e.g., Hasan et al., 1985; Georgopoulos, 1986; Hogan et al., 1987; Bullock and Grossberg, 1986; Gottlieb et al., 1989) and duplication is unnecessary. Nor do we intend to critically review or assess each theory that has been put forward regarding upper limb movement organization; the contributions within this section of chapters address these issues. Instead, we will try to place into context the class(es) of tasks to which a given theory is meant to apply and the relative importance of biomechanical considerations.

11.2 Relationships Between Tasks and Theories

The human upper limb is involved in a prodigious variety of tasks, ranging from pointing to eating to drawing to throwing. These natural movements tend to be graceful and usually involve many limb segments. Different tasks typically require quite different sequencing of limb motion and muscle activation and different information from sensors. Some involve dynamic interaction with the environment, others do not.

Multiple Muscle Systems: Biomechanics and Movement Organization
J.M. Winters and S.L-Y. Woo (eds.), © 1990 Springer-Verlag, New York

How are these movements organized? Perhaps one of the most fundamental questions about movement organization is *which muscles are used and in what pattern* (Bernstein, 1967). This question can be surprisingly difficult to answer experimentally. Part of the difficulty stems from the extreme adaptability and versatility of the upper limb. The task being performed profoundly influences the deployment and use of muscles, limbs and sensors and this makes the comparison of data obtained in different tasks problematical. For example, in Chpater 18 (Gielen et al.) it is shown that even the apparently minor difference between isometric and isotonic conditions can result in different patterns of muscle activation.

The reasonable tendency among researchers has been to create highly structured, well-controlled experiments to test specific hypotheses. Consequently, there is a close relationship between reported experimental data and the theories they were designed to test. The relation to other competing theories is typically less clear. In the following sections we will examine several types of theories and classes of tasks which have been investigated.

11.3 Optimization Theory

One widely-used mathematical tool is optimization theory. There are four components to an optimization problem: *i)* a task goal, defined mathematically via a performance criterion (cost function); *ii)* a system that is to be controlled (usually dynamic); *iii)* set of controls to this system that are available for modulation; and *iv)* an algorithm capable of finding an analytical or numerical solution. Optimization has the appealing feature that the cost function or performance criterion to be optimized is, in a well-defined sense, a mathematical model of the specific task to be performed, in some cases independent of the structures used to perform it. Nevertheless, the predicted output behavior will reflect any information about those structures which are used in the analysis, and the predicted control input provides insight regarding the optimal "strategy" for executing the mathematically-defined "task." If a mathematical model of the mechanics and dynamics of the structures used to perform a task is available, it is possible (though often computationally challenging) to use dynamic optimization theory to analyze the impact of the mechanics and

dynamics of the musculo-skeletal system on the performance that is achieved. Thus optimization theory is a valuable integrative and predictive tool for studying the interaction between the many complex factors which result in observable behavior.

One might argue that, given a model of musculo-skeletal mechanics, optimization theory merely re-maps Bernstein's problem of choosing among an infinity of possible patterns of muscle activation into an equivalent problem of choosing among an infinity of possible performance criteria [see also discussion in Chapter 14 (Gottlieb et al.)]. This is a legitimate objection, but it should be recognized that "merely" restating a question often makes the answer more obvious. The cost functions used to model *task goals* can be chosen to reflect physically meaningful concerns such as the need to minimize stress or energy consumption or plausible qualitative aspects of behavior such as maximizing smoothness. Combinations of such entities, i.e. subcriteria, are also possible, and in Chapter 19 (Seif-Naraghi and Winters) it is suggested that one of the advantages of the optimization approach lies in observing how predicted movement strategies change with changes in the relative weights between *competing* subcriteria. Certainly voluntary movements scale with ease simply by a change in conscious thought, and perhaps terms as "effort", "energy", "accuracy", "fatigue" or "smoothness" come closer to describing the higher levels of movement planning than terms such as "pulse width" or "percent overshoot." Thus, optimization theory serves to spell out the consequences of these broad generalizations in specific behavioral detail.

A second objection to optimization approaches is that many movements, especially those in everyday life, may be "*satisficing*" as opposed to "optimal". A conservative movement strategy that simply satisfies some basic, perhaps "fuzzy", objective may be adequate for many of the tasks of life. It is here, however, that the potential of optimization methods has yet to reach fruition. There has been a tendency in the study of movement for goals to be defined in kinematic terms; even terms such as "effort" have been given a kinematic flavor (Hasan, 1986). However, this certainly need not be the case — the distance runner doesn't win the race by sprinting the whole way. In Chapter 19 it is shown that without penalty for neural and/or muscle activity, predicted *EMG* signals tend to be much more aggressive than is ever seen experimentally, regardless of the form of the kinematic criteria. This suggests that some neuromuscular penalty could always be added to task-based subcriteria simply to mimic inherent tendencies of neuromotor circuitry to go back toward moderate nominal levels[1]. Interestingly, for single-joint systems a byproduct of increasing neuromuscular penalty is high levels of smoothness (Chapter 19).

Another aspect of the "satisficing" argument is related to the concept of desiring a certain margin of safety/stability — why "live of the edge" when not neccesary? It is true that certain tasks may be hard to specify by a performance criterion. Consider a popular task within this book, that of picking up and drinking a cup of hot coffee without spilling and without undue effort (Chapters 14, 17, 19). In the optimization approach, a margin of safety/stability can be approximated by assuming neural and/or environmental noise sources; the greater the applied noise, the greater the desired margin. As shown in Chapter 19, there are numerical techniques for solving such problems, with the predicted strategy a function of the noise level. In fact, such techniques have provided perhaps the most unbiased evidence to date regarding the positive use of nonlinear muscle properties, especially as related to impedance modulation.

11.4 Sensory-Motor Transformations

One viewpoint on the problem of movement organization is that it is fundamentally a process of transforming sensory perception into motor action. A prominent and highly controversial example of this perspective is the so-called "tensor" theory of sensory–motor transformation proposed by Pellionisz and Llinas in 1979 and elaborated in a series of subsequent papers. This exclusively geometric theory does not take the mechanics or

dynamics of the sensory-motor system into account. The central claim of the theory is that in order to relate the information from multiple non-orthogonal sensors to the action of multiple non-orthogonal actuators a metric tensor is required. The cerebellum (and certain other neural structures) was postulated to be the embodiment of that metric tensor. In fact, the mathematical aspects of this theory are seriously flawed [see Arbib and Amari (1985) for a critique] and recent experimental evidence (Kay et al. 1989a,b) indicates that sensory-motor processes are not based on a consistent underlying metric. Nevertheless, the fundamental importance of coordinate transformations in sensory-motor control cannot be denied.

11.5 Neural Networks

A related approach is the use of massively parallel distributed processing, the so-called "neural networks", for sensory-motor coordination. The foundation of this approach lies in two concepts: *i)* that the strength of synaptic connections (i.e., "gains") between cells and cell excitation thresholds can be modulated (e.g., due to synchronous activity); and *ii)* that the behavior of the overall intermeshed network can be considerably more complex than that of each individual cell. The particular appeal of the approach is that parallel distributed processors have been shown in several instances to exhibit significant learning and adaptive behavior. In addition, their structure bears at least a family resemblance to networks of neurons in the biological system.

At the time of writing, this field remains in a state of flux and valid generalizations are difficult to establish. With that caveat, parallel distributed processors may be considered to implement a general nonlinear mapping (usually "smooth" in some sense) between a multi-dimensional input and a multi-dimensional output. For example, Poggio and Girosi (1990) have established an equivalence between a class of three-layer networks and classical function approximation techniques such as regularization theory.

Applied to the problem of movement organization, neural network approaches typically attempt to establish a static (memory-less) functional relation between sensor input and motor output (e.g.

---

[1]An alternative to neuromuscular penalty may be redefine the "system" to include lower-level neurocircuitry which is wired to prevent excessive activity; similar to the *RC* and *RI* commands of Chapter 12 (Feldman et al.); the control inputs for optimization could then be the *R* and *C* commands of Chapter 12.

Kuperstein, 1987). Recent work has begun to extend these approaches to consider dynamics (e.g. Jordan, 1988, 1990; Kawato et al. 1987). Characteristic of these methods is that they treat the biological system somewhat as a "black box" and say little about its components, e.g. which structures are critical for which behavior.

In Chapter 20 (Denier van der Gon et al.) approaches are outlined for utilizing neural networks to: *i)* create internal representations of arm movements; *ii)* learn appropriate reflex responses for certain situations, and *iii)* generate sequences of activation patterns (a "motor program"). Although at present only applied to quite simple upper limb examples, this chapter provides a foundation for future exploration within neural networks, as relevant to neuromuscular systems.

One consequence of the neural network approach is that, unlike much of systems theory, neural networks are not bound to simple linearized biomechanical representations. In fact, as with numerical optimization, a neural network might learn to take advantage of nonlinear musculoskeletal properties (e.g. task-specific modulation of impedance). Also, as outlined in Chapter 20, the neural networks approach need not encompass a full sensory-to-motor mapping. Networks of subsystems, and especially the lower spinal circuitry, might prove fruitful. For instance, a neural network with a structure similar to that outlined in Chapter 10 (Loeb and Levine), could be developed, perhaps driven by both sensory inputs and a command structure similar to that outlined in Chapter 12 (Feldman et al.). In Chapter 10 an optimization approach is utilized to estimate appropriate task-specific feedback gains, while in Chapter 20 a neural network is considered. Which approach is better? This is an open question. However, it should be realized that neural networks and traditional optimization methods have certain aspects in common, and in one sense the neural network method is a form of ongoing optimization. Perhaps hybrid approaches will evolve in the future.

11.6 Hierarchical Organization

In detailing the nature of the transformation between sensory input and motor output, a multi-stage process seems plausible and consistent with known neural architectures (Bernstein, 1967; Arbib, 1972). Saltzman (1979) has argued for a multi-stage process, hierarchically organized with multiple levels of representation ranging from an abstract specification of task goals to a concrete specification of motoneuron activities.

Though there is little direct evidence for (or against) the existence of such a hierarchy, it is an implicit assumption in most of the work that has been reported. It is certainly compatible with our existing knowledge of brain structures, though aspects of the pyramidal tract may constitute an exception. For this reason, in comparing different theories and experimental results it is important to understand clearly *which level of the (assumed) hierarchy is addressed.* Many apparent contradictions may be resolvable by taking this into account.

11.7 Planning vs. Execution

One of the more common assumptions of hierarchical organization is that the production of motor behavior occurs in at least two stages: planning and execution. This begs the question: at what level of abstraction is the motor plan formulated? One possibility is that planning for most tasks takes place at the level of kinematics: the trajectory of the limb is planned, but the details of joint torques and muscle forces required to carry out that plan are not considered in the planning stage. They are assumed to be determined by the subsequent process of executing the motor plan (see review by Georgopoulos, 1986). This viewpoint may be cast into a mathematical framework by formulating a dynamic optimization problem in which the performance criterion is a function of kinematic variables only, for example a measure of the smoothness of the motion (Hogan 1982, 1984, Flash and Hogan 1985).

An alternative possibility is that the planning takes the details of movement execution into account. This may also be formulated as a dynamic optimization problem by choosing a performance criterion which depends on variables related to movement production (Uno et al. 1989; Kawato et al. 1990). These two possibilities are compared in Chapter 17 (Flash).

Of note is that execution of the plan, as prescribed by, say, an optimization algorithm, is a function of the system model and of the class of tasks. In Chapter 19 (Seif-Naraghi and Winters) it was found that for tasks which do *not* involve dynamic interaction with the environment, the basic *form* of the predicted strategy was relatively insensitive to the order of the model or whether it was nonlinear or linearized; details such as the *magnitude* of the excitation pulse, etc., were of course system-dependent. Perhaps this helps explain why the experimental data presented in Chapter 14 (Gottlieb et al.) and Chapter 15 (Corcos et al.) could be interpreted so easily — for their tasks muscle could be adequately viewed as a simple filter. In contrast, for tasks involving *environmental interaction* (e.g. isotonic or random loads), both the model structure and the assumed properties influenced the form of the control strategy.

11.8 Coordinate Frames

The viewpoint that planning takes place at the kinematic level leads naturally to the question of coordinate frames. A given arm motion admits many inter-related representations: as a trajectory of the hand, as a sequence of joint angles, as a time pattern of muscle lengths, and so forth. These may be considered as descriptions of the same behavior in alternative coordinate frames. A question which has received some attention is: Which one of these coordinate frames is the basis of motor planning?

A number of studies have been performed in which subjects performed self-paced point-to-point movements. One of the earliest investigations was that of Morasso (1981), who reported data showing that hand paths were roughly straight with unimodal tangential speed profiles. Compared with the corresponding joint trajectories, these features of the hand motion showed little variation with movement duration or location in the subject's workspace. From this Morasso suggested that the central command was formulated in terms of the hand motion, in spatial, body-centered coordinates.

In contrast, Soechting and Lacquaniti (1981) investigated sagittal (vertical) plane movements and reported a tight kinematic coupling between the angular speed of the shoulder and the elbow during the deceleration phase of movement which was independent of movement duration, target location and or load. From this they argued for trajectory planning in terms of joint coordinates. The issue of coordinate frames is discussed further in Chapter 17 (Flash) and need not be duplicated here. However, it is appropriate to note that the

presentation of the task (move between visually displayed points which are stationary in a laboratory reference frame) may bias the results and generalizations should be made with caution.

Soechting et al. (1986) also investigated three-dimensional drawing motions and reported that elbow and shoulder rotation were linearly related: both joint angles exhibited predominantly sinusoidal patterns with fixed phase relations between them. They suggested that linear shoulder-elbow relationships could be expressions of general neural constraints which facilitate the mapping of intrinsic and extrinsic coordinates.

It must be recognized that kinematic studies alone tell only a small part of the story of movement production, and may not be sufficient to resolve certain questions. Even if behavior is *planned* in terms of the kinematics of hand motion, the *execution* of that plan will be heavily influenced by the mechanics and dynamics of the peripheral musculo-skeletal system. Phase entrainment during oscillatory behavior may be due to the nonlinear neural and mechanical coupling which exists between degrees of freedom in the upper limb rather than a reflection of a higher-level computational processes. For example, Kelso (1984) and colleagues (e.g. Kay et al. 1987) have investigated human bimanual coordination. They have reported many of the phenomena typical of coupled nonlinear oscillators (e.g. frequency entrainment, phase resetting, critical slowing, etc) and have shown that their observations are well-described by a low-dimensional coupled nonlinear oscillator (Haken et al. 1985). Consequently the coupling reported by Soechting et al. (1986) may be peculiar to the cyclic movements they investigated and may not be preserved during terminated reaching movements of a more episodic nature.

11.9 Curved Motions

Many movements in life have inherently curved paths. Examples include handwriting and various sports activities. Are there certain salient features of such movements that can be identified? As outlined in Chapter 17 (Flash), one feature of such movements is that the trajectories of the hand in space are compatible with concept of maximal smoothness, i.e. minimal hand jerk (Viviani and Terzuolo, 1980; Morasso and Mussa-Ivaldi, 1982;

Schneider and Zernicke, R.F.; 1989). As discussed in Chapter 17, this high compatibility with theory need not imply that the *CNS* actually calculates minimum-jerk trajectories; rather, it suggest that the concept of maximum smoothness captures certain salient features of a variety of movements. As addressed in Chapter 19 (Seif-Naraghi and Winters), to some extent smooth trajectories are a byproduct of intrinsic muscle properties and of neuromuscular penalty.

11.10 Fast Tracking Movements

One-degree-of-freedom point-to-point movements have been a favorite of researchers (e.g. review by Gottlieb et al., 1989; Chapters 14–15). They allow well-defined experimental protocols and seem to challenge subjects. One area of interest has been the speed-accuracy trade-off during movements in which subjects are instructed to move as fast as possible and yet as accurately as possible. The appeal of this type of experiment is that the investigator has control over two task parameters (usually movement distance and target width) and measures a third variable (usually movement time). Many combinations are possible, movements of different systems and/or degrees of freedom can be explored and all variables can be determined without detailed measurement of movement trajectories. Empirical relations exist between these parameters (e.g., Fitts' law, 1954).

Others have focused on *EMG* burst behavior and its correlation to movement kinematics and various task parameters (e.g., Freund and Budingen, 1978). Still others have studied the effects of added inertial loads and/or variable speeds (e.g., Lestienne, 1979).

Recently Gottlieb et al. (1989) completed an extensive review of the work related to single-joint point-to-point movements and attempted to place these efforts within a unifying framework. Relations between task variables and *EMG*, between task and kinematic variables, and between *EMG* parameters and kinematic measures were documented across the literature. They assumed that observed *EMG* patterns may be the response of a first-order, low-pass system to an "excitation pulse" which can vary in height and/or width. They then identified two fundamental strategies (the "Dual Strategy" hythothesis): "*speed insensitive*" (excitation pulse and initial *EMG* slope constant, duration modulated) and "*speed sensitive*" (excitation pulse, initial *EMG* slope, and movement speed modulated, pulse duration constant). A wide variety of data seem fit into these categories, which makes the theory appealing, and it is the basis of a theory of motor skill acquisition [Chapter 15 (Corcos et al.). However, all of the tasks considered are uncommon outside the environment of controlled experiments. How this theory may be generalized to more natural multiple degree-of-freedom movements is not clear at this time.

11.11 Importance of Mechanics

For certain classes of tasks movement planning may appear to be represented and planned at a kinematic level. Examples include maximum smoothness (Flash and Hogan, 1985; see also Chapter 17) and the documentation of direction-sensitive cells (in world or visually-relevant coordinates) within the motor cortex [Georgopoulis, 1986; Schwartz et al, 1988; see also review in Chapter 16 (Karst and Hasan)]. However, in interpreting neuromotor experimental data the importance of mechanics and of mechanical principles should not be underestimated. Here we briefly outline a few examples.

For simple fast movements of one degree of freedom, it is important to note that many of the correlations that have been discovered experimentally [e.g. the "impulse-timing" theory of Wallace (1981)] are essentially expressions of expected behavior for the mechanical system, and indeed could not have been otherwise. This can be seen from simple optimization results for, say, second-order overdamped models (reviewed in Chapter 19; see also Nelson, 1983 for similar insights for a visco-inertial model). For instance, to move a greater distance, the agonist pulse height and/or width must increase. Increasing only pulse height increases velocity and movement distance, increasing only pulse width increases movement distance and movement time but not initial velocity, adding inertia while maintaining the same excitation pulses lowers peak velocities and accelerations and increases movement times, etc. ... – basic causal input-output mechanics.

As addressed in depth in Chapter 8 (Zajac and Winters), inertial dynamics introduce nonlinear coupling between body segments. This is due to the fact that the motion of one limb segment induces an acceleration of connected segments (the Coriolis and centrifugal effects). The effects of such coupling, which are a major focus in many of the chapters emphasizing the lower limb [reviewed in Chapter 35 (Mungiole and Winters)], can be quite complex and difficult to comprehend without simulation. However, most upper limb movements at *preferred* speeds are essentially quasi-static, and for such tasks gravitational and visco-elastic musculotendon effects tend to dominate over inertial effects.

Of note is that the inherent coupling between limb segments can potentially be used to advantage. This concept is the basis of an elegant method described in Chapter 21 (Meek et al.) for deducing the moments about the distal (artificial) joints of an upper-limb amputation prosthesis from measurements of *EMG* from muscles of proximal (natural) joints. However, it severely complicates the problem of associating muscle action with the resulting motion, to the point that the "obvious" flexor muscle of a joint may sometimes cause acceleration which would extend the joint [see review in Chapter 8 (Zajac and Winters)]. How does the biological system deal with this complexity?

11.12 Computation vs. Mechanics

Despite the manifest complexity of multi-link dynamics, even a young child is capable of a wide variety of effortless movements. How is this done?

One possible approach is to formulate a dynamic model (based on rigid-body inertial mechanics) of the relation between joint torques and the resulting limb motions, and use this model to compute "backwards" from a specified motion to the corresponding joint torques. This is the so-called "inverse dynamics" approach which has been championed by Hollerbach (1982). It implies that the central nervous system contains a detailed model of the kinematic and inertial properties of the skeletal system. However, the effective dynamic and mechanical behavior of the muscles and neural feedback circuits are not taken into account. It also implies that the central nervous system explicitly performs a computation that experience with a comparable problem in robotics has shown to be extremely demanding.

An alternative approach assumes a look-up table (Albus, 1975; Raibert, 1976). Here the computation problem is replaced by a memory storage/retrieval problem. Unfortunately, such tables may become very large very quickly. Although initially appealing and still under consideration (Atkeson, 1989), up to the present practical problems have made utilization of this approach difficult.

An alternative and somewhat simpler view is that the central nervous system takes advantage of the effective dynamic and mechanical behavior of the muscles and neural feedback circuits to circumvent most of the computational complexities of coordinating multi-joint actions. As reviewed in Chapter 7 (Hannaford and Winters), several aspects of muscle behavior are in sharp contrast to the behavior of the typical motors of modern machinery. Prominent among those differences is the fact that the steady-state force exerted by a muscle for constant neural input varies with muscle length, producing the so-called "spring-like" behavior [see Chapter 5 (Winters) for biological sources of this effect and Chapter 9 (Hogan) for mechanical implications]. Qualitatively similar behavior is observed with and without neural feedback. In the absence of external loads, the combined action of a group of muscles spanning a joint, including antagonists as well as agonists, will define an equilibrium posture for the joint. Changing the relative inputs to the muscles will (in general) define another equilibrium position and the joint will tend to move to the new equilibrium position. This equilibrium-point hypothesis was originally proposed by Feldman (1966) for single-joint motions and was subsequently pursued by other researchers (e.g. Bizzi et al. 1976, 1984). As outlined in Chapter 12 (Feldman et al.), it may be used to account for observed patterns of *EMG*.

The same idea, with a few modifications, may be generalized to account for the coordination of multiple degrees of freedom (Hogan 1985, Flash 1987, Feldman et al., 1986; Chapters 12 and 17). If the posture defined by the pattern of neural activity is moved continuously, the limb will tend to follow, driven by the action of the forces generated by the effective dynamic and mechanical behavior of the muscles and neural feedback circuits (e.g. the "spring-like" behavior, etc.). In general the position of the limb will differ from the neurally-defined "equilibrium" posture (and in the presence of external loads such as gravity may differ even at equilibrium) and hence the time-course of neurally-defined postures has been termed a virtual trajectory (Hogan 1982, 1984). A simulation study by Flash (1987, Chapter 17) has shown that this theory is competent to describe certain sets of observed movements to a remarkable degree.

Quite aside from its ability to reproduce certain experimental observations, one of the most important aspects of this theory is that it dramatically simplifies the computational complexity of multi-joint coordination. The effective mechanics of the neuro-muscular system account for the complex dynamics leaving only a much simpler static computation (albeit a non-trivial one) to be performed.

It must be stressed that neither of the above theories directly address the difficult problem of *which muscles are selected* to perform a multi-joint motion and how they are activated. This matter is taken up, at a simple level, in Chapter 16 (Karst and Hasan). Here initial muscle activity is documented for horizontal plane movements spanning a wide variety of directions and initial locations. Since different theories predict different initital muscle patterns, this approach potentially allows the various theories to be distinguished. Experimentally, the best fit of the data was with the angle defining the target direction with respect to the forearm, which is suggested not to be compatible with any of the aforementioned theories. However, it should be noted that an implicit assumption behind their "attractor" simplification of the *EP* theory is an isotropic stiffness; it turns out that experimentally measured anisotropy of the stiffness (as documented in Chapter 17) can profoundly affect the predictions.

11.13 Equilibrium, Mechanics and Neural Feedback

There has been some controversy regarding formulations of the two continually evolving "equilibrium point" hypotheses outlined above, the so-called (by Feldman) λ- and α-models. The λ-model was pioneered by Feldman (Feldman, 1966, 1986, Chapter 12), while the so-called α-model (as defined by Feldman) is based on the work of Bizzi and colleagues Bizzi (Bizzi et al, 1976, 1984; discussed in Chapter 17). The basic foundations

behind each model have been reviewed and contrasted elsewhere (e.g., Hasan et al., 1985). Here we briefly address some similarities and differences and then suggest that these are perhaps reconcilable.

Fundamental to both models is the concept that a muscle or a muscle–reflex system has spring-like properties, and that these properties, in conjunction with neural commands and any external load, set a steady-state equilibrium. *EP* models in general suggest that neuro-mechanical dynamics are so intertwined that they cannot be separated [see also Houk and Rymer (1981) and Chapter 13 (Wu et al.)]. However, it must also be recognized that there are qualitative similarities between the responses of intact and deafferented monkeys (Bizzi et al., 1984).

The assumed inputs to the models differ. In the λ-model, commands are related to motoneuron recruitment *thresholds*, defined mathematically as the offset of a spring. The *EMG* is considered an internal *consequence* of recruitment. In this foundation *EMG* activity is predicted via muscle activation areas (*MAAs*), which are defined by the ongoing muscle length and velocity along with the ongoing λ (see Figure 12.1 in Chapter 12). The further into an *MAA* (i.e. from the λ line), the greater the *EMG*. In past forms of the so-called α-model, the input has been activation, which in turn has modulated the *slope* of the force–length relation.

Of note is that both models tend to utilize slopes greater than *CE* tension–length properties could predict; values used for the α-model are often closer to *SE* stiffness, which is essentially what is experimentally estimated by perturbation studies [see review in Chapter 5 (Winters)], while values for the λ-model are based on experimental data generated by Feldman's group (reviewed in Feldman, 1986). Of note is that there tends to be a correlation between *SE* stiffness (which is nearly linearly related to muscle force) and activation, at least for equilibrium conditions and for small oscillatory perturbations [Chapter 5 (Winters)]. Furthermore, feedback gains tend to modulate with activation level (e.g., Lacquaniti and Soechting, 1981), and simulations with a nonlinear antagonistic muscle-joint model with delayed sensory feedback [e.g., Winters and Stark, 1985; see also Chapter 8 (Zajac and Winters)] showed that maintenance of stable yet effective position control was achieved by letting feedback gains modulate with activation. Interestingly, ongoing co-contraction levels compensate for feedback time delays [e.g., Chapter 7 (Hogan)], while feedback helps compensate for muscle yielding [Houk and Rymer, 1981; Cordo and Rymer, 1982; see also Chapter 13 (Wu et al.)]. Thus, the available mechanisms for predicting modulation of joint stiffness and threshold with antagonistic coactivation, some based on muscle mechanics and others on neurocircuitry, appear compatible. Chapter 12 develops additional neurally-based sources for these and related effects, based on *RI* (reciprocal inhibition) and *RC* (Rhenshaw cell modulation) commands. Careful consideration of each of the sources suggests that these are potentially complementary and in any case not conflicting.

Another difference between the models is the form of the assumed *virtual* (equilibrium) trajectory [defined here as a mechanical representation of the input that is applied so as to achieve a desired (*reference*) trajectory, which the *actual* trajectory tries to achieve]. To produce voluntary movements, Chapter 12 assumes ramp (constant velocity) trajectories, while Chapter 17 assumes bell-shaped velocity trajectories. Are these differences in shape incompatible? This remains an open question. Of note is that the mapping between the virtual trajectory and the *EMG* does differ in the two cases; in particular Chapter 12 emphasizes the *MAA* concept. Chapter 12 suggests that the assumption of smoothed virtual trajectories has a problem related to modification of ongoing movement plans, yet Chapter 17 provides evidence showing that observed corrective behavior that is seen with target changes in rapid progression is well approximated by the concept of independent, superimposed smooth virtual trajectories.

To help place the concept of virtual trajectories within a broader framework, in Chapter 19 (Seif-Naraghi and Winters) a slightly different form of a virtual trajectory is utilized. Here it is shown that when neuromuscular penalty is added within a performance criterion, the "reference" trajectory becomes a virtual (cost) trajectory that may represent general kinematic plans as opposed to an instant-to-instant desired trajectory. Because of neuromuscular dynamics, optimized solutions for a variety of tasks are insensitive to short-time transient details of such virtual trajectories, especially for fast movements (results for different forms, including ramps and various smoothed signals, are given in Seif-Naraghi, 1989). Thus, actual trajectories may be relatively insensitive to the exact form of the virtual trajectory that is utilized to initiate and terminate movements; interestingly, these are the regions where the approaches of Chapters 12 and 17 differ.

11.14 Contact Tasks

Much of the work on motor coordination has focused on unrestrained or minimally restrained motions, e.g. point-to-point reaching movements. Though clearly important, these motions are a small part of the repertoire of normal upper-limb behavior. Most functional uses of the upper limb require contact with objects.

In one sense, contact tasks might seem to be an order-of-magnitude more complex than unrestrained motions. The mechanics of contact are not adequately described by simply assuming an external load. The forces generated by contact depend on the motion of the limb (or vice versa) whereas an external load (e.g. gravity) need not. Thus, if an internal model of the musculo-skeletal system is postulated as necessary for the coordination of unrestrained motions, then contact tasks would appear to require a model of the object being wielded as well.

However, the effective dynamic and mechanical behavior of the muscles and neural feedback circuits may once again play a role in simplifying this problem. The "spring-like" muscle behavior which can be used to circumvent the "inverse dynamics" computations may also be used to control contact force. Furthermore, this approach has several distinct advantages. It circumvents the contact stability problem which has plagued attempts to develop robots capable of controlling the force they exert on objects [see Chapter 9 (Hogan)]. It also permits the same strategy (and perhaps even the same neural apparatus) to control contact tasks, unrestrained motions and the transitions between the two. The extraordinary effectiveness of this approach is demonstrated in Chapter 20 (Crago et al.), where stiffness regulation as the basis of a Functional Neuromuscular Stimulation (*FNS*) system for controlling grasp.

The usefulness of the effective dynamic and mechanical behavior of the muscles and neural feedback circuits is by no means restricted to the "spring-like" behavior. The force exerted by a muscle, with and without feedback, is a function of its speed of shortening. This imparts an effective viscous behavior to the muscle which may play an important functional role, e.g. in reducing oscillations or decelerating the limb. The effective viscous behavior need not be linear and indeed is not [e.g., see review in Chapter 5 (Winters)]. In Chapter 13 (Wu et al.) the effects of various forms of nonlinear viscous behavior are documented via phase plane analysis for isotonic and isokinetic cases of environmental coupling. These properties can be shown to facilitate adaptation to different external loads without the need for explicit identification of environmental parameters.

11.15 Conclusions and Future Directions

Proposing a perspective on the large body of research on upper-limb movement organization is no small undertaking. Compared to the vast repertoire of human upper-limb function and the remarkable adaptability of human sensory-motor behavior, the available experimental data is sparse. What has been learned does not constitute a "fabric of knowledge"; it is more like a few isolated knots of fact connected by a tenuous web of theory. We believe that many of the apparent contradictions and inconsistencies in the available data may be reconcilable by carefully considering the *experimental context in which the observations were made*. When evaluating alternative interpretations of the available evidence, it is imperative to clearly identify which *levels* of the motor control system (from cognitive processes to musculo-skeletal mechanics) the theories address and which levels are likely to be responsible for the data. We do not presume to draw any final conclusion but we believe that the beginnings of a consensus on at least some of the issues can be discerned.

Where does this lead? Here is small sample of a few areas that appear ripe for exploration:

1. Isn't this redundant? Most tasks can be performed in more than one way. While experimentally challenging, more data is needed for *freely* moving individuals performing reaching movements and common manipulation tasks. *3-D*

kinematic patterns need to be documented. For example, for what tasks does the elbow stay lower vertically than the hand? In conjunction with such experiments, dynamic optimization studies need to be performed which make kinematic predictions but include appropriate biomechanical models. By comparing optimization predictions to experimental data, it should be possible to illuminate some of the organizational strategies that solve problems in kinematic and muscle redundancy.

2. Why am I being perturbed? EMG and kinematic information on movement is often not fully illuminating — mechanics cannot be ignored. Perturbation studies can help extract such phenomena. However, anticipation of perturbation may (for better or worse) influence the strategy of the subject. Clever techniques for externally perturbing natural movements are needed. Such experiments would be especially appropriate for helping illuminate relations between: *i)* neuromusculoskeletal dynamics and kinematic *smoothness* and *ii) impedance modulation* over the course of various tasks.

3. Falling from equilibrium. The current evolution of *EP* theories toward multiple-link applications and more complex tasks needs to be continued. Additionally, foundations behind *EP* models need to be incorporated into Hill-based models, including some of the insights into neurocircuitry that are provided in Chapter 12 (Feldman et al.). *EP* models also need to be extended to the torso, which, because of fundamental concerns regarding postural stability, would provide a good (and perhaps even better) "testing ground" for continued theoretical development.

4. The arm has a body. With only a few exceptions [e.g. Chapter 22 (Meek et al.)], the arm has been considered as an isolated entity connected to a rigid base. In the past this has been partly due to our lack of understanding of the torso. As seen by Chapters 23-34, however, this has changed. Whenever a manipulation task is being performed, loads are transferred between the arm and back. Shoulder and back muscles are also heavily involved in fast arm tracking movements and in tasks such as throwing. When a relatively rigid base is desired for arm movements, how much of this is achieved simply by cocontracting torso muscles? How much by activity in specific muscles?

5. What's that muscle doing over there? In Chapter 18 (Gielen et al.) it is shown that since muscle work output is sensitive to velocity and length, multi-articular and uni-articular muscles may "work" differently. As with so many other recent insights regarding multi-articular muscles [e.g., Hogan et al., 1985; Chapters 17–18, 26, 41], this finding needs to be explored futher. However, the concept of muscle work efficiency is especially enticing because it can be tested via optimization studies.

6. So that's why I have two of them! Unless very astute or lucky, an uninformed creature reading a sampling of research papers involving upper limb movements may not realize that humans normally have two upper limbs. Many quite relevant tasks are performed with two arms. These tasks are often goal-directed, well-planned movements that are executed with extreme coordination. Understanding such movements is especially relevant for groups involved in upper-limb artificial limb or *FNS* research.

7. What's that baby doing now? The vast majority of experiments are performed on adults who are reasonably coordinated and who have previously practised performing the task under study. Such goal-directed tasks are fine for documenting skilled performance or for comparing to optimization results. However, with regards to understanding movement *organization*, investigation into the process involved in *developing* such organizational strategies may turn out to be an easier path. As mentioned in Chapter 15 (Corcos et al.), there has been somewhat of a schism between motor control and motor learning. This exists both in terms of theoretical issues and experimental techniques. Perhaps much could be learned by exploring more completely, with the best techniques the field has to offer, motor skill development in the child.

8. Do we need to network? As mentioned previously, the field of neural networks remains in a state of flux. However, with regard to neuromotor systems the real question is whether to "think big" or "think small". We suggest starting small, i.e. with reasonably-sized networks at the spinal levels, similar in concept to Chapter 20 (Denier

van der Gon et al.). Clearly neural networks have the potential to be complementary to other approaches such as optimization.

References

Albus, J.S. (1975) A new approach to manipulator control: the cerebellar model articulation controller (CMAC). *J. Dynamic. Sys., Meas. & Control*, **97**: 220-227.

Arbib, M.A. (1972) *The metaphorical brain: an introduction to artificial intelligence and brain theory.*, Interscience, New York.

Arbib, M. A. and S.-I. Amari (1985) Sensori-Motor Transformations in the Brain (with a Critique of the Tensor Theory of Cerebellum). *J. Theor. Biol.*, **112**: 123-155.

Atkeson, C.G. (1989) Learning arm kinematics and dynamics. *Ann. Rev. Neurosci.* **12**: 157-183.

Atkeson, C.G. and Hollerbach, J.M. (1985) Kinematic features of unrestrained vertical arm movements. *J. Neurosci.*, **5**: 2318- 2330.

Benati, M., Morasso, P. and Tagliasco, V. (1982) The inverse kinematic problem for anthropomorphic manipulator arms. *J. Biomech. Eng.*, **104**: 110-113.

Bernstein, N. (1967) *The co-ordination and regulation of movements.* Pergamon Press, New York.

Bizzi, E., A. Polit and P. Morasso (1976) "Mechanisms Underlying Achievement of Final Head Position", *J. Neurophysiol.*, **39**: 435-444.

Bizzi, E., N. Accornero, W. Chapple and N. Hogan (1984) "Posture Control and Trajectory Formation During Arm Movement", *J. Neurosci.*, **4**: 2738-2744.

Bullock, D. and Grossberg, S. (1986) Neural dynamics of planned arm movements: emergent invariants and speed-accuracy properties during trajectory formation. *Psychol. Rev.*, **95**: 49-90.

Cordo, P.J. and Rymer, W.Z. (1982) Contributions of motor-unit recruitment and rate modulation to compensation for muscle yielding. *J. Neurophys.*, **47**: 797- 809.

Feldman, A.G. (1966) Functional tuning of the nervous system with control of movement or maintenance of a steady posture, II: controllable parameters of the muscle. Biophysics 11, 565-578.

Feldman, A.G. (1986) Once more on the equilibrium-point hypothesis (λ model) for motor control. *J. Motor Behav.* **18**: 17-54.

Fitts, P.M. (1954) The information capacity of the human motor system in controlling the amplitude of the movement. *J. Exp. Psychol.*, **47**: 381-391.

Flash, T. (1987) The control of hand equilibrium trajectories in multi-joint arm movements. *Biol. Cybern.* **57**: 257-274.

Flash, T. and N. Hogan (1985) The Coordination of Arm Movements: An Experimentally Confirmed Mathematical Model. *J. Neuroscience*, **5:** 1688-1703.

Freund, H.-J. and Budingen, H.J. (1978) The relationship between speed and amplitude of the fastest voluntary contractions of human arm muscles. *Exper. Brain Res.* **31:** 1-12.

Gottlieb, G.L., Corcos, D.M. and Agarwal, G.C. (1989) Strategies for the control of voluntary movements with one mechanical degree of freedom. *Behav. & Brain Sci.*, **12:** 189-250.

Haken, H., J.A.S. Kelso and H. Bunz (1985) A theoretical model of phase transitions in human hand movements. *Biol. Cybern.* **51:** 347- 356.

Hasan, Z. (1986) Optimized movement trajectories and joint stiffness in unperturbed, inertially loaded movements. *Biol. Cybern.*, **53:** 373-382.

Hasan, Z., Enoka, R.M. and Stuart, D.G. (1985) The interface between biomechanics and neurophysiology in the study of movement: some recent approaches. *Exercise and Sport Sci. Rev.*, **13:** 169-234.

Hogan, N. (1982) Control and Coordination of Voluntary Arm Movements. in: *Proc. of 1982 Amer. Control Conf.*, M.J. Rabins and Y. Bar-Shalom (eds.) **2:** 522-527.

Hogan, N. (1984) An Organizing Principle for a Class of Voluntary Movements. *J. Neurosci.* **4:** 2745-2754.

Hogan, N. (1985) The mechanics of multi-joint posture and movement. *Biol. Cybern.*, **52:** 315-331.

Hogan, N., Bizzi, E., Mussa-Ivaldi, S. and Flash, T. (1987) Controlling multi-joint motor behavior. *Exerc. and Sport. Sci. Rev.*, **15:** 153-189.

Hollerbach, J.M. (1982) Computers, brains and the control of movement. *Trends in Neurosci.* **6:** 189-192.

Houk, J.C. and Rymer, W.Z. (1981) Neural control of muscle length and tension. *Handbook of Physiology - The Nervous System II*, Chapter 8, pp. 257-323.

Jordan, M.L. (1990) Indeterminate Motor Skill Learning Problems. In: M. Jeannerod (ed.), *Attention and Performance, XIII*, Lawrence Erlbaum, Hillsdale, NJ.

Jordan, M. (1988) Supervised learning and systems with excess degrees of freedom. Massachusetts Institute of Technology COINS Technical Report 88-27.

Kay, B.A., J.A.S. Kelso, E.L. Saltzman and G. Schoner (1987) Space-time behavior of single and bimanual rhythmical movements: data and limit cycle model. *J. Exp. Psych.: Human Percep. and Perform.*, **13:** 178-192.

Kay, B.A., N. Hogan, F.A. Mussa-Ivaldi and E.D. Fasse (1989a) Perceiving the Properties of Objects Using Arm Movements: Workspace Dependent Effects. *Proc. 11th Ann. Conf. IEEE EMBS*, Vol. 11, Part 5.

Kay, B.A., N. Hogan, F.A. Mussa-Ivaldi and E.D. Fasse (1989b) Perceived Properties of Objects Using Kinesthetic Sense Depend on Workspace Location. Society for Neuroscience Abstracts.

Kawato, M., K. Furukawa and R. Suzuki (1987) A hierarchical Neural-Network Model for Control and Learning of Voluntary Movement. *Biol. Cybern.*, **57:** 169-185.

Kawato, M., Y. Madea, Y. Uno and R. Suzuki (1990) Trajectory formation of arm movement by cascade neural network model based on the minimum torque change criterion. *ATR Technical Report*, TR-A-0056.

Kelso, J.A.S. (1984) Phase transitions and critical behavior in human bimanual coordination. *Amer. J. of Physiol.* **43:** 1183-1195.

Kuperstein, M. (1987) Adaptive visual-motor coordination in multi-joint robots using parallel architecture. *IEEE Conf. on Robotics and Automation*, pp. 1595-1602.

Lestienne, F. (1979) Effects of inertial load and velocity on the braking process of voluntary limb movements. *Exp. Brain Res.*, **35:** 407-418.

Lacquaniti, F., Licata, F. and Soechting, J.F. (1982) The mechanical behavior of the human forearm in response to transient perturbation. *Biol. Cybern.*, **44:** 35-46.

Marsden, C.D., Obeso, J.A. and Rothwell, J.C. (1983) The function of the antagonist muscle during fast limb movements in man. *J. Physiol.*, **335:** 1-13.

Morasso, P. (1981) Spatial control of arm movements. *Exp. Brain Res.*, **42:** 223-227.

Morasso, P. and Mussa-Ivaldi, F.A. (1982) Trajectory formation and handwriting: a computational model. *Biol. Cybern.* **45:** 131-142.

Mussa-Ivaldi, F.A. Hogan, N. and Bizzi, E. (1985) Neural, mechanical and geometric factors subserving arm posture in humans. *J. Neurosci.*, **5:** 2732-2743.

Nelson, W.L. (1983) Physical principles for economies of skilled movements. *Biol. Cybern.* **46:** 135-147.

Nichols, T.R. and Houk, J.C. (1976) Improvements in linearity and regulation of stiffness that results from actions of stretch reflex. *J. Neurophys.*, **39:** 119-142.

Pellionisz, A. and Llinás, R. (1979) Brain modeling by tensor network theory and computer simulation. The cerebellum: distributed processor for predictive coordination. *Neurosci.* **4:** 323-348.

Poggio, T. and F. Girosi (1990) Regularization algorithms for learning that are equivalent to multilayer networks. *Science*, **247:** 978-982.

Raibert, M.H. (1976) A state space model for sensorimotor control and learning. *MIT Artif. Intel. Memo No. 351*, January.

Saltzman, E. (1979) Levels of Sensorimotor Representation. *J. Math. Psych.*, **20:** 91-163.

Schmidt, R.A., Zelaznik, H., Hawkins, B., Frank, J.S. and Quinn, J.T. (1979) Motor output variability: a theory for the accuracy of rapid motor acts. *Psychol. Rev.*, **86:** 415-451.

Schwartz, A.B., Kettner, R.E. and Georgopoulos, A.P. (1988) Primate motor cortex and free arm movements to visual targets in three-dimensional space. I. Relations between single cell discharge and direction of movement. *J. Neurosci.*, **8:** 2913-2927.

Schneider, K. and Zernicke, R.F. (1989) Jerk-cost modulations during the practise of rapid arm movements. *Biol. Cybern.* **60:** 221-230.

Seif-Naraghi, A.H. (1989) Control of human arm movements via optimization. Ph.D. Dissertation, Arizona State University.

Soechting, J.F. (1982) Does position sense at the elbow reflect a sense of elbow joint angle or one of limb orientation? *Brain Res.*, **248:** 392-395.

Soechting, J.F. (1984) Effect of target size on spatial and temporal characteristics of a pointing movement in man. *Exp. Brain Res.*, **54:** 121-132.

Soechting, J.F. and Lacquaniti, F. (1981) Invariant characteristics of a pointing movement in man, *J. Neurosci.*, **1:** 710-720.

Soechting, J.F., Lacquaniti, F. and Terzuolo, C.A. (1986) Coordination of arm movements in three-dimensional space. Sensorimotor mapping during drawing movement. *Neurosci.*, **17:** 295-311.

Soechting, J.F., Dufrense, J.R. and Lacquaniti, F. (1981) Time-varying properties of myotatic response in man during some simple motor tasks. *J. Neurophys.*, **46:** 1226-1243.

Stein, R.B. (1982) What muscle variable(s) does the nervous system control in limb movements? *Behav. and Brain Sci.*, **5:** 535-577.

Uno, Y., M. Kawato, and R. Suzuki (1989) Formation and control of optimal trajectory in multijoint arm movement: minimum torque change model. *Biol. Cybern.* **61:** 89-101.

van Sonderen, J.F., Denier van der Gon, J.J. and Gielen, C.C.A.M. (1988) Conditions determining early modification of motor programmes in response to changes in target location. *Exp. Brain Res.*, **71:** 320-328.

Viviani, P. and Terzuolo, C. (1983) The organization of movement in handwriting and typing. *Language Production,* **2:** 103-146.

Wadman, W.J., Denier van der Gon, J.J. and Derksen, R.J.A. (1980) Muscle activation patterns for fast goal-directed arm movements. *J. Human Movem. Stud.*, **6:** 19-37.

Wallace, S.A. An impulse-timing theory for reciprocal muscular activity in rapid, discrete movements. *J. Motor Behav.*, **13:** 144-160.

Winters, J.M. and Stark, L. (1985) Task-specific second-order movement models are encompassed by an eighth-order nonlinear musculo-skeletal model. *Proc. IEEE Sys., Man & Cybern.*, pp. 1111-1115.

CHAPTER 12

The Origin of Electromyograms - Explanations Based on the Equilibrium Point Hypothesis

A. G. Feldman, S. V. Adamovich, D. J. Ostry and J. R. Flanagan

12.1 Introduction

In the present chapter, we review and further develop the equilibrium-point (*EP*) hypothesis or λ model for single and multi-joint movements (Feldman 1974, 1986; cf. Chapters 11, 13-22). A departure point is the notion of the measure of the central control signals underlying movement production. According to the *EP* hypothesis, central commands parameterize the threshold of motorneuron (*MN*) recruitment. The usual assumption that central signals are directly associated with muscle activation, i.e. recruitment of *MN*s and their firing frequencies, is rejected (see also Bernstein, 1967). This assumption ignores the role of muscle afferents in motor control as well as the non-linear threshold properties of *MN*s. In this chapter, we discuss electromyographic (*EMG*) patterns of single- and multi-joint movements in terms of the *EP* hypothesis. Reference will be made to the central control signals which set the inter-muscular interaction (cf. Nichols, 1989) as well as to the concept of muscle activation area which is essential for the explanation of the *EMG* patterns.

It is necessary to make an initial remark concerning the understanding of the *EP* hypothesis. A *static* arm position is associated with an equilibrium state of the corresponding spinal and supraspinal systems, including peripheral afferent systems, muscle properties, and external forces (loads). This static state is achieved by the dynamic interaction of these subsystems. The *EP* hypothesis suggests that the *CNS* can alter the equilibrium state at the spinal level by changing specific neurophysiological parameters. Subsequently, the *CNS* initiates **dynamic** processes which force the subsystems to find a new equilibrium. Changes in muscle activation and forces as well as limb movement itself are a reflection of the dynamic processes associated with the transfer of the neuro-biomechanical system from one equilibrium state to another. The final equilibrium state essentially depends on the parameters the *CNS* specifies as well as on the external load. It should be emphasized that muscle activation by way of the recruitment or derecruitment of motor units occurs as a **consequence** of the disturbance of the initial equilibrium state at the neuronal level. Neither the timing nor the magnitude of the muscle activation ("*EMG* bursts") have to be programmed for the execution of the movement, just as muscle elastic properties, forces and kinematic variables do not appear to be preplanned or calculated in the course of movement. Therefore, the *EP* hypothesis can be considered as an alternative to the notion that the *CNS* preplans movement kinematics and performs inverse-dynamic computation in the course of movement to generate appropriate muscle activation and forces (see Hollerbach, 1985).

The *EP* hypothesis suggests a more primary reason for biological movements than merely a change in muscle forces or, indeed, even muscle activation. In contrast, muscle activation is considered to be the basic mechanism in shifts of the limb equilibrium in an alternate version of the *EP* hypothesis [Bizzi, 1980; Hogan, 1984; for discussion see Chapters 11 (Hogan and Winters) and 17 (Flash)]. But, in essence, the cause and effect of movement production are inverted in this version of the original formulation (Feldman, 1974, 1986).

Multiple Muscle Systems: Biomechanics and Movement Organization
J.M. Winters and S.L.-Y. Woo (eds.), © 1990 Springer-Verlag, New York

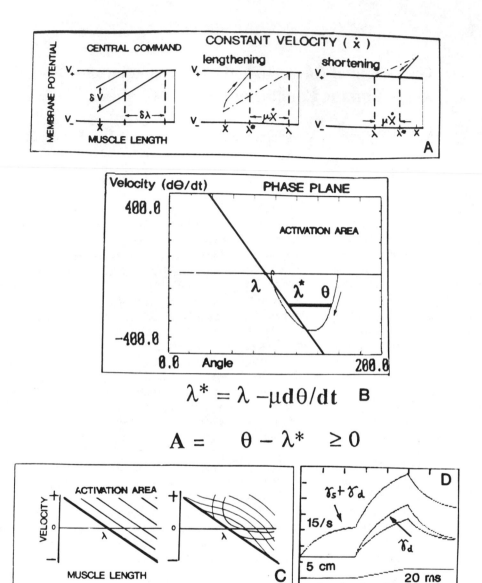

Figure 12.1: The muscle activation area and its properties. *A) Left panel:* Two equivalent measures (δV and $\delta\lambda$) of central control signals. V_-, V_+ are minimum and maximum threshold membrane potentials of the motoneuron. If control signals are constant, membrane potential (V) increases with muscle length, x (solid inclined lines). λ is the static threshold length for motoneuronal recruitment. *Right panels:* Dynamic threshold length (λ^*) decreases with velocity (dx/dt). Trajectories of the membrane potential are shown for statics (dashed-pointed lines) and dynamics (solid lines). μ is the damping coefficient for dx/dt. *B)* Muscle activation area in angular coordinates. The solid line is the border of the *MAA*. The horizontal bar is a measure of muscle activation, A. A simulated phase trajectory for fast active movements is shown. *C)* Two examples of the inner structure of the *MAA* with a stable (left) and a variable (right) order of the motoneuronal recruitment as a function of velocity. *D)* Modelled responses of muscle spindle afferents to a ramp muscle stretch and stimulation of gamma static (γ_s) and dynamic (γ_d) efferents.

12.2 Basic Concepts and Mathematical Equations

12.2.1 Muscle Activation Area (*MAA*)

A constituent part of the *EP* hypothesis is the concept of *MAA* (Feldman, 1974, 1986) which integrates, in a compact form, the non-linear properties of α *MNs*, the effects of afferent influences and descending central commands to α, β, γ *MNs*, and the effects of interneurons mediating afferent and efferent inflows to α *MNs*. The properties of specific phasic and tonic reflexes (the tendon reflex, unloading reflex, stretch reflex) are also integrated in the concept.

Statics

Consider a single α *MN* with intact afferent and efferent connections. Let *V* be an initial, subthreshold membrane potential of the *MN* at an initial muscle length, *x*, when descending control signals are fixed. Now let us suppose that the *CNS* specifies a new magnitude of the tonic control signal. The effect can primarily be measured by a decrement (δ*V*) in the membrane potential (Fig. 12.1a, left panel). The decrement results from both direct influences of the signals on the *MN* and indirect influences mediated by β and γ *MNs*, muscle spindle afferents, and interneurons. A quasi-static stretch of the muscle from the initial length results in increasing depolarization as a function of *x* because of the proprioceptive feedback. The threshold membrane potential (V_+), and consequently the recruitment of the *MN*, will be reached at a muscle length $λ_1$ if the central command is absent and at a muscle length $λ_2$ if the central command (δ*V*) is present. Thus, the command is expressed as a decrement (δλ) of the threshold muscle length at which the *MN* is recruited (Figure 12.1a). Motoneuronal activation in statics occurs when:

$$x - λ > 0 \qquad (12.1)$$

where *x* is associated with the actual muscle length and λ with the threshold muscle length. In the suprathreshold area, motoneuronal firing is an increasing function of $x - λ$.

Each *MN* has its own threshold $λ_i$, and all thresholds are interrelated so that:

$$λ_i = f_i(λ) \qquad (12.2)$$

where $λ = λ_1$ is the threshold of the *MN* which is recruited first. According to Burke et al. (1976), motor units are recruited in the order of *S*, *FR*, and *FF* where *S* are slow, fatigue-resistant motor units, *FR* are fast, also fatigue resistant, and *FF* are fast, fatiguable motor units. In the suprathreshold area (*Eq. 12.1*), muscle activation, *A* (i.e. the number of active *MNs* and their firing frequencies), is an increasing function of the difference between *x* and λ. For the purpose of simplicity, muscle activation will be directly associated with this difference if $x ≥ λ$ and 0 if $x < λ$:

$$A = x - λ \qquad (12.3)$$

Invariant Characteristics (ICs)

The static muscle torque, *T*, is a function of muscle activation such that $T = f(x - λ)$. When λ is constant, muscle torque depends only on the muscle length. We call this dependence the invariant characteristic (*IC*). It should be emphasised that the *IC* of the muscle including feedback is not equivalent to the torque-position function obtained when the muscle is deprived of feedback and stretched under constant level of activation. A given *IC* may be characterized by a single threshold; however, muscle activation varies as a function of length.

Dynamics

In dynamics, if the muscle is stretched at a speed, *dx/dt*, the muscle spindle afferents produce an additional speed-dependent component of activity which gives rise to an additional depolarization of the *MN*. As a result, the threshold V_+ will be reached at a muscle length (λ*) which is less than the static threshold length, λ (Figure 12.1A, middle panel). On the other hand, if the *MN* has been recruited, muscle shortening can lead to de-recruitment. In this case, the dynamic threshold length, λ*, also depends on the speed of shortening and is greater than the static threshold length (Figure 12.1A, right panel). On the whole, the threshold λ* is a decreasing function of speed *dx/dt*:

$$λ* = λ - dx/dt \qquad (12.4)$$

where μ is a coefficient having the dimension of time. Note that muscle shortening is considered negative. *Eq. 12.3* for muscle activation is modified for dynamic conditions correspondingly:

$$A = x - \lambda^* = x + dx/dt - \lambda \qquad (12.5)$$

The above relations can be represented in a simple graphical form (Figure 12A,B). The boundary condition $x - \lambda^* = 0$ represents a straight line in a phase plane (i.e. muscle length x versus speed dx/dt; see also Chapter 13 (Wu et al.). The *MAA* is the part of the plane to the right of the line. The position of the line (determined by the λ) can be modified by central commands as discussed above. The slope is associated with the coefficient μ of velocity (*Eq. 12.4*) and reflects the dynamic sensitivity of muscle spindle afferents (see below). The sensitivity is specified by activity of γ dynamic and β *MNs*. Consequently, it also can be modified by central commands. The level of muscle activation is measured by the horizontal distance between the threshold line and the point (x, dx/dt), which represents the current combination of the kinematic variables. The *MAA* can also be represented in corresponding angular variables (Figure 12.1B). In this case parameter λ is the threshold joint angle for recruitment of *MNs* of the corresponding muscle.

A single-joint movement can be represented as a trajectory on the phase plane. If the trajectory enters the *MAA*, muscle activity arises and increases as the trajectory goes deeper into the area. When the trajectory leaves the area, muscle activity disappears. In particular, Figure 12.1B shows a final position of the border of the flexor *MAA* and the trajectory for a fast flexor movement. It can be seen that the trajectory leaves and then reenters the *MAA*. Thus, flexor activation is predicted to be bi-phasic in this fast movement. The muscle activations and de-activations occur after time delays but in the present model these delays were not included.

Inside the *MAA* there are threshold lines for different *MNs* (Figure 12.1C) so that the area has a definite inner structure ("landscape") which allows, in principle, the prediction of the number of active motoneurons and their firing rates as a function of control and kinematic variables. Unfortunately, the structure of the *MAA* is not fully understood and further experimental work is required to map out the *MAA* in detail.

We may consider theoretical examples of the inner structure of the *MAA* with special reference to the problem of ordered or selective recruitment of MNs. In Fig. 12.1C (left panel), the threshold lines of *MNs* do not cross each other. In this case, the order of motoneuronal recruitment (*S, FR, FF*) remains the same irrespective of the method of muscle activation (changes in variables x and dx/dt or control parameters λ and μ). However, reversals in recruitment order are known to exist. In this case, individual threshold lines will cross each other. One possibility is shown in Figure 12.1C (right panel), in which the order of recruitment is speed dependent. The recruitment order of MNs also differs with respect to fatigue. This suggests that the structure of threshold lines in the MAA is non-stationary (i.e., time-dependent).

The above analysis shows that reversals in the order of recruitment of *MNs* are not necessarily associated with the use of specific central inputs to *MNs*. The intrinsic, synaptic organization of the motoneuronal pool, as well as the properties of MNs themselves, can give rise to the reversals. This is consistent with the idea that the activity of a motoneuronal pool is a function of one integral variable, $x - \lambda^*$, as has been suggested in *Eq. 12.5*.

12.2.2 Muscle Spindles

Static and dynamic properties of muscle spindles and their effects on *MNs* have been integrated in the concept of the *MAA*. Consider, however, muscle spindle properties in more detail (cf. Chap. 13), with the purpose of further developing the concept of the *MAA*. We suggest the following differential equation for the firing frequency (*S*) of muscle spindle afferents:

γ static γ dynamic & β *MNs*
$\downarrow$ $\downarrow$

$$S + c\, dS/dt = a(x - l) + b\, dx/dt \qquad (12.6)$$

The first term on the right side of the equation is a static length-dependent component of activity of muscle spindle afferents. The second term is the dynamic speed-dependent component. The arrows indicate the parameters controlled by γ static, γ dynamic, or β *MNs*. An increase in the tonic component of muscle spindle activity is associated with activation of γ static *MNs* and, as a result, with a decrease in the parameter l. The coefficient *b* represents the dynamic sensitivity of spindle afferents. It can be modified by γ dynamic and β MNs. The coefficient *a* is the spindle afferent positional sensitivity. The fact that the sensitivity is different for short and large lengthenings [e.g. see Chapter 13 (Wu et al.)] can be taken into account if we assume that *a* is an appropriate function of lengthening, δ*l*. The second term on the left-hand part of *Eq. 12.6* is associated with decay in spindle primary afferent activity after the end of muscle stretch which is presumed to be exponential with a time constant *c*. *Eq. 12.6* reproduces typical responses of muscle spindle afferents to ramp stretch as well as the effects of stimulation of γ static and γ dynamic *MNs* (Figure 12.1D).

During muscle stretch at a constant speed the speed–dependent component of the spindle primary response has the form $b \, dx/dt - c \, dS/dt = (b - c \, a) \, dx/dt$, where S is derived from *Eq. 12.6*. This component is less than the speed–dependent response $b \, dx/dt$ (*Eq.12.6*) in the absence of decay. Houk and Rymer (1981; see also Chapter 13 (Wu et al.)) have indicated that the gain of muscle spindle afferent responses to velocity can be low during constant velocity stretch. Our model shows that the gain is equal to $b - ca$, i.e. it can be low due to decay while the actual dynamic sensitivity of the afferents to velocity remains high. Thus, in contrast to Houk and Rymer (1981), we see no reason for postulating a complex non-linear dependence of muscle spindle activity on velocity.

Now let us use *Eq. 12.6* to further develop the concept of *MAA*. We assume that spindle afferent signals are transformed linearly to the motoneuronal membrane potential V:

$$V = g \, S + e \qquad (12.7)$$

where g is the gain of the transformation ("the weight" of synaptic transmission) and e is the component of membrane potential associated, in particular, with direct central inputs to *MNs* not depending on the spindle afferent transmission. Notice that the condition of motoneuronal recruitment, $V = V_+$, can be observed in dynamic conditions when $x = \lambda^*$ or in static conditions when $x = \lambda$ but dx/dt and dS/dt equal zero. It follows from *Eqs. 12.6* and *12.7* that:

$$\lambda^* = \lambda - \mu' \, dx/dt + (c/a) \, dS/dt \qquad (12.8)$$

where $\mu' = b/a$ is the ratio of dynamic to positional sensitivity for spindle afferents. It can be seen that the threshold of muscle activation has a time-dependent component due to the decay of spindle afferent activity. Geometrically, this suggests that the boundary line of *MAA* can be shifted by central control signals or by decay of spindle afferent discharge. Consequently, the boundary line can move even though the central control signals are fixed. *Eq. 12.5* for the magnitude of muscle activation remains but *Eq. 12.4* for λ* is replaced with *Eq. 12.8*.

Eq. 12.6 for muscle spindle firing must be modified to be consistent with the fact that muscle spindle afferent activity temporarily disappears during a twitch contraction of the muscle under isometric conditions — a standard test for spindle afferents. To reproduce this effect, it is necessary to take into account the interaction of the contractile and the series elastic components of the muscle. To do so, the length of the whole muscle (*x*) in *Eq. 12.6* must be replaced with the length of its contractile component (x_c). The same holds for all equations related to the *MAA*.

12.2.3 Angular Variables

While considering the co-ordination of activity of flexor and extensor muscles of a joint, it is convenient to use angular variables (joint angle θ, angular velocity $\omega = d\theta/dt$, etc.) instead of linear ones (*x*, *v*, etc.). Flexor length increases and extensor length decreases with increases of joint angle. The symbol λ will now refer to threshold angle: λ_1 for flexor muscles and λ_2 for extensors. In addition, invariant characteristics (*ICs*) will refer to muscle torque/angle functions associated with a constant value of threshold angle λ. The transformation from angular to linear variables may be linearly approximated by $x = m\theta + n$ where *m* is associated with the muscle moment arm about the joint. The sign of *m* is opposite for flexor and extensor muscles. Note that the form of *Eqs. 12.4, 12.6*, and *12.8* is unaffected by this substitution. The condition of activation for the flexor and extensor muscles in angular variables is given by:

$$A_1 = \theta - \lambda_1^* \geq 0; \quad A_2 = \lambda_2^* - \theta \geq 0 \quad (12.9)$$

where λ_1^* and λ_2^* are dynamic threshold angles for activation of the flexor and extensor muscles, respectively.

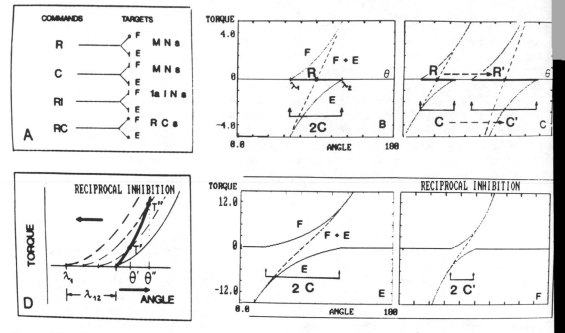

Figure 12.2: Central commands and intermusclar interactions in the λ model. *A)* neurophysiological schemes for the commands. *F MNs* and *E MNs* are flexor and extensor motoneurons; inhibitory synapses are marked with filled circles (see Sections 12.2 and 12.3). *B, C)* R and C commands in terms of shifts of the invariant torque/angle characteristics (ICs). *D-F)* Reciprocal inhibition of antagonist muscles increases the slope of the agonist *IC* and decreases the size of the coactivation area for flexor and extensor muscles.

12.2.4 Central Commands

It is suggested that high brain levels control flexor and extensor muscles as a coherent unit, but there exist a variety of coherent commands (Figure 12.2A) that allow the nervous system to set any combination of flexor-extensor activity (Feldman, 1980). According to the λ model, however, each central command is primarily expressed not in terms of muscle activations but in terms of λ_1 and λ_2. Nevertheless, the names of the commands reflect their typical (but not necessarily universal) effects on flexor and extensor activity: reciprocal (*R*), coactivation (*C*), reciprocal inhibition (*RI*), and Renshaw inhibition (*RC*) commands.

We first review the definition of the *R* and *C* commands (Feldman, 1980). The effects of the interaction of agonist and antagonist muscles mediated by specific afferents and spinal interneurons (e.g., those of reciprocal inhibition) will be considered later in this chapter.

C Command

If $\lambda_2 > \lambda_1$ (Figure 12.2B), the muscles work together in the range $\lambda_1 < \theta < \lambda_2$. Summation of the flexor and extensor torques in this coactivation range gives rise to the total *IC* of the joint (dashed line). The slope (stiffness) of the total *IC* exceeds that of either individual *IC*. The coactivation range expands with the difference between the thresholds. Thus, $C = (\lambda_2 - \lambda_1)/2$ provides a measure of the specific coactivation command. Its modification from *C* to *Cî* (Figure 12.2C) results in a greater distance between flexor and extensor *IC*s and gives rise to a change in the slope of the total *IC*. Note that the *C* and *R* commands are independent such that the coactivation area may vary without shifting the position *R*. For simplicity, we have assumed that the flexor and extensor *IC*'s are the same form. Thus, equal but opposite shifts of these *IC*s will not affect *R*. An additional effect of the command *C* is a linearization of the total *IC*. If $\lambda_1 > \lambda_2$, the operational range of joint angles attainable *in situ* has three zones in which either one of the muscles or none of them is active. The absence of coactivation range can be considered a negative coactivation: *C* < 0, in which case the term total *IC* may also be used. The total *IC* consists then of a flexor and an extensor *IC* situated apart.

R Command

Now consider a command $R = (\lambda_2 + \lambda_1)/2$. Since the form of flexor and extensor ICs is assumed to be identical, the R coincides with the joint angle at which the total IC crosses the θ-axis if $C > 0$ (Figure 12.2B). This command can be associated with the position of the total IC on a torque/angle plane. At the same time, if the external load at the joint is zero, the R coincides with the equilibrium position of the joint. To modify this position from R to $R\hat{I}$, the two individual ICs have to be shifted in the same direction (Figure 12.2C). Physiologically, this command may be elicited by descending central signals with reciprocal effects on flexor and extensor MNs (Figure 12.2A) and thus is called the reciprocal command (Feldman, 1980). Its modification shifts the total IC and the equilibrium position. Consequently, the limb moves to the new position. Damping of the system to avoid oscillations is provided by velocity dependent activity of muscle spindle afferents and the mechanism of muscle contraction [see below and Chapter 5 (Winters) on force–velocity relation].

12.3 Intermuscular Interactions

Muscle spindle afferents have mono- and polysynaptic connections with homonymous and heteronymous MNs. We will describe in terms of the λ model the intermuscular interactions between MNs of synergist, agonist and antagonist muscles mediated by muscle afferents.

12.3.1 Reciprocal Inhibition (RI) of Antagonist Muscles

The system of RI between flexor and extensor MNs has been studied in detail (Hultborn, 1972; Nichols, 1989). It is active during natural movements in man and animals (Feldman and Orlovsky, 1975; Baldissera et al., 1981). Ia interneurons (Ia INs) mediating RI are controlled by descending pathways (Grillner, 1975; Lundberg, 1982) and receive effective inhibitory inputs from Renshaw cells (Hultborn, 1972) and excitatory inputs from antagonist muscle spindle afferents.

From a theoretical point of view, it is important to find an adequate measure of RI. It is clear that the effects of Ia INs are not necessarily expressed in terms of a decrease in the activity of antagonist MNs: the inhibitory action of Ia INs can be subthreshold but not negligible, which can affect the timing and magnitude of future activity of antagonist MNs.

It has been shown that the stretch reflex threshold of the gastrocnemius muscle in the decerebrated cat increases if its antagonist lengthens (Feldman and Orlovsky, 1972). In other words, the RI effect can be measured by a shift in the threshold length of the extensor muscle, under the influence of spindle afferents of the flexor muscle and vice versa. Consequently, we hypothesize that the flexor reflex threshold angle λ^*_1 is modified by a value $\lambda^*_{12} > 0$ conditioned by extensor spindle afferent activity S_2 so that the new threshold angle is given by

$$\lambda^*_1 = \lambda^*_1 + \lambda^*_{12} \qquad (12.10)$$

where λ^*_{12} is an increasing function of S_2. The same effect occurs for the extensor muscle as the interaction between flexor and extensor MNs is mutually inhibitory (Hultborn, 1972). Note that the form of relationship also holds for statics.

Figure 12.2D shows that the RI gives rise to a change in the slope (stiffness) of the IC. The initial angle θ coincides with a threshold angle defined by $\lambda_1 + \lambda_{12}$. This threshold angle is composed of two components: λ_1 associated with central commands and λ_{12} with RI. Let the joint angle increase quasi-statically from θ to θ'' through θ^3. If the inhibitory effect (λ_{12}) were fixed, the flexor torque would increase according to the IC (thin solid line) specified by threshold $\lambda_1 + \lambda_{12}$. However, in fact, the inhibitory effect on the flexor MNs decreases as θ increases and the extensor shortens. This displaces the flexor IC to the left (Figure 12.2D, dashed lines). As a consequence, torques T^3 and T'' are generated. This results in a steeper muscle characteristic (thick solid line) than would occur in the absence of RI. The same effect occurs for the extensor muscle. Thus, RI acts to modulate the positional gain (stiffness) of the system as has been suggested (Nichols, 1989). Therefore, increases in stiffness need not be attributed to the autogenic stretch reflex. The influence of RI on stiffness under static conditions (i.e., $dx/dt = 0$) can also be shown analytically:

$$dA_1/d\theta = d(\theta - \lambda_1 - \lambda_{12})/d\theta$$
$$= 1 - d\lambda_{12}/d\theta \qquad (12.11)$$

where λ_{12} is an increasing function of antagonist spindle afferent activity S_2 and a decreasing function of θ. As a result $dA_1/d\theta > 1$, i.e., the gain of the system with RI exceeds that of the system without RI.

It is of interest to consider the role of RI when a C command acts. To do so, we used a simple model based on $Eq.\ 12.10$ for static conditions with linear dependencies of λ_{12} and λ_{21} on spindle afferent activity S_2 and S_1, respectively. Figure 12.2E and 2F show ICs when RI was either absent (2E) or present (2F). RI can influence the slope of the total IC by changing the size of the coactivation zone and the slope of the ICs of individual muscles. By comparison, without RI, the central C command affects the system's gain only by changing the size of the coactivation zone (Figure 12.2C).

The RI also affects reflex damping which characterises the ability of the system to change muscle activity as a function of velocity. In the absence of RI, this component of damping is equal to μ. When RI acts, the afferent component of damping is:

$$dA_1/d\omega = d(\theta - \lambda^*_1 - \lambda_{12})/d\omega$$

$$\qquad\qquad (12.12)$$

$$= \mu - d\lambda_{12}/d\omega > \mu$$

The inequality is justified by the fact that λ_{12} is an increasing function of extensor spindle afferent activity S_2 whereas S_2 is a decreasing function of angular velocity ω.

12.3.2 Renshaw Cells (RCs) in the
Inter-Muscular Interaction

Agonist RCs inhibit the Ia interneurons that inhibit antagonist MNs (Figure 12.2A). This disinhibitory effect will be expressed as a change in the antagonist λ. For example, the effect elicited by extensor RCs on flexor MNs is denoted by λ_{12}'. This shift for flexor muscles is negative, its absolute value increases if extensor activity A_2 increases. A_2, in turn, is a decreasing function of θ. The effect of RC on the individual and total ICs is opposite to the effect of RI. Using the same analysis described above for RI, it can be shown that RC reduces the slopes of the individual and total ICs. On the whole, both Ia INs and RCs are likely to belong to the system which establishes appropriate values of stiffness and damping.

12.3.3 Mutual Facilitation of Synergists

Ia afferents of muscle spindles terminate on both homonymous and heteronymous MNs. This gives rise to a steeper IC for each of the synergist muscles than would be the case in the absence of the facilitatory interaction. The considerations are similar to those used for the estimation of RI effects (Figure 2E). When a flexor muscle is stretched from a threshold angle, Ia afferents of synergist muscles give rise to a decrease in the flexor threshold angle. As a result, the IC which represents autogenic afferent effects shifts to the left so that the slope of the muscle torque-angle relationship is steeper than in the absence of the facilitatory interaction.

12.3.4 RI and RC Central Commands

Both Ia INs and RCs are effectively controlled by descending systems. We associate this action with the occurrence of independent central commands. How can these commands be expressed in terms of the λ model? By analogy with the MN, the threshold membrane potential of an Ia IN and its recruitment occur at a specific joint angle. An independent central action on Ia INs (we call this action the RI command) can be measured by a decrement or an increment of the threshold angle. Similarly, central RC commands specify the membrane potentials of Renshaw cells. As a result, the inhibitory action of RCs can be associated with a specific threshold level of muscle activity.

Thus, the RI command allows the nervous system to specify the angular range in which stiffness and damping will be enhanced whereas the RC command specifies the range of muscle activation in which the RI action is attenuated.

12.4 Muscle Torques:
Hill Force-Velocity Relation

The dependence of muscle force on velocity (e.g. see Chapters 1, 5) reflects the mechanism of attachment and detachment of actin-myosin cross-bridges (Descherevsky, 1977). The force-velocity relation has usually been studied under conditions of constant electrical stimulation of the muscle. It is a non-trivial problem to apply this relation to a muscle in vivo with varying activity. Within the framework of the λ model, we offer the following solution. Hill's equation will be represented in terms of muscle torque and angular velocity and refer only to the contractile component of the muscle. We represent the equation for torque (T) in the general form:

$$T = h \, (P_o, \omega_c) \qquad (12.13)$$

where ω_c is the velocity of the contractile (c) component, h is an increasing function of ω_c; P_o is isometric torque specified by muscle activation. Note that $h(P_o, 0) = P_o$ and shortening is considered negative. For a combined consideration of flexor and extensor muscles, it is convenient to transform muscle length into joint angle θ. Furthermore, θ can be decomposed into contractile (c) and series (s) elastic components such that $\theta = \theta_c + \theta_s$. Then we find $\omega_c = d\theta_c/dt$.

To solve *Eq. 12.13*, it is necessary to consider the series elastic component, which is assumed to have a spring-like characteristic:

$$T = g \, (\theta_s) \qquad (12.14)$$

In addition, it is necessary to specify the isometric torque P_o as a function of muscle activation, A. This will be done in several steps. Under static conditions, P_o represents the muscle torque. Thus, P_o is a function of muscle activation, $P_o = f(A)$, where the function f describes the muscle *IC* and $A = \theta - \lambda$. Strictly speaking, A in this equation is static tonic activation. However, the dynamic muscle activation, $A = \theta - \lambda^*$ is equivalent to the static activation, $A = \theta - \lambda\hat{I}$, if the static threshold, $\lambda\hat{I}$, is numerically equal to the dynamic threshold, λ^*. Consequently, the equation $P_o = f(A)$ represents isometric muscle torque irrespective of whether the muscle activation A is static or dynamic and *Eq. 12.9* and $P_o = f(A)$ can be used to find P_o.

Note that the active state, P_o, for a given level of activation, sets in gradually with a time constant. Thus we assume that an additional time-dependent transformation plays a role in the above relation so that:

$$D \, (P_o) = f \, (A) \qquad (12.15)$$

where D is assumed to be an operator of the second order:

$$D \, (P_o) = (1 + \tau_1 \, d/dt + \tau_2{}^2 \, d^2/dt^2) \, P_o \qquad (12.16)$$

The time constants, τ_1 and τ_2, characterize the calcium-dependent proceess of the excitation-contraction coupling.

To make the above relations complete, an equation of motion has to be specified. In addition, it is necessary to specify the timing of central commands. For single joint movements the equation of motion has the form:

$$I \, d\omega/dt = T_2 - T_1 + L \qquad (12.17)$$

where T_1 and T_2 are muscle torques, L is an external load (e.g., gravitational torque), and I is the inertia of the movable part of the system distal to the joint. In the double–joint model, Newton–Euler equations of motion were used (e.g., Hollerbach and Flash, 1982).

12.5 Timing of Central Commands for Single Joint Movements

It has been suggested in terms of the λ model (Feldman, 1979; Adamovich and Feldman, 1984) that the brain can specify the velocity and duration of changes in the R command and thus indirectly controls basic kinematic characteristics of movements (speed, duration and magnitude). This hypothesis has been corroborated for movements performed at moderate speeds (Abend et al., 1982)[1] and fast movements (Adamovich et al., 1984).

Figure 12.3A shows a scheme of the formation of the R command for stereotyped point-to-point single-joint movemements realized on a computer λ model. The R command is assumed to result from the summation of individual components each of which produces an elementary shift, R_i, in the equilibrium position. The movement distance is specified by the number of elementary components the brain issues. A priori, the components can be activated either successively (Figure 12.3A, input 1) or simultaneously (input 2). In the case of successive activation, the R command will shift gradually to its final value. The rate of shift

---

[1]Abend et al. (1982) concluded that there is a gradual shift in the equilibrium position for point-to-point movements. This conclusion is quite consistent with the λ model but not with the α model the authors suggest. In the α model, shifts in equilibrium are assumed to be a consequence of changes in muscle activation. Consequently, a gradual shift in equilibrium must be associated with a monotonic change in flexor and extensor *EMG*s. This is not consistent with experimental data showing that non-monotonic, three-burst *EMG* patterns are typical for these movements.

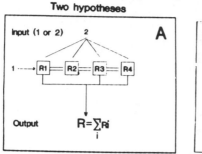

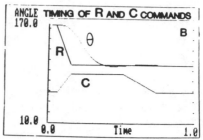

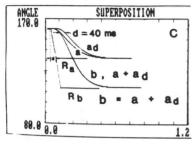

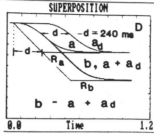

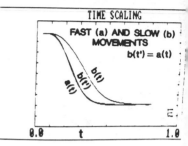

Figure 12.3: Properties of central commands and resulting movements in the λ model. *A)* Hypothetical components (R_i) of the R command. In scheme 1, the components are activated sequentially. In scheme 2, they are activated in parallel. Scheme 1 is consistent with experimental data and was used in the model. *B)* Typical form of R and C commands for fast active movements in the model (θ is joint position). *C, D)* Examples of verification of the principle of superposition for fast (*C*; $dR/dt = 600°/s$) and slow (*D*, 60°/s) movements. *E)* Demonstration of time scaling for movements of the same amplitude.

depends on the time interval between successive activations and can be centrally controlled. If the rate of shift is constant for a set of successively generated movements, but the duration varies, then the movement paths will initially be the same and then will diverge as a function of distance. However, if the rate of shift, and consequently torque, varies then the trajectories will deviate from each other from the very beginning. In constrast, in the case of simultaneous activation, only the amplitude of the shift, but not the rate of shift, can be controlled. The amplitude of shift will depend on the number of simultaneously activated units. The fact that the trajectories of single-joint fast movements coincide at an initial phase (Wadman et al., 1979; Adamovich et al., 1984) is consistent with the hypothesis of successive summation of command components. In addition, the successive model can account for experimental movements in which the paths diverge from the very beginning (cf. Chapter 14 (Gottlieb et al.)).

The C command and other central commands can be graded in a similar way and may be used either in combination with the R command or in isolation, depending on motor tasks. For example, the C command can be set before the movement initiation produced by the R command.

Figure 12.3B shows a hypothetical time course of the R and C commands used in the model for stereotyped one-joint point-to-point movements. The commands gradually change at constant speeds until the necessary final values are reached. The values as well as the speeds are controlled variables. The fastest movements are associated with the maximal values of rate of shift of central commands. After the end of the movement the final R command remains constant but the C command gradually diminishes: the first provides the final equilibrium position and has to be sustained whereas the higher stiffness of the joint the C command creates is necessary only during the movement and after its end the C command may be reduced. The magnitudes and speeds of the R and C commands may be correlated, especially for fast movements. However, this need not be the case. For example, a constant C command can be specified before the movement and remain during the movement.

12.6 The Principle of Superposition and Time Scaling

The hypothesis that gradual constant-rate control signals underlie stereotyped one-joint movements has numerous consequences. Figures 12.3C and D show a computer test of one of them, the principle of superposition. In 3C, two position-time functions, a and b, were elicited by R_a and R_b commands having the same speed (−600°/s) but R_b was twice the duration of R_a. The duration of their common path was 40 ms. Movement a was shifted to the right at time = 40 ms and denoted a_d. Summation of a and a_d resulted in the curve $a + a_d$ that coincides with movement b. The same is true for movements performed at a slow command speed (−60°/s, Figure 12.3D). In general, a single movement trajectory having distance $n\ a$ can be split into n identical trajectories of distance a and generated one after another with a time delay that is equal to the duration of the common path of the large and small movements. The only constraint is that the velocity of the R commands underlying the movements must be the same whereas the rest of the central commands may remain constant or vary as a single-valued function of the R command duration.

The principle of superposition has experimentally been verified for $n = 5$ and used to measure the rate (500–700°/s) of the R command for fast point-to-point movements (Adamovich et al., 1984; Abdusamatov et al., 1987).

Another advantage of constant velocity control signals is that a simple modification of the control velocity provides scaling of movements in time (Figure 12.3E). This is consistent with experimental data. In addition, R commands having an equal duration but different velocities give rise to movements that can be superimposed by scaling in amplitude. For discussions of movement scaling see Schmidt (1982), Chapter 14 (Gottlieb et al.) and Chapter 19 (Seif-Naraghi and Winters).

12.7 Movement Corrections

One more interesting consequence of the constant-velocity form of the control signals concerns corrections of movements in response to an unexpected shift of the target. Provided the new target position is not presented too late, the sweeping of the control signal can be stopped earlier or continued further depending on the new target position. The movement kinematics would be the same as a movement initially planned to the final target, and there will be no inflection point in the movement trajectory. This effect has been demonstrated for both arm and eye movements (Pellisson et al., 1986). (The simultaneous model discussed in Section 12.5 cannot account for these findings.) Note that the rate of the control signals in movements in which a final position must be reached very precisely may be slowed down while the arm is approaching the target.

12.8 Wave Command Generator

The neuronal organization of the brain structures underlying central commands is not known. The wave hypothesis proposed earlier (Adamovich et al., 1984) and described below is one simple possibility. The R command is assumed to be graded by a hypothetical segmental or suprasegmental neuronal ensemble arranged sequentially. Higher brain levels specify constant tonic influences on the ensemble during some time. The amplitude of the tonic signal specifies the rate of propagation of excitation along the neuronal ensemble. This causes new neurons of the ensemble to become tonically active, and these neurons discretely contribute to the R command. The number of neurons recruited is associated with the value of R and the final value of R will depend on the duration of the tonic control signal and its amplitude. The discrete neuronal activations result in individual movements which sum up successively according to the superposition principle. Similar wave structures may be associated with the C and other central commands.

12.9 Single-Joint Movements: EMGs and Kinematics

For simulations of single joint movements, we used λ models of different complexity. In a minimal model, the concepts of MAA, IC, R, and C central commands with appropriate timing, and Newton's equation of motion were used. However, this model did not include Hill's force–velocity relation and intermuscular interactions mediated by muscle afferents. It was essential to demonstrate that even such a simplified model qualitatively reproduces typical EMG patterns and basic kinematic characteristics of single joint movements including the three-burst EMG pattern typical of fast point-to-point

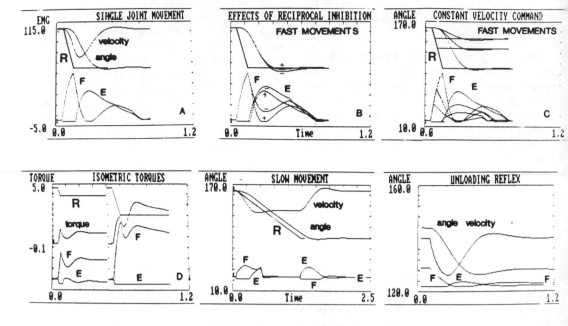

Figure 12.4: Flexor (*F*) and extensor (*E*) activation patterns ("*EMG*s") and kinematics in the model for single joint movements. (See text for details.)

movements. In more complex models, a linear approximation of the Hill force–velocity relation, series elastic and contractile muscle components, and reciprocal interaction between antagonist muscles were used. These additional properties of the system improve somewhat the characteristics of performance but do not change them radically. The value of the reflex damping parameter μ ranged between .05 and .12 *s*. This range encompassed both overdamped and underdamped motions.

12.9.1 Fast Active Movements

Figure 12.4A-C shows typical simulated *EMG* patterns for fast point–to–point movements. The calculations were made with the use of the minimal model (Figure 12.4B, curves with "−") and the model that included the reciprocal interaction between antagonist muscles and the Hill force–velocity relation (Figure 12.4B, curves with "+" and Figures 12.4A and C). A tri-phasic pattern of *EMG* is especially pronounced if there is a reciprocal interaction between antagonist muscles (compare "+" and "−" in B). This is independent of movement amplitude (Figure 12.4C). It should

be emphasised that this pattern results from a constant velocity control signal *R* (600°/s). The resulting movement has a bell-shaped velocity profile (Fig. 12.4A) which is typical for actual movements (Chapter 14 (Gottlieb et al.)). The dependence of the amplitude of the agonist and antagonist burst activity on movement amplitude (Figure 12.4C) is also consistent with experimental data (Brown and Cooke, 1981; see also Chapter 14).

12.9.2 Isometric Torques

In isometric conditions, the same central commands as in Figure 12.4A–C give rise to a fast increase in the agonist muscle torque (Fig. 12.4D). In this model, a series elastic component and a contractile muscle component with a linear dependence of force on velocity were included. For a fast torque production of moderate magnitude (left panel), a bi-phasic *EMG* pattern with reciprocal activation of agonist and antagonist muscles was observed (cf. Gordon and Ghez, 1984). The antagonist muscle activity was totally suppressed in the case of fast torque generations of large magnitudes (right panel).

12.9.3 Active Movements at Moderate Rates

The rate of change of the R command was diminished to about 60°/s to produce movements having a moderate speed. The C command was correspondingly attenuated. As a result, after a short period of acceleration the arm moves at a constant velocity (Figure 12.4E). The movement ends soon after the R command reaches its final value. In contrast, fast movements last relatively long after the end of control signals (cf. Figure 12.4A). During large movements at a moderate rate, the muscles are active in the initial and terminal phases of the movement whereas in the intermediate phase, when the velocity is constant, they are not active at all (Figure 12.4E) (Cooke and Brown, 1990).

12.9.4 Unloading Reflex

The effects of perturbations during posture and movements may be tested with the λ model. As an example, Figure 12.4F shows a simulation of *EMG*s and kinematics in the case of an abrupt unloading of the agonist muscle with constant R and C commands. The model reproduces characteristic features of the unloading reflex if the subject is instructed not to correct the deflections of posture (Feldman, 1986): a silent period in the agonist *EMG*, a transfer of the arm to a new position, and a lower level of tonic *EMG* corresponding to the residual load after the end of movement.

12.9.5 Rhythmic Movements

In the λ model, active rhythmical movements can be elicited in different ways. First, a central generator can produce rhythmical changes in the R command. The other commands can either remain constant or be changed in-phase with the R command. *EMG* can arise, indirectly, from changes in central commands. However, *EMG* responses may also arise rapidly from perturbations during movement while the form of the central commands remains unchanged. The phase of oscillations remains the same in spite of perturbations. Second, rhythmical movements can be received when central control signals remain constant and there is strong coactivation in combination with a high gain for the reciprocal interaction between antagonist muscles. This produces alternative activation of antagonist muscles. The movement amplitude of the limb can be controlled by the coactivation and reciprocal inhibition commands

and the midpoint of the movement range is specified by the R command. Activation and deactivation of antagonist muscles are locked to specific positions of the limb. The period of the oscillation depends on the inertia of the limb. If the limb is arrested, tonic activity in the agonist muscle is established and the oscillations cease. If the limb is released, the oscillations resume with a phase shift equal to the duration of the pause. Thus, the generator has some features reminiscent of that for slow walking (Grillner, 1975).

12.10 Double-Joint Movements: Control Signals, EMGs, and Kinematics

We have examined two–joint point-to-point arm movements using a double–joint model based on all the equations described above, where λ^* is given by *Eq. 12.4*. The model includes six muscles: single–joint flexor and extensor muscles for each joint and double–joint flexor and extensor muscles. Thus, the number of muscles is redundant. We consider first some properties of double–joint muscles in terms of the λ model and then suggest a strategy for the coordinative control of single– and double–joint muscles.

12.10.1 Double-Joint Muscles

The double-joint flexor muscle length is, to a first approximation, a linear function of the two joint angles, θ_1 and θ_2:

$$x = a\,\theta_1 + b\,\theta_2 + c \qquad (12.18)$$

where coefficients are positive. We define a weighted angle, θ:

$$\theta = p\,\theta_1 + q\,\theta_2 \qquad (12.19)$$

where $p + q = 1$ and $p = a/(a + b)$. There is a single-valued correspondence between this angle and the muscle length: $\theta = (x - c)/(a + b)$. The sense of this is that the length $x - c$ is considered as an arc of a circle having the radius $a + b$. The angle corresponding to this arc is θ. The same rule is used for transformation of the threshold length of each double–joint muscle into an angular one, λ. As a result, the condition of activation for the double–joint flexor muscles has the form:

$$\lambda - \theta \geq 0 \qquad (12.20)$$

which is similar to that for single joint flexor muscles.

For double–joint muscles, the R and C commands are defined as for single joint muscles but in terms of weighted thresholds. Thus, we consider commands R_1 and R_2 for the single and R for double–joint muscles. A given value of R is associated with an equilibrium value of weighted angle $\theta = R$. According to *Eq. 12.19*, a given value of the threshold angle can be achieved by different combinations of joint angles. It follows that central control signals to double–joint muscles provide a certain relation between equilibrium joint angles but not their specific values. Within the limits of this relation, the arm can go from one equilibrium combination to another in response to perturbations (indifferent equilibrium). In contrast, control signals R_1 and R_2 for single joint muscles provide a specific equilibrium configuration of the arm. For correspondence between the equilibrium positions specified by single– and double–joint muscles, the control signal R is established in accordance with *Eq. 12.19*:

$$R = p R_1 + q R_2 \qquad (12.21)$$

12.10.2 General Scheme of Performance

A hypothetical scheme for the performance of reaching movements may be described as follows. There is a neuronal analog of the subject's external space in the sense that activation of a neuronal population localized about a point in the neuronal structure is associated with a point (X,Y) of the external space. In particular, this point can coincide with the position of the arm endpoint. When a target is presented to the subject, the localisation of the neuronal activity changes, and, correspondingly, the equilibrium position of the arm endpoint shifts in the external space in the direction of the target. We assume that this shift in equilibrium, if there are no special constraints, occurs along a straight line at a constant velocity until the target is reached. This control signal is then transformed into commands for each joint (R_1, R_2 and the double–joint R command) to elicit an actual movement to the target.

12.10.3 Equilibrium Spaces

To make the above representations more precise, we define two neuronal spaces in the λ model. The first is the space of equilibrium end point positions corresponding to actual end point positions. We call this notion the equilibrium space (*ES*). The second is the space of equilibrium joint configurations which represents all possible combinations of control signals R_1 and R_2. Assuming equal link lengths, m, for the double–joint arm, the transformation between equilibrium joint space and *ES* is:

$$X' = m \sin R_1 - m \sin (R_1 + R_2)$$
$$Y' = m \cos R_1 - m \cos (R_1 + R_2) \qquad (12.22)$$

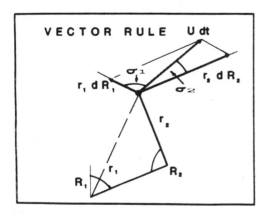
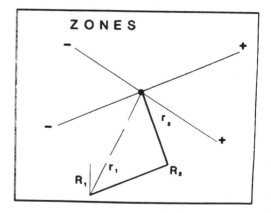

Figure 12.5: Relationship between the end point and joint control signals. Left panel: the end point displacement vector $U\,dt$ can be decomposed into endpoint displacements ($r_i\,dR_i$) due to rotations of individual joints. r_1 and r_2 are the radial vectors at the shoulder and elbow. R_1 and R_2 are the central commands for these joints. Right panel: four zones of inter-joint coordination. The boundary line (with negative slope) orthogonal to radial vector r_1 corresponds to shoulder motion. Flexion is $-$. The other boundary line (orthogonal to r_2) is associated with elbow motion.

Here X' and Y' are the coordinates of the end point of the limb in *ES*. Since R_1 and R_2 are specified by the brain and since the form of the above transformation also describes the relation between actual end point and joint coordinates, the *ES* is an isomorphic representation of external space. The brain produces movements by specifying a velocity vector, U, based on the *ES*. The effect is tantamount to a shift in the equilibrium position of the end point in the actual external space.

Given a goal-directed vector U, it is necessary to specify individual commands for each joint. They must be coordinated in such a way that the vector summation of the displacements of the end point elicited by rotations in each joint together give the desired vector U (Figure 12.5, left panel). The displacement of the end point due to rotation in the *i-th* joint equals the vector product of the radial vector r_i directed from the axis of the joint to the end point and the vector of rotation dR_i. Thus:

$$\Sigma\, r_i\, dR_i = U\, dt \qquad (12.23)$$

Eq. 12.23 applies to systems with any number of links as well as two- and three-dimensional motion. For double–joint limb and two-dimensional space the equation can be solved for R_i:

$$dR_1/dt = -a_1\, U \sin \sigma_2 \qquad (12.24)$$

R_2 is obtained by transposing subscripts *1* and *2*. In *Eq. 12.24*, U is the length of the velocity vector, σ_2 is the angle between the vector U and the displacement of the end point due to rotation of the second joint (Figure 12.5, left panel), $a_1 = 1/r_1 \sin(\sigma_1 - \sigma_2)$ depends only on the current configuration of the limb but not on the direction of the vector U. *Eq. 12.24* shows that the control signal for one joint depends on the parameters of the other joint. This strategy is different from that suggested by Berkinblit et al. (1986).

Thus, several levels of motor control are suggested in the λ model. At a high neuronal level, a constant-velocity vector U is specified that corresponds to the direction and rate of the shift in equilibrium position of the end point. The signal is then transformed into individual commands for each joint and the movement is realized according to the λ model. Indeed, both the magnitude and direction of U can be modified in the course of

movement to correct errors, to accelerate or decelerate the movement, to react to a sudden change in the target position, or to avoid obstacles. Otherwise, if there are no special constraints, the velocity vector remains constant until the target is reached.

The λ model gives a realization of the scheme of motor control suggested by Georgopoulos et al. (1988) [see also Chapter 16 (Hasan and Karst) and Chapter 17 (Flash)] in their study of motor cortex neuronal activity. We can associate the population vector in their study with vector U in the λ model. In addition, *Eq. 12.24* shows that, for any configuration, the control signal is maximal when $\sigma_2 = 90°$. The control signal will be a cosine function of the angle between this optimal direction and the actual movement direction. The corresponding dependence is characteristic of cortical neurons. This allows us to suggest that motor cortex neurons convey information about the target vector U as well as the individual central commands that produce shifs of the equilibrium for each joint.

12.10.4 Velocity Profiles and *EMGs*

Numerous studies have shown that the end point velocity profile in point-to-point arm movements is typically bell-shaped [e.g., see Chapter 17 (Flash)]. According to the λ model, movements are smooth because of the system's natural dynamics [see also Chapter 19 (Seif-Naraghi and Winters)]; the brain does nothing to produce a smooth movement [cf. Chapter 11 (Hogan), Chapter 17 (Flash)]. Figure 12.6 demonstrates that the λ model, with constant velocity shifts in the equilibrium position of the end point, is able to produce a bell-shaped velocity profile of the actual movement of the end point as well as the individual joints.

The magnitude and duration of *EMGs* in the model are a function of movement direction (Figure 12.6A,B). In the absence of joint movement reversals, agonist-antagonist patterns of *EMG* are tri-phasic.

Figure 12.5 (right panel) shows four zones relating movements of the end point to the directions of joint motion. Within each zone, joint motion directions (e.g., ++) remain constant. Joint reversals occur when the direction of U changes from one zone to another. The borders between zones are defined by the equation $\sin (\sigma_i) = 0$ (see

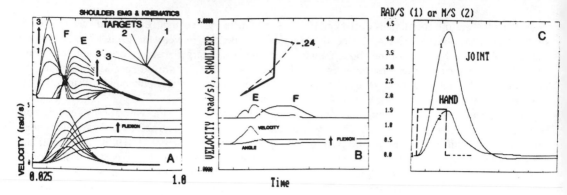

Figure 12.6: Shoulder EMG patterns and kinematics in the double-joint model. The numbers in A and B indicate movement direction (*rad*). *A, B)* Movements without joint reversals. *B)* A movement near to the border between zones shown in Figure 12.5, right panel. Note that the extensor is active first during flexor movement. *C)* Bell-shaped end point tangential and joint velocity profiles. Notice that a constant-velocity control signal (dashed line) underlies these movements.

Eq. 12.24; see also Chapter 16 (Karst and Hasan). Near the boundaries of the zones the model can produce *EMG* patterns that are counter-intuitive. For example, the beginning of flexor movement can be associated with an initial extensor *EMG* as has been shown experimentally (Hasan & Karst, 1989; Chapter 16). Bernstein (1967) has also described rather widespread cases when the movement direction is opposite to what we could predict based on the *EMGs*. Reactive forces and other dynamic effects in multi-joint systems may also affect the relations between *EMGs* and movement kinematics.

12.11 Double-Joint Movements: Corrections and the Principle of Superposition

The principle of superposition formulated above for one-dimensional control signals and resulting movements can be generalized to multi-dimensional performance. We have suggested that the control signal which is specified at the level of the *ES* is a vector. As a consequence, the vector can be decomposed in a sum of two or more components. For example, in case of an orthogonal decomposition we have:

$$U = U_x + U_y \qquad (12.25)$$

It is suggested that the corresponding decomposition is also possible for the resulting movement.

In this connection it is interesting to apply the principle of superposition to movement correc-

tions in response to a sudden shift of the target. This paradigm has been used by Flash (Chapter 17) with another model based on control signals with bell-shaped velocity profiles. In our model, there is a single control vector, *U*, which may be modified (in both magnitude and direction) in response to a shift in target position. On the other hand, we may assume that the old control signal is not interrupted and a new control signal is added. The new signal, *V*, is directed from the old target to the new target (Figure 12.7). Within the framework of the λ model the two strategies are formally equivalent and the resulting movements may coincide.

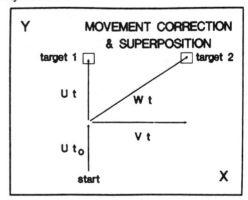

Figure 12.7: Movement corrections and the principle of superposition in a vector formulation of the control processes for multi-joint movements. (See text for details.)

In the model developed by Flash (Chapter 17), control signals are bell-shaped. To form them the nervous system has to estimate beforehand movement distances. In the case of movement correction, the old control signal continues simultaneously with the new correction signal. In constrast, our model assumes a single constant velocity control signal. Thus, the nervous system need not specify movement amplitude in order to initiate movements and reserves the possibility to specify or correct the final equilibrium position of the end point during the course of the movement depending on the current situation.

12.12 The Redundancy Problem

If the number of joints exceeds the dimension of external movement space and its inner, *ES* model, the solution of *Eq. 12.23* is ambiguous, i.e., the system is redundant (see also Chapters 6-11, 16-22). If there are no additional constraints (e.g., those limiting possible coordinations of control variables), the brain can apply one or another optimal control strategy to issue concrete commands. One solution to *Eq. 12.23* is to find commands dR_i/dt which give rise to a vector U' having a least square deflection from the desired vector U. With this constraint, a solution of *Eq. 12.23* can be found with the use of the method of the pseudo-inverse matrix (Gantmaher, 1966; cf. Mussa Ivaldi et al., 1988). Thus, the commands will be minimal in the sense of their "energy" (Gantmaher, 1966):

$$\S\, (dR_i/dt)^2 = \text{minimum} \qquad (12.26)$$

This solution of *Eq. 12.26* could, in principle, be found through dynamic neuronal interaction between and within the end point and joint equilibrium spaces. This interaction gives rise to control signals which minimize the movement deflection from the target vector U (*Eq. 12.23*) which, in turn, minimizes the "energy" of control signals in the sense of *Eq. 12.26* [see also Chapter 19 (Seif-Naraghi and Winters) for a similar approach based on dynamic optimization].

12.13 Conclusions and Future Directions

We have illustrated several ideas and conceptions which seem essential for the understanding of motor control and performance. First of all, it is necessary to find an adequate measure of the action of brain control structures on motoneuronal pools which is independent of kinematic variables characterizing the state of the periphery. This action is associated with the voluntary control of both posture and movement. In this analysis the threshold properties of *MN*s have been taken into account. The control signals are manifest in a change in the threshold length, λ, at which motoneurons become recruited. The λ is an experimentally measurable variable (Feldman, 1986). One more concept - the muscle activation area - is essential in the explanation of *EMG*s and the kinematics and dynamics of single- and double-joint movements. Intermuscular interactions have also been described in terms of the λ model. Importantly, according to the λ model, muscle activities, forces, and movement kinematics need not be iteratively calculated over the course of movement.

We have also illustrated the notion that each level of motor control may be associated with a specific invariant variable and performs specific motor functions (cf. Bernstein, 1967). In particular, a constant value of the parameter R is associated with the postural control of the joint. In this case, the system can generate responses to perturbations and a new equilibrium position will be established if the external load changes. The invariance of the target vector, U, produces an active directional movement of the arm endpoint. The level U does not exclude the behavior associated with the level R but may modify the behavior.

We offered various formulations of the principle of superposition both for control signals and resulting movements. Usually, principles of superposition are considered a characteristic of linear systems. The λ model is thus an example of linear behavior of the system in spite of nonlinearity of its single components. This does not mean that we can neglect nonlinearities.

It is a characteristic of the λ model that it combines both biomechanical and neurophysiological notions. The biomechanical aspects of the λ model can be developed by combining it with equations of the chemical kinetics of muscle contraction (Feldman, 1979). The neurophysiological aspects of the λ model allows us to explain different *EMG* patterns. Potentially, neurophysiological elements of the λ model can be modelled using neuronal nets (e.g., Chapter 20 (Denier van der Gon et al.). In this case, the λ model may be coordinated with experimental data concerning the activity of neurons in the motor cortex and other brain structures.

References

Abdusamatov R.M., Adamovich S.V., Berkinblit M.B., Chernavsky A.V. and Feldman A.G. (1988) Rapid one-joint movements: a qualitative model and its experimental verification. In *Stance and Motion: Facts and Concepts* (Eds. Gurfinkel V.S., Ioffe M.E., Massion J. and Roll J.P.), Plenum Press, New York, pp. 261-270.

Abdusamatov R.M., Adamovich S.V. and Feldman A.G. (1987) A model for one-joint motor control in man. In *Motor Control*. (Eds. Gantchev, G., Dimitrov, B. and Gatev, P.), Plenum Press, New York, pp. 183-188.

Abdusamatov R.M. and Feldman A.G. (1986) Description of electromyograms by a mathematical model of single joint movements. *Biofizika* 31: 503-505.

Adamovich S.V., Burlachkova N.I. and Feldman A.G. (1984) Wave nature of the central process of formation of the trajectories of change in joint angle in man. *Biophysics* 29: 130-134.

Abend W., Bizzi E. and Morasso P. (1982) Human arm trajectory formation. *Brain* 105: 331-348.

Adamovich S.V. and Feldman A.G. (1984) Model of central regulation of the parameters of motor trajectories. *Biophysics* 29: 338-342.

Baldissera F., Hultborn H. and Illert M. (1981) Integration in spinal neuronal systems. In *Handbook of Physiology, Sec. 1, Vol. II, The Nervous System: Motor Control, Part 1*, (Ed. Brooks, V.B.), Williams and Wilkins, Baltimore, pp. 509-595.

Bernstein N.A. (1967) *The Coordination and Regulation of Movements*. Pergamon Press, London.

Berkinblit M.B., Gelfand I.M. and Feldman A.G. (1986) A model for the control of multi-joint movements. *Biofizika* 31: 483-488.

Bizzi E. (1980) Central and peripheral mechanisms in motor control. In *Tutorial in Motor Behavior* (Eds. Stelmach G.E and Requin J.), North-Holland, Amsterdam, pp. 131-144

Brown S.H. and Cooke J.D. (1981) Amplitude- and instruction-dependent modulation of movement-related electromyogram activity in humans. *J. Physiol.* 316: 97-107.

Burke R.E., Rymer W.Z. and Walsh J.V. (1976) Relative strength of synaptic input from short-latency pathways to motor units of defined type in cat medial gastrocnemius. *J. Neurophysiol.* 39: 447-458.

Cooke J.D. & Brown S.H. (1990) Movement related phasic muscle activation. II: Generation and functional role of the tri-phasic pattern. J. Neurophysiol. 63: 465-472.

Descherevsky V.I. *(1977) Mathematical Models of Muscle Contraction.* Nauka, Moscow, pp. 1-160.

Feldman A.G. (1974) Control of the length of a muscle. *Biophysics* 19: 776-771.

Feldman A.G. (1979) *Central and Reflex Mechanisms in Motor Control.* Nauka, Moscow, pp. 1-184.

Feldman A.G. (1980) Superposition of motor programs. II. Rapid forearm flexion in man. *Neurosci.* 5: 91-95.

Feldman A.G. (1986) Once more on the equilibrium-point hypothesis (< model) for motor control. *J. Mot. Behavior* 18: 17-54.

Feldman A.G. and Orlovsky G.N. (1972) The influence of different descending systems on the tonic stretch reflex in the cat. *Exp. Neurol.* 37: 481-494.

Feldman A.G. and Orlovsky G.N. (1975) Activity of interneurones mediating reciprocal Ia inhibition during locomotion in cats. *Brain Res.* 84: 181-194.

Flash T. (1987) The control of hand equilibrium trajectories in multi-joint arm movements. *Biol. Cybern.* 57: 257-274

Gantmaher F.R. (1966) *The Theory of Matrices.* Nauka, Moscow, pp. 1-576.

Georgopoulos A.P., Kettner R.E. and Schwartz, A.B. (19XX) Primate motor cortex and free arm movements to visual targets in three-dimensional space. II. Coding of the direction of movement by a neuronal population. *J. Neurosci.* 8: 2928-2937

Gordon, J. and Ghez, C. (1984) EMG patterns in antagonist muscles during isometric contraction in man: Relations to response dynamics. *Exp. Brain Res.* 55: 167-171.

Grillner, S. (1975) Locomotion in vertebrates: central mechanisms and reflex interactions. Physiol. Rev. 55: 247-304.

Hasan Z. and Karst G.M. (1989) Muscle activity for initiation of planar, two-joint arm movements in different directions, *Exp. Brain Res.* 76: 651-655.

Hogan N. (1984) An organizing principle for a class of voluntary movements. *J. Neurosci.* 4: 2745-2754.

Hollerback, J.M. (1985) Computers, brains and the control of movements. *Trends in Neurosci.* 5: 189-192.

Houk J.C. and Rymer Z.W. (1981) Neural control of muscle length and tension. In *Handbook of Physiology, Sec. 1, Vol. II, The Nervous System: Motor Control, Part 1* (Ed. Brooks, V.B.), Williams and Wilkins, Baltimore, pp.257-323.

Hollerbach, J.M. and Flash, T. (1982) Dynamic interaction between limb segments during planr arm movement. *Biol. Cybern.* 44: 67-77.

Hultborn H. (1972) Convergence of interneurons in the reciprocal Ia inhibitory pathway to motoneurones. *Acta Physiol. Scand., Suppl.* 375: 1-42.

Lundberg A. (1975) Control of spinal mechanisms from the brain. In *The Nervous System, Vol. 2*, (Ed. Tower, D.B.), Raven Press, New York, pp. 253-265.

Mussa-Ivaldi F.A., Morasso P. and Zaccaria, R. (1988) Kinematic networks. A distributed model for representing and regulation of motor redundancy. *Biol. Cybern.* **60**: 1-16.

Nichols T.R. (1989) The organization of heterogenic reflexes among muscles crossing the ankle joint in the decerebrate cat. *J. Physiol.* **410**: 463-477.

Pellison D., Prablanc C., Goodale M.A. and Jeannerod M. (1986) Visual control of reaching movements without vision of the limb. II. Evidence of fast unconscious processes correcting the trajectory of the hand to the final position of a double-step stimulus.

Exp. Brain Res. **62**: 303-311.

Schmidt R.A. *(1982) Motor Control and Learning: A Behavioral Emphasis.* Human Kinetic Publishers. Champaign, IL, pp. 303-326.

Soechting J.F. and Lacquaniti F. (1981) Invariant characteristics of a pointing movement in man. *J. Neurosci.* **1**: 710-720.

Viviani P. and Terzuolo C.A. (1982) Trajectory determines movement dynamics. *Neurosci.* **7**: 431-437.

Wadman W.J., Danier van der Gon J.J., Geuze R.H. and Mol C.R. (1979) Control of fast goal-directed arm movements. *J. Hum. Mov. Studies* **5**: 3-17.

CHAPTER 13

Nonlinear Damping of Limb Motion

Chi-haur Wu, James C. Houk, Kuu-Young Young and Lee E. Miller

13.1 Introduction

The muscle–reflex mechanisms form a feedback system called the motor servo (Houk and Rymer, 1981; Gielen and Houk, 1987), which consists of a muscle, its spindle receptors, and the corresponding reflex pathways back to the muscle. This neuromuscular system mediates the stretch and unloading reflex of the muscle by the feedback. Motivated by a desire to understand the capabilities of biological arms, many researchers have studied limb movements (Asatryan and Feldman, 1965; Feldman, 1966; Freund and Budingen, 1978; Gottlieb et al., 1989a,b; Hasan, 1983,, 1985,, 1986; Hogan, 1984a,b; Polit and Bizzi, 1979; Stein, 1982) and findings have been applied to arm control and task planning (Atkeson and Hollerbach, 1985; Bizzi et al., 1978,, 1982,, 1984; Flash, 1987; Flash and Hogan, 1985; Hogan, 1985; Hollerbach, 1982; Hollerbach and Flash, 1982; Mussa–Ivaldi et al., 1985,, 1988; Uno et al., 1989). Many encouraging results benefitting robotic control have been found in these studies. However, most research efforts thus far have investigated linear dynamic aspects of the motor servo (Bawa et al., 1976; Oguztoreli and Stein, 1976; Rack, 1981; Stein, 1974). The nonlinear aspects of muscle-reflex mechanisms have for the most part been ignored. From the neurophysiological studies (Gielen and Houk, 1984,, 1987; Gielen et al., 1984; Houk et al., 1981a–c,, 1989; Rack et al., 1969; Zahalak, 1981), the results strongly suggested that nonlinear dynamics in the neuromuscular system may have a substantial functional significance for having the

superior adaptability and facile performance in limb control.

Recent experimental findings (Gielen et al., 1984a,b; Miller, 1984) further indicate that the motor servo has two salient nonlinear dynamical features, a stiffness enhancement at small signals and a fractional velocity-dependent viscosity. The presence of the short-range elasticity enhancement indicates that the stiffness of the neuromuscular system is not constant over the stretching range. The position-dependent mechanism will have more impact on the short-range response. As for the presence of the velocity-dependent viscous force response, it shows a nonlinear viscosity that has been postulated to dampen limb movements when variable loads are involved. According to an extensive body of experimental evidence, this nonlinear effect extracted from force–velocity data shows that the force response is proportional to a low fractional power, 0.17, of muscle's velocity. This unusual kind of damping property acts as a control function regulating the limb motion in such a manner that the muscle-reflex system will have a rapid response from the effect of low viscosity in the high-velocity range, but will have a very slow response from the effect of high viscosity in the low-velocity range. This nonlinear damping effect will help in terminating limb motion promptly when velocity decreases to the low-velocity range; similarly, it will enhance the rapid motion when high velocity is needed. These nonlinear dynamics also provide an important capability for the primate limb to adapt to the changes of different loads and motion constraints.

Multiple Muscle Systems: Biomechanics and Movement Organization
J.M. Winters and S.L-Y. Woo (eds.), © 1990 Springer-Verlag, New York

Preliminary results from this study were reported in Houk et al. (1989). The uniqueness of these properties motivates this study of nonlinear damping of the neuromuscular system.

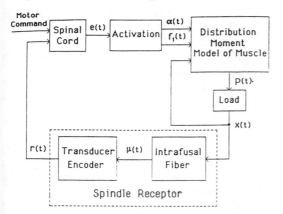

Figure 13.1: Muscle mechanisms.

One model of the motor servo that can replicate the muscle-reflex system is depicted in Figure 13.1 (Gielen and Houk, 1987). This model incorporates a nonlinear description of the behavior of muscle spindle receptors, a delay in the reflex loop via the spinal cord, and a nonlinear model describing muscle mechanical properties. From Hasan's study on muscle spindle receptors (Hasan, 1983), it was found that the response of muscle spindle receptors bears a fractional power relationship to velocity, and that the velocity sensitivity is multiplicative with the length sensitivity. From the experimental results in Gielen and Houk (1987), the reflex pathway through the spinal cord can be simply modeled as a time delay of 0.03 sec. As for the model of muscle mechanical properties, its force activation can be simulated through Zahalak's distribution-moment model (Zahalak, 1981). This model basically is a modification of Huxley's two-state sliding filament model in 1957 (Huxley, 1957). Before the nonlinear damping effect of the limb movement is studied, some of the properties and functions of the above muscle–reflex system will be briefly described and identified in the following manner.

13.1.1 Spindle Receptor Behaviors

Based on Hasan's results in, 1983, the model for producing reflex-induced electromyographic (EMG) activation of muscle (e(t) in Figure 13.1)

consists of three main components: intrafusal fiber, transducer–encoder, and spinal processing. Muscle length $x(t)$ is mechanically filtered by intrafusal muscle fibers in spindle receptors to produce the stretches of the sensory nerve endings $\mu(t)$. The nerve ending stretch is transformed into the discharge rate $r(t)$ of the afferent nerve fiber through the transducer–encoder process. This afferent nerve signal is conducted to the spinal cord where it is processed with the motor command and sent back to the muscle to produce an electromyographic signal $e(t)$ specifying the degree of the muscle activation.

According to Hasan's formulation, the length of the sensory zone $\mu(t)$ as a function of muscle length $x(t)$ is governed by a nonlinear first-order differential equation:

$$\frac{d\mu(t)}{dt} = \frac{dx(t)}{dt} + a\left(\frac{b\mu(t) - x(t) + c}{x(t) - \mu(t) + c}\right)^3 \quad (13.1)$$

where a and b are constants that affect the degree of nonlinearity in the differential equation, and c is a reference length specifying the lower bound of the position range where the muscle spindle responds to stimuli. Hasan also formulated the linear dynamics of the transducer-encoder process as the following equation:

$$r(t) = H\left(\mu(t) + \tau\frac{d\mu(t)}{dt}\right) \quad (13.2)$$

where H represents the small signal sensitivity of the receptors, and τ is the time constant of the encoder dynamics. As for the reflex pathway through the spinal cord, it can also be simply formulated as a time delay of 0.03 sec as follows (Gielen and Houk, 1987):

$$e(t) = r(t - 0.03) \quad (13.3)$$

Although no general solutions of the equations describing this reflex loop have been found, an analytic expression for the asymptotic response at constant stretch velocity, which is frequently adopted as target response in the experiments on the reflex-induced *EMG* activation of muscle, has been derived as follows (Hasan, 1983):

$$e(t) = \frac{H}{b}\left[1 + \left(\frac{dx/dt}{a}\right)^{1/3}\right]\left[x(t-0.03) - c + \tau\frac{dx(t-0.03)}{dt}\right]$$

$$(13.4)$$

From the above solution, it is shown that the reflex response is proportional to only a fractional power of muscle velocity. It is also indicated that the combined effect from the aforementioned non-linear velocity and the stretched length is multiplicative. The dependence of the nonlinear velocity in Hasan's model indicates that the spindle receptors will reflexly promote the operation of a nonlinear viscous damper. Therefore, to develop a model for emulating the muscle–reflex system, these nonlinear dynamical properties have to be included.

13.1.2 Muscle Mechanical Properties

From the study of muscle systems (Huxley, 1957; cf. McMahon, 1984), muscle fibers are divided into sarcomeres based on their striated bands and the sarcomere further consists of actin and myosin filaments (see Chapter 1 (Zahakak) for review).

Because of the complexity of Huxley's model on expressing the bond-distribution functions, Zahalak (1981) estimated the distribution functions with the low-order moments of the approximated distributions for computational simplification [see Chapter 1 (Zahalak) for more details]. Zahalak's model is included in Figure 1.1 for simulating muscle force responses $P(t)$. One input to this model is the muscle length $x(t)$, yielding the muscle mechanical response. The other input is $e(t)$, the simulated EMG response, yielding the neurally mediated component of the force response. The force of contraction $P(t)$ can be computed from the first-order moment of the distribution function $n(y,t)$, which describes the fraction of actin-myosin bonds with length $y(t)$ at time t, and formulated as:

$$P(t) = \alpha C \int y\, n(y,t)\, dy \qquad (13.5)$$

where α, here assumed proportional to $e(t)$, is the degree of muscle activation and C is a constant that depends on parameters such as bond stiffness, sarcomere spacing, cross-sectional area, and wrist geometry (Zahalak, 1981). As for $y(t)$, it is a function of muscle length $x(t)$ and muscle velocity (Gielen and Houk, 1987). In summary, the total effect described by Zahalak's model in *Eq. 13.5* represents a very nonlinear muscle stiffness mechanism.

13.1.3 Muscle-Reflex System

Many results about the muscle-reflex mechanisms have been found in the study of the motor servo. As shown in Figure 13.1, the muscle-reflex system basically consists of two major mechanisms, spindle receptors and the muscle stiffness mechanism. The mechanism of spindle receptors has a unique property of nonlinear damping effect proportional to a fractional power of velocity. The property of muscle stiffness mechanism described by Zahalak's model in *Eq. 13.5* also shows a dynamical effect of nonlinear stiffness. Although Zahalak simplified the mechanism from Huxley's model, it is still complicated and nonlinear. Combining this nonlinear muscle stiffness mechanism with the nonlinear spindle receptors, the complete neurophysiological model of the neuromuscular system is very complicated for applications such as limb motion emulations and robotic controls. Therefore, it is beneficial that some simplified analogous models can be developed for emulating the behaviors of the neuromuscular system.

To study the practicality of this idea, the nonlinear damping effect proportional to a fractional power of velocity presented by the muscle–reflex system will be analyzed first. As nonlinearity is involved in the analysis, the phase-plane approach (Ogada, 1970) will be employed for studying and exploring the characteristics of the system's behavior. Based on the phase-plane approach, a trace of the system's states in the space of velocity versus position will be generated. As time varies, a trajectory of states will represent the state history of the system. This trajectory is called a phase-plane trajectory that represents the system's dynamic behavior. By utilizing this phase-plane approach, the dynamical behavior of the nonlinear damping can then be evaluated.

In Section 13.2, the approach of phase-plane analysis will be introduced and a second-order system with the nonlinear damping effect will be analyzed. The dynamical effects of different loads, damping constants, and stiffnesses will be studied. To provide fitting "template" for developing an analogous muscle-reflex model, in Section 13.3 an experimental setup was designed for collecting different sets of data (Gielen et al., 1984a,b; Miller, 1984). Two types of involuntary movements, step disturbances and ramped stretches, were studied. Each type of movement will provide its unique features for identifying the functional capabilities of the neuromuscular system. The phase-plane analysis in Section 13.2 confirms that the muscle–reflex system is insensitive to variable loads and can adapt to different forces through its unique nonlinear damping effect. Based on this theoretical study, in Section 13.4 two basic models, referred to as the additive model and the product model, are proposed for emulating the response of this nonlinear damping behavior. To study the validity of the models, the responses of two types of involuntary movements, step disturbances and ramped stretches (discussed in Section 13.3) are used as templates for evaluating the models' performance. Through minimizing the mean-square errors in fitting these data, both models can replicate some of the salient features of the responses of the muscle–reflex system; especially the product model. Furthermore, from the effects of hysteresis and stiffness enhancement for small perturbations, a representation based on the product model is evolved for emulating the muscle–reflex system.

13.2 Analysis of a Nonlinear Damping System

From the brief description of the muscle-reflex system in the previous section, it is clear that the neuromuscular system has many nonlinear features that are unique. One specific feature is the nonlinear viscosity. Although the fractional power of velocity dependence indicated by *Eq. 13.4* is 1/3 , being coupled with the nonlinearity in muscle stiffness described by *Eq. 13.5*, a power of 0.17 is extracted from the experimental data (Gielen and Houk, 1987; Miller, 1984). This overall viscous damping effect indicates that the muscle-reflex system will express low viscosity at a high velocity but will transfer into high viscosity at a low velocity. In other words, the nonlinear damping effect will enhance the rapid motion when high velocity is needed; similarly it will help in terminating limb motion promptly when velocity decreases to the low-velocity range. This property shows prominent capability for adapting to different loads and forces. Therefore, before modeling any muscle-reflex system, the property of this nonlinear damping effect on different loads and forces needs to be investigated.

To simplify the initial analysis, the muscle stiffness of the muscle-reflex system is assumed to be a linear stiffness. Then based on the overall fractional power of velocity dependence, estimated to be 1/5 according the experimental data, a simplified model with the estimated nonlinear damping effect and a linear elastic stiffness can be simulated by the following simple second-order system:

$$J\ddot{x} + B\dot{x}^{1/5} + K(x - x_0) = F_a \tag{13.6}$$

where J represents the inertia of the load; B represents the damping coefficient of the system, K represents the linear stiffness of the muscle mechanism, x represents the muscle length, x_0 represents the equilibrium position of the muscle, and F_a represents the applied or observed force. To analyze the behavior of this nonlinear system, the phase-plane approach will be applied in the following manner.

13.2.1 Phase-Plane Analysis

To study the transient response of the simulated system in *Eq. 13.6*, the initial muscle length is assumed to be displaced from equilibrium by an amount $x(0) - x_o$ at position $x(0)$. Since the system is nonlinear, the phase-plane approach will be applied. For the phase-plane analysis, *Eq. 13.6* is rearranged into the following form:

$$\ddot{x} + \frac{K}{J}\dot{x} = -\frac{B}{J}\dot{x}^{1/5} + \frac{K x_0 + F_a}{J} \tag{13.7}$$

Then by defining a δ_t function as follows:

$$\delta_t = \left(-\frac{B}{J}\dot{x}^{1/5} + \frac{K x_0 + F_a}{J} \right) \frac{J}{K} \tag{13.8}$$

Equation 13.7 can be rewritten into the following standard form:

$$\ddot{x} + \omega^2 (x - \delta_t) = 0 \tag{13.9}$$

where $\omega = (K/J)^{1/2}$. Since $\ddot{x} = \dot{x}\, d\dot{x}/dx$, after integrating both sides of *Eq. 13.9* with respect to the position variable x, a circular trajectory in the space of $(x, \dot{x})$ can be generated as follows:

$$\left(\frac{\dot{x}}{\omega} \right)^2 + (x - \delta_t)^2 = R_t^2 \tag{13.10}$$

where R_t is the radius of the circle centered at

(δ_t,0). Based on *Eq. 13.10*, the relationship between $x(t)$ and $\dot{x}(t)$ is well defined at any time t, which depends on the radius $R_t(t)$ and the circle's center. In other words, the phase-plane trajectory of position versus velocity can be easily generated through computing the radius and the circle's center at any time t from the previous values of $x(t)$ and $\dot{x}(t)$. This approach can be implemented through a discrete method as the following equations:

$$\delta_t(t) = -\frac{B}{K}(\dot{x}(t-\Delta t))^{1/5} + \frac{K\,x_o + F_a(t)}{K} \quad (13.11)$$

$$R_t^2(t) = \left(\frac{\dot{x}(t-\Delta t)}{\omega}\right)^2 + (x(t-\Delta t) - \delta_t(t))^2 \quad (13.12)$$

where Δt is the sampling time. By knowing $x(t-\Delta t)$ and $\dot{x}(t-\Delta t)$, the phase-plane trajectory of $x(t)$ and $\dot{x}(t)$ can be generated from *Eq. 13.10* by estimating $\delta_t(t)$ and $R_t(t)$ from *Eqs. 13.11* and *13.12*.

For easily generating the phase-plane trajectory of position versus velocity, *Eq. 13.9* can be renormalized as follows:

$$\frac{d^2 x}{d\tau^2} + (x - \delta_t) = 0 \quad (13.13)$$

where τ is the normalized time and

$$\tau = \omega t \quad (13.14)$$

After integrating *Eq. 13.13* with respect to x, the relationship between x and $dx/d\tau$ is defined by the following circular equation:

$$\left(\frac{dx}{d\tau}\right)^2 + (x - \delta_t)^2 = R_t^2 \quad (13.15)$$

where

$$\delta_t(t) = -\left(\frac{B}{K}\right)\left(\frac{K}{J}\right)^{1/10}\left(\frac{dx}{d\tau}\right)^{1/5} + \frac{K\,x_o + F_a}{K} \quad (13.16)$$

To generate the phase-plane trajectory, the radius R_t at any τ can be evaluated by the values of $x(\tau - \Delta t)$ and $\frac{dx}{d\tau}(\tau - \Delta t)$ as follows:

$$R_t^2(t) = \left(\frac{dx}{d\tau}(t-\Delta\tau)\right)^2 + (x(t-\Delta t) - \delta_t(t))^2 \quad (13.17)$$

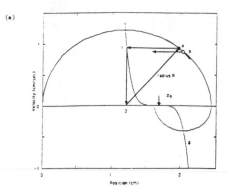

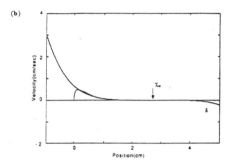

Figure 13.2: *a)* Phase-plane trajectory generation. *b)* Trajectory in low-velocity range.

where

$$\delta_t(t) = -\left(\frac{B}{K}\right)\left(\frac{K}{J}\right)^{1/10}\left(\frac{dx}{d\tau}(\tau - \Delta\tau)\right)^{1/5} + \frac{K\,x_o + F_a(\tau)}{K} \quad (13.18)$$

Figure 13.2a demonstrates how to utilize the phase-plane approach and the above equations for generating the position vs. velocity trajectory. This example assumed that no forces were applied to the system in *Eq. 13.6*, but the system was released from the initial zero position to the equilibrium position x_o. The figure shows how to find the new circle's center from the δ_t function using the previous location of position versus velocity, i.e., point a in Figure 13.2a. Then, based on the newly evaluated radius R_t, the next location in the space can be estimated at point b. After repeating the same procedure from the initial position $x(0)$ to the equilibrium position, the phase-plane trajectory of position vs. velocity is generated as shown.

13.2.2 Properties of Nonlinear Damping System

Observe from Figure 13.2a that the nonlinear function $\delta_t(\tau)$ acts as an asymptotic function regulating the behavior of the phase–plane trajectory of position vs. velocity. While the trajectory is in the large signal range of the $\delta_t(\tau)$ function, i.e. high-velocity range, it will overshoot a certain amount of position depending on where the trajectory intersects the $\delta_t(\tau)$ function. To demonstrate the response within the range of low velocity, another example of phase-plane trajectory was generated in Figure 13.2b. As this figure demonstrates, as long as the trajectory falls into the small signal range of $\delta_t(\tau)$ function, i.e. the low-velocity range, which is referred to as the *dead-band*, the system will take a very long time to reach the equilibrium position as if it is stopped. This property demonstrates that this nonlinear damping system will quickly respond to the desired motion if the phase–plane trajectory is in the large signal range. However, after the trajectory falls into the region of dead–band, the system will quickly damp the responses to stability.

Some other interesting properties can also be observed from *Eq. 13.18*. The damping coefficient B and stiffness K have dominant effects on the system's performance, because they change the region of dead-band governed by the $\delta_t(\tau)$ function. However, the effect of load J is very small due to the power of $1/10$ in the $\delta_t(\tau)$ function. This result means that a system with the specified nonlinear damping effect can adapt to a wide range of inertial loads without causing stability problems typically in a system with linear damping. To demonstrate the above properties, Figures 13.3a-b show the effect of different damping constants. The phase-plane plots in Figure 13.3a. show that a large damping constant B will induce a wider dead-band and make the system tend to undershoot, while the small damping will make the system tend to overshoot. This effect can be observed from Figure 13.3b showing the corresponding position-time plots of Figure 13.3a. As for the effect of different stiffnesses, it has the reverse effect on the system in comparison with the damping constant as indicated in *Eq. 13.18*.

The simulation results are demonstrated in Figures 13.4a,b. By setting the same initial position in the simulation, the simulated system with different stiffnesses will result in different equilibrium positions, as shown in Figure 13.4b. In comparison with a linear damping system, the most prominent effect of this nonlinear damping system is the property of load insensitivity. Figure 13.5a shows the phase–plane plots of this nonlinear damping system under different loads varied from 0.1 to 100, a 1000–fold range. The variables in Figure 13.5a are not normalized for the purpose of making the δ function unique in the phase plane; in other words, *Eq. 13.11* is used for generating the δ_t function. Figure 13.5b shows the time plots of the positions, which are well damped by the system. To demonstrate the load sensitivity of a linear damping effect, in comparison with the nonlinear system described by *Eq. 13.6*, a linear second-order system with the viscous force changing linearly with respect to the velocity is also simulated. The same 1000–fold range of loads used in the nonlinear system were also applied to this linear system. Figure 13.6a shows the counter part of Figure 13.5a for this linear damping system. Figure 13.6b shows the corresponding position–time plots. The results indicate that this linear damping system has the stability problem when the load is too large. Comparing the results, it is obvious that Figures 13.5 and 13.6 show the superiority of the nonlinear damping system over the linear one in the aspect of load insensitivity.

Concluding from the above results, a simple second-order system with muscle's nonlinear damping effect can capture the adaptability of the muscle–reflex system on different forces and loads in a manner that reduces the stability problem. Based on this analysis, we confirm that a nonlinear damping system can provide the adaptability that a muscle–reflex system has. Evolved from the theoretical study of this nonlinear damping property, in Section 13.4 we will propose three models to emulate the muscle's responses by actually fitting some sets of experimental data recorded from an experimental setup described in the following section.

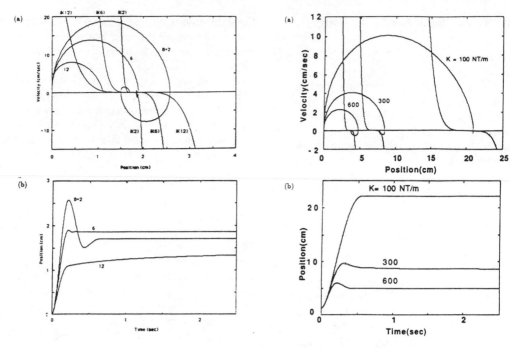

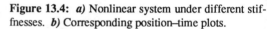

Figure 13.3: *a)* Nonlinear system under different dampings. *b)* Corresponding position–time plots.

Figure 13.4: *a)* Nonlinear system under different stiffnesses. *b)* Corresponding position–time plots.

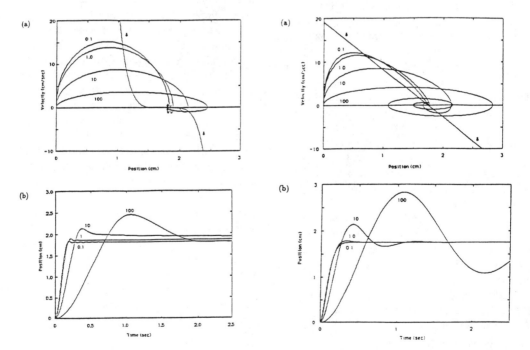

Figure 13.5: *a)* Nonlinear system under different loads. *b)* Corresponding position–time plots.

Figure 13.6: *a)* Linear system under different loads. *b)* Corresponding position–time plots.

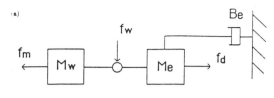

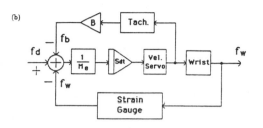

Figure 13.7: *a)* Device body diagram. *b)* System block diagram.

13.3 Experimental Data

13.3.1 Experimental Setup

To be able to provide some templates for developing simplified muscle–reflex models, the responses of two types of involuntary movements were collected from an experimental setup depicted in Figure 13.7 (Miller, 1984). The apparatus was designed to study wrist flexion and extension in the horizontal plane. The subject held a lightweight handle that had strain gauges installed for measuring wrist-force production. The forearm was supported in the horizontal plane by an arm brace, and the subject was seated so that the upper arm was approximately perpendicular to the forearm. This arrangement allowed comfortable and natural wrist flexion-extension motion. The handle was attached to a mechanical stimulus generator designed to operate in one of two modes controlling either position or force. The device was also capable of providing different external dampings for the stability concern of the whole system. Specifically, the parameters of this experimental setup shown in Figure 13.7 are defined as follows: f_d is the applied disturbance force; B_e is a linear damping constant provided by the

device for stability consideration; M_e is a mass simulated from the device; and M_w is the mass of the hand plus the handle connecting the forearm and the device. The measured wrist force f_w is the muscle force f_m plus the inertial force $M_w \ddot{x}_w$ where $\ddot{x}_w$ is the wrist acceleration.

Based on the above setup, two types of involuntary responses, using step force disturbances and ramped stretches, were recorded. The transient responses for force disturbances were recorded by applying four different force steps: 3.88, 8.64, 13.52, and 18.34 N. As for those of ramped stretches, they were recorded by imposing two ramped stretches, 3.62 and 0.92 cm/sec, respectively. The stretched length was 4.6 cm. To identify a subset consisting of the most stereotyped trials from a set of trials, a trial comparison technique (Gielen et al., 1984b; Miller, 1984) was applied. The technique will compare each trial of force trace to every other trial that was generated by the same stimulus condition in the experimental session. Each pair of trials is synchronized at stimulus onset and the difference between the trials is measured at each data point. If at any point, the difference exceeds a "divergence criterion" specified by the user, the trials are assumed to be dissimilar. If the divergence criterion is never exceeded, a "match" is counted for both trials. After all comparisons have been made, a summary table is compiled with the quantity of matches for each trial. From this table, the trials with the greatest number of matches are selected and the most highly stereotyped trials are assumed to represent the response of the motor servo without the addition of other modifying reactions. Naturally, the entries in the summary table will vary with the user's choice of the maximum difference. Experience with a large number of data sets collected under different conditions indicates that a useful divergence criterion is one that yields some trials with very few or no matches whereas other trials match 25-35% of the total number of trials. Under these conditions, taking the 10-20% of trials that match most frequently with the others generally yields a very stereotyped template set. There are some exceptions, however, since the procedure does not require that all trials be similar to a number of other trials. Occasionally, the template set will contain two classes, each showing good intraclass agreement. These may result from differing amounts of co-contraction. Template sets are always visually inspected and where multiple classes are apparent, the most frequent is chosen. Trials in a template set are averaged to produce a single template record.

Based on the above comparison technique, four sets of data displayed in Figure 13.8 were produced for the responses of these four step disturbance forces. Figures 13.8a–b display the corresponding position–time plots and the velocity–position plots representing the phase–plane trajectories. Similarly, two sets of data were produced for the responses of the two imposed ramped stretches. Figure 13.9a shows the force–position plots that emulate the system's elasticity. The stiffness enhancement effect for the small signal is clearly demonstrated in this figure. Furthermore, the force–time plots of the system displayed in Figure 13.9b demonstrate the slow decaying effect during the stretch-relaxation. This slow decaying effect indicates that the nonlinear damping mechanism dominates the response of stretch-relaxation. To understand these unique features from different responses generated through the muscle–reflex system, the properties of these two types of involuntary movements will be briefly described in the following subsections. Then, these data will be used as the templates for developing the proposed muscle–reflex models in Section 13.4.

13.3.2 Properties of Step Disturbance Responses

Enhancement of stiffness for small Perturbations

From the study of viscoelasticity of the wrist motor servo (Gielen et al., 1984b), it is found that if triggered reactions are eliminated, the responses of the neuromuscular system then contain inertial forces due to acceleration of the wrist, muscle mechanical forces, and forces of stretch-reflex origin. This result indicates that the elastic property of the human motor servo is not simply spring-like. A significant difference is a prominent static hysteresis in the position of the limb. This effect appears to arise as a result of a short-range enhancement of stiffness. Stiffness is high for small amplitude displacements about the operating point of the muscle. As the amplitude increases, the short-range effect gives way to

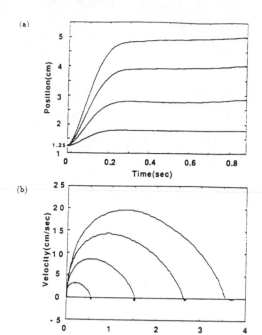

Figure 13.8: *a)* Position–time plots for step disturbance responses. *b)* Corresponding phase–plane trajectories.

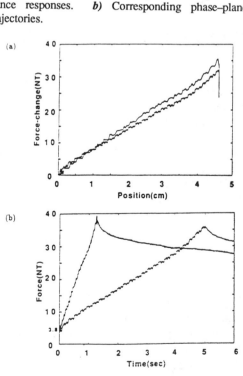

Figure 13.9: *a)* Force–position plots for ramp stretches. *b)* Corresponding force–time plots.

lower stiffness. Such short-range effects are not uncommon. Examples can be cited both for muscle's mechanical properties and for its reflex control. When restricted to approximately 2% of its physiological range of length change, a reflexive muscle exhibits a very linear, relatively high stiffness response to imposed length changes (Matthews, 1972; Nichols and Houk, 1976). Beyond this short range, the cross bridges between actin and myosin begin to pull apart and the muscle stiffness decreases abruptly. Upon termination of the stretch, a short length of time is required to reform these cross bridges and restore the higher stiffness.

Hysteresis

Because of the short-range stiffness enhancement, the limb position is dependent not only on force, but also on the direction to the final position as well. A directional dependence of this type implies hysteresis. The hysteresis in elastic properties can be studied more systematically by contrasting responses to loading and unloading. An experiment was set up for studying this behavior (Gielen et al., 1984b). The responses to loading used force steps that loaded the flexors from the 6 N flexor preload. Initial conditions for the unloading were set to the values for force and position, which were measured following the largest force step (to 25 N) in the loading. Force steps were delivered from that point to unload the flexors. The experimental results obtained in Gielen et al. (1984b) displayed two curves for loading and unloading, respectively. Both curves showed an initial high stiffness for movements less than 0.5 cm. Beyond this point, the curves asymptotically approached a lower stiffness for large amplitude movements. This short–range enhancement of stiffness created an opposite concavity in the two curves and gave rise to hysteresis in the elastic properties. The experiments indicated that the human motor servo behaves in a more complicated manner than does a simple mechanism. Disturbance responses of less than 0.5 cm appear to have up to 4 times higher stiffness than larger responses. This makes the response to loading different from that to unloading. The result is a significant hysteresis in the elastic properties of the system.

As a result of this elastic hysteresis in the motor servo, the task of the central nervous system in directing movements may become more complicated. In order to achieve accurate, swift placement of the limb, the calculation of the motor command may need to include the history of the limb as well as intended final position.

13.3.3 Properties of the Ramped Stretch Responses

It has been shown (Houk et al., 1981b) for primary and secondary endings in the decerebrate cat that their responses to ramped stretches can be described by the following relation:

$$r - r_0 = K (x - x_0) v^n \qquad (13.19)$$

where r is the firing rate of ending; r_0 is the steady-state firing rate at rest length; x_0 is the reference muscle length; v is the ramped velocity; n is the exponent, which is about 0.3 for primary and secondary endings; and K is the scale factor. In addition, it has also been shown that the responses of force, after correction for delay times in the reflex loops, are well described by a similar relation (Gielen and Houk, 1984), namely

$$F - F_0 = C_1 v^n (x - x_0) \qquad (13.20)$$

The above equation implies that for a ramped stretch the response should increase linearly as a function of muscle position. Furthermore, because of the power function v^n, the slope of the increasing response becomes steeper with the stretched velocity.

The aforementioned results show that, except for an early transient phase, responses of force during the ramped stretch can be adequately described by a product relationship of a position-dependent term and a low fractional power function of velocity. This indicates that the properties of the stretch reflex are essentially nonlinear and that any attempt to approximate the action of the stretch reflex by a linear model is restricted to a limited range of positions and velocities. The fact that the parameters in the product relationship tend to show only small variations for different subjects gives additional support to the characterization.

The characterization of the wrist motor servo presented here only applies to the region of the ramped stretch after the initial transient part of the response and prior to stretch plateau. *Equation 13.20* predicts no reflex-induced response at stretch plateau, when velocity is zero. Instead, the experimental records indicate that after stretch velocity becomes zero, force decays with a very slow time course as demonstrated in Figure 13.9b. Even after 5 sec a steady-state level was not reached, and this was found for all stretch velocities. In summary, the mechanism represented by *Eq. 13.20* must be combined with some other mechanisms to deliver the decaying responses.

13.4 Modeling Muscle-Reflex System and Simulations

The phase-plane analysis of a nonlinear damping system suggests that the nonlinear viscosity plays a dominant role in regulating the behavior of the muscle–reflex system. Therefore, to emulate the functional capabilities of the muscle–reflex system, a simplified model must have at least this nonlinear damping property. Inspired by this concept, two simplified models, referred to as the additive and the product model, will be introduced in Sections 13.4.1 and 13.4.2. Both models can capture some of the salient features of the muscle–reflex system. Through identifying the properties of stiffness enhancement for small perturbations, a more complete model referred to as the muscle–reflex model is proposed in Section 13.4.3. This model quite accurately emulates the responses of the neuromuscular system. In the end, the behaviors of the short-range stiffness enhancement and the hysteresis will also be emulated through the muscle–reflex model.

13.4.1. Additive Model

Modeling System

From the structure of the human wrist and forearm, the wrist is connected to the forearm muscle by a tendon which has an estimated stiffness in the range 8000 to 12,000 N/m. Since the tendon behaves like a strong spring, the combined stiffness of this tendon and forearm muscle is assumed here to be close to muscle stiffness [cf. Chapter 5 (Winters)]. By modeling the spindle mechanism as a nonlinear damper with the extracted nonlinear damping effect, the forearm muscle-reflex system can then be simplified as a spring in parallel with a nonlinear damper, both in

series with a strong spring representing the tendon. The mechanical formulation of this model is depicted in Figure 13.10. Under this simplified formulation, the effect of the reflex loop is implicitly incorporated through adjusting the nonlinear damping coefficient and stiffness. As the effects of nonlinear viscosity and parallel elasticity are additive, we refer to this system as the additive model.

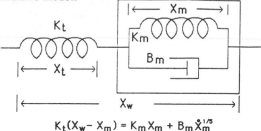

$$K_t(X_w - X_m) = K_m X_m + B_m \dot{X}_m^{1/5}$$

Figure 13.10: Additive model.

From the above formulation, the additive model that emulates the agonist-antagonist pair of muscle-reflex systems acting upon an inertial load at the wrist can be formulated by the following equations:

$$f_w - M\ddot{x}_w = K_t(x_w - x_m) \tag{13.21}$$

$$K_t(x_w - x_m) = K_m x_m + B_m \dot{x}_m^{1/5} \tag{13.22}$$

where x_w and $\ddot{x}_w$ represent the wrist position and acceleration; x_m and its derivative represent muscle position and velocity; M is the inertial load; B_m is the viscous coefficient of the nonlinear damping element; K_t and K_m are the elastic stiffnesses of the tendon and muscle, respectively; and f_w is the force at the wrist.

Simulations

The simulations here are designed to replicate the collected experimental data described in Section 13.3. Two types of involuntary responses, using step forces and ramped stretches, will be simulated. By using the additive model to emulate the muscle-reflex system, the aforementioned experimental system can be simulated with the following equations:

$$f_d - B_e \dot{x}_w - (M_e + M_w)\ddot{x}_w = K_t(x_w - x_m) \tag{13.23}$$

$$f_m = K_t(x_w - x_m) = K_m x_m + B_m \dot{x}_m^{1/5} \tag{13.24}$$

$$f_w = M_w \ddot{x}_w + K_t(x_w - x_m) \tag{13.25}$$

As described in the experimental setup, f_d is the applied disturbance force; B_e is the linear damping constant provided by the device for stability consideration, which is set to 50 N-s/m; M_e is the simulated mass from the device and is set to 2 kg; and M_w is the mass of the hand plus the handle connecting the forearm and the device, which is estimated as 0.65 kg. The measured wrist force f_w is the muscle force f_m plus the inertial force $M_w \ddot{x}_w$. To study the transient responses of step forces using the additive model, *Eqs. 13.23–24* will be utilized by applying a step force input of f_d. The simulation block for studying this type of response is shown in Figure 13.11a. As for studying the ramp-stretch responses, *Eqs. 11.24 and 25* will be the simulation equations by imposing a ramped time course for x_w at the wrist (Figure 13.11b). Since the ramp response of muscle movement has a constant velocity (after an initial start-up), the acceleration term is zero. Therefore, the equations can be simplified after a small time period to:

$$f_w = K_t (x_w - x_m) \tag{13.26}$$

$$K_t (x_w - x_m) = K_m x_m + B_m \dot{x}_m^{1/5} \tag{13.27}$$

where x_w is the imposed ramped stretch at the wrist.

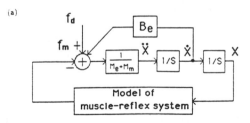

Simulation Block for Step Disturbances

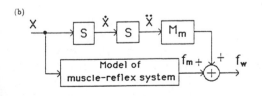

Simulation Block for Ramp Stretches

Figure 13.11: *a)* Simulation block for step disturbances. *b)* Simulation block for ramp stretches.

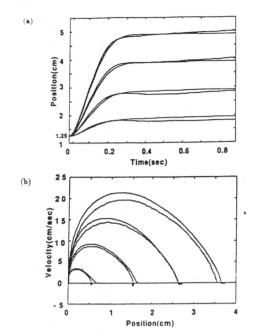

Figure 13.12: *a)* Additive model: position–time plots of simulated results and experimental data for step disturbance responses. *b)* Corresponding phase plane trajectories.

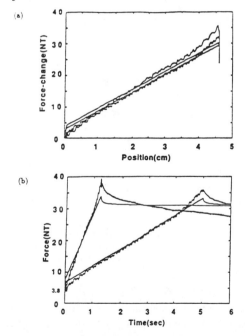

Figure 13.13: *a)* Additive model: force-position plots of simulated results and experimental data for ramp stretches. *b)* Corresponding force–time plots.

Using the additive model in the above setup to fit the experimental data, the model parameters in *Eqs. 13.21–22* will be optimized through minimizing the mean square errors between the simulation results and the experimental data. In the case of fitting step forces, a set of optimized parameters are obtained as follows: $K_t = 8000$ N/m, $K_m = 520$ N/m, and $B_m = 2.6$ N-(s/m)$^{1/5}$. The simulated results overlaying with the experimental data are displayed in Figure 13.12. In these figures the smooth curves are simulated results. Figure 13.12a shows the time plots of simulated and actual wrist positions and Figure 13.12b shows the corresponding phase-plane plots. These results show that the additive model captures the behavior of the transient responses to step disturbing forces quite well. However, the responses shown in Figure 13.12 also indicate that this model needs stiffness enhancement for small movements.

As for fitting the transient responses of ramped stretches, the optimized parameters are obtained as follows: $K_t = 8000$ N/m, $K_m = 620$ N, and $B_m = 8.2$ N-(s/m)$^{1/5}$ Figures 13.13a–b display the simulated results (smooth curves) overlaying with the experimental data. Figure 13.13a shows the stiffness plots of wrist forces versus positions. Figure 13.13b shows the corresponding time plots of wrist forces. Observed from these results, the additive model fits the experimental data reasonably well during the stretch but not well for the initial rising part and the final decaying part.

The results from the above study conclude that the modeling of the nonlinear damping effect in the additive model does capture muscle behavior to some extent. However, because of the incomplete muscle properties in the additive representation, the model does not fit the ramp stretches very well. In particular, it cannot fit the slow decaying behavior shown in the experimental data in Figure 13.13b. To capture this behavior, in the following another model referred to as the *product model* will be proposed.

13.4.2 Product Model

Modeling System

The experimental data of the force response from ramped stretches, the noisy curves in Figure

13.13b, showed two very distinct properties, a near-linear response for ramped stretch and a very slow decaying responses for relaxation. These two distinct properties indicate that during the stretch the muscle–reflex system responds like a linear spring; however, when the stretch ends, the system behaves as a nonlinear damping element to allow stress-relaxation to occur. These facts conclude that a model for the muscle-reflex system should at least be the combination of a linear spring in series with a nonlinear damping element. Specifically, this linear spring should be stiff enough for the near-linear behavior of the ramped stretches and then the nonlinear damping element allows the relaxation to occur. Additionally, in Section 13.3.3 we have described that the response of a ramped stretch can be represented by a product relation of a position-dependent term and a fractional power function of velocity. This product relation can also be observed from the spindle's properties, as described by *Eq. 13.4*. Motivated on the basis of this product relation, a new model referred to as the product model is proposed. The mechanical formulation of this model is depicted in Figure 13.14.

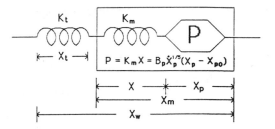

Figure 13.14: Product model.

In this product model, a linear spring is modeled for muscle stiffness to provide the stiffness necessary for describing the near-linear slope part of ramped stretches. Extracted from the experimental data, this stiffness ranges from 300 to 1000 N/m (Miller, 1984). To yield various decaying responses after different ramped stretches end, the nonlinear damping element is modeled as the product of a linear function of position change and a nonlinear viscous force that is proportional to $v^{1/5}$. From the above formulation, an agonist-antagonist pair of reflex-regulated muscles acting upon an inertial load at the wrist can be modeled by the product model with *Eq. 13.21* plus:

$$K_t(x_w - x_m) = K_m(x_m - x_p) = B_p \dot{x}_p^{1/5}(x_p - x_{po}) \quad (13.28)$$

where all terms are defined as in *Eqs. 13.21–22* and x_p and $\dot{x}_p$ are the internal position and velocity of the product model; x_{po} is the bias internal position of the product model; and B_p is the scaled damping coefficient of the product model.

Simulations

As in the additive model, simulations were designed for replicating the same experimental results recorded from four different step forces and two ramped stretches. After incorporating the product model into the experimental system described earlier, the simulation model for analyzing the transient responses of step disturbing forces (f_d) utilizes *Eqs. 13.21, 13.23,* and *13.28.* As for analyzing the transient responses of ramped stretches imposed at x_w, due to no accelerations in the ramped stretches, the simulated system is simplified by *Eq. 13.26* and *Eq. 13.28.* The simulation blocks are the same as in Figure 13.11, except using the product model for the muscle–reflex system.

Since the tendon represented as K_t is in series with the more compliant muscle represented as K_m, the combined stiffness will be merely represented by the neuromuscular stiffness. Therefore, in our simulation we have eliminated the tendon effect by assuming that $x_w = x_m$. In the sense of minimizing the mean square errors between the simulation results and the experimental data, after fitting the transient responses of four step forces, a set of parameters were obtained as follows: $K_m = 640\,\text{N/m}$, $B_p = 4200\,\text{N/m} \cdot (\text{s/m})^{1/5}$, $x_{po} = -0.004\,\text{m}$. The simulated results overlaying with the experimental data are displayed in Figures 13.15a–b. The simulation results show that the product model emulates the muscle behavior quite well in the case of larger step forces, but it still needs stiffness enhancement for smaller step movements. In comparison with the results in Figure 13.12a–b, simulated from the additive model, the product model can capture the final drifting toward the equilibrium position much better than the additive

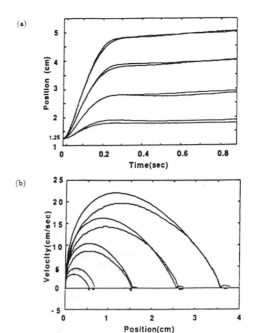

Figure 13.15: Product model. *a)* Position–time plots of simulated results and experimental data for step disturbance responses. *b)* Corresponding phase-plane trajectories.

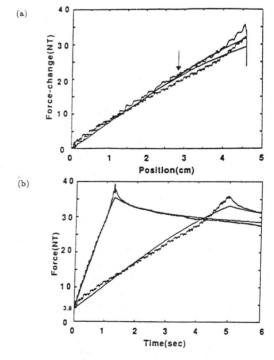

Figure 13.16: Product model. *a)* Force–position plots of simulated results and experimental data for ramp stretches. *b)* Corresponding force–time plots.

model. Quantitatively, the mean square errors for fitting all four curves in the product model are about 60% of the fitting errors in the additive model. This improvement over the additive model is quite substantial.

After applying the same optimization technique to fit the transient responses of ramped stretches, the parameters for the product model are obtained as follows: K_m = 740 N/m, B_p = 2600 N/m•(s/m)$^{1/5}$, x_{po} = –0.028 m. Figures 13.16a–b display the fitting results overlaying with the experimental data. In Figure 13.16b, the force–time plots indicate that the model fits the experimental data quite well in both the stretch and decaying parts, except the small signal range and the sharp force notches in the beginning of the stretch relaxation. In comparison with Figure 13.13, the results simulated from the additive model, the product model is much better in replicating the behavior of the transient responses of ramped stretches. Quantitatively, the mean square errors for fitting the two stretch responses in the product model are appreciably less than the additive model by a factor of 0.32. This improvement over the additive model is excellent.

In summary, the simulation results show that the product model is much better than the additive model in emulating muscle behavior, especially ramped stretches. The improvement over the additive model shows that the nonlinear product relation in the spindle mechanism is a very important function in modeling the muscle–reflex system in order to emulate the overall responses of ramped stretches. However, in order to replicate the overall responses through minimizing the mean square errors, the model could not produce enough stiffness enhancement for the small signal. This property of stiffness enhancement is clearly indicated by the experimental data, both in Figures 13.8 and 13.9. The experimental data of the four step disturbances in Figure 13.8 had different elastic stiffnesses as 705, 504, 486, and 486 N/m, respectively. The short-range enhancement was demonstrated through that the smallest response had the largest elastic stiffness and the largest response had the smallest one. As for the ramped stretches in Figure 13.9, the stiffness enhancement was demonstrated through the initial jump in force and the sharp force notch in the beginning of the stretch relaxation. Through our simulations, we also found that adjusting parameter x_{po} in *Eq. 13.28*

would have the effect of adjusting the point at which separation occurs for different ramped stretches. This point also demarcates the range of small signals for stiffness enhancement. The fitting results shown in Figure 13.16a demonstrate that the point of separation and the upper bound of stiffness enhancement range is around the position of selected x_{po} (2.8 cm in Figure 13.16a). The simulation results suggested that if the model satisfied the required stiffness enhancement for the small signal, the responses of large movements could not be replicated. To fit the overall responses in the case of ramped stretches, a desired short range for stiffness enhancement can then not be maintained. This effect explains why different values of x_{po} were obtained for fitting the step disturbances and ramped stretches.

In comparing the product model with the neuromuscular system shown in Figure 13.1, the differences are a separate mechanism for muscle mechanical stiffness and a natural input for motor commands from the brain. Without the input for motor commands, voluntary movements cannot be simulated, but it should not affect the involuntary movements requiring no motor programs. Therefore, we believe that the missing separate stiffness mechanism in the product model may be the reason why the model must produce less stiffness enhancement for the short-range signal in order to capture the overall features within a wide range of movements. Therefore, to compensate for the missing functions in the product model, in the following subsection we will propose a more complete model to emulate capabilities of the neuromuscular system.

13.4.3 Muscle-Reflex Model

Modeling System

The simulated results from the product model show that this model emulates the involuntary movements of the neuromuscular system quite well except for not producing enough stiffness enhancement for the small signal. Due to the properties of the nonlinear damping, the response of the product model should show a higher elastic stiffness for the short-range signal because of the dead-band discussed in Section 13.2; however, when the signal is out of the dead-band, the response should show a lower elastic stiffness because of the enhanced rapid motion. This property demonstrates that the product model can produce the stiffness enhancement for the small signal. However, through our simulations, if the system captured this stiffness enhancement, then the response for the large-signal range could not be replicated. Specifically, the stiffness for the large-signal range must be reduced to satisfy the stiffness enhancement at the small signal. As the muscle–reflex system depicted in Figure 13.1 has a separate muscle mechanism for stiffness effect, we believe that by modeling a separate muscle stiffness mechanism to compensate for the overall stiffnesses in the large-signal range, enough short-range stiffness enhancement can then be produced by the nonlinear damping to replicate the overall responses of different movements. Furthermore, the muscle–reflex system also accepts the motor commands sent from the brain that are responsible for voluntary movements of limbs. Therefore, to compensate for the missing muscle properties in the product model and to provide a natural input for motor commands, a model with an architecture similar to that of the neuromuscular system is proposed.

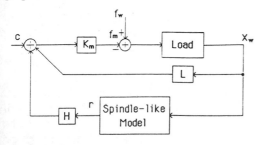

Figure 13.17: Muscle–reflex model.

On the basis of the aforementioned needs, a muscle–reflex model evolved from the product model is developed and depicted in Figure 13.17. The architecture of the proposed muscle-reflex model is similar to that of the neuromuscular system with a spindle-like model and a simple muscle stiffness mechanism. As demonstrated in the previous section, the product model can capture the overall performance of the neuromuscular system to a great extent. Therefore, it is reasonable to design the spindle-like model with the same functions as the product model dominating the response. To allow the nonlinear damping to produce enough short-range stiffness enhancement to replicate the overall responses of a wide-range

of movements, a linear muscle stiffness mechanism is added for compensating for the overall stiffness. Based on this discussion, the complicated muscle force-distribution model described in Section 13.1 is simplified by a simple stiffness K_m in Figure 13.17 representing the muscle mechanical stiffness. This mechanical stiffness is different from the total neuromuscular stiffness, defined in the additive and product models. The total neuromuscular stiffness is a composite elastic stiffness including both the muscle mechanical stiffness and the reflex stiffness in the spindle mechanism. In the previous two models, we have assumed the neuromuscular stiffness as this composite elastic stiffness because of no separate muscle stiffness mechanism. However, in order to produce enough short-range stiffness enhancement, in this new model the muscle mechanical stiffness and the reflex stiffness will be defined separately.

In this muscle–reflex model depicted in Figure 13.17, the discharge rate of the spindle-like model, r, scaled through a reflex gain coefficient, H, is combined with the motor command, denoted as c. The combined signal will produce a simulated, reflex-induced *EMG*. Since the muscle length will change with the load position, a linear feedback with a gain coefficient L is used to represent this linear effect of length change, such as from the muscle length–tension and the other unmodeled linear position feedback terms (Houk and Rymer, 1981; cf. McMahon, 1984). The combined signal of the simulated, reflex-induced *EMG* and the feedback signal of length change will then emulate a length change to produce the muscle force, f_m. The resultant muscle force combined with the disturbing forces, f_w, sensed from the load will then move the limb. Since the strong tendon stiffness can be ignored from our simulation as described in the product model, in this new model we will assume that the wrist position change, x_w, is the same as the muscle position change, x_m. Based on the above formulation, an agonist-antagonist pair of reflex-regulated muscles acting upon an inertial load at the wrist can be represented by the muscle–reflex model with the following equations:

$$f_w = f_m + M\ddot{x}_m \qquad (13.29)$$

$$r = K_r(x_w - x_p) = B_p \dot{x}^{1/5}(x_p - x_{po}) \qquad (13.30)$$

$$f_m = K_m(L x_w + H r - c) \qquad (13.31)$$

By ignoring the tendon stiffness, the wrist position change x_w is the same as the muscle position change. In *Eq. 13.30*, K_r represents the reflex stiffness in the spindle mechanism and K_m in *Eq. 13.31* represents the muscle mechanical stiffness. The other parameters are described earlier. Specifically, the above three equations represent, respectively, the load system model, the spindle-like model, and the spinal cord processing and the muscle stiffness mechanism. In summary, the proposed muscle-reflex model shown in Figure 13.17 is a simplified analogy of the neuromuscular system shown as in Figure 13.1.

In this new model, the effect of the muscle stiffness mechanism can be taken away by setting the value of parameter L to zero. In addition, by setting both K_m and H to 1, the muscle–reflex model is the same as the product model. By adding an extra linear stiffness to compensate for the overall stiffnesses for a wide range of movements, the reflex response of the spindle-like model can be adjusted to replicate the short-range stiffness enhancement. Although the linear feedback gain coefficient, L, can be used to adjust the overall stiffness of the system, when $L = 1$, K_m will represent the muscle mechanical stiffness that describes the linear effect of muscle length-tension. Therefore, to simplify our discussion and ignore other linear position feedback terms, we will set L to 1 and adjust all the other parameters to emulate the neuromuscular responses. Based on this formulation, in the following the proposed muscle-reflex model will be simulated to fit the experimental data.

Simulations

By setting L to 1, simulations are designed for replicating the same experimental data recorded from the involuntary movements of four step forces and two ramped stretches. Since the involuntary movements involve no motor commands, the model will be simplified by setting that $c = 0$. To analyze the transient responses of step disturbing forces, after incorporating the muscle-reflex model into the experimental system, the simulated model is *Eq. 13.30* plus the following:

$$f_d - f_m = (M_e + M_w)\ddot{x}_w + B_e \dot{x}_w \qquad (13.32)$$

$$f_m = K_m (x_w + H r) \qquad (13.33)$$

$$f_w = M_w \ddot{x}_w + f_m \qquad (13.34)$$

where f_d is the disturbing force applied from the experimental device to the wrist. For the same experimental setup, the values of parameters, B_e, M_e, and M_w are the same as those in the product model. As for analyzing the transient responses of ramped stretches, due to the constant velocity, the simulated model can be further simplified to *Eq. 13.30* plus the following:

$$f_w = f_m = K_m (x_w + H r) \qquad (13.35)$$

By imposing x_w to be ramped stretches, which means that $f_w = f_m$, the spindle-like model represented by *Eq. 13.30* will then respond to induce a reflex response. As for *Eq. 13.35*, its condition of zero acceleration forces x_w to be the imposed ramped stretches. In both simulations of step disturbing forces and ramped stretches, the parameters are optimized to fit the experimental data.

In the sense of minimizing the mean square errors between the simulated results and the experimental data, after fitting the transient responses of four step forces, a set of parameters were obtained as follows: $H = 0.00333$, $K_m = 138$ N/m, $K_r = 1100$ N/m, $B_p = 5700$ N/m•(s/m)$^{1/5}$, and $x_{po} = -0.004$ m. The simulated results overlaying with the experimental data are displayed in Figure 13.18. The short range for stiffness enhancement is achieved by fitting the parameter x_{po} to the desired range. In comparison with Figure 13.15 simulated using the product model, the simulated results from this muscle–reflex model emulate the step force responses very well. Quantitatively, the mean square errors for fitting all four curves are about the same as the fitting errors in the product model. Qualitatively however, this model demonstrates a certain degree of refinement in comparison with the product model. An interesting result is that the fitted value of K_m is in the range of muscle mechanical stiffness measured from the physiological experiments (Miller, 1984), in which three different muscle mechanical stiffnesses, 130, 160, and 230 N/m, were measured from three subjects, respectively.

After applying the same optimization technique to fit the transient responses of two ramped stretches, the parameters were obtained as follows: $H = 0.00333$, $K_m = 231$ N/m, $K_r = 1150$ N/m, $B_p = 2200$ N/m•(s/m)$^{1/5}$, $x_{po} = -0.008$ m. This result shows that K_m is still in the range of muscle mechanical stiffness. Since the step disturbances and ramped stretches were recorded from two dif-

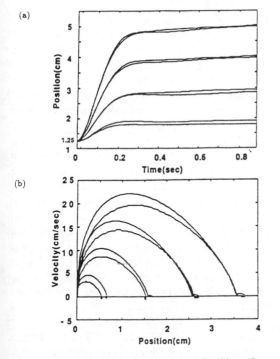

Figure 13.18: Muscle-reflex model. *a)* Position–time plots of simulated results and experimental data for step disturbance responses. *b)* Corresponding phase–plane trajectories.

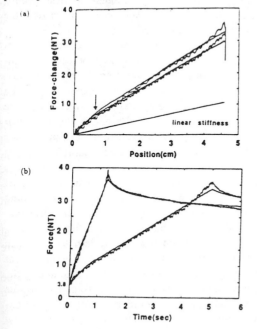

Figure 13.19: Muscle-reflex model: **a)** Force–position plots of simulated results and experimental data for ramp stretches. **b)** Corresponding force–time plots.

ferent subjects, different mechanical stiffnesses are expected. The short range for stiffness enhancement is also achieved by fitting the parameter x_{po} to a desired range (0.8 cm), which is very close to the value of 0.4 cm obtained in fitting the step disturbances. In Figure 13.19a, which displays the force–position plots, x_{po} demarcates the separation for different ramped stretches. In comparison with Figure 13.16a, where x_{po} was 2.8 cm fitted from the product model, this new model has provided good agreement in fitting both step disturbances and ramped stretches. The corresponding force–time plots are displayed in Figure 13.19b. In Figure 13.19a, the linear-stiffness line represents the effect of length–tension from K_m. As expected, the short-range stiffness enhancement is clearly demonstrated in Figure 13.19a. Besides the stiffness improvement at the small signal, the sharp force notch in the beginning of the stretch relaxation shown in Figure 13.19b has also been observed in comparison with Figure 13.16b. Quantitatively, the mean square errors for fitting the two stretch responses are about a fraction of 0.55 relative to the fitting errors in the product model. Such notable improvement is the result of adding a separate stiffness mechanism that represents the muscle length–tension effect.

In summary, the muscle–reflex model can accurately simulate the responses of the limb involuntary movements. In comparison with the neurophysiological model of the muscle–reflex system depicted in Figure 13.1, the proposed muscle–reflex model is much simpler for simulation or actual implementation. To conclude the modeling and simulations, we will use this model to discuss and emulate the important behavior of hysteresis in the neuromuscular system.

Short-Range Enhancement and Hysteresis

In comparison with the product model, a muscle stiffness mechanism is added into the muscle–reflex model for allowing the nonlinear damping to produce a desired stiffness enhancement for small signals. Because of this property of short-range enhancement, the loading and unloading of muscle movements will have a hysteresis behavior as described in Section 13.3.2. To demonstrate this behavior through the muscle–reflex model, the set of parameters' values obtained in the case of the ramped stretches were used to generate the loading and unloading curves for simulating the phenomenon of the hysteresis.

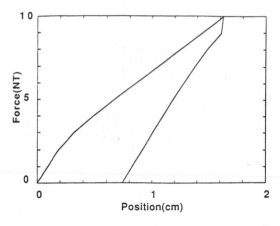

Figure 13.20: Simulated loading and unloading curves for short-range enhancement of stiffness and hysteresis.

To generate the loading curve that demonstrates the short-range stiffness enhancement, step disturbing forces ranging 0 to 10 N were applied to the system. The initial force step was 0.5 N for producing a short movement. To have a good resolution for the short-range stiffness enhancement, the sampling force step was 0.1 N from range 0.5 to 1 N. From range 1 to 10 N, the force step change was 1 N. Each knot point on the curve represents an applied step force and its corresponding response after 5.0 sec. The same range of force steps were unloaded from the largest force step by using the same set of parameters. Similarly, the step forces and the corresponding responses after 5.0 sec were recorded. Both curves of loading and unloading were displayed in Figure 13.20. The loading curve clearly demonstrates the phenomenon of short-range stiffness enhancement. When simulating the unloading, we found that the system needs to unload a certain amount of force to counterbalance the drifting forces due to the hysteresis. Therefore, when we simulated the first unloading force from the largest force step, 10 N, we had to unload at least 1 N in order to have the net reverse movement. This unloaded 1 N force caused a sharp drop as shown in the beginning of the unloading curve in Figure 13.20. Because of the effect of the hysteresis, there exists a zone between the loading and unloading curves. This zone clearly represents that the muscle-reflex model possesses the property of hysteresis presented by the neuromuscular system.

13.5 Discussion and Conclusion

The adaptability of limb movements in performing the constrained task is clearly superior to that of current robotic systems. To study the practicality of modeling the muscle-reflex mechanisms, the features of the muscle-reflex system are analyzed and then modeled in this chapter. Basically, the neuromuscular system consists of many nonlinear features but with two major mechanisms, the muscle stiffness mechanism and the spindle receptors. The response of the neuromuscular system demonstrates that the system has a unique nonlinear damping property that may be responsible for the limb's adaptability in performing the constrained task. Extracted from an extensive body of experimental evidence, this nonlinear damping property is proportional to a fractional power of muscle's velocity. On the basis of this finding, a phase-plane analysis was performed on a second-order system with such a nonlinear damping. The theoretical results show that such a system has the capability of adapting to various loads and different forces without changing the system's parameters. The encouraging results from this analysis suggest that it is possible to model the functional capability of the neuromuscular system. Initiated by these results, three models were developed to model the neuromuscular-like system.

In the physiological study, an experimental device was set up for recording two types of involuntary movements, step force disturbances and ramped stretches. The recorded experimental data were used as templates for modeling the neuromuscular system. Based on the nonlinear damping effect proportional to a fractional power of muscle's velocity, a simple model referred to as the additive model was formulated first. The simulation results showed that the additive model emulated the responses of the step disturbing forces well but not the ramped stretches. Nevertheless, this model captured the salient features of the neuromuscular system to a certain extent. This fact demonstrates that the nonlinear damping is a necessary property in modeling the muscle–reflex system. Comparing the results simulated from the additive model with the experimental data, we also found that the slow decaying behavior of stretch relaxation should be the product relation of position and velocity. This product relation can also be found in the property of the spindle receptors. On the basis of this finding, another model referred to as the product model was further introduced. The results simulated from this model showed great improvement over the additive model, especially for the responses of ramp stretches. Although the product model emulated the limb responses to a great extent, in order to replicate the overall responses of a wide range of movements, it could not produce enough short-range stiffness enhancement. Additionally, the product model also lacks a natural input for the motor command from the brain that is responsible for the voluntary movement. For these reasons, a muscle–reflex model that can accurately emulate the response of the limb involuntary movement was developed.

During the process of modeling, the model used for phase–plane analysis was eliminated from representing the neuromuscular system. The reason is that although this model is similar to the additive model, without the muscle tendon as an absorber for ramped stretches, the system will stop rigidly when a stretch ends without having any response associated with stretch relaxation. In both the additive and product models, we have modeled the parameter K_m as the total muscular stiffness that combines the muscle mechanical stiffness and the reflex stiffness. However, in the muscle–reflex model we have separated these two stiffnesses by defining K_m as the muscle mechanical stiffness and K_r as the reflex stiffness. Because the data collected for the step disturbances and ramped stretches were from two different subjects, two sets of values were optimized for the parameters in each model through fitting the data. In fitting the step disturbances through minimizing the mean-square errors, the fitting errors in the additive and product models were about 1.69 and 1.0 times, respectively, in comparison with the muscle–reflex model. As for fitting the ramped stretches, the mean-square errors in the additive and product models were about 5.69 and 1.82 times, respectively, in comparison with the muscle–reflex model. The improvement of the last model over the other two is impressive.

From the physiological study (Houk and Rymer, 1981), the results indicated that the linear stiffness term in the neuromuscular system has two potential origins: muscle length–tension effect and other linear position feedback terms. In our muscle–reflex model, to simplify the discussion, we have set the linear position feedback coefficient, L, to one so that the linear stiffness, K_m, will represent the muscle length–tension effect. Through our simulations in fitting both step disturbances and ramped stretches, the obtained values for K_m were within the measured range of muscle mechanical stiffness (Miller, 1984). The analysis of the other possible linear position feedback terms will be one of our future studies. As for the nonlinear term of the muscle–reflex system, the same study also indicated that it has two origins: spindle receptors and remnant nonlinear muscle properties (e.g. as described by Zahalak's model in *Eq. 13.5*). In our muscle–reflex model, the nonlinear damping properties of the spindle receptors are emulated through the spindle-like model. However, using a linear stiffness to represent the nonlinear muscle mechanical properties, the second origin contributing the nonlinear term is missing from our model. Therefore, understanding this remnant nonlinear term is important for improving the model.

In summary, through modeling the nonlinear damping properties of the neuromuscular system, the muscle–reflex model can accurately emulate the responses of involuntary movements. Without adjusting the values for the system's parameters, this neuromuscular-like model can adapt to a wide range of loads and disturbing forces without causing stability problems. In other words, this system has the advantage of adapting to any sudden change caused by the contact environment. As the constrained condition under robotic compliance control (Wu, 1988) is similar to that under the limb's involuntary movements, they both require that the limb be continuously in contact with the environment. The development of a neuromuscular-like model may be beneficial to robotic compliance control [see also Chapter 9 (Hogan)]. As for future research, besides the needed improvements for the models, to further understand the limb, movement studies will apply these models, particularly the muscle–reflex model, to the characterization of voluntary movements. Exploring the inverse relation from limb movement responses to motor commands is necessary for understanding how limbs are commanded.

Acknowledgment

This work is supported by the ONR contract N00014-88-K-0339.

References

Asatryan, D.G. and Feldman, A.G. (1985) "Functional Tuning of the Nervous System with Control of Movement or Maintenance of a Steady Posture # I. Mechanographic Analysis of the Work of the Joint on Execution of a Postural Task," *Biophysics USSR* **10**:925-935.

Atkeson, C.G. and Hollerbach, J.M. (1985) "Kinematic Features of Unrestrained Vertical Arm Movements," *J. Neuroscience* **5**: 2318-2330.

Bawa, P., Mannard, A., and Stein, R.B. (1976) "Predictions and Experimental Tests of a Visco-Elastic Muscle Model Using Elastic and Inertial Loads," *Biol. Cybern.* **22**:139-145.

Bizzi, E., Accornero, N., Chapple, W., and Hogan, N. (1984) "Posture Control and Trajectory Formation During Arm Movement," *J. Neuroscience* **4**:2738-2744.

Bizzi, E., Chapple, W., and Hogan, N. (1982) "Mechanical Properties of Muscles, Implication for Motor Control," *Trends Neurosci.* **5**:395-398.

Bizzi, E., Dev, P., Morasso, P., and Polit, A. (1978) "Effect of Load Disturbances During Centrally Initiated Movements," *J. Neurophys.* **41**:542:556.

Feldman, A.G. (1966) "Functional Tuning of the Nervous System with Control of Movement or Maintenance of a Steady Posture, II. Controllable Parameters of the Muscles," *Biophysics USSR* **11**:565:578.

Flash, T. (1987) "The Control of Hand Equilibrium Trajectories in Multi-Joint Arm Movements," *Biol. Cybern.* **57**:257:274.

Flash, T. and Hogan, N. (1985) "The Coordination of Arm Movements: An Experimentally Confirmed Mathematical Model," *J. Neuroscience* **5**:1688-1703.

Freund, H.J. and Budingen, H.J. (1978) "The Relationship Between Speed and Amplitude of the Fastest Voluntary Contractions of Human Arm Muscles," *Exp. Brain Res.* **31**:1-12.

Gielen, C.C.A.M. and Houk, J.C. (1984) "Nonlinear Viscosity of Human Wrist," *J. Neurophys.* **52**(3):553-569.

Gielen, C.C.A.M., Houk, J.C., Marcus, S.L., and Miller, L.E. (1984) "Viscoelastic Properties of the Wrist Motor Servo in Man," *Annals of Biomedical Engineering* **12**:599-620.

Gielen, C.C.A.M. and Houk, J.C. (1987) "A Model of the Motor Servo: Incorporating Nonlinear Spindle Receptor and Muscle Mechanical Properties," *Biol. Cybern.* **57**:217-231.

Gottlieb, G.L., Corcos, D.M., and Agarwal, G.C. (1989a) "Organizing Principles for Single-Joint Movements. I. A Speed-Insensitive Strategy," *J. Neurophysiology* **62**(3): 342-357.

Gottlieb, G.L., Corcos, D.M., and Agarwal, G.C.

(1989b) "Organizing Principles for Single-Joint Movements. II. A Speed-Sensitive Strategy," *J. Neurophysiology* **62**(3): 358-368.

Hasan, Z. (1983) "A Model of Spindle Afferent Response to Muscle Stretch," *J. Neurophys.* **49**:989-1006.

Hasan, Z. and Enoka, R.M. (1985) "Isometric Torque-Angle Relationship and Movement-Related Activity of Human Elbow Flexors: Implications for the Equilibrium-Point Hypothesis," *Exp. Brain Res.* **59**:441-450.

Hasan, Z. (1986) "Optimized Movement Trajectories and Joint Stiffness in Unperturbed, Inertially Loaded Movements," *Biol. Cybern.* **53**:373-382.

Hogan, N. (1984) "An Organizing Principle for a Class of Voluntary Movements," *J. Neuroscience* **4**:2745-2754.

Hogan, N. (1984) "Adaptive Control of Mechanical Impedance by Coactivation of Antagonist Muscles," *IEEE Trans. Automatic Control* AC-29, no. 8, pp. 681-690.

Hogan, N. (1985) "The Mechanics of Multi-Joint Posture and Movement Control," *Biol. Cybern.* **52**:315:331.

Hollerbach, J.M. (1982) "Computers, Brains, and the Control of Movement," *Trends Neurosci.* **5**:189-192.

Hollerbach, J.M. and Flash, T. (1982) "Dynamic Interactions Between Limb Segments During Planar Arm Movement," *Biol. Cybern.* **44**:67:77.

Houk, J.C., Crago, P.E., and Rymer, W.Z. (1981) "Function of the Spindle Dynamic Response in Stiffness Regulation - A Predicative Mechanism Provided by Nonlinear Feedback," *Muscle Receptors and Movement*, edited by Taylor, A. and Prochazka, A., Macmillan, London.

Houk, J.C., and Rymer, W.Z. (1981) "Neural Control of Muscle Length and Tension," *Handbook of Physiology # The Nervous System II*, Bethesda, MD, American Physiol. Soc. Sect. 1, Vol. II, Chap. 8, pp. 257-323.

Houk, J.C., Rymer, W.Z., and Crago, P.E. (1981) "Dependence of Dynamic Responses of Spindle Receptors on Muscle Length and Velocity," *J. Neurophysiology* **46**:143-166.

Houk, J.C., Wu, C.H., and Young, K.Y. (1989) "Nonlinear Damping of Limb Motion," *XXXI International Congress of Physiological Sciences*, Helsinki, Finland, July, Abs. 533.

Huxley, A.F. (1957) "Muscle Structure and Theories of Contraction," *Prog. Biophys. Chem.* **7**:257-318.

Matthews, P.B.C. (*1972*) *Mammalian Muscle Receptors and Their Central Actions*, Williams & Wilkins, Baltimore.

McMahon, A. T. (1984) *Muscles, Reflexes, and Locomotion*, University Park Press, Baltimore.

Miller, L.E. (1984) "Reflex Stiffness of the Human Wrist," M.S. Thesis, Department of Physiology, Northwestern University, Evanston, IL.

Mussa-Ivaldi, F.A., Hogan, N., and Bizzi, E. (1985) "Neural, Mechanical, and Geometric Factors Subserving Arm Posture in Humans," *J. Neuroscience* **5**: 2732-2743.

Mussa-Ivaldi, F.A., Morraso, P., and Zaccaria, R. (1988) "Kinematic Networks: A Distributed Model for Representing and Regularizing Motor Redundancy," *Biol. Cybern.* **60**:1-16.

Nichols, T.R. and Houk, J.C. (1976) "Improvement in Linearity and Regulation of Stiffness That Results From Actions of Stretch Reflex," *J. Neurophys.* **39**:119-142.

Ogada, K, (1970) *Modern Control Engineering*, Prentice-Hall Inc., Englewood Cliffs, NJ.

Oguztoreli, M.N. and Stein, R.B. (1976) "The Effects of Multiple Reflex Pathways on the Oscillations in Neuro-Muscular Systems," *J. Math. Biol.* **3**:87-101.

Polit, A. and Bizzi, E. (1979) "Characteristics of Motor Programs Underlying Arm Movements in Monkeys," *J. Neurophysiology* **42**:183-194.

Rack, P.M.H. (1981) "Limitations of Somatosensory Feedback in Control of Posture and Movement," *Handbook of Physiology # The Nervous System II*, Bethesda, MD, American Physiol. Soc. Sect. 1, Vol. II, pp. 229-259.

Rack, P.M.H. and Westbury, D.R. (1969) "The Effects of Length and Stimulus Rate on Tension in the Isometric Cat Soleus Muscle," *J. Physiol.* **204**:443:460.

Stein, R.B. (1982) "What Muscle Variable(s) Does the Nervous System Control in Limb Movements?" *Behav. Brain Sci.* **5**:535-577.

Stein, R.B. (1974) "The Peripheral Control of Movement," *Physiol. Rev.* **54**:215-243.

Uno, Y., Kawato, M., and Suzuki, R. (1989) "Formation and Control of Optimal Trajectory in Human Multijoint Arm Movement," *Biol. Cybern.* **61**:89-101.

Wu, C.H. (1988) "Compliance," *International Encyclopedia of Robotics: Application and Automation*, John Wiley and Sons, New York, Vol. I, pp., 192-202.

Zahalak, G.I. (1981) "A Distribution - Moment Approximation for Kinetic Theories of Muscular Contraction," *Math. Bioscience* **55**:89-114.

CHAPTER 14

Principles Underlying Single-Joint Movement Strategies

Gerald L. Gottlieb, Daniel M. Corcos, Gyan C. Agarwal and Mark L. Latash

14.1 Introduction

Humans make "rational" decisions in different ways. One is according to a plan that optimizes some consequence of the decision. Such a decision–making strategy is normative, leading to a behavior appropriate for achieving some desired goal. Another way is to use a heuristic or rule based approach that guides behavior on the basis of simpler criteria than those characterized by an "optimal" result. For example, optimal strategies for obtaining wealth from the stock market or the race track are sufficiently subtle that few have found them. On the other hand, "rules of thumb" are many, and perhaps almost as much money is made from the sale of such rules as from their application. In fact, people often apply such rules in the face of objective evidence that they are in fact not optimal.

These observations on general properties of human decision making have some bearing on how humans plan and perform voluntary movements. They could perform movements that are "optimal" in a sense that maximizes or minimizes some predefined goal or performance criterion important to the mover [Hogan, 1984; Nelson, 1983; Chapter 8 (Zajac and Winters); Chapter 19 (Winters and Seif-Naraghi)]. Alternatively, they could use simple rules that lead to movements that are not optimal in any unique sense, but accomplish a task that is loosely defined and achievable in a number of equivalent ways, none of which is "superior" to the others.

Consider the task of moving a cup of coffee from a table to your mouth. The overall goal is to make it possible to sip the coffee with added conditions that it neither spill nor get cold on the way. This goal provides few constraints on the movement's trajectory or speed yet it is one we can accomplish easily and without a moment's thought on just how to do it. Therefore, we presumably do not have to ponder and choose from an infinite number of motor programs that are capable of generating a similarly infinite number of trajectories. We promptly select and execute one program that gets us our hot coffee, keeps us dry, and does not require excessive effort. Is it uniquely "optimal" or is it merely "good enough"?

Let us note first that these are not mutually exclusive options. Surely the "best" trajectory is a member of the set of acceptable ones. Furthermore, a trajectory governed by a simple, suboptimal set of heuristically based rules might be indistinguishable from an optimal trajectory. It is our intention here to describe a set of such heuristically based rules that appear to be used to control a class of single–joint movements. In the discussion, we will consider whether these rules are compatible with possible optimization schemes that have been proposed.

We will consider fast, single-joint flexions of the elbow. "Fast" movements can be operationally defined as being associated with burst-like *EMG* patterns, at least in the agonist muscles (Wacholder and Altenburger, 1926; Wadman et al., 1979; Waters and Strick, 1981). This agonist burst is due to the sudden and strong activation of the muscles needed to produce the initial acceleration of the limb. The control of this burst, to produce the desired movement characteristics (e.g.

Multiple Muscle Systems: Biomechanics and Movement Organization
J.M. Winters and S.L-Y. Woo (eds.), © 1990 Springer-Verlag, New York

distance and speed for the given load), has been described by many authors in terms of what we will term a "pulse–timing" model (Freund and Bundingen, 1978; Ghez, 1979; Wallace, 1989). We too have found utility in such a model (Corcos et al., 1989; Gottlieb et al., 1989a,b, 1990), which proposes that the amplitudes and durations of the agonist and antagonist *EMG* bursts, and the onset latency of the antagonist *EMG* burst, are the parameters of the motor program that are selected to accomplish the desired movement.

Reducing the problem of control to one of determining a small number of relatively well–defined parameters allows some straightforward questions to be posed regarding how different movement tasks are performed. Can one (or more) parameters be found that relate to some movement feature? For example, Freund and Budingen (1978) proposed that to move rapidly to a target and return, burst duration is selected to control the movement's time and burst intensity to control its distance. However, the fact that people can make movements (and perform experiments) of similar or different durations over different or similar distances has not led to any consensus on how these different performance variations are accomplished. Furthermore, most of the rules that have been described are only applicable to a specific load, and some must be modified if the load changes. Do we have one rule for dealing with changes in movement distance, a second for changes in movement speed, a third for changes in limb load, and so on for each permutation and combination, ad infinitum? How might we combine rules when more than one feature of the planned movement changes? We have summarized this issue in a recent article (Gottlieb et al., 1989a). The notion of characterizing the control signal for fast movements as a pulse would have a compelling simplicity if we were able to provide a rational and small set of rules for specifying its parameters in terms of the physical task to be performed. These would be the kinds of "rules of thumb" about which we ruminated at the beginning of this article.

14.2 Strategies of Movement

The argument we shall develop is that rules for movement can be usefully formulated in terms of signals that drive motoneuron pools. However, this signal is not directly observable so we must infer it from the variables that we can measure. These are myoelectric (*EMG*), kinematic (angles and their derivatives), and kinetic (torques and their derivatives). We wish to demonstrate that the signal to the motoneuron pool can, for reasonably fast movements, be described as a rectangular pulse that we will call an "excitation pulse." The excitation pulse can be modulated in amplitude and in duration.

We will call the choice of modulation mode a "strategy" for movement control. If the movement task is defined simply as moving from point A to point B, subjects always have a choice of strategies in making a movement. They can choose to modulate pulse width or pulse height or perhaps both. This element of choice is important if the notion of strategies is to provide insight. A rock, dropped from the hand, has no choice but to fall under the influence of gravity. We would never say "the rock chooses to accelerate downward at 9.8 m/s^2."

What if the task is to move from point A to point B in 500 ms? Does the subject still have a choice under these more constrained kinematic conditions? Can the subject still chose a strategy for movement? The answer to these questions is "yes." This may be illustrated by the following analogy. A football coach always has a choice of strategies in playing a game. Pass or run, long or short, left or right. However, the goal is not just to play but to win the game. The primary goal of playing is augmented by a secondary goal of winning, and not all strategies may be compatible with both. (Clearly one cannot win if one does not play and one cannot move in 500 ms if one does not move at all.) However, the notion of primary and secondary goals is arbitrary, and to some what we might define as secondary from a kinematic perspective might actually be the primary or even the sole goal from another perspective (e.g. the "winning is The Only Thing" philosophy attributed to V. Lombardi of the Green Bay Packers (1964)). In such circumstances the coach retains the powers of choice, only now the consequences include winning or losing. To say that the coach must "first establish the ground game and then go to the pass, there is no choice" is a figure of speech, not a statement of fact. As many who have lost games might attest, a wrong choice after the fact was no less a choice before it! Returning to the more banal task of moving between two positions in a specified interval, the choice of movement strategy remains. One strategy may fail to reach the final position in a timely manner, but it is the problem of the mover, as it is the problem of the coach, to choose the right strategy to accomplish all aspects of the task. The possibility of error (like the possibility of "sin") is the basis for the existence of a meaningful choice.

What we intend to show is that certain classes of tasks are accomplished by pulse width modula-

tion of the excitation pulse while others are accomplished by pulse height modulation. A pattern of simultaneous modulation of both parameters, while in no way excluded from consideration, is unnecessary to describe any of the movements we will discuss here (cf. Gottlieb et al., 1989a, for further discussion of this point.) Because only two distinguishable patterns of control are needed for the movements we have examined, we will refer to this as a "dual strategy" hypothesis.

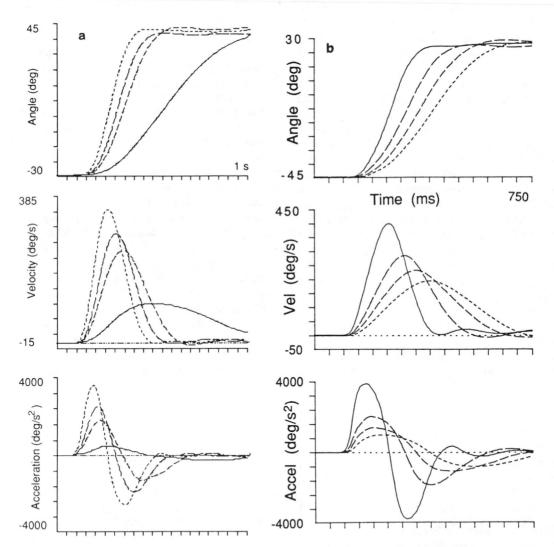

Figure 14.1: Angle, velocity and acceleration for two groups of four movements, performed at different speeds. *a)* Changes in speed were induced by four different instructions to the subject that affected speed (redrawn from Figure 4 of Gottlieb et al., 1990). *b)* Changes in speed resulted from changes in the inertial load moved by a subject who was trying to move quickly but accurately to a target (from Figure 3 of Gottlieb et al., 1989b). The traces are all averages of 10 movements. Reproduced with permission.

14.3 Some Illustrative Examples

Figure 14.1 shows two sets of 72° flexion movements. They were all made in the horizontal plane to a 9° target window. A video monitor provided continuous visual feedback of joint angle to the subject.

These two sets of movements look kinematically quite similar but are kinetically quite different. In Figure 14.2 we show the inertial joint torques generated by the muscles (the product of the moment of inertia and the joint acceleration, both of which were measured and the rectified, smoothed *EMGs* from a principal agonist and antagonist. Among the features that differ between the two data sets, we suggest that the onsets of the inertial torque and of the *EMG* bursts are most informative. On the left, the steepest rise of torque and *EMG* occurs for the shortest movement time while on the right, the rising phases of those variables are independent of movement time.

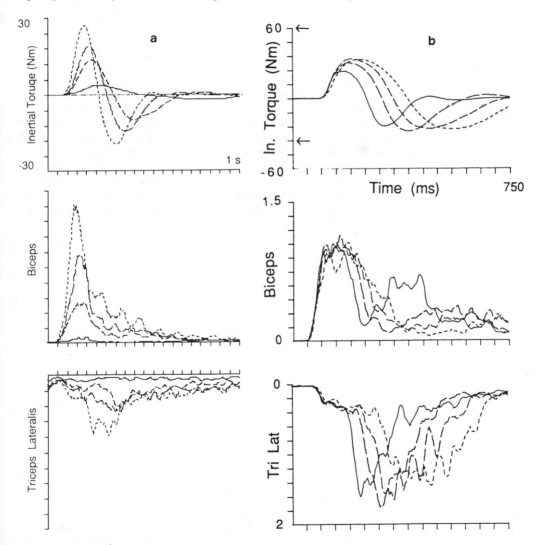

Figure 14.2: Inertial torque and rectified and smoothed EMGs from the principal agonist (biceps) and antagonist (Lat. Triceps) muscles for the movements shown in Figure 14.1. *a)* Peak inertial torque and the area of the EMG bursts increase with intentional changes in movement speed (from Figure 4 of Gottlieb et al., 1990). *b)* These measures increase with falling speeds that are consequent to added inertial loading (from Figure 3 of Gottlieb et al., 1989b). Reproduced with permission.

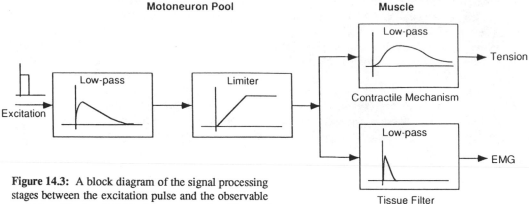

Figure 14.3: A block diagram of the signal processing stages between the excitation pulse and the observable variables. The two left blocks represent the effects of the motoneuron pool which transforms the graded potentials produced by synaptic excitation into trains of propagating action potentials. The limiter represents the bounds imposed by the number of motoneurons and their maximum firing frequencies. The two right blocks incorporate different aspects of muscle. The upper block embodies the dynamics of excitation, contraction and coupling mechanisms. The lower block represents the passive electrical properties of the tissue that affect the measurement of the *EMG*.

The origin of these differences is to be found in the nature of the movement task. The four sets of movements on the left were performed with four different instructions to the subject, each designed to affect movement speed (Gottlieb et al., 1990). The four sets of movements on the right were each performed with a different inertial load on the limb (Gottlieb et al., 1989b).

Such an operational distinction between behaviors offers little insight into why the difference exists. To look more deeply into this question, we have proposed a simple model for the relation between the signal that excites the motoneuron pool and the resulting *EMGs* and muscle forces. This model is illustrated for a single muscle in Figure 14.3. Its salient features are:

1. Motoneuron excitation (which is unmeasurable) can be characterized as a rectangular pulse. It generates trains of action potentials, modulated in frequency and in number of motor units. Both are proportional to the amplitude or intensity of the pulse. This signal is also, for most practical purposes, unmeasurable.

2. The signals from the motoneurons are low-pass filtered by two physically separate processes to produce the measurable variables, tension and *EMG*.

There are several implications of this model. The rectangular shape of the motoneuron excitation pulse allows only two controllable parameters for the agonist muscle, pulse width and pulse height. For the antagonist there are three, its height and width and also its latency with respect to the agonist onset. The low-pass filtering causes differences in excitation pulse height to be observable only as differences in the derivatives of the measured variables. This is demonstrated in Figure 14.4, which shows the effects of low-pass filtering on pulses that are modulated either in height (left) or width (right). This model realizes many of the significant properties of *EMG* modulation. Greater verisimilitude would require a more complex model and a more complex input waveform. This very simple model is adequate for the issues of concern here.

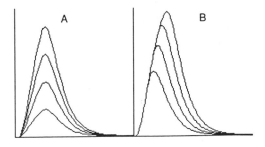

Figure 14.4: Simulated envelopes of *EMG* waveforms resulting from low-pass filtering of excitation pulses that are modulated either in amplitude (*a*) or duration (*b*). The simulation represents 500 ms and the filters (see Figure 3) are $1/(s+20)$ and $1/(s^2 + 100s + 2500)$. In *a)*, the pulse has a duration of 100 ms and varies in amplitude from .5 to 2 in equal steps. In *b)*, the amplitude is fixed at 1 and the duration varies from 50-125 ms in equal steps. The limiter element in Figure 3 has not been included in this simulation.

To apply this model to movement, we must have explicit rules for all five parameters. In order to account for data such as illustrated in Figures 14.1 and 14.2, we have proposed the need for two sets of rules or strategies which we call "speed-sensitive" (*SS*) and "speed-insensitive" (*SI*) (Gottlieb et al., 1989a). The rules for the five parameters are summarized in Table 14.1. The '+' symbols indicate a significant positive correlation between the parameter and the kinematic variable the subject is trying to control. A '−' indicates a significant negative correlation, and a zero means the parameter remains essentially invariant. This makes it clear that the *SS* strategy is one of pulse height modulation and the *SI* strategy is one of pulse width modulation. Note that both parameters will directly increase or decrease the total amount of muscle excitation in a similar manner. Note too that the latency of the antagonist reverses its relationship to the amount of muscle activation across the two strategies.

Table 14.1

Muscle	Modulation	SS strategy	SI strategy
Agonist	Pulse height	+	0
	Pulse width	0	+
Antagonist	Pulse height	+	0
	Pulse width	0	+
	Latency	−	+

Using these rules allows us to simulate a single–joint contraction. There are two identical and opposed muscles, each represented by a system such as Figure 14.3. The antagonist pulse is delayed relative to that of the agonist. The net joint torque is the difference between the two muscles. We illustrate this in Figure 14.5, showing the simulated agonist and antagonist *EMG*s

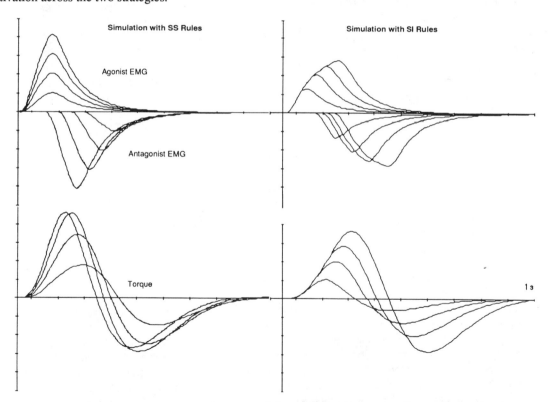

Figure 14.5: Simulation of agonist and antagonist EMG's and net joint Torques under the rules for each of the two strategies. *a)* SS rules apply. The excitation pulses to both muscles are equal in amplitude and the four traces differ in the pulse amplitude. The latency of the antagonist decreases as the amplitude of excitation increases. *b)* SI rules apply. Excitation pulses are of equal amplitude but differ in duration. The latency of the antagonist increases with pulse duration.

and the net torque under the rules for each of the two strategies. This is a highly simplified model, neglecting all the nonlinearities of muscle activation, contraction, and shortening dynamics which are described elsewhere in this book. Nevertheless, the results given SS rules (Figure 14.5a) and SI rules (Figure 14.5b) demonstrate many of the distinguishing features of the two modes of behavior.

We can apply this model to our interpretation of Figure 14.2. The left side of Figure 14.2, where EMG and torque slopes diverge from the onset of contraction, is consistent with pulse height modulation and antagonist latencies that decrease with increasing peak tension, or an SS strategy. The right side of the figure, where EMG and torque slopes are initially the same, is consistent with constant height pulses of varying duration (or pulse width modulation) and antagonist latencies that increase with peak tension or an SI strategy.

14.4 How Different Is "Different" and How Similar Is "Similar"?

The experimental distinction between the two strategies rests upon the assertion that for an SI strategy, measured variables (initial EMGs and joint torques) are similar for a period of time that depends upon the independently manipulated task variable. After this interval, they diverge and become different. In using an SS strategy to control movement speed as a task variable, the measured variables diverge from the outset of the contraction.

In our earlier studies, we have used the agonist EMG, integrated over the first 30 ms after its onset (Q_{30}), to distinguish between strategies (see Appendix 1 of Gottlieb et al., 1989b). In the SI strategy, Q_{30} is constant and independent of the task variable, while in the SS strategy it is proportional to it. This distinction can be seen in Figure 14.6. We have plotted Q_{30} for four different distances of movement under four sets of instructions regarding speed (Gottlieb et al., 1990) on the left (14.6A) and the total integration of the agonist burst (Q_{acc}) on the right (14.6B). Since speed increases monotonically on both of the base axes, the graphs show that when distance varies, Q_{30} is insensitive to speed but when instruction varies it is proportional. This is distinctly different from Q_{acc} which always scales with speed if the load is unchanged.

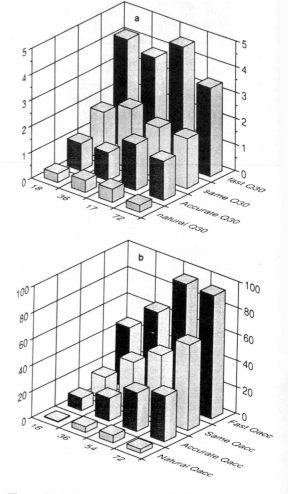

Figure 14.6. Integrated agonist EMG for movements of four different distances, performed under four different instructions to the subject affecting movement speed. At the left, the integration was performed over the first 30 ms of the agonist burst while at the right, it was performed over the whole burst, using the period of acceleration towards the target to define this interval.

The way in which strategy affects the similarities or differences in torque has previously been left to the eye of the observer to discern. To provide a more objective method, we have performed the following analysis. The data plots in Figure 14.2 show inertial torques four sets of movements, to the same target with the same load, performed at four different "chosen" speeds. For these four sets we have performed a one way repeated measures $ANOVA$ on the inertial torque,

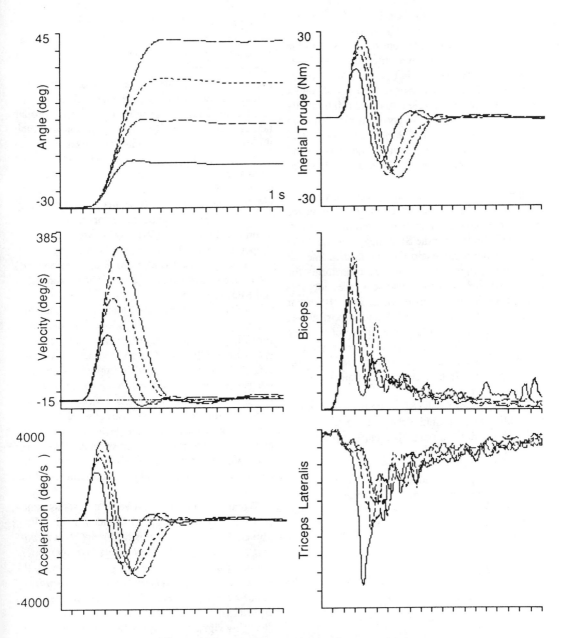

Figure 14.7: Movements of four distances with instruction to be as fast, but accurate, as possible. Each trace is the average of 10 movements. The data are redrawn from Figure 1 of Gottlieb, et. al., 1990a. Reproduced with permission.

measured every 4 ms, to determine when the sets become separable. Figure 14.7 shows for four sets of movements of four different distances with the same load and instruction influencing the choice of speed. Note that this is not the "same speed" since subjects cannot perform such a task under most conditions, when asked to move different distances. They always vary speed in proportion to distance, although that is not their conscious intent (as opposed to the experiment in Figure 14.2) and they may not even be aware of the change in speed (Gottlieb et al., 1990).

Figure 14.8 shows the F and p values as a function of time after the onset of inertial torque for each of the two cases that result from this analysis. The thin solid line for the SI strategy (Figure 14.7) represents the F value and the thin dashed line the probability p that the four torque trajectories for movements of different distances are not significantly different. It is only after over 60 ms that the probability falls below 0.5 and almost 80 ms before it is below .05. We interpret this as evidence that it is at least 80 ms before there is a statistically significant probability (at the .05 level) that the four acceleration trajectories in Figure 14.7 are different.

The thick lines for the SS strategy show that the probability that the four accelerations under different instructions are not different is vanishingly small from the first moments.

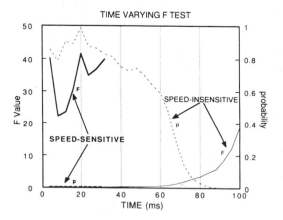

Figure 14.8: The graph plots the results of a one way repeated measures *ANOVA* (*F* and *p* values) for the inertial torques from two sets of four movements illustrated on the left side of Figures 14.1 and 14.2 and in Figure 14.7.

14.5 The Basis For Choosing Strategies

We are now able to address the issue of what elements of a movement task lead to the choice of one strategy over another. Some are physical constraints. For example, movements performed too quickly or too slowly require that the mover correct the joint torques in order to achieve a proper rate of acceleration. This is a condition in which the mover is concerned with the speed of movement (e.g., Figures 14.1a and 14.2a), or equivalently, the movement time, and therefore adopts an *SS* strategy. The strategy is required by the dynamical laws of Newtonian mechanics that specify the application of (larger or smaller) forces to obtain the desired (faster or slower) kinematic properties. Note that the choice is imposed, not by the movement itself (i.e., from point A to point B) but by the ***added constraint*** that it be performed at some desired speed.

Similar constraints are imposed by changing target size and emphasizing movement accuracy. Fitts' Law (Fitts, 1954) describes a reduction in speed for smaller targets, a speed-accuracy trade-off. This control of speed is implicit in that it is not part of the instructions given to the subjects, nor is it necessarily part of the subjects' conscious understanding of how to correctly perform the task. Nevertheless, the change in speed emerges as a way of controlling accuracy, and this change is accomplished by the same control strategy that is used to intentionally alter speed (Corcos et al., 1989).

When subjects alter movement distance, they also alter movement speed (Corcos et al., 1988; Milner, 1986). How is this different? *It is important to note that although there are dynamical reasons for why the SS strategy should exist, the existence of the SI strategy and the uniformity of initial torques and EMGs seems entirely gratuitous.* There is no obvious reason why diverging signals should not be seen in both cases, other than that this appears to be the choice made by the motor controller. There is nothing preventing the torques from being modulated for controlling distance in almost exactly the same manner that they are for controlling speed. There would have to be a change in the timing of the antagonist but that would not prevent the initial profiles of acceleration (and *EMG*) from resembling those of *SS* strategy controlled movements.

That this is not the case is clear from Figures 14.7 and 14.8. Instead, we observe uniform profiles that diverge much later in the trajectory. This behavior is consistent with our model for the *SI* strategy where uniform excitation is modulated in duration. Higher speeds are reached because acceleration is more prolonged rather than more steeply rising. Movement of different inertial loads (Figures 14.1b and 14.2b) is controlled in the same manner.

Since people usually make individually different movements, rather than repeat 10 or 20 similar movements at a time, how do these notions of strategy apply? To make one unique movement, a person may make it slowly enough to allow continuous correction of the trajectory using visual and proprioceptive feedback. This is not the kind of movement, however, that we are considering. To make a movement that is too fast for continuous correction, a person makes a prediction of the kinematic consequences of their motor commands and executes their best guess. In most cases, this is good enough. If it is not, they either can try again or suffer the consequences of their error. If they do try again, however, they will use one strategy or the other to improve. If the movement falls short, they will prolong muscle activation and delay the antagonist. If they are too slow, they will intensify muscle activation and recruit the antagonist earlier. If they erred in both regards, they will superimpose the strategies, prolonging and intensifying their muscle excitation and compromising on antagonist latency.

14.6 Equilibrium-Point Hypothesis

The development of our ideas has to this point been framed in terms of how the nervous system generates neural "programs" to produce muscle forces that lead to movement. This approach omits all consideration of afferent contributions from muscle and joint receptors that undoubtedly play a part in this process. It treats all the peripheral elements, the muscles and their associated reflexes, as if they were a force generator under the absolute control of descending signals from higher centers.

However, the peripheral neuromuscular system is not a length-independent force generator. Its spring-like and viscous properties have long been recognized [reviewed in Chapter 11 (Winters and Hogan), elaborated in Chapter 12 (Feldman et al.)]. These properties are entirely consistent with the two strategies we have described for force modulation. If we model the peripheral system as a controllable spring producing a force F at a length l:

$$F = k(l - \lambda) \qquad (14.1)$$

with a stiffness k and an equilibrium length λ, then we can obtain the desired forces by modulating either k or λ according to the chosen strategy.

This property of the peripheral motor system provides an alternative conceptual approach, based on spring–like muscle properties (Feldman, 1974; Bizzi et al., 1976) or, in the Sherringtonian tradition, on parameterization of muscle reflexes (Berkinblit et al., 1986). The λ-model of the equilibrium-point hypothesis, for example, treats the mechanism of limb control as modulating one parameter per muscle, identifying this parameter with the threshold of the muscle length-sensitive reflex [Feldman, 1986; Chapter 12 (Feldman et al.)]. In either case, central control can be described in terms of changing compliant characteristics of the peripheral apparatus, and the observable performance will depend on both central control signals and external conditions of movement execution. In this framework, the *EMG* patterns are determined by two equally important factors: the central command and reflex feedback.

One way to reconcile these views with the dual–strategy hypothesis is to presume that the higher centers predict afferent contribution and adjust the central commands to make the input to the α–motoneuron pools a rectangular excitation pulse (Figure 14.9, *C2*). This of course depends upon the ability of the higher centers to predict the shape of the afferent signal and generate an appropriately modified central command. If the movements are practiced, then this prediction can be firmly based on prior experience. If the movement is novel, then it may err because of the failure of the central controller to properly anticipate the pattern of afferent activity. However, since rapid novel movements are often not accurate, this does not seem to be a fundamental objection. This approach means abandoning the assumption of a rectangular excitation pulse as a central command to the α–motoneurons (see the "input signal" to the point of summation in *C2*, Figure 14.9). However, it does not mean rejection of the dual-strategy hypothesis.

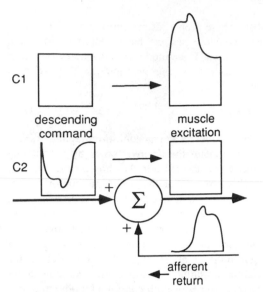

Figure 14.9: The pattern of muscle excitation will be influenced by a summation of descending and afferent signals at various levels of the central nervous system. According to the equilibrium-point hypothesis (*C1*), a central command in a "simple" (rectangular, in *C1*) feedback-independent form is summed with the afferent feedback signals and form a non-rectangular excitation pulse. If higher centers are able to predict the afferent pattern (*C2*), then they can achieve any arbitrary pattern of muscle excitation, including a rectangular one, by taking that afferent pattern into account in the generation of the descending command. Since afferent inputs reflecting peripheral events consequent to descending commands must be delayed by neuronal conduction times and peripheral system dynamics, the *initiation* of changes in muscle excitation, in both cases *C1* and *C2*, should reflect the descending rather than the afferent component. The point of summation is tentatively considered to be at the motoneuronal level.

The equilibrium-point hypothesis can be extended [Abdusamatov and Feldman, 1986; Adamovich and Feldman, 1989; Chapter 12 (Feldman et al.)] to produce the *EMG* patterns of either strategy by assuming simple standardized changes in λ for the agonist and antagonist muscles (λ_{ag} and λ_{ant}) and taking into account the contribution of the afferent reflex loops (Latash, 1989). One can assume that the deficiencies of a rectangular model on muscle excitation arise from the fact that the actual muscle excitation is in fact a more complex waveform, and reflects, in part, afferent contributions (*C1* in Figure 14.9). Models

based on such afferent contributions must be further developed, however, to account for the similarities in the *EMG* patterns that exist between free movements and those that are unexpectedly blocked (Wadman et al., 1979).

Thus, the equilibrium-point approach changes the site at which strategy based differences are implemented to a higher level in the motor control hierarchy than the motoneuron pool without changing the dual-strategy principle. In this framework, the speed-insensitive strategy is associated with a shift in λ_{ag} at a constant rate ω for different intervals (corresponding to our different pulse widths) while the speed-sensitive strategy is associated with changing λ_{ag} at different ω (corresponding to different pulse heights). The rules for choosing one strategy or the other do not differ from those described earlier.

During the first several tens of milliseconds, this approach is indistinguishable from the assumption of a rectangular excitation pulse since reflex loops need time to start affecting the *EMG*s. However, for the later events, including the antagonist burst, the predictions of the equilibrium point approach differ from those derived from assuming a rectangular excitation pulse. For example, the equilibrium-point approach leads to a possibly of non-monotonic changes in the antagonist *EMG* amplitude with movement amplitude under the *SI* strategy (Latash, 1989).

14.7 "Optimal" Strategies

A number of studies have investigated the optimality of simple movements according to various criteria. Hogan (1984) analyzed the problem of generating the smoothest trajectory using the following criterion.

$$J = \int_0^T \left[\frac{d^3 \phi}{dt^3} \right]^2 dt \qquad (14.2)$$

The goal was to calculate the trajectory ($\phi(t)$) for a movement of specified duration T and distance D that would minimize J, the squared third derivative of ϕ ("jerk"), integrated over T. A similar type of analysis was performed by Hasan (1986) using a different measure of performance, given by

$$J = \int_0^T k \left[\frac{d\lambda}{dt} \right]^2 dt \qquad (14.3)$$

The parameters k and λ are the same as in *Eq. 14.1*. This equation is treated as a measure of "effort" that the subject is presumed to minimize.

Since "strategies" are rules used by the nervous system to control the muscles, then certainly *Eq. 14.2* and *Eq. 14.3* qualify as implying specific strategies. There exist specific torques that must be generated by the muscles to produce those optimal trajectories and these require specific (although not unique) patterns of muscle activation. What is the relation between these optimal strategies and the simpler *SI* and *SS* strategies described above which deal only in pulse width and pulse amplitude modulation?

It must be emphasized that these methods require the *a priori* specification of movement time T before an optimal solution can be calculated. Given that time, each of these optimal criteria predicts a trajectory that is symmetric and can be characterized, in part, by two invariant relations. These are:

$$\left(\frac{d\phi}{dt}\right)_{max} = K_v \frac{D}{T} \qquad (14.4)$$

$$\left(\frac{d^2\phi}{dt^2}\right)_{max} = K_a \frac{D}{T^2} \qquad (14.5)$$

Hogan's optimal values for K_v and K_a are 1.88 and 5.77 respectively while Hasan's are 1.97 and 6.14. Figure 14.10 plots data from movements over four distances and with four different instructions concerning speed (see Figure 14.1 as well as Figures 1 and 4 from (Gottlieb et al., 1990). The solid lines are *Eq. 14.4* and *14.5* for the two sets of optimal coefficients. The dashed lines are the linear regression curves which give estimates of K_v and K_a as 1.73 and 7.28.

Since Figure 14.10 includes data from movements that encompass both strategies, we conclude that *SI* and *SS* strategies both result in trajectories such that parameters of movement like peak velocity or peak acceleration do not distinguish them from either of these optimal control strategies which minimize jerk or effort. However, the actual trajectories are not perfectly symmetrical (Nagasaki, 1989), a fact that is

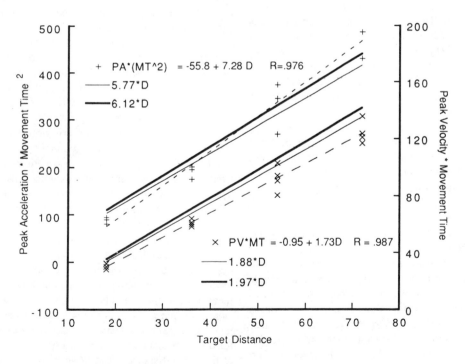

Figure 14.10: Plots based on *Eq. 14.4* (x) and *14.5* (+) for experimental data for movements of different distances and different speeds. The thin lines are the predicted relations for a minimum jerk trajectory (*Eq. 14.2*). The thick lines are for a minimum "effort" trajectory (*Eq. 14.3*)

problematic for the optimal control laws. This still leaves open the question of how a single optimal strategy, be it based on *Eq. 14.2* or *Eq. 14.3*, can be consistent with our requirement of two strategies, depending on the task.

Consider Hogan's minimum jerk trajectory for moving a distance D in a time T.

$$\phi = \phi_0 + D \left[10\left(\frac{t}{T}\right)^3 - 15\left(\frac{t}{T}\right)^4 + 6\left(\frac{t}{T}\right)^5 \right] \quad (14.6)$$

From this we can calculate the initial jerk as the third derivative of *Eq. 14.6* at $t = 0$.

$$\frac{d^3 \phi}{dt^3} \simeq \frac{60\ D}{T^3} \ ; \quad t \ll T \qquad (14.7)$$

The *SS* strategy predicts that the initial jerk will be proportional to D for movements of constant time and inversely proportional to speed for movements of constant distance. This prediction is obviously satisfied by *Eq. 14.7*. The *SI* strategy predicts that the initial jerk is insensitive to the distance moved. This implies from *Eq. 14.7* that for a minimum jerk trajectory, D/T^3 must be constant for movements of different distances.

In Figure 14.11 we have plotted this quantity for the same set of movements used above for Figure 14.10 and find that jerk increases for different instructions that affect speed (different symbols). Furthermore, for a single instruction and movements of different distances, T (which was not specified by the conditions of the task) varied with distance in such a way as to keep *Eq. 14.7* relatively constant. The significance of this is not yet entirely clear however since for trajectories such as those in Figure 14.7, although the quantity D/T^3 remains constant for different distances, it is not proportional to the initial jerk and *Eq. 14.7* is not satisfied.

Optimal strategies of the type we are considering describe trajectories that minimize their respective performance criteria for movements of a specified duration. They require that people "choose" movement time and then apply the optimal control law to calculate muscle activation commands. This *a priori* choosing of movement time is no more unreasonable than our earlier assumption of an *a priori* choice of the level of effort (intensity of the excitation pulse).

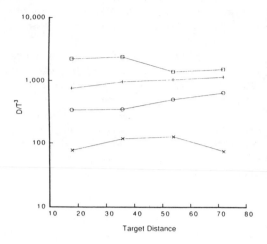

Figure 14.11: A plot of D/T^3 for movements of different distances. The different symbols correspond to different instructions to the subject concerning speed.

However, ***optimal strategies offer no guidance as to how to choose movement time***. Instead of choosing movement time *a priori*, one could leave it free to find the absolute minimum jerk or effort and the movement time at which it occurs. The solution to this problem is trivial. Zero effort and jerk (the absolute minimum of a non-negative function) will be achieved by applying zero force (or constant λ for *Eq. 14.3*) so that all the derivatives of angle are zero. This solution is unsatisfactory because movement time is infinite. Alternatively, one could use a criterion that sums jerk or effort with a monotonically increasing function of movement time. Such an approach does not offer any significant advantages over the *a priori* selection of movement time since it merely substitutes one arbitrary choice (of movement time) with another, equally arbitrary choice (of the weighting function of movement time).

The issue addressed by our dual strategy hypothesis, as it describes the kinetics and kinematics of movements, bears on how movement time might be chosen. The *SS* strategy emerges when movement times are chosen according to external criteria such as remaining constant for different distances or changing for movements of the same distance. The *SI* strategy emerges in the way that movement time is chosen to move different distances or inertial loads.

The dual strategy hypothesis also describes how the activation patterns of the motoneuron pools can be chosen to achieve the observed kinematics, in terms of pulse width or pulse height modulation of the excitation pulses. This aspect of the hypothesis is possibly incompatible with optimal rules of movement, but that is not yet determinable. It depends upon what rules for activating the motoneuron pools arise from dynamic optimization and how they compare with *EMG* data and the predictions of our pulse modulation model. [Some predictions of dynamic optimization are provided in Chapter 19 (Winters and Seif-Naraghi)]. It further remains to be determined whether the predictions of dynamic optimization will be sufficiently different from the pulse modulation model that various hypotheses can be distinguished from experimental data. Until that time, we conclude that optimal control laws and dual strategy rules of thumb are complementary principles that describe different aspects of the motor control problem.

14.8 Summary

We have reviewed the data that support the hypothesis that single-joint voluntary elbow movements may be usefully segregated into two classes. One class is performed under conditions which constrain movement speed. These conditions either are explicitly recognized by the mover or are implicit in what is called the speed–accuracy trade-off. This class of movements in controlled by a speed–sensitive strategy for modulating the intensity of motoneuron excitation. The other class of movements is performed without constraints on movement speed. These are governed by a speed–insensitive strategy that uses a uniform intensity of motoneuron excitation, in spite of distance and load dependent changes in movement speed. The accommodation to different distance and load conditions is achieved by this strategy by varying the duration of motoneuron excitation.

We have discussed this hypothesis in the context of two other classes of movement theories: the equilibrium point model and the optimal control model. We conclude that within the precision of the available data, the notion of dual strategies and its kinematic and myoelectrical manifestations are entirely compatible with both models.

The dual strategy hypothesis draws a distinction between classes of movements that needs to be recognized and incorporated in higher level theories of motor control.

Acknowledgments

We thank Mr. Om Paul for programming support and Mr. K. Lee for assistance in performing these experiments. This work was supported, in part, by NIH Grants NS 23593 and AR 33189.

References

Abdusamatov, R. M. and Feldman, A. G. (1986). Description of the electromyograms with the aid of a mathematical model for single joint movements. *Biofizika* 31: 503-505 (Translation 549-552).

Adamovich, S. V. and Feldman, A. G. (1989). The prerequisites for one-joint motor control theories. *Beh. Brain Sci.* 12: 210-211.

Berkinblit, M. B., Feldman, A. G. and Fukson, O. I. (1986). Adaptability of innate motor patterns and motor control mechanisms. *Beh. Brain Sci.* 9: 585-638.

Bizzi, E., Polit, A. and Morasso, P. (1976). Mechanisms underlying achievement of final head position. *J. Neurophysiol.* 39: 434-444.

Corcos, D. M., Gottlieb, G. L. and Agarwal, G. C. (1988). Accuracy constraints on rapid elbow movement. *J. Motor Behav.* 20: 255-272.

Corcos, D. M., Gottlieb, G. L. and Agarwal, G. C. (1989). Organizing principles for single joint movements: II - A speed-sensitive strategy. *J. Neurophysiol.* 62: 358-368.

Feldman, A. G. (1974). Control of the length of a muscle. *Biophysics* 19: 766-771.

Feldman, A. G. (1986). Once more on the equilibrium-point hypothesis (λ model) for motor control. *J Mot Beh.* 18: 17-54.

Fitts, P. M. (1954). The information capacity of the human motor system in controlling the amplitude of movement. *J. Exp. Psy.* 47: 381-391.

Freund, H. and Budingen, H. J. (1978). The relationship between speed and amplitude of the fastest voluntary contractions of human arm muscles. *Exp. Brain Res.* 31: 1-14.

Ghez, C. 1979. Contributions of central programs to rapid limb movment in the cat. In: *Integration in the nervous system* (edited by H. Asanuma and V. Wilson) pp. 305-319. Tokyo: Igaku-Shoin.

Gottlieb, G. L., Corcos, D. M. and Agarwal, G. C. (1989a). Strategies for the control of single mechanical degree of freedom voluntary movements. *Beh. Brain Sci.* 12(2): 189-210.

Gottlieb, G. L., Corcos, D. M. and Agarwal, G. C. (1989b). Organizing principles for single joint movements: I - A speed-insensitive strategy. *J. Neurophysiol.* 62(2): 342-357.

Gottlieb, G. L., Corcos, D. M., Agarwal, G. C. and Latash, L. M. (1990). Organizing principles for single joint movements: III - The speed-insensitive strategy as default. *J. Neurophysiol.* 63:625-636.

Hasan, Z. (1986). Optimized movement trajectories and joint stiffness in unperturbed, inertially loaded movements. *Biol. Cybernetics* **53**: 373-382.

Hogan, N. (1984). An organizing principal for a class of voluntary movements. *J. Neurosci.* **11**: 2745-2754.

Latash, M. L. (1989). Dynamic regulation of single-joint voluntary movements. Unpublished Ph.D. Dissertation, Rush University, Chicago.

Milner, T. E. (1986). Controlling velocity in rapid movements. *J. Mot. Beh.* **18**: 147-161.

Nagasaki, H. (1989). Asymmetric velocity and acceleration profiles of human arm movements. *Exp. Brain Res.* **74**: 319-326.

Nelson, W. L. (1983). Physical principles for economies of skilled movements. *Biol. Cyber.* **46**: 135-147.

Wacholder, K. and Altenburger, H. (1926). Beitrage zur physiologie der willkurlichen bewegung. X. Einzelbewegungen. *Pfl. Arch. Ges. Physiol.* **214**: 642-661.

Wadman, W. J., Denier van der Gon, J. J., Geuze, R. H. and Mol, C. R. (1979). Control of fast goal-directed arm movements. *J. Human Move. Studies* **5**: 3-17.

Wallace, S. A. (1981). An impulse-timing theory for reciprocal control of muscular activity in rapid, discrete movements. *J. Mot. Beh.* **13**: 144-1160.

Waters, P. and Strick, P. L. (1981). Influence of 'strategy' on muscle activity during ballistic movements. *Brain Res.* **207**: 189-194.

CHAPTER 15

Organizing Principles Underlying Motor Skill Acquisition

Daniel M. Corcos, Gerald L. Gottlieb,

Slobodan Jaric, Ronita L. Cromwell, and Gyan C. Agarwal

15.1 Introduction

A review of the literature on motor behavior suggests that there have been few attempts to understand motor learning and motor control from a common theoretical perspective. Both fields of study have a long history of research but with little overlap. This is unfortunate since the development of knowledge in each field must ultimately depend on the development of knowledge in the other. For example, many experimental results in the motor control literature are potentially confounded by the prior history of experimental subjects and the degree of learning that took place before and during data collection. Experimental paradigms requiring subjects "not to intervene" (Asatryan and Feldman, 1965; Gottlieb and Agarwal, 1988) usually require considerable subject practice and are seldom possible in untrained subjects. Observations such as the triphasic *EMG* pattern (Angel, 1974; Hannaford and Stark, 1985) are contingent on movement speed that is in turn affected by the skill of the subject at performing the particular task. Similarly, an understanding of the principles that underlie the control and regulation of movement will give insight into how movement skills are acquired and how control principles change as a function of learning.

This chapter will briefly summarize the Dual Strategy hypothesis of motor control that was presented in Chapter 14 (Gottlieb et al.). It will then discuss some of the findings on motor learning that can be found in the literature. Once this framework has been established, we will generate some predictions for the rules that underlie performance improvement. We will then present data to show how the Dual Strategy hypothesis can be used to interpret changes in motor performance resulting from prior experience in performing movement tasks. We will consider changes that arise from adaptation in typical and highly skilled individuals as well as learning and transfer.

15.2 Assessment of Pertinent Literature

In this section, we will first review the Dual Strategy hypothesis of motor control, then consider the processes of adaptation and learning and conclude with some predictions for learning that can be derived from the Dual Strategy hypothesis.

15.2.1 Dual Strategy Hypothesis

Individuals can generate movements over different distances, against different types of load and at different speeds. This wide repertoire of movements raises the question whether these movements are all controlled in the same manner, controlled in different manners or whether certain classes of movement tasks are controlled by one set of rules and other classes by other rules. We have shown that when a single joint movement task is to move different distances to a constant size target and movement time is not constrained, subjects adopt a "speed–insensitive" (*SI*) strategy (Gottlieb et al., 1989a). The term insensitive is used because this task can be performed "correctly" at a wide variety of speeds. This task can be contrasted with one in which a person is told to move "accurately" and to targets of different sizes. In most situations, subjects do not move at their fastest possible speed since this leads to inaccurate performance (Fitts, 1954; Fitts and Peterson, 1964). Instead they exert control over movement speed by slowing their movements in order to meet the accuracy requirements of the task. This movement strategy has been termed "speed-sensitive" (*SS*) (Corcos et al., 1989b). The same strategy is employed when individuals intentionally move at a specific speed.

We have used the term strategies to refer to a set of rules for the control of movement, of which we have described two (Gottlieb et al., 1989b). The rules refer to how the parameters of muscle excitation are modulated and how *EMG*s and kinematics vary as a consequence of this modulation. Motoneuron excitation refers to the total excitatory input that arrives at the motoneuron pool from all sources and, for the movements we are considering, will be assumed to be a rectangular pulse of controllable height or width (see Chapter 14 for a detailed explanation of the relationship of excitation to *EMG*s and kinematics).

Rules for the Speed-Insensitive Strategy:

SI rule for the excitation pulse: Intensity is constant and duration is modulated.

SI rule for the EMG: The initial pattern of motoneuron pool discharge is independent of the magnitude of the task variable. This generates *EMG*s which rise at the same rate, irrespective of changes in distance or load. Changes in the task affect the area and duration of the *EMG* burst.

SI rule for kinematics: Because initial motoneuron pool activation is insensitive to the magnitude of the task variable, initial muscle force rise is also unaffected. For constant inertial loads, this is reflected in constant initial rates of acceleration. For different inertial loads, acceleration scales inversely with load.

Rules for the Speed-Sensitive Strategy:

SS rule for the excitation pulse: Intensity is modulated. Duration is constant.

EMG rules: The initial rate of recruitment and firing rates of the alpha motoneurons are adjusted to adapt to changes in the task. This results in changes in the initial slope of the *EMG* and in the area of the agonist burst. The duration of the agonist burst will be nearly constant if the duration of the excitation pulse is constant.

Kinematic rules: The slope of the initial rise in muscle force (or joint torque) will scale with the intensity of the excitation pulse. For constant inertial loads, this implies that acceleration will be proportional to intensity (Gottlieb et al., 1989b, p. 196).

These rules state that the control of movement can be characterized by pulse height or pulse width modulation of a hypothetical excitation pulse that is determined by task demands.

15.2.2 Kinematic and Myoelectric Patterns of Adaptation/Learning

In the study of time-dependent changes in motor performance, it is important to distinguish changes that are relatively permanent from those that are temporary. Enhanced performance that occurs in the course of one experimental session has been termed adaptive to connote changes in behavior that are determined by practice but are not retained beyond the experimental session (Brooks and Watts, 1988; Ito, 1976, 1984). Motor learning, on the other hand, can be defined as "a set of processes associated with practice or experience leading to relatively permanent changes in skilled behavior" (adapted from Schmidt, 1988, p. 346). This distinction allows that there might be a difference in the processes that underlie adaptation and learning, a point we will consider in Section 15.4.3.

Several theories of motor learning have been proposed (Adams, 1971; Pew, 1966; Schmidt, 1975, 1988), but none are directly related to motor control models and mechanisms. Consequently, it is very difficult to draw conclusions from studies in the motor learning literature that have used myoelectric and kinematic variables. For example, increased muscle activation is one mechanism to enhance movement speed and increase *EMG* activity (McGrain, 1980; Schmidtbleicher et al., 1988). However, unpracticed movements are often characterized by greater activation than is required if the muscles were used more efficiently (Person, 1960). In such cases, improvements in performance might be characterized by decreases in muscle activation (Payton and Kelley, 1972). A third possibility is that the quantity (area) of myoelectric activity changes very little but that changes in the timing of muscle activity are responsible for performance changes (Hobart et al., 1975; Ludwig, 1982).

Such divergent findings are difficult to reconcile in the absence of specific rules relating myoelectric parameters to kinetic and kinematic parameters in the context of the specific task performed.

15.2.3 Predictions for Adaptation and Learning

Three defining characteristics of skilled performance are speed, accuracy, and consistency. It is possible to predict the way in which individuals will acquire movement skills in terms of the Dual Strategy hypothesis. Any time individuals repeat a task more quickly, *EMGs* and kinematics should change according to the speed-sensitive strategy. That is, the *EMGs* and kinematics should rise at progressively faster rates as a function of increases in movement speed.

The issue of how accuracy and consistency will improve is more difficult to predict than speed since the terms accuracy and consistency are difficult to define and the mechanisms that underlie accuracy and consistency are different. Movement accuracy can be defined either in terms of: *1)* the absolute value of the distance moved from the "target" (called absolute error, or *AE*), *2)* the signed value of the distance moved from the "target" (constant error, or *CE*), *3)* the standard deviation of the values of the final position (called variable error, or *VE*), or *4)* some composite value (*E*) (see Spray (1986) for a discussion of these measures). In this chapter we will refer to the second measure as movement accuracy and the third measure as movement consistency.

The Speed-Insensitive strategy proposes that movement distance is planned by selecting excitation intensity and then determining the appropriate values for duration and antagonist latency. If we define accuracy in terms of the signed value of the actual distance moved from the mid-point of the target, errors in accuracy will be due to incorrect specification of the duration of the excitation pulse and the time delay in activating the antagonist muscle. For example, longer durations and/or delayed antagonist activation would lead to movement overshoots. As such, improved accuracy would be related to refining the selection of these two parameters. However, the size of movement errors is indirectly related to the intensity of the excitation pulse. The smaller the value of intensity, the less the effect of errors in selecting values

excitation pulse. The smaller the value of intensity, the less the effect of errors in selecting values of duration (Gottlieb et al., 1989b). Therefore, individuals can become more accurate either by making slower movements (reducing intensity) or by using more appropriate values of duration and latency. The choice they make will depend heavily on experimental instructions and prior experience.

If we also consider accuracy in terms of reduced variability, then there will be greater consistency in selecting values for the excitation pulse from trial to trial. In such cases we predict that there will be a decrease in the variability of selected *EMG* measures. For example, the area of the *EMG* calculated over the first 30 ms [which is proportional to excitation intensity (Gottlieb et al., 1989a)] should become more consistent from trial to trial.

15.3 Methods

The majority of the data to be presented in this chapter are taken from a series of studies that are reported in detail in (Corcos et al., in preparation (a,b); Jaric et al., in preparation). In this section we will present only the experimental design for this series of studies and two other studies. The experimental protocols, mechanical measurements, and *EMG* analysis can be found in Gottlieb et al. (1989a).

15.3.1 Pre- and Post- Test — Sessions 0 & 8

A series of experiments consisting of nine experimental sessions was performed. Nine sessions numbered 0 to 8 were performed on alternate days with a day of rest in between. The experimental sessions 0 and 8 will be referred to as the "pretest" and the "post-test." In the pre-test, 10 subjects first performed three maximum voluntary isometric contractions (*MVCs*) in both elbow flexion and extension. They then performed series of flexion movements to a 3° target at distances of 18°, 36°, 54°, 72°, and 90°. They were instructed that when a computer-generated tone sounded, they should:

"Always be as fast as possible. I know that you will almost always miss the target. You may adjust your next movement on the basis of the previous undershoot or overshoot, but only under the condition that you not slow down. Do not adjust your final position."

The instructions also explained that reaction time was not important. For each experimental condition, subjects performed 11 movements in about 90 seconds and then had two minutes' rest before the next experimental block. The first movement of each condition was discarded from all subsequent analyses. Target distance for each block of movements was set in a randomized sequence with a brief rest between blocks. The data for this session will be reported in Section 15.4.4.

15.3.2 Early Phase — Session 1

Two days after the pre-test, five of the ten subjects first performed *MVCs* in both flexion and extension. They then made 200 elbow flexion movements, partitioned into 10 blocks of 20 movements, of 54° to a target 3° in width. The pause between movements was about 10 seconds and between blocks, two minutes. During the interblock interval, subjects were told the average movement time in the most recent block, and a comparison between it and the movement time in the previous block. If they had slowed, they were asked to try to increase their movement speed. The data from this experiment will be reported in Section 15.4.1. In addition, we also collected data from one college-level baseball player who performed exactly the same sequence of movements (in the flexion direction) and also performed movements in the extension direction. These data will be reported in Section 15.4.2.

15.3.3 Late Phase - Sessions 2-7

The five subjects who participated in the early phase (Session 1), repeated the same protocol for six more experimental sessions with a day of rest intervening between experimental sessions. Only the data from block 1 were recorded for further analysis for Sessions 3 to 6. The data from this experiment will be reported within Section 15.4.3.

15.3.4 Effects of Instruction

We also ran a set of experiments to investigate the effects of experimental instruction. Five subjects with no previous experience as experimental subjects generated 2,000 elbow flexion movements on 10 consecutive days (200 movements divided in 10 blocks over 54° distance to a 3° target). Three of the subjects were instructed to perform the fastest possible movements over the entire learning process, while the other two were

only allowed to increase movement velocity under the condition that at least 18 out of 20 movements in the previous block were accurate. The data from this set of experiments will be presented within Section 15.4.3.

15.4 Results and Discussion

In this section we will present data that illustrates the rules that underlie adaptation of normal and highly skilled individuals, learning, and transfer.

15.4.1 Adaptation Highly Skilled in Normal Individuals

The data in Figure 15.1 show the first three blocks of trials from one subject (S.4) in session 1. This subject increased movement speed over the first three blocks of trials. The question of interest is whether the "rules" used to increase speed are the same as those used when speed is intentionally varied [see Section 15.2.1 and Figures 14.1 and 14.2 of Chapter 14 (Gottlieb et al.)]. That they are can be seen from the fact that acceleration and both *EMGs* rise at a progressively faster rate from blocks 1 to 5 and the antagonist latency becomes shorter. These data suggest that improved performance is caused by increasing the intensity of the excitation pulse and activating the antagonist muscle earlier. These findings are consistent with the rules outlined in Section 15.2.1. These results are also in accord with studies that show increased coactivation during learning since the "speed-sensitive" strategy results in increased *EMG*, earlier antagonist activation, and consequently more temporal overlap in the activity of the two muscle groups.

Several other features of the performance are also interesting. The data in Figure 15.2 depict peak velocity and its standard deviation (*part a*) for one subject (S.3), the consistency of this subject (*part b*), and the ratio of acceleration time to deceleration time *part c*). The data show that peak velocity increases for about the first five blocks of trials (100 trials) and then decreases. The standard deviation of peak velocity decreases steadily throughout the ten blocks of trials. These data show that the subject became more consistent. The data in Figure 15.2 (*part c*) show that changes in total movement time were accomplished primarily by shortening the deceleration time.

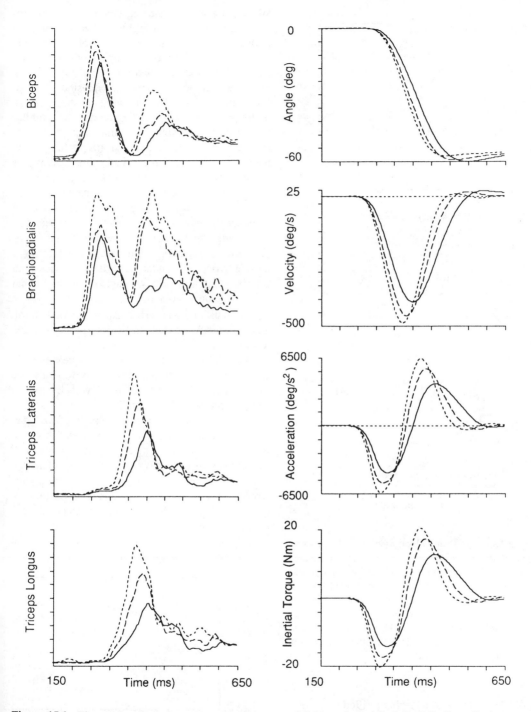

Figure 15.1: Time series plots of rectified and filtered *EMGs* of the biceps, brachioradialis, and lateral and long heads of the triceps. Data are averaged over the 20 trials for blocks 1 (solid line), 3 (long dashed line), and 5 (short dashed line) for 54° elbow flexion movements performed by subject 4. Angle is measured in degrees, velocity in deg/s, accleration in deg/s² and inertial torque in Newton meters.

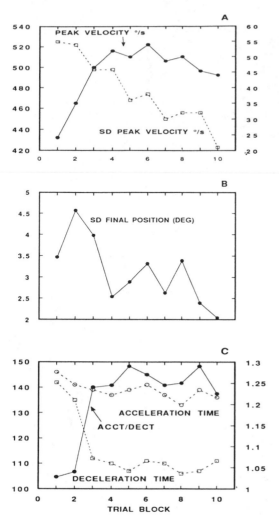

Figure 15.2: a) Peak velocity (solid circles) and the standard deviation of peak velocity (open squares) plotted as a function of trial block (20 trials per block). **b)** The standard deviation of final position plotted as a function of trial block (20 trials per block). **c)** Acceleration time (open circles), deceleration time (open squares), both measured in ms, and acceleration time divided by deceleration time plotted as a function of trial block (solid circles).

15.4.2 Adaptation of a Highly Skilled Individual

In the previous section we presented data from an extremely "simple" task in which considerable performance improvement was observed in one session over at least the first five blocks of trials.

We were also interested in whether such an improvement occurs in a highly skilled athlete who performs a task that uses some of the muscle groups involved in our "simple" elbow flexion task. Therefore we tested a baseball pitcher with movements in both flexion and extension. This allowed us to answer the following questions. First, is the performance in the flexion direction similar to that of normal subjects? Since flexors are not the agonist muscles involved in pitching, they have undergone no more extensive training as agonist muscles than the typical subjects tested. Second, is performance in the extension direction similar to that in the flexion direction? Since the extensor muscles are the principal elbow agonists in pitching, this might constitute special and extensive prior experience. Here, we are comparing a muscle group that undergoes extensive use as an agonist in pitching (elbow extensors) with one that does not (elbow flexors).

The data in Figure 15.3 depict the first, third, and fifth blocks of trials in flexion (*part a*) and extension (*part b*).

One difference from the data presented in Figure 15.1 is that the flexion movements are much faster. The average peak velocity of our five subjects was about per 500°/sec. The baseball pitcher was moving at a peak velocity of 720° per second as shown in Figure 15.4. In fact this subject is the fastest subject to have performed movements in our laboratory (n > 25). Clearly, this observation is not surprising since everyone would expect a pitcher to be able to move his arm quickly. However, the biceps muscles are not the agonist muscles for pitching, and there is frequently little transfer between tasks (cf Section 15.4.4). As such, why does he move so fast?

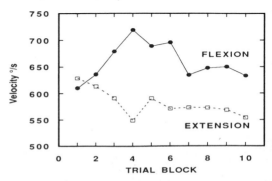

Figure 15.4: Peak velocity in flexion (solid circles) and extension (open squares) plotted as a function of trial block (20 trials per block).

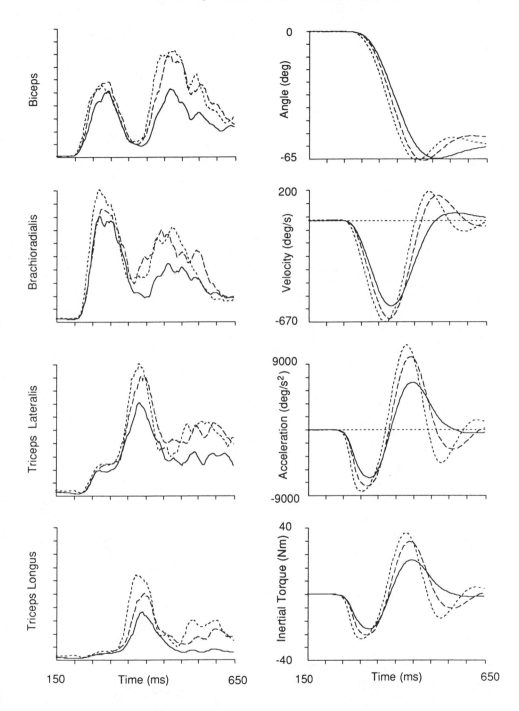

Figure 15.3: *a)* Time series plots of rectified and filtered *EMGs* of the biceps, brachioradialis, and lateral and long heads of the triceps. Data are averaged over 20 trials for blocks 1 (solid line), 3 (long dashed line), and 5 (short dashed line) for 54° elbow flexion movements to a 3° target performed by a college baseball pitcher. Angle is measured in degrees, velocity in deg/s, accleration in deg/s², and inertial torque in Newton meters. *b)* see next page.

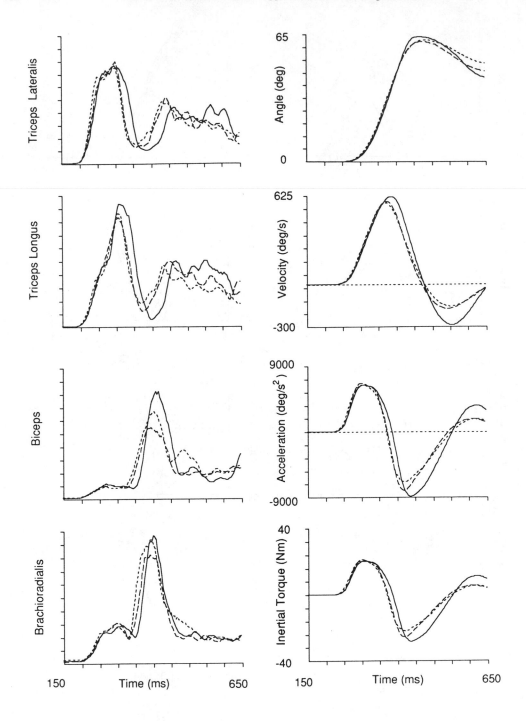

Figure 15.3: (cont.)   *b)* The same information as presented in *a)*, except that the movements are in the extension direction.

There are several possible reasons, and many of them will apply. They include a wide array of physiological and biomechanical properties alluded elsewhere [e.g. Chapter 19 (Seif-Naraghi and Winters)]. The reason we would like to consider here is antagonist muscle strength. This subject's MVC in flexion was about 70 Nm and about 48 Nm in extension. Colebatch et al. (1986) showed the average value of flexion torque to be about 62 Nm for males, but only about half that for extension torque. Therefore, the strength of the pitcher's flexors is slightly above the normal range, but his extensors are almost one and a half times as strong as the subjects in the study by Colebatch et al.

Several interesting points emerged from the extension movements. First, as already noted, his maximal voluntary contraction in extension is stronger than that of normal subjects. Second, he undergoes no increase in peak movement velocity in extension. This point can be seen in Figure 15.4. In this figure, the peak movement velocity of the 10 blocks of flexion and extension movements are depicted. For flexion movements, peak movement velocity improves over the first four blocks of trials and then falls. His performance in extension is somewhat different. His peak movement velocity starts to decline after the first block of trials and reaches a stable level at the sixth block. The main reason for the decline in peak movement velocity is that his movements were slightly long and he shortened them, as can be seen in Figure 15.3b. Longer movements are normally associated with higher movement velocities (Milner, 1986). If this argument is accepted, the most plausible explanation of this data set is that little or no change in performance occurred over the blocks of trials. Why might this be? One possible reason is that there is little room for improved performance in muscle groups that undergo extensive use. In other words, there has been positive transfer from what is learned in pitching to this elbow extension task.

15.4.3 Learning (Retention)

In this section we will first consider how different experimental instructions influence the rate at which a task is learned and then consider the degree to which learning differs from adaptation.

Influence of experimental instruction

The data in Figure 15.5a (solid lines) are from one of the three subjects in the experimental group that performed 10 blocks of 200 trials over 10 experimental sessions using the fast instruction. This subject increased his peak velocity at least 20%, and this increase extended over almost the entire learning process. The data of the other subject (dashed lines) are from one of the two subjects who practiced using the accuracy instruction and did not increase movement velocity.

The variability of peak movement velocity (Figure 15.5b) decreased threefold in the subject instructed to be fast despite the fact that only movement speed was stressed (cf Figure 15.2a).

The improvement in accuracy of the two subjects is impossible to compare using any of the standard measures because of the different instructions — the subject instructed to be accurate had to retain the same accuracy of the final position during the entire learning process. However, the damped oscillations around the final position (the standard deviation of the maximum movement amplitude during the terminal phase of the movement) offered an indirect accuracy measure for both subjects. The results show that both subjects notably improved their accuracy, but the subject given the accuracy instructions was approximately two times more accurate during the entire learning process (Figure 15.5c).

Some previous investigations have demonstrated more or less symmetrical, bell-shaped velocity profiles in single joint target movements (see Figure 15.2c, first block of 10 trials). Since changes in movement time are mainly based on extending the deceleration phase, this symmetry is sensitive to movement speed (Fisk and Goodale, 1989; Nagasaki, 1989). Figure 15.5d depicts changes in the symmetry of the velocity profile shape as a result of practicing movements under fast or accurate instructions. The two subjects showed opposite trends. The high movement speed achieved by the subject instructed to be fast was associated with a movement time decrease that was mainly attained by shortening the deceleration phase. However, the other subject did not significantly change movement speed or movement time but did prolong his deceleration phase. The reasons for

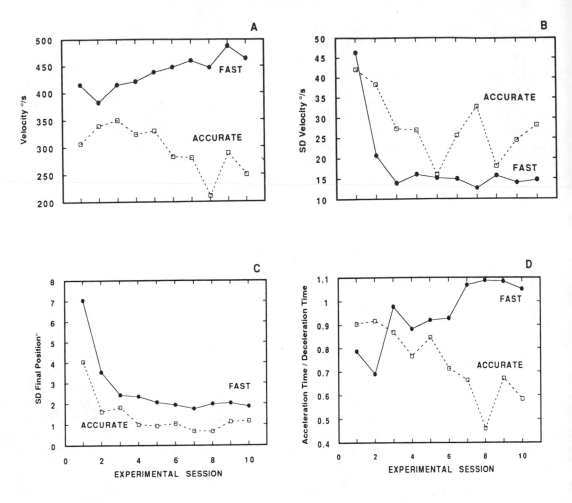

Figure 15.5: a) Peak velocity for a subject instructed to be fast (solid circles) and a subject instructed to be accurate (open squares) plotted as a function of experimental session. The data for each session are the averaged data for the block of twenty trials in each session. **b)** The standard deviation of peak velocity for a subject instructed to be fast (solid circles) and a subject instructed to be accurate (open squares) plotted as a function of experimental session. *c)* The standard deviation of peak excursion for a subject instructed to be fast (solid circles) and a subject instructed to be accurate (open squares) plotted as a function of experimental session. **d)** Acceleration time divided by deceleration time for a subject instructed to be fast (solid circles) and a subject instructed to be accurate (open squares) plotted as a function of experimental session.

these differences in symmetry are related to the different task demands. For example, in the case of the instructions to move accurately, the more time subjects have in the vicinity of the target, the more time they have to generate visually based corrections for the movement.

Learning versus adaptation

The data in Figure 15.6 are taken from days 1, 2, and 4 for one of the subjects (S.5) from the main experiment. He increased his movement speed significantly in spite of the fact that he did not enhance his maximum isometric strength in

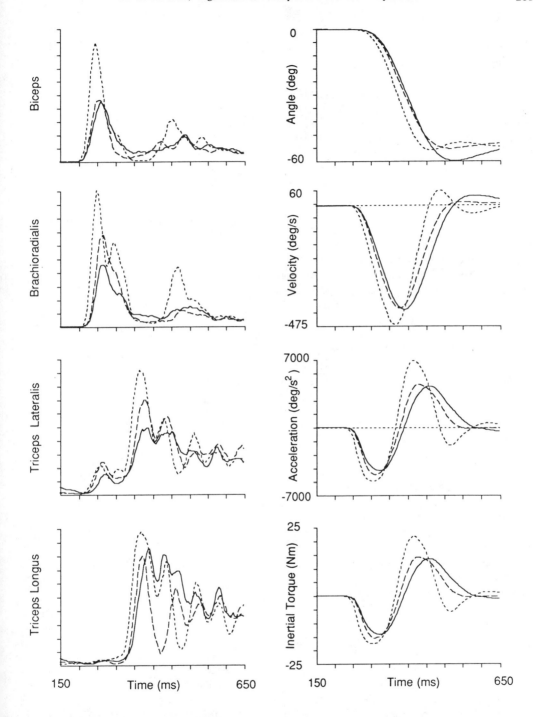

Figure 15.6: Time series plots of rectified and filtered *EMGs* of the biceps, brachioradialis, and lateral and long heads of the triceps. Data are averaged over the first 20 trials for day 1 (solid line), 2 (long dashed line), and 4 (short dashed line) for 54° elbow flexion movements performed by subject 5. Angle is measured in degrees, velocity in deg/s, acceleration in deg/s², and inertial torque in Newton-meters.

either elbow flexor or extensor muscles. Therefore, we conclude that the change is due to improvements in the neural activation pattern controlling the movement. Figure 15.6 presents the main changes in that pattern in parallel with the associated changes in angle, velocity, and acceleration. All three *EMG* bursts became enhanced, suggesting greater excitation of both agonist and antagonist muscles. In addition, the duration of the first agonist burst remained relatively unchanged, while antagonist burst onsets occurred earlier. These changes are in line with the SS strategy of motor control.

Given that the data of this subject conform to the "speed sensitive" strategy, we can ask whether there are any differences between adaptation and learning. The two main differences are that the rate of performance improvement decreases over time, as can be seen from the data of subject 4 that are plotted in Figure 15.7a). The first ten data points are the same as plotted in Figure 15.2a. Performance on the second day resumes at a higher level than at the end of day 1. The point that is new in these experiments is that the antagonist latency does not follow this same pattern. The data in Figure 15.7b show that the latency of the antagonist *EMG* falls to about 100 ms after the first experimental session and no further by the second session although velocity continues to increase. There is also some suggestion that towards the end, the latency begins to increase. This initial reduction in latency during adaptation and the subsequent increase during learning make sense if we postulate that two separate processes are at work. Over the first five blocks of trials velocity increases by the rules of the "speed-sensitive" strategy. The antagonist latency decreases. However, as subjects become more skilled, it is commonly observed that there is a decrease in coactivation. This would occur if individuals learn to delay the activation of the antagonist muscle. Learning such a pattern of muscular activation would lead to higher velocity movements and a more efficient movement pattern. Figure 15.7c is a prediction of how the antagonist delay is modulated to deal with the fact that it shortens for adaptation and lengthens for learning. That is, the time delay decreases within the course of a day but increases over days.

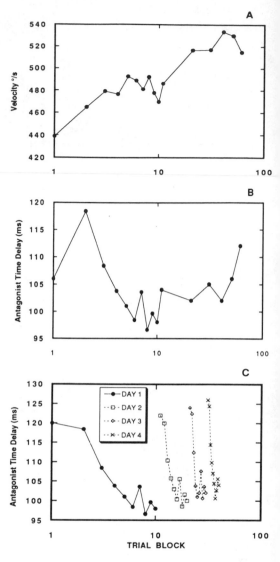

Figure 15.7: a) Peak velocity for subject 4. The data are all ten blocks on day 1 and the first of ten blocks on the next six days. The data are plotted on a logarithmic scale as a function of trial block. **b)** Antagonist time delay plotted in the same manner as in Figure 15.7a. **c)** A hypothetical plot of antagonist time delay as a function of trial block and day.

Anecdotal support for this suggestion can be seen in Figure 15.3a-b in which the latency of the antagonist muscles of the pitcher can be seen. In both directions, the muscles are not activated until well over 100 ms after the agonist, and there is very little coactivation.

15.4.4 Learning (Transfer)

The permanence of changes in performance that characterize learning can be determined in one of two ways. The first is by demonstrating the retention of changes in behavior in the task that is practiced. We have already shown this occurs by means of the rules that underlie the "speed–sensitive" strategy. The second way of demonstrating learning is by evaluating performance on a different task. This is referred to as either "transfer of training" or "generalizability." More specifically, transfer can be defined as "the gain (or loss) in the capability for responding on a transfer or criterion task as a function of practice

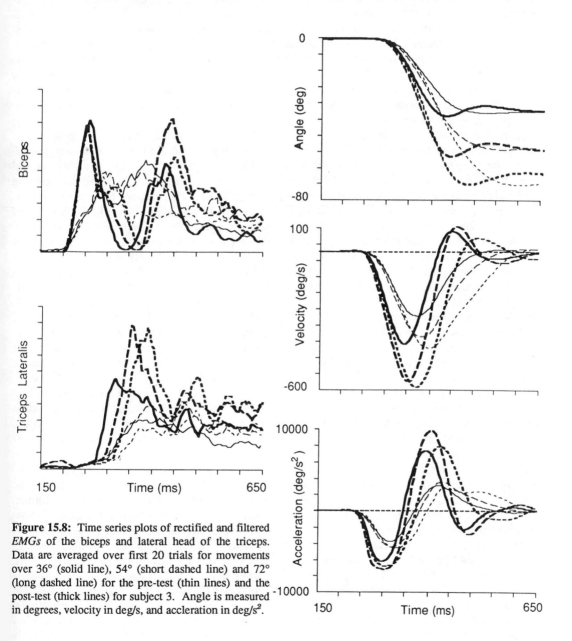

Figure 15.8: Time series plots of rectified and filtered *EMGs* of the biceps and lateral head of the triceps. Data are averaged over first 20 trials for movements over 36° (solid line), 54° (short dashed line) and 72° (long dashed line) for the pre-test (thin lines) and the post-test (thick lines) for subject 3. Angle is measured in degrees, velocity in deg/s, and accleration in deg/s².

or experience on a training task" (Schmidt and Young, 1987, p. 48). In our analysis of the highly skilled baseball pitcher, we have shown that transfer does or does not occur depending on movement direction. In the flexion direction (the antagonist muscle group in pitching), the pitcher improved his performance using the speed-sensitive rules. For extension movements, there was no increase in performance. The influence of pitching on the performance of the extension task is an example of positive transfer since this is the most reasonable interpretation of why there was no change in performance in this direction. In the flexion direction, the interpretation of whether any transfer occurs is somewhat more difficult since the pitcher clearly moves much faster than normal subjects, but there is still room for improvement.

Another way to demonstrate transfer is to have subjects make movements to different distances after practicing at one distance. The data in Figure 15.8 are taken from one of the subjects (S.3) making flexion movements over three different distances before and after practicing 1400 movements to the same distance. The first point we wish to make is that the degree of transfer differed markedly across the five subjects tested. The data we are presenting here are those of the subject who demonstrated the most transfer.

The data show clear evidence of learning and also transfer. The effect of learning can be seen most clearly in the velocity traces. The three lines corresponding to the pre-test rise at the same rate as would be expected for movements of different distances performed by means of the speed-insensitive strategy. If we look at the post-test movements, the 54° movement rises, as expected, more steeply than in the pre-test. The most interesting feature of this data set though is that the other two post-test distances also rise more steeply than in the pre-test, although not quite as rapidly as the 54° movement. These are therefore examples of positive but not complete transfer.

The way that learning and transfer are accomplished can be seen from an inspection of the electromyograms. The EMGs for the post-test rise more steeply than for the pre-test and occur earlier in time. Both these observations are consistent with the idea that this subject learned to increase the intensity of muscle excitation and decrease the latency at which the antagonist was activated.

15.5 Discussion and Conclusions

The studies we have presented show how the Dual Strategy hypothesis of motor control can be used to explain how performance improves over time and can reconcile many discrepant findings in the motor learning literature that have used myoelectric and kinematic variables. For example, the data in Figures 15.1, 15.3a, and 15.6 all show EMGs that rise at faster rates and have increased EMG quantities when movements are performed at faster speeds. Both these observations are consistent with the rules of the Dual Strategy hypothesis. The findings of increased EMG quantities are also consistent with studies by (Gottlieb et al., 1988; McGrain, 1980; Schmidtbleicher et al., 1988). We have shown this finding to be true for adaptation in typical individuals (Figure 15.1), in a highly skilled baseball player (Figure 15.3a) and for learning (Figure 15.6). The data we have presented also suggest that there is no change in the duration of the first agonist burst (Figures 15.1 and 15.3a). However, some studies have shown decreases in agonist burst duration (Kamon and Gormley, 1968; Hobart et al., 1975; Ludwig, 1982; Payton and Kelley, 1972). This observation could be associated with a reduction in movement time.

The movements made in these figures were all generated under instructions to be fast. Movements performed under these instructions are also characterized by decreases in the standard deviation of peak velocity (Figure 15.2a), consistency of final position (Figure 15.2b), and proportionally larger decreases in deceleration time than acceleration time (Figure 15.2c). These observations are, however, dependent on experimental instruction. When subjects are instructed to be accurate, they make their movements considerably more slowly and change the overall kinematic profile. Movements performed under instructions to be accurate usually show only improvements in movement accuracy and consistency (Figure 15.5b-c). Similar findings have been observed by Brooks and Watts (1988), Darling et al. (1988), Georgopoulos and Massey (1981), and Ludwig (1982). Movement time often does not change under instructions to be accurate, but the ratio of acceleration time to deceleration time does change; the deceleration phase is prolonged, as shown in Figure 15.5d and by

Hobart et al. (1975) and Payton and Kelley (1972). These changes in movement kinematics do not require larger forces (unlike increases in velocity) and will consequently be very difficult to detect in the electromyogram. We suspect this is the reason that no electromyographic changes were found in some of the studies that have emphasized movement accuracy (Brooks and Watts, 1988; Ludwig, 1982; Normand et al., 1982).

As pointed out by Schmidt and Young (1987), there has been very little interest in the transfer of motor control abilities. This is unfortunate since the study of transfer can contribute significantly to what is learned. Perhaps the most interesting finding in our set of studies concerns the observation that improving performance at one movement distance can lead to improvements in performance at other movement distances. This is accomplished by using an increased excitation pulse intensity and decreased antagonist latency (Figure 15.8).

The data on the baseball pitcher also imply that transfer occurs. The skill of pitching requires rapid, controlled elbow extension. We showed that the pitcher can perform a simple elbow extension task more rapidly than typical individuals but undergoes no performance improvement in the extension task. In the flexion task, the pitcher again moves much faster than typical individuals (some transfer) but undergoes performance enhancement. One interpretation of these data is simply that he has the ability to move quickly and would have performed the laboratory task as quickly before he became a pitcher. However, this explanation would not account for the observation that he could still undergo performance enhancement in flexion but not extension. We predict that he would have performed the flexion task at a higher initial speed if he had practiced an activity requiring elbow flexors and would not have undergone the same level of performance improvement.

In conclusion, the data presented in this chapter suggest the rules that underlie changes in movement speed are the same irrespective of whether the changes are induced by instruction (Corcos et al., 1989b), accuracy constraints (Corcos et al., 1989b), or adaptation (Corcos et al., in preparation, a). We have also suggested that the antagonist is activated relatively later as the task becomes learned (Jaric et al., in preparation).

15.6 Recommended Future Directions

At least two areas emerge from these studies for future research. The first is to refine our understanding of the processes that underlie the attainment of greater accuracy and consistency. The second question concerns the degree to which this theory can be used to explain how complex movements are learned. Clearly the study of multiple degree of freedom movements introduces many complexities that are not contained in the analysis of single degree of freedom movements that we have presented [cf. Corcos et al., 1989a and commentaries as well as Chapter 17 (Flash)]. Examples of such factors are the appropriate sequencing and coupling of proximal–distal musculoskeletal linkages [Bernstein, 1967; see also Chapter 39 (Chapman and Sanderson)] and how limb dynamics change as a function of practice (Schneider et al., in press).

Sometimes the findings from studies using complex movements are in accord with predictions from studies of single degree of freedom movements even though the "rules" might require elaboration. For example, Southard (1989) had subjects learn a ball-striking task under three sets of instructions: fast, accurate, and fast-accurate. He found that the movements were performed more slowly when using the accuracy instructions and that subjects did not learn how to transfer angular momentum when instructed to be accurate. Clearly, therefore, speed should be emphasized when learning single or multiple degree of freedom tasks. The way the rules should be changed to accommodate the sequencing of muscle activation is not so clear. However, the idea that antagonist muscles become activated progressively later as skill is acquired is a starting point. For example, McDonald et al. (1989) have shown that improved performance in a dart–throwing task is accompanied by an uncoupling of the degrees of freedom involved in the task. One way that degrees of freedom might be coupled is through simultaneous activation of synergistic and antagonistic muscle groups. As skill level increases, antagonist muscle activation can be delayed and the degrees of freedom uncoupled.

The study of movement organization and biomechanics will benefit from efforts to generate unified theories of motor control that can account for changes in motor behavior over time. The degree to which these theories will be task dependent is unclear. As Newell (1989) has pointed out, most of the theories, hypotheses, and models that have been developed are specific to the studied task and not general theories of motor control. There are several theories or models that can account for certain tasks or certain data sets (Feldman, 1986; Ghez and Gordon, 1987; Gottlieb et al., 1989b; Hogan, 1984; Schoner and Kelso, 1988), but none yet offer a unified theory of motor control or learning. We are optimistic that the considerable efforts being made to generalize these theories will be fruitful.

Acknowledgments

We thank Mr. Om Paul for programming support and Mr. K. Lee for assistance in performing these experiments. This work was supported, in part, by NIH Grants NS 23593 and AR 33189.

References

Adams, J. A. (1971). A closed-loop theory of motor learning. *J. Mot. Beh.* 3: 111-150.

Angel, R. W. (1974). Electromyography during voluntary movement: the two burst pattern. *Electroenceph. Clin. Neuro.* 36: 493-498.

Asatryan, D. G., Feldman, A. G. (1965). The functional tuning of nervous systems with control of movement or maintenance of a steady posture 1. Mechanographic analysis of the work of the joint on execution of a postural task. *Biophysics* 10: 925-935.

Bernstein, N. (1967). *The Coordination and Regulation of Movements.* Oxford: Pergamon.

Brooks, V. B., Watts, S. L. (1988). Adaptive programming of arm movements. *J. Mot. Beh.* 20(2): 117-132.

Colebatch, J. G., Gandevia, S. C., Spira, P. J. (1986). Voluntary muscle strength in hemiparesis: distribution of weakness at the elbow. *J. Neurol. Neurosurg. Psychiatry* 49: 1019-1024.

Corcos, D. M., Gottlieb, G. L., Agarwal, G. C. (1989a). Does constraining movements constrain the development of movement theories? *Behav. Brain Sci.* 12(2): 237-246.

Corcos, D. M., Gottlieb, G. L., Agarwal, G. C. (1989b). Organizing principles for single joint movements: II - A speed-sensitive strategy. *J. Neurophysiol.* 62: 358-368.

Corcos, D. M., Gottlieb, G. L., Jaric, S., Agarwal, G. C. (in preparation, a). Principles for learning single joint movements: I - Rules for enhanced motor performance.

Corcos, D. M., Jaric, S., Agarwal, G. C., Gottlieb, G. L. (in preparation, b). Principles for learning single joint movements: III - Generalizability.

Darling, W. G., Cole, K. J., Abbs, J. H. (1988). Kinematic variability of grasp movements as a function of practice and movement speed. *Exp. Brain Res.* 73: 225-235.

Feldman, A. G. (1986). Once more on the equilibrium-point hypothesis (λ model) for motor control. *J. Mot. Behav.* 18: 17-54.

Fisk, J. D., Goodale, M. A. (1989). The effects of instructions to subjects on the programming of visually directed reaching movements. *J. Mot. Beh.* 21(1): 5-19.

Fitts, P. M. (1954). The information capacity of the human motor system in controlling the amplitude of movement. *J. Exp. Psy.* 47: 381-391.

Fitts, P. M., Peterson, J. R. (1964). Information capacity of discrete motor responses. *J. Exp. Psy.* 67: 103-112.

Georgopoulos, A. P., Kalaska, J. F., Massey, J. T. (1981). Spatial trajectories and reaction times of aimed movements: Effects of practice, uncertainty, and change in target location. *J. Neurophysiol.* 46: 725-743.

Ghez, C., Gordon, J. (1987). Trajectory control in targeted force impulses. I. Role of opposing muscles. *Exp. Brain Res.* 67: 225-240.

Gottlieb, G. L., Agarwal, G. C. (1988). Compliance of single joints: Elastic and plastic characteristics. *J. Neurophysiol.* 59: 937-951.

Gottlieb, G. L., Corcos, D. M., Agarwal, G. C. (1989a). Organizing principles for single joint movements: I - A speed-insensitive strategy. *J. Neurophysiol.* 62(2): 342-357.

Gottlieb, G. L., Corcos, D. M., Agarwal, G. C. (1989b). Strategies for the control of single mechanical degree of freedom voluntary movements. *Behav. Brain Sci.* 12(2): 189-210.

Gottlieb, G. L., Corcos, D. M., Jaric, S., Agarwal, G. C. (1988). Practice improves even the simplest movements. *Exp. Br. Res.* 73: 436-440.

Hannaford, B., Stark, L. (1985). Roles of the elements of the triphasic control signal. *Exp. Neurol.* 90: 619-634.

Hobart, D. J., Kelley, D. L., Bradley, L. S. (1975). Modifications occurring during acquisition of a novel throwing task. *Am. J. Phys. Med.* 54: 1-24.

Hogan, N. (1984). An organizing principal for a class of voluntary movements. *J. Neurosci.* 11: 2745-2754.

Ito, M. 1976. Adaptive control of reflexes by the cerebellum. In: *Understanding the Stretch Reflex* (edited by S. Homma), Progress in Brain Research 44:435-443.

Ito, M. 1984. *The Cerebellum and Neural Control.* New York: Raven Press.

Jaric, S., Corcos, D. M., Agarwal, G. C., Gottlieb, G. L. (in preparation.) Principles for learning single joint movements: II - Adaptation and learning.

Kamon, E., Gormley, J. (1968). Muscular activity pattern for skilled performance and during learning of a horizontal bar exercise. *Ergonomics* 11(4): 345-357.

Ludwig, D. A. (1982). *EMG* changes during acquisition of a motor skill. *Am. J. Phys. Med.* 61: 229-243.

McDonald, P. V., van Emmerik, R. E. A., Newell, K. M. (1989). The effects of practice on limb kinematics in a throwing task. *J. Mot. Beh.* 21(3): 245-264.

McGrain, P. (1980). Trends in selected kinematic and myoelectric variables associated with learning a novel motor task. *Res. Q. Ex. Sp.* 51: 509-520.

Milner, T. E. (1986). Controlling velocity in rapid movements. *J. Mot. Beh.* 18: 147-161.

Nagasaki, H. (1989). Asymmetric velocity and acceleration profiles of human arm movements. *Exp. Br. Res.* 74: 319-326.

Newell, K. M. (1989). On task and theory specificity. *J. Mot. Beh.* 21(1): 92-96.

Normand, M. C., Lagasse, P. P., Rouillard, C. A., Tremblay, L. E. (1982). Modifications occurring in motor programs during learning of a complex task in man. *Br. Res.* 241: 87-93.

Payton, O. D., Kelley, D. L. (1972). Electromyographic evidence of the acquisition of a motor skill: a pilot study. *Phys. Ther.* 52: 261-266.

Person, R. S. (1960). Studies of human movements when elaborating motor habits. *Proceedings of Third International Federation for Medical Electronics.* London.

Pew, R. W. (1966). Acquisition of hierarchical control over the temporal organization of a skill. *J. Exp. Psychol.* 71: 764-771.

Schmidt, R. A. (1975). A schema theory of discrete motor skill motor learning. *Psy. Rev.* 82: 225-260.

Schmidt, R. A. 1988. *Motor Control and Learning: A Behavioral Emphasis.* Champaign, Illinois: Human Kinetics Publishers, Inc.

Schmidt, R. A., Young, D. E. 1987. *Transfer of Movement Control in Motor Skill Learning.* Edited by S. M. Cormier and J. D. Hagman. *Transfer of Learning.* Orlando, FL: Academic Press.

Schmidtbleicher, D., Gollhofer, A., Frick, U. 1988. *Effects of a stretch-shortening typed training on the performance capability and innervation characteristics of leg extensor muscles.* Edited by G. Groot, A. P. Hollander, P. A. Huijing, and V. I. Schenau. Biomechanics XI-A. Amsterdam, Netherlands: Free University Press.

Schneider, K., Zernicke, R. F., Schmidt, R. A., Hart, T. J. (in press.). Changes in limb dynamics during the practice of rapid arm movements. *J. Biom.*

Schoner, G., Kelso, J.A.S. (1988). Dynamic pattern generation in behavioral and neural systems. *Science* 239: 1513-1520.

Southard, D. (1989). Changes in limb striking pattern: effects of speed and accuracy. *Res. Quart. Exer. Sci.* 60(4): 348-356.

Spray, J. A. (1986). Absolute error revisited: an accuracy indicator in disguise. *J. Motor Behav.* 18(2): 225-238.

CHAPTER 16

Direction-Dependent Strategy for Control of Multi-Joint Arm Movements

G.M. Karst and Z. Hasan

16.1 Introduction

In order to manipulate objects with our hands, we routinely use coordinated rotations about the shoulder and elbow to convey the hand from point to point within the surrounding workspace. The seeming ease with which the nervous system is able to generate muscle activation patterns appropriate for pushing an elevator button or reaching for the telephone appear to belie the well-documented complexity of the mechanics underlying such movements. In order to delineate rules by which the *CNS* might choose suitable patterns of muscle activation, and to identify the movement parameters most important in implementing those rules, we have investigated the electromyographic (*EMG*) activity of shoulder and elbow muscles associated with a variety of two-joint pointing movements in the horizontal plane (Hasan and Karst, 1989; Karst, 1989). In this chapter, we will consider data from those studies, as well as from studies of movement-related neuronal activity at higher levels of the *CNS*, which suggest that an important parameter of such movements is the spatial direction of hand movement.

16.2 Direction-Related Neuronal Activity in the Brain

In recent years, recordings obtained during the performance of target-directed, multijoint reaching tasks have demonstrated that modulation of neuronal activity in several regions of the primate brain is correlated with the spatial direction of hand movement during the task. Evidence of direction-dependent activity has been obtained from neurons in the motor cortex (Georgopoulos et al., 1982), *area 5* of the parietal cortex (Kalaska et al., 1983), and cerebellum (Fortier et al., 1989) during multijoint reaching tasks. Since many of these findings have been detailed in recent reviews (Georgopoulos 1988a, 1988b; Kalaska, 1988), this section includes only a brief synopsis of the experimental support for direction-dependent neuronal activity, with the primary emphasis on results characterizing direction-dependent neuronal activity in motor cortex.

16.2.1 Directional Coding in the Motor Cortex

Many of the findings discussed below are based on an experimental paradigm (Georgopoulos et al., 1982) in which extracellular recordings from single motor cortical neurons in rhesus monkeys were obtained during the performance of target-directed, reaching movements. These volitional movements were initiated in response to the lighting of visual targets, and the movement-related activity of individual neurons was quantified in terms of the change in firing rate (relative to the rate before the target appeared) during both the reaction time (target onset until movement onset) and throughout the movement. Data were collected only from those neurons which showed activity related to movement of the proximal joints of the contralateral arm. Unlike the more constrained, single-joint tasks used in most earlier studies of cortical activity, these multijoint movements encompassed a variety of directions in 2-dimensional (Georgopoulos et al., 1982) or 3-

dimensional space (Schwartz et al., 1988). Furthermore, this paradigm has also been modified to include movements initiated from different regions within the workspace (Caminiti et al., 1988; Kettner et al., 1988) and to examine the effects of various external forces applied to the hand. (Kalaska and Hyde, 1985; Kalaska et al., 1989).

Single-Cell Activity

Movement related changes in the discharge rate of task-related motor cortical cells were typically observed 80–100 ms prior to movement onset. Studies in which movement direction was varied in two-dimensional or three-dimensional space yielded similar basic findings, which may be summarized as follows:

1) Of the task-related motor cortical cells studied, 70–80% exhibited changes in discharge rate that varied systematically with the spatial direction of hand movement.

2) These cells typically exhibited a gradual, continuous modulation of discharge frequency over a broad range of movement directions. This modulation of discharge was modeled as a cosine function of the difference between the direction of movement and the cell's "preferred direction," the movement direction eliciting the largest response.

3) Although baseline activity of motor cortical cells varied with the spatial location of the hand, movement-related *changes* in discharge were independent of final hand position, and thus, similar for parallel movements in different regions of the workspace.

These findings indicate that, for many motor cortex cells related to shoulder function, changes in discharge rate appear more closely linked to the spatial direction of hand movement than to movement distance or final hand position. Likewise, the movement does not appear to be represented at the motor cortical level in terms of intrinsic coordinates, such as joint angles. The directional representation need not, however, be strictly kinematic in nature, as some cells also demonstrated variations in discharge in response to constant external loads applied to the hand (Kalaska et al., 1989), indicating sensitivity to kinetic, as well as kinematic, parameters of movement. Nevertheless, cells sensitive to kinetics modulated their activity in relation to the direction of external loading such that the increase in discharge rate was maximum when the external load was in a direction opposite the preferred movement direction for that cell. Thus, it appears that the preferred movement direction remains an important functional characteristic of the cell, whether that cell is modulated primarily in relation to kinematic or kinetic parameters of the movement.

Neuronal Population Activity

Though the broad tuning of motor cortical cells generally precludes unambiguous coding of directional information by the output of single motor cortical cells, Georgopoulos and colleagues have shown that directional information could be coded unambiguously by the activity of populations of cortical cells. If the activity of individual cells is represented by a vector whose orientation is parallel to the cell's preferred direction and whose magnitude is proportional to the movement-related change in its activity, the "population vector" may be derived as vectorial sum of those individual cell vectors. This population vector corresponds to the spatial direction of hand movement for multijoint reaching movements in either two-dimensional or three-dimensional space (Georgopoulos et al., 1986; 1988). Furthermore, when calculated for each 20 ms interval, the population vector appears to predict the direction of movement over time, since it begins to point in the direction of the subsequent movement and to increase in magnitude approximately 100 ms prior to movement onset. Finally, for parallel movements in the same direction, but varying in distance and location within the workspace, the congruency of the population vector with the actual hand trajectory remains relatively constant (Caminiti et al., 1988; Kettner et al., 1988), demonstrating that movement-related changes in motor cortical activity correspond to the spatial direction of hand movement, as opposed to the final hand position. Thus, like the changes observed in single-cell activity, it appears that modulation of the activity represented by neuronal population vectors corresponds most closely to the spatial direction of the upcoming hand movement, and shows much less dependence on movement distance or location within the workspace.

16.2.2 Directional Coding in Other Regions of the Brain

Direction-related changes in neuronal activity have been reported in other regions of the brain, as well (Fortier et al., 1989; Kalaska et al., 1983). These findings have led to suggestions that the movement may be represented at multiple levels of the *CNS*, with directional information serving as a common code to enhance communication between various levels of the sensorimotor system, or as a means of facilitating coordinate transformations (Kalaska, 1988). Moreover, as exemplified by the comparison of direction-dependent activity in motor cortex and *area 5* of parietal cortex which follows, the distinctions between directional representations at different levels of the *CNS* may provide important clues about the organization of the sensorimotor system.

Cells in *area 5* of the parietal cortex show direction-related changes in activity that are, in many ways, similar to those observed in motor cortical cells. That is, single-cell discharge modulation shows broad directional tuning, and population vectors representing *area 5* activity correspond to the spatial direction of hand movement (Kalaska et al., 1983). And, though cells in *area 5* tend to be activated 50–80 ms later than those in motor cortex, changes in activity still precede movement onset, suggesting a role in planning, as well as ongoing guidance of the movement. On the other hand, the *area 5* cells appear to differ from many motor cortical cells in their response to external loads. Whereas modulation of activity in motor cortical cells frequently reflects an additive effect related to the external load, movement-related modulation of neuronal activity in *area 5* is typically unaffected by external loading (Kalaska et al., 1989). Based on these findings, Kalaska and colleagues have suggested that movement-related modulation of *area 5* cells encodes directional information in strictly kinematic terms, whereas the activity of motor cortical cells may relate primarily to dynamic variables (e.g. direction of forces required to counteract external loads). Such a purely kinematic representation might provide accurate information about limb configuration for both planning and execution of reaching movements.

Yet another example of direction-related neural activity has recently been reported by Fortier et al. (1989). They found that many cerebellar neurons, including both cortical and nuclear cells, demonstrate direction-dependent changes in activity associated with reaching movements. These cells also typically display the broad directional tuning observed in motor and parietal cortical cells, once again suggesting the possibility of a common code based on spatial direction of movement.

All-in-all, these results suggest that the encoding of directional information by changes in neural discharge rates is a recurrent theme in many regions of the brain participating in the control of movement, whereas there is little indication of similar neural coding of alternative movement representations such as final hand position or joint angle changes (although a possible alternative interpretation of motor cortical activity in terms of muscle state variables has been recently suggested by Mussa-Ivaldi [1988]). As such, these findings lend support to the premise that spatial direction of movement plays an important role in formulating the motor output for initiation of multijoint reaching movements. In subsequent sections, we will examine some direction-related characteristics of the muscle activation patterns associated with such movements, and consider how those features might relate to directional coding at the motor cortical level.

16.3 Direction-Related Features of Electromyographic Activity

By serving as an indicator of the actual output of spinal motoneuron pools, *EMG* data obtained during the performance of multijoint arm movements can provide information supplementary to that derived from kinematic and kinetic analyses. We have analyzed initial *EMG* activity in muscles acting at the shoulder and elbow in order to test proposed strategies for the planning and execution of multijoint limb movements (Hasan and Karst, 1989, Karst and Hasan, 1990a).

Subjects performed self-paced movements to visual targets, with the arm supported in the horizontal plane and relaxed prior to each trial (Figure 16.1). It is important to note that these movements encompassed a broad range of initial positions, distances and directions, as illustrated in Figures 16.2A, B, which portray the initial and final hand positions in relation to the reachable workspace within the horizontal plane. For each movement, *EMG* records were analyzed *qualitatively* to determine which muscle groups (i.e., flexors or extensors) were the first to be activated at each joint, and *quantitatively* in terms of onset time and magnitude of that initial *EMG* activity. In agreement with previous reports (Accornero et al., 1984; Wadman et al., 1980) we observed that the pattern of muscle activity at each joint was typically characterized by alternating flexor and extensor bursts, comparable to the two- or three-burst *EMG* patterns frequently associated with rapid, single-joint movements involving the elbow (Hallet et al., 1975; Karst and Hasan, 1987) or shoulder (Angel, 1974; 1977; Pantaleo, et al., 1988). This pattern of *EMG* activity was particularly consistent at the shoulder, where determination of the "*agonist*" muscle group (defined as the first muscle group activated at that joint) was possible for 95% of the movements trials, whereas the agonist muscle group at the elbow could be clearly identified for 89% of the trials.

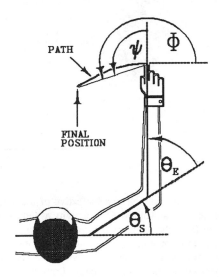

Figure 16.1: Conventions for defining joint angles and the spatial direction of hand movement from initial to final position. Note that ψ is defined as the target direction relative to the initial orientation of the forearm, while ϕ is the absolute target direction. Thus, $\phi = \psi + \theta_S^{init} + \theta_E^{init}$.

Three proposed initiation rules were tested; two are based only on statics, while the third rule takes into account intersegmental dynamics. Since the latter two of the proposed rules (described below)

can be formulated in terms of the same two movement parameters, ψ and θ_E^{init} (see Figure 16.1), it follows that each of those rules predicts the partitioning of the four combinations of shoulder and elbow agonists (*FF*, *FE*, *EF*, or *EE*) into specific regions of the ψ and θ_E^{init} plane. Thus, by plotting the data on axes corresponding to ψ and θ_E^{init}, using different symbol types to indicate the observed combination of shoulder and elbow agonists, the predictive power of each rule was evaluated based on comparison of the predicted and observed partitioning of symbol types (Karst, 1989; Karst and Hasan, 1990a). Results for all three rules may be summarized as follows:

1) The first rule tested predicts that the sign of initial muscle activity at each joint should correspond to the direction in which that joint must rotate in order to reach the final position (e.g., initial elbow flexor activity should accompany all elbow flexion movements). Though this rule would suffice for single-joint movements in the horizontal plane, it is clearly contradicted for these two-joint movements, particularly with respect to initial elbow muscle activity.

2) The second rule predicts that the sign of initial muscle activity at each joint is chosen such that the distal tip of the limb (hand) exerts an *initial force* in the direction of the final tip position. Observed initial shoulder muscle activity clearly contradicted this proposed rule, which is based on the idea that the *CNS* controls the movement by specifying the equilibrium position of the hand.

3) The third rule predicts that the sign of initial muscle activity at each joint is chosen such that the *initial acceleration* of the hand is directed toward the final position. This rule, based on the notion that the *CNS* plans a straight-line hand trajectory, then uses inverse kinematics and inverse dynamics to determine appropriate joint trajectories and joint torques, was also contradicted by the observed shoulder muscle activity for certain movements. Furthermore, the effects of added inertial loads predicted by this rule were not observed for trials in which a 1.8 kg mass was attached to the hand.

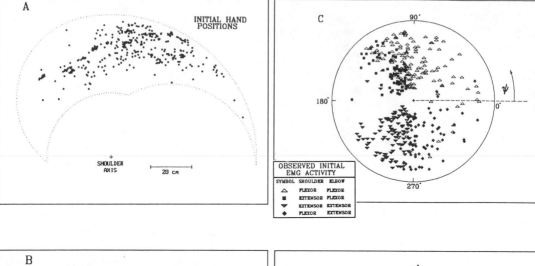

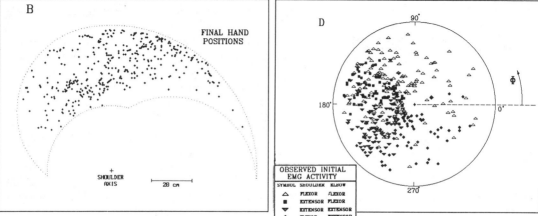

Figure 16.2: Illustration of the wide range of initial and final hand positions associated with the movements depicted in subsequent figures. Each symbol represents a single movement (data from 8 subjects; 390 movements) and depicts the initial (*A*) or final (*B*) hand position in relation to the reachable workspace within the horizontal plane. (In order to combine data from several subjects., segment lengths were normalized across subjects., and ranges of joint motion were assumed to be –10° to 100° of horizontal flexion at the shoulder., and 0° to 130° at the elbow for all subjects.) *C–D* depict the observed sign (flexor or extensor) of initial *EMG* activity at the shoulder and elbow as a function of spatial direction defined in terms of ψ (*C*) and of φ (*D*). Each symbol represents a single move-ment., with the symbol type indicative of the sign of initial shoulder and elbow *EMG* activity as indicated in the key. In *C*, the value of ψ for each movement corresponds to the central angle formed by the initial line and a radius passing through the symbol. There is no significance to the radial distance from the origin., which was assigned randomly in order to reduce overlapping of symbols with similar ψ values. The tendency for the different symbol types to be partitioned into different sectors of *C* indicates that ψ is, by itself, a relatively good indicator of the initial muscle activations chosen by the *CNS*, and contrasts with the much greater overlap of different symbol types seen in *D*, which is identical except that the data are plotted as a function of φ.

16.3.1 Selection of Agonists at the Shoulder and Elbow

Although the analyses described above did not support any of the specific initiation rules tested, they demonstrated that the observed initial muscle activity at both joints was well predicted by combinations of movement variables which included ψ, the *direction of movement relative to the forearm* (see Figure 16.1). In fact, though certain other variables (e.g. initial elbow angle and movement distance) appear to have a slight influence on the observed choice of agonists (Karst and Hasan, 1988), the data indicate that ψ is, by itself, a good predictor of the observed choice of agonist muscle groups at each joint. Furthermore, this relationship has proven consistent across subjects, for movements encompassing nearly the entire reachable workspace in the horizontal plane, and despite altering the mechanical characteristics of the limb by the addition of inertial loads of 1.8 kg to the hand (Karst, 1989; Karst and Hasan, 1990a).

The degree to which ψ accurately predicts the shoulder and elbow agonists is illustrated in Figure 16.2C, which depicts a total of 390 individual movements performed by eight subjects. Each symbol represents a single movement, and the symbol type indicates the observed combination of initial *EMG* activity at the shoulder and elbow. Figures 16.2A and 16.2B characterize the broad scope of the movements studied by denoting, for each movement, the location of the initial (Figure 16.2A) and final (Figure 16.2B) hand positions relative to the reachable workspace within the horizontal plane. The same movements are portrayed in the polar plot of Figure 16.2C, where the central angle corresponding to each symbol indicates the value of ψ for that movement. (To emphasize the partitioning due to ψ alone, there is no significance to the symbol's radial distance from the origin, which was varied randomly simply to limit overlapping of symbols with similar ψ values.) The tendency for the different symbol types to occupy relatively distinct sectors of Fig. 16.2C supports the link between ψ and the sign of initial muscle activity. In contrast, there is a much greater overlap of different symbol types in Fig. 16.2D, where the same movements are plotted as a function of *absolute* movement direction (ϕ). (Comparison of ψ and ϕ is discussed in greater detail in Section 16.4.)

16.3.2 Magnitude and Timing of Initial Shoulder and Elbow EMG

The fact that some overlap of symbol types does occur in Figure 16.2C indicates that ψ alone does not fully account for the choice of shoulder and elbow agonists made by the *CNS*. Nevertheless, ψ appears to be unique in that, by itself, it is a very robust predictor of initial agonist choice throughout the workspace. Furthermore, for each of the muscle groups studied, the magnitude of initial *EMG* activity (quantified in terms of either peak amplitude or area of the rectified, filtered *EMG* during the initial 100 ms of activity in each muscle) varied systematically with respect to ψ.

Each muscle group's contribution to the overall pattern of muscle activation is related to ψ, in terms of both the magnitude of initial activation and the role (i.e., agonist or antagonist) of the muscle group during the execution of a given movement. This is illustrated in Fig. 16.3 by plotting the magnitude of the initial *EMG* activity as a positive value if the muscle group acted as the agonist, and as a negative value for movements in which it acted as the antagonist. Data from individual movements are shown in this manner in Figure 16.3A, where initial *EMG* magnitudes recorded from the pectoralis major (clavicular portion) are plotted against ψ for a total of 355 movements performed by eight subjects, thus revealing the systematic variation of initial agonist magnitude with respect to ψ. Likewise, each of the other muscle groups studied demonstrated distinct patterns of direction-related changes in initial *EMG* magnitude that were consistent enough to remain apparent even when averaged across subjects and for movements from different initial positions. This direction-related modulation is depicted for all five muscles in Figures 16.3B-F, where for each muscle, the initial *EMG* magnitude of all quantifiable trials has been bin-averaged over 20 deg intervals of ψ, so that each symbol represents the mean ($\pm$ S.E.) initial EMG magnitude ("+" for agonist; "-" for antagonist) for 5 to 60 individual movements. For comparison of averaged results with individual movement results, note that Figure 16.3B is the bin-averaged pectoralis (*PEC*) *EMG* data for the 355 trials depicted individually in Figure 16.3A.

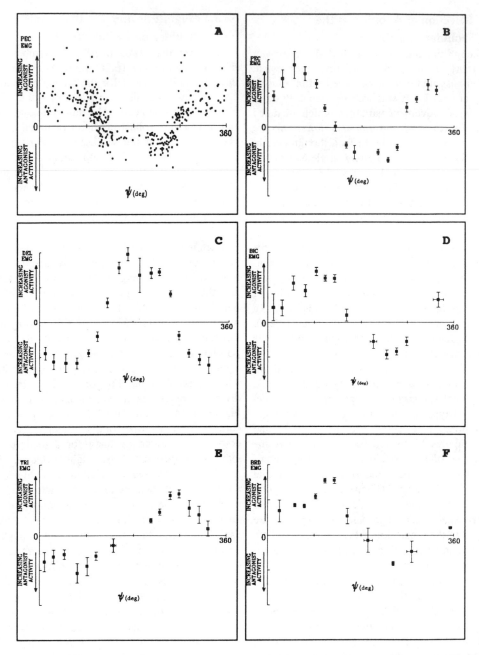

Figure 16.3: Characterization of the direction-dependent variations of muscle function (i.e. agonist vs antagonist) and magnitude of initial *EMG* activity for individual shoulder and elbow muscles. In each panel, the magnitude of the initial *EMG* activity is represented as a positive value if the muscle group acted as the agonist, and as a negative value for movements in which it acted as the antagonist. Initial pectoralis major *EMG* magnitude is shown in relation to ψ in *A* and *B*, where data from 355 movements (8 subjects) are represented individually (in *A*) and as bin-averaged means (in *B*). Bin-averaged initial *EMG* data for the posterior deltoid, biceps brachii, triceps brachii, and brachioradialis, respectively, are shown in *C–F*, where each symbol represents the mean (± S.E.) initial *EMG* magnitude ("+" for agonist; "–" for antagonist) and direction (ψ) for each bin. (Bin width = 20°, except where adjacent bins were collapsed so that $n > 5$ for each bin; horizontal error bars extend beyond the symbols only for collapsed bins).

Presenting the data in this format illustrates that the absolute magnitude of *EMG* activity tends toward a minimum for movement directions near the agonist–antagonist transition for that muscle group (e.g., near $\psi = 120°$ for the *PEC*), and, though the minima are similar for anatomical antagonists at each joint, they differ across joints. As a result, when the movement direction is near the agonist–antagonist transition for one of the joints, the net torque acting at that joint depends more on the torque due to dynamical interactions and less on the muscular torque at that joint, thus making the choice of agonists somewhat less critical near those transitions. Though the initial kinematics would differ depending on the choice of agonist, such differences may be very slight, since the magnitude of the muscle torque at that joint would be small, regardless of sign. For example, Figure 16.3A indicates that, of the movements for which $\psi = 120°$, some were initiated by *PEC* activity and others by deltoid (*DEL*) activity, yet the overall direction of movement was identical.

The data shown in Figure 16.3 support the possibility of ψ playing a significant role in coordinating the activity of various muscle groups at the *same* joint. In order to initiate multijoint movements, however, it is apparent that there must be coordination of the initial muscle activation patterns *across* joints as well. Using data from the same movements depicted above, we examined both the relative timing and the relative magnitude of initial agonist *EMG* activity at the shoulder and elbow. For all trials in which the time of agonist onset at each joint could be clearly identified, we determined the agonist onset time difference, and quantified the relative magnitude of shoulder and elbow initial agonist activity as the $\log_{10}$ ratio of shoulder/elbow *EMG* magnitude during the first 100 ms of activity. The results of those analyses (Karst, 1989; Karst and Hasan, 1990b) are illustrated in Figure 16.4 and briefly summarized below.

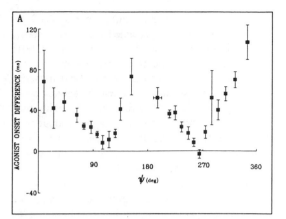

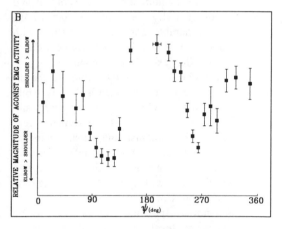

Figure 16.4: *A)* This illustrates that the agonist onset difference varies systematically with ψ. Bin-averaged values for 345 movements; 8 subjects. (Positive values indicate that shoulder agonist activity preceded elbow agonist onset.) Each square symbol and vertical error bar indicate the mean (± S.E.) of the agonist onset difference for the movements within that bin. Bin width is 10 deg., except where adjacent bins have been collapsed to insure that $n \geq 5$ for all bins. (Collapsed bins are those with visible horizontal error bars indicating the S.E. for ψ within the bin., while., for the 10 degree bin widths., the horizontal error bars are not visible., since they are narrower than the symbols.) *B)* Illustration showing that the shoulder–elbow agonist magnitude ratio also varies systematically with ψ. Conventions as in *A*, except that the vertical axis now represents the relative magnitude of the initial 100 ms of agonist *EMG* activity at the shoulder and elbow. Increasing *Y* values indicate increasing shoulder *EMG* activity and/or decreasing elbow *EMG* activity. Data from the same movements as in *A*.

For most movements, initial agonist *EMG* activity at the shoulder preceded that at the elbow, with a mean onset difference of 26 ms ($p < 0.001$; $n = 345$), of which less than 3 ms was attributable to axonal conduction to the more distal muscles. Figure 16.4A also illustrates that the agonist onset varies systematically as a function of ψ, with minima corresponding to the agonist–antagonist transitions for the shoulder muscles and maxima near the analogous transitions for the elbow muscles. As a result, there is a nearly linear change in the relative timing of shoulder and elbow agonist onsets over each of the four ranges of ψ occupied by the four possible combinations of shoulder and elbow agonists. Furthermore, plotting the relative magnitude of initial shoulder and elbow agonist activity, as seen in Figure 16.4B, reveals a comparable pattern of variation with respect to ψ. Consequently, for movement directions near the shoulder agonist-antagonist transitions, initial agonist *EMG* activity at the elbow tends to be relatively large and early while shoulder agonist activity is relatively small and late; the converse being true for movement directions near the elbow agonist-antagonist transitions.

The manner in which both the relative timing and the relative magnitude of initial shoulder and elbow activity vary with target direction suggests that the temporal aspects of the motor output pattern might not be specified independently at the pre-motoneuronal level. Instead, once the relative magnitude of the shoulder and elbow motor command is specified, the observed variations in relative timing could simply emerge from an inverse relationship between the magnitude of the excitatory input to a motoneuron pool and the time required for those motoneurons to reach the threshold of activation. If so, temporally stereotyped motor commands, in which the input to the shoulder agonist motoneurons precedes that to the elbow motoneurons by a fixed interval (perhaps 25–30 ms), could still result in the observed range of agonist onset differences.

In summary, we have demonstrated that, for planar, two-joint, reaching movements, the main features of the initial muscle activation patterns at each joint show consistent correlations with ψ, the spatial direction of hand movement relative to the forearm. For movements throughout the reachable workspace in the horizontal plane, ψ may be used to predict which muscle groups will be initially activated at each joint, as well as the relative timing and magnitude of those initial activations. These observations suggest that relatively simple rules, based on directional information such as that encoded in motor cortical output, may play a primary role in determining the appropriate muscle activation patterns for the initiation of multijoint reaching movements.

16.4 Comparison of EMG Activity and Motor Cortical Output

Though it appears that the directional information encoded in motor cortical activity exerts an important influence on the *EMG* pattern, that influence is far from direct. For proximal muscle groups, such as those we have studied, motor cortical neurons do not project directly to the spinal motoneuron pools. Instead, they terminate primarily on interneurons of the *C3–C4* propriospinal system. There is considerable divergence of motor cortical output, due in part to axonal branching within the corticospinal tract itself, as well as to connectivity within the *C3–C4* propriospinal system. This system, which receives a variety of descending and peripheral inputs, projects directly to spinal motoneurons, and appears to play an important role in shaping the final motor output associated with reaching movements (Alstermark et al., 1981). As a result, the output of a single motor cortical neuron affects multiple spinal motoneuron pools, and, conversely, a given motoneuron is influenced by converging input from a variety of sources, including multiple motor cortical cells.

How the directional information encoded in the motor cortical output is actually used to specify muscle activation patterns is still unknown, though at least one possibility has been suggested. A general scheme, proposed by Georgopoulos and colleagues (Georgopoulos, 1988; Schwartz et al., 1988), incorporates the divergence of motor cortical output from a single cell to multiple motoneuron pools, and includes the following characteristics:

1) Individual motor cortical cells exert a weighted influence on the activity of motoneuron pools in several muscles.

2) The combination of muscles and weighting factors (a "muscle field") determines the preferred movement direction, for which that motor cortical cell is maximally active.

3) Thus, each muscle field acts as an independent functional unit, the activation of which varies as a function of the difference between the preferred direction and the specified movement direction.

Given the broad directional tuning of single cells, it follows that movement would result from the graded co-activation of many muscle groups. Our observations regarding *EMG* activity associated with planar, two-joint movements are not incompatible with the basic features of this scheme, although some of our findings suggest that the mechanism outlined above understates the role of spinal circuitry in mediating the transfer of cortical motor output to the motoneuron pools.

Having considered the assertions in preceding sections that the observed activity at both the motor cortical and *EMG* levels is related to spatial direction, the reader may well ask: direction relative to what? The evidence presented in Section 16.2 indicates that modulation of cortical neuronal activity is related to φ, the **absolute** direction of hand movement (i.e. relative to subject's fixed trunk), while our data demonstrate variations of initial *EMG* activity in relation to ψ, the direction of hand movement ***relative to the forearm*** (Figure 16.1). For movements initiated from the same position, φ differs from ψ only by a constant which describes the initial absolute orientation of the forearm in space (since $\phi = \psi + \theta_S^{i\,nit} + \theta_E^{i\,nit}$). Thus, for movements which share the same initial hand position, choosing between ψ and φ as the independent variable changes only the phase, not the form of a given relationship. Similarly, ψ and φ tend to covary for nearby starting positions. Since, in our studies of initial *EMG* activity, we attempted to maximize the range of initial hand positions (see Figure 16.2A), φ and ψ are no longer simply related, as indicated by the different distributions of the data in Figures 16.2C and 16.2D. Furthermore, comparison of the partitioning of symbol types in Figures 16.2C and 16.2D reveal that ψ is clearly a better predictor of the combination of shoulder and elbow agonists used for initiating these movements.

On the other hand, the studies of motor cortical activity reviewed in Section 16.2 showed that activity to be related to the absolute movement direction (φ). Even when those experiments included sets of parallel movements with varying initial and final positions (Georgopoulos et al.,

1984; Kettner et al., 1988), the population vectors corresponded to observed movement trajectories to a similar degree for parallel movements. Still, comparison of single-cell activity between pairs of parallel movements revealed that significant differences were observed in nearly one-third of those paired comparisons (Kettner et al., 1988). And, since it is unclear whether the range of initial positions in those experiments was comparable to that which we have examined, we cannot rule out the possibility that the apparent difference in coordinate systems used to represent direction at the motor cortical and *EMG* levels is due to methodological differences.

If further investigation should confirm that movement direction is, indeed, represented in different coordinates at the two levels, it would imply that the transformation between ψ and φ, which depends on the absolute orientation of the forearm, takes place subcortically. That scenario seems feasible, since, as we have already noted, a good deal of the requisite neural integration for these movements does occur at the spinal level. In fact, the *C3–C4* propriospinal system, which mediates the effect of motor cortical output on the spinal motoneuron pools, receives input from a variety of forelimb afferents as well as from the motor cortex, and thus would appear well-suited to effect such a transformation.

Another contrast between the output at the motor cortical and spinal motoneuron levels lies in the structure of the *EMG* patterns which we and others (Accornero et al., 1984; Wadman et al., 1980) have observed during the execution of planar, two-joint reaching movements. Those data give the impression: *1)* that there is a strong tendency toward the generation of an alternating pattern of agonist and antagonist activity at each joint, as has been shown for many single-joint movements; and *2)* that the patterns at the shoulder and elbow tend to be roughly in-phase; but *3)* that the modulation of that pattern may be much more comprehensive than has generally been observed in relation to movements constrained to a single joint. For example, Figure 16.5 depicts four individual movements from approximately the same initial position. The direction of these movements varies sequentially from ψ = 303° in Figure 16.5A to ψ = 248° in 16.5D, thus spanning a range of ψ which encompasses one of the agonist-antagonist transitions for the shoulder musculature. The switch of shoulder agonist from pectoralis in Figure 16.5B to posterior deltoid in Figure 16.5C illustrates that, even for movement directions very near this transition, an alternating pattern is seen in the muscles about the shoulder, though the magnitude of activity is greatly reduced.

Although we have argued that the directional information from motor cortex could specify some of the features of the initial *EMG* pattern, this does not mean that the basic structure of that pattern is represented at the motor cortical level. In general, discharge patterns of single motor cortical cells are not characterized by the sort of temporal modulation which might serve as the basis for the typical alternating *EMG* pattern (Murphy et al., 1985), nor is there a reversal of the population vector which might account for antagonist activation (Georgopoulos, 1988). [There is evidence, however, that the proximal to distal sequencing of agonist onsets may be reflected at the cortical level (Murphy et al., 1985).] Thus, the characteristic structure of this pattern seems likely to emerge at the spinal level. Although there have been previous suggestions of a stretch reflex origin for the antagonist burst (Angel, 1977; Ghez and Martin, 1982), our data provide clear evidence that antagonist activation is not dependent on stretch, since, in some cases, the entire three-burst pattern is expressed even though the antagonist is actually shortening throughout the movement (e.g. the *PEC* in Figure 16.5C). Nevertheless, afferent input to the *C3–C4* propriospinal system appears to play a significant role in reaching movements (Alstermark et al., 1981; 1986). At present, it seems reasonable to assume that pattern generating circuitry at the spinal level, activated by descending commands and modulated by afferent input, is responsible for the characteristic alternating *EMG* pattern observed during these movements.

Finally, our observations about the structure and modulation of these *EMG* patterns have implications as to how the directional information encoded in the motor cortical output might be incorporated into the generation of the *EMG* output. As noted in Section 16.2, directional information appears to be represented as a continuous variable in the output of the motor cortex. If a similar representation existed at the muscle activation level (as suggest in the "weighted influence" scheme suggested by Georgopoulos and colleagues), one would expect agonist–antagonist co-activation to be much more prominent than we or others have observed for the type of movements discussed here (Accornero et al., 1984; Hasan and Karst, 1989; Karst, 1989; Wadman et al., 1980). Instead, our observations suggest that directional information at the *EMG* level is used in two ways, being represented first as a discrete variable (specifying one of four combinations of shoulder and elbow agonists), then, within each of those subgroups, as a continuous variable (specifying relative magnitude and/or timing of initial agonist activity). Using directional information in this manner would seem to simplify the planning of initial muscle activations since the first step allows for the use of a stereotyped muscle activation pattern by simply "plugging in" the appropriate agonists, while the second step provides flexibility by modulating parameters within that pattern. Furthermore, this two-step use of directional information would explain the observed *EMG* activity near the agonist–antagonist transitions, such as that depicted for the shoulder in Figure 16.5.

To summarize, comparison of motor cortical and *EMG* activity supports the notion that directional information encoded in the activity of motor cortical cells plays a major role in determining the observed muscle activation patterns, but indicates that the previously proposed "weighted influence" scheme does not, by itself, completely account for the direction-related characteristics of *EMG* activity that we have observed. Our data suggest more extensive neural integration at the spinal level.

16.5 Summary and Conclusions

The main thrust of this chapter is the assertion that, in spite of the complexity of the mechanics, multijoint reaching movements might ensue from a strategy in which muscle activations appropriate for initiating a movement are determined by simple rules based primarily on the spatial direction of that movement. The experimental evidence offered in support of this premise includes our observations regarding the initial shoulder and elbow *EMG* activity for movements throughout the reachable workspace in the horizontal plane, as well as observations regarding the activity of motor cortical neurons in the rhesus monkey during a variety of volitional multijoint reaching movements. The former observations indicate that: *1)* the initiation of these movements is characterized by relatively simple patterns of *EMG*

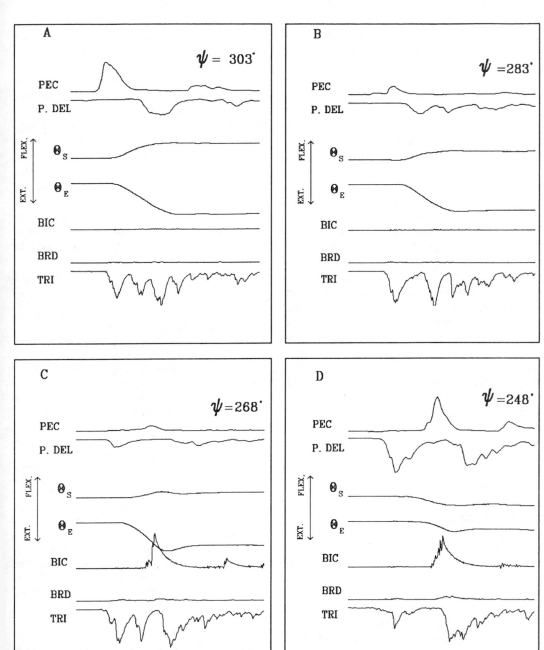

Figure 16.5: Representation of digitized *EMG* and kinematic data for movements in different directions. Each of the four records depicts a single movement performed by the same subject, initiated from approximately the same position (initial joint angles were $22° \pm 4°$ [mean ± S.D.] for θ_S^{init} and $57° \pm 5°$ for θ_E^{init}), but differing in movement direction, as indicated by the values of ψ. Angular displacements at the shoulder and elbow are indicated by θ_S and θ_E, respectively., with upward deflections indicating flexion. Each record also depicts the rectified and filtered *EMG*s of five shoulder and elbow muscles, with increasing activity indicated by an upward deflection of the *EMG* trace for flexor muscles (biceps, brachioradialis and pectoralis major) and a downward deflection for extensor muscles (triceps and posterior deltoid).

activity at each joint; *2)* modification of a few parameters of this stereotyped *EMG* pattern provides the flexibility to initiate movement in any direction; and *3)* appropriate specification of each of those parameters may be based primarily on the spatial direction of the movement. Moreover, the ready availability of such directional information is supported by the latter group of experimental observations, which indicate that the spatial direction of movement is encoded in the output of motor cortex, and, in fact, appears to be prominently represented at multiple levels within the sensorimotor system.

This proposed initiation rule does not explain the movement strategy as a whole, but would function to propel the hand in the approximate direction of the intended final position, with additional mechanisms required for ongoing correction and braking. Despite these potential drawbacks, the simplicity of this initiation rule, whereby the initial muscle activations could derive from a single parameter (which is known to be represented in the *CNS*), is inherently appealing. This simplicity contrasts sharply with the complexity of a strategy in which the *CNS* plans the initial hand trajectory, then uses inverse kinematics (plus additional constraints to insure a unique solution) to determine initial joint trajectories, then employs inverse dynamics (and precise knowledge of mass and inertial parameters) to determine initial joint torque requirements, and finally (by some unknown mechanism) chooses appropriate initial muscle activations based on those torque requirements.

Finally, in addition to its appealing simplicity and the extent of the experimental evidence which appears to support it, the concept of a simple, direction-dependent initiation rule has testable implications. Since a direction-based initiation rule, as proposed here, does not explicitly account for the effects of dynamic interactions between limb segments, some initial deviation from the intended direction of movement is an inherent characteristic. Thus, it seems likely that a strategy built around this rule would also exhibit the type of rapid, ongoing modulation of motor output that has been observed in other multijoint movement tasks (Cole et al., 1984). Experiments currently in progress will assess the effects of unpredictable perturbations on the *EMG* output during the initiation phase of multijoint reaching movements. Similarly, ongoing correction for the inherent

directional deviations would result in curved hand paths, with the extent and direction of that curvature dependent on the initial configuration of the limb and the direction of the movement. As more diverse movements have been studied, it has become apparent that this is, indeed, the case (Atkeson and Hollerbach, 1985; Karst, 1989). Whether or not the observed deviations from a straight-line hand path are fully explained by this initiation rule has not yet been determined. Finally, though the particular direction-based initiation rule discussed here applies only to planar movements, the search for an analogous rule for the initiation of three-dimensional movements is an obvious, albeit technically difficult, step in evaluating this concept.

References

Accornero, N., Berardelli, A., Argenta, M. and Manfredi, M. (1984) Two joints ballistic arm movements. *Neurosci. Lett.* **46**: 91-95.

Alstermark, B., Lundberg, A., Norrsell, U. and Sybirska, E. (1981) Integration in descending motor pathways controlling the forelimb in the cat. 9. Differential behavioral defects after spinal cord lesions interrupting defined pathways from higher centres to motoneurones. *Exp. Brain. Res.* **42**: 299-318.

Alstermark, B., Gorska,T., Johannisson,T. and Lundberg, A. (1986) Hypermetria in forelimb target-reaching after interruption of the inhibitory pathway from forelimb afferents to C3-C4 propriospinal neurones. *Neurosci. Res.* **3**: 457-461.

Angel, R.W. (1974) Electromyography during voluntary movement: The two-burst pattern. *Electroenceph Clin Neurophysiol* **36**: 493-498.

Angel, R.W. (1977) Antagonist muscle activity during rapid arm movements: central versus proprioceptive influences. *J Neurol Neurosurg Psych* **40**: 683-686.

Atkeson, C.G. and Hollerbach, J.M. (1985) Kinematic features of unrestrained vertical arm movements. *J Neurosci* **5**: 2318-2330.

Caminiti, R., Johnson, P.B., Pastore, R.A. and Urbano, A. (1988) Spatial coding of movement direction in frontal cortical areas. *Soc Neurosci Abstr* **14**: 342.

Cole, K.J., Gracco, V.L. and Abbs, J.H. (1984) Autogenic and non-autogenic sensorimotor actions in the control of multiarticulate hand movements. *Exp Brain Res* **56**: 582-585.

Fortier, P.A., Kalaska, J.F. and Smith, A.M. (1989) Cerebellar neuronal activity related to whole-arm reaching movements in the monkey. *J Neurophysiol* **62**:198-211.

Georgopoulos, A.P. (1988a) Spatial coding of visually guided arm movements in primate motor cortex.

Can J Physiol Pharmacol **66**: 518-526.

Georgopoulos, A.P. (1988b) Neural integration of movement: role of motor cortex in reaching. *FASEB J* **2**: 2849-2857.

Georgopoulos, A.P., Kalaska, J.F., Caminiti, R. and Massey, J.T. (1982) On the relations between the direction of two-dimensional arm movements and cell discharge in primate motor cortex. *J Neurosci* **2**: 1527-1537.

Georgopoulos, A.P., Kalaska, J.F., Crutcher, M.D., Caminiti, R. and Massey, J.T. (1984) The representation of movement direction in the motor cortex: Single cell and population studies. In: Edelman, G.M., Cowan, W.M. and Gall, W.E. (eds) *Dynamic aspects of neocortical function*. Wiley, New York pp 501-524.

Georgopoulos, A.P., Schwartz, A.B. and Kettner, R.E. (1986) Neuronal population coding of movement direction. *Science* **233**: 1416-1419.

Ghez, C. and Martin, J.H. (1982) The control of rapid limb movement in the cat: III. Agonist-antagonist coupling. *Exp Brain Res* **45**: 115-125.

Hallett, M., Shahani, B.T. and Young, R.R. (1975) EMG analysis of stereotyped movements in man. *J Neurol Neurosurg Psych* **38**: 1163-1169.

Hasan, Z. and Karst, G.M. (1989) Muscle activity for initiation of planar, two-joint arm movements in different directions. *Exp Brain Res* **76**: 651-655.

Kalaska, J.F. (1988) The representation of arm movements in postcentral and parietal cortex. *Can J Physiol Pharmacol* **66**: 455-463.

Kalaska, J.F., Caminiti, R. and Georgopoulos, A.P. (1983) Cortical mechanisms related to the direction of two-dimensional arm movements: relations in parietal area 5 and comparison with motor cortex. *Exp Brain Res* **51**: 247-260.

Kalaska, J.F., Cohen, D.A.D., Hyde, M.L. and Prud'homme, M. (1989) A comparison of movement direction-related versus load direction-related activity in primate motor cortex, using a two-dimensional reaching task. *J Neurosci* **9**: 2080-2102.

Kalaska, J.F. and Hyde, M.L. (1985) Area 4 and area 5: differences between the load direction-dependent discharge variability of cells during active postural fixation. *Exp Brain Res* **59**: 197-202.

Karst, G.M. (1989) Multijoint arm movements: Predictions and observations regarding initial muscle activity at the shoulder and elbow. Doctoral Dissertation, University of Arizona.

Karst, G.M. and Hasan, Z. (1987) Antagonist muscle activity during human forearm movements under varying kinematic and loading conditions. *Exp Brain Res* **67**: 391-401.

Karst, G.M. and Hasan, Z. (1988) Two parameters determine which muscles initiate planar, two-joint arm movements. *Soc Neurosci Abstr* **14**: 951.

Karst, G.M. and Hasan, Z. (1989) Timing and magnitude of EMG activity at the shoulder and elbow for initiation of planar arm movements. *Soc Neurosci Abstr* **15**: 48.

Karst, G.M. and Hasan, Z. (1990a) Initiation rules for planar, two-joint arm movements: agonist selection for movements throughout the workspace. (Submitted)

Karst, G.M. and Hasan, Z. (1990b) Timing and magnitude of electromyographic activity for two-joint arm movements in different directions. (Submitted)

Kettner, R.E., Schwartz, A.B. and Georgopoulos, A.P. (1988) Primate motor cortex and free arm movements to visual targets in three-dimensional space. III. Positional gradients and population coding of movement direction from various movement origins. *J Neurosci* **8**: 2938-2947.

Murphy, J.T., Wong, Y.C. and Kwan, H.C. (1985) Sequential activation of neurons in primate motor cortex during unrestrained forelimb movement. *J Neurophysiol* **53**: 435-445.

Mussa-Ivaldi, F.A. (1988) Do neurons in the motor cortex encode movement direction? An alternative hypothesis. *Neurosci Lett* **91**:106-111.

Pantaleo, T., Benvenuti, F., Bandinelli, S., Mencarelli, M.A. and Baroni, A. (1988) Effects of expected perturbations on the velocity control of fast arm abduction movements. *Exp Neurol* **101**: 313-326.

Schwartz, A.B., Kettner, R.E. and Georgopoulos, A.P. (1988) Primate motor cortex and free arm movements to visual targets in three-dimensional space. I. Relations between single cell discharge and direction of movement. *J Neurosci* **8**: 2913- 2927.

Wadman, W.J., Denier van der Gon, J.J. and Derksen, R.J.A. (1980) Muscle activation patterns for fast goal-directed arm movements. *J Human Mvmt Studies* **6**: 19-37.

CHAPTER 17

The Organization of Human Arm Trajectory Control

Tamar Flash

17.1 Introduction

Traditionally, studies of human and animal movements have focused on systems composed of a single muscle, or a single joint. However, most natural human actions such as reaching, walking, writing, etc., require coordination among a large number of muscles and joints. Although the excess degrees of freedom problem, also known as Bernstein's problem (Bernstein, 1967), arises even in the context of single-joint movements, it becomes especially complicated to resolve in the multi-joint case.

In the strict kinematic sense, degrees of freedom are "the least number of independent coordinates required to specify the position of the system elements without violating any kinematic constraints" (Saltzman, 1979). To describe the location and orientation of the hand in space, six independent coordinates are required. However, to uniquely characterize the configuration of the upper arm, at least seven coordinates are necessary. Given the kinematic redundancy of the human arm, the inverse kinematics problem, i.e., finding what joint angles correspond to a given hand location and orientation in space, does not have a unique solution. This is but one example of a large number of ill-posed problems that arise in the control of arm posture and movement. A problem is well-posed when a solution exists, is unique, and depends continuously on the input information (Tikhonov and Arsenin, 1977). Ill-posed problems fail to satisfy one or more of these criteria. Most motor problems are ill-posed in the sense that the kinematic solution to the problem is not unique. Since most limb segments are

operated upon by a much larger number of muscles than are strictly necessary from mechanical considerations, the problem of torque distribution among redundant muscles in another example of an ill-posed problem.

A third and perhaps more fundamental ill-posed problem arises at the task level. Any behavioral goal can be acheived in many different ways. Thus, for example, a cup of coffee might be reached while moving the hand along many different paths. Likewise, in drawing an ellipse, or writing the letter z, we may generate the same geometrical form, while using each time a totally different law of motion for the time course of pen position along the path. Again, how does the brain determine in what way to perform a given motor task in order to achieve the desired behavioral goals?

The excess degrees of freedom problem, therefore, poses a real challenge to any motor control theory. How does the motor system select specific solutions to any of these ill-posed problems? How does the system manage to handle the large number of controlled parameters available at all levels of sensorimotor representation? The two solutions that were offered to this problem were: *a)* the motor system is hierarchically organized; *b)* the system makes use of coordinative constraints.

Instead of trying to directly bridge the gap between the behavioral goals of any motor act and the neural input to muscles needed to achieve this goal, this gap is progressively narrowed down by using a hierarchy of motor control levels (Bernstein, 1967; Gelfand et al., 1971; Saltzman, 1979). Higher levels translate the motor problem

Multiple Muscle Systems: Biomechanics and Movement Organization
J.M. Winters and S.L-Y. Woo (eds.), © 1990 Springer-Verlag, New York

into terms that are more suitable for the lower level to handle, all the way down to the muscles. Associated with this idea is the concept of internal representations of motor actions (e.g., Bernstein, 1967; Saltzman, 1979; Keele, 1981). Many papers dealing with the problem of motor organization have stressed the need to distinguish between long-term and short-term, or working representation of action (for a review see Saltzman, 1979). While the working plan was postulated to be quite specific and to depend on the current task demands and the current state of the musculoskeletal system and of the environment, the long-term representation was postulated to be more abstract and to relate to the general features of any motor action. Two considerations led to this hypothesis. One is the limitation on the memory storage capacity of the system. The other is derived from the similarities that exist between the characteristics of movements produced by different effectors (e.g., handwriting produced by wrist and finger movements or by upper limb movements while writing on a blackboard). Since it would have been extremely inefficient for the system to store all possible variations of any single movement, it was argued that there should exist more abstract and general representations of movement. The actual performance of every motor act was therefore claimed to incorporate both abstract information derived from long-term memory, and particular information about the current task demands and the physical state of the environment and of the musculoskeletal system.

It was also argued that in order to simplify motor control, it is essential to reduce the number of controlled parameters and the amount of information needed to be analyzed in the performance of any motor act (Gelfand et al., 1971). Coordinative constraints, or the so-called basic motion synergies, were claimed to play an important role in establishing such working conditions for the higher motor levels. Gelfand et al. (1971) have defined synergies as "those classes of movements that have similar kinematic characteristics, coinciding active muscle groups and conducting types of afferentation." Such synergies, then, form the basic alphabet from which more complicated motor acts can be generated.

This chapter focuses on the organizing principles that underlie the generation of multi-joint arm movements. In discussing these principles, we will refer to the view that the motor system is hierarchically organized by addressing the question of what possible aspects of movement generation are dealt with at the various levels of sensorimotor representation. Section 17.2 deals with the topic of trajectory planning and reviews evidence from behavioral and modelling studies which supports the notion that higher levels deal with more abstract aspects of movement generation. In particular, a theory of motor coordination which offers a solution to the task-level redundancy problem is presented. Section 17.3 deals with the problem of motor execution. It focusses on the role played by the mechanical properties of muscles in simplifying the motor control problems arising in multi-joint posture and movement. This section also summarizes evidence in support of the view that arm trajectory control is a step-by-step process, whereby higher levels deal with motion planning while lower levels deal with the more concrete aspects of motor execution. Section 17.4 deals with possible strategies for trajectory modification. An experimentally confirmed model is presented which suggests that the modification of aimed arm movements in response to sudden changes in target location might involve the superposition of basic trajectory primitives. Next, the principles discussed in the former sections are summarized and an integrative view of arm movement organization is presented. Finally, we conclude by discussing several recommendations for future research directions.

17.2 Trajectory Planning

In recent years, the planning and control of multi-joint movements in general, and arm movements in particular, has attracted the attention of an increasingly large number of studies. The objective of many of these studies was to identify common kinematic features or sterotyped patterns of muscle activations which characterize intact motor behavior, and based on these observations to form new ideas and hypotheses about movement organization.

Motor behavior is fundamentally multi-dimensional. Hence, movements can be alternatively represented at the muscle, joint or task levels. This raises the two following fundamental questions: In what space(s) or coordinate frame(s) does the brain represent movement?

What rules govern the selection of specific trajectories among the infinite number of possible ones? One effective way to investigate these issues is to experimentally observe human movements. By looking for patterns of invariance in the observed behavior, certain hypotheses concerning the underlying organizing principles can be formulated leading to quantitative and testable theories of motor control. In this context, distinguishing between movement path and trajectory may provide a better understanding of the problems involved in multi-joint movement generation. *Path* is defined as the geometrical curve that the hand follows in space, while the term *trajectory* refers also to the velocity of movement along the path.

Based on experimental observations, several investigators (e.g., Soechting and Lacquaniti, 1981; Hollerbach and Atkeson, 1987; Flanagan and Ostry, 1990) have suggested that trajectories are planned in joint variables while others have argued that simplicity of motor control (at least for higher-level *CNS* planning) is achieved by planning hand trajectories in extracorporeal space; joint rotations are then tailored to produce the desired hand movements (Bernstein, 1967). This view gained support from several behavioral and neurophysiological studies of human and monkey reaching movements (Abend et al., 1982; Georgopoulos et al., 1981; Morasso, 1981; Flash and Hogan, 1985; Georgopoulos, 1986). When moving the hand between pairs of targets in the horizontal plane, subjects tended to generate roughly straight hand paths with single-peaked, bell-shaped speed profiles; this behavior was independent of the part of the work-space in which the reaching movement was performed. Because these common invariant features were only evident in the extracorporeal coordinates of the hand and not in the movements of individual limb segments, these results provided strong indication that planning takes place in terms of hand motion through external space and not in terms of joint rotations (Morasso, 1981).

This conclusion was also generalized to more complex movements, where kinematic invariances were again only present in the hand and not in joint movements. When subjects were instructed to generate curved, or obstacle-avoidance movements, the single-peaked velocity profiles were not preserved. Although the hand paths appeared smooth, movement curvature was not uniform and

the trajectories displayed two or more curvature maxima. The hand velocity profiles also had two or more peaks and the minima between adjacent velocity peaks temporally corresponded to the maxima in curvature (Abend et al., 1982). To account for the observed kinematic featuress of straight and curved hand trajectories a mathematical model of the organization of voluntary arm movements was formulated (Flash and Hogan, 1985). This model, which was based on dynamic optimization theory, enabled us to describe an assumed goal of this class of movements, by a relatively simple formula, and to derive from that formula detailed predictions of the kinematics of a large number of specific motions.

17.2.1 The Maximum-Smoothness Theory for Arm Movements

Natural movements are characteristically smooth and graceful. This observation can be expressed by a mathematically concise model of motor coordination by postulating that voluntary movements are made, at least in the absence of any other overriding concerns (such as the minimization of movement duration), to be as smooth as possible under the circumstances. Mean squared magnitude of hand jerk (rate of change of hand acceleration, or equivalently, the third derivative of position) integrated over movement time was used as a measure of smoothness (Hogan, 1984; Flash and Hogan, 1985). Using dynamic optimization theory, the unique trajectory (among the infinite number of possible ones) that minimizes this performance measure was determined. More precisely, the smoothest motion is acheived by the trajectory that minimizes the following objective function:

$$C_T = \frac{1}{2} \int_0^{t_f} \left(\frac{d^3 r}{dt^3} \right)^2 dt \qquad (17.1)$$

where $r(t)$ is the vector of hand position and t_f is the movement duration. Based on this definition the actual movement had to be worked out, its details depending on the conditions assumed at the beginning and end of the movement.

The above model has been initially derived for single-joint movements (Hogan, 1984) but most natural movements are multi-dimensional, i.e. they involve simultaneous rotations about several joints. In extending the maximum smoothness

principle to the multi-joint case, however, a crucial question is what coordinates should be used in
order to define the above measure of smoothness.
For example, the vector r may relate to the position coordinates of the hand in space, or to the
angular coordinates of the different joints (e.g.,
shoulder and elbow). A basic postulate of this
theory is that the objective of motor coordination
should be expressed in the coordinate system in
which movement planning is assumed to occur.
Since, as discussed above, experimental observations have indicated that arm movements are
planned in terms of hand trajectories, the vector r
was chosen to be expressed in terms of the
Cartesian coordinates of the hand. The minimization of this cost resulted in analytic expressions for
the description of both point-to-point and curved
trajectories. For point-to-point movements, starting and ending at rest, the expressions derived for
the description of $x(t)$ and $y(t)$, the hand position
coordinates were:

$$x(t) = x_o + (x_f - x_o)\ (6(t/t_f)^5 - 15(t/t_f)^4 + 10(t/t_f)^3 \quad (17.2)$$

$$y(t) = y_o + (y_f - y_o)\ (6(t/t_f)^5 - 15(t/t_f)^4 + 10(t/t_f)^3$$

where (x_o, y_o), (x_f, y_f) are, respectively, the initial
and final hand positions, t is the time and t_f is the
movement duration.

These predicted movements were shown to
have the following characteristics:

1) The trajectories of the hand follow straight
   paths.

2) The velocity profile for moving along that path
   is smooth and unimodal.

3) The shape of the trajectory is invariant under
   translation, rotation, amplitude, and speed
   scaling.

Experimental studies of planar horizontal arm
movements have shown that these predictions
agree quite well with the experimental observations. Figure 17.1 shows theoretical predictions
and experimental observations for two typical
point-to-point movements.

Several studies indeed have reported that
during point-to-point motions the path of the hand
through extrapersonal space is essentially straight
and the velocity profile is basically bell-shaped independently of end-point locations (Morasso,
1981; Flash and Hogan, 1985), thus satisfying the
prediction that the trajectory is invariant under
translation and rotation. It is both true for large
and small movements and for movements at different speeds, thus, satisfying the predictions that
the trajectories are invariant under amplitude and
speed scaling.

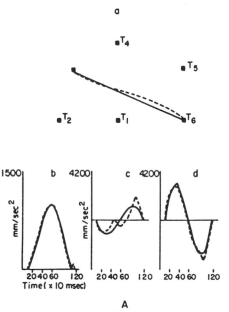

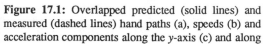

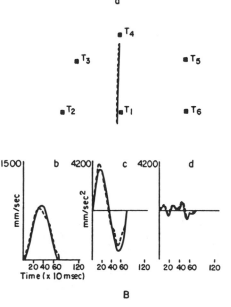

Figure 17.1: Overlapped predicted (solid lines) and
measured (dashed lines) hand paths (a), speeds (b) and
acceleration components along the y-axis (c) and along
the x-axis (d) for two point-to-point movements.
(From Flash and Hogan, 1985; reproduced with permission.)

To account for the kinematic features of curved movements, again the maximum smoothness model was applied assuming that curved motions are generated by specifying a small number of accuracy ("via") points along the trajectory. The hand is then required to pass through these points on its way between the initial and final positions. The time at which the hand passes through these points, or the velocity at that time, need not be *a priori* specified. These are predictions of the model. As before, limb motion was expressed in terms of hand coordinates. Taking the simplest case of one via point between the initial and final positions, the theory yielded explicit mathematical expressions for the description of curved motions (see in Flash and Hogan, 1985) as well as the following predictions:

1) The hand velocity profile exhibits two peaks with a valley between them.

2) The depth of this valley increases with the lateral deviation of the via point from the straight line joining the initial and final positions.

3) The hand path exhibits a single curvature peak.

4) The peak in curvature temporally coincides with the valley in tangential velocity.

5) The shape of the trajectory is invariant under translation, rotation amplitude and time scaling.

6) The durations of the motions from the initial position to the via point, and from the via point to the final position are almost always equal, except for cases in which the via point is located very close to either one of the movement end-points. This behavior will be referred to below as the *isochrony principle* (Viviani and Terzuolo, 1982).

Each of these predictions was corroborated by observation. Figure 17.2 compares between theoretical predictions and experimental observations for two typical curved movements. Peak in curvature, valleys in velocity profiles, and the time coincidence between them were reported both for curved and obstacle–avoidance movements (Abend et al., 1982) and for handwriting (Viviani and Terzuolo, 1980; Edelman and Flash, 1987). Likewise, for drawing movements it was observed that the same law of motion applies whether the movements are large or small, slow or fast (for a review see Lacquaniti, 1989). The additional prediction corroborated by experimental data is the isochrony principle, namely, the phenomenon that within a given trajectory, movement durations for large and small segments are roughly equal.

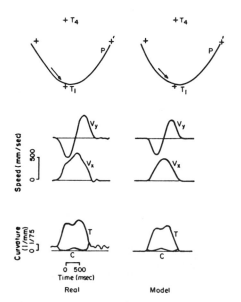

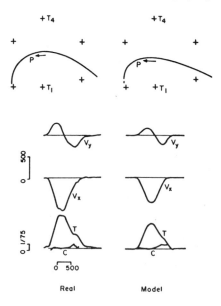

Figure 17.2: Two examples of measured (real, left columns) and predicted (model, right columns) curved trajectories. Displayed are the hand paths, P, and plots of hand speed, T, curvature, C and velocity components, V_x and V_y, versus time. (Reproduced with permission from Flash and Hogan, 1985.)

Since this theory was based on the hypothesis that arm movements are planned or coordinated based on hand motions in external space, the success of this theory in accounting for the observed behavior provided further support for this hypothesis. How does this view agree with other observations of arm movements? It has been reported that in straight pointing movements performed in the vertical plane the ratio between joint velocities approaches a constant value towards the end of the movement. This finding served as an argument in favour of the idea that arm movements are planned in terms of joint coordinates (Soechting and Lacquaniti, 1981). The motions, recorded in these studies, were directed away from the body and ended near the boundaries of the workspace. Based purely on mathematical grounds, Hollerbach and Atkeson (1987) have shown, however, that if the hand follows a straight path, the ratio of joint velocities is bounded to converge to a constant as the hand approaches the workspace boundaries. Thus, such movements cannot provide evidence for joint–based planning.

Not all point-to-point movements are ideally straight. Several researchers presented experimental observations showing that although some pointing movements, performed in the vertical plane, follow straight paths, others are characteristically quite curved (Atkeson and Hollerbach, 1985; Flanagan and Ostry, 1990). To account for the curvature of some of these movements, Hollerbach and Atkeson (1987) have proposed a joint-based strategy which they called *staggered joint interpolation.* The suggestion was that the velocity profiles of all joints have the same kinematic form. The times at which the joints begin moving, however, might be delayed or staggered with respect to one another. These velocity profiles are also appropriately scaled with respect to amplitude and duration. Choosing appropriate parameters, this strategy can produce essentially straight lines in certain parts of the workspace. This strategy, however, does not allow joint reversals during the performance of any movement. Hence, Hollerbach and Atkeson (1987) have argued that in regions where the performance of straight paths requires joint reversals, the movements are bound to be curved.

The trajectories observed in many studies of point-to-point movements do display, however, joint reversals and are also very close to being straight (Hogan, 1988; see also Figure 4 in Flash and Hogan, 1985). The predictions of the staggered-joint interpolation strategy, therefore, are incompatible with experimental observations, at least in the case of horizontal planar movements. Why pointing motions are more curved is not yet clear. However, as will be shown in the next section, it has been possible to account for the deviations of horizontal planar point-to-point movements from straight paths by considering how desired motion plans are possibly executed by the neuromuscular system. Whether this conclusion can be extended to movements in the vertical plane remains to be seen.

The maximum-smoothness objective, as expressed by the above theory, is independent of limb kinematics and the neuromuscular dynamics. This theory is, therefore, completely consistent with the notion that there exists an abstract representation of movement that is independent of the mechanical effectors used in the performance of the motor task (Bernstein, 1967). This theory is also consistent with the notion that movement generation is heirarchically organized, whereby at higher levels only the more general features of movement, i.e. those that remain invariant under changes in the temporal and spatial scales of the movement, are represented (Keele, 1981). Specific motion parameters, such as end-point or via-point positions and movement durations, defined on the basis of the current task demands, are then used to obtain a more detailed specification of any particular movement (Flash and Hogan, 1985).

Although hand jerk was shown to decrease with practice (Schneider and Zernicke, 1989), it should be stressed that nowhere does this theory hypothesize or suggest that hand jerk is actually sensed, or that minimum-jerk trajectories are computationally derived by the nervous system. Thus, one possible physiological interpretation for the success of this theory in accounting for the observed behavior is that the central nervous system employs a trajectory planning strategy that is captured by this model. Another possible explanation is that the smoothness of motion is an outcome of the intrinsic properties of the neural and musculoskeletal hardware [Flash and Hogan, 1985; Chapter 19 (Seif-Naraghi and Winters)]. The possible rationale for maximizing smoothness might be the wish to maximize the predictability of the trajectory, which is consistent with minimizing its higher time-derivatives (Hogan, 1984; Flash and Hogan, 1985), or the wish to minimize the amount of information that needs to be presented and proccessed in the planning and execution of motor tasks.

Similar optimization principles may also apply in the generation of more complicated movements. In handwriting or drawing, for example, the system need not internally represent all possible letters or figural forms, but may use, instead, a limited set of basic primitives or strokes which can then be concatenated to form more complicated figural shapes (Morasso and Mussa-Ivaldi, 1982; Lacquaniti, 1989). These strokes, themselves, might be internally represented, based on a limited set of position and shape parameters, and generated according to motion planning rules, similar to the ones described above (Edelman and Flash, 1987). These internal constraints may again give rise to the observed coupling between motion speed and curvature.

In naturally executed drawing movements, angular velocity was found to decrease with increasing curvature and to be proportional to the two-thirds power of the latter (Lacquaniti et al., 1983). The gain factor of this relationship was demonstrated to be piecewise constant and to be determined by the linear extent of each individual segment (Viviani and Cenzato, 1985). These observations were interpreted to suggest that in spite of the apparent continuity of drawing movements, they are, in fact, intrinsicly discontinuous and are constructed of individual segments or strokes (Morasso and Mussa-Ivaldi, 1982; Viviani and Cenzato, 1985). It has been suggested that hand velocity during any movement need not be explicitly coded but is automatically derived from the coupling between speed and curvature expressed by the two-third power law (Lacquaniti, 1989). It is not yet known how this relationship between geometrical form and velocity comes about in the natural execution of movement, but it is possible that optimization theories of the type presented above may offer a possible explanation. (See also Nelson, 1983; Stein et al., 1985; Wann et al., 1988.)

17.3 Trajectory Execution

In the previous section we have focused on the topic of motion planning. To execute a desired

motion plan, however, appropriate joint torques and muscle activation patterns must be genereteed. How are these trajectory plans realized by the system? One possible way that this can be achieved (Hollerbach, 1982) is by transforming hand trajectory plans into joint rotations. The required joint torques can then be derived by solving the second-order nonlinear dynamic equations of motion (the so called "inverse dynamics" problem). For multi-jointed arms, there exist inertial dynamic interactions between the moving skeletal segments and several muscles pull across more than one joint [e.g., Chapter 18 (Gielen et al.); Chapter 8 (Zajac and Winters)]. As indicated by a study of point-to-point movements (Hollerbach and Flash, 1982), all interaction torques, including Coriolis and centripetal forces, are significant when the entire course of the movement is considered even at slow movement speeds. Moreover, the latter forces completely dominate arm dynamics at movement midpoint, when all the other torque terms which depend on joint accelerations go through zero as the hand switches between acceleration and decceleration. Also, joint torque profiles, required for the generation of one particular point-to-point movement, cannot be used in the generation of a kinematically similar movement between a different pair of end–points. Since the motor system seems to be capable of executing quite complicated multi–joint movements, this implies that it must have devised some means for computing or appropriately compensating for the interaction forces. It has been suggested, however, that the time–scaling property of human movement may simplify dynamic computations. Since for arm trajectories which simply scale with speed, the rate–dependent torque terms also simply scale with speed, this may indicate that the CNS has developed a trajectory formation strategy which reduces some of the complexities involved in motor execution (Hollerbach and Flash, 1982).

If indeed the system explicitly computes or, alternatively, derives the necessary joint torques from look-up tables (e.g. Atkeson, 1989), these torques must still be distributed among a considerable number of muscles. Moreover, having internal models of arm dynamics is not sufficient, since it does not guarantee robustness of the behavior in the face of unpredictable external disturbances or errors in internal models. Such robustness can be guaranteed, however, by the utilization of the viscoelastic properties of muscles [see Chapter 9 (Hogan)].

Is the derivation or computation of joint torques the only possible scheme for executing limb movements? As the theory presented below will suggest, there exists an alternative possible scheme that may allow the system to circumvent the need to solve the complicated inverse dynamics problem. Instead, the brain may only transform the desired hand trajectory plans into hand equilibrium trajectories. The forces needed to track the equilibrium trajectory may then be automatically generated as a consequence of the mechanical properties of muscles [Hogan, 1985; Flash, 1987; Chapter 12 (Feldman et al.), Chapter 9 (Hogan)].

The force exerted by a muscle on the limb increases as the muscle is stretched. The magnitude of this force depends on both the muscle stiffness and rest length, which are specified by the level of the neural activation of the muscle (for a review see Houk and Rymer, 1982). Consequently, several investigators have proposed that postural control is achieved by the motor system through the choice of a particular pair of torque–angle curves for the agonist–antagonist muscle pairs acting on the limb (Bizzi et al., 1976; Feldman, 1986; Chapter 12). This choice will determine the equilibrium position for the limb and the stiffness about the joint. According to the *final position* hypothesis for movement generation, motion towards a specified final position can be achieved without explicit planning of the trajectory of the limb, but merely on the basis of a pulse-like shift of the equilibrium point to the final position (Bizzi et al., 1976; Feldman, 1986). Recent observations of single–joint elbow movements in monkeys have indicated, however, that the *CNS* generates a control signal which defines a series of equilibrium positions and not merely the final position (Bizzi et al., 1984).

17.3.1 The Control of Arm Posture

The mechanical properties of muscles may also play an important role in the control of multi-joint posture and movement [Hogan, 1985; Berkenblit et al., 1986; Chapter 9 (Hogan)]. In particular, the actions of individual arm muscles ultimately combine to produce the overall mechanical behavior of the hand during posture. The net spring-like behavior of the hand was recently characterized by

measuring the elastic field at a number of locations within the horizontal plane (Mussa–Ivaldi et al., 1985). The hand was displaced from equilibrium and the resulting restoring forces were measured at the displaced positions at steady state and before the onset of voluntary reactions. The field of forces, measured at the hand during posture, indicated that this field is mechanically conservative and can be completely described, in the vicinity of each hand position, by the stiffness matrix which relates force to displacement vectors. The hand stiffness matrices, obtained from these measurements, were graphically represented as ellipses, characterized by three parameters (see also Chapter 9): size (the area of the ellipse), shape (the ratio between the lengths of the major and minor axis) and orientation (the direction of the ellipse major axis with respect to a laboratory fixed coordinate system).

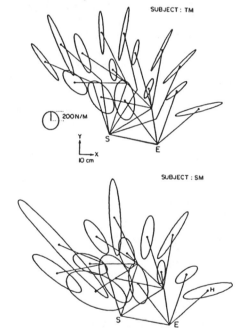

Figure 17.3: Hand stiffness ellipses obtained from two subjects during the postural task. Each ellipse was derived by regression on about 60 force and displacement vectors. The upper arm and forearm are represented by the two line segments (*S–E*) and (*E–H*), respectively, and the ellipses are placed on the hand (*H*). The calibration for the stiffness is provided by the circle to the left, which represents an isotropic hand stiffness of 200 N/m. (Reproduced with permission from Flash, 1987.)

The experimental findings indicated the existence of a strong and systematic dependence of the shape and orientation of the stiffness ellipse on the location of the hand in the horizontal plane. In particular, the results shown in Figure 17.3 indicate that the major axis of the hand stiffness ellipse at any location was observed to be nearly coaligned with the direction of the radial axis of a polar coordinate system located at the shoulder, where this radial axis is defined by the line connecting the hand with the shoulder. As far as shape is concerned, the ellipses became more elongated (i.e. hand stiffness was less isotropic) as the hand approached the distal boundary of the workspace, the major axis being in the proximal-distal direction. These patterns of stiffness shape and orientation were observed to be the same in all the subjects participating in the study and to remain invariant over time. In contrast, the values obtained for the size parameter varied substantially among subjects and in the same subject among experimental sessions. Even when a disturbance force in a well-specified direction was imposed on the hand, Mussa-Ivaldi et al. (1985) have found that only minor changes could be seen in the orientation and shape of the stiffness field.

In a subsequent study, the underlying causes for the observed spatial pattern of variation of the hand stiffness ellipse were investigated (Flash and Mussa-Ivaldi, 1990). Three possible factors that could have contributed to the observed characteristics of the stiffness field were considered. First, for a given vector of joint torques the magnitude and direction of the net force experienced at the hand depends on the configuration of the arm (see Chapters 9 and 11). Consequently, even if the values of all joint stiffnesses were to remain constant throughout the workspace, the geometrical parameters of the hand elastic field would be expected to change with hand position. Second, the contribution of each muscle to the resultant joint stiffness depends on the muscle moment arm. Since the lengths of muscle moment arms change with joint angles, the stiffness contributions of muscles are also expected to change with arm configuration. Third, muscle stiffness depends on muscle length and activation through the length-tension relationships. Hence, the spatial features of hand stiffness at different positions of the workspace are also affected by neural activation as well as by the muscle spring-like properties.

Examining the effects of these factors on the characteristics of the hand stiffness field and on the patterns of variations of the joint stiffness matrix with hand position, it was found that arm configuration alone can not be the mere cause for the experimental observations. Using anatomical data derived from the study by Wood et al. (1989) and considering the effects that muscle cross-sections and changes in muscle moment arms have on the joint stiffness matrix, we found that these anatomical factors are also not sufficient to account for the observed pattern of variation of joint stiffnesses in the workspace. However, based on the mathematical analysis of the relation between hand and joint stiffness matrices, it was found that in order for the stiffness ellipse to have the observed characteristics, the shoulder stiffness must covary in the workspace with the stiffness component provided by the two-joint muscles (Flash, 1987; Flash and Mussa-Ivaldi, 1990).

This condition was indeed found to be satisfied by the measured joint stiffness components. Figure 17.4 shows stiffness surfaces describing the variations of the net shoulder, net elbow and two-joint stiffnesses with elbow and shoulder angles. Thus, as these results indicate, the two-joint and the net shoulder stiffnesses do covary with arm configuration while the net elbow stiffness surface displays a different pattern of variation. We then examined whether the coupling between shoulder and two-joint stiffnesses may result from the coactivation of the muscles contributing to these stiffnesses. EMG signals were recorded from shoulder, elbow, and two-joint muscles. However, our results indicated that, while some muscle coactivation may indeed exist, it can be found for only some of the muscles and in only part of the workspace.

Nonetheless, these results have indicated that functional muscle synergies can be identified. In this context, the concept of a synergy implies that in spite of the excessive number of arm muscles, there is nearly a fixed relationship between the stiffnesses contributed by different muscle groups. This coupling finds its manifestation in the net mechanical response of the human arm to external disturbances. At least in principle (Hogan, 1985), the redundant number of arm muscles offers the CNS the possibility of selectively tuning the hand stiffness field according to task requirements. The indicated limited capability of the system to significantly modify the characteristics of the stiffness field may, therefore, indicate the operation of coordinative constraints which might also be present during movement generation.

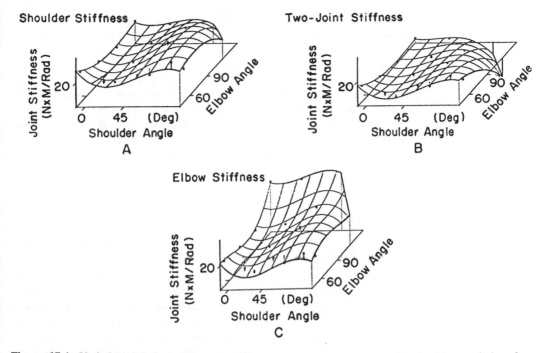

Figure 17.4: Variations of the measured joint stiffnesses with shoulder and elbow angles. *a)* Net shoulder stiffness. *b)* Two-joint stiffness. *c)* Net elbow stiffness. (Reproduced with permission from Flash and Mussa-Ivaldi, 1990.)

17.3.2 The "Equilibrium Trajectory" Control Theory

The above section has concentrated on the importance of the mechanical properties of muscles in the control of multi–joint arm posture. Do these properties play any role in the execution of multi–joint movement? As postulated by the *equilibrium trajectory* theory discussed below, the motor system may use muscle "spring-like" properties to obviate the need for dynamic computations. Furthermore, as indicated by a recent simulation study, the control of arm trajectories, at least in the context of kinematically unconstrained movements, might not be fundamentally different from the control of posture (Flash, 1987).

According to the equilibrium trajectory hypothesis, limb movements are achieved by gradually shifting the hand equilibrium position between the movement end-points (Hogan, 1985; Flash, 1987). The magnitude and direction of the forces acting on the arm, at any point in time, are determined by the magnitude and direction of the displacement vector of the hand from equilibrium and by the hand elastic and viscosity fields about the equilibrium positions.

The equilibrium position and trajectory and the characteristics of the stiffness and viscosity fields are determined by the neural activations of arm muscles. Given the "spring-like" properties of muscles, an equilibrium position and trajectory for the limb can always be defined. Are they, however, explicitly controlled or coded for by the brain? As the model presented below suggests, the system might use the equilibrium trajectory control scheme as a vehicle in the realization of desired motion plans. To test the validity of this hypothesis, an explicit model of arm trajectory control was developed (Flash, 1987). The model was based on the notion that the execution of reaching movements involves explicit planning of straight hand equilibrium trajectories which are invariant under translation, rotation, amplitude, and time scaling. A simple mathematical description of the musculoskeletal system was constructed and the suggested control scheme was implemented in computer simulations. The stiffness values used in these simulations were derived from the experimentally measured static stiffness values (Mussa–Ivaldi et al., 1985), under the assumption that the orientation and shape of the stiffness field, when the arm moves through any workspace location, are similar to those of the static field at that location. The simulations were used to derive hand equilibrium trajectories from measured movements. These equilibrium trajectories were found to follow straighter paths than the actual movements. Forward computations were also performed. An equilibrium trajectory, derived from one representative movement, was used (after suitable amplitude and time scaling, translation, and rotation) to simulate actual movements of different amplitudes, directions, and speeds.

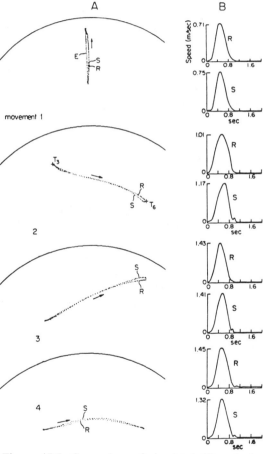

Figure 17.5: Comparison of simulated (*S*) and observed (*R*) point-to-point movements in the horizontal plane. Column A shows hand paths. Column B shows profiles of hand speed along the path. A virtual trajectory (marked *E*) was derived from observations of movement 1 (top panel) and used with appropriate scaling and rotation to simulate movements 2, 3 and 4. (Reproduced with permission from Flash, 1987.)

As seen in Figure 17.5, the predicted trajectories captured the observed features of the measured movements down to fine details of kinematics. In particular, the curvatures of the measured movements were successfully accounted for, as well as the presence of small hooks or movement overshoots as the hand approaches the final target and the kinematic differences observed between movements generated between the same pair of targets but in opposite directions.

On the basis of the success of this model in accounting for the fine details of movement curvature, we may argue that the characteristic deviations from ideally straight paths observed in actual data may reflect the combined effects of arm inertia, centrifugal and interaction torques, and the local characteristics of the arm stiffness and viscosity fields. Were joint torques explicitly computed by the brain and then distributed among the muscles, there would have been no apparent reason why the desired straight hand paths could not be generated. The success of this model in capturing the observed behavior therefore suggests that at least in the case of point-to-point movements, the inverse dynamics problem is not explicitly solved by the brain. The system, therefore, has seemed to adopt a simpler control strategy even at the price of movement accuracy. Moreover, since the stiffness fields used in these simulations were assumed to have similar characteristics to the static ones, this may indicate that similar coordinative constraints operate during both posture and movement.

17.3.3 An Alternative Model

The view presented in this chapter is that the transformation of behavioral goals into actual movements involves a step-by-step process. The first step is only concerned with movement kinematics and deals with the generation of motion plans. The second step involves the execution of these plans by taking advantage of the physiological properties of the neuromuscular system. Recently, an alternative view to the one presented here was suggested (Uno et al., 1989, Kawato et al., 1990). Again, it was postulated that motor commands are directly calculated from the goal of the movement, represented by some performance index. However, in contrast to the maximum-smoothness objective which was expressed in terms of world coordinates, Uno et al.

(1989) have postulated that the underlying criterion for the selection of specific motions from the infinite number of possible ones can be defined as follows:

$$C_T = \frac{1}{2} \int_0^{t_f} \sum_{i=1}^{n} \left(\frac{dz_i}{dt} \right)^2 dt \qquad (17.3)$$

where z_i is the joint torque for actuator i. Hence the proposed performance index is the sum of square of the rate of change of joint torques, integrated over the entire movement. This movement objective, unlike the one proposed by Flash and Hogan (1985), is critically dependent on the dynamics of the musculoskeletal system.

The results obtained from the minimization of C_T were found to be in good agreement both with the observed behavior and with the predictions of the minimum-jerk model. For certain movements, however (e.g., movements starting when the hand is stretched away from the midline), the trajectories predicted by this model were more curved than the ones predicted by the minimum-jerk model and in better agreement with the data. This was also found to be the case for certain observations made with respect to curved movements.

Independently from the above study, computationally efficient trajectory planners based on the multi-grid approach were developed and were applied to this problem of finding the unique trajectories which minimize C_T for a two-joint arm (Ben-Zvi, 1987; Flash, 1988). In this latter study,

however, different predictions were obtained from the ones obtained by Uno et al. (1989). These differences do not seem to result from the differences between the methods used to arrive at the optimal solution, but to reflect differences in the values of the inertial parameters used to model the upper limb. Since, as stressed above, C_T strictly depends on arm dynamics, the results obtained from the minimization of this cost are highly sensitive to the values used for these parameters.

As was explained above, however, the curvature seen in measured point-to-point movements can be accounted for by considering the strategy used by the system to execute the desired motion plans (see Figure 17.5). Hence, the actual movements produced on the basis of the equilibrium trajectory control scheme are more curved than the ones that are assumed to be planned based on the minimization of hand jerk. This is indicated in Figure 17.6, where the trajectories predicted by the minimum-jerk and the equilibrium trajectory models are displayed side by side. However, when using the same inertial parameters as the ones used in the equilibrium trajectory model to predict the trajectories that would result from the minimization of C_T, the resulting movements failed to match the measured ones. These findings indicate that multi-joint movement generation may indeed involve a step-by-step trajectory control process and that arm dynamics may not be internally represented as postulated by the minimum-torque change model (Uno et al., 1989; Kawato et al., 1990).

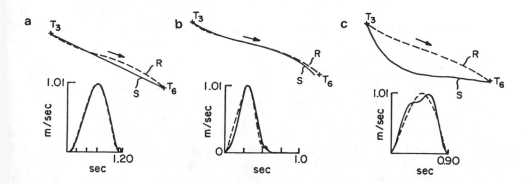

Figure 17.6: Comparison of measured (dashed line) and predicted (solid line) movements. *a)* The minimum-jerk model. *b)* The equilibrium-trajectory model. *c)* The minimum–torque rate of change model. (Reproduced with permission from Flash, © 1988 *IEEE.)*

17.4 Arm Trajectory Modification

So far we have discussed the topics of motion planning and execution, focussing mainly on the generation of point-to-point and curved movements. During many daily activities, however, we do not simply generate motions towards static visual objects, but must also actively interact with the environment or respond efficiently and quickly to dynamic changes in the locations of external objects. Reaching towards visual targets in space involves a series of events and sensorimotor transformations leading from the neural encoding of target location, derived from vision or memory, to the performance of aimed movements. As was discussed in the section on trajectory planning, the temporal scales of any specific movement, i.e. motion amplitude and duration, are used to transform the general motion plans into a detailed trajectory for the hand through space. When, however, the target towards which we intend to move our limb suddenly changes its location, our motor response must be planned and executed while the arm is no longer at rest. Even if the target is displaced during the reaction time, it can be assumed that the first planning process has already begun. How, therefore, does the nervous system modify an ongoing motion or update an ongoing planning process?

The characteristics of aimed movements when the target changes location either during the reaction or movement time have been described by several investigators (e.g. Georgopoulos et al., 1981; Gielen et al., 1984). Basically, these studies indicated that the change in target location elicits a graded movement towards the first target, followed by a change in movement direction and a subsequent motion towards the second target. Occasionally, for a short interstimulus interval (*ISI*; the time elapsing between target presentation and its displacement), the hand moved directly to the second target or in between the two targets. The duration of the initial motor response to the first target was found to be a linear function of the *ISI* and no delays were found beyond the normal reaction time, i.e., there was no appreciable psychological refractory period. The orderly modifications of movements produced in response to double target displacements suggested that the aimed motor commands are emitted in a continuous fashion as a real time process that can be interrupted at any time (Georgopoulos et al., 1981).

It is not yet clear, however, what strategy is used by the brain in the modification of ongoing motion plans.

To investigate this issue, human horizontal planar movements were recorded using the double-step target displacement paradigm. The experiments were performed in a darkened room, eliminating any visual feedback from the moving limb (Henis and Flash, 1989; Flash and Henis, in preparation). The target was displaced once or twice along the *x* or *y* axes or obliquely, using various directions and amplitudes of target displacements and ISI's ranging between 50 and 700 ms. In general, our findings were consistent with previously reported results.

One plausible strategy for trajectory modification may involve aborting the rest of the initially planned trajectory and replacing it by a new movement between the location of the hand (at the modification time) and the final target position. For the two movement parts to be smoothly joined together, since motion planning preceeds its execution, this strategy would require of the brain to predict, or derive, information about the kinematic state of the hand at the switching time.

In this section, we summarize evidence that the system may use an alternative and simpler modification scheme than the one described above. Instead of aborting the rest of the initial trajectory plan, it is suggested here that this plan continues unmodified until its intended completion and is vectorially summed with a second, time-shifted point-to-point hand trajectory plan for moving between the initial and new target locations. Thus, both trajectory units start and end at rest and each trajectory is separately planned regardless of the other one. Since the initial trajectory is neither aborted nor modified when the second part is injected, this guarantees that the second target location will always be reached independently of the modification time. However, the exact details of the combined motion plan, and consequently of the actual movement executed by the system, do strongly depend on the time shift between the initiation of the two superimposed trajectory plans and on their durations.

In this context, it should be mentioned that the idea that more complicated movements may emerge from the superposition of basic trajectory primitives was postulated in the context of speech (Munhall and Lofqvist, 1987), locomotion (Flashner et al., 1988), and both single (Adamovitch and Feldman, 1984) and multi–joint arm movements (Morasso and Mussa–Ivaldi, 1982). The validity of this hypothesis was also tested for movements recorded in double–step target displacement trials (Massey et al., 1986). However, the underlying assumption made in that study was that the superimposed trajectories should have the same durations as control point-to-point movements separately performed by the same subject when moving between the two target pairs. Since in that study, the simulated movements that were predicted by this scheme failed to match the measured ones, it was concluded that arm trajectory modification cannot possibly involve the superposition strategy.

In our work (Henis and Flash, 1989), no *a priori* assumption was made with respect to the durations of the superimposed trajectory plans. Instead, we wished to test whether the modified movements may emerge from the vectorial summation of "control-like" point-to-point trajectories, i.e., movements that have the same kinematic form as simple reaching motions, whatever their durations might be. To assess the success of this scheme in accounting for the data, the following analysis was separately applied to the x and y components of each individual measured modified movement. The simulated x and y components were then combined to predict the entire measured trajectory:

a) Time scaled velocity profiles corresponding to a point-to-point minimum jerk trajectory between the initial and first targets were superimposed on the initial part of the movement.

b) The time of the first detectable deviation of the measured speed profile from that of the superimposed control movement (t_s) was extracted.

c) From this time on, the following fifth-order polynomial was added to the first polynomial.

$$x(t) = (x_2 - x_1)(a_3 t_r^3 + a_4 t_r^4 + a_5 t_r^5) \quad (17.4)$$

where $t_r = (t - t_s)/(t_f - t_s)$, t_f is the total duration of the modified movement, derived from data,

a_3, a_4 and a_5 are coefficients to be calculated, and $(x_2 - x_1)$ is the x component of the displacement vector between the first and second target locations. The expression for $y(t)$ was analogous to *Eq. 17.4*. This family of trajectories has zero initial positions as well as zero initial (but not final) velocities and accelerations.

d) The values of the three unspecified coefficients of this polynomial were determined using a least-square fitting method based on the position error between the simulated and measured movements.

For the modified motions, the analysis showed that point-to-point minimum–jerk trajectories between the first and second targets provided the best fit for the second trajectory units. Statistical tests further showed that the coefficients of the added trajectory units are not significantly different from the measured ones. Thus, the modified movements were found to result from the vectorial summation of the initial unmodified trajectories with point-to-point trajectories between the first and second target locations which have the same kinematic form as simple point-to-point movements. Following the first detectible deviation, the motions resulting from the superposition scheme were found to be in good agreement with the measured ones.

The recorded hand paths and velocity profiles for all target configurations and all *ISI*s were successfully accounted for. Figure 17.7 shows two examples of measured modified movements, the corresponding trajectories predicted by the superposition scheme, and the superimposed trajectory units. The exact kinematic details of any modified movement were found to depend on the specific durations of the superimposed trajectories and on the time delays between their initiation. The alternative strategy for trajectory modification which involves aborting the rest of the initial trajectory and replacing it by a new trajectory plan was also mathematically modelled. However, the movements predicted by this scheme failed to match the measured ones. Hence, this scheme was significantly less successful than the superposition model in accounting for the observed behavior.

Our results were therefore consistent with the assumptions of the superposition model that the second target presentation activates a separate planning process while the first process continues

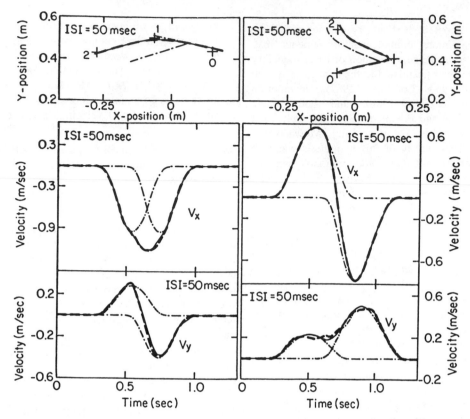

Figure 17.7: Comparison of measured modified movements (*ISI* = 50 ms) and movements resulting from the superposition scheme. The initial hand positions are marked by 0. The first and second target locations are marked by 1 and 2, respectively. Hand paths are shown in the top panels. The *x* and *y* components of velocity, V_x and V_y, respectively, are shown in the two bottom panels. Measured trajectories are marked by solid lines, superimposed initial and added trajectory units by alternating dots and dashes and their vectorial sums by dashed lines. In the top panels the added trajectories are shown to begin at the locations corresponding to the modification times.

unmodified. Using information about the first and second target locations, derived from vision, this process generates a trajectory plan that obeys the same relationship between movement amplitude, duration and speed as a single reaching movement starting and ending at rest. When this trajectory plan is ready, it is continuously added to the first plan to yield the combined motion. For long *ISIs*, the initial response is ready before the second process is activated, giving rise to movements initially directed towards the first target. For short *ISIs*, however, it was often observed that the hand either directly moves towards the second target, or along some intermediate direction [the "averaging" phenomenon (Van Sonderen et al., 1989)]. It remains therefore to be seen whether in this case, too, the modified trajectories result from the superposition of the initial trajectory plan with

a second trajectory, or whether, for short *ISIs*, the initial trajectory is already directed towards some intermediate target located in between the first and second target locations. The simplicity of the proposed superposition scheme may explain the observed lack of a psychological refractory period. Furthermore, our findings may indicate the presence of parallel planning of trajectory primitives.

17.5 Conclusions

The studies reviewed in this chapter have indicated the existence of hierarchical control levels for arm trajectory generation. Higher levels are concerned with the planning of trajectories that maximize the smoothness of hand motion in extracorporeal space. These desired motions are then executed by taking advantage of the mechanical properties of muscles. An important consequence of the proposed hierarchical scheme is to make computational control of multi–joint arm movements much simpler. Since a hand location on the equilibrium trajectory can be directly associated with a co-contraction pattern of muscles, the joint torque profiles, required for the realization of the intended movements, are automatically generated whenever the actual hand position deviates from the instantaneous equilibrium position, thus eliminating the need for inverse dynamics computations.

The above scheme assumes two separate levels of trajectory planning and execution. One possible alternative scheme, which assumes that the cost that is minimized in movement execution is the rate of change of joint torques, does not adhere to this step-by-step transformation from the behavioral goal of movement to the neural activations to muscles. When comparing the predictions of these two alternative schemes, our results indicated that, in contrast to the findings presented by Uno et al. (1989), only the predictions of the step-by-step scheme were compatible with experimental observations. Hence, the studies reviewed here show that the observed kinematic data are completely consistent with the hypothesis that: *a)* simple movements are first planned in terms of desired hand motions in external space; *b)* are expressed in terms of equilibrium trajectories; and *c)* are executed via the mechanics of the neuromuscular system, which acts to keep the actual path of the limb reasonably close to the equilibrium trajectory.

In dealing with the modification of ongoing movements, the success of the superposition model in accounting for the observed behavior suggests that the system may have adopted a simple movement modification strategy that eliminates the need to rely upon efference copies of past motor commands or on kinesthetic information in order to derive or predict the expected hand position at the switching time. Instead, independently and possibly in parallel the system plans a second point-to-point trajectory, using information derived from vision about the first and second target locations. The two trajectory plans are then continuously summed together to give the combined plan for the modified movement. This combined plan can then be transformed into a combined hand equilibrium trajectory and ex-

ecuted in the same way as a separate point-to-point movement by relying upon the mechanical propertes of muscles. Since the superimposed primitives were shown to be hand trajectory plans, this provides an additional support for the hierarchical organization of the trajectory control processes and suggests that such an organization might be especially beneficial when coping with unpredictable disturbances or with dynamic motor tasks.

Given that the superimposed movements were shown to have the same kinematic form, as though derived from a common template, this indicates that more complicated motor acts and motion sequences might be constructed from the superposition of more elementary trajectory primitives.

17.6 Recommended Future Directions

The issues discussed in the previous sections have dealt with several aspects of arm trajectory control. Of course, there are many other aspects of multi–joint arm movement generation that were not addressed here and many questions remain. Here, we will discuss only several recommendations for future research directions. Some of these recommendations address issues that arise mainly in the context of three-dimensional arm movements and movements in the vertical plane, while others address more general problems.

Of the three ill-posed motor problems that were discussed in the introduction to this chapter, we, as well as other investigators, have focussed mainly on the question of how the system resolves the task-level redundancy. There is a strong evidence to support the hypothesis that, indeed, arm movements are internally represented in terms of task-level coordinates. It is not yet clear, however, whether and how the system transforms these task-level motion plans into joint coordinates. Here we have focussed mainly on two-joint arm movements in the horizontal plane. An additional issue that does not arise in this context is the issue of kinematic redundancy. Based on the kinematic analysis of three-dimensional drawing movements, Soechting and Terzuolo (1986) have suggested that the brain is endowed with an explicit representation of movements in joint space and an algorithm for embedding a world 'space trajectory into joint rotations. It was also suggested that the internal representation of actual movement in joint space is not an exact reproduction of the intended hand movement, but is one that simplifies the transformation from hand to joint coordinates. Another consideration, for resolving redundancies, may involve, for example, the maximization of smoothness. Based on theoretical grounds, Jordan and Rosenbaum (1989) have suggested that a smoothness constraint in articulatory space, which implies that targets nearby in time are achieved using similar limb configurations, might be used to resolve kinematic redundancies. Thus, a greater research effort should be directed at investigating whether and how hand trajectory plans are transformed into joint movements. Nonetheless, for practical reasons, most of the studies addressing this and similar questions have focused on well-defined motions that are easy to study from within a laboratory setting (e.g., pointing movements, drawing ellipses, etc.). There is therefore a great need to examine what principles underlie the generation of more natural and perhaps less confined movements.

Another recommended direction for future studies has to do with the characterization of arm impedance, i.e. hand and joint stiffnesses and viscosities during movement, especially when gravity must be taken into account. As was suggested above, the observed curvature of some pointing movements in the vertical plane may result from the way by which the nervous system executes desired motion plans. As long as arm impedance cannot be measured and characterized during movement, the validity of this hypothesis, as well as the validity of the equilibrium trajectory hypothesis in the context of planar horizontal reaching movements, cannot be experimentally tested. Thus, it is strongly recommended that research efforts should be directed, as indeed they currently seem to be (e.g., Xu et al., 1989) at developing technical means that will enable us to measure arm stiffness and viscosity during movement.

It is also not yet clear how the system solves the ill-posed problem of muscle activation, i.e., selecting what muscles to activate, in what order, and to what level of activation. In many studies attempts were made to determine the rules, according to which the motor system specifies the pattern of activation of individual muscles. The minimization of various performance costs, such as the

sum of muscle forces, stresses, etc. were hypothesized [e.g., see Patriarco et al., 1981 or review in Chapter 8 (Zajac and Winters)]. However, the force distributions predicted in these studies were found to depend more strongly on the muscle attachment geometry than on the types of cost functions being minimized. Furthermore, these methods failed to predict the coactivation of muscle synergists and antagonists, usually seen in experimental data [however, see Chapter 19 (Winters and Seif-Naraghi)]. In a recent paper assessing the question of the existence of particular muscle synergies during load perturbations or intentional arm movements, it was concluded that the concept of a muscle synergy is not appropriate to characterize, in any economical fashion, the activities of muscles involved in upper limb movements or the response of the muscles to applied load perturbations (Soechting and Lacquaniti, 1989). Instead, it was suggested that each muscle is related in a unique and different manner to the kinematic and dynamic variables of the motor task. Thus, it was argued that the simplification of the problem of controlling a large number of degrees of freedom does take place in the sense suggested by Bernstein (1967), but that this occurs at the level of limb kinematics and does not manifest itself in terms of fixed patterns of activation among different muscles. As indicated, however, by the findings reported by Hasan and Karst (1989, Chapter 16), it does seem that during planar horizontal two–joint point-to-point movements, the order of muscle activation, i.e. which muscles are first activated, is reproducible over trials and among subjects and depends on the initial limb configuration. Another promising line of investigation is reported in Chapter 18 (Gielen et al.). Hence, although some progress has been made with respect to this issue, further efforts should be directed at deciphering the rules that dictate what muscles to activate during multi-joint arm posture and movement.

Nothing was said in this chapter about the relevance of the studies reviewed here to neurophysiology or to the analysis of motor disorders. However, although very often neurophysiological findings can be interpreted in many different ways and do not always lead to unequivocal answers to theoretical questions, they may provide insight into the ways by which the brain solves some of the computational problems discussed above. For example, the hypothesis that arm movements are internally represented in terms of hand coordinates was supported by recent neurophysiological findings (for a review see Georgopoulos, 1986). These series of studies indicated that the motor cortex is a key area in the control of spatial aspects of hand trajectories. The direction of the population vector, which reflects the activity of a large population of neurons, was found to provide unique information concerning the direction of movement of the hand in 2-D or 3-D space. A signal related to the instantaneous hand velocity was also described. Although it has been argued that these findings might also be interpreted in other ways (Mussa–Ivaldi, 1988), the similarities that were shown to exist between the patterns of activities of cortical cell populations when moving the hand in the same direction but from different starting positions (Kettner et al., 1988) did support the idea of task–level representation of multi–joint arm movements. Further studies are therefore required either to establish, or to refute, this conclusion as well as additional studies designed to investigate whether and how motion plans are transformed into joint and muscle coordinates.

Another recommended direction for future studies is in the area of motor disorders. Thus, for example, the kinematic analysis of motion in basal ganglia or cerebellar disorders may offer a new opportunity to shed new light on the still mysterious role of these brain areas in motor planning and execution. Recently, for example, we have investigated the kinematic properties of planar horizontal point-to-point and curved movements in Parkinsonian patients (Flash et al., 1988). Pertinent to our discussion here are the findings related to the production of curved movements which showed that unlike healthy subjects, patients often stopped at points of maximum curvature before changing movement direction. Moreover, the isochrony principle (Viviani and Terzuolo, 1982) did not hold in patient movements: movement duration increased proportionally to the length of the movement segment. Thus, since it has been indicated that the performance of complex, i.e. sequential or simultaneous movements is especially impaired in Parkinson's disease (Benecke et al., 1986), the study of such motor tasks may provide new insight into the functional roles of these structures in neuromotor planning.

Finally, more research studies should focus on the areas of motor learning and adaptation. Such studies may offer us new opportunities to resolve the questions of what it is that the system learns during skill acquisition and when adapting to new loads or external disturbances, what performance costs does the system wish to minimize or optimize during skill acquisition, etc. Atkeson (1989) as well as Kawato et al. (1990) have claimed that the equilibrium trajectory scheme does not allow efficient learning from practice, but no evidence was provided to support this claim. In contrast, as indicated by a recent study based on the use of recurrent neural networks, motor learning based on the use of the equilibrium control scheme is at least feasible (Jordan, 1990). Thus, much more effort should be directed at investigating how it is that the motor system allows us to learn new skills or to improve motor performance on the basis of practice. Finally, efforts should be directed at developing theories that will account for the phenomenon of motor equivalence (e.g., in Berkenblit et al., 1986), which implies that even if external conditions do not vary, the system may generate a set of different solutions for one and the same motor task.

Acknowledgments

Tamar Flash is an Incumbent of the Corinne S. Koshland Career Development Chair. This research was partially supported by grants No. 85-00395 and 88-00141 from the United States–Israel Binational Science Foundation (BSF), Jerusalem, Israel.

References

Abend, W.K., Bizzi, E. and Morasso, P. (1982) Human arm trajectory formation, *Brain* **105**: 331-348.

Adamovitch, S.H. and Feldman, A.G. (1984) Model of the central regulation of the parameters of motor trajectories. *Biofizika* **29**: 306-309 (English Translation 338-342).

Atkeson, C.G. (1989) Learning arm kinematics and dynamics. *Ann. Rev. Neurosci.* **12**: 157-183.

Atkeson, C.G. and Hollerbach, J.M. (1985) Kinematic features of unrestrained vertical arm movements. *J. Neurosci.* **5**: 2318-2330.

Benecke, R., Rothwell, J.C., Dick, J.P.R., Day, B.L. and Marsden, C.D. (1986) Performance of simultaneous movements in patients with Parkinson's disease. *Brain* **109**: 739-757.

Ben-Zvì, I. (1987) Optimal trajectory planning for robotic manipulators: a multi-grid approach. *M.Sc. Thesis*, Dept. of Appl. Math. and Computer Sci., The Weizmann Institute of Science, Rehovot, Israel.

Berkenblit, M.B., Feldman, A.G. and Fukson, O.Z. (1986) Adaptability of innate motor patterns and motor control mechanisms. *Behavioral Brain Sci.* 9: 585-638.

Bernstein, N. (1967) *The Coordination and Regulation of Movements*. Pergamon Press, Oxford.

Bizzi, E., Accornero, N., Chapple, W. and Hogan, N. (1984) Posture control and trajectory formation during arm movement. *J. Neurosci.* 4: 2738-2744.

Bizzi, E., Polit, A. and Morasso, P. (1976) Mechanisms underlying achievement of final head position. *J. Neurophysiol.* 39: 435-444.

Edelman, S. and Flash, T. (1987) A model of handwriting. *Biol. Cybern.* 57: 25-36.

Feldman, A.G. (1986) Once more on the equilibrium-point hypothesis for motor control. *J. Motor Behavior* 18: 17-54.

Flash, T. (1987) The control of hand equilibrium trajectories in multi-joint arm movements. *Biol. Cybern.* 57: 257-274.

Flash, T. (1988) Models of human arm trajectory control. *Proc. IEEE Eng. in Medicine and Biology, 10th Ann. Int. Conf.*, New Orleans.

Flash, T. and Hogan, N. (1985) The coordination of arm movements: an experimentally confirmed mathematical model. *J. Neurosci.* 7: 1688-1703.

Flash, T. and Mussa-Ivaldi, F.A. (1990) Human arm stiffness characteristics during the maintenance of posture. *Exp. Brain Res.*, In press.

Flash, T., Inzelberg, R. and Korczyn, A.D. (1990) Quantitative methods for the assessment of motor performance in Parkinson's disease. *Methodological Problems of Clinical Trials in Parkinson's Disease*. Demos Publ., New York, In press.

Flashner, H., Beuter, A. and Arabyan, A. (1988) Fitting mathematical functions to joint kinematics during stepping: implication for motor control. *Biol. Cybern.* 58: 91-99.

Flanagan, J.R. and Ostry, D.J. (1990) Trajectories of human multi-joint arm movements: evidence of joint level planning. *Experimental Robotics* (Edited by Hayward, V.). Springer–Verlag, New York.

Gelfand, L.M., Gurfinkel, V.S., Tsetlin, M.L. and Shik, M.L. (1971) Some problems in the analysis of movements. *Models of the Structural-Functional Organization of Certain Biological Systems* (Edited by Gurfinkel, L.M., Gurfinkel, V.S., Fomin, S.V. and Tsetlin, M.L.). MIT Press, Cambridge, MA.

Georgopoulos, A.P. (1986) On reaching. *Ann. Rev. Neurosci.* 9: 147-170.

Georgopoulos, A.P., Kalaska, J.F. and Massey, J.T. (1981) Spatial trajectories and reaction times of aimed movements: Effects of practice, uncertainty, and change in target location. *J. Neurophysiol.* 46: 725-743.

Gielen, C.C.A.M., Van der Heuvel, P.J.M. and Van der Gon, D.J.J. (1984) Modification of muscle activation patterns in fast goal-directed arm movements. *J. Motor Behavior* 16: 2-19.

Hasan, Z. and Karst, G.M. (1989) Muscle activity for initiation of planar, two-joint arm movements in different directions. *Exp. Brain Res.* 3: 651-655.

Henis, E. and Flash, T. (1989) Mechanisms subserving arm trajectory modification. *Perception* 18: 495.

Hogan, N. (1984) An organizing principle for a class of voluntary movements. *J. Neurosci.* 4: 2745-2754.

Hogan, N. (1985) The mechanics of multi-joint posture and movement. *Biol. Cybern.* 52: 315-331.

Hogan, N. (1988) Planning and execution of multi-joint movements. *Canadian J. Physiol. Pharmacol.* 66: 508-517

Hollerbach, J.M. (1982) Computers, brains and the control of movement. *Trends Neurosci.* 5: 189-192.

Hollerbach, J.M. and Atkeson, C.G. (1987) Deducing planning variables from experimental arm trajectories pitfalls and possibilities. *Biol. Cybern.* 56: 279-292.

Hollerbach, J.M. and Flash, T. (1982) Dynamic interactions between limb segments during planar arm movement. *Biol. Cybern.* 44: 67-77.

Houk, J.C. and Rymer, W.Z. (1982) Neural control of muscle length and tension. *Handbook of Physiology Vol. 2: Motor Control, Section 1: The Nervous System* (Edited by Brooks, V.B.), pp. 257-323, Williams and Wilkins, Baltimore.

Jordan, M.I. (1990) Indeterminate motor skill learning problems. *Attention and Performance*, XIII. (Edited by Jeannerod, M.). Lawrence Erlbaum, Hillsdale, NJ.

Jordan, M.I. and Rosenbaum, D.A. (1989) Action. *Foundations of Cognitive Science* (Edited by Posner, M.I.). MIT Press, Cambridge, MA.

Kawato, M., Madea, Y., Uno, Y. and Suzuki, R. (1990) Trajectory formation of arm movement by cascade neural network model based on the minimum torque change criterion. ATR Technical Report, TR-A-0056.

Keele, S.W. (1981) Behavioral analysis of movement. *Handbook of Physiology, Vol. 2: Motor Control, Section 1: The Nervous System* (Edited by Brooks, V.B.), pp. 225-260. Williams and Wilkins, Baltimore.

Kettner, R.E., Schwartz, A.B. and Georgopoulos, A.P. (1988) Primate motor cortex and free arm movements to visual targets in three dimensional space. III. Positional gradients and population coding of movement direction from various movement origins. *J. Neurosci.* **8**: 2938-2947.

Lacquaniti, F. (1989) Central representations of human limb movement as revealed by studies of drawing and handwriting. *Trends in Neurosci.* **12**: 287-291.

Lacquaniti, F., Terzuolo, C. and Viviani, P. (1983) The law relating the kinematics and figureal aspects of drawing movements. *Acta Psychol.* **54**: 115-130.

Massey, J.T., Schwartz, A.B. and Georgopoulos, A.P. (1986) On information processing and performance a movement sequence. *Exp. Brain Res. Suppl.* **15**: 242-251.

Morasso, P. (1981) Spatial control of arm movements, *Exp. Brain Res.* **42**: 223-227.

Morasso, P. and Mussa-Ivaldi, F.A. (1982) Trajectory formation and handwriting: a computational model. *Biol. Cybern.* **45**: 131-142.

Munhall, K. and Lofquist, A. (1987) Gestural aggregation in speech. *PAW Rev.* **2**: 13-17.

Mussa-Ivaldi, F.A. (1988) Do neurons in the motor cortex encode movement directions? An alternative hypothesis. *Neurosci. Lett.* **91**: 106-111.

Mussa-Ivaldi, F.A., Hogan, N. and Bizzi, E. (1985) Neural, mechanical and geometric factors subserving arm posture in humans. *J. Neurosci.* **5**: 2732-2743.

Nelson, W.L. (1983) Physical principles of economics of skilled movements. *Biol. Cybern.* **46**: 135-147.

Patriarco, A.G., Mann, R.W., Simon, S.R. and Mansom, J.M. (1981) An evaluation of the approaches of optimization models in the prediction of muscle forces during human goal. *J. Biomechanics* **14**: 513-525.

Saltzman, E. (1979) Levels of sensorimotor representation, *J. Math. Psychol.* **20**: 96-1063.

Schneider, K. and Zernicke, R.F. (1989) Jerk-cost modulations during the practice of rapid arm movements. *Biol. Cybern.* **60**: 221-230.

Soechting, J.F. and Lacquaniti, F. (1981) Invariant characteristics of a pointing movement in man. *J. Neurosci.* **1**: 710-720.

Soechting, J.F. and Lacquaniti, F. (1989) An assessment of the existence of muscle synergies during load perturbation and intentional movements of the human arm. *Exp. Brain Res.* **3**: 535-548.

Soechting, J.F. and Terzuolo, C.A. (1986) An algorithm for the generation of curvilinear wrist motion in an arbitrary plane in three dimensional space. *Neurosci.* **19**: 1393-1405.

Stein, R.B., Oguztoreli, M.N. and Capaday, C. (1985) What is optimized in muscular movements? *Human Muscle Performance: Factors Underlying Maximal Performance* (Edited by Jones, N.L., McCartney, N. and McComas, A.J.). Kinetic Publishers, New York.

Tikhonov, A.N. and Arsenin, V.Y. (1977) *Solutions of Ill-Posed Problems*, W.H. Winstron, Washington, D.C.

Uno, Y., Kawato, M. and Suzuki, R. (1989) Formation and control of optimal trajectory in multijoint arm movement: minimum torque-change model. *Biol. Cybern.* **61**: 89-101.

Van Sonderen, J.F., Van der Gon, J.J.D. and Gielen, C.C.A.M. (1989) Conditions determining early modifications of motor programs in response to changes in target location. *Exp. Brain Res.* **71**: 320-328.

Viviani, P. and Cenzato, M. (1985) Segmentation and coupling in complex movements. *J. Exp. Psychol.: Human Perception and performance* **11**: 828-845.

Viviani, P. and Terzuolo, C. (1980) Space-time in motor skills. In *Tutorials in Motor Behavior* (Edited by Stelmach, G.E. and Requin, J.), pp. 525-533. Elsevier North-Holland Publishing Co., Amsterdam.

Viviani, P. and Terzuolo C. (1982) Trajectory determines movement dynamics. *J. Neurosci.* **7**: 431-437.

Wann, J.P., Nimmo-Smith, I. and Wing, A.M. (1988) Relation between velocity and curvature in movement: equivalence and divergence between a power law and the minimum-jerk model. *J. Exp. Psycol.: Human Perception and Performance* **14**: 622-637.

Wood, J.E., Meeke, S.G. and Jacobson, S.C. (1989) Quantitation of human shoulder anatomy for prosthesis arm control. Part II. Anatomy matrices. *J. Biomechanics* **22**: 309-326.

Xu, Y., Bennet, D.J., Hollerbach, J.M. and Hunter, I.W. (1989) Wrist-airjet system for identification of joint mechanical properties of the unconstrained human arm. *Soc. Neurosci. Abstr.* **15**: 396.

CHAPTER 18

The Activation of Mono- and Bi-Articular Muscles
in Multi-Joint Movements

Stan Gielen, Gerrit-Jan van Ingen Schenau, Toine Tax, Marc Theeuwen

18.1 Introduction

Biological limbs are designed in a way that gives them large flexibility to make a wide variety of complex movements. This flexibility is provided by the relatively large number of joints and of mono- and poly-articular muscles. This large set of joints and muscles has raised the issue of coordination for simple movements, for which the neuro-muscular system is redundant with respect to the number of degrees of freedom required for these movements. Several suggestions have been proposed in the literature. Most of these suggestions are based on the notion of imposing additional constraints. Among these are the minimization of total muscle torque (Yeo, 1976) and the minimization of fatigue (Dul, 1984a, 1984b). It has been shown, however, that most of these criteria cannot explain the observed activation of arm muscles in man (Gielen et al., 1988a). A more detailed review of alternative hypotheses on the control of redundant neuromuscular systems and a comparison between theoretically predicted and observed activation patterns can be found in Gielen et al. (1988a) and in van Zuylen et al. (1988).

18.2 Statement of the Problem

In this chapter we will focus on the contribution of mono- and bi-articular muscles in the control of force and position. The main issue of this chapter is illustrated in Figure 18.1. Let us assume that the subject's arm in Figure 18.1 is moving in the horizontal plane by rotations of elbow and shoulder joints only. Since torque is defined as the vector cross-product between the

lever arm and the force vector there is a unique relationship between the external force at the hand and the torque in each of the joints. An isometric force in the direction indicated in Figure 18.1 requires a flexion torque in both elbow and shoulder.

For a subject to make a movement producing a force against the external force in Figure 18.1, an extension of the elbow and a movement in anteflexion direction of the shoulder is required. The point is that by making an isotonic movement against the external force the change in elbow joint angle (extension) is opposite to the direction of the torque in the elbow joint. This implies that the elbow joint is dissipating work, rather than contributing work. Notice that the total work done by the hand is the sum of the work done at the elbow and at the shoulder:

$$\Delta W = T_{el}\, \Delta\phi_2 + T_{sh}\, \Delta\phi_1 \qquad (18.1)$$

Consequently, because of the negative work delivered in the elbow, the work done at the shoulder exceeds the work done at the hand. For some movement directions the work done in one joint may be about twice as large as the work done by the hand (Gielen and van Ingen Schenau, 1990). This example shows that if biological limbs would be supplied with mono-articular muscles only, movement control would be rather inefficient since, for a large number of movements, some mono-articular muscles would dissipate a large part of the work delivered by other muscles. In this chapter we will show that with bi-articular muscles the negative work in a joint can be

Multiple Muscle Systems: Biomechanics and Movement Organization
J.M. Winters and S.L-Y. Woo (ed.), © 1990 Springer-Verlag, New York

avoided, thus making movement coordination more efficient.

The different constraints on the direction of torque and change in joint angle can be understood easily. Let us define a Cartesian x-y coordinate system with the origin at the shoulder (see Figure 18.1). The relationship between hand position (x,y) and joint angles (ϕ_1, ϕ_2) in the shoulder and elbow, respectively, is then given by

$$\begin{bmatrix} x \\ y \end{bmatrix} = \begin{bmatrix} l_1 \cos(\phi_1) + l_2 \cos(\phi_1 + \phi_2) \\ l_1 \sin(\phi_1) + l_2 \sin(\phi_1 + \phi_2) \end{bmatrix} \quad (16.2)$$

In this equation the length of the upper arm and forearm is represented by l_1 and l_2. For small displacements the relation between changes in joint angles $\Delta\phi_1$ and $\Delta\phi_2$ is related to the displacement $(\Delta x, \Delta y)$ in world space by the equation

$$\begin{bmatrix} \Delta x \\ \Delta y \end{bmatrix} = J \begin{bmatrix} \phi_1 \\ \phi_2 \end{bmatrix} \quad (18.3)$$

where J is the Jacobian given by

$$J = \begin{bmatrix} -l_1 \sin(\phi_1) - l_2 \sin(\phi_1 + \phi_2) & -l_2 \sin(\phi_1 + \phi_2) \\ l_1 \cos(\phi_1) + l_2 \cos(\phi_1 + \phi_2) & l_2 \cos(\phi_1 + \phi_2) \end{bmatrix}$$

$$(18.4)$$

The relationship between a change in torque ΔT in the joint space and a change in force ΔF at the hand is given by the relation

$$\Delta T = J^T \Delta F \quad (18.5)$$

where J^T is the transpose of the matrix J. Note that the Jacobian (Eq. 18.4) is not symmetric in general. Therefore, J and its transpose may be different. Also note that even when the displacement Δx and the change in force ΔF are in the same direction, the resulting change in joint angle $\Delta\phi$ and change in torque ΔT may be in different directions. Even the sign of the respective components may be different, such as for the elbow in the example presented in Figure 18.1.

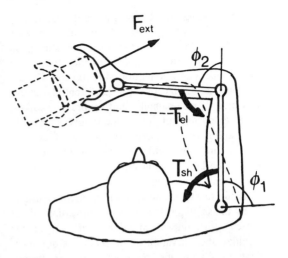

Figure 18.1: Illustration of torques and changes in joint angles for isotonic contractions against an external force in a horizontal plane. Note that the direction of torque and change in joint angle are not the same for the elbow joint. T_{el} and T_{sh} refer to the flexion torques in elbow and shoulder joints, respectively, counteracting an external force F_{ext}; ϕ_1 and ϕ_2 refer to the joint angles at the shoulder and elbow, respectively.

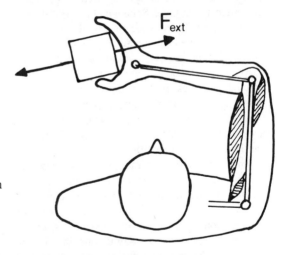

Figure 18.2: This illustrates how mono- and bi-articular muscles may be activated for movement in a particular direction acting against an external force F_{ext}, in a more efficient way than is possible with mono-articular muscles only. Note, that the mono-articular elbow flexors, which are lenthening during the movement, should not be activated in order to prevent negative work (energy dissipation!) by these muscles.

The dissipation of energy could be easily remedied by incorporating bi-articular muscles (see Figure 18.2). For the example shown in Figure 18.2 the flexion torque in the elbow joint can be obtained by activation of m. biceps. The activation of m.biceps is preferential above activation of m. brachialis and m. brachioradialis, since extension of the elbow gives rise to lengthening of the mono-articular m. brachialis and m. brachioradialis, thereby causing negative work done by these muscles. On the other hand, lengthening of biceps, which would otherwise arise due to the elbow extension, is compensated for by the shoulder flexion. Since the amplitude of the elbow torque is smaller than that at the shoulder, the activation of m. biceps is not sufficient to obtain the required shoulder torque. Therefore, the contribution of a shoulder flexor is required to make up the total shoulder torque. The mono-articular elbow extensor (possibly the medial or lateral part of m. triceps) may be necessary to cause the elbow extension. In this way the same work is done by the hand in a much more efficient way. Such a role for mono- and bi-articular muscles was previously reported for cycling movements (van Ingen Schenau et al., 1989).

This particular role for bi-articular muscles is helpful for movements when lengthening of active mono-articular muscles would occur. For other movement directions, when the active mono-articular muscles are shortening the bi-articular muscles may be lengthening and in these conditions the same reasoning argues against the activation of bi-articular muscles (see Gielen and van Ingen Schenau, 1990).

This chapter focuses mainly on two questions concerning the coordination of mono- and bi-articular muscles. The first issue concerns inter-muscle coordination and relates to the coordination between different muscles in isometric and isotonic contractions in different directions. The second issue concerns the activation of motor units within a muscle in isometric and isotonic contractions, and in particular, the contribution of recruitment and firing rate in isometric and isotonic contractions.

18.3 Methods

Muscle activation was studied in isometric and isotonic contractions with surface electrodes and with intramuscular electrodes. The directions of both force and movement were always in the horizontal plane. In all conditions force in the vertical direction (upwards and downwards) was measured and was kept zero. This is important since m. biceps contributes to torque in exorotation at the shoulder and since the activation of m. biceps depends on the amount of torque in both exorotation and endorotation of the humerus (Buchanan et al., 1986; ter Haar Romeny et al., 1984).

The relative activation of muscles for isometric and isotonic contractions in different direction was investigated with surface electrodes. *EMG* activity was recorded from the mono-articular muscles m. brachialis and m. brachioradialis and from the bi-articular m. biceps (caput longum). In order to measure the muscle activation in isometric conditions the wrist of the subject was fixated to a device which measured force with a resolution of 0.1 N. For isotonic movements the wrist was fixated to a rope running over a pulley. In this way a constant force was applied to the wrist of the subject. Movement velocity was constant and low (e.g. about 5 cm/s). As a result inertial force components were negligibly small. In these experiments the joint angles in shoulder and elbow (see Figure 18.1) were both 90°.

In addition, the activation of mono- and bi-articular arm muscles in isometric contractions was investigated with intramuscular electrodes. The comparison of surface *EMG* data with motor-unit recruitment thresholds can be done with some reasonable assumptions. The procedure is explained in the following.

With intramuscular electrodes the recruitment threshold of motor-units was determined for force in various directions. It is generally accepted that a motor-unit is recruited when the total synaptic input to the motoneuron exceeds a particular threshold (Henneman, 1957, 1981). This idea is related to the "Size Principle" (Henneman, 1957) which assumes a homogeneous input to all motoneurons of a particular motoneuron pool and which states that the recruitment order in this motoneuron pool is related to the size of the motoneuron soma. As a consequence, any combination of forces in the horizontal plane that are present when a motor-unit is recruited can be considered to reflect a particular, constant input to the motoneuron pool. Using surface electrodes, the amount of surface *EMG* activity is measured for constant forces in various directions. The difference between these approaches is that with intramuscular electrodes a ***constant output*** signal of the motoneurons of a particular muscle is obtained for various forces, whereas with surface electrodes a ***variable output*** activity is obtained for a constant force at the wrist. Therefore, the comparison of recruitment data of motor units with surface *EMG* activity requires some explanation. The force direction θ_{min}, where a motor unit reveals the lowest recruitment threshold F_{min}, corresponds to the force direction of maximal input, which is the direction with the largest surface *EMG* activity for constant forces in different directions. Assuming that the *EMG*–force relationship is more or less linear, which is a good approximation for forces in the range below 30% of maximum voluntary contraction (*MVC*), the recruitment threshold $F(\theta)$ can be compared with surface *EMG* activity for the direction θ by comparing the ratio of $F_{min}/F(\theta)$ with the ratio of *EMG* activity relative to the maximum *EMG* activity for that force for all directions.

The activation of motor units in isometric and isotonic contractions was investigated with intramuscular wire electrodes inserted in m. biceps and m. brachioradialis. Intramuscular motor-unit recordings were obtained after insertion of nylon-coated fine wire electrodes (diameter 25 μm, California Fine Wire Company) with a hollow needle (diameter 0.4 mm) which was extracted after insertion, leaving the wires in the muscle. A detailed analysis of the experimental procedures can be found in van Zuylen et al. (1988) and Tax et al. (1989). In the motor-unit experiments the joint angles at shoulder and elbow (see Figure 18.1) were 0° and 100°, respectively. Motor-unit behavior was studied in isometric contractions and in isotonic contractions. In the latter conditions both lengthening and shortening contractions were studied. In order to eliminate any effects of the force–velocity relationship, the movements were made as slowly as possible. Usually the velocity was 3 deg/s, which corresponds to a velocity of about 1.3% of the rest length per second. At these velocities the effect of the force–velocity relation-

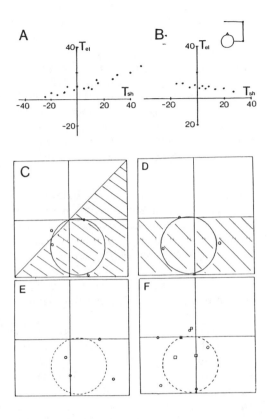

Figure 18.3: A and B show recruitment thresholds of a motor unit in m. brachioradialis (A) and m. biceps (B). Units along the axis refer to flexion (positive) and extension (negative) torque (Nm) in the elbow T_{el} and shoulder T_{sh}. When the data points in A and B are fitted by a straight line, the transformation of this line in torque space to the force space gives the curves in C and D. The open circles refer to data obtained with surface *EMG* during isometric contractions. The hatched areas in C and D correspond to regions where the muscle would be shortening (see text). Data in Figure 18.3E and F refer to *EMG*-activity measured during isotonic movements in the same (squares) or opposite direction (circles) as the external force. In all parts torque and force refer to the torque and force exerted by the limb as a result of muscle contraction. Therefore, the external force (torque) acting on the limb is pointing in opposite direction. Also note that scaling of surface *EMG* is done for m. biceps and m. brachioradialis separately, such that a quantitative comparison of the relative activation of these two muscles in different conditions cannot be done with the data shown here. The inset shows the position of the subject in the experimental set-up.

ship is very small.

In all motor-unit experiments the right arm of the subject was 80° abducted. Movements were in the range between 90° and 110° elbow flexion. All data were collected when the arm moved at constant velocity through the position of 100° elbow flexion. In this way any artefacts in our results due to the force–length relationship of muscles could be eliminated. In these experiments the shoulder of the subject was fixated eliminating any movements of the upper arm.

18.4 Results

The motor-unit recruitment behavior for isometric forces in various directions is shown in Figure 18.3. Recruitment thresholds of motor units in m. brachioradialis and m. biceps are shown in Figures 18.3A and B, respectively. As expected, motor units in these muscles are recruited mainly for flexion torques in the elbow joint. These data are very similar to those reported earlier by Jongen (1989).

Using *Eq. 18.5* the recruitment data in Figure 18.3A and B were plotted as a function of force at the hand. Moreover, the data were converted such that they could be compared with data obtained with surface electrodes using the procedures outlined in Section 18.3. In this conversion, it was assumed that the length of upper arm and forearm was the same. The results are shown in Figure 18.3C–D. Since different motor units do have different recruitment thresholds the amplitude of the response in Figure 18.3C–D was normalized to 100% for the activation at the direction with the lowest recruitment threshold. The recruitment data in Figure 18.3A–B for brachioradialis and biceps seem to fall along straight lines with a slightly different slope. This difference in slope becomes visible in Figure 18.3C-D in the different orientation of the curves (circles).

Also shown in Figures 18.3C-D are data obtained with surface electrodes. For the direction for isometric force with the largest *EMG* activity, the *EMG* activity was normalized to 100%. Considering that the motor-unit data and surface-*EMG* data were obtained in different experiments (from the same subject), both data are in good agreement. Notice that the activation of m. brachioradialis is mainly for force directions that would give rise to muscle *shortening* if the arm was allowed to move. These directions correspond with forces to the lower-right of the diagonal in Figure 18.3C. A significant part of the activation of m. biceps is for forces that would produce elbow flexion and shoulder flexion. If the lever arms of m. biceps at the elbow and shoulder would be the same, the force directions with a negative F_y component in Figure 18.3D would correspond to directions where m. biceps is shortening.

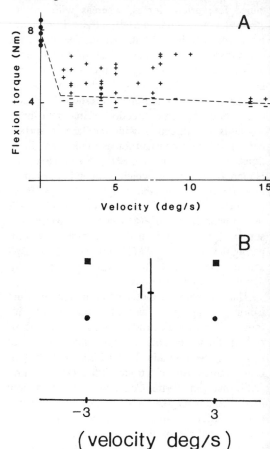

Figure 18.4: *Part A shows the absence ('−') or presence ('+') of motor-unit activity for different torques at the elbow as a function of shortening velocity. Part B shows averaged recruitment thresholds for motor units in m. biceps (circles) and m. brachioradialis (squares) for isometric contractions (v = 0) and for voluntary isotonic flexion (v = 3 deg/s) and extension (v = −3 deg/s) movements. (Part A reprinted with permission from Tax et al., 1989.)*

EMG activity recorded during isotonic contractions against a constant preload (20 N at the hand) is shown in Figure 18.3E–F. Comparing the data with those in Figure 18.3C–D shows that the main orientation of the data in E and F corresponds roughly with those in C and D, suggesting that the relative activation of m. brachioradialis and m. biceps is approximately the same in isometric and isotonic contractions. The variability in *EMG* activity did not allow to make explicit statements on whether the data, obtained for isotonic contractions, deviated significantly from the data obtained with isometric contractions. However, the data in Figure 18.4 and Figure 18.5 clearly indicate that there is a different activation in the two conditions (see Section 18.5 for discussion).

The data in Figure 18.3E–F suggest that the relative activation of m. biceps and m. brachioradialis during movements in the same direction and opposite to the external force may be different. The circles in Figure 18.3E–F refer to the relative amount of activation during movements in a direction opposite to the external force. During movements in the same direction as the external force the amount of *EMG* activity in both muscles decreased considerably in comparison to the amount of *EMG* activity for movements in the direction opposite to the external force. For m. biceps the relative variation of *EMG* activity as a function of movement direction (or with other words, the orientation of the curves in Figure 18.3D) was not different for movements opposite to or in the same direction as the external force. In m. brachioradialis small variations in *EMG* activity could be observed for movements in the same direction as the external force, but they were too small to make accurate estimates of the *EMG* activity. Presumably, the decreased *EMG* activity at the same force can be explained with the force–velocity relationship, which gives a larger force for small lengthening contractions at constant activation for the velocities used in this experiment (about 5 cm/s, corresponding to about 5% of the rest length per second).

Recruitment data obtained in m. biceps and m. brachioradialis are shown in Figure 18.4. The main finding was that the recruitment thresholds of motor units is not the same for isometric and isotonic contractions. For isotonic contractions the recruitment threshold of motor units in m. biceps is decreased (see Figure 18.4A). It is decreased considerably even for slow movements. It appeared to be almost impossible to make voluntary movements at constant velocities below 3 deg/s. However, recruitment thresholds were determined at velocities exceeding 3 deg/s. For higher velocities the recruitment threshold decreased very slowly, quantitatively in agreement with the force–velocity relationship for arm flexor muscles (Jorgensen, 1976). Extrapolation of the recruitment threshold to velocity zero suggests a clear discontinuity with respect to the isometric recruitment threshold. This sudden decrease of recruitment threshold at slow velocities indicates a discontinuity at velocity zero (see Tax et al., 1989). For motor units in m. brachioradialis the recruitment threshold was higher for isotonic contractions than it was for isometric contractions. Just as for motor units in m. biceps, all data for m. brachioradialis suggested a discontinuity at velocity zero (see Tax et al., 1990).

Since the absolute changes in recruitment threshold were proportional to the isometric recruitment threshold (Tax et al., 1989) we have plotted average values of the recruitment threshold for movements relative to the isometric recruitment threshold. Figure 18.4B clearly shows that during isometric contractions the recruitment threshold of motor units in m. biceps is higher than that during voluntary movements. The recruitment threshold for voluntary flexion and voluntary extension was not significantly different. For the recruitment behavior of motor units in m. brachioradialis just the opposite was found. Here the recruitment threshold for isometric contractions is lower than for voluntary movements. Again, there was no clear difference between the recruitment threshold for flexion or extension movements. On the average the recruitment threshold for movements is decreased by a factor of 0.77 (standard error = 0.02) for m. biceps and is increased by a factor of 1.29 (standard error =0.03) for m. brachioradialis and m. brachialis. These data are averaged for three subjects. It should be mentioned that there are small quantitative differences for data from different subjects (see Tax et al., 1989, 1990).

Because of the different effect of movement on the recruitment threshold of motor units in m. biceps and m. brachioradialis, motor units in different muscles may show reversal of recruitment order in isometric and isotonic contractions.

However, the recruitment order of motor units within the same motor-unit population was the same both in isometric and isotonic conditions.

The firing-frequency behavior of motor units during isometric and both isotonic flexion and extension movements is shown in Figure 18.5. In Figure 18.5A it is shown that the brachioradialis motor unit is recruited at lower torque levels

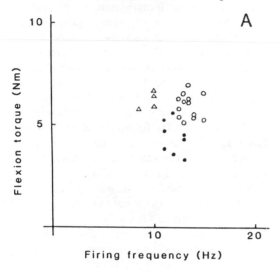

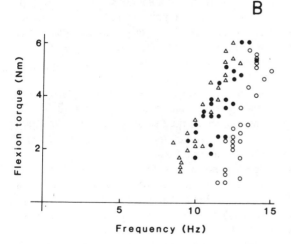

Figure 18.5: This shows the relationship between flexion torque in the elbow and firing frequency of a motor unit in m. brachioradialis (A) and in m. biceps (B). Filled circles refer to data obtained at isometric contractions, open circles refer to extension movements, and squares refer to flexion movements. (Figure 18.5B reprinted with permission from Tax et al., 1989.)

during isometric contractions than during isotonic movements, in agreement with the data in Figure 18.4. In addition, the firing frequency is slightly higher during isotonic flexion movements and lower during isotonic extension movements than it is during isometric contractions.

Figure 18.5B shows the data for a motor unit in m. biceps. Clearly the motor unit is recruited at higher torques during isometric contractions than for isotonic contractions, in agreement with the conclusions from Figure 18.4. The main point of Figure 18.5B is that at the same torque the firing rate is different for isotonic flexion movements and for isometric contractions. At the same torque the motor unit has a higher firing rate for flexion movements compared to the firing rate at the same isometric torque.

The data in Figure 18.4 and Figure 18.5 show that the activation of the mono-articular brachioradialis and the bi-articular m. biceps varies in a different way for isometric and isotonic contractions. Moreover, they show that the firing rate of motor units can be modulated independently from the recruitment such that the gradation of force by recruitment and firing rate is different in isometric and isotonic contractions.

18.5 Discussion

The main finding of this paper is that mono- and bi-articular muscles in the human arm have a different role in isometric and isotonic contractions and that the relative activation of mono- and bi-articular muscles can, at least qualitatively, be understood based on some simple theoretical considerations regarding the efficiency of multi-limb movements. The changes of motor-unit recruitment threshold and firing frequency at recruitment related to isometric or isotonic contractions are as yet not well understood.

As explained in the results, the variability in *EMG* activity did not allow us to decide whether the data obtained for isometric and isotonic contractions in Figure 18.3C-F deviated significantly. However, the motor-unit data, obtained in Figure 18.4 show that the recruitment threshold of motor units in the mono-articular brachioradialis and the bi-articular biceps changes in a different way for isometric and isotonic contractions. Owing to mechanical restrictions in our set-up we could investigate motor-unit behavior in isometric and isotonic contractions in one direction only (in flexion/extension direction of the elbow). Since for this direction the relative activation of m. brachioradialis and m. biceps changes in a different way for isometric and isotonic contraction, we believe that this reflects a more general phenomenon and we speculate that the activation for both m. brachioradialis and m. biceps is different in isometric and isotonic contractions. We are collecting more data to verify this prediction.

Some caution is warranted with respect to the interpretation of the data obtained with surface EMG electrodes. In the first place, it has been shown that within a muscle there are several groups of motor units (ter Haar Romeny et al., 1982), each with a different activation. Since surface electrodes measure the weighted activity of a relatively large group of motor units, one cannot determine how different motor-unit populations contribute to the surface *EMG* activity. Based on the location of the different populations of motor units (see ter Haar Romeny et al., 1984; van Zuylen et al., 1988), we are confident that the *EMG* activity measured with the *EMG* surface electrodes reflects the activity of the largest motor-unit populations (see van Zuylen et al., 1988), which is the same type of motor units that was recorded from with the intramuscular wire electrodes.

A second methodological issue, which should be raised with respect to the interpretation of surface *EMG* data as a measure of muscle activation and muscle force, concerns the fact that the two force-grading mechanisms (i.e. recruitment and firing frequency of motor units) do not contribute equally to the amplitude of the surface *EMG*. It is well known that the surface *EMG* activity is, for the most part, determined by recruitment. Therefore, if recruitment thresholds and firing frequency of motor units can be modified independently, as shown by Tax et al. (1989, 1990; see also Conway et al., 1988; Crone et al., 1988) it becomes very difficult to interpret the surface *EMG* activity quantitatively in terms of activation of the motoneuron pool or of the muscle. However, despite these problems, our data show that mono- and bi-articular muscles do have a different role in movements and that the relative activation of mono- and bi-articular muscles is different in isometric and isotonic (both shortening and lengthening) contractions, as discussed previously.

18.5.1 Comparison With Other Experimental Observations

The important role of bi-articular muscles has been described earlier for jumping, skating and bicycling by van Ingen Schenau et al. (1987), Bobbert and van Ingen Schenau (1988), van Ingen Schenau (1989), and Chapter 41 (van Ingen Schenau et al.)). A review of the different role of mono- and bi-articular muscles can be found in a recent target article by van Ingen Schenau (1989) and the commentaries included in that volume. The observation that mono- and bi-articular muscles may receive a task-dependent activation has been reported before by others as well. For example, Nardone and Schieppati (1988) found that during active lengthening of the triceps surae, as when gradually yielding to a dorsiflexing load, the soleus muscle is silenced whereas the gastrocnemius is active. During active shortening contractions against an external load both muscles are active. This observation fits nicely with the ideas and the analysis presented in this chapter.

A need for bi-articular muscles has been suggested earlier by Hogan (1985). According to his ideas, bi-articular muscles are necessary in order to obtain an isotropic end-point stiffness in the workspace (instead of in the jointspace). This idea is not in conflict with the ideas outlined in this chapter. Instead, it puts another constraint on the activation of mono- and bi-articular muscles. The constraint that follows from our results follows from the efficient activation of mono- and bi-articular muscles in movements against an external force. The constraint predicted by Hogan (1985; Chapter 9) is related to end-point stiffness. This constraint should be considered for isometric contractions. Whether it is valid for movements as well is an open question since it is still unclear whether the same joint stiffness is found for moving and stationary limbs. End-point stiffness has been investigated mainly in static conditions (e.g. see Mussa-Ivaldi et al., 1985; Vincken et al., 1983); Flash has estimated stiffness for simple dynamic movements (Chapter 17). It has been postulated that the same principles underlie the generation of movements (see Feldman, 1986, and Chapter 12 (Feldman et al.)). However, experimental data about stiffness regulation during movements are scarce. Our motor-unit data (see Figures 18.4 and 18.5), which show that the activation of mono- and bi-articular muscles is different in isometric and isotonic contractions, suggest that stiffness may be different.

As was explicitly mentioned before, all movements were made in a horizontal plane and movement velocities were relatively low. This was done in order to be able to avoid variations in gravitational force. In a study by Karst and Hasan (1989) *EMG* activity was recorded from flexor and extensor muscles in the shoulder and elbow joint [see also Chapter 16 (Karst and Hasan)]. The main result of their investigations was that the activation pattern which initiated arm movements could be predicted based neither by the changes in joint angles nor the (isometric) force that was necessary to initiate these movements. Unfortunately, these authors pooled all flexor muscles (both mono- and bi-articular muscles), which makes it impossible to interpret their data in the present context. Moreover, the movement speed in their experiments was such that inertial components could not be neglected, which requires some further extension of the ideas presented in this chapter. A way to do this may be to incorporate the ideas suggested by Zajac and Gordon (1989) [see also Chapter 8 (Zajac and Winters)].

An interesting issue for further research concerns the stage in the functional hierarchy of motor programming where the coordination of mono- and bi-articular muscle activation takes place. Recently, it has been demonstrated that the relative activation of muscles in response to external perturbations is different in the short-latency and long-latency reflex components (Lacquaniti and Soechting, 1986; Gielen et al., 1988b). The short-latency reflex was found only in muscles that were stretched, which is compatible with the notion that the short-latency reflex prevents muscle-yielding due to the sudden break of actin-myosin bonds (Allum and Mauritz, 1984). The long-latency reflex component (latency about 50 ms in elbow flexors) gives rise to the activation of a wider set of muscles, some of which are not stretched by the perturbation. However, the activation of these additional muscles is necessary for the coordinated response to the perturbation (Gielen et al., 1988b). This result, in our view, suggests that the coordination center should be found at a relatively peripheral level in the motor-programming scheme.

Acknowledgements

Part of this research was supported by the European Community in the ESPRIT 2, BASIC RESEARCH Program MUCOM, project number 3149.

References

Allum, J.H.J. and Mauritz, K.-H. (1984) Compensation for intrinsic muscle stiffness by short-latency reflexes in human triceps surae muscles. *J. Neurophysiol.* **52**: 797-816.

Bobbert M.F. and van Ingen Schenau G.J. (1988) Coordination in vertical jumping. *J. Biomech.*, **21**: 249-262.

Buchanan, T.S., Almdale D.P.J., Lewis J.L., Rymer W.Z. (1986) Characteristics of synergistic relations during isometric contractions of human elbow muscles. *J. Neurophysiol.* **56**: 1225-1241.

Conway, B.A., Hultborn,H., Kiehn, O. and Mintz,I. (1988) plateau potentials in α-motoneurones induced by intravenous injection of L-DOPA and clonidine in the spinal cat. *J. Physiol.* **405**: 369-384.

Crone C., Hultborn,H., Kiehn,O., Mazieres,L. and Wistrom,H. (1988) Maintained changes in motoroneuroneal excitability by shortlasting synaptic inputs in the spinal cat. *J. Physiol.* **405**: 321-343.

Dul, J., Johnson J.E., Shiavi, R. and Townsend, M.A. (1984a) Muscular synergisM. -I. On the criteria for load sharing between synergistic muscles. *J. Biomech.* **17**: 663-673.

Dul, J., Johnson J.E., Shiavi, R. and Townsend, M.A. (1984b) Muscular synergisM.- II A minimum fatigue criterion for load sharing between synergistic muscles. *J. Biomech.* **17**: 675-684.

Feldman, A.G. (1986) Once more on the Equilibrium-point hypothesis (λ-model) for motor control. *J. Motor Behav.* **18**: 17-54.

Gielen C.C.A.M., van den Oosten, K. and Pull ter Gunne, F. (1985) Relation between EMG activation patterns and kinematic properties of aimed arm movements. *J. Motor Behav.*, **17**: 421-442.

Gielen S.C.A.M., van Zuylen E.J., Denier van der Gon J.J. (1988a) Coordination of arm muscles in simple motor tasks. In: *Int. Series on Biomechanics,* Vol. 7A. Eds.: G. de Groot, A.P. Hollander, P.A. Huying, G.J. van Ingen Schenau. Free University Press, Amsterdam, pp. 155-166.

Gielen C.C.A.M., Ramaekers,L. and van Zuylen,E.J. (1988b) Long-latency stretch reflexes as co-ordinated functional responses in man. *J. Physiol.* **407**: 275-292.

Gielen, C.C.A.M. and van Ingen Schenau G.J. (1990) The constrained control of force and position by multi-link manipulators. *IEEE Trans. Syst., Man, and Cybern.*, submitted for publication.

Haar Romeny, B.M. ter, Denier van der Gon, J.J. and Gielen C.C.A.M. (1982) Changes in recruitment order of motor units in the human biceps muscle. *Exp. Neurology* **78**: 360-368.

Haar Romeny, B.M. ter, Denier van der Gon J.J. and Gielen C.C.A.M. (1984) Relation of the location of a motor unit in human biceps muscle and its critical firing levels for different tasks. *Exp. Neurol.* **85**: 631-650.

Hasan, Z. and Karst G.M. (1989) Muscle activity for initiation of planar, two-joint arm movements in different directions. *Exp. Brain Res.* **76**: 651-655.

Henneman, E. (1957) Relation between size of neurons and their susceptibility to discharge. *Science* **126**: 1345-1347.

Henneman, E. (1981) Recruitment of motoneurons: the size principle. In: *Motor Unit Types, Recruitment, and Plasticity in Health and Disease. Progress in Clinical Neurophysiology*, **Vol. 9**: pp. 26-60. Karger, Basel.

Hogan, N (1985) The mechanics of multi-joint posture and movement control. *Biol. Cybern.* **52**: 315-331.

Jongen, H.A.H. (1989) Theories and experiments on muscle coordination during isometric contractions. Thesis, University of Utrecht, The Netherlands.

Jorgensen, K. (1976) Force-velocity relationship in human elbow flexors and extensors. In: Komi (Ed) *Intern. Series on Biomechanics*, **Vo. 1A**, University Park Press, Baltimore.

Lacquaniti,F. and Soechting J.F. (1986) Responses of mono- and bi-articular muscles to load perturbations of the human arm. *Exp. Brain Res.* **65**: 135-144.

Loeb, G.E. (1985) Motoneurone task groups: coping with kinematic heterogeneity. *J. Exp. Biol.* **115**: 137-146.

Mussa-Ivaldi, F.A., Hogan N., and Bizzi E. (1985) Neural, mechanical, and geometrical factors subserving arm posture in humans. *J. of Neurosci.* **5**: 2732-2743.

Nardone, A. and Schieppati,M. (1988) Postural adjustments with voluntary contraction of leg muscles in standing man. *Exp. Brain Res.*, **69**: 469-480.

Tax, A.A.M., Denier van der Gon J.J., Gielen C.C.A.M., and van den Tempel C.M.M. (1989) Differences in the activation of m. biceps brachii in the control of slow isotonic movements and isometric contractions. *Exp. Brain Res.*, **76**: 55-63.

Tax, A.A.M., Denier van der Gon, J.J., Erkelens. C.J. (1990) Differences in coordination of elbow flexor muscles in force tasks and in movement tasks. *Exp. Brain Res.*, accepted for publication.

van Ingen Schenau, G.J., Bobbert M.F. and Rozendal R.H. (1987) The unique action of bi-articular muscles in complex movements. *J. Anatomy*, **155**: 1-5.

van Ingen Schenau, G.J. (1989) From rotation to translation: Constraints on multi-joint movements and the unique action of bi-articular muscles. *Hum. Mov. Sci.* **8**: 301-337.

van Zuylen, E.J., Gielen, C.C.A.M. and Denier van der Gon J.J. (1988) Coordination and inhomogeneous activation of human arm muscles during isometric torques. *J. Neurophysiol.* **60**: 1523-1548.

Vincken M.H., Gielen,C.C.A.M. and Denier van der Gon, J.J. (1983) Intrinsic and afferent components in apparent muscle stiffness in man. *Neuroscience* **9**: 529-534.

Yeo, B.P. (1976) Investigations concerning the principle of minimal total muscular force. *J. Biomech.* **9**: 413-416.

Zajac, F.E. and Gordon, M.E. (1989) Determining muscle's force and action in multi-articular movements. *Exerc. Sport Sci. Rev.* **17**: 187-230.

CHAPTER 19

Optimized Strategies for Scaling Goal-Directed Dynamic Limb Movements

Amir H. Seif-Naraghi and Jack M. Winters

19.1 Introduction

A recurring theme of the previous five chapters has been a consideration of control strategies for goal–directed movements involving one of the upper limbs. In each chapter specific classes of tasks were considered. For each task, specific instructions were given to adult subjects who presumably complied as well as possible with the stated task goals. Their movement strategies were a function of a specific neuromusculoskeletal apparatus which had been tuned through years of constant use. Their performance, however, was also a function of many hard-to-define variables, including "mood", perception of instructions, the amount of practice, the level of mental and/or physical fatigue, and choice of limb. These factors contribute to intra- and intersubject variability. As pointed out in Chapter 34 (Pedotti and Crenna), however, such variability may be related to different *strategies* as opposed to "noise" about a single strategy. An issue, not directly addressed in these previous chapters, is how subtle changes in movement instructions or subjective goals affect the movement strategy. Within the context of understanding movement strategy, we suggest, this is of fundamental importance. Tradeoffs exist between meeting task-related *goals* (e.g. movement to a specified target) and the "*effort*" level of the individual, especially for everyday movements performed. In the laboratory setting, movements can also scale due to explicit changes in task requirements such as speed, magnitude or direction.

In this chapter dynamic optimization is utilized to address basic issues related to strategies for scaling simple goal–directed upper–limb movements. The use of a mathematical modeling framework requires an explicitly defined biomechanical system and its goals. In optimization studies "goals" are specified by a scalar performance criterion ("cost function", "penalty function", "performance index") which is to be minimized or maximized [see also Chapter 8 (Zajac and Winters)]. Varying the system structure and/or parameter values while maintaining the same performance criterion allows investigation of the sensitivity of the control strategy to biomechanical assumptions. Varying relative weights between competing subcriteria within a generalized performance criterion while using the same biomechanical system yields information regarding the organizational strategies for scaling movements.

An implicit assumption in this study is that obtaining optimized solutions for a biomechanical system, without consideration of neurocircuitry, is a worthwhile and potentially rewarding endeavor. The rationale behind this approach is that organizational strategies underlying movement should reflect the fact that we are inherently goal–directed, biocybernetic creatures that move within a physical world subservient to mechanical laws. Movement goals can vary widely; upper limb movements can range from the practical task of picking up a full cup of coffee and bringing it to the mouth (preferably with minimal exertion or conscious "effort") to very skilled tasks such as accurately throwing a ball or playing piano. Each of these tasks can be accomplished *many different ways* by very subtle changes in desire.

Multiple Muscle Systems: Biomechanics and Movement Organization
J.M. Winters and S.L-Y. Woo (eds.), © 1990 Springer-Verlag, New York

Variations in performance due to subtle changes in a subjective goal may yield insight to many aspects of movement organization. Yet there has been little investigation along these lines. Indeed, the concept tends to be at odds with the traditional experimental approach, which usually requires a well-controlled, highly constrained experiment designed to test an explicit hypothesis. In this chapter the emphasis is on a synthesis of our current understanding regarding how changes in goals, and especially subtle changes, affect predictions for optimal movement strategies.

This chapter is divided into four sections. Section 19.2 summarizes closed–form optimal control strategies for scaling movements in very simple systems. Surprisingly, existing closed–form results for such systems have not been reviewed previously within the motor control field, nor are these findings developed in any other part of this book. Consequently, this section, which emphasizes results for simple parallel mass–dashpot–spring (JBK) models of a single joint (Figure 19.1a), is included. Section 19.3 considers how strategies are modulated when the input to the "joint" comes from one or two muscle "filters" (Figures 19.1b–c). Although results for simple JBK models are informative and capable of providing basic organizational principles, one must consider the limitations of their predictions. In Section 19.4 we show that control strategies often take advantage of higher-order model structure and/or nonlinear muscle properties, and that scaling of movements can be well approximated by changes in the relative weights among performance subcriteria. Finally, Section 19.5 summarizes our findings and Section 19.6 identifies areas for future exploration of upper limb movement organization which we believe are likely to be fruitful, especially as related to extensions to multi-articular systems.

19.2 Closed Form Solutions to JBK Models

Closed form analytical solutions have a significant advantage over numerical solutions in that *explicit relationships* are developed that relate model structure, model parameter magnitudes, variation in performance criterion, and optimal control inputs. Thus, one sees immediately the interrelationships between changes in system parameters and the control strategy. This is in contrast to the consideration of cases where

numerical search-based optimization methods are necessary, in which many separate optimization solutions are necessary to investigate the effects of variation in even a single model parameter. Unfortunately, however, closed–form solutions are only possible for very simple systems, typically involving movement of a single joint with crude (and even nonexistant) muscle properties. Nevertheless, an understanding of the basic *form* of the solution will turn out to be quite valuable.

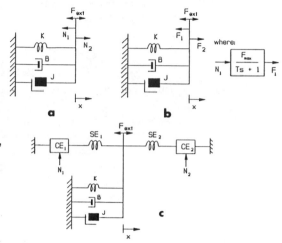

Figure 19.1: Basic model structures under consideration here. *a)* Second-order "spring–mass–dashpot" (JBK) system, with the dashpot (B) and spring (K) containing potentially nonlinear properties, driven by neural inputs ($0 < N_1 < N_{max}$; $0 < N_2 < N_{max}$, but pulling in opposite directions), and potentially an external load (F_{ext}). In all cases we assume J, B and K are greater than zero. Special cases of practical interest are: "visco-elastic" ($J = 0$; first-order system), "pure inertial" ($K = B = 0$), "visco-inertial ($K = 0$), "overdamped" ($B^2 > 4 J K$), "underdamped" ($B^2 < 4 J K$). *b)* Higher–order system in which the neural inputs work through one or two first– or second–order linear muscle "filters" (first-order shown to right), with the output(s) of the filter(s) driving a passive JBK model. *c)* Antagonistic muscle-joint model, where muscle properties are nonlinear and a function of both neural input and joint kinematics [see also Chapter 8 (Zajac and Winters)]. Here F may refer to force or moment, x represents changes in length or angle.

The goal here is to present optimal control inputs that satisfy some explicitly defined goal. We will not develop nor utilize the mathematical foundation, whether based in calculus of variations, Pontryagin's maximum principle (Pontryagin et al., 1962)), or geometric techniques for determining switching functions. In fact, from the present perspective, the mathematical tools utilized to find solutions are irrelevant — certainly the brain does not solve these problems in a way that resembles our (relatively simplistic) analytical techniques. Rather, we present the form of results and briefly discuss the consequences as relevant to motor control. Our closed-form examples are divided into two categories: *i)* second-order (or simpler) models (Section 19.2); and *ii)* linear models where muscle-like filters drive a second-order model (Section 19.3).

The following generalized performance criterion (C_j) is used throughout this section:

$$C_j = w_s[x_r(T)\text{-}x_i(T)] + \int_0^T \{\, w_t + w_x[x_r(t)\text{-}x(t)]^2 +$$

$$w_{u_f}|u(t)| + w_{u_e}u^2(t) + w_j\dddot{x} + w_e"E(t)" \,\} \, dt \qquad (19.1)$$

where w_i are weighing functions, x_r is a "reference" (desired) position (in deg), x is position (in deg), u is the control input, T is the movement time (which may be fixed, free or bounded), and the five summed expressions go by the names: final error (typically forced to zero), minimum time (time optimality), integrated state error, minimum input "effort" ("fuel"), minimum input "energy", and an expression representing various "metabolic/mechanical energy" considerations.

For each case of interest, we must specifiy: *i)* a system to be controlled; *ii)* movement goals (as specified by performance criterion which are subsets of *Eq. 19.1*); *iii)* constraints on the inputs or states (e.g. physiological relevant ranges); and *iv)* boundary conditions (e.g. initial and final states and perhaps the movement time).

The form of the second-order differential equation of Figure 19.1a is:

$$J\frac{d^2x}{dt} + B\frac{dx}{dt} + K(x-x_o) = N_1 - N_2 + F_{ext} \qquad (19.2)$$

where $x = x(t)$ all terms are defined in Figure 19.1a. Additionally, T is defined as the time interval of the movement and D is the distance moved [i.e., $x(T) - x(0)$]. When appropriate, t_s is the time of switching between muscles. Notice that the neural inputs and the external load F_{ext} structurally enter at the same location. An implicit assumption is that the zero position, essentially the mid-operating range, is the resting length of the spring, i.e. $x(0) = x_o$. In terms of boundary condi-

tions, the most common case has been to prescribe initial position and velocity (the latter usually zero) and a desired final position and velocity (often zero). For some cases specification of acceleration conditions are also relevant.

In order of increasing complexity, the following systems will be considered: *i)* simple first-order linear ("viscoelastic") systems $(J = 0)$; *ii)* second-order viscoinertial systems $(K = 0)$, including one case with a nonlinear viscosity described by Hill's equation plus an isotonic load; and *iii)* full *JBK* systems, with an emphasis on the overdamped case.

1. Minimum Time ($w_t > 0$, all others zero).

It turns out that the optimal control structure will be "*bang-bang*", i.e. maximal level input signals, for all cases to be considered here.

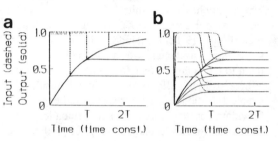

Figure 19.2: Minimum time solutions for first-order "viscoelastic" system. *a)* Overplot of optimal control pulses (dashed) and first-order system responses (solid) as the movement distance increases, assuming $K = 1$. *b)* Control pulses (dashed, shown nonidealized) and system responses (solid) for case where pulse height is constrained when movement magnitudes are small.

First-Order "Viscoelastic" System $(J = 0)$. Consider the classical case of performing movement of magnitude D, with the initital velocity and the desired final velocity both zero. Here a nonzero input holds the initital position $x(0)$ (e.g. if $x(0) > 0$ then $N_1 = K x(0)$). The optimal solution is quite intuitive: for $D > 0$, make N_1 maximal until $x(t_{cr}) = x_{desired}(t)$. The required duration of the pulse is

$$t_{sw} = t_{mt} = -\frac{1}{N_{1_{max}}} \frac{B}{K} \ln(1 - \frac{D}{K}) \qquad (19.3)$$

In order to hold the new position, allow $n_i = K x(t_{sw}) = K x(T)$. This could be called a "*bang-hold*" strategy. Such a "hold" is necessary whenever either K, the final position, or the final (isotonic) level F_{ext} is nonzero. The relationship between t_{sw} and D is apparent from Figure 19.2a. Notice that: *i)* the pulse width equals the movement time; *ii)* smaller magnitude movements require a smaller pulse width (*PW*); *iii)* a "hold" is necessary that is only a function of K and the desired position; *iv)* coactivation serves no useful purpose and would never be predicted in the solution; and *v)* for a given D, the pulse width increases linearly in magnitude with increasing viscosity B or decreasing maximal strength N_1.

Obtaining a maximal neural input with small pulse widths may not be physiologically realizable. Figure 19.2b considers the practical concept of "bing-hold" control, i.e. the pulse height is constrained by the movement magnitude D until some maximal magnitude, x_{1max}, is reached (Figure 19.2b). For the regions of "bing-hold" control, the pulse width can be found by replacing N_{1-max} in Eq. 19.3 by n_1, and similarly for n_2, where n_i are the available pulse height. Notice that this causes mild increases in pulse widths (e.g., Figure 19.2b).

Pure Inertial System. ($B = K = 0$).

Assuming the classical case of zero initial and final velocity, we simply list results presented by various authors, primarily Nelson (1983), where he additionally assumed that $|N_{1-max}| = |N_{2-max}|$:

$$u_{opt} = \begin{matrix} N_{1-max} & 0 \le t < t_s \\ N_{2-max} & t_s < t \le T \end{matrix} \qquad (19.4)$$

where t_s, the switching time, and D, the movement distance, are

$$t_s = \left[2 \frac{D}{N_{1-max}} \frac{N_{2-max}}{N_{1-max} + N_{2-max}} \right]^{1/2} \qquad (19.5)$$

$$T = t_s \frac{N_{1-max} + N_{2-max}}{N_{2-max}} \qquad (19.6)$$

Notice that if $|N_{1-max}| = |N_{2-max}|$, $t_s = T/2$, i.e. the switch occurs at the half-way point — acceleration followed by an equal deceleration. This is the limiting case in Figure 19.3a as the magnitude of J continues to increase relative to K and B.

Visco-Inertial Model. The minimum time solution for this model is also "bang–bang", with the switching time past the mid-point $T/2$ (Nelson, 1983; see also Figure 19.3b as B becomes large or Figure 19.3c as K approaches zero):

$$t_s = \frac{T}{2} + \frac{b^2}{2} \frac{D}{U} > \frac{T}{2} \qquad (19.7)$$

For a given system inertia and control bounds, as the viscous drag increases, the movement time increases and the peak velocity decreases (see Appendix A of Nelson, 1983 for details). In terms of state variables, the optimal switching time is (Athans and Falb, 1966)

$$t_s = \frac{\theta}{|\theta|} (1 - e^{(B/J)\theta}) - \omega) \qquad (19.8)$$

Second-order, nonlinear visco–inertial. In FitzHugh (1977) idealized inertial and isotonic loads were controlled by a Hill-based muscle model (which functioned essentially as a nonlinear viscous element) with a single control input. A "*bang–bang–hold*" strategy was found which was a function of distance, inertia and the magnitude of the isotonic load, with the final step level that required to hold the isotonic load. The optimal pulse width differed for shortening and lengthening muscle. For shortening muscle, the bang–bang pattern is based on driving the system velocity to be optimal for maximal muscle power generation, assuming the classical Hill equation for the force–velocity relationship.

Second-Order, Overdamped ($B^2 > 4JK$). The time optimal control is "*bang–bang–hold*" (Figures 19.3 and 19.4). The bang-bang switching times, as a function of variation in J, B and K, are presented in Figure 19.3. Notice that as B increases (i.e., the system becomes progressively more overdamped), the movement time increases and the switching time becomes essentially the same as the movement time. Similarily, when J is small, the normalized switching time is near one, i.e., the system approaches first-order; conversely, as J becomes progressively larger, the switching time approaches that of a pure inertial system. Finally, notice that the optimal switching time is relatively independent of K until it has a very large value. Of note is that whether the switching and movement time increases or decreases with increasing K depends on the initital position and movement magnitude. Athans and Falb (1966)

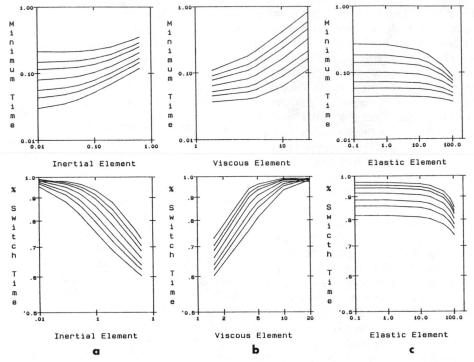

Figure 19.3: Minimum time optimal solutions for an overdamped *JBK* model as a function of model parameters *J* (part *a*), *B* (part *b*), and *K* (part *c*). In each case both movement times (top) and switching times as a percentage of movement times (bottom) are presented. The constant values of *J*, *B*, & *K* are 0.06 $(Nm/rad/s^2)$, 5.0 (Nm/rad/s), 10.0 (Nm/rad). The traces in each plot are obtained by varying the maximal Torque input from 10 to 90 Nm.

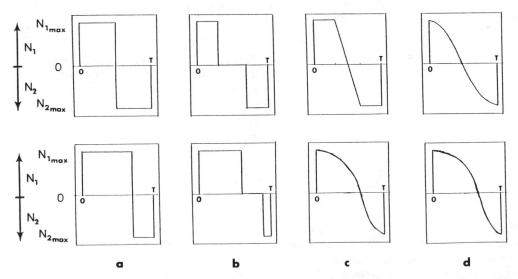

Figure 19.4: Conceptual overview of the basic forms of the optimal control of point-to-point movements for four classical isolated criteria: *a)* minimum time (time optimal); *b)* minimum neural "effort" (absolute input); *c)* minimum neural "energy" (squared input); and *d)* "jerk" (maximum smoothness of output). In each case control input forms are presented for purely inertial (*J*) system (top) and visco-inertial (*J-B*) system (bottom).

provide switching curves as functions of state variables.

Second-Order, Underdamped ($B^2 < 4JK$). This is a more difficult problem. While the solution is "bang-bang", it may include many switches, with the number of switches tending to increase with distance. Also, the maximum time at a given control level is π/ω, where $\omega = \sqrt{K/J}$ is the natural frequency (see Athans and Falb, 1966 for details). The suboptimal, yet practically clever control strategy known as "Posicast control" (e.g., see Takahashi et al., 1972), is of special interest. This engineering solution is based on the concept that if, during the underdamped response to a step, a second step of strategic magnitude can be added at the time of the peak overshoot due to the initial step, an approximate cancellation of the two superimposed responses can result. The net result is a rapid rise and reasonably fast clamping at the new value.

2. Minimum State Error (i.e. $w_x = 1$, relative here to a steady reference level).

This criterion differs subtlely from minimum time since it also penalizes the path that is traversed to arrive at the reference position. As the power to which the state error is raised grows, large errors in particular are agressively eliminated with small errors near the reference position of less relative importance.

First-order "viscoelastic" system. The solution is the same as for the minimum time case.

Second-order "inertial" system. It turns out that the control is bang–bang, with multiple switches and oscillation. Indeed, as m grows greater "chattering" (and initial overshoot) occurs about the reference point (Takahashi et al., 1972). The optimal position trajectory crosses the "final" position sooner than for the classical time optimal case. However, from a practical viewpoint, as an isolated criterion, nonphysiological behavior is apparent, therefore this criterion will not be considered any further. However, later it will be shown to be quite effective as a *subcriteria* that "competes" with measures of effort.

3. Minimum "Effort" (Minimum "Impulse"; $w_u = 1$).

This criterion has been especially well studied in control systems literature within the framework of minimizing fuel consumption. The form of the optimal control solution is of "*bang-zero-bang*" type for the operating ranges of interest (see Figure 19.4b).

First-order "viscoelastic" model. The solution for this case, given that $x(0)$ and $x(T)$ have the same sign and latter has a greater magnitude is given by

$$u(t) = \text{sgn}\,\{x(T)\},$$

and

$$T = \frac{1}{a}\,\log\frac{x(0) - \text{sgn}\,\{x(T)\}}{x(T) - \text{sgn}\,\{x(T)\}} \tag{19.9}$$

Second-order "inertial" system. The optimal solution turns out to be equivalent to minimizing the maximum velocity while satisfying specified movement time considerations (Figure 19.4b). If there were no bounds on the input, the optimal control is $u_{opt} = (2DM/T)((T/2)-t)$, which describes a ramp; this allows the maximum velocity to be the average velocity. Often, however, the desired movement time T is low enough such that the control saturates. It turns out that the optimal control is "*bang-zero-bang*" strategy, and thus there are two switches (see Figure 19.3b). If we assume for simplicity that $N = N_{1\text{-}max} = N_{2\text{-}max}$, the switching times, for the practical regions of the solution space of interest, are defined by (Athans and Falb (1966):

$$t_{S1} = \frac{1}{2}[\,T - (\,T^2 + 4D\,)^{1/2}\,]$$
$$t_{S2} = \frac{1}{2}[\,T + (\,T^2 + 4D\,)^{1/2}\,] \tag{19.10}$$

where T is the maximum allowable time (i.e., the optimal solution will **utilize all available time**). As an approximation to a "movement turnaround", which is of great importance in motor control, if an initial velocity exists that opposes the intended direction, the optimal control will decrease the "coasting" time and produce an acceleration pulse that is wider than the deceleration pulse (Athans and Falb, 1966):

$$t_{S1} = \frac{1}{2}\{\,T - v_o - [(T - v_o)^2 + 4D - 2v_o^2]^{1/2}\,\}$$
$$t_{S2} = \frac{1}{2}\{\,T - v_o + [(T - v_o)^2 + 4D - 2v_o^2]^{1/2}\,\} \tag{19.11}$$

For an initial velocity in the direction of the movement, the solution is not unique for this idealized system. However uniqueness may be achieved by penalizing movement time as well as "effort"; this important case will be considered later.

Second-order "visco-inertial" system. Here we again see a "*bang–zero–bang*" strategy, with the switching times now more complex, with numerical solution to nonlinear equations necessary (Nelson, 1983).

4. Minimum Energy ($w_{ue} > 0$, all others zero).

Another possibility for control is to move a certain distance D in a specified time T while minimizing the quadratic of neural drive. This popular type of quadratic index has been related to input "energy" (Athans and Falb, 1966).

First-order "viscoelastic" model. If the input is unbounded (i.e. doesn't saturate), the continuous optimal control turns out to be:

$$n = \frac{2a(x_f e^{-aT} - x_o)}{(1 - e^{-2aT})} e^{at} = A e^{at} \qquad (19.12)$$

where $a = K/B$ and $D = x_f - x_o$. Notice that we start with a control magnitude of A, and decreases *exponentially* until the final time. If the desired movement time is on the order of the system time constant and the movement magnitude is over the primary range, *saturations* occurs. While for greater movement time, the exponentially-shaped input signal progressively decreases (e.g., see Takahaski et al., 1972).

Second-order "inertial" system. A continuous ramp-like optimal control signal results if not saturated [i.e. $T \geq (6D/U)^{1/2}$, (Nelson, 1983)]:

$$u_{opt} = \frac{6D}{T^2} (1 - 2(t/T)) \qquad (19.13)$$

It turns out that the peak velocity exceeds the average velocity by 50%. If saturation is reached, i.e. $T < (6D/U)$, the form of the control input is "*bang–ramp–bang*", as shown in Figure 19.4.

Second-order visco-inertial system. The optimal control is, assuming no saturation (Nelson, 1983):

$$u(t) = C - 2 e^{(B/J)t} \qquad (19.14)$$

where

$$C = \frac{B^2}{J} \frac{e^{bT} D}{(BT-2J)e^{bT} + BT + 2J} + J \qquad (19.15)$$

As in the first-order case we have a exponential relation unless saturation is reached.

5. Minimum Jerk (i.e. $W_j = 1$, all others zero).

Here the optimal control has a graded form, [Nelson, 1983, assuming no saturation; see also Chapter 14 (Gottlieb et al.) and Chapter 17 (Flash)]:

$$U_{opt} = \frac{5D}{T^2} (1 - 6(t^2/T^2) + 3(t^3/T^3)) \qquad (19.16)$$

The maximum input appears at the initital time, then scales downward.

1&3. Minimum Time and Minimum Effort (i.e. w_t & $w_{uf} > 0$, T not prespecified).

Second-order "inertial" system. Here tradeoffs between a task parameter (minimum time with no final error) and the control effort are considered. The solution, which is unique, turns out to be "bang–zero–bang", with the switching times set by regions of the phase plane and the relative weight w between the competing subcriteria (Athans and Falb, 1966; see also Figure 19.5)

Second-order "visco-inertial" system. This situation also produces a unique solution with "*bang–zero–bang*" behavior, with the switching times, which again can be viewed from within the phase plane, given in Athans and Falb (1966). The continuum of optimal strategies obtained by increasing the relative weight of minimum time subcriterion (Figure 19.5) shows a transition from bang–zero–bang to a near bang–bang. Notice that the width of the negative (third) pulse increases (relative to total movement time) with decreasing viscosity. With small relative weight given to minimum time, the coasting period dominates the movement strategy, especially when viscosity is low (left column of Figure 19.5). Also notice that increased N_{max} equally scales the timing of all three pulses when large weights are given to minimum time.

An example developed by Athans and Falb (1966) in which the viscous drag was a function of the squared velocity. The optimal control law took forms such as "*bang–bing–zero–bang*", showing explicitly that *submaximal* activation may indeed emerge as an optimal strategy for *nonlinear* systems.

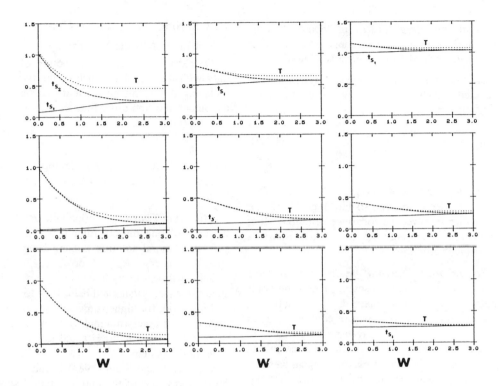

Figure 19.5: General forms for scaling input pulses as a function of the relative weights between minimum effort and minimum time (w) subcriteria. A *J-B* model with *J* = 0.1 Nm/rad/s and *B* = 0.1, 1.0, 2.0 Nm/rad/s (left to right columns) and the maximum torque N_{max} = 1.0, 5.0, 10.0 Nm (top to bottom rows) is considered. The horizontal axes represent the $\log_{10}w$. Compare case of B=.1, N_{max} =1 with case of B=1, N_{max}=10 to see the effects of reduced *J*. Shown: switching time from maximum to zero (ts_1, solid), from zero to negative maximum (ts_2, dashed), and movement time (*T*, dotted).

2&3: Minimum State Error and Minimum Energy.

The solution for the case of penalizing the square of both state error and input is especially appropriate for analytical analysis. The resulting optimal control law can be formulated in terms of the state variables, and thus state feedback can be utilized to realize the control. For our purposes, of note is that: *i)* the solution depends on the relative weight *w*, being more aggressive if *w* >> 1; *ii)* quadratic terms in the performance criterion suggest disproportionately high penalty for large error/input; *iii)* the form of the solution is exponential; *iv)* the solution depends on the movement time (if not specified, the optimal control will involve a decaying exponential that does not converge until *t* = ∞); and *v)* the solution may depend on whether or not the idealized control would hit the saturation boundary N. We are in fact more interested in cases of finite time and bounded control. For finite time and a final position as mid-range, the control is:

$$u = \frac{x_o}{e^{cT}}(e^{-c(T-t)} - c\,e^{c(T-t)}) \qquad (19.17)$$

where $c = w + a^2$

Notice that the optimal control involves a combination of falling and rising exponentials, and can be shown to be quite intuitive (e.g., see Takahashi et al. (1972), p. 649). Of note is that the optimal feedback gain is a function of time if the movement is to be made in finite time (e.g. see Takahashi et al., 1972 for details).

Final State Error and Input "Energy". Here we consider briefly what happens when some "slack" in the final position error is allowed.

Second-order visco-inertial system. For specified final time T, the optimal control input turns out to be:

$$u_{opt} = \frac{-2wx_o}{(2+w)e^{2\,aT}-1}\,e^{at} \qquad (19.18)$$

6. Minimal Expended Energy.

The reader is referred to FitzHugh (1977) for this rather involved performance criterion. It is considered here because it is in essence a "combined" criterion.

Second-order "inertial" system with isotonic load. For lengthening muscle a "bang-bang-hold" strategy was always obtained. For shortening muscle, however, two classes of control strategies were revealed: *i)* bang–bang–hold; and *ii)* bang–level–bang–hold. In each case the initital pulse width of the initital maximal signal was that necessary to achieve the critical velocity for optimum muscle power generation. However, for certain regions, as defined by the magnitude of the movement distance, the inertia, and the isotonic load level, a "level hold" was required which allowed efficient muscle power production to continue for a certain time period.

Second-Order, Oscillatory, With Variable (Nonlinear) K. With $B = 0$ and $K = u(t)$ ($0<u<1$), the inertial system can be driven from any initial position (zero initial velocity) to origin with a bang–zero input (Takahaski et al., 1972). The maximum input results in maximum force at each dynamic position and therefore the highest accelerations the system is capable of producing. The subsequent zero spring constant results in a zero acceleration (i.e., constant velocity) coasting period which brings the inertial system to the origin. This transfer to the equilibrium (i.e. origin) through stiffness modulation is of special interest since Hogan (1984) suggest movement may be carried out by a continuous shift of equilibrium.

19.3 Second-Order Plants Controlled by Linear Muscle Filters

Consider the equation of motion of *Eq. 19.2*, only with the signal for each of the two muscles replacing the neural inputs, and this signal being the output of a first-order filter of the form

$$\tau F_i + F_i = N_{1_i} - N_{2_i} \qquad (19.19)$$

This could represent, for example, a simple muscle filter acting on an inertial object such as a limb.

Minimum Time ($w_{mt} > 0$, all others zero). For the case of one muscle filter that can push and pull, the general solution is bang–bang with *two switches*, i.e. $N_1 \to N_2 \to N_1$ or $N_2 \to N_1 \to N_2$, with switching times set by a switching surface in *3-D* state space (see Athans and Falb, 1966 for details).

In general, for higher-order linear systems with negative real poles (**normal** systems to be exact), the number of switches is equal to one less than order of the system.

Oguztoreli and Stein (1983) formulated closed form expressions of the optimal control strategies for a model consisting of a pair of linear, first-order antagonistic muscles with the Hill model structure that could be driven directly by a neural input or by the output of a first-order filter representing muscle activation. In both cases the optimal control was found to be bang-bang, with the switching times complex functions of the model parameters.

Other Performance Criteria. Using the linear two-muscle model described above, Oguztoreli and Stein (1983) also considered several other performance criteria, including oscillation around final position, an involved expression representing total energy, and neural input. As might be expected, the form of all neural strategies were found to be bang–bang or bang–zero–bang except for the minimum energy case. Thus there exists a natural progression from simpler systems to more structurally complex systems.

An important point made by the authors in an earlier paper (Ouztoreli and Stein, 1982) is that two *linear* antagonist muscles may be combined into one since co-activation does not vary any of the system characteristics and is not utilized.

Combined Criteria. The need for a weighted sum of some or all the above subcriteria was suggested by Oguztorelli and Stein (1983); however, the numerical solution was feared to be "quite difficult". Corraborating this observation, our own attempts at such solutions have proved unsuccessful (Seif-Naraghi, 1989). To our knowledge no closed-form solutions exist for muscle–filter–joint–plant systems with a generalized criteria that include both measures of task performance and neuromuscular penalty. Consequently, numerical solution becomes a necessity. [Of note is that solutions for optimal feedback gains, given a performance criteria that includes both penalty for both states and inputs, can be formulated for certain classes of regulator problems – see Chapter 10 (Loeb and Levine)].

19.4 Higher-Order Nonlinear Movement System

Mathematical representations describing the same physical phenomenon are often of varying degrees of complexity. The proper choice of model is of great importance since what can be learned from the study is often a direct function of the modeling choice. Use of extremely simplified, lower-order models leads to predictions that may not be generalized beyond the specific conditions under which the model is valid. Extremely detailed models with numerous parameters may lead to little insight into the overall system behavior. Having considered the simple models of movement in Section 19.2 and reached the limits for closed-form solutions in Section 19.3, we now consider a more realistic nonlinear muscle–joint model. The following subsections discuss inherent scaling of movement strategies that occur naturally through systematic changes in the movement goal, as described by the optimization criterion. The results suggest that the system utilizes nonlinear muscle properties, such as ability to rapidly modulate internal dynamics, to perform tasks more efficiently. Predictions made by simple second-order models are modified to take advantage of these properties.

19.4.1 Modeling/Optimization Foundations

Muscle Model. The classical Hill model structure consists of a contractile element (CE) in series with a viscoelastic element (SE). The CE is represented by the product relationship of a bell-shaped length–tension relationship and a modified Hill force–velocity relationship [e.g., see Chapter 5 (Winters)]. The SE force–extension relationship is approximated by an exponential relating force and extension, using peak force and peak extension as parameters. When coupled, the CE and SE can be represented with one nonlinear ordinary differential equation. An input to this first-order system is activation/attachment, which is itself the output of two cascaded first-order unicausal filters approximating neural dynamics underlying excitation (E) and activation (A) processes, respectively, with an idealized neural signal (N) being the input to the neuromuscular unit. The intermediate signal between the excitation and activation processes, which is one of the state variables, resembles a linear enveloped electromyographic (EMG) signal. Each of these is developed in detail in Winters and Stark (1985) and in Chapter 5 (Winters).

Muscle-Joint Model. Two antagonistic equivalent muscles coupled to a joint is utilized in this modeling approach; this is the simplest structure capable of producing basic muscle-joint dynamic phenomena (Winters aned Stark, 1985). As discussed in Chapter 8 (Zajac and Winters), the muscle parallel elasticity is incorporated into the second-order parallel plant which consists of lumped elastic (K_p) and viscous (B_p) elements and link inertia (J_p). Joint acceleration is produced by the summation of the opposing muscle torques [e.g. for flexor (l) and extensor (e) muscles] plus a possible torque term due to the external environment:

$$J_p \ddot{\theta} + B_p \dot{\theta} + K_p \theta = F_f(N,\theta,\dot{\theta}) - F_e(N,\theta,\dot{\theta}) + M_{ext}$$

(19.20)

where the muscle torques are a function of not only neural activation but also system dynamics ("bicausal" muscle filters). The full model is therefore nonlinear, eighth-order, with three inputs (i.e., neural inputs to two muscles plus external load) and one output (i.e., joint position).

Optimization Criterion. We favor a generalized performance criterion which is a function of state and input variables. In addition to joint position output, other available variables include: joint velocity, acceleration and jerk, each muscle force, the resulting net torque, and the dynamic force "lost" due to the viscous properties of the system.

Various measures of energy storage (across springs) or dissipation (across dashpots) can be obtained. The higher-order model allows inclusion of these variables, alone or in tandem, within the optimization criterion. The generalized performance criterion that we will consider here is a natural extension of *Eq. 19.1* and is a subset of the more generalized form developed in Chapter 8 (Zajac and Winters):

$$J_c = \int ITO^n + REF^n + JK^n + \Sigma \,(NE^n + MS^n + MD^n)$$
(19.22)

Optimization Technique

Theoretical optimal control, when applied to a high-order system and a general optimization criterion, involves solving complicated differential equations with split boundary conditions by numerical means. Two general methods often used are *gradient-based*, which does not guarantee conversion to a solution, and random search, which are computionally inefficient. Many of the more recent additions are essentially hybrid methods which attempt to incorporate the best of each approach. The modified Bremermann optimizer, which will be used here, combines a random search method with a gradient-like feature in order to provide an optimized solution within reasonable computation time. A multivariate type search along a random direction is carried out in order to obtain five points on the optimization cost function. A one-dimensional fourth-order polynomial fit through these points is then differentiated to find its minimum. This point in the parameter space is mapped to its actual cost. The minimum of these six points on the cost surface (one previous minima, one calculated new minima, and the four perturbed points) is the new minimum cost and the corresponding point in the parameter space represents the control parameter set.

Since the optimization criterion is digitally integerated over the time course of *whole task*, the performance criterion may include equality constraints such as final position and/or movement time (defined as the time to arrive at and stay within a neighborhood of the final position). Also, a finite number of control parameters are necessary, in contrast to the possibility of continuous control signals for closed–form solutions. This points out the need to assume a *structure* for the control signal. Here a sequence of pulses is assumed, with pulse heights and widths designated

as the control parameters. Such a multi–pulse structure is in agreement with experimental findings as well as the results of the lower-order systems. Our own experimentation with input structure has consistently shown that only three or four pulse heights and widths for each muscle will be utilized by the algorithm (minimum jerk is an exception). Nevertheless, it should be realized that what is obtained is an "optimized" solution rather than the "optimal" solution.

19.4.2 Strategies for Scaling Movements

The goal of this section is to systematically vary the optimization criterion and observe the scaling of the movement strategies. The "base" movement is a very fast, nearly time optimal movement, followed by a variety of combined criteria.

Scaling Via Changes in Task Kinematic.

1. Movement Time and Position Error. Fast time optimal movements may be obtained by a direct minimization of the time it takes the model to arrive and stay within a described neighborhood of the final position. It turns out that forcing the model to follow a very fast reference trajectory (also may be refered to as virtual cost trajectory) gives very similar results (Seif-Naraghi, 1989). Optimization convergence results made it clear that a criterion including both movement time and position reference trajectory subcriteria is computationally most efficient. In fact, movement speed may be scaled by either specifying the desired movement time or the speed of the reference model without significant changes in the optimized strategy (Seif-Naraghi, 1989). The solution, a near bang-bang control strategy with a longer acceleration phase (agonist pulse width) than deceleration phase (antagonist pulse width), is similar in basic form to that obtained in the slightly overdamped second-order case; the larger agonist pulse is in part due to viscous properties of the joint. However, the nonlinear properties of muscle, rather than the neural input, shape the developed muscle forces so as to cause significant overlap during the second half of the deceleration phase. These increased muscle forces, although not accelerative in either direction, help to increase the stiffness of the muscle-joint system, which in turn helps bring it to a quick stop [Seif-Naraghi, 1989; see also Chapter 5 (Winters)].

Also, the peak antagonist force and peak deceleration are larger than peak agonist force and peak acceleration, respectively. The small peak negative velocity is an indication of the overshoot of the position trace. The amount of overshoot is a function of the width of the allowable neighborhood — thus, in line with the insights from Chapter 34 (Pedotti and Crenna), subtle changes in goal will cause subtle changes in control strategy.

2. Scaling By Virtual Cost Trajectory.
Movements of various magnitude and/or speed may be obtained with scaling of the amplitude or the "speed" (i.e., the poles of the assumed second order form) of the virtual cost (or reference) trajectory.

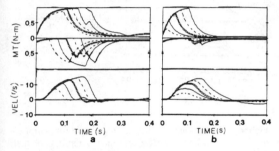

Figure 19.6: Optimized solutions for various magnitudes of reference trajectories. *a)* *EMGs* for both agonist and antagonist muscles (flexion upward, and extension downward) and velocity profiles for optimized 20°, 40°, 60°, 80°, and 100° elbow movements with position error cost subcriterion active, while a fast reference model and the appropriate input step magnitude provide the reference trajectory. Notice the decreased onset of the antagonist muscle *EMG* burst, overlaping the agonist activity, with decreased movement magnitude. *b)* Same as *a)* except with neuromuscular cost *(NE)* added. Notice the smooth and more symmetrical velocity profiles and diminishing antagonist activity.

For *amplitude* scaling, reasonable control signals may be obtained for these movements if the optimization criterion also includes some measure of neuromuscular penalty (Figure 19.6a and b). As seen in Figure 19.6b for 20°, 40° and 60° movements, smooth symmetrical velocity profiles occur when any of the neuromuscular cost subcriteria are active. The higher magnitude movements have nearly saturated velocities, which is partly due to the nonlinear *CE* force–velocity relation.

Scaling movements by varying the *speed* of reference trajectory, essentially the optimization equivalent of the "inverse dynamics" calculation approach, predicts neural signals that are overly aggressive (Figure 19.7a). However, inclusion of neuromusclar penalty reduces the overlaping *EMGs* and results in much smoother pursuit of both fast and slow trajectories (Figure 19.7b). As to be presented below, we suggest that speed scaling can be achieved in a more natural manner by scaling relative weights between kinematic and neuromuscular penalty subcriteria.

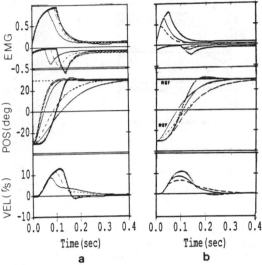

Figure 19.7: Scaling by the reference trajectory speed. *a)* *EMGs* for both agonist and antagonist muscles (i.e. flexion in the positive, and extension in the negative direction), position, and velocity profiles for five 60° elbow flexions, optimized for position error between joint position and five reference trajectories: *r1, r2, r3, r4,* and *r5* that are the output of step responses from various second-order models. Position plots include these reference trajectories, which in all cases start faster than the biomechanical model. *b)* Same optimization as 3 of the 5 runs (i.e., *r1, r4,* and *r5*) in *a)*, only with a 10% relative weight given to neural effort subcriterion, which results in smoother trajectories, lower *EMG* pulse heights, and less coactivation.

3. Kinematic Scaling Via Minimum Jerk. As seen in Figure 19.8 (small fig), this highly publicized criterion requires a smoother structure for the controller. In other words, it seems to deviate from the pulse-like neural input that is assumed as a default; unlike strategies for most of combinations of subcriteria, the optimal strategy will take advantage of additional pulses, in a graded fashion, if they are provided. This, of course, is similar to the case of the second-order system (Hogan, 1984). Another interesting characteristic of minimum jerk continuum of movements is the reduced delay in the onset of the antagonist burst, which helps smooth the acceleration trace as the relative weight of jerk subcriterion increases. In general, minimizing jerk tends to cause more pulse overlap and, if weighted heavily, a "pull-pull" strategy that includes cocontraction, that is not necessarily an efficient use of neuromuscular effort (note slower rate of decrease of m and n costs in Figure 19.8b).

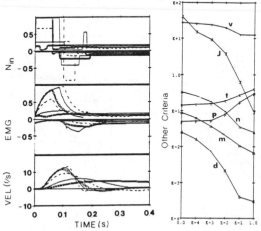

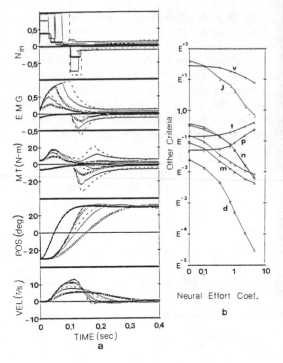

Figure 19.8: Continuum for scaling movement via *minimum jerk. a)* Neural inputs and (enveloped) *EMGs* for both agonist and antagonist muscles (i.e. flexion in the positive, and extension in the negative direction) and velocity profiles for various relative weights of minimum jerk subcriterion (w_{JK}) of 0 (dash), .0001 (single), .001 (double), .01 (dash), .1 (single), and 1 (double). *b)* Final costs of all active and inactive sbcriterion as a function of the relative weight of minimm jerk subcriterion. The codes on the traces are *J (JK), t (ITO), p (PE), n (NE), m (MS), d (MD),* and *v (JV).*

Scaling Via Addition of Neuromuscular Penalty

We shall now discuss the effects of adding various forms of simple neuromuscular penalty to the time optimal plus squared position error subcriterion. (Notice that for dynamic optimzation using a forward dyanmic model, unlike static optimization with an inverse dynamic model, some "task" subcriteria must be specified in addition to neuromuscular penalty.) In each case we will see that both control strategy and movement kinematics scale as a continuum with changes in relative weights; however, the form of the scaling changes with the form of the neuromuscular penalty added to the optimization criterion.

Figure 19.9: Continuum when scaling weight for *neural effort. a)* Neuro-inputs, *EMGs,* and muscle torques for both agonist (i.e. flexor in the positive direction) and antagonist (i.e. extensor in the negative direction), plus position and velocity profiles, and position trajectories for elbow movement task. Position plots include a reference trajectory (*r3*) in thick line. *ITO* target size intervals were ±3° and ±600 deg/sec. *NE* relative weights (w_{NE}) are 0 (dash), .01 (single), .5 (dash), 1.0 (single), and 5 (double). To discern between similar line types note that increasing weights corresponds to smaller pulse heights or widths, decreasing peak muscular activity, increased movement time, and reduced peak velocity and peak acceleration. *b)* Final costs of all active and inactive sbcriterion as a function of the relative weight of neural effort subcriterion.

1. Neural Effort. While the addition of very small relative weights for neural effort increases the efficiency of the movement use of neural input, it does little to the dynamics of the position trajectory (Figure 19.9a). Initially, the agonist pusle width is modulated and the pulse height remains at 100%. However, increased *NE* relative weight results in reduction of agonist pulse height (i.e., initial rate of change of *EMG*). Using the terminology of Chapter 14 (Gottlieb et al.), this subcriterion seems to switch from "speed–insensitive" to "speed–sensitive" strategy as the moment time increases. The second agonist muscle force burst appears to diminish as a consequence of this efficiency measure. This, together with decreased antagonist neural pulse height and

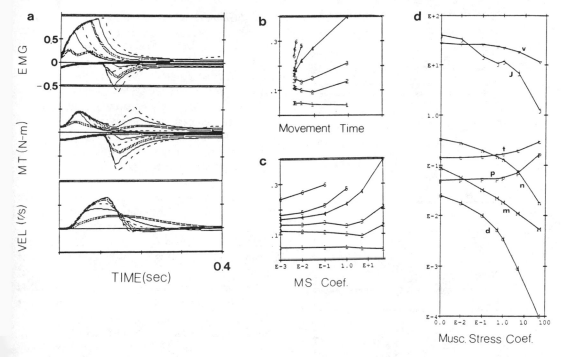

Figure 19.10: Scaling continuum with changes in the relative weight for *muscle stress*. a) EMGs and muscle torques for both agonist and antagonist muscles, and velocity profiles for 60° elbow movement task. Criterion includes *PE* relative to *r3* reference model, 1% *ITO* with a target size of ±3° and ±600 deg/sec, and various weights for to *MS*: 0 (dash), .01 (single), .1 (double), 1.0 (dash), 5 (single), and 50 (double). To discern between similar line types refer to Figure 19.9a. b&c. Timing plots of six kinematic points as a function of movement time (b) and relative weight of *MS* (c). The six kinematic points are: *1*) peak acceleration, *2*) peak velocity, *3*) maximum decceleration or minimum acceleration, *4*) peak position, *5*) minimum velocity, *6*) second zero velocity. Note the linear scale of b and log scale of c. d) Final costs of all active and inactive subcriteria as a function of the relative weight of neural effort subcriterion (codes given in Figure 19.8b).

a corresponding decrease in muscle torque, reduces the stiffness during the deceleration phase. One major finding, depicted in Figure 19.9b, is that all other inactive subcriteria fall as the *NE* is reduced, at the expense of slight increases in movement time and position error. The net torque and acceleration plots (not illustrated) show a softening of the transition between the peaks with increasing *NE* penalty. This is consistent with the "minimum fuel" results suggested by the second-order system.

2. Muscle Stress. This subcriterion results in slightly different neuroinput strategy, though similar general changes in trajectory dynamics (Figure 19.10). Pulse width modulation is the dominating mode of scaling movement. While second and third muscle force peaks are reduced, the timing of these peaks, which are tightly coupled to movement time, hardly changes. Note that the pulses representing the envelope of *EMG* signals modulate in height and width in both *NE* and *MS* cases. This is an interesting observation considering the fact that in the experimental studies *EMG* signals are often measured and analyzed. It is clear that *EMG* is neither the input to the system nor proportional to muscle force or stress. The latter is in part due to *CE* as well as the *SE* of the muscle.

With *MS* relative weight below 1.0, the timing of important kinematic points such as peak acceleration and velocity is hardly modified. This observation, coupled with reduced muscle activity for the same movements, suggests that the *MS* subcriterion facilitates movement efficiency. As with *NE* variation, all inactive subcriteria are reduced sharply with increased relative weight of the *MS* subcriterion (Figure 19.10d).

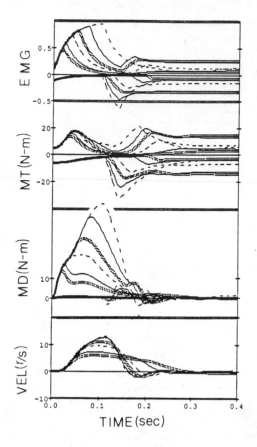

TIME (sec)

Figure 19.11: Scaling continuum with changes in weight of *muscle dissipation* subcriterion. *EMGs*, and muscle torques, dissipated torques for both agonist and antagonist muscles and velocity profiles for *PE* and *ITO* relative weights similar to Figure 19.9a, with *MD* subcriterion relative weights of: 0 (dash), .1 (single), 1 (double), 5 (dash), 50 (single), and 200 (double). To discern between similar line types refer to Figure 19.9a.

3. Muscle Dissipation. Addition of this subcriterion lowers the dynamic losses in the muscle model. This is accomplished by reducing muscle torque production during high velocity regions (Figure 19.11). Additionally, there is an increase in use of muscle torque in the static holding of the position, persisting cocontraction, and a tendency toward triphasic behavior. The neural strategy for *MD* is pulse width modulation of the first agonist pulse and reduced muscle torque peaks. The early behavior is similar to that seen when employing the *MS* subcriterion.

4. "Generalized" Neuromuscular Penalty

The three previous continnum of results all differ in subtle ways. These differences are general (i.e. not model specific), and were seen for elbow, wrist and head models. Which is "better"? We suggest that each has certain beneficial attributes.

In addition to simply penalizing control inputs for traditional reasons, *NE* may represent certain inherent features of neurocircuitry (e.g., reciprocal inhibition), which when functioning normally, does in fact seem to deter sustained high levels of motoneuronal activity.

MD alone fails since it does not penalize force production during quasi-static conditions due to being only a dynamic measure; hence cocontraction, which in fact costs muscular energy, is not penalized unless the movement velocity is significant.

MS does penalize such co-contraction. Also, as the relative weight of *MS* increases, all of the other inactive subcriteria (*NE*, *MD*, *JK*, or even *JV*) follow decreases in *MS* cost by also displaying sharp decreases of value. This behavior suggests that by minimizing *MS*, other criteria of interest also move toward their minimum simply as a byproduct of the "active" *MS* subcriteria – an indication of an effective general criterion. This is also apparent from results of Table 19.1, where a sensitivity matrix is presented for all considered subcriteria to variation in the value of the active subcriterion. Using *MS* as an example, numbers larger than 1.0 in the *MS2* row indicates lowering of other criteria as *MS* cost decreases. Also, smaller negative numbers in that row under *PE2* and *TO1* show the efficiency of this criterion in lowering neuromuscular expenditure with smaller adverse effect on the joint trajectory.

Although the *PE–MS* combination appears to be the best of the pairs considered, the combination of larger numbers of active criteria, including *PE* and *MS* with *NE*, *MD* or *JK*, is suggested to be a better approximation of reality. The *PE–MS–NE* combination helps prevent early maximal neural activity, which *MS* alone does not really discourage during the first 50 ms since the muscle torque pulse during active shortening is never incredibly high (this is due to nonlinear muscle dynamics). In fact, constraints on maximal activity are likely to be neural in origin since each motoneuron receives a wealth of converging information – it is unlikely that every excitatory connection is maximally active and every inhibitory connection silent at any given time, even

Table 19.1: Optimization Sensitivity Matrix.

	NE2	MS2	MD2	JK2	JV2	PE2	TO1
ELBOW:							
NE2	+1.00	+0.63	+1.63	+0.97	+0.33	-0.44	-0.28
MS2	+1.12	+1.00	+2.24	+1.28	+0.35	-0.46	-0.30
MD2	+0.44	+0.58	+1.00	+0.45	+0.17	-0.16	-0.13
JK2	+0.45	+0.40	+0.78	+1.00	+0.15	-0.35	-0.19
JV2	+0.87	+0.81	+2.63	+0.90	+1.00	-0.96	-0.69
WRIST:							
NE2	+1.00	+0.63	+1.86	+0.89	+0.35	-0.35	-0.26
MS2	+1.12	+1.00	+2.53	+1.23	+0.28	-0.21	-0.23
MD2	+0.36	+0.32	+1.00	+0.40	+0.18	-0.72	-0.19
JK2	+0.42	+0.43	+0.76	+1.00	+0.17	-0.34	-0.23
JV2	+0.54	+0.42	+1.93	+0.60	+1.00	-0.68	-0.23
HEAD:							
NE2	+1.00	+0.41	+1.60	+0.88	+0.37	-0.41	-0.33
MS2	+1.93	+1.00	+2.76	+2.05	+0.66	-0.79	-0.59
MD2	+0.43	+0.33	+1.00	+0.61	+0.20	-0.16	-0.19
JK2	+0.49	+0.32	+0.97	+1.00	+0.23	-0.34	-0.30
JV2	+1.68	+2.29	+3.96	+2.29	+1.00	-0.18	-1.26

in a motivated human. The *PE–MS–MD* combination seems especially appropriate — together *MS–MD* provide an estimation of the sum of energy storage and loss in muscle (Oguztorelli and Stein 1983), with *MS* providing a measure of both series elastic storage and resting metabolic energy loss and *MD* estimating dynamic energy (power) loss across the contractile element. Furthermore, they are highly complimentary, with *MS* preventing excessive co-activation and *MD* providing a tendency toward tri-phasic neural behavior.

Considering that the inactive jerk criterion is reduced when neural effort or muscle stress are penalized and that the opposite is *not* necessarily the case, it is more reasonable to optimize effort and obtain a reduction of jerk as an added advantage. At least those experiments linking practise of a task to reduction of jerk are no proof that it is jerk that is being minimized. Nearly symmetrical bell-shaped velocity profiles, so commonly seen in experiments, are also observed as relative weights for neuromuscular penalty subcriteria are increased.

We suggest that *all of these* are relevant, and suggest the following form for a generalized performance criterion:

$$J_c = J_{\text{"task"}} + W * J_{neuromuscular\ penalty} \qquad (19.xx)$$

where W is the relative weight. Notice that this is a natural extension of the second-order case (e.g. see Figure 19.4). In order to combine features of various neuromuscular subcriteria into one criterion, one needs to consider the absolute sizes of each of the cost subcriteria. A rough renormalization of the sizes of these costs should result in a neuromuscular criterion which include the basic advantages of each one of them. We suggest the following relative weights within the neuromuscular penalty subcriterion (Seif-Naraghi and Winters, 1989a):

$$J_{np} = .02\,NE + .10\,MS + 1.0\,MD \qquad (19.21)$$

This combined neuromuscular penalty subcriteria, which we consider most representative of physiological reality, will be utilized in a few of the subsequent simulations that are presented.

19.4.4 Optimization in the Presence of External Loading

Many everyday movements involve interaction with the environment. Such interaction may range from steady applied load (isotonic and inertial) to unpredictable transient perturbation. However, second-order linear *JBK* models of movement *cannot distinguish* between external forces acting on the joint and the modeled neuro-input changes or *EMG* levels, and therefore they *are not suitable for study of most loaded movements*. Higher-order structural models, on the other hand, may be used to simulate various types of loading and their optimal strategies. For brievity, the results is limited to select few cases.

1. Pseudo-Random Perturbation. Random load perturbation allows simulation of movements within an unpredictable environment, as is often encountered in daily movements. Since optimization by nature involves predictions that should be compared to tasks which have been practiced many times under conditions which are fixed and well known, random perturbation findings may also be relevant to strategies applied for novel, unpractised tasks. For example, holding a child while making a movement requires strategies which account for sudden changes in magnitude and direction of the load due to the child.

Increased neural input, specifically as *co-contracting antagonist activity*, results from inclusion of pseudo-random joint torques during a point-to-point movement [Figure 19.12a; see also Seif-Naraghi and Winters (1988) and Winters et al. (1988)]. This increases the muscle force levels holding the final position and causes oscillation of the position trajectory due to increased stiffness. The modulation of mechanical impedance in the presence of noise was also reported by Hogan (1982, 1984) in a second-order nonlinear model capable of *K* modulation. Note that co-contraction will remain although the performance criterion includes moderate neuromuscular penalty. Importantly, if the *model is linearized, the resulting optimal strategy does not employ* cocontraction (Seif-Naraghi, 1989). Also of note is that the stiffness that is being modulated is that due to *SE* as opposed to the *CE* tension-length relation. In addition, the average "viscosity" of the Hill relation increases with co-contraction [see also Chapter 5 (Winters)].

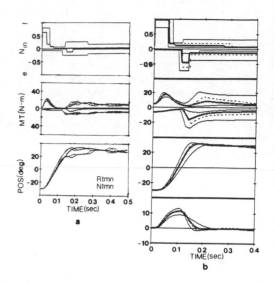

Figure 19.12: Effects of random loading on control strategy and neural effort continuum. *a)* Neural inputs and muscle torques for both agonist and antagonist muscles and velocity and position trajectories for four perturbed and one free optimized 60° elbow movement. In order to investigate optimized movement strategies in an unknown environment special optimization criterion is formed. A pseudo-random sequence of +1, 0, and -1's, with a zero mean, together with a scaling factor (6 Nm in **part a** and 8 Nm in **part b**)) was randomized in 20 ms bins, defined one of four external torque sequences. Another one is obtained by a shift in time of one sample, and two more by inverting the first two sequences. The four external torque sequences generated in this manner have a zero mean within each one and across them. At each optimization step, in order to obtain an average-sense optimized movement, the subcosts for each of the four runs of the system with the four pseudo-random external torque sequences are summed. Note that this results in four times as many runs as unperturbed optimization, which increases the computational cost accordingly. Although this is a deterministic loading of the muscle-joint system since the four sequences are fixed, the resulting optimized control strategy has to account for changes in perturbation direction of the load as well as its timing. The cost criterion utilized here includes **PE** relative to **r2** reference model and an additional 1.0 times **NE** plus 1.0 times **MS** subcriteria. Increased muscle activity mainly in co-activation helps resist the external perturbations. The four runs for the final optimized control signal, plus the average of the four runs, is overplotted. *b)* Increased weight of **NE** subcriterion added to the performance criterion, here with only the average trajectory of the four runs plotted. **NE** relative weights (w_{NE}) are: 0 (dash), .01 (single), .1 (double), and 1 (single with lower activity and speed).

This is an explicit (and quite important) example of how the neuromotor system can take advantage of nonlinear muscle properties.

Another interesting finding involves the scaling of the strategies in the presence of random perturbation as the relative weight of each subcriterion is systematically varied. Results indicate that the modulation of the *agonist* pulse in height and width seems to be a function of the chosen criterion and not affected by presence of the perturabation (Seif-Naraghi, 1989).

2. Inertial Loading. Increased inertial loading of the elbow joint results in increased neuromuscular activity in the optimized control strategy (Figure 19.13). Very large co-contraction occurs if there is no penalty for neuromuscular effort (Figure 19.13a). As a result, the position traces show oscillatory overshoot, which becomes progressively less as the relative weight for the neuromuscular criterion increases (Figure 19.13b). Notice that a silent period is followed by an increase in height of the antagonist pulse while the agonist pulse width increases. With increased inertial loading, the system behaves more and more like a pure inertial system. This is apparent from the bang-zero-bang strategy which occurs with the addition of neuromuscular cost − a natural extension of the second-order case. Additionally, if the plots were redrawn against normalized time, position traces would be nearly identical, which is another characteristic observed in the case of pure inertial system.

The increased neural activity, together with lower velocities, results in muscle torque trajectories of increased magnitude (Figure 19.13a). Reduced peak acceleration and deceleration, as well as peak velocity, scale in their timing proportional to movement with added inertial load in a manner that is in agreement with experimental data (Gottlieb et al., 1989).

When the model is linearized based on the optimal run and then re-optimized with the linear model, one sees decreased agonist activity, increased muscle torque, lower initial acceleration, and higher peak velocities (Seif-Naraghi and Winters, 1989b).

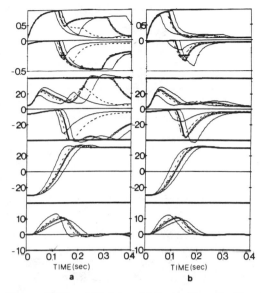

Figure 19.13: Added inertial load with and without neuromuscular criterion *a) EMGs* and muscle torques for both agonist and antagonist muscles, velocity profiles and position trajectories for the 400 ms elbow movement task. Performance criterion include 1.0 times *PE* and .01 times *ITO* under various added inertial loads: 0 (single, fastest with lowest activity), .06 (dash), .10(double) and .20 Kgm^2/s (single, slowest with most activity). *b)* Same as a except for .01 times *NE* and .02 times *MS* added to the performance criterion.

3. Isotonic Loading. Consider 60° elbow movements made in the presence of constant external torque inputs (isotonic loading) of varying magnitude which may be directed "with" or "against" the intended movement. In the case of only *PE* and *ITO* subcriteria being active (i.e. no neuromuscular penalty), the result is so much activity that the graphical representation of the result is at best "messy" (Seif-Naraghi and Winters, 1989b). However, a substantial reduction in neuromuscular activity and more moderate velocities result from added neuromuscular penalty (Figure 19.14). For the case of the highest opposing isotonic torque in Figure 19.14, a nearly trapezoidal velocity profile is accompanied by a flat muscle torque to overcome the external torque. One may expect an increase in agonist activity for the opposing external torques, but since the movement time is increased and there is a cost associated with neuromuscular activity, the best use of available input is to spread the movement over time for a coasting period.

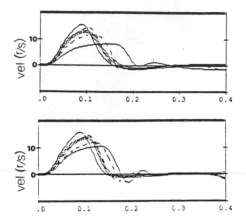

Figure 19.14: Added isotonic load with and without active neuromuscular subcriteria. **a.** Velocity profiles for optimized elbow movement task. Performance criterion includes 1.0 times *PE* and .01 times *ITO* under various added isotonic loads: +8 (single, fastest), +4 (dash, faster), 0 (double), –4 (dash, slower), –8 Nm (single, slowest). Note that + direction is the direction of the movement and – opposes the movement. *b)* Same as a except for .01 times *NE* and .02 times *MS* added to the performance criterion.

Notice that the velocity profiles (and also the muscle torque trajectories) are very different in the case of isotonic loading from those of inertial loading. The marked increase in muscle torque peaks which was the case with inertial loading is not present in isotonic loading, and the velocity profiles with inertial loading are much more bell-shaped than with isotonic loading. In addition the initial muscle torque levels are adjusted in isotonic loading to allow the initial joint position to be held in steady-state; this is automatically done for the final position by the optimization of the final neural excitation levels.

Despite considerable reduction of agonist and antagonist activity with added weight for the neuromuscular penalty subcriteria, a sufficiently large agonist muscle torque level is maintained to oppose the large external torque. The linearized and re-optimized movements lose all co-activation and exhibit reduced frequency of oscillation and smooth velocity trajectories since impedance modulation is no longer possible (Seif-Naraghi and Winters, 1989a). Additionally, higher peak velocity and lower initial acceleration of the linearized model demonstrate the ability of the nonlinear model to modulate the viscosity of the

CE through the neural input. The absence of co-contraction as a means of increasing impedance in the linearized case once again points to the fact that the optimal control solution takes advantage of the nonlinear series elastic relation.

19.5 Observations and Conclusions

1. To help place our findings in perspective, notice that none of the models that have been presented here allow for inclusion of neurocircuitry. Idealized, optimized open loop neurocontrol pulses to the muscle are obtained without constraints of delayed sensory feeback. The optimized neuromotor control strategies are simply solutions to a dynamic optimization problem in which the **goal** is to get a **biomechanical model** to perform an **explicit task** optimally. The goal of the movement is defined by the performance criterion, and the optimization algorithm performs many iterations (or trials) to converge toward (or learn) the solution (or strategy). Consequently, in all cases except for tasks with added psuedorandom perturbations our results **should be compared to performance in well-practiced, skilled subjects.** Note that even in the case of the pseudorandom loading the nature and peak magnitude of the loading would be well known through practise. However keep in mind that the biomechanical system has ample chance to practise simple everyday tasks.

2. For **unloaded** point-to-point movements involving a single joint, there is a remarkably consistent natural progression between optimal solutions ("strategies") for low- and high-order models. In other words, for this class of tasks, the form of the strategy is most sensitive to the **form of the performance criterion.** Subtle details, such as the timing of neural pulses, are governed by the form of the biomechanical model and the magnitudes of the parameters within the model. For instance, "minimum time" solutions tend to be "bang-bang" in all cases, while adding "effort"-related penalty results in strategic scaling of pulses in height and/or width (depending on form of subcriteria), often with "zero" input coasting regions. Of note is that the simple penalty functions for the second-order model (e.g. $|u|$, u^2) become expanded into a more generalized "neuromuscular" penalty subcriteria which includes inputs and some state variables.

3. As addressed in Chapter 12 (Feldman et al.) and Chapter 14 (Gottlieb et al.), a major issue in movement control is how (typically unloaded) movements are *scaled* in speed and magnitude. The conceptual approaches in Chapters 12–17 tend to be based on *kinematic* foundations. Using optimization methods we find that scaling movements via kinematic criteria alone produces *overly aggressive* control signals – this general trend was true for criteria as diverse as a virtual cost reference model and measures of smoothness (jerk). A highly refined reference trajectory is an awkward approach for scaling, as s the concept of explicit inverse dynamic calculation. In general kinematically-forced scaling results in aggressive "pull–pull" strategies that seem to be uncharacteristic of available experimental observation (e.g., data presented in Chapters 12, 14, 15). Adding penalty for jerk does not change this situation. Solutions for runs which include neuromuscular penalty become relatively insensitive to the subtle differences in the form of the virtual trajectories [compare Chapter 12 (Feldman et al.) to Chapter 17 (Flash)] as long as this trajectory is faster than (i.e., leads) the model. Our results encompass elements of both "speed sensitive" and "speed insensitive" strategies presented in Chapter 14 (Gottlieb et al.). In general, higher neuromuscular penalty results in slower movements that are more compatible with the speed-sensitive strategy. However, when *NE* penalty is higher, pulse-height modulation is more prevalent, and when *MS* penalty is higher, pulse-width modulation is more likely.   Perhaps kinematic-based optimization could dominate for certain sets of instructions to the subject. However, in our opinion, in most everyday tasks, the subject's perception of "effort" strongly influences both movement planning performance and neuromotor execution of the plan. These observations hold true for all models checked: elbow, wrist and head, and thus seem to represent a general finding (Seif-Naraghi, 1989).

4. It is recommended that neuromuscular penalty be included for all dynamic optimization work (other than perhaps certain short-term sports activities performed by elite athletes). We also suggest that criteria such as *NE*, *MS* and *MD* can (and perhaps should) be employed as a unit. This recommendation contrasts the norm in upper-limb movement organization studies, where theories on movement organization tend to be kinematically based. Examples of this include the minimum jerk hypothesis (Hogan, 1984; see also Chapter 17 (Flash) and the "effort" hypothesis of Hasan (1984). Interestingly, we find that jerk is dramatically lowered as a natural byproduct of increasing the neuromuscular penalty. One rationale behind the minimum jerk hypothesis is that bell-shaped velocity trajectories are experimentally common, as predicted by this hypothesis. Notice that the nominal "very fast" movement is not bell shaped, but that as any of the weights for neuromuscular penalty grow, the velocity profile becomes progressively more symmetrical (Figures 19.9–19.14).

5. Added *inertial loading* had distinct effects on control strategies.   Neural activity, movement times, and muscle torques all increased, velocities and accelerations decreased, and velocity profiles become symmetric. Increasing *J* has the effect of making the model appear more "second-order", and consequently the control strategies becomes progressively more similar to those found for a simple second-order model.

6. For *interaction with the environment* (i.e. external loading other than simply adding *J*, *B,* or *K*), second-order linear models become useless. Furthermore, nonlinear properties become of great importance. Movements under external perturbation show *modulation of the system impedance* through co-contraction directed at minimizing position error. They also prove the generality of some of the features of different neuromuscular criteria; relative to each other, the *NE*, *MS*, and *MD* subcriteria show similar characteristics both with and without the application of external perturbation. Furthermore, as with unloaded movements, the same trends are seen for elbow, wrist and head models (Seif-Naraghi, 1989).

7. Isotonic loading of the joint results in increased activity when opposing the direction of the movement. However, the velocity profiles are more trapezoidal and adding neuromuscular criterion to the performance index results in a zero-acceleration (i.e. coasting) period which is accomplished by sustained neural activity (to oppose the load) at lower than maximum levels. These results generally favor inclusion of some neuromuscular penalty performance criterion of daily movements. They also make predictions which can be tested experimentally.

8. In Section 14.7, Gottlieb et al. discuss optimization. It is stated that optimization methods "... require the *a priori* specification of movement time T before an optimal solution can be calculated ... ." This is not true for many of the second-order closed-form solutions which we have reviewed and for *all* of our solutions with the higher-order model. Movement time was not prespecified; conversely, it was merely a byproduct of the optimization process. Their comments are only valid for overly simple, single-component optimization criterion which the authors chose to critique. Optimization criterion which include multiple (competing) subcriteria will by nature have variable movement times.

9. There is no basis for the common perception that finding *numerical* solutions for nonlinear models is more difficult than for linearized models; in fact, we have found that the converse is often the case. The neural inputs can indirectly modify the internal parameters of the nonlinear model and therefore provide flexibility and richness to the control strategy. For instance, with nonlinear models, impedance modulation (of *both* stiffness and viscosity) via cocontraction becomes a useful option. Although not emphasized in this chapter, this is especially true for tasks involving movement turn-arounds (Seif-Naraghi, 1989). Interestingly, the utilized optimization algorithm has a more difficult time arriving at a solution when there is no neuromuscular penalty. This may be due to such penalty sculpturing the otherwise rugged cost–control-parameter space; perhaps this also holds true for higher neuromotor structures as well. It also has a harder time finding solutions for simple linear second-order models than for complex nonlinear, higher-order models. There is also no basis for the common perception that performance criteria with fewer terms may be easier to optimize. Perhaps by searching for relatively simplistic rule-based organizational strategies, researchers miss the opportunities to explore the underlying processes.

10. The above observations have ramifications regarding neural network research. We suggest that such researchers should utilize nonlinear muscle models so that the neural network can learn to take advantage of beneficial nonlinear muscle properties (i.e., a "mechanically smart" periphery).

11. What experiments are most likely to help uncover movement organizational strategies? Conventional wisdom (and pragmatic experimental common sense) suggest that well-controlled tasks with little variation should be studied. We find that insights do not come from isolated optimization results but rather from considerations of how *strategies change as goals change* within a certain class of movements. We also often find "valleys" within the cost–control space where there are a spectrum of solutions that are within a small neighborhood of the optimal solution. Our results suggest that fundamental organizational strategies are likely to be revealed by considering the subtle *differences* that occur when task goals are changed, preferably by the subject as opposed to the experimenter. Furthermore, as discussed in Chapter 34 (Pedotti and Crenna), when tasks are not highly constrained, a number of significantly different strategies may be utilized that are all nearly optimal.

19.6 Future Directions

19.6.1 Needed Experiments

Mathematical modeling and dynamic optimization can be a useful tool for making predictions about the goal-directed movements. However, difficult experimental designs are needed to test these predictions.

The issue of perception of instructions, which often preceed a performed task, is also worthy of consideration for future experiments. These commands to the subject, in our view, help shape the relative weights in the performance criterion and therefore decide the objective goal of the movements. Experiments linking the verbal instruction to the movement startegies and possibly to the performance criterion will help test the predictions made by the model.

19.6.2 Optimizing "Above" Neurocircuitry

Up to the present, with only a few exceptions where the control space included feedback gains (Seif-Naraghi and Winters, 1988), our concentration has been on open-loop control at the motoneuronal level. Another possibility, suggested by insights within Chapter 12 (Feldman et al.) and Chapter 16 (Karst and Hasan), is to include basic neural connections within the "system" and then optimize control parameters that could include terms such as the C, R and RI commands of Chapter 12.

19.6.3 Multi-Joint Movements

An interesting question, which we are currently addressing, is whether the insights from one joint extend to multi-joint systems. Observations in favor of this include the fact that for certain combinations of active performance subcriteria, certain *forms* of solutions were seen irrespective of the details of the model; however, the model structure and parameters did dramatically affect the actual signals. Conversely, however, when external loads were added, higher-order, nonlinear models became crutial. As discussed in Chapter 8 (Zajac and Winters), a given joint doesn't distinguish between external loads and dynamic coupling between links. This suggests that for multi-link systems we should immediately start with higher-order nonlinear muscle models. Within this context we believe that conceptual foundations found for the higher-order, single-joint model will extend to multi-joint cases.

However, the problem is difficult because of the wealth of possibilities. Multi-joint movements can scale not only in speed and magnitude but also direction. They also can take place within different regions of the workspace. Furthermore, external loads have direction as well as magnitude. Finally, tasks other that point-to-point are of great interest. Unlike for cyclic and propulsive movements involving the lower limb, one cannot define certain stereotyped tasks for consideration.

As discussed earlier, we believe insight into movement strategies comes not from any one optimization run but rather from a consideration of how optimal solutions changes as the goals for a certain class of tasks change. Consider a minimal set of point-to-point movements. With only two movement magnitudes, six movement directions spaced 60° apart, and three different regions within the workspace result in 36 distinct tasks. Furthermore, for each task we would want to employ three models: a linearized Hill model with single-joint muscles at shoulder and elbow, a nonlinear Hill-based model with only single-joint muscles, and a nonlinear model that also includes bi-articular muscles (i.e. 108 optimization runs). Additionally, for each of these, we can immediately identify 7 basic performance criteria which would be representative: *i)* small "target" criteria (a circle of a radius equal to 5% of distance); *ii)* large "target" criteria (a circle of a radius equal to 10% of distance); *iii)* straight-line virtual cost trajectory (very fast, similar to "step" change in position), *iv)* slow virtual trajectory; *v)* fast virtual trajectory with jerk added; *vi)* also with small nerumuscular penalty; and *vii)* large neuromuscular penalty. If we were to perform all 7 of these fundamental possibilities for each of the 108 cases, we would require **756 optimizations** just for point-to-point movements! This appears overwhelming. However, we are exploring certain strategic areas of the workspace — for instance, the strategic directions described in Chapter 16 (Karst and Hasan).

Four other classes of experiments are of high priority: *i)* the curved-path experiments of Flash (e.g. see Chapter 17); *ii)* the curved-path experiments of Schneider et al. (1989); *iii)* rapid voluntary elbow flexion-extension oscillatory movements in which the shoulder is to remain relatively fixed (which should help establish relationships between posture and movement, including a role for cocontraction of a proximal "base"); and *iv)* maximal velocity throwing experiments (which may identify optimal strategies for utilizing the stretch-shortening cycle).

Another future direction of great practical and theoretical relevance is to consider arm-torso interaction in the seated position. This would essentially be an extension of the types of studies presented in Chapter 29 (Ramos and Stark) and Chapter 31 (Ong et al.), where postural stability and volitional movement must both be addressed. The minimal model for addressing fundamental issues would appear to be a four-link system including a 2-link arm, a link between the shoulder and thoracic spine, and a link between the pelvis and the thoracic spine.

In closing let us re-emphasize that there are many possible applications for dynamic optimization in the study of goal-directed movement strategies. These predicted strategies are byprodcts of a stated goal within the performance criterion that can then be tested experimentally.

References

Athans, M. and Falb, P.L. (1966) Optimal control: an introduction to the theory and its applications. McGraw-Hill, New York.

Bremermann, H. (1970) A method of unconstrained global optimization. *Math Biosci.*, **9**: 1-15.

Burges, D. and Graham, A. (1980) Introduction to control theory including optimal control. Ellis Horwood, Chister, England.

Corcos, D.M., Gottlieb, G.L., Agarwal, G.C. (1989) Accuracy constraints upon rapid movements. *J. Motor Behavior.*

Dul, J., Townsend, M.A., Shiavi, R. and Johnson, G.D. (1984) Muscular synergism - I. On criteria for load sharing between synergistic muscles. *J. Biomech.*, **17**: 663-673.

Georgopoulos, A.P.: On reaching. *Ann. Rev. Neurosci.*, **9**, 147-170, (1986).

Hannaford, B., Stark, L.: Roles of the elements of the triphasic control signal. *Exper. Neurol.*, **90**, 619-634, (1985).

Hasan, Z., Enoka, R.M., Stuart, D.G. (1985) The interface between biomechanics and neurophysiology in the study of movement: some recent approaches. *Exerc. & Sport Sci. Rev.*, **13**: 169-234.

Hogan, N. (1984) Adaptive control of mechanical impedance by coactivation of antagonistic muscles. *IEEE Trans. on Autom. Control,* **AC-29**: 681-690.

Hogan, N., Bizzi, E., Mussa-Ivaldi, F.A., Flash, T. (1987) Controlling multijoint motor behavior. *Exerc. & Sport Sci. Rev.*, **15**, 153-190, (1987).

Lehman, S., Stark, L. (1982) Three algorithms for interpreting models consisting of ordinary differential equations: sensitivity coefficients, sensitivity functions, global optimization. *Math Biosci.*, **62**, 107-122.

Lestienne, F. (1979) Effects of inertial load and velocity on the braking process of voluntary limb movements. *Exp. Brain Res.*, **35**, 407-418, (1979).

Levine, W.S., Zajac, F.E., Belzer, M.R., Zomlefer, M.R. (1983) Ankle controls that produce a maximum vertical jump when other joints are locked. *IEEE Trans. Autom. Control* **AC-28**, 1008-1016.

Nelson, W.L.: Physical principles for economies of skilled movements. *Biol. Cybern.* **46**, 135-147, (1983).

Oguztorelli, M.N. and Stein, R.B. (1983) Optimal control of antagonistic muscles. *Biol. Cybern.*, **48**: 91-99.

Pontryagin, L.S., Boltyanskii, V.G., Gamkredligze, R.W., Mishchenko, E.F. (1962) *The mathematical theory of optimal processes.* John Wiley, New York.

Seif-Naraghi, A.H. (1989) Control of human arm movement via optimization. Ph.D. Dissertation, Arizona State University.

Seif-Naraghi, A.H., Winters, J.M.: Fast movements in unknown environments: tradeoffs. (1987) *Proc. IEEE Engng. Med. & Biol.*, pp. 268-269, Boston.

Seif-Naraghi, A.H. and Winters, J.M. (1989a) Effects of task-specific linearization on musculoskeletal system control strategies. *ASME Biomech. Symp.*, **AMD-98**: 347-350.

Seif-Naraghi, A.H. and Winters, J.M. (1989b) Changes in musculoskeletal control strategies with loading: inertial, isotonic, random. *ASME Biomech. Symp.*, **AMD-98**: 355-358.

Takahashi, Y., Rabins, M.J. and Auslander, D.M. *Control and Dynamical Systems.* Addison-Wesley Publ. Co., Reading, Mass.

Wadman, W.J., van der Gon, J.J., Derkson, R.J.A. (1979) Muscle activation patterns for fast goal-directed arm movements. *J. Human Movem. Stud.*, **5**, 3-17.

Winters, J.M., Seif-Naraghi, A.H. Optimized neural strategy for dynamic movements: I. Basic concepts and findings. *Biol. Cybern.*, submitted.

Winters, J.M., Stark, L. (1985) Analysis of fundamental movement patterns through the use of in-depth antagonistic muscle models. *IEEE Trans. Biomed. Engng.*, **BME-32**: 826-839, (1985).

Winters, J.M., Stark, L. (1987) Muscle models: what is gained and what is lost by varying model complexity. *Biol. Cybern.* **55**: 403-420, (1987).

Winters, J.M., Stark, L. and Seif-naraghi, A.H. (1988) An analysis of sources of muscle-joint system impedance. *J. Biomech.*, **12**: 1011-1025.

CHAPTER 20

Self-Organizing Neural Mechanisms Possibly Responsible for Muscle Coordination

J.J. Denier van der Gon, A.C.C. Coolen, C.J. Erkelens, and H.J.J. Jonker

20.1 The Motor System

In this chapter we will discuss certain mechanisms that may play an important role in the organization and the learning of motor control. These mechanisms are based on the generally accepted notion that the strength of synaptic connections between neurons in the central nervous system can be modified under the influence of synchronous activity of these neurons. Thus, neural signals not only contain current information, they also contribute to the long-term organization of neural structures.

Two streams of neural information are essential in motor control. One, afferent stream of signals contains a manifold of sensory information originating from all the different types of sensory systems. These systems provide the brain with partially supplementary and partially overlapping information. With signals from each of these systems, the brain can build up an internal representation of a specific physical aspect of the state of the body and/or of the environment. Simultaneous activity of the sensory systems enables the brain to form a complete internal representation of the state of the body in relation to its environment.

The other, efferent stream of information consists of control signals for the effector systems (α-activation), as well as for the sensory systems (for instance γ-activation). Such signals initiate contractions of muscle fibers which enable man and animal species to navigate through their environment. Moreover, muscle contractions generate motion that is important for the creation of internal representations because motion initiates simultaneous activity in the various sensory systems.

An oversimplified, hierarchically structured, functional model of the motor system is shown schematically in Figure 20.1. General motor plans are formed at the highest level of control. Target selection, choice of the speed of movement, choice of the effector system, considerations with respect to constraints of movement (obstacle avoidance) are ingredients of the general motor plan. This intriguing, but extremely difficult to model, level of control will not be the subject of further discussion. At the next lower level of control we find two functions which are thought to be localized in the sensorimotor cortex (Brooks, 1986). In this structure an internal representation (*IR*) is formed of the position (and movement) of the selected target in relation to the position of the body. The creation of a specific motor plan is supposed to take place in the same part of the brain. This plan is a lay out of a distinct pathway or trajectory (van Sonderen et al., 1989, 1990) for the intended movement. Information for this plan is obtained from the *IR* and restrictions are set by the general motor plan. At a still lower level, movement trajectories are translated into motor programs for those effectors that will be involved in the action. These motor programs contain the codes of the activation patterns for the muscles. Finally, at the level of the spinal cord, these patterns are distributed among the appropriate muscles. Adjustments can be made by means of regulating the gains and thresholds of spinal reflexes mediated by the proprioceptive system.

Multiple Muscle Systems: Biomechanics and Movement Organization
J.M. Winters and S.L-Y. Woo (eds.), © 1990 Springer-Verlag, New York

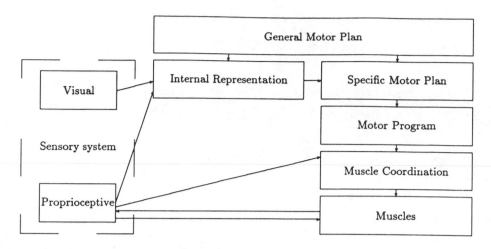

Figure 20.1: Functional model of the motor system. The terminology has been adopted from Brooks (1986).

In this chapter attention will focus on possible mechanisms underlying the learning of muscle coordination and the generation of motor programs. In a first example we will show how in principle an internal representation of elbow movement can be created on the basis of proprioceptive information obtained from muscle spindles. In a second example, a method is presented by which reflexes in response to random disturbances can be made more adequate. To achieve this result, groups of excitatory and inhibitory interneurons learn to contribute to these reflexes on the basis of activity of homonymous connections. Finally, we will show how neural systems may generate sequences of activation patterns, which together may make up a motor program.

20.2 Learning in Neural Networks

Neural networks are best viewed as consisting of two types of variables: fast variables and slow variables. The states of the neurons are the fast variables; firing states can vary on a timescale of milliseconds. The neural connection strengths (or synaptic efficacies) and the neural firing thresholds are the slow variables (their values are modified on a much larger timescale). In the values of the latter variables all information is stored, both concerning the outside world and concerning the function (or program) of the network. In neural network theory it is generally assumed that a neuron i will fire if

$$\sum_j J_{ij} s_j > \theta_i \qquad (20.1)$$

The left-hand side of *Eq. 20.1* is the post-synaptic potential of neuron i; θ_i is the threshold for excitation of neuron i. J_{ij} denotes the strength of the synaptic connection between the axon of neuron j and the dendrite or cell-body of neuron i. If $J_{ij} > 0$, neuron j has an excitatory effect on neuron i; if $J_{ij} < 0$, the effect is inhibitory. Finally, s_i represents the firing frequency of neuron i. In a physicist's model s_i is usually taken to be either -1 (neuron i is at rest) or $+1$ (neuron i fires with maximum frequency), a so called *Ising spin*.

The learning process amounts to modifying connections J_{ij} and thresholds θ_i, based on the current activity $s = (s_1,...,s_N)$ of the N neurons in the network. The rule used for modification of connections and thresholds (the 'learning rule') will be called *local* if it is based solely on the information available at the physical location of the variable that undergoes the modification. If a synapse J_{ij} is modified, then ΔJ_{ij} can only depend on s_i, s_j and $\sum J_{ij} s_j$ (the state of the post-synaptic neuron, the state of the pre-synaptic neuron and the post-synaptic potential, respectively). If a threshold θ_i is modified, then $\Delta \theta_i$ can only depend on s_i and $\sum J_{ij} s_j$. Only models which apply local learning rules can be considered realistic from a neurophysiological point of view. We will restrict ourselves accordingly.

In 1982 Hopfield (1982) introduced a model in which he used a version of Hebb's (1949) learning rule: if the network is in state $\vec{s}$ all synapses are modified according to

$$\Delta J_{ij} = \frac{1}{N} s_i s_j \qquad (20.2)$$

All thresholds are zero. If this system undergoes a learning phase, in which a number of activity states $\vec{\xi}^{(\mu)}$ are enforced upon the system, the final connection matrix is such that the states $\vec{\xi}^{(\mu)}$ are the stable activity states of the network (they are said to be 'stored'). If a network state resembles one of the stored patterns, say pattern μ, then the subsequent autonomous evolution of the neural activity will be a relaxation towards equilibrium state $\vec{\xi}^{(\mu)}$ ('associative recall'). If the network is divided into an input layer and an output layer, then the very same procedure allows one to store input-output mappings. Finally, if one takes into account transmission delays of neural signals, systems of the Hopfield type can store and generate sequences of activity patterns as well.

One of the prominent features of these types of models is that they allow for analytical solutions. As a result of their resemblance to statistical mechanical models for magnetic substances, one can apply the same (rich) toolbox. The *microscopic* level of description (the states of single neurons) is abandoned in favor of a level of well-chosen *macroscopic* features. For these features one generally chooses the correlations between the actual microscopic network state $\vec{s}$ and the patterns $\vec{\xi}^{(\mu)}$. In this way the equilibrium properties (Amit et al., 1985, 1987; van Hemmen and Kühn 1986) as well as the dynamical properties (Derrida et al., 1987; Coolen and Ruijgrok, 1988) of these models can be studied analytically.

Another class of neural network models are the so-called competitive systems. They consist of two or more separate interacting layers. In an input-layer input patterns are presented; in a second layer (not directly accessible from the outside world) competition takes place, such that the neurons in this layer will tend to respond to only one specific input (or one specific type of input) of the input-layer. In other words, *feature detectors* will emerge. One specific model proposed by

Kohonen (1982) is capable of creating topologically correct feature maps of the input. One can create topological maps of the outside world, if this world is observed by sensors whose outputs are fed into the input–layer of Kohonen's model. The topological maps can then be used as internal representations of the outside world. The only restriction is that the sensors must preserve the world's topology, i.e. small changes in observed events (limb positions, target positions, sound frequencies etc.) must lead to small changes in the corresponding sensory signals. One disadvantage of Kohonen's original model is that during the competition process a "supervisor" is needed who decides which neuron is the winner of each round of the competition and who can tell which neurons are neighbors. However, it can be shown that the original model can be modified in such a way that Hebb's rule will suffice to generate the desired order without a supervisor.

We believe that the neural mechanisms just discussed (all of which have the advantage of using autonomous local learning rules) in principle seem sufficient to (learn to) perform the tasks that are encountered in human movement coordination and sensor-motor interaction. In the following sections we will illustrate this by studying some specific examples in more detail.

20.3 Internal Representations Created from Spindle Outputs

As an illustration of the construction of an internal representation we consider the elbow system. This system has two degrees of freedom, namely flexion-extension and supination-pronation of the forearm. Movements resulting from rotations in the elbow joint or moments generated by muscles that act over the elbow joint can be drawn in a diagram like Figure 20.2. The diagram shows the directions of the moments exerted by the main muscles.

In this example we will confine ourselves to directions and disregard magnitudes of displacements or forces. These directions can be mapped topologically onto a one-dimensional closed array of neurons, e.g. a circular array. According to Kohonen's theorem this mapping can be brought about by a sensory system that has directional sensitivity, for instance the spindles in the muscles that act over the elbow joint. A disturbance of the position of the forearm will only cause activity in the spindles of those muscles that are stretched. Each spindle is assumed to project onto each neuron of the circular array (Figure 20.3).

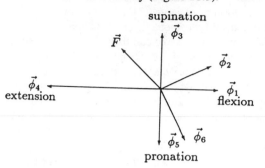

Figure 20.2: Moments ϕ of the main muscles, acting over the elbow joint. F denotes an external disturbance which leads to activity of the spindles in the muscles that are stretched. Meaning of the symbols ϕ_1: m. brachialis/brachioradialis; ϕ_2: m. biceps; ϕ_3: m. supinator; ϕ_4: m. triceps; ϕ_5: m. pronator; ϕ_6: m. pronator teres; F: direction of disturbance.

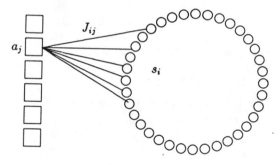

Figure 20.3: Spindle sensors a_j and a circular array of feature detectors s_i. The connections between sensors and detectors are modified according to Kohonen's organization principle.

For simplicity we replace the spindles within a muscle by just one sensor that signals the total activity a_j of the spindles of muscle j. A neuron of the circular array now receives an input equal to

$$h_i = \sum_j J_{ij} a_j \qquad (20.3)$$

where J_{ij} denotes the synaptic weight of spindle group j onto neuron i in the 'ring'. Initially the values J_{ij} are chosen randomly. A learning trial consists of applying a disturbance in a random direction to the elbow joint and determining which

of the neurons in the ring is activated most according to *Eq. 20.3* (say neuron number k). The next step is to update synaptic weights according to the following procedure: we let all J_{kj} increase in proportion to the spindle activity a_j. Next the new values of J_{kj} are rescaled in order to keep the total synaptic weight $\sum_j J_{kj}$ a constant.

$$J'_{kj} = \frac{J_{kj} + \alpha a_j}{\sum_l (J_{kl} + \alpha a_l)} \qquad (20.4)$$

where α is a constant. Following Kohonen we assume that this process is not restricted to neuron k but that it also applies to neurons that are, up to certain degree, neighbors of k (see also Bonhoeffer et al., 1989). Initially the neurons that are activated most upon disturbances in the elbow joint are randomly distributed over the ring. However, after a number of trials, in which the weights are updated according to the process described above, these neurons reflect topologically the direction of the disturbance applied. A result is shown in Figure 20.4. For the central nervous system the elbow joint is part of the outside world and an internal representation of this part is formed. If the system generates disturbances itself it is obviously able to develop actively internal representations.

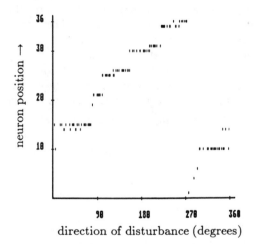

Figure 20.4: The result of the learning process in Kohonen's model: the position of the neuron which is activated most as a function of the angle of the disturbance.

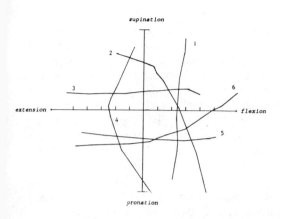

Figure 20.5: Voluntary exerted moments that give rise to a constant activation of the muscles. The numbers refer to the muscles as listed in Figure 20.2.

At the end of the learning phase the neuron that is activated most receives its input from a few mostly synergistic muscles. If the weights J_{ij} with which the muscle sensors contribute to the activity of a certain neuron are, in turn, used as weights to activate the muscles, coordinated muscle activity results which shows a striking overall resemblance with actually observed coordinated muscle activity. Figure 20.5 shows an example. The lines in the diagram represent exerted moments for which the muscles are activated proportionally to the weights J_{ij}. The level of activation is chosen such that for each line a muscle has a constant activation. These lines can be compared directly to critical firing level lines (see e.g. van Zuylen et al., 1989). Average measured slopes show a fair agreement with the computations.

Kohonen's model suggests an efficient way of creating an internal representation of the outside world as 'viewed' by the muscle spindles. It is clear, however, that from a neurophysiological point of view there are non-realistic ingredients. There is the supervisor who decides which neuron is activated most. Also, a mechanism is needed to restrict the synapse modification to connections belonging only to this very neuron and its neighbors.

In an alternative model we assume that all connections are modified according to a Hebbian rule:

$$\Delta J_{ij} = s_i (a_j - <a_j>) - \varepsilon J_{ij}$$ (20.5)

From the sensory signals a_j we have subtracted the average value (this procedure is necessary if the average is not equal to zero (Amit et al., 1987)). A decay term is introduced to replace the rescaling of the connections. In the second layer (the 'ring') we introduce two types of neural interactions: a long-range inhibition J and a short-range excitation Γ. The local input h_i of a neuron in the ring will now consist of two contributions:

$$h_i = h_i^{ext} + h_i^{int}$$ (20.6)

The external contribution (coming from the input layer) is again given by Eq. 20.3. In addition we have:

$$h_i^{int} = -J \sum_j s_j + \Gamma \sum_{j \in N_i} s_j$$ (20.7)

where N_i contains the neurons near i that contribute to Γ. Owing to their mutual interactions, the activity of the neurons in the ring will no longer be a simple function of the sensor signals a_j. If, for some given sensory input $\{a_j\}$, all neurons in the ring update their activity in random order according to the value of the input h_i minus the neural threshold θ, then one will observe a relaxation process in which some equilibrium configuration of the activity in the ring will be approached (Figure 20.6).

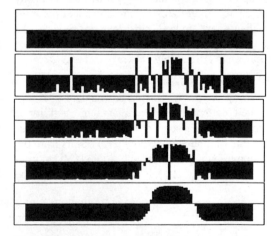

Figure 20.6: Development of activity of the neurons in the ring due to the presence of long-range inhibition, short-range excitation and a positive threshold. The upper panel shows the initial states of neural activity (all neurons at rest). The bottom panel shows the equilibrium states of neural activity (active neurons are clustered).

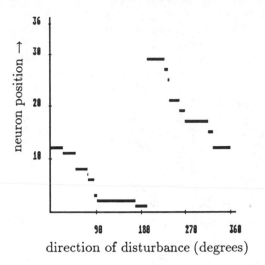

direction of disturbance (degrees)

Figure 20.7: The result of the learning process in the non-supervised version of Kohonen's model: the position of the neuron which is actived most as a function of the angle of the disturbance.

By choosing J and θ, one can control the average activity level of the regime of possible equilibrium states. Finally the short-range excitation Γ will further restrict the set of possible equilibrium states in the ring to those which are the same shape as the equilibrium state shown in Figure 20.6. The actual position of the active group of neurons in the ring will be determined by the input. Figure 20.7 demonstrates that these ingredients are indeed sufficient to replace the supervisor in the standard Kohonen model (if the parameters are suitably chosen).

20.4 Reflex Coordination

When a limb is displaced suddenly, a segmented pattern of electromyographic activity is often observed in some of the muscles. In this pattern, the first burst of activity (*M1*) results from the homonymous, monosynaptic loops. The next burst of activity (*M2*) is often referred to as the

functional stretch reflex because of its longer duration and often higher amplitude (Marsden et al., 1976). Additionally, the force generated by this part of the reflex counteracts the disturbing force much more accurately than does the early reflex (Gielen et al., 1988). In this section we present a mechanism that may play a role in the organization of this long latency reflex and in the shaping of polysynaptic, heteronymous connections.

In the model muscle spindle activity during stretch is related to the relative change in muscle length $\Delta L/L_o$, where ΔL represents the amount of stretch and L_o the initial length of the muscle (Jongen et al., 1989). Anatomical data concerning the elbow joint are used in our model to compute the muscle spindle activity (Jongen et al. 1989).

In the model (Figure 20.8), groups of muscle spindles project monosynaptically onto the motoneuron pool of the homonymous muscle through fixed connections. In addition, the groups of muscle spindles project onto corresponding groups of inhibitory interneurons and onto groups of excitatory interneurons. Each group of inhibitory interneurons projects onto all groups of excitatory interneurons, with modifiable connections (J^-). The groups of excitatory interneurons project onto the motoneuron pools of all muscles, also with connections with modifiable weights (J^+). All neurons obey Dale's law: excitatory neurons can only have excitatory connections whereas inhibitory neurons can only have inhibitory connections.

A possible criterion for an adequate reflex is that the direction of the total, generated force is opposite to the direction of the disturbance vector $\vec{F}$. Thus, if Θ_{refl} represents the direction of the generated reflex and Θ_{dist} the direction of the disturbance, the direction of an 'ideal' reflex is described by:

homonymous monosynaptic projections

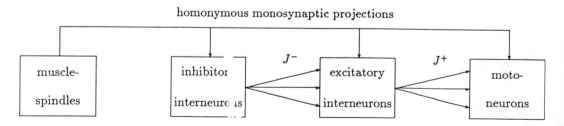

Figure 20.8: Architecture of the network.

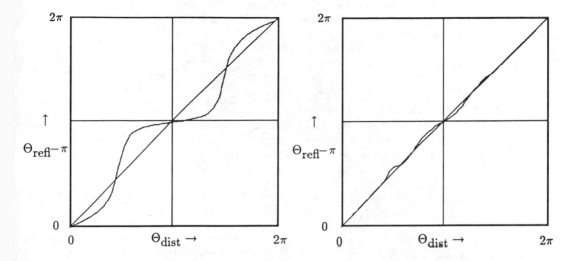

Figure 20.9: Direction of the *M1* reflex as a function of the direction of disturbance. The straight line represents the direction of the 'ideal' reflex, i.e.: opposite to the direction of disturbance.

Figure 20.10: Direction of the *M2* reflex as a function of the direction of disturbance. The straight line indicates the direction of the 'ideal' reflex.

$$\Theta_{refl} = \Theta_{dist} + \pi \qquad (20.8)$$

The *M1* reflex, caused by homonymous, monosynaptic connections, does not perform well with respect to the chosen criterion. Figure 20.9 clearly shows that the line indicating the direction of the *M1* reflex deviates largely from the straight line representing the direction of the 'ideal' reflex.

In order to obtain an *M2* reflex which is more adequate than the *M1* reflex, the modifiable connections in the neural network of Figure 20.8 are changed. During the learning process, randomly chosen disturbances $\vec{F}$ are applied to the elbow joint. Consequently certain muscles are stretched and their spindles are activated. As a result of the connections, activity spreads out through the network. Both J^+ and J^- are modified according to Hebbian rules, that is, J^+ according to *Eq. 20.2* but J^- according to an alternative Hebbian rule applied to inhibitory neurons (see Jonker et al., 1989). The latter rule assumes that inhibitory connections become more inhibitory if pre- and post-synaptic neurons are active simultaneously. Figure 20.10 shows the adequacy of the reflex after a learning period. Comparison of this figure with Figure 20.9 clearly shows that the adequacy of the reflex has improved dramatically.

20.5 Storage and Processing of Motor Programs

Motor programs can be thought of as sequences of activity patterns which are generated by the central nervous system and which are ultimately sent to the muscle fibers. Since the activity patterns run ahead of the movement, we looked for mechanisms that could store and generate sequences of patterns. A possible candidate is a network of the Hopfield type with transmission delays. If this network has experienced a learning phase during which a sequence of patterns is presented, it will end up having connections in which the correlations between bits of subsequent patterns are stored (Hopfield, 1982; Sompolinski and Kanter, 1986):

$$J_{ij} = \frac{1}{N} \sum_{\mu} \xi_i^{(\mu+1)} \xi_j^{(\mu)} \qquad (20.9)$$

If this network is forced into the first state $\vec{\xi}^{(1)}$ of the sequence, it will subsequently generate all the following states of this sequence:

$$\vec{\xi}^{(1)} \rightarrow \vec{\xi}^{(2)} \rightarrow \vec{\xi}^{(3)} \rightarrow \dots \qquad (20.10)$$

In this way motor programs may be stored and processed.

Furthermore, since it is rather unlikely that each of all the possible motor programs has been stored, a neural network must be able to perform interpolations between the activity patterns that actually have been learned. The question of how interpolations can be performed in neural networks is more difficult to answer. Recently a model was proposed that consists of two layers: an input layer of excitatory neurons and an output layer of mutually interacting inhibitory neurons. As a learning rule the same rule as introduced in Section 20.4 was used in the model. It was shown that this two-layer system is capable of performing interpolations between the input–output relations that have been learned (Jonker et al., 1989).

20.6 Concluding Remarks

If one is interested in obtaining a basic understanding of complex motor functions, it seems worthwhile reflecting upon their neural substrates and the organizing mechanisms of the neural machinery. After all, many motor functions are learned and are adaptable to changing circumstances (e.g. due from growth). Experimental exploration of parts of the nervous system is often an endless task unless one knows what kind of mechanisms and organization one should look for. Neural network theory suggests a number of possible learning- and self-organizing mechanisms that may be used by the central nervous system. In this contribution we discussed a few of these mechanisms and we have speculated on where they might play a role in the motor organization. Although we do not claim that these mechanism are indeed used in the way we suggested, considerations like those presented give rise to hypotheses that can be tested. We have confined ourselves to muscle- and reflex- coordination and to a few aspects of motor programs, since much data are available on these subjects. The examples are still far remote from the real motor system. However, we think that further modelling based on neural principles could turn out to be a rewarding investment.

Acknowledgment

We gratefully acknowledge the contributions of D.J. Glastra, G.J. Meijer and P.B. Nederkoorn, who carried out the simulations.

References

Amit, D.J., Gutfreund, H. and Sompolinsky, H. (1985) Spin-glass models of neural networks. *Phys. Rev.* A32: 1007-1020.

Amit, D.J., Gutfreund, H. and Sompolinsky, H. (1987) Information storage in neural networks with low levels of activity. *Phys. Rev.,* A35: 2293-2303.

Bonhoeffer, T., Staiger, V. and Aertsen, A. (1989) Synaptic plasticity in rat hippocampal slice cultures: Local 'Hebbian' conjunction of pre- and postsynaptic stimulation leads to distributed synaptic enhancement. *Proc. Natl. Acad. Sci.,* 86: 8113-8117.

Brooks, V.B. (1986) *The Neural Basis of Motor Control.* Oxford University Press, Oxford.

Coolen, A.C.C. and Ruijgrok, Th.W. (1988) Image evolution in Hopfield networks. *Phys. Rev.* A38: 4253-4255.

Derrida, B., Gardner, E. and Zippelius, A. (1987) An exactly solvable asymmetric neural network model. *Europhys. Lett.,* 4: 167-173.

Gielen, C.C.A.M., Ramaekers, L. and Zuylen, E.J. van (1988) Long-latency stretch reflexes as co-ordinated functional responses in man. *J. Physiol.* 407: 275-292.

Hebb, D.O. (1949) *The Organization of Behaviour.* Wiley, New York.

Hemmen, J.L. van and Kühn, R. (1986) Nonlinear neural networks. *Phys. Rev. Lett.* 57: 913-916.

Hopfield, J.J. (1982) Neural networks and physical systems with emergent collective computational abilities. *Proc. Natl. Acad. Sci.* 79: 2554-2558.

Jongen, H.A.H., Gielen, C.C.A.M. and Denier van der Gon, J.J. (1989) Activation of human arm muscles during flexion/extension and supination/pronation Tasks: A theory on muscle coordination. *Biol. Cybern.* 61: 1-9.

Jonker, H.J.J., Coolen, A.C.C. and Denier van der Gon, J.J. (1989) Linear interpolation with binary neurons. *Proc. Artifciul Neural Networks.* IEE, London.

Kohonen., T. (1982) Self-organized formation of topologically-correct feature maps. *Biol. Cybern.,* 43: 59-69.

Marsden., C.D., Merton, P.A. and Morton., H.B. (1976) Stretch reflex and servo action in a variety of human muscles. *J. Physiol.* 259: 531-560.

Sompolinsky, H. and Kanter, I. (1986) Temporal association in asymmetric neural networks. *Phys. Rev. Lett.* 57: 2861-2864.

Sonderen, J.F. van, Gielen, C.C.A.M. and Denier van der Gon, J.J. (1989) Motor programmes for goal-directed movements are continuously adjusted according to changes in target location. *Exp. Brain Res.,* 78: 139-146.

Sonderen, J.F. van and Denier van der Gon, J.J. (1990) A simulation study of a programme generator for centrally programmed fast two-joint arm movements: responses to single- and double-step target displacements. *Biol. Cybern.* 63: 35-44.

Zuylen, E.J. van, Gielen, C.C.A.M. and Denier van der Gon, J.J. (1988) Coordination and inhomogeneous activation of human arm muscles during isometric torques. *J. Neurophysiol.* 60: 1523-1548.

CHAPTER 21

External Control of Limb Movements Involving Environmental Interactions

Patrick E. Crago, Michel A. Lemay, and Like Liu

21.1 Introduction

External control of both upper and lower limbs can be achieved by electrical stimulation of muscles, allowing paralyzed patients to regain useful function. External control by Functional Neuromuscular Stimulation (*FNS*) has been achieved for restoration of hand grasp (Peckham et al., 1988; Keith et al., 1989; Handa et al., 1986), arm and elbow control (Allin and Inbar, 1986; Hoshimiya et al., 1989; Miller et al., 1989), standing (Kralj et al., 1986; Jaeger et al., 1989), and locomotion (Kralj et al., 1983; Marsolais and Kobetic, 1986). All of these systems are open loop in the sense that no automatic corrections are made for errors in the output. It is well recognized that incorporation of feedback control into these systems can improve the repeatability of performance by automatically compensating for internal disturbances (particularly the nonlinear and time varying properties of muscles) as well as external disturbances due to interaction with objects in the environment. In both upper and lower extremity applications, the limbs interact with external loads that are constantly changing, complicating the design of single variable (e.g. force or position alone) control systems. In contrast, stiffness regulation (achieved by combining force and position feedback from external sensors) shows potential advantages in *FNS* applications, since it can achieve good compensation for internal disturbances even in the presence of changing external loads.

In this chapter, we will first briefly review *FNS* with special attention to control issues, and then present work that we have done with active stiff-

ness regulation, in both laboratory experiments and in simulation. We have employed stiffness regulation to regulate the input-output properties of a hand grasp neuroprosthesis. We will present experimental results obtained in spinal cord injury patients that document the performance of the system under varying loading conditions. Analytical and simulation results will then be presented that elucidate the special role of the value of regulated stiffness in determining performance under different loading conditions.

21.2 Functional Neuromuscular Stimulation

In muscles that have been paralyzed by an upper motor neuron lesion, the motor axons innervating the muscle remain intact and can be activated by electrical current pulses. This type of paralysis occurs both in stroke and in spinal cord injury (*SCI*). Gradation of contraction strength is important in most applications, yet is technologically difficult to realize in a well-controlled manner. The strength of the contraction can be modulated by varying the parameters of stimulation. The number of motor axons activated by each pulse can be varied by adjusting the amplitude or duration of individual pulses, allowing recruitment modulation of contraction force (Crago et al., 1980). The frequency of pulses can be used to vary the overlap of successive twitches, allowing modulation of force by temporal summation. Thus, neuroprostheses can employ the same mechanisms of force modulation that are used by the nervous system, except that all motor axons that are activated by a single stimulus pulse are forced to fire in synchrony, and the order of recruitment is generally not the same.

Multiple Muscle Systems: Biomechanics and Movement Organization
J.M. Winters and S.L-Y. Woo (eds), © 1990 Springer-Verlag

The type of stimulating electrode affects the selectivity, reproducibility, and slope of recruitment modulation. Stimuli can be delivered via electrodes placed within the muscle (intramuscular electrodes, e.g. Peckham, 1987), sewn onto the surface of the muscle (epimysial electrodes, e.g. Grandjean and Mortimer, 1986), located next to the nerve (epineural electrodes, e.g. Holle et al., 1984), in cuffs surrounding the nerve (nerve cuff electrodes, e.g. Naples et al., 1989), or on the skin surface overlying the muscle or nerve (surface electrodes, e.g. Kralj et al., 1983). Surface electrodes are convenient to apply, especially for short term studies in human subjects, yet have poor selectivity and poor reproducibility. Nerve electrodes have good reproducibility and are also selective if they are placed on a nerve innervating only one muscle. Epimysial electrodes offer reasonable selectivity and do not require direct access to the muscle nerve. Coiled wire intramuscular electrodes have selectivity similar to the epimysial electrodes, and can be inserted percutaneously via a hypodermic needle, unlike nerve and epimysial electrodes that require surgical placement. Intramuscular electrodes also offer reasonable long-term stability, and are therefore convenient for the development of clinical systems, even if they are to be replaced by surgically implanted epimysial or nerve electrodes in the final version of a neuroprosthesis.

Almost all motor system neuroprostheses modulate contraction strength by recruitment modulation with the stimulus frequency fixed to the lowest rate that gives an adequately fused response. This strategy is taken to minimize the rate of fatigue, which occurs more slowly at low than at high stimulus frequencies (Mortimer, 1981). The relationship between the force of contraction and the stimulus pulse width or amplitude is referred to as the recruitment characteristic (Crago et al., 1980). In general, this relationship is nonlinear, with a deadband below threshold, saturation at full recruitment, and variable slope between these two extremes, regardless of the type of electrode (Crago et al., 1980; Grandjean and Mortimer, 1986; Durfee and MacLean 1989; Vodovnik et al., 1967). With all but nerve cuff electrodes, the fraction of the muscle that is recruited at a particular stimulus level is also a function of the muscle length (and thus, the joint geometry), a property referred to as length dependent recruitment (Crago et al., 1980). It is felt that length dependent recruitment results from changes in the location of the nerve axons within the field produced by the stimulus pulse, as the muscle moves. This same dependence on geometry makes the recruitment characteristic extremely variable from electrode to electrode, necessitating initial calibration, and recalibration if electrodes are changed or moved.

The simplest functional systems are those for open loop standing in *SCI* patients (Kralj et al., 1986; Jaeger et al., 1989). Fixed stimulation patterns are turned on and off by commands from the patient. In two channel systems, the quadriceps of each leg are stimulated to raise the body from a seated position and to lock the knees in extension. The patient uses his arms in conjunction with balance aids to get into a posture that forces the center of gravity of the head, arms and trunk to be behind the hips. This eliminates the need for active hip extension. The arms also provide stability at the ankles, which are otherwise free to move. Additional channels can be added to activate muscles at the hip to provide active extension and increased stiffness in the coronal plane (Chizeck et al., 1988). Hybrid systems, combining stimulation and orthotics, provide both the active muscle power (with stimulation) required for standing up, sitting down, and walking, and passive stabilization (without stimulation) at the knee and ankle during quiet standing to reduce the amount of time that muscles must be activated (Andrews et al., 1988).

Walking is achieved by activating muscles with temporally varying patterns of stimulation that a patient triggers by means of switches (see reviews by Cybulski et al., 1984; Chizeck et al., 1988). Each stored sequence of stimuli takes the patient from one stable postural state to the next. The simplest systems use two channels of stimulation per leg, one to activate the quadriceps to provide active extension, and another to stimulate the peroneal nerve to activate a flexor withdrawal reflex for swing (Kralj and Bajd, 1983). More complex systems use more electrodes and stimulus channels. Holle et al. (1984) employ a multichannel implantable nerve stimulation system to directly activate hip and knee extensors, providing reciprocating gait. Marsolais and Kobetic (1987) employ as many as 16 stimulus channels per leg to provide hip flexion, extension and abduction, as well as flexion and extension at the knee and ankle joints.

Neuroprostheses for hand grasp have been developed for patients with spinal cord injuries at the fifth and sixth cervical levels (Peckham et al., 1988; Handa et al., 1986). These patients can position their hand voluntarily in a reasonable working space in front of them, and achieve opening and closing of the hand by controlled electrical stimulation of finger and thumb muscles. Eight channels of stimulation provide two modes of grasp, as well as a single channel of cutaneous stimulation for sensory feedback (Keith et al., 1989). Typically, a single signal (command), derived from voluntary movement of the opposite shoulder, provides coordinated stimulation of thumb and finger muscles by means of pulse width maps. At one extreme of the command range, the fingers and thumb are fully extended. As the signal is graded toward the other extreme, the stimulus pulse widths to the extensors are decreased and those to the flexors are increased (as programmed in the pulse-width maps), providing gradual closing of the hand and exertion of grasp force. By altering the choice of muscles and the pattern of stimulus modulation, either palmar grasp (tip pinch) or lateral grasp (key grip) is achieved. The system enables patients to function independently at home, school, and work since they can pick up and hold ordinary objects such as pencils, papers, books, and eating utensils.

In all of these systems the patterns of stimulation specified in the pulse width maps are developed by an iterative process, starting from conventional knowledge about which muscles provide each of the functions that are needed, and from studies of how muscles are activated normally. This is supplemented by measurements of function when each electrode is stimulated separately (e.g. Kilgore et al., 1989). The need for iterative tuning is a result of poor predictability of the electrically elicited responses for individual electrodes, the dynamic coupling of torques at different joints because of gravitational, inertial and Coriolis effects, and the complex functional anatomy of the muscles and limbs [Chapter 8 (Zajac and Winters)]. Because of the anatomical structure, individual muscles frequently have more than one action at a single joint (i.e. they do more than just flex or extend) and produce coupled torques at all of the joints that they cross (e.g. the rectus femoris flexes the hip at the same time that it extends the knee). The result is that there are not one-to-one relationships between stimulus channels and degrees of mechanical freedom, and this makes tuning of the system more difficult. The clinical tuning process would be facilitated greatly by the use of accurate models of the extremities, including both skeletal and muscle mechanical properties; certainly this book is timely in this regard.

Once stimulus patterns have been established in an open loop system, they remain fixed until they need to be retuned, which frequently necessitates a return visit to the laboratory. Satisfactory performance depends on the ability of the chosen patterns to provide the desired motor output in the face of internal disturbances (changing muscle properties) and external disturbances (changing interactions with the environment). The input-output properties of stimulated muscle are time dependent, as described below, making the motor output for a given stimulus pattern nonrepeatable. The stimulus patterns are set up for expected external loading conditions. For example, the knee extensors are stimulated at a certain point in time during the swing phase, in the anticipation that heel contact will be made. The time of contact can not be predicted accurately, even under the best circumstances. In standing, significant postural disturbances are generated by functional use of the upper extremities. These are obviously impossible to predict, and restricting the use of the arms would eliminate an important functional benefit of standing. Similarly, in hand grasp, pulse width maps are set up to control position and force for a certain size and compliance load. With different loads, the control properties can change substantially.

Muscle gain changes produce substantial internal disturbances in *FNS* systems. They arise from changes in the slope of the recruitment characteristic as discussed above, and from changes in the contractile strength of muscle (the force produced by a given set of active fibers). Changes in muscle contractile strength arise from potentiation (Burke et al., 1976), fatigue (Jones et al., 1979), length–tension properties (Rack and Westbury, 1969), and the trophic effects of exercise (Lieber 1986) or disuse (Mayer et al., 1984). In principle, length–tension effects and length dependent recruitment can be characterized experimentally, and accounted for in the specification of open–loop stimulation patterns. Since potentiation takes place relatively rapidly, it may be less important in tonic activities, where the effects of fatigue will predominate. Changes in force due to exercise or disuse, can not be predicted accurately, and can cause such large changes in strength that open loop stimulus patterns need to be altered in the clinic.

The incorporation of closed loop control systems into neuroprostheses is expected to provide several significant improvements. Perhaps the most important benefit is to compensate for internal disturbances: the time varying properties of muscle, length dependent recruitment, and muscular or dynamic coupling across several joints or across degrees of freedom at a single joint (i.e. feedback can decouple interacting degrees of freedom [see also discussion in Chapter 10 (Loeb and Levine)]. This would provide substantial improvement in predictability of the input-output properties of the system, making it possible to specify movements or forces directly in terms of physical variables, rather than indirectly in terms of stimulus parameters. Feedback correction for internal disturbances may also reduce the need for frequent retuning, but only if the closed loop system's performance itself is robust, ie relatively insensitive to the internal disturbances. If the closed loop system is not robust, the need to retune stimulus parameters may just be replaced by a need to retune the feedback controller.

The second benefit of feedback control is stabilization of an otherwise unstable system, such as upright posture (Cybulski et al., 1984; Chizeck et al., 1988). In open loop standing systems, upright posture is stabilized in two ways: *i)* by voluntarily controlling the upper extremities, and *ii)* by stimulating the quadriceps to drive the knee joint into hyperextension where it is stabilized by passive structures such as ligaments. High levels of quadriceps stimulation ensure that the knees remain stable in the face of disturbances. In the long term, it is felt that hyperextension will cause joint degeneration. With feedback control, the stimulus level is continuously adjusted to stabilize the knee by active muscle forces without driving the knee into hyperextension. Feedback control can also stabilize the ankle, a task that can not be accomplished with open loop techniques since the passive stiffness is so low in the middle of the range of motion, where stability is desired.

The third advantage of feedback control is that endurance during submaximal activities could be increased in comparison with that obtained using open loop systems. Closed loop systems adjust the stimulus levels to those just needed to accomplish the task, and as the muscles fatigue, the stimulus levels are increased automatically. Open loop systems, on the other hand, overdrive the muscles to ensure that there is enough force even in the presence of fatigue.

When patients have voluntary control of the stimulus gradation, as in the neuroprostheses for hand grasp, they can make corrections for disturbances by increasing or decreasing the voluntary command. Patients with spinal injury generally have major sensory as well as motor deficits, forcing them to visually monitor their performance to detect errors. Incorporation of substitute sensory feedback to augment a patient's perception of errors should improve performance (Riso et al., 1989), but changes in command still have to be made by the patient, thus adding an attention demand.

Closed loop systems for regulating joint position, muscle force, or muscle length, have been studied in several laboratories, with the result that there are now several different types of controllers available to be used in clinical studies. Controllers may be divided into two groups: fixed parameter controllers that are first tuned but then operate with a fixed set of values (e.g. Chizeck et al., 1988; Durfee, 1989); and adaptive controllers that continually monitor the input-output properties of the system and adjust the controller parameter values to optimize performance as the properties of the system change (e.g. Allin and Inbar, 1986; Bernotas et al., 1987). Fixed parameter controllers offer the advantage of computational simplicity, while adaptive controllers have the potential advantage of being able to adjust to changes in the system. While the adaptive ability is attractive for *FNS* because of the time varying plant, adaptive controllers are potentially unstable, work best when the changes in the system are slow, and show oscillatory behavior when the system changes rapidly. Detailed review of closed loop controller design is beyond the scope of this chapter, but it should be pointed out that nearly all controllers have been designed on an ad-hoc basis, without detailed consideration of muscle properties or the biomechanics of the systems to be controlled. A notable exception to this is a computer simulation model of feedback control of *FNS* standing, developed by Khang and Zajac (1989 a,b). The simulation was used to study the recovery of upright posture from perturbed initial conditions and the maintenance of upright posture during external disturbances induced by movements of the upper extremities. The control system performed well in simulation, but has yet to be tested experimentally.

21.3 Active Stiffness Regulation

Most of the controller studies have looked at regulation of a single variable, either force or position, under loading conditions where only that one variable is changing. Single variable control systems such as these will not allow compensation for internal disturbances when the loading conditions switch. In contrast, regulating the relation between force and position, specifically stiffness, has several functional advantages, including the ability to compensate for internal disturbances under varying loading conditions. Stiffness can be regulated by the feedback control system shown in Figure 21.1. The input to the system is the static equilibrium position, and will be referred to as the virtual position, P_V. The difference between the virtual position and the actual position, P, as measured by a sensor, forms a position error. The position difference is multiplied by the stiffness coefficient, K_C, to determine the desired contact force. A force error, calculated as the difference between the desired and measured contact forces, is processed by a feedback controller, and the output is used to specify how the antagonistic muscles should be stimulated. The following equations describe the stiffness regulator and the activation controller.

$$E_k = K_C (P_{Vk} - P_k) - F_k^i \qquad (21.1)$$

$$D_k = D_{k-1} + G E_k + m G E_{k-1} \qquad (21.2)$$

where the subscript k is the time index, T is the stimulus period (time equals kT), E_k is the force error (input to the activation controller), P_{Vk} is the virtual position, P_k is the actual position, F_k^i is the actual interaction (contact) force with any external object, D_k is the controller output, G is the controller gain, and m is the controller zero location. The operation of the stiffness regulator and the feedback controller given in *Eq. 21.2* are described in previous publications (Chizeck et al., 1988; Crago et al., 1990).

If the controller is functioning properly the error should be zero. For our implementation, the controller has integrator action, ensuring zero steady-state error for maintained inputs. Under this assumption,

$$F_k^i = K_C (P_{Vk} - P_k) \qquad (21.3)$$

which is the equation of a spring, and

$$P_{Vk} = P_k + \frac{F_k^i}{K_C} \qquad (21.4)$$

which indicates that the input, under conditions of zero error (i.e. at steady state), is equal to the sum of the position and the force scaled by the stiffness, rather than a single variable. Note that the convention assumed here is that an increase in force in the direction of flexion causes a decrease in position. Thus, the sum of position and scaled force can be viewed as the single output of a stiffness regulator.

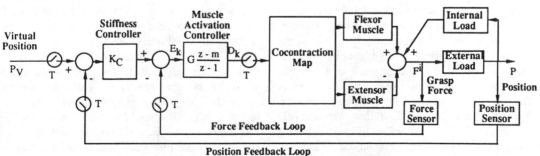

Figure 21.1: Block diagram of an active stiffness regulator controlling joint position and contact force by electrical stimulation of antagonist muscles.

In the absence of an external load, the contact force will be zero, and the system reduces to a single feedback loop, position regulation system. In the presence of a rigid load, the position feedback signal remains constant at a fixed offset with respect to the virtual position, and the system becomes a single feedback loop, force regulation system.   With loads that are compliant, or have viscous or inertial properties, both the position and force loops are significant in the regulatory action.

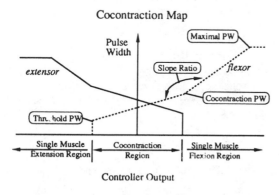

Cocontraction Map

Figure 21.2: Cocontraction map relating the output of the controller to the pulse widths of the stimuli applied to the antagonistic muscles.

The muscles are stimulated according to a map (cocontraction map, Figure 21.2) that relates the output of the activation controller to the stimulus pulse width applied to each muscle. The controller output is divided into three activation regions, allowing both cocontraction and single muscle contractions. Antagonists are cocontracted in the central region and activated alone in the regions on either side. Coactivation in the central region increases the inherent stiffness during unloaded position control or when exerting low forces, and also prevents deadbands in control when switching from stimulation of the flexor to stimulation of the extensor and vice versa. The flexor and extensor are activated alone when the stiffness of the system is known to be high, ie during extension near the limit of the range of motion, and during moderate to high force contractions. The two muscles are activated reciprocally in the cocontraction region. An increase in the controller output (to the right on the abscissa) increases the

pulse width sent to the flexor, and decreases the pulse width sent to the extensor.

The cocontraction pulse widths and the thresholds define the conditions for cocontraction. Since the abscissa is controller output, and can be scaled by changing controller gain, the controller outputs delimiting the cocontraction region can be chosen arbitrarily and the overall gain tuned in the controller. At the extreme ends of the cocontraction region, one or the other of the two muscles is inactive, since its pulse width is just below threshold. Therefore, the pulse width sent to the other muscle can be chosen to define the mechanical output of the system at that point (for a given load). The slope of the line specifying the pulse width to each muscle in its single muscle contraction region, is specified relative to the slope for that muscle inside the cocontraction region (ratios r_f and r_e).

21.4 Experimental Methods

Our experimental studies of stiffness regulation have focussed on applications for neuroprosthetic control of hand grasp. The hand grasp neuroprosthesis provides two modes of grasp: palmar prehension, where the thumb is held in opposition and grasp is controlled by flexion and extension of the fingers; and lateral prehension (key grip) where the fingers are held in flexion and the thumb is flexed to hold objects against the lateral aspect of the index finger (Peckham et al., 1989; Keith et al., 1989). Application of stiffness regulation to hand grasp involves controlling the digit that moves and exerts force, rather than the digit that is fixed.   Thus in lateral prehension, the thumb is controlled by the stiffness regulator, with the finger fixed in flexion by constant stimulation (Cabrera et al., 1987).

Some of the subjects that have received hand grasp neuroprostheses (Peckham et al., 1989; Keith et al., 1989) have also participated in the studies on feedback regulation of hand grasp stiffness.   The electrodes and the implantable stimulator system used by the particular subject whose data are shown in this chapter have been reported in detail elsewhere (Keith et al., 1989). The experimental methods for closed loop stiffness studies are also described elsewhere (Crago et al., 1990), and will only be summarized briefly here.   For tests on tuning and measurement of input-output properties, the subject's hand and arm are placed in a cast, fixed to a stereotaxic table. Only the digit (index finger or thumb) that is under feedback control is free to move. A rigid force transducer is placed at a position where the digit would normally come into contact with opposing digit during flexion. A position transducer is attached to measure the displacement of the controlled digit when it is extending. Position is taken as zero when the digit is in contact with the force transducer, and positive for extension. Force is taken as zero when the digit is not touching the transducer, and positive during flexion. The stiffness controller is implemented by a digital laboratory computer that samples the transducer outputs just prior to each stimulus pulse, performs the controller calculations and cocontraction mapping, and then stimulates each of the muscles via a multichannel computer controlled stimulator. In the examples reported in this chapter, the virtual positions were all generated by the computer.

Tuning the stiffness regulator consists of defining the cocontraction map and choosing parameters for the regulated stiffness and the activation controller. The pulse widths in the cocontraction map are defined on the basis of steady state measurements of the force versus pulse width for the flexor, and the position versus pulse width for the extensor. The thresholds are chosen as the maximal values that produce no activation. The maximal pulse widths are chosen as the smallest values that produce maximal activation without exceeding safe pulse width limits or activation of adjacent muscles. In hand grasp, the cocontraction pulse widths are chosen to limit cocontraction at approximately 20% of maximal grasp force, and 80% of maximal grasp opening when stimulating the flexor or extensor alone.

The controller structure and the loading conditions encountered during hand grasp are used to simplify tuning of the stiffness regulator, by eliminating interactions between parameters. The system loop gain is tuned to perform satisfactorily under at least four unique combinations of loads and activation regions that are encountered in hand grasp (see below). There are four gain terms (controller gain, regulated stiffness, flexor and extensor pulse width slope ratios) that are tuned to adjust the loop gain of the system so that performance is satisfactory under each condition. The

tuning criteria are rise time less than one second and overshoot less than 30% for step inputs. The final criterion is that the response should remain steady with less than 10% oscillation for maintained inputs. The parameters are adjusted from trial to trial by factors of 2.

The loop gain is a function of both feedback loops, and is therefore also a function of the external loading conditions. For a thin load, contact takes place within the cocontraction region. If this load is rigid, the point of contact divides the cocontraction region into two parts, one where the digit is unloaded, and another where the digit is isometric. As pointed out above, the position feedback loop is effectively opened when the load is rigid, making the loop gain independent of the regulated stiffness, which under these conditions only scales the input to the force regulation loop. Therefore, the controller gain (G) can be adjusted by itself to obtain satisfactory performance under isometric cocontraction conditions. Increasing the desired force moves the system from the cocontraction region to the single muscle flexion region. The flexor slope ratio (r_f) is then adjusted to provide satisfactory performance under this condition.

Under unloaded conditions, the force feedback is eliminated and the system loop gain depends on the product of controller gain and regulated stiffness. With the digit extended just enough to be lifted off the surface of the force transducer, the digit is operating under conditions of unloaded cocontraction. Leaving the controller gain fixed at the value obtained for isometric loads, the value of regulated stiffness (k_C) is adjusted to give satisfactory performance under these conditions. Increasing the position command moves the system out of cocontraction into single muscle extension, where the extensor slope ratio (r_e) is adjusted to give satisfactory performance. To summarize, controller gain (G) is adjusted under isometric cocontraction conditions, the flexor slope ratio (r_f) is adjusted with isometric loads in the single muscle flexion region, regulated stiffness (K_C) is adjusted under unloaded cocontraction conditions, and the extensor slope ratio (r_e) is adjusted during unloaded position control in the single muscle extension region.

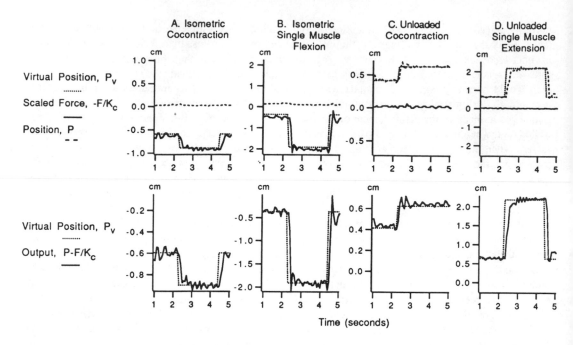

Figure 21.3: Step responses of the stiffness regulation system in all four combinations of loading and activation that are used in tuning. Force, position, and virtual position are shown at the top, and the combined output is shown at the bottom. The step responses show similar rise time and overshoot under all conditions, with a single set of system parameters.

21.5 Experimental Results and Discussion

We have found that the stiffness regulation system produces good performance under the variety of loading conditions that are encountered in hand grasp. This is evidenced by the fact that the step responses obtained in each of the loading and activation regions during tuning, have acceptable rise time and overshoot, and have a stable output. Furthermore, the responses are also acceptable when the system is tested with other inputs and other loads.

Example step responses obtained when tuning the system are shown in Figure 21.3 for one subject. The steps beginning at approximately two seconds were obtained for the following conditions: *a)* isometric cocontraction, *b)* isometric single muscle flexion, *c)* unloaded cocontraction, and *d)* unloaded single muscle extension. Forces (solid lines), positions (dashed lines), and virtual positions (dotted lines) are shown as functions of time in the top panels, and the combined outputs $(P - F^i/K_C)$ are shown in the bottom panels. All of the step responses show rise times less than 0.5 sec and overshoots less than 30% of the step amplitude, except for the response in unloaded

cocontraction, where the overshoot was about 40%. Since rise time and overshoot were assessed visually in these tests, and since there is some variability in responses, the response criteria were not always strictly met in individual trials.

Transitions between loading conditions and between activation regions were evaluated during smoothly graded contractions (ramps) from extension to flexion and vice versa. Three loads were used in these tests: the zero thickness rigid load that was used during tuning, a large rigid load consisting of a wooden block placed on the force transducer, and a compliant load consisting of a foam rubber block placed on the force transducer. The ramp rate was kept low to approximate static conditions. With the zero thickness rigid load (Figure 21.4a), virtual position was constant for a few seconds before the ramp began. In this period there was a steady state error, due to saturation of the extensor. That is, maximal activation of the extensor could not achieve a grasp opening as large as requested. As the virtual position ramped down, the steady state error was reduced and position began to track the virtual position. Force stayed at zero until contact was made. After contact, the position remained nearly constant and force began to track the virtual position. Good tracking was evident in the reverse direction also. With the large rigid object (Figure 21.4b), two differences are evident. First, position decreases only about one cm before contact is made. Second, force starts increasing earlier and reaches a higher value, with the result that force is offset from the virtual position. The data in Figure 21.4c were obtained with the compliant load. Contact was made at a larger virtual position and force and position changed simultaneously over a larger range of virtual positions than with the rigid objects. The stiffness of the load increased with compression, with the result that the force and position relationships were curved rather than straight lines as would be expected with linear compliant objects. Transitions from one activation region to another, or from one loading region to another, were all accomplished smoothly.

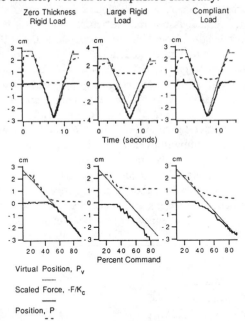

Virtual Position, P_v

Scaled Force, $-F/K_c$

Position, P

Figure 21.4: Force and position as functions of time (top) for virtual position ramps that traverse the range of available positions and forces, for three different kinds of load, as indicated at the top of each panel. The dependence of input–output properties on load are shown at the bottom, in plots of force and position versus virtual position.

Virtual position, position and scaled force are shown as functions of the normalized virtual position (Command) below each time plot in Figure

21.4. The command represents the patient's input to the stiffness regulator, assuming that there is a straight line relationship between the command and the virtual position, and that the slope relating command to position is equal to the slope relating command to scaled force. That is, the control of unloaded position and of scaled isometric force have the same sensitivity to changes in the patient's command. As expected, the relationship of position to command is unaltered prior to contact. The size and compliance of the load alter the relationship between command and position. The size of an object causes a vertical shift and the compliance of an object causes a change in slope of the command to force relationship.

These relationships are what would be expected with stiffness regulation: if the endpoint of the digit moves as if it was being controlled by changing the slack length of a spring of constant stiffness. Several functional benefits can be expected from this achievement. First, the system controls either position, force, or both with a single input command, and without switching control strategies between regulation of force, position, or both. The single input recognizes the physical impossibility of simultaneously regulating force and position to independent values. Position and force must be related by the properties of the load and the control system. The type of regulation is achieved automatically by simply summing both the position and the force feedback signals. In application to hand grasp, the system offers predictable and repeatable input–output properties. In unloaded conditions, the position is proportional to the command signal, and with a load, the force is proportional to the difference between the desired (virtual) and the actual positions. This should reduce the number of corrections that patients would need to make in their command signals, and reduce the attention demands in controlling their hand.

21.6 A Simulation Study of Active Stiffness Regulation

Stiffness regulation with electrically stimulated antagonist muscles has been studied in simulation as an aid in understanding the operation of the system, and to study the sensitivity of the performance to system parameter values. The simulation incorporates a nonlinear model of electrically stimulated muscle (Figure 21.5), including activation dynamics, length-tension properties, and asymmetrical force-velocity properties, similar to that described in detail in Chapter 5 (Winters). Two loads are included, an internal one representing passive mechanical properties, and an external one representing the mechanical properties of objects in the environment. The limb is allowed to make and break contact with the external load during movements. A constant load, linearized version of the model is used to derive conditions for making performance independent of the external load. These conditions are confirmed by simulation with the nonlinear system.

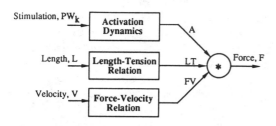

Figure 21.5: Block diagram of the muscle model employed in the simulation. Force is the product of three factors representing activation, force–velocity properties, and length–tension properties.

A discrete time model of isometric muscle that is suitable for control purposes has been developed by Bernotas et al. (1986). This model has recently been expanded to incorporate length–tension and force–velocity properties so that it could be used under nonisometric loading conditions. Muscle force is calculated as the product of three components: the activation by stimulation, and two terms representing the force–velocity and the length–tension properties of muscle. Methods of parameter estimation have recently been developed by Chizeck and Geng (unpublished). The model assumes constant frequency, variable pulse width stimulation.

Since both the flexor and extensor are modeled the same way, only the equations describing a general muscle are given here. The force produced by a muscle is modeled as the product of an activation dependence factor, a length dependence factor, and a velocity dependence factor:

$$A_{k+1} = a_1 A_k + a_2 A_{k-1} + b PW_k \qquad (21.5)$$

$$A(kT+t) = A_k + (t/T)(A_{k+1} - A_k) \qquad (21.6)$$

$$LT = e(L - L_o) \qquad (21.7)$$

$$FV = 1 + h_1 T V U(V) - h_2 T V U(-V) \qquad (21.8)$$

$$F = A\ LT\ FV \qquad (21.9)$$

The activation dependence factor, A_k, is a linear, second order, discrete time model of the dynamics of isometric force modulation, where a_1, a_2 are autoregressive activation parameters, b is the linear recruitment gain, and PW_k is the pulse width (in normalized units) representing the recruitment level ($0 \le PW_k \le 1$). The second equation linearly interpolates between successive values of A_k, giving a continuous time representation of the isometric force. The length dependence factor, LT, is a linear approximation of the length tension properties, where e is the slope of LT at maximal activation, L is the muscle length, and L_o is the slack length. Note that the product be is the stiffness at maximal activation. The velocity dependence factor, FV, is a piecewise linear approximation of the force–velocity properties, with separate slopes h_1 and h_2 for lengthening and shortening respectively. V is the velocity, and $U(\cdot)$ is the unit step function. $A, L, V, LT, FV,$ and F are all continuous functions of time. The equation for F calculates the active muscle force as a product of three factors: the force–velocity factor FV, the length–tension factor LT, and the activation A. In isometric conditions and at the discrete times kT, this reduces to the model of Bernotas et al. (1986).

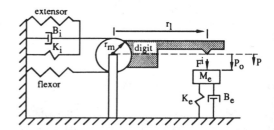

Figure 21.6: Schematic of the biomechanical system employed in simulation of stiffness regulation by stimulated antagonists.

The muscles are incorporated into a system mimicking the control of a single digit interacting with an external load (Figure 21.6). The two muscles act as antagonists (one flexor and one extensor) with constant, identical moment arms (r_m). The flexor is assumed to shorten with increasing joint position, P, while the extensor increases its length by an equal amount. The digit position is measured as the arc length that the contact point of the digit moves through when the joint rotates. The point of contact with the external load is assumed to be at radius r_l. At rest, the two muscles are midway between their minimal, L_0, and maximal, L_{max}, physiological lengths. That is,

$$L = \frac{(L_0 + L_{max})}{2} \pm \frac{r_m}{r_l} P \qquad (21.10)$$

where the positive sign is for the extensor and the negative sign is for the flexor.

Two loads are included: an internal load representing the passive properties at the joint, and an external load representing an object in the environment. Both are assumed to be second order linear loads, with the following differential equations.

$$M_i \ddot{P} + B_i \dot{P} + K_i P = \frac{r_l}{r_m} (F^f - F^e) - F^i \qquad (21.11)$$

$$M_e \ddot{P}_o + B_e \dot{P}_o + K_e (P_o - P_{oe}) = F^i \qquad (21.12)$$

M, B, and K represent the mass, viscosity, and stiffness of the loads. The subscripts i and e are used for the internal and external loads respectively, P is the position of the digit, P_o is the position of the external load, P_{oe} is the equilibrium (unloaded) position of the external load, F^i is the interaction force between the digit and the external load at a radius r_l, F^f and F^e are the forces exerted by the flexor and extensor respectively. The dots above the variables indicate derivatives with respect to time.

The simulation is broken down into two components. The controller output is calculated at discrete times corresponding to multiples of the stimulus period, T. Between each stimulus period, the force acting on the mechanical load is integrated by an adaptive step-size, fourth order Runge-Kutta method (Press et al., 1986). The digit (including the two muscles and the internal

load) is allowed to make and break contact with the external load. When the digit is not contacting the load, the two systems are integrated separately and when they are in contact, they are integrated together. The two systems are joined if the positions begin to overlap, and they are separated if the force exerted by the digit on the load becomes negative. If the boundary conditions differ by more than a certain tolerance at the end of each integration step, the time step is reduced and the system is reintegrated.

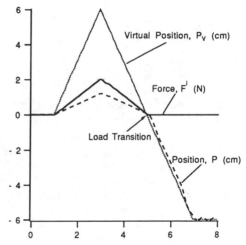

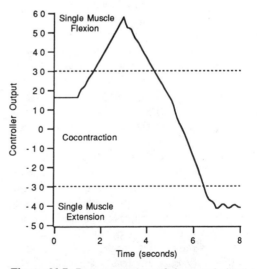

Figure 21.7: Ramp responses of the tuned simulated system for a compliant load. In the positive direction, the load is contacted and force and position change in parallel. In the negative direction, when the load is not in contact, force remains at zero and position tracks the virtual position. The controller output, shown at the bottom, indicates the activation region transitions.

Table 22.1: Model Parameters

Muscle and Controller Parameters		Biomechanical and Load Parameters	
a_1	0.19	r_l	7.5 cm
a_2	0.0094	r_m	7.5 mm
be	0.46 N/mm	K_i	0.512 N/cm
L_0	-26 mm	B_i	0.0752 Ns/cm
L_{max}	0 mm	M_i	0.00276 Ns²/cm
h_1	0.002 s/mm	K_e	1.7 N/cm
h_2	-0.0128 s/mm	B_e	0.076 Ns/cm
m	0.6	M_e	0.00042 Ns²/cm
G	20	P_{oe}	0 cm
K_C	0.5 N/cm		
r_f	4.4		
r_e	4.0		

The simulated system is tuned with a procedure similar to that described above for the clinical system. An example of the simulation response to ramp inputs is shown in Figure 21.7, using muscle parameters (Table 21.1) estimated from an experiment in cat soleus (unpublished results). Force, position and virtual position as functions of time are shown at the top for ramp changes in virtual position traversing all of the activation regions, and with a load transition. At the start, the virtual position is just at the point of contact, and the controller output, shown below, is in the cocontraction region. As the virtual position increases, the contact force and position increase in proportion, and there is a smooth transition from cocontraction to single muscle flexion at about 1.6 s, as seen in the controller output. The reverse transition is also made smoothly during the decreasing ramp. As the virtual position passes zero, there is a transition from simultaneous control of force and position to unloaded position control. Although there is a change in the slope of the controller output at this time, the mechanical transition is made smoothly. Continued decrease in the virtual position requires a transition from cocontraction to single muscle extension. There is a slight change in the slope of position versus time at this point. Overall, the results are qualitatively similar to results obtained in patients.

21.7 The Dependence of Performance on External Load

The loop gain of a feedback control system is an important determinant of the system's ability to regulate the output in the face of internal disturbances. An analytical expression for the loop gain is not feasible for the complete system described above, due to the nonlinearities. However, we will derive an expression for the loop gain of the linearized system under steady state conditions during cocontraction with a constant stiffness external load. This equation offers insight into the relative importance of the three stiffnesses in the model: the regulated stiffness, the internal stiffness, and the stiffness of the external load. In particular, we will show that if the regulated stiffness is equal to the internal stiffness (passive stiffness and intrinsic active muscle stiffness), then the loop gain becomes insensitive to the stiffness of the external load. We then use the simulation to show that the result holds for the nonlinear system as well.

Under steady state conditions, the equation for the force produced by one muscle is

$$F = \frac{be}{a_1 + a_2} PW \left[\frac{L_{max} - L_0}{2} \pm \frac{r_m}{r_1} P \right] \quad (21.13)$$

This is linearized with respect to PW and P by taking the first order Taylor series expansion about a fixed PW (which will be designated PW_a) and $P = 0$ (unstretched external stiffness). The linearized model is given by

$$F = F_0 + B \Delta PW \pm K_{ia} \Delta P \quad (21.14)$$

$$B = \frac{be}{a_1 + a_2} \left(\frac{L_{max} - L_0}{2} \right) \quad (21.15)$$

$$K_{ia} = \frac{be}{a_1 + a_2} PW_a \frac{r_m}{r_l} \quad (21.16)$$

and

$$F_0 = B PW_a \quad (21.17)$$

where B is the linearized gain for stimulation, K_{ia} is the linearized internal active stiffness, and F_0 is the force produced at the point of linearization. We further assume that the muscles are coactivated at the same pulse width, and have the same slope in the coactivation map. Then, the net muscle force is the difference between the two muscles, and is given by

$$F^f - F^e = 2 B \, \Delta PW - 2 K_{ia} \Delta P \qquad (21.18)$$

The second term represents the contribution of the intrinsic active stiffness of the stimulated muscles, and this can be added to the internal passive stiffness (K_i) to form a total internal stiffness, K_i .

$$K_i' = K_i + 2 K_{ia} \qquad (21.19)$$

The system block diagram under these conditions is shown in Figure 21.8, and the loop gain, calculated as the product of all the gains around the loop is given by

$$LG = G_c \frac{K_e + K_C}{K_e + K_i'} \qquad (21.20)$$

$$G_c = 2 B G s (1 - m) \qquad (21.21)$$

where G_c is a gain term that is independent of load, and s is the slope of pulse width versus controller output in the cocontraction map. Taking the derivative of the loop gain with respect to K_e shows how the performance would be expected to change with external load stiffness.

$$\frac{d(LG)}{d(K_e)} = G_c \frac{K_i' - K_C}{(K_e + K_i')^2} \qquad (21.22)$$

If the internal stiffness, K_i', is equal to the regulated stiffness, then the loop gain is independent of the stiffness of the external load. If K_i' is greater than the regulated stiffness, then loop gain will increase with the external stiffness, and if it is less than the regulated stiffness, the loop gain will decrease with the external stiffness. Note that this derivation does not depend on the specific form of the muscle model, only that the model must include an intrinsic active stiffness term.

This behavior was verified in the nonlinear system (the simulated system) by analyzing step responses with varying external load stiffness for three combinations of internal stiffness and regulated stiffness (Figure 21.9). The internal stiffness was adjusted by varying the level of cocontraction, keeping the net muscle force equal to zero, for a zero input. Since changing the cocontraction level also changes the muscle gain for pulse width inputs, the product of controller gain and cocontraction map slope was kept constant. Unit step responses starting from $P_V = 0$ were quantified in terms of rise time, overshoot and percent RMS error. When the internal stiffness was less than the regulated stiffness, rise time and RMS error increased, and overshoot decreased with increasing external load stiffness. This behavior would be expected if loop gain decreased with increasing external stiffness. Reversing the stiffness relationship also reversed the behavior of the step responses. In the last case, the internal and regulated stiffnesses were made equal, and the dependence of the step response on external stiffness was greatly reduced (Figure 21.9, bottom left).

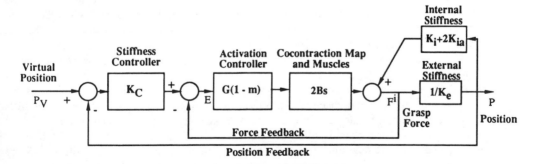

Figure 21.8: Block diagram of the linearized stiffness controller.

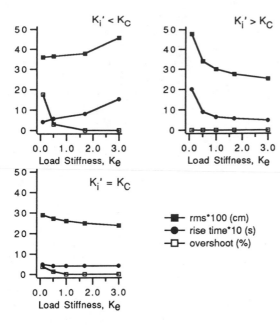

Figure 21.9: Dependence of step response parameters on the stiffness of the external load, for three relationships between internal stiffness and regulated stiffness. In all cases, the internal stiffness, K_i', was 0.512 N/cm. When the regulated stiffness (3.0 N/cm) was greater than the internal stiffness (upper left), the responses behaved as they would if the loop gain decreased with increasing external stiffness. The opposite occurred when the regulated stiffness (0.2 N/cm) was less than the internal stiffness (upper right). The dependence on external load stiffness was much smaller when the internal and regulated stiffnesses were equal (lower left).

21.8 Conclusions

Active stiffness regulation, achieved by combining feedback of contact force and limb position, is a useful and simple method of providing controlled interactions with the environment [see also Chapter 9 (Hogan), Chapter 11 (Winters and Hogan), and Chapter 12 (Feldman et al.)]. It provides compensation for internal disturbances under a wide range of external loading conditions. The relative contributions of force and position to the feedback compensation are determined by the properties of the load, rather than by a discrete switch within the controller between control of force or of position. Logical switches between controlled variables would be prone to errors, and could cause oscillations. Tuning the system is straightforward, eliminating interactions between

variables by taking advantage of the dependence of loading conditions and performance on individual parameters or combinations of parameters. Once tuned, the system performs well under the full range of loading conditions encountered in hand grasp.

The input–output relationship of the system is simple. In the absence of a load, the actual position tracks the virtual position. When a load is present, the force is proportional to the difference between the virtual and actual positions. In a neuroprosthesis, the command (generated voluntarily by the patient or automatically by the hardware) determines the virtual position, and the value of regulated stiffness determines the relative sensitivities for controlling unloaded position or isometric force. The fact that stiffness is at an intermediate value (i.e. neither overly stiff nor overly compliant) is important for patient control. It is the difference between the actual and virtual positions that determines the interaction force, and this difference can occur because of either an external disturbance or an error in virtual position. The change in interaction force with either type of error is kept low if stiffness is low, and the force opposing displacement is high if stiffness is high. An intermediate value of stiffness provides a compromise between these goals.

In the experimental studies, the value of stiffness is determined by tuning the system under conditions of unloaded position control. Stiffness is adjusted with a fixed controller gain so that position is stable, with a fast rise time to step inputs. Controller gain is set similarly to adjust performance under isometric conditions. The chosen values of stiffness and controller gain make performance nearly the same under the two extremes of load stiffness. With these values, there is no guarantee that performance will remain the same for loads with intermediate values of stiffness. The analytical and simulation results, however, indicate that a particular value of stiffness makes the loop gain independent of load stiffness. This value is the sum of the intrinsic muscle stiffnesses and the passive stiffness. Since intrinsic muscle stiffness is proportional to activation level (force), and passive stiffness depends on joint angle, it follows that it would be advantageous to have the regulated stiffness be variable, rather than fixed as it is in the present control system. Since the intrinsic muscle stiffness is more significant than the passive stiffness over most of the range of joint movement, and since intrinsic stiffness may be predictable from the recruitment characteristics and the cocontraction map (with some initial calibration), it may be relatively simple to incorporate this nonlinearity into the control system. There is at least one possible disadvantage to having variable stiffness: the relationship between command and isometric force becomes nonlinear, raising the possibility that it may be more difficult for a patient to control force.

21.9 Recommended Future Directions

Stiffness regulation is one aspect of impedance control, which is viewed as a general method of controlling limbs that is especially important when environmental interaction is required [Hogan, 1985a,b,c; Chapter 9 (Hogan)]. Development of variable stiffness controllers and the addition of damping are obvious extensions to the work presented above. Damping is relatively unimportant in hand grasp since inertial loads (in compression) are very unusual. However, in applications to arm control, walking or standing, inertial loads are extremely important, and if the limb behaved as a pure spring, oscillations would die out very slowly. Just as there are choices to be made for stiffness, the type of damping for limb control must be investigated. Damping in the physiological system is nonlinear [Chapter 13 (Wu et al.); Gielen and Houk, 1984], with the viscous component being relatively more important at low velocities than at high velocities [e.g. see Chapter 5 (Winters)]. One functional advantage of this type of damping may be to terminate slow movements quickly, without overly impeding rapid movements.

Stiffness regulation must also be considered in other, more general applications in neuroprostheses. It is rare that only a single limb segment with a single degree of freedom is controlled. The more general case is a multisegment, multidegree of freedom limb. End-point stiffness regulation can be regarded as a general paradigm for limb control since virtually all movements involve environmental interactions, and these interactions occur at the end-point. Work towards this goal has already begun, by extending the stiffness regulator to control a two segment, two degree of freedom limb (Lan et al., 1989). Feedback regula-

tion of the magnitude and orientation of the end-point force vector has been achieved by stimulating knee and ankle flexors and extensors in the cat hindlimb. More work has to be done in the area of controlling end-point position, load transitions, and limb trajectories.

Acknowledgements

The authors gratefully acknowledge the support of the NIH-NINDS Neuroprosthesis program, under contract NO1-NS-6-2303.

References

Allin, J. and Inbar, G.F. (1986). FNS control schemes for the upper limb. *IEEE Trans. Biomed. Eng.* **33**: 818-828.

Andrews, B.J., Baxendale, R.H., Barnett, R.H., Phillips, G.F., Yamazaki, T., Paul, J.P. and Freeman, P.A. (1988). Hybrid FES orthosis incorporating closed loop control and sensory feedback. *J. Biomed. Eng.* **110**: 189-195.

Bernotas, L.A., Crago, P.E. and Chizeck, H.J. (1986). A discrete-time model of electrically stimulated muscle. *IEEE Trans. Biomed. Eng.*, **33**: 829-838.

Bernotas, L.A., Crago, P.E. and Chizeck, H.J. (1987). Adaptive control of electrically stimulated muscle. *IEEE Trans. Biomed. Eng.* **34**: 140-147.

Burke, R.E., Rudomin, P. and Zajac, F.E. III (1976). The effect of activation history on tension production by individual muscle units. *Brain Res.* **109**: 515-529.

Cabrera, K.T., Crago, P.E. and O'Malley, J.M. (1987). A combined open and closed loop system for a functional neuromuscular stimulation hand grasp, *Proc. 9th Ann. Conf. IEEE Eng. in Med. and Biol. Soc.*, Boston.

Chizeck, H.J., Crago, P.E. and Kofman, L. (1988). Robust closed-loop control of isometric muscle force using pulse width modulation, *IEEE Trans. Biomed. Eng.* **35**: 510-517.

Chizeck, H.J., Kobetic, R., Marsolais, E.B., Abbas, J.J., Donner, I.H. and Simon, E. (1988). Control of functional neuromuscular stimulation systems for standing and locomotion in paraplegics. *Proc. IEEE* **76**: 1155-1165.

Crago, P.E., Nakai, R.J. and Chizeck, H.J. (1990). Feedback regulation of hand grasp opening and contact force during stimulation of paralyzed muscle. accepted for publication, *IEEE Trans. Biomed. Eng.*

Crago, P.E., Peckham, P.H. and Thrope, G.B. (1980). Modulation of muscle force by recruitment during intramuscular stimulation. *IEEE Trans. Biomed.*

Eng. **27**: 679-684 .

Durfee, W.K. (1989). Task based methods for evaluating electrically stimulated antagonist muscle controllers. *IEEE Trans. Biomed. Eng.* **36**: 309-321.

Durfee, W.K. and MacLean, K.E. (1989). Methods for estimating isometric recruitment curves of electrically stimulated muscle. *IEEE Trans. Biomed. Eng.* **36**: 654-667.

Edstrom, L. (1970). Selective changes in the sizes of red and white muscle fibers in upper motor neuron lesions and Parkinsonism. *J. Neurol. Sci.* **11**: 537-550.

Gielen, C.C.A.M and Houk, J.C. (1984). Nonlinear viscosity of human wrist. *J. Neurophysiol.* **52**: 553-569.

Grandjean, P. and Mortimer, J.T. (1986). Recruitment properties of monopolar and bipolar epimysial electrodes. *Ann. Biomed. Eng.* **14**: 53-66.

Handa, Y., Handa, T. and Hoshimiya, N. (1986). A portable FNS system for the paralyzed upper extremities, *Proc. 8th Annual Conference of the IEEE Eng. in Med. and Biol. Soc.*, Fort Worth, TX, 65-67.

Hogan, N. (1985a). Impedance control: An approach to manipulation: Part I-Theory. *J. Dyn. Sys. Meas. Cont.*, **107**: 1-7.

Hogan, N. (1985b). Impedance control: An approach to manipulation: Part II-Implementation. *J. Dyn. Sys. Meas. Cont.* **107**: 8-16.

Hogan, N. (1985c). Impedance control: An approach to manipulation: Part III-Applications., *J. Dyn. Sys. Meas. Cont.* **107**: 17-24.

Holle, J., Frey, M., Gruber, H., Kern, H., Stohr, H. and Thoma, H. (1984). Functional electrostimulation of paraplegics-Experimental investigations and first clinical experience with an implantable stimulation device. *Orthop.* **7**: 1145-1160 .

Hoshimiya, N., Naito, A., Yajima, M. and Handa, Y. (1989). A multichannel FES system for the restoration of motor functions in high spinal cord injury patients: a respiration-controlled system for multijoint upper extremity. *IEEE Trans. Biomed. Eng.* **36**: 754-760.

Jaeger, R.J., Yarkony, G.M. and Smith, R.M. (1989). Standing the spinal cord injured patient by electrical stimulation: refinement of a protocol for clinical use, *IEEE Trans. Biomed. Eng.* **36**: 720-728.

Jones, D.A., Bigland-Ritchie, B. and Edwards, R.H.T. (1979). Excitation frequency and muscle fatigue: mechanical responses during voluntary and stimulated contractions. *Exper. Neurol.* **64**: 401-413.

Keith, M.W., Peckham, P.H., Thrope, G.B., Stroh, K.C., Smith, B., Buckett, J.R., Kilgore, K.L. and Jatich, J.W. (1989). Implantable functional neuromuscular stimulation in the tetraplegic hand. *J. Hand Surg.* **14A**: 524-530.

Khang, G. and Zajac, F.E. (1989 a). Paraplegic standing controlled by functional neuromuscular stimulation: Part I- Computer model and control system design. *IEEE Trans. Biomed. Eng.* **36**: 873-884 .

Khang, G. and Zajac, F.E. (1989 b). Paraplegic standing controlled by functional neuromuscular stimulation: Part II- Computer simulation studies. *IEEE Trans. Biomed. Eng.* **36**: 885-894.

Kilgore, K.L., Peckham, P.H., Thrope, G.B., Keith, M.W. and Gallaher-Stone, K.A. (1989). Synthesis of hand grasp using functional neuromuscular stimulation. *IEEE Trans. Biomed. Eng.* **36**: 761-770.

Krajl, A., Bajd, T., Turk, R. and Benko, H. (1986). Posture switching for prolonging functional electrical stimulation standing in paraplegic patients. *Paraplegia.* **24**: 221-230.

Krajl, A., Bajd, T., Turk, R., Krajnik, J. and Benko, H. (1983). Gait restoration in paraplegic patients: a feasibility demonstration using multichannel surface electrode FES. *J. Rehab. Res. Develop.* **20**: 3-20.

Lan, N., Crago, P.E. and Chizeck, H.J. (1987). An adaptive controller for regulation of joint stiffness by coactivation of antagonist muscles. *Proc. 9th Annual Conference of the IEEE Eng. in Med. and Biol. Soc.*

Lan, N., Crago, P.E. and Chizeck, H.J. (1989). Feedback control of stiffness in a multi-joint limb by FNS. *Proc. 11th Ann. Int. Conf. IEEE Eng. Med. Biol. Soc.*, Seattle, 971-972 .

Lieber, R.L. (1986). Skeletal muscle adaptability. III: Muscle properties following chronic electrical stimulation. *Dev. Med. Child Neurol.* **28**: 662-670.

Marsolais, E.B. and Kobetic, R. (1987). Functional electrical stimulation for walking in paraplegia. *J. Bone Joint Surg.* **69A**: 728-733.

Mayer, R.F., Burke, R.E., Troop, J., Walmsley, B. and Hodgson, J.A. (1984). The effect of spinal cord transection on motor units in cat medial gastrocnemius muscles. *Muscle and Nerve* **7**: 23-31.

Miller, L.J., Peckham, P.H. and Keith, M.W. (1989). Elbow extension in the C5 quadriplegic using functional neuromuscular stimulation. *IEEE Trans. Biomed. Eng.* **36**: 771-780 .

Mortimer, J.T. (1981). Motor Prostheses. In: *Handbook of Physiology. The Nervous System. Motor Control.* Bethesda, MD, Am. Physiol. Soc. sect. 1, vol. II, part 1, chapter 5, 155-187.

Naples, G.N., Mortimer, J.T. and Yuen, T.G.H. (1990). Overview of peripheral nerve electrode design and implantation. In: Agnew, W.F. and McCreery, D.B. (eds.)., *Neural Prostheses*, Prentice Hall, Englewood Cliffs, New Jersey, 107-146.

Peckham, P.H. (1987). Functional electrical stimulation: Current status and future prospects of applications to the neuromuscular system in spinal cord injury. *Paraplegia* **25**: 279-288.

Peckham, P.H., Keith, M.W. and Freehafer, A.A. (1988). Restoration of functional control by electrical stimulation in the upper extremity of the quadriplegic patient. *J. Bone Joint Surg.* **70-A**: 144-148.

Press, W.H., Flannery, B.P., Teukolsky, S.A. and Vetterling, W.T. (1986). *Numerical Recipes: The Art of Scientific Computing.* Cambridge University Press, Cambridge.

Riso, R.R., Ignagni, A.R. and Keith, M.W. (1989). Electrocutaneous sensations elicited using subdermally located electrodes. *Automedica* **11**: 25-42.

Stanic, U. and Trnkoczy, A. (1974). Closed-loop positioning of hemiplegic patients' joint by means of functional electrical stimulation, *IEEE Trans. Biomed. Eng. BME* **22**: 365-370.

Vodovnik, L., Crochetiere, W.J, and Reswick, J.B. (1967). Control of a skeletal joint by electrical stimulation of antagonists. *Med. Biol. Eng.* **5**: 97-109.

CHAPTER 22

Model-Based, Multi-Muscle EMG Control of Upper-Extremity Prostheses

Sanford G. Meek, John E. Wood, and Stephen C. Jacobsen

22.1 Introduction

Mathematical modelling of natural limb motion and actuation can greatly facilitate understanding of the biomechanics and control of a human limb, and can be used in the design of controllers for multi–axis prosthetic arms or the design of functional neuro–stimulators (*FNS*) for paralyzed limbs. The motion of a human limb involves the simultaneous control of each muscle of the limb. This simultaneous control provides stability, linkage stiffness, and force balance of the entire limb system, if not the entire body, in addition to the primary action of the limb. This idea of synergy of the entire system suggests that a prosthetic limb or a neuro–stimulated paralyzed limb should not be considered as an autonomous system but rather as an integral, dynamically coupled part of the person. Such a control scheme should free functionally sound parts of the body from controlling or actuating prosthetic or stimulated limbs, thus minimizing the conscience effort on the part of the person. Jacobsen (1973) proposed such a scheme, called Postulate Control, which is a unifying model involving the dynamics of the natural and prosthetic system and the natural and prosthetic efferent and afferent signal paths. While we have investigated afferent pathways (Meek et al., 1989) for improved prosthesis control, in this chapter we discuss only our work in investigating the efferent pathways.

Components of the Postulate Control model include: *i)* the kinematic and dynamic equations of motion of the natural and prosthetic limb links; *ii)* the instantaneous moment arms of all of the rem-

nant muscles about all of the remnant joints of the arm (or leg); *iii)* the recruitment patterns of the muscles; *iv)* models of the muscle forces as functions of muscle length, velocity, and fatigue; and *v)* the relationships between muscle forces and the cutaneous myoelectric signals. We, herein, present the component models, and the results of our research to date, noting that quantification of some of the models is not yet complete. The models are separated into two approaches, analytical and empirical. The analytical approach is the more detailed approach, modelling each subcomponent separately, providing the foundations of the total neuro-musculo-skeletal system model. The empirical approach is the more pragmatic (and computationally simpler) approach, which lumps the natural and prosthetic system in an overall manner into a phenomenological set of equations so that practical controllers can be developed for existing prosthetic limbs. While we have concentrated our efforts towards controllers for prosthetic arms, the methods and models presented herein should have utility for others interested in the biomechanics and control of intact human limb motion.

22.2 Postulate Control

Dynamic equations of motion (e.g. Lagrange's) are the starting point for the description of the Postulate Controller. The governing dynamic system can be described mathematically by the following matrix equation:

$$\begin{bmatrix} M_{nat} \\ M_{pros} \end{bmatrix} = P(\theta)\alpha + Q(\theta,\omega) + R(\theta) \qquad (22.1)$$

Multiple Muscle Systems: Biomechanics and Movement Organization
J.M. Winters and S.L-Y. Woo (eds.), © 1990 Springer-Verlag

where θ, ω, and α are seven-element vectors representing angular position, velocity, and acceleration respectively. Of the 11 degrees of freedom of the upper arm, from the clavicle to the forearm supination/pronation, only seven degrees of freedom are significant for an upper-extremity prosthesis controller. These degrees of freedom, spanning both the natural (*nat*) and prosthetic (*pros*) degrees of freedom, are: clavicular abduction/adduction, clavicular flexion/extension, humeral abduction/adduction, humeral flexion/extension, humeral rotation, elbow flexion/extension, and wrist pronation/supination. For an above-the-elbow amputation, humeral rotation, elbow flexion/extension, and wrist pronation/supination are the controlled degrees of freedom of the prosthesis. Terminal device prehension is not included in this analysis as it is not dynamically coupled sufficiently to be controlled well by a Postulate Controller. M_{nat} is a five element subvector representing the torques applied by the remnant musculature. In a prosthetic limb controller, these are estimated from cutaneous electromyographic (*EMG*) signals. Humeral rotation torque is included in this vector. It must also be a controlled motion because, with an above-the-elbow amputee, insufficient torque and motion of the humerus are transmitted through the soft tissues of the remnant limb for effective "direct" use. Therefore, humeral rotation is "directly" controlled by the *EMG*-estimated torque M_{nat} and not by the Postulate Controller, though the humeral rotation torque is also an input via M_{nat} to the Postulate Controller. M_{pros} represents the two remaining command torques, elbow flexion/extension and wrist supination/pronation, applied to the prosthesis. The P matrix represents the position dependent coefficients (the generalized impedance) of the angular acceleration vector. The centrifugal and Corriolis terms are contained in the Q matrix. The gravity terms are contained in the R matrix. The arm positions and velocities are measured by goniometers attached to the amputee and by transducers in the prosthesis (Meek, 1982), leaving nine unknowns for the seven dynamic equations - the seven angular accelerations (α) and the two prosthetic torques (M_{pros}). This imbalance of unknowns and equations can be resolved by postulating a constraint between some of the degrees of freedom of the linkage. We have linearly related the two degrees of freedom of the clavicle with the remaining five

degrees of freedom of the humerus and forearm, as the clavicle typically has the smallest motion of the upper extremity and is the most difficult to monitor with a goniometer:

$$\theta_{constrained} = \eta \; \theta_{free} \qquad (22.2)$$

where η represents an experimentally determined 2x5 element constraint matrix. Using the constraint equation and its appropriate derivatives, the dynamic equations can be partitioned thus:

$$\begin{bmatrix} M_{nat} \\ M_{pros} \end{bmatrix} = [\,P_1\,P_2\,] \begin{bmatrix} \alpha_{constrained} \\ \alpha_{free} \end{bmatrix} + Q + R \qquad (22.3)$$

The equations can be rearranged, solving for the seven unknowns, α_{free} and M_{pros}:

$$\begin{bmatrix} M_{nat} \\ M_{pros} \end{bmatrix} = \left[\, P_1\eta + P_2 \begin{array}{c} 0 \\ \hline -1 \end{array} \right]^{-1} [\, M_{nat} - Q - R \,] \qquad (22.4)$$

The command signal to the prosthesis torque servo is M_{pros}; the α_{free} are not directly required for the servo.

22.3 Derivation of the Controller Equations

The complete derivation of the controller equations can be found in the references (Jacobsen, 1973; Jerard, 1976), with further discussion in Fullmer (1983) and Fullmer et al. (1985). Some general points of interest for the development of a model-based prosthetic controller are discussed below.

22.3.1 Kinematics

The combined natural and prosthetic skeletal system is considered to be a set of rigid links interconnected with revolute joints, starting from the clavicular abduction/adduction joint and ending at the wrist supination/pronation joint. This enables the use of successive rotation matrices (Fu et al., 1987), from the clavicle to the wrist, to describe the kinematics.

22.3.2 Dynamics

The dynamic equations represent the major computational burden for the controller. Efficient recursive inverse dynamic algorithms now exist for robotic controllers [Hollerbach, 1980; reviewed briefly in Chapter 8 (Zajac and Winters)]. However, these are not directly applicable to the **partitioned** form of the controller equations. An exact analytical form of the P

matrix was determined using a symbolic manipulation program (Jerard, 1980; Fullmer, 1985). The values of the Q matrix (which were not partitioned) were found by Hollerbach's algorithm (Hollerbach, 1980; Fullmer et al., 1985).

Of primary interest is whether the centripetal and Coriolis accelerations contribute significantly to the moments on the linkage. It has been found from our laboratory experiments that with an amputee performing tasks of daily living with his prosthesis, that the Q matrix could be ignored without noticeably degrading performance (Fullmer, 1983; Meek et al., 1989).

22.4 Estimation of the Joint Moments – Analytical Approach

The Postulate Controller requires that we determine the vector M_{nat} which represents the moments about the remnant joints due to muscle forces. For a prosthetic controller, these moments can be estimated from the cutaneous myoelectric (*EMG*) signals of the remnant musculature, as represented by the following matrix equation:

$$M_{nat} = G E + N \qquad (22.5)$$

where E is the vector of *EMG* signals, G is a matrix of position and velocity dependent coefficients, and N is a vector of moments due to nonactive elements about the joint. We call this equation the vectormyogram (*VMG*) equation.

Four biomechanical components form the above matrix transduction of *EMG*'s to muscle moment estimates. These components are: *1)* an *EMG transmission* matrix which accounts for attenuation, dispersion and cross-talk of the *EMG* signals as they are transmitted from their fiber-depolarization sources to the electrode sites on the skin; *2) muscle modelling* matrices which modify the *EMG*-based force estimate of each muscle as a function of its length, velocity, and fatigue; *3) muscle recruitment* matrices which correlate the activity of synergistic muscles; and *4)* an *anatomy* matrix (for musculature) which provides the geometric factors for converting the muscle forces into joint torques. If we could completely understand these components, we could analytically find the relationship between the joint torques and the cutaneous *EMG* signals. Each of these components is discussed below from analytical and empirical bases, with varying levels of completeness of the component models.

22.4.1 *EMG* Transmission Matrix

Cutaneous *EMG* signals are typically used for practical prosthesis control because their noninvasive nature is better suited for daily use of prostheses. The cutaneous signals indicate more of the overall activity of the muscle, and the skin and tissues provide some integration (smoothing) of the individual motor unit depolarizations. The relationship between the cutaneous *EMG* signals of the accessible muscles and their subcutaneous *EMG* signals can be given in matrix form:

$$E_{(subcutaneous, \, accessible)} = T E_{(cutaneous, \, accessible)} \qquad (22.6)$$

The off-diagonal terms of the transmission matrix include the *EMG* cross-talk terms. We have not specifically investigated the relationships between subcutaneous and cutaneous *EMG* signals, since, at least for empirical *VMG* and practical prosthesis control, we always are interested in cutaneous *EMG* signals. Consequently, in the analysis, we have lumped the transmission matrix into the muscle modeling matrix.

22.4.2 Muscle Modeling Matrix

The amplitude of the subcutaneous *EMG* signal is used as an estimate of the muscle force. The relationship between the *EMG* level and the muscle force can be given by:

$$F_{muscles} = H_1(L,V,f) \, E_{subcutaneous} + H_2(L,V) \qquad (22.7)$$

where H_1 is the "active" muscle matrix and H_2 is the vector of passive muscle forces. The force-to-*EMG*-amplitude relationship is nonlinear, depending upon the length (L), velocity (V), and the fatigue level (f) of the muscle. This relationship has been studied or modelled by many researchers [Heckathorne and Childress, 1981; Wood and Mann, 1981; Patla et al., 1982; see also review in Chapter 5 (Winters)].

We have estimated the muscle length dependency of the cutaneous *EMG* amplitude for an intact elbow. Thus, we have combined the *EMG* transmission matrix, T, with the H matrices, yielding H'. A third order model of *EMG*s vs. length was used for H' with the constant terms (H_2) forced to be zero. The primary origin of this dependency is the change in the filament overlap within the muscle (see *Eq. 22.8*). A secondary consideration, though unmodelled, might be the movement of the muscle belly with respect to the cutaneous electrodes with limb movement.

The estimation procedure was as follows. Filtered cutaneous *EMG* signals of the biceps, triceps, and brachialis muscles were recorded as a subject moved his elbow over a 40° range. Torques were applied about the elbow in a sinusoidal manner with a ±3 ft-lb amplitude. Figure 22.1 shows the results where the measured and estimated moments about the elbow are displayed, along with the elbow position and the *EMG* amplitude of each muscle.

ing the *EMG* level to muscle force for the biceps muscle:

$$F_{muscle} = H'(L) E_{cutaneous} \qquad (22.8)$$

where $E_{cutaneous}$ is the cutaneous *EMG* signal amplitude, F_{muscle} is the muscle force, and $H'(L)$ is the length–dependent coefficient which has the transmission matrix and modelling matrix coefficients nested within.

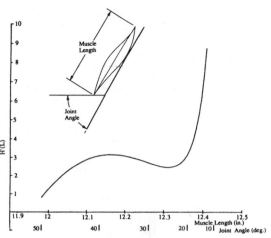

Figure 22.2: The length-dependent coefficient, $H'(L)$ relating the cutaneous *EMG* signal, E, with the muscle force, F. Determined by regression.

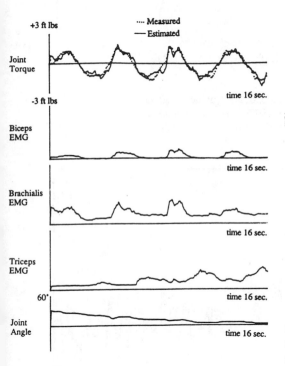

Figure 22.1: Actual and estimated via *VMG* elbow torques with a varying angle of the elbow. Filtered cutaneous *EMG* signals of the biceps, brachialis, and triceps also shown.

A regression analysis was performed on the measured *EMG* signals, measured torques, and elbow angle. The length of each muscle and the moment arms of each muscle were estimated from measurements of the muscles about the elbow joint of a cadaver (Section 22.4.4). Figure 22.2 shows the length-dependent coefficient $H'(L)$ relat-

At the relatively slow contraction rates associated with prosthesis control in laboratory experiments, the velocity dependent relationships of H_1 do not seem critical. What is very critical to prosthesis control is the change in the *EMG* signal amplitude with fatigue of the muscle which would change the command signal to the prosthesis. Figure 22.3 shows this effect.

Experimentally, this effect was determined by applying a constant load to the muscle with a weight. As can be seen, the amplitude of the filtered *EMG* signal increases with time. Figure 22.4 shows that the median frequency of the *EMG* spectrum decreases under the same conditions. This suggests the possibility of measuring fatigue level and theoretically, to compensate the *EMG*/torque relationship for the fatigue level.

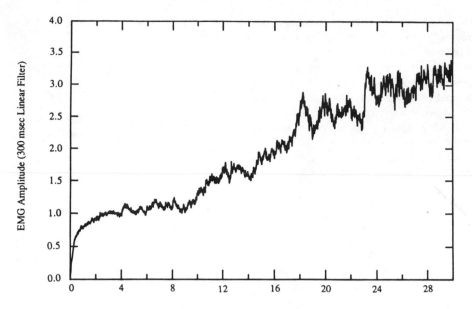

Figure 22.3: The amplitude of the cutaneous *EMG* signal for a constant force (80% maximum voluntary contraction (*MVC*)) on the biceps showing the increase of the signal due to fatigue.

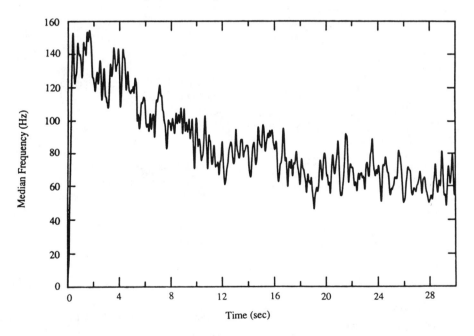

Figure 22.4: The median frequency of the cutaneous *EMG* (same *EMG* as in Figure 22.3, constant force at 80% *MVC*), showing the descrease due to fatigue.

Current research at the CED is investigating relationships between the median frequency and the amplitude change in order that a fatigue-compensated *EMG* processing system could be built. Real-time measurement of fatigue level has been investigated by several researchers (Deluca, 1984; Kramer et al., 1987) and is a critical issue for prosthetics, neuro–stimulation, and biomechanics research.

22.4.3 Muscle Recruitment Matrix

There are typically several muscles which cause any given action, as well as several muscles which oppose the motion and/or provide joint stability and bone stress reduction. For example, the biceps, brachialis, and brachioradialis all flex the elbow. Each muscle provides a different proportion of the torque across the joint. At different positions of the joint, the relative proportions of the total torque from each muscle will change (Basmajian and Latif, 1957; Crago et al., 1980; Hof and van den Berg, 1977; Dul et al., 1984a,b). The contraction force of each of the muscles with respect to forces of the other muscles acting about the joint is known as the recruitment pattern. The forces are then mapped to torques about each joint by the Anatomy Matrix (Section 22.4.4). Jacobsen (1973) described the recruitment relationships between groups of muscles in terms of the *EMGs* by the following matrix equation assuming linear synergism

$$E_{(subcutaneous,\ unaccessible)}$$
$$= \alpha_1 E_{(subcutaneous,\ accessible)} + \alpha_2 \tag{22.9}$$

An understanding of the recruitment patterns is helpful for artificial arm control as well as necessary for basic biomechanics knowledge. For an artificial arm which is reliable and convenient for the amputee to use, it is desired to minimize the number of muscles and electrodes needed to control the prosthetic arm. Moreover, many of the muscles are not accessible from surface *EMG* electrodes. For a practical "take–home" artificial arm, it is desirable to use only surface electrodes and to minimize the number of electrodes for the comfort and convenience of the amputee. The muscle recruitment patterns describe how the muscles work together for given motions, from which the minimum number of muscles needed for the control of a prosthesis could be determined and the activity of the unaccessible muscles could

be estimated. Defining the recruitment patterns of the muscles of a complex joint such as the shoulder is a formidable task, so a method of grouping the muscles according to their actions is needed. The principal components techniques (Massey, 1965) group the variables according to their correlations with each other. The principal components are the eigenvalue-weighted eigenvectors of the correlation matrix of *EMG* vectors (EE^t). For example, Table 22.1 shows the second and third components of the set of principal components for 10 muscles about the scapulo-humeral joint while the three simultaneous torques, humeral flexion/extension, humeral abduction/adduction, and humeral rotation, were applied to that joint. The groups can be identified by knowing the general actions of the muscles and seeing when the muscles involved with a certain action have significant principal component coefficients.

Table 22.1: Principal component groupings of muscle actions about the shoulder during simultaneous humeral flexion/extension, abduction/adduction, and rotation. (The first component does not group to any particular degree of freedom.)

Muscle	2nd Principal Component	3rd Principal Component
Pectoralis Major - Abdominal part	.1128	-.2694
Pectoralis Major - Sternocostal part	.0869	-.2701
Pectoralis Major - Clavicular part	.2996	-.6427
Deltoid - Clavicular part, medial	.1477	.1313
Deltoid - Clavicular part, lateral	-.0121	.2162
Deltoid - Acromial part	-.7873	.0760
Deltoid - Scapular part	.0022	.0133
Infraspinatus	.4233	.5760
Teres Major	.2369	.1989
Latissimus Dorsi	.1135	-.0133

In this example, the second principal component weights the *EMGs* according to their action with humeral abduction. It shows, for instance, that the mid deltoid (*PC#2* = –0.7873) is the only significant abductor (negative direction) of the humerus. The third principal component groups the humeral rotators. It indicates that the most significant indicator of medial (negative direction) rotation is the clavicular pectoralis major (*PC#3* = –0.6427) and the most significant indicator of lateral (positive direction) rotation is the infraspinatus (*PC#3* = 0.5760). This implies that only these two muscles would possibly be needed to estimate humeral abduction/adduction torque. Similar analyses, including the minimum number of muscles required for the *VMG* equation, could be performed for the other degrees of freedom.

22.4.4 Anatomy Matrix

The conversion of muscle forces to joint torques requires the determination of "effective" geometric moment arms. This can be described by the following relationship:

$$M_{nat} = A_m(\theta) \, F_m \tag{22.10}$$

where A_m is the anatomy matrix containing the instantaneous moment arms of the muscles about the joints. In the simplest cases, the muscle has a linear line-of-action which remains linear with limb displacements. In more complicated cases, the muscle usually begins with a curved line of action which then transitions to a linear line-of-action extending from a point of tangency on one body segment to a point of insertion on another segment. In such cases, with limb displacement, there is motion of the *effective* origin and/or insertion points, relative to their respective anatomical attachments to rigid bones, thus causing changes in the *effective* anatomy matrix coefficients. Regardless of the problems, the nature of the geometric data comprising the anatomical conversion factors remains similar in form.

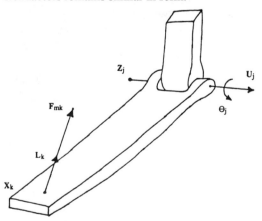

Figure 22.5: Muscle acting about a revolute joint. The force of muscle k, F_{mk}, acts at point X_k along unit vector L_k. Unit vector U_j defines the direction of the joint, j, while Z_j locates the joint in space.

Quantitation of Moment Arms

Each A_m-matrix coefficient can be found by the triple-scalar-product:

$$A_{m(jk)} = (X_k - Z_j) \times L_k \cdot U_j \tag{22.11}$$

where: X_k is the point of attachment of the *k-th* muscle, Z_j locates in space the unit vector of revolution U_j of the *j-th* joint, and L_k is a unit vector of the *effective* line-of-action of the *k-th* muscle. Figure 22.5 shows the notation. Spherical joints, such as at the shoulder, are defined by three successive revolute displacements. The term $A_{m(jk)}$ is thus a scalar, consisting solely of geometric information (all of which is a function of the skeletal linkage configuration), and is thus the geometric factor (or moment arm) which converts the *k-th* scalar force $F_{m(k)}$ into a scalar moment about the *j-th* joint (U_j, Z_j). If we have several muscle forces acting about joint *j*, then we would have the total net moment for musculature about joint *j* given by the scalar summation over all muscles

$$M_{nat(j)} = \sum_k A_{m(jk)} \, F_{m(k)} \, . \tag{22.12}$$

The muscle force could be the force of a single fiber, a motor unit, a gross muscle bundle with a tendon which attaches at a fairly discrete point, or the resultant force vector of a muscle with a distributed attachment pattern. The line-of-action of the muscle force is the line from the insertion to the origin of the muscle in the linear line-of-action case. For curvilinear muscles, the line-of-action is tangent to the centroid line of the muscle from the insertion to the origin. A_m-matrix coefficients were calculated along the length of the centroid of each muscle. The centroid line of each muscle was calculated from *3-D* locations of points on the surface of each muscle. Mathematical cross sections (perpendicular to the muscle fibers) were found for each muscle along the length of the muscle. For each cross section, the centroid and area (projected area of the muscle) were calculated. The tangents of the centroid line were calculated by the set of line segments cross section centroid.

Quantitatively, we determined the elements of each $A_{m(jk)}$ for 30 muscles about the shoulder and elbow joints (eleven degrees of freedom total) of a cadaver mounted inside a rigid measuring frame at the Center for Engineering Design (Wood, 1989a,b). A 3-degree-of-freedom digitizer was then used to collect *3-D* coordinate data for muscle trajectories, muscle and bone surfaces, and joint kinematics.

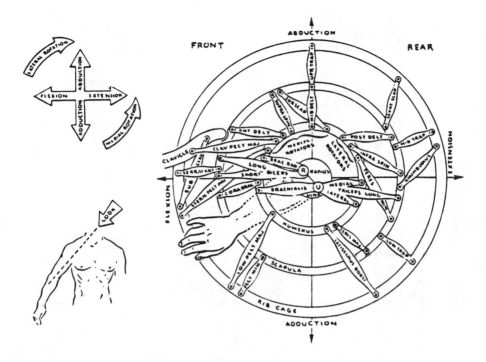

Figure 22.6: Original Circle Diagram by Jacobsen (1973) of the muscles acting about the shoulder.

The results are shown in Table 22.2. The "*effective*" moment arm is given for each muscle. The choice of the "*effective*" moment arm is based upon whether the best estimate of the line of tangency of the trajectory of the line-of-action of the muscle occurs at the origin (*O*) or at the insertion (*I*). Also indicated is whether a linear trajectory (*L*) is sufficient to model the line of action of the muscle (e.g. a straight "string" model from the origin to the insertion is adequate) or whether a curved trajectory (*C*) is required.

Circle Diagram

A graphical method of showing the moment arms about several joints is by the Circle Diagram (Jacobsen and Mann, 1974). Jacobsen originally based his anatomy coefficients on the primary actions as described in anatomy texts (Figure 22.6). Figure 22.7 shows a computer-generated circle diagram based upon measured anatomy coefficients from our dissection for the same primary actions (Wood et al., 1989b). For example, we might consider the primary actions of the biceps to be elbow flexion and wrist supination. A secon-

dary action would be humeral flexion. Thus, the Circle Diagram is a graphical representation of the mechanical influence of the muscles about several degrees of freedom in this case about the joints of the upper limb. Three degrees of freedom (*DF*) can be shown for each muscle on a single diagram. The degrees of freedom are, typically, a flexion/extension, an abduction/adduction, and a rotation of a given joint. Another diagram would be needed to represent the moment arms of those muscles about another joint (or set of *DF*s). Explicitly, the vertical and horizontal axes represent the abduction/adduction (*A/A*) and flexion/extension (*F/E*) directions, respectively. The angle (θ) that a muscle is drawn with respect to these axes indicates the relative lengths of its *A/A* and *F/E* moment arms. The width (*w*) of the muscle body indicates the relative magnitude of the "hypotenuse" (*a*) of the *A/A* and *F/E* moment arms, times the effective cross-sectional area of the muscle, thus giving a measure of the total moment capacity of the muscle. The perpendicular distance (*d*) from the muscle "line-of-action" (as

Table 22.2: A-matrix coefficients: Coefficients are in units of inches

I— point of tangency is at the insertion,   O— point of tangency is at the origin

No.[†] Name	Linear(L) or Curvilinear (C) Trajectory	1 Clavicular Rotation	2 Clavicular Flexion/ Extension	3 Clavicular Abduction/ Adduction	4 Scapular Rotation	5 Scapular Flexion/ Extension	6 Scapular Abduction/ Adduction	7 Humeral Flexion/ Extension	8 Humeral Rotation	9 Humeral Abduction/ Adduction	10 Elbow Flexion/ Extension	11 Wrist Supination/ Pronation
1. Biceps-long head	L							O 1.1	O 0.0	O 0.3	I 1.7	I 0.6
2. Biceps-short head	L							O 1.2	O 0.1	O -0.7	I 1.7	I 0.6
3. Brachioradialis	L										O 1.3	I 0.5
4. Brachialis-lower head	L										I 1.1	
5. Brachialis-upper head	L										I 1.0	
6. Coracobrachialis	L							O 1.1	O -0.2	O -0.9		
7. Deltoid-acromial part	C							O -0.7	O 0.0	O 0.8		
8. Deltoid-clavicular part	C							I 1.3	I -0.1	I 0.6		
9. Deltoid-scapular part	C				O 0.2	O 0.8	O -0.5	I -3.1	I 0.3	I -0.1		
10. Infraspinatus	L							I 0.0	I 0.9	I 0.2		
11. Latissimus dorsi	C	I 1.3	I -5.9	I -4.2	I -2.4	I -0.9	I -2.2	I -1.6	I -0.4	I -1.4		
12. Pectoralis minor	L	I 1.3	I 2.0	I -2.9	I -0.2	I -0.6	I 0.8					
13. Pect. major-abdominal	C	I 2.1	I 3.5	I -3.3	I 2.5	I 2.0	I -3.0	I 2.2	I -0.7	I -2.1		
14. Pect. major-clavicular	C	I 1.2	I 1.8	I -2.3	I 2.1	I 1.6	I -1.8	I 2.1	I -0.5	I -1.0		
15. Pect. major-sternocostal	C	I -4.6	I -4.3	I -0.1	I 1.8	I 1.7	I -3.1	I 1.7	I -0.8	I -2.0		
16. Rhomboid major	C	I -2.6	I -4.6	I 0.9	I -1.6	I -2.7	I -5.3					
17. Rhomboid minor	C	I 2.0	I 6.2	I 3.7	I -0.4	I -3.4	I -3.1					
18. Serratus anterior-lower	C	I -1.2	I 5.2	I 0.5	I 4.8	I 0.2	I 3.3					
19. Serratus anterior-upper	L				I 0.8	I -0.4	I 1.1					
20. Supraspinatus	L							I 0.2	I 0.1	I 0.7		
21. Subclavius	C	L 0.3	L -0.4	L 0.0								
22. Subscapularis	C							I -0.1	I -1.0	I 0.1		
23. Teres major	L							I -1.8	I -0.4	I -1.5		
24. Teres minor	L							I -0.4	I 0.8	I -0.5		
25. Triceps-lateral head	L										I -0.8	
26. Triceps-medial head	L										I -0.8	
27. Triceps-long head	L							O -1.0	O 0.5	O -1.2	I -0.9	
28. Trapezius-lower part	L	I 5.9	I -4.1	I -1.7	I 2.6	I -1.5	I 1.4					
29. Trapezius-middle part	L	I 0.5	I -6.1	I 1.5	I 0.2	I -2.5	I -0.2					
30. Trapezius-upper part	L	I -0.2	I -4.2	I 2.4	I 0.0	I -1.0	I 0.2					

† Number corresponds to the muscle in the circle diagram, Figure 22.7

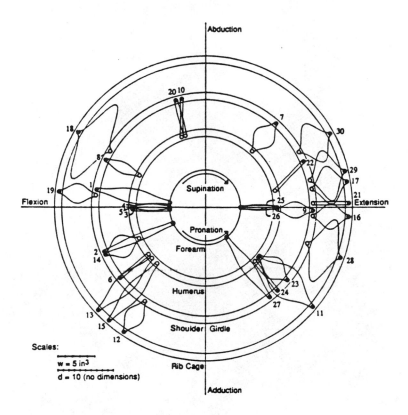

Figure 22.7: Computer-generated Circle Diagram for the muscles acting about the shoulder. Numbers correspond to muscles listed in Table 22.2. (From Wood et al., 1989b; reprinted with permission.)

drawn) to the center of the circles is proportional to the normalized (by a) rotational moment arm. For example, suppose that a muscle (m) is generating a stress and it affects the humeral degrees of freedom (it might also be acting about other DFs, but they must necessarily be plotted on another diagram). The humeral A/A moment (HAA), for orientation θ with respect to axis F/E, would then be calculated by:

$$M_{HAA(m)} = (\sigma)\,(\text{width})\,(\sin\theta)$$

$$= \sigma\,(A_c \cdot a)\,\sin\theta \qquad (22.13)$$

$$= F_m\,a_{HAA}$$

where

$$a = \sqrt{(a_{HAA}^2 + a_{HFE}^2)}, \quad F_m = \sigma A_c \text{ and}$$

$$a_{HAA} = a \sin\theta$$

Likewise, the humeral flexion/extension moment ($M_{HFE(m)}$) would be calculated using $\cot\theta$, or, alternatively,

$$M_{HFE(m)} = M_{HAA(m)} \cos\theta \qquad (22.14)$$

The humeral rotation (HR) moment would be calculated by:

$$M_{HR(m)} = (\sigma)\,(\text{width})\,(d)$$

$$= \sigma\,(A_c \cdot a)\,(a_{HR}/a) \qquad (22.15)$$

$$= F_m \cdot a_{HR}$$

Similar analyses can be made for muscles affecting movement of the shoulder girdle bones (clavicle and scapula) or forearm or any other limb (e.g. Seireg and Arvikar, 1973, 1975).

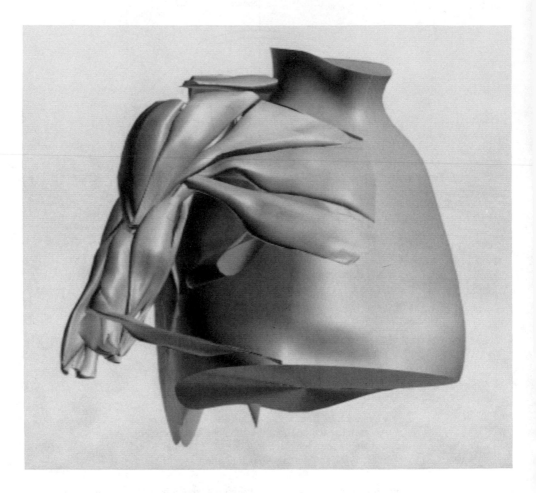

Figure 22.8: Alpha-1 generated image of surface topology reconstruction of the muscles about the shoulder (from Wood et al., 1989a)

Surface Modelling

Mathematically continuous surface models of the thirty muscles acting about the shoulder and upper limb were produced using Coons' surface (Coons, 1967) methods. This allowed us to produce a computer–generated image of any or all of the muscles and bones. Figure 22.8 shows an Alpha_1 (Cohen and Riesenfeld, 1982) solid model reconstruction of all of the muscles.

Volumetric data can be calculated from these data. The results of the maximum projected cross sectional area and volume calculations are given in Table 22.3 along with the *physiologic cross section* areas (Crowninshield, 1981), which is the volume divided by the length of the muscle. The reader can make the appropriate comparisons of the maximum "projected" cross-sectional area and the physiologic cross-sectional area, which are surprisingly close in most cases.

Table 22.3:

Lengths (L), Surface Areas (S)[*], Volumes (V), Cross-Sectional
Areas (A_c)[**], and Physiological Areas (A_p)[***] of Muscles

No.	Muscle Name	L(in.)	S(in.2)	V(in.3)	A$_c$(in.2)	A$_p$(in.2)
1.	Biceps-long head	12.3	27.9	4.3	1.1	0.3
2.	Biceps-short head	13.1	23.4	2.3	0.5	0.2
3.	Brachioradialis	11.3	26.0	2.5	0.4	0.2
4.	Brachialis-lower head	3.0	11.8	2.1	0.8	0.7
5.	Brachialis-upper head	5.9	22.2	4.0	0.9	0.7
6.	Coracobrachialis	7.3	12.8	1.3	0.2	0.2
7.	Deltoid-acromial part	6.5	36.7	13.6	2.8	2.1
8.	Deltoid-clavicular part	7.6	27.6	5.4	1.0	0.7
9.	Deltoid-scapular part	7.5	25.0	4.6	0.7	0.6
10.	Infraspinatus	5.9	26.0	5.2	1.3	0.9
11.	Latissimus dorsi	10.6	125.7	20.7	2.0	2.0
12.	Pectoralis minor	5.3	24.6	3.2	0.7	0.6
13.	Pectoralis major-abdominal part	8.6	33.3	4.8	0.8	0.6
14.	Pectoralis major-clavicular part	7.3	29.5	5.7	1.3	0.8
15.	Pectoralis major-sternocostal part	8.5	39.1	5.9	0.8	0.7
16.	Rhomboid major	4.5	17.3	2.5	0.8	0.6
17.	Rhomboid minor	3.1	10.2	1.7	1.1	0.5
18.	Serratus anterior-lower	6.8	73.4	9.9	1.6	1.5
19.	Serratus anterior-upper	4.1	28.6	1.9	0.5	0.5
20.	Supraspinatus	3.4	11.1	2.4	0.8	0.7
21.	Subclavius	3.0	-	-	-	-
22.	Subscapularis	4.8	33.2	7.4	2.1	1.5
23.	Teres major	4.8	19.8	4.3	1.3	0.9
24.	Teres minor	4.1	10.0	1.5	0.4	0.4
25.	Trieps-lateral head	9.7	33.9	6.6	0.9	0.7
26.	Trieps-medial head	8.4	23.1	3.8	0.6	0.5
27.	Triceps-long head	12.3	38.4	7.8	1.1	0.6
28.	Trapezius-lower part	8.2	73.4	12.2	2.0	1.5
29.	Trapezius-middle part	4.9	31.7	5.5	1.1	1.1
30.	Trapezius-upper part	4.7	24.0	4.9	1.2	1.0

[*] Surface areas do not include areas of ends of the muscles (typically small)
[**] Cross-sectional areas are maximum projected areas (normal to line-of-action of muscle force)
[***] Physiological areas are defined as: $A_p = V/L$

22.4.5 Summary of Analytical Methods

The *EMG* transmission, muscle modelling, muscle recruitment, and anatomy matrices represent the fundamental components needed to estimate the natural moments of the muscles acting about a joint. The mathematical models of these components could be further refined. Even as is, without further refinement they are not fully quantitated. Much work remains in the basic understanding of the complete system. In order to bypass the incompleteness of the "analytical" data, empirical methods (described below) were used to get lumped-parameter data.

22.5 Estimation of Joint Moments –
Empirical Methods

In the next two sections, we present our methods of empirically measuring the relationships between joint torques of the shoulder and arm and cutaneous *EMG* signals of the muscles about the shoulder and arm.

22.5.1 Myoelectric Signal Processing -
Adaptive Filtering

In order to have accurate estimates of muscle forces from *EMG* signals, a processing system with high signal-to-noise ratio as well as short rise time is required. Many filtering schemes have been investigated, including linear filters, averaging filters, spatial, pre-whitening filters, and adaptive filters. Adaptive filters have been shown to provide the best results (Meek et al., 1987).

The *EMG* signal, when used for proportional control, is usually treated as an amplitude-modulated signal. The mean amplitude of the cutaneous myoelectric signal is used as the control signal and is the desired output from the myoelectric processor. A multiplicative signal with frequencies from below 10 Hz to over 1000 Hz (the *EMG* spectrum) functions as the "carrier" of the amplitude modulated system (Deluca, 1979). However, the use of myoelectric signals as control signals poses a fundamental problem: the *EMG* spectrum of frequencies overlaps the frequencies of desired control (Jacobsen et al., 1984; Fullmer, 1985). This makes the separation of the desired control signal (*EMG* amplitude) from the noise (*EMG* carrier) difficult. We observed a fundamental filtering paradox whereby, with a stationary filter, it is possible to have either a fast response or a high signal-to-noise ratio, but not both (Jacobsen et al., 1984).

The basic idea behind the adaptive filtering method in overcoming the filtering paradox is to vary the time constant of the filter depending upon the rate of change of the signal. The rate of change of the signal is determined by the derivative of the smoothed signal. When the signal is changing rapidly (high derivative), the time constant is low, allowing fast response but with high noise. When the signal is steady (derivative is low or zero), the time constant is high, allowing high signal-to-noise ratio but slow response. The assumption is that when operating a prosthesis, an amputee can tolerate noise when rapidly moving the prosthesis but will not tolerate delays in con-

trol. When holding the prosthesis steady or performing slow, dexterous tasks such as threading a needle, the amputee will tolerate slow response as long as there is low noise. The response time and the signal-to-noise ratio depend, then, not upon the time constant but rather upon the adaption logic. The adaption logic is defined by:

$$\tau_{nl} = \frac{\tau_l - \tau_s}{a\dot{Z}^2 + 1} + \tau_s; \qquad Y = \tau_{nl}\dot{E} + E \quad (22.16)$$

where:
τ_{nl} is the adaptive time constant
τ_l is the max. time constant of the adaptive filter
τ_s is the min. time constant of the adaptive filter
$\dot{Z}$ is the derivative of the smoothed *EMG* signal
E is the rectified unsmoothed *EMG* signal
Y is the output (control) signal
a is a gain factor for the adaption.

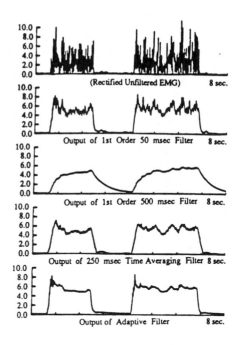

Figure 22.9: Comparison of the response of a 50 ms linear filter, a 500 ms linear filter, a 250 ms averaging filter, and the adaptive filter.

The ability of the adaptive filter to produce high signal-to-noise ratios as well as allowing fast response can be demonstrated in Figure 22.9. The original rectified raw *EMG* signal is shown with the responses of a 500 ms linear filter, a 50 ms linear filter, an averaging filter of 250 ms averaging time, and the adaptive filter. Note that the response of the adaptive filter has the same rise time as that of the 50 ms linear filter and that it has the same signal-to-noise ratio as that of the 500 ms filter.

22.5.2 Empirical Controller

Vector Myogram Equation

For practical artificial arm use using the postulate controller, the muscle forces are estimated from *EMG* signals. A relationship between joint torques and *EMG* signals can be found empirically by measuring known torques about a particular joint and associated *EMG* signals. This relationship is known as the vectormyogram (*VMG*).

VMG Data Collection

The empirical measurement of the *VMG* must begin with the data collection. Simultaneous joint torque and *EMG* data are collected from a subject by one of two methods (Meek, 1984).

The first method employs the Force Loading Joystick (*FLJ*), which is a three degree-of-freedom force sensor. The shoulder joint torques initiated by the subject in three directions are measured as forces at the distal end of the limb. Since the distances can easily be measured from the *FLJ* to the center rotation of each joint, the torques are easily computed. The torque applied by the subject, as well as the simultaneous *EMG* signals, are recorded by the computer for later processing.

In order to insure that all ranges and directions of the torques about all the degrees of freedom are achieved, the subject is directed by the computer to specified combinations of torque levels. A target is presented to the subject on a computer graphics screen, with the actual torque that the subject is applying to the *FLJ* also displayed. When the applied torque matches the target, the simultaneous torque and *EMG* data are recorded by the computer.

The targets are set at specified levels and are presented in a random order to the subject. This is to prevent fatiguing of any one group of muscles. As an additional means to remove fatigue as a factor, the subject is paced to the target level at a specified rate.

There are several limitations of the Force Loading Joystick. Firstly, only static torques can be applied. Secondly, since the subject must initiate three dimensions of torque and the computer graphics display is two dimensional, controlling

more than three torques at specified levels becomes exceedingly difficult for the subject. For these reasons, another data collection system was devised, the Dynamic Skeletal Loading System (*DSLS*).

The *DSLS* is an apparatus consisting of several fully gimbled, computer-controlled pushers which can apply forces to the arm. The gimbles are instrumented to measure angular positions at each joint of the gimble. The pusher rod is also instrumented to measure the position, velocity, and force at the end of the rod. The motor can push or pull the rod with a maximum force of 10 pounds (B.K. Hanover, personal communication).

If three pushers are connected together at one point, then the force in any direction can be controlled. Also, since the pushers are gimbled, the limb is no longer constrained to be in a fixed position. Another advantage of the *DSLS* is that since the force can now be computer-controlled to push or pull on the subject's limb, the subject is no longer required to initiate the torques and follow complicated target displays on a computer graphics system. The subject merely is required to hold his limb in position, resisting the applied torques. This enables the *DSLS* to be capable of data collection in more than 3 degrees of freedom, as well as collecting simultaneous torque/*EMG* data on a moving limb. Experiments have been performed with up to 5 degrees of freedom of torques.

Humeral Rotation

In the experiments involving both the *FLJ* and the *DSLS*, the humeral rotation torques are not readily obtained from above-elbow amputees. The amputee's muscles apply rotary torques about the humerus and the humerus rotates; however, the humeral torque is not transmitted to the outside skin since the humeral bone remnant is rotating inside the soft tissues of the stump and these tissues are incapable of transmitting the full torque to the outside. This is not a problem for the non-amputee since a loadcell could simply be placed on the forearm.

In order to determine the humeral torques on an amputee, an assumption must be made. Whenever a bone moves in a joint, certain resistances due to passive ligament, tendon, and muscle tensions as well as inherent joint friction must be overcome. When the amputee's humeral bone rotates inside the soft tissues of the stump, some resisting torque must be present. This resisting torque may also be dependent upon the position of the bone in the joint, since the ligaments will be stretched different amounts in different positions. We assumed that the resisting torque and, therefore the applied muscular torque, vary linearly with position. The angular humeral rotation position measurement can be accomplished with the use of a goniometer applied to the end of the amputee's stump. The goniometer moves with very little resisting torque so that the motion of the soft tissues can turn it. Enough rotation transmits out through the soft tissues, if the stump is long and the tissues are not too loose, to allow a torque estimate.

Data Analysis

The simultaneous joint torque and muscle *EMG* relationship is estimated via a multivariant regression. Since the muscle actions are redundant, the corresponding *EMG* signals can be highly correlated. This leads to an instability when ordinary least squares regression is used to calculate the equation coefficients. In order to circumvent this problem, two other regression techniques have been investigated, specifically *ridge regression* and *principal components regression* (Meek et al., 1984). The complete details of the regression analysis will not be presented here. However, a cursory explanation is helpful. If we partition the *VMG* equation as:

$$M = [\, G \vdots N \,] \left[\frac{E}{1} \right] \qquad (22.17)$$

then the least squares formulation results in:

$$[\, G \vdots N \,] = M\, E^T\, (E\, E^T)^{-1} \qquad (22.18)$$

The covariance matrix, $E\, E^T$, is of interest for investigating the correlation problem. If there is a high correlation between the independent variables, the *EMG*'s, then this matrix tends to be singular, causing unstable regression coefficients. *Ridge regression* (Marquart, 1975) and *principal components regression* (Massy, 1965) are two methods to reduce the singularity of the covariance matrix.

Ridge regression reduces the singularity of the covariance matrix by adding a small constant value to the diagonal terms of the matrix. This increases the determinant of the matrix, making the inverted matrix, and therefore the regression coef-

ficients, less sensitive to small changes in the values of the variables. The regression coefficients, therefore, become more stable, that is to say, they have a smaller mean-squared-error. The mean-squared error is lowered at the sacrifice of a small biasing of the regression coefficients from their "true" values.

Principal components regression treats the singularity problem by removing the source of the singularity. The procedure to do this is straightforward. First, the *EMG* vectors are rotated by the eigenvectors of the covariance matrix. These rotated *EMG* vectors are ranked according to their eigenvalues from largest to smallest. The smallest eigenvalues represent the more singular data and are eliminated. The eigenvectors representing the highest eigenvalues are called the principal components. The regression is then performed between the remaining rotated *EMG* vectors and the torque vectors. The resulting regression coefficients are then rotated back to the original space by their eigenvectors. Thus, the singularity is reduced at a small sacrifice of some lost information (some of the *EMG* data is thrown away).

22.5.3 Results of the Empirical *VMG* Experiments

The correlation between the actual measured torques and the *VMG* estimated torques are typically in the range of 0.90 to 0.96. Correlations as high as .98 have been achieved (Meek, 1982). Figure 22.10 shows the results of a typical 5 degrees of freedom *VMG* experiment performed on an amputee.

Notice that the humeral torques are fairly accurate. The humeral rotation estimate is showing the effects of the nonlinearity between the *EMG*'s and the position (what is actually recorded).

In general, the clavicular torques are not estimated as well as the humeral torques. This is due to several reasons: *1)* the muscles providing clavicular flexion are difficult to access with surface *EMG* electrodes; and *2)* it is difficult to apply torques to the clavicle. The techniques and results are, however, quite adequate for arm control in the laboratory setting.

22.5.4 Arm Control With *VMG* Estimates

The purpose of the *VMG* is to estimate the natural joint torques, which become the inputs to the Postulate-based multiple degree-of-freedom artificial arm controller. The ultimate criterion of

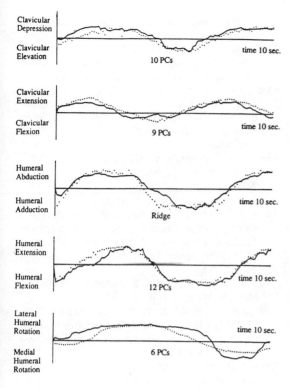

Clavicular Depression
Clavicular Elevation
10 PCs
time 10 sec.

Clavicular Extension
Clavicular Flexion
9 PCs
time 10 sec.

Humeral Abduction
Humeral Adduction
Ridge
time 10 sec.

Humeral Extension
Humeral Flexion
12 PCs
time 10 sec.

Lateral Humeral Rotation
Medial Humeral Rotation
6 PCs
time 10 sec.

Figure 22.10: Comparison of the measured and *VMG*-estimated joint torques about the degrees of freedom of the shoulder.

success of our ability to find the "empirical" *VMG* is the amputee's ability to control the arm. These techniques have been used successfully for real-time, multiple degree-of-freedom, Postulate-based arm control with above-elbow amputees. The subjects have been able to simultaneously and proportionally control elbow flexion, humeral rotation, and wrist rotation. Evaluating the controllability is difficult. Both subjective (amputee opinion) and objective evaluation methods have been used. Assessing the amputee's opinion is a formidable task (Jerard, 1976; Sears, 1983) and is not discussed here. Objective methods of evaluating control have been used such as trajectory tracking tests and object manipulation tests. The results of objective tests rarely have a high correlation with the amputees' opinions (Jerard, 1976).

22.6 Conclusions

We have found from our experiences with multi-degree-of-freedom arm control experiments, and from our fitting of over 500 Utah Arms [from Iomed, Inc., UT] outside the laboratory environment, that the major impediments to improvements in prosthetic arms are better command signals (*EMG* signal processing) and better electro-mechanical hardware (lighter, stronger, more reliable, more efficient). In this sense, biomechanical modelling does not help us design better, cheaper, stronger, or more rugged artificial arms. However, such modelling does help us in the basic understanding of human limb motion. Specifically, better kinematic and anatomic models could help in the design of better harnessing systems for artificial arms, while generally, anatomical and muscle modelling would be helpful in the area of ergonomics and safety, sports and exercise biomechanics, and surgical orthopedics.

Acknowledgments
Primary support for this work came from the National Institute of General Medical Sciences, Grant 5-R0I-GM23499-05 and in part by Air Force contract, F33615-83-D-0603.

References
Basmajian, J.V. and Latif, A. (1957) Integrated Actions and Functions of the Chief Flexors of the Elbow. *J. Bone and Joint Surgery*, **39-A:** 1106-1118.

Cohen, E. and Riesenfield, R. F. (1982) General Matrix Representations for Bezier and B-Spline Curves. *Computers in Industry*, **3:** 9-15.

Coons, S.A. (1967) Surfaces for Computer-Aided Design of Space Forms. *MAC-TR-41*, M.I.T.

Crago, P.E., Peckham, P.H. and Thrope, G.B. (1970) Modulation of Muscle Force by Recruitment During Intramuscular Stimulation. *IEEE Trans. on Biomed. Eng.*, **BME-27:** 679-684.

Crowninshield, R.D. and Brand, R.A. (1981) A Physiologically Based Criterion of Muscle Force Prediction in Locomotion. *J. Biomech.* **14:** 793-801.

DeLuca, C.J. (1979) Physiology and Mathematics of Myoelectric Signals. *IEEE Trans. on Biomed. Engng.* **BME-26:**

DeLuca, C.J. (1984) Myoelectric Manifestations of Localized Muscular Fatigue in Humans. CRC *Crit. Rev. in Biomed. Engng.*, Vol. 11, Issue 4.

Dul, J., Townsend, M.A., Shiavi, R. and Johnson, G.E. (1984a) Muscular Synergism — I. On Criteria for Load Sharing Between Synergistic Muscles. *J. Biomech.* **17:** 663-673.

Dul, J., Johnson, G.E., Shiavi, R. and Townsend, M.A. (1984b) Muscular Synergism — II. A Minimum-fatigue Criterion for Load Sharing Between Synergistic Muscles. *J. Biomech.* **17**: 675-684.

Fu, K.S., Gonzalez, R.C. and Lee, C.S.G. (1987) *Robotics Control, Sensing, Vision, and Intelligence.* McGraw-Hill.

Fullmer, R.R. (1983) Generation of Postulate-based Controller Equations for a Seven Degree of Freedom Prosthetic Arm. M.S. Thesis, Dept. of M.E., University of Utah, August.

Fullmer, R.R., Meek, S.G. and Jacobsen, S.C. (1984) Optimization of an Adaptive Myoelectric Filter. *6th Conference IEEE/Eng. in Med. and Biol. Soc.*

Fullmer, R.R., Meek, S.G., Jacobsen, S.C. (1985) Generation of the 7 Degree of Freedom Controller Equations for a Prosthetic Arm. *Conf. on Applied Motion Control*, Univ. of Minnesota, Minneapolis.

Heckathorne, M.S. and Childress, D.S. (1981) Relationhips of the Surface Electromyogram to the Force, Length, Velocity, and Contraction rate of the Cineplastic Human Biceps. *Amer. Jour. of Phys. Med.*, **60**: 1-19.

Hof, A.L. and Van Den Berg, J.W. (1977) Linearity Between the Weighted Sum of the *EMG*s of the Human Tricepes Surae and the Total Torque. *J. Biomech.*, **10**: 529-539.

Hollerbach, J.M. (1980) A Recursive Lagrangian Formulation of Manipulator Dynamics and a Comparative Study of Dynamics Formulation Complexity. *IEEE Trans. on Sys. Man. and Cybern.*, **SMC-10**: 730-736.

Jacobsen, S.C. (1973) Control Systems for Artificial Arms. Doctoral Dissertation, M.I.T., January.

Jacobsen, S.C. and Mann, R.W. (1974) Graphical Representation of the Functional Musculoskeletal Anatomy of the Shoulder and Arm. *27th ACEMB*, Philadelphia, PA, 6-10 October.

Jacobsen, S.C., Meek, S.G. and Fullmer, R.R. (1984) An Adaptive Myoelectric Filter. *6th Conf. IEEE/Eng. in Med. and Biol. Soc.*

Jerard, R.B. (1976) Application of a Unified Theory for Simultaneous Multiple Axis Artificial Arm Control, Ph.D. Dissertation, University of Utah.

Jerard, R.B. and Jacobsen, S.C. (1980) Laboratory Evaluation of a Unified Theory for multaneous Multiple Axis Artificial Arm Control. *Trans. of the ASME, J. Biomech. Engrg.* **102**(3):199-207.

Kramer, C.G.S., Hagg, T. and Kemp, B (1987) Real-time Measurement of Muscle Fatigue Related Changes in Surface *EMG*. *Medical and Biological Engineering and Computing* **25**:627-630.

Marquart, D.W. and Snee, R.D. (1975) Ridge Regression in Practice. *American Statistician* **12**(3):591-612.

Massy, W.F. (1965) Principal Components Regression in Exploratory Statistical Research. *American Statistical Association Journal*, pp. 234-256, March.

Meek, S.G. (1982) Command Inputs for a Multiple Degree of Freedom Artificial Arm Controller. Ph.D. Dissertation, University of Utah, December.

Meek, S.G., Fetherston, S., Schoenberg, A. and Milne, K. (1987) Multi-Channel, Adaptive Myoelectric Filtering. *24th Annual Meeting of the Society of Engineering Science*, Salt Lake City, Utah, September.

Meek, S.G., Fullmer, R.R., and Jacobsen, S.C. (1984) Control Inputs to a Multiple Degree of Freedom Artificial Arm. *IEEE Conference on Engineering in Medicine and Biology*, Los Angeles.

Meek, S.G., Jacobsen, S.C. and Goulding, P.P. (1989) Extended Physiologic Taction _ A Proportional Terminal Device Force Feedback System. *Journal of Rehabilitation Research and Development* **26**(3):53-62.

Morrison, D.F. (1976) Multivariate Statistical Methods, McGraw-Hill.

Patla, A.E., Hudgins, B.S., Parker, P.A. and Scott, R.N. (1982) Myoelectric Signal as a Quantative Measure of Mechanical Output. *Medical Biological Engineer Computers* **20**:319-328.

Sears, H.H. (1983) Evaluation and Development of a New Hook-Type Terminal Device. Ph.D. Dissertation, University of Utah, June.

Seireg, A. and Arvikar, R.J. (1973) A Mathematical Model for Evaluation of Forces in Lower Extremities of the Musculo-Skeletal System. *J. Biomech.* **6**:313-326.

Seireg, A. and Arvikar, R.J. (1975) The Prediction of Muscular Load Sharing and Joint Forces in the Lower Extremities During Walking. *J. Biomech.* **8**:89-102.

Wood, J.E. and Mann, R.W (1981) A Sliding-Filament Cross-Bridge Ensemble Model of Muscle Contract for Mechanical Transients. *Math. Biosci.* **57**:211-263.

Wood, J.E., Meek, S.G. and Jacobsen, S.C. (1989) Quantitation of Human Shoulder Anatomy for Prosthetic Control, Part I. *J. of Biomechanics* **22**(3):273-292.

Wood, J.E., Meek, S.G. and Jacobsen, S.C. (1989) Quantitation of Human Shoulder Anatomy for Prosthetic Control, Part II. *J. of Biomechanics* **22**(4):309-326.

CHAPTER 23

Role of Muscle in Postural Tasks: Spinal Loading and Postural Stability

Gunnar B. J. Andersson and Jack M. Winters

23.1 Introduction

The major issues of interest in Section IV can be illuminated by considering a simple, everyday task: picking up an object (say of medium-weight) off the floor and placing it in a new location (say at a specific spot on a tabletop). Of note is that robots have a surprisingly difficult time performing such "simple" tasks. Let's assume that our goal is quite practical: to successfully complete this task without undue effort. One fundamental issue is related to tissue loading: clearly high loads in various passive and muscular tissues are undesirable. Since tissue loading is likely to be a function of how the task is completed, this concern may influence movement strategy. Another fundamental consideration in completing this task is to not have any part of the body buckle or the whole body fall: local mechanical stability (e.g. of the spine) and overall dynamic stability (i.e. "good balance") must be satisfied. Furthermore, some margin of safety is desired, first because of inherent neural and mechanical "noise" and second in anticipation of the unforeseen (e.g., the object weight is different than expected). It turns out that minimizing tissue loading and maintaining stability are sometimes competing criteria. A third consideration involves visual tracking of: *i)* the object; *ii)* perhaps one's hands as they reach for the object; *iii)* the path the object must pass, to avoid any obstacles; and *iv)* the new location at which the object is to be placed. This involves eye, head and perhaps torso movements, and needs to be done with fairly minimal effort. A fourth issue, related to the others (especially stability and visual orientation), is to perform this voluntary task in a coordinated, smooth, "easy" fashion. This holds whether the task is performed slowly or quickly, and in our example involves the whole body. It also usually involves the vestibular system, which essentially measures head orientation and rotation.

This chapter starts with a brief summary of the classes of models that are used to study posture. The rest of the chapter is then organized along the lines of these four basic issues. We have two interrelated goals. First, we seek to provide a synopsis of our current understanding of the role of muscles in posture and balance within the context of a global view of tissue loading, balance, posture and orientation. Second, we attempt to synthesize how the chapters within Section IV address these basic practical and conceptual issues. Our emphasis will be on addressing postural adjustment, balance, and head/visual orientation during movement tasks of interest within daily living.

23.2 Mathematical Modeling Foundations

Because of the biomechanical complexity of the torso and of whole-body musculoskeletal systems in general, mathematical modeling becomes a integral part of the process of gaining improved understanding. Biomechanical models provide predictions which can be compared to the limited experimental variables that usually can be obtained, such as *EMG* measurement of some strategic subset of muscles, movement kinematics (typically as estimated from markers coupled to the skin at strategic landmarks), perhaps loading at locations of contact with the environment (e.g. force platforms), and/or (within certain laboratory

Multiple Muscle Systems: Biomechanics and Movement Organization
J.M. Winters and S.L-Y. Woo (eds.), © 1990 Springer-Verlag, New York

settings) disk pressure measurements. Mathematical modeling techniques for the study of human limb movements in general are reviewed extensively in Chapter 8 (Zajac and Winters). The classes of models employed for torso and whole-body posture, however, span an even greater range. No single model provides all answers. It turns out that each approach has its place; hence the reason for this brief review, designed to place the various modeling approaches in perspective.

Models can be classified as follows:

1. "Rigid body" quasi-static models. Here a free body diagram cut is made, typically through the lumbar region, with mechanically relevant muscle and passive tissue forces represented by vectors. Typically some type of static optimization algorithm is used to solve the tissue redundancy problem [e.g. Andersson et al., 1980; Schultz et al., 1983; Chapter 24 (Ladin); Chapter 25 (Gracovetsky)]. This class of model has been utilized to estimate lumbar spinal loading during a wide variety of tasks and condition, with emphasis on ergonomic considerations (e.g. Chaffin and Andersson (1984)) or rehabilitation (e.g. Chapters 24–25).

2. Multi-link lumped-parameter models. Here bones are considered to be rigid bodies, ligaments and cartilage are represented by assorted spring and (sometimes) dashpot elements, and muscle is represented by either a passive spring, an idealized force generator, or a dynamic muscle model. Lumped-parameter models can be further divided into three classes: *i)* those emphasizing rigid body dynamics [e.g., impact studies (reviewed in Huston and Perrone, 1978; Winters, 1988)]; *ii)* quasi-static spring-mass models emphasizing tissue loading in musculo-skeletal disease [e.g. Closkey and Schultz, 1988; Ghista et al., 1988; Chapter 27 (Deitrich et al.) or principles related to posture and stability (Bergmark, 1987; Chapters 26–27); and *iii)* lumped parameter models complete with muscle dynamic properties [e.g. Chapter 30 (Ramos and Stark)].

3. Analytical continuum mechanics models. In a number of studies the spine has been represented as a beam (reviewed in Yoganandan et al., 1987). These types of models, often used to study responses to impact, are less popular today because of the wide availability of finite element models.

4. Finite element models. Here tissues are structurally broken down into (usually deformable) elements in mechanical contact with each other (e.g., revew by Yoganandan et al., 1987; Chapter 27 (Dietrick et al.). Although used in Chapter 27 for a complete torso system, the primary uses for this class of model have been to estimate tissue load distribution [e.g., within intervertebral disk (Yoganandan et al., 1987)]. Typically finite element models are only used to investigate quasi-static load distribution. Notice that lumped-parameter models are a (more computationally efficient) subset of finite element models which are typically more appropriate for dynamic simulations.

Of note is that all of these classes of models are useful and will likely continue to be useful. In contrast to models of the upper and lower extremities, as presented elsewhere in this book, it is interesting to note that quasi-static models are most common for the torso; this is a reflection of the complexity of the system.

23.3 Trunk Muscle Activity During Various Tasks

The activities of the trunk muscles cannot be measured directly in mechanical terms but have been estimated indirectly, using electromyography. Although the relationship between electric output of muscles and force is often nonlinear, for quasi-static tasks it is monotonic. An appropriate calibration procedure, therefore, will result in reasonably accurate estimates of force from *EMG* data.

The relationship between torque and myoelectric signal amplitude for the trunk muscles in flexion-extension was studied by Stokes et al. (1987). They found a linear relationship for the *ESM* and extension moment, while a quadratic regression better described the relationship of abdominal muscle activity and flexion torque. Seroussi and Pope (1987) found linear relationships for isometric extension efforts and for the difference between the left and right erector spinae myoelectric activity and the lateral bending moment. Vink et al. (1988) placed twelve surface electrodes over the lumbar spine at *L1*, *L3* and *L5* levels to record myoelectric activity over the multifidus, longissimus and iliocostalis muscles. When performing isometric extension efforts while standing, a different relationship of force to myoelectric activity was found for the three muscle groups. The multifidi showed a linear relationship, the longissimus and iliocostalis a curvilinear.

McGill (1990) recorded muscle activities during static and dynamic twisting efforts. Peak muscle activity during maximal twisting efforts was low, compared to the maximum activity of all trunk muscles. Although generally a monotonic relationship occurred between torque and activity, the relationship was not consistently linear or nonlinear. This supports previous studies by Pope et al. (1986).

To summarize, the relationship of torque and trunk muscle myoelectric activity during quasi-static tasks, while monotonic, appears to differ between groups of muscles, and to be influenced by the specific loading condition. Of course, as shown in many locations throughout this book [e.g., Chapter 8 (Zajac and Winters)], this relationship becomes more complicated which motion occurs. As an example of relevance here, Marras and Mirka (1990) performed an experiment to determine the myoelectric response of the trunk muscles to trunk velocity, trunk position (both forward bending and asymmetric angle) and trunk force exertion level. Subjects produced constant torque about the lumbro-sacral junction while moving the trunk under constant velocity conditions. Significant reactions to velocity, force level and unique combinations of trunk angle and velocity were seen in all muscles of the trunk.

23.3.1 Myoelectric Activity and Posture: Standing and Sitting

The myoelectric activity of the trunk muscles in erect standing and in sitting postures has been studied extensively. Much of this work has been reviewed previously (Andersson, 1974, 1982); Ortengren and Andersson, 1977; Hosea et al., 1986; Andersson et al., 1989).

Standing

In studies of standing, a slight myoelectric activity is typically observed in the paraspinal muscles; with higher levels in the thoracic than in the lumbar and cervical regions (Ortengren and Andersson, 1989). Asmussen and Klaussen (1962) found slight activity in erect standing posture in either the posterior back muscles or the abdominal muscles, but not in both. This may result form the normal presence of postural sway. Studies of the rectus abdominis and the external

and internal oblique muscles reveal slight activity during relaxed standing, particularly of the internal oblique muscles (Floyd and Silver, 1950). There is also slight activity in the vertebral portion of the psoas major muscle (Andersson et al., 1974).

Sitting

The activity of the lumbar paraspinal muscles is similar in standing and in unsupported sitting, while there is a somewhat higher level of activity in the thoracic region during sitting (Floyd and Silver, 1950; Andersson and Ortengren, 1974; Andersson et al., 1974a,b). Carlsoo (1963) recorded slight activity in the anterior oblique muscles, but he did not investigate the rectus abdominis and transverse muscles. His results agreed with those of Schultz et al. (1982) and Andersson et al. (1980), who studied several supported sitting postures. They found slight activity in the rectus and oblique abdominal muscles, even through the posture was sagitally symmetric. This suggests mild cocontraction. The iliopsoas muscle is also slightly active when a sitting posture is assumed (Andersson et al., 1974; Floyd and Silver, 1955).

Andersson (1986, 1987) summarized several studies of supported sitting and of work activities in sitting postures. These studies showed that the myoelectric activity of the trunk muscles was influenced by the posture of the seated subject, by supports incorporated into the chair, and by the specific work activities performed. The use of backrests was particularly important, with muscle activity strongly influenced by the angle between the seat and the backrest (Andersson and Ortengren, 1974; Andersson et al., 1974,a,b). Levels of activity were quite low in all trunk muscles when the trunk is adequately supported by a reclining back rest (Figure 23.1.)

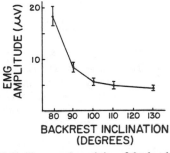

Figure 23.1: The muscle activity of the lumbar erector spinae muscles decreases as the backrest inclination increases (adapted from Andersson et al., 1974).

The importance of the slope of the seat pan has also been studied to define optimal seating in the office setting. Bendix et al. (1985) found no difference in activity. Soderberg et al. (1986), on the other hand, recorded lower levels of activity found with forward inclination in the lumbar as well as cervical regions of the back. Myoelectric activities directly measured during office work as well as model calculations of trunk forces suggest that, in general, the myoelectric signals are quite low and marginally influenced by table and chair adjustments (Andersson, 1986, 1987; Andersson et al., 1986).

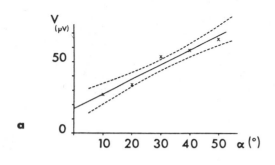

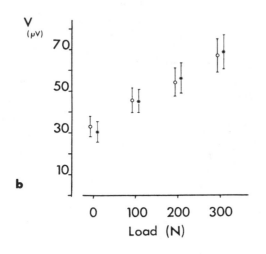

Figure 23.2: The myoelectric activity of the lumbar back muscles. *a)* Activity increases linearly with the angle of flexion of the trunk. *b)* Activity increases as the load held by the hands increases (measurements at 30° of trunk flexion). (Adapted from Andersson et al., 1977).

23.3.2 Myoelectric Activity in Flexed Postures and During Forward Flexion

Studies of flexed postures have revealed an increase in the myoelectric activity of the back muscles, both when the angle of flexion is increased and when external loading is increased at a fixed angle of flexion [Figure 23.2; Schultz et al., 1982; Andersson et al., 1977; Chapter 24 (Ladin)]. In attempted flexion resisted by external forces, on the other hand, the abdominal muscles are strongly active while there are only low levels of activity in the lumbar part of the erector spinae muscle (Jonsson, 1970; Schultz et al., 1983, 1987; Zetterberg et al., 1987).

Forward flexion of the trunk is a combined movement of the spine and pelvis. The muscles of the trunk, pelvis and thighs combine to control the pattern of motion. During the first portion of a flexion movement, strong activity is found in the gluteus maximus, the gluteus medius, and the hamstring muscles; this muscle action locks the pelvis and prevents motion at the hip joints. As flexion progresses, the increasing trunk moment is balanced by a corresponding increase in back muscle activity. In the fully flexed position, however, myoelectric activity decreases and often ceases almost completely. Floyd and Silver (1951; 1955) called this the flexion-relaxation phenomenon of the back muscles and hypothesized that it results from stretch reflex inhibition [see also discussion in Chapter 25 (Gracovetsky)]. Other possible explanations are that in the fully flexed posture the trunk moment is resisted by passive support structures such as the ligaments, thoraco-lumbar fascia and facet joints (see also Chapter 25). Certainly the stretched extensor muscles also contribute. Valencia and Monroe (1985), using wire electrodes, found a decrease in activity in full flexion only in subjects who reached their pre-experimental degree of trunk flexion.

The main reason for the interest in the flexion-relaxation phenomenon is the potentially harmful effects of performing lifting and other work activities in flexed postures. It has been postulated that the inactivity of muscles leave the spine unprotected, and increase the forces inside the spine because the moment arm acting on the ligaments is shorter. Kippers and Parker (1984) determined the relationship between degree of flexion and decrease in myoelectric activity, and also discussed a variety of possible explanations for the phenomenon. Schultz and Associates (1985) estimated the tissue tensions resulting from the flexion-relaxation phenomenon. They also determined that the back muscles immediately become electrically quite active when exertions are performed in the fully flexed posture.

23.3.3 Myoelectric Activity During Extension

When raising the trunk from the flexed to the upright posture, the sequence of muscular activity is the reverse of that when bending forward. The gluteus maximus, along with the hamstrings, is active early and initiates extension by a posterior rotation of the pelvis. The paraspinal muscles then become active and then increase their activity until the upright posture is reached. Extensor muscle activity is typically greater when the trunk is being raised than when it is being lowered, although in neither case is it close to its maximum. The direction of the movement in relation to the weight forces of body segments is obviously important. Concentric contractions are known to require more muscle force than eccentric.

When the trunk is extended from the upright position, myoelectric back muscle activity is strong early on during the initial phase, but again only after the gluteus maximum has become active. Both muscle groups are active in the position of full extension, while between these two extreme postures there is only slight activity. The abdominal muscles, particularly the rectus abdominis, show increasing activity throughout the extension movement. Extension of the trunk against resistance results in a marked increase in the activity of the muscles of the lumbar region of the back (Jonsson, 1970; Schultz et al., 1983, 1987; Zetterberg et al., 1987). In fact, this is the activity in which the back muscles show their maximum activity.

23.3.4 Myoelectric Activity During Lateral Flexion and Twisting

When the trunk is flexed laterally, the myoelectric activity increases in the posterior back muscles on both sides of the spine. The main increase in activity in the lumbar region is on the side contralateral to the direction of lateral bend. This is also the case when the trunk is loaded in lateral flexion, where comparatively higher levels of activity are found on the contralateral side of the lumbar region [see also Chapter 24 (Ladin)]. Muscle activity in the lumbar region is typically higher in the sacrospinal than in the transversospinal muscles (Andersson et al., 1977). Jonsson (1970) found the sacrospinal muscles to be active in lateral flexion, whereas the multifidi muscles, which are closer to the spine, were usually inactive. Intuitively, this observation makes sense. The abdominal muscles are active in lateral flexion both ipsilaterally and contralaterally, with higher activity on the contralateral side. Carlsoo (1961) recorded strong activity in the gluteus medius and the tensor fasciae latae muscles on the ipsilateral side, which reflects the force necessary to rotate the pelvis.

Raftopoulos et al. (1989) studied whether the flexion-silence phenomenon of back muscles this is observed in full flexion also exists in lateral flexion. A relaxation phenomenon does seem to occur in the fully laterally bent trunk posture, but only in the trunk extensor muscles. The oblique abdominal muscles remain active.

Pope and associates (1986, 1987) studied the myoelectric activity of trunk muscles when twisting was attempted, both with and without prerotation of the trunk. In general, a linear relationship was established between force output and myoelectric activity (Figure 23.3). However, high levels of antagonistic activity were found in both abdominal and posterior back muscles. In some muscles, prerotation increased the antagonistic activity. In both these experiments, the highest activity levels were found in the erector spinae and external oblique abdominal muscles. Ladin et al. (1989; see also Chapter 24) modeled a lumbar cross-section as a loading plane to which external moments were applied in different combinations. This loading plane was assumed to have a horizontal axis to which flexion-extension moments were applied, and a vertical axis for lateral bending moments. Any point in the plane was viewed as a loading point describing any combination of bending moments. Using this scheme, the myoelectric activities of lumbar trunk muscles were well predicted for various moment applications.

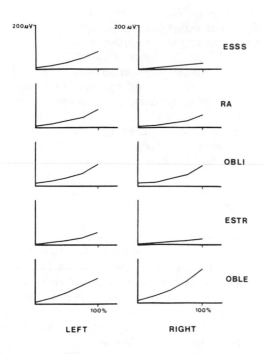

Figure 23.3: Average myoelectric activity versus contraction level for the five muscle pairs. *ESSS*, sacrospinal; *RA*, rectus abdominus; *OBLI*, oblique internus; *ESTR*, transversospinal; *OBLE*, oblique externus. (Adapted from Pope et al., 1986.)

23.3.5 Myoelectric Activity During Lifting

Because of the common occurrence of low back strain with lifting, this topic has been of particular interest. The back muscles, the muscles of the buttocks, and the hamstring muscles are all myoelectrically active during a lift. The abdominal muscles are also active, but to a lesser degree. The levels of activity in these various muscles are directly related to the external moment and are, therefore, influenced by the weight lifted, the body posture, the location of the mass center of the weight and the speed of the lift (Andersson et al., 1976, 1980; Schultz et al., 1982; McGill and Norman, 1986; Gracovetsky, 1988). At a practical level the question of whether the spine should be flexed or straight during the lift has been addressed using *EMG* measurements. Generally, in these studies, activity of the back muscles is similar in a leg lift (spine held straight) and a back life (spine flexed) or, sometimes, is greater in the back lift.

Bobet and Norman (1984) measured *EMG* activity from the lumbar *ESM* when subjects carried a backpack placed with just below the midback or just above the shoulders. The high load placement resulted in significantly higher levels of muscle activity. Cook and Neumann (1987) studied the effect of different load placement (anterior, posterior, and lateral) when subjects were walking. The lowest levels of activity occurred when carrying the load in a backpack. Carrying the load anteriorly resulted in significant increased activity levels. Further, in this carrying mode, significantly higher levels of activity were found for women than for men. Side-carrying resulted in high contralateral muscle activities. The conclusion derived from these studies is that the lifting mode is less important than load placement and lifting speed.

3.9 Myoelectric Activity During Walking

A few studies report on the myoelectric activity of the trunk muscles during walking [Thorstenson et al., 1982; see also Chapter 33 (Winter et al.)]. Battey and Joseph (1966) recorded short periods of activity over the lateral part of the *ESM*; one at the start of stancephase, a second at the end. Using indwelling electrodes Waters and Morris (1972), on the other hand, report multifidus activity of heelstrike and *ESM* activity at the time of contralateral heelstrike. More recently, Dotterhof and Vink (1985) confirmed those findings, recording short bursts of activity just before left and right heelstrike over both the multifidus and iliocostalis lumborum muscles. They also found that load carrying altered the pattern of activity corresponding to the external moment resulting from load and its location.

There are a number of possible explanations for the observed lumbar muscle activity. Perhaps the posterior back muscles are responding to a flexion moment occurring when the lower part of the body at heelstrike is decelerated. The theory of the "spinal engine" (Gracovetsky, 1988), presented within Chapter 25 (Gracovetsky), provides another possible explanation. Here it is suggested that the spinal musculature, along with the unique spinal linkage, plays an integral role in the production of gait. In Chapter 33 (Winter et al.) posterior back muscle activity is implicated as being important for maintenance of whole body stability and in minimizing horizontal and vertical fluctuations of

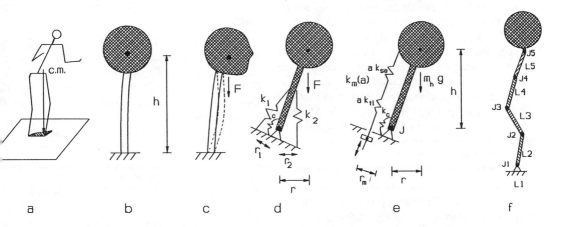

a   b   c   d   e   f

Figure 23.4: Inverted pendulum models of stability. *a)* classical whole-body view. The center of mass must reside within the base of support. If not, the body will have a tendency to fall. Dynamic stability can be maintained if the base of support can change before it's too late. *b)* Stability of an elastic column under load (see text). *c)* Elastic beam with an eccentric load. *d)* Inverted pendulum controlled by muscle-like springs. Muscle activation (a; $0 \leq a \leq 1$) is assumed to scale the stiffness of both the *CE* tension-length (= $a\, k_{tl}$) and series elastic (= $a\, k_{se}$) springs, plus shift the offset ("slack" length) of the muscle (shown conceptually by shifting "lock" at base). *e)* Multi-link inverted pendulum, which requires spring-like muscle-reflex actuators (not shown) to maintain stability.

the head (and thus of the vestibular and visual systems). We now consider issues related to postural stability.

22.4 The Problem of Postural Stability

It is interesting that the basic approach of the previous section, which emphasizes the concept of muscles as force generators, does not explicitly address the issue of stability. Rather, a kinematic configuration of interest is assumed to already be stable. Since analyses proceed from experimentally documented initial kinematics, this represents a logical approach. Stability, however, is a real issue, and furthermore an important one.

Whole Body Stability, Given Base of Support. The standing person can be thought of as an inverted pendulum structure (Gurfinkel and Osovets, 1972). Statically, for a given orientation of the whole body to be stable, the center of mass must be located within the base of support at the feet (see Figure 23.4a). With regards to stability, typically a margin of safety is desired. Hayes (1982), summarizing conventional thought regarding whole body stability, states that the degree of

stability is: *i)* proportional to the base of support; *ii)* related to the horizontal distance between the center of gravity and the edge of support; *iii)* inversely related to the height of the center of gravity relative to the base of support; and *iv)* proportional to the weight of the body (true for rigid body systems that "tip" but, as shown below, not true in general). Implicit in the above statements is the assumption of no moment between the foot and the ground. Stabilometry, the measurement and recording via force platforms of the continuous oscillation of the body center of mass, helps document this level of relative stability (Terekhov, 1976).

For human movements what really matters is dynamic stability. During tasks such as gait, certain inherently "unstable" periods of time are not only tolerated but also desired since they are part of the process of propulsion [Chapter 33 (Winter et al.)]. However, to maintain dynamic stability either the base of support must change (i.e. step to new position) or there must be appropriate dynamic modulation between body parts (e.g., the child with arms out while walking on balance beam).

Stability Within the Body. For individual parts of the body, if the superior aspect is not within a base of support (e.g. upper body center of mass anterior to the spinal column or the head anterior to the neck), appropriate moments necessary to balance this eccentric loading must exist. However, such conditions are *necessary but not sufficient* [Bergmark, 1987; Chapter 9 (Hogan); Chapter 26 (Crisco and Pajabi)]. In addition, a *stable* static equilibrium requires that, in the event of a perturbation (e.g. "noise"), the body will *return back* to equilibrium. Further complicating the situation is the fact that for musculoskeletal systems, which are inherently nonlinear, stability may depend on the size of the perturbation.

Such issues are difficult to grasp when employing inverse (position to force) mechanical analyses [(although see Chapter 33 (Winter et al.)], and require either a different analytical model or a forward (direct) dynamic simulation [see Chapter 8 (Zajac and Winters) for discussion]. Bergmark (1987) in particular has recently shown that stability is a major issue that can influence predictions of mechanical loads, etc. In Chapter 9 (Hogan) it is shown that an inititally stable system may become unstable when in contact with the environment (e.g. picking on an object of significant mass). This is referred to as *contact instability* and remains one of the biggest challenges within robotics and prosthetics [see also Chapter 11 (Hogan and Winters)]. It also turns out that for stable equilibrium the neuromusculoskeletal system must possess the attributes of springs, and that for stability the springs need to exceed a certain critical stiffness. The passive, relaxed person is inherently instable at many levels — when someone faints, the head falls if the shoulders are supported, the torso falls if the body is supported at the pelvis, and the whole body falls if the person is freely standing. Certainly the central nervous system is partly responsible for this behavior, but the primary mode of its influence can be considered to be adjusting muscle "spring-like" behavior [Chapter 9 (Hogan), Chapter 26 (Crisco and Panjabi)].

The functioning neuromusculoskeletal system handles stability so remarkably well that it is hard sometimes to even identify it as a problem. Perhaps this helps explain why it has been addressed so rarely in movement and occupational biomechanics [Chapter 26 (Crisco and Panjabi)].

Interestly, Chapters 26–33 explicity address the issue of stability.

Basic Mechanical Principles and the Spine. An appropriate starting point is to consider mechanical energy conservation [Chapter 9 (Hogan), Chapter 26 (Crisco and Panjabi); Chapter 32 (Ong et al.)]. Connective tissues such as ligaments store potential energy when stretched. The energy stored is the area under the load-extension curve. A system of connective tissue, such as the spinal column, may be though of as a rotational (torsional) spring which stores potential energy. If a loaded spring is released, the stored energy may be transferred to, say, a mass. Masses can store potential energy by virtue of their height and kinetic energy by virtue of their speed of movement.

To help establish fundamental principles as related to system stability, let's consider a few simple examples. As a first example, consider the simple elastic column of Figure 23.4b. For such elastic columns, a crucial concern is buckling. The force causing bucking failure in such an idealized linear elastic column is (e.g. Shigley, 1977):

$$F_{cr} = n\frac{\pi^2 EI}{L^2} = n\frac{\pi^2\ EA}{(L/c)^2} = n\ \pi^2 c(c/L)K_a \quad (23.1)$$

where F_{cr} is the critical (peak) load, n is a constant that depends on the boundary conditions [e.g., 1/4 if bottom is fixed and top is free (most useful case, shown in Figure 23.4b) and 2 if one end fixed, the other pin-jointed], E is Young's modulus, I is the area moment of inertia, A is the cross-sectional area, c is the radius of gyration, (L/c) is called the slenderness ratio, and K_a is the axial stiffness (EA/L). Although overly simplistic, it provides a conceptual foundation: elastic beams which are *long*, *thin*, *flexible* and *free at one end* are more likely to be inherently unstable. Other relations exist for the more common case of eccentric loading shown in Figure 23.4c, but these are more complex and add little to our conceptual foundation. Of note is that a nominally curved beam with a convavity as shown by the dashed line in Figure 23.4c is more likely to be stable along its length; interestingly, both the cervical and lumbar spines normally have lordotic curves which, relative to the typical anteriorly-placed eccentric loads, are in the direction as shown in the Figure.

Passive stiffness values have been fairly well

documented for various levels of the spine (reviewed in Ashton-Miller and Schultz, 1988). There are six components: compression, anterior-posterior shear, lateral shear, bending in flexion-extension, lateral bending, and torsion. Because of the structural arrangement of spinal units stacked in series, extensions add and the overall stiffness for a given mode is less than the lowest individual component:

$$K_{eq} = \frac{1}{\sum_i (1/K_i)} \qquad (23.2)$$

As can be seen from the review by Ashton-Miller and Schultz (1988), it turns out that stiffness is least (compliance is highest) in the cervical spine for all modes of bending, with the $C1$-$C2$ primarily responsible for torsional rotation, the range of flexion-extension relatively uniform along the cervical column, and lateral bending most significant in the mid- and lower-cervical column. Also, the thoracic spine is reasonably compliant in torsion while the lumbar spine capable of considerable flexion and to a lesser extent extension and lateral flexion.

Let us now consider a second conceptual view for assessing system stability which emphasizes the effects of tissues such as muscle (see Figure 23.4d). The potential energy of a single-joint system with 2 linear springs (k_1 and k_2) is (Bergmark, 1987):

$$\Delta V = \{ \delta_1 F_1 + \delta_2 F_2 - \delta m g \} +$$
$$\{ \frac{1}{2}k_1 \delta_1{}^2 + \frac{1}{2}k_2 \delta_2{}^2 + \frac{1}{2}k_\phi \delta\phi^2 \} \quad (23.3)$$
$$= \Delta V_1 + \Delta V_2$$

where we have assumed that for small angular perturbations from equilibrium $\delta\phi \approx \sin(\phi)$ and $\cos(\phi) = 1.0$. For equilibrium, $\partial\Delta V_1/\partial\Delta\phi = 0$ for $\Delta\phi = 0$, and assuming $\delta_i = \phi\, r_i$, we have:

$$F_1 r_1 + F_2 r_2 + M_\phi = m g r_m \qquad (23.4)$$

where the sign of the moment arms r_i would differ if the springs were antagonistic. For stable equilibrium the potential energy must grow as we are displaced from equilibrium [see also Chapter 26 (Crisco and Panjabi], and thus:

$$k_1 r_1{}^2 + k_2 r_2{}^2 + K_\phi > m g h \qquad (23.5)$$

or, noting that torsional stiffness is related to translational stiffness, to a first approximation, by $K_i = r_i^2 k_i$, for stable equilibrium we have:

$$K_1 + K_2 + K_\phi = K_{eq} > m g h \cong K_{crit} \quad (23.6)$$

Thus, the passive system is stable if the torsional stiffness is higher than some **critical value** that depends on the mass and the height of the column. For more detailed analyses see Bergmark (1987, 1989). This simplistic analysis shows also that stability is more of a challenge for inverted pendulum systems. Notice that for static equilibrium any moment due to the antagonist would **subtract** from the agonist moment, while for stability, the antagonist and agonist stiffnesses **add**. Since the stiffness increases with activation, stiffness and thus system stability can be modulated via co-contraction.

Considering posture, three possible systems are immediately evident: head-neck; torso-back, and body-ankles. Taking crude estimates of masses and heights, we estimate that critical bending stiffness values are on the order of:

$$K_{cr\text{-}neck} \sim 10 \text{ N-m/rad}$$
$$K_{cr\text{-}low\ back} \sim 100 \text{ N-m/rad}; \qquad (23.7)$$
$$K_{cr\text{-}ankle} \sim 250 \text{ N-m/rad (per ankle)}$$

where here we have assumed an idealized joint at the base of the segment of interest and the height to be that from the base to the center of mass. The ankle value comes from Bergmark (1987). Of interest is that passive stiffnesses are much lower than these values. Now let's consider the order of magnitude of the extensor moments, assumed due to muscle activity, that are necessary to maintain static equilibrium ($M = m g r$):

$$M_{neck} \sim 4 \text{ N-m}; \quad M_{low\ back} \sim 40 \text{ M-m};$$
$$M_{ankle} \sim 17 \text{ N-m (per ankle)} \qquad (23.8)$$

These values are approximately 15%, 25% and 10% of the maximum possible moments via isometric muscle contraction. To maintain a stable equilibrium, not only must the muscle tissue provide the moment necessary to balance the system but at the same time must provide appropriate stiffness. As described in Chapter 5 (Winters) and shown in Figure 23.4e, it turns out that muscle displays a few different types of mechanical stiffness, one related to the series elastic (*SE*) component (a "transient, or "dynamic" stiffness) and one related to static stiffness [contractile element (*CE*) tension–length curve]. Furthermore, because of "tonic" reflex activity the neuromuscular unit can be thought of a possessing a "reflex" stiffness [Chapter 12 (Feldman et al.); Chapter 13 (Wu et al.); Chapter 35 (McMahon)]. It also turns out that the muscle series element (transient) stiffness and *CE* tension–length are, to first approximations, directly proportional to force:

$$k_i = z_i(F/F_{max}) \text{ or } K_i = z_i r^2 (M/M_{max}) \qquad (23.9)$$

where z is a constant, F is muscle force, M is the corresponding moment, and the subscript i is an index. Based on the dimensionless constitutive equations presented in Chapter 5 (Winters), and assuming a maximal muscle stress 0.4 MPa, a peak series element extension of 4%, and a *CE* tension–length slope of 20% of σ_{max} for a 10% change in muscle fiber length over the nominal operating range (perhaps high), we have:

$$z_{se} \sim 10 \text{ MPa} \frac{A_m}{L_{mt}} \text{ and } z_{tl} \sim 2.0 \text{ MPa} \frac{A_m}{L_{mt}} \qquad (23.10)$$

where A_m is the physiological cross-sectional area, L_{mt} is the length of the musculotendinous unit and r is the idealized moment arm.

Based on these crude assumptions, the estimated stiffnesses that would occur as a byproduct of obtaining the required moments are on the order of:

$$K_{se\text{-}neck} \sim 50 \text{ N-m/rad}; \quad K_{tl\text{-}neck} \sim 10 \text{ N-m/rad}$$

$$K_{se\text{-}low\,back} \sim 300 \text{ N-m/rad}; \quad K_{tl\text{-}low\,back} \sim 60 \text{ N-m/rad}$$
$$\qquad\qquad (23.11)$$
$$K_{se\text{-}ankle} \sim 400 \text{ N-m/rad}; \quad K_{tl\text{-}ankle} \sim 80 \text{ N-m/rad}$$

Interestingly, the series stiffnesses are higher than the estimated "critical" values for stable equilibrium, yet the *CE* stiffnesses may be lower. Due to the series arrangement, as in *Eq. 23.2* the "equivalent" stiffness is lower that the lower value; thus, some cocontraction and/or reflex enhancement of stiffness may be necessary to maintain stability. Assisting stability is the fact that the slope of the *CE* force–velocity relation, which serves as a "soft ground" for the series element, is high (i.e. quite viscous) near zero velocity.

The above analysis, although illuminating with regards to capturing the essense of stability, is clearly an oversimplification. The simplest realistic model for addressing biomechanical stability is that shown in Figure 22.4f. This addition of a second joint has a number of important implications. First, in allows one to address relative stiffness variation between different links. Second, it allows one to consider dynamic interaction between links, which as many chapters within this book point out, is of great importance. Third, allows to system to become unstable by either of the classic two possibilities: "collapsing" as outlined initially by the Euler beam analogy and "tipping". Finally, it allows issues such as separate roles for uni- and multiarticular muscles to be explicitly addressed [Chapter 26 (Crisco and Panjabi)]. Chapters 27–34 all address stability concerns from within the context of either conceptual (Chapters 29, 31, 33, 34) or mathematical (Chapters 27, 28, 30, 32, 34) with two or more joints.

Posture and Stability: a Synthesis of the Contributions Within Part IV.

In Chapter 26 Crisco and Panjabi open by developing the concepts of clinical and mechanical stability, especially as it applies to the spine. They then address fundamental issues related to uniarticular versus multiarticular skeletal muscles. One observation was that any vertebrae devoid of musculature are inherently unstable, irrespective of the muscular architecture. Of special importance was the finding that longer, multiarticular muscles are advantageous in terms of system stability, with the efficiency increasing with the number of vertebrae spanned. Such insights have far-reaching consequences, in particular as related to traditional thought which often regards the deep internal muscles as "stabilizers" and the larger superficial muscles as "prime movers". Their observations can be made even more conclusive by realizing that, given a certain moment arm r, the moment scales linearly with force ($M = r F$) while the torsional stiffness scales with the square of r ($K = r^2 k$). Thus, since larger superficial muscles have larger moment arms, such muscles are especially appropriate for efficient stiffness modulation. Crisco and Panjabi also point out that while antagonistic muscle coactivation is, at least for some tasks, a normal physiological phenomenon, most past modeling efforts do not predict such coactivation and in fact stability considerations have traditionally not entered into the method of solution.

Finally, Figure 26.5 brings out an important consequence of nonlinear systems such as the spinal column — the concept of multiple equilibrium positions. In particular, the nonlinear, concave upward force–extension curve of soft connective tissues allow the passive stiffness to increase with extension. With disruption of one tissue, relative vertebra rotation or translation will occur until resisted some other tissue(s). Because of the lack of normal load-sharing, this secondary tissue is likely to be extended more than normal, and thus in its high-stiffness operating. Consequently, in the absence of severe injury, the injured spinal column will tend to find a secondary (yet nonideal) equilibrium rather than completely collapsing. Thus, from purely a stability perspective, nonlinear, concave-upward force–extension behavior is highly desirable. This has implications toward conditions such as scoliosis, which is addressed from a biomechanical perspective in Chapter 27 (Dietrich et al.).

In contrast to (yet complimentary with) the approach of Chapter 26, Chapter 27 utilizes a considerably more complex model and explores relations between stability, mild muscle imbalances, and idopathic scoliosis. Using a static optimization criterion related to the minimum of elastic energy to distribute forces, they show that with even small muscle asymmetry, the spine can become inherently unstable, especially in lateral bending. Both small local instabilities and gross curve changes can be predicted, with the form of the curvature a function of the external loading.

In Chapter 28 (Winters and Peles) it is seen that models which include only the 10 major superficial muscles of the neck region tend to, ironically, be inadequate at maintaining upright postures of certain orientation. Small suboccipital muscles, at minimum, appear essential for maintenance stable head orientation. Additionally, because of the natural lordotic curve, it was found that contraction of the large posterior muscles crossing the cervical spine creates a tendency toward buckling of the spine; perhaps this is one reason why muscles such as the longus colli have been found to be active during a variety of movements (Vitti

et al., 1974). Another interesting finding was the documentation of "release and hold" mechanisms for certain ranges of head movement: head extensor activity can control head flexion movements and, for upward head extension positions, the sternoceidomastoids can play a major role in controlling head extension. Such findings show that for multi-articular inverse pendulum systems, the distinction between "agonist" and "antagonist" depends on the operating range.

Starting with Chapter 29, the emphasis shifts to postural stability as related to the whole body. Chapter 29 (Keshner and Allum) considers whole-body stability from the perspective of documenting neuromuscular responses to rotational or translational *perturbations* at the foot. Their work represents an extension of the classical studies by Nashner and colleagues (e.g., Nashner, 1976) which had emphasized the role of ankle muscles in system stability. Of special interest are relative timing between muscular response at various levels such as the ankle, hips and neck, which is shown to be a function of the direction and type of perturbation. Such data helps document and catalog the growing body of knowledge regarding neuromuscular responses, and in particular relations between simple reflex loops, triggered responses, and voluntary responses. Could these responses be predicted based purely on mechanical principles, as applied to a multi-link inverted pendulum? Simulation studies appear to be necessary to address such issues.

Simulations of experimental data, using two-joint inverted pendulum models of the body, are presented in Chapter 31 (Ramos and Stark). Here, however, what is of interest are "anticipatory" postural adjustments (*APA*) made before or during rapid *voluntary* movements [Chapter 30 (Bouisset and Zatarra)]. Such activity, first documented by Belenkii et al (1967), have turned out to be a general phenomena (Cordo and Nashner, 1982; Chapter 30). Are such adjustments solely related to inverted pendulum mechanics? Is selective cocontraction a major mode of modulation? Might the primary reason for such adjustments be to "set" the system for neurosensory collection? Such issues are addressed in these chapters. Insight is facilitated by measuring link accelerations as well as collecting the usual *EMG* and force platform data. It is shown that *APA* are preprogrammed movements that are part of the motor program and serve to counterbalance anticipated perturbation to balance. It is also emphasized that the *APA* are sensitive to the initial body configuration and to the dynamic asymmetric of the task. Simulation insights in Chapter 31 suggest that there are multiple reasons for anticipatory postural adjustments, with the relative importance of these possibilities being a function of the type of task. In particular, for tasks involving dynamic arm movements, neuromotor pulses of moderate magnitude appear sufficient; cocontraction also appears to be useful. For dynamic movements involving the torso as well, inertial dynamics are of greater relative importance, and stability requires well-timed control pulses to strategic muscles. Finally, it is shown that the appropriate postural responses may change as the task progresses.

Chapter 32 (Ong et al.) also considers voluntary tasks, and in particular the task of sitting on a chair. Their analysis of this task involves forward dynamics simulation, complete with analytical methods for guaranteeing stability and minimizing performance errors. Although correlation between experiment and simulation was only moderate, perhaps due to extensive linearization, this study provides an indication of the type of path that further numerical studies involving dynamic stability will likely take. Of special interest is their development of a Lyapunov function for stability analysis which is based on energy considerations. Also of interest was the prediction of cocontraction and the need to divide certain task into constrained and unconstrained regions — clearly the contact between the chair and the body affects stability. It is confirmed that there are certain critical feedback and stiffness values for certain tasks that are necessary to maintain stability throughout the task.

A fascinating aspect of human movement is the of maintenance of dynamic stability during tasks such as walking. In fact, the primary control-related problems with *FNS* walking systems is not producing the basic sagittal-plane gait-producing muscle patterns but rather maintaining stability! As pointed out in Chapter 33 (Winter et al.), during the 80% of the gait cycle in which only one foot is on the ground, the system is dynamically unstable. This very instability helps facilitate forward propulsion. Chapter 33 provides insights into how stability is maintained. In particular, the

hip extensor muscles are primarily responsible for sagittal plane stability (see also Cappozzo, 1983), while the primary determinant of frontal plane stability is medio-lateral foot placement, which in turn is controlled primarily by hip abductor activity.

Chapters 29 and 33 emphasize the importance of the *vestibular system* in the maintenance of balance during gait and other tasks. From a neuromotor control perspective, two points seem especially appropriate: *i)* great effort is made to mimimize head movement oscillations, thus facilitating high quality signals for sensors measuring head and eye movement; and *ii)* the vestibular system is ideally situated for control of stability and balance, being situated within the top link of inverted pendulum structure and in tight proximity to the brain.

Another important point raised in Chapter 33 is the concept that *variability* in the gait cycle can be related to unique patterns designed to solve stability patterns, as opposed to "noise" around a single strategy. This concept is expanded in Chapter 34 (Pedotti and Crenna), where it is pointed out via numerous examples that for normal, self-paced whole-body movement tasks, multiple, distinct, strategies are the rule rather than the exception. Thus, one must be careful to distinguish between "noise" and variability in task strategy and execution; both *EMG* patterns and movement kinematics may be affected.

Tradeoffs Between Stability and Movement. Based on the above analyses, it might be assumed that, because of stability concerns, a "stiff" system is always better. This is not true. In particular, one of the primary functions of the spinal column, and especially the cervical and lumbar regions, is to rotate. To do so, a compliant spine is desirable for many tasks, including that described at the beginning of this chapter. A healthy spine is quite mobile, and in fact range of motion is one measure of health. Furthermore, muscle activity comes at metabolic cost. Even more importantly, muscle cocontraction causes high tissue loads, especially in axial compression of disk tissue. Thus, muscle activity should be used selectively. Our point, however, is that determination of appropriate muscle activity requires more than just an assessment of static or dynamic equilibrium; stability needs to also be addressed. Whether or not the results of past optimization studies, especially involving the lower back, would have been influenced by stability concerns remains an open question.

22.4 Posture, Sensors, and Movement

In the above discussion emphasized biomechanical aspects of postural stability, with only modeate discussion of the crucial role of sensory feedback. Important modes of sensory information include proprioceptive, vestibular, and visual. For appropriate postural control, these separate modes of information must be *integrated together*. Sensory information may play many roles in movement regulation, including: *i)* aiding task planning; *ii)* helping provide initial conditions for "motor programs"; *iii)* initiating "triggered responses", *iv)* serving as part of traditional real-time feedback loops; and *v)* providing information that may be used "off line" to assist neural learning structures involved in future applications of the task.

Proprioceptive information may come from throughout the body. However, there are certain regions that appear to have special relevance with regards to postural regulation. For instance, neck muscles have a massive supply of sensory receptors (Richmond, 1988), and furthermore have well-documented neural connections to the vestibular and occulomotor systems. Vestibular receptors essentially measure the orientation and movement of the head in space. Vision provides geometric information regarding the relative spatial location of objects in the environment and of parts of the body, as seen by the eyes. Extensive visual information processing only occurs within a narrow range of approximately 2° of arc at the fovea. This primary direction of view is termed the gaze. The orientation of gaze with respect to some world reference frame is

$$\theta_g = \theta_{e/h} + \theta_{h/t} + \theta_{t/w} \qquad (23.12)$$

where the three terms on the right are the eye orientation with respect to the head, the head with respect to the torso, and the torso with respect to the world frame, respectively. It turns out that there are only a few stereotyped classes of normal eye movements: *i)* saccades (very fast movements, completed within 20–60 ms); *ii)* smooth pursuit (moderate- and low-speed movements which track moving targets); *iii)* the vestibulo-ocular reflex (*VOR*, where the eyes move in a direction oppos-

ing ongoing head rotation); and *iv)* vergence (where the eyes move in opposite directions so as to provide depth perception). These movements have long fascinated bioengineers, with many groups investigating how such eye movements interact with each other (e.g., Young and Stark, 1963; Winters et al. Stark, 1984) and with movements of the head (Berthoz, 1985; Guitton, 1988) and hand (e.g., Prablanc et al., 1979). It is known, for instance, that there is tight coupling between voluntary eye and head movements, both in terms of static orientation (Andre-Deshays et al., 1988; Roucoux and Crommelinck, 1988) and dynamic movement (e.g., reviewed in Guitton, 1988).

Of special interest here is the strong relationship between vision and standing posture. Postural sway is well known to increase when the eyes are closed, and furthermore a moving visual surround is known to cause changes in the body pitch (Soetching and Berthoz, 1978; Berthoz, 1979). In addition to such automatic, subconscious coupling, visual orientation is crucial for the successful completion of most movement tasks, including that which was described in the opening paragraph of this chapter.

Of note is that sensors not only affect performance but can also be the driving force behind a postural adjustment. For instance, a postural adjustment may occur simply to obtain a proper orientation for effective vision. Such adjustments may involve the eyes alone, the eyes and the head, or in some cases the entire body. Within the work environment, considerable time may be spent sitting with a given body orientation, for example viewing a video terminal and typing at a keyboard. Thus, we see that body orientation and posture are often intricately linked.

22.5 Posture/Movement Strategies: What is Being Optimized?

Most movements are performed with some sense of purpose in mind. This may be as simple as to stand without falling, to pick up an object, or to set one's gaze on a certain object of interest. As such movement is goal-directed. Biocybernetic principles suggest that goal-directed systems should move toward optimized solutions. As outlined in Chapter 34 (Pedotti and Crenna), there are typically many ways to complete a task. One consideration is kinematic redundancy — for instance, in the absence of explicit instructions, different people will use different kinematic patterns to pick up the same object. Another level of redundancy, repeatedly addressed throughout this book, involves muscle force distribution. Also entering into the picture are measures such as "effort" and "fatigue", each of which has multiple levels (e.g., mental and muscular). All of these considerations, and undoubtedly many others, influence how the task described at the beginning of this chapter is executed. Given all these possibilities, how does one choose between the available options?

The goal of this section is to address this topic from within the framework of mathematically-based optimization, with special reference to tasks involving the torso. Although each postural task differs, there do seem to be certain salient features that cross between tasks which can be emphasized, and whcih distinguish these tasks from those considered in other parts of the book.

As discussed in detail in Chapter 8 (Zajac and Winters), there are two fundamental classes of optimization investigation: "static" and "dynamic". The former tends to be the approach of choice for postural biomechanical studies that involve the torso. Typically a certain posture is assumed and the magnitudes of the muscle forces crossing the region of interest are to be determined. A common approach for the lumbar spine region has been to assume that tissue stress should in some sense be minimized. This may involve muscle stresses, axial compressive stressses, or shear stresses [see also the appendix of Chapter 25 (Gracovetsky)]. Minimization of such criteria result in low tissue stresses, which seems to be an appropriate strategy for minimizing injury. Furthermore, the muscle stress, raised to a power of, say, 2 or 3, has been correlated to minimization of muscle fatigue (Crowninshield and Brand, 1981; Dul et al., 1985).

A simple, computationally efficient variation of the muscle stress criterion, used in Chapter 24 (Ladin), is to minimize the maximal muscle stress (Schultz et al., 1983; Bean et al., 1988). As outlined in Chapter 24, this approach is currently only applicable for relatively unright postures. In Chapter 24 the basic approach is generalized, with the result being muscle activity surfaces which provide predictions of muscle activity for arbitrary combinations of external moments. Such 3-D surfaces thus predict certain mapping between muscle activity and load. Switching curves, which

defines the boundary between whether or not a given muscle will be active, can also be identified and utilized. Comparison to EMG records suggests that this optimization approach is reasonably effective. However, as would be expected, cocontraction is never predicted. Our previous stability analysis suggested that cocontraction may at times be necessary. Does it occur? To some extent. For instance, the correlation between model results and EMG is better for contralateral muscles (primary agonists for this inverted pendulum system) than ipsilateral muscles (muscles potentially cocontracting).

In Chapter 25 (Gracovetsky) a two-stage optimization strategy is employed. First, a static optimization criterion is used which minimizes a linear combination of muscle stresses, spinal compression, and spinal shear, summed across five lumbar joints. These tend to be complimentary (as opposed to competing) subcriteria. Next, an iteration scheme is used to minimize the distribution of load variation along the column while allowing mild kinematic changes.

Chapter 27 (Deitrich et al.) employ a lumped parameter model and utilize a criterion related to the stored energy across the muscle springs. This criterion predicted that muscles capable of large torques (with moments arms) are most active; smaller muscles join in for larger loads when reasonable constraints are placed on larger muscles. Of interest is the comment within this chapter that, for "reasonable" performance criteria, the optimal solutions are relatively insensitive to the form of the criterion. This may be true in part but must be applied with caution.

In Chapter 28 (Winters and Peles) a "forward static" analysis is employed, i.e. the task is quasi-static but the inputs to the system are muscle forces and the outputs are changes in the kinematic configuration. Chapter 27 may also have used this approach. By using this input-output causality, the optimization criterion can include kinematic parameters as well as parameters related to muscle or joint stress. Specifically, Chapter 28 employs 3-D screw axis parameters in this regard. "Reference" axis of rotation parameters are specified based on experimental data or a purposeful reason to want to see what muscles are necessary to cause a certain unique head rotation axis. Muscle stress is also penalized. Unlike the method of Chapter 25, where muscle stress and joint stresses are *complimentary* subcriteria, these kinematic and muscle subcriteria are *competing*, similar to was is common for dynamic optimization studies [e.g. Chapter 19 (Seif-Naraghi and Winters), Chapter 43 (Yamaguchi)]. Consequently, different *strategies* can be investigated by varying the relative weights between these subcriteria. Of interest is that for the 10-muscle scheme discussed within Chapter 28, it was not possible to find solutions for certain physiological combinations of head orientations and head axis of rotation. Clearly small internal muscles must play major roles during voluntary head movements.

Chapter 32 (Ong et al.) employs dynamic optimization to determine appropriate control parameters for their cycling and sitting tasks. A few different criteria are used. Of interest is one related to stiffness, which led to predictions of cocontractions.

How can optimization techniques be best utilized in the future? First, the type of optimization approach will depend on the *goal of the analysis*. For cases where predictions of tissue loading is of primary interest, static optimization methods may suffice. These methods have the important advantage of being computationally efficient, to the point were solutions can be obtained quite quickly [Chapter 24 (Ladin)]. However, such efforts are likely to yield a lower limit on tissue loading since cocontraction is not predicted. If cocontraction is seen via *EMG* measurement, then these methods will either need to be supplemented by heuristic methods forcing some cocontraction or else be set aside in favor over other methods.

If postural or movement *organizational strategy* is of interest, static optimization methods, which prespecify movement kinematics, are less likely to be effective. Notice that a number of chapters, and especially Chapter 34 (Pedotti and Crenna), have pointed out that there are typically multiple strategies for solving a given task that may be kinematically distinct. Dynamic optimization, as outlined in Chapter 8 (Zajac and Winters), provides a natural causality for the neuromuscular system, with movement kinematics becoming a system output. For postural systems, this implies that the optimization algorithm must truely solve the inverted pendulum problem. This method further allows tradeoffs between "effort-related" subcriteria and "task-related" subcriteria [Chapter 19 (Seif-Naraghi and Winters)].

However, traditional dynamic optimization approaches may not be sufficient, especially for systems with a small base of support. Sabilometry measurements on standing humans repeatedly show that there is significant sway of a oscillatory nature. Either this is a purposeful act of the nervous system or there is noise in the system. Incorporation of "noise", whether due to the nervous system or external perturbation, turns a deterministic dynamic optimization problem into a stochastic optimization problem. Although in some cases such problems can be solved [e.g., see Chapter 19 (Seif-Naraghi and Winters) for one approach], solutions are difficult. However, such an optimization approach would allow movement strategy to be investigated within the natural context of goal-directed task completion, combined with maintenance of stability within a specified margin of safety.

23.6 Future Directions

The biomechanical systems considered within this chapter are quite complex. Furthermore, the tasks considered here are by and large difficult to understand.

One basic area of great need is quantitative musculotendon data for trunk muscles. In the Appendix to this book (Yamaguchi et al.), the *quantitative* musculotendon data, from various authors, for the musculature of the lower limb spanned 24 pages; that for the trunk only 2 pages. Yet the trunk includes many, many muscles. Such lack of data discourages modeling efforts. Better data on muscle physiological cross-sectional areas, muscle fiber lengths, fiber compositions, etc., which would facilitate existing anatomical investigations [e.g. that of Bogduk (1980) and Bogduk and MacIntosch (1984)], is needed.

Considerable work remains to further improve our understanding of posture and movement. Specifically, we feel that efforts are needed to address the issue of postural stability, muscle coordination in asymmetric activities, and effects of fatigue, disease and injury on the functions of the trunk muscles.

As discussed throughout this book, the ability to predict muscle force is a central problem in biomechanics. Because the problem is usually statically indeterminate, static optimization techniques are frequently employed. Often muscles are grouped together, and co-contractions of antagonist muscles are not predicted. Yet, we know from electromyographic experiments that co-contractions are common, particularly in situations of asymmetry. For example, in the experiments by Pope et al. (1986, 1987) considerable antagonistic activity was recorded when performing static twisting activities while standing. Thus the conventional assumption of reciprocal innervation, which would postulate a complementary pattern of activity between agonist and antagonist, would seem to be refuted by the experimental data. The reasons of this co-contraction require further exploration. However, our crude analyses shows that co-contraction may be quite important from the point of view of mechanical stability, providing appropriate stiffness to the trunk and likely to other regions of the body as well. This concepts require further clarification.

While posture obviously is critical to understand, further complexity will occur with movements. Here the relationship of muscle activity to external moments is poorly understood. Clearly, muscles work in harmony to provide motion. Agonistic and antagonistic activities are well synchronized, but the precise relationships are unknown. Insights in this area will require additional experimental investigations complemented by simulation studies.

As emphasized in Chapters 28-34, the trunk is intricately coupled to the limbs. Relations between the spine, pelvis and lower limbs for various tasks is now, quite appropriately, an area of active investigation (e.g. Chapters 25, 32-34). A variety of muscles attach between the spine and the shoulder girdle. Consequently, muscles in the trunk are involved in shoulder stabilization and in arm movement. Although some studies have explored basic mechanical coupling between the spine and the arm (e.g., Andersson and Schultz, 1979), this is a virtually untapped area.

We also need to better understand how fatigue, disease and injury influence muscle activity and movement in general. There is experimental evidence that the technique of lifting is influenced by fatigue of back and knee muscles. Is this a major factor contributing to overloading and injury? Given a specific type of injury, can we predict how muscle coordination will adapt? Can observations of task performance be used to determine abnormalities in the use of muscles? These are clinically important questions that require answers.

References

Aggahyan, R.V. and Pal'tsev, Y.I. (1975) Reproduction of certain special aspects of the dynamics of the maintanance of the vertical posture by man using a mathematical model. *Biofizika*, **20**: 137-142.

Andersson, G.B.J. (1974) On Myoelectric Back Muscle Activity and Lumbar Disc Pressure in Sitting Postures, Gotab, thesis. University of Goteborg.

Andersson, G.B.J. (1982) Measurements of Loads on the Lumbar Spine. In *Symposium on Idiopathic Low Back Pain*, ed. A.A. White III, S.L. Gordon, pp. 220-251, Mosby, St. Louis.

Andersson, G.B.J. (1986) Loads on the Spine During Sitting. In *The Ergonomics of Working Postures* (Ed. Corlett, N., Wilson, J., Manenica, T.) Taylor and Francis, London, pp. 309-318.

Andersson, G.B.J. (1987) Biomechanical Aspects of Sitting: An Application to VDT Terminals. In *Behavior and Information Technology*, Taylor and Francis, London **6**:257-269.

Andersson, G.B.J., Bogduk, N., DeLuca, C. et al. (1989) Muscle: Clinical Perspective. In *New Perspectives on Low Back Pain* (Eds. J.W. Frymoyer, S.L. Gordon). Amer. Acad. Ortho. Surg., Park Ridge, IL, pp. 293-334.

Andersson, G.B.J., Herberts, P. and Ortengren, R. (1976) Myoelectric back muscle activity in standardized lifting postures, in Komi P.V. (ed): *Biomechanics 5-A*. Baltimore, University Park Press, pp. 520-529.

Andersson, G.B.J., Jonsson, B., Ortengren, R. (1974) Myoelectric activity in individual erector spinae muscles in sitting. *Scan. J. Rehabil. Med. Suppl.* **3**:91-108.

Andersson, G.B.J., Ortengren R. (1974) Myoelectric back muscle activity during sitting. *Scand. J. Rehabil. Med.* **3**(supple):73.

Andersson, G.B.J., Ortengren, R. and Herberts, P. (1977a) Quantitative electromyographic studies of back muscle activity related to posture and loading. *Orthop. Clin. North Am.* **8**:85-96.

Andersson, G.B.J., Ortengren, R., Machemson, A. et al. (1974) Lumbar disc pressure and myoelectric back muscle activity during sitting: I. Studies on an experimental chair. *Scand. J. Rehabil. Med.* **6**:104-114.

Andersson, G.B.J., Ortengren, R. and Schultz, A.B. (1977b) Analysis and measurement of the loads on the lumbar spine during work at a table. *J. Biomech,,* **13**: 513-520.

Andersson, G.B.J., Ortengren, R. and Schultz, A. (1980) Analysis measurement of the loads on the lumbar spine during work at a table. *J. Biomech.* **13**:513-520.

Andersson, G.B.J. and Schultz, A.B. (1979)

Transmission of moments across the elbow joint and the lumbar spine. *J. Biomech.* **12:** 747-755.

Andersson, G.B.J., Schultz, A.B., Ortengren, R. (1986) Trunk muscle forces during desk work. *Ergonomics* **29:**1113-1117.

Andre-Deshays, C., Berthoz, A. and Revel, M. (1988) Eye-head coupling in humans. I. Simultaneous recordings of isolated motor units in dorsal neck muscles and horizontal eye movements. *Exp. Brain Res.,* **69:** 399-406.

Ashton-Miller, J.A and Schultz, A.B. (1988) Biomechanics of the human spine and trunk. *Exercise & Sport Sci. Rev.,* **16:** 169-204.

Asmussen, D. and Klaussen, K. (1962) Form and function of the erect human spine. *Clin. Orthop.* **25:**55.

Battey, C.K., and Joseph, J. (1966) An investigation by telemetering of the activity of some muscles in walking. *Med. Biol. Eng.* **4:**125-135.

Bean, J.C., Chaffin, D.B. and Schultz, A.B. (1988) Biomechanical model calculation of muscle contraction forces: a double linear programming method. *J. Biomech.* **21:** 59-66.

Belenkii, Y.Y., Gurfinkel, V. and Paltsev, Y.I. (1967) Element of control of voluntary movements. *Biofizika,* **12:** 135-141.

Bendix, T., Winkel, J., Jensen, F. (1985) Comparison of office chairs with fixed forwards and backwards inclining, or tiltable seats. *Eur. J. Appl. Physiol.* **54:**378-385.

Bergmark, A. (1987) *Mechanical stability of the human lumbar spine.* Ph.D. Dissertation, Lund Inst. of Technol., Lund, Sweden.

Bernstein, N. (1967) *The co-ordination and regulation of movements.* Pergamon Press, Oxford.

Berthoz, A. (1985) Adaptive mechanisms in eye-head coordination, Chapter 12, *Adaptive mechanisms in gaze control, facts and theories.,* (eds: Berthoz, A. and Melville Jones), Chapter 12, pp. 177-201.

Berthoz, A., Lacour, M., Soechting, J.F and Vidal, P.P. (1979) The role of vision in the control of posture during linear motion. In *Reflex cotrol of posture and movement,* (Granit, R. and Pompeiano, O., eds.), pp. 197-209.

Bizzi, E. and Polit, (1978) Effect of load disturbance during centrally inititated movement. *J. of Neurophys.,* **41:** 542-556.

Bobet, J. and Norman, R.W. (1984) Effects of load placement on back muscle activity in load carriage. *Eur. J. Appl. Physiol.* **53:**71-75.

Bogduk, N.A. (1980) A reappraisal of the anatomy of the human lumbar erector spinae. *J. Anat.,* **131:** 525-540.

Bogduk, N. and MacIntosch, J. (1984) The applied anatomy of the thoracolumbar fascia. *Spine* **9:** 164-170.

Black, F.O., O'Leary, D.P., Wall, C. and Furman, J. (1977) The vestibulospinal stability test: normal limits. *Trans. Am. Acad. Opthal. Otolaryngol.* **84:** 549-560.

Carlsoo, S. (1961) The static muscle load in different work positions: An electromyographic study. *Ergonomics* **4:**193.

Carlsoo, S. (1963) *Table, Chair and Work Posture.* Folksam and Facit. (Swedish).

Chaffin, D.B. and Andersson, G.B.J. (1984) *Occupational Biomechanics.,* John Wiley and Sons.

Clement, G., Gurfinkel, V.S., Lestienne, F. Lipshits, M.I. and Popov, K.E. (1984) Adaption of postural control to weightlessness. *Exp. Brain Res.,* **57:** 61-72.

Cook, J.M. and Neumann, D.A. (1987) The effects of loac placment on the EMG activity of the low back muscles during load carrying by men and women. *Ergonomics* **30:**1413-1423.

Cordo, P.J. and Nashner, L.M. (1982) Properties of postural adjustments associated with rapid arm movements. *J. Neurophys.,* **47:** 287-302.

Crowninshield, R.D. and Brand, R.A. (1981) A physiologically based criterion of muscle force prediction in locomotion. *J. Biomech.* **14:** 793-801.

Dotterhof, A.S.M. and Vink, P. (1985) The stabilizing function of the mm. iliocostales and the mm. multifidi during walking. *J. Anat.* **140:**329-336.

Dul, J., Townsend, M.A., Shiavi, R. and Johnson, G.E. (1984) Muscular synergism - I. On criteria for load sharing between synergistic muscles. *J. Biomech.* **17:** 663-673.

Floyd, W.F. and Silver, P.H.S. (1950) Electromyographic study of patterns of activity of the anterior abdominal wall muscles in man. *J. Anat.* **84:**132.

Floyd, W.F. and Silver, P.H.S. (1951) Function of erectores spinae in flexion of the trunk. *Lancet* **1:**133.

Floyd, W.F. and Silver, P.H.S. (1955) The function of the erectores spinae muscles in certain movements and postures in man. *J. Physiol.* (London) **121:**184.

Goel, V.K., Clark, C.R., McGoman, D. and Goyal, S. (1984) An in-vitro study of the kinematics of the normal, injured and stabilized cervical spine. *J. Biomech.,* **17:** 363-376.

Gracovetsky, S. (1988) *The spinal engine.* Springer-Verlag, Wien, New York.

Guitton, D. (1988) Eye-head coordination in gaze control. In *Control of Head Movements* (Peterson, B.W. and Richmond, F.J., eds.), Oxford Press, New York.

Gurfinkel, V.S., and Osovets, S.M. (1972) Dynamics of equilibrium of the vertical posture in man. *Biofizika,* **17:** 478-485.

Harms-Ringdahl, K., Ekholm, J., Schuldt, K., Nemeth, G. and Arborelius, U.P. (1986) Load moments and myoelectric activity when the cervical spine is held in full flexion and extension. *Eggonomics,* **12:** 1539-1552.

Hayes, K.C. (1982) Biomechanics of postural control. *Exer. and Sprot Sci. Rev.*, **10**: 363-391.

Hayes, K.C., Weiss, P.L. and Darling, W.G. (1980) Kinetics of postural control in man. *J. Biomech.*, **13**: 198.

Hayes, K.C., Winter, D.A. and Norman, R.W. (1976) Energetics of bipedal locomotion. *Yearbook. Phys. Anthropol.*, **20**: 481-490.

Hosea, T.M., Simon, S.R., Delatizky, J., Wong, M.A., Hsieh, C.G. (1986) Myoelectric analysis of the paraspinal musculature in relation to automobile driving. *Spine* 11:928-936.

Huston, R.L. and Perrone, N. (1978) Dynamic response and protection of the human body and skull in impact situations. In *Perspectives in Biomechanics, Vol. I*, (ed. Ghista et al.), pp. 531-472, Harwood Acad., New York.

Jonsson, B. (1970) Topography of the lumbar part of the erector spinae muscle. *Anat. Entwicklunsgesch* 130:77.

Kippers, V., and Parker, A.W. (1984) Posture related to myoelectric silence of erectores spinae during trunk flexion. *Spine* 9:740-745.

Koozekanani, S.H., Stockwell, R.B. McGhee, R.B. and Firoozmand, F. (1980) *IEEE Trans. Biomed. Engng.* BME-27: 605-609.

Ladin, Z., Murthy, K.R., DeLuca, C.J. (1989) Mechanical recruitment of low back muscles. *Spine* 14:927-938.

Marras, W.S. and Mirka, G.A. (1990) A comprehensive evaluation of trunk response to asymmetric trunk motion. Manuscript.

McGill, S.M. (1990) Electromyographic activity of the abdominal and low back musculature during the generation of isometric and dynamic axial trunk torque. Manuscript, Univ. of Waterloo.

McGill, S.M. and Norman, R.W. (1986) Partitioning of the L4/L5 dynamic moment into disc, ligamentous and muscular components during lifting. *Spine* 11:666-678.

Nachemson, A.L. (1981) Disc pressure measurements. *Spine*, **6**: 93-97.

Nashner, L.M. (1976) Adapting reflexes controlling the human posture. *Exp. Brain Res.*, **26**: 59-72.

Nashner, L.M. and Cordo, P.J. (1981) Relation of automatic postural responses and reaction-time voluntary movements of human leg muscles. *Exp. Brain Res.* **43**: 395-405.

Nashner, L.M. and McCollum, G. (1985) The organization of human postural movements: A formal basis and experimental synthesis. *Brain. Behav.*, **8**: 135-172.

Ortengren, R. and Andersson, G.B.J. (1977) Electromyographic studies of trunk muscles with special reference to the functional anatomy of the lumbar spine. *Spine* 2:44-52.

Ortengren, R. and Andersson, G.B.J. (1989)

Pope, M.H., Andersson, G.B.J., Broman, H., Svensson, M. and Zetterberg, C. (1986) Electromyographic studies of the lumbat trunk musculature during the development of azial torques. *J. Orthop. Res.*, **4**: 288-297.

Pope, M.H., Andersson, G.B.J., Broman, H., Svensson, M., Zetterberg, C. (1986) Electromyographic studies of the lumbar trunk musculature during the development of axial torques. *J. Orthop. Res.* 4:288-297.

Prablanc, C., Echallier, J.F., Komilis, E. and Jeannerod, M. (1979) Optimal response of eye and heand motor systems in pointing at a visual target. *Biol. Cybern.*, **35**: 113-124.

Raftopoulos, D.D. Rafco, M.C., Green, M., et al. (1989) Relaxation phenomena in lumbar trunk muscles during lateral bending. *Spine*, accepted for publication.

Richmond, F.J., Baker, D.A., Stacy, M.J. (1988) The sensorium: receptors of neck muscles and joints. In: *Control of head movements*, (ed: Peterson, B.P. and Richmond, F.J.), Oxford Univ. Press, New York.

Richmond, F.J.R. and Vidal, P.P. (1988) The motor system: joints and muscles of the neck. In *Control of Head Movements*, (eds: Peterson, B.W. and Richmond, F.J.), Chapter 1, pp. 1-21.

Rondot, P. (1988) Clinical disorders of head movement. In: *Control of head movements*, (ed: Peterson, B.P. and Richmond, F.J.), Oxford Univ. Press, New York.

Sances, A., Myklebust, J.B., Maiman, D.J., Larson, S.J., Cusick, J.F. and Jodat, R.F. (1984) The biomechanics of spinal injuries. *CRC Crit. Rev. Bio. Engng.*, **11**: 1-xx.

Schor, R.H., Kearney, R.E. and Dieringer, N. (1988) Reflex stabilization of the head. In: *Control of head movements*, (ed: Peterson, B.P. and Richmond, F.J.), Oxford Univ. Press, New York.

Schuldt, K., Ekholm, J., Harms-Ringdahl, K., Nemeth, G. and Arborelius, U.P. (1986) Effects of changes in sitting work posture on static neck and shoulder muscle activity. *Ergonomics*, **12**: 1525-1537.

Schultz, A.B. and Andersson, G.B.J. (1981) Analysis of loads on the lumbar spine. *Spine*, **6**: 76-82.

Schultz, A.B., Andersson, G.B.J., Haderspeck, K., Ortegren, R., Nordin, M. and Bjork, R. (1982) Analysis and measurement of lumbar trunk loads in tasks involving bends and twists. *J. Biomech.*, **15**: 669-675.

Schultz, A.B., Andersson, G.B.J., Ortengren, R., Bjork, R. and Nordin, M. (1982) Analysis and quantitative myoelectric measurements of loads on the lumbar trunk muscles in isometric performance of mechanically complex standing tasks. *J. Orthop. Research*, **1**: 77-91.

Schultz, A.B., Andersson, G.B.J., Ortengren, R., et al. (1982) Analysis and quantitative myoelectric measurements of loads on the lumbar spine when holding weights in standing postures. *Spine* **7**:390-397.

Schultz, A.B., Cromwell, R., Warwick, D., Andersson, G. (1987) Lumbar trunk muscle use in standing isometric heavy exertions. *J. Orthop. Res.* **3**:320-329.

Schultz, A.B., Haderspeck-Grib, K., Sinkora, G. et al. (1985) Quantitative studies of the flexion-relaxation phenomenon in back muscles. *J. Orthop. Res.* **3**:189-197.

Schultz, A.B., Haderspeck-Grib, K., Warwick, D., Portillo, D. (1983) Use of lumbar trunk muscles in isometric performance of mechanically complex standing tasks. *J. Orthop. Res.* **1**:77-91.

Seroussi, R.E., Pope, M.H. (1987) The relationship between trunk muscle electromyography and lifting moments in the sagittal and frontal planes. *J. Biomech.* **20**:135-146.

Shigley, J.E. (1977) *Mechanical Engineering Design.* McGraw-Hill, New York.

Soetching, J.F. and Berthoz, A. (1979) Dynamic role of vision in the control of posture in man. *Exp. Brain Res.* **36**: 551-561.

Soderberg, G.L., Blanco, M.K., Cosentino, T.L., Kurdelmeier, C.A. (1986) An EMG analysis of posterior trunk musculature during flat and anteriorly inclined seating. *Human Factors* **28**:483-491.

Stark, L. (1968) *Neurological Control Systems.* Prenum Press, New York.

Stokes, I.A.F., Rush, S., Moffroid, M., Johnson, G.B., Haugh, L.D. (1987) Trunk Extensor EMG-Torque Relationship. *Spine* **12**:770-776.

Thorstenson, A., Carlson, H., Zomlefer, M.R., Nilsson, J. (1982) Lumbar back muscle activity in relation to trunk movements during locomotion in man. *Acta Physiol. Scand.* **116**:13-20.

Valencia, F.P. and Monroe, R.R. (1985) An electromyographic study of the lumbar multifidus in man. *Electromyogr. Clin. Neurophysiol.* **25**:205-221.

Vink, P. Van der Velde, E.A., Verbout, A.J. (1988) A functional subdivision of the lumbar extensor musculature. *Electromyogr. Clin. Neurophysiol.* **28**:517-525.

Vitti, M. and Basmajian, J.V. (1974) The functions of semispinalis capitis and splenius capitis muscles: an electromyographic study. Atat. Rec., 179: 477-480.

Vitti, M., Fujiwara, M., Basmajian, J.V. and Iida, M. (1974) The integrated roles of longus colli and sternocleidomastoid muscles: an electromyographic study. *Anat. Rec.*, 177: 471-484.

Waters and Morris (1972)

Williams, J.L. and Belytschko, T.B. (1987) A three-dimensional model of the human cervical spine for impact simulation. *J. Biomech. Engng.*, **105**: 321-330.

Winters, J.M. (1988) Biomechanical modeling of the human head and neck. In: *Control of head movements*, (ed: Peterson, B.P. and Richmond, F.J.), Oxford Univ. Press, New York.

Winters, J.M., Nam, M.H. and Stark, L. (1984) Modeling dynamical interaction between fast and slow movement: fast saccadic eye movement behavior in the presence of the slower VOR. *Math. Biosci.*, **68**: 159-187.

Yoganandan, N., Myklebust, .B., Ray, G. and Sances, A. (1987) Mathematical and finite element analysis of spine injuries. *CRC Crit. Rev Biomed. Engng.*, **13**: 29-93.

Young, L.R. and Stark, L. (1963) Variable feedback experiments tesing a sampled data model for eye tracking movements. *IEEE Trans. Human Fact. Elec.*, **HFE-4**: 38-51.

Zangemeister, W.H. and Stark, L. (1982) Gaze latency: variable interaction of head and eye latency. *Exp. Neurol.*, **75**: 389-406.

Zetterberg, C., Andersson, G.B.J., Schultz, A.B.z (1987) The activity of individual trunk muscles during heavy physical loading. *Spine* **12**:1035-1040.

CHAPTER 24

Use of Musculoskeletal Models in the Diagnosis and Treatment of Low Back Pain

Zvi Ladin

24.1 Background

Recent epidemiological studies document the continued impact of low back pain (*LBP*) on our society. *LBP* occurs at some time in over 60% of the population in industrialized nations (Andersson, 1981; Deyo, 1983), and each year 5% of American adults experience an episode of low back pain (Frymoyer and Cats-Baril, 1987). As many as 75 million Americans currently suffer from back problems (Kelsey et al., 1979; Kelsey and White, 1980), and each year seven million new cases develop. Of these new cases, five million will become partially disabled and the remaining two million will be unable to work at all. The decrease in work time associated with lower-back pain has been estimated to be 93 million work days per year, second only to the common cold (Nordby, 1981). The long–range projections of this problem are even more serious than the above numbers suggest because the rate of disability resulting from *LBP* increased at 14 times the rate of population growth between 1977 and 1981 (Cats-Baril et al., 1987). Although the vast majority of patients recover within six months (Cats-Baril et al., 1987), a few studies point out that up to 90% of the costs are related to the treatment of only 10% of the patients who are disabled for more than six months (Benn and Wood, 1975; Cats-Baril and Roth, 1985; Pheasant, 1977). As the debilitating condition persists, the potential for recovery decreases. Only 20–40% of patients return to work after a disability of 1 year, and none after a 2–year disability (Frymoyer and Cats-Baril, 1987).

At the present time, most clinicians approach the treatment of back disorders as a biomechanical problem (Farfan, 1973). This approach is based on the supposition that the onset of lower back pain is related to an imbalance in the mechanical components of the back. In recent years there have been a few studies that suggest that physical therapy exercises improve the prognosis of *LBP* patients. Nachemson (1983), in a comprehensive review of diagnostic and therapeutic approaches to low back pain, suggests that "early but moderate and gradual motion and loading improve healing in all the structures that build the back." The philosophy behind this approach gave rise to the development of a simple set of guidelines that are intended to rehabilitate and educate low back patients, as summarized and taught at "back schools." The improved prognosis of patients that have gone through such training and educational courses (Moffett et al., 1986) no doubt contributed to the popularity of such schools in recent years.

The high success rate in treating acute cases of low back pain (Wiesel et al., 1984) can therefore be contrasted with the low success rate in treating chronic low back patients. One is left to wonder why is it not possible to extend the same kind of educational and therapeutic approach to the portion of the patient population that is clearly in great need for it. The high human and economic costs that are involved in the care for the chronically disabled low back patients should not dictate abandonment of the physical therapy approach. Rather, what appears to be lacking is a comprehensive understanding of the role muscles play in the

Multiple Muscle Systems: Biomechanics and Movement Organization
J.M. Winters and S. L-Y. Woo (eds), © 1990 Springer-Verlag

control of trunk motion and posture. The development of such understanding could lead to assessment techniques for determining *quantitatively* and *objectively* both the diagnostic nature of a given muscular deficiency and the optimal therapeutic regimen to improve such a deficiency. Assessment techniques that provide an outcome measure for treatment exist for the skeletal system (X-rays), the soft tissue (CAT scans), and neurological structures (myelograms). For muscles, the assessment of the efficacy of a specific physical therapy regimen is usually based on indirect measures such as muscle strength and a joint range of motion. Such measures have been successfuly used in the evaluation of post–surgical rehabilitation of knee surgery where the individual role of a single muscle in the control of a given joint is much simpler than in the back. The large number of muscles that intersect the lumbar region, the mechanical complexity of the posture and movement in that area, and the lack of a comprehensive framework to understand the role of the muscles in a variety of physical tasks make the design of rehabilitation exercises to improve the muscular function of the lower back a seemingly intractable problem.

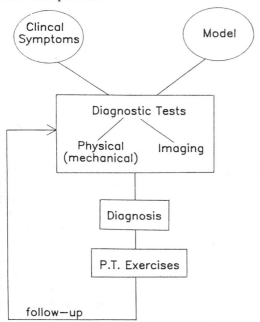

Figure 24.1: A conceptual framework for the integration of the biomechanical model into a clinical decision-making process.

A valid biomechanical model of the lower back could serve as an invaluable link in the sequence of diagnosis, prescription, and evaluation that exists in every clinical problem. The role of the model is schematically depicted in Figure 24.1. The presentation of clinical symptoms, coupled with the biomechanical model, leads to the specification of diagnostic tests that could combine imaging and mechanical loading tests. The imaging tests are intended to reveal basic structural deficits such as disk or vertebral problems. The mechanical loading exercises are intended to selectively activate or deactivate specific muscles. The patterns of muscular activity thus detected could identify muscular deficits. The diagnosis of those deficits will then lead to the prescription of individualized exercises that are intended to selectively activate or deactivate those muscles. The prescription of such exercises requires a basic understanding of the role of the different muscles in response to externally applied loads. The biomechanical model is intended to summarize and provide such an understanding.

24.2 Introduction

The lower back is a complex system of muscles, vertebral bodies, intervertebral disks, ligaments, and layers of fascia. The basic structure is the vertebral column that is made up of multiple vertebrae. This configuration provides the upper body with the required mechanical strength to support the weight of the trunk and head as well as any externally applied loads, while maintaining the flexibility necessary to provide flexion, lateral bending, and torsional motion. The vertebral bodies are separated by fibrocartilaginous intervertebral disks. All the vertebral bodies have bony prominences called processes: some serve as attachment sites for the muscles and ligaments, while others serve as articular surfaces during the intervertebral motion. The articular surfaces of two adjacent vertebral bodies are covered with hyaline cartilage and form a synovial joint (Snell, 1973). Hence there are two kinds of intervertebral joints: the synovial joints formed by the articular surfaces, and the fibrocartilaginous joints formed by the intervertebral disks. The ligaments are attached longitudinally to the anterior and posterior surfaces of the vertebrae and serve to restrict the amount of intervertebral motion that can be sustained by the vertebrae.

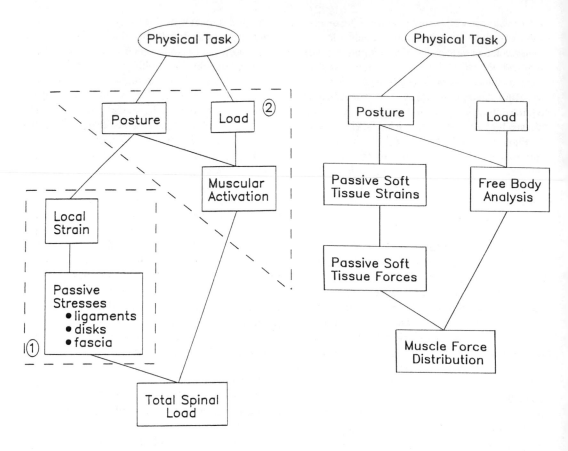

Figure 24.2: A flow-chart for the determination of total spinal load arising from a given physical task (*a*) and the calculation of muscle force distribution in response to a loading task (*b*).

The muscles of the lumbar region can be divided into three groups: the anterolateral group, the posterior group, and the deep paravertebral group. The anterolateral group contains the rectus abdominus and the internal and external oblique; the posterior group includes the erector spinae group (the multifidus, iliocostalis, and longissimus) and the latissimus dorsi; the deep paravertebral group includes the psoas and the quadratus. The thoracolumbar fascia (also referred to as the posterior ligamentous system) (Bogduk, 1984) gives rise to the upper fibers of the internal oblique and to the transversus muscles, and is attached medially to the transverse processes of the lumbar vertebrae.

The mechanical properties of the lumbar region are determined by tissues that have only passive mechanical properties (vertebral bodies, ligaments, disks, and fascia) and tissues that have both passive and active properties (muscles). The loads sustained by the different elements will therefore depend on the anatomic posture (i.e. the geometry) that determines the strains (and therefore the stresses) of the passive elements, as well as on the activation level of the different muscles. Such a hierarchical relationship is depicted in Figure 24.2a. Any physical task can be described by two indepenedent determinants: the body posture and the external load. The posture determines the local strains and therefore the resulting stresses in the passive tissues, whereas the muscle force distribution is determined by both the posture *and* the external loads. The total sum of the stresses in all the soft tissues determines the overall internal force and moment that need to equilibrate the externally applied loads to maintain the static equilibrium or the dynamic motion required by the physical task.

Figure 24.2a also provides a convenient framework to examine previous biomechanical models. Block 1 in the figure represents the studies that were aimed at finding the constitutive equations (i.e. the stress/strain relationship) of the passive soft tissues. They include studies that were aimed at quantifying the motion of vertebral segments (Kippers and Parker, 1989), the development of phenomenological models for the disk (Panagiotacopulos et al., 1987; Hickey and Huckey, 1980), and the facet joints (Shirazi-Adl and Drouin, 1987). Block 2 in the figure describes studies that addressed the relationship of physical tasks and muscle force distribution, for example Yettram and Jackman (1980), Schultz et al. (1983), Pope et al. (1986), Seroussi and Pope (1987), and others [see also Chapter 23 (Andersson and Winters) and Chapter 25 (Gracovetsky)]. Studies that analyzed the force distribution in the lumbar muscles were limited to the erect posture: a subject was standing erect while holding (or resisting) the forces and moments exerted by weights (Schultz et al., 1983, 1982; Pope et al., 1986; Seroussi and Pope, 1987). A biomechanical analysis of physical tasks that involve a different posture increases significantly the complexity of the problem, since the internal forces arising from the local strains imposed on the passive tissues will have to be accounted for. This is the reason why attempts at analyzing different postures were limited to the study of overall ground reaction forces (Kromodihardjo and Mital, 1987; Freivalds et al., 1984), spinal joint loads (Garg and Herrin, 1979; Bejjani et al., 1984), or continuous models of the spinal column such as the passive beam model by Lindbeck (1987) and the arch model by Aspden (1989).

The mechanical determination of the lumbar muscle force distribution can be schematically described as a variation of the above process, and is depicted in Figure 24.2b. The physical task determined by the combination of external loads and anatomic posture defines the equilibrium equations obtained by a "free body analysis" of any imaginary cross section in the lumbar region. For example, consider an imaginary transverse cross section performed through the lumbar region on a weight lifter standing in an erect posture. Six equilibrium equations can be written, three for the force components and three for the moment components, representing the six degrees of freedom of the virtual joint through which the cross section was conducted. The number of unknowns in these equations corresponds to the number of soft tissue components that can generate force, i.e., the passive and active elements listed above. Another set of equations can be incorporated by calculating the local strains arising from the anatomic posture, and then using the constitutive equations of the passive soft tissues to calculate the forces contributed by them. These forces can then be substituted into the equilibrium equations, leaving the muscle forces as the only unknowns in these equations. Since the number of muscles crossing the lumbar level exceeds the number of degrees of freedom of the transverse cross section, the problem becomes mechanically indeterminate.

Two approaches have been taken to resolve the mathematical redundancy in determining the muscle force distribution: functional grouping and optimization. In functional grouping, muscles that were believed to work in consort were grouped together, thereby reducing the number of unknowns to match the number of equations (Seroussi and Pope, 1987; Morris et al., 1961). The optimization approach assumed the existence of a cost function defined by the muscle forces, and searched for a solution that optimized the value of the cost function (Schultz et al., 1983; Bean et al., 1988; see also Chapters 8, 23, 25-27). Some authors tried to avoid the use of optimization and still maintain some degree of anatomic accuracy by invoking physiological assumptions; i.e. if one muscle group is active, then an opposing muscle group will not be active (Schultz et al., 1982). It is clear that each of the above approaches carries its own benefits and penalties: the mathematical simplification that is achieved by reducing the number of unknowns using physiological assumptions (one can view the functional grouping as a physiological assumption that constrains the grouped muscles to always act together) reduces the ability to study *individual* muscle activation patterns. Therefore, the detailed study of individual lumbar muscles requires at this point the use of optimization techniques. This approach will be illustrated through the rest of this chapter.

24.3 The Muscle Activity Surfaces

The vertebaral bodies that make up the spinal column create a multi–joint, multi–link system that is held together by a series of passive elements such as the ligaments and by the muscles that have both passive and active properties. The external loads that are applied at any level of the spinal column include moments and forces that can be separated into six components: two shear forces and a compressive force, and two bending moments and an axial torsion moment. Andersson et al. (1980) suggested that the moments generated by the muscles in the lumbar region balance the overall moments that are applied to a given lumbar cross section. This suggestion was followed in later works by Schultz et al. (1982), Seroussi and Pope (1987), and Bean et al. (1988). This hypothesis can be stated as a physiological decoupling assumption: in response to the external loads, *the muscles balance the external moments*; the other vertebral and paravertebral structures equilibrate the external forces and the balance of the muscle forces. This analysis does not take into account the moments contributed by all the non-muscular soft tissues in the cross section of interest. Such a simplification is probably appropriate for the erect posture, where one could argue that the ligaments and the fascia are in their slack, rest position and therefore not generating any forces. Those forces will clearly have to be taken into account for any posture that strains those structures.

By limiting the class of physical tasks under consideration to external loads applied to the upper body in the erect posture condition, it is clear from the above assumption that the external moment combination will determine the muscle–force distribution. If indeed the muscle forces are uniquely determined by such a combination, one can describe a three-dimensional space called the *"loading space"* whose axes are the three independent components of the external moment loading, i.e. the two bending moments and the axial (or torsional) moment. The muscle force distribution

in *all* the lumbar muscles is therefore determined by a point in the "loading space": the coordinates of that point will describe the external moment components that dictate the forces in all the lumbar muscles.

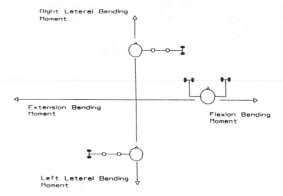

Figure 24.3: *The loading plane.* The stick figures represent characteristic loading tasks that correspond to points along the axes. (Reproduced from Ladin et al. (1989), with permission.)

The "loading space" describes all the different moment combinations that could be externally applied, and they include in the most general case, two bending moments, and an axial (or a torsion) moment. The analysis could be initially limited to loading conditions that have no torsion component. In this case, the "loading space" becomes a "loading plane" and is described in Figure 24.3. The horizontal axis is the flexion–extension bending moment, and the vertical axis is the lateral bending axis. The stick figures on the axes represent typical loading conditions that are described by points along the axes. For example: holding equal weights in both hands, while the arms are symmetrically extended in front of the body, generates a pure flexion bending moment and no lateral bending moment, and therefore will be represented by a point *on* the flexion bending moment axis. In a similar fashion, holding a weight in the right hand while the arm is being extended laterally to the right side produces a pure right lateral bending moment and no flexion–extension moment. Such a task is therefore represented by a point *on* the right lateral bending moment axis. Since most daily activities involve holding weights in front of the body, the analysis will be limited to the right half of the "loading plane", representing combinations of flexion *and* lateral bending moments. It is important to note that any physical task can be described by a single point on the "loading plane", provided that it does not have an externally applied torsion moment and that it is executed in an erect posture. A typical exercise could be moving the upper extremities while holding weights and keeping an erect posture. The moment combination arising from holding the weights in any position will translate into a resultant lateral bending moment and a resultant flexion moment, thereby defining a single point on the "loading plane". Moving the weights in the process of performing the exercise will therefore map into a curve in the loading plane. The inverse calculation cannot be generally done without invoking additional assumptions, since a single point in the loading plane could correspond to many physical tasks.

The physiological decoupling assumption listed above, and the identification of the loading plane as the single determinant of the lumbar muscle force distribution, open the door to the description of the muscular response to different loading conditions. This process can be executed by discretizing the loading plane, i.e. creating a discrete grid on the plane, and solving the muscle force distribution problem repeatedly for different points on the grid, thus covering a range of loading conditions. The predicted muscle forces in a single muscle for the given range of the loading plane can be stored as a three-dimensional surface called the *muscle activity surface* (*MAS*). Such a surface provides us with a predicted activity map of the muscle: given the external moment loading generated by a specific task, the expected muscular force can be immediately determined.

The biomechanical model discussed in this chapter is based on the model presented by Schultz et al. (1983), and is described in detail by Murthy (1990). The moment equilibrium equations of the lumbar cross section at the *L3* level were written for a given loading condition represented by a single point on the loading plane. These equations included the unknown muscle forces applied in the directions of the muscle fibers, and the locations of the centroids of the cross sectional areas of those muscles. A cost function that described the muscular spinal compression, i.e., the total sum of the axial components of the muscle tensile forces, was

defined. The goal of the solution process was to optimize the muscle force distribution, i.e. to find a distribution of muscle forces that minimized the cost function while satisfying the moment equality constraints. The solution space was bound by adding inequality constraints that prevented the individual muscle stresses (calculated as the muscle forces divided by the appropriate cross-sectional areas) from exceeding a given upper bound. By gradually increasing the value of this bound, the linear programming algorithm applied in this problem searched for *the smallest possible muscle forces that could satisfy the moment equilibrium equations and minimize the muscular spinal compression force.*

The *MAS* of all the 22 lumbar muscles were calculated by discretizing the loading plane using a step-size of 10 Nm, and covering the range of 0–100 Nm along the flexion bending moment, and 0–100 Nm along the lateral bending moment for both the right and left sides. The grid was made up of a total of 231 points. The surfaces were then stored as three-dimensional arrays representing the muscular response to external loading for any given loading condition described by a point that lies in the specified range of the loading plane. The procedure to find the predicted muscle force in response to a given physical task can now be summarized in the following steps:

1. Calculate the external moment combination arising from the physical task.

2. Use the moment combination to determine the loading point on the "loading plane."

3. Find the predicted muscle force from the muscle activity surface (*MAS*) of that muscle.

The *MAS* of all the lumbar muscles provide us with a summarized biomechanical model of the lumbar region. They capture the muscle force distribution in response to an *arbitrary* loading of the lumbar region. Such a loading could come about by performing weight-holding exercises using the upper extermities, or by directly loading the trunk in a controlled experiment. As long as the basic assumptions used in the process of calculating the *MAS* are not violated, the surfaces could be used as our best estimates of the momentary muscle force distribution. The graphic display of the muscle activity surface enables us to visually scan the

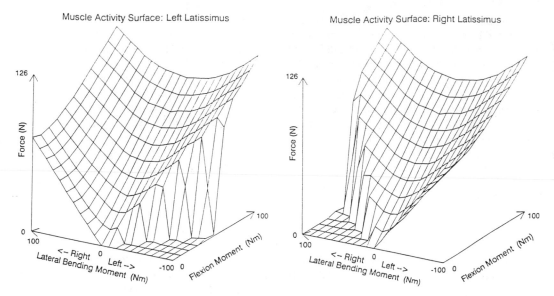

Figure 24.4: Muscle activity surfaces of the left *(a)* and right *(b)* latissimus dorsi.

predicted response of the muscle to different loading conditions. It can show whether the muscle is more sensitive to load increases in one direction or another and it can show if there are moment combinations that predict no activation of the muscle. The comparison of the *MAS* for different muscles can teach us about muscle recruitment schemes, i.e., moment combinations that require the activation of different muscle groups. The biomechanical model can therefore become a useful integral tool in the physiological study of the relationship between the external loading tasks and the lumbar muscular response.

The concept of *MAS* as a summarized map of muscular activity could be easily extended to handle arbitrary moment combinations. The most general loading condition in the erect posture involves a torsion moment in addition to the two bending moments. The loading space in this case is a three-dimensional (*3-D*) space, and therefore the *MAS* becomes a four-dimensional (*4-D*) surface. Such a surface can be stored as a *4-D* array in the computer memory, but could not be visualized in its entirety due to the limitations of the *3-D* geometric space we live in. The reduced order *MAS* can still be studied by observing the surface that is defined by combinations of two out of the three independent moment components. The above discussion addressed the *MAS* that is defined by the two bending moments, with no torsion. One can similarly study the *MAS* defined by a combination of one bending moment and the tor-

sion moment with a fixed value of the other bending moments. The geometric display of such surfaces may help explain the overall relationship between the muscular force and the external loading conditions.

24.4 Physiological Applications of the Lumbar Model

The precalculated values of the *MAS* of all the lumbar muscles enable us to study the response of the lumbar muscles to arbitrary loading conditions. One of the most intriguing results was the prediction of muscular switching curves. The role that the curves play in predicting the activation state of a given muscle, and the physiological tests that were conducted to test those predictions, will be described in the following sections. Finally, a *real-time* simulator of the effect of upper–limb loading on the lumbar muscle force distribution will be described.

24.4.1 Muscle Switching Curves

The *MAS* of the left latissimus muscle is shown in Fig. 24.4a. The axes of the loading plane are labeled according to the different bending moments that determine the muscle force distribution. The surface shows a monotonic increase in the direction of increasing the flexion bending moment, predicting that the muscle force will increase with an increase of the flexion bending moment. Its dependence on the lateral bending moment is more complex. It monotonically in-

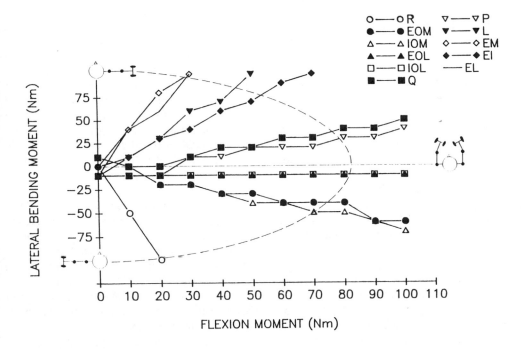

Figure 24.5: The switching curves of the right-side lumbar muscles: R-rectus abdominus, EOM-medial portion of external oblique, IOM-medial portion of internal oblique, EOL-lateral portion of external oblique, IOL-lateral portion of internal oblique, Q-quadratus, P-psoas, L-latissimus, EM-multifidus, EI-iliocostalis, EL-longissimus; activation direction is counterclockwise. The dashed line represents the exercise discussed in the text.

creases with the right lateral bending moment and it crosses the loading plane for some combinations of a left lateral and flexion bending moments. The *MAS* also has a local maximum for moment combinations that include the left lateral bending moment and a non-zero flexion moment. From the shape of the surface it can be seen that it has a local minimum along the line that represents no lateral bending moments (i.e. a value of lateral bending moment of zero). One can draw a practical conclusion from such an observation, namely that the optimal position for holding a weight in a fixed distance in front of the body (a task that creates a fixed flexion bending moment) is to hold it in both hands right in front of the body. Holding the weight in this fashion will produce no lateral bending moment, and therefore will minimize the latissimus muscle force.

The intersection of the *MAS* and the loading plane occurs along a curve that separates loading combinations that will not activate the muscle (on one side of the *MAS*) from those that will activate the muscle (on the other side of the curve). This curve was therefore named the *switching curve*. The switching curves of all the different muscles that comprise the lumbar region can be plotted on the same loading plane, since all the surfaces are defined by the same plane. Such a plot (seen in Figure 24.5) describes the activation pattern of the eleven muscles on the right side of the body in response to an arbitrary external bending moment loading. The muscles are activated clockwise: in response to a load held in the right hand while it is extended laterally to the right side, thereby exerting a pure right lateral bending moment, all the muscles on the right side are predicted to be inactive. As the right hand rotates forward, decreasing the lateral bending moment and increasing the flexion bending moment, the erector spinae group is first activated, the deep paravertebral group is activated next, as the load is moved in front of the body and held by the two hands (creating a symmetric loading condition by exerting only a flexion bending moment with no lateral bending component), the lateral portion of the oblique muscles is activated. As the load is moved to the left by the left hand, the left lateral bending moment increases while the flexion moment decreases, causing the activation of the medial portion of the obliques, and finally the activation of the right rectus abdominus. A similar plot was made for the muscles on the left side of the body. Because of the symmetry of the anatomic information used in the model, it is not surprising that the *MAS* and the switching curves of the muscles on the left side of the body are mirror images of the corresponding ones on right side of the body.

The model can thus be used as a predictive tool: a given external moment combination defines a point on the loading plane. Using Fig. 24.5 one can predict what muscles will be active and what muscles will not be active in order to oppose the external load. An exercise that creates such a loading combination could be performed, and the monitored *EMG* signal could be used as a validation test of the model. Ladin et al. (1989) performed such a study by monitoring the surface *EMG* activity of six muscles at the *L3* level, in the process of performing weight-holding tasks. Eight subjects were tested during the performance of six different tasks. The prediction success rate ranged from 60% in the worst case (for the ipsilateral erector spinae) to 100% (for the contralateral erector spinae). The laterality was defined in terms of the externally held weight: for a load held in the right hand, the left muscles were termed contralateral while the right muscles were the ipsilateral. The model predictions for all the muscles in all the subjects were correlated with the measured activity patterns, and a chi-square analysis showed that the results in one series of tests were significant at $p < 0.005$. Only one muscle in one series of tests showed a smaller confidence level ($p < 0.1$). The model was more successful in predicting the activity of the contralateral muscles (92–100% success rate) than those of the ipsilateral muscles (60–75% success rate).

The concept of the switching curves as describing a loading threshold needed to activate a specific muscle is supported by other studies:

Seroussi and Pope (1987) reported the existence of a threshold level of the lateral bending moment required to activate the centralized external oblique and that the threshold level is independent of the flexion moment. An examination of the switching curves of the lateral oblique muscles in Fig. 24.5 does show that the curves are parallel to the flexion moment axis, reflecting the independence of the threshold level and the flexion bending moment. Bean et al. (1988) described the activation of the left external and internal obliques achieved by a task that involved shifting a hand-held load by 30° to the right of the mid-sagittal plane. Mapping such a task on the loading plane and examining the switching curves of the appropriate muscles reveals indeed that such an exercise is expected to activate the lateral portions of the obliques, though the medial portion is not expected to be activated. Bean et al. (1988) further reported that a 90° rotation of the load (i.e. creating a pure lateral bending moment) will activate the ipsilateral obliques and the rectus abdominus. Our model predicts the activation of the ipsilateral rectus abdominus, but does not predict the activation of the obliques under those conditions.

Schultz et al. (1982) reported that pure flexion bending moments did not increase the activity of the rectus abdominus muscles above their rest levels. Such tasks are represented by points along the flexion moment axis, a line that lies entirely in the *inactive* region of the rectus abdominus, and therefore are predicted not to activate that muscle. The activity of the contralateral erector spinae muscles required to balance a pure lateral bending moment (e.g. the activity of the left erector spinae required to balance a right lateral bending moment created by a load held laterally by the right hand) has been reported by many researchers, including Floyd and Silver (1955), Jonsson (1970), Andersson et al. (1980), and Seroussi and Pope (1987). As the pure lateral bending moment axis is in the *active* region of the contralateral erector spinae, it is predicted by this model as well. Jonsson (1970) reported that such a task did not activate the ipsilateral erector spinae, a result that is predicted by this model, since the a pure lateral moment represents a task that is in the inactive region of the ipsilateral erector spinae.

In summary, the model created a framework that enabled to examine previously reported studies in a comprehensive manner. The model suggests that electromyographic activity patterns that have been reported in the literature for different tasks and by different researchers generally agree with the model predictions. Such a model could therefore be used as an aid in studying the predicted effects of external loading conditions on the distribution of muscle forces in the lumbar region. As the following section will illustrate, it can also be used to design loading exercises that are intended to activate a particular muscle or group of muscles.

24.4.2 The Lumbarator

The biomechanical model described above served as the basis for the development of a real-time simulation program that graphically describes the muscle force distribution of *all* the lumbar muscles in response to a given weight–holding exercise. The program is composed of two elements that represent the two stages of the simulation program: the exercise generator and the load distribution display. The screen display of the program (dubbed the Lumbarator) was implemented on a Silicon Graphics 4D system and is shown in Fig. 24.6. It is composed of two screens: The first one is the *exercise design screen* and is described in Figure 24.6a. The second screen displays the lumbar muscle force distribution and is shown in Figure 24.6b. Both screens serve as the user interface as well as the information display.

The exercise design screen shows three projections and a perspective image of a person holding weights and moving the upper extremities in a horizontal plane. The exercise design begins by specifying the weights held in both hands. An exercise is then prescribed for each arm by using the generic *angle/angle plane* in the center of the screen. The axes of this plane are the elbow flexion angle (the horizontal axis) and the shoulder internal rotation angle (the vertical axis). By constraining the movement of the upper limbs to a horizontal plane, a mapping between the angle/angle plane and the orientation of the upper limb was obtained. Therefore, the movement of the cursor on the angle/angle plane is mapped into a spatial trajectory that determines the motion of the left or right upper extremity. The motion is generated in real time in the central window of the screen. The specification of the loads and the trajectory of both arms complete the exercise design stage, and the program proceeds to display the second screen.

The second screen, shown in Fig. 24.6b, describes the prescribed exercise and the resulting load distribution. The person in the top left portion of the screen shows the actual exercise, the window at the bottom of the left part of the screen shows the loading plane and the movement of the loading point resulting from the prescribed exercise. The schematic figure in the right portion of the screen describes the *L3* cross section. It shows the vertebral body (the bottom of the figure is the direction of the back, or the posterior direction), and all 22 of the different muscles that cross that level. Each muscle is represented by a circle whose area is proportional to the cross sectional area of the muscle. The circles are centered around the corresponding locations of the centroids of the cross sectional areas. The table at the bottom represents all the different muscles and is used to plot the switching curve of a given muscle on the loading plane. As the figure starts exercising in the top left window, the moment loading combination resulting from the momentary posture is calculated and displayed on the loading plane. The loading point is used to intersect the 22 different muscle activity surfaces and determine the corresponding muscle force distribution. The predicted forces in the individual muscles are displayed *in real time* according to a color code in the lumbar cross-section window.

The Lumbarator can be used in two modes: as a teaching tool and as an aid in prescribing physical therapy exercises. The simulator can be used as a tool that describes the lumbar muscle force distribution in response to a given exercise. Using the momentary location of the loading point on the loading plane and the switching–curve for the muscle under study, an exact determination can be made as to the posture (for the given hand-held weights) that will cause the activation of that particular muscle. Such a mode of use can begin to develop an insight to the activation pattern of different muscles in response to specific loading conditions. The model could be used in an opposite mode as an exercise prescription tool: by specifying a goal of exercising a specific muscle by a given amount, an exercise that includes a set of weights and a sequence of different postures that achieves such a loading condition can be designed. The sequence can then be printed out and used as a guide for the patient during the in-home exercise.

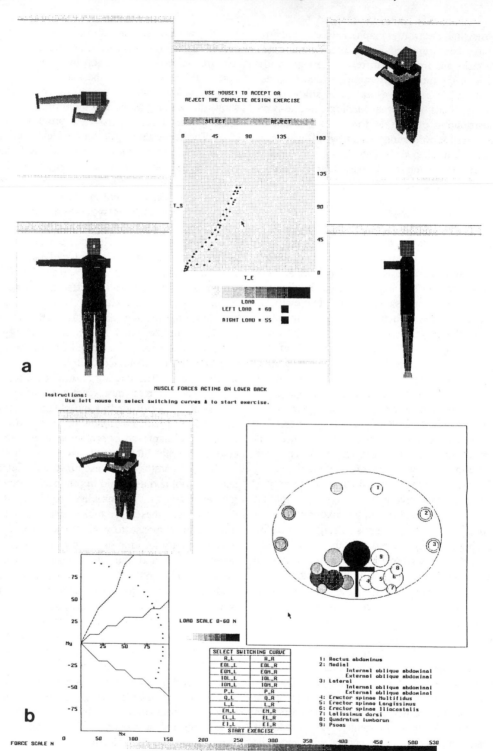

Figure 24.6: The screen layout of the Lumbarator: the exercise design screen *(a)* and the lumbar force distribution screen *(b)*.

24.5 Summary and Future Work

The biomechanical model presented in this chapter was designed as a practical tool that links a physical loading task and the resulting lumbar muscle force distribution in a consistent and comprehensive manner. An underlying question that is still unresolved concerns the physiological accuracy of the model predictions. The ultimate test of the model is clearly a study that will compare specific muscle force predictions to physical measurements of such forces. The limitations of current technology preclude such a validation scheme, leaving the *EMG* signals generated by the muscle as our only avenue for comparing muscle force predictions and activity levels. Comparisons between muscle force predictions and *EMG* measurements have been reported by many authors, and an extensive comparison of the assumed relationships can be found in Basmajian and De Luca (1985). There is broad agreement that such a relationship is monotonic, though its exact nature and the question of whether such a general relationship (independent of the particular muscle) exists is still an open research question.

The theoretical prediction of the switching curves as determining the activity state of a given muscle provided a new vehicle to test some of the basic predictions of the model. The study of the switching curves eliminates the need to assume a certain force–*EMG* relationship, and as such has the potential of being a general tool for the study of muscle activation, independent of the specific muscle under examination. As reproted by Ladin et al. (1989), the model has shown a good correlation between its predictions and the measured activity states of the monitored muscles in a group of healthy subjects. One of the most intriguing questions that needs to be addressed is the nature of the switching curves of the lumbar muscles in low back patients. Changes in the switching curves could suggest physiological variations that are correlated with low back pain. The long-range monitoring of the switching curves could also help in providing a quantitative tool to assess the efficacy of physical therapy exercises.

The usefulness of biomechanical models for the lumbar region was demonstrated by the Lumbarator program. Such a model is a prerequisite for the establishment of a link between a physical task that involves a given posture and loading, and the resulting muscle force distribution. As discussed in this chapter, the biomechanical model presented here is quite successful in providing a single framework that explains previously conducted biomechanical studies and the observed muscle activity patterns. A successful biomechanical model could therefore serve as an integral link in the diagnostic and therapeutic processes that take place in the context of treating low back pain.

Much work remains to be done before we get to the point of actually utilizing the biomechanical model as a fully integrated clinical tool. The model described in this chapter is clearly limited to the erect posture, and assumes that the moments arising from the muscular forces equilibrate the external moments. A change in the posture will clearly activate the ligaments and other passive tissues, which then contribute to the overall moments. The development of the constitutive equations for those tissues in a variety of postures is therefore necessary for extending the model to deal with those postures. Since the model equations are based on the cross-sectional anatomy of the lumbar region, obtaining accurate and reliable information that will enable extraction of geometric properties of the muscles is necessary if the model is to be integrated into the clinical decision–making process. Finally, further research is needed to ascertain the physiological assumptions on which the model is based. The model presented here assumes that the muscle force distribution minimizes the spinal compression. Whether this is really the case is still an open question, though the more relevant issue is finding a model that will be able to predict muscle activation patterns that will correlate well with measured patterns under a variety of physical tasks. Only when we reach that stage could we view the biomechanical model as a complete and helpful tool in clinical and physiological studies.

Acknowledgment

This research was supported by a grant from the Veterans Administration Research and Development Service of the Department of Medicine. My appreciation to Dr. Carlo J. De Luca for the helpful discussions leading to some of the ideas expressed in this work, and to Ramesh Murthy, Sujoy Guha, and Kristin Neff for their assistance on this study.

References

Andersson, G.B.J., Ortengren, R. and Schultz, A. (1980) Analysis and Measurement of the Loads on the Lumbar Spine During Work at a Table. *J. of Biomech.* **13**:513-520.

Andersson, G.B. (1981) Epidemiologic Aspects of Low-Back Pain in Industry. *Spine* **6**:53-60.

Aspden, R.M. (1989) The Spine as an Arch. *Spine* **14**: 266-274.

Basmajian, J.V. and De Luca, C.J. (1985) *Muscles Alive.* Williams and Wilkins, Baltimore.

Bean, J.C. and Chaffin, D.B. and Schultz, A.B. (1988) Biomechancal Model Calculation of Muscle Contraction Forces: A Double Linear Programming Method. *J. of Biomech.* **21**: 59-66.

Bejjani, F.J., Gross, C.M. and Pugh, J.W. (1984) Model for Static Lifting: Relationship of Loads on the Spine and the Knee. *J. of Biomech.* **17**:281-286.

Bendix, T., Krohn, L. and Flemming J. (1985) Trunk Posture and Trapezius Muscle Load While Working in Standing, Supported-Standing, and Sitting Positions. *Spine* **10(5)**:433-439.

Benn, R.T. and Wood, P.H. (1975) Pain the Back; An Attempt to Estimate the Size of the Problem, *Rheumatol. Rehabil.* **14**:121.

Bogduk, N. (1984) The Applied Anatomy of the Thoracolumbar Fascia. *Spine* **9**:164-170.

Cats-Baril, W.L., Tessaglia, D.L. and Haugh, L.D. (1987) Low Back Pain Disability: Comparing Physicians' Predictions to Expert and Empirical Models. *Proc. of 10th Ann. Conf., RESNA*: 296-298.

Cats-Baril, W. and Roth, W.A. (1985) System Dynamics Model to Determine the Effect of Research, Education and Industrial Change on Societal Costs Due to Low Back Pain. *13th Intern. Conf. on Modeling and Simulation,* June, Switzerland.

Deyo, R. (1983) Conservative Treatment for Low Back Pain. *JAMA* **250(8)**:1057-1061.

Farfan, H.F. (1973) *Mechanical Disorders of the Low Back.* Lea and Febiger, Philadelphia.

Floyd, W.F. and Silver, P.H.S. (1955) The Function of the Erectores Spinae Muscles in Certain Movements and Postures in Man. *J. of Physiol.* **129**:184-203.

Freivalds, A.V., Chaffin, D.B. and Garg, A. and Lee, K.S. (1984) A Dynamic Biomechanical Evaluation of Lifting Maximum Acceptable Loads. *J. of Biomech.* **17**:251-262.

Frymoyer, J. W. and Cats-Baril, W. (1987) Predictors of Low Back Pain Disability. *Clin. Orthop. and Related Res.* **221**:89-98.

Garg, A. and Herrin, G.D. (1979) Stoop or Squat: A Biomechanical and Metabolic Evaluation, *AIIE Trans.* **11(4)**:293-302.

Gracovetsky, S. (1985) An Hypothesis for the Role of the Spine in Human Locomotion: A Challenge to Current Thinking. *J. of Biomed. Engin.* **7**:205-216.

Gracovetsky, S. (1986) Determination of Safe Load, *British J. of Indust. Med.* **43(2)**:120-133.

Hickey, D. and Huckey, D. (1980) Relation Between the Structure of the Annulus Fibrosus and the Fracture and Failure of the Intervertebral Disc. *Spine* **5**:106-116.

Hutton, W. C. and Adams, M. A. (1982) Can the Lumbar Spine Be Crushed in Heavy Lifting? *Spine* **7(6)**:586-590.

Jonsson, B. (1970) The Functions of Individual Muscles in the Lumbar Part of the Spinae Muscle, *Electromyography* **1**:5-21.

Kelsey, J.L. and White, A.A. III, Pastides, H. and Bisbee, G.E. Jr. (1979) The Impact of Musculoskeletal Disorders on the Population of the United States. *J. of Bone and Joint Surgery* **61-A**:959-963.

Kelsey, J. L. and White A. A. III (1980) Epidemiology and Impact of Low-Back Pain. *Spine* **5**:133-148.

Kippers, V. and Parker, A.W. (1989) Validation of Single-Segment and Three Segment Spinal Models Used to Represent Lumbar Flexion. *J. of Biomech.* **22**:67-75.

Kromodihardjo, S. and Mital, A. (1987) Biomechanical Analysis of Manual Lifting Tasks, *J. of Biomech. Engineering* **109**:132-138.

Ladin, Z., Murthy, K.R. and De Luca, C.J. (1989) Mechanical Recruitment of Low-Back Muscles - Theoretical Predictions and Experimental Validation. *Spine* **14**:927-938.

Lindbeck, L. (1987) Analysis of the Asymmetrically Loaded Spine by Means of a Continuum Beam Model. *J. of Biomech.* **20(8)**:753-765.

McGill, S.M., Patt, N. and Norman, R.W. (1988) Measurement of the Trunk Musculature of Active Males Using CT Scan Radiography: Implications for Force and Moment Generating Capacity About the L4/L5 Joint. *J. of Biomech.* **21(4)**:329-341.

Moffett, J.A., Klaber, Chase, S.M., Portek, I. and Ennis, J.R. (1986) A Controlled, Prospoective Study to Evaluate the Effectiveness of a Back School in the Relief of Chronic Low Back Pain. *Spine* **11(2)**:12-122.

Morris, J.M., Lucas, D.B. and Bressler, B. (1961) Role of the Trunk in the Stability of the Spine, *J. of Bone and Joint Surgery* **43-A(3)**:327-351.

Murthy, K.R. (1990) External Loading and Its Effect on Muscle Force Distribution in the Lower Back. Boston University, Biomedical Engineering Department, *M.JSc. Thesis,* January.

Nachemson, A. (1983) Work for All. *Clinical Orthopaedics and Related Research* **179**: 77-85.

Nordby, E. J. (1981) Epidemiology and Diagnosis in Low Back Injury. *Occup. Health and Safety:* 38-42.

Ortengren, R. and Andersson, G. B. J. (1977) Electromyographic Studies of Trunk Muscles, With

Special Reference to the Functional Anatomy of the Lumbar Spine, *Spine* **2**:44-51.

Panagiotacopulos, N.D. and Pope, M.H., Krag, M.H. and Bloch, R.A. (1987) Mechanical Model for the Human Intervertebral Disc. *J. of Biomech.* **20**:839-850.

Pheasant, H.C. (1977) The Problem Back. *Curr. Pract. Orthop. Surg.* **7**:89.

Pope, M.H., Andersson, G.B.J., Broman, H., Svensson, M. and Zetterberg, C. (1986) Electromyographic Studies of the Lumbar Trunk Musculature During the Development of Axial Torques. *J. of Orthop. Res.* **4**:288-297.

Schultz, A., Andersson, G.B.J., Ortengren, R., Bjork, R. and Nordin, M. (1982) Analysis and Quantitative Myoelectric Measurements of Loads on the Lumbar Spine when Holding Weights in Standing Postures. *Spine* **7**:390-397.

Schultz, A., Haderspeck, K., Warwick, D. and Portillo, D. (1983) Use of Lumbar Trunk Muscles in Isometric Performance of Mechanically Complex Standing Tasks. *J. of Orthop. Res.* **1**:77-91.

Seroussi, R.E. and Pope, M.H. (1987) The Relationship Between Trunk Muscle Electromyography and Lifting Moments in the Sagittal and Frontal Planes. *J. of Biomech.* **20**:135-146.

Shirazi-Adl, A. and Drouin, G. (1987) Load-Bearing Role of Facets in a Lumbar Segment Under Sagittal Plane Loadings. *J. of Biomech.* **20(6)**:601-613.

Snell, R.S. (1973) *Clinical Anatomy for Medical Students*. Little, Brown and Company, Boston.

Wiesel, S.W., Leffer, H.L. and Rothman, R.H. (1984) Industrial Low-Back Pain - A Prospective Evaluation of a Standardized Diagnostic and Treatment Protocol. *Spine* **9**:199-203.

Yettram, A. L. and Jackman, M. J. (1980) Equilibrium Analysis for the Forces in the Human Spinal Column and Its Musculature. *Spine* **5**:402-411.

CHAPTER 25

Musculoskeletal Function of the Spine

Serge Gracovetsky

25.1 The Problem of Low Back Pain

Low back pain is an enormous problem in North America today. It is the most common disability in persons under the age of 45; in those over 45 it is third only after arthritis and heart disease.

The impact of spinal dysfunction both on the lives of individuals affected and on the health care system motivated the Quebec Workers' Compensation Board to set up a task force headed by Dr. W. Spitzer (1987) to review the state of available treatment procedures. The task force considered and classified the work contained in over 7,000 publications and as such the report produced by the Task Force is an invaluable source of information. One important conclusion reached by the task force is that despite the intense research activity of the past 50 years, low back pain is still an unknown phenomenon, requiring further research.

How can we hope to restore normalcy to the injured and design efficient rehabilitation protocols if we do not know what is the function to be restored? It seems that the study of the spine either clinically or via in vitro experimentation is conducted without first answering a basic question: What is the function of the spine? A purely in vivo experimental investigation into that question is unacceptable for obvious ethical reasons. One is left mainly with a combination of in vitro experimental analysis supplemented by mathematical modeling and simulation. To structure a mathematical formulation requires a detailed knowledge of the anatomy and some hypotheses as to the probable function of the vertebrate spine.

25.1.1 The Optimum Human Spine

The theory of evolution and natural selection implies the survival of the fittest and the elimination of the weak from the pool of genetic information available for the continuity of a species. What is the meaning of "fit for survival," and what ranking system decides that the weak are, indeed, weaker than some ideal norm? One can start with the hypothesis that survival means that the animal will not self-annihilate. This implies that, at any given time, the level of mechanical stress within the musculoskeletal system cannot exceed some ultimate value. Hence, regardless of the task being accomplished, the central nervous system will activate appropriate muscles to prevent such an event. Since the various components of the musculoskeletal system have different mechanical characteristics, it is not unreasonable to speculate that the level of stress in all its components during the execution of a task will be proportional to the ultimate limit of each individual component. For example, if, for an arbitrary task, the stress within the bone reaches 2/3 of its ultimate, then the stress within the ligaments will also be 2/3 of its own ultimate.

This idea has a number of implications. As each vertebra is composed of the same material, the stress within each should therefore be equal. This is another way of saying that an efficient animal will execute a task in such a way that the stress within the spine is equalized at its lowest level.

But what is the task for which the spine has been designed? Although there is no way of knowing for sure what that particular task might be, we propose that the most important activity for mem-

Multiple Muscle Systems: Biomechanics and Movement Organization
J.M. Winters and S.L-Y. Woo (eds.), © 1990 Springer-Verlag, New York

bers of the vertebrate species is locomotion. It is therefore appropriate to examine the stages of development of our ancestors' locomotive ability. In so doing, the spine and its surrounding tissues emerge as the pervasive element, the primary engine of locomotion in animals such as ourselves.

It is suggested that the spine of a fish and its surrounding tissues represent the primary engine which the animal uses for locomotion. To this day, gait analysis is essentially the analysis of the motion of the legs. The legs are certainly useful, but are they essential? The answer is definitely no. Human bipedal gait can be demonstrated not to require the presence of any extremities. In retrospect, it was evident that the primary function of the spine, so obvious in the fish, was never transferred to any of our extremities during the long evolutionary journey.

Locomotion is but one of the many tasks that the human spinal engine is asked to perform. In many ways, tasks such as weight-lifting require the use of the spinal engine for applications for which it has not been optimally designed. The analysis of this wide variety of possible tasks is beyond the intent of this chapter. We will therefore restrict ourselves to the study of flexion in the sagittal plane and locomotion.

25.2 Pathology and the Mechanical Etiology of Most Spinal Injuries

The first step in understanding the function of the spine is to appreciate the nature of the forces which it has to support. The study of the pathology of the intervertebral joint generates facts on the natural outcome of these forces. Since the early 1900s, a considerable number of distinguished researchers have attempted to reproduce in the laboratory the damages on the intervertebral joint noted by pathologists by subjecting the spine to a variety of excessive forces. At least three types of loading appear to be responsible for serious damage when combined in specific ways. These are axial compression, axial torque, and lateral bending. However, lateral bending may also induce an axial rotation because of the coupled motion of the spine. Hence, it could be argued that there are really only two basic loading configurations responsible for a large number of pathological observations: axial compression and torsion. In this regard it is interesting to note that Hirsch and Schajowicz (1953) reported two dis-

tinct early injuries in the annulus as well as at the central portion of the disc.

The type of injury is determined by the actual distribution of stress on the components of the lumbar spine. Thus, one of the key factors in determining the effect of the stress is the geometric arrangement of the individual spinal segments. From the geometric features seen in simple radiographs, it is sometimes possible to locate the particular joint that is at risk. This is particularly helpful either when the radiographs of the joints show no injury or where other indications point to multiple possibilities. In the lumbar spine, 95% of all injuries occur at the *L4/5* and *L5/S1* levels. Of the remaining 5%, problems at *L3/4* appear to be far more common than at the other two levels (Farfan and Kirkaldy-Willis, 1981). The shape and size of the disc, as well as its location relative to the pelvis, have considerable influence on both the type of injury induced and its consequences for the anatomy and the function of the joint. Disc problems cannot be amalgamated into a single entity.

Farfan (1973) proposed that the existing data on pathology could be explained by the existence of two basic sequences of pathological processes resulting from the application of excessive compression and excessive torsion to the intervertebral joint. Each sequence behaves in its own clinical manner and responds to different types of treatment.

Sequence 1: Compression Injury: This injury results in central damage to the disc with an end-plate fracture of varying magnitude, followed by an ingrowth of vascular granulation tissue through the fracture into the disc nucleus. As a result, the avascular nucleus and inner annulus are gradually destroyed. In the early stages, the facet joints are not greatly affected. At a later stage, with continued loss of volume of the nucleus, the height of the disc is reduced and the facet joints dislocate. Arthritic changes appear but are rarely severe. The outer annulus survives and is gradually pushed out from between the end-plates.

External violence may cause intervertebral joint injury and subsequent degenerative change. This is generally accompanied by obvious damage to the bone. There are, however, special cases involving certain types of accidental falls on the backside. In these instances, high axial load is applied when neither the ligamentous nor the muscular system is set to restrain the motion; the result is an explosive rupture of the disc.

Sequence 2: Torsion Injury: This causes peripheral damage to both the disc and facet joints. The annulus is avulsed from the end-plate, and its laminae separate. However, the central portion of the disc and end-plate remain intact. At a later stage, the annulus develops radial fissures, while the nucleus remains relatively untouched. Because the changes in the facet joints are severe, the intervertebral joints may become unstable.

The reluctance to consider torsion as an important element in the etiology of spinal disorders is based upon the interpretation of many in vitro experiments such as those of Adams et al. (1982). These experiments involving bending (such as hyperflexion) did produce annular damage and were believed to represent a plausible explanation for disc prolapse. There are, however, a number of reasons to be cautious. First, there is no clear-cut relationship between the data collected and an *in vivo* situation, and second, the initial conditions of the discs are unknown. For example, it is conceivable that discs already damaged by torsion were included in the sample. Hence, the prolapse subsequently observed may have been due to the combination of the preceding injury and the hyperflexion. Furthermore, the level of facet asymmetry was not noted, even though it is bound to have an effect as it may result in a net torque being applied to the annulus. Thus it could be argued that Adams et al. observed annular failure from a peculiar combination of compression and torsion.

The simple distinction made between injuries resulting from the application of an excessive axial compression and/or torsion is attractive but certainly incomplete; it is also far from being universally accepted. Yet, the usefulness of the concept can be appreciated in the consideration of two important but basic types of spinal motion: *1)* flexion in the sagittal plane, in which the main type of forces is axial compression; and *2)* locomotion, in which the main type of forces is axial torsion and compression.

25.3 The Role of the Spinal Musculature in Flexion

The clarification of the role of muscles in flexion, and the resulting axial compression applied to the spine, are largely the result of mathematical simulation with appropriate *in vivo* validation experiments.

One would expect that the study of the anatomy would precede the development of mathematical models. Regrettably, this has not been the case.

Although mathematical modeling was a highly popular exercise in the 1960s and 1970s, these approaches were invariably based on the description of the ill-defined back musculature found in Gray's Anatomy, which dates from the beginning of the century. Surprisingly, few individuals challenged these fuzzy descriptions, and it was not until 1984 that the correct anatomical description of the back musculature was finally established by Bogduk and Macintosh (1984). For the last two decades researchers have performed an impressive number of computer simulations using models based upon an erroneous anatomical representation of the spine which included the following two basic mistakes: *1)* The back musculature was misrepresented as an imprecise mass called "erectores spinae" with few hard facts as to the layout of muscular fibers. *2)* The role of the posterior ligamentous system (*PLS*), especially that of the thick layers of collagen fibers connecting the ilium to the spinous processes, was nearly totally overlooked.

The consequence of such errors can be appreciated when considering the role of the spinal musculature as the spine flexes. During the last 50 years, it has been frequently argued in the literature that the back muscles are responsible for lifting loads in the sagittal plane. In 1955, Floyd and Silver studied a series of sagittal lifts performed by 150 individuals (Figure 25.1). To their surprise, they noted that the electromyographic activity of the erectores spinae decreased at the time when they are most needed, that is, when the load is just lifted off the ground. This muscle relaxation phenomenon presented quite a challenge to the school which promotes the role of the lumbar musculature in lifting. To resolve the difficulty, Bartelink (1957) proposed that the internal abdominal pressure would rise and supplement the action of the back muscles (Figure 25.2). Theoretical and experimental considerations refuted this suggestion (Gracovetsky, 1977, 1981, 1988). There is no doubt that the maximum moment which those muscles are able to generate at *L5/S1* is about 250 Nm. This is sufficient to handle 50 kg or so but not much more (Figure 25.3). Hence, the question: If the erectores are too small to lift significant loads and do not fire at the very moment they are needed most, then where does the necessary lifting force come from?

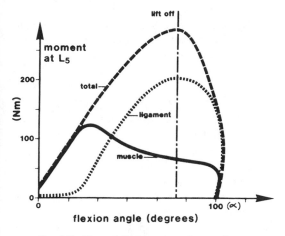

Contribution of ligament and muscle

Figure 25.1: The *EMG* data of the erectores spinae demonstrate an almost total relaxation at the time the volunteer lifts a load from the floor. This phenomenon, first reported by Floyd and Silver (1955), is difficult to explain if one believes that the low back muscles generate the moment necessary to balance the load on the spine.

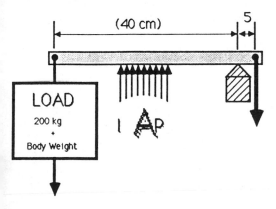

Figure 25.2: The concept of internal abdominal pressure (*IAP*) was introduced to explain lifts in excess of 50 *kg*. Indeed, there are simply not enough back muscle mass to balance external loads in excess of 50 *kg*. The *IAP* is presumably pushing up the diaphragm which assists the erectores in their action. The difficulty with this concept is that the *IAP* will have to rise well above any physiological limit to be of any significant assistance to the muscles. This concept must therefore be abandoned.

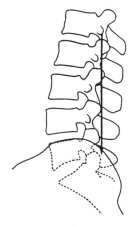

Figure 25.3: The anatomical arrangement of longissimus and iliocostalis lumborum is such that none of them are well placed to exert the maximum moment to balance loads in sagittal plane movement.

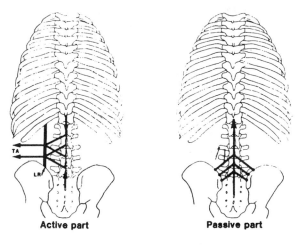

Figure 25.4: Bogduk's dissections demonstrated the corrections in the mathematical prediction of two basic anatomical structures which form the thick lumbodorsal fascia (*LDF*) linking the ilium to the spinous processes: *a*) a network of deep collagen fibers connecting the ilium to the tips of the spinous processes. This is the passive part of the *LDF* because it requires a change in spinal geometry to be tightened; *b*) a superficial layer of collagen fibers connecting the lateral raphe of the fascia to the transversus abdominis. When the transversus contracts, the collagen geometry forces the spinous processes to draw closer, thereby inducing a moment supporting the activity of the erectores. This moment is not significant (Bogduk and Twomey, 1987). The whole assembly integrates neatly into a network of muscles and ligament fibers. Proper load transmission from shoulders to ground requires that all parts of the fibers (muscles and ligaments) be tightened. In particular, the passive part of the *LDF* must be tightened. This can be achieved through lordosis changes. The control of lordosis is of fundamental importance both in lifting and walking, and it is suggested that this is the primary responsibility of the erectores spinae and psoas muscle.

Gracovetsky (1977) suggested an answer for this question. Briefly, the force necessary to perform a lift is generated by the hip extensors, which are attached to the ilium. This force must then be transmitted from the ilium to the upper extremities where it is required (Figure 25.4). It was hypothesized that the posterior ligamentous system (*PLS*) comprising the ligamentum flavum, the capsular ligaments of the facets, and the thick lumbodorsal fascia is the main structure channeling these forces. Anatomical studies (Bogduk and McIntosh, 1984; Bogduk and Twomey, 1987) showed that the lumbodorsal fascia portion of the *PLS* is comprised of two major sub-structures which must share the transmission of forces:

1. The upper sub-structure, comprised of the latissimus dorsi and other thoracic muscles, channels the force from the tips of the spinous processes to the shoulders and arms. Because it is a muscular component, it can be activated regardless of posture.

2. The lower sub-structure, comprised of the back muscles and the *PLS*, links the ilium to the tips of the spinous processes. It is crucial to determine the division of the force between the back muscles and the *PLS*. Mathematical modeling of the problem using appropriate anatomy indicates that the back muscles can transmit about 1/5 of the force: the remaining 4/5 of the force is transmitted by the *PLS*.

It is therefore apparent that the *PLS* plays a critical role in transmitting the forces generated by the hip extensors from the ilium to the tips of the spinous processes. To do this, the *PLS* must be taut. Since it is made of passive collagen fibers, the *PLS* cannot be tightened until the lordosis is sufficiently reduced. We suggest that the control of lordosis is precisely the role of the lumbar musculature, and that lordosis is a vitally important parameter describing the ability of the spine to support loads (examined in more detail later). Indirect support comes from study of Triano et al. (1987), where *EMG* activity of the erectores was less for normals than the injured. This demonstrates that the normal must use a nonmuscular component in the lifting process, and that this nonmuscular component may be what is disabled by injury.

The suggestion that the *PLS* is the main component in the transmission of force was greeted with considerable skepticism. To clarify the role of the *PLS* requires a review of the relationship between the musculature and the other soft tissues of the spine, especially the ligaments.

25.3.1 The Nonmuscular Soft Tissues of the Spine

Muscles are attached to bone by strong collagen fibers called tendons (or, occasionally, ligaments). Bone can be attached to bone by purely ligamentous tissues. In general, ligaments do not restrict movement unless the range of motion exceeds physiological limits. Hence, the popular representations of the muscle/ligament relationship, whereby muscles will contract to move bones, and the whole movement will be maintained within limits by the ligaments [e.g. Chapter 8 (Zajac and Winters)]. This representation is certainly true for the cruciate ligaments of the knee. Is it also true for the spinal motion? If it were, then there would be less need to account for the presence of ligamentous tissue in the mathematical modeling of spinal function until near the limits of the physiological range.

Virtually all mathematical models of the spine and its surrounding soft tissues use some form of optimization technique to overcome the basic problem of indetermination created by the excess of control parameters (muscles) over constraints; for the spine, this has been static optimization [reviewed in Chapter 8 (Zajac and Winters) and Chapter 23 (Andersson and Winters)]. The performance criteria used by the optimization algorithm have been expressed in the form of minimizing some functional, such as muscular activity, energy expenditure, compressive stress at *L3*, etc. An example of such modeling is given at the end of this chapter in which the calculated muscle firing sequence results in the minimization and equalization of stress throughout the lumbar spine.

Although the mathematical solution representing muscle activity depends upon the specific criteria used, it must be realized that this type of approach has a fundamental limitation independent of the criteria chosen. That is, the model of a purely muscle driven spine generates a muscle firing sequence which is unlikely account for the muscle relaxation phenomenon noted by Floyd and Silver (1955). Furthermore, a matrix relating the activity of the muscles to the stress on the joint was found to be positive semi-definite (Gracovetsky, 1988). It is known that such a matrix is also representative of the energy in the system, and that a quadratic form with a positive semi-definite matrix describes a system with an infinite number of states that have the same minimum energy associated with them. This type of system is unstable, for it can change between these minimum energy states. Such behavior would be clearly nonphysiological. To stabilize the system, the matrix in the quadratic form describing the energy in the system must be positive definite.

To convert a positive semi-definite matrix into a positive-definite matrix, it is sufficient to introduce a mathematical term in each diagonal element. With this correction, the muscle relaxation phenomenon is now accounted for. However, the model is supposed to be a faithful representation of the anatomy. The need to correct the model suggests that a major structural component of the spine has been overlooked in the anatomical dissections; the more recent work of Bogduk's group (Bogduk and Twomey, 1987) identified the previously neglected *PLS* as the "missing" structural component predicted by the mathematical model.

It can now be appreciated that if the *PLS* cannot be correctly tightened, the available moment at *L5* will have to be supplied mostly by the erectores. This will severely limit the ability to lift. Furthermore, the lever arm of the erectores is approximately 50% shorter than those of the *PLS*, since the muscles are attached to the transverse processes and not the spinous processes of the vertebrae. Therefore, using these muscles rather than the *PLS* will always result in a higher compression stress on the disk.

An injury to the disk annulus or the capsular ligaments of the facet will restrict the motion of the joint. Consequently, the erectores cannot adjust the correct spinal lordosis and the *PLS* remains slack. There has been a reluctance to relinquish the tradition of assigning the erectores the major role in sagittal lift. For instance, McGill et al. (1986, 1988) propose a model based on an acceptable representation of the anatomy, including ligamentous tissue. Yet their conclusion was that the *PLS* could not possibly intervene during the lift and therefore the erectores still represented the

major contributors to the lift. How can this be explained? McGill et al. did not recognize the importance of the passive part of the lumbodorsal fascia, which was assigned the role of limiter of range of motion, and as such was implicitly removed from the equations. Specifically, the elimination of the ligamentous contribution is based entirely on the idea that the ligaments remain slack at up to 14° of flexion of the *L4/5* joint; based on data such as Adams et al. (1980), who measured an average flexion angle at *L4/5* of 10 ° or so, it is suggested that the value of 14° is too extreme. It is thus not surprising that McGill et al. did not find any ligament contributions within the normal spinal range of motion. Introducing the value of the 10 deg in McGill et al.'s equations will result in a reversal of the conclusion: the *PLS* becomes now the major force transmission structure (McGill 1986, p. 142).

It is noteworthy that McGill et al. (1986) did not attempt to test their model for the type of lifts which are common among weight lifters. Such tests can be expected to give results similar to those of Granhed et al. (1987), who analyzed the performance of lifters in the 300 Kg lift range, using an equivalent muscle model proposed by Schultz et al. (1982). He found that the calculated disc compression (36,000 N) was three times higher than the ultimate (12,000 N) determined experimentally by Adams et al. (1980)! This high compression, if truly representative of the *in vivo* situation, would be severely disabling. It is entirely based on the premise that the back muscles are responsible for lifting. In the opinion of the author, this school of thought cannot explain simple observations, no longer helps clarify the function of the spine, and thus should be abandoned.

As the calculations unfolded, another surprising feature of the spine emerged: the spine and its surrounding soft tissues behaved like a gigantic muscle. A proper description of this phenomenon requires a reexamination of the notion of muscle.

25.3.2 The Synthesis of Muscular Response

A muscle is usually defined by its material composition as an ensemble of individual fibers whose length can be changed within certain physiological limits. This feature is described by an equation relating the force applied to the muscular fiber to the resulting elongation (Figure 25.5a). Defining a muscle by its physical appearance is convenient but restrictive.

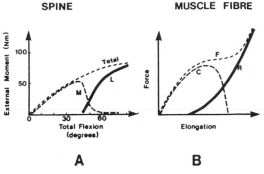

A   **B**

Figure 25.5: *a)* Distribution of moments between muscles (*M*) and ligaments (*L*) as a function of the angle of forward flexion; *b)* response of an individual muscle fiber to elongation. *F* is the total tension force, *C* the muscle contractile force, *R* the elastic resistance (Frost 1973). The overall response of the spine during forward flexion has a pattern similar to that of a single muscle "fiber." This may not be coincidental; the spine and its surrounding soft tissues behave like a very large muscle.

We propose enlarging this definition by introducing the concept of 'mega muscle.' A mega muscle is defined to be anything which will be described by the same force/elongation relationship as a muscle fiber. Hence, a mega muscle is defined by its function, and not by the way it looks. Note that this definition certainly includes what are generally termed muscles. The key point is that it also includes any complex arrangement of components (not necessarily made of muscular tissue), which exhibit the same force/elongation transfer function.

The reason for enlarging the definition of a muscle is due to its physical constraints in bulk and size. To eliminate these physical constraints and still keep the same behavior we must use components other than muscle fibers.

25.3.3 The Spine as a Mega Muscle

It is remarkable that the calculated response of the spine, for sagittal lifts, exhibits the same characteristic response as that of a single muscle fiber (Fig. 25.5b; Gracovetsky, 1988). Therefore, according to our previous definition, the complex system represented by the spine and its soft tissues must be called a mega muscle. The freedom in the

design of mega muscles can be appreciated when considering the need to house the abdominal content and other vital organs. It seems difficult to imagine how a regular biceps could function with a "hole" in its center. Hence, the spine, as a mega muscle, represents a feasible solution to the problem of synthesizing the muscle force/elongation transfer function with a hole in its center (e.g. the abdominal cavity). There is no obvious alternative: if the moment at *L5* necessary to lift heavy loads were to be generated by the low back musculature, then these muscles will have to be enlarged from their current size to almost fill out the abdominal cavity.

The study of the function of the spine in flexion/extension revealed that the most important type of stress on the intervertebral joint is of a compressive nature. Accordingly, a flurry of *in vitro* experiments tried to reproduce the pathology of an injured joint by subjecting intact specimens to excessive compressive stress. A good correlation was expected from such simple experiments. Unfortunately, the data stated otherwise. Virgin (1951) subjected intervertebral joints to excessive compression *in vitro* and demonstrated that the annulus fibrosus is rarely damaged by compression. Numerous other researchers have repeated this experiment with the same results. The calculations made by Shirazi–Adl et al. (1984) suggest that when the end-plate ruptures, the collagen fibers of the annulus fibrosus are only stretched by about 4%, well below their estimated limit of 20–30%. In short, the annulus cannot be damaged easily by compression and therefore disc prolapse cannot be a consequence of the application of pure compressive forces to the joint.

Pathology showed that axial torsion is a source of crippling spinal injuries. This in itself is paradoxical: why some mechanism did not evolve to prevent the spine from being injured by torsion? There had to be some fundamental advantage in exposing the spine to such potentially damaging types of forces. This problem was resolved when we came to understand that the spine was not a passive structure, but rather an engine capable of driving the pelvis that has adapted to our bipedal mode of locomotion. Torsion thus emerged as perhaps the most important motion of the spine.

However, a problem remained. If torsion was indeed necessary to ensure pelvic rotation, how could any axial torque be generated from the more or less longitudinal pull of the erectores? It is conceivable that the obliques could contribute to axial rotation during locomotion; the problem is that they do not (Basmadjian, 1979). Somehow the spine converts the primitive lateral bend of our fish-like ancestors into an axial torque. How this is done was discovered by Lovett (1903) and is at the very heart of the spinal engine theory.

25.4 The Role of the Spinal Musculature in Locomotion: The Theory of the Spinal Engine

The theory of the spinal engine is an attempt to put under a single conceptual roof the many apparent paradoxes and experimental contradictions which have surfaced over the years (Gracovetsky, 1988). The theory states that the locomotive function of the spine, so obvious in the fish, was never transferred to any extremities. In other words, the spine is not a supporting column passively transported by the legs but rather the primary engine driving the pelvis: the legs follow and amplify the movement. This theory represents a radical departure from the current representations of human gait, which assigns a predominant role to the lower extremities, even thought it is realized that fully 2/3 of our body mass is located above the legs. Conventional representation of gait does not assign any particular important role to this mass, which is believed to be passively carried away at a significant energy cost. Such a simplified model may explain the considerable difficulties encountered by many in the study of pathological gait.

To be credible, a theory of human bipedal locomotion cannot be limited to man. It has to explain the features of other vertebrates in the context of evolution.

25.4.1 Evolution of the Spine

The main steps in the evolutionary sequence hypothesized here to lead to our present day anatomy are are outlined in Gracovetsky (1988). Mechanically, the type of movement of interest is illustrated by movement in the lizard, in which the locomotive process demands the alternate lifting of the legs to clear obstacles as the animal advances. Hence, at each step, the spine is bent laterally to one side, while the shoulder of the other side is lifted. This motion applies an axial rotation to a spine already flexed laterally. What happens next is a straight-forward problem in physics: Any homogeneous rod bent in the horizontal plane, while an axial torque is applied, induces another motion off the horizontal plane. This phenomenon has been known for a long time (Lovett, 1903). It corresponds to the so called coupled motion of the spine which will be discussed elsewhere. Hence, axial rotation in the presence of a lateral bend induces a motion of flexion/extension in the sagittal plane.

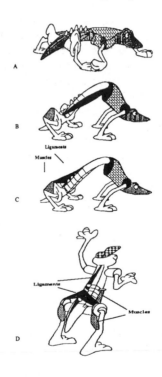

Figure 25.6: *a)* Motion by lateral bend requires the alternate firing of paraxial muscles. The power available for locomotion is half the installed power; *b)* Motion by flexion (gallop) requires the firing of both paraxial muscles. The power available is double; *c)* Flexion can be controlled by muscles placed outside the pelvis/shoulder complex. Hence, the growth of the power source (hip extensors) is no longer hindered by the abdominal content, provided this power is returned to the spinal engine (*PLS*); *d)* It was then a matter of time before the animal could develop enough hip extension power to lift himself off the ground in the erect stance.

The induction of flexion in the sagittal plane opened up new avenues in the design of the animal (Figure 25.6). Flexion/extension permitted a transition from the slow, low-powered crawl of the lizard, to the rapid, high-powered gallop of the cheetah. The muscular power source had to switch from the alternate firing of the left and right erectores to the simultaneous firing of all the erectores. The increased demand that a gallop puts on muscular power could not be met by the space constrained erectores; this problem was resolved by developing muscular power sources (hip extensors) outside of the abdominal cavity, together with the means of returning this power to the spine (*PLS*). Without the constraint in size and volume imposed by the need to house the abdominal content, the hip extensors increased in size and power. It was therefore inevitable that the corresponding increase in moment in the sagittal plane generated by these muscles would permit the animal to switch abruptly from 4 feet to 2.

From this point on, bipedal locomotion becomes a problem of axially rotating the pelvis using a muscular mass (erectores spinae and hip extensors), which is essentially parallel to the axis of rotation. As described below, the solution to this problem requires the existence of a gravitational field.

25.4.2 Rotating the Pelvis

Axial rotation of the pelvis cannot efficiently be accomplished by direct action of the hip extensors. An indirect route must be taken. The sequence of steps necessary to rotate the pelvis is as follows:

1. The hip extensors lift the body as the spine extends. This is particularly obvious in a runner as both feet leave the ground. Muscular energy is now stored in potential form as the center of gravity rises in the earth's gravitational field.

2. As the center of gravity rises and falls, the effect of the changing vertical velocity is to reduce the load on the spine. To prepare for step 3, the spine must be positioned for impact. Appropriate corrections in lordosis and lateral bending postures are made by the trunk musculature, and in particular iliocostalis, longissimus and psoas. These muscles, once again, act as controllers of the spinal curvature, and need not have the power to locomote the subject.

3. When the heel strikes, the trunk deceleration converts the potential energy stored in the gravitational field into kinetic form. The spine, already positioned during step 2, is further bent laterally with force, in the presence of lordosis. The coupled motion converts this forced lateral bend into an axial torque. This effect is enhanced at heel strike by the axial compressive pulse of propagating upward. Compressing the spine increases the axial torque strength of its joints and permits a relatively large axial torque to be generated from small displacements.

4. The pelvis and the shoulders are forced to counter-rotate by the spine and its surrounding tissues. The legs follow and amplify the pelvic motion.

Locomotion is the product of the spine oscillating in the gravitational field. There are two resonating frequencies for this engine, depending on whether or not the posterior ligamentous system plays a major role in the exchange of energy between the gravitational field and the animal. Each of these frequencies corresponds to a different style of movement: walking, which is a muscle predominant strategy, and running, which is a ligament predominant strategy.

A parallel can be made with the previous study of lifting, in which there were also two basic types of lifting. For small loads below 50 kg (walking) the power demand of the spine can be met by the spinal musculature, and there is no need to tighten the *PLS* to directly transmit further forces from the ilium to the tips of the spinous processes. When large loads are lifted, (running), the power demand of the spine cannot be met by the spinal musculature alone and therefore the *PLS* must be taut.

Finally it must be noted that the shape of the compressive pulse returned to the spine by the heel strike impact cannot be arbitrary if maximum axial torque is to be generated at minimum energy expenditure. This implies that the combination of track surface and shoes must be very specific. This has been known for some time by athletes. An extreme case of poor match is felt by anyone walking or running over soft sand: the compressive pulse is so damped that the energy recovered by the spine from the descending trunk is inadequate. To rotate the pelvis, additional power sources acting directly on the spine (spinal and trunk musculature) must be recruited. The inefficiency of this solution is obvious in that one becomes tired very quickly. This demonstrates that the spinal musculature is not designed to supply the power to locomote, any more that it is designed to supply the power to lift. In both cases, it is suggested that the role of the spinal musculature is to modify the spinal curvature to control the use of power generated by the hips extensors. This representation of spinal function explains the anatomical arrangement of longissimus, iliocostalis lumborum, multifidus and psoas (Figure 25.7).

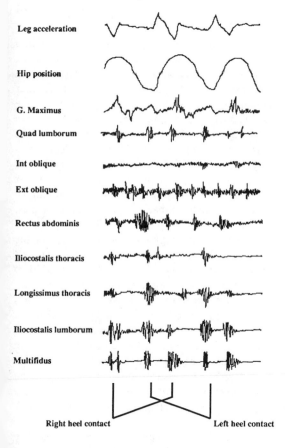

Figure 25.7: Right side *EMG* measurements of a subject walking. Redrawn from Liberson (1965) and Waters (1972).

A few observations are appropriate:

1. A contradiction. If the legs were to axially rotate the pelvis, then the conservation of angular momentum would require that an equal an opposite torque be transferred to the ground. This has not been measured.

2. The use of the gravitational field. Nachemson (1966) proposed that the psoas muscle is a controller of spinal posture. Dempster (1955) noted that "whenever the body exerts forces on its environment, forming a closed chain system of forces, limb and trunk muscles do not directly exert pull forces; instead they maintain joint postures which permits body weight to exert an effective moment." Hence the idea that the gravitational field ought to play a role in locomotion has been in the literature for quite some time.

3. Minimizing lateral displacements. To minimize lateral displacement of the pelvis and the shoulder, it is advantageous to construct the spinal engine in two sections, corresponding to the lumbar lordosis and thoracic kyphosis. The reversal in curvature cancels the lateral displacement of each section and that of pelvis and shoulders, without affecting the generation of axial torque.

4. Stabilizing the head. The need to stabilize the head as a platform for eyes/ears sensors, demands that the rotation of the shoulders be decoupled from the head movement. This is achieved by using the coupled motion of the cervical spine. It is suggested that the cervical lordosis converts the lateral cervical spine motion into an axial torque, which cancels the torque transmitted to the cervical spine by the rotating shoulders. In short, the cervical spine stabilizes the head by de-rotating the motion of the shoulders.

5. Bipedalism and human gait. Bipedalism has been around for a very long time: dinosaurs walked on two legs and so do today's birds. Walking on two feet does not require the peculiarities of the human spine. The spine of apes and monkeys does not have the lordosis of the human spine, and although they are basically quadrupeds, they can walk fairly well on two feet for some time. The distinctive form of human bipedalism is very specific. People with a fused spine or disabled spine can be expected to walk on two feet, although they will not achieve the normal level of efficiency (Inman, 1966). The characteristic counter rotation of pelvis and shoulders will be affected and so will energy exchanges with the gravitational field. It is therefore relevant to study the theoretical predictions of power flow through the spine and compare these calculations with experimental data.

25.4.3 Power Flow in the Spinal Engine

To better appreciate the intricate details of the energy transfers which take place within the spine, we reviewed the available experimental data on spinal motion during gait, together with the in vitro mechanical response of the intervertebral (*IV*) joint to compression and torsion (Gracovetsky, 1988). As the spinal engine is a collection of stacked intervertebral joints, the dynamic response of the joint must be clarified by using data from in vitro experiments and modeling.

1. The torque needed to drive the trunk and pelvis must rotate the individual *IV* joints and overcome the inertia of the surrounding soft tissues. The torsional response of the joint can be determined from the data first collected by Farfan and co-workers (Farfan, 1969, Farfan et al., 1970) and later on by Liu et al. (1985).

2. These data were used to derive a semi-empirical nonlinear first order differential equation describing the joints. This equation relates the axial torque necessary to rotate the joint, to the axial angular displacement and velocity and to the applied axial compression.

3. The torque needed to overcome the inertia of the trunk mass was calculated by Cappozzo (1983) based on observations of volunteers walking at different speeds. From these data, the total torque, which is the sum of the torque required to overcome the inertia of the trunk mass plus the torque required to overcome the resistance of the intervertebral joint, can be calculated. This total torque can then be decomposed into the components supported by the annulus and facets.

4. Once the total torque and angular displacement are known, the mechanical energy expended and the power used by the joint can be calculated. To evaluate the entire contribution of the spine, it is necessary to estimate the amount of power shared between both thoracic and lumbar spines. The lumbar contribution can be readily evaluated, assuming that the lumbar joint rotates by one-fifth of the total displacement measured by Thurston and Harris (1983). The thoracic contribution has been evaluated from the data generated by Gregerson and Lucas (1967). Within the limitations of this experimental knowledge and the validity of the many assumption made in the calculation, the power available at the pelvis may be estimated to reach about ten times that available at the joint, with a peak of up to 30 W.

We could not find any data from the literature to directly verify the validity of these predictions, and therefore used an indirect approach based on the experimental data published by Winter (1983a,b), who considered the kinematics of the legs and calculated the power available at the hip joint regardless of the detailed motion of the masses above the pelvis. The rationale for this validation is as follows: if the spine does indeed drive the pelvis, then the calculated power delivered at the hip joint by the spinal engine should be similar to that measured (e.g. by Winter, 1983a; see also Gracovetsky, 1988).

Although the magnitude of the power at the hips is different from that calculated at the *L4* level, its pattern should nevertheless yield some indications of the reasonableness of our approach. The superimposition of power patterns shows striking similarities. Hence, it appears that the proposed theory can offer a rational link between the respective motions of the spine and legs during locomotion. The apparent success of this correlation does not necessarily prove the theory to be true; rather, it simply demonstrates its usefulness in putting pieces together from a wide range of different sources.

This theoretical analysis relies on the validity of equations which have been derived from experiments conducted under artificial conditions. Proper experiments will have to be carried out in order to analyze the response of the intervertebral joint under more physiological conditions. It is possible to compile a list of some testable consequences of the spinal engine theory and compare them with those of the pedestrian theories of locomotion (Table 25.1).

It can be appreciated that the coupled motion of the spine is a fundamental feature tuned to the variations in spinal curvature and the mechanical properties of collagen. Hence, the unique S shape of our spine is of fundamental importance for human gait.

From a clinical and rehabilitation point of view, the theory highlights the fundamental importance of axial torsion in the motion of the spine, coupled with a critically timed axial compression. In the formulation of a diagnosis, the clinician should remember the critical role of torsion as a source of injuries to the ligamentous structures of the joint. The current difficulty in ascertaining the extent of ligamentous damage may explain the hesitation of many to accept the predominant impact of axial torsion on the etiology of spinal disorders. The verification and validation of the proposed theory is certainly a formidable task. Yet we should not be deterred from using it as a qualitative guide for lack of better alternative.

Table 25.1: Some testable consequences of the spinal engine and classical theories of locomotion.

| Situation | Predictions | |
	Classical Theories	Spinal Engine
Shortened or otherwise disabled legs.	*Reduced amplitude of spinal and pelvic oscillations.*	*Increased amplitude of spinal and pelvic oscillations.*
Less spinal mobility (i.e. spinal fusion)	*Walking ability unimpeded. Pelvic oscillations greater from theories which see the role of spinal oscillations to dampen pelvic oscillations, or unchanged from models which ignore trunk entirely.*	*Pelvic oscillations and walking ability impaired.*
Complete restraint of spine by cast or harness.	*Walking ability unimpaired; perhaps some difficulties in maintaining balance.*	*Inability to walk or extremely modified gait; or, inability of harness to restrain the spine.*
Loss of control of psoas.	*Minimum effect.*	*Major consequences as lordosis control is lost.*

25.5 The Coupled Motion of the Spine

The ability of the spine to convert a lateral bend into an axial torque has been known for some time. The earliest reference we could find is that of Lovett (1903). Formal studies involving radiology and or *in vitro* laboratory tests were made by Tanz (1953), Roaf (1958), Miles and Sullivan (1961), Gregerson and Lucas (1967), Pope et al. (1977), White and Panjabi (1978), Frymoyer et al. (1979), Schultz et al. (1979), Pearcy and Tibrewal (1984), and Panjabi (1985).

The results are variable; they range from "there is little or no coupling" (Schultz et al., 1979) to "coupling is essential to the understanding of the lumbar spine kinematics" (White and Panjabi, 1978). To explain these wide differences in opinion it was proposed that the direction and magnitude of coupling is a function of the relative position of the center of rotation of the joint and the disc pathology (Gracovetsky, 1988). In short, in the normal spine, given a lateral bend to one side, the induced axial torque can be either clockwise, counter clockwise, or neutral. Hence, it is suggested that the difficulty in controlling the position of the center of rotation may be responsible for the paradoxical results of apparently similar experiments.

When the disc is injured, the coupled motion of the spine is affected, the induced axial torque may become negligible, and consequently the patient may lose the ability to locomote. To maintain locomotion in the presence of an injury, the induced axial torque must be restored to an acceptable level. This compensation mechanism can be calculated. It requires the facets to supplement for the loss of disc strength. To enhance the function of the facets, it is necessary for the disc to collapse (i.e. the disc space is reduced); if that is not sufficient, the facet may be reoriented. Hence, the theory may explain the frequent deformations noted on *CT* scans called 'facet arthropathy' or 'facet remodeling'.

25.5.1 Theoretical Method for Detection of Spinal Injuries

As seen earlier the mechanical failure of the joint may occur by a combination of excessive torsion and compression. Unless the injury is fresh and clean, the clinician is faced with complicated detective work because *CT, MRI,* and *X-rays* cannot give a clear image of the precise damage to the collagen of the annulus fibrosus, and injection of contrast material generally reveals the existence of multiple anatomical anomalies (Heslin, 1987). In fact, it is known that 30% of asymptomatic patients exhibit some kind of radiological anomaly (Boden et al., 1990). Since the correlation between anatomy and function is tenuous at best, the question remains: What anomaly, among the many visible on the radiographic film, is the one responsible for the patient's problem?

To answer such a question, the response of the joint must be tested in vivo under the two basic modes of loading: an increase in axial compression and an increase in axial torsion. The increase in compression is achieved by requesting the patient to bend forward in full flexion. This increase is mainly supported by the anterior part of the intervertebral joints, and in particular, the disc. The torsional strength of the joint (and the ability of the disc and the facets to transmit it) is tested by asking the patient to bend laterally. If a patient is asked to bend to the left, then the pelvis will be driven in a clockwise direction (when looking from above) by the coupled motion - i.e., the left inferior facet of a vertebral body will push against the left superior facet of the one immediately below. If the joint is healthy, the resulting motion of the spinous process will be smooth. The whole spine will bend evenly to one side.

But if the joint has been previously injured by excessive torsion in the clockwise direction, then executing a lateral bend to the left will force the joint to rotate in a clockwise direction, that is, in the same direction which caused the patient's injury. The patient will unconsciously refuse to do this. The lower spine will go to the right first (to axially derotate the injured level) and then from a certain level up go to the left, as requested. This phenomenon is well documented in the literature where it is referred to as "the lateral bending sign" (Weitz, 1982) or "protective scoliosis" (Caillet, 1974).

Torsional injuries deform the joint. Restoring normal alignment of the vertebral bodies requires derotation. In principle, spinal derotation could be achieved by external manipulation, or by asking the patient to use appropriate muscles such as multifidus. But using muscles is tiring and inefficient because spinal muscles are more or less aligned with the spine. The patient left to himself uses his own body weight to derotate his spine through a combination of protective scoliosis and lordosis in order to reduce the pain. The patient's list betrays his torsional injury. Note that the protective scoliosis disappears when the patient lies supine on a table.

Care must be exercised to avoid confusing protective and idiopathic scoliosis. Rare are asymptomatic spines which do not have some degree of idiopathic scoliosis. Following torsional injury, the protective scoliosis adds its deformation to the preexisting idiopathic scoliosis.

Therefore, the detection of a torsional injury is facilitated by conceptually separating the spinal scoliosis into its primary components: the permanent deformation (idiopathic scoliosis) and the temporary, compensatory deformation (protective scoliosis) which characterizes the injury. It would be beneficial to design a machine capable of distinguishing between compression injuries (by analyzing the motion of the spine flexing in the sagittal plane) and torsional injuries (by distinguishing between idiopathic and protective scoliosis as the spine bends laterally). The Spinoscope (described in Section 25.6) is one such machine; it opens a new field for assessing the function of the spine in a non-invasive manner.

25.5.2 The Problem of Assessing Spinal Function

The difficulties involved in objectively assessing the function of the spine are rather formidable. Besides the fact that the spine consists of 24 individual intervertebral joints (all of which are mobile, and any one of which can be the site of an injury), the whole assembly is mounted on the pelvis, which itself moves. The strong connective tissues that attach the vertebrae directly to the ilium ensure that the motions of the spine and pelvis are inextricably linked (Figure 25.8).

This is why the various isokinetic, isotonic, isoinertial, and related machines intended to measure muscular trunk strength do not do justice to the biomechanical complexity of the spine. By attempting to mechanically isolate the spine from the pelvis using harnesses and belts (Figure 25.8), and allowing its movement around a fixed axis of rotation, these machines interfere substantially with the normal functioning of the spine. What they measure is in fact the disturbance of the spine's function by the machine. While this disturbance may be repeatable, it is not necessarily relevant. After all, most workplace tasks are not performed with the worker's pelvis restrained by a belt, nor are they performed isokinetically, isoinertially, or in any other such artificial way. Berkson et al. (1977) compared the muscle strength of male patients (those healthy and those with acute low back pain) and found that acute back syndrome patients have considerable physical ability under many circumstances. As a rule, either patients were unable to assume a given position, or else their performance was close to the norm in all exercises in that position. The positions which could not be assumed were generally those involving substantial trunk twisting, or a combination of bending and twisting.

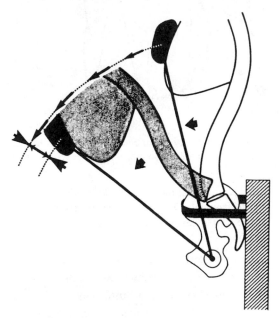

Figure 25.8: Harnessing the pelvis to isolate the spine is a common procedure. However, spinal function is inextricably linked to the pelvis.

A spinal injury is usually a ligamentous or a collagenous injury. Damage to ligamentous tissues cannot be readily detected by radiology, and there exists no evidence in the literature sustaining the view that machines which measure trunk muscle strength can characterize this injury either. Finally, to complicate these issues, the ligaments are visco-elastic — that is, their response is time dependent. Meaningful data on a spinal in-

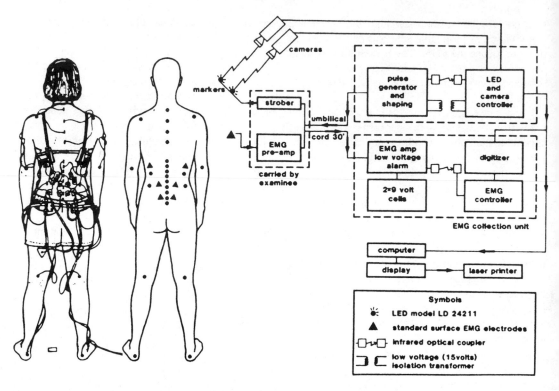

Figure 25.9: Overview of Spinoscope. Note the 12 markers placed above the mid line of the spine, while two more are placed above the iliac crest and two above the Achilles' tendons.

jury requires that the contribution of the ligaments be determined. This task must be done dynamically, not quasi-statically as the spine moves freely.

25.6 Assessing Spinal Function and Injuries: A Practical Solution: The Principles of Spinoscopy

To verify the application of the theory of the spinal engine in explaining human locomotion, it was first necessary to collect data on individuals with no lower extremities. The motion of the spine during locomotion was approximated by tracking the movement of markers placed on the skin, while monitoring the *EMG* activity of selected muscle groups. The precision necessary to measure variations in lordosis and intersegmental mobility precluded the use of *TV* based motion analysis systems; the *Spinoscope* (Figure 25.9) uses infrared cameras with a resolution 64 times better.

The *Spinoscope* obtains geometric data from the patient's spine using a dual camera system, while the activity of multifidus is recorded using skin surface *EMG* electrodes. By mathematical analysis, detailed information about the coordination between spine, pelvis and muscles can be deduced. For a normal individual, this coordination is task–specific, and can be used as a reference. When a patient with an unknown condition is tested, his spine and pelvis coordination can be compared with the reference obtained from the normal individual. Any discrepancy between them is interpreted as a loss of spinal function.

The kinematics of the spine are obtained by tracking the motion of markers placed on the skin with adhesive tape. These markers are small light–emitting diodes. Twelve of them are distributed along the spine, only two of which need to be located with precision: the first one must be above the spinous process of *C7*, and the ninth below must be above *L4*. With this accomplished, the other markers can be placed approximately, because their relation to the underlying vertebrae can be calculated using anthropomorphic tables. In addition, two more markers are placed above the iliac crest, and two more on the Achilles' tendons. The *EMG* electrodes are placed above multifidus at *L5*. In this region, multifidus stands alone and the electrical signal is not disturbed by the activities of the iliocostalis and the longissimus.

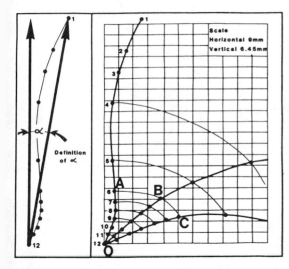

Figure 25.10: Subject in flexion with the reconstructed motion of the markers in the sagittal plane.

During data collection, each marker emits a short pulse of infrared light. At full speed, each marker flashes about 5,000 times per second. This allows the reconstruction of spinal motion at a rate of 180 images per second.

When the patient bends forward, the motion of the markers in the sagittal plane contains information on the intersegmental mobility and lordosis reduction (Figure 25.10). The reconstructed motion can be played back for detailed analysis of the kinematics of the spine. This information, however, is distorted by skin motion, which introduces an error into the measurement process.

25.6.1 The Skin Motion Problem

The clinical data generated by the Spinoscope is derived from the tracking of markers placed on the skin. As the patient moves, the spine and the skin move. It is pertinent to examine to what extent it may be possible to obtain information on the spine movement from the kinematics of the skin markers. This idea is not new (Bryant, 1989) and has been used to determine vertebral body positions.

Our basic protocol for this investigation consists in taping 3-mm steel balls over the spinous processes and the iliac crest of volunteers to be X-rayed. To control for image magnification of the X-ray machine, a one inch diameter steel ring is taped above multifidus as close to *L3* as possible.

This ring will appear as an ellipse on the film and will permit scaling the image until the shadow of the largest ellipse is exactly one inch. By taking several *AP* and lateral radiographs of the subject so instrumented, one can measure the relationship between the position of the markers and the vertebrae for a wide variety of postures. From these data, it is possible to derive some important physiological parameters.

25.6.2 The Measure of Lordosis

Lordosis refers to the curvature of the lumbar spine. For the purpose of this study, lordosis is defined as being the lumbosacral angle, which is the angle between the bisectors of the *T12/L1* and *L5/S1* discs. The lumbosacral angle derived from direct vertebrae measurements is said to be "true." In addition to the shape of the vertebrae, the radiograph clearly shows the corresponding location of the steel balls, which define a curve. The perpendiculars to the curve at the level of *T12/L1* and *L5/S1* also define an angle. This lumbosacral angle derived from the steel balls measurements is termed "effective." The correlation between the "true" and "effective" angles has been studied in Gracovetsky (1988). Essentially, the relationship is monotonic; in fact, it is well approximated by a straight line (Figure 25.11). This has been verified on 30 individuals, and is consistent with the findings of others.

The correlation in variation in lumbosacral angle, calculated from the kinematics of the *LED* markers and those of the steel balls with the true motion of the vertebrae, is also consistent for a given individual. But what about inter-subject variation? Although the patterns of motion are similar, an absolute inter-subject comparison using only marker data is not possible because of the variable thickness of the soft tissues above the spine. Hence, the correlation curve for an individual must be obtained before the absolute true lumbosacral angle can be deduced from the data collected on skin markers. In practice, this restriction is not severe because the relation between the "effective" and "true" lordosis is at least monotonic and therefore the clinician can compare the patterns of responses of different individuals. In addition, it is clinically important to compare the data collected on the same individual during the performance of different tasks.

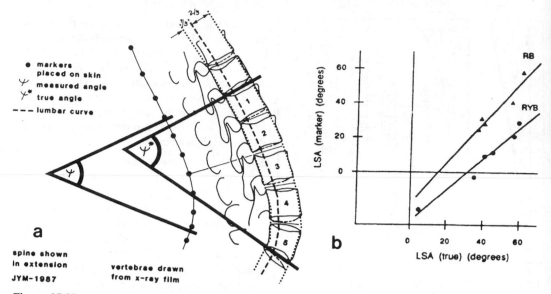

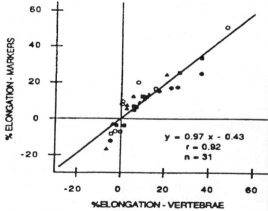

Figure 25.11: *a)* Lateral view showing the relative position of vertebrae and steel balls. These define the lumbosacral angle Y* as the angle between *L5/1* and *T12/L1* and the angle *Y* as the estimate of calculated from the positions of the steel balls, *b)* The correlation between Y* and Y is fairly linear, indicating that the motion of external markers placed on the skin does contain information relevant to the estimation of the true lumbosacral angle.

25.6.3 The Schober Test

A very popular test performed by clinicians consists in measuring the variation in distance between two marks placed on the skin above *S1* and *L1*. As the patient bends forwards, the lordosis is reduced and the distance between the two marks increases and is well correlated with the true relative elongation of the distance between the corresponding spinous processes of *S1* and *L1* (Gracovetsky, 1988) (Figure 25.12).

Note that in addition to any inter-trial variability, at least part of the variance between the measurements using the *LEDs* and steel balls is due to the fact that the *LEDs* could be tracked dynamically, while the radiographs of the steel balls were taken with the subject in quasi-static positions. The visco-elastic ligamentous structures of the spine may creep while the radiographs are taken. This may account for the fact that the dynamic measures made from the *LEDs* yield a consistently lower calculated response than the response obtained from the quasi-static data collected from radiographs of the steel balls. In the normal subjects exhibiting the muscle relaxation phenomenon, the elongation at relaxation was approximately 30%.

25.6.4 The Non-Invasive Evaluation of True Intersegmental Mobility

The accuracy of the Spinoscope camera system 1 part in 500) makes it easy to track gross movements such as range of motion. It is, however, much more difficult to determine the true intersegmental mobility of the various joints of the spine, because the magnitude of the errors due to skin motion becomes comparable to the relative displacement of one joint above another. This is not to say that the skin movement introduces a level of uncertainty which reduces to zero the clinical sig-

Figure 25.13: Cumulative data on seven individuals showing the correlation between the variation in the distance between the *L5* and *T12* vertebrae and the variation in distance between markers placed above the spinous processes of *L5* and *T12*.

nificance of the intersegmental mobility calculated from the steel balls or from the *LED*s; skin motion cannot be random and must contain information characterizing both the spine and its surrounding soft tissues. Since both are affected by pathology, it must follow that skin motion is also affected by the presence of spinal pathology. Therefore the clinical relevance of the calculated "effective" intersegmental mobility (calculated from the kinematics of the skin markers) must be separated from the clinical relevance of the "true" intersegmental mobility (calculated from the kinematics of the vertebral bodies) because each contains different amounts and types of information. The usefulness of the effective intersegmental mobility to the clinician is beyond the scope of this chapter (see Gracovetsky, 1990). Briefly, the clinician wants to know if there exists a relationship between effective intersegmental mobility and pathology, and to what extent this relationship will help him or her better appreciate the condition of the patient. In this regard, the correlation between the effective and true intersegmental mobility may be considered of secondary importance, although it would clearly be an advantage to obtain reliable data on the actual motion of the vertebrae from non-invasive sources. The relationship between the kinematics of the skin markers and true intersegmental mobility of the spine, for a population of 24 subjects, is documented elsewhere (Gracovetsky, 1990).

Table 25.2 summarizes the experimental results by providing the statistical error between true and effective intersegmental mobility. These data suggest that it is not unreasonable to use skin markers to approximate the motion of the spine. Of course some precautions must be taken and judgment must be exercised. For instance, very fat people will not be good candidates for such an analysis.

Table 25.2: Average error (in degrees) between true and effective intersegmental mobility.

	FLEXION		L. BEND		R. BEND	
	Average	SD	Average	SD	Average	SD
L1	2.59	1.32	2.00	1.54	0.71	0.78
L2	1.58	1.78	1.79	1.36	1.63	1.19
L3	2.91	2.26	1.25	0.84	1.38	0.83
L4	3.65	1.49	2.33	2.15	1.21	1.56
L5	2.60	1.84	3.00	2.94	3.23	2.16

At the very least, it is suggested that this type of non-invasive measurements should be considered as an acceptable source of information on the hard-to-get dynamic response of the unrestrained spine. This is desirable in view of the limited alternatives available to identify spinal dysfunction. The visco-elastic nature of collagen requires that functional analysis of the spine be performed at speed. This dynamic information has important clinical implications. Yet the issue at hand is a larger one: we do not know how to integrate fully the data collected from the variety of non-invasive sources available to us.

25.7 Clinical Application of Spinoscopy

25.7.1 Experimental Muscles and Ligaments Response as the Load Lifted Increases

A subject who exhibits a near normal response when carrying no load might demonstrate a markedly different response when handling a certain amount of weight. It is therefore desirable to study the variations in performance of a patient as he is asked to manipulate increasing levels of weights. As the load and therefore the stresses within the spine increase, there must be a point at which the spine reaches its maximum ability to handle that load. The spine will "brace" itself, thereby manifesting slightly different kinematics which can be objectively detected by the Spinoscope (e.g. Figure 25.13). Notice that the pattern of lifting was remarkably similar (see Figure 25.15).

There is a limit to what the collagen network of the lumbodorsal fascia can transmit. As the load exceeds a limit (which varies according to the anthropomorphic characteristics of each individual), the maximum trunk velocity begins to decrease. When this occurs, the collagen must stretch more to maintain its ability to transmit an increasingly high level of force. This stretching cannot continue indefinitely since the soft tissues of the spine will be damaged. Thus a decrease in peak trunk velocity indicates that the lifter is approaching his or her physiological limits. This observation suggests a measure for evaluating the safety of a load. It is proposed that the load is safe as long as it is in a range during which the maximum trunk velocity increases. The value of the load at which the peak velocity begins to decrease would then be considered as the maximum for safe

loading. In Figure 25.15, the safe load is estimated to be about 50 kg.

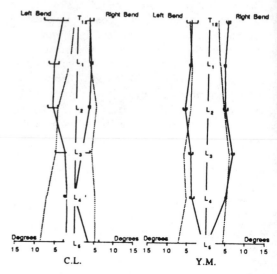

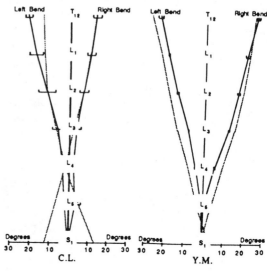

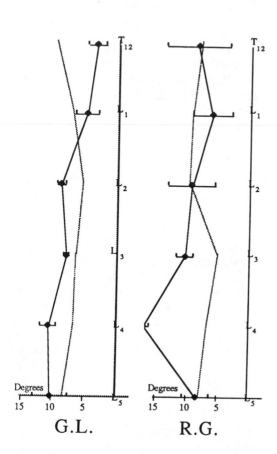

G.L.   R.G.

Figure 25.13: Data for ten healthy volunteers between
the ages of 23 and 32 who were requested to lift a
series of increasingly heavier loads using whatever
posture they felt most comfortable and to stop whenever they felt the lift would endanger their spine. The
data from several lifts of increasing loads are overlaid.
Notice the similarity of the responses. The data suggest that the coordination of spine and pelvis is
consistent regardless of the load lifted, as long as the
spine can handle the load. See also Figure 25.15.

The net decrease in lordosis is defined by the
change in lordosis as the subject bends down unloaded to pick up the load. The net increase in
lordosis is the variation in lordosis as the patient,
after having picked up the load, returns to the erect
stance (Figure 25.15a).

Figure 25.14: For each load lifted the peak trunk
velocity changes. The velocity peaks for a certain
angle of forward flexion. Note that the maximum
velocity first increases with the load to be lifted, and
then begins to decrease. It is suggested that the load at
which the peak velocity begins to decrease reflects the
limit of ligaments stretching, and as such, represents a
reasonable index of the maximum safe load.

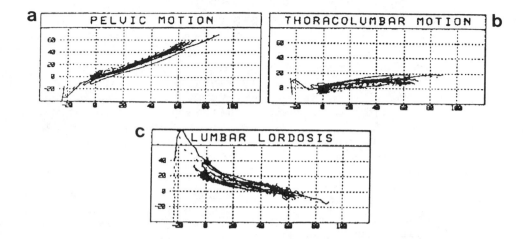

Figure 25.15: *a)* The variation in lordosis is given as a function of the load lifted. Note the similarity between the net increase and decrease in lordosis, that is whether or not the patient is loaded. It seems that the subject anticipates the posture to be adopted to lift the weight and adjusts his lordosis accordingly. Note that lordosis decreases as the weight lifted increases. This is consistent with a tightening of the posterior ligamentous system.   *b)* The dynamic Schober Test demonstrates a reduction in the elongation of the soft tissues of the spine because the back flattens when the load increases. What is not represented by this measurement is the impact of the pelvis rotating around the acetabulum as the knees flex. *c)* As the load lifted increases, the contribution of the pelvis to the reduction of lordosis increases to compensate for the flattening of the back. Indeed, the back must be flattened to minimize the external moment which must be supported by *L5*.

It is apparent from Figure 25.15b that the elongation of the soft tissues reduces from 50% (no load with the spine flexing to its maximum) to 25% (100 kg with "flat back"). Can we conclude that lifting with a "flat back" implies a lifting with a back more lordotic than that of a flexed spine? Notice that the initial lordosis also changes with loads; a 50% of reduction of the initial lordosis with no loads is equal to a 25% of reduction of the initial lordosis with a 100 kg load, which explains why the final lordosis is the same in both cases. Hence, a "flat back" posture does not necessarily imply a lordotic spine because the accompanying flexion of the knees permits a significant rotation of the acetabulum so that the pelvis may now compensate for the reduced spinal action. This permits a level of lordosis reduction sufficient to tighten the *PLS*. Note that the action of the erectores is necessary to control the reduction of lordosis. Hence, the observed multifidus activity is related to the need to adjust posture (a control action shared with psoas which requires little power, Nachemson, 1966) and not to a need to deliver the power for the lift. A more complete discussion on the theoretical implications of the role of the *PLS* is given in Gracovetsky (1988).

This demonstrates the considerable impact of pelvis rotation on the mechanics of the spine, and why it is unreasonable to restrain the pelvis in any way when testing patients. This is dramatically illustrated in Figure 25.15c in which the relative contribution of the spine and pelvis are plotted versus the load lifted.

These data suggest a method for determining a safe level of load. The relationship between the ligament elongation, velocity and load lifted is very particular, and similar patterns can be found in all healthy subjects we examined. The importance of the ligaments in the lifting process makes it imperative that they function in an optimal manner. The visco-elastic nature of collagen suggests that speed is an important parameter which must be maintained if the strength of the collagen is to be sustained. Hence, the beginning of trunk velocity reduction as shown in Figure 25.14 may be indicative of the asymptomatic moment at which the spine begins to acknowledge the stresses created by the load. It is therefore suggested that a "safe load" be defined as the load at which the maximum trunk velocity begins to decrease.

25.7.2 Range of Motion Versus Coordination of Motion

There is an important distinction to make between range of motion and coordination. Range of motion is under voluntary control and can be easily altered. Coordination is harder to change because it is, generally speaking, not controlled consciously. For example, it is virtually impossible to rotate at will the *L4* vertebrae over *L5*. The coordination of joint motion is a characteristic of the health of the spine and represents a desirable signature which can be measured.

In fact, the lifting capacity of an asymptomatic spine can also be determined by detecting the load at which normal coordination breaks down. This is possible because the coordination between spine and pelvis is not affected by increasing loads, as long as the subject is normal. Moreover, there is little change in spine/pelvis coordination when a lift is performed with the back appearing either flat or flexed. This may be surprising because the various styles of lifting appear quite different. Yet, the spine/pelvis coordination of an individual is relatively consistent throughout the range of weight lifted, as long as the spine can safely support the load.

25.7.3 The Unique Optimum Coordination Between Spine and Pelvis

The theory of the spinal engine predicts that the coordination between spine and pelvis is determined by the need to minimize the level of stress which the spine has to support, regardless of the load to be lifted. This unique optimum coordination will be affected by injury, and the resulting changes can be readily detected. That the coordination between spine and pelvis is invariable for a healthy spine should not really surprise us, since the human body has other invariants such as temperature. For example, no one would expect a large individual to have a high temperature, because a specific body temperature is characteristic of the species. Similarly, the coordination between spine and pelvis is something which is characteristic of the human species, not just the individual, and as such represents a highly desirable parameter to be quantified by Spinoscopy.

25.8 Conclusion

The theory of the spinal engine has been able to explain in rational form a considerable body of experimental facts. Nevertheless, this success is not a validation. It is simply a question of time before new evidence forces a review of the theory which will then have to be amended or discarded. In the mean time, it is hoped that our work will be seen as a useful contribution to the solution of the eternal back problem.

25.9 Future Directions

The theory of the spinal engine assigns an important role to the *PLS* and the coupled motion of the spine. Therefore, our recommendations are directed towards encouraging research in three important areas:

1. *The properties of the PLS.* This study can be done in vitro by analyzing the response of the collagen, while maintained at body temperature, under various regimes of elongation, loads and velocities.

2. *The precise role of the PLS in locomotion.* This is a theoretical approach involving the redesign of the current mathematical models of the spine. The spine and its surrounding tissues must oscillate in the earth's gravitational field, using the sagittal flexion of the trunk lifted by the hips extensor as input. The result of the sequence in energy transfer is the axial rotation of the pelvis. The thick passive part of the lumbodorsal fascia linking the ilium to the tip of the spinous processes and the tendonous attachment of latissimus dorsi should be carefully modeled. They are expected to play a crucial role in the determination of the resonating frequency; as such, the energy storage property of the *PLS* should be investigated.

3. *The coupled motion of the spine.* Until now, experiments designed to evaluate the coupled motion of the spine did not reproduce the conditions which can be calculated during walking or running. That is, as the joint rotates around a moving axis of rotation, the intervertebral joint is subjected to high compressive pulses. The design of an apparatus capable of applying such high variable loads cyclically would certainly represent a challenge.

Appendix: Three-Dimensional Model of the Human Spine

The development of a physiological *3-D* model of the spine can be viewed as the consolidation of four separate tasks. The first task is to acquire an accurate description of the relevant anatomy and quantify the appropriate components. The second task is to develop a mathematical method to generate different spinal postures, using the anatomical description provided by completion of the first task. The third task, given the above, is to generate equations capable of describing the moments, shears and compressions induced on the spine by the various muscles and ligaments. The fourth and final task involves the application of systems theory to take these equations, along with information about the external loads on the spine in a given posture, and generate a solution that minimizes and equalizes joint stress as a function of muscle activity, ligament tension and spinal posture. It is this latter task that is of greatest interest in this book, and thus the first three tasks will only be briefly outlined (see Gracovetsky (1988) for details.

25.A.1 The First Task - Anatomical Details

The anatomy relevant to a spinal model must be determined and then described mathematically. Our model includes the following anatomical groups: the thoracic spine (including the shoulder girdle and arms), the lumbar spine, the pelvis, the thoracic back muscles, the abdominal muscles, the lumbar back muscles, and the thoracolumbar ligamentous system. The cervical anatomy (including the head) has not been incorporated, but is a natural extension. Within this context, twelve muscles and four ligaments have been incorporated into the model.

The muscles have been further broken down into their fascicular construction, with the identification of the points of origin and insertions of the various fascicles on the bones. A similar procedure has been applied to the various ligament strands. The bones with muscle/ligament attachments have been given numbers, and each individual attachment point on a particular bone is also numbered. Thus, a fascicle or strand can be identified by a co-ordinate pair (bone number, attachment point number). Each muscle and ligamentous structure is then identified by a group of fascicles and strands respectively.

This numerical coding of the anatomy is used to construct the anatomical database, which is simply a list of the muscles and ligaments with their respective fascicle and strand co-ordinate pairs. Designing the database in this fashion allows for simple modification of the anatomy. Muscle fascicles or ligament strands can be added, deleted or displaced by modifying the appropriate co-ordinate pair. This numbering is a "generic" representation of each bone. Determination of *3-D* bone sizes, and scaling for subjects, is discussed further in Gracovetsky (1988).

25.A.2 Second Task - Spinal Posture Details

Generation of spinal postures in *3-D* allows for the simulation of almost any real life activity as a sequence of connected spinal postures. To generate spinal postures requires an understanding of how the spine moves in space. Studying *3-D* motion can be confusing, so the problem is broken down into three sub-problems: motion in the sagittal plane, motion in the coronal plane, and motion in the transverse plane. See Gracovetsky (1988, Appendix I) for details of this process.

25.A.3 The Third Task - Mathematical Details

The muscles and ligaments can induce forces and moments on the various spinal components. However, for the purposes of the *3-D* model, only the forces and moments on the lumbar vertebrae are required. When the cervical spine is included, then the forces and moments on both the cervical and lumbar vertebrae will be required.

Given the spinal posture and anatomical database, it is possible to generate force vectors for each muscle fascicle and ligament strand. Knowing the force vectors, it is possible to compute the moment induced at each lumbar vertebra.

Automatic determination of which vertebrae are spanned by a given fascicle or strand is accomplished by building a sequence table for the 57 bones. The sequence table defines the order of the bones, thus allowing for the identification of the vertebrae spanned by any muscle fascicle or ligament strand (see Gracovetsky, 1988) for further details on this algorithm.

25.A.4 The Fourth Task - System Details

The *3-D* model is used to predict patterns of muscle activity and ligament tension for an individual during a load-bearing activity. This activity is defined by a sequence of spinal postures. Execution of the first three tasks yields sets of equations that give the compression forces, the shear forces, and the moments induced at the vertebral bodies for given spinal postures as functions of muscle activity and ligament tension. Given the load that is borne during an activity, the forces and moments generated by the load at the vertebral bodies are computed for each spinal posture. Equilibrium conditions dictate that the resultant forces and moments at each vertebral body must be zero for each spinal posture.

Equilibrium Conditions. Static equilibrium of the lumbar spine is accomplished by balancing all the moments and loads at each of the five lumbar levels. The intervertebral discs are assumed to be able to support shear and compressive loads, but little or no moments (just like a ball-and-socket joint). This assumption leads to the equality constraints imposed on the system: all net external moments at all levels must be zero. It is convenient to separate the moment at each intervertebral disc into its X, Y, and Z components. Thus, we have three sets of equality constraints (one for each axis) and each set has five equations (one for each lumbar level). The individual sets of equations can be represented in matrix form as:

$$MLM_x X + EM_x = 0$$

$$MLM_y X + EM_y = 0 \qquad (25.1)$$

$$MLM_z X + EM_z = 0$$

where N is the total number of independent muscle fascicles and ligament strands, MLM_i is the muscle and ligament tension to $i = x,y,z$ direction moment scaling (5xN) matrix, and EM_i is the $i=x,y,z$ (5x1) vector direction component of the moment induced by external loads and body weight.

Physiological Constraints. Muscles only exert force when contracting. They cannot push. A positive muscle activity represents a pull whereas a negative muscle activity represents a push. Thus, muscle activity values must be greater than or equal to zero. Ligaments can be considered to act as cables, as they can only support tensions greater than or equal to zero. The mathematical constraints expressing the above properties of muscles and ligaments are represented by a family of inequalities:

$$X_i \geq 0 \; ; \; i = 1 \text{ to } N \qquad (25.2)$$

Constraints to Reflect a Given Pathology. An injury to the back can manifest itself in many ways and may involve muscles, ligaments, bones, or joints. To prevent injuries we must impose constraints on the maximum and minimum shear or compression to which any given intervertebral joint is subjected, *2)* the maximum muscle activity a given muscle can have, and *3)* the maximum tension a given ligament can support. The muscle and ligament constraints are implemented by a family of equations:

$$X_i \leq X_{max_i}; \; i = 1 \text{ to } N \qquad (25.3)$$

where X_{max} is an N vector with the maximum values permitted for the individual muscles and ligaments. The equations used to compute the shears and compressions are as follows:

$$
\begin{aligned}
\text{Shear } X \text{ direction} &= S_x X + Se_x \\
\text{Shear } Y \text{ direction} &= S_y X + Se_y \qquad (25.4) \\
\text{Compression} &= CX + C_e
\end{aligned}
$$

where S_x and S_y are the muscle and ligament tension to X and Y direction shear scaling (5xN) matrices, respectively; Se_x and Se_y are the muscle and ligament tension to compression scaling (5xN) matrices, respectively; and Ce is the (5x1) vector compression induced by external loads and body weight. Each expression is subject to constraint inequalities that set minimum and maximum intervals.

Formation of the Objective Function. The control hypothesis employed to formulate the objective function states that the human body will perform a given task in such a manner as to minimize the stress induced on the body while expending the minimum amount of energy. This hypothesis is formulated mathematically as follows:

$$OF(X) = P_1^2 \sum_1 [S_x X + Se_x]_i^2 2 + P_2^2 \sum_2 [S_y X + Se_y]_i^2 +$$

$$P_3^2 \sum_3 [CX + Ce]_i^2 + P_4^2 \sum_4 [X_i]^2 + P_5^2 \sum_5 [X_j]^2$$

$$(25.5)$$

where $\sum_1, \sum_2, \sum_3$ are on the index $i = 1, 5$; $\sum_4$ has the index $i=1, NMUS$; $\sum_5$ has the index $i=NMUS+1, N$; $P_i, i = 1$ to 5 are scaling factors weighting the different components ($P_i \geq 0$); *NMUS* is the number of muscle fascicles; $N - NMUS$ is the number of ligament strands. This equation is a quadratic function. With some algebraic manipulation, it can be written in the form:

$$OF(X) = 1/2 \, X'G' \, X + B \, X + A \qquad (25.6)$$

where G is an (NxN) symmetric positive definite matrix, B is an (N) vector and A is a scalar.

The implications of positive definiteness for the matrix G are important from a system-theoretic point of view. If this constraint is not made, i.e. if G is only a positive semi-definite matrix, then there exists some non-zero X for which $X'G' \, X = 0$. If it is assumed that the objective function is a measure of the energy within the system, then having G as a positive semi-definite matrix means that there exists some non-zero state within the system (ie. some non-zero X) for which the energy is zero. If the solution X is not unique, this implies that the system is unstable and can exist in an infinite number of states for which the energy is zero. We can force the matrix G to be positive definite by adding the identity matrix to G. This is equivalent to adding a ligament to the model. Therefore this requirement of positive definiteness suggests that modeling the spinal system without including the posterior ligamentous system is questionable.

Minimization of the Objective Function. The objective function is minimized subject to all constraints. All the constraints are linear. The objective function itself is quadratic with G being positive definite. The algorithm used to minimize the objective function is outlined in the next section. The objective function presented here is strictly for the minimization of stress. Stress equalization is dealt with in a later section.

The Optimization Algorithm. The control criterion described in the previous section is the minimization of musculoskeletal stress. We expressed this criterion mathematically as a quadratic function of muscle activity and ligament tension. The desired result is the minimization of this function with respect to muscle activity and ligament tension. Certain inequality constraints are placed on the range of feasible solutions, such as positive muscle activity, positive ligament tension and ligament tension less than or equal to some prescribed maximum. These constraints are all linear functions of the muscle activity and liga-

ment tension. The equality constraints imposed are that the resultant moment at each level be zero. An outline of the algorithm is given in the lower part of Figure 25.18.

The Nature of the Objective Function. The human lumbar spine is modeled using linear functions of muscle activity and ligament tension, while performing tasks such as lifting and/or twisting. Resultant shear and compression at the various lumbar levels are expressed as linear functions of muscle activity and ligament tension. These linear functions follow directly from a classical mechanical analysis of the loads and forces acting on the lumbar spine. The task being performed induces external loads on the spine which are balanced by internal forces generated by muscle activity and ligament tension, as well as by reaction forces at the intervertebral joints. This balancing action is modeled by equality constraints. The equality constraints reflect the fact that the moments and forces must be balanced at each intervertebral joint to assure that the spine is in equilibrium.

Muscle activity is modeled in such a way that any given muscle produces force only when contracting. Similarly, ligaments can only support tensile loads, not compressive loads, and are modeled as such. These two conditions lead to inequality constraints on the muscle activity and ligament tension which force them to be greater than or equal to zero.

Modeling the lumbar spine in this manner allows one to compute the resultant forces at each intervertebral joint as well as the moments and forces induced individually by the muscle activity, ligament tension and external load. The equality and inequality constraints form a space which must contain muscle activities and ligament tensions. This space contains an infinite number of solutions. The question we ask is: Is there only one best solution or a number of them?

It is apparent from this question that some criterion must be formed with which to compare one solution against another. It is reasonable to assume that the system will perform the task (i.e. choose a solution) while expending the minimum amount of energy. From a systems point of view, having a system that functions at minimum energy implies stability of that system.

Energy in this spinal system can be represented as a function of the stress in the various spinal components. The stress can be approximated by taking the Euclidian norm of the shear, compression, muscle activity, and ligament tension. The energy can then be represented by a quadratic form. The matrix of this quadratic form is always positive definite provided that the square of the muscle activity and the square of the ligament tension are included in the approximation. It can be mathematically proven that if only the Euclidian norm of the shear and compression is used, the matrix of the quadratic form is positive semi-definite. As explained elsewhere, the need to make this matrix positive definite lead to the realization that an important structure was forgotten (the *PLS*). Thus, the function for the spinal system must include the muscle activity squared plus the ligament tension squared. Adding the square of the shears and the square of the compression gives a more complete representation of the system.

The System as a Function of Time. In the strictest sense of systems theory, our objective function should be dependent on time as well as muscle activity and ligament tension (see Chapter 8 (Zajac and Winters) for a discussion of dynamic optimization). We should be minimizing an integral of the form:

$$OF(X(t)) = \int [1/2\, X(t)'\, G(t)'\, X(t) +$$
$$B(t)'T\, X\,(t) + A(t)]\; dt \qquad (25.7)$$

What we are actually minimizing is an individual sequence of static functions of the form:

$$OF_i\,(X(i)) = 1/2\, X(i)'T'\, G(i)\, X(i) +$$
$$B(i)'T\, X(i) + A(i) \qquad (25.8)$$

where i is the index of the flexion angle. Each $OF_i(X(i))$ is a quadratic function whose minimum value is 0. In effect, we are minimizing individual quadratic functions at each flexion angle. This can be written as:

$$\sum 1/2\, [\, X(i)'T\,'G(i)\, X(i) +$$
$$B(i)'T\, X(i) + A(i)]\; \Delta i \qquad (25.9)$$

If this equation is written so that the index i runs from 0 to 7 in steps of n ($n>0$) instead of 1, then its reduces to Riemann sum. Thus, our way of modeling the system can be viewed as a discrete-

time approach. The flexion angle can be viewed as a representation of time because it is estimated as a linear function of time. When we obtain solutions for $X(i)$ and then plot them for i going from 0 to 7, we observe points that make up smooth, continuous curves.

Total Stress Minimization and Joint Stress Equalization. Stress in the lumbar spine can be readily computed as a function of external load, spinal geometry, muscle activity and ligament tension. Holding the external load and spinal geometry constant then allows for the stress to be minimized as a function of muscle activity and ligament tension. Stress equalization has been achieved by modifying the geometry through perturbation of the positions of the centers of reaction independently of the stress minimization procedure. Once a new geometry is chosen, it is held constant for the minimization process. This two-stage procedure is used to obtain a solution in terms of muscle activity, ligament tension and geometry. This solution is one for which stress is minimized and equalized. It should be noted that stress equalization as a function of muscle activity and ligament tension alone has not been attempted.

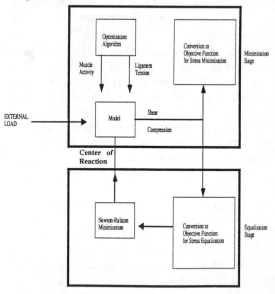

Figure 25.18: *Two-stage procedure for stress minimization and equalization. 1)* The minimization stage is carried out for a given geometry. *2)* Stress is examined to see if equalization has been achieved. If yes, stop. If no, go to step 3. *3)* Modify the geometry. *4)* Go to step 1.

A block diagram of this two-stage procedure, plus the outline of the optimization algorithm, is illustrated in Figure 25.18. Note that the method for determining the perturbation of the centers of rotation is not as sophisticated as the method used to solve for muscle activity and ligament tension. It is desirable to consolidate both stages and thus use only one optimization algorithm to determine the muscle activity, ligament tension, and geometry that give minimized and equalized stress. To achieve this goal, the stress must be written as a function of muscle activity, ligament tension and geometry. Because of the nature of the system, several complications arise, as shown by the following argument:

a) The resultant moment must be zero. For example, in the *X*-direction:

$$MLM_x X + EM_x = 0 \qquad (25.10)$$

where EM_x is the external moment; MLM_x is the muscle activity or ligament tension to moment scaling matrix; and X the independent variable (muscle activity or ligament tension).

b) Expansion of the moment scaling factors shows their dependence on geometry.

$$MLM_{i\,j} = [(X_j - CX_i, Y_j - CY_i, Z_j - CZ_i) \times (u_j)]_x ; \qquad (25.11)$$

where u_j is the unit vector for the line of action of muscle or ligament j; (X_j, Y_j, Z_j) is the coordinate of the point of application of muscle or ligament j; (CX_i, CY_i, CZ_i) is the co-ordinate of the center of rotation at level i.

c) The cross-product can then be expanded. The result then contains terms of $CX_i\,u_j$, $CYi_i\,u_j$ and $CZi_i\,u_j$. If we make u_j, CX_i, CY_i, and CZ_i independent variables, then the equality constraints are nonlinear.

Changing the optimization algorithm has severe implications. What would be required is a general optimization algorithm that handles nonlinear constraints (e.g. see Chapter 19 (Seif-Naraghi and Winters)). These types of algorithms tend to be computationally slow. The advantages of this algorithm is that it converges in at most N steps where N is the dimension of the problem. If the initial point $x(0)$ is fully constrained (i.e. N inequality constraints active), the algorithm will find a solution in at most N iterations. This is due to the fact that every iteration reduces the solution range by one dimension.

If all the K-T multipliers are positive and N constraints are active, then the solution has been found. If all the K-T multipliers are positive and less than N constraints are active, then a line search is done to eliminate the possibility of finding a solution with positive K-T results that can still be minimized.

References

Adams, M.A., Hutton, W.C. (1980) The role of the apophyseal joints in resisting intervertebral compressive force. *J Bone Joint Surg [Br]* **62**: 358-362.

Adams, M.A., Hutton, W.C. (1982) Prolapsed intervertebral disc: A hyperflexion injury. *Spine* **7(3)**: 184-191.

Bartelink, D.L. (1957) The role of abdominal pressure in relieving the pressure on the lumbar intervertebral disc. *J Bone Joint Surg [Br]* **39B**: 718-725.

Basmadjian, J. (1979) *Muscles Alive: Their Function Revealed by Electromyography.* Baltimore, Williams and Wilkins.

Berkson, M., Schultz, A., Nachemson, A., Andersson, G. (1977) Voluntary strengths of male adults with acute low back syndromes. *Clin Ortho Rel Res* **129**: 84-95.

Boden, S.D., Davis, D.O., Dina, T.S., Patronas, N.J., Wiesel, S.W. (1990) Abnormal magnetic-resonance scans of the lumbar spine in asymptomatic subject. *J. Bone Joint Surg* **72**:403-408.

Bogduk, N., Macintosh, J. (1984) The applied anatomy of the thoracolumbar fascia. *Spine* **9**: 64-170.

Bogduk, N., Twomey, L.T. (1987) *Clinical Anatomy of the Lumbar Spine.* New York, Churchill Livingstone.

Bryant, J., Reid, J., Smith, B., Stevenson, J. (1989) Method for determining vertebral body position in the sagittal plane using skin markers. *Spine* **14(3)**: 258-265.

Cappozzo, A. (1983) The forces and couples in the human trunk during level walking. *J Biomech* **16**: 265-277.

Caillet, R. (1981) *Low Back Pain Syndrome*, 3rd Ed. Philadelphia, F.A.Davis, 230 pages.

Dempster, W.T. (1955) Space requirements of the seated operator. *WADC Technical Report* **55**: 159 pages.

Farfan, H. F. (1969) Effects of torsion on the intervertebral joints. *Can J Surg* **12**: 336-341.

Farfan, H. F., Cossette, J.W., Robertson, G.H., Wells, R.V., Kraus, H. (1970) The effects of torsion on the lumbar intervertebral joints: The role of torsion in the production of disc degeneration. *J Bone Joint*

Surg [Am] **52A(3)**: 468-497.

Farfan, H.F. (1973) *Mechanical Disorders of the Low Back*. Philadelphia, Lea and Febiger.

Farfan, H.F., Kirkaldy-Willis, W.H. (1981) The present status of spinal fusion in the treatment of lumbar intervertebral joint disorders. *Clin Orthop* **158**: 198-214.

Floyd, W.F., Silver, P.H.S. (1955) The function of erector spinae muscles in certain movements and postures in man. *J Physiol (Lond)* **129**: 184-203.

Frost, H. (1973) *Orthopedic Biomechanics. Vol. V*. Springfield, Ill., C.C. Thomas.

Frymoyer, J.W., Frymoyer, W.W., Wilder, D.G., Pope, M.H. (1979) The mechanical and kinematic analysis of the lumbar spine in normal living human subjects in vivo. *J Biomech* **12**: 165-172.

Gracovetsky, S., Farfan, H., Lamy, C. (1981) The mechanism of the lumbar spine. *Spine* **6**: 249-262.

Gracovetsky, S., Farfan, H.F., Lamy, C. (1977) A mathematical model of the lumbar spine using an optimization system to control muscles and ligaments. *Orthop Clin North Am* **8(1)**: 135-154.

Gracovetsky, S.A., Newman, N.M., Asselina, S. (1990) The problem of non-invasive assessment of spinal function. *Spine* (Paper submitted.)

Granhed, H., Jonson, R., Hansson, T. (1987) The loads on the lumbar spine during extreme weight lifting. *Spine* **12(2)**: 146-149.

Gregerson, G.G., Lucas, D.B. (1967) An in vivo study of the axial rotation of the human thoracolumbar spine. *J Bone Joint Surg [Am]* **49Az(2)**:247-262.

Heslin, D.J., Saplys, R.J., Brown, W. (1987) Frequency of asymptomatic abnormalities in the lumbosacral spine. *Modern Med Canada* **42(6)**: 469-472.

Hirsch, C., Schajowicz, F. (1953) Studies on the structural changes in the lumbar annulus fibrosus. *Acta Orthop Scand* **22**: 184-230.

Inman, V.T. (1966) Human locomotion. *Can Med Assoc J* **94**:1047-1057.

Liberson, W.T. (1965) Biomechanics of gait: A method of study. *Arch Phys Med* **46**: 37-48.

Liu, Y.K., Goel, V.K., Dejong, A., Njus, G., Nishiyama, K., Buckwalter, J. (1985) Torsional fatigue of the lumbar intervertebral joints. *Spine* **10(10)**: 894-900.

Lovett, A.W. (1903) A contribution to the study of the mechanics of the spine. *Am J Anat* **2**: 457-462.

McGill, S.M. (1986) Partitioning of the L4/L5 dynamic movement into muscular, ligamentous and disc components for calculation of tissue loads during lifting. *Ph.D. thesis*, University of Waterloo, Canada.

McGill, S.M, Norman, R.W. (1986) Partitioning of the L4-L5 dynamic moment into disc, ligamentous, and muscular components during lifting. *Spine* **11**: 666-678.

McGill, S.M., Norman, R.W. (1988) Potential of lumbodorsal fascia forces to generate back extension moments during squat lifts. *J Biomed Eng* **10**: 312-318.

McIntosh, J.E., Bogduk, N., Gracovetsky, S.A. (1987) The biomechanics of the thoracolumbar fascia. *Clin Biom* **2**:78-83.

Miles, M., Sullivan, W.E. (1961) Lateral bending at the lumbar and lumbosacral joints. *Anat Rec* **139**: 387-398.

Nachemson, A. (1966) Electromyographic studies on the vertebral portion of the psoas muscle. *Acta Ortho Scand* **37**: 177-190.

Panjabi, M.M. (1985) The human spine: Story of its biomechanical functions. In: *Proceedings of the Ninth International Congress of Biomechanics*. Waterloo, Canada, Human Kinetics Publishers.

Pearcy, M.J., Tibrewal, S.B. (1984) Axial rotation and lateral bending in the normal lumbar spine measured by three-dimensional radiography. *Spine* **9(6)**: 582-587.

Pope, M.H., Wilder, D.G., Buturla, E., Matteri, R., Frymoyer, W.W., Frymoyer, J. W. (1977) Radiographic and biomechanical studies of the human spine. *Report of the Air Force Office of Scientific Research (AFOSR)*, USAF, AFOSR-74-2738.

Pope, M.H., Wilder, D.G., Frymoyer, J.W., Buturla, E.M. (1977) In vivo load-deflection studies of the lumbar spine. Presented at the *Fourth Meeting of the International Society for the Study of the Lumbar Spine*, Utrecht, Netherlands, May 5, 1977.

Pope, M.H., Wilder, D.G., Matteri, R.E., Frymoyer, J.W. (1977) Experimental measurements of vertebral motion under load. *Orthop Clin North Am* **8(1)**: 155-167.

Roaf, R., (1958) Rotation movements of the spine with special reference to scoliosis. *J Bone Joint Surg [Br]* **40B**: 312-332.

Rolander, S.D. (1966) Motion of the lumbar spine with special reference to the stabilizing effect of posterior fusion. An experimental study on autopsy specimens. *Acta Orthop Scand [Suppl]* **90**: 1-44.

Schultz, A.B., Andersson, G.B.J., Haderspeck, K., Nachemson, A. (1982) Loads on the lumbar spine: validation of biomechanical analysis by measurement of intradiscal pressures and myoelectric signals. *J Bone Joint Surg (Am)* **64A**: 713-720.

Schultz, A.B., Warwick, D.N., Berkson, M.H., Nachemson, A.L. (1979) Mechanical properties of the human lumbar spine motion segments. Part 1: Responses in flexion, extension, lateral bending and torsion. *J Biomech Eng* **101**: 46-52.

Shirazi-Adl, S.A., Shrivastavi, S.C., Ahmed, A. (1984)

Stress analysis of the lumbar disc-body unit in compression: 3-dimensional non-linear finite element study. *Spine* **9**: 120-133.

Spitzer, W.O., Leblanc, F.E., Dupuis, M. (1987) Scientific approach to the assessment and management of activity-related spinal disorders - A monograph for clinicians. *Spine* **12** (7S), S1-S59.

Tanz, S.S. (1953) Motion of the lumbar spine. A roentgenologic study. *Am J Roentgenol Radium Ther Nucl Med* **69**: 399-412.

Thurston, A.J., Harris, J.D. (1983) Normal kinematics of the lumbar spine and pelvis. *Spine* **8**: 199-205.

Triano, J.J., Schultz, A.B. (1987) Correlation of objective measure of trunk motion and muscle function with low-back disability ratings. *Spine* **12(6)**: 561-565.

Virgin, W. (1951) Experimental investigations into the physical properties of the intervertebral disc. *J Bone Joint Surg (Br)* **33B (4)**: 607-611.

Waters, R.L., Morris, J.M. (1972) Electrical activity of muscles of the trunk during walking. *J Anat* **111(2)**: 191-199.

Weitz, E.M. (1981) The lateral bending sign. *Spine* **6(4)**: 388-397.

White, A.A., Panjabi, M.M. (1978) *Clinical Biomechanics of the Spine.* Philadelphia, Lippincott.

Winter, D. (1983a) Biomechanical motor patterns in normal walking. *J Motor Behavior* **15**: 302-330.

Winter, D. (1983b) Moment of force and mechanical power in jogging. *J Biomech* **16**: 91-97.

CHAPTER 26

Postural Biomechanical Stability and Gross Muscular Architecture in the Spine

Joseph John Crisco III, and Manohar M. Panjabi

26.1 Introduction

Consider a ligamentous thoracolumbar spine specimen, sacrum-$T1$, fixed at the sacrum and carrying a load at $T1$. It has been determined experimentally that such a spine buckles when the load reaches a critical value of about 20 N (Lucas and Bresler, 1961). Or consider a lumbar spine specimen, sacrum-$T1$, its critical load is less than 90 N (Crisco, 1989; Crisco, et al., 1990). On the other hand, we have a world-class weight lifter who can carry more than 3000 N without damaging the spine (Granhed et al., 1987). Simply stated, the difference between the two behaviors is the spinal muscles. The same is true in another example. Compare a healthy normal person and a polio patient. The latter, with the back muscles paralyzed, cannot even hold his/her own trunk in the upright position. The important role of the muscles as the stabilizers of the spine is unquestioned and essential for its function. A deficiency in either muscle function or bony-ligamentous function will eventually lead to disabling clinical problems.

In this chapter we will first briefly review the working definitions of stability in the clinical and engineering professions, concentrating on spinal and mechanical definitions. We will define and differentiate the principles of equilibrium and stability for elastic mechanical structures. These principles will then be applied to a simple, single functional spinal unit to demonstrate that stiffness regulation through bilateral cocontraction of muscle is necessary and sufficient for stability. Finally, our simple model is extended to the whole lumbar spine. With this model we examine how gross muscular architecture (defined as a pattern of positions and orientations) influences the muscle's ability to stabilize a multiple joint structure. Though we have used the spine in our model, the concepts presented and the findings reported in this chapter apply, within the assumptions and limitations given here, to all jointed structures. Our underlying assumption was that the neuromuscular system controls posture with spring–like behavior. Thus, this chapter is limited to postural stability.

26.2 Spinal Instability

The case of a polio patient described above illustrates two aspects of a spinal instability problem. First, the ligamentous spine, without functional muscles, is incapable of carrying any substantial loads before it buckles. We may call this the mechanical or biomechanical instability problem. Second, the mechanical instability, owing to lack of muscle function, manifests itself as a clinical problem of a serious nature. This we may call the clinical instability problem. In the following paragraphs we look at these two aspects of the spinal instability in some detail.

26.2.1 Definitions of Clinical Instability

Clinical instability arises either from acute trauma and surgical procedures or from cumulative degenerative changes in the spine. Acute trauma and surgery result in the instantaneous failure or destruction of ligamentous and/or osseous structures which may lead to acute instability. Degenerative changes in the spine

Multiple Muscle Systems: Biomechanics and Movement Organization
J.M. Winters and S.L-Y. Woo (ed.), © 1990 Springer-Verlag

over time may also produce cumulative damage in the same structures. We may call this cumulative instability. Either of these two modes may result in clinical changes acutely or over time in the patient. This is clinical instability.

For the diagnosis of clinical spinal instability due mainly to trauma, a two-column model of the ligamentous spine (the anterior column loaded in compression and the posterior column loaded in tension) was proposed by Holdsworth (1970), Whitesides (1977), and Bradford (1980). More recently Denis (1983) proposed a three-column model. The posterior column of the three-column model was essentially the same as that of the two-column model: the posterior bony arch and the posterior ligamentous complex (supra/interspinous ligaments, facet capsules, and ligamentum flavum). The middle column was the posterior longitudinal ligament, the posterior portion of the annulus fibrosus, and the posterior vertebral wall. The anterior column consisted of the anterior longitudinal ligament, the anterior portion of the annulus fibrosus, and the anterior wall of the vertebral body. The advantage of these stability models is their simplicity and ease of clinical application; if any two of the three columns are determined by imaging techniques to have failed, then the spine is considered clinically unstable. However, none of these classifications are based on any experimentally determined data. Instead, they represent collective clinical experience.

In the case of recurring damage to the spine, Kirkaldy-Willis and Farfan (1982) proposed to divide the temporal process of cumulative damage and degeneration into three stages: (1) temporary dysfunction; (2) unstable phase; and (3) restabilization phase. They went on to define clinical instability as the clinical status of the patient with back problems, with the least provocation steps from the mildly symptomatic to the severe episode, i.e., from stage (1) to stage (2). In an attempt to explicitly define the step into the realm of instability, the authors used catastrophic theory and manifolds. In this representation, a patient's progress is displayed as a path on the catastrophe surface, defined by "increased pathology and abnormal motion." Instability arises when there is a discontinuous or catastrophic increase in the patient's "symptoms." The terms "increased pathology," "abnormal motion," and "symptoms" were not rigorously defined.

The most encompassing clinical definition of instability was proposed by White and Panjabi (1978): "Clinical instability is defined as the loss of the ability of the spine under physiological loads to maintain relationships between vertebrae in such a way that there is neither damage nor subsequent irritation to the spinal cord or nerve roots, and, in addition, there is no development of incapacitating deformity or pain from structural changes." Based upon this definition and in vitro experimental data, specific checklists for the determination of spinal instability for the cervical, thoracic, and lumbar and lumbosacral regions were proposed (White et al., 1975; Panjabi et al., 1981; Posner et al., 1982).

As an example, the checklists proposed for the lumbar and lumbosacral regions are presented in Table 26.1. The use of these checklists is straightforward. If a clinician determines that there are any positive findings in the element category, then appropriate point values are assigned and summed up. A total of five or more represents clinical instability.

These and other clinical definitions of stability were reviewed in more detail by Panjabi et al. (1988) and Frymoyer et al. (1988). The role of the musculature was not specifically addressed in any of these sources.

Table 26.1: Checklists by White et al. (1981) for the diagnosis of clinical instability in the lumbar, *L1–L5*, and lumbosacral, *L5–S1*, spines. It is assumed clinically unstable if the total is greater than 5.

Element	Point	L1–L2	L5–S1
Cauda equina damage	3		
Relative flexion translation	2	>16%	>25%
Relative extension transl.	2	>12%	>12%
Relative flexion rotation	2	>11°	>19°
Anterior elements destroyed	2		
Posterior elements destroyed	2		
Dangerous loading anticipated	1		

Figure 26.1: A ball rolling down an incline is not in static equilibrium, and therefore its condition of stability is not defined.

26.2.2 Definitions of Mechanical Stability

In elastic mechanical systems, *equilibrium is necessary but not sufficient for stability.* As an example, let us consider a ball on a flat surface. If the surface is inclined, the ball rolls down (Figure 26.1). Hence the ball is not in static equilibrium. In Figure 26.2 a ball is at rest on three different surface shapes. Although the ball is in static equilibrium on each surface, it is not stable in all cases. The ball is considered to be stable if, after a perturbation from its equilibrium position, it returns to that same equilibrium position. In Figures 26.2a and 26.2b, when the ball is perturbed, it will not return to its equilibrium position and hence it is unstable. Only the ball at the the bottom of the valley is stable (Figure 26.2c), and will return to its equilibrium position when perturbed.

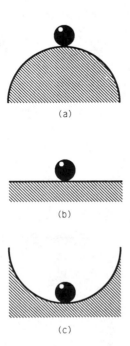

(a)

(b)

(c)

Figure 26.2: Even though the ball is in equilibrium in *(a)*, *(b)*, and *(c)*, it is unstable in *(a)* and *(b)*, and stable only in *(c)*.

An elastic system is characterized by its potential energy V. We define the potential energy of a system to be the total work, by both internal and external loads, required to perturb the system from its equilibrium position. The potential energy will be referred to as the potential hence forth in this chapter. From the potential, the equilibrium position and stability conditions can be derived. For a single degree of freedom elastic system whose position is given by θ, the equilibrium position(s) are the minimum(s) and maximum(s) of the potential, i.e. the zeros of the derivative of the potential with respect to displacement,

$$\frac{dV}{d\theta} = 0 \qquad (26.1)$$

We define the initial equilibrium position as the reference position, $\theta = 0$. The stability condition is determined by the shape of the potential at the equilibrium position, i.e. the sign of the second derivative of the potential with respect to the displacement. If

$$\frac{d^2V}{d\theta^2} < 0 \qquad (26.2)$$

or

$$\frac{d^2V}{d\theta^2} = 0 \qquad (26.3)$$

then the system is *unstable*, and if

$$\frac{d^2V}{d\theta^2} > 0 \qquad (26.4)$$

the system is *stable*. Equations 26.2, 26.3, and 26.4 mathematically define the stability conditions represented in Figures 26.2a, 26.2b, and 26.2c, respectively.

26.2.3 Definitions of Biomechanical Stability

Presently, our biomechanical definition of stability is that of the mechanical definition applied to biological systems. Previously, a simple biomechanical definition of spinal stability was given by Pope and Panjabi (1985). In that definition the stiffness of the spine was suggested as an indicator of its stability.

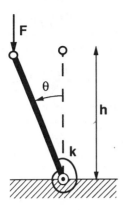

Figure 26.3: A single degree of freedom column, height h and hinged at its base with torsional stiffness k, is perturbed from the vertical reference position ($q = 0$) by the vertical force F.

The first definition of biomechanical stability that incorporated muscles was presented by Bergmark (1987, 1989). By approximating the mechanics of the neuromuscular system by elastic springs (Broberg, 1981), Bergmark was able to derive an elastic potential energy function for his muscular lumbar spine model, which investigated the role of global and local muscular systems in maintaining stability as a function of load and posture.

26.3 The Muscular Euler Column

A column subjected to a vertical compressive load will not fail until the stresses in the column exceed the failure stress of the material. However, a slender column will become unstable and buckle at a load significantly smaller than its failure load. This is the classical example of structural mechanical instability known as the Euler column.

Consider a single bone and joint as a single degree of freedom Euler column. The bone is a rigid column of height h, connected to a fixed base by a hinge joint (Figure 26.3). The passive ligamentous tissue of the joint are approximated as a linear torsional spring whose stiffness coefficient is k. The position of this ligamentous column is given by θ. The reference position is the vertical position, $\theta = 0$.

If a vertical load F is applied to the top of the limb, the potential energy of the column is the sum of the potentials due to the applied load and the torsional spring,

$$V = \frac{1}{2}k^2\,\theta \,-\, F\,h\,(1-\cos(\theta)) \qquad (26.5)$$

The equilibrium equation of the column is derived from *Eq. 26.1*

$$\frac{dV}{d\theta} = k\,\theta \,-\, F\,h\,\sin(\theta) = 0 \qquad (26.6)$$

and is graphed in Figure 26.4 by plotting the position (θ) as a function of the magnitude of the load F. Throughout these illustrative examples we have set the height of the column to 1 m and the stiffness of the spring to 1 Nm/rad. The equilibrium equation shown in Figure 26.4 is not a simple function. As the load is increased from zero, the column may remain in equilibrium at $\theta = 0$ for any given load. In addition, at 1 N the equilibrium path bifurcates. That is, for loads above the bifurcation point there are two equilibrium positions other than the reference position.

The stability condition for the column is determined from the second derivative of the potential energy,

$$\frac{d^2V}{d\theta^2} = k \,-\, F\,h\cos(\theta) \qquad (26.7)$$

With $\theta = 0$, the reference position, the above equation becomes

$$\frac{d^2V}{d\theta^2} = k \,-\, F\,h \qquad (26.8)$$

By applying *Eqs. 26.2-26.4*, we find that the column is **unstable** when

$$\frac{k}{h} \leq F \qquad (26.9)$$

or

$$\frac{k}{h} = F \qquad (26.10)$$

and **stable** when

$$\frac{k}{h} > F \qquad (26.11)$$

From Figure 26.4 we see that *Eq. 26.11* is true below the bifurcation point and *Eq. 26.9* is true above the bifurcations point. Therefore, we define the load at the bifurcation point to be the **critical load** (F_{cr}). For loads greater than or equal to the critical load the column is unstable and for loads less than the critical load the column is stable at θ

= 0. *Eq. 26.10* gives critical load explicitly,

$$F_{cr} = \frac{k}{h} \qquad (26.12)$$

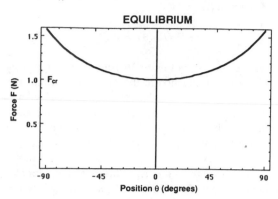

EQUILIBRIUM

Figure 26.4: The equilibrium paths for the Euler column of Figure 26.3 as a function of position and load magnitude. The Euler phenomena is represented by the bifurcation at the critical load (F_{cr}), 1 N in this example. The height and stiffness were 1 m and 1 Nm/rad, and the reference postion is $\theta = 0$.

The shape of the potential curves allows us to comprehend graphically the phenomena of the Euler column. In Figure 26.5 the potential energy (*Eq. 26.5*) is plotted for two loads: one less than and the other greater than the critical load. For loads less than the critical load the shape of the potential energy curve is that of the valley ball of Figure 26.2c. If the column, when loaded with a load less than the critical load, is perturbed, the potential increases and the column returns to the reference position. When the column is supporting a load greater than the critical load, and there are no perturbations, the column may remain at equilibrium in the reference position, as shown by the equilibrium path at the zero position in Figure 26.4. However, the column is not stable, the potential decreases to both sides as the hill of Figure 26.2a, and when perturbed infinitesimally the column will **buckle**. In our example, we see from the potential curve that there are two local minimums. Thus the buckled column moves from the reference position to a new equilibrium position. The shape of the potential curves varies for each applied load but can be visualized by the evolution of the curves with load: from the deepest valley, at a zero load, to being nearly flat at the critical load, and above the critical load, the hill

centered at the reference position continues to evolve to a steeper and steeper hill as the two local valleys become deeper and move outward.

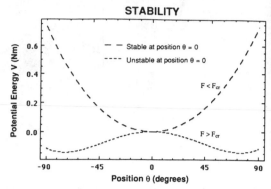

STABILITY

Figure 26.5: These potential energy curves give the stability conditions of the Euler column at two specific loads. For the load less then the critical load, the potential curve resembles a valley, and the column is stable. For the load greater than the critical load, the column is unstable when vertical, $q = 0$. Therefore, the perturbed column will buckle and, in this example, equilibrate at one of the two new equilibrium positions.

This simple column model may be extended to the ligamentous spine, where the column of Figure 26.3 approximates a single vertebral body, and the hinge joint simulates the intervertebral joint. In an *in vitro* study, Wilder (1985) found that single vertebral segments of human cadavers had a critical load of 500 N. Since healthy humans do not present with a buckled spine, the *in vivo* critical load must be higher than the ligamentous critical load. From *Eq. 26.12*, we see that one can increase the critical load by either decreasing the height of the vertebrae or increasing the stiffness of the joint. Since one cannot regulate vertebral height, one must regulate intervertebral stiffness to be stable *in vivo*.

Muscles regulate intervertebral joint stiffness. Therefore, returning to our model, we construct a muscular Euler column with muscles that originate from the base and insert on each side of the limb (Figure 26.6). If we are to model this column as a mechanical structure, we must approximate the behavior of the muscles by classical mechanical elements.

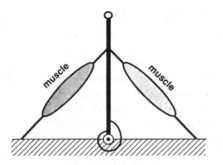

Figure 26.6: The muscular Euler column. This is the same simple column of Figure 26.3 supplemented with bilateral muscles, approximated as springs.

Though the regulated muscular variables may depend upon numerous, as yet unknown factors (Stein, 1982), the regulation of stiffness has been observed experimentally (e.g. Houk, 1979; Greene and McMahon, 1979), and found to be an elegant and useful characteristic for approximation in theoretical work (McMahon and Greene, 1978; Bergmark, 1989; see also Chapters 9, 11–12, 36–38). Therefore, we shall assume that the muscles of Figure 26.6 act as springs. Moreover, their stiffness is variable and proportional to an increase in muscular activity. Setting aside – until the following section – the influence of gross muscular architecture on the Euler column, the muscles in this section are considered to generate a pure torque about the joint. Furthermore, each muscle is capable of generating a torque in only its contracting direction. Therefore, we represent their bilateral contribution to the potential energy as a torsional spring of stiffness k_m, and the total potential energy of the muscular Euler column is then

$$V = \frac{1}{2}(k + k_m)\,\theta^2 - F\,h\,(1 - \cos(\theta)) \qquad (26.13)$$

In vivo loads vary, and if the load increases above the critical load of the ligamentous column, the column will buckle. In order to stabilize the column, the muscular stiffness must increase to increase the joint stiffness, which raises the critical load of the column. Moreover, to maintain the reference position of the column, this muscular activity must increase bilaterally. For a given applied load that is greater than the passive or ligamentous critical load, there is a minimal muscular stiffness that stabilizes the column. This muscular stiffness is the ***critical muscular stiffness***. This simple model predicts coactivation of

bilateral muscles when the critical load is larger than the critical load of the passive ligamentous structures.

The concept of mechanical stability is not new [it is generally attributed to the German scientist Kirchhoff in 1850), nor is the concept of the neuromuscular reflex system regulating stiffness, first attributed to the experimental work by Nichols and Houk (1976). Hogan and colleagues have further developed a theoretical foundation for stiffness–impedance modulation [e.g. Chapter 9 (Hogan), Chapter 17 (Flash)]. However, it is believed that Bergmark (1987, 1989) first united these ideas in his theoretical study of the stability of the lumbar spine.

The cervical, thoracic, and lumbar vertebral columns may be modeled as multiple Euler columns of Figure 26.3 stacked upon one another. The ligamentous thoracolumbar spine and the ligamentous lumbar spine buckled at critical loads of 19 N and 88 N, respectively, in experimental studies on cadaveric specimens by Lucas and Bresler (1960) and Crisco (1989). By stacking the muscular Euler column of Figure 26.6, we shall investigate in the next section how muscular architecture may influence the muscles' ability to stabilize the multiple joints of a spine.

26.4 Postural Stability and Gross Muscular Architecture in the Lumbar Spine

Leonardo da Vinci, depicting the vertebral muscles of the neck as ropes, was the first investigator of mechanical spinal stability (Figure 26.7). He hypothesized that the muscles of the cervical spine stabilized the neck analogous to the guy ropes of a ship's mast. In reference to the influence of muscular architecture, da Vinci stated, "the more central muscles stabilize, the more lateral bend the neck" (Keele, 1983). In modern anatomical texts, the multifidus muscles are referred to as the stabilizers of the vertebral column. Bogduk and Twomey (1987) have suggested the role of the deep muscles, including the multifidus, is to prevent consequential motion produced by the more superficial muscles as they move the thoracic cage and pelvis. Unfortunately, there is no scientific evidence for any of these statements.

Bergmark (1987) was one of the first to mathematically analyze the muscular stability of the spine. He grouped the muscles of the back into two groups: local (vertebral insertions and origins) and global (pelvic origins and thoracic insertions), and studied the three-dimensional stability of the lumbar spine as a function of posture. His spinal model was stable if the potential energy, which incorporated the ligamentous and muscular stiffnesses, was a minimum.

The work in this chapter is a first attempt at applying the principles of potential energy to the investigation of the stability of the spine as influenced by gross muscular architecture. Gross muscular architecture refers to a specific pattern of muscle positions and orientations, i.e. intersegmental, multisegmental, etc. The objective was to compare the lateral stabilizing potential as a function of the gross muscular architecture of the lumbar spine.

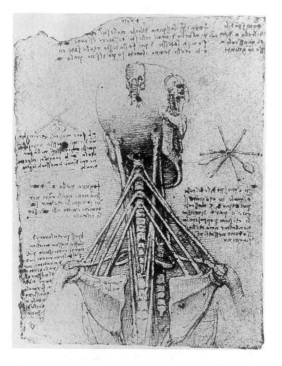

Figure 26.7: Sketch by Leonardo da Vinci depicting the lines of actions of the cervical spine muscles. da Vinci hypothesized that the muscles stabilize the spine like the guy ropes of a ship's mast, and the more central muscles stabilize the spine and the more lateral muscles move the spine. (From Keele, 1983.)

26.4.1 The Model

The model was composed of two elastic systems: one represented the ligamentous spine and the other represented the passive and active properties of the muscles and their stretch reflex. The ligamentous spine was modeled as a discrete elastic column: the vertebral bodies were simulated by rigid bodies and all elastic interbody tissues were simulated by torsional elastic springs. The loads were chosen to be vertical and compressive, simulating the weight of the body-section at each vertebral body. Motion was restrained to the frontal plane.

The active and passive properties of the muscles and their reflex activity were simulated by variable stiffness springs, the stiffness of the muscle being assumed proportional to the level of muscular activation [see also Chapter 5 (Winters) and Chapter 9 (Hogan)]. It was assumed that the neuromuscular system regulates muscular stiffness in postural control. This muscular stiffness was assumed to be linear, proportional to muscle activation, and equal for all muscles. The critical muscular stiffness was the minimal muscular stiffness required to stabilize the spine model, under the given loading. The stabilizing efficiency of various muscular architectures was compared by their critical stiffness: the lower the critical muscular stiffness, the more efficient the given muscular architecture was at stabilizing the model. Although the overall length of each muscle varied in the models, it was assumed that the contractile components that regulate the muscular stiffness were of equal length. Due to these assumptions, displacements were limited to be small and, therefore, only the critical load in upright posture was studied. In this study we defined the inferior attachment of a muscle to be the origin.

Table 26.2: The insertions and origins for each morphological model of muscular architecture.

Insertions	ARCH 1	ARCH 2 Origins	ARCH 3	ARCH 4
L1	L2	L3	Pelvis	Pelvis
L2	L3	L4	Pelvis	Pelvis
L3	L4	L5	Pelvis	None
L4	L5	Pelvis	Pelvis	Pelvis
L5	Pelvis	Pelvis	Pelvis	Pelvis

The Muscular Spine Models

Four simple morphological models were constructed to study the effect of muscular architecture on the critical muscular stiffness. The models, labeled *ARCH 1* through *ARCH 4*, are schematically drawn in Figure 26.8, and defined by the origins and insertions listed in Table 26.2.

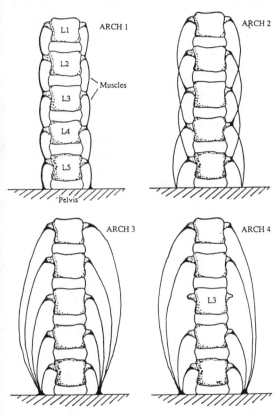

ARCH 2, the muscles were defined to span two joints. In *ARCH 3*, each muscle was defined to span the maximum number of segments, with each muscle originating from the pelvis. The lack of musculature was investigated in *ARCH 4* by using the same architecture of *ARCH 3* and removing the muscles from a single vertebrae, in this model from *L3*.

The details of the mathematics are given elsewhere (Crisco, 1989). Essentially, the potential energy function was a function of the positions of muscle origin and insertion, and was derived for each muscular architecture model. The potential energy was then formulated as a generalized eigenvalue problem. The smallest eigenvalue was the critical muscular stiffness for each muscular architecture model.

26.4.2 Results

For each muscular architecture, the critical muscular stiffness required to stabilize the intact ligamentous spine is listed in Table 26.3. These results show that the intersegmental muscular architecture (*ARCH 1*), consisting of muscles which originated from adjacent vertebrae, required the highest muscular stiffness for stability. This stiffness decreased by 70% when the muscles skipped a vertebra and originated on the vertebrae inferior to the adjacent vertebra (*ARCH 2*). As the number of vertebrae that were spanned by the multisegmental muscles increased, so did the efficiency of stabilization. In *ARCH 3* the muscles originated from the pelvis, an architecture that permitted the largest possible number of vertebrae to be spanned for each muscle. The result was a 90% increase in the efficiency of stabilization, relative to *ARCH 1*. We also found the efficiency of any muscle increased the more lateral the muscle's origin and insertion.

Figure 26.8: A schematic representation of the insertion and origin architecture of the four spinal models. *ARCH 1* models unisegmental muscles that span one intervertebral joint. *ARCH 2* models mutlisegmental muscles that span two intervertebral joints. *ARCH 3* models multisegmental muscles that span the maximum number of intervertebral joints, with each muscle originating from the pelvis. *ARCH 4* is identical to *ARCH 3*, but devoid of muscle at *L3*. The muscular lines of action are renderings only.

The muscles of the models do not simulate specific muscles but rather approximate possible muscular architectures. The first model, *ARCH 1*, approximated the smaller intervertebral muscles that span only a single intervertebral joint. In

Table 26.3: Critical muscular stiffness (kN/m) for the various muscular architectures.

ARCH 1	ARCH 2	ARCH 3	ARCH 4
129.5	37.8	10.5	oo

ARCH 4 was identical to that of *ARCH 3*, except the muscles of *L3* were removed. It was determined that such a model was unstable, regardless of the muscular stiffness. We found that this instability occurred for any vertebrae devoid of musculature, irrespective of the muscular architecture.

26.4.3 Discussion

Critical Muscular Stiffness

Contrary to what da Vinci hypothesized and what has been generally accepted throughout the anatomical literature, our model found that the deep intervertebral muscles (ARCH 1) were the least efficient at laterally stabilizing the spine. The model demonstrated that the efficiency of the multisegmental muscles increased with the number of vertebrae spanned, and that the most efficient architecture (ARCH 3) consisted of muscles that attached to the pelvis − spanning the maximum number of vertebrae. The significant difference between the critical muscular stiffness of ARCH 1 and ARCH 3 can be visualized by considering the following example: Assume the vertical model (at equilibrium) is rotated only at L5, with no other relative motions. In the intersegmental model (ARCH 1), only the muscles spanning L5 are elongated and increase the potential energy of the system. The other vertebrae are still in equilibrium since their relative motion is zero. However, in ARCH 3, all muscles are elongated, and all contribute to the increase in potential energy, forcing the system back to equilibrium. This example also demonstrates why multisegmental muscle efficiency increases with the number of joints spanned: the more muscles that span a joint, the greater the potential for restoring the equilibrium position of that joint.

When the intervertebral joint stiffness decreased − as with ligamentous injury − an increase in muscular stiffness attained spinal stability. Since muscular stiffness was assumed proportional to activation, the temporal stability of the injured ligamentous spine would be limited by muscular fatigue. Therefore, the model suggests that chronic instability due to ligamentous deficiencies may result from muscular overload, indicated by spasm and pain.

When a vertebral level was devoid of muscle (ARCH 4), the compressive load at that level was supported solely by the ligamentous spine, capable of supporting 88 N before becoming laterally unstable. Therefore, the muscular spine was unstable for any loads greater than the critical load of the ligamentous spine, regardless of the remaining spinal musculature.

Assumptions and Limitations

In our model we assumed that muscular stiffness was constant throughout each model so that we could specifically address the issue of muscular architecture. This assumption was based upon the concept of material optimization in biological systems: stresses, which in muscle are proportional to activation levels, are regulated to be equivalent throughout the material. Our assumption here is a further simplification. The cross-sectional area of the spinal muscles was also not taken into account. However, because the intervertebral muscles are smaller than the multisegmental muscles, the findings reported here would only be accentuated if cross-sectional areas were incorporated.

In the formation of our model we assumed that the active muscle length was constant and independent of the distance between the origin and insertion. For common engineering materials the stiffness of the material is proportional to its length. Although the relation between muscle stiffness and length has not been determined, we reformulated our model by assuming muscle stiffness was proportional to muscle length. We found, as was expected, that the critical muscle stiffness increased for each muscle architecture. However, the relative efficiencies of the architectures were unchanged.

Hoffer and Andreassen (1981), investigating the muscle mechanics of the hind limbs of decerebrated cats, showed that the stiffness of the muscles increased with force up to 25% of the maximal force, but remained within ±15% thereafter. The stiffness decreased slightly as the force approached tetanic contraction. Additionally, with the reflex present, they found that the stiffness varied little with operating length. For the noninnervated muscle (i.e. reflex absent), the muscular stiffness was linearly proportional to the force (exponential elastic behavior). Many other groups have shown similar findings (e.g. see Chapters 1, 3, 6, 34-36). Therefore, the assumption made here, that muscular stiffness is proportional to muscle activation, is in agreement with the findings of Hoffer and Andreassen (1981) for forces one-fourth of maximal contraction. It is reasonable to assume that posture is maintained with a level of muscular activation that produces forces less than one-fourth of maximal contraction.

EMG and Bilateral Cocontraction

Since the work of Sherrington (1909), the subject of coactivation of antagonistic muscles has received much attention and still remains an issue which is far from being resolved. We shall only discuss, and briefly, the work that is directed towards the coactivation of back muscles during isometric contracts, which is synonymous with our limitations of small angles and postural control. Note that since the spine is only subjected to axial loads in our model, there is by definition neither an agonist muscle nor an antagonistic muscle. Thus in this chapter, we are concerned only with cocontraction or coactivity that occurs bilaterally, and this bilateral activity must occur symmetrically for the spine to maintain its upright position.

Floyd and Silver (1955) were the first to study the electromyograms (EMG) activity of back muscles. They showed that the back muscles were active in the relaxed upright posture. Under static body weight, the ligamentous spine was unstable, and our model suggests that coactivation of back muscles are necessary to laterally stabilize the spine, as Floyd and Silver (1955) found.

Numerous researchers, including Schultz et al. (1987), Andersson et al. (1977), Serroussi and Pope (1987), and Basmajian and Deluca (1982), have since used EMGs to record antagonistic activity of the back muscles [see also Chapter 24 (Ladin) and Chapter 25 (Gracovetsky)]. Most researchers now agree that coactivation is a normal physiological phenomena. However, of the noteworthy biomechanical models to date for the spine, none predict coactivation (Hughes and Chaffin, 1988). Most muscular biomechanical models have used some form of an objective function, based indirectly on muscle activity. Thus coactivation, the expenditure of muscular energy without mechanical work, would not be a possible solution [however, cf. Chapter 19 (Seif-Naraghi and Winters)].

Let us assume the following: 1) critical muscle stiffness is a physiological concept neurologically regulated to the minimal value for maintaining upright posture; 2) the nervous system can sense an increase in the applied compressive loads on the spine. Then the model predicts an increase in coactivation of bilateral muscles as the compressive load on the spine increases. Cocontraction increases the stiffness of the spine, raising the critical load above the applied load and stabilizing the spine.

Serroussi and Pope (1987) proposed and verified that the difference between the EMGs of the contra and ipsilateral erector spinae muscle group was correlated with an induced frontal moment. These frontal plane moments were developed with constant dead weights in the outstretched arms of subjects. The moments were varied by varying the effective moment arm. The load on the spinal column, due to the extended dead weight, can be simply decomposed in to an axial compressive force and a moment (or couple). For the spinal column to support the increase in compressive force, the critical muscle stiffness must be raised; therefore, cocontraction is indicated by our model. The moment would be supported by the additional activation of the contralateral muscle, as Serroussi and Pope (1987) found. Since Serroussi and Pope (1987) reported results with only a single load, direct verification of the proposed coactivation role cannot be done.

Floyd and Silver (1955) recorded the EMG signals while a subject was standing with weights in each hand. They state: "With an equal weight in each hand there was no asymmetry of erectores spinae action, and usually very little increase in [EMG] activity over resting [standing relaxed]...." Unfortunately, the weights were not reported, nor was the frequency of the increase in EMG activity. All other published data on EMG studies have been directed at determining the specific role of muscles and their levels of activity for various postures and loads [see also review in Chapter 23 (Andersson and Winters)]. The simplest case of supporting increasing weights while maintaining upright posture, which would allow a direct verification of our model prediction of a bilateral increase in activity with increasing load, has not yet been thoroughly investigated. Floyd and Silver (1955) did report some increase in activity with increasing load, which is predicted by our model and cannot be predicted by equilibrium based models.

The assumption that the nervous system can sense an increase in the compressive loads on the spine is a difficult one to address. The nervous system can only sense displacement. With an increase in the compressive load on the spine, displacements could develop as lateral perturbations or as changes in the curvature of the spine in the sagittal plane. However, the magnitude or type of displacement needed to evoke an efferent signal is not known. Furthermore, the source of such signals is not known. However, two studies have recently touched upon this topic. Hunter and Kearney (1983) showed, when monitoring the motion of the human ankle, that dynamic muscle stiffness was invariant with fatigue. They were not able, however, to determine if this phenomena of muscle stiffness was a result of an invariance in the associated muscles' mechanical properties, or achieved via an unknown nervous regulatory mechanism. Skinner et al. (1986) showed that joint position sense decreased with fatigue. They associated this finding with a decrease in the efficiency of the muscle receptors, and concluded that the capsular receptors play a secondary role in position sense, and were unaffected by fatigue.

Muscle Stiffness and Muscle Force

In order to correlate muscular stiffness with the better understood variable of muscle force (or stress), Bergmark (1987) proposed a simple linear relationship between skeletal muscular stiffness and force in his stability model. Using the cross-bridge model for muscular contraction (Huxley, 1974), this relationship was based upon the work by Broberg (1981).

Bergmark (1987) proposed that the muscular stiffness (k_m) was proportional to the muscle force (F), and inversely proportional to the muscle length (l_m). The constant of proportionality was designated as q, and this relationship is written as

$$k_m = q\frac{F}{l_m} \qquad (26.14)$$

From the work of Morgan (1977) and Hunter and Kearney (1983), Bergmark (1987) calculated an average value of 40 for q (q is dimensionless).

In order to determine q more thoroughly, calculations were performed using the data from an extensive review of the literature: Morgan (1977), Rack and Westbury (1969), Joyce et. al. (1969), Agarwal and Gottlieb (1977), Hoffer and Andreassen (1981), Cannon and Zahalak (1982), An et. al. (1981), van Ingen Schenau et. al. (1988) and Woittiez et. al. (1987). From these calculations, a mean value of 10 was determined for q. The values ranged from 0.5 to 42. In addition to the difficulty with defining l_m, we conclude that the present knowledge of gross muscle mechanics is insufficient to endorse such a relationship as proposed by Bergmark (1987). This relationship, however, certainly deserves further consideration, since it is a noteworthy and simple concept.

Supportive Studies

In a histological study, Nitz and Peck (1986) determined the percentage volume of muscle spindles from various back muscles. They found that the rotators brevis spindle percentage volumes ranged form 4.58 to 7.30 times higher than those of the multifidus and semispinalis. The differences were significant ($p < 0.0001$). Based primarily upon these findings, Nitz and Peck (1986) hypothesized that the smaller intervertebral muscles play more of a sensory role than a mechanical role in stabilizing the spine, and that the mechanical requirements for stability were being fulfilled by the larger multisegmental muscles.

Thus, from a mechanical potential energy analysis and from a histological perspective, it appears that the intersegmental muscles do not play a primary role in the mechanical stability of the spine. However, simply classifying the roles of the intersegmental muscles as sensory and multisegmental muscles as mechanical is a simplistic view that should prove incomplete as our understanding increases.

The predictions of this model would also apply to any structure approximated as elastic. For example consider some spinal ligaments: If the ligaments of the spine were approximated as springs, then by analogy with the model, the ligaments would be more efficient, in a material sense, if they spanned several intervertebral joints. Recently, Bogduk and Twomey (1987) showed that the origins and insertions of the anterior and posterior longitudinal ligament spanned several vertebral joints.

26.5 Summary

Even though no rigorous definition of clinical stability exists, the spine is certainly dependent upon both its bony-ligamentous and muscular systems for maintaining stability [see also Chapter 25 (Gracovetsky) and Chapter 27 (Dietrich et al)]. The understanding of the mechanical stability of these systems should constitute the foundation for a better clinical understanding. In this chapter we have attempted to demonstrate the difference between equilibrium and stability in elastic mechanical systems and the difference between equilibrium-based and stability-based biomechanical models. By applying these definitions, with the given assumptions, to the spine modeled as a muscular Euler column we have:

1. Demonstrated that the multisegmental muscles of the spine were more efficient than the intersegmental muscles at stabilizing the spine in the frontal plane, with their efficiency increasing with the number of intervertebral joints spanned. The pelvic-originating muscles, which spanned the maximum number of joints, were 90% more efficient at laterally stabilizing the spine than the least efficient architecture of intersegmental muscles.

2. Found that the efficiency of stability increased the more lateral the muscular positions of attachment. This was due to the increasing moment arm of the muscular line of action.

3. Determined that the spine was unstable when a vertebral body was devoid of muscle, indicating instability if intervertebral stiffness is not increased when a vertebra is resected of its musculature.

4. Predicted that when the compressive load on the spine was increased, bilateral cocontraction of the back muscles occurs in order to stiffen the spine and raise the critical load.

Acknowledgments

This research was supported in part by NIH Grants AR30361 and AR39209, and CDC Injury Grant R49/CCR103551. We would also like to thank Ms. Cathy Nelson for her assistance in the preparation of this manuscript.

References

An, K.N., Hui, F.C., Morrey, B.F., Linscheid, R.L. and Chao, E.Y. (1981) Muscles across the elbow joint: a biomechanical analysis. *J. Biomech.* **14**: 659-669.

Andersson, G., Ortengren, R. and Nachemson, A. (1977) Intradiskal pressure, intra-abdominal pressure and myoelectric back muscle activity related to posture and loading. *Clin. Ortho. Res.* **129**: 156-164.

Agarwal, G.C. and Gottlieb, G.L. (1977) Oscillation of the human ankle joint in response to applied sinusoidal torque on the foot. *J. Physiol. Lond.*, **268**: 151-176.

Basmajian, J.V. and Deluca, C. (1982) *Muscles Alive: Their Function Revealed by Electromyography*. Williams and Wilkins, Baltimore.

Bergmark, A. (1987) Mechanical stability of the human lumbar spine. Doctoral Dissertation, Lund Institute of Technology, Department of Solid Mechanics, Lund, Sweeden.

Bergmark, A. (1989) Stability of the lumbar spine. A study in mechanical engineering. *Acta Ortho.*

Scand. No.230, Vol. 60, 1989 Bogduk N and Twomey LT (1987) *Clinical Anatomy of the Lumbar Spine*. Churchill Livingston, New York.

Bradford, D.S. (1980) Spinal stability: orthopaedic perspective and prevention. *Clin. Neurosurg.* **27**: 591-610.

Broberg, K.B. (1981) The mechanical behavior of the spinal system. Report from the division of solid mechanics, Lund Institute of Technology, Lund, Sweeden.

Cannon, S.C. and Zahalak, G.I. (1982) The mechanical behavior of active human skeletal muscle in small oscillations. *J. Biomech.* **15**: 111-121.

Crisco, J.J., Panjabi, M.M., Yamamoto, I., Oxland, T.R. (1990) The Euler stability of the lumbar spine: Part II. Experiment (in preparation).

Crisco, J.J. (1989) The biomechanical stability of the human lumbar spine: experimental and theoretical investigations. Doctoral Dissertation, Department of Mechanical Engineering, Yale University, New Haven, Conn.

Floyd, W.F. and Silver, P.H.S. (1955) The function of the erectores spinae muscles in certain movements and postures in man. *J. Physiol.* **129**:184-203.

Frymoyer, J.W., Pope, M.H., and Wilder, D.G. (1988) *Segmental Instability. The Lumbar Spine, ISLS*, WB Saunders, Orlando, FL.

Granhed H., Johnson, R., Hansson, T. (1987) The loads on the lumbar spine during extreme weight lifting. *Spine* **12**:146-149.

Greene, P.R. and McMahon, T.A. (1979) Reflex stiffness of man's anti-gravity muscles during kneebends while carrying extra weights. *J. Biomech.* **12**:881-891.

Hoffer, J.S. and Andreassen (1981) Regulation of soleus muscle stiffness in premammillary cats: intrinsic and reflex components. *J. Neurophys.* **45**:267-285.

Holdsworth, F. (1970) Fractures, dislocations, and fracture-dislocations of the spine. *J. Bone and Joint Surg.* **52-A**:1534-1551.

Houk, J.C. (1979) Regulation of stiffness by skeletomotor reflexes. *Ann. Rev. Physiol.* **41**:99-114.

Hughes, R.E. and Chaffin, D.B. (1988) Conditions under which optimization models will not predict coactivation of antagonist muscles. *Proc. Am. Soc. Biomech.* University of Illinois, Urbana.

Hunter, I.W. and Kearney, R.E. (1983) Invariance of ankle dynamic stiffness during fatiguing muscle contractions. *J. Biomech.* **16**:985-991.

Joyce, G.C., Rack, P.M.H. and Westbury, D.R. (1969) The mechanical properties of cat soleus muscle during controlled lengthening and shortening movements. *J. Physiol.* **204**:461-474.

Keele, K.D. (1983) *Leonardo da Vinci's Elements of*

the Science of Man. Academic Press, New York.

Kirkaldy-Willis, W.H. and Farfan, H.F. (1982) Instability of the lumbar spine. *Clin. Ortho.* **165**:110-123.

Lucas, D.B. and Bresler, B. (1960) Stability of the ligamentous spine. *Technical Report ser. 11 No. 40*, Biomechanics Laboratory, Univ. California, San Francisco.

McMahon, T.A. and Greene, P.R. (1978) Fast running tracks. *Sci. Amer.* **239**:148-163.

Morgan, D.L. (1977) Separation of active and passive components of short-range stiffness of muscle. *Am. J. Physiol.* **232**:45-49.

Nichols, T.R. and Houk J.C. (1976) Improvement in linearity and regulation of stiffness that results from action of the stretch reflex. *J. Neurophy.* **39**:119-142.

Nitz, A.J. and Peck, D. (1986) Comparison of muscle spindle concentrations in large and small human epaxial muscles acting in parallel combinations. *American Surgeon* **52**:273-277.

Panjabi, M.M., Thibodeau, L.L., Crisco, J.J. and White, A.A. (1988) What constitutes spinal instability? *Clin. Neurosurg.* **34**:313-339.

Panjabi, M.M., Hausfeld, J.N., White, A.A. (1981) A biomechanical study of the ligamentous stability of the thoracic spine in man. *Acta Orthop. Scand.* **52**:315-326.

Pope, M.H. and Panjabi, M.M. (1985) Biomechanical definitions of spinal instability. *Spine* **10**:255-256. (Symposium of Instability of the Lumbar Spine during the International Society for the Lumbar Spine in Cambridge, England, April, 1983.)

Posner, I., White, A., Edwards, T. and Hayes, W. (1982) A biomechanical analysis of the clinical stability of the lumbar and lumbosacral spine. *Spine* **7**:374-389.

Rack, P.M.H. and Westbury, D.R. (1969) The effects of length and stimulus rate on tension in the isometric cat soleus muscle. *J Physiol.* **204**:443-460.

Schultz, A., Cromwell, R., Warwick, D. and Andersson, G. (1987) Lumbar trunk muscle use in standing isometric heavy exertions. *J. Ortho. Res.* **5**:320-329.

Serroussi, R. and Pope, M. (1987) The relationship between trunk muscle electromyograhy and lifting moments in the sagittal and frontal planes. *J. Biomech.* **20**:135-146.

Sherrington, C.S. (1909) On reciprocal innervation. In D. Denny-Brown (ed), *Selected Writtings of Sir Charles Sherrington*, P.B. Hoeber, New York, 1940, pp 237-313.

Skinner, H.B., Wyatt, M.P., Hodgdon, J.A., Conard, D.W. and Barrack, R.L. (1986) Effect of fatique on joint position sense of the knee. *J. Ortho. Res.* **4**:112-118.

Stein, R.B. (1982) What muscle variable(s) does the nervous system control in limb movements? *Behav. Brain Sci.* **5**:535-577.

Van Ingen Schenau, G.J., Bobbert, M.F., Ettema, G.J., De Graaf, J.B. and Huijing, P.A. (1988) A simulation of rat edl force output based on intrinsic muscle properties. *J. Biomech.* **21**:815-824.

White, A.A., Johnson, R.M., Panjabi, M.M. and Southwick, W.O. (1975) Biomechanical analysis of clinical stability in the cervical spine. *Clin. Ortho.* **109**:85-96.

White, A.A. and Panjabi, M.M. (1978) *Clinical Biomechanics of the Spine*. JB Lippincott Co., Philadelphia.

Whiteside, T.E. (1977) Traumatic kyphosis of the thoracolumbar spine. *Clin. Ortho. Related Res.* **128**:78-92.

Wilder, D. (1985) On loading of the Human Lumbar Intervertebral Motion Segment. Ph.D. Dissertation, Civil and Mechanical Engineering Dept, Univ of Vermont.

Woittiez, R.D., Brand, C., Haan, de A., Hollander, A.P., Huijing, P.A., Tak, R. van der and Rijnsburger, W.H. (1987) A multipurpose muscle ergometer. *J. Biomech.* **20**:215-218.

CHAPTER 27

Modeling of Muscle Action and Stability of the Human Spine

Marek Dietrich, Krzysztof Kedzior and Tomasz Zagrajek

27.1 Introduction

Biomechanical investigations of the human spine are nowadays rarely taken up for purely cognitive reasons. Usually they are stimulated by needs of various areas of contemporary civilization and directed toward applying their results in medicine (etiology of diseases and defects of the spine, rehabilitation, occupational medicine), technology (e.g. influence of a vehicle upon the spine of its human operator in normal or emergency conditions), and sports (influence of overloads due to sport training upon the spine).

Muscles play a very important role in the spine system owing to their active influence upon bone elements, which in turn allows the position of the human body to be controlled, even while under external loads, both in static conditions and during motion. The bones and joints in the spine system seem, at least at first glance, to be rather clear, which is contrary to the system of muscles, which seem to be rather complex. This impression is due to the large number of muscles, their complicated and varied shapes, and their multi-functionality.

It seems that the progress in the spine system biomechanical research, especially concerning the principles underlying spine movement organization, depends largely on discovering the principles of operation and cooperation of muscles belonging to the system.

27.2 Assessment of Pertinent Literature

27.2.1 General Remarks

The range of *in situ* experimental research of the human spine system is very restricted. Such research may produce only limited information, and in certain cases [e.g. measurement of pressures in intervertebral disks (Nachemson, 1981)] may be rather risky to the subject's health. The present state of knowledge, however, does not allow for quantitative measurement of tensions and deformations in elements of the system, and forces exerted by each of the muscles in particular. Even obtaining reliable data representing geometrical dimensions of the system for a given subject is impossible without employing methods such as X-ray radiography.

For the above-mentioned reasons most investigations of the spine systems are conducted with the help of either material or mathematical models. The first of the two methods resolves itself to investigating specimens from corpses (Yang and King, 1984), replicas made of photo-elastic resins (Dietrich and Kurowski, 1983), or replicas made of composite materials representing bones, ligaments, and muscles [Chapter 28 (Winters et al.)]. Such investigations are, however, toilsome and expensive, and these methods are limited in terms of investigating the operation of muscles in the spine system. For the method using photo-elastic resins, this is due to the fact that it is impossible to model a system

Multiple Muscle Systems: Biomechanics and Movement Organization
J.M. Winters and S.L-Y. Woo (eds), © 1990 Springer-Verlag

composed of both rigid (bones) and elastic (muscles) elements and to investigate tension distribution in both kind of the elements simultaneously. The second of the two methods - mathematical modeling aided with computer simulation - is free of such limitations and therefore is now used most frequently. However, this does not mean that using that method it is easy to model muscles. Generally, one is faced here with the following difficulties:

1. The necessity to account in one model for a large number of muscles differing in size, shape, number of attachments, and directions of muscle fibers.

2. The necessity to account for the double function of each muscle. One of the two functions is developing passive force. In this aspect a muscle not stimulated with a nervous signal can be treated as an elastic element of the system built of a material (muscular tissue) of anisotropic, non-linear mechanical properties. The other function is exerting active force, the direction of which is the same with the direction of muscle fibers (a muscle can only pull), and its value depends on three parameters: stimulation, length, and speed of contraction (in case of concentric work) or extension (in case of eccentric work) of the muscle.

3. The necessity to formulate a criterion describing the way muscles are controlled by the nervous system during the considered motion act.

Now let us consider the above–mentioned problems.

27.2.2 Geometrical Modeling of the Muscular System

During mathematical modeling of the musculoskeletal system it is usually assumed that whole muscles or their parts are one-dimensional strings, the axis of which forms a straight or a broken line. This type of simplification is commonly used while modeling the muscles of human limbs, for which the string model is an adequate approximation (Seireg and Arvikar, 1976; Dabrowska and Kedzior, 1981). This model, however, is also employed while describing the spine system, where its use is better justified for modeling muscles of small diameters and small, almost

point-like attachments to bones (e.g. muscles: rotatory, multifidus, interspinal), but in case of such muscles as erector spinae, of large diameter and large number of attachments located on the lateral surface of the muscle, this type of a model is a very rough approximation of reality. And for abdominal muscles (transverse, oblique, diaphragm), which play a very important role in mechanical operation of the human spine system by producing pressure in the abdominal cavity, this model is totally unsuitable. What is more, the string model does not account for the fact that muscles when contracting change their shape and act upon neighboring muscles, pressing them with their lateral surfaces. Therefore the real muscular force operation is different from that determined with the string model. It seems that a way to overcome the above mentioned difficulties is by employing the Finite Element Method (*FEM*) to model complex muscular systems. The method allows for effective, three-dimensional modeling of all elements of the human spine system, which was mentioned in earlier works (Dietrich et al., 1989; Dietrich and Zagrajek, 1989).

27.2.3 Modeling of Muscles as Force Generators

The second of the mentioned problems – passive and active role of the muscles – may seem to be easier to account for in the spine model. Modeling of separate skeletal muscles as dynamic force generators is one of the more developed areas of contemporary biomechanics, and a great deal of information concerning the subject has been gathered [Hatze, 1980; Chapter 8 (Zajac and Winters)]. Usually the proposed mathematical models are in the form of an ordinary differential equation or set of equations (Kedzior, 1973; Soechting and Roberts, 1975; Biezanowska and Kedzior, 1981) and coefficients of the equations are usually functions of the muscle length (Dabrowska and Kedzior, 1985). In mathematical modeling it is not very important whether the equations are derived based on considerations of the visco-elastic properties of the muscle, or treated as an object of regulation. The main problem here is that the coefficients (time constants) occurring in the equations are different for different muscles (Winter, 1979), and sometimes for a different muscle group a different degree of the differential equation is proposed

(Dabrowska and Kedzior, 1985). Furthermore, muscle force depends not only on neural input but also on the muscle length history (Chapter 1 (Zahalak and Winters)). Therefore, prior to modeling of the spine system, one should identify the parameters of the mathematical model for each of the muscles in the system. Nowadays it is not possible, which, of course, decreases the reliability of the obtained results.

In some of the older models (Andriacchi et al., 1974; Belytschko, et al., 1973; Orne and King Liu, 1971; Schultz and Galante, 1970; Schultz et al., 1973) active components of muscular forces are omitted, and approximate passive components (elasticity) were accounted for. Those models served mainly for determining external forces in the spinal column during high accelerations acting upon the human body (car crash, catapulting of airplane pilot), i.e. in extreme situations not occurring during normal human activity. In the modern models of the human operator during emergency situations a simplified way of accounting for the active component of muscular forces is employed which involves one equation (constructed on the basis of the visco-elastic model) for the whole group of muscles operating at a given joint (Laananen et al., 1982; Laananen, 1987).

There are also models which account only for the active component of muscle or muscle groups (e.g. Seireg and Arvikar, 1976; Yettram and Jackman, 1980). Both the active and the passive roles of muscles have been accounted for in a number of models designed throughout the last decade [Dietrich and Kurowski, 1985; Bergmark, 1987; Deng and Goldsmith, 1987; Williams and Belytschko, 1983; Chapter 28 (Winters et al.)].

Omitting the passive component (elasticity) of muscular forces is an acceptable simplification in static or quasi–static analysis of the human locomotion system (including the spine model) for large loads, i.e. when active forces developed by muscles are considerably greater than the passive ones. In investigations considering dynamics and stability, however, the passive component should be accounted for since its role in the overall elasticity and damping of the system is considerable.

27.2.4 Modeling of Muscle Control

The way the nervous system, by muscle stimulation, acts upon the bone elements of the spine system can be regarded as execution of an algo-rithm (program) of control. The question is according to what "program" the nervous system acts. This is, of course, a task for neurophysiology, and in regard to the spine system it still requires serious research work. In biomechanics the problem is called the "muscle cooperation problem" or the "general distribution problem" (GDP). Lacking the appropriate neurophysiological information, the research workers try to give an approximate description of the muscle system operation. For the description some general observations are used, and it is assumed that the control is done in a "logical, optimum" way. The literature gives many examples of solutions of optimization problems based on the analysis of the human body mechanics [see Chapter 8 (Zajac and Winters) for review]. Arvikar and Seireg (1978), modeling intervertebral disks under sudden accelerations, minimized the linear combination of forces and torques. Coefficients of objective function were selected so that the results were consistent with quantitative observations in situ. Gracovetsky et al. (1977; see also Chapter 25), modeling the lumbar spine, accepted as the optimization criterion the minimum sum of the shearing forces' squared in five intervertebral disks. Crowninshield et al. (1978) investigated mechanics of the hip, using a linear objective function in the form of the sum of quotients of all muscle forces and physiological cross sections of respective muscles. Yettram and Jackman (1980), investigating the lumbar section of spine, minimized the sum of forces in the muscles. The same was done by Dietrich and Kurowski (1985), and they minimized also the sum of intervertebral forces and minimized the maximum shearing force in lumbar disks.

Looking at the above examples, the conclusion drawn by Dietrich and Kurowski (1985) and confirmed by Aruin et al. (1989) is interesting and encouraging. They proved that for different "logical" criteria of muscle control, similar distributions of muscular forces are usually obtained. It should also be mentioned that most of the tasks presented so far of the GDP type consider the operation of the musculo–skeletal system in steady–state conditions of all muscles involved (tetanic contraction-stimulation does not change; muscular force depends on the length and speed of the contraction but does not explicitly depend on time). Such tasks can be solved with parametric optimization methods. In those few tasks which dealt with the contraction accounting for the transient processes in muscles (when during a motion separate muscles join in or out of the work), there were also attempts to achieve a mathematical model allowing use of parametric optimization (Hatze 1975; Biezanowska, 1984). Solving the optimization task of a multi-dimensional, non–linear dynamic process, although theoretically possible, is numerically difficult to perform. It refers mainly to investigations of the human spine system, where the number of cooperating muscles is large.

27.3 Method

To investigate the operation of the multi-muscular system serving the human spine we have used an improved mathematical model designed with the help of *FEM* (described in Dietrich et al., 1989; Dietrich and Zagrajek, 1989). This model accounts for those muscles which considerably influence the spine system operation: longissimus, iliocostal, spinal, interspinal, intertransverse, semi-spinal, multifidus, rotatory, straight m. of abdomen, oblique internal m. of abdomen, oblique external m. of abdomen, transverse m. of abdomen, diaphragm, quadratus, psoas, levators of ribs. Those muscles have been divided into 600 three-dimensional, 20-node finite elements (Zienkiewicz, 1977). The *FEM* mesh has been done so that every element would be a functional unit of the muscle with parallel fibers. Assumed were linear, anisotropic mechanical properties of the muscle tissue (passive component), described in a local cartesian reference system, one axis of which runs along the muscular fibers direction. The system changes its orientation in space owing to change of the muscular fibers direction. It has been assumed that the finite elements in the muscles contract along muscular fibers under neural stimulation. Contraction of the elements modeling muscles produces forces (active components) acting upon the bone system of the spine. In the described model, 302 independent contracting muscle zones have been distinguished (some finite elements form groups contracting together). The division mesh of some muscles into finite elements is shown in Figure 27.1.

The dependence between the tensions, the deformations, and stimulations has been assumed in the following form (steady-state tetanic contraction, small changes of muscles length):

$$\sigma = C\,\varepsilon + C\,K\,p \qquad (27.1)$$

where: σ = tension tensor, ε = deformation tensor, C = matrix of material constants, K = matrix of constant coefficients, p = vector of stimulations of individual muscular zones (groups of elements).

The first term in the sum of *Eq. 27.1* describes the passive components, the second one the active components. The material data of muscles, bones, cartilage, and tendons were taken on the basis of literature (Obrastsov, 1988; Skalak and Shu Chien, 1987). Because of the wide scatter of values cited in the literature, averaged data have been used. For example, for the not stimulated muscular tissue the Young's modulus along muscular fibers has been accepted as $E = 0.4$ MPa, and cross muscular fibers as $E = 0.2$ MPa. The geometrical and mass data of the model elements have been gathered on the basis of our own measurements of spine specimens.

The abdominal cavity has been treated as an incompressible (with constant volume) body despite the fact that tension of the muscles (mainly abdominal muscles and the diaphragm) causes pressure in it, which decreases the loading of the spine column in its lumbar section. The assumption would be precise if the abdominal cavity were filled with liquid. However, there are also gases in it. Considering both liquids and gases would be difficult because of the lack of appropriate data in literature, and it would make the model much more complex. Moreover, a slight pressure change influences the abdominal cavity volume very little. It can be estimated that pressure change $\Delta p = 0.02$ MPa, with 10% of the gas component in the abdominal cavity causing the change of its volume less than 2%.

The spine system model was loaded with the body weight of 300 N, which was substituted by a system of concentrated forces applied at nodes of the model finite elements (~20,000 nodes). The external load, modeling a weight of 400 N held in hands, was applied in the form of concentrated forces to nodes in the zone of scapulas, symmetrically to the sagittal plane.

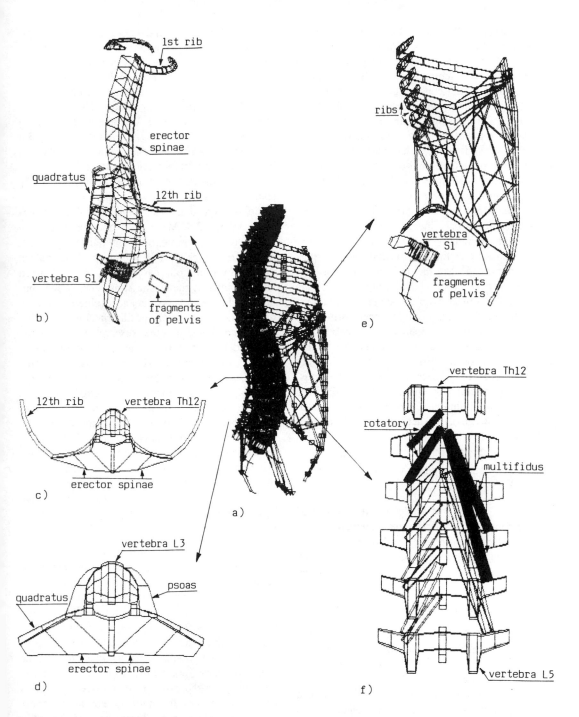

Figure 27.1: *FEM* mesh: *a)* general (lateral) view of the spine system model in the straightened position (muscles, bones, ligaments, disks); *b)* erector spinae (right part) and quadratus muscle (left part); *c)* attachments of the erector spinae to the bone system; *d)* muscles: erector spinae, quadratus, and psoas in cross section between vertebrae L3-L4; *e)* abdominal muscles; *f)* fragments of multifidus muscle (right part) and short rotatory muscles (left part).

The human spine system model employed in our investigation consists of 2640 finite elements – rigid (bones) and elastic (muscles, ligaments, cartilages, intervertebral disks). Its respective mathematical model consists of 13,107 generally non–linear equations, which are linearized for the considered spine position. Using computer simulation we have investigated two cases of work of the muscles. In the first case it has been assumed that all muscles participate in balancing the external load; in the second case that the external load is balanced only by muscles which are close to the spine: multifidus and interspinal. Stability of the spine system has been investigated as well, determining spine deflection shapes due to instability.

The tensions in the muscles have been determined assuming that the task of the muscular system in both cases is balancing the weights (body and external), with the spine position kept straightened. The condition has been formulated assuming that all components of the pelvis displacement and the horizontal component in the sagittal plane of the first thoracic vertebra are equal to zero. It has been assumed that the muscular system performs the task with minimum work expense (minimum elastic energy accumulated in muscles). That criterion has been formulated is the following form:

$$\sum_{m=1}^{k} \int_{V_m} \frac{N_m^2}{E_m} \, dV_m = \min, \quad N \geq 0 \qquad (27.2)$$

where: k = number of independently stimulated muscle zones, V_m = muscle zone volume, N_m = tension along muscle fibers, E_m = Young's modulus (the passive and active component together) along muscle fibers (different values for different muscles).

27.4 Results and Discussion

27.4.1 Operation of the Spine System Muscles

Predicted tension distributions along muscle fibers in exemplary muscles are presented in Figure 27.2 (first case - all muscles work). Similar tension distributions have been obtained for all other muscles of the system. The obtained values of tensions (about 0.2 MPa) seem to be correct (Obrastsov, 1988). The pressure in the abdominal cavity, which in this case equals 0.0047 MPa, is small in comparison with the predicted

values (Chaffin and Andersson, 1984). Analysis of the tension distribution configuration in the whole spine system has proven that the external load is in this case balanced mainly by the tension of muscles with large diameters and operating on large radii, like the erector spinae. The tensions in short muscles with small diameters operating on small radii, e.g. rotatory, are relatively small. But their tension is also necessary, since they play the important role in stabilization of the spine.

Tension distribution for the second case (only some muscles work) proved to be not reliable, as could be expected. The obtained values of tensions (about 3.5 MPa) with the external load relatively small more than three times exceeds maximum values cited in literature. The tension distributions in both cases have been obtained with minimizing the elastic strain energy accumulated in the muscular system according to criterion of Eq. 27.2. That criterion, like most optimization criteria employed by other researchers, is suitable for modeling of muscle control in static conditions. As a result of its use for small external loads, mainly those muscles work that attain large torques with respect to the joint rotation axes. The other muscles join in the work for large loads when constraints built in the optimization algorithm in the form of maximum acceptable tensions in muscular tissues are active. Thus a new criterion should be found which, simultaneously to the increase of the spine load, would cause the increase of tension in short muscles. Quantitative formulation of such a criterion requires further investigation.

27.4.2 Stability of the Spine System

Idiopathic scoliosis is the most common type of scoliosis. Medical publications present various hypotheses concerning the causes of the disease. A new hypothesis, however, can be formulated: idiopathic scoliosis is due to the fact that a symmetrically built and symmetrically loaded spine can be instable. Stability here is understood in the Lapunov sense, commonly used in mechanics. Depending on the external load value and values of muscular forces, the spine system can be stable or instable. The spinal column can be deflected in different forms because of instability. Nonlinearity of the system is the reason that every form of deflection may have its respective form of local equilibrium. All these forms are asymmetri-

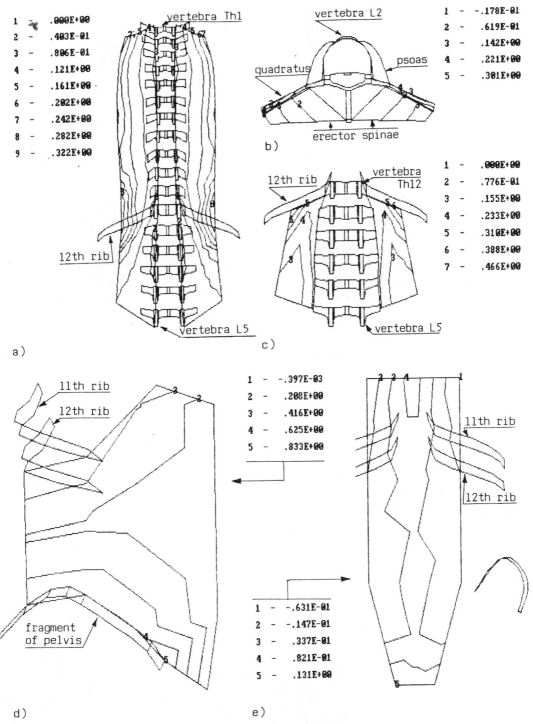

Figure 27.2: Tension distribution along muscle fibers: *a)* erector spinae; *b)* muscles: erector spinae, quadratus, and psoas, cross-section in transverse plane on the level of vertebra L2; *c)* quadratus muscle; *d)* transverse muscle of abdomen; *e)* straight muscle of abdomen. In tables there are given tension values in MPa respective to numbers of isolines.

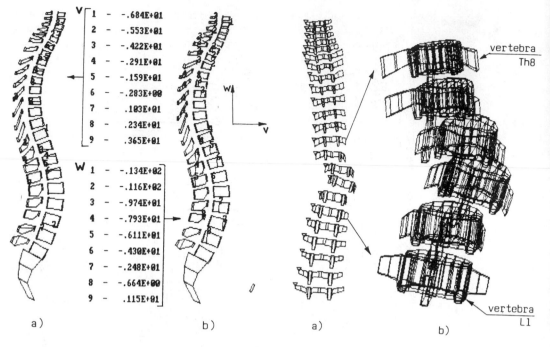

V	1	− −.684E+01
	2	− −.553E+01
	3	− −.422E+01
	4	− −.291E+01
	5	− −.159E+01
	6	− −.283E+00
	7	− .103E+01
	8	− .234E+01
	9	− .365E+01
W	1	− −.134E+02
	2	− −.116E+02
	3	− −.974E+01
	4	− −.793E+01
	5	− −.611E+01
	6	− −.430E+01
	7	− −.248E+01
	8	− −.664E+00
	9	− .115E+01

a)   b)   a)   b)

vertebra Th8

vertebra L1

Figure 27.3: Spine equilibrium in the state of stability (side view): *a)* marked isolines of the transverse component of displacement in sagittal plane *v*; *b)* marked isolines of the vertical component of displacement in sagittal plane *w*. Tables give numbers of the isolines of displacement values in millimeters.

Figure 27.4: First form of deflection due to instability of the spine (back view): *a)* the whole spine; *b)* vertebrae *T8-L1*.

cal and are characterized by lateral bending and rotation of vertebrae, despite the aforementioned fact that the original spine system and the external loads were fully symmetrical.

Figure 27.3 presents the spine system displacements in the state of stability (only vertebrae are shown) deformed by load (in the considered case all muscles work). Because of the symmetry of both the investigated system and the applied load in respect to the sagittal plane, displacements in the frontal plane equal zero. It can be noticed that, in comparison with Figure 27.1a, the load causes an increase of the spine curves (kyphosis and lordosis).

When the load is increased by about 20% the spine system loses its stability. A form of spinal deflection is shown in Figure 27.4. Because of linearization of the equations, only the proportions of the displacements and rotation angles of individual vertebrae are known, not their absolute values. The form shown in Figure 27.4 is charac-

terized by large displacements of vertebrae *T10* and *T11* and simultaneous rotation of all vertebrae. The rotation angle of vertebrae, beginning at the *L5*, increases in the lumbar section and attains its maximum value at the *T11*. Then it decreases in the thoracic section. The relative rotation axes of the vertebrae are beyond the spinal column. The spine deformation visible from behind is mostly due to the fact that during the rotation the location of the natural spine curves changes (lordosis and kyphosis are now in different planes).

Increase of the external load of the system may cause a few further forms of deflection to appear (at the increase by about 38%, 60%, and 180%, respectively). In order to state which of those forms can occur, calculations taking into account the nonlinearity of the system would have to be done.

Simulations have also been performed for the second case, when only some muscles work. It turns out that in this case the spine system is instable even for the load that is smaller by 15% than the original external load.

Among the investigated forms of spine deflection (due to instability), none have been found that are characterized by bending in the sagittal plane. Considering the system mechanics, this fact can be explained by the existence of the almost incompressible volume of the abdominal cavity and strong abdominal muscles and the diaphragm. When the spine is bent forward, in the abdominal cavity additional pressure must appear which causes tension in the abdominal muscles and the diaphragm; otherwise the abdominal cavity's constant volume would not be preserved. In that situation the tension of the abdominal muscles and the diaphragm indirectly opposes the forward bending of the spine, increasing the system stiffness in that direction (Tesh et al., 1987). Similar, although probably to a lesser degree, is the influence of the incompressibility areas of nuclei pulposi of the intervertebral disks.

27.5 Conclusions

The conducted research indicated the necessity of three-dimensional modeling of the muscular system while investigating complex biomechanical objects such as the human spine system.

The presented results of computer simulation allow formulation of a hypothesis concerning the cause of idiopathic scoliosis: the cause is the loss of stability of the spine system due to incorrect operation of the muscular system. The operation is incorrect when some muscle groups are too weak in comparison with other groups. The phenomenon does not have to be asymmetrical (which has been assumed so far), since even with full symmetry of the spine system and of the acting loads, there appear asymmetrical forms of deflections due to the instability. Those forms are characterized by lateral (frontal) bending and vertebra rotation. That deflected state can be fixed by remodeling of the bones, which leads to permanent deformation of the spine (Stokes et al., 1981). Local deformations of the spine due to instability can be the reason of further ailments popularly called discopathy.

27.6 Recommended Future Directions

The present work proves that for fully symmetrical (in respect to sagittal plane) geometry, distribution of masses, and external loads, asymmetrical positions of spinal column can occur. The problem requires further qualitative and quan-

titative investigations of the system treated as a dynamic one (i.e. with inertia and damping forces). Because of its multi-dimensionality and nonlinearity, it is practically impossible to employ the known methods of stability investigation in order to state whether (and in what state parameter zones) a given point of equilibrium is asymptotically stable. Therefore further investigations can be conducted by trial method only. For various symmetrical and asymmetrical initial configurations and different external loads, the search can be done in order to investigate more closely these selected zones of the system.

In conjunction with investigations of the whole spine system, it seems advisable to conduct simulations concerning its fragments. Models of the subsystem can be more detailed and can represent more precisely the geometry, material properties, pathologies (e.g. osteoporosis), the liquid flow between the system elements, the nonlinearity of parameters, etc. External loads of those fragments can be determined on the basis of investigation of the whole system. Results of such investigations can be helpful for explaining the underlying biomechanical reasons for certain ailments and diseases (e.g. discopathy, vertebra fractures, etc.). Such investigations have already been initiated (Dietrich et al., 1988).

References

Aruin A., Zatsiorsky V., Prilutsky B. (1989). Decision of distributional problem using "logical" and "illogical" optimization criteria. In: Gregor R.J., Zernicke R.F., Whiting W.C. (eds), *Proceedings of the XII Int. Congr. of Biomech.*, Univ. of California, Los Angeles, 431.

Andriacchi T., Schultz A.B., Belytschko T., Galante J. (1974). A model for studies of mechanical interactions between the human spine and rib cage. *J. Biomech.* 7: 497-507.

Arvikar R.J., Seireg A. (1978). Distribution of spinal disc pressure in the seated posture subjected to impact. *Aviation, Space and Environmental Medicine* 49(1): 166-169.

Belytschko T.B., Andriacchi T.P., Schultz A.B., Galante J.D. (1973). Analog studies of forces in the human spine. *J. Biomech.* 6: 361-371.

Bergmark A. (1987). Mechanical stability of the human lumbar spine. Ph.D.Thesis, Lund Institute of Technology, Sweden.

Biezanowska E. (1984). modeling of muscle coaction under dynamic conditions. *Biology of Sport* 1: No.3/4, PWN, Warsaw, 199-208.

Biezanowska E., Kedzior K. (1981). Simulation approach to modeling and investigation of static and dynamic properties of skeletal muscles. In: Morecki A. et al. (eds), *Biomechanics VIIA*, University Park Press - Baltimore, PWN, Warsaw, 208-214.

Chaffin D.B., Andersson G.B. (1984). *Occupational Biomechanics*. John Willey and Sons, New York.

Crowninshield R.D., Johston R.C. Andrews J.G., Brand R.A. (1978). A biomechanical investigation of the human hip. *J.Biomech.* **11**: 75-85.

Dabrowska A., Kedzior K. (1981). Cooperation of muscles under dynamic conditions. In: Morecki A. et al (eds), *Biomechanics VIIA*, University Park Press, Baltimore; PWN, Warsaw, 215-222.

Dabrowska A., Kedzior K. (1985). Investigation and modeling of relationship between integrated surface EMG and muscle tension. In: Winter D.A. et al. (eds), *Biomechanics IXA*, Human Kinetics Pub., Champaign, Illinois, 308-312.

Deng Y.C., Goldsmith W. (1987). Response of a human head/neck/upper-torso replica to dynamic loading - II. Analytical/numerical model. *J. Biomech.* **20**: 487-497.

Dietrich M., Kedzior K., Zagrajek T. (1988). Finite element method analysis of human spine segment. In: de Groot et al. (eds), *Biomechanics XIA*, Free University Press, Amsterdam, 333-337.

Dietrich M, Kedzior K., Zagrajek T. (1989). Model of human spine system. In: Gregor R.J., Zernicke R.F., Whiting W.C. (eds), *Proceedings of the XII Int. Congr. of Biomech.*, Univ. of California, Los Angeles, 381.

Dietrich M., Kurowski P. (1983). Model of the human lumbar spine. *Proc. of Sixth IFToMM Congress on Theory of Machines and Mechanisms*, Indian Institute of Technology, Delhi, vol.2, 1386-1389.

Dietrich M., Kurowski P. (1985). The importance of mechanical factors in the etiology of spondylolysis. A model analysis of loads and stresses in human lumbar spine. *Spine* **10(6)**: 532-542.

Dietrich M., Zagrajek T. (1989). Simulation of human spine system. *Second Int. Symp. on Computer Simulation in Biomechanics*, Univ. of California, Davis, 32-33.

Gracovetsky S., Farfan H.F., Lamy C. (1977). A mathematical model of the lumbar spine using an optimized system to control muscles and ligaments. *Orthop. Clin. N.A.* **8(1)**:135-153.

Hatze H. (1975). A control model of skeletal muscle and its application to a time optimal biomotion. Ph.D. thesis, Univ. of South Africa, Pretoria.

Hatze H. (1980). Neuromusculoskeletal control systems modeling - a critical survey of recent developments. *IEEE Transactions on Automatic Control*, **AC-25(3)**: 375-385.

Kedzior K. (1973). Investigation of dynamic properties of isolated skeletal muscles. *Archive of Mechanical Engineering* **XX(2)**, Polish Scientific Pub., Warsaw, 219-238.

Laananen D.H. (1987). Passenger response in transport aircraft accidents - a simulation. *Soma - Engineering for the Human Body*, **2 (1)**: 18-25.

Laananen D.H., Bolukbasi A.D., Coltman J. (1982). Computer simulation of an aircraft seat and occupant in a crash environment (final report), Simula Inc., Tempe, Arizona.

Nachemson A.L. (1981). Disc pressure measurements. *Spine* **6**: 93-97.

Obrastsov I.F. (ed) (1988). *Strength problems in biomechanics*. Wysschaya Schola, Moscow (in Russian).

Orne D., King Liu Y. (1971). A mathematical model of spinal response to impact. *J. Biomech.* **4**: 49-71.

Schultz A.B., Belytschko T.B., Andriacchi T.P. (1973). Analog studies of forces in the human spine: mechanical properties and motion segment behaviour. *J. Biomech.* **6**: 373-383.

Schultz A.B., Galante J. (1970). A mathematical model for the mechanics of the human vertebral column. *J. Biomech.* **3**: 405-416.

Seireg A., Arvikar R.J. (1976). A mathematical model for evaluation of forces in the lower extremities of the musculoskeletal system. *J. Biomech.* **6**: 313-326.

Skalak R., Shu Chien (eds), (1987). *Handbook of Bioengineering*. McGraw-Hill Book Co., New York.

Soechting J.F., Roberts W.J. (1975). Transfer characteristics between EMG activity and muscle tension under isometric conditions in man. *J. Physiology Paris* **70**: 779-793.

Stokes I.A.F., Bigalow L.C., Moreland M.S. (1981). Three dimensional spinal curvature in idiopathic scoliosis. *J. Orthop. Research* **5**: 102-113.

Tesh K.M., Dunn S.J., Evans J.H. (1987). The abdominal muscles and vertebral stability. *Spine* **12(5)**: 501-508.

Williams J.L., Belytschko T.B. (1983). A three-dimensional model of the human cervical spine for impact simulation. *J. Biomechanical Engineering* **105**: 321-330.

Winter D.A. (1979). *Biomechanics of Human Movement*. John Wiley and Sons, New York.

Yang K.H., King A.I. (1984). Mechanism of facet load transmission as a hypothesis for low-back pain. *Spine* **9(6)**: 557- 565.

Yettram A.L., Jackman M.J. (1980). Equilibrium analysis for the forces in the human spinal column and its musculature. *Spine* **5(5)**: 402-411.

Zienkiewicz D.C. (1977). *The Finite Element Method in Engineering Science*. 3rd edition, McGraw-Hill, London.

CHAPTER 28

Neck Muscle Activity and 3-D Head Kinematics During Quasi-Static and Dynamic Tracking Movements

Jack M. Winters and Joseph D. Peles

28.1 Introduction

The head-neck system is one of the least understood (and most complex) neuromechanical systems of the body. The head rests on the top of the cervical spinal column; it can be considered the final link of an open kinematic chain that also includes the seven cervical and upper thoracic vertebrae. Each vertebra is connected in series to its neighbors by multiple joints and is capable of *3-D* relative rotation. This capacity for rotation is a function of the level within the spine, with the top two (anatomically unique) cervical vertebrae causing especially unique kinematic and mechanical features. The flexible cervical column, surrounded by passive and active tissue, must function at three levels: *i)* generation of appropriate head movements in three dimensional space; *ii)* maintanance of mechanical stability of the head-neck system at a given orientation; *iii)* distribution of loads within local neck tissues.

Consideration of the neuromotor aspects of head function adds to the mystery. Head movements are closely coupled to movements of the eyes, assisting in visual stabilization of images (see reviews by Berthoz (1985) and Guitton (1988)), and to vestibular function (e.g. vestibulocollic reflex), assisting in postural concerns that often involve the entire body (Peterson, 1988). Head muscles are also richly innervated with sensors, which provide information both for local muscle reflexes and for postural orientation/stability of the entire body (Richmond et al., 1988). The tie between neck biomechanics and head movement control is tight. For instance, biomechanical injury to neck tissues can cause changes in neuromotor strategy (Albright et al., 1984; Bohlman et al., 1982). Conversely, neuromotor disease conditions not only affect voluntary movement but can also cause deformity (Rondot, 1988).

Our goal here is gain insight into the role of neck muscles in head movement organization. To maximize insight, we have been simultaneously using three complimentary approaches: *i)* experiments on human subjects (Peles, 1990); *ii)* construction and testing of an advanced anthrobotic head-neck replica (Liang, 1989); and *iii)* computer simulations using a *3-D* neuromusculoskeletal model (Daru, 1989). The focus of this chapter, in line with the goals of this book, is to search for fundamental patterns related to the organization of head movements, in particular as related to goal-directed tracking movements and postural orientation. Our emphasis will be "natural" movements made at comfortable speeds in many directions and with different initital orientations. It will turn out that many fundamental issues, such as muscle redundancy and the concept of what constitutes a "synergist", "agonist", or "antagonist" will emerge. Experimental results on humans will be emphasized for a simple practical reason: the amount of available data for the human movement system is quite small, suggesting that for this particular system the greatest need is for presenting a breadth of basic experimental data. The two modeling approaches will be utilized in a supporting capacity to help illuminate and interpret various experimental findings.

Multiple Muscle Systems: Biomechanics and Movement Organization
J.M. Winters and S.L-Y. Woo (eds.), © 1990 Springer-Verlag, New York

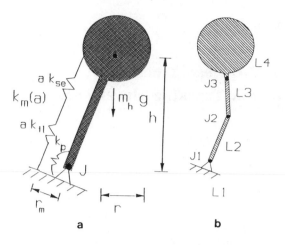

Figure 28.1: Conceptual view of the head-neck system as an inverted pendulum (of mass m_h in gravitational field g). *a)* Inverted pendulum with a single joint used to represent the cervical column, a torsional spring representing "passive" column rotational stiffness (k_p), and a translational spring used to represent "passive" and "active" musculotendon stiffness (k_m). The latter is assumed here to be a linear function of activation a ($0 < a < 1$) for both the "tension–length" (k_{tl}) and "series elastic" (k_{se}) relations, and since these springs are in series and $k_{se} \gg k_{tl}$, $k_m \simeq a\, k_{tl}$. *b)* Four–link, three–joint inverted pendulum model to be used here, minus all springs. Link 1 lumps vertebrae *T4–C7,* link 2 lumps *C6–C3,* link 3 lumps *C2–OC* (*OC* is occiput (skull)).

28.2 Basic Head-Neck Neuro-Biomechanics

The basic biomechanics of the head and neck are outlined elsewhere (e.g., Winters, 1988; Sances et al., 1984). Here we emphasize the implications of this biomechanical structure. The head can be approximated as an inverted pendulum connected to the cervical vertebral column (Figure 28.1a). Two important characteristics of this system are: *i)* a relatively large head mass (approx. 4-5 kg) and inertia (approx. 0.02 kg-m² axial rotation, 0.03-0.04 kg-m² frontal and lateral flexion); and *ii)* a cervical vertebral column that can be bent fairly easily in every direction of rotation. Because the center of mass of the head is anterior to the neck, if all neck muscles were to relax the head would fall forward, rotating about an axis of rotation located somewhere between the first cervical and first thoracic vertebrae. Voluntary head movements also rotate about axes located within this interval. Of interest here is the causal relation between activation of the more than 40 neck muscles which cross through the neck region and head kinematics.

28.2.1 Passive Column Biomechanics

Passive resistance to neck bending is attributed to soft tissue (ligaments and intervertebral disks) and bony (vertebral facets) constraints. Ligaments help limit the relative rotation between vertebrae by generating force when extended. These tissues display a classical concave upward force-deflection behavior over the physiological operating range of interest. A recent study of spinal ligament tensile properties (Myklebust et al., 1988) suggests that all ligaments in this region, and not just the ligamentum flavum and nuchae, are quite extensible; based on their force-extension data we suggest typical peak strains of 20-60% for most ligaments. Intervertebral disks, composed of a fluid (nucleus pulposus) surrounded by cartilage (annulus fibrosis), also exhibit the classical concave upward load-deflection behavior, and act as viscoelastic "shock absorbers", reaching peak extension and compression strains on the order of 50% (Myers and Mow, 1983).

Since the cervical vertebrae are essentially stacked on each other, the structure is mechanically in series (see Figure 28.2). Consequently, for a given applied moment (or force), applicable rotational (and translational) components add, and thus the overall rotation (and translation) is greater than that between any two vertebrae. The overall stiffness (compliance) is thus less than (greater than) that of individual components. The picture that emerges is a passive spinal column that is inititally quite compliant, but progressively stiffer as rotation increases; studies of passive vertebra–soft tissue units (Goel et al., 1984, 1988, Moroney et al., 1988, and Panjabi et al., 1986 for cervical spine) support this view.

Conventional thought is that for most tasks less load is transmitted through the facets than the disks (e.g. Winters, 1988). However, the relative contribution of each mode varies with the type of loading and magnitude of rotation. The cervical facets, spatially orientated at approximately a 45° angle to the vertical, play a special role in kinematically coupling lateral bending and axial rotation (see White and Panjabi (1978) for explanation and Scholten and Veldhuizen (1985) for relevant computer simulations).

The above observations show that the passive cervical spine has the mechanical properties of a compliant viscoelastic beam in which the stiffness increases with increasing rotation and in which cross-coupling stiffness terms exist. Given the relatively large mass of the head, an important emerging concern (too often ignored in the past) is system stability [see also Chapter 26 (Crisco and Panjabi)]. For an inverted pendulum, Bergmark (1987) has shown that it is not enough to simply balance the moments – the system must also be mechanically stable (i.e. the potential energy at equilibrium must represent a minima). Using his "T–shaped" model case (Figure 28.1a), it turns out that the conditions for sagittal stable mechanical equlibrium are (Bergmark, 1987):

$$F_m r_m + M_p = m_h g r \tag{28.1}$$

$$k_m r_m^2 + k_p \geq m_h g h \tag{28.2}$$

where k_m is the translational muscle stiffness, k_p is the passive column rotational stiffness and F_m is muscle force (see Figure 28.1). Here Eq. 28.1 is necessary for equilibrium, while Eq. 28.2 ensures stable equilibrium. As a ballpark example, if we let $m_h = 4$ kg, $h = 0.16$ m, $r = 0.06$ m, $r_m = 0.03$ m, assume a (negligible) value of 1 N-m/rad for lumped passive stiffness (K_p), and assume a negligible passive moment M_p at this resting orientation, we find from Eq. 28.1 that for equilibrium without co-contraction a force of about 80 N is necessary. However, from Eq. 28.2 we see that for **stable** equilibrium a muscle stiffness of at least 70 N/cm (or angular stiffness of at least 6-7 N-m/rad) is necessary. Assuming an "equivalent" head extensor muscle with a physiological cross-sectional area of 20 cm^2 [obtained from adding individual muscle cross-sections estimated in Daru (1989) for extensor muscles crossing $C7$ (other than trapezius)], the muscle stress is on the order of 0.04 MPa, or about 10% of maximum. Assuming an average muscle fiber length of 0.1 m (some muscles connect to skull, some don't), and a positive CE tension–length slope with a crude value of $a F_{max}/L_{rest}$ (the best possible case), where a is the relative activation, the predicted tension–length stiffness is on the order of 80 N/cm, or about the same value. Since if anything this is an overestimate, this suggests that maintenance of stability may require co-contraction [see also Chapter 23 (Andersson and Winters)].

Assessment of stability for the real system is more difficult. The greatest rotation tends to occur in the upper spine, and thus at minimum a useful model must include a second joint. An additional confounding factor is that the normal cervical spine has a lordotic curve. Thus, the model of Figure 28.1b appears to be the simplest model that encompasses the basic features of the head-neck system; this is the model we will use here for our similations. Interestingly, we have found regions of inherent instability within our default 10-muscle, 3-joint model, especially in the neighborhood of upright posture (discussed in Section 28.4.1).

28.2.2 Kinematics and Redundancy

The estimated ranges of motion for flexion-extension, lateral bending and axial rotation, due to static equilibrium between voluntary muscle contraction and (primarily) passive tissues, are shown in Figure 28.2. Notice that with the exception of axial rotation, where $C1$–$C2$ is especially important, there is fairly uniform relative rotation. Of note is that mid-range head orientations, which are of much greater importance in everyday movements, can be performed with multiple combinations of rotation (e.g., see Schuldt et al. (1986)). For example, with a little practice it is possible to perform low magnitude head movements (e.g. ±20° flexion–extension) about an axis near $C1$ (mainly OC–$C1$ rotation), an axis about the base of the neck, or via an axis between these two extremes (the normal case). Muscles must be causing this difference.

28.2.3 Neck Muscles

The above observations lead to questions regarding neck muscle activity and head-neck kinematics. As seen in Figure 28.2, there are many more neck muscles capable of head rotation than would seem to be necessary. Traditionally, the deep muscles have been thought of as stabilizers while the long outer muscles connecting the skull to the shoulder girdle and vertebral column have been considered the prime movers. For reasons distinct from (yet complimentary to) those described in Chapter 26 (Crisco and Panjabi), we will suggest in this chapter that this is an oversimplification.

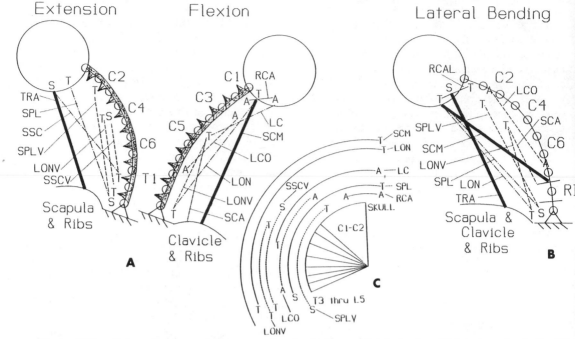

Figure 28.2: Schematic summarizing conventional wisdom with regards to relative contributions to the typical range of motion seen for *a)* flexion and extension ("pitch"); *b)* lateral flexion ("yaw"); and *c)* axial rotation ("roll"; displayed as a *pie chart* with relative rotations adding). Also shown (by strings in *a)* and *b)* and by arcs in *c)*) are selected muscles conventionally considered to cause these motions, here with "average" locations for origins and insertions (many attach to multiple vertebrae). An *S* near origin-insertion site indicates connection to the spinous process or to the occiput *(OC)* posterior to the vertebral foramen *(VF)*, *T* represents connection to transverse process or lateral to *VF*, and *A* represents connection to the anterior portion of vertebral body or anterior of *VF*. **Muscle key:** Muscles are classified into six groups: *i) Thick lines:* large, superficial muscles which connect the skull with the shoulder girdle (sternocleidomastoid *(SCM)*, trapezius *(TRA)*); *ii) Long Dash - Dot lines:* long, often large muscles linking the skull with the vertebral column (semispinalis capitis *(SSC)*, splenius capitis *(SPL)*, longus capitis *(LC)*, longissimus capitis *(LON)*); *iii) Single lines:* short suboccipital muscles and ventral muscles (rectus capitis anterior *(RCA)*, rectus capitis lateralis *(RCAL)*,); *iv) Dashed lines:* Long, deep muscles between cervical and thoracic vertebrae (or ribs (Longus Colli *(LCO)*, Splenius cervicis *(SPLV)*, Semispinalis Cervicis *(SSCV)*, Longissimus Cervicis *(LONV)*) and *v) Short Dash-Dot:* muscles connecting cervical vertebrae to ribs, scapula or clavical (Scalenius (anterior, medius, posterior) *(SCA-)*, parts of trapezius *(TRCV)*)

28.3 Past Investigations

Available experimental methods for assessing neck muscle function and neck kinematics are quite limiting. Kinematic and mechanical investigation of cadaveric spines have helped yield basic information regarding passive spine properties and the relative contribution of tissues such as ligaments (e.g. Panjabi et al., 1975, 1988, Goel et al., 1984, 1988), and as such are important for studies of injury biomechanics and for model parameter estimation. However, such studies do not yield direct information regarding the roles of muscles or movement organization.

Radiographic analysis can be utilized to estimate relative rotations of vertebrae. A few graphical/numerical methods have emerged which attempt to use sequences of low-dose x-rays to es-timate joint angles and axes of rotation (e.g. Gonon et al., 1987; Breen et al., 1988). Additionally, Moffat and Schulz (1979) have estimated relative vertebral centers of rotation during 16 sequential radiographs made while the subjects rotated from full flexion to full extension. They found reasonably good correlation between the measured head path and the predicted head path obtained from a pinned link model of the vertebrae. They also provide evidence of non-uniform relative rotation between vertebrae. In general, however, these methods are crude at best and furthermore yield little information regarding muscle function. Also, such invasive methods could not be ethically used to study a variety of tracking tasks.

Recently, Chao et al. (1989) used a *3-D* motion analysis system to study *3-D* head ranges of motion, employing Euler angles to show explicitly the coupling between lateral bending and axial rotation. This approach is complimentary to that which is used here, which will emphasize calculation of axes of rotation parameters during tracking tasks.

EMG measurement has been utilized in a few cases, often in conjunction with kinematic measurement. Some studies have assessed activation patterns in specific muscles during various simple, relaxed tasks. Needle electrodes have been placed into longus colli (*LCO)* and longissimus cervicis (*LONV*) muscles (Fountain et a., 1966; one subject), the *LCO* and *SCM* muscles (Vitti et al., 1973), and *SSC* and *SPL* (Takebe et al, 1974). Vitti et al. (1973) found that the *LCO* and sternocliedomastoid (*SCM*) were inactive during relaxed sitting, were synergists during both free flexion and forward flexion against applied loads. During lateral bending, they worked homolaterally, while during axial rotation, say to the right, the right *LCO* worked with the left *SCM*. The *LCO* in particular was found to be very active during talking, coughing and swallowing, presumably to help prevent excessive increases in the cervical lordotic curve during pharyngeal contractions (Basmajian, 1979). Of interest is that the *SCM* are incapable of providing this function The *LONV* was found to be primarily an extensor of the spine, with little lateral bending effect, and marked activity during axial rotation (Fountain et al., 1966). The semispinalis capitis (*SSC*) was found to be limited to head extension, but was relatively inactive during erect postures (Takebe et al., 1974), while the splenius capitis (*SPL*) was of importance for both axial rotation and extension.

A group in Sweden has utilized surface *EMG* electrodes and head orienation measurements to help identify correlations between the work environment and levels of fatigue/injury in females involved in assembly work (Ekholm et al., 1986; Harms-Ringdahl, 1986). They found that working with the head flexed resulted in especially high activity in the extensor muscles near the skull (likely in large part to *SSC*), with activity in the posterior group of the lower neck most influenced by the seat inclination. It was concluded that a slanted work table (e.g., 35°) and a slightly backwardly inclined chair backrest (e.g., 10–15°) were ideal from the point of view of minimizing muscle activity (Ekholm et al., 1986). However, based on kinematic inspection and simple biomechanical analyses of likely moments at *OC–C1* and *C7–T1*, especially at movement extremes, it was pointed out that different subjects will employ *different strategies* for various postures, and will adapt postures as appropriate for the necessary movements and changes in orientation required on the job (Harms-Ringdahl, 1986).

Using surface electrodes, Vorro and Johnston (1987) have studied *EMG* timing (time to initiation, duration) during various neck bending tasks in subjects diagnosed (via range of motion during lateral bending) as having "symmetric" or "asymmetric" neck function. Interestingly, they have found that relative to movement times (provided as percentage of total time), there were significant differences between symmetric and aysmmetric populations, with *EMG*'s for the symmetric group starting sooner, peaking sooner and lasting longer. There were also right-left differences. The authors' postulate that these results are due to dysfunction of the neuromotor reflex control system.

In contrast to the above study of postural muscle activity, Stark and colleagues have investigated *EMG* activity and rotational head kinematics during very fast horizontal tracking movements and eye–head coordination during such movements. Inititial studies showed fairly similar *SCM* and *SPL* activity (Zangemeister et al., 1982); subsequent studies emphasized *SPL* muscles. In general, tri-phasic agonist–antagonist–agonist patterns were seen. As the movement magnitude increased, the *EMG* pulses increased in height and width, the peak velocity and accelerations increased, and the time duration increased (Zangemeiser et al., 1982; Hannaford et al., 1984). The effects of added viscous, inertial and spring loads were also investigated, with viscous loads causing greater agonist over a longer time period and less antagonist activity, inertial loads increasing the pulse widths, and elastic loads causing a change primarily in bias *EMG* levels (Hannaford and Stark, 1984; Hannaford et al., 1985). Another study found that there were significant deviations from straight-line paths during oblique movements (Hannaford et al., 1983). Phenomenological antagonistic muscle–joint models of the horizontal component, employing the classical Hill model structure outlined in Chapter 5 (Winters) for "equivalent" axial rotators, have been shown to be able to predict each of the basic trends seen experimentally (Zangemeister et al., 1982; Hannaford and Stark, 1985; Winters, 1985), although in general less second and third pulse activity has been necessary in the simulations. Also, optimal control results with such model structures predict more overlap between the second and third pulses than is seen experimentally (Winters et al., 1987).

Neural control studies in animals and the human have shown repeatedly that head movement organization is strongly coupled both to eye movements and to postural orientation. A recent book volume (edited by Peterson and Richmond, 1988) considers such coupling in great detail [see also review in Chapter 23 (Andersson and Winters)]. Of note is that neck muscles have an especially plentify supply of muscle spindles and Golgi tendon organs (Richmond et al., 1988). Thus, sensory reasons alone suggest an important role for neck muscle activity in global strategies involving orientation and posture [e.g., see Chapter 28 (Keshner and Allum)]. On the motor side, neck muscles receive converging information from a variety of sources. Certain tight connections are well documented: the vestibulo-collicular reflex (Schor et al., 1988); the "tonic neck reflex", which generally implicated tight coupling between the head and posture (Wilson, 1988); and nearly identical head and eye agonist drive signals during visual orientation changes of sufficient magnitude (Zangemeister et al., 1982; Guitton et al., 1988; with the superior colliculus usually implicated (Roucoux and Crommelinck, 1988)). Finally, Pellionisz and Peterson (1988) have used tensor methods to address the challenging question of how geometric mapping between sensory activity and muscle activity might occur.

Finally, of note is that disease or injury conditions can affect neck neuromechanics (Sances et al., 1984). For example, *CNS* brain lesions may selectively affect the base drive to various muscles. The result can be head orientations that are off the midline (e.g. torticolis, where there is axial rotation). If sustained, permanent (and often progressive) biomechanical deformity and degradation of head movement function can result (Rondot, 1988; Stark et al., 1988). Similarly, biomechanical injury may affect not only the biomechanical integrity of the spine but neuromotor control of neck muscles.

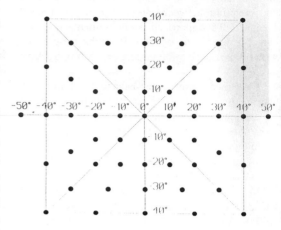

Figure 28.3: Target grid on the spherical dome, as would be seen by the subject. Dashed lines outline target patterns used for the four primary classes of tracking tasks: "horizontal" (*A*), "vertical" (*B*), "oblique" (*C*), and "box" (*D*). (Ranges shown are for "normal" subjects; injured "whiplash" subjects, of less interest here, typically tracked a subset of these ranges due to limited range of motion in various directions.)

28.4 Methods

28.4.1 Human Experiments

For the human tracking experiments the subject was placed on a chair facing a dome arc with a dark blue background which contained a grid of 59 lamps of 2.5 cm dia. The subject was placed at the approximate center of the dome, which had a radius of 1.61 m. The targets were arranged systematically as shown in Figure 28.3. Each target could be controlled separately by a digital computer (PC-AT 286). A light-weight helmet, which included five reflective markers on the left side of the mid-sagittal line and a laser pointer connected to the top, was fit snugly to the subject's head (Figure 28.4). The five markers were in view of two synchronized video cameras. Eight *EMG* electrodes were placed on the subject in the positions shown in Figure 28.4; these placement locations were chosen based on a survey of the literature and a number of trial runs with various placements.

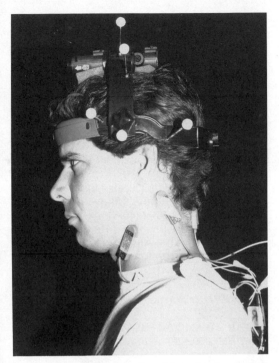

Figure 28.4: Photograph of a subject, showing the helmet with five attached markers and a laser, and typical *EMG* placement for measurement of the *SCM*'s (located by palpation), *SPL*'s (located laterally and superiorly near the mastoid process so as to lessen crosstalk due to the trapezius and other posterior musculature, *UC* (over upper dorsal cervical spine, orientated vertically, about 1 cm from the mid-sagittal plane, and *LC* (lower cervical spine orientated vertically and 1 cm lateral). We assume that *UC* includes *SSC* plus deep *OC–C1–C2* paravertebral muscles and trapezius, and that *LC* includes the three cervicis muscles (semispinalis, longissimus and splenius) which originate on the upper thoracic vertebrae and insert on the upper cervical vertebrae, and possibly rhomboid activity or activity from intervertebral muscles.

EMG information was rectified, lightly smoothed (20 ms double time constant), sampled via a Data Translation 2801A board, and then collected onto a *PC-AT* computer via the *Labtech Notebook* (Laboratory Technologies) data acquisition software package. All data were inititaly reduced and plotted using command files from within the *DaDisp* (DSP Devel. Co.) software environment. Selected final plots were produced using the *Grapher* (2-D) software package (Golden Software, Inc.).

Video information, sychronized to *EMG* collection and to target sequencing, was analyzed off-line by the *3-D Expert Vision* (Motion Analysis, Inc.) system. The resulting centroidal coordinates for the five markers were then utilized by a software algorithm designed to find "finite" and "instantaneous" screw axis parameters. To accomplish this, the data were separated into "on target" and "moving" aspects based on a velocity threshold. The finite screw axis parameters (see Figure 28.5) were determined by using averaged data between two subsequent "on target" locations and then applying the Spoor-Veldpaus method (Spoor and Veldpaus, 1981), as formulated by Woltring et al. (1985) and documented in Daru

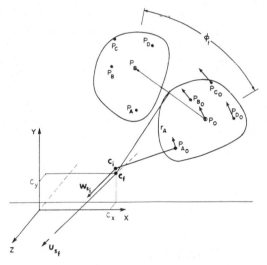

Figure 28.5: Graphical overview of *3-D* Kinematic Analysis Methods. *Finite screw axis parameters* [finite rotation vector direction (u_f), the coordinates of a point on this axis (c_f), the rotation angle (ϕ_f), and sliding along the axis] are calculated from estimation of marker locations *before and after* a certain *finite rotation*. In contrast, *instantaneous screw axis parameters* (magnitude ω and direction (u_i) of angular velocity vector and coordinates of a point (c_i) on this axis) are calculated from estimation of the *position and velocity* of the markers at *each instant* that is sampled (shown for right-most body). Significant marker velocities are required for the latter method to provide useful data. Displacement and acceleration are then calculated from the smoothed angular velocity. For simple cases (e.g. nearly planar rotations that stop at targets), the two methods can be checked against each other by comparing integrated rotation over time predicted by the *ISAP* method with the finite rotations predicted by the *FSAP* method.

(1989). We have found this algorithm to be less sensitive to noise than the classical techniques described in Suh and Radcliffe (1978; see also Kinsel (1972) and Panjabi et al. (1981)); it also has the advantage of including all five markers instead of combinations of any four. During the "moving" phases, the instantaneous screw axis algorithm in Angeles (1982), modified by light pre- and post-application smoothing, was used (see Figure 28.5). Data was plotted in *3-D* using the *Surfer* plotting package (Golden Software).

Testing Protocol

Testing has been performed on populations of young normal controls ($n = 9$) and a group of subjects with "whiplash" symptoms who are undergoing conservative chiropractic treatment ($n = 12$). For the "normal" subjects of primary interest here, targets changed every 2 sec. Tracking sequences could be divided into horizontal, vertical, oblique (45° angle from horizontal and vertical) and a "square box" (Figure 28.3). Additionally, two types of instructions were given to the subject: *i)* "track the lit target with the head (laser) pointer at a comfortable pace" (*COMF*); and *ii)* "follow the target quickly", i.e. "always keep the difference between the head pointer and target small" (*FAST*).

The entire protocol, including *EMG* placement, took approximately 2 hours. Under 20% of this time actually involved testing; subjects were free to relax or read between tasks. The first task, which will not be discussed further here, involved natural tracking of targets without a head pointer or instructions of any kind — this was done first so that the subject wouldn't be biased by previous head tracking tasks. It was felt that *FAST* movements, which are less natural and spontaneous and involve significant mental load, might bias *COMF* movements; thus *COMF* tasks were performed next, followed by the *FAST* movement tasks. The session concluded with the self-paced "slow", "moderate" and "fast" horizontal and vertical oscillation movements of ±20° and ±40° (not discussed further here).

For *horizontal COMF* movements, the pattern was:

a 0→10→20→30→40→50→40→30→20→10→0→−10→
     −20→−30→−40→−50→−40→−30→−20→−10→0.

Vertical COMF movements were similar, but with the target sequence only reaching 40°.

For both horizontal and vertical *FAST* movements the sequences were:

b −20→20→−20 (repeated six times, last four stored)

c 0→40→0→−40→0 (repeated six times, last four stored)

d Unpredictable pattern between −40←→0←→40.

For *oblique* movements, only the following predictable patterns were used:

(0,0)→(10,10)→(20,20)→(30,30)→(40,40)→
(30,30)→(20,20)→(10,10)→(0,0)→(−10,−10)→
(−20,−20)→(−30,−30)→(−40,−40)→(−30,−30)→
(−20,−20)→(−10,−10)→(0,0)→(−10,10)→
(−20,20)→(−30,30)→(−40,40)→(−30,30)→(−20,20)→
(−10,10)→(0,0)→(10,−10)→(20,−20)→(30,−30)→
(40,−40)→(30,−30)→(20,−20)→(10,−10)→(0,0)

for the *COMF* instruction and

(−20,−20)→(20,20)→(−20,−20)
(−20,20)→(20,−20)→(−20,20)

for the *FAST* instruction. Our primary goal for these tasks was simply document what muscles were active and what kinematic patterns were seen (e.g. locations of axis of rotation; type of path taken by the subject).

For the **"square box"** protocol, the following predictable target sequence was employed for the *COMF* instruction only:

(−40,40)→(−20,40)→(0,40)→(20,40)→(40,40)→
(40,20)→(40,0)→(40,−20)→(40,−40)→(20,−40)→
(0,−40)→(−20,−40)→(−40,−40)→(−40,−20)→
(−40,0)→(−40,20)→(−40,40)

reversed direction box:

(−40,40)→(−40,20)→(−40,0)→(−40,−20)(−40,−40)→
(−20,−40)→(0,−40)→(20,−40)→(40,−40)→
(40,−20)→(40,0)→(40,20)(40,40)→(20,40)→
(0,40)→(−20,40)→(−40,40)

The goals of the "box" data were to: *i)* examine *EMG* and kinematic behavior for a variety of orientations; *ii)* examine the effects of the movement operating range on movements of the same magnitude and in the same direction; and *iii)* examine the effects of direction of approach on how a given orientation is held. In particular, the horizontal movements made while holding flexion could be compared to those with no vertical component and those made while holding extension. Similarly, the same type of comparison could be done for vertical movements.

Figure 28.6: Photograph of the anthromorphic replica, which includes 10 artificial muscles. Head is darkened to minimize reflection during motion analysis.

28.4.2 Anthropomorphic Model

Results from an anthropomorphic replica of the human head, neck and upper torso (Liang, 1989; see Figure 28.6) will be utilized to help interpret the human experimental data, especially as related to kinematic effects of muscle contraction. This model included: *i)* fiberglass "bones" made from molds of actual bones (from *T5* upward, including upper ribs and clavicle); *ii)* artificial ligaments with nonlinear force–extension properties very similar to the various neck ligaments (anterior and posterior longitudinal, flavum, transverse and supraspinous were modeled; each also including embedded displacement gages not of interest here), *iii)* intervertebral discs (silicone rubber disks with embedded fabric which surrounded a central pocket filled with water); and *iv)* pneumatically-powered (and computer controlled) braided artificial "McKibbon-like" muscle actuators (described in Chapter 7; used for the following muscles: 2 *SCM*, 2 *SPL*, 2 *SCA*, 2 *TRA*, 2 *SSC*).

Kinematic orientation and motion could be measured via *3-D* motion analysis techniques similar to that mentioned earlier for the human system. The same algorithm as described previously was utilized to determine screw axis parameters. A triaxial accelerometer was mounted to the head.

The following classes of tasks have been studied: *i)* static external loading to the head, applied from a variety of directions; *ii)* contraction of the artifical muscles, both isolated, in pairs, and in "strategiec" combinations (all flexors, all extensors, all muscles on one side, all axial rotators); and *iii)* the model response to head impacts, especially with regards to the role of initital muscle co-contraction (ongoing). Only the second class of tasks will be of interest here.

28.4.3 Computer Simulation Models

Of primary interest here is the *3-D* quasi-static model of Daru (1989), as used in conjunction with static optimization. In order to investigate fundamental issues related to relations between muscle activity, head orientation, and head screw axis parameters (*HSAP*), the model configuration of Figure 28.1b was utilized. This appeared to the simplest model which had wide potential for variable *HSAP*. Each joint was surrounded by a nonlinear (concave upward exponential) torsional spring that represented the lumped passive properties of the cervical column. Given the series arrangement of stacked vertebrae, passive torque–angle properties were obtained by lumping deflections from *OC–C2* into the top joint, *C3–C6* into the middle joint, and *C7–T4* into the base joint. The indiviual passive properties were estimated based on data such as Goel et al. (1984, 1988). Of note is that the bottom "joint" was considerably stiffer in all directions, with the middle "joint" most compliant for flexion–extension and lateral bending, and the top "joint" initially very compliant for axial rotation (Daru, 1989). Three different sets of muscles were developed: a "minimal" set (8 large superficial muscles), a "default" set (12 muscles), and a "large" set (36 muscles). Our goal was to first fully understand the potential and limitations of the smaller sets. The method utilized to obtain parameter values was based primarily on cervical spine segmental data such as Goel et al. (1984,1988), as detailed in Daru (1989) [see also the appendix to this book (Yamaguchi et al.)].

Of interest here are two classes of tasks: *i)* tasks documenting kinematic sensitivity to systematic activation of individual muscles and combinations of muscles; and *ii)* tasks using static optimization in which the performance criterion is of the form:

$$J = K_1 \sum_i^3 (HAX_{r_i} - HAX_{m_i})^2 + K_2 (\theta_r - \theta_m)^2$$

$$+ K_3 \sum MS^2 \qquad (28.3)$$

where HAX_r and HAX_m represent the "reference" (e.g., experimental) and "model" head finite axes of rotation, respectively; θ_r and θ_m the respective head orientations of primary interest; MS muscle stress (force per area, essentially normalized force), and K_i's are relative weights between subcriteria. By changing the task (specified here by the "reference" signals), the relative weights between subcriteria, the assumed number of muscles, and/or model parameters for active or passive tissue, a wide variety of possibilities, numbering in the thousands, are possible — we have explored an "interesting" subset of these.

28.4 Results and Discussion

In conjunction with the general focus of the chapters within **Part IV** of this book, we will emphasize issues related to orientation and posture; thus we will concentrate on the *COMF* and off-center *FAST* tasks. We will start by briefly summarizing some insights from sensitivity analysis studies of various muscle contraction sequences. Our approach will then be to outline salient *EMG* and kinematic results, discuss possible causal relations between these findings, and use modeling results to help interpret experimental findings.

28.4.1 Overview of Modeling Insights

Stability. One interesting computer simulation finding is that when only the large muscles are employed there are regions of local instability, especially for orientations near the upright posture. Slight changes in muscle activation often resulted in "sudden" rotational sensitivity to muscle activity. Also, a given muscle pattern might have multiple equilibrium positions. Often a solution could only be found with the aid of the optimization algorithm, and even then only if "muscle stress" was not excessively penalized. There was also a general tendency for strong con-

tractions to cause some buckling of the link structure; for instance, multilink muscles on the opposite side of the column from contracting muscles would sometimes shorten in length. Stability was improved either by increasing the passive column stiffness or changing either the rest length or slope of the muscle tension-length relation. Interestingly, for the anthro-model, in part to minimize the magnitude of this problem, we had purposely "tightened" the spine by artificially increasing the stress of ligaments, especially those near the *OC–C1–C2* area; our rationale was that this helped represent deep muscle activity in the "alert" human (Liang, 1989).

High C1–C2 Sensitivity to Axial Rotation. One of the most interesting findings from systematic contractions of the artificial muscles within the anthropomorphic model was accidental: because of slight fabrication asymmetries in muscle properties, at low "symmetric" contraction levels in a single pair of muscles there was often a tendency toward some initial axial rotation that was clearly occuring at *C1–C2*. This suggests a high sensitivity to axial rotation in muscles such as *SCM* and *SPL*, which really only have moderate vectorial components in axial rotation. It also suggests that tight control near the mid-range requires activity in the *deeper muscles* within the *OC–C1–C2* region.

Sensitivity to Muscle Activity. In general, simulation sensitivity analysis results tended to confirm traditional views regarding the larger muscles (Daru, 1989). Also, with few exceptions (e.g. *SCM*), the change in the rotation angle (about the axis of rotation) became *less pronounced with increasing activation*. Interesting findings regarding specific muscles included: *i)* SCM contraction causes insignificant flexion-extension at the top joint due to producing a negligible moment, and thus high SCM activity is indicative of a *low* axis of rotation; *ii)* SCA contraction most significantly affects lateral bending of the lower joint; *iii)* SPL contraction causes all three rotation components, with axial rotation and lateral bending highest and about equal, and furthermore influences the top and middle cervical parts of the column about equally; and *iv)* due to gravitational coupling, muscles such as SPL and SSC can mildly influence the "joint" at the base of the neck without crossing this region.

Optimization. Some interesting general trends were: *i)* the relative contribution between synergists was dependent on the penalty for muscle stress — for instance, for 20° flexion, *SCA* activity was much higher than either *SCM* or *LC* when K_{ms} = 0 but decreased dramatically for even small K_{ms} levels; *ii)* the relative contribution between synergists was causally related to the location of the axis of rotation — for instance, *SCM* was a preferred muscle for flexion movements about *C7* but not for higher axes; and *iii)* pure lateral bending required the significant use of muscles on **both** sides of the mid-sagittal plane.

28.4.2 *EMG* Activity During *FAST* Movements

Horizontal

For *task b* (–20°↔→20°) the primary "driving" agonists were, as expected based on previous studies, the ipsilateral *SPL* and the contralateral *SCM*. *SPL* was usually a stronger antagonist than *SCM*. We usually did not see as dramatic of a triphasic bursting pattern as was seen by Hannaford et al. (1984), but did see agonist activity sustained through what would be the second agonist burst.

Interestingly, we saw significant activity in both *UC* and *LC*, especially for the ipsilateral side, with 60% of subjects showing *UC* pulses with over 25% of maximal activity. It was usually difficult to ascertain whether the activity was that of an "agonist" or an "antagonist"; there was a trend toward *ipsi UC* firing with agonists (70%) and *contra LC* being antagonists (70%). Of note is that pulse shapes consistently suggested that *UC* and *LC* cross-talk with *SPL* was minimal — certainly *SPL* doesn't affect *LC*. The sources of this activity may be large superficial dorsal muscles (*SSC*, *LON*, *LONV*, *SSCV*), but we suggest that small internal muscles may be a primary source. Four reasons for this activity can be postulated: *i)* small muscles with diagonal fibers functioning as agonists-antagonists (e.g. the suboccipital *Obliquus capitis inferior* for the *UC* signal, and the *rotatores spinae* for the *LC* signal); *ii)* co-contraction may be utilized to help stabilize the spinal column by increasing the general stiffness; *iii)* a selective increase in the vertical stiffness via co-contraction would help cause a straight horizontal trajectory, or *iv)* subtle movement coupling dynamics may cause a need for muscle activity that we as yet don't understand.

For *off-center FAST* horizontal movements (*task c*), the classical agonists (*ipsi SPL* and *contra SCM*) are both very strong when going **away** from the center line, with large pulse widths; when going **toward** the center line both displayed narrowing pulses, and *SCM* in particular is not as strong (see Figure 28.7). Antagonists (*contra SPL* and *ipsi SCM*) showed more activity (and wider pulses) for movements **away** from the center point than towards zero. In fact, virtually no antagonistic clamping was seen during "toward" tasks, and in most cases the antagonists (and especially *SCM*) exhibited initial "release" behavior. How are these fast "toward center" movements clamped? We suggest that deep muscles must be involved because: *i)* movement dynamics suggest that clamping must occur; *ii)* computer simulation insights show that resolution of position is difficult for the straight-ahead position when only the large muscles are active (Daru, 1989); and *iii)* this view is also compatible with the general kinematic finding that the first 30° of rotation occur almost exclusively at *C1–C2* (White and Panjabi, 1978).

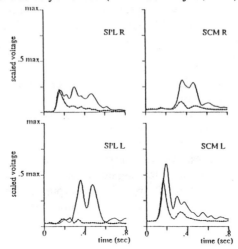

Figure 28.7: Overplot of ensemble-averaged *SPL* and *SCM EMGs* during off-center horizontal (axial) *FAST* movements made to the right (Subject N4). *a)* Toward centerline (with overplotted stars). *b)* Away from zero.

Another interesting finding is that when going away from the center line, both *UC* and *LC* on the *ipsi* side tended to be agonists, while the *contra* cervical muscles tended to be antagonistic. However, when going toward the center line the timing in all of these muscles was that of an agonist.

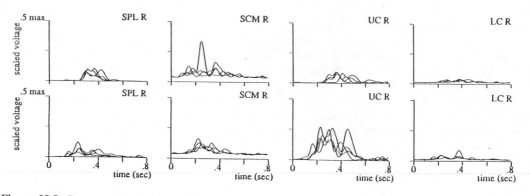

Figure 28.8: Representative *EMG* activity for *FAST* vertical flexion (top) and extension (bottom) movements between −20° and 20° (Subject N5). Trials overplotted based on kinematic threshold. Notice lack of maximal activity for these moderate-magnitude movements. Only right side *EMG*s displayed (left very similar).

Vertical

For *FAST* extension movements over the mid-range (−20° ↔ 20°), the dorsal musculature, most notably *UC*, were consistently active yet rarely at a maximal level (e.g. Figure 28.8). *SPL* and especially *LC* were clearly lower magnitude than *UC*. Interestingly, based on timing, in some subjects the *SCM* showed little sign of antagonistic activity, rather co-contracting with agonists, as is the case in Figure 28.8.

For *FAST* flexion movements, *SCM* was distinct yet never close to maximal, with a small pulse width that served to help initiate the movement. Figure 28.8 shows that multiple strategies can exist for flexion movements, as one case shows a distinct single pulse while the others show small double pulses. Many subjects had difficultly performing these movements quickly. In some cases "controlled release" by extensors appeared to provide part of the necessary drive toward flexion.

Antagonistic "clamping" by *UC* was often larger in magnitude and pulse width than seen in Figure 28.8.

The most fascinating finding regarding *off-center* vertical *FAST* extension movements was that strong *LC* (and often *UC*) pulses were seen when the movement was made *away* from the center line, in extension as an agonist and in flexion as an antagonist (Figure 28.9). We also found greater *SCM* activity for movements between 0°↔40° than between 0°↔−40°, often with steady *SCM* "holding" activity at 40°. Also, in comparing head flexion movements, *SCM* was more active for "toward" movements (40°→0°) than "away" (0→40°). Of interest is that the "antagonist" is necessary for "holding" the new position. Finally, although not the case for the subject shown in Figure 28.9, *LC* could display a holding pattern at either movement extreme.

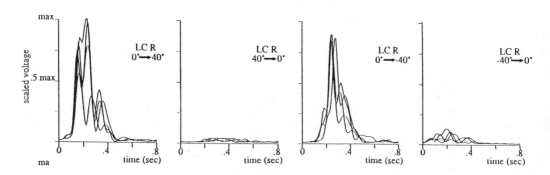

Figure 28.9: *EMG* activity in *LC* during *FAST* vertical movements both away from and toward the center position (Subject N5).

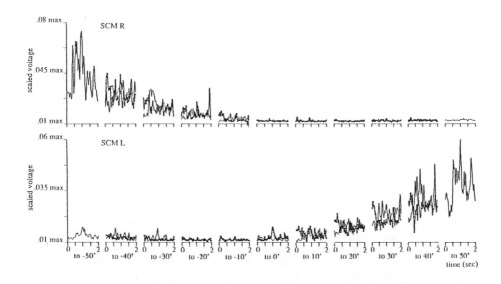

Figure 28.10: Example of *EMG* activity (*SCM*) during *COMF* horizontal tracking movements as a function of operating range (Subject N3). Original plot (over 40 sec of sampling) was split up by times of target changes (every 2 sec) and overplotted so that *EMGs* for movements ending at the same position could be compared. Traces with overplotted small stars are for movements approaching the given position from a larger angle (notice that these *EMGs* usually start with a higher value).

FAST Oblique

In general, *FAST* oblique movements contained many of the features seen in *FAST* **horizontal** movements, with the *ipsi SPL* and *contra SCM* always agonists and the *contra SPL* and *ipsi SCM* antagonists. However, the *UC* and *LC* activity was more clear cut than for horizontal movements. For **upward** oblique movements, both *UC*'s were clearly agonists, while *LC* pulse activity timing was usually closer to what would be expected for an antagonist. For **downward** oblique movements, *contra SCM* antagonist was not as prevalent, while *SPL* took on the classical patterns expected for the horizontal component of the movement. However, both *UC* and *LC* activity, which was quite major, was usually better classified as antagonistic.

28.4.2 Operating Range Sensitivity for COMF Tracking

Horizontal

Variation in FSAP. In general, with the exception of the 40°–50° rotations, the head axis of rotation for horizontal *COMF* tracking was quite vertical. Furthermore, this data shows that during horizontal movements the axis stays within a fairly tight region. However, a consistent trend that emerged was that the axis shifted slightly laterally (about 1–2 cm) for movements to the same side. Based on a rank-ordering the *x*- and *y*-components crossing *C5* for each subject, we concluded that once this minor shift occurs, the axis stays about the same for rotations between 10° and 40°. For rotations between 40°↔50°, a significant lateral bending component emerged in most subjects that had the effect of raising the ipsilateral ear and lowering the contralateral ear.

EMG Trends and Possible Causal Relations. As horizontal movements for *task a* proceeded outward in 10° increments, steady *EMG* activity levels in agonists (*ipsi SPL* and *contra SCM*) increased. The *SCM* in particular showed a steady rise throughout, often with a disproportionally greater increase toward the extreme (e.g. Figure 28.10). Antagonist *contra SPL* and *ipsi SCM* activity was generally light, with most of that seen occuring near extreme orientations. The most interesting finding was that *ipsi UC* activity often increased as the extremes of motion were approached, presumably due to deep musculature. This trend is compatible with computer and anthropomorphic modeling studies, where it is apparent that the large superficial muscles (*SCM* and *SPL*) are incapable of providing the horizontal force vector component required for rotations over about 30°–40° (Daru, 1989; Liang, 1989). The agonist *SCM* cannot cause such rotations because

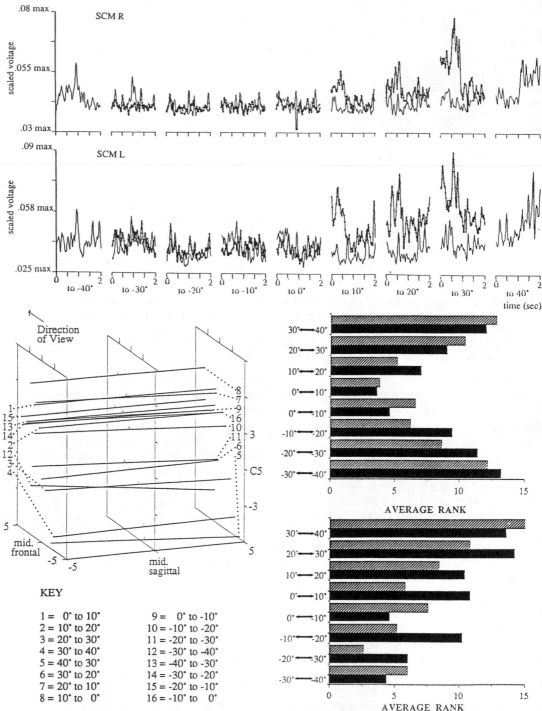

Figure 28.11: Example of *COMF* vertical tracking movements (10° magnitudes) as a function of operating range (Subject N8). *a)* Selected muscle activity (*SCM*), with movement location changing every 2 sec. Two movements are overplotted, with the traces with stars inititally from a higher angle. *b)* Finite axis of rotation variation across the neck region for 10° incremental rotations ranging between ±40°. *c)* Bar graphs of average of ranking of *y*- (vertical; top) and *x*-(anterior-posterior, with anterior positive; bottom) coordinate crossing locations in the mid-sagittal plane. For each subject, a rank of 1 was given for the highest (top) or most anterior (bottom) crossing location, 16 for the lowest. Solid bars are for movements going away from the center line, striped toward.

the muscle becomes nearly vertical; the *SPL* provides less force because it becomes too short (< *70%* of rest length). Furthermore, the maximal possible axial rotation at *C1–C2* is about 40° (e.g. White and Panjabi, 1978). Thus, the last 10° must include rotation along the *whole* cervical spine. Interestingly, this is the same operating range where the axis of rotation "tilt" from vertical was seen! Chao et al. (1989) appear to show a similar finding. We suggest that this is due to a combination of three sources: *i)* intervertebral muscle contraction; *ii)* passive coupling between axial rotation and lateral bending; and *iii)* a mechanically ineffective *contra SCM* "agonist" contraction that now serves to laterally tilt the head.

Vertical COMF Movements

Variations in FSAP. For pure vertical tracking movements the axis of rotation vector was, with few exceptions, directed horizontally, orthogonal to the mid-sagittal plane (e.g. Figure 28.11b). The axis of rotation was found to be highest for 0°↔10°, typically reaching the *C2* level. For 10°↔20°, the level was typically *C3*. Interestingly, it lowered dramatically both for movements between 30° and 40° and for the four movements between –20° and –40° (Figure 28.11b). The lowest level reached was typically about *C7*, with this occuring most commonly for rotations between –30° and –40°. For most subjects ventral–dorsal (front–back) variation in the axis was small, reaching only a few centimeters. However, it did tend to be more posterior for movements above the mid-line and more anterior for movements below; the only consistent trend was that movements between 30° and 40° crossed posteriorly by about 2 cm.

EMG Trends and Possible Causal Relations. For extension movements, *SPL* and *UC* were the primary agonists, with *SPL* showing an increase in steady-state *EMG* level from 0° to 40° (usually not large until extremes), and a corresponding fall from 40° to 0°. *UC* (likely due to both large and small muscles) showed large movement pulses from 0° to 40° but little activity while moving back to the center. In half of the subjects *LC* often exhibited increased activity at the movement extreme (40°); this may be correlated to the lower axis of rotation that occurred as extreme upward orientations were reached and perhaps also with the posterior shift of the axis. For some subjects the steady-state *SCM* activity was a clear function of the previous direction of approach (Figure 28.11a). The *SCM*'s also had a slight rise in steady-state *EMG* level while tracking from 20° to 40°. Also interesting was that the *EMG* levels were significantly higher for some subjects during "holding" aspects while going from 40° through 20°, during which the *SCM*'s were prime movers for the initital movement phase. The lack of significant *SCM* for movements between 0° and 20° seems consistent with the high kinematic axis during these intervals *OC–C1–C2* muscles appear primarily responsible for these rotations. In fact, computer simulation studies with the three-joint model of Figure 28.1b show that paired *SCM* contractions cause virtually no moment nor rotation in the upper cervical spine, with the highest moment at the base of the neck and most of the rotation in the mid- and lower cervical spine. This interpretation is further supported by the fact that for movements between 30° and 40° the *SCM* increases in activity while for the same movements the axis lowers (Figure 28.11).

Finally, although we could not measure this activity, we suggest that the scaleni (*SCA*) muscles, which rotate the lower cervical column, likely assist in lowering the axis during flexion.

COMF Oblique Movements

In contrast to *FAST* oblique movements, *COMF* diagonal oblique movements possessed *EMG* features that were more similar to vertical *COMF* movements. This suggests that the dominant effect of static orientation behavior is gravity.

28.4.3 Effect of Bias Orientations on Movements: "Square" Tracking

The *COMF* "square box" tracking protocol provides a unique opportunity to assess the effects of bias orientations orthogonal to the movement direction. Also, a byproduct of traversing the "square" both clockwise and counterclockwise is that history-dependent effects can also be investigated.

Axis of Rotation Effects

Figure 28.12 summarizes behavior of the axis of rotation vector. Before assessing this data it is important to realize that horizontal and vertical movements are coupled via the orientations at the corner locations.

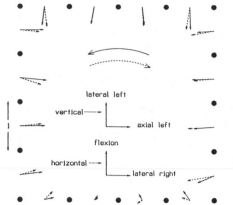

Figure 28.12: Averaged axis of rotation unit vector components (relative to body coordinates) in the off-axis directions during "square box" tracking (*n* = 4). Vectors originate at the half-way point of the movement between two consecutive targets (solid circles). The solid lines are for counterclockwise rotation , dashed for clockwise. Notice that the smaller the line magnitude, the more "on axis" the data.

Notice that downward *vertical* off-line tracking results in an axial component of 0.5, or 30°. This is less than the 40° rotation that would be expected if the subject was making pure flexion-extension movements in the new (gaze-centered) plane, yet much greater than the 0° that would be expected if the rotation occured in the (body-centered) initial coordinate system of the torso — clearly a *compromise* emerges. Notice also that the slight lateral bending component shifts sign. Rising off-

center extension movements display a 25° axial component, again less than the 40° gaze-centered case.

There was great asymmetry between *horizontal* rotations while looking upward and horizontal rotations while looking downward (Figure 28.12). The former case displayed the same type of "compromise" axis as seen for vertical movements, with a *F–E* component of *20–25°*. However, for horizontal rotation while looking downward, we see that the *F–E* component was under *10°*! Thus, *body-centered* axial rotation is preserved. Furthermore, when rotating toward the mid-line the *L–B* component rotates so as to eliminate the pre-existing mild *ipsi* ear up posture, then rotates so as the raise the *ipsi* (formerly *contra*) while going away from the midline! This fascinating asymmetry was documented for "squares" in either direction and is a clear sign of direction- and/or history-dependence.

EMG Activity and Potential Explanations for Axis Vector Data

For downward off-center vertical tracking, the *ipsi SCM* was inactive while the *contra SCM*, which had a virtually vertical orientation, was quite active in the extended positions, but showed little activity otherwise (Figure 28.13a). The posterior cervical musculature (*UC, LC*) exhibited a "hold-release" pattern, with activity especially high at –40°V.

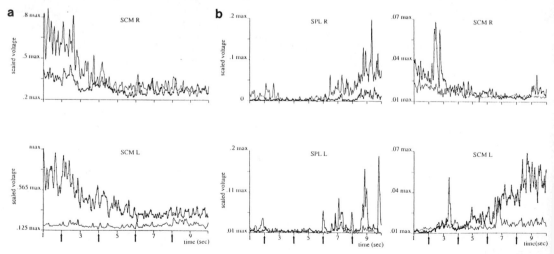

Figure 28.13: *EMG* behavior for selected muscles while subject performs clockwise and counterclockwise "box" patterns. *a) SCM* behavior during downward vertical movements, with cases of right (with overplotted stars) and left horizontal biases over-

plotted (Subject N1). *b) SPL* and *SCM* behavior during horizontal movements to the right, overplotted with different downward and upward (with overplotted stars) vertical biases (Subject N3).

For horizontal movements with downward vertical components, the *LC* musculature is now low and steady – this is compatible with the concept of "high axis" rotation. *SCM* activity is marginal or nonexistant. Mechanically, we might have expected greater *SCM* activity, especially for the "away" aspect where *L–B* was seen that could have been due to the *contra SCM*. *SPL* agonist activity is low from 40° to 0°, then becomes a clear agonist from 0°H to 40°H.

For upward off-center vertical, the *SPL* and *UC* show little activity, while the *LC* increases progressively. It is likely some of the smaller vertebral muscles with an axial component are partly responsible for observed behavior. The *ipsi SCM* appears to be involved in holding the top corner positions.

As seen in Figure 28.13b, for the upward horizontal sequence, the *SCM* opposite the head orienation was quite active, with the *SPL* inactive and the *UC* and *LC* involved in "hold–release" patterns.

28.5 Summary

We have seen that the head-neck system plays an important role in visual orientation and whole body postural control. In this chapter we go beyond traditional range-of-motion or isometric contraction studies to explore muscle organization and head kinematics during a variety of head tracking tasks. We supplement our experimental findings, some of which were quite novel, with insights from modeling studies.

Some of the more interesting findings include: *i)* asymmetries in *FAST* movements made in the same direction only over different *operating ranges*; *ii)* a situation (vertical fast movements) in which the "agonists" and "antagonists" for the "fast" portion of the movement differ dramatically from those necessary for "steady" step levels – this is a consequence of controlling an inverted pendulum with an eccentric mass; *iii)* indications that *FAST oblique* movements are closer in organizational strategy to horizontal movements, yet that *COMF* movements and steady orientations in general are dominated by features found in vertical movements; *iv)* indications of dramatic extensor activity during *FAST* horizontal movements, suggesting both that small internal muscles play a significant role and that the stiffness in the direction orthogonal to the movement (i.e.

vertical) is increased, perhaps to aid straight-line path production; *v)* indications that extensor activity during *FAST* horizontal movements is a function of the operating range, with movements "toward" the center showing bilateral symmetry and "away" movements showing initial *ipsi* muscle activity; *vi)* cases where *COMF* movements are made by a "hold–release antagonist", with minimal agonist acitivity (the "agonist" being gravity); *vii)* cases where *EMG* activity for a given orientation is a function of the direction of approach; *viii)* "compromise" kinematic behavior for off-center vertical *COMF* movements, where the axis of rotation was between the "gaze-centered" and "body-centered" but closer to the former – additionally, *EMG* activity in muscle pairs such as *SCM*, which normally for vertical movements are synergists, was markedly asymmetric; *ix)* dramatic asymmetric sensitivity of *COMF* horizontal movements to the vertical component of orientation, with movements with a downward component having a "body-centered" rotation yet movements with an upward component being a compromise; *x)* indications of history dependent kinematic redundancy, in which the axis of rotation is influenced by the recent tracking history; *xi)* a general trend toward activity in stretched muscles; *xii)* computer modeling results show explicitly that 12-muscle schemes emphasizing the large superficial musculature are unable to capture certain fundamental features of the real system; and *xiii)* perhaps the most fundamental conclusion of our study is that the internal paraspinal musculature is very important during voluntary movements; in fact, the large, multilink superficial muscles with larger moments arms may be more effective "stabilizers" [see Chapter 26 (Crisco and Panjabi)].

28.6 Future Directions

Our knowledge of head–neck biomechanics and movement organization remains incomplete. In large part this is due to the enormous complexity of the system. However, as outlined in Winters (1988), it is also due to there being three very different disciplines involved in head-neck research with very different goals and little cross-fertilization: *i)* neuroscientists using rudimentary biomechanical models and highly evolved neuromotor concepts; *ii)* engineers studying impact–injury mechanisms using advanced models of passive biomechanics, with little or no consideration of neuromuscular factors; and *iii)* medically-based studies related to head-neck rehabilitation from injury (perhaps further classified as orthopedic-based or neurologically-based).

The greatest need seems to be in *rehabilitation*, which has long been the domain of medical professionals and more recently an area of active pursuit by orthopedic biomechanists. An area largely untapped are the relationships between muscles, voluntary movement organizational strategies, and the head-neck rehabilitation process. We suggest that this is the most pressing area.

There is also a great need for new *diagnostic tools* for functional assessment of the head-neck system. Current imaging techniques provide useful clues regarding skeletal or neurological function, but really say little about the neuromusculskeletal system. Classical range-of-motion and palpation techniques also do not really address the functioning system. Because of the strong kinematic link between the head and neck, screw axis parameters may provide a useful "window" for assessing head–neck neuromechanical function. We are currently testing this hypothesis on a population of subjects with "whiplash"–type symptoms. Preliminary results have been encouraging, yet have pointed out that a multitude of problems may exist within the cervical spine. For instance, we found one subject who always rotated about the *C1–C2* area, yet another who tended to have very low (and wildly varying) axes of rotations. While we often have an intuitive feel for what changes in head kinematics mean clinically, interpretation of results must be done cautiously. In particular, a more detailed model of the cervical spine is necessary; we are currently developing such models.

Our study, in conjunction with Chapters 23 and 26-27, point out the need for a better understanding of system stability, and especially the role of muscle properties and muscle activation level to such stability. As outlined in these chapters, biomechanical assessment of equilibrium conditions and of dynamic interaction is not sufficient. Unfortunately, we find that the head–neck system is too complex for effective analytical stability analysis, at least for the questions we seek to address. Clearly computer models are necessary. Yet our modeling studies of the head–neck system

suggest that 12-muscle models are not sufficient, with at least 24 and perhaps more necessary to correctly simulate basic head kinematics during voluntary movements. Furthermore, determination of appropriate muscle activation patterns is very difficult, due to both the complex anatomy and the "inverted pendulum" structure. Consequently, to find solutions, optimization methods, both static and dynamic, become essential supporting tools. However, as discussed in Bergmark (1987), static optimization, and even deterministic dynamic optimization, may fail to provide insight into issues related to stability. Thus, we need to consider stochastic optimization approaches, a very difficult adventure.

References

Albright, J.P., van Gilder, J., el Khoury, G., Crowley, E. and Foster, D. (1984) Head and neck injuries in sports, In *Principles of Sports Medicine*, (eds: Scott, W.N., Nisonson, B. and Nicholas, J.A.), Williams & Wilkins, Baltimore.

Angeles, J. (1982) *Spatial Kinematic Chains*. Springer-Verlag, New York.

Basmajian, J.V. (1978) *Muscles Alive. Their functions revealed by electromyography*. Williams & Wilkins Co., Baltimore.

Bergmark, A. (1987) *Mechanical stability of the human lumbar spine*. Ph.D. Dissertation, Lund Inst. of Technol., Lund, Sweden.

Berthoz, A. (1985) Adaptive mechanisms in eye-head coordination. In: *Adaptive mechanisms in Gaze Control: Facts and theories*, (eds: Berthoz, and Melville Jones), pp. 177-201, Elsevier Sci.

Bizzi, E. and Polit, (1978) Effect of load disturbance during centrally inititated movement. *J. of Neurophys.*, **41**: 542-556.

Bohlman, H.H., Ducker, T.B. and Lucas, J.T. (1982) Spine and spinal cord injuries. In *The Spine*, (eds: Rothman, R.H. and Simeone, F.A.), W.B. Saunders, Philadelphia.

Breen, A., Allen, R., Morris, A. (1988) An image processing method for spine kinematics - preliminary studies. *Clin. Biomech.*, **3**: 5-10.

Chao, E.Y.S., Tanaka, S., Korinek, S. and Cahalan, T. (1989) Measurement of neck range and pattern of movement. *Proc. XII Intern. Congr. Biomech.*, Abstract 319, UCLA, Los Angeles.

Daru, K.M. (1989) *Computer simulation and static analysis of the human head, neck and upper torso*, M.S. Thesis, Arizona State University.

Fountain, F.P., Minear, W.L. and Allison, R.D. (1966) Function of longus colli and longissimus cervicis muscles in man.

Goel, V.K., Clark, C.R., McGoman, D. and Goyal, S.

(1984) An in-vitro study of the kinematics of the normal, injured and stabilized cervical spine. *J. Biomech.*, **17**: 363-376.

Goel, V.K., Clark, C.R., Gallaes, K. and Liu, Y.K. (1988) Moment-rotation relationships of the ligamentous occipito-atlanto-axial complex. *J. Biomech.*, **21**: 673-680.

Gonon, J.P., Deschamps, G., Dimnet, J. and Fisher, L.P. (1987) Kinematic study of the inferior cervical spine in sagittal plane. In: *Cervical Spine I*, (eds: Kehr, P. and Weidner, A.), pp. 32-37, Springer-Verlag, New York.

Harms-Ringdahl, K., Ekholm, J., Schuldt, K., Nemeth, G. and Arborelius, U.P. (1986) Load moments and myoelectric activity when the cervical spine is held in full flexion and extension. *Ergonomics*, **12**: 1539-1552.

Hannaford, B., Maduell, R., Nam, M.-H., Lakshminarayanan, V. and Stark, L. (1983) Effects of loads on time optimal head movements: EMG, oblique, & main sequence relationships. *Proc. 19th Ann. Conf. Manual Control*, pp. 483-499, Cambridge.

Hannaford, B., Nam, M.H. and Stark, L. (1984) Electromyographic evidence of neurological controller signals with viscous load. *J. Motor Behav.*, **16**: 255-274.

Hannaford, B. and Stark, L. (1985) Roles of the elements of the tri-phasic control signal. *Exp. Neurol.*, **90**: 619-634.

Kinzel, G.L., Hall, A.S. and Hillberry, B.M. (1972) Measurement of the total motion between two body segments - I. *J. Biomech.*, **5**: 93-105.

Liang, D. (1989) *Mechanical response of an anthropomorphic head-neck system to external loading and muscle contraction.* M.S. Thesis, Arizona State University.

Moffat, E.A. and Schulz, A.M. (1979) X-ray study of the human neck during voluntary motion. *Soc. Autom. Eng.*, SP-790134, pp. 31-36.

Moroney, S.P., Schultz, A.B., Miller, J.A. and Andersson, G.B. (1988) Load-displacement properties of lower cervical spine motion segments. *J. Biomech.*, **9**: 185-192.

Myers, E.R. and Mow, V.C. (1983) Biomechanics of cartilage and its response to biomechanical stimuli. In: *Cartilage, Volume I*, Academic Press.

Myklebust, J.B., Pintar, F., Yoganandan, N., Cusick, J.F., Maiman, D., Myers, T.J. and Sances, A. (1987) Tensile strength of spinal ligaments. *Spine*, **13**: 526-531.

Panjabi, M.M., Krag, M.H. and Goel, V.K. (1981) A technique for measurement and description of three-dimensional six degree-of-freedom motion of a body joint with an application to the human spine. *J. Biomech.*, **14**: 447-460.

Panjabi, M., Dvorak, J., Duranceau, J., Yamamoto, I., Gerber, M., Rauschning, W. and Ulrich Bueff, H. (1988) Three-dimensional movements of the upper cervical spine. *Spine*, **13**: 726-730.

Panjabi, M.M., Summers, D.J., Pelker, R.R., Videman, T., Friedlaender, G.E. and Southwick, W.O. (1986) Three-demensional load-displacement curves due to forces on the cervical spine. *J. Orthop. Res.*, **4**: 152-161.

Panjabi, M.M., White, A.A. and Johnson, R.M. (1975) Cervical spine mechanics as a function of a function of transection of components. *J. Biomech.*, **8**: 327-336.

Peles, J. (1990) *Relations between neck muscle activity and three dimensional head kinematics during quasi-static and dynamic head tracking tasks.* M.S. Thesis, Arizona State University.

Pellionisz, A.J. and Peterson, B.W. (1988) A tensoral model of neck motor activation. In: *Control of head movements*, (ed: Peterson, B.P. and Richmond, F.J.), Oxford Univ. Press, New York.

Peterson, B.W. (1988) Cervicocollic and cervicocular reflexes. In: *Control of head movements*, (ed: Peterson, B.P. and Richmond, F.J.), Oxford Univ. Press, New York.

Peterson, B.W. and Richmond, F.J., eds., (1988) *Control of head movement.*, Oxford Univ. Press, New York.

Richmond, F.J., Baker, D.A., Stacy, M.J. (1988) The sensorium: receptors of neck muscles and joints. In: *Control of head movements*, (ed: Peterson, B.P. and Richmond, F.J.), Oxford Univ. Press, New York.

Roucoux, A. and Crommelinck, M. (1988) Control of head movement during visual orientation. In: *Control of head movements*, (ed: Peterson, B.P. and Richmond, F.J.), Oxford Univ. Press, New York.

Rondot, P. (1988) Clinical disorders of head movement. In: *Control of head movements*, (ed: Peterson, B.P. and Richmond, F.J.), Oxford Univ. Press, New York.

Sances, A., Mylebust, J.B., Maiman, D.J., Larson, S.J., Cusick, J.F. and Jodat, R.W. (1984) The biomechanics of spinal injuries. *CRC Crit. Rev. Biomed. Engng.*, **11**: 1-76.

Schor, R.H., Kearney, R.E. and Dieringer, N. (1988) Reflex stabilization of the head. In: *Control of head movements*, (ed: Peterson, B.P. and Richmond, F.J.), Oxford Univ. Press, New York.

Scholten, P.J.M. and Veldhuizen, A.G. (1985) The influence of spine geometry on the coupling between lateral bending and axial rotation. *Engng. in Med.*, **14**: 167-171.

Schuldt, K., Ekholm, J., Harms-Ringdahl, K., Nemeth, G. and Arborelius, U.P. (1986) Effects of changes in sitting work posture on static neck and shoulder muscle activity. *Ergonomics*, **12**: 1525-1537.

Spoor, C.W. and Veldpaus, F.E. (1980) Rigid body motion calculated from spatial co-ordinates of markers. *J. Biomech.*, **13:** 391-393.

Stark, L., Zangemeister, W.H. and Hannaford, B. (1988) Head movement models, optimal control theory, and clinical application. In: *Control of head movements*, (ed: Peterson, B.P. and Richmond, F.J.), Oxford Univ. Press, New York.

Takebe, K., Vitti, M. and Basmajian, J.V. (1974) The functions of semispinalis capitis and splenius capitis muscles: an electromyographic study. Atat. Rec., 179: 477-480.

Vitti, M., Fujiwara, M., Basmajian, J.V. and Iida, M. (1974) The integrated roles of longus colli and sternocleidomastoid muscles: an electromyographic study. *Anat. Rec.*, 177: 471-484.

Vorro, J. and Johnston, W.L. (1987) Clinical biomechanic correlates for cervical function: Part II. A myoelectric study. *JAOA*, **87:** 353-367.

White, A.A. and Panjabi, M.M. (1978) *Clinical biomechanics of the spine.* J.B. Lippincott, Philadelphia.

Winters, J.M. (1988) Biomechanical modeling of the human head and neck. In: *Control of head movements*, (ed: Peterson, B.P. and Richmond, F.J.), Oxford Univ. Press, New York.

Winters, J.M., Liang, D. and Daru, K.R. (1988) Effect of loading on head "axis of rotation": a comparison of approaches. *Adv. in Bioengng.*, ASME, **BED-8:** 95-98, Chicago.

Winters, J.M., Radhakrishnan, S. and Seif-Naraghi, A.H. (1987) Analysis of complex, multi-directional head movement tasks. *Proc. IEEE Engng. Med. & Biol.*, pp. 270-271, Boston.

Woltring, H.J., Huiskes, R., De Lange, A. and Veldpaus, F.E. (1985) Finite centroid and helical axis estimation from noisy landmark measurements in the study of human joint kinematics. *J. Biomech.*, **18:** 379-389.

Zangemeister, W.H. and Stark, L. (1982) Gaze latency: variable interaction of head and eye latency. *Exp. Neurol.*, **75:** 389-406.

Zangemeister, W.H., Lehman, S. and Stark, L. (1982) Simulation of head movemnt trajectories: model and fit to main sequence. *Biol. Cybern.*, **41:** 19-32.

CHAPTER 29

Muscle Activation Patterns Coordinating Postural Stability from Head to Foot

Emily A. Keshner and John H. J. Allum

29.1 Characteristics of Human Equilibrating Movements: The Problem of Redundancy

In order to truly understand how the central nervous system plans and produces movements to match environmental demands, one needs to take into account the many muscle activation patterns and movable links available to the human body when a movement is generated. The human musculoskeletal system, because it has the potential to move in multiple directions and has more muscles surrounding its joints than appear necessary to produce the functional range of movements (Pellionisz and Peterson, 1988), is a unique redundant system. Unfortunately, human movement systems, particularly for the study of postural control, are frequently reduced to 1-degree-of-freedom, single-axis, rigid bodies in order to simplify the gathering, analysis, and interpretation of data. The problem with this approach is that the results of such conceptual assumptions may lead to conclusions about sensory signals and resulting movement control that do not match the reality of the tasks presented to the human central nervous system [e.g. Chapter 8 (Zajac and Winters); Chapter 10 (Loeb and Levine); Chapter 34 (Pedotti and Crenna)]. In fact, during standing the *CNS* is controlling a multi–link system with several degrees of freedom. Thus the first aspect of the redundancy problem which should be investigated is the redundant modes of movement of the multi-link structure of the human body.

A second aspect of the redundancy problem concerns the number of muscles acting at a joint. Consider, for example, the neck "joint" (kinematically actually at least two "joints"; Vidal et al., 1986). Twenty-three different muscles directly link the skull on either side of midline to the vertebral skeleton (Sherk and Park, 1983). The multiple muscle attachments might not be so surprising if the head were involved in the fine motor control and variety of motions found in the hand and fingers. Motions of the head relative to the trunk, however, are primarily directed toward orienting and stabilizing the position of the eyes and head in space [Outerbridge and Jones, 1971; Chapter 28 (Winters et al.)]. Even for such a relatively "simple" task the redundancy problem of which muscle is selected and its activation pattern has to be examined in order to understand how a movement is generated. Although these muscle activation patterns are often described in terms of agonist muscle activity, or as a balance between agonist and antagonist forces, there is the potential for muscles acting on the joint to join in varying combinations of activation and still produce the same directional force output. Examples of such systems include not only the head-neck motor system (Peterson et al., 1985), but also the shoulder (Flanders and Soechting, 1989) and the elbow joint (Buchanan et al., 1989).

The final aspect of the redundancy problem concerns the type of motor program that generates a movement. Thus a multi-muscle system can potentially switch its control operations between

Multiple Muscle Systems: Biomechanics and Movement Organization
J.M. Winters and S.L-Y. Woo (eds.), © 1990 Springer-Verlag, New York

reflexes, voluntary mechanisms, and biomechanical resonant properties in order to achieve the appropriate response. Biomechanical dynamics are easily described [e.g. see Chapter 8 (Winters and Zajac)]. These include the viscoelastic and inertial characteristics of each segment and act almost instantaneously. However, these forces must be taken into consideration by later acting neural responses in order to produce muscle activation patterns that match the externally applied forces. Definitions of neural responses controlling posture are usually based on three properties of the response: latency, task dependence, and invariance of motor output. Reflexes are assumed to have the shortest latencies (less than 70 ms), to be independent of task demands, and to be organized as an invariant output to a specific pattern of sensory inputs. Voluntary responses are believed to have the longest latencies (greater than 150 ms), are dependent upon specific task parameters, and are highly variable in their response organization. For some movements that are well learned and predictable (postural movements fall into this class), these response definitions can be confounded. Voluntary responses are capable of occurring at latencies as early as those of reflexes (Rothwell, 1988), and triggered reactions, that can be modified by the demands of the task, appear at latencies longer than reflexes (80–120 ms) (Crago et al., 1976).

In this chapter our primary goal is to lay the framework for an appreciation of human postural control based on these redundancy considerations. On the problem of multi–link redundancy we will suggest that a given environmental task is usually achieved by only one multi–link mode totally dependent on a whole body muscle synergy. Our arguments in support of this hypothesis will be developed by considering how muscle activation patterns at the ankle, hip, trunk, and neck are linked together. Nested within this multi-joint synergy concept is the idea that the number of muscles and activation patterns that could be involved at a single joint when a stabilizing movement occurs are in fact less redundant than at first sight because of preprogrammed optimization and reduction of possible patterns by the CNS. Finally, in our consideration of the link and muscle pattern redundancy, a by-product will be an understanding of the role of reflex and voluntary responses in equilibrating movements.

29.2 Maintaining Multisegmental Body Equilibrium

29.2.1 Describing Whole-Body Stabilizing Strategies

When upright stance is disrupted, equilibrating reactions could be generated by the peripheral and central nervous system in a number of ways. One extreme possibility is that movements at various joints are solely dependent on local reflex responses to perturbations at each joint. Only reflexes would have controlled the timing and strength of muscle activity, and thus joint angle changes would be practically independent of each other, except for biomechanical coupling. The other extreme is to assume that both muscular activation and movement coordination are part of a centrally preprogrammed strategy (cf. Horak and Nashner, 1986) which is triggered by the sensory inputs signalling the onset and direction of destabilizing forces. A centrally preprogrammed response would presumably involve coordinated patterns of muscle activity or synergies between and across joints. Regardless of how the resulting movement coordination pattern is generated, it is generally termed a strategy. Through an investigation of the strategies and synergies comprising stabilizing reactions, one can develop a descriptive solution to the aforementioned redundancy problem of human posture.

The idea that preprogrammed or stereotypical automatic responses to postural disturbances initiated at the base of support were responsible for stabilizing upright stance was first introduced by Nashner (1976). His experimental technique had subjects stand on a platform that could be translated in an anterior or posterior direction, or rotated into plantar or dorsiflexion within limited velocity ranges. Once the velocity ranges of this technique were extended, it was recognized as an extremely useful approach to studying postural control. Early studies concentrated only on muscle responses in the lower limbs, particularly those at the ankle joint. Logically, but inaccurately, assumptions about the organization of stabilizing responses and early models of human posture were based on a single–link inverted pendulum movement strategy. As has already been emphasized here, the body is multisegmental with multiple joints within some segments such as the neck and back [reviewed in Chapter 23

(Andersson and Winters); Chapter 46 (Brooke and McIlroy)]. The validity of rigid body approximation was challenged following observations of accelerations of the center of gravity (Kodde et al., 1979) and of individual body segments (Valk-Fai, 1973) during quiet standing that differed from those of the inverted pendulum model. Crucial to this point were the findings of Stockwell and his collaborators (1981), who developed a four-link model composed of a foot, shank, thigh, torso, and head. They found significant movement at all measured body joints, indicating that at least four degrees of freedom would be required to adequately describe postural sway in the sagittal plane. Not surprisingly, Nashner and his colleagues have begun to question their original hypothesis and to suggest that body posture is a function of the position of several different joints (Nashner and McCollum, 1985) and multiple sensory inputs (Nashner et al., 1989).

Once the multi–link nature of human sway was widely accepted, the next step involved determining whether postural reactions to one type of destabilization always adopted a unique multi–link strategy fitting the force requirement of the task or even the immediate sensory inputs evoked by support surface movements. One way to investigate this question is to compare the response strategy to rearward translation of the support surface with that induced by a toe-up rotation which mimics the ankle dorsiflexion occurring during translation. Specifically, one is asking two questions. Firstly, are the modes of multi–link movement strategy inherently different for rotations and translations? And, if not, is the sensory input at the ankle joint the cause of the common response strategy?

It turned out that rearward translation caused, from the very first translation, a characteristic set of leg, trunk, and head movements different from those induced by rotation despite an equal angle of imposed ankle dorsiflexion. The differences are shown as biomechanical recordings in Figure 29.1 and reduced to a schema in Figure 29.2. Those for translation may be best understood by assuming that the hips remain stationary in space for the first 100 ms; that is, are thrust forward relative to the support base. An ankle dorsiflexion and hip extension is thereby induced. As the hips are thrust forward, the head and trunk are accelerated backwards.

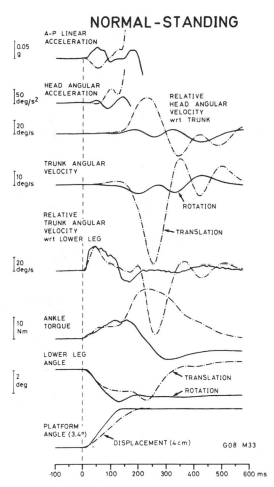

Figure 29.1: Biomechanical changes and equilibrating movements induced by rotating (solid line) and translating (interrupted line) the support surface. All traces are from one normal subject. An upward deflection of the head, trunk, and lower leg traces in the figure indicates backwards pitching. Increasing plantar flexion torque on the support surface is represented by an upward deflection of the ankle torque curve. (From Allum et al., 1989. Reprinted with permission.)

Compensatory action opposing the stimulus induced body sway commences at approximately 150 ms from stimulus onset and is also different for translation and rotation. For translation the changes in ankle torque and shear force resulting from the hips being thrust back coincide with fast trunk motion pitching forwards and a return of lower leg angle to upright. As illustrated in Figures 29.1 and 29.2, an overshoot, past upright, of the initial lower leg inclination almost always occurs. Simultaneously, the head pitches back-

Support Surface Rotation elicits a Stiffening Strategy

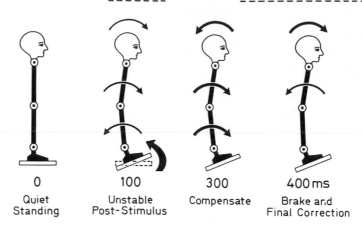

0	100	300	400 ms
Quiet Standing	Unstable Post-Stimulus	Compensate	Brake and Final Correction

Support Surface Translation elicits a Multi-Link-Strategy

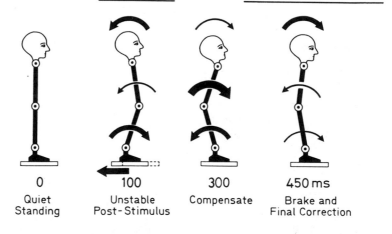

0	100	300	450 ms
Quiet Standing	Unstable Post-Stimulus	Compensate	Brake and Final Correction

Figure 29.2: Schema of stabilizing strategies for translation and rotation support surface perturbations. Each of four stages of the response are illustrated with faster angular velocities of body segments represented by broader arrows.

wards on the trunk, thus maintaining gaze direction, though the head is not stable in space. Because the equilibrating action compensating for support surface translation involves major rotations about the ankle, hip, and neck joints, it has been termed a "multi–link" strategy (Allum et al., 1989) The term "ankle–hip" strategy (Horak and Nashner, 1986) is almost as appropriate.

Rotation of the support surface elicits a completely different movement strategy (see Figures 29.1 and 29.2), which has been termed a "stiffening" multi–link strategy (Allum et al., 1989). Despite an equal amount of ankle dorsiflexion and equal, if oppositely directed head accelerations being imposed for rotation, the compensatory rotational velocities of body segments are of an order of magnitude less than those for translation. Essentially the rotation of the support surface is absorbed at the ankle joint though it is evident from the traces in Figure 29.1 that the hips are thrust backwards by the toe-up rotation. Thus, the two movement strategies, multi–link and stiffening, are very different despite comparable changes at the ankle joint. In short, the appropriate strategy appears to be dependent on the direction of trunk and head acceleration, and is executed as if preprogrammed from the very first response.

29.2.2 Describing Whole-Body Muscle Synergies

The above description of stabilizing movement strategies left three crucial issues untouched. The first issue concerns the type of muscle response synergies underlying movement strategies. These synergies will be described here in order to provide the reader with an overview of muscle activation patterns that restabilize an unstable posture. The second issue concerns how the activation patterns of muscles that act as synergists along the ventral or flexing surface and along the dorsal or extending surface of the body are correlated with the movement strategy, in order to reduce joint and muscle redundancy by locking degrees of freedom together. Finally, the dependence of muscle responses on incoming sensory signals deserves consideration.

Muscle activation patterns recorded in the trunk and leg muscles are consistent with the intersegmental movements caused by support surface perturbation and the compensation strategy utilized to reestablish upright posture. Figure 29.3 shows that in response to posterior translations, soleus (*SOL*), hamstrings, and abdominal muscles were activated earlier and more powerfully than their antagonists at the lower leg, thigh, and trunk, respectively. When the hips were thrust backward by toe-up rotation, tibialis anterior (*TA*), quadriceps, and paraspinal muscles provided the earliest and most predominant responses. By comparing Figures 29.1 and 29.3 it is evident that these different muscle responses are well correlated in timing and amplitude (*EMG* area) with differently directed torques exerted on the support surface, and with the overall different movement strategies exhibited by tracings of trunk and head angular velocity. Excepted from these observations on response differences are the short latency (*SL*) stretch reflex responses in soleus. As Figure 29.3 illustrates, the soleus *SL* responses, and for that matter gastrocnemius responses, are common to rotation and translation.

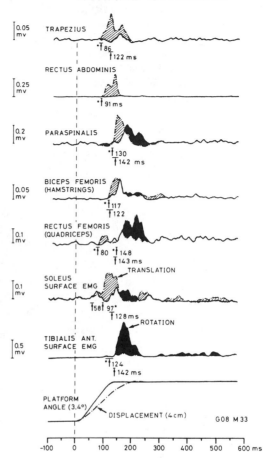

Figure 29.3: Muscle response patterns of a normal subject to rotations and translations (same subject as in Figure 29.1). The traces are average *EMG* responses for the 2nd 8 trials. Differences between the means of the *EMG* curves are shown as a filled area when the rotation response was larger, as a hatched area when the translation response was larger. The zero latency reference is given by a vertical dashed line, and corresponds to the first inflexion of ankle torque (see Figure 29.1). Average values of *EMG* burst-onset latencies are marked by a vertical arrow, one standard deviation by a horizontal bar above the arrow. The upper set of latencies (marked with an asterisk) below each trace correspond to translation responses, the lower set for rotation responses, except for *SL* onset in soleus which was equal in latency for rotation and translation. (From Allum et al., 1989; reprinted with permission.)

It is noticeable that response onsets in Figure 29.3 lie between 90 and 150 ms and do not fall into the neat categories of either a reflex response (onset prior to 80 ms) or voluntary response (greater than 150 ms onset; see also Figure 29.4). A rather elegant way to account for those onset delays would be to suggest that the proprioceptive input at the ankle joint, specifically muscle stretch receptors from triceps surae muscles, excited first a stretch reflex in the lower legs which then radiated upwards along the body reaching the neck muscles last. With appropriate pre-programming of intersegmental neuronal circuits, one would conceive of an ascending synergy triggered by the sensory input from the ankle joint. In essence, this idea is embodied in Horak and Nashner's (1986) paper on stabilizing strategies. There are arguments, however, that make this otherwise reasonable hypothesis untenable.

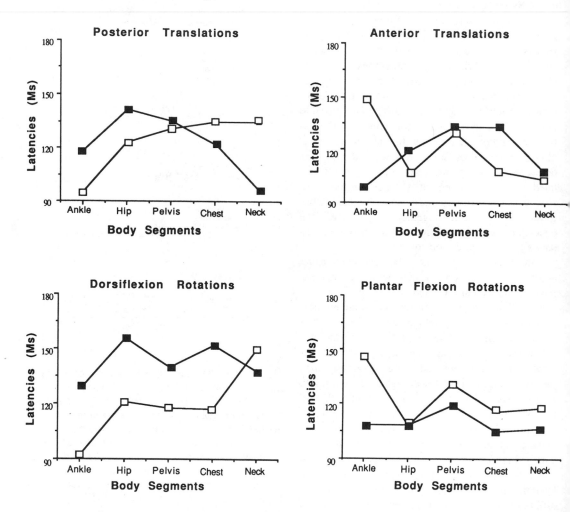

Figure 29.4: Mean medium latency *EMG* responses of muscles to four directions of platform perturbation. Muscles on the ventral surface (open squares) and dorsal surface (filled squares) of the body are plotted on each graph to illustrate the pattern of muscle activation. Muscles tested at each segment include: soleus and tibialis anterior at the ankle, hamstrings and quadriceps at the hip, lumbar paraspinals and abdominals at the pelvis, thoracic paraspinals and pectorals at the chest, and neck extensor and flexor at the neck. (From Keshner et al., 1988; reprinted with permission.)

First, the rotation of the support surface in Figure 29.1 mimics the dorsiflexion ankle rotation occurring during rearward support surface translation. Nonetheless, the movement strategy in Figure 29.1 and muscle activation patterns in Figure 29.3 are so fundamentally different for the two types of perturbations that neither muscle spindle receptors in the lower leg, ankle joint receptors, nor pressure receptors in the sole of the foot appear to trigger the muscle activation patterns or to be responsible for the movement strategy. In other words, the *SL* reflex response in triceps surae muscles may well be an epiphenomenon to the movement strategy.

Second, Keshner et al. (1988) recorded from 11 leg, trunk, and neck muscles. The timing of postural muscle responses within and between body segments was employed in order to determine whether they maintained a consistent temporal relationship under translational and rotational platform movement paradigms. The results did not support a strict ascending pattern of activation. In response to posterior platform translations (Figure 29.4), an ascending pattern of muscle responses along the ventral surface of the body was observed. In addition, responses elicited in the neck flexor and abdominal muscles occurred as early as those of the stretched ankle muscles (*SOL*), suggesting a simultaneous descending pattern of activation. The crossed ascending/descending pattern also existed for anterior platform displacements, where early neck extensor muscle responses were observed at the same time as stretched ankle flexor (*TA*) activation (Figure 29.4). For platform rotations there was an even greater tendency for all muscle responses to occur simultaneously, although earlier on one side of the body.

The conclusions that can be drawn from these studies are twofold. The fundamental aspect is that the two movement strategies, multi–link rotations for translation, stiffening for support surface rotations, appear to result from two types of muscle activation synergies each of which is a different solution to the redundancy problem and comprises an attempt to link movements of body segments together. In response to translation, a multi–link stabilizing strategy is generated by an ascending–descending muscle response pattern; ascending on the ventral surface when body sway is induced forward, on the dorsal surface when induced sway is backwards. For rotation of the support surface, the resulting stiffening strategy involves a simultaneous activation of muscles at all segments with stronger activation on the dorsal body surface for dorsi-flexion rotation and vice versa for plantar flexion rotations. The auxiliary aspect to these intersegmental studies is that sensory inputs from the ankle joint are not instrumental in triggering the different strategies and the consequent reduction in muscle redundancy. In fact, as the next section will document, studies of coactivation patterns acting across a joint have indicated that vestibular and proprioceptive systems in the upper body play a role in both initiating and modulating stabilizing muscle responses.

29.2.3 Coactivation Patterns at the Ankle and Hip Joints During Equilibrating Reactions

One optimizing strategy for reducing redundant linkages is to coactivate agonist–antagonist muscle pairs across a joint within the movement synergies. To generate appreciable movement about each joint during the multi–link strategy, activation of antagonist muscles acting at a joint must be delayed in onset but coactivated at intensities relative to one another. The stiffening strategy can occur if muscles acting there have simultaneous onset of correlated activation [e.g. see Chapter 5 (Winters) and Chapter 9 (Hogan)]. For hinge joints such as the ankle and hip, a consideration of this solution is easier because movements are basically restricted to flexion and extension when stance is two-legged. At the neck, where the joint structure is complex, solutions to the redundancy problem have a matching complexity [see Chapter 28 (Winters and Peles)]. For this reason, neck muscle synergies will be described separately in a later section (29.3).

Figure 29.5 documents, for all subjects tested, the activation patterns observed in reaction to rotation and translation as medium latency (*ML* onset at 90 to 150 ms) responses in lower leg muscles. The upper part of the figure plots the onset of the tibialis anterior *ML* response against that of soleus. The lower part of Figure 29.5 plots the *ML* response areas measured under the first 80 ms of each *EMG* response. The separation of data points for rotation and translation in Figure 29.5 indicates significant differences with respect to coactivation at the ankle joint for these stimuli. Translation responses consist of a sequential activation of soleus followed by tibialis anterior after 40 ms. There is no significant covariation between response amplitudes as measured by *EMG ML* areas. Rotation responses, in contrast, appear to be coactivated because both onset latencies are within one standard deviation of each other, and a significant linear correlation exists between response areas.

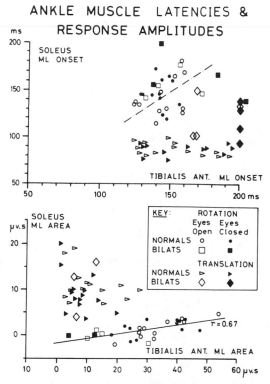

Figure 29.5: Response amplitudes and latencies for soleus (*SOL*) and tibialis anterior (*TA*) muscles following a rotation or a translation perturbation to stance. In the upper part of the figure *SOL ML* onset is plotted against *TA ML* onset, equal onset is marked by the dash-dot line. Several *TA* responses to translation for the vestibular deficit patients had onset latencies over 200 ms. The lower part of the figure shows plots of the area of the first 80 ms of the response. Rotation response areas had a significant linear correlation. The plots illustrate an earlier larger response in *SOL* for translation, a coactivated action of *SOL* and *TA* for rotation with *TA* the larger response. Inset indicates visual conditions and subject category associated with the data point symbol. (From Allum et al., 1989; reprinted with permission.)

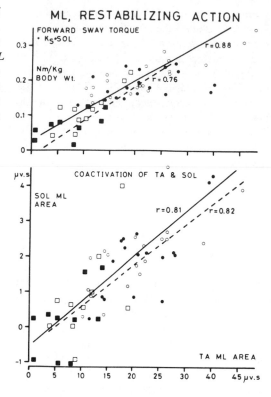

Figure 29.6: Coupled activity of *SOL* and *TA* and their joint influence on ankle torque at medium latencies. Lower half of figure: average areas of *TA* activity of each normal subject (circles) and patient (squares) plotted against *SOL*, and regression line drawn for eyes open (solid line) and closed (broken line) data. Respective correlation coefficient (*r* value) printed next to each line. Upper half: multivariate regression of *TA* and *SOL* on ankle torque. Average activity of the weaker predictor is multiplied by its regression coefficient and added to the torque values (e.g., $Ks*SOL$) (From Keshner et al., 1987; reprinted with permission.)

A similar coactivation effect, albeit with subjects initially leaning back slightly prior to onset of the support surface rotation, has been described previously (Keshner et al., 1987). These authors studied the correlation of kinematic parameters with neural components of the response to platform perturbations in standing adults. That is, the magnitude of restabilizing ankle torque was examined for temporal concurrence and significant correlation with *EMG* response areas of medium latency (approx. 120 ms) and long latency (75 ms later) activity in *SOL* and *TA*. The results confirmed the hypothesis of functional coactivation of these muscles (see lower half of Figure 29.6) consistent with current theories suggesting that muscles around a joint are coactivated in a range of limb orientations in order to produce appropriate limb movement (Buchanan et al., 1989; Baker et al., 1985; Peterson et al., 1989). At the ankle joint, interrelationships between the kinematic measure (ankle torque) and *EMG* activity were examined through a multivariate regression, and revealed that each ankle muscle alone was poorly correlated with resultant ankle torque. When, however, the coactivated action of both ankle muscles was considered, high significant correlations ($p < 0.0005$) appeared between ankle torque and muscle activity at medium and long latencies (upper half of Figure 29.6). The motor control picture emerging from this analysis suggested that at medium latencies, the action of *TA* was to increase forward sway torque about the ankle joint to counteract the rearward thrust from the platform. Longer latency torque was best predicted by *SOL* which produced a negative sloping regression line, suggesting a braking torque action on the forward sway produced at medium latencies.

Interestingly, the coactivation picture at the ankle joint for support surface rotation, and its absence for translation, is completely reversed for the antagonist abdominal and lumbar paraspinal muscles that act to rotate the trunk about the hip joint. Figure 29.7 presents the onset delays and coactivation between these muscles and demonstrates that the trunk muscle synergy is constant across subjects. Although abdominals respond some 40 ms prior to paraspinals during translation, the range of onset differences between subjects is small, suggesting a fixed coactivation delay. The concept of coactivation in these muscles is supported by the linear correlation between response magnitudes as shown in the lower part of Figure 29.7; such a covariation was conspicuously absent from the ankle muscle responses to translation (Figure 29.5) and from the trunk muscle responses to rotation. For rotations, no responses were observed in abdominal muscles (see Figure 29.3). Returning to Figure 29.1, it is apparent that the large pulse of forward pitching trunk acceleration thrusting the hips backward is a major component of the restabilizing strategy in response to translation. Thus, it would be useful

to determine, in a similar way attempted for rotation response at the ankle joint (see Figure 29.6), whether the coactivation pattern at the trunk muscles is correlated with torque exerted about the hip joint. Since there is not a simple way of measuring this torque, studies of patient responses might provide the necessary insights because clinical studies are a means of exploring where control components might lie for certain motor functions. The temporal organization and magnitude of EMG responses, as well as resultant biomechanical changes, have been shown to be significant indicators of balance disorders (Black et al., 1983; Diener et al., 1984; Allum et al., 1988).

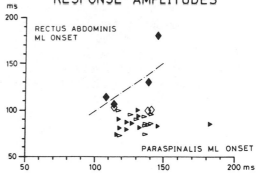

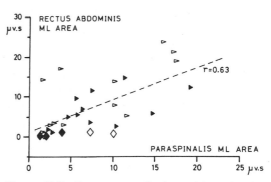

Figure 29.7: Response amplitudes and latencies for rectus abdominous and paraspinal muscles after a support surface translation. See Figure 29.5 for explanation of symbols. (From Allum et al., 1989; reprinted with permission.)

29.2.4 Sensory Inputs Underlying the Organization of Equlibrating Strategies

Differences noted in muscle synergies and coactivation patterns for the strategies elicited by translation and rotation perturbations provide the strongest evidence for the hypothesis that sensory inputs other than those in the lower legs determine the organization of the equilibrating strategy. If one can exclude proprioceptive inputs from the lower legs, what then are the necessary and sufficient sensory inputs to trigger and/or modulate the appropriate synergy and movement strategy? The most effective receptors must rapidly detect the occurrence and direction of body movements. Thus, sensory systems registering differences in polarity and having early onset would be the best candidates. Reviewing Figure 29.1, it appears that the vestibular system registering head angular accelerations is an ideal candidate because vestibular afferents have an early onset (20 ms).

Studies comparing the responses of normals and vestibular deficient subjects have suggested that vestibular afferents excited by head angular accelerations play a major role in the control and organization of equilibrating reactions to support surface rotation (Allum and Pfaltz, 1985; Keshner et al., 1987). Areas under the *EMG* bursts in the ankle muscles, soleus (*SOL*) and tibialis anterior (*TA*), and ankle torque recordings of patients with bilateral labyrinthine deficits were found to be significantly diminished when compared with normal subjects ($p < 0.05$) with both eyes open and closed (see Figure 29.6). The magnitude of *EMG* activation correlated with the extent of peripheral vestibular deficit (Allum et al., 1988), suggesting that the amplitude of lower limb postural reflexes is under the control of the vestibulospinal system. Whether or not these latter results can be extended to translation responses requires further investigation. The number of vestibular deficient subjects examined to date is too small to permit definitive conclusions. However, a tendency for reduced abdominal responses has been observed (see Figure 29.7) in these subjects.

Head and trunk accelerations could also excite a number of proprioceptive sensory systems which would be appropriate in generating suitable balance reactions. Among the sensory systems influenced by these accelerations are muscle stretch receptors in the hip, trunk, and neck muscles, as well as joint receptors in the spinal column. At this point it would be useful to review experiments distinguishing between the sensory inputs excited by head and trunk accelerations, and to consider the complex redundancy problem posed for the neck muscles.

29.3 Towards an Understanding of Head Stabilization Strategies

The troika of solutions suggested above for solving the redundancy problem of human postural stabilization could be examined purely through the head movement system which is a microcosm of a multi–link, multi–muscle system. Investigations to date have not provided evidence, however, indicating that opposing neck muscles are coactivated in the same way as leg and trunk muscles to reduce degrees of freedom during equilibrating reactions. Unlike the lower limb, difficulties exist in examining the head and neck motor system because of its intrinsic complexity, including overlapping muscles, multiple insertions on the intervertebral joints, and several central control mechanisms. A confounding factor is that the multiple muscles acting on the head, including many small, deep muscles, demonstrate well-specified directions of preferred activation [Baker et al., 1985; Keshner et al., 1988; see also Chapter 27 (Winters and Peles)]. It may well be that the failure to find coactivation between antagonistic neck muscles may simply be the result of incorrectly defined recording locations with respect to the stabilization task.

The second difficulty is that the afferent origin of neck muscle responses noted during postural stabilization has not been clearly identified (Keshner et al., 1987), even though the predominant role of vestibulocollic and cervicocollic reflexes has been postulated for some time in humans. In the cat, head stabilization for pitch rotations is dominated by short latency vestibular and neck reflexes (Peterson et al., 1985). Evidence for short latency effects in man are seen in vestibularly deficient subjects, but only as a possible enhancement of cervicocollic reflexes (Allum and Pfaltz, 1985). There is some evidence of short latency mechanisms controlling head stability at high frequencies (2-3 Hz) of horizontal head rotations (Keshner and Peterson, 1988b), but this frequency range has not yet been correlated with functional activity. Resonant characteristics of the head were found to predominate at higher frequencies (> 3 Hz), matching the head angular acceleration frequency characteristics during equilibrating reactions to pitch platform rotations (Keshner et al., 1987). It appears that man tends to rely more heavily on voluntary motor commands

(Guitton et al.,1986; Kasai and Zee, 1978) and passive mechanics of the head–neck system (Viviani and Berthoz, 1975; Allum et al., 1990) to stabilize the head. Aspects of recent studies attempting to ameliorate the two difficulties in exploring solutions to redundancy in the head–neck system, will be highlighted in the following sections.

29.3.1 Muscle Activation Patterns During Isometric Head Stabilization

In order to obtain a basic understanding of the types of correlated activation that might be expected during dynamic head stabilization, Keshner et al. (1989) studied directional preferences of human neck muscles during voluntary, isometric head stabilization. In total, four neck muscles, including semispinalis capitis (*SEMI*), splenius capitis (*SPL*), trapezius (*TRAP*), and sternocleidomastoid (*SCM*) were recorded with surface electrodes at placements later verified with bipolar intramuscular electrodes. The polar plots of directional preferences for these muscles shown in Figure 29.8a may be more easily understood by referring to the adjacent schema. For example, an applied isometric force at zero degrees was taken in the direction requiring neck extension (pitch extension) torques to stabilize the head. A force at 90° required right lateral flexion (right roll) torques of the head, and -90° required left roll. Applying the force directly behind the subject resulted in pure forward flexion (pitch flexion) torques of the head (180°).

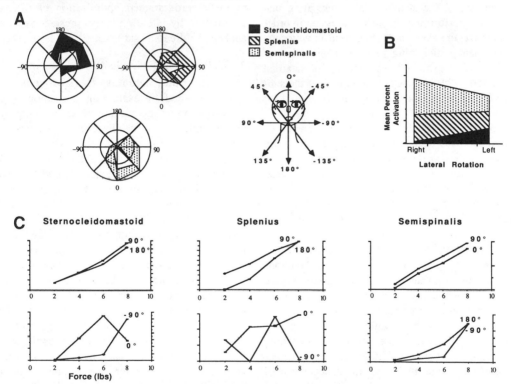

Figure 29.8: *a)* Mean percent of maximum *EMG* activation plus and minus one standard deviation for the group. A narrow shaded area indicates less variability among the subjects; the point closest to the circumference is the maximal activation direction for that muscle. Diagram to the right demonstrates the direction of resistive torques at each head orientation in the frontal plane. *b)* Mean percentage of *EMG* activation plotted for the two directions of lateral rotation. *c)* *EMG* responses of one subject with incrementing weights. Preferred activation directions for each muscle are in the top row, non-preferred directions in the bottom row. Numbers to the right of each line indicate orientation of the head. Amplitude of muscle electrical activity has been normalized so that the largest response in the four orientations was given a value of 1. [Modified from Keshner et al., 1989 and Keshner (in press), reprinted with permission.]

Referring to Figure 29.8a, it can be seen that each muscle has a restricted range of excitation with a consistent direction of maximal activation indicative of that muscle's potential dynamic responsiveness. Zero, or null, responses of each muscle in directions opposite to the maximum indicate that muscles are reciprocally activated during this task. *SEMI* worked primarily in pitch extension with some lateral rotation (0° to 45°), and therefore is the muscle of choice to record from during posturography investigations inducing pitch sway. Superior fibers of trapezius presented a large area of variation in all of the tested directions and low levels of activation during isolated movements of the head. Intramuscular *EMG* responses of the trapezius muscle were greatest when the subject was asked to perform isolated movements of the shoulder joint, indicating that *TRAP* was performing scapular depression in order to stabilize the scapular during head movements, rather than participating in pitch stabilization of the head. *SPL* exhibited two directions of maximal activation. This muscle was used by half the subjects for roll with flexion, and by the other half for roll with extension of the head.

Robustness of the response to directional parameters was tested by increasing the isometric forces applied to the head. Such attempts at increasing the force requirements for head stabilization resulted not only in a change in the *EMG* activation patterns, but also in a shift from a pattern of reciprocal activation to one of cocontraction. Accompanying this change in motor program was a non-linear shift in *EMG* output with increased force output in the directions that previously produced no *EMG* response (Figure 29.8c). Non-linearity in the directional parameters of the response raises some question about how the *CNS* programs coactivation. If, as suggested above, an essential element of the stiffening strategy is that the coactivation and reciprocal activation commands are issued simultaneously at all times, non-linear increasing activation of agonist and antagonists with increased loading on the neck joint should not only alter the stiffness of the joint, but cause a change in head position. With linear changes in activation levels one would expect a stable equilibrium to be maintained, as predicted by Feldman (1980; see also Chapter 12). The sudden jump from a null to a strongly activated response in the neck muscles (see lower trace of

each muscle in Figure 29.8c) would, therefore, require an equal increase in the activation level of whichever muscle was acting in its preferred direction if no movement was required. It would appear that at some critical force level there is a change to an alternative program of muscle response which is suitably matched to the stiffening equilibrating strategy, rather than a simple shifting in the tension characteristics of the muscles as found during reciprocal activation.

The results of these isometric studies suggest that muscle patterns involved in head stabilization may be quite localized and therefore difficult to investigate unless recordings are obtained from a number of neck muscles. In addition, differences between the muscle response patterns in preferred versus non-preferred muscle pulling directions due to the preprogrammed optimization of different parameters by the *CNS* may present a further analysis hurdle to be surmounted.

29.3.2 Patterns of Muscle Activity During Reflex and Voluntary Head Movements

One method of examining the complex optimization of muscle pulling directions employed during human head stabilization is to investigate stabilization responses in more simple animal models. Animal models have the advantage that the experimenter can hold external conditions constant enabling responses to be examined over a range of stimuli. In the decerebrate cat, for example, neck muscles maximally excited by one direction of head rotation (stimulation of the vestibulocollic reflex, *VCR*) are consistent over time and over animals (Baker et al., 1985). Of course, the optimal response of the neck muscles may be very different in both the freely moving animal or in an alert animal whose head is stabilized. These differences, however, can provide insight into how the *CNS* employs one sensory mode, vestibular signals from the semicircular canals, as compared to voluntary commands that are organized on the basis of somatosensory, vestibular, visual, and cortical inputs in order to select muscle pulling directions.

As a working hypothesis for their experiments Peterson and his colleagues (Keshner et al., 1986; Keshner and Peterson, 1988a) expected that reflex responses obtained when the *VCR* was stimulated (by placing the animal on a platform and rotating it in the dark while the head was fixed in relation

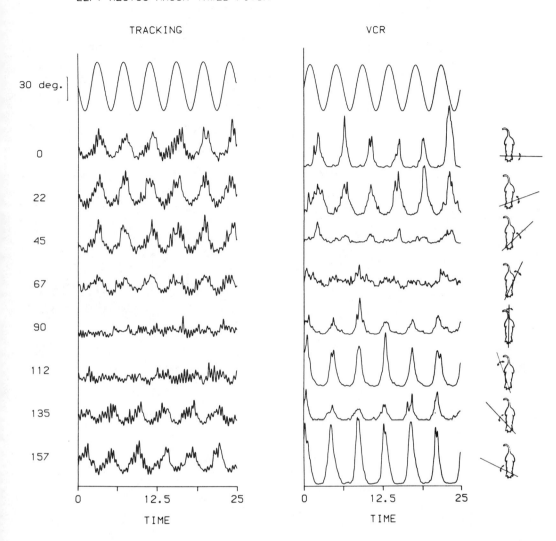

LEFT RECTUS MAJOR YAWED PITCH

Figure 29.9: Activation of the left rectus major muscle in an alert cat during sinusoidal whole body rotation (right traces) and sinusoidal head tracking with body fixed (left traces). On the far right of the figure, the solid lines with curved arrows indicate the axis of rotation for that set of body or head rotation records. This series of rotations includes pitch (0°) and roll (90°). *EMG* responses are sinusoidally modulated in both tasks, but the null plane where the response grows small and reverses direction clearly differs for muscle activation during body rotation (*VCR*) and head tracking (from Keshner and Peterson, 1988a; reprinted with permission).

to the body) would occur in the opposite direction to those obtained for voluntary head motion about the same axis. The responses of the left rectus major muscle during the voluntary and reflex tasks are compared in Figure 29.9, where it can be seen that differences in the reflex (*VCR*) and voluntary (tracking) muscle activation patterns exist both in the orientation of maximum and minimum activation, and in the phase of the response.

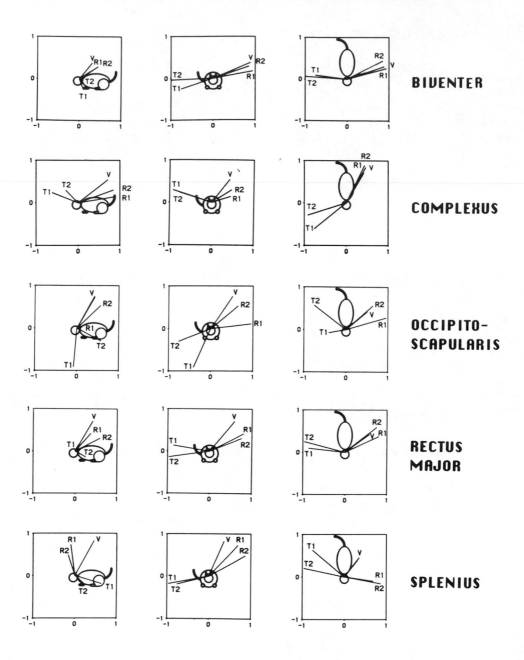

Figure 29.10: Averaged response vectors of five muscles during voluntary tracking in two alert cats (*T1* and *T2*) and whole-body rotations in alert (*R1* and *R2*) and decerebrate (*V*) cats. Left muscles are inverted to appear as though they are on the right side of the body. Each vector represents the axis of the rotation that would maximally excite the muscle. Maximal activation directions can be visualized from the vectors in the figure by placing one's right thumb along the vector and noting the direction in which the fingers curl around the vector axis. (From Keshner and Peterson, 1988a , reprinted with permission.)

In fact, for this muscle and for many others (see Figure 29.10) reflex and voluntary responses followed a more orthogonal relationship. As seen in Figure 29.10, only in the biventer cervicis muscle was the hypothesis of oppositely directed responses supported. It is important to note for the purpose of this comparison of head stabilization strategies in man and cat that maximal activation was not in the direction expected from the anatomical position of the muscle (Peterson et al., 1989). For example, right complexus produced its major response component while the head was rolling to the left rather than assisting the head in turning to the right as would be expected for a right sided muscle during voluntary head movement. In some human subjects, splenius was found to act primarily in roll with flexion rather than extension (see Figure 29.8). Although voluntary muscle patterns were quite consistent for an individual cat over several months of testing, they differed much more from animal to animal than the reflex muscle patterns, suggesting that the *CNS* chooses a single solution to generate the *VCR* from among the many possible motor patterns that could be used, but voluntary responses are selected on the basis of multiple parameters and strategies.

Thus, the solution for the head stabilization synergy from the viewpoint of an individual human *CNS* may vary from person to person. As a result, the concept with which this paper was started, namely that solutions of the muscle and joint redundancy problem for postural control are common for a given (patient or normal) population may well be poorly conceived for head stabilization. No matter how excellent an explanation synergy and strategy approaches are for all lower body segments, these concepts require rethinking before they can be applied to head stabilization. Future questions must first address whether neck muscle activation patterns are programmed to match frequency requirements as well as spatial and force constraints. Second, the role of the deep versus superficial muscles and their relation with various joint arrangements and the frequency characteristics of the system should be examined. Third, effects on the spatial characteristics of the neck muscle activation patterns at different frequencies of head rotation need to be investigated, especially given that a different spatial organization exists for the reflex and voluntary response patterns at low frequencies.

29.4 Future Directions

One of the main difficulties in studying postural control concerns the complexity of the task involved: stability of an inverted multi–linked pendulum, held upright by muscle torques generated at each link, is sensed by receptor systems at the ankle joint and head. The first approach of many modellers would be to reduce the system to a single inverted pendulum in order to better understand human postural control. This drastic solution to the redundancy problem of several links and numerous muscles acting at each joint undoubtably leads to conclusions that do not match the reality of the task presented to the central nervous sytem. The question that must be faced if one is to argue that a given stabilization task is achieved by one multi–link mode totally dependent on a whole muscle synergy, is the extent to which muscle synergies are modulated by incoming sensory signals on a continuous scale, or whether a switching between different movement strategies is dependent on the original biomechanical response to the body's perturbation. Thus future research could be targeted towards examining the number of centrally organized movement strategies, and the extent to which vestibular and upper body proprioceptive systems participate in the selection of the associated muscle synergies.

Crucial to the question of switching between different movement strategies is a deeper understanding of the body's biomechanics. One might ask whether for a given type of body perturbation a particular synergy is chosen by the *CNS* because sensory information indicated that either passive mechanical forces were nearly appropriate to compensate for imposed forces or that only one type of synergy was quick or strong enough to stabilize the body. Considerable progress has been made in recent years examining normal and pathological muscle action during stance stablization. This research has spotlighted the latter course of action by *CNS* command centers. Very little progress has been made, however, in understanding the effect that stance perturbations have on passive biomechanical trajectories of body segments and the modifications of these trajectories by active muscle forces. It was when head stabilization strategies were considered that the separate contributions of active muscle and passive biomechanical forces assume an important function in the environmental task.

Consideration of head–neck responses under a variety of environmental conditions suggests that a single functional movement pattern can be under the control of different motor programs and, very possibly, utilize different control mechanisms, yet still attain the final behavior. Thus the system can potentially switch or coordinate its control operations between proprioceptive reflexes, vestibulocollic reflexes, voluntary mechanisms, and mechanical properties of the system to achieve the appropriate response. Yet it is for the neck muscles that experimental identification of muscle synergies has proved to be most elusive, possibly because difficulty exists, in man, in discriminating between the activities of nearby muscles and in monitoring the function of deep muscles. Despite this drawback, it is for the head–neck system that greater progress in understanding the basic biomechanics has occurred using optimization techniques (Pellionisz and Peterson, 1988). Hopefully applying similar techniques to a multi–degree of freedom model of upright stance will improve knowledge on one aspect of the redundancy problem discussed in this chapter, namely the distinction between a movement strategy and a purely passive biomechanical response to an unstable posture.

References

Allum, J.H.J. and Pfaltz, C.R. (1985) Visual and vestibular contributions to pitch sway stabilization in the ankle muscles of normals and patients with bilateral peripheral vestibular deficits. *Exp. Brain Res.* **58**: 82-94.

Allum, J.H.J., Honegger, F. and Pfaltz, C.R. (1989) The role of stretch and vestibulospinal reflexes in the generation of human equilibrating reactions. Afferent control of posture and locomotion. In *Progress in Brain Research* (edited by J.H.J. Allum and M. Hulliger), Elsevier, Amsterdam, **80**:399-409.

Allum, J.H.J., Honegger, F., and Keshner, E.A. (1990) Head-trunk coordination in man: Is trunk angular velocity elicited by a support surface movement the only factor influencing head stabilization. In *The Head-Neck Sensory-Motor System: Evolution, Development, Disorders, and Neuronal Mechanisms* (edited by A. Berthoz, W. Graf, and P.P. Vidal), Oxford University Press, London (in press).

Allum, J.H.J., Keshner, E.A., Honegger, F. and Pfaltz, C.R. (1988) Indicators of the influence a peripheral vestibular deficit has on vestibulo-spinal reflex responses controlling postural stability. *Acta*

Otlaryngol. (Stockh.) **106**: 252-263.

Baker, J., Goldberg, J. and Peterson, B. (1985) Spatial and temporal response properties of the vestibulocollic reflex in decerebrate cats. *J. Neurophysiol.* **54**: 735-756.

Black, O.F., Wall, C., and Nashner, L.M. (1983) Effects of visual and support surface orientation references upon postural control in vestibular deficient subjects. *Acta Otolaryngol.* **95**:199-210.

Buchanan, T.S., Rovai, G.P.and Rymer, W.Z. (1989) Strategies for muscle activation during isometric torque generation at the human elbow. *J. Neurophysiol.* **62**: 1201-1212.

Crago, P.E., Houk, J.C. and Hasan, A. (1976) Regulatory actions of human stretch reflex. *J. Neurophysiol.* **39**: 925-935.

Diener, F., Dichgans, J., Bacher, M., and Guschlbauer, B. (1984) Characteristic alterations of long-loop "reflexes" in patients with Friedrieich's disease and late atrophy of the cerebellar anterior lobe. *Neurol. Neurosurg. Psychiatr.* **47**:679-685.

Feldman, A.G. (1980). Superposition of motor programs. II. Rapid forearm flexion in man. *Neurosci.* **5**: 91-95.

Flanders, M. and Soechting, J. (1989) Patterns of muscle activity in isometric arm postures. *Soc. Neurosci. Abstr.* **15**: 51.

Guitton, D., Kearney, R.E., Wereley, N., and Peterson, B.W. (1986) Visual, vestibular and voluntary contributions to human head stabilization. *Exp. Brain Res.* **64**:59-69.

Horak, F.B. and Nashner, L.M. (1986) Central program of postural movements: Adaptation to altered support-surface configurations. *J. Neurophysiol.* **55**: 1369-1381.

Kasai, T. and Zee, D. (1978) Eye-head coordination in labyrinthine-defective human beings. *Brain Res.* **144**:123-141.

Keshner, E.A. (1990) Controlling stability of a complex movement system. *Phys. Ther.* (accepted for publication).

Keshner, E.A., Allum, J.H.J., Pfaltz, C.R. (1987) Postural coactivation and adaptation in the sway stabilizing responses of normals and patients with bilateral peripheral vestibular deficit. *Exp. Brain Res.* **69**: 66-72.

Keshner, E.A. and Peterson, B.W. (1988a) Motor control strategies underlying head stabilization and voluntary head movements in humans and cats. Vestibulospinal Control of Posture and Movement. In *Progress in Brain Research, vol. 76*, (Eds. O. Pompeiano and J.H.J. Allum), pp. 329-339, Elsevier, Amsterdam.

Keshner, E.A. and Peterson, B.W. (1988b) Mechanisms of human head stabilization during random sinusoidal rotations. *Soc. Neurosci. Abstr.* **14**:1235.

Keshner, E.A., Woollacott, M.H. and Debu, B. (1988) Neck and trunk muscle responses during postural perturbations in humans. *Exp. Brain Res.* **71**: 455-466.

Keshner, E.A., Campbell, D., Katz, R., Peterson, B.W. (1989). Neck muscle activation patterns in humans during isometric head stabilization. *Exp Brain Res.* **75**: 335-344.

Keshner, E.A., Baker, J., Banovetz, J., Peterson, B.W., Wickland, C., Robinson, F.R., and Tomko, D.L. (1986) Neck muscles demonstrate preferential activation during voluntary and reflex head movements in the cat. *Soc. Neurosci. Abstr.* **12**: 684.

Kodde, L., Geursen, J.B., Venema, E.P., and Massen, C.H. (1979) A critique of stabilograms. *J. Biomed. Eng.* **1**: 123-124.

Nashner, L.M. (1976) Adapting reflexes controlling human posture. *Exp. Brain Res.* **26**: 59-72.

Nashner, L.M. and McCollum, G. (1985) The organization of human postural movements: A formal basis and experimental synthesis. *Brain Behav.* **8**: 135-172.

Nashner, L.M., Shupert, C.L., Horak, F.B., and Black, F.O. (1989) Organization of posture controls: An analysis of sensory and mechanical constraints. Afferent control of posture and locomotion. In *Progress in Brain Research.* (edited by J.H.J. Allum and M. Hulliger) Elsevier, Amsterdam. **80**: 411-418.

Outerbridge, J.S. and Melvill Jones, G. (1971) Reflex vestibular control of head movements in man. *Aerospace Med.* **42**: 935-940.

Pellionisz, A. and Peterson, B.W. (1988) A tensorial model of neck motor activation. In *Control of Head Movement* (edited by B.W. Peterson and F.J. Richmond), pp 178-186, Oxford, New York.

Peterson, B.W., Goldberg, J., Bilotto, G., and Fuller, J.H. (1985) Cervicocollic reflex: Its dynamic properties and interaction with vestibular reflexes. *J. Neurophysiol.* **54**: 90-109.

Peterson, B.W., Pellionisz, A.J., Baker, J.F., and Keshner, E.A. (1989) Functional morphology and neural control of neck muscles in mammals. *Amer. Zool.* **29**: 139-149.

Rothwell, J.C. (1987) Set-dependent modifiability of human long-latency stretch reflexes in the muscles of the arm. In *Higher Brain Functions* (edited by S.P. Wise), pp 113-132. Wiley, New York.

Sherk, H.H. and Parke, W.W. (1983) Normal adult anatomy. In *The Cervical Spine.* Cervical Spine Research Society, pp 8-22, Lippincott, New York.

Stockwell, C.W., Koozekanani, S.H., Barin, K. (1981) A physical model of human postural dynamics. *Ann. N.Y. Acad. Sci.* **374**:722-730.

Valk-Fai, T. (1974) Analysis of the dynamical behaviour of the body whilst "standing still." *J. Ky. Med. Assoc.* **72**: 21-25.

Vidal, P.P., Graf, W., Berthoz, A. (1986) The orientation of the cervical vertebral column in unrestrained awake animals. I. Resting position. *Exp. Brain Res.* **61**: 549-559.

Viviani, P. and Berthoz, A. (1975) Dynamics of the head-neck system in response to small perturbations: Analysis and modelling in the frequency domain. *Biol. Cybern.* **19**: 19-37.

CHAPTER 30

Segmental Movement as a Perturbation to Balance?
Facts and Concepts

Simon Bouisset and Maurice Zattara

30.1 Introduction

That postural reactions are associated with voluntary movement has probably been known in the medical field at least since Babinski's observations (1899). Since that time, many kinesiologic studies have been carried out with the aim of describing the muscular patterns which underlay maintenance of a given posture, slow changes in posture, and movements performed at a significant speed. They have included segmental movements performed on a stable underpropping area, as well as gross movement involving body progression, such as gait. It has been observed that postural muscular activities can precede, accompany, and also follow intentional movement.

As far as the postural activities which precede the onset of intentional movement, usually referred to as anticipatory postural adjustments (*APA*), are concerned, their approach in a neurophysiological context started with the pioneering work of Belenkii et al. (1967). They were based on the hypothesis proposed by Gelfand et al. (1966) in the line of Bernstein's general theories (1935), according to which programming of a motor act includes two components: one which is in relation to the intentional movement itself and the other which is related to the postural support. Since that time, diverse types of segmental movements have been considered, mainly involving the upper limb (Lee, 1980; Cordo and Nashner, 1982; Clement et al., 1984; Brown and Frank, 1987) and trunk (Crenna et al., 1987).

The aim of this chapter is to put forward some ideas about the postural roles played by muscles in relation to segmental movement. These ideas are based on the study of anticipatory postural activities.

More precisely, it has been assumed that anticipatory postural adjustments depend on two main factors: *1)* a factor which integrates the effect of movement to body structure, i.e., not only the parameters of the intentional movement, but also its location with respect to the body's axes of symmetry; *2)* the actual functional state of the sensori-motor and musculo-skeletal systems (impairment in disease or trauma, fatigue, conditioning, etc.). The originality of the present research may be situated within this context.

APA associated with intentional movement will be studied in both bilateral and unilateral movements, performed without and with an added inertia. It will be shown that *APA* are specific to the characteristics of the forthcoming intentional movement. They will be considered "preprogrammed". It will be assumed that postural adjustments constitute a part of the motor program that they tend to create inertial forces which, when the time comes, will counterbalance the disturbance of balance due to the forthcoming intentional movement. In this context, the intentional movement is considered as a perturbation to balance. The concept of *dynamic asymmetry is proposed to characterize this factor, which depends not only on the parameters of movement but also on its location with respect to the body's axes of symmetry*.

In Parkinsonian patients it will be shown that there was a lack of APA and that movement velocity decreased. In order to take into account the ability to react efficiently to the forthcoming perturbation, the concept of *posturo-kinetic capacity is proposed, which is supposed to depend on the actual state of the sensori-motor system*.

Multiple Muscle Systems: Biomechanics and Movement Organization
J.M. Winters and S.L-Y. Woo (eds.), © 1990 Springer-Verlag, New York

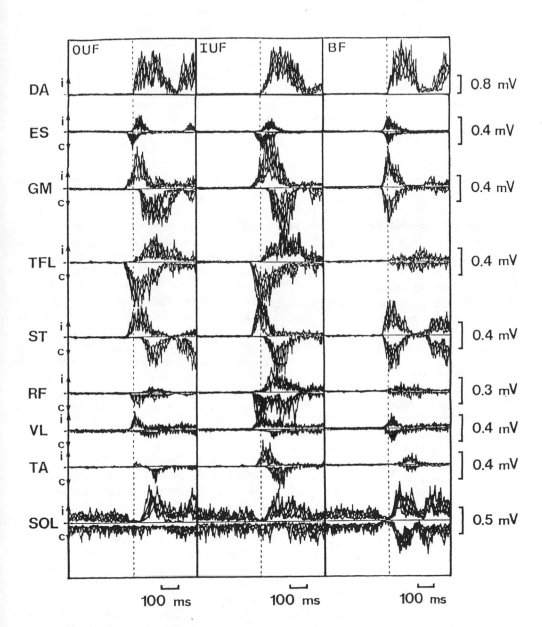

Figure 30.1: *EMG activities of lower limbs and pelvis recorded for the three experimental conditions. EMG* activities are reported from one subject performing the three types of movements. For each type of movement, rectified and smooth *EMG* activities of five trials were superimposed by synchronizing records on the onset of DAi activities (dotted line). *From left to right: OUF,* unilateral flexions with no additional inertia; *IUF,* unilateral flexions with an additional inertia; *BF,* bilateral flexions. *From top to bottom:* activities of anterior portion of Deltoideus (*DA*) Erectores Spinae (*ES*), Gluteus Maximus (*GM*), Tensor Fasciae Latae (*TFL*), Semi-tendinosus (*ST*), Rectus Femoris (*RF*), Vastus Lateralis (*VL*), Tibialis Anterior (*TA*), Soleus (*SOL*). Ipsilateral (*i*) and contralateral (*c*) activities with respect to the moving limb are displayed in opposite directions.

30.2 Methods

Subjects stood normally and had to execute, according to a simple reaction time paradigm, at a maximal speed, unilateral upper limb movements (*UF*), without (*OUF*) and with (*IUF*) an inertial load, and bilateral upper limb movements (*BF*). The inertial load was a lead bracelet (1 kg) fastened to the wrist.

This paradigm has several advantages and, in particular, it offers the possibility of dividing the body into two parts: the upper limb(s), which are voluntarily moved, and the rest of the body.

Local movements were studied by electromyography (*EMG*) and accelerometry (*ACG*). General body movements were studied by means of a force platform, which gave the resultant forces and the resultant torque about the vertical axis as well.

30.3 Results and Discussion

30.3.1 Study of Local Movements

This study includes a systematic description of APA, mostly from a temporal point of view (Bouisset and Zattara, 1981, 1983, 1987; Zattara, 1982; Zattara and Bouisset, 1986, 1988).

EMG anticipatory activities

The pattern of *EMG* postural activities was specific to the intentional movement (Figure 30.1). Synchrony of homonymous muscle activities was the rule for *BF*, contrarily to *UF* for which asynchrony increased when an inertial load was added to the upper limb.

As far as *APA* were concerned, there was a given sequence of disactivations (–) and activations (+): $SOLi(-)$, $TFLc(+)$ and $RFc(+)$, $STi(+)$ and $GMi(+)$, $ESc(+)$ for *UF* and both $SOL(-)$, $ST(+)/GM(+)$ and $ES(+)$ for *BF*. There were no consistent anticipatory *EMG* modifications either in foot or in neck muscles. The *APA* increased from *BF* to *OUF* and from *OUF* to *IUF*. Indeed, mean *APA* ranged from 25-30 ms for *BF*, 40-45 ms for *OUF* and 60-70 ms for *IUF*: they almost doubled from *BF* to *IUF*.

ACG anticipatory activities

The *ACG* activities concern all the body segments, and their pattern was also specific to the forthcoming movement (Figure 30.2). The existence of accelerations indicated that anticipatory activities are *movements*.

The *ACG* presented the same main features as the *EMG*: *a)* identity of APA at a given level was the rule for *BF*, contrarily to *UF* and, *b)* duration of *APA* increased from *BF* to *OUF* and from *OUF* to *IUF*; they almost doubled from *BF* to *IUF*, showing values similar to *EMG*s. There was a given order in anticipatory *ACG*: first contralateral lower limbs and hip, and then ipsilateral ones for *UF*, and both sides simultaneously for *BF*. Therefore, the *ACG* results were consistent with the *EMG* ones and the electrical muscular activities could be related to local accelerations: the local movements are not passive movements.

Moreover, it has been shown that the peak amplitude of anticipatory *ACG* increased in a quadratic manner with the peak velocity of the intentional movement. Besides, for a given velocity, the peak amplitude is higher for inertial loaded movements (Zattara and Bouisset, 1983).

From these *EMG* and *ACG* results, it must be stressed that *APA* are specific not only to parameters of the forthcoming intentional movement, such as its velocity, but also to its location with respect to the body's axes of symmetry. Their combination is called **dynamic asymmetry**. Therefore, as *APA* occur by definition before the onset of voluntary movement, it can be pointed out that dynamic asymmetry is "pre-programmed": it is not equivalent, in terms of posturo-kinetic programming, to raise both upper limbs simultaneously, one upper limb or the other.

30.3.2 Study of General Body Movements

The general shape of kinetic curves (*Aw*) was reproducible according to a given experimental condition. Upper limb movement was always diphasic: an acceleration was followed by a deceleration phase (Figure 30.3). As far as the resultant force or the resultant moment was concerned, the following was observed: during *BF*, ΔR_z and R_x were diphasic, while the variations of R_y and M_z were nil or negligible. On the contrary, for *UF*, resultant force and moment showed consistent variations for all the considered variables, but these were more complex than for *BF* (Bouisset and Zattara, 1987).

As can be seen in Figure 30.3, ΔR_z and R_x variations began before the onset of *Aw*; these accelerations were positive, i.e. corresponded to upward and forward accelerations of the body center of gravity. As far as they were concerned, R_y

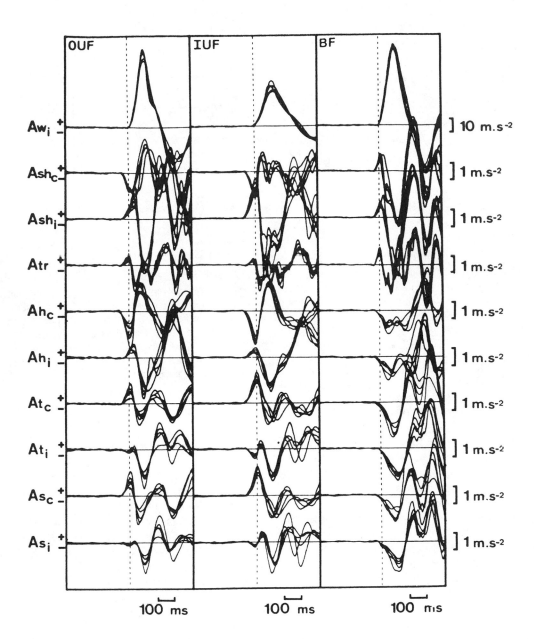

Figure 30.2: *Local accelerations recorded for the three experimental conditions.* Antero-posterior local accelerations are reported from one subject performing the three types of movement. For each type of movement accelerations of five trials were superimposed by synchronizing records on the onset of *Awi* (dotted line). *From left to right: OUF*, unilateral flexions with no additional inertia; *IUF*, unilateral flexions with an additional inertia; *BF*, bilateral flexions. *From top to bottom: Awi*, tangential acceleration of the upper limb measured at wrist level (positive sign corresponds to the acceleration phase of the movement); *Ash, Atr, Ah, At* and *As*, antero-posterior accelerations measured at level of shoulders, trunk, hips, thighs, and shanks (positive sign corresponds to forward accelerations); *i* and *c*, ipsilateral and contralateral accelerations with respect to the moving limb.

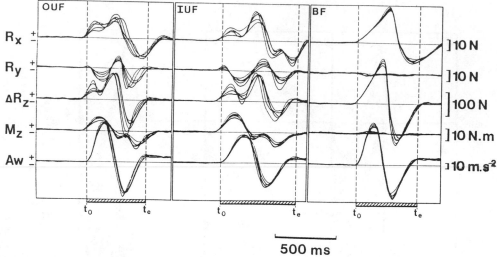

Figure 30.3: *Force platform data recorded for the three experimental conditions.* Force platform data recorded from one subject, performing upper limb(s) movements according to the three experimental conditions. For each type of movement, records of five trials were superimposed by synchronizing them on the onset of Aw_i. *From left to right: OUF,* unilateral flexions with no additional inertia; *IUF,* unilateral flexions with an additional inertia; *BF,* bilateral flexion. *From top to bottom: R_x, R_y* and *ΔR_z,* antero-posterior, lateral and vertical components of the resultant force ($\Delta R_z = R_z - P$, with P being the weight of the subject); positive sign corresponds respectively to forward, right-to-left and upward forces); M_z, resultant moment about the vertical axis (positive sign corresponds to a moment which tends to rotate the body from the right side to the left one); Aw_i, tangential acceleration of the upper limb recorded at wrist level; t_o and t_e correspond to the onset and the end of the upper limb movement.

and M_z were positive for *UF*: these accelerations corresponded to a lateral acceleration of the body center of gravity directed toward the contralateral side and induced a moment directed also toward the contralateral side.

Duration of *APA* were longer for ΔR_z than for the other variables. Moreover, they increased from *BF* (29 ms) to *OUF* (50 ms) and from *OUF* to *IUF* (56 ms): they doubled in *IUF* as compared to *BF*, which was consistent with the data reported above. Concurrently the peak velocity of intentional movement decreased from 7.9 rad/sec for *BF* (at right wrist level, classically less at the left one, summing up at about 15.5 rad/sec) to a slightly lower value (7.5 rad/sec) for *OUF* and to 5.8 rad/sec for *IUF*.

The finality of *APA* may be proposed on the basis of an analysis of forces acting at shoulder level and at the body center of gravity (Figure 30.4), whose direction was opposite. From this analysis, it can be assumed that *APA* tend to create inertial forces which, when the time comes, will counterbalance the disturbance to postural equilibrium due to the intentional forthcoming movement (Bouisset and Zattara, 1984).

According to this hypothesis, the intentional movement constitutes a perturbation which depends not only on the parameters of the movement, but also on its location with respect to the body's axes of symmetry, i.e. to its dynamic asymmetry. Hence, it is tempting to consider that the postural perturbation associated with voluntary movement might result form two factors, which correspond to the two vectors to which the perturbation force system may be reduced, i.e., its resultant and its moment (in reference to the body center of gravity, for example). One is linear, the other is rotational. Both may include three orthogonal components.

BF   UF

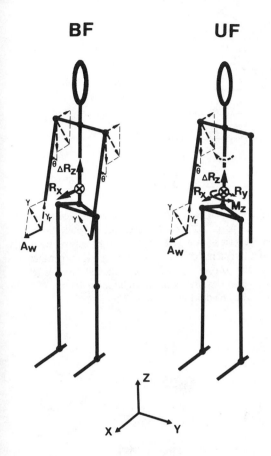

Figure 30.4: *Interpretation of the finality of anticipatory postural adjustments.* The filled arrows correspond to the actual recorded biomechanical data, the interrupted arrows correspond to theoretical parameters. θ, angular displacement of the upper limb(s). Aw, γ_r and γ, tangential, radial and total upper limb acceleration. R_x and ΔR_z, antero-posterior and vertical acceleration of the body center of gravity, G. M_z, resultant moment about the vertical axis crossing G.

30.3.3. Study of Reaction Time

The previous results tended to prove that dynamic asymmetry was "pre-programmed". A direct argument may be brought by a chronometric study (Zattara and Bouisset, 1986).

The simple reaction time was studied in the same three conditions. The results showed that *RT* varied from one condition to an other (Figure 30.5). But if *RT* was divided into two parts, motor latency (*ML*) and postural anticipation (*PA*), it could be seen that *ML* was constant whereas *PA*

varied. More precisely, *PA* increased with the dynamic asymmetry, which was, in fact, not surprising because it corresponded to the *APA*.

By definition, as is well known, simple *RT* cannot vary according to the parameters of a voluntary movement. Therefore, *ML* was equivalent to the "true *RT*" in that type of motor activity, and it can be assumed that the motor activity started with *APA*. Furthermore, as *APA* are triggered by a feed-forward command and are specific to the characteristics of the forthcoming movement, they have to be determined from previous knowledge of the disturbing effect of forthcoming voluntary movement. Finally, a question can be raised: could intentional movement be initiated only when convenient postural dynamics have been developed? In other words, without suitable *APA*, does intentional movement have to be slower? This problem has been studied in Parkinsonians (Bazalgette et al., 1986) and in paraplegics (Do et al., 1985).

30.3.4 Anticipatory Postural Adjustments and Parkinson's Disease

Data were collected from 18 Parkinsonian patients (patients with severe tremor or marked L-Dopa–induced dyskinesia were excluded), who ranged from 39 to 73 years of age (mean: 59.6 years) and who were hospitalized for treatment. Data were also collected from five volunteer subjects in the same age range, who were free from neurological or other severe diseases. With the aim of comparing healthy subjects and patients, electromyographic and accelerometric data were recorded in the same way as in the studies reported above.

The main results concerned the movement duration and the *APA*. In accordance with the classical data, the movement time increased in a highly significant way ($p < 0.01$) in Parkinsonian patients, ($m = 752 \pm 42$ ms, as compared to $m = 457 \pm 23$ ms) i.e. the velocity decreased (from about 8 rad/sec to 4.5 rad/sec). Furthermore, the *APA* were usually absent and the postural adjustments which occured during the movement course were not specific to the voluntary movement. Indeed, contrarily to healthy subjects, in whom early postural adjustments associated with unilateral movements have been reported as asymmetric (*Asi* directed backward, *Asc* forward), Parkinsonian patients displayed symmetrical pos-

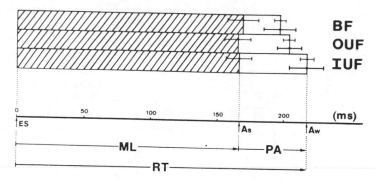

Figure 30.5: *Chronometric indexes for the three experimental conditions.* On the lower part of the figure, meanings of the chronometric indexes are illustrated (*RT*, reaction time; *ML* motor latency; *PA*, postural anticipation) (*RS* being the response signal, *As* and *Aw* indicating respectively the onset of the earlier shank acceleration and the onset of the wrist acceleration). On the upper part of the figure, mean values and standard deviations of *RT*, *ML* (hatched areas), and *PA* (non-hatched areas) are given for five subjects performing ten movements of each type during two experimental sessions. *From top to bottom:* bilateral flexions (*BF*), unilateral flexions without additionnal inertia (*OUF*) and with an additional inertia (*IUF*).

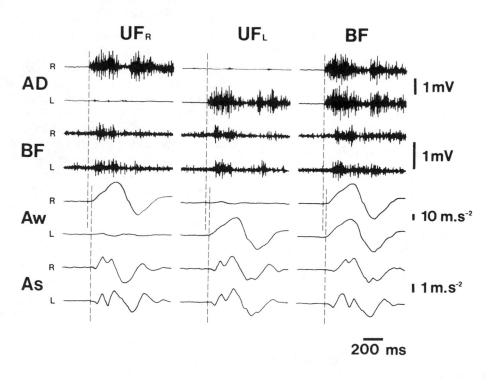

Figure 30.6: *Electromyographical and accelerographical recordings in a Parkinsonian patient. From left to right: UFR,* right unilateral shoulder flexion; *UFL,* left unilateral shoulder flexion; *BF,* bilateral shoulder flexion. From top to bottom: *AD,* EMG of anterior portion of right (*R*) and left (*L*) deltoideus; *BF,* EMG of right (*R*) and left (*L*) biceps femoris; *Aw,* tangential acceleration of the right (*R*) and left (*L*) upper limbs, measured at wrist level; *As,* antero-posterior acceleration of the right (*R*) and left (*L*) shanks. The dotted vertical lines indicated the onset of prime mover activity and the onset of the upper limb acceleration.

tural movements of both legs (always directed forward) in unilateral as well as in bilateral movements (Figure 30.6.)

These results raised up a fundamental question: was bradykinesia in Parkinson's disease due to pathological postural adjustments or postural adjustments reduced in relation to bradykinesia? Some of the present data are in agreement with the first hypothesis. Indeed, *APA* have not been observed in healthy subjects, for submaximal movements, above 4 rad/sec (Zattara and Bouisset, 1983), which could correspond to a threshold below which the inertial forces due to the intentional movement became negligible. The upper limit for the movement velocity in Parkinsonian patients could be explained by their inability to counterbalance these perturbing forces. The results, obtained in paraplegics, could also be interpreted in favor of the same hypothesis.

In order to take into account the ability to react to the forthcoming perturbation, the concept of posturo-kinetic capacity has been proposed. This is supposed to depend on the actual state of the sensori-motor and musculo-skeletal systems (impairment in trauma or disease, conditioning, fatigue, etc.).

30.4 Conclusions

Dynamic asymmetry has been shown to be a discriminative experimental factor for the study of *APA*. Inasmuch as it integrates the effect of movement parameters to body's structure, it allows a more comprehensive approach to *APA*. It might be interesting to test the hypothesis according to which *APA* result from the arrangement of two biomechanical factors. These factors correspond to two vectors to which the perturbing force system may be reduced, i.e., its resultant and its moment (in reference to the body center of gravity, for example).

The concept of posturo-kinetic capacity has been proposed to characterize the ability to react to the perturbation due to intentional movement. The interest of this concept has been proved on disabled subjects. For healthy subjects, on the other hand, this posturo-kinetic capacity could also depend on the configuration of postural basis. Preliminary results (Zattara and Bouisset, 1987) seem to support this assumption.

References

Babinski, J. (1899) De l'asynergie cerebelleuse. *Rev. Neurol.* **7**:806-816.

Bazalgette, D., Zattara, M., Bathien, N., Bouisset, S. and Rondot, P. (1986) Postural adjustments associated with rapid voluntary arm movements in patients with Parkinson's disease. In Yarr, P.D. and Bergmann, K.J.: *Adv. Neurol.* **45**:371-374, Raven Press, New York.

Belenkii, Y.Y., Gurfinkel, V. S. and Paltsev, Y.I. (1967) Element of control of voluntary movements. *Biofizika* **12**:135-141.

Bernstein, N. (1935) *Coordination and regulation of movements*. New-York:Pergamon Press (Amer. translation, 1967).

Bouisset, S. and Zattara, M. (1981) A sequence of postural movements precedes voluntary movement. *Neurosci. Lett.* **22**:263-270.

Bouisset, S. and Zattara, M. (1983) Anticipatory postural movements related to a voluntary movement. In *Space Physiol.*, Toulouse, Cepadues Pubs., pp. 137-141.

Bouisset, S. and Zattara, M. (1987) Biomechanical study of the programming of anticipatory postural adjustments associated with voluntary movement. *J. Biomech.* **20**:735-742.

Brown, J.E. and Franck, F.S. (1987) Influence of event anticipation on postural actions accompaying voluntary movement. *Exp. Brain Res.* **67**:645-650.

Clement, G., Gurkinkel, V.S., Lestienne, F., Lipshits, M.I. and Popov, K.E. (1984) Adaptation of postural control to weightlessness. *Exp. Brain Res.* **57**:61-72.

Cordo, P.J. and Nashner, L.M. (1982) Properties of postural movements related to a voluntary movement. *J. Neurophysiol.* **47**:287-303.

Crenna, P., Frigo, C., Massion, J. and Pedotti, A. (1987) Forward and backward axial synergies in man. *Exp. Brain Res.* **65**:538-548.

Do, M.C., Bouisset, S. and Moynot, C. (1985) Are paraplegics handicapped in the execution of a manual task? *Ergonomics* **28**:1363-1375.

Gelfand, I.M., Gurfinkel, V.S., Tsetlin, M.L. and Shik, M.L. (1966) Problems in analysis of movements. In:Gelfand, I.M., Gurfinkel, V.S., Fomin, S.V. and Tsetlin, M.L. (Eds); *Models of the structural functional organization of certain biological systems* (Amer. translation, 1971), M.I.T. Press, Cambridge, MA, pp. 330-345.

Lee, W.A. (1980) Anticipatory control of posture and task muscles during rapid arm flexion. *J. Mot. Behav.* **12**:185-196.

Zattara, M. (1982) Ajustements posturaux anticipateurs associes au mouvement du membre superieur. These de 3e cycle, Orsay, France.

Zattara, M. and Bouisset, S. (1983) Influence de la vitesse d'execution du mouvement volontaire sur les accelerations locales anticipatrices.   Huitieme Congres de la Societe de Biomecanique, Lyon, in resume des communications, pp. 113-114.

Zattara, M. and Bouisset, S. (1986) Chronometric analysis of the posturo-kinetic programming of the voluntary movement. *J. of Motor Behavior*, **18**: 215-223.

Zattara, M. and Bouisset, S. (1987) Posturo-kinetic organization during upper limb movement performed from varied foot positions. *Neuroscience Letters*, suppl. 22, 1969.

Zattara, M. and Bouisset, S. (1988) Posturo-kinetic organization during the early phase of voluntary upper limb movement. 1. Normal subjects. *J. of Neurology, Neurosurgery and Psychiatry*, **51**: 956-965.

CHAPTER 31

Simulation Experiments can Shed Light on the Functional Aspects of Postural Adjustments Related to Voluntary Movements

Constance F. Ramos and Lawrence W. Stark

31.1 Introduction

Voluntary movements of the upper body are accompanied by anticipatory postural adjustments in the lower body in a standing subject. The experimental studies described in Chapter 30 (Bouisset and Zattara) support the hypothesis that these anticipatory activities initiate an opposing acceleration of the body which counteracts the oncoming perturbation to the center of gravity caused by the voluntary movement. This idea is centered around the physical principle of "Action = Reaction" and is a compelling biomechanical explanation. However, it may also be that the anticipatory activity serves a *neurophysiological* purpose, for example, related to setting proprioceptive feedback gains in the postural muscles.

Computer simulation experiments can shed light on the functional aspects of the postural adjustments that are related to voluntary movements. This chapter proposes a generic modeling scheme for studying the neurological control of multi-joint movements, and presents simulation examples of how the model can be used to elucidate different aspects of the postural control processes during movement.

As described below, a stretch reflex model that has been used previously to investigate aspects of single-joint movements (Ramos and Stark, 1987; Ramos et al., 1990) is expanded into a double-joint system. Each joint is controlled by its own agonist-antagonist excitation inputs and proprioceptive feedback. With this model, computer simulations can fully explore the control

aspects of the postural activities associated with voluntary movements, those which anticipate the voluntary muscle activity as well as those which follow.

31.2 Empirical Evidence

Studies on anticipatory postural adjustments have considered a number of rather complex movement paradigms, including push-pull experiments (Cordo and Nashner, 1982) and head movements in quadrupeds (Debu and Gahery, 1988). In this chapter, the simulation studies focus on two movements for which a standing subject can be regarded as a simple two-link inverted pendulum; these are rapid upward arm raising and fast forward bending.

31.2.1 Rapid Upward Arm Swings

Several different groups have chosen to look at the postural activities that accompany a rapid upward arm swing, including experiments conducted under the zero-gravity conditions of orbital space flight (Clement et al., 1984); this particular paradigm lends itself well to the simulation methodology.

Three important experimental series are exemplary. The first group of experiments has been conducted by Friedli and colleagues (1984, 1988). Their experiments revealed the clear distal-to-proximal order of the postural muscle activations in the gastrocnemius, biceps femoris, and erector spinae muscles, as well as reciprocal activation between agonist/antagonist muscle pairs. Overall, their studies suggested that whereas these anticipatory activities may be preprogrammed, the

Multiple Muscle Systems: Biomechanics and Movement Organization
J.M. Winters and S. L-Y. Woo (eds.), © 1990 Springer-Verlag, New York

postural adjustments stabilizing the body during the voluntary movement and after it has ended are probably resulting from reflex control.

The second group of experiments, conducted by Lee, Buchanan, and Rogers (1987), have considered the effects of different arm accelerations on the postural perturbation. They attempted to find a correlation between the anticipatory postural muscle activity (in this case, the ratio of mean rectified *EMG* amplitudes) and the acceleration of the voluntary movement. Their experiments showed that although such a correlation might be made for each individual subject, no general rule could be made for data across several subjects. Furthermore, they found no evidence that the anticipatory period (anywhere from 7 ms to 86 ms) depended upon the arm acceleration. From their experiments, it might be speculated that anticipatory postural activities associated with voluntary movements may not serve a purely biomechanical purpose.

The third group of experiments, conducted by Bouisset and Zattara (1981, 1987; Zattara and Bouisset, 1988), and described in Chapter 30, primarily focussed on the anticipatory period in the postural activity. Their data included electromyographic recordings from shank, thigh, hip, and trunk muscles (flexors and extensors) on both right and left sides of the body; locally recorded accelerations for these body segments; ground reaction forces; the anterior deltoid *EMG*; and wrist acceleration. The experimental conditions in their studies included both unilateral and bilateral flexions, with and without an added 1 kg mass to the wrist. The model simulations of the present study are qualitatively compared to their empirical observations.

Important aspects of their electromyographic data include the following: *1)* both the semitendinosus and soleus muscles exhibit anticipatory activities; the semitendinosus becomes active approximately 30 ms prior to the onset of the voluntary activity in the anterior deltoid, whereas the soleus becomes silent about 60 ms prior to the voluntary activity; *2)* the tensor faciae latae exhibits no anticipation, but seems to increase its activity during periods of quiescence in the semitendinosus (a reciprocal activation pattern).

These initial, anticipatory postural activities (but not the soleus silencing) would seem to correspond to an initial backward motion of the lower body and, indeed, as recorded by accelerometers, the postural sway preceding the voluntary movement is clearly in the backwards direction for the hips and lower limbs.

31.2.2 Fast Forward Bending

Postural activities associated with fast forward bending show similarites to those associated with rapid arm movements. Experiments conducted by Crenna et al. (1987, 1988) have shown that fast forward bending movements are accompanied by a backwards motion of the hips and lower limbs. Motion analysis of the full body kinematics revealed that the voluntary movement is generally preceded by activity at the level of the ankle (and also manifested in a slight but functionally insignificant flexion at the knee). Electromyographic evidence was reported to show that anticipatory postural muscle activity of the lower limbs occurs in the soleus (early silencing) and tibialis anterior (early activation). *EMG* activity in the vastus medialis was reported to appear approximately simultaneously with the rectus abdominis (voluntary) activity. Their complex interpretation of these experimental results was that postural activity produced a necessary backwards motion in a feed forward manner, which served to minimize a destabilizing forward acceleration of the center of gravity caused by the voluntary movement.

31.3 The Modeling Scheme

31.3.1 The Biomechanical System

The mechanical system used in mathematically describing both the upward arm swing and the forward bending movements is the simple, two–link inverted pendulum (Figure 31.1). The equations of motion for the inertial torques on this system can be easily derived using Lagrangian methods. In the example of the arm swing paradigm, the biomechanical coupling between the two links is described by:

$$M_{body} = -(J_2 + m_2 D_2{}^2)\ddot{\theta}_2 - m_2 g(D_2 \sin(\theta_2) - L\sin(\theta_1))$$

$$+ m_2 D_2 L (\ddot{\theta}_1 + \ddot{\theta}_2)\cos(\theta_2 + \theta_1) \qquad (31.1)$$

$$- m_2 D_2 L (\dot{\theta}_2 - \dot{\theta}_1)^2 \sin(\theta_2 + \theta_1)$$

$$M_{arm} \simeq 0.0$$

(since the motion of the arm is overwhelmingly dominated by the muscle torques)

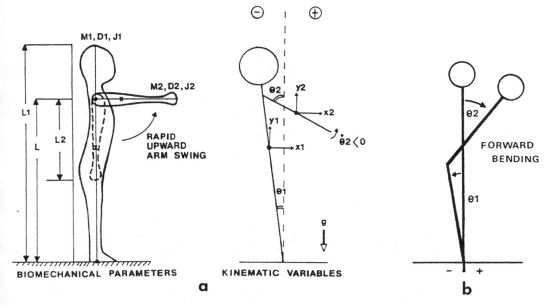

Figure 31.1: *a)* A two link simplification of a standing subject for modeling postural maintenance during a rapid upward arm swing. *Left:* The body is modeled as one rigid link and the arms (together) as the other link. The values of the biomechanical parameters are listed in Table 31.1; *Right:* Definition of the kinematic variables with which the biomechanical coupling equations are derived. *b)* Equivalent inverted pendulum model for the forward bending paradigm.

where M are moments, J is the inertia, m is mass, and the distances are defined in Figure 31.1. Notice that the total torque at each joint is a combination of this inertial component, the gravitational torque and the muscle activity, and summarized by the second order equation:

$$J_p \alpha = M_{link} - M_{ext} + M_{flex} + M_{grav\text{-}link} - B_p \omega - K_p \theta$$

$$(31.2)$$

where

$$J_{p\text{-}body} = J_1 + m_2 L^2 + m_1 D_1^2$$

and

$$J_{p\text{-}arm} = J_2 + m_2 D_2^2$$

The motion of the center of gravity of the *system*, that is, body plus arm links, is also computed for each simulation. Center of gravity accelerations may be compared to experimentally recorded ground reaction forces (biomechanical parameters for the model are summarized in Table 31.1). A similar set of equations describes the forward bending paradigm.

Table 31.1: Biomechanical parameter values as estimated from the scheme proposed by Zatsiorsky and Selayanov (1983) (subject's mass = 75 kg; subject's height = 175 cm)

		Arm Swing	Bending
Lower Link Mass	m_1	65.55 kg	27.92 kg
Upper Link Mass	m_2	7.40 kg	45.03 kg
Lower Link Length	L_1	1.75 m	0.92 m
Upper Link Length	L_2	0.68 m	0.82 m
Lower Link C.O.G.	D_1	0.96 m	0.58 m
Upper Link C.O.G.	D_2	0.24 m	0.36 m
Lower Prin. Inertia	J_1	0.71 kg-m²	3.93 kg-m²
Upper Prin. Inertia	J_2	0.23 kg-m²	10.62 kg-m²

for the upward arm swing paradigm:

Arm(s) Inser. Pt	L	1.50 m

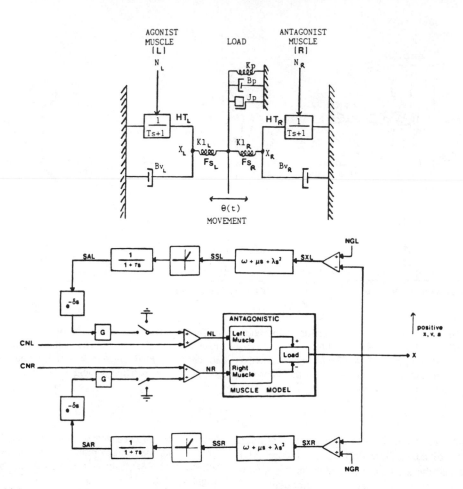

Figure 31.2: *a)* Block diagram of the single-joint, neurological control model that is used to independently control the movement of each joint ("shoulder" or "hip" and "ankle"). Two muscles work as an antagonistic pair in moving their common load. Feedback control from the proprioceptive system is not depicted (see Figure 31.2b). The two model inputs, *NL* and *NR*, are comparable to experimental recordings of electromyographic activities, whereas the model output is directly comparable to the recorded kinematics. *b)* The complete stretch reflex model. The single-joint, neurological control system can now function in either open-loop (neurologically "preprogrammed") or closed-loop (reflexive) modes.

31.3.2 The Neurological Control System

In addition to the inertial coupling, the motion of each link in this biomechanical system is independently controlled by an eighth-order, nonlinear model for neuromuscular control (Hannaford and Stark, 1985, Ramos and Stark, 1987; the system state equations can be found in Ramos et al., 1990). This model consists of an antagonist pair of muscles and their common load (Figure 31.2a). Each muscle is constructed from a force generation element, a parallel viscosity that incorporates the well-known, nonlinear force-velocity relationship ($(F+a)(v+b) = constant$), and a series elasticity, which is also nonlinear and implemented as a

"hard" spring [see Chapter 5 (Winters) for further discussion of the Hill-based muscle model]. The muscles are identical in their mechanical properties and act symmetrically on their common load. The load itself is a second-order plant described by an inertia (J_p), viscosity (B_p), and elasticity (K_p). The model also incorporates proprioceptive feedback from the muscles spindles (Figure 31.2b). Earlier simulations (Ramos, 1989) have determined a set optimal parameter values for both the feed-forward and feedback components of the model, which are also adopted for the present simulation study (Table 31.2).

Table 31.2: Muscle, load and feedback parameters for the neurological control model.

Muscle Parameters		
electromechanicl delay		25 ms
activation time constant	T	50 ms
apparent viscosity factor	Bh	350 deg/s
series elasticity (linear)	k_1	0.006 Nm
series elasticity (exp)	k_2	0.010 $(deg)^{-1}$
Load Parameters		
parallel elasticity	k_p	1.0 Nm/deg
parallel viscosity	B_p	1.0 Nms/deg
Feedback Parameters		
position feedback		1
velocity feedback		1.94
acceleration feedback		1.96
lag time constant		100 ms
transmission delay		8 ms
spinal gain		G (variable)

Both the joints in the biomechanical system (shoulder and ankle in the case of the arm swing experiments, hip and ankle in the forward bending experiments) are independently controlled by such a neurological control system. Input signals for each joint represent the degree of neuromuscular excitation for the agonist and antagonist muscles and are expressed in terms of muscle tension level. The muscle–load model therefore has two inputs that can be compared to experimentally recorded electromyograms in their impulse patterns and overall time course. The model output is the joint angle and derivatives, which can also be compared to the experimental kinematics.

31.3.3 The Simulation Scheme

The simulation method requires four neurological inputs: two input signals specifying the muscle activation levels for the voluntary (arm) movement, and two input signals specifying the postural activities. Bursts in the neurological control signal are modeled by Gaussian functions (Ramos and Stark, 1987). The model outputs are the joint angles, velocities and accelerations for the joint involved in the voluntary movement (shoulder or hip), and the ankle joint, which shows the postural sway.

In the first simulated paradigm, the upward arm swing is simulated for a set of control signals that are determined by matching the kinematic output to the quantitative data taken from Zattara and Bouisset (1988). In the second paradigm, the fast forward bending movement of the "trunk" link is simulated to a final, forward position of 40°. The movement time and peak velocity approximately correspond to what has been observed in human experiments (Ramos, Frigo, and Pedotti, unpublished data). Thus, by specifying the kinematics for each of the voluntary movements, the input signals for the upper joint are found by an inverse modeling technique.

The postural adjustments associated with these voluntary movements are found by taking a set of input signals for the descending neuromuscular control at the ankle joint (for example, a constant cocontraction) *and* the angular kinematics of the voluntary movement, which together comprise the simulation input for the computer model.

31.4 Simulation Experiments

The following section describes the results of simulation experiments for the two movement paradigms, a rapid upward arm swing and a fast forward bend, both in the absence and then in the presence of anticipatory postural activities.

31.4.1 The Upward Arm Swing

For the first paradigm, the arm movement simulations determine that the upward arm swing is controlled by two independent, successive bursts (separated by a brief silent period) in the anterior deltoid muscle; next follows a steady level contraction, which maintains the arm horizontally. It is interesting to note that this double burst pattern of the simulated input control signal fits into the triphasic *EMG* pattern for fast movements (cf. Wadman et al., 1979; Hannaford and Stark, 1985). The absence of *pB* (the antagonist braking pulse, second in the triphasic signal) is due, in the model simulation, to the adequacy of gravity in carrying out this function. In zero-gravity conditions (see Clement et al., 1984), the model thus predicts the appearance of *pB*. Experimental evidence confirming this prediction should be forthcoming.

Kinematically, the resulting simulated movement shows a slight overshoot of about 3° before coming to rest at 89° to 92°. This characteristic is also seen in human subject experiments (personal

BODY   KINEMATICS

(A)

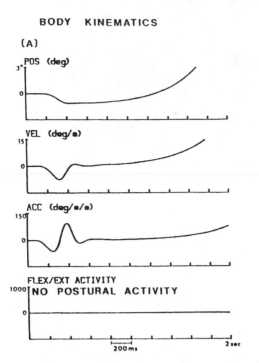

FLEX/EXT ACTIVITY
NO POSTURAL ACTIVITY

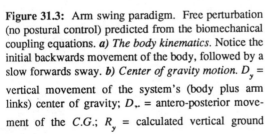

CENTER OF GRAVITY MOTION

(B)

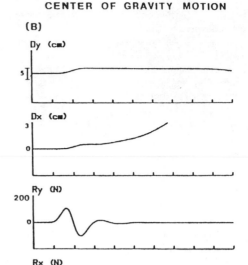

Figure 31.3: Arm swing paradigm. Free perturbation (no postural control) predicted from the biomechanical coupling equations. *a) The body kinematics.* Notice the initial backwards movement of the body, followed by a slow forwards sway. *b) Center of gravity motion.* D_y = vertical movement of the system's (body plus arm links) center of gravity; D_{x} = antero-posterior movement of the *C.G.*; R_y = calculated vertical ground reaction force (equal to the acceleration of the center of gravity, multiplied by the total body mass); R_x = calculated antero-posterior ground reaction force. Notice that although the body link is initially thrust backwards, the motion of the system's center of gravity is always forwards. Thus, the upwards arm swing clearly dominates the center of gravity motion in the absence of any postural control activities.

observaton). The total movement time is 527 ms. The peak acceleration is 3100°/s^2 and the peak velocity is 443°/s.

In the absence of any postural activities in the lower limbs and trunk (other than a minimal cocontraction at the ankle of 80 N (about 1% of maximally possible contraction, necessary for state equation stability)), the predicted postural sway is initially about 1° *backward* (Figure 31.3). This corresponds to approximately 16 mm of body link's center of mass, and is due to the backwards reaction of the body to the forward muscular thrust on the arms. The peak backwards acceleration of the body is -65.7 deg/s^2, and the peak backwards velocity is -7.5 deg/s. The motion then slowly changes to the *forward* direction, caused by the large, upward angular momentum of the arms; the body continues to move forward at a speed of about 6.5 deg/s (with a peak forward acceleration

of 95.6 deg/s^2). Since there are no postural activities, the body continues falling forward until in the limit it would fall flat on the floor.

This two-phased postural sway (first backwards, then forwards) results solely from the biomechanical coupling between the two segments (arm and body) in this idealized system. Since normal human subjects cannot suppress their postural activities, the model simulation is invaluable in clarifying this "reduced" trajectory. Evidently, the backwards perturbation to the body resulting from the muscular thrust at the shoulder is quite small.

With regard to the movement of the center of mass for the complete, two- link system, the motion is always upwards (2 cm) and forwards (0.5 cm, initially, then increasing with the continuing forward sway).

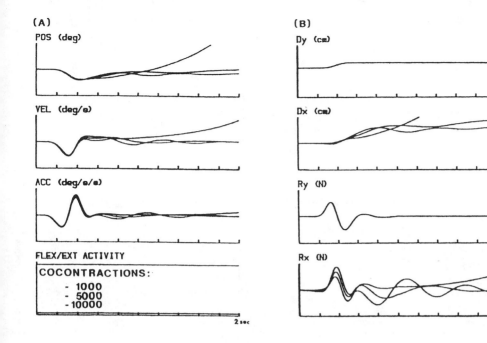

(A)

POS (deg)

VEL (deg/s)

ACC (deg/s/s)

FLEX/EXT ACTIVITY

COCONTRACTIONS:
- 1000
- 5000
- 10000

2 sec

(B)

Dy (cm)

Dx (cm)

Ry (N)

Rx (N)

2 sec

Figure 31.4: Arm swing paradigm. Effects of anticipatory cocontractions on the postural sway. Cocontraction levels in the flexor/extensor postural muscle pair are considered at 1000N (~20% maximal contraction), 5000 N (~100%), and 10,000 N (~200%). Although cocontraction does tend to suppress the forward component of the postural sway, it also introduces an oscillation that increases in frequency with increasing cocontraction level. *a)* Body kinematics. *b)* Center of gravity motion. (Scales are as in Figure 31, pos = + 3 ; vel = + 15°/sec; acc = + 150°/sec^2; flex/ext = 1000 N; $x = + 50$ N; $R_y = + 200$ N; $D_x = + 3$ cm; $D_y = 85$ cm - 100 cm).

The effects of increased levels of cocontraction at the ankle have also been considered (Figure 31.4). Simulations have tested a preset level of cocontraction at the ankle joint for 1000 N, 5000 N, and 10,000 N. These values would reasonably correspond to about 20%, 100%, and 200% of a maximal cocontraction level for the ankle flexion/extension system. No stretch reflex activity is present. The results show that increasing the cocontraction corrects the postural perturbation by returning the body to 0°, or straight up (note that 1000 N is too low to achieve a final upright posture, even though it is about 20% of maximal force). However, cocontraction also has the effect of introducing a marked oscillation, the frequency of which augments from 1 Hz at a cocontraction level of 5000 N, to 3 Hz at 10,000 N.

31.4.2 Fast Forward Bending

For the second movement paradigm, simulating the fast forward bending of the trunk link determines a voluntary movement of the upper body that is controlled by two symmetric, reciprocally active bursts, first in the rectus abdominis to initiate the movement, then in the erector spinae to brake the movement. Next follows a steady level contraction in the erector spinae, which maintains the trunk at 40° from the vertical (Figure 31.3 Left, lower panel; dashed trace). The resulting simulated movement shows a smooth rise and hold from zero to 40°, with a total movement time (as measured from the acceleration trace) of about 580 ms. The peak acceleration is 400 deg/s^2 and the peak velocity is 87 deg/s.

BODY KINEMATICS

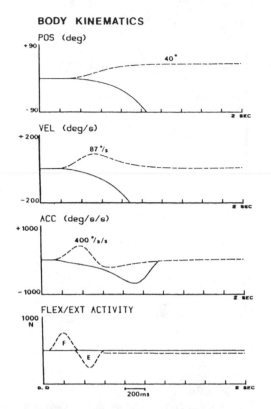

CENTER OF GRAVITY MOTION

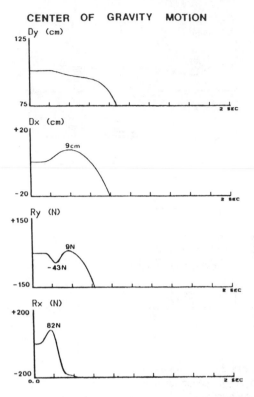

Figure 31.5: Forward bending paradigm. Postural perturbation in the absence of postural activity. *Left:* The dashed trace is the voluntary, fast forward bending movement, controlled by a single burst (*F*) in the flexor (rectus abdominis: amplitude = 600 N; width = 200 ms; peak 200 ms), followed by a burst (*E*) in the extensor (erector spinae; amplitude 600 N; width 200 ms; peak 450 ms) and a steady level contraction (170

N) to maintain the final position at 40°. The solid line displays the trajectory of the lower body, which completely falls over backwards within 200 ms after the voluntary movement has stopped. *Right:* the motion of the center of gravity is first forwards, then backwards, in the absence of any postural maintenance activity in the hips and lower limbs.

Preliminary simulations have demonstrated that for purely reflexive postural activity, increasing levels of reflex gains progressively diminish the perturbation to the lower body link but have no perceptible effect on the large initially forward excursion of the center of gravity of the system (Ramos and Stark, submitted). Since the experimental evidence shows that the center of gravity moves only about 1 cm during fast forward bending, this result implies that feedback alone is not sufficient for postural maintenance during the movement. A simulation experiment using an initial burst of descending activity in the anterior muscles of the lower body, followed by stretch reflex activity in these muscles, gives a rough idea of how feedback control coupled with an initial descending signal significantly improves the overall stability of the full body movement (Figure 31.6).

In the absence of any neuromuscular activity in the hips and lower limbs, the simulation experiment clearly shows a perturbation to the lower body link which is **backward**, but slightly delayed with respect to the voluntary movement and also slower (Figure 31.5, left; solid trace). The motion of the system's center of gravity is biphasic (Figure 31.5): first about 9 cm *forward*, then reversing direction until the body falls completely over on its back about 800 ms after the onset of the fast forward bend (i.e. about 200 ms after the voluntary movement has stopped).

This motion of the center of gravity (first forwards, then backwards) results solely from the activity at the "hip" joint and from biomechanical coupling between the two segments (upper and lower body links) in this idealized system. The backwards perturbation to the body results from a significantly large momentum transfer from the voluntary movement.

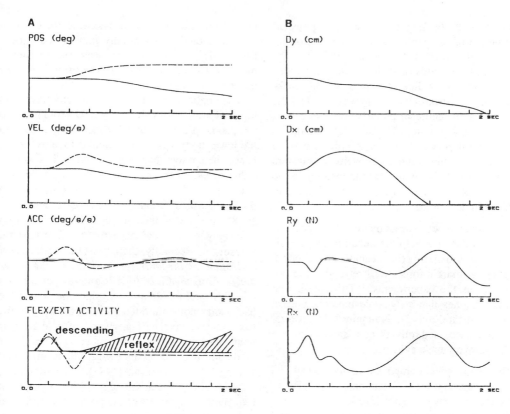

Figure 31.6: Forward bending paradigm. Effects of stretch reflex feedback on the postural perturbation. *Left:* A descending burst of activity in the anterior muscles of the lower body is followed by a proprioctive feedback signal (G = 3) which serves to stabilize the total body movement. Without this initial descending burst, the effect of feedback alone (not shown) can diminish the backwards motion of the hips and lower limbs but not the center of gravity (dashed line = voluntary forward bending; solid line = lower body response). *Right:* Center of gravity motion.

31.5 Summary and Conclusions

Clearly, from a biomechanical point of view, certain postural commands must exist simultaneously with voluntary motor commands in order to counterbalance the reaction forces accompanying the primary movement. What is not entirely clear is the reason for the *anticipatory* activities in the postural system. Does the anticipation serve a biomechanical purpose in producing a postural sway before the onset of the voluntary movement, or is there a neurophysiological reason for activating the postural muscles before the "prime mover"? The simulation studies presented in this chapter have considered the effects of removing postural activities that normally accompany voluntary movements. The perturbations to balance that result from an upward arm swing and a fast forward bend in the absence of any postural muscle activities can be regarded as the "reduced trajectory" of the body posture response. The simulations have also taken a preliminary look at possible control strategies for anticipatory and synchronous postural activities; they include cocontractions at the ankle for the arm swing paradigm, and a mixture of descending and reflex control for the forward bending paradigm. Important observations from these simulation experiments are described below.

31.5.1 On the Upward Arm Swing Simulations

For the upward arm swing paradigm, the control aspects investigated by the simulation study can be summarized as follows:

1) The biomechanical component of the model predicts the physical propogation of the postural perturbation that is a consequence of the upward arm swing. The initial backward

thrust is to be expected from the physical principle of mechanics "Action = Reaction" — that is, the force at the shoulder throwing the arm link upwards and forwards must also throw the body link downwards and backwards. Unexpectedly, however, is the evidently insignificant perturbation this causes to the body posture. On the other hand, the subsequent *forward* motion, predictable from the fact that the voluntary movement transfers a significantly large angular momentum to the body link seems to introduce the important destabilizing torque.

2) A sustained cocontraction of the postural muscles does not help the uncontrolled, physical perturbation (biomechanical model), except at muscle forces well beyond those observed in the experiments. This simulation result implies that a dynamic bursting pattern in the postural muscles is required for postural maintenance, and this is consistent with the electromyographic evidence.

The neurological component of the model can describe consequences of alternate strategies to cocontractions, such as stretch reflex activity, and anticipatory and synchronous descending postural activities (or combinations thereof). This ability to compare aspects of various neurological strategies is an indispensable feature of the model, which may show that the anticipatory activites are primarily of neurophysiological importance, not necessarily used for a biomechanical advantage; the anticipatory activities may be used as a means of setting a certain "motor rhythm" or as a means of preparing the reflex pathways (cf. Wollacott et al., 1984).

31.5.2 On the Forward Bending Simulations

For the fast forward bending paradigms, human subject experiments have shown that the maximal excursion of the center of gravity during fast forward bending is very small and on the order of 1 cm (Crenna et al., 1987). The simulation experiments have shown that in the absence of an active, backwards movement of the hips and lower limbs, the center of gravity displaces itself forward by approximately 9 cm. This is because the *passive* backwards movement resulting from the physical coupling of the body links is not quick enough to initially compensate for the fast forward voluntary

movement. However, that backward perturbation *is* large enough to eventually pull the body over backwards within 200 ms after the voluntary movement stops, *even though the trunk is held in the forward position*.

The implication is that postural activities in the hips and lower limbs should be a two-fold process: first, some preprogrammed, descending control to the lower body would be required to actively enhance the passive, backwards motion, in this way reducing the otherwise large center of gravity displacement. This prediction is in accordance with the hypothesis of Crenna and his colleagues (1987, 1988), in spite of the fact that they do not find electromyographic evidence for this first postural adjustment. Secondly, there must be a *subsequent activation in the anterior muscles of the lower body (vastus medialis, tibialis anterior, etc.) to arrest this backwards motion*, since otherwise the backwards momentum would carry the body to the floor in less than half a second after the upper body movement has terminated. This latter prediction is in accordance with the experimental findings of Crenna et al. (1987). The simulation experiments have clarified these two opposite and sequential components of the postural perturbation from the "reduced trajectory."

31.6 Future Directions

A variety of neuromuscular activation patterns at the ankle joint, both descending and reflex derived, can be explored by changing the neurological input and feedback parameters of the system, and these hypothetical activities are directly comparable to the relevant electromyographic recordings. Effects of stretch reflex feedback activity and anticipatory activation/deactivation of the postural muscles can then be observed in simulations with a fixed voluntary movement. Such simulation experiments would determine, with a theoretical method, the optimal combinations of higher level and feedback control for postural maintenance during movement.

In addition, further simulation studies may include variations in the voluntary movement kinematics with the same paradigm, such as increasing acceleration as in the experiments of Lee and colleagues (1987), to see what quantitative effects this may have on the postural perturbation, under different neurological control strategies and

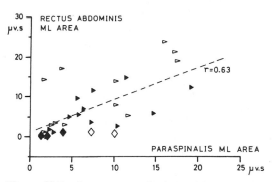

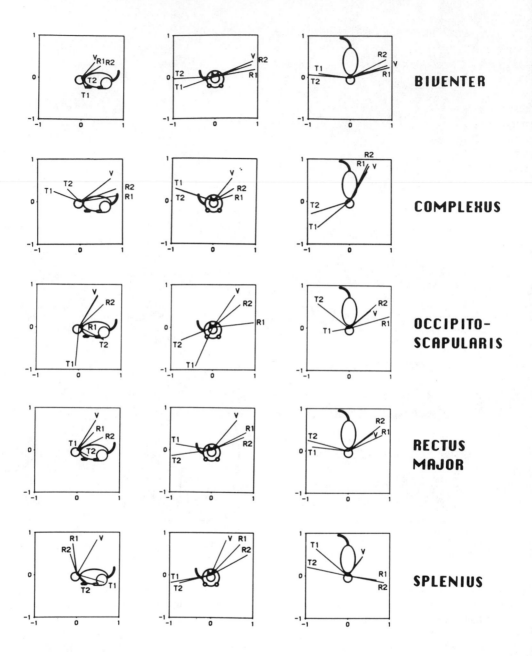

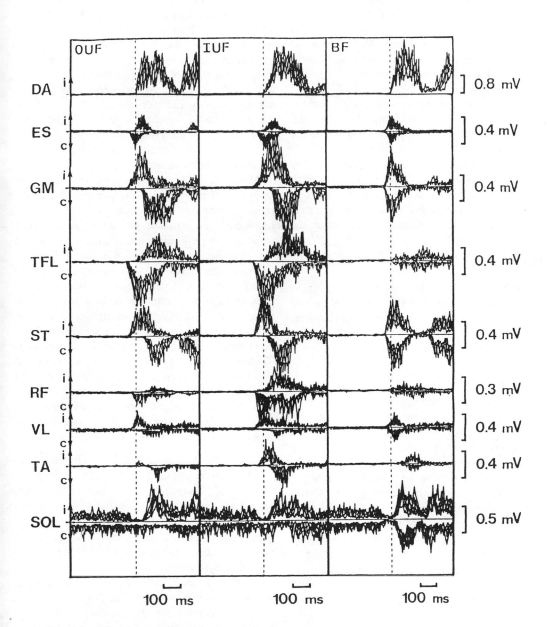

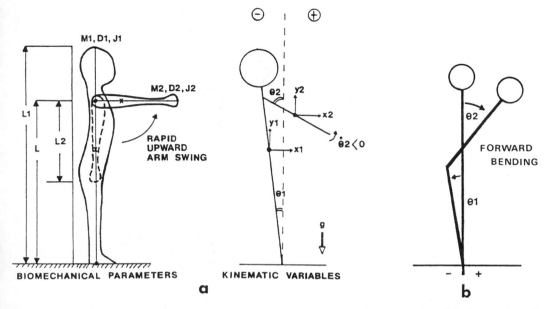

Figure 31.1: *a)* A two link simplification of a standing subject for modeling postural maintenance during a rapid upward arm swing. *Left:* The body is modeled as one rigid link and the arms (together) as the other link. The values of the biomechanical parameters are listed in Table 31.1; *Right:* Definition of the kinematic variables with which the biomechanical coupling equations are derived. *b)* Equivalent inverted pendulum model for the forward bending paradigm.

where M are moments, J is the inertia, m is mass, and the distances are defined in Figure 31.1. Notice that the total torque at each joint is a combination of this inertial component, the gravitational torque and the muscle activity, and summarized by the second order equation:

$$J_p \alpha = M_{link} - M_{ext} + M_{flex} + M_{grav\text{-}link} - B_p \omega - K_p \theta$$

$$(31.2)$$

where

$$J_{p\text{-}body} = J_1 + m_2 L^2 + m_1 D_1^2$$

and

$$J_{p\text{-}arm} = J_2 + m_2 D_2^2$$

The motion of the center of gravity of the *system*, that is, body plus arm links, is also computed for each simulation. Center of gravity accelerations may be compared to experimentally recorded ground reaction forces (biomechanical parameters for the model are summarized in Table 31.1). A similar set of equations describes the forward bending paradigm.

Table 31.1: Biomechanical parameter values as estimated from the scheme proposed by Zatsiorsky and Selayanov (1983) (subject's mass = 75 kg; subject's height = 175 cm)

		Arm Swing	Bending
Lower Link Mass	m_1	65.55 kg	27.92 kg
Upper Link Mass	m_2	7.40 kg	45.03 kg
Lower Link Length	L_1	1.75 m	0.92 m
Upper Link Length	L_2	0.68 m	0.82 m
Lower Link C.O.G.	D_1	0.96 m	0.58 m
Upper Link C.O.G.	D_2	0.24 m	0.36 m
Lower Prin. Inertia	J_1	0.71 kg-m^2	3.93 kg-m^2
Upper Prin. Inertia	J_2	0.23 kg-m^2	10.62 kg-m^2

for the upward arm swing paradigm:

Arm(s) Inser. Pt	L	1.50 m

BODY   KINEMATICS

(A)

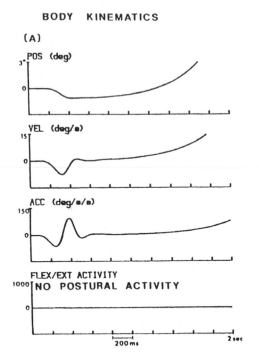

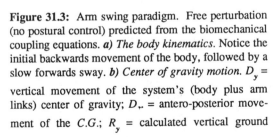

FLEX/EXT ACTIVITY
1000 NO   POSTURAL   ACTIVITY

|__200 ms__| 2 sec

CENTER OF GRAVITY MOTION

(B)

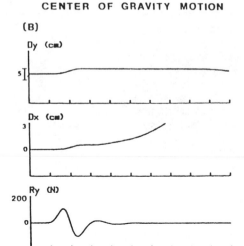

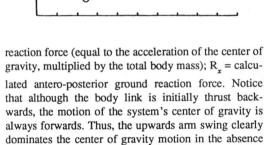

Figure 31.3: Arm swing paradigm. Free perturbation (no postural control) predicted from the biomechanical coupling equations. *a) The body kinematics.* Notice the initial backwards movement of the body, followed by a slow forwards sway. *b) Center of gravity motion.* D_y = vertical movement of the system's (body plus arm links) center of gravity; D_x = antero-posterior movement of the *C.G.*; R_y = calculated vertical ground reaction force (equal to the acceleration of the center of gravity, multiplied by the total body mass); R_x = calculated antero-posterior ground reaction force. Notice that although the body link is initially thrust backwards, the motion of the system's center of gravity is always forwards. Thus, the upwards arm swing clearly dominates the center of gravity motion in the absence of any postural control activities.

observaton). The total movement time is 527 ms. The peak acceleration is 3100°/s² and the peak velocity is 443°/s.

In the absence of any postural activities in the lower limbs and trunk (other than a minimal cocontraction at the ankle of 80 N (about 1% of maximally possible contraction, necessary for state equation stability)), the predicted postural sway is initially about 1° *backward* (Figure 31.3). This corresponds to approximately 16 mm of body link's center of mass, and is due to the backwards reaction of the body to the forward muscular thrust on the arms. The peak backwards acceleration of the body is -65.7 deg/s², and the peak backwards velocity is -7.5 deg/s. The motion then slowly changes to the *forward* direction, caused by the large, upward angular momentum of the arms; the body continues to move forward at a speed of about 6.5 deg/s (with a peak forward acceleration

of 95.6 deg/s²). Since there are no postural activities, the body continues falling forward until in the limit it would fall flat on the floor.

This two-phased postural sway (first backwards, then forwards) results solely from the biomechanical coupling between the two segments (arm and body) in this idealized system. Since normal human subjects cannot suppress their postural activities, the model simulation is invaluable in clarifying this "reduced" trajectory. Evidently, the backwards perturbation to the body resulting from the muscular thrust at the shoulder is quite small.

With regard to the movement of the center of mass for the complete, two- link system, the motion is always upwards (2 cm) and forwards (0.5 cm, initially, then increasing with the continuing forward sway).

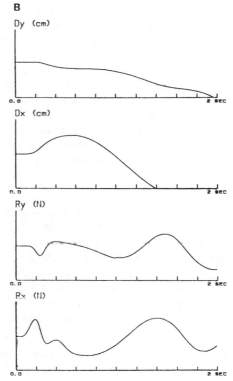

to be expected if satisfactory performance level is to be maintained. In fact, differences in the *EMG* pattern between control and narrow-support conditions were actually detected. These did not in general involve the latency of the synergistic muscles, since both in untrained subjects and in gymnasts, no systematic changes affected the order of muscle activation. Rather, the most striking changes involved the amplitude of the *EMG* activity in the calf muscles, and most interestingly, were detected exclusively in the gymnasts group. In these subjects, on moving from control to narrow-support conditions, the early *EMG* burst typically observed on calf muscles was consistently reduced in amplitude from the very first trials, and in three out of four subjects completely disappeared in the majority of trials (see Figure 34.5).

34.3 Conclusions

The motor tasks considered here, which were performed self-paced by freely behaving subjects, belong to the repertoire of everyday life and are characterized by a great level of complexity deriving from the interplay of many body segments.

One of the main findings which emerges from the data presented is the existence of well defined individual patterns of muscle recruitment, associated with relevant biomechanical variables in the execution of the same motor task, in the *same* experimental condition. The observed variables (several muscles, mechanical moments, trajectories, *CG* displacements) constitute only a relatively small subset of the high number of factors involved in such complex movements. However, they indicate the presence of individual motor strategies as general schemes of organization of a motor action, which result in the *prevalence of a criterion of optimization* (e.g., velocity, force, equilibrium, precision, energy expenditure, elegance) *in order to obtain a given goal*.

The redundancy of the motor organization is one of the basis of such individual strategies, which on the other hand can be related to specific characteristics of each subject, such as anthropometric parameters, individual levels of training, and, possibly, inborn predisposition.

References

Baldissera, F., Hultbom, H. and Illert, M. (1987) Integration in spinal neuronal systems. In V.B. Brooks: *Handbook of Physiology*, 2, I, Nervous System, Motor Control, I, American Physiological Society, Bethesda, 509-595.

Bouisset, S. and Zattara, M. (1987) Biomechanical study of the programming of anticipatory postural adjustments associated with voluntary movement. *J. Biomech.* 20:735-742.

Cole, K.J. and Abbs. J.H. (1986) Coordination of three-joint digit movements for rapid finger-thumb grasp. *J. Neurophysiol.* 55:1407-1423.

Crenna, P., Frigo, C., Massion, J., Pedotti, A., (1987) Forward and backward axial synergies in man. *Exp. Brain Res.* 65:538-548.

Ferrigno, G. and Pedotti, A. (1985) ELITE: a digital dedicated hardware system for movement analysis via real time TV processing. *IEEE Trans. Biomed. Eng. BME-32* 11:456-461.

Kuypers, H. and Huisman, A.M. (1982) The new anatomy of the descending brain pathways. In Sjolund, B. and Bjorkllund, A.: *Brain stem control of spinal mechanisms*, 2, Elsevier, Amsterdam, 29-55.

Lemon, R. (1988) The output map of the primate motor cortex. *Trends Neurosci.* 11:501-505.

Lundberg, A., Malmgren, K. and Schomburg E.D. (1987) Reflex pathways from group II muscle afferents. 3. Secondary spindle afferents and the FRH; a new hypothesis. *Exp. Brain Res.* 65:294-306.

Pedotti, A. (1977) A study of motor coordination and neuromuscular activities in human locomotion. *Biolog. Cybernet.* 26:53-62.

Pedotti, A., Krishnan V.V. and Stark, L. (1978) Optimization of muscle-force sequencing in human locomotion. *Math. Biosci.* 38:57-76.

Pedotti, A. and Ghista, D.N. (1981) Human locomotion analysis in: *Orthopaedic Mechanics*, Vol. II, Ed. D.N. Ghista, R. Roaf, Academic Press, pp. 111-174.

Pedotti, A., Crenna P., Deat A., Frigo, C. and Massion J. (1989) Postural synergies in axial movements: short and long-term adaptation. *Exp. Brain Res.* 74:3-10.

CHAPTER 35

Overview: Influence of Muscle on Cyclic and Propulsive Movements Involving the Lower Limb

Michael Mungiole and Jack M. Winters

35.1 Introduction

We are one of the few living animals who normally locomotes bipedally. Because of this, our lower limbs have features and requirements that are somewhat unique in the animal kingdom. We are endowed with relatively strong leg extensor muscles and long lower limbs, characteristics well suited to our locomotion and postural requirements. These features, however, also result in different loading than would normally be encountered by the upper limbs.

When comparing the functional requirements of the upper and lower limbs, some contrasting characteristics are evident. The upper limbs generally are used as an open kinematic chain in which the trunk serves as a base while the hands undergo various curvilinear paths. Typically, the loads acting at the hands are small compared to the weight of the body. In contrast, the lower limbs are often expected to both support and move the mass of the body. Because one or both feet are usually in contact with the environment, the musculoskeletal system experiences high external loads during continuous or cyclic tasks such as posture and locomotion. Furthermore, a closed kinematic chain is encountered whenever both feet are in contact with the ground, resulting in a kinematically and dynamically difficult indeterminate problem.

The muscles involved in movements of the lower limb have specific features that are compatible with their functional requirements. Architectural features include pennate fibers and relatively large cross-sectional areas, both of which help in allowing these muscles to develop large forces. The more distal muscles have long tendons that are useful for storing elastic energy. The relatively short and proximally-located muscle fibers help to minimize the limb mass at the distal aspect of the segment (Chapter 36 (Alexander and Ker)). This characteristic is functionally important in terms of reducing metabolic energy requirements during locomotion (Chapters 36, 38). The control and organization of movement is often considered differently for arm-dominated versus leg-dominated movements. As outlined in Chapter 11 (Hogan and Winters) and exemplified by Chapters 12–17, the arms have been looked at more from a perspective of identifying factors involved in the movement trajectory. Theories of motor control that are developed using arm movement paradigms do not directly apply to movements of the lower limb, which have been primarily investigated during cyclic tasks such as locomotion. These cyclic movements, which are much more common for the lower limbs, are often considered to be produced by muscle activation patterns resulting from some type of central pattern generator (Grillner, 1975).

Investigations of lower limb movement organization have frequently focused on the consideration of mechanisms influencing movement performance. These include such issues as the storage and utilization of elastic energy, the role of biarticular muscles in producing and controlling movement, and energy transfer between and within segments. These topics are of particular relevance to several chapters in this section of the book and will be reviewed in Section 35.4.

Multiple Muscle Systems: Biomechanics and Movement Organization
J.M. Winters and S.L-Y. Woo (eds.), © 1990 Springer-Verlag

35.2 Features of Volitional Movements Involving the Lower Limb

35.2.1 Interaction with the Environment

The movement constraints placed on the lower limbs are somewhat unique since the feet are often dynamically interacting with the environment. This may involve contact with either the ground or some moving object (e.g., bicycle pedal) which provides resistance. The torso, which is attached to the proximal end of each limb, represents the base of support during bicycling but not during most forms of locomotion.

In the review of upper limb movement [Chapter 11 (Hogan and Winters)], it was shown that dynamic interaction with the environment can have profound effects on the musculoskeletal system, and that the central nervous system has the capacity to modulate this dynamic interaction [see also Hogan et al., 1987 and Chapter 9 (Hogan)]. These interactions and resulting effects also occur for the lower limb, which usually is in contact with the environment. Ground contact essentially provides a position input to the neuromusculoskeletal system with the ground reaction force represented as an output [e.g., Chapter 43 (Yamaguchi)]. This provides additional constraints on movement which have implications in the motor control requirements during posture and locomotion. Issues related to posture and stability are the focus of Section III of this book and are reviewed in Chapter 23 (Andersson and Winters); here we emphasize cyclic and propulsive movements.

35.2.2 Types and Magnitudes of Loading Encountered

Because the legs are usually involved in supporting the rest of the body, large forces are often developed. When the feet are in contact with the ground during a postural or movement task, the required support of body weight often results in significant ground reaction forces, especially during periods where only one foot is on the ground (Miller and Nissinen, 1987). Furthermore, because large muscle forces are often necessary to provide appropriate joint torques, joint contact (bone on bone) forces can be large (Burdett, 1982). Table 35.1 gives typical estimates and direct measures (obtained from the research literature) of forces and moments developed during several common activities, athletic movements, or stationary positions. The estimates were based primarily on either simple reductionist models of joint load distribution or on static optimization methods [for details regarding such methods see Section 8.4 of Chapter 8 (Zajac and Winters)] and are quite sensitive to modeling assumptions (e.g., estimated moment arms). Also, most of the estimated values are likely to underestimate the true forces or moments because muscle co-contractions were usually not assumed.

A consequence of the relatively high forces being exerted by the leg muscles is that, in the absence of dynamic coupling between links, generated joint angular velocities tend not to be as high as those often encountered during arm movements. One contributing factor is the contractile element (CE) force–velocity property, where for a particular level of activation, higher forces result in reduced CE shortening velocity. Additionally, because of unique skeletal features (e.g., calcaneus, patella, greater trochanter, the shape of the pelvis), the major muscles of the lower limb have relatively large moment arms. This facilitates joint torque generation but at a cost to joint angular velocity. Finally, inertial components tend to scale with size at a greater rate than joint moments (McMahon, 1984). Thus, in proportion to limb inertia, larger moments can be developed about joints of the upper limb as compared to the lower limb. This would result in proportionately larger accelerations and, hence, larger velocities for the upper limb segments. We will see throughout this sequence of chapters, however, that dynamic coupling and elastic energy storage can be used to one's advantage to help increase the velocities of the lower limb segments.

35.2.3 Importance of Cyclic Movements

For most cyclic movements, the active muscles are continuously undergoing an alternating stretching and shortening, known as the stretch-shortening cycle (Norman and Komi, 1979). Several rhythmic movements have been investigated with respect to their particular relevance to stretch-shortening, including running (Alexander and Bennet-Clark, 1977; Cavagna and Kaneko, 1977; Ito et al., 1983), walking (Hof et al., 1983), and cross-country skiing (Komi and Norman, 1987). There is some controversy, however, regarding whether muscles are actively stretched

Table 35.1: Representative summary of peak forces and moments experienced by the lower limb for various movements and positions. Shown are vertical ground reaction forces, bone-on-bone forces, musculotendon and ligament forces, and joint moments. All forces are in multiples of body weight and joint moments are in Nm.

load/anatomical location	movement/ position	average peak value	reference
vertical ground reaction force:			
	walking @ 1.34 m/s	1.2	Larish et al. (1988)
	walking @ 1.21 m/s	1.1	Chao et al. (1983)
	running @ 4.5 m/s	2.8	Cavanagh & Lafortune (1980)
	sprinting @ 7 m/s	3.0	Hamill et al. (1983)
	hopping	4.2	Fukashiro & Komi (1987)
	V/B spike (landing)	4.8	Adrian & Laughlin (1983)
	bkwd. somersault (tkoff.)	4.8	Bruggemann (1985)
	running fwd. somersault (lndg.)	13.6	Miller & Nissinen (1987)
bone on bone forces:			
ankle	running @ 4.5 m/s	9.0	Harrison et al. (1986)
hip	walking @ 1.2 m/s	5.2	Crowninshield et al. (1978a)
hip	walking @ 1.0 m/s	4.3	Crowninshield et al. (1978b)
hip	ascending stairs @ 0.5 m/s	5.5	Crowninshield et al. (1978b)
knee	landing from 1 m height	24.4	Smith (1975)
ankle	landing from 1 m height	7.4	Smith (1975)
ankle	running @ 4.47 m/s	10.8	Burdett (1982)
ankle	ballet (2 foot spring to toes)	10.0	Galea & Norman (1985)
hip	walking @ 1.3 m/s	2.5*	Rydell (1965)
hip	walking @ 0.9 m/s	1.8*	Rydell (1965)
hip	walking @ 0.65 m/s	1.3*	Rydell (1965)
hip	walking @ 1.3 m/s (swing phase)	1.3*	Rydell (1965)
hip	standing on one leg	2.5*	Rydell (1965)
hip	walking	2.0*	Kilvington & Goodman (1981)
hip	single legged stance	2.2*	Kilvington & Goodman (1981)
knee	sideways bend (double support)	2.4	Fuller & Winters (1988)
hip	sideways bend (double support)	2.1	Fuller & Winters (1988)
knee & hip	marching (single support)	3.1	Fuller & Winters (1988)
knee	side kick (single support)	2.8	Fuller & Winters (1988)
hip	side kick (single support)	2.5	Fuller & Winters (1988)
knee	back kick (single support)	2.7	Fuller & Winters (1988)
hip	back kick (single support)	3.0	Fuller & Winters (1988)
hip	single legged stance	2.2	Kilvington & Goodman (1981)
knee	squatting (fast descent)	5.6	Dahlkvist et al. (1982)
patellar/fem.	squatting (fast descent)	7.6	Dahlkvist et al. (1982)
musculotendon or ligament force:			
ach. tendon	walking @ 1.8 m/s	3.0	Alexander & Vernon (1975)
patellar lig.	running @ 3.9 m/s	7.4	Alexander & Vernon (1975)
quadriceps	standing long jump (tkoff.)	7.8	Alexander & Vernon (1975)
gastrocnemius	running @ 4.5 m/s	7.0	Harrison et al. (1986)
hamstrings	running @ 4.5 m/s	3.1	Harrison et al. (1986)
ach. tendon	bicycling @ 90 rpm (1 kp)	0.6*	Gregor et al. (1987)
ach. tendon	bicycling @ 60 & 90 rpm (3 kp)	0.8*	Gregor et al. (1987)
ach. tendon	bk. somersault (takeoff)	14.1	Bruggemann (1985)
ach. tendon	running @ 4.47 m/s	7.0	Burdett (1982)
patellar lig.	landing from 1 m height	16.6	Smith (1975)
gastrocnemius	landing from 1 m height	6.1	Smith (1975)
quadriceps	squatting (fast descent)	6.9	Dahlkvist et al. (1982)
hamstrings	squatting (fast descent)	2.2	Dahlkvist et al. (1982)
gastrocnemius	squatting (fast descent)	1.2	Dahlkvist et al. (1982)
joint moment:			
ankle	running @ 3.9 m/s	200	Alexander & Vernon (1975)
ankle	walking @ 1.8 m/s	96	Alexander & Vernon (1975)
knee	standing long jump	125	Alexander & Vernon (1975)
ankle	isometric	134	Nistor et al. (1982)
ankle	isokinetic @ 30 deg/s	101	Nistor et al. (1982)
ankle	bk. somersault (takeoff)	282	Bruggemann (1985)
knee	running @ 4.5 m/s	188	Harrison et al. (1986)
hip	running @ 4.5 m/s	98	Harrison et al. (1986)
ankle	running @ 4.5 m/s	174	Harrison et al. (1986)

* direct measure obtained.

during bicycling although recent research indicates that this does occur for some muscles [Gregor et al., 1987; Chapter 40 (Hull and Hawkins)].

Cyclic movements also provide the foundation behind the concept of the central pattern generator, which is considered by many to be primarily responsible for rhythmic type movements (Grillner, 1975). From this viewpoint it is believed that central pattern generators likely control functional groupings of muscles (synergies) during cyclic movements, with local control over individual segments of the leg (Brooks, 1986). Past research addressing this concept, primarily conducted on cats, have centered on studies where mechanical or electrical perturbations were applied. The purpose of such studies has been to determine the relative influence of feedback pathways versus spinal and higher brain centers [see Grillner, 1975 and Chapter 45 (Brooke and McIlroy)].

35.2.4 Greater Importance of Spring-Mass Characteristics

There is a good bit of evidence [reviewed in McMahon (Chapter 37)] supporting the spring-like characteristics of muscles during movements such as running and bouncing. The bouncing studies (Cavagna, 1970; Greene and McMahon, 1979) show the importance of muscle mechanical properties in determining the output characteristics of the movement. Bach et al. (1983) found that subjects can obtain a resonant frequency of bouncing due partly to the mechanical properties of their ankle plantarflexor muscles. Each of these studies assumed the body to be represented by mass and viscoelastic components, with a force generator also included in the Bach et al. study. Greene and McMahon found that mass had a minor effect on leg stiffness measurements while geometric factors (such as lower limb joint angles) influenced stiffness to a much greater extent (see also Chapter 37).

The spring-like characteristics of muscle have traditionally been considered to be the primary reason for enhanced performance (above that obtained when there is no prior muscle stretch) during the concentric phase of the stretch-shortening cycle. Specifically, the early studies (see Komi, 1984 for review) which have investigated the stretch-shortening cycle for jumping

and hopping movements have generally concluded that enhanced performance is primarily due to the storage and utilization of elastic energy (see also Chapter 38 (Hof)). In contrast, van Ingen Schenau (1984) has questioned the amount of elastic energy that is capable of being stored and utilized based primarily on measures of musculotendon stiffness and changes in tendon length during walking, running, and jumping [see also Chapter 39 (Chapman and Sanderson)].

35.2.5 Greater Interest in Energy and Power Transfer and Biarticular Muscles

There has been a good deal of research, primarily investigating walking and running, that has considered the role of energy transfer and its relationship to the efficiency of movement (see Winter, 1984a for review). The individual energy transfer assumptions inherent in mechanical energy analyses, however, have resulted in a large range of mechanical work, power, and efficiency measures during movement (Williams and Cavanagh, 1983; Winter, 1984a). These assumptions encompass the various possibilities of segment energy transfer, ranging from insignificant exchange (Norman et al., 1976) to complete exchange between nonadjacent segments. The kinetic approach, however, which employs a mechanical power analysis (Elftman, 1939; Robertson and Winter, 1980) eliminates the need to make any energy transfer assumptions. Also, Winter has indicated that this analysis can provide information on where energy generation, absorption, and transfer take place. Although ground reaction force measures are required, outputs obtained using this analysis include the individual joint reaction forces and moments and the instantaneous power measures attributed to them.

Zajac and Gordon (1989) recently published a review in which they consider dynamic interactions that occur between segments for multijoint movements [see also Chapter 8 (Zajac and Winters) and Chapter 42 (Pandy)]. They emphasize that an active muscle connecting two adjacent segments can potentially contribute to movements of other segments due to dynamic coupling and suggest that one of the primary roles of biarticular muscles is to transfer power among the various body segments (van Ingen Schenau et al., 1987). Wells (1988) has shown support for this role of biarticular muscles during walking. Using a reductionist algorithm, he found that there was a reduced mechanical energy cost when the joint moments were partitioned among both monoarticular and biarticular muscles, as compared to the case when only monoarticular muscles were active. An interesting theoretical discussion on the overall role of biarticular muscles was recently given in a target article by van Ingen Schenau (1989).

35.2.6 Some Unique Functional Features

The lower limb has at least two unique features that are not present in the upper limb. The first of these relates to the structure of the knee joint and, in particular, the function of the patella. It is generally considered that the main purpose of the patella is to increase the moment arm of the knee extensor muscles, allowing this muscle group to develop larger moments than could be obtained if the patella were not present. Since stiffness scales with the square of the moment arm [Chapter 9 (Hogan)], the patella also serves to increase the knee joint angular stiffness.

A second unique characteristic of the lower limb is the high stiffness of the ankle joint. This feature is primarily related to the large moment arm of muscles crossing the ankle joint (see Winters and Stark, 1988). Ankle joint stiffness appears to serve functional requirements during volitional tasks such as locomotion, where it helps offset the high equivalent mass so as to provide appropriate spring-mass behavior. The high stiffness is also significant with regards to postural adjustments provided by muscles spanning the ankle joint. Grillner (1972) indicated that joint stiffness provides a load compensation mechanism in controlling movement for which the mechanical response to perturbations is instantaneous, in contrast to the inherent time delay that occurs with feedback response (see McMahon, 1984).

35.3 Current Methodologies

35.3.1 Experimental

Both hardware and software have evolved considerably in recent years. Capabilities in the area of imaging (motion analysis) have also been greatly expanded. The most popular types of imaging systems include cinematography, videography, and optoelectric systems. With an improvement in the quality and resolution of its

image in recent years, video has gained in popularity. One is usually restricted, however, to a maximum sampling rate of 60 frames/s, which is adequate for gait analysis. Video has the advantages of overall ease of handling, the ability to obtain immediate feedback, and the retention of a permanent record of the movement being investigated.

Kinetic analyses involving the lower limb are usually conducted while using a force platform to obtain the ground reaction forces imparted to the foot. Pressure sensing mats (Hennig et al., 1982), a more recent biomechanical tool, provides the researcher with the added capability of measuring the pressure distribution beneath the foot or shoe area contacting the mat. A variation of the pressure mat is the discrete force sensor, which can be placed inside the shoe to obtain force measures at particular anatomical locations (Gross and Bunch, 1988). Load cells and strain gages have also been used to measure forces either applied to the body or those imparted to athletic equipment coupled to the body (Hay et al., 1979).

One of the most important requirements in biomechanical analyses is the determination of the individual forces produced by the active muscles. Methods for force estimation are presented in Section 8.4 of Chapter 8 (Zajac and Winters). For lower limb movements, Hof and Van den Berg (1981) used electromyography (EMG) to estimate muscle force while using a three element Hill-based model and inputs of joint angle and velocity. Active state was assumed to be a function of the filtered EMG. They obtained fairly good agreement with experimental results for an ankle plantarflexion movement. For walking, Olney and Winter (1985) used EMG and kinematic information to validate a deterministic model which calculated moments at the ankle and knee joints.

An unfortunate limitation for each external measuring method (including the more invasive procedure of using EMG wire electrodes) is that no direct measure of muscle force is made. Thus, one is often forced to make questionable assumptions to obtain the forces produced by individual muscles. Recently, buckle transducers have been temporarily implanted around the achilles tendon, allowing for direct measures of tendon forces. Some researchers have used a similar device on the cat hindlimb muscles (Walmsley et al., 1978;

Gregor et al., 1988) and the elbow flexor muscles in monkeys (Landjerit et al., 1988), while Komi and associates (Komi et al., 1987; Gregor et al., 1987) have had success in implanting of a buckle transducer on the achilles tendon of humans. Norman (1989) considers the determination of individual muscle and ligament forces noninvasively as the most important problem faced by researchers involved in human movement studies. Although a direct measure of musculotendon length has been rarely attempted, Whiting et al. (1984) and Abraham and Loeb (1985) have obtained this measure for some hindlimb muscles in cats.

35.3.2 Modeling

Since muscle modeling has been considered in depth elsewhere [e.g., Chapter 5 (Winters)], it will not be reviewed in this section. (It should be indicated that Hill-based muscle models are used in Chapters 38, 39, 40, 42, and 43 in this section of the book.) Here, we will briefly address modeling, as related to multiarticular movement dynamics and control of the lower limb. There are two main approaches that have been considered in determining the kinematic and kinetic aspects of movement, inverse and forward (direct) dynamics [Chapter 8 (Zajac and Winters)]. Inverse dynamics is a simpler approach which can be readily applied to both two and three dimensional problems. By measuring the joint and/or segmental kinematics along with the externally applied loads (e.g., ground reaction forces), one can systematically obtain the joint moments and reaction forces by progressing in a distal to proximal sequence while solving algebraic equations only (Bresler and Frankel, 1950). Inverse dynamics has been used more often than forward dynamics in kinetic analyses, probably because of the reduced complexity of the former method. Much of the published research on locomotion and other complex movements has employed this method.

In many instances it is desirable to obtain the loading contributions of the various tissues crossing a joint by a distribution of the net joint moment. Unfortunately, many combinations of individual muscle forces are capable of producing similar multisegmental movements. In an attempt to overcome this indeterminate problem in inverse dynamic analyses, one is forced to either use heuristic reductionist approaches or static op-

timization techniques to solve for individual muscle forces (described in Chapter 8). The inherent difficulties in experimental validation of the several performance criteria used in static optimization, however, continues to hamper the researcher. A related confounding influence is exemplified in a gait study conducted by Winter (1984b), who found large variations in the knee and hip torques despite fairly constant lower limb kinematics across walking trials.

Inverse dynamics has additional limitations as well. As outlined in Chapter 8, the researcher utilizing such methods may not even need to be aware of fundamental issues concerning neural control strategies. Consider the common example of human gait. To a first approximation, the upright human torso system can be thought of as an inverted pendulum [see Chapter 37 (McMahon)]. Such systems are very difficult to stabilize, yet maintaining stability [reviewed in Chapter 23 (Andersson and Winters)] represents one of the primary neuromotor control issues of the lower limb. With inverse dynamics, stability becomes a moot point since the kinematics represent the input to the model. Thus, it is not surprising that a review of the inverse dynamics literature finds little mention of stability requirements. In contrast, when forward (direct) dynamic simulations of gait are performed [Chapter 43 (Yamaguchi)], it becomes a vital question because the model will literally "fall" if this issue is not appropriately addressed. Even here, however, deterministic models may not allow a full investigation of stability issues since such models may not be set up to assess the model response to perturbations (see also Chapter 23).

As another example, the concept of goal-directed behavior is related to a task that unfolds over time. Such tasks cannot be adequately treated from within the framework of inverse dynamics since each instant in time is mathematically distinct from any other instant. This is especially evident in static optimization, where there is no mechanism available for incorporating the performance criterion which is a function of the entire temporal task. In reality, neural control strategies at one instant are applied for the purpose of affecting future instants in time, based on trying to meet the goals of some task (e.g., Chapter 42 (Pandy)).

In contrast to inverse dynamics, the direct

(forward) dynamics approach represents the natural causality of physical systems. This approach offers the important advantage of forcing the researcher (or computer) to solve the physical problem in a manner similar to the functioning neuromotor system, i.e., the neural drives to muscles become inputs to the musculoskeletal system [see Chapter 8 (Zajac and Winters)]. In this case, state equations must be integrated over time which results in the solution of differential equations. Forward dynamics has often been utilized in conjunction with lower limb dynamic optimization methods [e.g., Hatze, 1978; Chapter 43 (Yamaguchi); Chapter 42 (Pandy)]. In Chapter 8 it is shown that direct dynamics, used in conjunction with dynamic optimization, provide the greatest potential for gaining insight into neuromotor control strategies employed during movement.

Both inverse and direct dynamics modeling approaches, at least as utilized to date, make no *a priori* assumptions regarding neural connections between muscles. It is hopeful that future applications of dynamic optimization will include assumptions on the neural connections which, upon excitation, would have the ability to produce inherent coupling between muscles. Control parameters to be optimized need not only be motoneuronal input parameters; feedback gains or tuning parameters for "hard-wired" neural circuitry may also be optimized for specific (stereotyped) tasks. Eventually, the modeler may consider if an optimization algorithm (or perhaps a neural network) can produce a central pattern generator. Cyclic lower limb movements would provide the ideal paradigm for such investigations.

35.4 Major Conceptual Issues and Current Status

35.4.1 Basic Implications of Elastic Energy Storage and Utilization

Elastic energy and its biomechanical (and neuromotor) implications has been the focus of much investigation, especially as they relate to muscles in the lower limb. Yet the source and relative importance of elastic energy storage and utilization remain controversial. In several chapters (particularly Chapters 36–38) within this section emphasizing the lower limb, recognized leaders provide a unique focus on some important

conceptual and practical issues regarding elastic energy.

Implications of Elastic Properties in Musculoskeletal System Design

Chapter 36 (Alexander and Ker) provides a thoughtful commentary on the relationship between architectural features of limb muscles and tendons in various animals and the capacity for elastic energy storage during locomotion. This has been an ongoing topic of interest by Alexander and colleagues (e.g. Alexander, 1988), and in conjunction with Chapter 37 (McMahon) and Chapter 38 (Hof) provides a unique global perspective on spring-like musculotendinous structures. The authors indicate how these structures have evolved in animals to optimally meet the competing needs of task performance (e.g., locomotion speed) versus metabolic energy requirements. Assuming typical stress–strain material behavior for tendon [see review in Chapter 5 (Winters)], Alexander and Ker provide a convincing argument that various tendons have dramatically different force–extension properties which serve different purposes. Based on architectural features, they distinguish between three major types of musculotendinous units and, from a teleological perspective, propose reasons for these differences. These three types of musculotendinous units are:

1. Muscles with **long fascicles and relatively short tendons**. These muscles tend to be located proximally on the limb and are often large in size. They are capable of producing a large amount of work, while accelerating a limb segment over a significant range of motion. When it is necessary to help decelerate the limb, these muscles are also capable of absorbing energy.

2. Muscles with **long and relatively thick tendons** which only extend up to 2% of their length. These tendons do not store a significant amount of strain energy and would thus appear to be overdesigned. Building on the observations of Rack and Ross (1984), Alexander and Ker suggest that the tendons may serve the purpose of **remote operation**. These musculotendinous units often have long tendons and distal insertion sites which allow the muscle mass to be proximally located and help in minimizing limb inertia. Due to their thickness, these tendons have the stiffness necessary to meet the important functional requirement of position control, especially appropriate for prehensile muscles of the fingers and thumb. Because of their long

length, however, the tendons also aid in the control of force and regulation of stiffness. Since the operating range of these tendons are within the "toe" region (see Fung, 1981, p. 210), their stiffness increases dramatically with an increase in muscle force. Thus, the theoretical foundation for impedance modulation described by Hogan (Chapter 9), based on the capacity to modulate stiffness, would still apply. In addition, these properties circumvent two disadvantages of compliant tendons stated in Chapters 36 and 38. The first of these is represented by the higher metabolic cost and more sluggish nature of voluntarily-induced isometric or concentric muscle action. This is especially true when muscle action is initiated from a low force level, where the tendon compliance is higher. Under this condition, a relatively large amount of CE shortening accompanies a given change in muscle force. The second disadvantage is concerned with potential neuromuscular control difficulties which are due to: *i)* the muscle spindles not measuring absolute musculotendinous length; *ii)* a lack of precise position regulation; and *iii)* the requirement that the central nervous system determine the appropriate time course for a concerted contraction (Chapter 38) that would result in reduced metabolic energy consumption.

. Musculotendinous units with relatively *short muscle fascicles and long and slender tendons* (e.g., ankle plantarflexors). These units are represented by the anti-gravity muscles located in the distal segments. When large forces are present in these muscles, the tendons undergo large extensions (up to 5%) enabling them to store significant amounts of elastic energy. As indicated in Chapters 36 and 38, such tendons also save energy because: *i)* for a given force level, less metabolic energy is required due to the relatively small muscle mass; *ii)* the tendon acts as a spring, storing energy that can later be released with little viscous energy loss; and *iii)* the contractile machinery does not experience as much variation in length as the entire musculotendinous unit, allowing the CE to remain within a narrower and more efficient region of its tension–length and force–velocity relations [see also Chapter 39 (Chapman and Sanderson)]. Hof (Chapter 38) provides quantitative documentation of points *ii)* and *iii)* while using a Hill–based model of the ankle plantarflexors. While considering the positive and negative work of the CE and the entire musculotendinous unit along with the elastic energy stored, he shows the various sequences of energy flow during slow, moderate, and fast speed walking, and running.

Fundamental Global Effects of Spring-Like Muscles

The findings of Chapter 37 (McMahon) complement the above observations related to architecture while treating the gross limb musculature as a single telescoping spring. Here, simple spring-mass models are considered while addressing elastic energy during movements, particularly the "elastic bounce" seen during running and walking. The physical source of spring-like behavior is not directly addressed because McMahon considers stiffness from a global perspective in which control-modulated reflex stiffness works in conjunction with passive musculotendinous elasticity. Conceptual issues related to the advantageous use of spring-like elements in animals and robots are emphasized in his chapter. He documents experimental observation of macroscopic spring behavior, showing the relatively small change in global leg stiffness over a large range in force and the more significant variation in stiffness due to kinematics (e.g., knee angle). Also, he makes a distinction between vertical and leg stiffnesses. The telescoping spring leg is utilized in considering the theoretical benefits of spring-mass systems during running as related to the tradeoffs between stiffness, stride length, and forward velocity. Finally, McMahon indicates that model predictions are in reasonably good agreement with the empirical relationship between stride length and forward velocity (both variables normalized to leg length) obtained by Alexander (1976) for a wide range of animals.

35.4.2 Stretch-Shortening: Controversy and Significance

Bicycling represents a movement in which there has been controversy as to whether the stretch–shortening cycle exists to a significant extent. When electromechanical delay (Norman and Komi, 1979) is taken into account, Hull and Hawkins (Chapter 40) conclude that elastic energy is stored and utilized in cycling but may be dependent on methodological factors such as pedalling speed and whether toeclips are used. They report that their subject pedalled at a relatively high speed (85 rpm) while using toeclips and cleats. These results are consistent with those obtained by Andrews (1987) and Gregor et al. (1987) who also show evidence that the stretch–shortening cycle exists. In contrast, previous studies which concluded that elastic energy was not stored and utilized, had subjects pedaling at approximately 60 rpm and without toeclips.

Force potentiation induced by the stretch–shortening cycle has been documented for a variety of movements (reviewed in Komi, 1984). The relative contribution and importance of elastic energy, however, remains controversial. An interesting discussion on this issue was addressed by van Ingen Schenau (1984) in which he suggested that elastic energy storage may be overestimated. In general, Chapman and Sanderson (Chapter 39) seem to agree with this assessment and state that the real advantage of the stretch–shortening cycle is to allow time for muscles to develop force. This allows a high force level to be achieved by the time the muscle starts to shorten, resulting in improved performance during the concentric phase.

The importance of this high force level for movement performance [Chapman and Sanderson (Chapter 39)] is consistent with results obtained by Bosco et al. (1981) at movement turnaround, where the direction of joint(s) rotation is reversed. Bosco et al. found that vertical jump performance was strongly affected by the brief delay period (coupling time) at turnaround during which the knee joint was maintained at a constant angle. They have shown that an increase in coupling time resulted in a reduction in the vertical ground reaction force at turnaround along with a corresponding reduction in movement performance (maximum height jumped). Also, an increase in coupling time corresponded to reduced angular acceleration of the joints at turnaround.

The importance of joint angular acceleration (and ongoing muscle force and *CE* power) at the initiation of the concentric phase of the stretch–shortening cycle was addressed in a recent experimental and simulation study for ankle dorsi/plantarflexion movements (Mungiole, 1990). Computer simulations, using a model with Hill-based muscle properties, were obtained in which ankle joint angular accelerations at movement turn-around were compared by varying the neural input to the plantarflexor (agonist) muscles during the eccentric phase of the movement. Mechanical output variables were compared during the first 200 ms of ankle plantarflexion. Figure 35.1 shows plots of simulation predictions for joint angle, torque, and power and *CE* power for the three levels of angular acceleration at turnaround.

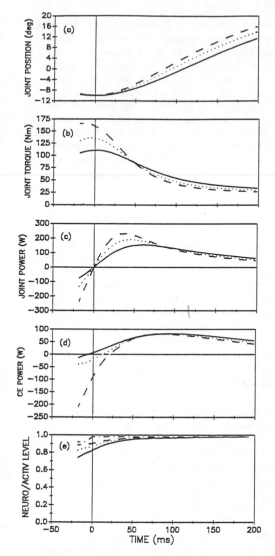

Figure 35.1: Ankle dorsi/plantarflexion simulation of experimental protocol (Mungiole, 1990) obtained from a muscle model (Winters and Stark, 1985) showing the effect of minipulating neural signals on ankle plantarflexion (time > 0). Solid, dotted, and dashed lines represent ankle joint angular accelerations of 41.5, 59.2, and 76.7 rad/s^2, respectively, at movement turnaround (time = 0). *a)* Joint position. Anatomical position (relaxed standing) is at 0°. Positive values represent plantarflexion. *b)* Joint torque. Positive values represent plantarflexor torque. At turn-around, torque values are 111, 136, and 161 *Nm*, respectively. *c)* Joint power (torque * angular velocity). Positive values indicate joint is plantarflexing. *d)* Contractile element (*CE*) power (muscle force * *CE* shortening velocity). Positive values indicate *CE* is shortening. *e)* Neural excitation and activation (simulated *EMG*) levels. Values can range between 0 and 1. The dashed-dotted line represents the neural excitation input.

Movements were simulated while a resistive load of 36% of maximum isometric plantarflexion strength was applied. The neural input values were varied in a manner that produced equal neural excitation and nearly equal activation levels (Figure 35.1e) at turnaround (time = 0) and during the concentric phase. Joint excursion, torque, and power results agreed reasonably well with experimental data in which subjects lowered and raised weights while using their ankle plantarflexor muscles (Mungiole, 1990). Considering the early part of the concentric phase, it is apparent that the joint power and *CE* power (Figures 35.1c and d, respectively) follow different trends across the three simulation runs. As angular acceleration at turnaround increases, the joint power increases while the *CE* power decreases. This trend difference indicates that, as turnaround acceleration increases, the series elastic component (*SE*) contributes to a greater extent both in utilizing stored elastic energy and in setting *CE* force and velocity operating ranges. What is important is not **how much** elastic energy is stored but rather **when** it is stored and how this affects the system state variables. In particular, the difference between the instantanous *CE* power and the joint power is a direct consequence of series element behavior.

These simulation results are compatible with Chapman and Sanderson (Chapter 39) who indicate that the large muscle torque developed early in the concentric phase results in the joint angular velocity increasing faster than would occur for a movement starting from rest. Chapman (1985) also reached a similar conclusion in his comprehensive review article on muscle mechanics in which he compared model results of a stretch–shortening cycle with a movement starting from rest undergoing a concentric phase only. Thus, the main reason for improved performance in the stretch–shortening cycle is the high torque developed at movement turnaround. Our simulation experience with other joints suggest that although the details may differ, the basic trends extend to most joints throughout the body.

35.4.3 Distinguishing Between Stretch-
Shortening Phenomena

Before proceeding, we would like to distinguish between the classical definition of the stretch– shortening cycle and "elastic bounce".

For the stretch–shortening cycle (reviewed in Chapters 39–41), eccentric muscle action due to increased muscle activation facilitates the strength and speed of the ensuing concentric action. With this movement, changes in *voluntary* muscle action are responsible for inducing the stretch–shortening cycle. "Elastic bounce" exists when a sudden (and often prolonged) impact *load* is applied to the musculoskeletal system, inducing muscle stretching. Prior to impact, the muscle(s) soon to be stretched is not typically lengthening. In fact, there is often shortening, with the ongoing activation level increasing in anticipation of the impact. The increase in activity is partly responsible for producing muscle recoil sometime after impact. Our simulation results of the muscle-activation–induced stretch–shortening cycle and external-load–induced "elastic bounce" indicate that they are different situations, with muscle receiving different types of inputs [see also Chapter 5 (Winters)]. Sensitivity analysis trajectories indicate that the classical stretch–shortening cycle against a constant load predicts less of a role for elastic energy storage than does an "elastic bounce" simulation. This perhaps explains why elastic energy is of greater concern in chapters emphasizing running (Chapters 36–38), a movement indicative of an "elastic bounce" effect. It is downplayed and of lesser importance, however, in situations where the windup aspects of throwing, striking, and kicking are addressed [Chapman and Sanderson (Chapter 39)], movements closely identified with the stretch–shortening cycle.

35.4.4 Revisiting Responses to Impact

In Chapter 38, Hof considers the contribution of ankle plantarflexors to the energetics of walking and running and the relative role of elastic energy storage at different speeds. Using a Hill-based muscle model, he shows that that elastic energy is utilized more effectively during normal walking as opposed to slow walking, and that this is due to muscle mechanical properties. It should be noted that these simulations assume no muscle yielding [see Chapter 1 (Zahalak)]. This assumption is compatible with the findings of Cordo and Rymer (1983), who showed that when additional fibers are being recruited (a typical occurrence when impact is anticipated), yielding is insignificant. As further support, strong reflex activity due to muscle lengthening (see Chapter 13 (Wu et al.) and tendon compliance also tend to help mini-mize a yielding effect. These issues are discussed in more detail in Chapter 5 (Winters).

For most impact-induced "elastic bounce" tasks (particularly, cyclic movements such as running), the initial conditions at impact can be strategically set by neural activity. As an example, the initial speed and rate of change of force of the ankle plantarflexor muscles at the time of impact can be set by prior neural activation. For both force and velocity to be independently set, however, either temporary antagonistic co-contraction or strategic multi-articular coordination may be required. Co-contraction changes the system stiffness/impedance [Chapter 5 (Winters); Chapter 9 (Hogan)]. McMahon (Chapter 37) suggests that there is an optimal leg stiffness, of intermediate value, which allows a cyclical task to be repeated without variation. This optimal stiffness would depend on the goal of the movement along with the mass and inertial dynamics of the system [see also discussions in Chapter 38 (Hof) and Chapter 39 (Chapman and Sanderson)]. It may also depend on the magnitude and duration of the im-pact. Consider the example of kicking a ball of moderate mass as far as possible. The impact time may be brief and the impulse and energy imparted to the rotating joint(s) are considerably less than that which occurs when the foot impacts the ground and bears the full weight of the body. In fact, an eccentric phase may not even occur. In this case the optimal stiffness might be a value which allows elastic recoil to act on the object (and within the object) during the impact time period. Thus, for such a striking task, maximal performance depends on the magnitude and direc-tion of the contact velocity [Chapter 39 (Chapman and Sanderson)], segment geometry at impact, and the compliance of the object being struck. If con-tact time is relatively long (i.e., the object has significant compliance), manipulation of the CE and SE operating ranges takes on greater sig-nificance.

35.4.5 Intersegmental Coupling, Muscle, and Optimal Sports Performance

To fully understand elastic energy storage and most other aspects of lower limb neuromusculo-skeletal function, we need to consider dynamic interactions between limb segments and the unique roles of multiarticular muscles. The con-cept of multi-link dynamic coupling represents a

major issue in Chapters 39 and 41–45, which con-sider the movements of throwing and kicking (Chapter 39), jumping (Chapters 41–42), and walking (Chapters 43–45). Inverse dynamic models provide one avenue for exploration and have been utilized extensively for study of energy and power transfer (see Section 35.4.7). These types of approaches essentially document dynami-cal interaction effects, and as such, provide valuable information. For example, in Chapter 45 (Hinrichs) helps to identify a major role for arm swing during running — stabilization of the axial angular momentum.

Direct (forward) dynamic methods, however, can provide insights into intersegmental coupling that are not possible with the inverse dynamic ap-proach (Zajac and Gordon, 1989). For instance, Zajac and Gordon indicate that the contraction of a muscle crossing the ankle joint affects not only the ankle joint but also accelerates and transfers energy to more proximal joints. Such findings force reassessment of fundamental concepts, such as muscle synergy (see also Chapter 8 (Zajac and Winters)). In addition, one of the primary ad-vantages of forward dynamic simulations is that sensitivity analysis techniques can be more fully utilized since changing model (or control) parameters also changes movement kinematics.

Chapters 39–43 collectively consider the effects of intersegmental dynamics and muscle properties as related to optimal performance (e.g., in sports activities). Until recently, there have been few ex-plicit ties made between dynamic coupling, muscle properties, and performance. Considering a basic issue, Chapman and Sanderson (Chapter 39) and Hof (Chapter 38) suggest that optimal cadence during cyclic tasks may in part be set by considerations of muscle properties and the operat-ing range of the CE.

In Chapter 39, Chapman and Sanderson indic-ate that studies in sports biomechanics rarely consider how muscle mechanical properties may predispose athletes toward adopting specific tech-niques. The authors attempt to remedy this by providing numerous insights on how muscle mechanical factors fundamentally influence move-ment strategy. (Given our bias towards the importance of muscle properties, we suggest that this chapter should be read by all sports biomechanists.) The primary focus of this chapter rests with the stretch– shortening cycle, as ex-tended to multi-link systems. They use the inverse dynamics approach to document the proximal to distal nature of throwing and kicking tasks and then utilize muscle simulation experience to consider the implications as they relate to muscle mechanics. When muscle action produces the rotation of a proximal segment, the adjacent distal segment in the open kinematic chain tends to initially experience backward rotation. This, in turn, results in an efficient stretch–shortening cycle at the joint between these segments due to subsequent activity of the muscles crossing this joint. The stretch–shortening cycle can then occur at some later time at the next more distal joint. With appropriate neuromotor timing, this process could be an effective strategy which takes advantage of muscle properties.

Chapter 39 closes by pointing out that one of the goals of training is to change muscle properties and consequently, performance. Also, they indicate that extensive sensitivity analysis studies (using muscle describing parameters) will be crucial if biomechanical modeling is to make an impact on sports performance. We fully agree, and would further suggest that as far as sports training is concerned, exact simulations would not be a requirement. Rather, what the coach and athlete need to know is the *direction* they need to take to improve performance. This may involve changing either the neuromotor control strategy or the biomechanical system (with training). Typically, neurocontroller parameter sensitivity would help determine the appropriate direction for modifications in *task execution* (i.e., the goals for an athlete's neuromotor system), while sensitivity to muscle (and perhaps joint) mechanical parameters would help determine how to best *train*.

Related to this, Pandy (Chapter 42) shows the interdependence of the neural and mechanical systems. For jumping, he indicates that as the muscle properties change, the optimal neural pattern may also change. Muscle strength, contraction speed, and tendon compliance are varied to determine their effects on maximal jumping height. His results indicate that variations in strength have the largest effect on height while the control-related aspects show that as strength is increased, the importance of muscular coordination (or the timing of proximal to distal muscle activity) decreases. Finally, he found that the optimal strategy for a jump changes dramatically as the mechanical parameter of peak unloaded CE velocity is varied.

Chapter 41 (van Ingen Schenau et al.) and Chapter 42 (Pandy), both of which consider energy transfer and the role of biarticular muscles in jumping, appear to give conflicting results. Although van Ingen Schenau et al. state that the transport of energy refers to the power transported due to the action of biarticular muscles, one would normally need to indicate in what context this term is being used. Traditionally, energy transfer has usually referred to the flow of energy between various segments, or within a segment when a transfer between kinetic and potential energy occurs. Van Ingen Schenau et al. indicate that they are referring to an energy balance summed over all segments while also reporting that biarticular muscles are responsible for a transfer of power from the proximal to distal joints during jumping. Contradicting these results, Pandy shows that all lower limb muscles, including biarticular muscles, are involved in transferring power in a proximal direction, primarily to accelerate upwards the relatively heavy trunk. A possible reason for the contradictory results, as indicated by van Ingen Schenau et al., may be that Pandy's interpretation of the definition of energy transfer is different from that used by the former authors.

35.4.6 Insights Into The Roles of Biarticular Muscles

Based on the several early references mentioned in Chapter 41 (van Ingen Schenau et al.), it is evident that a basic understanding of possible uses for multiarticular muscles has been around for a surprisingly long time. In Chapter 41, modeling approaches are used to gain insight into cycling and two explosive push-off activities — jumping and speed skating. The authors develop a number of important reasons for the positive use of biarticular muscles: *i)* coupling of joint movements via tendon-like biarticular leg muscles; *ii)* capacity for biarticular muscles to contract within the relatively efficient low velocity region of the CE force–velocity relation; *iii)* transport of energy via coupling actions; *iv)* use of coactivation of a monoarticular agonist and a biarticular antagonist to modify and stabilize movements; *v)* capacity for more energy-efficient use of muscles (documented by the same group in Chapter 18 (Gielen et al.)) and *vi)* capacity to modulate the direction of force application of the contact location at the distal segment. The last reason appears related to another possibility indicated by Hogan (Chapter 9) — the capacity to modulate end-point stiffness. *Point iv* requires some elaboration. Here, monoarticular agonist and biarticular "antagonist" muscles work together, with the latter acting as a relatively stiff elastic band. It is suggested that this muscle coactivation represents synergistic rather than antagonistic action because it makes it possible to "redistribute joint moments and joint powers in such a way that specific task demands are met" (p. xxx). Specifically, this "could serve the purpose of transporting energy from joints where it can no longer be used effectively, to joints where it still can" (p. xxx).

It is interesting that monoarticular muscles are found in Chapter 42 (Pandy) to be more important in terms of power generation. Furthermore, in Chapter 39 (Chapman and Sanderson), the proximal to distal timing was viewed as a mechanism for placing monoarticular muscles into efficient *CE* operating ranges of their *CE* curves. Perhaps biarticular muscles, as suggested in Chapters 40-42, are not as much the main power generators as they are the subtle, efficient modulators of performance. This role for the biarticular muscles is consistent with the results obtained by Winter (1984b), who attributed a major portion of the variability in gait kinetics to these muscles. This increased task-specific variability of activation for biarticular muscles (as compared to monoarticular muscles) may indicate that these muscles are based on different organizing principles, at this time not well understood.

Of note are two other uses for multiarticular muscles that have been suggested in other parts of this book. In Chapter 9, Hogan points out that selective modulation of multiarticular muscles may help the overall system appear "passive", and consequently remain stable when in contact with other passive objects. In Chapter 26, Crisco and Panjabi point out that multiarticular muscles may inherently facilitate system stability, especially for multilink inverted pendulum structures.

35.4.7 Walking Into the Adapting Future

Up to this point we have emphasized tasks involving rapid lower limb movements that, in general, have fairly well defined goals. We now consider a type of task that is perhaps one of the most difficult to understand - bipedal walking. The difficulty arises because: *i)* walking involves both single and double support periods and *ii)* one cannot easily define the goal of the task. This common task does not tax the limits of the musculoskeletal system. Several research studies have

been conducted that have utilized various static optimization schemes for walking (reviewed in Berme et al., (1985); Zajac and Gordon, 1989). The results of these studies, in trying to identify the strategies underlying the control of gait, have not been particularily satisfying (Patriarco et al., 1981; Winter, 1984a). Because of the high visibility of these studies, optimization techniques (under the guise of the more limiting static optimization approach) have at times been given a bad name. However, the problem rests more with the type of task than optimization *per se.* Furthermore, as documented by Winter (1984b), quite different inverse dynamics predictions of joint torque can produce nearly the same movement kinematics. This suggests that walking may not have a specific optimization strategy, at least in terms of kinetic criteria. Perhaps this concept is reflected by the increased variability in the net muscle torques for the hip and knee joints that are crossed by several biarticular muscles during gait (Winter, 1984b).

Chapter 43 (Yamaguchi) and Chapter 44 (Winter et al.) represent quite nicely the directions toward which experimental and modeling studies of gait analysis are moving. These chapters (as well as Chapter 33 (Winter et al.) and Chapter 45 (Hinrichs), indicate the need for improved modeling of the torso and arms and a better understanding of gait as an integrated activity involving the entire body. Additionally, both Chapters 43 and 44 emphasize the use of computer models for pathological gait, with Chapter 43 using the direct dynamics approach, Chapter 44 inverse dynamics. Chapter 44 discusses the assessment of gait and makes an important distinction between adaptations to gait pathologies by the intact neuromuscular system as compared to pathology itself. Because of the inherent plasticity of the central nervous system, atypical adaptations will likely disappear or be reduced after successful surgical intervention followed by appropriate therapy. By utilizing direct dynamic simulation techniques, Chapter 43 emphasizes the predictive capacity of gait simulations, especially as they relate to likely effects of surgical intervention (e.g., tendon transfers) or orthotic treatment. This latter technique can also be utilized to explore possible neuromotor adaptations and issues related to stability. These two fundamentally different modeling approaches are both needed in pathological gait study and, together, provide a vision of what the future will offer.

35.4.8 *EMG* and Neural Control Strategies

Whether the direct or inverse dynamics approach is utilized in biomechanical analyses, it is useful to have information on the muscle activation patterns because it will provide some insight into motor control requirements. From a modeling perspective, the rectified and filtered *EMG* patterns [see Chapter 40 (Hull and Hawkins) for a review of techniques] during various movements have been used: *i)* in *EMG*-to-force processing [Chapter 38 (Hof)]; *ii)* to help elucidate regions of positive and negative muscle work (Chapter 40); and *iii)* to compare and potentially validate experimental results with those predicted by using static or dynamic optimization. Review of classical and/or especially interesting *EMG* patterns for various lower limb tasks is beyond the scope of this chapter. It should be pointed out, however, that muscle activation patterns are the "final common pathway", and thus they include both open-loop inputs (e.g., central pattern generator) and closed-loop sensory feedback. The usual way to distinguish between reflex contributions and central inputs is to perturb the mechanical system and then study the *EMG* and mechanical effects of this perturbation (reviewed in Chapter 46 (Brooke and McIlroy)). In recent years perturbations have applied to complex movements involving multiple limb segments, with the results suggesting that a significant portion of the response is due to centrally-triggered (pre-structured) "programs". In Chapter 46 the effects of such disturbances are documented for a cycling task. It is seen that corrective movements involve the whole limb.

35.5. Future Directions

35.5.1 Expanded Utilization of Advanced Modeling Tools

It was suggested in Chapter 8 (Zajac and Winters) that, as computer technology continues to advance, the future lies in combining experimental investigations with direct dynamic simulation and dynamic optimization. Another useful tool that is likely to find expanded use is sensitivity analysis. Our objective here, however, is to suggest some more tangible areas that are more directly related to understanding lower limb biomechanics and neuromotor movement organization.

35.5.2 Bearing Good Fruit Depends on One's Vision: Visions Worth Exploring

Many tasks have been considered in this set of chapters. Often, somewhat opposing views have emerged. When viewed in context, however, we find that most of the studies within this group of chapters are compatible with one another. For instance, the relative contribution of elastic energy to movement is a function of the characteristics of the task, and can change throughout the task. If one point emerges from the assortment of chapters, it is that the techniques utilized to assess data help define the visions that are possible. What techniques and approaches are most fruitful? From the overall point of view of the movement biomechanics community, the application of multiple approaches appears to be quite advantageous, while also making meetings more enjoyable. From our viewpoint, however, certain approaches do appear most likely to bear better fruit. Here is one of the many possible examples. In Chapter 41 van Ingen Schenau et al., suggesting that "nature never works against herself" (Pettigrew, 1873, as cited by Tilney and Pike, 1925), conclude that since co-contraction consists of muscles producing forces that cancel and moreover comes at metabolic cost, co-contraction would violate this adage. However, this is perhaps a limited vision. The standing human has certain characteristics of an inverted pendulum, which is inherently unstable [Chapter 23 (Andersson and Winters)]. Stable equilibrium requires that a certain "critical" stiffness is met. Because of nonlinear muscle properties and reflex feedback gain modulation, co-contraction causes joint stiffness modulation. Thus, co-contraction can often stabilize an otherwise unstable system, and, more importantly, can provide a margin of safety to a system that might otherwise just barely be stable. Nature may possibly want to allow herself a margin of safety, especially in an unpredictable environment. Thus, even co-contraction could at times be compatible with this adage. As shown in Chapter 19 (Seif-Naraghi and Winters) for dynamic optimization studies of simple arm movements, optimal solutions will in fact predict co-contraction while movements must be made in an unpredictable environment.

The above problem required that direct dynamics would need to be employed. This may not be enough, however. To study stability in complex nonlinear systems, sensitivity to "noise" perturbations needs to be addressed, and thus a *stochastic model* is needed. This adds considerable complexity to the problem, especially to optimization studies, but perhaps it will be worth the effort.

35.5.3 How Do Tasks Scale?

One can walk and run at different speeds, jump to different heights, throw different distances, cycle at different average power levels. Athletes typically use different strategies during warm-up than during competition. One hundred percent effort is the exception, not the rule. Tasks scale whenever the person's goal changes. Can we describe such scaling of goals? The concept of a generalized optimization criterion, as proposed in Chapter 8 (Zajac and Winters), suggests that we can for many tasks of interest (e.g., in sports and rehabilitation). By changing the relative penalty between measures of task performance and measures of neuromuscular effort, optimal solutions scale quite naturally [e.g., Chapter 19 (Seif-Naraghi and Winters)]. We suggest that as the movement biomechanics field matures, increased effort will be placed in the experimental and modeling investigations of how movements scale. In turn, this will produce a subtle but real change in focus toward neuromuscular strategy. It will also, in conjunction with sensitivity analysis studies, provide an avenue for investigation of athletic performance and training techniques.

35.5.4 Incorporating Torso and Arms into Lower Limb Models

In the past, most lower limb models have assumed that the torso is just a large mass that needed to be supported. In the fourth section of this book, several chapters (Chapters 24, 28–31) were concerned with integrating torso movements to lower and upper limb movements. In line with Chapters 35 and 43–45, we suggest that better models of body segments above the pelvis are necessary to better integrate these segmental movements.

35.5.5 What is The Muscle Doing?

As observed by Chapter 39 (Chapman and Sanderson), there have been surprisingly few studies which have seriously looked into how muscle mechanical properties influence performance. Compared to man-made motors, muscle properties are quite unique (Chapter 7 (Hannaford

and Winters)). Chapters 36, 38, and 39 show that the muscle model structure (both in terms of architecture and the *SE* and *CE* series arrangement) can be advantageous. Chapter 5 (Winters) points out a number of situations in which muscle nonlinear properties are advantageous. We simply need to understand the effects of muscle structure and nonlinear properties better. One should be cautious in interpreting results when linearized muscle models are used; they can limit one's view and, worse yet, may provide misinformation. This is especially true for muscles of the lower limb since muscle forces and velocities can vary over a wide range.

35.5.6 Roles for Multi-Articular Muscles

The recent exciting work regarding possible fundamental roles for lower limb multi-articular muscles will continue to progress. Differences of opinion, such as those expressed by the authors of Chapters 41 and 42 need to be resolved. We need to document task-specific features more accurately and develop a better understanding of the roles of multiarticular muscles for various skilled and stereotyped movements. Improved understanding of neurosensory integration of the information from multiarticular muscles will take a substantial effort. Advanced animal studies, such as the approach taken by Loeb's group (Chapter 10), are likely to become quite important.

35.5.7 A Move Toward Clinical Relevance

The movement biomechanist needs to be able to share information with not only research colleagues but also with other professionals interested in certain aspects of movement. It is expected that the biomechanist will also tend to converge toward areas of pursuit that can be funded. The area that appears most likely to experience growth, at least in the United States, is rehabilitation biomechanics. To meet these goals, the biomechanist will have to be able to provide information graphically and also bring professionals (such as the orthopedist, therapist, or coach) into the research process. It follows that these professionals will need to see causal results from their interaction, such as the effects produced due to a change in treatment or training procedure. This is another area where sensitivity analysis and particularly optimization performance criterion can be quite useful.

References

Abraham, L. D. and Loeb, G. E. (1985) The distal hindlimb musculature of the cat. *Exp. Brain Res.* 58: 580-593.

Adrian, M. J. and Laughlin, C. K. (1983) Magnitude of ground reaction forces while performing volleyball skills. *Biomechanics VIII-B* (Edited by Matsui, H. and Kobayashi, K.), pp. 903-914. Human Kinetics Publishers, Champaign, IL.

Alexander, R. McN. (1976) Estimates of speeds of dinosaurs. *Nature* 261: 129-130.

Alexander, R. McN. (1988) *Elastic Mechanisms in Animal Movement.* Cambridge Univ. Press, Cambridge.

Alexander, R. McN. and Bennet-Clark, H. C. (1977) Storage of elastic strain energy in muscles and other tissues. *Nature* 265: 114-117.

Alexander, R. McN. and Vernon, A. (1975) The dimensions of knee and ankle muscles and the forces they exert. *J. Hum. Movmt. Studies* 1: 115-123.

Andrews, J. G. (1987) The functional roles of the hamstrings and quadriceps during cycling: Lombard's Paradox revisited. *J. Biomech.* 20: 565-575.

Bach, T. M., Chapman, A. E. and Calvert, T. W. (1983) Mechanical resonance of the human body during voluntary oscillations about the ankle joint. *J. Biomech.* 16: 85-90.

Berme, N., Heydinger, G. and Cappozzo, A. (1987) Calculation of loads transmitted at the anatomical joints. *Biomech. of Engng.: Modelling, Simulation, Control* (Edited by Morecki, A.), pp. 89-131. Springer-Verlag, Wien.

Bosco, C., Komi, P. V. and Ito, A. (1981) Prestretch potentiation of human skeletal muscle during ballistic movement. *Acta physiol. Scand* 111: 135-140.

Bresler, B. and Frankel, J. P. (1950) The forces and moments in the leg during level walking. *Trans. Am. Soc. Mech. Engrs.* 72: 27-36.

Brooks, V. B. (1986) *The Neural Basis of Motor Control.* Oxford University Press, New York.

Bruggemann, P. (1985) Mechanical load on the achilles tendon during rapid dynamic sport movements. *Biomechanics: Current Interdisciplinary Research* (Edited by Perren, S. M. and Schneider, E.), pp. 669-674. Martinus Nijhoff, Boston.

Burdett, R. G. (1982) Forces predicted at the ankle during running. *Med. Sci. Sports Exercise* 14: 308-316.

Cavagna, G. A. (1970) Elastic bounce of the body. *J. appl. Physiol.* 29: 279-282.

Cavagna, G. A. and Kaneko, M. (1977) Mechanical work and efficiency in level walking and running. *J. Physiol.* 268: 467-481.

Cavanagh, P. R. and Lafortune, M. A. (1980) Ground reaction forces in distance running. *J. Biomech.* 13: 397-406.

Chao, E. Y., Laughman, R. K., Schneider, E. and Stauffer, R. N. (1983) Normative data of knee joint motion and ground reaction forces in adult level walking. *J. Biomech.* 16: 219-233.

Chapman, A. E. (1985) The mechanical properties of human muscle. *Exercise and Sport Sciences Reviews* (Edited by Terjung, R. L.), Vol. 13, pp. 443-501. Macmillan Publishing Company, New York.

Cordo, P.J. and Rymer, W. Zev. (1983) Contributions of motor-unit recruitment and rate modulation to compensation for muscle yielding. *J. Neurophys.* 47: 797-809.

Crowninshield, R. D., Brand, R. A. and Johnston, R. C. (1978a) The effects of walking velocity and age on hip kinematics and kinetics. *Clin. Orthop.* 132: 140-144.

Crowninshield, R. D., Johnston, R. C., Andrews, J. G. and Brand, R. A. (1978b) A biomechanical investigation of the human hip. *J. Biomech.* 11: 75-85.

Dahlkvist, N. J., Mayo, P. and Seedhom, B. B. (1982) Forces during squatting and rising from a deep squat. *Engng. Med.* 11: 69-76.

Elftman, H. (1939) Forces and energy changes in the leg during walking. *Am. J. Physiol.* 125: 339-356.

Fukashiro, S. and Komi, P. V. (1987) Joint moment and mechanical power flow of the lower limb during vertical jump. *Int. J. Sports Med.* 8: 15-21.

Fuller, J. J. and Winters, J. M. (1988) Estimated joint loading during exercises recommended by the arthritis foundation. *1988 Adv. in Bioengineering* (Edited by Miller, G. R.), pp. 159-162. American Society of Mechanical Engineers, New York.

Fung, Y. C. (1981) *Biomechanics: Mechanical Properties of Living Tissues.* Springer-Verlag, New York.

Galea, V. and Norman, R. W. (1985) Bone-on-bone forces at the ankle joint during a rapid dynamic movement. *Biomechanics IX-A* (Edited by Winter, D. A., Norman, R. W., Wells, R. P., Hayes, K. C. and Patla, A. E.), pp. 71-76. Human Kinetics Publishers, Champaign, IL.

Greene, P. R. and McMahon, T. A. (1979) Reflex stiffness of man's anti-gravity muscles during kneebends while carrying extra weights. *J. Biomech.* 12: 881-891.

Gregor, R. J., Komi, P. V. and Jarvinen, M. (1987) Achilles tendon forces during cycling. *Int. J. Sports Med.* 8: 9-14.

Gregor, R. J., Roy, R. R., Whiting, W. C., Lovely, R. G., Hodgson, J. A. and Edgerton, V. R. (1988) Mechanical output of the cat soleus during treadmill locomotion: In vivo vs in situ characteristics. *J. Biomech.* 21: 721-732.

Grillner, S. (1972) The role of muscle stiffness in meeting the changing postural and locomotor requirements for force development by the ankle extensors. *Acta Physiol. Scand.* **86:** 92-108.

Grillner, S. (1975) Locomotion in vertebrates: Central mechanisms and reflex interaction. *Physiol. Rev.* **55:** 247-304.

Gross, T. S. and Bunch, R. P. (1988) Measurement of discrete vertical in-shoe stress with piezoelectric transducers. *J. Biomed. Engrg.* **10:** 261-265.

Hamill, J., Bates, B. T., Knutzen, K. M. and Sawhill, J. A. (1983) Variations in ground reaction force parameters at different running speeds. *Hum. Movmt. Sci.* **2:** 47-56.

Harrison, R. N., Lees, A., McCullagh, P. J. J. and Rowe, W. B. (1986) A bioengineering analysis of human muscle and joint forces in the lower limbs during running. *J. Sports Sci.* **4:** 201-218.

Hatze H. (1978) A general myocybernetic control model of skeletal muscle. *Biol. Cybern.* **28:** 143-157.

Hay, J. G., Putnam, C. A. and Wilson, B. D. (1979) Forces exerted during exercises on the uneven bars. *Med. Sci. Sports* **11:** 123-130.

Hennig, E. M., Cavanagh, P. R., Albert, H. T. and Macmillan, N. H. (1982) A piezoelectric method of measuring the vertical contact stress beneath the human foot. *J. Biomed. Engrg.* **4:** 213-222.

Hof, A. L. and Van den Berg, Jw. (1981) EMG to force processing I: An electrical analogue of the Hill muscle model. *J. Biomech.* **14:** 747-758.

Hof, A. L., Geelen, B. A. and Van den Berg, Jw. (1983) Calf muscle moment, work and efficiency in level walking: Role of series elasticity. *J. Biomech.* **16:** 523-537.

Hogan, N., Bizzi, E., Mussa-Ivaldi, F. A. and Flash, T. (1987) Controlling multijoint motor behavior. *Exercise and Sport Sciences Reviews* (Edited by Pandolf, K. B.), Vol. 15, pp. 153-190. Macmillan Publishing Co., New York.

Ito, A., Komi, P. V., Sjodin, B., Bosco, C. and Karlsson, J. (1983) Mechanical efficiency of positive work in running at different speeds. *Med. Sci. Sports Exercise* **15:** 299-308.

Kilvington, M. and Goodman, R. M. F. (1981) In vivo hip joint forces recorded on a strain gauged "English" prosthesis using an implanted transmitter. *Engng. Med.* **10:** 175-187.

Komi, P. V. (1984) Physiological and biomechanical correlates of muscle function: effects of muscle structure and stretch-shortening cycle on force and speed. *Exercise and Sport Sciences Reviews* (Edited by Terjung, R. L.), Vol. 12, pp. 81-121. The Collamore Press, Lexington, MA.

Komi, P. V. and Norman, R. W. (1987) Preloading of the thrust phase in cross- country skiing. *Int. J.*

Sports Med. **8:** (Suppl. 1), 48-54.

Komi, P. V., Salonen, M., Jarvinen, M. and Kokko, O. (1987) In vivo registration of achilles tendon forces in man. I. Methodological development. *Int. J. Sports Med.* **8:** (Suppl. 1), 3-8.

Landjerit, B., Maton, B. and Peres, G. (1988) In vivo muscular force analysis during the isometric flexion on a monkey's elbow. *J. Biomech.* **21:** 577-584.

Larish, D. D., Martin, P. E. and Mungiole, M. (1988) Characteristic patterns of gait in the healthy old. *Ann. N. Y. Acad. Sci.* **515:** 18-32.

McMahon, T. A. (1984) *Muscles, Reflexes, and Locomotion.* Princeton University Press, Princeton, NJ.

Miller, D. I. and Nissinen, M. A. (1987) Critical examination of ground reaction force in the running forward somersault. *Int. J. Sport Biomech.* **3:** 189-206.

Mungiole, M. (1990) Factors influencing the mechanical output of the ankle plantar flexor muscles during concentric action, with and without prior stretching. Ph.D. Dissertation. Arizona State University.

Nistor, L., Markhede, G. and Grimby, G. (1982) A technique for measurements of plantar flexion torque with the Cybex II dynanometer. *Scand J. Rehab. Med.* **14:** 163-166.

Norman, R. W., Sharratt, M. T., Pezzack, J. C. and Noble, E. G. (1976) Reexamination of the mechanical efficiency of horizontal treadmill running. *Biomechanics V-B* (Edited by Komi, P. V.), pp. 87-93. University Park Press, Baltimore.

Norman, R. W. and Komi, P. V. (1979) Electromechanical delay in skeletal muscle under normal movement conditions. *Acta Physiol. Scand.* **106:** 241-248.

Norman, R. W. (1989) A barrier to understanding human motion mechanisms: A commentary. *Future Directions in Exercise and Sport Science Research* (Edited by Skinner, J. S., Corbin, C. B., Landers, D. M., Martin, P. E. and Wells, C. L.), pp. 151-161. Human Kinetics, Champaign, IL.

Olney, S. J. and Winter, D. A. (1985) Predictions of knee and ankle moments of force in walking from EMG and kinematic data. *J. Biomech.* **18:** 9-20.

Patriarco, A. G., Mann, R. W., Simon, S. R. and Mansour, J. M. (1981) An evaluation of the approaches of optimization models in the prediction of muscle forces during human gait. *J. Biomech.* **14:** 513-525.

Rack, P. M. H. and Ross, H. F. (1984) The tendon of flexor pollicis longus: its effects on the muscular control of force and position at the human thumb. *J. Physiol.* **351:** 99-110.

Robertson, D. G. E. and Winter, D. A. (1980) Mechanical energy generation, absorption and transfer amongst segments during walking. *J. Biomech.*

13: 845-854.

Rydell, N. (1965) Forces in the hip joint. Part (II) Intravital measurements. *Biomechanics and Related Bioengineering Topics* (Edited by Kenedi, R. M.), pp. 351-357. Pergamon Press, Oxford.

Smith, A. J. (1975) Estimates of muscle and joint forces at the knee and ankle during a jumping activity. *J. Hum. Movmt. Studies* 1: 78-86.

Tilney, F. and Pike, F.H. (1925) Muscular coordination experimentally studied in its relation to the cerebellum. *Archives of Neurology and Psychiatry* 13: 289-334.

van Ingen Schenau, G. J. (1984) An alternative view of the concept of utilisation of elastic energy in human movement. *Hum. Movmt. Sci.* 3: 301-336.

van Ingen Schenau, G. J. (1989) From rotation to translation: Constraints on multi-joint movements and the unique action of bi-articular muscles. *Hum. Movmt. Sci.* 8: 301-337.

van Ingen Schenau, G. J., Bobbert, M. F. and Rozendal, R. H. (1987) The unique action of bi-articular muscles in complex movements. *J. Anat.* 155: 1-5.

Walmsley, B., Hodgson, J. A. and Burke, R. E. (1978) Forces produced by medial gastrocnemius and soleus muscles during locomotion in freely moving cats. *J. Neurophysiol.* 41: 1203-1216.

Wells, R. P. (1988) Mechanical energy costs of human movement: an approach to evaluating the transfer possibilities of two-joint muscles. *J. Biomech.* 21: 955-964.

Whiting, W. C., Gregor, R. J., Roy, R. R. and Edgerton, V. R. (1984) A technique for estimating mechanical work of individual muscles in the cat during treadmill locomotion. *J. Biomech.* 17: 685-694.

Williams, K. R. and Cavanagh, P. R. (1983) A model for the calculation of mechanical power during distance running. *J. Biomech.* 16: 115-128.

Winter, D. A. (1984a) Biomechanics of human movement with applications to the study of human locomotion. *CRC Crit. Rev. Biomed. Engng.* 9: 287-314.

Winter, D. A. (1984b) Kinematic and kinetic patterns in human gait: variability and compensating effects. *Hum. Movmt. Sci.* 3: 51-76.

Winters, J. M. and Stark, L. (1985) Analysis of fundamental movement patterns through the use of indepth antagonistic muscle models. *IEEE Trans. biomed. Engrg.* BME-32: 826-839.

Winters, J. M. and Stark, L. (1988) Estimated mechanical properties of synergistic muscles involved in movements of a variety of human joints. *J. Biomech.* 21: 1027-1041.

Zajac, F. E. and Gordon, M. E. (1989) Determining muscle's force and action in multi-articular movement. *Exercise and Sport Sciences Reviews* (Edited by Pandolf, K. B.), Vol. 17, pp. 187-230. Williams and Wilkins, Baltimore.

CHAPTER 36

The Architecture of Leg Muscles

R. McN. Alexander and R. F. Ker

36.1 Introduction

Each of the four muscles shown in Figure 36.1 (a to d) consists of muscle fascicles (bundles of muscle fibers) connected at either end to tendons, but they show striking differences of architecture. Most authors would describe (a) and (c) as pennate, but (b) and (d) as parallel–fibered. It often seems convenient to use these adjectives, but the distinction that they make is not a sharp one: it is easy to imagine a continuous series of intermediates between (a) and (b) or between (c) and (d). It is sometimes suggested that the diagnostic feature of a pennate muscle is that its fascicles attach obliquely to the tendons. However, the cross-sectional areas of the tendons are always much less than the total of the cross-sectional areas of the muscle fascicles, so geometry requires that the attachment be oblique even in muscles such as (b) and (d) that would generally be described as parallel–fibered. Muscle (e) has fascicles that attach at one end directly to a bone rather than to a tendon.

The significant difference between the muscles of Figure 36.1 is that (a), (c), and (e) have short fascicles and relatively long tendons whereas (b) and (d) have long fascicles and short tendons. (We refer to fascicle length rather than fiber length because individual muscle fibers may not extend the whole length of the fascicles (Loeb et al., 1987). In this chapter we discuss the functional significance of this difference.

We also discuss the cross–sectional areas of tendons. We will show that some tendons are much stronger than seems necessary to transmit the forces exerted by their muscles, and will inquire why this should be.

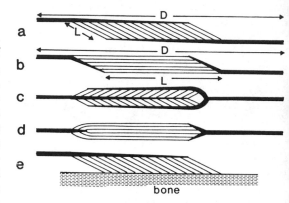

Figure 36.1: Diagrams of four muscles showing their tendons (thick lines) and fascicles.

36.2 Methods

The anatomical data of this paper come from dissections of about 40 species of mammal obtained in Kenya, Britain and (in a few cases) the USA. They range in size from small rodents of about 0.1 kg body mass to a 2500 kg elephant. Some were obtained from the wild, some had died in captivity, and a few were domestic animals. Leg muscles were dissected out and weighed individually, and fascicle lengths were measured. Weighted harmonic mean fascicle lengths were calculated for groups of muscles (Alexander et al., 1981). The total of the cross-sectional areas of the fascicles of each muscle, and the cross-sectional areas of tendons, were determined gravimetrically, and the stresses likely to act in tendons were estimated by assuming a stress of 0.3 MPa in the muscle fascicles (Ker et al., 1988). The effective length of tendon for each muscle with tendon at both ends (Figure 36.1, a to d) was estimated as

---

Multiple Muscle Systems, Biomechanics and Movement Organization
J.M. Winters and S.L-Y. Woo (eds.), © 1990 Springer-Verlag, New York

the difference between the overall length D (Figure 36.1 a,b) and the fascicle length L. If the angle between fascicles and tendon is less than about 30° (as it generally is), this is approximately the total length of tendon in series with each fascicle (note that cos 30° = 0.87). Rack and Westbury (1984) have shown for cat soleus muscle that the elastic compliance of the entire tendinous component could be calculated with acceptable accuracy by multiplying the compliance per unit length of the external part of the tendon by (D–L). This estimate of effective tendon length is not appropriate to pennate muscles that attach directly to bone (Figure 36.1e). [See also Section 4.3 of Chapter 4 (Ettema and Huijing).]

Further anatomical measurements were made on single species only. Freshly killed rabbits (*Oryctolagus cuniculus*) were arranged in positions matching selected frames from cine films of galloping. Rigor mortis was allowed to develop (to ensure that the muscles were taut), and sarcomere lengths were measured by diffraction of light (Dimery, 1985). Experiments were performed on the legs of dead horses (*Equus caballus*) to discover the relationships between muscle lengths and joint angles, so that the muscle length changes that occur during running could be calculated from measurements on films (Dimery et al., 1986).

Dynamic tensile tests were performed on various tendons from ten species of mammal, using an Instron servo-hydraulic dynamic testing machine (Bennett et al., 1986).

36.3 Results and Discussion

36.3.1 Fascicle Lengths

Figure 36.2 shows the principal muscles of the hind leg of a typical mammal. It shows the attachments of the muscles and gives a general indication of the differences of fascicle length, which are the subject of this subsection.

It will be convenient to group some of these muscles together in presenting results. We will group the hamstrings (biceps femoris, semi–tendinosus, semimembranosus, and gracilis) with the other large muscle of the back of the thigh, the adductor femoris. The quadriceps group consists of the rectus femoris and vasti. By "ankle extensors" we mean gastrocnemius, soleus, and plantaris, and when we refer to "toe flexors" we exclude the plantaris although it flexes the digits as well as extending the ankle.

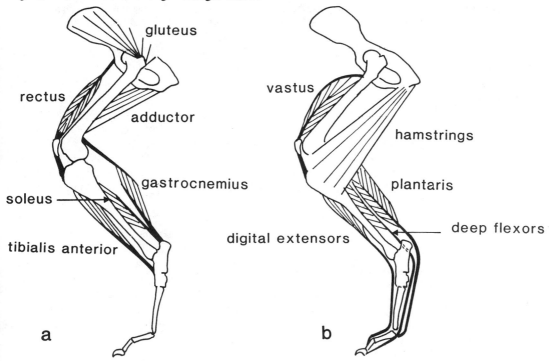

Figure 36.2: Diagrams of the skeleton and some of the muscles of the hind leg of a typical mammal.

Table 36.1: Relative fascicle lengths (mean ± standard deviation) for major muscles and muscle groups in mammals investigated by Alexander et al. (1981). Fascicle lengths have been divided by femur length (hind leg muscles) or humerus length (foreleg muscles). The muscle groups are defined in the text.

	Primates (6 species)	Fissipedia (8 species)	Bovidae (8 species)
Gluteus superficialis	0.28 ± 0.07	0.42 ± 0.15	0.44 ± 0.07
Hamstrings & adductor	0.41 ± 0.12	0.58 ± 0.09	0.62 ± 0.09
Quadriceps	0.17 ± 0.03	0.21 ± 0.06	0.25 ± 0.05
Ankle extensors	0.12 ± 0.02	0.13 ± 0.07	0.06 ± 0.01
Toe flexors	0.13 ± 0.05	0.11 ± 0.05	0.09 ± 0.02
Triceps	0.23 ± 0.06	0.28 ± 0.05	0.40 ± 0.06
Wrist flexors	0.18 ± 0.01	0.09 ± 0.04	0.05 ± 0.01

Muscle fascicles lengthen and shorten as the animal moves its legs but we will show in the next subsection that these length changes are generally small compared to the differences that we will demonstrate between different muscles in the same leg. For many muscles of the thigh, the overall length D (Figure 36.1) is approximately equal to the length of the femur. For some muscles of the lower leg, the overall length is approximately equal to the length of the tibia, which in all but one of the species for which we present data was between 0.82 and 1.32 times the femur length. (The exception is the elephant *Loxodonta* for which the ratio was 0.58.) We therefore give fascicle lengths of hind limb muscles as fractions of femur length: we will call this fraction the ***relative fascicle length***.

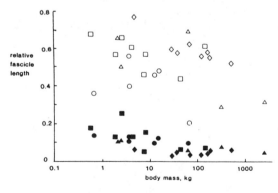

Figure 36.3: Relative fascicle lengths of the hamstring and adductor muscles (open symbols) and of the ankle extensors (filled symbols), plotted against body mass, for mammals studied by Alexander et al. (1981): ○ ● Primates; □ ■ Fissipedia; ◇ ◆ Bovidae; △ ▲ others.

Figure 36.3 shows relative fascicle lengths for two muscle groups, plotted against body mass. Several points seem clear. First, the hamstrings and adductor invariably have much longer fascicles than the ankle extensors. Secondly, there are differences between major groups of mammals: most notably, the ankle extensors have smaller relative fascicle lengths in antelopes, etc. (Bovidae) than in primates. Finally, there is no marked dependence of relative fascicle length on body mass. Homologous muscle groups generally have approximately equal relative fascicle lengths in related mammals of different sizes.

Table 36.1 includes data for more muscle groups. It shows that gluteus superficialis and the hamstrings and adductor generally have longer fascicles than quadriceps, which in turn has longer fascicles than the ankle extensors and toe flexors. The ankle extensor fascicles are especially short in Bovidae.

Table 36.1 also includes some data for muscles of the foreleg. For these, humerus length rather than femur length has been used to calculate relative fascicle length. The table shows that the triceps (the extensor muscle of the elbow) has longer fascicles than the "wrist flexors," in which group we include the digital flexors as well as carpal flexors. The wrist flexors have especially short fascicles in Bovidae.

Some other ungulate mammals as well as Bovidae have very short fascicles in some leg muscles. The most extreme examples are found in horses and camals (Dimery et al., 1986).

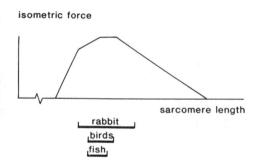

isometric force

sarcomere length

rabbit

birds

fish

Figure 36.4: A schematic graph of isometric force against sarcomere length for vertebrate striated muscle. Corresponding points on the graph would occur at different sarcomere lengths, in different species. Bars below the graph indicate the ranges used by leg muscles of galloping rabbits (Dimery, 1985), wing muscles of birds (Cutts, 1986), and red swimming muscles of carp (*Cyprinus carpio*; Rome et al., 1988).

36.3.2 Fascicle Length Changes

A vertebrate striated muscle fiber can contract to short lengths or be stretched to long ones, but can exert large forces only over a limited range of lengths. A short muscle fascicle (i.e. one composed of a small number of sarcomeres in series), required to work over a wide range of length, could exert only small forces in parts of the range, but a longer fascicle could exert large forces over the whole range. It will help us to understand the differing fascicle lengths of different muscles, if we know which parts of their force–length curves muscles actually use.

Dimery (1985) measured sarcomere lengths in leg muscles, in rabbit carcases arranged in positions imitating galloping. She concluded that most of the muscles worked in the sarcomere length range 1.7 to 2.7 µm, the range in which 80% or more of maximum force can be exerted. This result, and the results of similar studies of bird flight muscles and fish swimming muscles are shown in Figure 36.4. They give the impression that muscles generally work mainly on the ascending limb and plateau of the force-length curve. Poliacu Prosé's (1985) work on cat leg muscles gives a similar impression.

All these measurements are subject to possible error if the forces in the muscles were different from those that would act in the same limb position in locomotion, because the forces cause

elastic extension of tendons [e.g. Chapter 5 (Winters); Chapter 4 (Ettema and Huijing)]. This problem is unlikely to be important in muscles such as the hamstrings, which have long fascicles and short tendons, but Loeb et al. (1987) have suggested that length changes of the short fibers of such muscles may not be proportional to the length changes of the fascicles if fibers of the same motor unit are not connected directly in series. In that case, elastic extension and recoil of the epimysial matrix might contribute appreciably to fascicle length changes. This seems unlikely to be important in strenuous activities if most or all of the motor units are recruited so that only short lengths of epimysium are interpolated between active units.

Hoffer et al. (1989) avoided the problem of tendon stretching in their study of the cat gastrocnemius by using implanted piezo-electric crystals to measure fascicle length changes as the animal walked. In a typical experiment they found that a fascicle fluctuated in length between about 15 and 22 mm during each stride.

Griffiths (1989) fitted buckle transducers to the medial part of the gastrocnemius tendon of wallabies (*Thylogale billardierii*) so that he could record the force exerted by the medial gastrocnemius muscle during hopping. He also filmed the wallabies hopping and analyzed the film to determine the changes in overall length of the muscle with its tendon. He made tensile tests on excised tendons to discover how much the tendons would stretch under the forces indicated by the buckle transducer. Finally, he subtracted the tendon length changes from the overall length changes to determine the length changes of the fascicles. He found that the length of the fascicles fluctuated, during the part of the stride in which they were active, through a range of about 6 mm. Their length at the minimum of the range was probably about 19 mm (Morgan et al., 1978).

These data suggest that, in the locomotion of mammals, the fascicles of limb muscles generally work over ranges in which the minimum length is about 0.7 times the maximum.

36.3.3 Tendon Length Changes

Mechanical tests on tendons show no obvious systematic differences in properties between mammalian species or anatomical sites (Bennett et al., 1988). The ultimate tensile strength is about 100

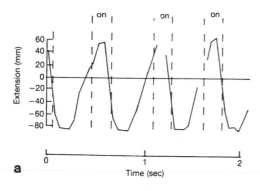

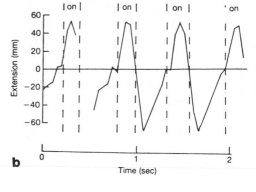

Figure 36.5: Graphs showing how the lengths of (a) the plantaris muscle and (b) the superficial flexor of the digit of the foreleg of a horse fluctuate during galloping. From Dimery et al. (1986).

MPa (or possibly a little higher) and tendons stretch by about 8% before breaking. [See also Chapter 5 (Winters); Chapter 4 (Ettema and Huijing).]

The remainder of this subsection concerns tendons that suffer substantial strains in normal activities. We will show later that many tendons suffer much smaller maximum strains.

Alexander and Vernon (1975a) made force plate records and films of a wallaby (*Macropus rufogriseus*) hopping, and calculated the forces in the principle leg muscles. Ker et al. (1986) reanalysed their data and made tensile tests on the tendons in question. We calculated that peak stresses in the gastrocnemius, plantaris, and deep digital flexor tendons were about 41, 35 and 15 MPa, in slow hopping. These would stretch the tendons by about 4, 3 and 2% respectively, of their lengths (see Figure 1.6 of Alexander, 1988). Griffiths' (1989) records indicate that the medial gastrocnemius tendon of *Thylogale* stretches by 3.2% in slow hopping and 4.4% at a higher speed.

Some of the leg muscles of horses and camels (*Camelus dromedarius*) have long tendons and exceedingly short muscle fascicles. Camp and Smith (1942) realized that some of these tendons must stretch substantially and recoil in each running stride. Dimery et al. (1986) measured the changes in overall length (D, Figure 36.1) of some of these muscles during walking, trotting and galloping. Figure 36.5 shows some of our results. The muscle belly of the plantaris is vestigial, with fascicles only 1–2 mm long, and that of the superficial digital flexor has 3 mm fascicles. Length changes of these short fascicles can have contributed very little to the length changes shown in Figure 36.5. In these graphs, zero extension

means that the tendon was only just taut. The negative extensions of up to –80 mm mean that during part of the stride, when the foot was off the ground, the tendons were slack and must have been folded. While the foot was on the ground the plantaris tendon stretched by 50 mm (6% of its length) and the superficial flexor by 60 mm (9%). Nine per cent seems improbably high, as tendons generally break in tensile tests at about 8% strain, and may perhaps have been enlarged by an unidentified error. The tendons extend less in slower gaits. Dimery et al. (1986) showed that these estimates of tendon strain were generally consistent with measurements of tendon properties and estimates of tendon stresses reported in earlier papers.

The peak force in the Achilles tendon of a 70 kg man running at a middle-distance speed is about 4700 N, and the cross-sectional area of the tendon is about 89 mm^2 (Ker et al. 1987). Thus the peak stress is about 53 MPa, and the tendon must be stretched by about 5% (refer again to Figure 1.6 of Alexander, 1988).

These studies suggest that some leg tendons of many mammals may be stretched by around 5% during running. However, we will show in Section 36.3.5 that many other muscles are incapable of exerting the forces required to stretch their tendons so much.

This and the previous subsection show that many muscle fascicles shorten and lengthen by about 30% during running and that some tendons stretch and recoil by about 5%. If the fascicles are less than about one sixth as long as the tendons, their length changes may contribute less to the movements of the joints than do the length changes of the tendons. Many leg muscles have fascicles as short as this: many of the muscles in Table 36.1 have fascicles less than 0.15 times as long as the femur or humerus although their overall lengths (including tendons) are at least equal to the lengths of these bones.

36.3.4 Tendon Compliance Saves Energy

An animal running steadily on level ground needs very little net work in each stride, merely enough to overcome aerodynamic drag, friction in the joints and tissue viscosity. However, the leg muscles of mammals do much larger quantities of work at some stages of the stride and act as brakes, degrading mechanical energy to heat, at other stages (e.g. see Alexander and Vernon, 1975a). Tendons that stretch (removing kinetic energy from the body) at one stage of the stride and recoil (returning it) at another can largely perform the tasks that would otherwise be required of muscle fascicles. (See also Chapter 37 (McMahon) and Chapter 38 (Hof).)

Muscles use metabolic power whenever they are active, exerting tension. They use more metabolic power when they are shortening (doing work) and less when they are lengthening (acting as brakes) under the same force (Woledge et al., 1985; Heglund and Cavagna, 1987). Because work and braking almost balance, we can ignore this difference and assume that the metabolic energy consumed during a stride is approximately proportional to the time integral of muscle force. However, more metabolic power is needed to maintain the same force in long muscle fascicles than in short ones, and in fast fascicles than in slow ones. Tendons that contribute to the length changes required of muscles, by stretching and recoiling, do not alter the forces that the muscles have to exert but may enable the animal to make do with shorter or slower fascicles, and so reduce the metabolic power required for running.

If large energy savings are to be made, the tendons must stretch to large strains, as the following argument shows. The radius of the articular surfaces in a joint sets a lower limit to the moment arms of muscles working the joint. The mid-shaft radii of the long bones of mammals are typically about 4% of bone length (Alexander et al., 1979), and the radii of the articular surfaces at the distal ends of the bones are probably about as large. Thus the moment arms of muscles are unlikely to

be less than 4% of the lengths of the bones alongside which they lie. Some muscles, such as the gastrocnemius at the ankle, have much larger moment arms (Alexander et al., 1981). The tendons of muscles that have short fascicles are generally almost as long as the bones alongside which they run. Thus if tendon stretching is to allow joint angle changes of around one radian, such as occur commonly in running (for example Goslow et al., 1973), the tendons must stretch by at least 4%. This requires stresses that are large fractions of the ultimate tensile stress.

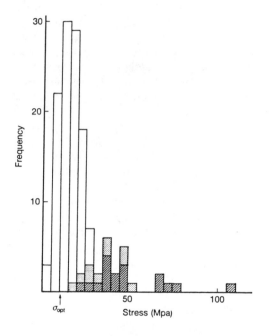

Figure 36.6: The distribution of maximum stresses among limb tendons of mammals. Stippling and hatching indicate tendons that would be stretched by more than one quarter and one half of the length of the muscle fascicles, respectively, when these stresses acted. (From Ker et al. (1988); reprinted with permission.)

36.3.5 Tendon Thickness

Ker et al. (1988) measured various limb muscles and their tendons from ten diverse species of mammal. We calculated the stresses that would act in the tendons when the muscles exerted their maximum isometric forces. Figure 36.6 shows that for most of the tendons this stress was 5-25 MPa but for some (including those discussed in

Section 36.3.3) it was much larger. The ultimate tensile stress of tendon is 100 MPa or more (see above), so the lightly stressed tendons seem very much thicker than is necessary to transmit the forces that their muscles can exert. Why are they so thick?

The highly stressed tendons are the ones that seem to serve as energy–saving springs, as discussed in Section 36.3.4, where we argued that such tendons *must* be highly stressed. Shading in Figure 36.6 indicates tendons that would be stretched (by maximum isometric forces) by more than one quarter of the length of the muscle fascicles. The information on fascicle length changes in Section 36.3.2 suggests that for most of these muscles, tendon stretching is likely to contribute more than fascicle length changes to movement at the joints. These are the muscles that seem adapted to save metabolic energy by serving as springs.

The most thoroughly studied example of a lightly stressed tendon is the tendon of insertion of the human flexor pollicis longus (Rack and Ross, 1984). The muscle belly is in the forearm but it serves to bend the interphalangeal joint of the thumb, so its force has to be transmitted by a long tendon. It can exert forces up to 140 N (calculated from Figure 3D of Brown et al., 1982), which imposes a stress of 15 MPa on the tendon. This is only about one sixth of the ultimate tensile stress but is enough to stretch the tendon by 1.7%. The muscle originates directly on bone as in Figure 36.1e so (D-L) is not a good estimate of the effective length of its tendon: 170 mm seems a reasonable estimate. Thus the tendon may be stretched by about 2.9 mm, enough to allow 21° movement at the thumb joint. If the tendon were more highly stressed and so stretched more, it would be very difficult to control the position of the joint when fluctuating forces acted on the thumb. On the other hand, if the tendon did not stretch it would be difficult to control the force exerted by the thumb (Rack and Ross, 1984).

It is not obvious what the best compromise will be between the requirements of position control and force control, but a different approach suggests optimum thicknesses for tendons (Ker et al., 1988). Suppose that a typical task for a muscle is to exert maximum isometric force and move a joint to a particular position. A thinner tendon would stretch more than a thick one and require

the muscle fascicles to shorten more to take up this stretch. If the fascicles are required to shorten more they should be longer, and so have greater mass. Thus reduction of tendon mass requires increase of fascicle mass, and vice versa. We argued that the combined mass of tendon plus fascicles would be minimized when the tendon had a particular cross-sectional area. The stress in the tendon, when the muscle exerted its maximum isometric force, would then be about 10 MPa. It may be argued that tendons should be thicker than this, and exposed to even lower stresses, because a given mass of tendon uses less metabolic energy than an equal mass of fascicles. However, the mode in Figure 36.6 is about 13 MPa, reasonably close to the prediction of the simple theory.

36.3.6 Location and Architecture

The data of this paper suggest that we should distinguish three main types of limb muscle. Typical examples of the three types are very different from each other and tend to be found in different parts of the limb, but intermediates occur, and lack of data makes it difficult to place some muscles in the classification.

Type (i) muscles have long fascicles and relatively short tendons: the hamstring muscles are good examples. They include the largest muscles of the limb which, because of their volume of muscle fascicles, can do more work than smaller muscles. Remember that the work that a fascicle can do is the product of the force it exerts (proportional to its cross-sectional area) and the distance it shortens (proportional to its length), so this work is proportional to its volume. When an animal accelerates or jumps, its muscles perform net work, which must be done largely by these muscles. When it decelerates it also requires large muscles to degrade kinetic energy to heat. The large *type (i)* muscles are found in the proximal segments of limbs where their mass adds less to the limb's moment of inertia, than if they were located more distally. The greater the moment of inertia of the limb about its proximal end, the larger the forces that muscles must exert to accelerate and decelerate it as it swings forward and back. This seems bound to increase the metabolic energy cost of locomotion, although Taylor et al. (1974) were unable to demonstrate the effect in their comparison of cheetahs, gazelles and goats.

Type (ii) muscles have relatively thick, lightly stressed tendons that are longer than the muscle fascicles. The human flexor pollicis longus is one good example.   Others are the digital extensor muscles of both fore and hind limbs, of various mammals (Ker et al., 1988). These muscles are generally not concerned in supporting the weight of the body, so the compliance of their tendons cannot save energy in the manner envisaged in Section 36.3.4. Many of these muscles operate joints that are remote from the muscle belly, for example muscles in the human forearm that operate the hand. If the bellies of these muscles were in the hand, the moment of inertia of the arm would be higher and the hand would be inconveniently bulky. Remote operation requires long tendons, which would stretch by large amounts if they were highly stressed. We discussed the advantages of relatively thick tendons for such situations in Section 36.3.5.

Type (iii) muscles have relatively slender, highly stressed tendons that are much longer than the muscle fascicles. In extreme cases such as the horse plantaris the fascicles may be rudimentary. The anti-gravity muscles of the lower leg and forearm of ungulate mammals all belong to *type (iii)*, but homologous muscles of some other mammals do not.   For example, the plantaris and superficial and deep digital flexors have highly stressed tendons and so belong to *type (iii)* in horses and sheep but have lightly stressed tendons, and so are better placed in *type (ii)*, in monkeys (Ker et al., 1988). *Type (iii)* muscles are the ones involved in energy saving by tendon compliance (Section 36.3.4). They are generally more distally placed than *type (i)* muscles but because of their smaller mass do not add unduly to the moment of inertia of the limb. Many *type (iii)* muscles have fascicles *less than one fifth* as long as *type (i)* muscles in the same limb (compare the ankle extensors and hamstrings of Bovidae, Table 1), so the same force can be exerted by a muscle whose belly has only one fifth the mass.

There is another reason, in addition to that of moment of inertia, why *type (i)* muscles should be proximal and *type (ii)* muscles distal. When mammals run at constant speed, the forces on the feet remain approximately in line with the legs (Figure 36.7a), so their moments about proximal limb joints are relatively small. When they accelerate or take off for a jump, however, the forces on the hind feet must have large forward components and so must exert large moments about the hip (Figure 36.7b). *Type (i)* hip extensor muscles (the hamstrings and adductor) are well placed to do the work of accelerating the animal but need not exert large forces in steady running.

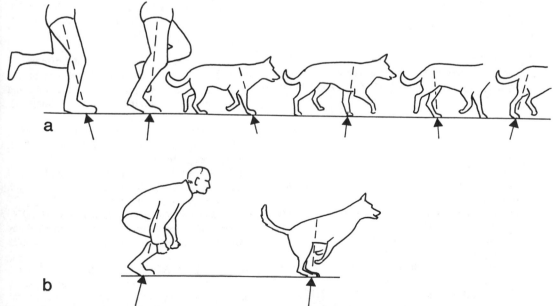

Figure 36.7: The directions of forces on the feet of a dog and a man during *(a)* running and *(b)* takeoff for a standing jump. Based on force plate records and simultaneous films made by Alexander (1974) and Alexander and Vernon (1975b).

36.4 Future Directions

Our information about the ranges of sarcomere length that are used in normal activities is fragmentary, and we have no quantitative theory of the trade-offs that determine the optimum range. Plainly, if a very narrow range were used, the muscle fascicles would have to be long and the metabolic cost of exerting tension would be high, whereas if a wide range were used the fascicles could be short, but little force could be exerted near the extremes of the range, but we have no theory to predict just what the optimum range should be. Our knowledge of the strains that occur in tendons in normal use is limited to a few cases, and some of these depend on calculations based on doubtful assumptions. We would like to have more and better data. In particular, we would like more data about tendon strain in muscles such as the triceps and vastus that have long aponeuroses but only short external tendons. It also seems desirable to extend our studies in a newer direction. Limb muscles are composed of fascicles of different intrinsic speeds that are recruited in sequence as the animal performs increasingly strenuous activities, but we do not know how the spectrum of speeds in any particular limb muscle is matched to that muscle's normal range of activities. Rome et al. (1988) showed how the slow red fibers and the fast white ones of fish swimming muscle both have intrinsic speeds adapted to their tasks, but there has been no similar study of limb muscles.

References

Alexander, R.McN. (1974) The mechanics of jumping by a dog (*Canis familiaris*). *J. Zool., Lond.*, **173**: 549-573.

Alexander, R.McN. (1988) *Elastic Mechanisms in Animal Movement*. Cambridge University Press, Cambridge, England.

Alexander, R.McN., Jayes, A.S., Maloiy, G.M.O. and Wathuta, E.M. (1979) Allometry of the limb bones of mammals from shrews (*Sorex*) to elephant (*Loxodonta*). *J. Zool., Lond.*, **189**: 305-314.

Alexander, R.McN., Jayes, A.S., Maloiy, G.M.O. and Wathuta, E.M. (1981) Allometry of the leg muscles of mammals. *J. Zool., Lond.*, **194**: 539-552.

Alexander, R.McN. and Vernon, A. (1975a) The mechanics of hopping by kangaroos (Macropodidae). *J. Zool., Lond.*, **177**: 265-303.

Alexander, R.McN. and Vernon, A. (1975b) The dimensions of knee and ankle muscles and the

forces they exert. *J. Human Movt. Stud.*, **1**: 115-123.

Bennett, M.B., Ker, R.F., Dimery, N.J. and Alexander, R.McN. (1986) Mechanical properties of various mammalian tendons. *J. Zool, Lond. (A)*, **209**: 537-548.

Brown, T.I.H., Rack, P.M.H. and Ross, H.F. (1982) Forces generated at the thumb interphalangeal joint during imposed sinusoidal movements. *J. Physiol.*, **332**: 69-85.

Camp, C.L. and Smith, N. (1942) Phylogeny and functions of the digital ligaments of the horse. *Mem. Univ. Calif.*, **13**: 69-124.

Cutts, A. (1986) Sarcomere length changes in the wing muscles during the wing beat cycle of two bird species. *J. Zool., Lond. (A)*, **209**: 183-185.

Dimery, N.J. (1985) Muscle and sarcomere lengths in the hind limb of the rabbit (*Oryctolagus cuniculus*) during a galloping stride. *J. Zool. Lond. (A)*, **205**: 373-383.

Dimery, N.J., Alexander, R.McN. and Ker, R.F. (1986) Elastic extension of leg tendons in the locomotion of horses (*Equus caballus*). *J. Zool., Lond. (A)*, **210**: 415-425.

Goslow, G.E., Reinking, R.M. and Stuart, D.G. (1973) The cat step cycle: hind limb joint angles and muscle lengths during unrestrained locomotion. *J. Morph.*, **141**: 1-42.

Griffiths, R.I. (1989) The mechanics of the medial gastrocnemius muscle in the freely hopping wallaby (*Thylogale billardierii*). *J. Exp. Biol.*, **147**: 439-456.

Heglund, N.C. and Cavagna, G.A. (1987) Mechanical work, oxygen consumption and efficiency in isolated frog and rat striated muscle. *Am. J. Physiol.*, **253**: C22-C29.

Hoffer, J.A., Caputi, A.A., Pose, I.E. and Griffiths, R.I. (1989) Roles of muscle activity and load on the relationship between muscle spindle length and whole muscle length in the freely walking cat. *Progr. Brain Res.* **80**: 75-85.

Ker, R.F., Alexander, R.McN. and Bennett, M.B. (1988) Why are mammalian tendons so thick? *J. Zool., Lond.*, **216**: 309-324.

Ker, R.F., Bennett, M.B., Bibby, S.R., Kester, R.C. and Alexander, R.McN. (1987) The spring in the arch of the human foot. *Nature*, **325**: 147-149.

Ker, R.F., Dimery, N.J. and Alexander, R.McN. (1986) The role of tendon elasticity in hopping in a wallaby (*Macropus rufogriseus*). *J. Zool., Lond. (A)*, **208**: 417-428.

Loeb, G.E., Pratt, C.A., Chanaud, C.M. and Richmond, F.J.R. (1987) Distribution and innervation of short, interdigitated muscle fibers in parallel-fibered muscles of the cat hindlimb. *J. Morph.* **191**: 1-15.

Morgan, D.L., Proske, U. and Warren, D. (1978) Measurements of muscle stiffness and the mechanism of elastic storage of energy in hopping kangaroos. *J. Physiol.,* **282:** 253-261.

Poliacu Prosé, L. (1986) De functionele stabiliteit van de knie van de kat. Thesis, Free University of Brussels, Belgium.

Rack, P.M.H. and Ross, H.F. (1984) The tendon of flexor pollicis longus: its effects on the muscular control of force and position at the human thumb. *J. Physiol.,* **351:** 99-110.

Rack, P.M.H. and Westbury, D.R. (1984) Elastic properties of the cat soleus muscle and their functional importance. *J. Physiol.,* **347:** 479-495.

Rome, L.C., Funke, R.P., Alexander, R. McN., Lutz, G., Aldridge, H., Scott, F. and Freadman, M. (1988) Why animals have different muscle fiber types. *Nature,* **335:** 824-827.

Taylor, C.R., Shkolnik, A., Dmi'el, R., Baharav, D. and Borut, A. (1974) Running in cheetahs, gazelles and goats: energy costs and limb configuration. *Am. J. Physiol.,* **227:** 848-850.

Woledge, R.C., Curtin, N.A. and Homsher, E. (1985) *Energetic Aspects of Muscle Contraction.* Academic Press, London.

CHAPTER 37

Spring-Like Properties of Muscles and Reflexes in Running

Thomas A. McMahon

37.1. Introduction

37.1.1 Early Evidence for Spring-Like Events in Running

The idea that running is essentially bouncing may be an ancient one, but the first good quantitative evidence for it was presented by Cavagna et al. (1976). Force–plate measurements of steady-speed human walking and running were used to calculate fluctuations of the kinetic energy and the gravitational potential energy of the body's center of mass. In walking, the calculations showed that changes in forward kinetic energy were out of phase with changes in gravitational potential energy, so that the total mechanical energy of the center of mass was nearly steady throughout a walking step. In running, the kinetic and gravitational energy fluctuations were found to be in phase, so that large changes in the sum of the two took place during a step. A simple way to put this is to say that in walking, the body is highest at the moment the forward kinetic is least, in mid-step (when the hips pass over the ankles). In running, as may be seen in the strobe photograph in Figure 37.1, the body is lowest in mid-step, when the forward kinetic energy is least.

Cavagna and his collaborators pointed out that in walking, the exchange of energy between potential and kinetic forms means that not all the energy required to lift and accelerate the body has to come from the muscles. In running, however, gravitational and kinetic energy exchange is not possible because the center of mass is losing both height and speed at the same time. Thus, if potential energy is to be stored and re-used during the running cycle, it must be stored in an elastic, rather than a gravitational form.

Figure 37.1: Strobe photograph of a subject running at constant speed. The strobe frequency was 8.4 Hz. The white lines on the floor are 1.0 m apart. (From McMahon et al. (1987); reprinted with permission.)

37.1.2 Measurement of Spring-Like Properties in Muscles and Reflexes

1. Tension-length curves in isolated muscle

A fundamental property of muscle is that it is capable of bearing a greater force at longer lengths than at shorter lengths. Furthermore, a muscle stimulated to either partial or fully fused tetanus can bear a higher force than an unstimulated (passive) muscle at the same length. The developed tension is the difference between the tension measured while a muscle is active and the tension measured while it is passive at a particular length. The fact that the developed tension has a maximum near the length a muscle is found in the body (the rest length) and declines at longer lengths has been used as a point of evidence in favor of the sliding–filament model of muscle contraction (Gordon et al., 1966).

Multiple Muscle Systems: Biomechanics and Movement Organization
J.M. Winters and S.L-Y. Woo (ed.), © 1990 Springer-Verlag

2. Reflex stiffness

The skeletal muscles of the limbs contain specialized structures known as stretch receptors that are known to report information to the spinal cord concerning the magnitude and rate of local length changes taking place in a muscle. In addition, Golgi tendon organs at the junction between a muscle and its tendon report force information to the spinal cord. Since the discovery and description of these organs, physiologists have wondered what roles they might play in the control of motor movements. Results from several classes of experiments have suggested that the motor servo including both types of receptors regulates and maintains the ratio of force change to length change, rather than controlling either the force or length separately (Houk, 1979).

Experiments done in cats decerebrated at the premammillary level corroborated the idea that stiffness is maintained nearly constant by the action of spinal reflexes (Hoffer and Andreassen, 1978, 1981). In decerebrate preparations like the one used by Hoffer and Andreassen, the force level in the various muscles changes spontaneously over a wide range during the several hours the preparation is viable, but stays at a nearly constant level for many seconds at a time. The investigators used a torque motor to apply small length perturbations (1 mm) for short durations (200–500 ms) in order to evaluate the stiffness of the muscle by dividing the change in force by the change in length. The results (Figure 37.2) show that reflex effects combine with intrinsic properties to give a stiffness that increases rapidly with force at low levels, then reaches a plateau that varies by less than 15% when the operating force ranges between 25% and 100% of its maximum value. By contrast, when the muscle was isolated by cutting the soleus nerve and applying an electrical stimulation to the cut end, the stiffness measured by the torque motor technique over the same range of force levels was lower than with the reflex intact.

3. Cavagna's "elastic bounce of the body"

Giovanni Cavagna was able to demonstrate the reflex stiffness of calf muscles in an elegant way (Cavagna, 1970). Subjects jumped on a force platform, landing on the balls of the feet (M–P joints) while keeping the knees fully extended. The force record showed a ringing, damped trace that was used to extract a spring constant and damping ratio. The values reported for the damping ratio (average = 0.20) illustrate that, at least in this exercise, the reflex stiffness of the muscles behaves as a fairly lightly-damped spring.

4. Board bouncing experiments

Running involves more muscles than just those acting to plantar-flex the ankle. In an effort to characterize the reflex stiffness of all the antigravity muscles of the leg, Greene and McMahon (1979) measured the vertical vibratory motions of subjects standing on wooden springboards of

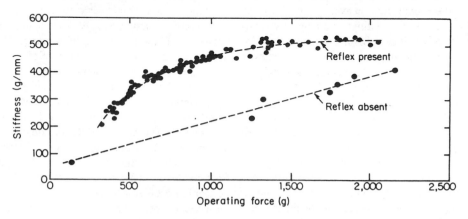

Figure 37.2: Measurements of stiffness vs. force in the soleus muscle of a decerebrate cat. From Hoffer and Andreassen (1978).

known spring stiffness. The subjects held their knees fixed at certain angles while executing up-and-down bouncing motions of small amplitude. It was found that shrugging the shoulders or making small flexions and extensions of the back was enough to excite bouncing deflections of the springboard of an inch or so in amplitude, providing that the excitation was carried out at one particular frequency; excitation at other frequencies had little effect. The subjects carried barbell weights on their shoulders in order that the steady force level on the legs could be varied. When the subjects were not carrying weights, they folded their arms rigidly across the chest to keep their arms from becoming an additional sprung mass. It was possible to excite either the parallel mode of oscillation, where both the subject and the springboard were moving downward at the same time, or the anti-parallel mode where the subject was moving down as the board was moving up (and vice versa), but it was not possible to excite both modes at one time.

The frequencies of both the parallel and anti-parallel modes were measured, and the results were analyzed using a simple theoretical model assuming a separate lumped mass for the board and the man, an undamped spring for the board and a damped spring for the man. The calculations showed that the spring stiffness of the leg was about 60.2 kN/m when the angle of the thigh with respect to the horizontal was about 70°, a value representative of the midstance posture during running. The stiffness depended very much on the angle of the knee; it fell, on average, to 37.6 kN/m when the angle of the thigh with the horizontal was 45°. A significant finding was that the stiffness increased, but only by a very little when extra weights were added to the subject's shoulders. One man increased the weight on his shoulders by 200 lbs without changing his stiffness. On average, the stiffness increased by less than 5% as subjects put an additional weight equal to their body weight on their shoulders.

The stiffness for bouncing on one leg was found, on average, to be 19% less than for bouncing on two. This finding will be important later as we attempt to understand the role of the stiffness of the trunk as well as that of the legs in running.

5. Preliminary conclusions concerning reflex stiffness

The experimental evidence presented above demonstrates how a single muscle or group of muscles can behave like a spring. Furthermore, it shows how reflexes (most importantly the stretch reflex) act to increase muscle stiffness and maintain that stiffness at a level nearly independent of the force at moderate and high force levels. The board bouncing experiments showed that although the leg spring stiffness is nearly independent of the force, it depends on the knee angle because the mechanical advantage of the ground reaction force acting about the knee changes with knee angle.

37.2. Leg Springs and Robots

37.2.1 The McMahonimal

In 1976, an undergraduate, Michael Jacker, wanted a design project. I suggested he build a machine that would run. The final product is shown in Figure 37.3. It had four legs, each of which had an elbow or ankle with a spring acting to keep it extended. The rotary motions of the hips and shoulders were coupled using large gears so that moving the rear legs backward caused the front legs to move forward and vice versa. A microswitch was positioned to close when the ankle joints dorsiflexed past a certain angle; this caused a solenoid to open, sending pressurized nitrogen to an air cylinder that pulled a tendon, thereby plantar-flexing the ankles and extending the hips. The machine was mounted on the end of a counterweighted boom. As long as the machine was held off the ground, it did nothing. When it was dropped on its feet, it ran in a circle around the central support. Without asking me if I liked the idea, Jacker named it the McMahonimal.

It worked, but I never thought it was a very good simulation of an animal. It had only one forward speed. We had to fiddle for some time with the placement of the microswitch and the nose-down angle of the body before the jumping motions produced forward running. It was a relaxation oscillator; once it was running, we had no control over it beyond turning the gas supply on and off.

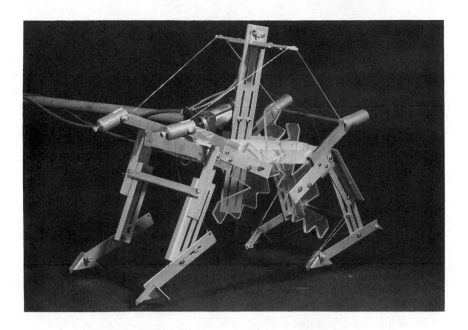

Figure 37.3: The McMahonimal, a running robot. The length of the body is 33 cm.

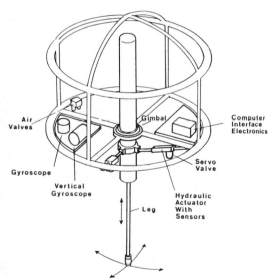

Figure 37.4: Raibert's one-legged hopping machine. The length of the leg is about 0.5 m. (From Raibert (1986); reprinted with permission.)

37.2.2 Raibert's Robots

Marc Raibert and his students have built running machines with one, two, and four legs, all under computer control, all with pneumatic leg springs enclosing a fixed volume of air (Raibert, 1986). The most extraordinary thing about these machines is that they balance themselves while running.

A diagram of Raibert's one–legged hopping machine is reproduced in Figure 37.4. The body of the machine is an aluminum frame containing hydraulic actuators for flexion–extension motions of the hip joint in two planes, air valves, gyroscopes, and computer interface electronics. The large computer that actually controls the machine sits in its own air-conditioned room many feet away; it communicates with the machine via an umbilical cord that dangles from the ceiling. The umbilical also brings in hydraulic power.

The driver stands somewhere out of harm's way and controls the machine by moving a joy–stick that commands changes in velocity. Computer algorithms determine where to put the foot in order to obtain the proper acceleration of the center of mass and to stabilize the body in pitch and roll. A famous demonstration of the machine's ability to balance itself begins as one of the machine's builders (they are called "wranglers" when they do this) steps forward and deliberately pushes the body sideways as the machine is hopping in place. After two or three steps of wobbling, it rights itself and hops back to the starting position, all without any human assistance.

37.2.3 McGeer's Theoretical Robot

Tad McGeer (1989) became interested in a different but related issue concerning the stability of running robots. He wondered whether it was theoretically possible for a bipedal robot with leg springs to operate in a stable limit cycle while running forward. He also considered the equivalent question for a stiff-legged walking machine (McGeer, 1990).

McGeer's bipedal robot model had two telescoping legs with linear springs. The legs were connected at the hip by a linear torsional spring. He found that some choices of parameters led to stable limit cycles and others to unstable ones; he pointed out that even the unstable limit cycles could be stabilized by a simple control law. A particularly interesting set of solutions required zero input of energy from the outside.

McGeer pointed out the curious fact that a theoretical running machine could go through a complete cycle without any energy input, but a passive walking machine loses energy in every step and therefore has to walk down a shallow grade in order to maintain its speed. A stiff–legged walking machine conserves momentum but loses energy when the rigid legs strike the ground. Thus the difference between compliant and rigid legs can make the difference between modes of locomotion that are potentially free of loss and those that inevitably require an input of power.

37.3. Comparing the Gaits: Why are Some Faster?

37.3.1 Alexander's Introduction of the Froude Number

In 1976, R. McNeill Alexander introduced a useful idea to the study of locomotion (Alexander, 1976). He wished to estimate the speed of a dinosaur from a knowledge of the distance between its footprints. The question led him to discover the remarkable correlation reproduced in Figure 37.5, where a dimensionless parameter

called relative stride length is plotted against another dimensionless parameter, the Froude number. The main advantage of using dimensionless parameters is that they allow data representing observations on animals of a wide range of body size to be compared on an equivalent basis.

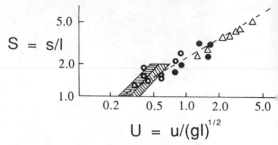

Figure 37.5: Relative stride length $S = s/l$ vs. $U = u/(gl)^{1/2}$. Animals represented include horses, other mammals, and an ostrich. Open circles = walk; filled circles = run, trot, or rack; triangles = canter or gallop. The shaded range shows the predictions of the ballistic walking model. The broken line shows Alexander's correlation $S = 2.3 U^{0.6}$. Adapted from Alexander (1977) plus Mochon and McMahon (1980); reprinted with permission.

The stride length s is the distance between footprints of the same foot; it is normalized by the leg length l to give relative stride length $S = s/l$. The Froude number is a group originally used in studies of ship hydrodynamics. When it is given in the form $u/(gl)^{1/2}$, (where u is forward velocity and g is the acceleration due to gravity), the Froude number is a dimensionless velocity. The term Froude number has also been applied to the group u^2/gl, which may be recognized as a dimensionless kinetic energy. Alexander gave $S = 2.3U^{0.6}$ for the best-fitting line in this log–log plot, where $U = u/(gl)^{1/2}$.

37.3.2 Ballistic Walking

Motivated by Cavagna's discovery that the kinetic plus potential energy of the body changes little during a walking step, Mochon and McMahon (1980) proposed a mathematical model of walking that coupled a compound pendulum representing the swing leg to an inverted pendulum representing the stance leg. The model was restricted to describing the swing period when one foot is on the ground and the other is moving forward. It was assumed that no torques act at the hip or knee throughout the swing period. Mochon and McMahon termed this motion "ballistic walking." The condition that the foot of the swing leg should not strike the ground during the swing period results in a particular relationship between the relative stride length S and the Froude number $u/(gl)^{1/2}$. When $u/(gl)^{1/2}$ is between 0.3 and 0.7, the relationship is approximated well by $S = 2.9\,u/(gl)^{1/2}$. The model also predicts that the vertical ground reaction force will approach zero in the middle of the step if $u/(gl)^{1/2}$ approaches 1.0. An inverted pendulum has a similar property: its weight will be zero when the centripetal acceleration u^2/l balances the gravitational acceleration g, where l is the length of the pendulum and u is its speed when it passes over the top. Thus we expect a transition from walking to running before $u/(gl)^{1/2}$ reaches 1.0, and this is what is found. Alexander (1977) found evidence that the walk-run transition occurs when $u/(gl)^{1/2}$ is near 0.8 in cats, men, and horses.

The shaded region in Figure 37.5 shows the domain in S vs. $u/(gl)^{1/2}$ space where ballistic walking would be expected. The right-hand boundary of the shaded region is set by the condition that the toe of the swing leg must not strike the ground. The left-hand boundary gives the condition where the maximum flexion angle of the knee is 125°, an angle not normally exceeded as any mammal walks. The middle line defines the condition where the maximum flexion angle of the knee is 90°. The theoretical range and the experimental points are in fair agreement, showing that the ballistic walking model can determine the neighborhood of the low-speed part of the figure relevant for walking. The high-speed part of the figure will be the topic of a later section of this chapter.

37.3.3 Comparing Trotting and Galloping

As quadrupedal animals move faster, they walk, then trot, then gallop. A simple mass-spring model separating the vertical and horizontal motions of the body may be used to predict that galloping is much faster than trotting (McMahon, 1985).

The model makes many assumptions in order to reduce the calculations required to use it. The total body mass is lumped in a single mass M sitting upon a massless spring K representing the effective vertical stiffness of the load path between the center of mass and the foot. The foot is assumed to be a circular arc so that the reaction forces on the center of mass are vertical. When feet are

placed on the ground in sequence, as they are in galloping, it is assumed that an unspecified energy exchange mechanism involving the inertial mass of the body allows the transfer of elastic energy from one leg spring to another as one foot is lifted from the ground and another is placed simultaneously. The rebound time (time from the end of one aerial phase until the beginning of the next) is assumed to be a given fraction (for convenience, one half) of a full cycle of vibration of the mass–spring system. The stiffness of the vertical spring when two feet are on the ground at once is assumed to be twice what it is when one foot is on the ground. The step length (distance moved forward when one foot is on the ground) is assumed to be a certain distance L, whatever the gait or running speed. The conclusions of the model are not changed in any important way when more realistic assumptions replace many of these above (McMahon, 1985, pp. 277-279).

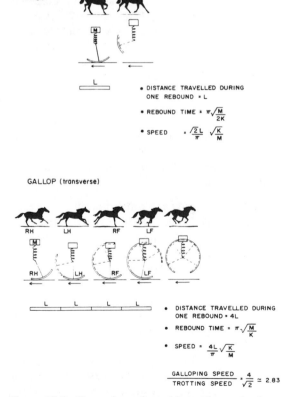

Figure 37.6: Comparison of speed in trotting and galloping using a mass–spring model separating vertical and horizontal motions. From McMahon (1985).

A result comparing trotting and galloping is shown in Figure 37.6. In trotting, since two legs are down at once, the vertical stiffness is $2K$ and the rebound time is proportional to $(M/2K)^{1/2}$. The trotting speed is given by the step length divided by the rebound time. In galloping, each of the four legs is placed in sequence, so that the total distance covered during rebound is $4L$. Because only one leg is on the ground at a time, the rebound time is proportional to $(M/K)^{1/2}$. The final line in the bottom panel of the figure shows that the ratio of the galloping speed to the trotting speed is predicted to be $4/(2)^{1/2}$, or 2.83.

Experimental comparisons have been in reasonable agreement with the above assumptions and predictions. For example, analysis of data presented by Cavagna et al. (1977) show that the contact period is the same fraction of a cycle of resonant vibration at low trotting and low galloping speeds in both small and large dogs, in agreement with the assumptions of the model, and the lowest galloping speed was about 2.4 times the lowest trotting speed (McMahon, 1987). A recent study of sixteen species of wild and domestic quadrupeds ranging in size from 30 g mice to 200 kg horses (including 3.5 kg suni, 4.35 kg dik-diks, 200 kg goats, 90 kg wildebeest, and 213 kg elands) concluded, among other things, that the lowest galloping speed was about 2.6 times the lowest trotting speed in all the animals studied (Heglund and Taylor, 1988). Furthermore, the preferred speeds naturally chosen by the animals for galloping and trotting were in approximately the same ratio.

37.4. An Apparent Contradiction

Until now in this chapter, there has been an implicit assumption that the vertical motion in running can be considered separately from the horizontal motion. The K in Figure 37.6 represents the effective vertical stiffness when a single foot is on the ground at a time. One way to define such a stiffness is to measure the vertical ground reaction force at any instant and divide by the compression of the spring, which may be measured by the downward vertical displacement of the body from the beginning of the contact phase. The vertical stiffness is a logical concept to propose because, together with the body mass, it determines the natural frequency for vertical motions of the body and hence it is the relevant stiffness governing the duration of the contact phase.

Figure 37.7: Groucho running. The subject was running at the same speed as in Figure 37.1. The strobe frequency was again 8.4 Hz. (From McMahon et al. (1987); reprinted with permission.)

37.4.1 Groucho Running

What sets the vertical stiffness for a given animal? Certainly, the intrinsic properties of muscles and reflexes must be important. Another variable must be posture, as the board-bouncing experiments implied (Section 37.1.2, part 4). In a study designed to investigate the effects of posture on vertical stiffness, subjects were asked to run at constant speed on a treadmill and over a force platform while deliberately flexing the knees more than usual (McMahon et al., 1987). The authors called this exercise "Groucho running."

A strobe photograph of a subject Groucho running is shown in Figure 37.7. Comparing Figure 37.7 with a strobe photo of the same subject running normally at the same speed (Figure 37.1) reveals that the contact time is longer and the aerial time is shorter in Groucho running. As subjects bent their legs, they softened their vertical springs, increasing the contact time until, as happened in the deepest postures, the aerial time disappeared entirely. Because there was no flight phase, deep Groucho running looked superficially like walking. The force–plate records showed, however, that it was actually running by Cavagna's criterion, because the body was lower in mid-step, just as in ordinary running.

The Groucho running studies made use of a new dimensionless group, $v\omega_o/g$, designated the Groucho number. In this group, v is the vertical landing velocity, $\omega_o = (k_{vert}/m)^{1/2}$ is the natural frequency for vertical vibration, k_{vert} is the effective vertical stiffness, m is the total body mass, and g is the acceleration due to gravity. For normal running, this group is near unity, and the contact phase occupies 3/4 of a full cycle of vibration of the mass–spring system comprising m and k_{vert}. In Groucho running, k_{vert} goes down and v diminishes nearly to zero (because the aerial phase disappears). As a consequence, the Groucho number goes toward zero and the contact phase occupies nearly a full cycle of vibration.

In another part of the Groucho studies, subjects ran normally at a range of speeds from 2.5 to 5 m/s. The effective vertical stiffness was found to increase by a factor of 2 or more over this speed range.

37.4.2 Direct Measurements of the Vertical Spring

Cavagna et al. (1988) analyzed the vertical motions of the center of mass in running humans and birds; trotting dogs, monkeys, and rams; and hopping kangaroos and springhares. They presented figures showing the vertical acceleration of the center of mass against the vertical displacement, where both variables were calculated from force-plate records. The figures generally showed a nearly linear relationship between vertical acceleration (hence vertical force) and vertical displacement during the contact phase, making it possible to obtain the vertical stiffness directly from the figure.

These authors also introduced a new method for calculating the vertical stiffness. They pointed out that the interval between the time the force is equal to body weight when the force is rising and the time it is again equal to body weight when the force is falling can be considered a half-period of vibration of the vertical mass-spring system. When the vertical spring is an ideal, linear spring, this method gives the same value for vertical stiffness that would be obtained from the slope of the vertical force-vertical displacement curve. The results for all the animals showed a vertical stiffness that increased by a factor of 2 or 3 as speed doubled.

37.4.3 Posing the Apparent Contradiction

Both the Groucho running results and the direct measurements of Cavagna et al. in running animals showed that k_{vert} increases by a large fac-

tor as speed increases. These studies also showed that the peak vertical ground reaction force measured in the middle of the running step increases with speed, although not quite so dramatically. Perhaps an explanation for both increases is that, as an animal runs faster, more muscle fibers are recruited into activity. Additional active muscle fibers in parallel would be expected to increase both the force–generating capacity and the stiffness of the muscles of the leg.

The problem with this explanation is that it doesn't agree with either the measurements of reflex stiffness in the hindlimbs of cats or the measurements of short–range leg stiffness in the board-bouncing experiments with human subjects (37.1.2). Both of these studies found that the stiffness of the leg increased very little with increasing force over the moderate to high force range encountered in running. A conclusion of these studies was that the leg stiffness was determined not by the number of muscle fibers active but by negative-feedback servo loops including at least the stretch reflex and probably other reflexes as well.

37.5 Resolution of the Contradiction

37.5.1 There Are Two Stiffnesses

Suppose a ball bounces vertically on a force platform. The force–plate record may be used to determine the vertical stiffness k_{vert} either by the direct method (obtaining the slope of the force–displacement relation) or the half-period method introduced by Cavagna et al. (1988). Now suppose the ball moves forward at a steady speed while bouncing (assume that it also rolls at a steady angular velocity in order to avoid skidding on the ground.) Again the force–plate records may be used to calculate k_{vert}, and the answer turns out the same as it did for vertical bouncing.

When the ball is replaced by a linear leg spring with a mass on the top, the situation changes. As long as the motion is confined to hopping in place, force–plate records analyzed by either of the two methods will calculate a k_{vert} equal to the linear spring constant of the leg spring, k_{leg}. When forward motion is included without changing the stiffness of the leg spring, calculations for k_{vert} will give a value higher than k_{leg}.

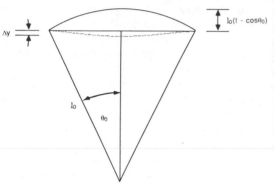

is dimensionless because it is the ratio of two lengths, an arc length to a radius. The dimensionless parameters U and V are Froude numbers based on the horizontal and vertical landing velocities. K_{LEG} is a dimensionless stiffness. It is also the ratio of two forces, the maximum force available from the spring when it is fully compressed divided by the total body weight. Thus, K_{LEG} must take a value greater than one to prevent the body mass from falling through the floor even under conditions of motionless standing.

Figure 37.8: Diagram showing the trajectory of the center of mass during a running step of the McMahon and Cheng model (broken curve). (From McMahon and Cheng (1990); reprinted with permission.)

Figure 37.8 may be used to explain why this is true. The broken curve shows the trajectory of the center of mass during the contact phase of a running stride. The initial (and final) angle of the leg with respect to the vertical is θ_o, and the initial (and final) length of the leg is l_o. The downward vertical displacement from the moment of contact until mid-step is Δy. If we call the ground reaction force at mid-step F_{max}, then the vertical stiffness k_{vert} is $F_{max}/\Delta y$. The stiffness of the linear leg spring is given by F_{max} divided by its entire change in length at mid-step, so that $k_{leg} = F_{max}/[l_o(1-\cos\theta_o) + \Delta y]$. Since the denominator of k_{vert} is smaller than that of k_{leg}, k_{vert} is greater than k_{leg} for nonzero values of θ_o.

37.5.2 The McMahon and Cheng Model

A comprehensive set of solutions for forward running using a linear leg spring have been given by McMahon and Cheng (1990). The parameters of the model include the total body mass m, the initial (zero-force) length of the leg l_o, the initial angle of the leg with respect to the vertical θ_o, the stiffness of the leg spring k_{leg}, the vertical landing velocity $-v$, the horizontal landing velocity u, and the acceleration due to gravity g. When the differential equations describing the motion are written using dimensionless variables $X = x/l_o$, $Y = y/l_o$, $L = l/l_o$, and $T = t(g/l_o)^{1/2}$ where t is time, the problem is specified in terms of four dimensionless groups of variables: θ_o, $U = u/(gl_o)^{1/2}$, $V = v/(gl_o)^{1/2}$, and $K_{LEG} = k_{leg}l_o/mg$. The initial angle θ_o

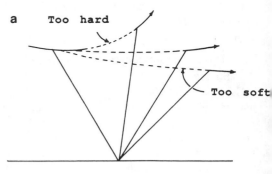

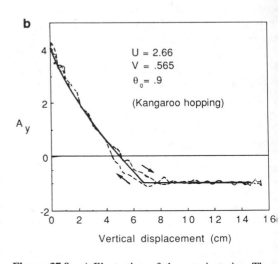

Figure 37.9: *a)* Illustration of three trajectories. The correct stiffness results in a symmetric trajectory that reverses the downward velocity and preserves the forward velocity. *b)* Result showing the normalized vertical acceleration $A_y = f_y/mg - 1$ versus the vertical displacement for a 21 kg kangaroo. Here, f_y is the vertical force. The broken curve (from Cavagna et al. 1988) shows experimental measurements. (From McMahon and Cheng, 1990; reprinted with permission.)

The method for obtaining solutions of the spring–leg running problem is illustrated in Figure 37.9a. Three of the four dimensionless parameters must be specified, and the differential equation model provides the fourth. We elected to specify θ_o, U, and V, so that the model gave K_{LEG} as an output. Our method of solution began by assuming a value for K_{LEG} and integrating the equations forward in time. The leg compressed, then extended, and the integration was stopped when the leg returned to its original length. If the value of K_{LEG} was too large (too hard), the final value of the horizontal velocity was less than the starting value and the magnitude of the final vertical velocity was greater than the starting value, as shown schematically in Figure 37.9a. If the value of K_{LEG} was too low, the final horizontal velocity was too large and the vertical velocity too small. Iterations were performed on K_{LEG} until the final and initial angles and horizontal velocities matched and the final vertical velocity was equal and opposite the starting value. These conditions ensure that the model can run forward without gaining or losing forward speed or vertical height.

A representative solution obtained with the model is shown in Figure 37.9b. The input conditions U, V, and θ_o are appropriate for simulating a 21 kg kangaroo of leg length 0.5 m hopping forward at 21.1 km/hr. The solid curve shows the predictions of the model when the dimensionless vertical acceleration of the center of mass $A_y = f_y/mg - 1$ is plotted against the vertical displacement in cm. The horizontal portions of the curve at $A_y = -1$ corresponds to the flight phase. After a flight phase, A_y rises as the vertical displacement falls during the stance phase. The broken curve shows the results of an experiment in which a 21 kg kangaroo ran at 21.1 km/hr across a force platform (Cavagna et al. 1988). The theoretical and experimental curves are in agreement, even to the point of showing a curved, rather than a linear force–displacement characteristic. The changing slope shows that the vertical spring is almost twice as stiff at the highest force levels as it is at the lowest. The fact that the leg was taken to be a linear spring in the model demonstrates that it is a geometric effect, not one involving the intrinsic or reflex properties of the muscles, that makes the kangaroo's effective vertical stiffness a nonlinear spring. The kangaroo, by the way, gave results that were not typical of the other animals

analyzed. Using the model, McMahon and Cheng showed that the vertical stiffness was nearly linear in running men and dogs.

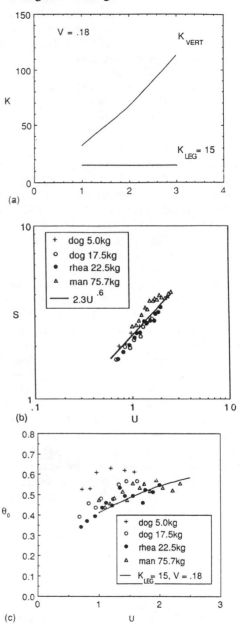

(a)

(b)

(c)

Figure 37.10: Results from the McMahon and Cheng model (solid curves). The animal points are calculated from experimental results given by Cavagna et al. (1988). *a)* Dimensionless stiffnesses K_{LEG} and K_{VERT} vs $U = u/(gl)^{1/2}$. *b)* Relative stride length $S = s/l$ vs. U. *c)* Initial leg angle θ_o vs. U. (From McMahon and Cheng, 1990; reprinted with permission.)

37.5.3 Running with Legs of Constant Stiffness

Earlier we saw how the ballistic walking model was able to make a fair prediction of the way in which relative stride length S increases with forward Froude number $u/(gl)^{1/2}$ in walking (Figure 37.5). Is there anything simple to be said about the rest of Figure 37.5, the running part?

The answer is yes. In Figure 37.10, the McMahon and Cheng model has been specified to have a leg of constant dimensionless stiffness K_{LEG} = 15. Additionally, we required that the relationship between the relative stride length $S = s/l$ and the horizontal Froude number $U = u/(gl)^{1/2}$ be the one given by Alexander as representative of all animals: $S = 2.3U^{0.6}$. The outputs of the model include K_{VERT} and θ_0; these are shown as solid curves in Figures 37.10a and 37.10c. The points are not from Alexander; they were calculated by McMahon and Cheng from data published by Cavagna et al. (1977).

Looking at Figure 37.10b, the first point worthy of note is that Alexander's line works just about as well to correlate these data on two dogs of different sizes, a 22.5 kg rhea, and the average of ten men (mean body weight 75.7 kg) as it did for the variety of animals Alexander used in its formulation. Figure 37.10c shows that all the animals took larger steps to go faster, and the solid curve from the model prediction is in generally good agreement with the animal data. Note that if one looked only at the open triangles describing the men in Figure 37.10c, the trend would show that θ_0 does not increase much with speed at moderate to high speeds, a fact noted by Cavagna et al. (1976) and McMahon and Greene (1979). Another trend is that the smaller the animal, the more its data points tend to lie above the predicted curve. A regular feature of animal scaling is that smaller animals tend to swing their legs through larger excursion angles, as discussed by McMahon (1975).

The conclusion available from Figure 37.10 is that a model for running, based on the assumption that the leg is a linear spring of constant stiffness, can make a realistic prediction of the way in which both stride length and leg excursion angles increase with running speed in vertebrate animals. In the model, the vertical stiffness increases markedly with running speed, in agreement with experimental observations (McMahon et al., 1987; Cavagna et al., 1988). The increase in K_{VERT} with forward speed is obtained in the model by increasing θ_0, not by increasing K_{LEG}. Coincidentally, Raibert's robots work similarly: forward speed is increased by taking larger steps and thereby increasing θ_0, not by making the leg spring stiffer.

37.6. Future Directions

37.6.1 Theoretical

The theoretical models discussed in this chapter provided a conceptual framework for interpreting various known experimental facts. They also pointed the direction for new experimental investigations that might not have been undertaken otherwise. Theoretical models are often developed in a hierarchical fashion, with the simplest ones presented first, in the hope that the most important principles are also the easiest ones to analyze.

Sometimes it is not sufficient to stop with the simplest model. For example, Mochon and McMahon (1980) considered first just an inverted pendulum as a model for walking. We rejected that model because an inverted pendulum alone specifies too broad a range of period for the swing leg, all the way from a period approaching infinity (when the pendulum nearly stops at the top of its travel) to the period corresponding to $u/(gl)^{1/2} = 1$ (when the pendulum is weightless in mid-step). We tried coupling a rigid stance leg to a pole-like rigid swing leg, and also found predictions that were in poor agreement with experiments. The model assuming a rigid stance leg with a foot coupled to a swing leg including a knee appeared to be the least complicated model that provided predictions for the swing period that were in good agreement with experimental observations.

The McMahon and Cheng model, with its single telescoping spring leg, has been successful in simulating many observations about animal running. It has its limitations, however. For one thing, it does not have a swing leg at all, and therefore assumes that there will always be enough time available during one stance and one aerial phase for the swing leg to move forward into the correct position for the next rebound. Cavagna et al. (1988) have objected to this limitation of spring models of running including only one mass. McGeer's model for bipedal running does include two legs, and it is valuable additionally because it investigates the stability, as well as the equilibrium dynamics, of running with springs (McGeer, 1989). A limitation of McGeer's model so far is that it does not have a knee on the swing leg, and this raises a problem about how the swing leg clears the ground. It would be interesting to see how a bipedal running model including a knee would work.

Another important contribution would be made by a quadrupedal model for running that could trot, canter, and gallop. The scheme outlined in Figure 37.6 was limited to vertical motions, but the coupling between vertical and horizontal motions that was a major theme of the McMahon and Cheng single-leg model must be even more important in quadrupeds. For example, the transfer of energy from one leg spring to another that was assumed in the galloping paradigm in Figure 37.6 might be incorporated naturally into a quadrupedal model with spring legs. As one leg is placed before another is lifted, the leg which is about to be lifted would recoil (extend), transferring its elastic energy partly into the kinetic energy of the body and partly into compression of the recently placed leg. Alexander (1989) has wondered how such a mechanism might work; it would be valuable to see.

37.6.2 Experimental

The satisfactory agreement between theory and experiment in Figure 37.10 leads one to ask: What physiological mechanisms are at work which might explain why leg-spring stiffness apparently does not change with running speed? The principle, whatever its explanation, appears to apply quite generally to vertebrate animals. Alexander (1988) has shown by calculations that when the length of the Achilles tendon is much longer than the length of muscle fibers in series with the tendon, the stiffness of the leg may be dominated by the stiffness of the tendon [see also Chapter 36 (Alexander)]. It is likely that many more investigations, including those centered on the mechanisms of control of the limb by the central nervous system, will be needed to understand the result.

The McMahon and Cheng model can make predictions of how all the dynamic variables would change as certain circumstances of running change, including the angle of inclination of gravity (running up and down hill) and the mag-

nitude of gravity. It would be useful to have experimental evidence to compare with these predictions.

Finally, the construction of robots that can trot and gallop represents a further fascinating possibility for studying the mechanical principles of animal gait. What should be the sequence of placement of the legs in galloping, and how should the robot make a transition from one gait to another? Frequently the attempt to synthesize something that occurs in nature can lead to deeper analytical questions and therefore a more comprehensive understanding than any straight analytical study. The mechanics of locomotion has been of interest to scientists and inventors for many hundreds of years, but its period of rapid flowering may have just begun.

References

Alexander, R. M. (1976) Estimates of the speeds of dinosaurs. *Nature (Lond.)* **261**:129-130.

Alexander, R. M. (1977) Terrestrial locomotion. In: *Mechanics and energetics of animal locomotion*, R. M. Alexander and G. Goldspink, eds. John Wiley and Sons, New York.

Alexander, R. M. (1988) *Elastic mechanisms in animal movement*. Cambridge Univ. Press, Cambridge.

Alexander, R. M. (1989) Optimization and gaits in the locomotion of vertebrates. *Physiol. Reviews* **69**(4): 1199-1227.

Cavagna, G. A. (1970) Elastic bounce of the body. *J. Appl. Physiol.* **29**: 279-282.

Cavagna, G. A., H. Thys, and A. Zamboni. (1976) The sources of external work in level walking and running. *J. Physiol.* **262**: 639-657.

Cavagna, G. A., N. C. Heglund, and C. R. Taylor. (1977) Mechanical work in terrestrial locomotion: two basic mechanisms for minimizing energy expenditure. *Am. J. Physiol.* **233**(5): R243-R261.

Cavagna, G. A., P. Franzetti, N.C. Heglund, and P. Willems. (1988) The determinants of the step frequency in running, trotting and hopping man and other vertebrates. *J. Physiol.* **399**: 81-92.

Gordon, A. M., A. F. Huxley, and F. J. Julian. (1966) The variation in isometric tension with sarcomere length in vertebrate muscle fibers. *J. Physiol.* **184**: 170-192.

Greene, P.R. and T.A. McMahon. (1979) Reflex stiffness of man's anti-gravity muscles during kneebends while carrying extra weights. *J. Biomech.* **12**: 881-891.

Heglund, N. C. and C. R. Taylor. (1988) Speed, stride frequency and energy cost per stride: How do they change with body size and gait? *J. Exp. Biol.* **138**:

301-318.

Hoffer, J. A. and S. Andreassen. (1978) Factors affecting the gain of the stretch reflex and soleus muscle stiffness in premammillary cats. *Soc. Neurosci. Abstr.* **4**: 937.

Hoffer, J. A. and S. Andreassen. (1981) Regulation of soleus muscle stiffness in premammillary cats: intrinsic and reflex components. *J. Neurophysiol.* **45**: 267-285.

Houk, J. C. (1979) Regulation of stiffness by skeletomotor reflexes. *Annu. Rev. Physiol.,* **41**: 99-114.

McGeer, T. (1989) *Passive bipedal running.* CSS-IS TR 89-02. Simon Fraser University, Centre for Systems Science, Burnaby, B.C., Canada.

McGeer, T. (1990) Passive dynamic walking. *Int. J. Robotics Research* **9**:62-82.

McMahon, T. A. (1975) Using body size to understand the structural design of animals: quadrupedal locomotion. *J. Appl. Physiol.* **39**: 619-627.

McMahon, T. A. (1985) The role of compliance in mammalian running gaits. *J. Exp. Biol.* **115**: 263-282.

McMahon, T.A. (1987) Compliance and gravity in running. In: *Biomechanics of normal and prosthetic gait,* J. L. Stein, ed. Am. Soc. of Mech. Engrs. Book no. G00410, New York.

McMahon, T. A. and P. R. Greene. (1979) The influence of track compliance on running. *J. Biomech.* **12**: 893-904.

McMahon, T. A., G. Valiant, and E. C. Frederick. (1987) Groucho running. *J. Appl. Physiol.* **62(6)**: 2326-2337.

McMahon, T. A. and G. C. Cheng. (1990) The mechanics of running: How does stiffness couple with speed? *J. Biomech.* (in press).

Mochon, S. and T. A. McMahon. (1980) Ballistic walking. *J. Biomech.* **13**: 49-57.

Raibert, M.H. (1986) *Legged robots that balance.* MIT Press, Cambridge, 233 pp.

CHAPTER 38

Effects of Muscle Elasticity in Walking and Running

At L. Hof

38.1 Introduction

38.1.1 Statement of the Problem

A person S of 78.5 kg weight who jogs at a speed of 2 m/s (7.2 km/h) makes up-and-down movements with his trunk of 12 cm, twice every stride of 0.88 s. These vertical movements represent a change in potential energy of $mgh = 60$ J. At the same time he decelerates and accelerates his horizontal velocity, so that the kinetic energy of his trunk fluctuates over 30 J, in phase (for running) with the 60 J of potential energy, thus making 90 J of trunk energy variations. His legs are active as well and reach 60 J of kinetic energy in midswing. Our subject, whose data were borrowed from Hof, Struwe and Nauta (1989) and of whom we will report more in this chapter, thus delivers mechanical work at a rate of 2 (90+60)/0.88 = 340 W. For this work he has available a metabolic power, measured from oxygen consumption, equivalent to 4.2 J/kgm, in all 660 W (Cavagna and Kaneko, 1977a). One would conclude from this that the efficiency of his muscles is over 50%. The efficiency of a single muscle, however, has been measured a great number of times, and it is commonly assumed that it cannot be higher than 25%, at most 30%.

On the above rough calculation several comments are possible (Williams and Cavanagh, 1983) but this does not invalidate the general conclusion: the apparent efficiency in human running is much higher than can be understood from muscle properties alone. In fact, the efficiency in the above example is less excessive than those reported for human sprint running (Cavagna and

Kaneko, 1977) and kangaroo hopping (Cavagna et al., 1977b), both of which can be over 80%.

To explain the above anomalies, generally the concept of elastic energy storage is advocated: the decrease of potential and kinetic energy in the first stage of stance is accumulated in strain energy of muscles and (mainly) tendons and released at a later time to propel the trunk again forward and upward. Essentially the same phenomena are seen in a bouncing ball, hence the "bouncing ball model" for running [see also Chapter 36 (Alexander and Ker) and Chapter 37 (McMahon)].

Some other phenomena are also considered to be due to storage of elastic energy. You can jump higher when the upward movement is preceded by one downward and still higher when you start from a platform (Asmussen and Bonde-Petersen, 1974). Such a countermovement is apparent in many more activities like throwing and kicking [discussed in Chapter 39 (Chapman and Sanderson)].

Is all this true, and is it so simple, can any superfluous amount of energy just be "stored" for some time until it is needed again? What are the complications for the control of muscles, which are to work with compliant tendons? We will try to address a few of these questions in this chapter. The method will be, very globally, that the interaction between contractile muscle fibers and elastic tendon material is reconstructed on the base of, on one hand, recordings of muscle force and length and, on the other hand, the estimated force-elongation curve of the muscle series elasticity. The activities are walking and running in man.

Multiple Muscle Systems: Biomechanics and Movement Organization
J.M. Winters and S.L-Y. Woo (eds.), © 1990, Springer-Verlag, New York

38.1.2 Literature

The subject of elastic energy storage has a strong appeal to many researchers; witness the number of relevant papers. Fortunately, a number of excellent reviews are available and for the purpose of this chapter we may confine ourselves to a short review of reviews.

A good name to begin with is A.V. Hill, the godfather of muscle mechanics. His many papers, reviewed in Hill (1965), as well as his book "First and last experiments in muscle mechanics" (Hill, 1970), contain a treasure of ideas on the measurement of elasticity and its function in movement.

A general introduction, with examples from all over the animal kingdom, is given in the book Alexander (1988a) "Elastic mechanisms in animal movement". More specifically related to human running is a paper of the same year (Alexander, 1988b) which, short as it is, makes an excellent and quite complete introduction to the present subject.

The links between muscle physiology, exercise physiology and biomechanics are expounded in Cavagna (1977). A contemporary paper (Cavagna et al. 1977) provides the vista of Noah's ark: dogs, ostrichs, hares, kangaroos, monkeys walking, hopping and galloping over the force plate. This work has later been extended to even more exotic animals (Taylor et al., 1981, Fedak et al., 1981, Heglund et al., 1981a,b).

The book of McMahon (1984) addresses elasticity effects, among many other subjects, in a very broad framework. Additions to his work in this field are a paper on "Groucho Running" (McMahon et al., 1987) and Chapter 37 in this book.

Although their main objective is to provide a review of methods of elasticity measurement in animal experiments, Proske and Morgan (1987) also address some points of wider interest, e.g. the control problems for a muscle that is connected to the outer world with a compliant tendon. The same problem is also dealt with by Rack (1981).

The review of Shorten (1987) is, like that of Alexander (1988b), well suited to serve as a general introduction. It has an emphasis on methods of elasticity measurement and data on elasticity effects in "real" movements. His results in jumping were markedly expanded shortly afterwards by Bobbert et al., (1986a,b, 1988). It turned out that, although the elastic properties of muscle are indispensable for explaining the observed power output, elastic energy storage yields only part of the work needed to move the body upward. Another important mechanism turned out to be the action of biarticular muscles (see Chapter 41 (Van Ingen Schenau et al.)). Finally, the simple fact that muscle action lasts longer in countermovement jumps relative to squatting jumps may be one more explanation for the observed differences in jumping height. Our own papers (Hof et al., 1983, Hof and Van den Berg, 1986) on the elastic effects of elasticity in walking belong to the same line of thought: What are the effects of *SEC* elasticity on the internal behavior of the muscle components in a natural movement, and how efficiently do they exploit the contractile properties? The present chapter can be considered as a follow up and extension (with regards to running) of these papers.

38.1.3 Assessment of Series Elasticity in Man

Elastic components should be included in any model of muscle next to the contractile properties. The simplest models contain three elements (see Figure 38.1). We will adopt the version *a*, which is easier to understand and to calculate with. Jewell and Wilkie (1958) have shown that it is impossible to decide between the two models on the basis of measurements alone because as a rule the parallel elastic component (*PEC*) is much more compliant than the series-elastic component (*SEC*). For convenience we will use angular units, moment of force (Nm) and joint rotation (deg or rad) instead of muscle force and length.

As is apparent from Figure 38.1a, the *PEC* represents all elasticity that remains when the *CC* does not exert force, i.e. when the muscle is passive. This comprises properties of muscle filaments (mainly short range elasticity, D.K. Hill, 1968, Boon et al., 1973), of sarcolemma, fasciae and — in musculoskeletal models — joint ligaments. Measuring the *PEC* moment–angle relation is a straightforward affair, although there are some complications in the form of visco-elastic effects and hysteresis. As regards the major leg joints, data are available for the hip (Yoon and Mansour, 1982, Walsh and Wright, 1988), the knee (Heerkens et al., 1985) and the ankle (Tardieu et al., 1981, Hof and Van den Berg, 1981b). The *PEC* moment is usually very low over most of the

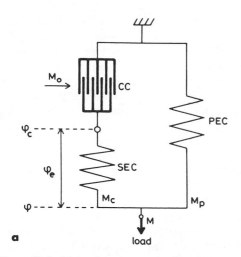

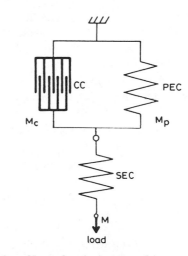

Figure 38.1: Three element muscle models. CC = contractile component, SEC = series elastic component, PEC = parallel elastic component. *a)* Maxwell model, *b)* Voight model. Model *a)* has been adopted in this study. Muscle length has been expressed in ankle angle

range of the joint, increasing only near the limits of this range. It has a function in constraining joint motion and damping limb movement in situations with no muscles active. In some muscles with very short fibers, as in the camel (Alexander et al., 1982) and the horse (Van den Bogert et al., 1989) the *PEC* accounts for most of the muscle force, also during active movement. It will be shown below that it has only minor importance for the ankle moment in human walking and running.

The *SEC* comprises the elasticity in the crossbridges, in the thick and thin filaments, in the tendinous material in series which each fiber, the aponeuroses and the tendons. Alexander and Bennet–Clark (1977) calculated from the dimensions of muscle fibers and tendons in many muscles that the *SEC* can be thought located for the major part in aponeurosis and tendon. Exceptions are only very long parallel fibered muscles like frog sartorius, the 'pet muscle' of physiologists.

To determine the elasticity curve for the *SEC* in the intact human is far less easy than for the *PEC*, because the *SEC* cannot be disconnected from the *CC*. The curve can be described by a relation $M_c(\phi_e)$ between M_c and the elastic stretch ϕ_e. The derivative at a certain M_c is the elasticity or stiffness $K'(M)$:

$$K'(M_c) = \left(\frac{dM_c(\phi_e)}{d\phi_e} \right)_{M_c} \qquad (38.1)$$

ϕ for the calf muscles. An increase of ϕ corresponds to plantarflexion and muscle shortening. *CC* length ϕ_c and *SEC* length ϕ_e have been expressed in corresponding angular units.

From animal experiments it is known that $M_c(\phi_e)$ is a nonlinear relation, the elasticity K' (or the compliance $1/K'$) is therefore not constant. All muscles of interest are connected to limbs with a considerable moment of inertia I, with respect to the joint rotation axis. In combination with a damping factor B, the origin of which need not be gone into at this point, this yields a second order differential equation for the moment externally applied to the joint

$$M_{ext} = M_c + I\ddot{\phi} + B\dot{\phi} + K'\phi \qquad (38.2)$$

in which M_c is the *CC* moment. For the assessment of K' four methods are available [see also Chapter 5 (Winters)]:

a) The complete $M(\phi_e)$ curve is determined from the course of the isometric moment, for example at the onset of a tetanus. This might be done by comparing real isometric tetanus with one in which a spring of known stiffness is interposed (MacPherson, 1953) or by curve fitting when the force–velocity relation of the *CC* is known. Our experience (Hof and Van den Berg, 1981b) is that a good model fit can be obtained in this way, but that the parameter sensitivity is not sufficient to get reasonably accurate *SEC* parameter values.

b) The angle ϕ is varied and the amplitude of the

fluctuations in M_{ext} is measured (Goubel and Pertuzon, 1973, Asmussen and Bonde-Petersen, 1976, Evans et al., 1983, Ma and Zahalak, 1985, Weiss et al., 1988).

c) Vice-versa: M_{ext} is changed and the resulting rotation is measured (Gottlieb and Agarwal, 1978).

d) Methods relying on determination of the frequency of natural oscillation

$$ f_o = \frac{1}{2\pi}\sqrt{\frac{K'}{I}} \tag{38.3} $$

The attractive point in these methods is their simplicity. In the elastic bounce method (Cavagna, 1970) the subject has to make a small jump and the period of the resulting damped oscillation is determined. In oscillating methods (Greene and McMahon, 1979; Hof and Van den Berg, 1981c; Bach et al., 1983) the subject is asked to make oscillations in his 'easiest' rhythm and it is assumed that this corresponds to the resonant frequency. The inertia I in the resonance experiments (for the ankle) is the one due to the total body weight, or half of it, with respect to the ankle, when standing on one or two legs respectively. This gives thus the stiffness K' at two values of M. A greater range of K' values was obtained by Shorten (1987) who loaded the bent knee of the sitting subject with a series of weights.

As far as the published data can be compared, and that is not always easy because of the various mathematical representations of $M(\phi_e)$ and $K'(M)$, they agree reasonably well with each other. Interindividual differences are probably the major source of disagreement. There is also a satisfactory agreement with published data on tendon stress–strain relations measured in vitro (see the discussion in Section 38.4.2).

All methods b, c, d, however, have two major drawbacks. The first is that they all measure the elasticity $K'(M)$ at only a few points of the curve and not the complete recoil of the SEC in a single fast release. Such a fast release would be really representative for the SEC function as an elastic energy buffer.

Usually it is assumed that the SEC has one unique $M(\phi_e)$ relation, which can be found by fitting the derivative of this function to the measured stiffness values. It is not certain that this supposition is true; maybe the recruitment of more muscle fibers at higher forces also recruits more of the small elastic elements that can be thought in series with each muscle fiber [cf. Shorten, 1977 and Chapter 4 (Ettema and Huijing)]. In that case the fast recoil of the SEC would not follow the same curve for different initial values of the muscle moment [see also Chapter 5 (Winters)]. It should be admitted that the differences in resultant SEC stretch are probably not large, as apparent from in vitro experiments (Cavagna et al., 1981, Goubel, 1987), and that at any rate the differences in elastic energy are minor.

The second drawback is the uncertainty whether M_c, the moment due to the CC, remains constant during the stiffness measurement. This is by no means certain. The stretch reflex, for example, can make the muscle force increase as a result of stretch, and this manifests itself as an apparent elasticity. In order to cover this eventuality, McMahon (1984 and Chapter 37) systematically speaks of 'reflex stiffness'. It should be understood, however, that the part of the apparent muscle stiffness that is due to reflex activity does *not* participate in the energy saving by storage of elastic energy, because it is related to active CC contractions which consume their usual metabolic energy.

38.2 Methods of Experimental Work

38.2.1 EMG to Force Processing

The results in this chapter on moment and work of the calf muscles have been obtained by EMG to force processing (Hof and Van den Berg, 1981a; Hof, 1984; reviewed in Chapter 8 (Zajac and Winters)). This method relies on a muscle model, for which the three-component Hill muscle model has been chosen, being an attractive compromise between model complexity and model realism (Winters and Stark 1987, Chapter 5 (Winters)). The arrangement of the elements is given in Figure 38.1a. The action of the CC is described with a force–length–velocity relation according to Hill (Abbott and Wilkie, 1953, Winters and Stark, 1985) driven by an 'active state' independent of muscle force or lengthening. Model equations and parameters have been summarized in Table 38.1.

For reasons of convenience the moment around the ankle M and the ankle angle ϕ are used instead of muscle force and length, respectively.

Surface *EMG* signals (in our case of mm. soleus and gastrocnemius) are rectified, summed and processed in such a way that the resulting signal has the major properties of the muscle active state: a fast rise, a plateau and a slow decay. This active state signal is input to the *CC* of a Hill model, together with the muscle length, which in our case can be measured by recording the ankle angle with a goniometer. It has been shown that the muscle moment thus obtained corresponds very well with the ankle moment determined with inverse dynamics (Hof et al., 1987), in spite of the fact that the *EMG* is intrinsically a strongly fluctuating 'noisy' signal. The effect is related to the low-pass properties of the muscle model due to the *SEC–CC* interaction.

The method by which the triceps surae moment has been obtained has in fact little relevance to the results to be presented. An inverse dynamics approach, with due consideration of antagonist muscle activity, would have been equally useful. On the other hand, it might not have induced us to investigate the interaction between series-elastic and contractile component.

Table 38.1: Formulae and parameters relating to the muscle model. In the last column 'S' means that the value of the parameter is specific to subject S, 'A' means that the parameter value is generally valid for adult human subjects. For details and parameter estimation see Hof and Van den Berg (1981a,b,c).

Formula	Parameters	A/S
EMG to active state		
$M_0(t) = \max \begin{cases} U(t-\Delta t) & (0 \le \Delta t \le \tau_2) \\ U(t-\Delta t).\exp\left[-\dfrac{\Delta t - \tau_2}{\tau_3}\right] & (\tau_2 \le \Delta t) \end{cases}$	$\tau_2 = 30$ ms	A
	$\tau_3 = 60$ ms	A
where $U(t)$ is the rectified EMG, smoothed over $\tau_1 = 25$ ms.		
Moment–angle relation (force–length relation)		
$f(\phi_c) = \begin{cases} 1 & (\phi_c < \phi_2) \\ \dfrac{\phi_1 - \phi_c}{\phi_1 - \phi_2} & (\phi_2 < \phi_c < \phi_1) \\ 0 & (\phi_c > \phi_1) \end{cases}$	$\phi_1 = 150°$	S
	$\phi_2 = 45°$	S
Moment–velocity relation (Hill relation)		
	$b = 1.2$ rad/s	A
$M_c = M_0 \dfrac{f(\phi_c) - n\dot{\phi}_c/b}{1 + \dot{\phi}_c/b}$	$n = 0.12$	A
$M_c \le (1+c)M_0 f(\phi_c)$	$c = 0$	A
Series Elastic Component		
	$\beta = 15$ rad^{-1}	S
$\phi_e = \dfrac{1}{\beta}\ln\dfrac{M_c + M_s}{M_s} + \dfrac{M_c}{K}$	$K = 740$ Nm/rad	S
	$M_s = 1$ Nm	A
Parallel Elastic Component		
$M_p = M_{p0} \exp\left[\dfrac{\phi - 90°}{\phi_p}\right]$	$M_{p0} = 10$ Nm	S
	$\phi_p = 38°$	S

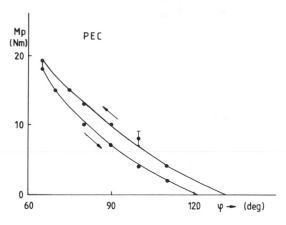

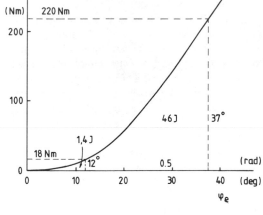

Figure 38.2: *a)* Elasticity curve of the Parallel Elastic Component for subject *S*, measured by passively moving the foot and recording the ankle moment at no muscle (*EMG*) activity. The area under the curve equals the energy stored in the *PEC* (see Table 38.2). There is some hysteresis, but the associated energy loss is very small (less than 2 J per cycle). *b)* Elasticity curve of the Series Elastic Component for subject *S*, measured by the free resonance method at moments of 50 and 100 Nm. The curve follows *Eq. 38.4* with β = 15 rad^{-1} and $K = 740$ Nm/rad. The dashed lines refer to the 'taking up the slack' effect (see Section 38.4.7).

38.2.2 Parameters of the Muscle Model

A description of the model and of the methods by which the parameters have been determined can be found elsewhere (Hof and Van den Berg, 1981a,b,c). In short, great effort has been made to devise different types of contractions, each specific to as few parameters as possible. Most parameters were individually assessed for every subject and showed considerable interindividual differences. Table 38.1 gives a summary.

Of special interest for this paper is the *SEC* load-extension relation. In the model it was

$$\phi_c = \frac{1}{\beta} \ln \left(\frac{M_c + M_s}{M_s} \right) + \frac{M_c}{K} \qquad (38.4)$$

This form has been chosen, because then the compliance $1/K$ as a function of M_c has a simple form:

$$\frac{1}{K'} = \frac{d\phi_e}{dM_c} = \frac{1}{\beta M_c} + \frac{1}{K} \qquad (38.5)$$

The parameters β and K were determined by the 'free oscillation' method (see 38.1.3), in two conditions, viz. standing on one and on two legs. This yields two values for the stiffness K' at two values of M ($\sim M_c$), from which β and K could be obtained. M_s had to be chosen arbitrarily. This makes that the total value of ϕ_e has not a well-founded experimental base. However, the subject of this study are mainly differences in ϕ_e at different values of M_c and these are hardly affected by the choice of M_s. Only the position of the moment angle relation, parameter ϕ_1, is dependent on M_c (Table 38.1).

38.2.3 Experiments, Subject

The results to be reported have been taken from a more extensive study on the relation between calf muscle work and segment energy changes in walking and running (Nauta and Hof, 1989; Hof, Struwe and Nauta, 1989). Data on one subject (*S*, male, age 28 yr, weight 78.5 kg, stature 1.82 m) will be given. He walked on a treadmill at constant speeds between 0.5 and 2.0 m/s, with increments of 0.25 m/s and ran at 2.0 and 2.75 m/s. (The treadmill did not allow higher speeds.) The results on walking can be seen as an extension of those in Hof et al., (1983); those on running are new.

Figure 38.2a gives the *PEC* curve and Figure 38.2b the *SEC* curve of subject *S*.

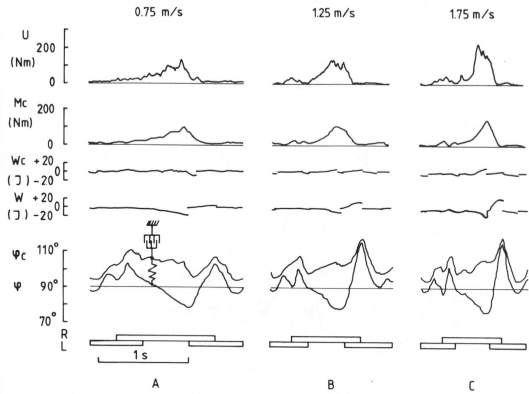

Figure 38.3: Calf muscle action in walking at *a)* 0.75 m/s (2.7 km/h), *b)* 1.25 m/s (4.5 km/h) and *c)* 1.75 m/s (6.3 km/h), respectively. From top to bottom: U = sum of rectified *EMG*s of soleus and gastrocnemius, smoothed with a time constant of 25 ms. The scale is calibrated in Nm isometric ankle moment. M_c = moment developed by the *CC*. The total moment M is at most 10 Nm higher. W_c = work done by the *CC*, running integral of *CC* power. W = work done by the whole muscle-tendon complex. ϕ_c = 'length' of *CC*, expressed in angular units, see Figure 38.1a, ϕ = angle of the ankle, plantarflexion positive, neutral standing position = 90°. Below: foot contact pattern and time scale.

38.3 Results

Figure 38.3 a,b,c gives recordings of walking at 0.75, 1.25 and 1.75 m/s respectively, and Figure 38.4 a,b those for running at 2.0 and 2.75 m/s. Next to the rectified *EMG* (sum of both muscles), triceps surae moment M_c and the ankle angle ϕ, three derived data have been given:

W, the work done by the calf muscle muscle-tendon complex. It is presented as the integral

$$W(t) = \int_{t_o}^{t} M \ \dot{\phi} \ dt \qquad (38.6)$$

and is reset at times t_o when $\dot{\phi} = d\phi/dt = 0$. This yields seperate branches related to, first, the negative work W^- and, next, the positive work W^+. As has been shown earlier (Hof and Van den Berg, 1983), W^- is in walking not very dependent on speed, amounting to –16, –12 and –16 J in Figure 38.3 a,b,c respectively. W^+ increases strongly with

speed (in fact with the freely chosen steplength), in Fig. 38.3 a,b,c, it amounts to 4, 12 and 24 J respectively. In fact the presented figures have been chosen in such a way that W^+ is less, equal to and more than W^-. In running both W^- and W^+ are considerably higher than in walking, in Figure 38.4a $W^- = -44$ J and $W^+ = +56$ J, in Figure 38.4b $W^- = -48$ J, $W^+ = +60$ J.

The *CC*–length ϕ_c has been calculated, according to Figure 38.1a, as

$$\phi_c = \phi + \phi_e \qquad (38.7)$$

in which ϕ is measured and ϕ_e derived from M_c according to *Eq. 38.4*. Because of the definition of ϕ and ϕ_c, plantarflexion positive, an increase of ϕ_c means a shortening of the *CC*. The diagram in Figure 38.3a, a two-component Hill model with the *PEC* left out, may illustrate the relation.

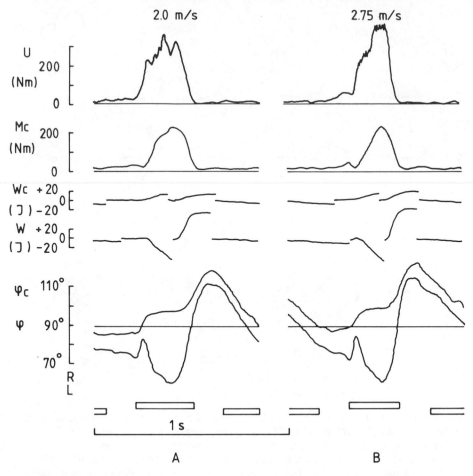

Figure 38.4: Calf muscle action in running at *a)* 2.00 m/s (7.2 km/h) and 2.75 m/s (10 km/h). For explana- tion see legend of Figure 38.3. Time scale is twice as large as in Figure 38.3.

The *CC* work W_c is defined analogous to *Eq. 38.6* as

$$W_c(t) = \int_{t_o}^{t} M_c \; \phi_c \; dt \qquad (38.8)$$

In the same way as W it is reset when $\dot{\phi}_c = 0$.

Immediately obvious from the Figures 38.3 and 38.4 are the essential differences between the ankle angle ϕ, corresponding to the length changes of the total muscle–tendon complex, and ϕ_c, which represents the length of the contractile component alone. The pattern for ϕ, both in walking and run- ning, is eccentric–concentric: stretch first, then shorten. The *CC* 'length' ϕ_c, on the other hand, is shortening in most of the contractions during most of the time; when force is developed, only short intervals and small amounts of lengthening are in- terposed between the gradual concentric action. Only in Figure 38.3a, where the amount of nega- tive work is more than the positive work, is lengthening of the *CC* is substantial.

This is the overall picture, details will be dis- cussed in the next section.

38.4 Discussion

38.4.1 Function of Negative and Positive Calf Muscle Work

The action of the calf muscles, the work done on them (W^-) or by them (W^+), must have, of course, some function in the complex movements of walking and running. In running negative and positive calf muscle work are directly related to the decrease and increase of potential and kinetic energy of the trunk moving down and up during stance (Hof, Struwe and Nauta, 1989). In walking positive calf muscle work serves to initiate leg swing. The origin of the negative work is as yet less clear (Nauta and Hof, 1988). From a mechanical point of view, one would expect that it would serve to decelerate the movement of the trunk over the ankle, but a corresponding decrease of trunk energy is not present. Maybe this is due to a simultaneous acceleration of the trunk due to some other force.

38.4.2 Shortcomings of the Method

There are a few aspects of the *EMG* to force processing method, that should be in mind when evaluating the results. It has the noisy rectified *EMG* as input. This results in some ripples in the output muscle moment M_c (cf. Section 38.2.1) and these in their turn may cause a few spurious deflections of ϕ_c and W^-_c. The general form and magnitude of the model output can be considered correct, however.

The resonance method for measuring *SEC* stiffness is not very accurate and may contain a reflex component (38.2.2). Some justification may be obtained from the fact that the measured parameter values agree with data on Young's modulus of tendon material of a wide range of muscles and animals, as given by Bennett et al., (1986). An exact match can be obtained for subject S if we assume a tendon length of 40 cm, a moment arm of 5 cm and a tendon cross-section of 0.5 cm^2, constant over the whole length. It is impossible to verify these measures accurately, but they are certainly of the correct order of magnitude.

Triceps surae is composed of three muscles, two of which — the medial and lateral heads of gastrocnemius — are biarticular. It would have been nice if all three muscles could have been modelled separately, to show (possibly) additional effects of biarticular muscles [e.g. see Chapter 18 (Gielen et al.) and Chapter 41 (van Ingen Schenau et al.)] and of any differences in activity between the three muscles. Unfortunately we have not yet available at present equipment and a parameter estimation procedure to be able to make a reliable model of the three separate muscles. The present muscle model consists therefore of one monoarticular muscle, even if two *EMG*'s are used.

38.4.3 Elastic Energy

In the above we defined the *CC* work W_c and the *SEC* elastic energy E_e, in addition to the muscle

work W. There is an interesting relation between these quantities. The *CC* and the *SEC* are connected in series (Figure 38.1a) and therefore with the sign convention adopted here.

$$\phi_c = \phi + \phi_e \qquad (38.7)$$

and

$$\dot{\phi}_c = \dot{\phi} + \dot{\phi}_e \qquad (38.9)$$

The *PEC* is parallel to the *CC* and the *SEC*, thus

$$M = M_c + M_p \qquad (38.10)$$

From (38.9 and 38.10) it follows that

$$\int M_c \dot{\phi}_c \, dt = \int M \dot{\phi} \, dt + \int M_c \dot{\phi}_e \, dt - \int M_p \dot{\phi} \, dt$$

or

$$\qquad (38.11)$$

$$W_c = W + E_e + E_p$$

provided that all integrals are taken over the same time interval. E_e and E_p are the elastic energy stored in the *SEC* and the *PEC*, respectively. The change of sign in E_p is because an increase of ϕ, muscle shortening, corresponds to a decrease of E_p. *Eq. 38.11* holds at any time in the movement cycle. It represents a relation between, on one hand, *CC* work W_c and muscle–plus–tendon work W, done up to a certain time t, and on the other hand, the elastic energy present at that time in *SEC* and *PEC*. We will now estimate the magnitude of E_p and E_e.

The energy E_p stored in the *PEC* can simply be found as the area under the curve of Figure 38.2a. It is only dependent on the ankle angle ϕ. Table 38.2 gives the maximal value of E_p, which occurs when ϕ is minimum.

The energy E_e stored in the *SEC* can be found from its moment–angle characteristic, Figure 38.2b and *Eq. 38.4*. This yields, neglecting a minor term:

$$E_e = \frac{M_c}{\beta} + \frac{M_c^2}{2K} \qquad (38.12)$$

which shows that E_e is only dependent on M_c, as it should be. Maximal values for M_c and E_e are also given in Table 38.2.

With the parameters found, we see that E_e reaches substantial values: of the same order as the negative work W^-. In comparison E_p is considerably smaller, although not completely negligible. In the following we will be mainly concerned with E_e and the *SEC-CC* interaction.

Table 38.2: From left to right: Speed of walking (W) or running (R), maximal values of *PEC* moment M_p and elastic energy E_p, which depends on the minimum value of ϕ. Maximal value of *SEC* elastic energy E_e, which depends on the maximum of M_c. Negative and positive work, W^- and W^+, done by the calf muscles on the ankle.

speed	ϕ_{min}	M_p	E_p	$M_c(max)$	E_e	W^-	W^+
(m/s)	(deg)	(Nm)	(J)	(Nm)	(J)	(J)	(J)
0.75 W	78	14	4.8	100	13.4	−16	+ 4
1.25 W	78	14	4.8	110	15.5	−12	+12
1.75 W	76	15	5.3	140	22.6	−16	+24
2.0 R	62	20	9.3	220	47.4	−44	+56
2.75 R	62	20	9.3	230	51.0	−48	+60

38.4.4 Concerted Contraction

Splitting W and W_c in phases of positive and negative work, W^+, W_c^+, W^- and W_c^- respectively (cf. Figures 38.3 and 4), we can evaluate *Eq. 38.11* to get the total work done in a complete contraction. Before and after a contraction $M_c = 0$ and thus $E_e = 0$ as well. E_p need not exactly to be zero at those times, but it turns out to be very small (2-3 J) and will not be considered here. For such a complete contraction thus holds

$$W_c^- + W_c^+ \simeq W^- + W^+ \qquad (38.13)$$

We see from Table 38.2 that in walking the sum ($W^- + W^+$) could be negative (0.75 m/s), zero (1.25 m/s) or positive (1.75 m/s). In running it was positive at both speeds.

W_c^+, positive or concentric *CC* work requires metabolic energy, at an efficiency of 20-30%. Negative muscle work W_c^- may also cost some energy, to maintain the muscle tension, but to all available evidence this amount seems to be small (Asmussen, 1953, Hill, 1960). It is uneconomical, nevertheless, to waste more W_c^- than necessary, because then an amount of W_c^+ has to be done to keep the sum equal to the right side of (38.13). We will consider the three possible cases in some detail.

38.4.5 Findings in Walking and Running

a) $-W^- = W^+$ (moderate speed of walking, as in Figure 38.3b)

The most efficient contraction in this case would be one in which both W_c^- and W_c^+ are zero. The only way to achieve this, is to keep ϕ_c constant all the time during which the active moment M_c has a significant value. This requires a muscle force which closely matches the muscle plus tendon lengthening and shortening, because ϕ_c — which should be kept constant — is the sum of muscle 'length' ϕ and the *SEC* stretch ϕ_e, which is wholly determined by M_c. Figure 38.5 gives a diagram with a two-component muscle model in such a contraction, for which we previously have proposed the term 'concerted contraction' (Hof and Van den Berg, 1983).

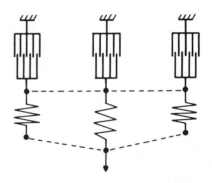

Figure 38.5: Diagram of a two component muscle model (as in Figure 38.1 with *PEC* left out) in a concerted contraction without shortening or lengthening. Externally muscle and tendon are stretched and shortened, but the muscle force is at each instant just thus high that the stretch of the *SEC* is equal to the external length change. The *CC* does not change in length in such a contraction.

For an ideal concerted contraction will hold as one of the consequences that the minimum of ϕ, which is also the borderline between W^- and W^+, should be simultaneous with the maximum of M_c, and thus with the maximum in the elastic energy. For this maximal elastic energy then should hold, from *Eq. 38.11*:

$$-W^- = E_e \,(max) \qquad (38.14)$$

In Figure 38.3b we see that this ideal is approximated quite reasonably; ϕ_c stays within 6 deg, while ϕ changes about 12 deg during the period that M_c is over ca. 40 Nm. Before and after that period larger lengthening and shortening movements of ϕ_c are seen, but then the associated W_c^- and M_c^+ are low, because M_c is low (see *Eq. 38.4.7*). Over the complete contraction we measure a W_c^+ of 6 J with a measurement error of about 4 J, while W^- and W^+ are both –12 and +12 J. The peak E_e is amply sufficient to fulfill *Eq. 38.14*. The main energy flow in this contraction is, to summarize:

$$W^- \to E_e \to W^+ \tag{38.15}$$

One might thus say that in a concerted contraction the muscle 'merely acts as a spring' and one is reminded of McMahon's (1984) 'spring plus rack–and–pinion' concept of muscle, with the rack and pinion stationary [see also Chapter 37 (McMahon)]. It should be kept in mind, however, that a concerted contraction requires a close match between muscle force and extension, and therefore a very special control.

b) $-W_c^- > W_c^+$ (slow walking, as in Figure 38.3a).

The most efficient *CC* action in this case, when eccentric action preponderates, is one in which ϕ_c is either stationary or decreasing (i.e. the *CC* is lengthening). In that case no W_c^+, with the associated energy consumption, is done. We see that this indeed happens in Fig. 38.3a; ϕ_c is seen to be constant — at most very slightly increasing — for most of the time, with short periods of abrupt decrease in between. (The number of these 'slips' may be slightly overrated due to excess fluctuations in M_c (see Sections 38.2.1 and 38.4.2).) The energy flow can be represented as:

$$W^- \to E_e \begin{array}{c} \nearrow W^+ \\ \searrow W_c^- \end{array} \tag{38.16}$$

c) $-W^- < W^+$ (fast walking, running, Figures 38.3c and 4a,b).

The most efficient muscle action in this case, when net concentric work is done, is one in which ϕ_c is increasing all the time, a 'concerted contraction with shortening'. W_c^- is then zero and W_c^+ has the minimal value, equal to the sum of W^- and W^+ (38.13). The energy flow can be represented as:

$$\begin{array}{c} W^- \searrow \\ W_c^+ \nearrow \end{array} E_e \to W^+ \tag{38.17}$$

In Figure 38.3c we see that this is roughly what happens in fast walking. There can be short 'slips' in between, e.g., when a peak in M_c does not exactly coincide with a minimum of ϕ, but the associated W_c^- is small. The results presented earlier (Hof and Van den Berg, 1983, Figures 3 and 4) were slightly less favourable in this respect. In running (Figure 38.4a,b) the ideal scenario is wholly followed. When running, therefore, of the large amounts of negative and positive work, –44 and +56 J in Figure 38.4a, only the sum, $56 - 44 = 12$ J needs to be delivered by concentric *CC* action, with the associated energy consumption. In McMahon's model: the rack–and–pinion is gradually cranked up all the while, but the biggest energy exchange is the stretch and rebound of the elastic component.

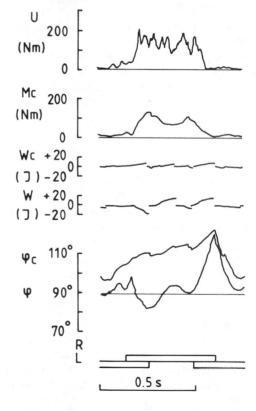

Figure 38.6: Calf muscle action in walking (1.5 m/s, 5.4 km/h) with exaggerated up and down movement, see text. For explanation see legend of Figure 38.3. Fast time scale as in Figure 38.4.

A peculiar case, from the "department of silly walks", can be seen in Figure 38.6. It concerns walking with an exaggerated up and down movement halfway stance, we heard it called once the 'salesman walk'. In the ankle angle we see an additional plantar- and dorsiflexion, which corresponds to phases of positive and negative work which, in their turn, can be found back in an up and down movement of the trunk. One would expect that such a needless movement, with the associated work and energy changes, would be very inefficient. We see, however, that the *SEC* stretch accounts for the whole excursion, while ϕ_c follows a smoothy increasing course. It may be that the total positive work done (ca. 30 J) is higher than normal, but the muscle does this work in an efficient way, without undue losses.

38.4.6 Consequences for Muscle Efficiency

The major energy saving mechanism has now been discussed:

1) Elastic energy storage. By this effect negative work is not wasted but made available for the positive work W^+. The *CC* work W_c^+, which determines the metabolic demands, can thus be lower than W^+.

There are, however, more effects of *SEC* elasticity.

2) Moment-angle relation. We have seen that during stance ϕ_c changes over a smaller range than ϕ. It is remarkable that this range of ϕ_c is close to the optimum muscle length in the moment-angle (M_c - ϕ_c) relation. In Figures 38.3 and 4 ϕ_c is always between 95° and 110° when M_c is over 40 *Nm*. The moment-angle relation has values between 1.0 and 0.85 in this interval (cf. Table 38.1). To put it another way, because of elastic stretch muscles with quite short fibers are adequate for eccentric–concentric actions involving large length changes. This point has been extensively discussed in Chapter 36 (Alexander and Ker).

3) Moment-velocity relation. The third advantage of *SEC* elasticity is that the *CC* works in the most efficient range of speeds. This is most dramatically illustrated in running, e.g. Figure 38.4b. Ankle plantarflexion reaches in that case an angular velocity up to 12 rad/sec, which is more than the maximal shortening speed of the muscle fibers ($\phi_o = b/n = 10$ rad/sec for the calf muscles, Table 38.1). In contrast, the shortening of ϕ_c occurs at rates not over 1.5 rad/sec as long as $M_c > 40$ *Nm*.

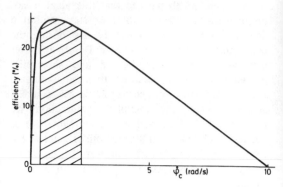

Figure 38.7: Efficiency of muscle contraction, inclusive recovery, as a function of positive *CC* shortening velocity ϕ_c. Calculated from formulae of Hill (1938) with parameters for the human calf muscles ($b = 1.2$ rad/sec, $n = 0.12$, optimum efficiency 25%). Shaded area: interval of ϕ_c where the efficiency is 90% or more of the maximum. (From Hof and Van den Berg (1983); reprinted with permission.)

In Figure 38.7 the efficiency of muscle contraction is given as a function of *CC* shortening velocity calculated from the moment-velocity relation with parameters from Table 38.1. We see that the curve has a flat maximum, such that the efficiency is over 90% of the optimum for *CC* shortening rates between 0.4 and 2.0 rad/s. The *CC* speeds in both walking and running are all in this range except for the fact that ϕ_c often shows alternating periods of no motion with periods of moderately fast shortening. Rough calculations, which will not be detailed here, show that the overall efficiency in this case is hardly different from the case that the muscle would shorten the same distance at a steady low speed. The *SEC–CC* interaction can thus ensure that the positive *CC* work W_c^+ is delivered at an efficiency close to the optimum 25%.

4) Smoothing of power peaks, catapult action. This effect is most nicely demonstrated in running (Figure 38.4b). In push-off the muscle-tendon complex delivers its work W^+ at rates of no less than 1100 watt. The rate of W_c^+ is much less, up to 300 watt, better in accordance with the physiological limit of muscle power production, estimated at 150–250 watt per kg of muscle. This is a manifestation of *SEC* 'catapult action' (Alexander and Bennet–Clark, 1977); the *SEC* acts as an energy buffer in which energy is accumulated at a rate efficient to the muscle fibers, but can be released almost instantaneously. We think that this effect is important in many 'explosive' movements as throwing, kicking and jumping etc.

Jumping without countermovement is relevant as an example of an activity in which negative work plays a minor role. 'Storage of elastic energy' is thus not a significant mechanism for achieving a large work output. The catapult mechanism, on the other hand, plays a major role by concentrating the accumulated muscle work for the very short time in which it is to be delivered at top speed. Admittedly, in man (Bobbert et al., 1988) this effect is less dramatic than in the flea (Alexander, 1988a).

One comment on the above discussion has to be made. We have considered the ankle angle and moment as given entitiles and deduced the behavior of the contractile element. In reality there is a close interaction between muscle active state, CC and SEC properties, and mechanical load. For example, the statement that CC speed is mostly between 0.4 and 2 rad/s, as long as M_c has a substantial value, may also be reversed: as soon as the CC reaches higher velocities, M_c decreases rapidly because of the force–velocity relation. An example can be seen in the final phase of push-off in running, Figure 38.8. This remark does not invalidate, in our opinion, the relevance of the four energy saving mechanisms, but questions how they are achieved in real-world movements.

38.4.7 Taking Up The Slack

In the recordings of running (Figures 38.4a,b) we see that at the onset of the contraction ϕ_c shows a steep rise. This is due to both an increase of M_c and the swift plantarflexion at heel strike. M_c is still low at that time, however, and the associated W_c^+ is very small. We can interpret this as the CC giving a slight pretension to the SEC, so that the 'toe' region of the SEC curve is left unused and the angular excursions can be kept smaller during the negative work phase (see Figures 38.4a and 38.2b). From the total 37° of SEC stretch, which corresponds to a moment of 220 Nm, the first 12° are delivered by the CC. This requires a low moment of 18 Nm and a W_c^+ of only 1.4 J for the elastic energy. The further 47.4 - 1.4 = 46 J of elastic energy are mainly provided by W^- (44 J) with an ankle dorsiflexion of no more than 22°. This effect was foreseen by A.V. Hill (1970) and

called 'taking up the slack'. For a further discussion see Hof and Van den Berg (1986). A further advantage is a high stiffness at heel strike, which provides better stability for the leg.

It should be noted, however, that the reverse can also occur, giving some extra slack. We see it in the walking recordings (Figures 38.3 a,b,c) of this subject.

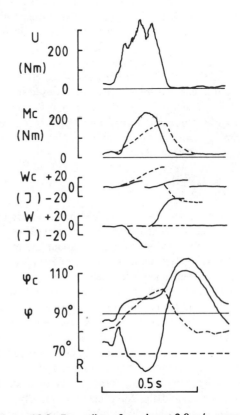

Figure 38.8: Recording of running at 2.0 m/s, same as in Figure 38.4a. Dashed lines: muscle moment, work, CC length etc., according to the muscle model in a hypothetical isometric contraction (ϕ fixed at 70°) with the same *EMG* input. The initial delay and the lower peak value are due to the long time constant of the CC–SEC combination, see text. The earlier drop of M_c in running, compared to the 'isometric' case, is due to the high CC shortening velocity.

38.4.8 Disadvantages of a Compliant Tendon

A compliant tendon gives important advantages to a muscle that has to perform eccentric-concentric or explosive actions. For other tasks it may be disadvantageous. To these belong actions that begin with an isometric phase. In an isometric action the total muscle plus tendon length is constant, but internally there is at the onset a shortening of the CC at the expense of a stretch of the SEC. This costs W_c^+ and metabolic energy, which cannot be regained in the decrescent phase, when the CC is lengthened again. Furthermore this process makes the development of force slower. The combination of the Hill-type CC and linear plus logarithmic SEC as used in our muscle model (Table 38.1) behaves in an isometric step response approximately as a first order system with time constant τ_m (cf. Hill, 1970, p. 106; Vit, 1978) in which

$$\tau_m = \frac{1+n}{b} \left(\frac{1}{\beta} + \frac{M_o}{K} \right) \tag{38.18}$$

For our parameters and $M_o = 220$ Nm this yields $\tau_m = 340$ ms, which is rather long compared to the duration of the activity in running, about 250 ms.

Figure 38.8 gives a recording of running on which is superimposed (dashed line) a hypothetical isometric contraction (ϕ constant at 70°) with the same activation. Figures like these are easy to make with a muscle model. The 'isometric' contraction is very sluggish compared to the real one and W_c^+ is at least equally high. In the real case the SEC is swiftly stretched by the external movement (the initial dorsiflexion), there is very little slack to overcome and the force can rise rapidly. It reminds one of the 'controlled stretch' experiments, in which A.V. Hill (1970) exposed the initial fast rise of the active state.

Next to this sluggish response is another problem in the control of muscles with compliant tendons. Immediately obvious in Figures 38.3 and 38.4 is the big difference in time course between ϕ (= muscle plus tendon length) and ϕ_c (= muscle fiber length). The muscle spindles are in parallel with the muscle fibers and can therefore not sense total muscle length, closely connected to joint rotation, but only a sum of muscle length and muscle force (translated into tendon stretch). This problem has been discussed by Rack (1981).

A third problem, in our probably incomplete list, is how to control a concerted contraction, with or without added stretch. To achieve the possible energy saving, force and length of the muscle must follow a closely connected time course (Section 38.4.4). At present we have only vague ideas how this might be performed. Feedback action by muscle spindles may be relevant, as the spindles sense CC length (see above). Also can we imagine that the interaction of a muscle CC plus SEC with a combination of a gravitational and inertial load, together representing the action of quadriceps or calf muscles in running, would more or less automatically produce a kind of concerted contraction, because of the inbuilt feedback due to the force–velocity relation.

Compliant tendons in muscles have their pros and cons. They are useful for saving energy in eccentric–concentric actions, they are a disadvantage in precise and fast movements and of no use in concentric movements requiring a great deal of muscle fiber shortening. It seems that a specialization along these lines indeed exists between muscles. In Chapter 36, Alexander and Ker found striking anatomical differences between muscles such as triceps surae, suited to eccentric-concentric work, with short muscle fibers and relatively compliant tendons, and other muscles, with longer fibers and relatively thick and stiff tendons.

38.5 Future Directions

It will be obvious that the above discussion contains sufficient open ends and gaps in the argumentation to provide research subjects for some time to come. We may name a few.

a) The present methods to access *in vivo* series elastic properties of human muscle are all rather indirect, not very accurate and they rely on a number of unproven assumptions (see Sections 38.1.3 and 38.4.2). A new method, against which the simple resonance method might be checked and which would separate purely passive elasticity from active reflex components, would be welcome.

b) Animal experiments with concerted contractions might provide a useful addition to the available data on the effects of muscle elasticity. Most experiments on elastic energy storage have been performed with tetanic stimulation and a prescribed lengthening and shortening of muscle plus tendon. Concerted contractions might be achieved with controlled lengthening and shortening, e.g. by means of a servo-controlled actuator, so that the external stretch counteracts the internal lengthening of the *SEC* (cf. Hill, 1970), with controlled stimulation (Rack and Westbury, 1969; Zhou et al., 1987) or with both. In such experiments it might be possible to demonstrate in vitro contractions in which eccentric work would be fully regained in later positive work, more convincingly than Morgan et al., (1978) could do.

c) Modeling studies to find possible rules of motor control for achieving concerted contractions in human activities, might clear up a few of the questions posed in Section 38.4.8.

38.6 References

Alexander, R.McN. and Bennett-Clark, H.C. (1977). Storage of elastic strain energy in muscle and other tissues. *Nature* **265**: 114-117.

Alexander, R.McN., Maloyi, G.M.O., Njau, R. and Jayes, A.S. (1982). The role of tendon elasticity in the locomotion of the camel (camelus dromedarius). *J. of Zoology* **198**: 293-313.

Alexander, R.McN. (1988a). Elastic mechanisms in animal movement. Cambridge UP,Cambridge

Alexander, R.McN. (1988b). The spring in your step: the role of elastic mechanisms in human running. In: G. de Groot et al., (eds.). *Biomechanics* **XI-A**: 17-25. Free University Press, Amsterdam.

Asmussen, E. (1953). Positive and negative muscular work. *Acta Physiol. Scand.* **28**: 364- 382.

Asmussen, E. and Bonde-Petersen, F. (1974). Storage of elastic energy in skeletal muscles in man. *Acta Physiol. Scand.* **91**: 385-392.

Bach, T.M., Chapman, A.E. and Calvert, T.W. (1983). Mechanical resonance of the human body during voluntary oscillations about the ankle joint. *J. Biomech.* **16**: 85-90.

Bennett, M.B., Ker, R.F., Dimery, N.J. and Alexander, R.McN. (1986). Mechanical properties of various mammalian tendons. *J. Zoology, London A* **209**: 537-548.

Bobbert, M.F., Huijing,P.A. and Van Ingen Schenau, G.J. (1986a). A model of the human triceps surae muscle-tendon complex applied to jumping. *J. Biomech.* **19**: 887-898.

Bobbert, M.F., Huijing, P.A. and Van Ingen Schenau, G.J. (1986b). An estimation of the power output and

work done by the human triceps surae muscle-tendon complex in jumping. *J. Biomech.* **19**: 899-906.

Bobbert, M.F. and Van Ingen Schenau, G.J. (1988). Coordination in vertical jumping. *J. Biomech.* **21**: 249-262.

Bogert, A. van den, Hartman, W., Schamhardt, H. and Sauren, A. (1988). In vivo relationship between force, EMG and length change in the deep digital flexor muscle in the horse. In: G. de Groot (eds.). *Biomechanics* **XI**, pp. 68-74, Free University Press, Amsterdam.

Boon, K.L., Hof, A.L. and Wallinga-De Jonge, W. (1973). The mechanical behaviour of the passive arm. In: E. Jokl (ed.). *Medicine and Sport,* **8**: *Biomech. III*, 243-248. Karger, Basel.

Cavagna, G.A. (1970). Elastic bounce of the body. *J. Appl. Physiol.* **29**: 279-282.

Cavagna, G.A. and Kaneko, M. (1977a). Mechanical work and efficiency in level walking and running. *J. Physiol.* **268**: 467-481.

Cavagna, G.A., Heglund, N.C. and Taylor, C.R. (1977b). Mechanical work in terrestrial locomotion: two basic mechanisms for minimizing energy expenditure. *Am. J. Physiol.* **233**: R243-R261.

Cavagna, G.A. (1977c). Storage and utilization of elastic energy in skeletal muscle. *Exercise and Sports Sci. Rev.* **5**: 89-129.

Evans, C.M., Fellows, S.J., Rack, P.M.H., Ross, H.F. and Walters, D.K.W. (1983). Response of the normal human ankle joint to imposed sinusoidal movements. *J. Physiol.* **344**: 483-502.

Fedak, M.A., Heglund, N.C. and Taylor, C.R. (1982). Energetics and mechanics of terristrial locomotion. II Kinetic energy changes of the limbs and body as a function of speed and body size in birds and mammals. *J. Exp. Biol.* **97**: 23-?0.

Gottlieb, G.L. and Agarwal, G.C. (1978). Dependence of human ankle compliance on joint angle. *J. Biomech.* **11**: 177-181.

Goubel, F. (1987). Muscle mechanics. In: Marconnet, P. and Komi, P.V. (eds.). *Muscular Function in Exercise and Training.* 24-35, Karger, Basel.

Greene, P.R. and McMahon, T.A. (1979). Reflex stiffness of man's antigravity muscles during kneebends while carrying extra weights. *J. Biomech.* **12**: 881-891.

Heerkens, Y.F., Woittiez, R.D. and Huijing, P.A., Huson, A. and Van Ingen Schenau, G.J. (1985). Inter-individual differences in passive resistance of the human knee. *Human Movement Science*, **4**: 167-188.

Heglund, N.C., Fedak, M.A., Taylor, C.R. and Cavagna, G.A. (1982). Energetics and mechanics of terrestrial locomotion. IV Total mechanical energy changes as a function of speed and body size in birds and mammals. *J. Exp. Biol.* **97**: 57-66.

Heglund, N.C., Cavagna, G.A. and Taylor, C.R. (1982). Energetics and mechanics of terrestrial locomotion. III Energy changes of the centre of mass as a function of speed and body size in birds and mammals. *J. Exp. Biol.* **97**: 41-56.

Hill, D.K. (1968). Tension due to the interaction between the sliding filaments in resting striated muscle, the effect of stimulation. *J. Physiol.* **199**: 639-681.

Hill, A.V. (1938). The heat of shortening and the dynamic constants of muscle. *Proc. Roy. Soc.* **B76**: 136-195.

Hill, A.V. (1960). Production and absorption of work by muscle. *Science* **131**: 897-903.

Hill, A.V.(1970). *First and Last Experiments in Muscle Mechanics*. University Press, Cambridge.

Hof, A.L. and Van den Berg, Jw. (1981a). EMG to force processing. I An electrical analogue of the Hill muscle model. *J. Biomech.* **14**: 747-792.

Hof, A.L. and Van den Berg, Jw. (1981b). EMG to force processing. II Estimation of parameters of the Hill muscle model for the human triceps surae by means of a calf ergometer. *J. Biomech.* **14**: 759-770.

Hof, A.L. and Van den Berg, Jw. (1981c). EMG to force processing. III Estimation of model parameters for the human triceps surae muscle and assessment of the accuracy by means of a torque plate. *J. Biomech.* **14**: 771-785.

Hof, A.L. and Van den Berg, Jw. (1981d). EMG to force processing. IV Eccentric-concentric contractions on a spring-flywheel set up. *J. Biomech.* **14**: 787-792.

Hof, A.L., Geelen, B.A. and Van den Berg, Jw. (1983). Calf muscle moment, work and efficiency in level walking; role of series elasticity. *J. Biomech.* **16**: 523-537.

Hof, A.L. and Van den Berg, Jw. (1986). How much energy can be stored in human muscle elasticity? (Letter to the editor). *Human Movem. Sci.* **5**: 107-114.

Hof, A.L., Pronk, C.N.A. and Van Best, J.A. (1987). Comparison between EMG to force processing and kinetic analysis for the calf muscle moment in walking and stepping. *J. Biomech.* **20**: 167-178.

Hof, A.L., Struwe, D.P. and Nauta, J. (1989). Calf muscle work and segment energy changes in running. *Proc. XII Congress of Biomech.*, Los Angeles, 1989: p.322.

Ingen Schenau, G.J. van (1989). From rotation to translation: constraints on multi-joint movements and the unique action of bi-articular muscles. *Human Movem. Sci.* **8**: 301-337.

Jewell, B.R. and Wilkie, D.R. (1958). An analysis of

the mechanical components of frog's striated muscle. *J. Physiol.* **143**: 515-540.

Ma, S.P. and Zahalak, G.I. (1985). The mechanical response of the active human triceps brachii muscle to very rapid stretch and shortening. *J. Biomech.* **18**: 585-598.

McMahon, Th.A. (1984). *Muscles, Reflexes and Locomotion*. Princeton Univ. Press, Princeton, N.J.

McMahon, Th.A., Valiant, G. and Frederick, E.C. (1987). Groucho running. *J. Appl. Physiol.* **62**: 2326-2337.

MacPherson, L. (1953). A method of determining the force-velocity relation of muscle from two isometric contractions. *J. Physiology* **122**: 172-177.

Morgan, D.L., Proske, U. and Warren, D. (1978). Measurements of muscle stiffness and the mechanism of elastic storage of energy in hopping kangaroos. *J. Physiol.* **282**: 253-261.

Nauta, J. and Hof, A.L. (1989). Calf muscle work and segment energy changes in human treadmill walking. In: G. de Groot (eds.). *Biomechanics* **XI-A**, 38-42: Free Univ. Press Amsterdam.

Proske, U. and Morgan, D.L. (1987). Tendon stiffness: methods of measurement and significance for the control of movement. A review. *J. Biomech.* **20**: 75-82.

Rack, P.M.H. and Westbury, D.R. (1969). The effect of length and stimulus rate on tension in the isometric cat soleus muscle. *J. Physiol.* **204**: 443.

Rack, P.M.H.(1981). Limitations of somatosensory feedback in control of posture and movement. In: *Handbook of Physiology. Section 1. The nervous system*. **Vol. 2** *Motor Control* (ed. by Brooks, V.B.). p. 229-256.

Shorten, M.R. (1987). Muscle elasticity and human performance. In: B. van Gheluwe and J. Atha (eds.). Current research in Sports Biomechanics (*Medicine and Sports Science*, **25**: p. 1-18), Karger, Basel.

Tardieu, C., Colbeau-Justin, P., Bret, M.D., Lespargot, A. and Tardieu, G. (1981). Effects on torque-angle curve of differences between the recorded tibia-calcaneal angle and the true anatomical angle. *Eur. J. Appl. Physiol.* **46**: 41-45.

Taylor, C.R., Heglund, N.C. and Maloiy, G.M.D. (1982). Energetics and mechanics of terrestrial locomotion. I. Metabolic energy consumption as a function of speed and body size in birds and mammals. *J. Exp. Biol.* **97**: 1-21.

Vit, K. (1978). Analytical solution to isometric mechanogram of Hill's model of striated muscle. *Bull. Math. Biol.* **40**: 359-368.

Walsh, E.G. and Wright, G.W. (1988). Postural thixotropy at the human hip. *Quart. J. Exp. Physiol.* **73**: 369-377.

Weiss, P.L., Hunter, I.W. and Kearney, R.E. (1988).

Human ankle joint stiffness over the full range of muscle activation levels. *J. Biomech.* **21**: 539-544.

Williams, K.R. and Cavanagh, P.R. (1983). A model for the calculation of mechanical power during distance running. *J. Biomech.* **16**: 115-128.

Winters, J.M. and Stark, L. (1985). Analysis of fundamental human movement patterns through the use of in-depth antagonistic muscle models. *IEEE Trans. BME* **32**: 826-839.

Winters, J.M. and Stark, L. (1987). Muscle models: what is gained and what is lost by varying model complexity. *Biol. Cybernetics* **55**: 403.

Yoon, Y.S. and Mansour, J.M. (1982). The passive elastic moment at the hip. *J. Biomech.* **15**: 905-910.

Zhou, B.H., Baratta, R. and Solomonov, M. (1987). Manipulation of muscle force with various firing rate and recruitment control strategies. *IEEE Trans. Biomed. Eng. BME* **34**: 128-139.

CHAPTER 39

Muscular Coordination in Sporting Skills

Arthur E. Chapman, David J. Sanderson

39.1 Introduction

It is our opinion that while research on muscle has blossomed since the early 1900s, as illustrated by the size of this book, we have hardly begun to apply knowledge of muscle to the understanding of athletic technique and its modification. For example, a review of the papers given at the 12 meetings of the International Society for Biomechanics reveals much work concerned with sports, yet relatively few are concerned with if and how mechanical properties of muscle predispose the athlete to adoption of a specific technique. This chapter is concerned with muscular coordination in sporting techniques, what information is currently available and where this area of research may lead.

Information on appropriate muscular coordination is buried deeply within the central nervous system of skillful individuals. Undoubtedly the nervous system has knowledge of the mechanical characteristics of muscles, and if not specific knowledge of characteristics, knowledge of the mechanical outcome of activating muscle. The process of execution of a skill begins with a specific aim from a given set of initial conditions which may or may not be entirely under the control of the athlete (e.g. the contrast between throwing a baseball from rest versus cross-country running over variable terrain). Consequently muscular coordination can range from an exact temporal sequence which would be ideal to repeat on each occasion (the throw) to variations on a basic temporal pattern according to conditions (the run). In either case there is an assumption that there is an ideal sequence of muscular coordina-tion which allows achievement of a given aim by means of a system of muscles acting upon a skeletal structure. Just as the engineer needs to know the characteristics of motors to design a control system (see Chapter 7 (Hannaford and Winters)), so the athlete's nervous system and the sports biomechanist need to understand the human motor. In this way the biomechanist may be able to aid the athlete's nervous system in control of muscle.

Muscular coordination can be investigated by a number of techniques. Electromyography gives a direct measure of the recruitment of muscles provided that caution is undertaken in the treatment and interpretation of the *EMG*. Kinematics provide a description of motion only. Kinetics gives joint forces and torques. Joint forces are a convenient biomechanical resultant of muscular, ligamentous and bone-on-bone forces. Torques are the net result of action of a number of muscular protagonists and antagonists. In this sense the torque is a result of activity of a single equivalent muscle. In combination with kinematics, torques can be identified as resulting from either concentric, isometric or eccentric muscular activity. Such information is useful in describing how muscular activity is coordinated in relation to the kinematic state of the muscles when activity begins. Therefore the use of such beneficial phenomena as the stretch–shortening cycle (*SSC*) of muscle can be identified. By means of models of individual muscles, the inverse dynamic approach has been used to solve the general distribution problem (*GDP*). The aim here is to identify the timing and magnitude of the contribu-

Multiple Muscle Systems: Biomechanics and Movement Organization
J.M. Winters and S.L-Y. Woo (ed.), © 1990 Springer-Verlag

tion of individual muscles to a specific task. The question of application of this technique is whether such information can be used by a performer. For example, would the athlete be at an advantage if he/she were to know that m. supinator was relatively more active than m. biceps brachii in a complex striking skill involving supinator torque? A final approach, and the only one with predictive qualities, is that of computer simulation. This approach allows the searching for an optimal solution for the skill in question. Armed with average values for segmental parameters and mechanical characteristics of single equivalent muscles, the simulator can identify general strategies for coordination of these muscles. A model representing a specific individual might lead to fine tuning of the general strategy. This work is in its infancy.

That the same segmented body is used for all athletic skills may be too obvious to state. Yet the aims within the spectrum of athletic skills are so varied that the same segmented body must adopt different strategies. For this reason the subsequent discourse on muscular coordination is divided into two basic classes of sporting actions; namely throwing (and striking and kicking) and cycling. These two activities are so different that we might expect significantly different principles of coordination to be adopted.

39.2 Muscle Mechanics

This section is not a lengthy discourse on the numerous, complex properties exhibited by skeletal muscle. Readers are directed to Chapman (1985) for references in this area and Chapters 1-5 for more detailed information. The aim in this section is to provide sufficient information to illustrate why a knowledge of muscle mechanics is a necessary precursor to interpreting experimental evidence on the role of muscular coordination in sporting skills.

While muscle comprises numerous complex structural elements involved in both the development and transmission of contractile force, a convenient functional model comprises at least two components. The contractile component (*CC*) is that which develops force under influence of neural control and the series elastic component (*SEC*) transmits this force to the bone.

Contractile Component. The *CC* develops force as a function of the intensity of neural

stimulation (activation). At a given level of activation the isometric, steady–state force produced by the *CC* is a non-linear function of its length. Force is low at a short *CC* length; it increases as *CC* length increases and it subsequently decreases with further increases in length. Thus the force of the *CC* is greatest at some intermediate length. For a given level of activation and length the force in the *CC* is a non-linear function of velocity. As shortening velocity increases (concentric contraction) force decreases, while an increase in lengthening velocity (eccentric contraction) produces increased force. There is therefore a four-dimensional relationship between force, length, velocity and activation [see also Chapter 5 (Winters)]. Of current, general interest is the fact that at any instant *CC* force is enhanced if the muscle has previously been stretched and it is depressed following prior shortening. These last two phenomena represent history-dependent properties of the *CC*. In muscle *in situ* activation provides the greatest variation in *CC* force followed in decreasing importance by velocity, length and history-dependent properties.

Series Elastic Component. The *SEC* is a component which is continually revealing increased complexity. The basic property is that the *SEC* transmits force to the bone as a non-linear function of its length such that its stiffness increases with increased length. As some of the series elasticity resides in the sarcomeres, *SEC* stiffness is dependent upon muscle activation and length. In developing a complex phenomenological model of muscle all of the above properties require incorporation. At the current level of understanding of muscular control in skill it is suggested that the simple recognition of the presence of a *SEC* is sufficient.

Chapman (1985) has described the interaction of the *CC* and *SEC* in typical types of muscular contraction; stretch followed by shortening (*SSC*) and repetitive cyclical motion of a load against gravity. The *SSC* has received much attention from a sporting standpoint. The basic story is that either external load or antagonist muscles move a limb segment in a direction opposite to the required final motion (See Figure 39.1). The protagonist muscles then begin activation in an eccentric state. Consequently force rises rapidly because the *CC* is working either on the eccentric part of the force–velocity relationship or at a small velocity on the concentric part. By the time the eccentric motion of the limb segment is arrested the force applied to that segment is high. Therefore subsequent concentric contraction begins from this high level. Figure 39.1 (S-R) demonstrates how the torque in the concentric phase of forearm supination, which follows muscle stretch (S), is much greater than that in a concentric contraction from rest (R). Although torque from rest eventually exceed that from the SSC, the early high torque in S produces the most rapid accumulation of angular impulse and therefore a faster rise in load angular velocity. It can be seen that the concentric torque following an isometric development of tension (I) is also greater than that from rest (R), but not as beneficial as that from the SSC. Simulation has shown that this benefit of the SSC is realized with a muscle model having only a CC (see Figure 2b in Chapman, 1988). The more realistic incorporation of the SEC also shows beneficial effects, and the relative benefit depends on the relative properties of the CC (rate of activation, FV relationship) and SEC (stiffness) (see Figure 2 in Chapman, 1988). The benefit of the SSC has been ascribed to storage and release of strain energy in elastic structures. While this view is convenient to indicate a saving of energy through use of a spring, it is the current authors' view that this explanation is misleading when applied to a single contraction. Without activation of the CC the SEC will store no energy. In fact the gain in velocity of a load is obtained at the expense of a greater amount of work done by the CC in a single SSC compared with that in a single contraction from rest (see Figure 2f in Chapman, 1988). Thus it appears that the advantage of the SSC is to allow greater muscular work to be done to stretch the SEC in the eccentric phase. It appears that storage and release of energy is a consequence of the SSC and not a reason for its use. The presence of the SEC modifies the kinematics of the CC, which allows the latter to work on a favorable part of the FV relationship. Many investigators may disagree with aspects of this explanation [e.g. see Chapter 38 (Hof)]. Reflex potentiation of activation has also been cited as a benefit of stretch. Yet there is evidence that reflexes do act (Aura and Komi, 1988) and do not act in maximal muscular effort (Thomson and Chapman, 1988).

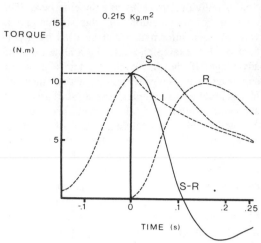

Figure 39.1: Torque versus time in forearm supination against the inertia shown. R, I and S refer respectively to contractions from Rest, a release from prior Isometric development of tension and from Stretch, where the supinators contracted against an eccentrically moving load. S-R is a measure of how much the torque in S exceeds that in R. (Reproduced with permission of the Free University Press, Amsterdam, The Netherlands.)

The story in cyclical contraction against gravity, such as in running, is simple (Chapman, 1985). At a specific cyclical frequency most of the external motion is accommodated by SEC stretch and recoil [see also Chapter 37 (McMahon)]. This allows the CC to remain in an almost isometric state which avoids the extra metabolic cost of excursions of the CC. At other than this resonant frequency the CC has to shorten and lengthen. Consequently cyclical SSC's will show different gains (e.g. force/activation or load kinetic energy/CC work) at different cyclical frequencies (Bach et al. 1983). In this repetitive type of contraction the concept of storage and release of strain energy in the SEC is a reasonable explanation of energy saving, but only in the sense that CC work is minimized but not obviated completely. When gravity does not play a significant part as in cycling, muscular contraction would appear to be entirely concentric. However, the work of Ryan et al. (1989) indicates that there are eccentric modes of contraction. In this case the benefits of the SSC may be present.

39.3 Kicking, Striking and Throwing

Striking and throwing have the common aim of projecting an object with various degrees of accuracy and speed. Conventional wisdom through observation of experts dictates that the sequence of progression of segmental involvement is from proximal to distal (*P–D*). Research in muscular coordination of throwing and striking has centered largely upon describing patterns of motion while few researchers have considered why the *P–D* progression appears best. That experts in these actions become experts by copying other experts suggests that research in this area is merely an academic exercise with little practical value. But until these skills are understood completely it can not be assumed that copying is the best method of learning. For example, Hatze (1976) demonstrated an improvement in learning of a kicking task by presenting a subject with kinematic information obtained from optimization of a myocybernetic model of the subject. This work suggests that the fine tuning of coordination can be determined for an individual to produce a better result than if the individual is left to their own devices. Unfortunately the work of Hatze (1976) has not as yet been built upon significantly by other researchers.

The reasons for using the *P–D* sequence have been investigated by Herring and Chapman (1988, 1989). A simple, three–segment, sagittal-plane, overarm throwing task was simulated using constant joint torques. A search field was used in which the temporal onset of the three joint torques was varied to find the throw yielding maximal velocity and maximal range. All throws began from the same joint angles and angular velocities (negative to replicate a backswing). Both ball velocity and range were maximized only if the sequence of onset of torques was initially shoulder extension followed by elbow extension and finally wrist flexion. Different sequences of onset produced poor results. From this work it can be concluded that the *P–D* sequence is desirable simply because of the linked segmental nature of the model (limb) irrespective of any impact which muscle properties may have on the action. A similar simulation by King and Huston (1989) confirmed the suitability of the *P–D* sequence.

A further observation by Herring (1989) was that ball range and velocity would be enhanced by reversing torques, again in a proximal to distal sequence. This type of antagonist activity has been suggested by others as being beneficial largely through observation of changes in segmental angular velocities. Evidence of reduced and reversed torques in human throwing is present but not universally observed in all types of throwing, striking and kicking actions. A study of the tennis serve Bahamonde (1989) indicated that those players who demonstrated a change from extensor to flexor elbow torque produced the greatest ball velocity. Furthermore Feltner (1989) stated that rapid elbow extension in a baseball pitch resulted from reversal of trunk rotation and not from the action of elbow extensors. That the muscles responsible were those of the trunk and legs demonstrates torque reversal of a very proximal set of muscles. Strangely Feltner and Dapena (1986) provide evidence for the production of elbow extensor torque in the same skill. Presumably the statement by Feltner (1989) implies that the elbow extensor torque was insufficient to account entirely for the observed kinematics of elbow extension. Although the outcome of a reversal of a given joint torque in a multisegmental body is dependent upon the kinematics of the linked segments, the reversal of a proximal torque will generally tend to decelerate the proximal segment. However, changes in a distal torque can also lead to deceleration of the proximal segment. For example, Dunn and Putnam (1988) and Putnam (1983) showed deceleration of the thigh in punting to be the result of shank kinetics. A flexor hip moment remained throughout the whole kick and was never reversed in sign (Figure 39.2). Putnam and Dunn (1987) also analyzed kicking off the ground and described similar kinematic patterns to the punt. However, the pattern of joint torques, while following the *P–D* sequence, showed significant differences. In the same type of kick Robertson and Mosher (1985) showed hip flexor torque to drop dramatically, while Zernicke and Roberts (1976) showed hip torque to become extensor prior to ball contact. Alternatively Roberts et al. (1974) show no evidence of torque reversal. An interesting modification of the *P–D* sequence was shown by Feltner (1989) in the baseball pitch. Peak angular velocity of elbow extension preceded the more proximal peak angular velocity of upper-arm internal rotation.

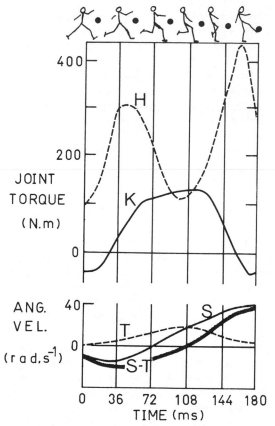

Figure 39.2: Experimental results from a punt. *H* and *K* refer repectively to torques about the hip (flexor is positive) and knee (extensor is positive) joints. *T* and *S* refer respectively to absolute angular velocities of the thigh and shank (counterclockwise is positive). *S–T* represents angular velocity of knee joint extension. Redrawn from Putnam (1983). (Reproduced with permission of Human Kinetic Publishers, Champaign, IL, and the author.)

Many investigators have described the *P–D* sequence on the basis of measurements other than joint torques. Wilson et al. (1989) showed that peak segmental angular velocities occurred in a *P–D* sequence when moving loads as fast as possible with the hands. Elliott and Chivas (1988) described a *P–D* sequence of peak segmental endpoint velocities in hitting a hockey ball. In this case the hockey stick behaves in much the same manner as any bodily segment. The same *P–D* sequence of peak segmental angular velocities was described by Luhtanen (1988) in experienced volleyball players. Less experienced players who

produced lower ball velocities did not always show this sequence. Similarly, in the tennis serve the maximal linear velocities of segment endpoints demonstrate a *P–D* sequence (Van Gheluwe and Hebbelinck, 1985).

Whichever criteria are used to assess the use of *P–D* sequence, it seems reasonable to conclude that the motion is caused by a *P–D* sequence of onset of protagonist torque. Undoubtedly the general *P–D* sequence of muscular coordination is a basis of execution which must meet the demands of varying initial conditions and varying specific aims of the skill. Whether deceleration of a proximal segment is caused by proximal torque reversal or by distal torque increase appears to depend upon the nature of the multisegmental task.

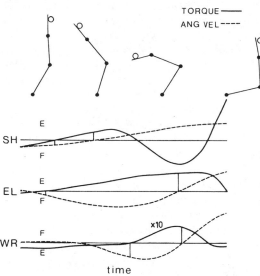

Figure 39.3: Experimental results from an overhand throw using the arm in sagittal-plane motion. *SM*, *EL* and *WR* refer respectively to the Shoulder, Elbow and Wrist joints. Extension (*E*) and Flexion (*F*) are indicated. For each joint torque the vertical line shows that torque is zero when the joint is moving in opposition to the throwing direction. The second vertical line shows how torque in the throwing direction is well established when the joint is at zero angular velocity (from Chapman, 1985, 1988; reproduced with permission of Free University Press, Amsterdam, The Netherlands).

Having established that the linked-segmental nature of a limb predisposes the system to *P–D* torque onset, the question remains as to the manner in which muscular properties are implicated. In this context the *SSC* is paramount. In throwing (Feltner and Dapena, 1986) and kicking (Putnam, 1983), the presence of a backswing is clearly evident. The immediate conclusion is that the protagonist muscles begin activity in an eccentric mode. As discussed in Section 39.2, the magnitudes of joint moments will be high at the onset of the concentric phase. For example, Figure 39.3 demonstrates a well established shoulder extensor torque when the shoulder joint is at zero angular velocity, having undergone a period of eccentric contraction. Similarly Figure 39.2 shows a hip flexor torque of 100 Nm when hip joint angular velocity is zero in kicking. Clearly the subsequent angular impulses applied to the upper arm in throwing (Figure 39.3) and the thigh in kicking (Figure 39.2) will be enhanced by the torques beginning above zero. The effect of these proximal protagonistic torques will be to induce a joint force at the proximal end of the next more distal segment. With reference to Figure 39.3, such a force will tend to induce and/or increase backward rotation of the forearm. Similarly in Figure 39.2 the shank backward rotation is aided, in this case by a knee flexor torque. The consequence of these effects is that the forearm (Figure 39.3) and shank (Figure 39.2) show increases in negative angular velocity during the initial period when the upper arm and thigh show increasing positive angular velocity. An examination of both elbow extensor torque and knee extensor torque shows that the extensor muscles are active in eccentric contraction. Eccentric contraction serves to reduce angular velocities of elbow and knee flexion such that when these velocities reach zero, both elbow and knee extensor torques are large. Again this represents clear evidence of use of the *SSC*, since the greater part of the elbow and knee extensor angular impulses serve to arrest backward motion of both forearm and shank respectively. For the throw in Figure 39.3 it can be seen that the wrist flexor moment begins eccentrically to arrest wrist extension and is high when zero angular velocity of the wrist is achieved. Such segmental end-point forces have been shown to induce external rotation of the upper arm prior to the required internal rotation in baseball pitching (Feltner and Dapena, 1986). The more proximal muscles producing horizontal abduction are responsible for this effect, and they result in putting the internal rotators into stretch. The use of successive *P–D* muscular recruitment requires appropriate timing for success, and the complexity would seem to increase with the greater number of segments involved. For example, Elliott et al. (1989) observed two types of forehand drive in tennis, one of which involved substantial intersegmental motion in the arm, and one in which the whole arm appeared to act as a single unit. In fact the segmental end-point velocities in the direction of the stroke showed a similar sequence of peaking despite the apparent observed difference in technique. This indirect evidence suggests use of a fundamental pattern of muscular recruitment. However, the greatest ball velocity shown by the multisegmental group suggests indirectly greater relative involvement of *SSC*'s to enhance total angular impulse.

In certain kicks the hip flexor torque is variable; however, it never becomes extensor in sign. This appears to negate the possibility that distal segment velocities can be enhanced by antagonistic activity of muscles crossing the most proximal joint, i.e. the use of a mechanism to aid transfer of angular momentum from proximal to distal segments. Yet Figure 39.3 shows an antagonistic torque at the shoulder for a period of time when both elbow and wrist torques are protagonistic. Although the throw shown in Figure 39.3 may not have been the best, the possibility still remains (as yet unconfirmed) that *P–D* antagonism following *P–D* protagonism may be a feature of control which enhances ball velocity in throwing.

The certain pattern which emerges is the successive *P–D* use of *SSC*'s in protagonist muscles of throwing, kicking and hitting. Use of the *P–D* sequence is therefore beneficial from the point of view of both the physical nature of the linked segments and the beneficial effects of the *SSC* of muscle. In the simulations by Herring and Chapman (1988, 1989), which involved constant muscular torques, the *SSC* is seen to occur in a *P–D* sequence. This begs the question as to how important are the muscular characteristics. It is tempting to suggest that muscular *FV* characteristics evolved to take advantage of the fact that a *SSC* is used in optimal throwing. Yet the specific action of throwing in the form which we understand it is confined to humans, and other species have muscles which exhibit similar dynamic muscular characteristics.

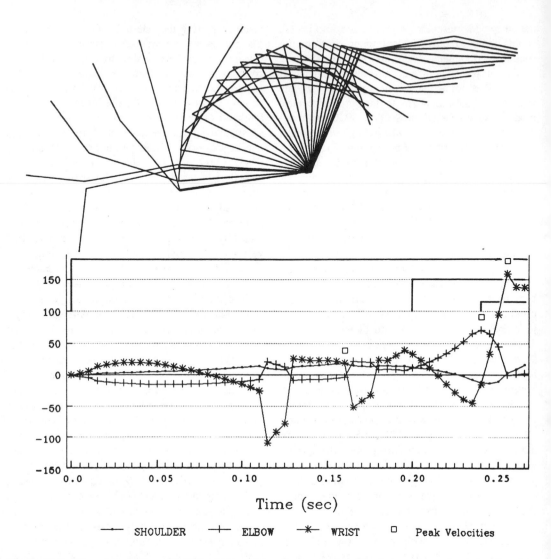

Figure 39.4: The upper diagram illustrates kinematics of a simulated throw to the left for maximal end-point velocity. The bottom diagram shows angular velocities (rad/s) of the joints indicated where a positive value is in the direction of the throw (counterclockwise). Square symbols illustrate peak angular velocities. The step functions are, in order of time, torque inputs at the shoulder (extension, 40 Nm), elbow (extension, 25 Nm) and wrist (flexion, 7.5 Nm). (From Herring (1989); reproduced with permission of the author.)

Muscular control in throwing depends upon the aim of the throw. The simulations of Herring and Chapman (1988, 1989) that are shown in Figures 39.4 and 39.5 illustrate best kinematics and temporal onset of torques for maximal velocity (Figure 39.4) and maximal ball range (Figure 39.5). Where maximal velocity is the aim (Figure 39.4), the shoulder extensor torque is turned on at time zero while elbow extensor and wrist flexor torques turn on at 200 ms and 240 ms, respectively. In contrast, maximal ball range resulted from *P–D* torque onsets of 120 ms, 220 ms and 260 ms, respectively. The early onset of shoulder torque in the maximal velocity throw presumably indicates the necessity for use of maximal angular impulse. Although the greatest total angular impulse could be achieved with simultaneous onset of all torques at zero time, the resulting pattern of

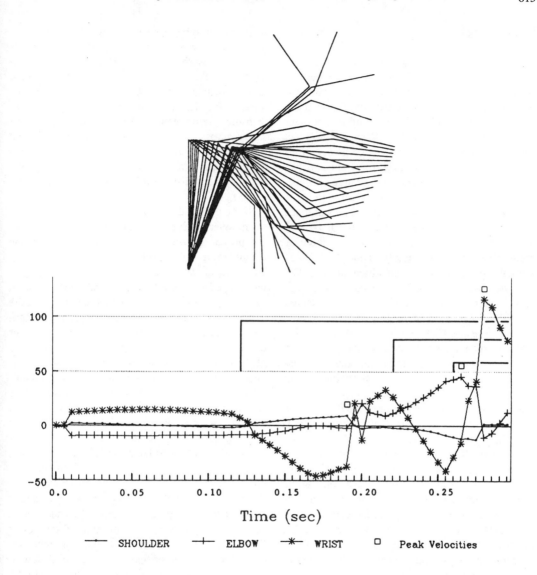

Figure 39.5: The kinematics of a simulated throw for maximal ball range. See caption to Figure 39.4 for explanation of symbols. (From Herring (1989); reproduced with permission of the author.)

motion proved inappropriate due to too early achievement of full wrist flexion. The aim here was to achieve maximal velocity irrespective of its direction and therefore segmental orientation. Had a directional requirement been imposed upon the ball velocity vector, different times of onset of torques would have been necessary. The extent of such a modification can be seen in the throw for maximal ball range (Figure 39.5). In this use the best combination of velocity and angle of release is required. The substantial delay in onset of shoulder torque (120 *ms*) enables considerable flexion of the elbow to be achieved under gravitational influence. Final elbow extension and wrist flexion therefore occur in a manner appropriate to achievement of the correct ball velocity vector. The *P-D* sequence of onset of torques and peak angular velocities is retained as in the case of the maximal velocity aim.

39.4 Cycling

Cycling, like walking and running, is an activity that is characterized by a repeating pattern of coordinated muscular effort. When the situation demands prolonged effort the pattern of movement is consistent and exhibits little variation. For example, in normal paced walking Winter (1987) has shown coefficients of variation of lower limb kinematics to be 8% and 21% for knee and hip joints respectively. Further, over a range of walking speeds there is relatively little change in these joint kinematics. Cycling represents an even more constrained task. The constraints of seated cyclists are such that there have been a number of authors (Hull et al. 1988; Redfield and Hull, 1985) who have presented predicted kinematic data for the lower limb based upon five–bar linked segment models. These models have presented reasonable kinematics of cycling from analytical solutions [see also Chapter 40 (Hull et al.)].

One of the interesting characteristics of cycling is that there appears to be an optimal rate at which it is performed. "Optimal" has been interpreted from a variety of perspectives. When the movement task is one dictated by prolonged effort the performer will select a rate which is consistent, repeatable, and relatively invariant for a given level of power output. When the terrain is predictable, as in a gymnasium or on a stationary bicycle, the control of this rate appears to be automatic, or at least at some level below the conscious task of on-line programming. However, the pattern is not so entrained that it cannot be consciously varied (e.g. McMahon et al. 1987) or indeed trained to be different (Sanderson and Cavanagh, 1990). In cycling the optimal rate or cadence has been shown to be dependent upon power output (Coast and Welch, 1985) and indirectly on the level of performance.

It seems reasonable to expect that the rate of pattern replication would reflect some match between required neuromuscular control and the mechanical characteristics of skeletal muscle. In other words, there is some level of optimization of resources which results in the particular frequency of the movement pattern. There have been a number of published works that have examined this phenomenon in cycling. The apparent goal has been to provide information to assist training of cyclists so as to achieve a desired pedaling rate

which is felt to be best for successful performance. That is, one wants to be able to make statements like "If you wish to ride at 200 watts then pedal at 87 rpm". To this end there is an abundance of data, primarily physiological in nature, which addresses the issue of minimal energy cost as measured by oxygen uptake. These data are consistent across running, walking, and cycling in that they show a clear parabolic relationship between oxygen uptake and cadence at particular power outputs with a distinct minimum associated with a given rate.

If we accept that oxygen uptake is an indicator of the overall level of muscular effort, and if we accept that the primary contribution to this value comes from the muscles of the legs, then we would conclude that the selection of a rate of pattern repetition exhibits minimal energy cost. Further, it would seem likely that the design of the lower limb musculature is such that each component muscle would develop maximum power output consistent with the power output demands of the task. One avenue to explore in the task of identifying the particulars of the muscle mechanical characteristics would be to query whether the rate of pattern repetition was similar for different tasks using the same muscles. For example, if the primary controlling factor is the force–velocity relationship, as postulated by Chapman (1985), then for similar power outputs in walking and cycling there should be a similar optimal rate. Published data on the limb kinematics (Cavanagh and Sanderson, 1986; Winter, 1987) indicate a similarity in the range of motion of the lower limbs in these two tasks. The external forces required, on the other hand, are quite different. However, walking is a weight–bearing activity, while cycling is not. Nonetheless there are an infinite number of force–velocity curves; when the factor of activation is accounted for it is not impossible to consider a combination that would result in a similar frequency.

We are suggesting that regardless of the length, or rather assuming that length change remains within the middle of the force-length relationship, the FV relationship dictates the most desirable rate. Hill (1922) showed that for a given FV relationship a power velocity relationship could be developed and that this showed a peak power output for the muscle. Cavanagh and Kram (1985) have argued that this relationship leads to the conclusion that these very mechanical properties are the main properties that are used in the optimizing process. In other words, the rate at which power output is maximized for a given demand determines the desired rate of movement. As this is a peak in the power output–velocity curve, the constraints on the system would then be to attempt to maintain the linear velocities of shortening near the same value. In walking at a normal pace the range of knee joint angular velocities is within $\pm$ 5 rad/s while in cycling at 60 rpm the ranges are similar (Nordeen, 1976).

Kroon (1983) presented an assessment of optimal cadence and argued that there indeed were similarities in the optimal frequencies of walking and cycling that indicated the same underlying controlling mechanism, namely the mechanical properties of skeletal muscle. This hypothesis suffers a risk of oversimplification, however. In the action of the lower limbs there are numerous muscles involved in moving the limbs. *EMG* examination indicates that there are activities in the one–joint muscles as well as the two–joint muscles. To find a simple relationship between force and velocity which would predict the exact optimal cadence will be difficult indeed. Given the variations in the geometrical relationship between the muscles and the joints, whether they be one or two–joint muscles, and the consequent variations in the lengths of the muscles during the ranges of motion it would require a very sophisticated development system to match. Even if the perfect match among the muscles is not made, there will be an optimal overall relationship even though some muscles would be working over ranges that are less than optimal. The issue may not be to minimize energy cost for individual muscles but for the whole system. The fact that one can ride without a particular muscle dropping out indicates some degree of success. Yet a pattern of recruitment which overloads a particular muscle is unlikely to be used, even if the energy cost of the whole system is small. Ryan et al. (1989) have presented data which show that there is indeed a *SSC* in the muscles of the lower limb during steady–state cycling. Specifically, the gastrocnemius showed active lengthening during the first 90° of crank motion after top dead center. Gregor et al. (1985) have presented data which show a delicate balance between the simultaneous contributions of the one- and two-joint muscles

during the pedaling cycle. While they did not have data on muscle lengths, their data would suggest that force and velocity characteristics in association with moment arm lengths are very important for effective cycling.

One issue in the literature that is controversial is the apparent lack of agreement on whether the optimal rate reflects muscle properties or some other level of control. For example, Pandolf and Noble (1973) have presented data which suggest that optimal cadence is not associated with a feeling of metabolic ease. Rather, a higher rate is more desirable. This perspective argues that the desired rate is one which gives the perception that the forces within the legs are minimized. As this occurs at a higher cadence, oxygen uptake is elevated. Kram (1987) has argued that for a shift in pedaling rate, the apparent advantage of reduced pedal forces and associated muscle moments outweigh the detrimental effects of increased metabolic cost. Merril and White (1985) have concluded from a different perspective that the higher pedaling rates of competitive cyclists are a reflection of peer pressure and training. This leads to an interesting question of whether the optimal cadence is based on mechanical properties of muscles, perception of metabolic effort, perception of perceived exertion or training using conventional wisdom as a guide.

39.5 Summary and Future Directions

The underlying presumption has been that it is beneficial to know all of the properties of all muscles potentially involved in a skill. If this were the case, how could the information be used to identify appropriate muscular coordination in a skill? The enormous amount of information would preclude a solution by means of thought alone. A reasonable solution appears to be through modelling and simulation of the task according to some objective function or optimization scheme (see also Chapter 8 (Zajac and Winters)). Providing that information can be given to the athlete in a usable form, the simulation could be considered successful in producing the most desirable performance. Further useful application would result from modifying muscular characteristics with a view to prescribing what and how to train. Incorporating fatigue functions would allow investigation of the possibility of varying muscular coordination to get the best from a muscular system which is subject to change. These applications are either in their infancy or unborn, but in the context of throwing and cycling they may take the following forms.

How much variation in muscular coordination in throwing is appropriate in a group of individuals showing different relative segmental inertias and dimensions? In other words, can we be sufficiently accurate in simulations to customize the simulation to an individual? Even if we can, is it necessary? Do we simply provide the basic pattern of coordination and let the nervous system deal with fine-tuning of the muscular system? These questions will be answered by the use of sensitivity analysis: varying segmental dimensions and inertias and muscular characteristics in our simulation [see also Chapter 8 (Zajac and Winters)]. One major advantage of this approach is to allow prediction of necessary adjustments in coordination when muscular characteristics change because of training.

To achieve maximal overground velocity in cycling, the angular impulse applied to the front chain ring needs to be maximized. This would be achieved by a particular pattern of force application to the pedal, which in turn would result from a specific sequence of contractions of single equivalent muscles of the lower limb in a simulation. The aim would be to obtain the required pedal-force profiles with a sequence of contractions requiring minimal activation. Therefore the initial need in determining desirable muscular coordination in cycling is to identify appropriate pedal-force profiles.

Should our simulations be successful, the final overriding problem is how to communicate muscular coordination to the performer. Only when this problem is solved will we be able to assess the success of our efforts to improve muscular coordination in sports.

References

Aura, O. and Komi, P.V. (1988) The mechanical efficiency of isolated leg extensor activities at different work intensity levels. *Biomechanics XI-A* (edited by de Groot, G., Hollander, A. P., Huijing, P.A., and van Ingen Schenau, G.J.), pp. 48-51. Free University Press, Amsterdam.

Bach, T.M., Chapman, A.E. and Calvert, T.W. (1983) Mechanical resonance of the human body during voluntary oscillations about the ankle joint. *J.Biomechanics* 16: 85-90.

Bahamonde, R.E. (1989) Kinetic analysis of the serving arm during the performance of the tennis serve. *Proc. 12th. Int. Cong. Biomechanics* (edited by Gregor, R.J., Zernicke, R.F. and Whiting, W.C.), Abstract #99, University of California, Los Angeles.

Cavanagh, P.R. and Kram, R. (1985) Mechanical and muscular factors affecting the efficiency of human movement. *Med. Sci. Sport. Exercise* 17: 326-331.

Cavanagh, P.R. and Sanderson, D.J. (1986) The biomechanics of cycling. Studies of the pedaling mechanics of elite pursuit riders. In *Science of Cycling* (edited by Burke, E.), Human Kinetics Publishers Inc., Champaign, Ill.

Chapman, A.E. (1985) The mechanical properties of human muscle. *Exercise and Sport Sciences Reviews* 13: 443-501.

Chapman, A.E. (1988) How muscular mechanical properties govern technique in Sports. *Biomechanics XI-B* (edited by de Groot, G., Hollander, A.P., Huijing, P.A. and van Ingen Schenau, G.J.), pp. 545-554. Free University Press, Amsterdam.

Chapman, A.E. (1988) An understanding of muscle is the cornerstone of sport biomechanics. *Proc. 5th. Biennial Conf. Can. Soc. Biomechanics* (edited by Cotton, C.E., Lamontagne, M., Robertson, D.G.E. and Stothart, J.P.), pp. 2-4, Spodym Publishers, London, Ontario.

Coast, J.R. and Welch, H.G. (1985) Linear increase in optimal pedal rate with increased power output in cycle ergometry. *Eur. J. Appl. Physiol.* 53: 339-342.

Dunn, E.G. and Putnam, C.A. (1988) The influence of lower leg motion on thigh deceleration in kicking. *Biomechanics XI-B* (edited by de Groot G., Hollander, A.P., Huijing, P.A. and van Ingen Schenau, G.J.), pp. 787-790. Free University Press, Amsterdam.

Elliott, B.C. and Chivers, L. (1988) A three dimensional cinematographic analysis of the penalty corner hit in field hockey. *Biomechanics XI-B* (edited by de Groot G., Hollander, A.P., Huijing, P.A. and van Ingen Schenau, G.J.), pp. 791-797. Free University Press, Amsterdam.

Elliott, B., Marsh, T. and Overheu, P. (1989) A biomechanical comparison of the multisegmental and single unit topspin forehand drives in tennis. *Int. J. Sport Biomechanics* 5: 350-364.

Feltner, M.E. (1989) Three-dimensional interactions in a two-segment kinetic chain. Part II: application to the throwing arm in baseball pitching. *Int. J. Sport Biomechanics* 5: 420-450.

Feltner, M.E. and Dapena, J. (1986) Dynamics of the shoulder and elbow joints of the throwing arm during a baseball pitch. *Int. J. Sport Biomechanics*

2: 235-259.

Gregor, R.J., Cavanagh, P.R. and Lafortune, M. (1985) Knee flexor moments during propulsion in cycling - a creative solution to Lombard's Paradox. *J. Biomechanics* **18**: 307-316.

Hatze, H. (1976) Biomechanical aspects of a successful motion optimization. *Biomechanics V-B* (edited by Komi, P.V.), pp. 5-12, University Park Press, Baltimore.

Herring, R.M. (1989) Computer simulation of throwing: the influence of segmental parameters on the progression of movement associated with maximal effort throws. *M.Sc. Thesis*, Simon Fraser University, B.C., Canada.

Herring, R.M. and Chapman, A.E. (1988) Computer simulation of throwing: optimization of endpoint velocity and projectile displacement. *Proc. 5th. Biennial Conf. of Can. Soc. Biomechanics* (edited by Cotton, C.E., Lamontagne, M., Robertson, D.G.E. and Stothart, J.P.), pp. 76-77. Spodym Publishers, London, Ontario.

Herring, R.M. and Chapman, A.E. (1989) Computer simulation of throwing: effect of upper limb parameters on progression of movement. *Proc. 13th. Ann. Meeting of Amer. Soc. Biomechanics* Vermont. pp. 96-97.

Hill, A.V. (1922) The maximum work and mechanical efficiency of human muscles, and their most economical speed. *J. Physiol. (Lond.)* **56**: 19-41.

Hull, M.L., Gonzales, H.K. and Redfield, R. (1988) Optimization of pedalling rate in cycling using a muscle stress-based objective function. *Int. J. Sports Biomechanics* **4**: 1-20.

King, T.P. and Huston, R.L. (1989) Recent advances in the analysis of throwing. *Proc. 12th. Int. Cong. Biomechanics* (edited by Gregor R.J., Zernicke, R.F. and Whiting, W.C.) Abstract #40. University of California, Los Angeles.

Kram, R. (1987) Optimal pedaling frequency selection. *Proc. Amer. Society of Mech. Engineers Symp. on Biomech. in Sport.*

Kroon, H. (1983) The optimum pedaling rate. *Bike Tech* **2**: 1-5.

Luhtanen, P. (1988) Kinematics and kinetics of serve in volleyball at different age levels. *Biomechanics XI-B* (edited by de Groot G., Hollander, A.P., Huijing, P.A. and van Ingen Schenau, G.J.), pp. 815-819. Free University Press, Amsterdam.

Merril, E.G. and White, J.A. (1985) Physiological efficiency of constant power output at various pedal rates. *J. Sports Science* **2**: 25-34.

McMahon T A., Valiant, G. and Frederick E.C. (1987) Groucho running. *J. Appl. Physiol.* **62**: 2326-2337.

Nordeen, K.S. (1976) The effect of bicycle seat height variation upon oxygen consumption and both ex-perimental and simulated lower limb kinematics. *M.Sc. Thesis*, Penn. State University, State College, Pa.

Pandolf, K.B. and Noble, B.J. (1973) The effect of pedalling speed and resistance changes on perceived exertion for equivalent power outputs on the bicycle ergometer. *Med. Sci. Sports* **5**: 132-136.

Putnam, C.A. (1983) Interaction between segments during a kicking motion. *Biomechanics VIII-B* (edited by Matsui, H. and Kobayashi, K.), pp. 688-694. Human Kinetics Publishers Inc., Champaign, Ill.

Putnam, C.A. and Dunn, E.G. (1987) Performance variations in rapid swinging motions: effects on segment interaction and resultant joint moments. *Biomechanics X-B* (edited by Jonsson, B.), pp. 661-665. Human Kinetics Publishers Inc., Champaign, Ill.

Redfield, R. and Hull, M.L. (1985) On the relation between joint moments and pedalling rates at constant power in bicycling. *J. Biomechanics* **19**: 317-329.

Robertson, D.G.E. and Mosher, R.E. (1985) Work and power of the leg muscles in soccer kicking. *Biomechanics IX-B* (edited by Winter, D.A., Norman, R.W., Wells, R.P., Hayes, K.C. and Patla, A.E.), pp. 533-538. Human Kinetics Publishers Inc., Champaign, IL.

Roberts, E.M., Zernicke, R.F., Youm, Y. and Huang, T.C. (1974) Kinetic parameters of kicking. *Biomechanics IV.* (edited by Nelson, R.C. and Morehouse, C.A.), pp. 157-162. University Park Press, Baltimore.

Ryan, M., Gregor, R.J., Whiting, W.C. and Rugg, S.G. (1989) Length change and EMG patterns of lower extremity muscles during cycling. *Proc. First IOC World Congress on Sport Sciences*, (edited by Dillman, C.J., Nelson, R.C., Nigg, B.N., Voy, R.O., and Newsom, M.M.), pp. 277-278.

Sanderson, D.J. and Cavanagh, P.R. (1990) Use of augmented feedback for the modification of the pedaling mechanics of cyclists. *Can. J. Sport Sciences*, in press.

Thomson, D.B. and Chapman, A.E. (1988) The mechanical response of active human muscle during and after stretch. *Eur.J.Appl.Physiol*, **57**: 691-697.

Van Gheluwe, B. and Hebbelinck, M. (1985) The kinematics of the service movement in tennis: a three-dimensional cinematographical approach. *Biomechanics IX-B* (edited by Winter, D.A., Norman, R.W., Wells, R.P., Hayes, K.C. and Patla, A.E.), pp. 521-526. Human Kinetics Publishers, Inc., Champaign, IL.

Wilson, B.D., Howick, I.A. and Putnam, C.A. (1989) Segment timing and interaction in arm movement for maximum velocity under different load conditions. *Proc. 12th. Int. Cong. Biomechanics* (edited by Gregor, R.J., Zernicke, R.F. and Whiting, W.C.), Abstract #351.   University of California, Los Angeles.

Winter, D.A. (1987) *The Biomechanics and Motor Control of Human Gait*. University of Waterloo Press, Waterloo, Ontario.

Zernicke, R.F. and Roberts, E.M. (1976) Human lower extremity kinetic relationships during systematic vriations in resultant limb velocity. *Biomechanics V-B* (edited by Komi, P.), pp. 20-25. University Park Press, Baltimore.

CHAPTER 40

Analysis of Muscular Work in Multisegmental Movements: Application to Cycling

M. L. Hull and D.A. Hawkins

40.1 Introduction

In many athletic activities, efficiency and/or power are the keys to superior performance. For example, in activities which are primarily aerobic (e.g. distance running, cross-country skiing), the ability to perform the activity with high efficiency is one important factor in realizing superior performance. On the other hand, in activities which are primarily anaerobic (e.g. jumping, sprint running, power lifting), the ability to develop high power is an important ingredient in the recipe for superior achievement.

Although the ability to perform an athletic activity with either high efficiency or high power is dependent on many factors (Cavanagh and Kram, 1985; Heglund and Cavagna, 1985; Ekblom, 1987; Komi, 1987), one important factor is neuromuscular coordination (Cavagna, 1977; Cavanagh and Kram, 1985). Neuromuscular coordination involves not only the activation of different muscles in relation to one another but also the activation of individual muscles in relation to the kinematics of that muscle. By coordinating activation and kinematics, it is possible to develop a "stretch-shortening cycle," which consists of a period of active lengthening followed by active shortening. Developing a stretch-shortening cycle has the potential of increasing both efficiency and power substantially. To appreciate this phenomenon more fully, it is useful to briefly review the literature relevant to muscle efficiency, power, and the stretch-shortening cycle.

Before embarking on the literature review, some of the terminology surrounding muscular work will be clarified. Muscle has the capability of responding to external limb loads through two kinds of work – positive or concentric, and negative or eccentric. In positive work, the muscle shortens while under contraction, whereas in negative work, the muscle lengthens during contraction. Thus, in positive work the muscle does work on the environment whereas in negative work the environment does work on the muscle. Note that the word "contraction" as used herein means only that the muscle is generating force but not necessarily changing length. Isometric contraction refers to force generation at constant length whereas concentric and eccentric contractions are defined as above.

The efficiency with which muscle performs positive work is dependent on a variety of factors. Positive work efficiency is a complex function of muscle fiber type, with efficiency being higher for slower muscles (Davies, 1965). Positive work efficiency is also strongly influenced by muscle kinematics and especially contraction velocity (White, 1977); at the extremes of the force–velocity curve, efficiency of positive work is zero. Finally, Whipp and Wasserman (1969) have observed that efficiency is also dependent on temperature with efficiency decreasing as temperature increases. Assuming that none of these factors are near extremes, the gross efficiency of whole muscle is about 30% (Whipp and Wasserman, 1969).

Similar to positive work, the "efficiency" (defined here as the work divided by the energy expended by muscle) of negative work also depends on a number of factors but the efficiency

Multiple Muscle Systems: Biomechanics and Movement Organization
J.M. Winters and S.L-Y. Woo (ed.), © 1990 Springer-Verlag

of negative work is much higher than positive work (Abbot et al., 1952; Asmussen, 1953; Stainsby, 1976). The explanation as to this disparity is traced in part to the force–generating capability of muscle which is higher in eccentric than in concentric contractions. Thus, fewer fibers need be stimulated in eccentric contraction to develop a given level of force (Komi, 1973). The explanation is also traced in part to the chemical processes accompanying forcible cross-bridge detachment in eccentric contractions [Curtin and Davies, 1975; White, 1977; see also Chapter 1 (Zahalak)].

As the positive work efficiency of a muscle depends on many factors, so does its ability to generate power. Similar to efficiency, muscle power is a function of the muscle's kinematics. At the extremes of the force–velocity curve, the power is zero, whereas maximum power is obtained at shortening velocities near 30% of the muscle's maximal velocity (Edgerton et al., 1986). Muscle architecture is another important factor affecting power production. Muscles of similar mass and composition but with varying arrangements of sarcomeres (i.e. arranged in either series or parallel) may develop the same maximum power, but at different absolute shortening velocities (Edgerton et al., 1986). The fiber composition of a muscle also influences the power generated. Fast twitch fibers generate about four times greater maximal power than slow twitch fibers (Green, 1986; Faulkner et al., 1986).

The mechanical performance of muscle may be altered remarkably when the muscle experiences a stretch–shortening cycle (Bosco et al., 1982; Goubel, 1987; see also Chapters 3–5, 35, 38–39). The force following an active stretch to a specified length is higher than the corresponding isometric force (Cavagna et al., 1968; Bosco et al., 1982; Goubel, 1987). Consequently the concentric work (and hence the power) is also higher when the muscle is allowed to shorten (Bosco and Komi, 1979). Further, the excess work increases with both the stretch velocity and the length change (Cavagna et al., 1968; Bosco et al., 1982). Any delay in initiating the concentric phase, however, will diminish the increased force and hence work following the eccentric phase (Cavagna et al., 1985). The increased force and work accompanying the stretch–shortening cycle appears due to two separate mechanisms, elastic energy stored in the series elastic component during the active stretch and "potentiation" (increased force) of the contractile component (Cavagna et al., 1968, 1985; see also Chapters 1-5, 33-39). Accompanying the remarkable increase in force and work of muscle as a result of the stretch–shortening cycle is an increase in efficiency of concentric work (Thys et al., 1972; Asmussen and Bonde-Petersen, 1974; Bosco et al., 1982; Aura and Komi, 1986; Goubel, 1987).

The fact that the performance of negative work immediately followed by positive work has the potential to substantially affect the overall efficiency of movement has caused evaluation of the importance of this mechanism in various activities [see also Chapter 39 (Chapman)]. The activities which have had the greatest attention are walking, running, and cycling. Discussion here will be restricted to running and cycling. Many studies have measured the overall efficiency of both running (Lloyd and Zacks, 1972; Zacks, 1973; Asmussen and Bonde-Petersen, 1974; Cavagna and Kaneko, 1977) and cycling (Whipp and Wasserman, 1969; Zacks, 1973; Asmussen and Bonde-Petersen, 1974). Depending on the study, the apparent efficiency of running ranged between 33% (Zacks, 1973) and 80% (Cavagna and Kaneko, 1977), whereas the efficiency of cycling was always below 30%. Because the measured efficiency of running was greater than the gross efficiency of muscle undergoing positive work sans prestretch effects, the conclusion was that the effects of prestretch play an important role in this activity. In cycling, on the other hand, the result that efficiency was less than the gross efficiency of muscle led many to conclude that prestretch plays no role in this activity and that negative work does not occur (Zacks, 1973; Asmussen and Bonde-Petersen, 1974; van Ingen Schenau, 1984). However, more recent studies in cycling biomechanics suggest that inferring a lack of negative work based on efficiency measurements may be faulty reasoning. The existence of stretch–shortening cycles in both the hamstrings and gastrocnemius muscles is evident in the work of Andrews (1987) and Gregor et al. (1987), respectively.

Inasmuch as the development of stretch-shortening cycles in muscle has the potential to impact both efficiency and power and hence the performance level of an activity, it is useful to develop a methodology for detailed examination of this phenomenon in multisegmental movements. One objective of this article is to present a general procedure for analyzing this phenomenon through identification of regions of positive and negative work in individual muscles. Because of conflicting claims which have surrounded the presence of negative work in cycling, a second objective is to apply this procedure to the study of this activity.

40.2 Methods

In order to identify regions of positive and negative work in individual muscles, two types of data must be available. The first is knowledge of the time course of muscle kinematics (i.e. length and velocity) and the second is the time course of the muscle contraction. By synchronizing these two types of data, regions of positive and negative work and hence the presence of stretch–shortening cycles may be readily identified.

40.2.1 Muscle–Tendon Kinematics

In order to provide the first type of data described above, kinematics of individual muscles must be determined for the activity under study. However, it is not practical to directly measure muscle kinematics from living subjects. To do so would require invasive techniques that would be time consuming, not well tolerated, and more than likely inhibit normal movements. To circumvent the need for a direct approach for measuring muscle kinematics, an indirect approach has been developed. This approach is based on the measurement of joint angles and the use of these angles in conjunction with a musculo-skeletal system model to determine muscle–tendon lengths. The generic methodology for determining muscle–tendon kinematics is described followed by a specific example that determines lower extremity muscle-tendon kinematics during cycling.

The methodology presented here is based on the idea that muscle–tendon length may be expressed as a function of the absolute location of its bony attachments (i.e. origin/insertion locations). Several steps are taken in the calculation of muscle–tendon length. In the first step, local (i.e. segmental) coordinate reference frames are defined for body segments to which muscles of interest are attached and muscle origin and insertion cites are located in terms of appropriate local coordinates. Usually, the local coordinate reference

frames are fixed to points which offer ready alignment of coordinate axes with anatomical landmarks and are easily identified visually. Points which offer these features are typically the joint centers. With the origin at the joint center, the coordinate axes are developed by aligning one along a line connecting the proximal and distal joint centers of the segment. A second axis is directed anteriorly and perpendicular to the first while the third is mutually orthogonal. In the second step, segment orientations and joint center locations are determined in a global coordinate system during the activity under study. In the final step, the coordinates of the muscle origin and insertion locations are determined in the global coordinate system and the length is computed.

The procedures described above are illustrated schematically in Figure 40.1 for a two-segment model and a single-joint muscle. Two local reference frames are shown, one for each body segment, $(x_1, y_1, z_1$ and $x_2, y_2, z_2)$ along with a third reference frame representing a global coordinate system (X, Y, Z). The location of the muscle origin (O_{x1}, O_{y1}, O_{z1}) is defined in terms of the local reference frame of *segment 1* while the location of the insertion (I_{x2}, I_{y2}, I_{z2}) is defined in terms of the local reference frame of *segment 2*. Segment orientation and joint center locations are defined in terms of the global reference frame. Given this information, origin and insertion locations may be transformed from local coordinates into global coordinates by multiplying local position vectors by the appropriate transformation matrix (see Figure 40.1). The transformation matrix includes both a translation and a rotation. Once both attachment sites are defined in terms of the global reference system, muscle–tendon length is calculated as the distance between these attachment sites (see Figure 40.1).

To fully describe muscle–tendon kinematics, the muscle–tendon velocity must be determined as well as the length. If muscle–tendon lengths can be expressed as analytical functions of the independent variables (i.e. joint angles) and these variables can be expressed as functions of time, then velocity may be obtained by simply differentiating the analytical functions. As is often the case, however, obtaining analytical functions of muscle lengths is impractical due to the complexity of the movement under study. Consequently, muscle tendon lengths must be determined from measurements of joint angles, requiring that muscle–tendon velocities be determined using finite difference techniques. A common differentiation technique of this type involves forward difference at the beginning of time series data, backward difference at the end, and central difference for everything in between (Hornbeck, 1975).

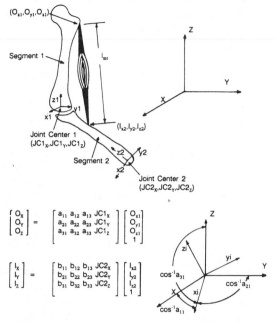

where: a_{ij} and b_{ij} are direction cosines for respective local coordinate frames.

Muscle-tendon length, $l_{mt} = \sqrt{(O_X - l_X)^2 + (O_Y - l_Y)^2 + (O_Z - l_Z)^2}$

Figure 40.1: Multistep procedure for determining muscle–tendon length. The first step involves defining local coordinate systems attached to individual segments and then locating muscle origin and insertion sites in these coordinate systems. The next step requires that local coordinate system origins and orientations be determined relative to the global coordinate system during the activity under study. In the final step, coordinates of muscle origin and insertion sites are determined in the global coordinate system and length is computed.

With the use of finite difference derivatives comes the problem of noise in the angle measurements being amplified in the difference process. To combat this problem, the usual approach is to digitally filter both the length data prior to computing finite differences and the resulting velocity data. Several filtering techniques have been employed including splines, Fourier series approximations, and digital filters. For a discussion of each of these techniques refer to the papers by Zernicke et al. (1976), Soudan and Dierckx (1979), Vaughn (1982), and Wood (1982).

A variety of techniques are available for measuring segment positions, thus enabling the quantification of both local coordinate system origins and segment orientations. A review of these techniques is a major topic in its own right so that only a brief cataloging of various techniques will be given here. Techniques include cinematography (e.g. Walton, 1981), videography (Hawkins et al., 1986; Kadaba et al., 1987; Andriacchi, 1987; Miller, 1987), goniometry (e.g. Chao, 1980), and linkage models (Hull and Jorge, 1985).

40.2.2 Time Course of Muscle Force

The second type of data necessary to examine regions of positive and negative muscular work in detail is the time course of muscle force. Historically there have been two basic approaches utilized to predict muscle force *in vivo* aside from attaching force transducers directly to tendons. The first approach is direct and relies on measurements of specific muscle parameters (e.g. activation levels, kinematics, architecture) and a suitable mathematical muscle model to compute individual muscle forces (e.g. Pell, 1972; Stern, 1974; Hof and van den Berg, 1981; Baildon and Chapman, 1983). The second approach is indirect, and involves solving the inverse dynamics problem first so as to determine intersegmental forces and moments, followed by distributing these loads (i.e. forces and moments) among the various structures of the joint (i.e. bones, ligaments, and muscles) to satisfy some musculoskeletal model. The distribution problem is generally solved using one of three methods: *1)* reduction methods (Morrison, 1968); *2)* optimization methods (Seireg and Arvikar, 1973; Penrod et al., 1974; Pedotti et al., 1978; Crowninshield, 1978; Crowninshield and Brand, 1981; Dul et al., 1984; Davy and Audu, 1987), or *3)* control model methods (Hatze, 1981). A detailed review of these various methods is beyond the scope of this chapter. Refer to the references cited above and to Chapter 8 (Zajac and Winters) for more information regarding each of these techniques.

If the regions of positive and negative work are only of interest rather than quantification of the actual work accomplished in those regions, then it is not necessary to quantify the force level. In these situations, the direct approach is advantageous because the time course of the muscle force is dynamically related to the muscle activation, which may be estimated experimentally by measuring muscle action potentials. Because the regions rather than the actual work are of interest here, subsequent discussion will be devoted to identifying these regions using the direct approach.

The direct approach involves the use of electrodes to measure muscle action potentials (*EMGs*). As outlined by Loeb and Gans (1986) the many types of electrodes available can be categorized as either invasive or non-invasive. By far the more widely used for kinesiological studies are the non-invasive type surface electrodes. Typically these electrodes are mounted in groups of three on the skin covering the muscle of interest and are oriented along the long axis of the muscle. To both avoid "cross-talk" from other muscles and insure a strong signal, the electrodes must be carefully placed. Anatomical guides such as that by Delagi et al. (1975) describe procedures for electrode placement on individual muscles.

Two of the three electrodes provide a signal which is amplified differentially while the third electrode provides a reference signal. Differential amplification is desirable to eliminate common mode signals which may be developed in the body acting as an antenna picking up electromagnetic radiation. Typically, the most troublesome radiation comes from commercially supplied power. In addition to amplification, other signal conditioning includes band pass filtering. The high pass characteristic is necessary to eliminate movement artifact (Basmajian et al., 1975) whereas the low pass characteristic reduces high frequency noise. Low and high frequency cutoffs are typically 5–10 Hz and 500–1000 Hz, respectively.

Since the raw *EMG* signal appears as band limited noise, usually the raw signal is further processed to develop a time varying signal, the value of which is used as an indication of the instantaneous level of activation. Processing may be done either in the analog domain or in the digital domain. With processing operations implemented via computer software, the digital domain offers considerably greater versatility. In either case, a variety of techniques has been developed such as integrated rectified *EMG*, average rectified *EMG* and true *RMS* (Basmajian, 1974; Basmajian et al., 1975). To provide a single instantaneous index of activity, a rectangular window is convolved with either the rectified or squared signal and then the integration is performed over the window duration. Yet another technique is the smoothed, rectified *EMG*. Note that if the data are processed in the digital domain, then it is possible to perform the smoothing operation (i.e. low-pass filtering) such that no phase shift is introduced. This requires that the smoothing be done twice, both forwards and backwards in time.

To accurately identify regions of positive and negative muscular work, the relation between the time when force is actually developed by muscle and the initiation and termination of the *EMG* signal must be considered. Komi and coworkers have studied experimentally the effect of various factors on the delay between time of activation and onset of muscle force, termed electromechanical delay (*EMD*). Cavanagh and Komi (1979) studied the dependence of *EMD* in the human elbow flexor group on contraction type and found that *EMD* was significantly different depending on whether the muscle was contracting eccentrically, concentrically or isometrically. The longest *EMD* of 55.4 ms was found for concentric contraction and the shortest *EMD* of 49.4 ms for eccentric contraction. Norman and Komi (1979) explored the effects of both velocity and muscle type (i.e. slow or fast) in addition to contraction type on the *EMD*. Similar to Cavanagh and Komi, the major muscles crossing the elbow were studied, with the triceps being faster than the biceps. Norman and Komi reported that the *EMD* was shorter for the faster muscle and that *EMD* was independent of velocity during concentric contraction. During eccentric contraction, however, *EMD* decreased with higher velocity but only for the slower muscle. The dependence of *EMD* on fiber type was subsequently confirmed by Viitasalo and Komi (1981), who correlated *EMD* to muscle fiber type as determined from muscle biopsies. For the vastus lateralis muscle, *EMD* increased from 25 ms for low percentages of slow twitch fibers to 55 ms for high percentages.

In addition to the factors mentioned above, the effect of *EMD* on muscle length has also been examined. In their experiments, Norman and Komi (1979) also included two ranges of motion. They found no significant differences between the *EMD* for the two ranges of motion. In contrast to this result is that of Muro and Nagato (1985) who studied *EMD* in the triceps surae under isometric contraction for different ankle flexion angles with the knee angle constant. Rather than having subjects voluntarily stimulate muscles, stimulation was imposed electrically. They found that *EMD* varied inversely with length; the delay at greatest length was 7.0 ms while at the shortest length was 11.7 ms.

In summary, *EMD* appears to be a complex result of muscle morphology (i.e. fiber composition), kinematics (length and velocity) and type of contraction (isometric, eccentric, concentric). Accordingly, because *EMD* values may be relatively long in relation to the duration of the activity, it appears desirable to measure these delays specific to the muscles of interest in both the athletes of interest and the kinematics of the particular activity. In the event that this is not possible, then data available from the literature must suffice. As is evident from the preceding discussion, data are available for a number of muscles with varying kinematics.

In the development of muscle force subsequent to activation, the electromechanical delay is only one consideration because once development commences, the muscle force does not track the *EMG* signal [Gottlieb and Agarwal, 1971; Chapter 5 (Winters)]. While the development of muscle force, once initiated, is clearly important to the quantitative assessment of positive and negative work, it is not necessary to appreciate the force level if only regions of positive and negative work are of interest. Since these regions occupy attention here, the methodology includes no provision for treating the time course of muscle force other than the electromechanical delay.

40.2.3 Application to Cycling

The complete formulation of the steps of this methodology, as they may be applied to the determination of lower extremity muscle-tendon lengths in cycling, will be described. A general set of local reference frames for the lower extremity is defined as shown schematically in Figure 40.2. The location of local coordinate system origins together with the definition of axes in

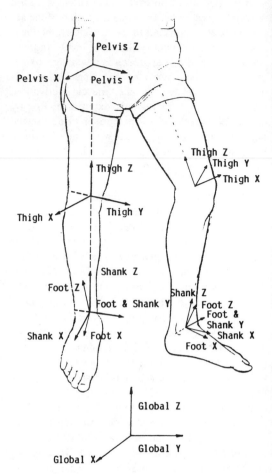

Figure 40.2: Local coordinate systems for the four segments used in the cycling analysis. See Table 40.3 for descriptions of these systems.

relation to anatomical landmarks is given in Table 40.1.

A variety of techniques has been used to estimate the locations of origins and insertions of lower limb muscles. Seireg and Arvikar (1973) estimated muscle attachment sites from a figure of a skeleton in a textbook. Crowninshield et al. (1978) determined origin and insertion locations from a single dry cadaver specimen. Dostal and Andrews (1982) estimated hip muscle origins and insertions from one side of seven dry cadaver specimens; they also examined three cadavers to obtain muscle attachment data for other lower extremity muscles.

Table 40.1:

Segment	Local reference frame origin	Definition of coordinate axes		
		x	y	z
Pelvis	Center of acetabulum.	Parallel to a vector created by taking the cross product of the y-axis with the z-axis.	Perpendicular to and directed medially from the plane of motion.	Parallel to a vector directed along the long axis of the torso and in the plane of motion.
Thigh	Midpoint between medial and lateral femoral epicondyles.	Parallel to a vector created by taking the cross product of the y-axis with the z-axis.	Perpendicular to and directed medially from the plane of motion.	Parallel to a vector from the the knee joint center to the hip joint center.
Shank	Midpoint between the medial and lateral malleoli.	Parallel to a vector created by taking the cross product of the y-axis with the z-axis.	Perpendicular to and directed medially from the plane of motion.	Parallel to a vector from the the ankle joint center to the knee joint center.
Foot	Midpoint between the medial and lateral malleolus.	Parallel to the component of a vector from the lateral aspect of the calcaneus to the lateral aspect of the 5th meta-tarsal in the plane of motion.	Perpendicular to and directed medially from the plane of motion.	Parallel to a vector created by taking the cross product of the x-axis with the y-axis.

Table 40.2:

| Segment | Normalizing factors for local axes | | |
	x	y	z
Pelvis	Anterior/posterior distance from the pelvic frontal plane to the sciatic notch	Medial/lateral distance from the anterior superior iliac spine to the hip center	Sagittal plane distance from the iliac crest to the ischial tuberosity
Thigh	Femoral width at condyles	Femoral width at condyles	Thigh length (greater trochanter to lateral femoral condyle)
Shank	Tibial plateau width	Tibial plateau width	Shank length (lateral femoral epicondyle to lateral malleolus)
Foot	Tibial plateau width	Tibial plateau width	Shank length (lateral femoral epicondyle to lateral malleolus)

Table 40.3:

| Muscle Name | Origin segment | Insertion segment | Origin Location* | | | Insertion Location* | | |
			x	y	z	x	y	z
Upper 1/3 of Gluteus Maximus	pelvis	thigh	-48.9	35.1	65.4	-20.2	-71.7	104.1
Biceps Femoris	pelvis	shank	-62.1	18.9	-23.7	-58.8	-68.6	107.4
Rectus Femoris	pelvis	shank	48.5	-49.4	16.4	5.91	0.84	130.4
Semimembranosus	pelvis	shank	-56.7	16.1	-22.5	-90.2	8.86	105.2
Medial Gastrocnemius	thigh	foot	-26.4	19.2	1.86	-56.5	-6.03	-13.7
Lateral Gastrocnemius	thigh	foot	-24.8	-28.9	1.16	-56.7	-5.94	-13.7
Vastus lateralis	thigh	shank	1.48	-68.6	53.9	13.8	-24.2	129.3
Vastus medialis	thigh	shank	4.83	-17.3	47.8	-13.2	22.4	127.5
Tibialis Anterior	shank	foot	-9.73	-21.3	76.2	35.5	31.2	4.29
Soleus	shank	foot	-45.9	-1.00	78.3	-55.6	-10.4	-13.9

* All values are expressed as a percent of the normalizing factor.

Probably the most extensively referenced muscle origin and insertion location data are those reported by Brand et al. (1982). They determined three-dimensional attachment cite locations for most muscles of the lower extremity [see Appendix (Yamaguchi et al.) for complete overview]. With the exception of the foot coordinate system, they defined attachment coordinates in terms of the local coordinate frames illustrated in Figure 40.2. Their data were determined from three fresh cadavers (six limbs). Muscles with either broad origins or broad insertions were marked as a single point if variations in the attachment location did not affect moment arm lengths (vastus intermedius, soleus), and multiple points if attachment location variations did greatly affect these lengths (gluteus maximus, gluteus minimus, gluteus medius, adductor brevis, adductor magnus). For muscles that do not act in a straight line, either the origin (for illiacus, psoas, obdurator internus) or insertion (for sartorius, gracilis, semitendinosus, tibialis posterior, flexor digitorum communis, flexor hallucis longus, peroneus tertius, peroneus brevis, peroneus longus) was marked by a single "effective" point where either the muscle or tendon crossed a bony prominence. In addition to examining muscle attachment locations, they also developed a scaling scheme which allows estimations to be made of origins and insertions in living subjects based on anthropometric measurements. They reported normalized muscle origin and insertion coordinate data for 47 muscles or equivalent muscles. Table 40.2 provides a list of muscle–tendon normalizing factors. Given in Table 40.3 is a list of normalized muscle tendon attachment locations for the ten muscles of the right leg studied in the cycling application here. Three dimensional coordinates for each attachment location are expressed as a percent of the normalizing factor in the direction of appropriate local coordinate axes. Illustrated in Figure 40.3 are the muscles listed in Table 40.3 (except for medial gastrocnemius).

Joint center locations for the ankle, knee and hip are determined during cycling using a closed loop five-bar linkage model as shown in Figure 40.4. The five links in this model represent the thigh, shank, foot, crank arm, and crank axis to hip center. In this model it is assumed that leg motion

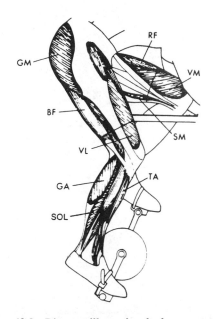

Figure 40.3: Diagram illustrating the lower extremity muscles considered in the analysis. *GM*—gluteus maximus, *BF*—biceps femoris (longhead), *SM*—semimembranous, *RF*—rectus femoris, *VM*—vastus medialis, *VL*—vastus lateralis, *TA*—tibialis anterior, *GA*—gastrocnemius (lateral head), *SOL*—soleus. Not labeled is gastrocnemius (medial head).

occurs in a single plane and that there is no relative motion between the pelvis and the bicycle seat. It is also assumed that the axes of rotation do not shift, and thus that the lengths of the links remain constant. Knowing linkage lengths (from bicycle setup information and subject anthropometrics) and the orientation of these links enables the hip, knee, and ankle joint centers, representative of local coordinate origins, to be determined throughout the crank cycle. Inasmuch as the model is a five-bar linkage, it has two degrees of freedom so that the orientation of all links is uniquely specified by knowing the orientation of any two links. Rotational potentiometers are used to measure the angular position of both the crank arm and pedal (i.e. foot link) throughout the crank cycle. See Hull and Davis (1981) for further details on the hardware used to make these measurements.

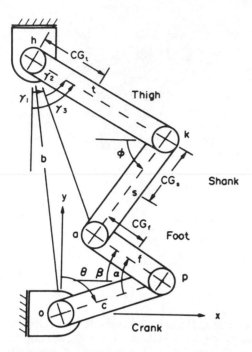

Figure 40.4: Five-bar linkage model of the lower limb in cycling. The five links are the thigh, shank, foot, crank, and the bicycle seat tube which is a fixed link connecting the crank and hip axes. Linkage movement is assumed to be confined to the sagittal plane. Also the link lengths are assumed fixed. Crank and pedal angles are measured to determine the linkage configuration throughout a crank cycle.

With the positions of local coordinate system origins, the orientation of these coordinate systems, and the muscle origin/insertion locations within these local coordinate systems known, the lengths of individual muscles over the complete crank cycle are determined using the procedures outlined in Figure 40.1. Muscle–tendon velocities are then computed using finite difference techniques and the results smoothed using a digital low-pass filter (Vaughn, 1982).

As described above, muscle activation, suitably processed, is used as an indicator of muscle force. Muscle activation information is recorded using three silver-silver/chloride surface electrodes mounted on the skin covering the muscle of interest and oriented serially along the long axis of the muscle fibers. The electrodes have an outer diameter of 12.5 mm, with a sensor diameter of 6 mm. The electrodes are applied to the skin with a center-to-center spacing of approximately 2.5 cm.

Muscle action potentials are amplified and high-pass filtered with a small signal conditioning circuit located within 10 cm of the electrode sensing elements. The signal conditioning circuit incorporates a single operational amplifier with a gain of approximately 850, and an RC high-pass filter with a cutoff frequency set to 34 Hz. The signal conditioning circuit is illustrated in Figure 40.5.

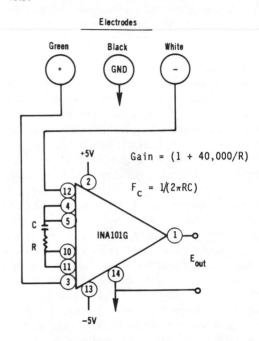

Figure 40.5: *EMG* signal conditioning circuit with high-pass filter. F_c is high-pass cutoff frequency.

EMG data are collected at 1000 Hz utilizing a Compaq computer with Metrabyte *A/D* converter. Further processing of the *EMG* signal includes rectification, smoothing, and normalization.

The smoothing routine uses a 2nd-order Butterworth digital low-pass filter (Vaughn, 1982). The filtering routine processes the data in both the forward and reverse directions to remove any phase shift. The cutoff frequency is set to between 4 and 6 Hz. Finally, the *EMG* data of each muscle are normalized with respect to the maximum levels developed for that muscle during a crank cycle.

To account for the electromechanical delay between muscle activation and the commencement of muscle force, it is assumed that the *EMD* is the same for all lower limb muscles. Also it is assumed that the delay time is independent of both muscle kinematics and type of contraction. The *EMD* value selected is 40 ms, which is the average value reported by Viitasalo and Komi (1981) for the vastus lateralis.

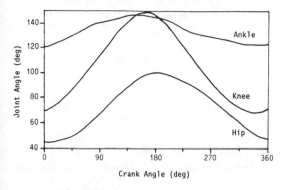

Figure 40.6: Joint angles of the leg over a complete crank cycle. The hip angle is the angle between the pelvis and the thigh (full hip extension = 180°), the knee angle is the angle between the shank and thigh (full knee extension = 180°), and the ankle angle is the angle between the shank and a line connecting the ankle joint axis to the pedal spindle axis. Extreme values do not coincide.

40.3 Results

The comparison of hip, knee, and ankle angles, which are plotted as a function of crank angle in Figure 40.6, highlights the differences in movement of these joints. The hip and knee angles are the included angles between the pelvis and thigh and thigh and shank repectively when the joint is flexed. Thus, full joint extension corresponds to an angle of 180° for both joints. Note that the angle labeled "ankle" in the plot is not the actual ankle angle but rather the angle of the foot link relative to the shank (see Figure 40.4). Because the foot link connects the ankle axis to the pedal axis, this angle is greater than 90° (see Figure 40.3). These plots indicate that the range of motion of the knee (75°) is considerably greater than either the hip (55°) or the ankle (25°). Further, the joints do not reach their maximum and minimum values simultaneously; the extremes of knee motion occur about 20° earlier in the crank cycle than the hip and about 20° later than the ankle. Finally, while the knee plot is approximated closely by a sinusoid, the hip and ankle plots are not.

To illustrate muscle-tendon kinematic profiles (i.e. length and velocity) as a function of crank angle, sample plots for four of the five biarticular muscles studied are presented in Figure 40.7. The four muscles are gastrocnemius medial head, rectus femoris, biceps femoris (long head) and semitendinosus. The gastrocnemius lateral head plots for both length and velocity are not shown because they track those of the medial head almost identically. In comparing the muscle length plots in Figure 40.7 to the joint angle plots in Figure 40.6, it is noticed that extreme values of angles and lengths do not coincide. This observation highlights the complex relation between joint angles and kinematics of biarticular muscles.

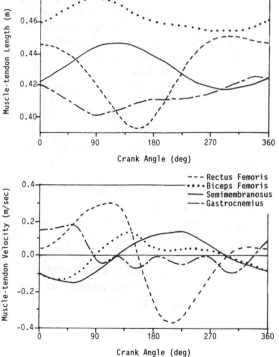

Figure 40.7: Kinematics of biarticular muscle–tendon complexes over a crank cycle. The kinematics of the medial head of gastrocnemius are representative of the lateral head. Extreme values of lengths do not coincide with extreme values of joint angles, indicating that lengths of biarticular muscles are complex functions of joint angles. None of the four muscles exhibits similar patterns of length change, despite the fact that two of the muscles (biceps femoris and semimembranosus) are in the same group. Regions of positive velocity for both knee flexors and extensors overlap.

Unlike the biarticular muscles, the kinematics of the uniarticular muscles can be appreciated to a large degree from the joint angle plots in Figure 40.6. In particular, the patterns of length change of individual muscles closely track the corresponding joint angle plot in Figure 40.6 with the phase adjusted depending on whether the muscle is either a flexor or an extensor. From the patterns of length change, the patterns of velocity change follow directly by noting extreme values (points of zero velocity) in conjunction with regions of increasing length (negative velocity) and decreasing length (positive velocity).

For the purposes of this article, the primary utility of the muscle length and velocity data is to identify possible regions of positive and negative work. With these regions identified it then becomes possible to evaluate the potential role that specific muscles might play in developing the propulsive power in cycling. According to the velocity plots, the gastrocnemius can only provide positive work in the range 340°–85°. For the rectus femoris, it is capable of positive work in the range 300°–160°. Finally, the positive work ranges are 90°–300° and 120°–300° for the biceps femoris and semitendinosus respectively. These regions together with those of the other muscles are presented in Figure 40.8.

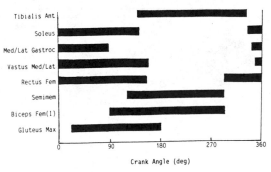

Figure 40.8: Regions of possible positive work for ten lower limb muscles over a crank cycle. Possible synergy exists between muscles crossing the opposite sides of the hip and knee joints.

Examining Figure 40.8 leads to several interesting observations regarding the synergy of muscles which cross joints on opposite sides. First, notice the tibialis anterior and muscles of the triceps surae group. For this subject, the positive work regions of these muscles do not overlap, in-

dicating no possible synergy. This is contrasted to muscles in the quadriceps and hamstrings groups. In the second half of the downstroke region of the crank cycle, the positive work regions of these muscles overlap. Accordingly, although these muscles are considered antagonistic at the knee, the overlap in positive work regions indicates the potential for the muscles in these groups to work synergistically in extending the leg in cycling. A similar observation holds for the gluteus maximus/rectus femoris as well as the gastrocnemius/quadriceps. Notice that there is no possibility of synergy between muscles crossing opposite sides of the same joint during the upstroke region (180°–360°). Inasmuch as the maximum power region in cycling is centered at about 100°, the potential synergy in the downstroke appears well suited to meeting this demand.

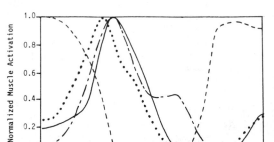

Figure 40.9: Plots of smoothed, rectified EMG's for four muscles over a crank cycle. Smoothing is performed with a zero phase shift digital low-pass filter. Smoothed EMG signals are normalized to highest levels recorded for corresponding muscles over the crank cycle. The cadence of the tests is about 85–90 RPM and the power level is 175–200 W.

The next results to be presented are those of muscle activation. Sample plots of the smoothed, rectified EMGs for the four muscles of Figure 40.7 are illustrated together in Figure 40.9. Each EMG signal is normalized to the highest EMG signal recorded for a particular muscle over the crank cycle.

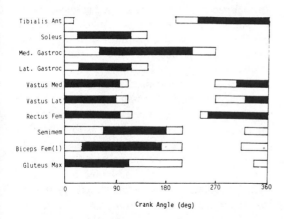

Figure 40.10: *EMG* activity regions for ten lower limb muscles over a crank cycle. White regions correspond to activity below 30% of the maximum in the same crank cycle for that muscle while dark regions indicate activity above the 30% threshold. Overall, the *EMG* activity picture is one of little activity in the upstroke (180°-360°) and relatively higher activity in the downstroke (0°-180°), where positive power is generated.

From plots such as those in Figure 40.9, the regions of activity for all muscles may be identified as illustrated in Figure 40.10. In this figure the activity region of each muscle is indicated by white and black regions. The black region corresponds to activity greater than 30% of the maximum level in a complete crank cycle whereas the white region indicates activity less than 30%. As is clear from the figure, there is little activity during the upstroke region. Beginning at about 240°, the tibialis anterior reaches its 30% level. Subsequently, the quadriceps become active with the rectus femoris preceding the vastii. At the top dead center position of the crank, every muscle is active to some degree with the relative activity being greatest in the quadriceps and gluteus maximus. Relative levels of activity greater than 30% in both the gluteus maximus and quadriceps are sustained throughout the first half of the downstroke region. At about 25°, the muscles of the triceps surae group reach 30% activity level as does biceps femoris (long head). Subsequently, the semimembranosus reaches its 30% level. At 90° past top dead center, every muscle except tibialis anterior exhibits activity greater than the 30% threshold. Over the second half of the

downstroke, the quadriceps become inactive whereas the hamstrings remain active. The gluteus also remains active but the activity drops below 30% over the majority of the second half of the downstroke region. Overall the *EMG* picture is one in which there is little activity in the upstroke region where either little or no positive power is generated. Conversely, relatively high levels of activity occur in all muscles except tibialis anterior during the downstroke region where positive power is generated over the entire region (Hull and Davis, 1981).

Combining the muscle activation results in Figure 40.10 with the positive velocity regions in Figure 40.8 enables assessment of the actual synergy between muscles as well as identification of stretch–shortening cycles. The actual regions of positive and negative work resulting from this combination process are depicted in Figure 40.11. Note that in identifying these regions only the activity region above 30% is used. Omitting the activity region below 30% is equivalent to a time delay of 40 *ms* or greater for all muscles.

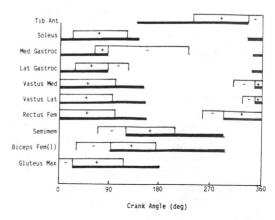

Figure 40.11: Actual regions of positive and negative work for ten muscles over a crank cycle. Work regions are developed using only the region of *EMG* activity greater than the 30% threshold. Omitting the region less than 30% is equivalent to a time delay of about 40 *ms* or greater for all muscles. All muscles except those crossing the ankle joint exhibit a stretch–shortening cycle.

Considering first the issue of synergy, not all of the potential synergy in the downstroke region evident from Figure 40.8 is realized. In particular, positive work regions for the quadriceps and hamstrings do not overlap. However, there is substantial overlap in positive work regions for the other two potentially synergistic muscle pairs, gluteus maximus/rectus femoris and quadriceps/gastrocnemius.

Considering next the issue of stretch–shortening cycles, notice that all muscles except those crossing the ankle joint exhibit negative work regions immediately followed by positive work regions. The durations of the negative work regions vary considerably, however. The regions are smallest for the vastus lateralis and gluteus maximus, occupying about 25° of the crank cycle (45–50 ms at 85 RPM). The vastii exhibit comparable negative work regions of about 40° (75 ms) whereas the negative work regions for the hamstrings are the largest, being 55° (105 ms) and 65° (125 ms) for semimembranosus and biceps femoris, respectively. With these levels of duration for negative work regions comes the possibility of developing substantial levels of force during stretch. Consequently, the stretch-shortening cycle with its concomitant increase in both efficiency and power appears to be an important mechanism in cycling mechanics.

40.4 Discussion

The presence of stretch–shortening cycles in both the hip and knee muscles indicates that these muscles perform negative work which is contrary to the claim of Zacks (1973), Asmussen and Bonde-Petersen (1974), and van Ingen Schenau (1984). It should be noted, however, that the experimental conditions herein are different from those of either Zacks or Asmussen and Bonde-Petersen. In particular, the pedalling rate of 85 RPM used here is notably higher than the 50 RPM and 65 RPM rate used by Zacks and Asmussen and Bonde-Petersen respectively. Also, the test subject who provided the data here pedalled with toeclips and cleats whereas these pedalling aids were not used by Zacks and Asmussen and Bonde-Petersen. Suzuki et al (1982) have shown that pedalling rate influences the timing of EMG signals during the crank cycle with higher pedalling rates generally advancing the phase of the biarticular muscles. Retarding the phase of EMG envelopes would decrease the negative work regions evident in Figure 40.11. Also, Ericson et al. (1985) and Jorge and Hull (1986) have shown that the foot–pedal connection influences EMG patterns as well. Consequently both these variables

impact pedalling mechanics to a degree where it is possible that the muscle use in the experiments performed here is fundamentally different than the muscle use in the experiments of Zacks (1973) and Asmussen and Bonde-Petersen (1974).

Because the electromechanical delay is known to be dependent on a number of factors and the delay in the analysis was assumed constant at 40 ms, the question naturally arises as to what effect this assumption might have on results. As mentioned previously the 40 ms value is the average of the values obtained by Viitasalo and Komi (1981). Recall that the delay is included here by omitting the region of EMG activity below 30% and that this is conservative in the sense that the region of EMG activity below 30% consumes at least 40 ms at the beginning of the activity period. Thus, reducing the EMD would not impact the negative work regions in Figure 40.11 providing that these regions are delineated by the 30% threshold. On the other hand, increasing the EMD would impact the negative work regions in Figure 40.11. Examining Figure 40.10 indicates that the EMG signal for the rectus femoris reaches the 30% level in the shortest time. This time is right at 40 ms. The next shortest time to the 30% level is about 50 ms for the gluteus maximus. This is followed by the vastus lateralis (75 ms). Consequently, increasing the EMD value to the longest time (55 ms) reported by Viitasalo and Komi would result in slight decrease of the negative work regions for only two muscles, gluteus maximus and rectus femoris.

Assuming that all the cycling variables (e.g. pedalling rate) are constant, the factor which will certainly impact the regions of positive and negative work is the test subject. The results presented here pertain to a single test subject and his corresponding patterns of muscle activation and joint angles. However, patterns of muscle activation are subject dependent (Jorge and Hull, 1986), as are joint angles. In fact joint angles not only vary between subjects (due to differences in anthropometry and pedalling mechanics) but also may vary between the legs of one subject due to asymmetry in pedalling mechanics (Lafortune et al., 1983). Because of the varying patterns of muscle activity and joint angles both between and within subjects, the positive and negative work regions may vary markedly from those shown in Figure 40.11.

Recognizing the possible variability in positive and negative work regions between subjects and the impact that the stretch–shortening phenomenon has on human performance, it is interesting to consider whether diagrams such as that shown in Figure 40.11 might serve to evaluate the potential of athletes to excel in cycling as well as other sports. In order to take this step, some means of interpreting the information in this type of diagram is necessary. To provide this means, one possibility is to develop these diagrams for both elite athletes and non-elite athletes who are comparable physiologically and then try to identify features which distinguish the two groups. Another possibility would be to quantitate the actual force produced by the individual muscles and then use these data in conjunction with a suitable biomechanical model to assess performance. This second possibility is attractive because it could lead to a fundamental understanding of how neurological and biomechanical factors blend in determining performance. While this second possibility is attractive, the implementation is difficult primarily because the actual muscle force cannot be quantitated with a great degree of certainty in dynamic contractions involving biarticular muscles. This would be a problem common to most, if not all, sports.

Beyond serving as a vehicle for evaluating athletic performance, the methodology presented herein serves to elucidate muscle kinesiology. It has been believed traditionally that muscles crossing opposite sides of a joint form agonistic/antagonistic pairs. Accordingly, simultaneous activity of these muscle pairs has been viewed as paradoxical. Probably the best known paradox is Lombard's which surrounds the coactivation of the hamstrings and quadriceps during knee extension (Lombard, 1903). The results in Figure 40.8, however, indicate the possibility of synergy between these muscle groups while being coactivated. The possibility of synergy results because the hamstrings muscles cross two joints, the angles of which both change. During leg extension in cycling, the way in which these joint angles change simultaneously allows the hamstrings muscles to shorten even though the knee is extending. Consequently, to accurately evaluate the functional role of biarticular muscles in a given activity, a detailed analysis of muscle length change in conjunction with stimulation according to the methods presented herein is warranted.

40.5 Summary

Because of the importance of stretch–shortening cycles of muscles in affecting human performance, it is of interest to develop a general methodology for identifying positive and negative work regions in individual muscles in complex movements. Thus it is the objective of this work to present such a method and then illustrate its application to the lower limb muscles in cycling. The method involves combining information surrounding muscle–tendon kinematics (i.e. length and velocity) with information concerning the time course of force developed by the muscle. Muscle–tendon kinematics are obtained via a biomechanical model in conjunction with experimental data which quantify both the positions and orientations of body segments during the activity of interest. The general procedure for developing a biomechanical model which serves to compute muscle kinematics, given these experimental data, is outlined. By examining velocity polarity of muscles, possible regions of positive and negative work are readily identified. To identify actual regions of positive and negative work, the muscle velocity must be considered together with the time course of muscle force. This time course is indicated by measurement and suitable processing of muscle electromyograms (EMG) during the activity of interest. The methods of making these measurements and techniques of processing are described briefly.

The methodology is applied to cycling. The biomechanical model consists of a five-bar linkage restricted to plane motion. The five links are the thigh, shank, foot, crank and bicycle (fixed link). The segment positions and orientations are determined throughout the crank cycle by measuring the angles of the crank and the foot links. Relying on previously published data which allow muscle origin and insertion sites to be determined relative to each link, the lengths and velocities of ten muscles are computed over a crank cycle. Electromyograms measured simultaneously with the linkage angle data allow the time course of muscle force to be synchronized with the muscle kinematics. This reveals stretch–shortening cycles for all muscles considered in the thigh. Consequently, the stretch–shortening cycle appears to be a potentially important mechanism in cycling mechanics.

References

Abbott, B. C., Bigland, B., and Ritchie, J. M. (1952), "The Physiological Cost of Negative Work," *J. of Physiol.* **117**: 380-390.

Andrews, J. G. (1987), "The Functional Roles of the Hamstrings and Quadriceps During Cycling: Lombard's Paradox Revisited," *J. of Biomech.* **20**: 565-575.

Andriacchi, T. P. (1987), "Clinical Applications of the SELSPOT System," *Proc. of the Biomech. Symp.*, AMD-Vol. 84, edited by D. L. Butler and P. A. Torzilli, American Society of Mechanical Engineers, New York, pp. 339-342.

Asmussen, E. (1953), "Positive and Negative Muscular Work," *Acta Physiol. Scand.* **28**: 364-382.

Asmussen, E., and Bonde-Petersen, F. (1974), "Apparent Efficiency and Storage of Elastic Energy in Human Muscles During Exercise," *Acta Physiol. Scand.* **92**: 537-545.

Aura, O. and Komi, R. V. (1986), "Effects of Prestretch Intensity on Mechanical Efficiency of Positive Work and on Elastic Behavior of Skeletal Muscle in Stretch-Shortening Cycle Exercise," *Intern. J. of Sports Med.* **7**: 137-143.

Baildon, R.W.A. and Chapman, A.E. (1983), "A New Approach to the Human Muscle Model," *J. of Biomech.* **16**: 803-809.

Basmajian, J. V. (1974), *Muscles Alive*, Williams and Wilkins Co., Baltimore.

Basmajian, J. V., Clifford, N. C., McLeod, W. D., and Nunnally, H. N. (1975), *Computers in Electromyogr.*, Butterworths, Boston.

Bosco, C., Ito, A., Komi, P. V., Luhtanen, P., Rahkila, P., Rusko, H. and Viitasalo, J. T. (1982), "Neuromuscular Function and Mechanical Efficiency of Human Leg Extensor Muscles During Jumping Exercises," *Acta Physiol. Scand.* **114**: 543-550.

Bosco, C. and Komi, P. V. (1979), Potentiation of the Mechanical Behavior of the Human Skeletal Muscle Through Prestretching," *Acta Physiol. Scand.* **106**: 467-472.

Brand, R. A., Crowninshield, R. D., Wittstock, C. E., Petersen, D. R., Clark, C. R. and van Krieken, F. M. (1982), "A Model of Lower Extremity Muscular Anatomy," *J. of Biomech. Engineering* **104**: 304-310.

Cavagna, G. A. (1977), "Storage and Utilization of Elastic Energy in Skeletal Muscle," *Exercise and Sports Sciences Reviews*, **5**:89-129.

Cavagna, G. A., Dusman, B., and Margaria, R. (1968), "Positive Work Done by a Previously Stretched Muscle," *J. of Appl. Physiol.* **24**: 21-32.

Cavagna, G. A. and Kaneko, M. (1977), "Mechanical Work and Efficiency in Level Walking and Running," *J. of Physiol.* **268**: 467-481.

Cavagna, G. A., Mazzanti, M., Heglund, N. C., and Citterio, G. (1985), "Storage and Release of Mechanical Energy by Active Muscle: A Non-Elastic Mechanism?," *J. of Exper. Biol. Design and Performance of Muscular Systems*, **115**:79-87.

Cavanagh, P. R. and Kram, R. (1985), "Mechanical and Muscular Factors Affecting the Efficiency of Human Movement," *Medicine and Science in Sports and Exercise* **17**: 326-331.

Cavanagh, P. R. and Komi, P. V. (1979), "Electromechanical Delay in Human Skeletal Muscle Under Concentric and Eccentric Contractions," *Eur. J. of Appl. Physiol.* **42**: 159-163.

Chao, E. Y. (1980), "Justification of the Triaxial Goniometer for the Measurement of Joint Rotation," *J. of Biomech.* **13**: 989-1006.

Crowninshield, R. D. (1978), "Use of Optimization Techniques to Predict Muscle Forces," *J. of Biomech. Engineering* **100**: 88-92.

Crowninshield, R. D. and Brand, R. A. (1981), "A Physiologically Based Criterion of Muscle Force Prediction in Locomotion," *J. of Biomechanics* **14**: 793-801.

Curtin, N. A. and Davies, R. E. (1975), "Very High Tension with Very Little ATP Breakdown by Active Skeletal Muscle," *J. of Mechanochemistry and Cell Motility* **3**: 147-154.

Davies, R. E. (1965), "Bioenergetics of Muscular Contraction," in *Control of Energy Metabolism*, edited by B. Chance, R. W. Estabrook and J. R. Williamson, Academic Press, New York, pp. 383-392.

Davy, D. T. and Audu, M. L. (1987), "A Dynamic Optimization Technique for Predicting Muscle Forces in the Swing Phase of Gait," *J. of Biomech.* **20**: 187-201.

Delagi, E. F., Perotta, A., Iazetti, J., and Morrison, D. (1975), *Anatomic Guide for the Electromyographs*, Charles C. Thomas, Springfield, Illinois.

Dostal, W. F. and Andrews, J. G. (1981), "A Three Dimensional Biomechanics Model of Hip Musculature," *J. of Biomech.* **14**: 803- 812.

Dul, J., Townsend, M. A., Shiaui, R. and Johnson, G. E. (1984) "Muscular Synergism. I. On Criteria for Load Sharing Between Synergistic Muscles," *J. of Biomech.* **17**: 663-673.

Edgerton, V. R., Roy, R. R., Gregor, R. J. and Rugg, S. (1986), "Morphological Basis of Skeletal Muscle Power Output," in *Human Muscle Power*, edited by N. L. Jones and A. L. McComas, Human Kinetics Publishers, Champaign, Illinois, pp. 43-58.

Ekblom, B. (1987), "External and Internal Factors Influencing Physical Performance," in *Medicine and Sports Science: Muscular Function in Exercise and Training*, edited by P. Marconnet and P. V. Komi,

Karger, New York, Vol. 26, pp. 90-97.

Ericson, M. O., Nisell, R., Arborelius, U. P. and Ekholm, J., (1985), "Muscular Activity During Ergometer Cycling," *Scand. J. of Rehabilitative Medicine* 17: 53-61.

Faulkner, J. A., Claflin, D. R. and McCully, K. K. (1986), "Power Output of Fast and Slow Fibres from Human Skeletal Muscles," in *Human Muscle Power*, edited by N. L. Jones and A. L. McComas, Human Kinetics Publishers, Champaign, Illinois, pp. 81-96.

Gottlieb, G. L. and Agarwal, G. C. (1971), "Dynamic Relationship Between Isometric Muscle Tension and the Electromyogram in Man," *J. of Appl. Physiol.* 30: 345-351.

Goubel, F. (1987), "Muscle Mechanics: Fundamental Concepts in Stretch- Shortening Cycle," in *Medicine Sport Science: Muscular Function in Exercise and Training*, edited by P. Marconnet and P. V. Komi, Karger, New York, Vol. 26, pp. 24-35.

Green, H. J. (1986), "Muscle Power: Fibre Type Recruitment Metabolism, Fatigue," in *Human Muscle Power*, edited by N. L. Jones and A. L. McComas, Human Kinetics Publishers, Champaign, Illinois, pp. 65-79.

Gregor, R. J., Komi, P. V., and Jarvinen, M. (1987), "Achilles Tendon Forces During Cycling," *Intern. J. of Sports Medicine* 8: 9-14.

Hatze, H. (1981), *Myocybernetic Control Models of Skeletal Muscles*, University of South Africa, Muckleneuk, Pretoria.

Hawkins, D. A., Hawthorne, D. L., De Lozier, G. S., Campbell, K. R. and Grabiner, M. D. (1987), "The Use of Videography for Three Dimensional Motion Analysis," in *High Speed Photography, Videography, and Photonics V*, edited by H. C. Johnson, International Society for Optical Engineering, Bellingham, Washington, pp. 42-45.

Heglund, N. C. and Cavagna, G. A. (1985), "Efficiency of Vertebrate Locomotory Muscles," *J. of Exper. Biology: Design and Perf. of Muscular Systems*, 115:283-292.

Hof, A. L. and van den Berg, J. W. (1981), "EMG to Force Processing I: An Electrical Analog of the Hill Muscle Model," *J. of Biomech.* 14: 747-758.

Hornbeck, R.W. (1975), *Numerical Methods*, Prentice-Hall, Inc., Englewood Cliffs, New Jersey, pp. 16-23.

Hull, M.L. and Davis, R.R. (1981), "Measurement of Pedal Loads in Bicycling: I. Instrumentation," *J. of Biomech.* 14: 843-855.

Hull, M. L. and Jorge, M. (1985), "A Method for Biomechanical Analysis of Bicycle Pedalling," *J. of Biomech.* 18: 631-644.

Jorge, M. and Hull, M. L. (1986), "Analysis of EMG Measurements During Bicycling," *J. of Biomech.* 19: 683-694.

Kadaba, M. P., Wotten, M. E., Ramarkrishnan, H. K., Hurwitz, D. and Cochran, G.V.B. (1987), "Assessment of Human Motion With VICON," *Proc. of the Biomechanics Symposium, AMD-Vol. 84*, edited by D. L. Butler and P. A. Torzilli, American Society of Mechanical Engineers, New York, pp. 335-338.

Komi, P.V. (1973), "Relationship Between Muscle Tension, EMG, and Velocity of Contraction Under Concentric and Eccentric Work," in *New Devel. in Electromyography and Clinical Neurophysiology*, edited by J. E. Desmedt, Karger, Basel, Vol. 1, pp. 596-606.

Komi, P. V. (1987), "Neuromuscular Factors Related to Physical Performance", in *Medicine and Sport Sciences: Muscular Functions in Exercise and Training*, edited by P. Marconnet and P. V. Komi, Karger, New York, Vol. 26, pp. 48-66.

Lafortune, M., Cavanagh, P. R., Valient, G. A. and Burke, E. R. (1983), "A Study of the Riding Mechanics of Elite Cyclists," *Medicine and Science in Sports and Exercise* 15: 113.

Lloyd, B. B. and Zacks, R. M. (1972), "The Mechanical Efficiency of Treadmill Running Against a Horizontal Impeding Force," *J. of Physiol.* 223: 355-363.

Loeb, G. E. and Gans, C. (1986), *Electromyography for Experimentalists*, University of Chicago Press, Chicago, Illinois.

Lombard, W. P. (1903), "The Action of Two-Joint Muscles," *Amer. Physical Education Review* 8: 141-145.

Miller, J.A.A. (1987), "Motion Analysis Using the 2 Camera CODA-3 Measurement System," in *Proc. of the Biomech. Symp., AMD-Vol. 84*, edited by D.L. Butler and P.A. Torzilli, American Society of Mechanical Engineers, New York, pp. 343-344.

Morrison, J. B. (1968), "Bioengineering Analysis of Force Actions Transmitted by the Knee Joint," *Biomed. Engineering* 3: 164-170.

Muro, M. and Nagato, A. (1985), "The Effects of Electromechanical Delay of Muscle Stretch of the Human Triceps Surae," *Biomechanics IX-A*, edited by D. A. Winter, R. W. Norman, R. P. Wells, K. C. Hayes and A. E. Patla, Human Kinetics Publishers, Champaign, Illinois, pp. 86-90.

Norman, R. W. and Komi, P. V. (1979), "Electromechanical Delay in Skeletal Muscle Under Normal Movement Conditions," *Acta Physiol. Scand.* 106: 241-248.

Pedotti, A., Krishnan, V. V. and Stark, L. (1978), "Optimization of Muscle- Force Sequencing in Human Locomotion," *Math. Biosci.* 38: 57-76.

Pell, K. M. and Stanfield, J. W. (1972), "Mechanical Model of Skeletal Muscle," *Amer. J. of Physical Med.* 51: 23-38.

Penrod, D. D., Davy, D. T. and Singh, D. P. (1974), "An Optimization Approach to Tendon Force Analysis," *J. of Biomech.* **7**: 123-129.

Seireg, A. and Arvikar, A. (1973), "A Mathematical Model for the Evaluation of Forces in Lower Extremities of the Musculo-Skeletal System," *J. of Biomech.* **6**: 313-326.

Soudan, K. and Dierckx, P. (1979), "Calculation of Derivatives and Fourier Coefficients of Human Motion Data While Using Spline Functions," *J. of Biomech.* **12**: 21-26.

Stainsby, W. N. (1976), "Oxygen Uptake for Negative Work, Stretching Contractions by In Situ Dog Skeletal Muscle," *Amer. J. of Physiol.* **230**: 1013-1017.

Stern, J. T. (1974), "Computer Modeling of Gross Muscle Dynamics," *J. of Biomech.* **7**: 411-428.

Suzuki, S., Watanabe, S. and Hamma, S., (1982), "EMG Activity and Kinematics of Cycling Movements at Different Constant Velocities," *Brain Research* **240**: 245-258.

Thys, H., Faraggiana, T. and Margaria, R. (1972), "Utilization of Muscle Elasticity in Exercise," *J. of Appl. Physiol.* **32**: 491-494.

van Ingen Schenau, G. J. (1984), "An Alternative View of the Concept of Utilization of Elastic Energy in Human Movement," *Human Movem. Science* **3**: 301-336.

Vaughn, C. L. (1982), "Smoothing and Differentiation of Displacement-time Data: An Application of Splines and Digital Filtering," *Intern. J. of Biomed. Computing* **13**: 375-386.

Viitasalo, J. T. and Komi, P. V. (1981), "Interrelationships Between Electromyographic, Mechanical, Muscle Structure, and Reflex Time Measurements in Man," *Acta Physiol. Scand.* **111**: 97-103.

Walton, J. S. (1981), Close Range Cine-Photogrammetry: A Generalized Technique for Quantifying Gross Human Movement, Doctoral Dissertation, Pennsylvania State University, University Park, Pennsylvania.

Whipp, B. J. and Wasserman, K. (1969), "Efficiency of Muscular Work," *J. of Appl. Physiol.* **26**: 644-648.

White, D.C.S. (1977), "Muscle Mechanics," in *Mechanics and Energetics of Animal Locomotion*, edited by R. McN. Alexander and G. Goldspink, Chapman and Hall, London, pp. 23-56.

Wood, G. A. (1982) "Data Smoothing and Differentiation Procedures in Biomechanics," *Exercise and Sports Science Reviews*, **10**:308-361.

Zacks, R. M. (1973), "The Mechanical Efficiencies of Running and Cycling Against a Horizontal Impeding Force," *Internationale Zeitschrift fur Angewandte Physiologie Einschlieblich Arbeitsphysiologie* **31**: 249-258.

Zernicke, R. F., Caldwell G. and Roberts, E. M. (1976), "Fitting Biomechanical Data with Cubic Spline Functions," *Res. Quart.* **47**: 9-19.

CHAPTER 41

The Unique Action of Bi-Articular Muscles in Leg Extensions

Gerrit Jan van Ingen Schenau, Maarten F. Bobbert, and Arthur J. van Soest

41.1 Introduction

In textbooks on the anatomy of the musculo–skeletal system, both muscles crossing only one joint (mono–articular muscles) and muscles crossing more than one joint (multi–articular muscles) are classified according to the location of their line of action relative to joint axes of rotation (e.g. Williams and Warwick, 1980). For instance, the line of action of the mono–articular vastus medialis passes anterior to the flexion/extension axis of the knee joint, and therefore the muscle is classified as a knee extensor. Similarly, the bi–articular gastrocnemius is classified as a knee flexor and ankle plantar flexor. As such, the gastrocnemius is considered to be an antagonist of the vasti at the knee joint.

This classification method, which is focussed on joint displacements, underlies the majority of contemporary descriptions of muscle actions required to perform a task. For instance, in jumping, hip extension and knee extension occur. Thus, the hip extensor and knee extensor muscles are expected to be active. Unfortunately, if we focus on joint displacements it is difficult to understand why the body is supplied with muscles crossing more than one joint; it seems that such muscles could well have been replaced with sets of mono–articular muscles. Arguing against the classification of muscles as described above, many authors have suggested that muscle actions can only be understood if their effects are studied in a natural environment, taking into account the actions of other muscles, forces on the environment, inertial forces and gravity (e.g. Fisher, 1902; Bernstein, 1967). Following this suggestion, we

have studied jumping, speed skating and cycling using an inverse dynamical approach, and we have identified a number of constraints in the transformation of rotations in joints into the desired translation of the body center of gravity relative to the foot (in jumping and speed skating) or the pedal trajectory relative to the trunk (in cycling). In dealing with these constraints, bi–articular muscles appear to play a unique role by distributing net joint moments and joint powers over the joints.

Both in this chapter and in Chapter 18 (Gielen et al.), this unique role, which often requires co-activation of mono–articular muscles and their bi-articular antagonists, will be explained in the terminology of multi–link models. Before describing our approach, and results obtained using it, some attention is paid to other approaches used in the literature to study the function of bi–articular muscles and co-activation. Also, in order to prevent misunderstandings, attention is paid to some of the concepts used in our approach.

41.2 Possible Actions of Bi-Articular Muscles

Since Borelli (1685) showed that the force development in the knee joint is influenced by the position of the hip joint, many researchers have advanced ideas about possible actions of muscles crossing more than one joint. The possible action of bi–articular muscles in the lower extremity has in particular been the subject of a lot of speculation over the past century (e.g. Cleland, 1867; Fick, 1879; Hering, 1897; Lombard, 1903; Baeyer, 1921; Fenn, 1938; Markee et al, 1955; Molbech, 1966; Wells and Evans, 1987; Ingen Schenau,

Multiple Muscle Systems: Biomechanics and Movement Organization
J.M. Winters and S.L-Y. Woo (eds.), © 1990 Springer-Verlag, New York

1989). When restricted to bi–articular muscles and actions which cannot be performed by an alternative set of two mono–articular muscles, the following (in part overlapping) functions and advantages of bi–articular muscles have been proposed:

* **Coupling of Joint Movements**

Many authors have stressed the fact that activation of bi–articular muscles leads to interdependency of the movements in both joints that are crossed (Cleland, 1867; Huter, 1863,1869; Fick, 1879; Langer, 1879; Elftman, 1939; Markee et al. 1955; Landsmeer, 1961; Winter, 1984; Hogan, 1985; Wells, 1988). If, for example, the hip is extended by the mono–articular hip extensors and the rectus femoris does not elongate, hip extension must be accompanied by knee extension. In a similar way, knee extension can be coupled to plantar flexion via the gastrocnemius muscle. This coupling is known as ligamentous or tendinous action (Cleland, 1867). Especially in animals such as the horse, a number of the bi–articular muscles have only a limited shortening capacity and can to a large extent be regarded as tendons (Bogert et al., 1989). These tendinous muscles allow the more proximally located mono–articular muscles to have indirect actions on joints which they do not pass (Cleland, 1867). As indicated by Cleland (1867) and by Fick (1879), coupling of joint movements by tendinous action of bi–articular muscles has the advantage that most of the muscle mass can be located close to the trunk, thus leaving the distal segments relatively free of muscle bulk. Other proposed advantages of these couplings are the ease of control of multi–joint movements (Hogan, 1985) and the transport of energy from one joint to a more distal joint (Cleland, 1867; Gregoire et al. 1984; see below).

* **Low Contraction Velocity**

This concept can be seen in the discussions of Cleland (1867), Fick (1879), Duchenne (1867), Fenn (1938), and Gregoire (1984). If hip extension and knee extension occur simultaneously, the shortening velocity of the bi–articular hamstring muscles is lower than that of the mono–articular hip extensors and the shortening velocity of the bi–articular rectus femoris is lower than that of the mono–articular knee extensors. Similarly, the shortening velocity of gastrocnemius is lower than that of mono–articular plantar flexors when knee extension is combined with plantar flexion. At this lower contraction velocity, the muscles are operating in a more favorable region of their force–velocity relationship compared to a situation where origin and insertion are not moving in the same direction. Baeyer (1921) used the term "concurrent movements" to define simultaneous movements in adjacent joints causing origins and insertions of bi-articular muscles to move in the same direction; for the opposite movements he used the term "counter-current movements."

* **Transport of Energy**

In the above-mentioned simultaneous hip and knee extension it might be said that the mono–articular hip extensors are doing work in extending the knee. In our recent work we defined this process as transport of energy, a concept coupled to our applied multi–link models. Expressed in other words, this transport mechanism was already proposed more than a century ago: Langer indicated in 1879 that the gluteus maximus can support plantar flexion in a leg extension by coupling actions of rectus femoris and gastrocnemius. The same idea was expressed at the same time by Fick (1879) and by other authors later on (Lombard, 1903; Fenn, 1938; Gregoire et al, 1984).

In addition to the above mentioned actions of bi-articular muscles, a number of other actions have been proposed, such as joint stabilization (Markee et al, 1955). However, such actions cannot be judged as unique for bi-articular muscles.

41.3 Co-Activation of Antagonists

From the point of view of joint displacements required in performing a particular task, it seems inefficient to activate antagonists since the force (and work) contribution of the agonists appear to be cancelled out by the antagonists. This apparent inefficiency may have led many to the opinion that such co-contractions should not (or do not) occur in voluntary movements. In this context, it has been stated:

"Nature never works against herself."
(Pettigrew, 1873, cited by Tilney and Pike, 1925)

Several authors have attempted to identify organizational principles which could prevent co-contraction of agonists and antagonists. Descartes (1662) was the first to describe some type of reciprocal inhibition (controlled by "vital spirits") and there have been many supporters since then; especially since Sherrington (e.g. Sherrington, 1909) published his series of papers on this subject (e.g. Fujiwara and Basmajian, 1975; Suzuki et al., 1982; Kumamoto, 1984; Yamashita, 1988; see Smith (1981) for more references).

Since Winslow's work in 1776 (cited by Tilney and Pike, 1925) many have opposed these views. According to Tilney and Pike (1925), Duchenne described co-activations of mono–articular agonists and bi–articular antagonists in 1857 as "Harmonie des Antagonistes": co-activations needed to modify and stabilize the movements. Much experimental evidence has since then been published to show that co-activations of antagonists indeed occur (see Tilney and Pike, 1925 and Smith, 1981 for further arguments and references on this controversy).

In fact, many results of studies of multi–joint movements (such as running, jumping, cycling and standing up from a chair) indicate that co-contractions of mono–articular agonists and their bi–articular antagonists are common rather than exceptional (Andrews, 1987; Elftman, 1939a,b; Gregor et al., 1985; Winter, 1984; Gregoire et al., 1984). In the remaining part of this chapter, as well as in Chapter 18 (Gielen et al.), it will be shown that these co-activations are not in conflict with Pettigrew's statement that "nature never works against herself."

41.4 Constraints in the Transfer of Rotation to Translation

41.4.1 Geometrical Constraints

Since the translational range of motion in human joints is very small, translations of hand or foot relative to the trunk have to be realized by rotations in joints. In the transformation of rotations of segments into translation of segmental end points, constraints are present. Because of these constraints, a particular pattern of coordination of mono– and bi–articular muscles is needed to prevent inefficient utilization of metabolic energy. One of the constraints is that the force exerted on the environment not only needs a magnitude but

also a direction [see also Chapter 9 (Hogan)]. Assuring a particular direction of the external force requires a certain distribution of net moments in the joints. This phenomenon was identified in an analysis of cycling (Ingen Schenau, 1989) and further elaborated for arm tasks in Chapter 18 (Gielen et al.). A second constraint is that joint angles influence the transfer of angular velocities into linear velocities, and the transfer of angular accelerations into linear accelerations. This constraint plays an important role in explosive ballistic movements where the aim of the movement is to obtain a velocity as high as possible in projecting the body center of gravity or an object. Examples are vertical jumping, pushoff in speed skating, and overarm throwing. The constraint was originally identified in the speed skating pushoff (Ingen Schenau et al, 1985) and further elaborated in an analysis of the vertical jump (Ingen Schenau et al, 1987; Bobbert and Ingen Schenau, 1988). It will be illustrated with the help of a simplified example of a push off as outlined in Figure 41.1.

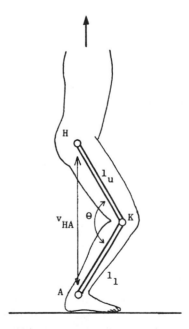

Figure 41.1: The velocity difference between hip and ankle is not only determined by the angular velocity $d\theta/dt$ but also by θ. The more the knee approaches full extension, the smaller is the transfer of $d\theta/dt$ to v_{HA}.

Imagine a pushoff where the trunk is to be accelerated by extending the knee joint without changing the position of the trunk or the foot. Such type of pushoff is required in speed skating: increasing the trunk angle would cause an increase in air friction, and plantar flexion in the leg that pushes off would increase ice friction as the skate of this leg glides forward (Ingen Schenau et al., 1985; de Koning et al., 1989). By taking the time derivative of the distance between hip and ankle, we obtain the vertical velocity of the hip relative to the ankle:

$$v_{HA} = \left[\frac{l_l \, l_u \sin\theta}{[(l_l^2 + l_u^2 - 2l_l \, l_u \cos\theta)]^{1/2}} \right] d\theta/dt$$

(41.1)

where l_u is upper leg length, l_l is lower leg length, and θ is knee angle. The expression between brackets may be regarded as a transfer function describing the transformation of knee angular velocity into the required translational velocity. This transfer function goes to zero when the knee joint reaches full extension. Thus, irrespective of the knee extension velocity, the translational velocity v_{HA} also goes to zero. Thus, dv_{HA}/dt becomes negative before the knee joint reaches full extension. It follows that the skate will loose contact with the ice before the knee is fully extended, at approximately the instant that dv_{HA}/dt reaches a value of -9.8 m/s^2. This indeed occurs in speed skating where the skate was found to lose contact with the ice at a knee angle of about 150° (Ingen Schenau et al., 1985). The reason is that at the instant that the decreasing transfer function begins to dominate the (still increasing) angular velocity of knee extension, the relatively heavy trunk pulls the lower leg, foot and skate from the ice since it has already obtained a velocity larger than the decreasing velocity v_{HA} at the last part of knee extension.

Needless to say, the same constraint, referred to as *geometrical constraint* (Ingen Schenau et al., 1987), is present in jumping. The only difference is that the transfer function is more complex because rotation of the trunk and foot are allowed. As a matter of fact, an early loss of contact was demonstrated by Alexander (1989) in a simulation of vertical jumps.

It is important to realize that when the pushoff ends with the knee still flexed, the knee extensor muscles have not shortened fully, and their capacity to do work has not been used fully for the pushoff. If the muscles remain active after takeoff, the work performed by them over the remaining shortening range will be used for a useless increase in rotational energy of segments.

41.4.2 Anatomical Constraint

In addition to the geometrical constraint imposed by the transfer function of equation (41.1), a second constraint is present during explosive pushoffs. This is due to the fact that the angular velocity needs to be reduced prior to full extension. In actual jumping, the knee angular velocity can reach values up to 17 rad/s. If this angular velocity were not actively decelerated to zero, the knee joint could be damaged. To preserve structural integrity of the knee joint, knee flexor activity is needed. This "anatomical constraint" (Ingen Schenau et al., 1987) should be accounted for in protocols used to simulate vertical jumping (see below and Chapter 42 (Pandy)). If Alexander (1989) would have incorporated this constraint in his simulations, he would have found smaller knee angles at the end of the pushoff even in the hypothetical jumps of his model with massless legs.

41.5 Possibilities of Dealing with Constraints

41.5.1 Co-Activation of Mono-Articular and Bi-Articular Muscles

In an actual vertical jump where the jumper is allowed to perform a plantar flexion, the geometrical constraint imposed by the transfer function as well as the anatomical constraint mentioned above can be dealt with effectively by activation of the bi–articular gastrocnemius. This is outlined in Figure 41.2a. The resulting knee flexing moment caused by this muscle reduces the angular acceleration in the knee joint and, due to the tendinous action described earlier, knee extension is now to a certain extent coupled to plantar flexion. It can be said that knee extensors pull on the calcaneus and thus support plantar flexion.

The importance of this tendinous action was demonstrated by a simple physical model as outlined in Figure 41.2b (Bobbert et al., 1987). The mono–articular knee extensors are modelled by a

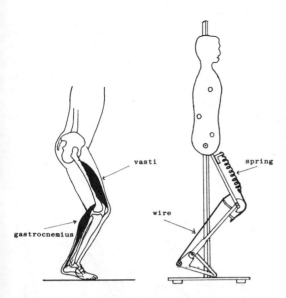

Figure 41.2: *a)* As soon as the velocity difference between hip and ankle can no longer be increased, the gastrocnemius muscle couples a further knee extension to plantar flexion. *b)* The effect of this coupling was demonstrated by a mechanical model ("jumping Jack") where the gastrocnemius is represented by a wire (see text and Bobbert et al., 1987).

spring, which is loaded with a certain amount of potential energy by flexing the knee (pushing the model downwards). The tendinous action of the gastrocnemius is modelled by a stiff wire. The length of this wire can be adjusted in order to simulate variation in timing of the coupling of knee extension and plantar flexion. It was found that when the wire became taut during the pushoff, jumping height was greater than when it remained loose. Moreover, it was found that an optimum occurred in wire length; with wire length adjusted to this optimum (representing optimal timing of the coupling of knee extension and plantar flexion), the model jumped almost twice as high as when the wire remained loose. Also, we performed computer simulations of this type of jumps (Soest et al., 1989) and found that the optimal timing results in a compromise between loosing ground contact too early (which occurs when coupling occurs early in the push off) and increasing the rotational energy in the lower and upper legs uselessly (which occurs when coupling occurs late during the pushoff). These examples suggest that co-activation of antagonists (in this case co-activation of knee extensors and the bi–articular gastrocnemius) can be highly effective in ex-

plosive pushoffs. In an electromyographic analysis of vertical jumping we found that human subjects indeed show co-activation of vasti and gastrocnemius. With these basic principles in mind it is now possible to explain the temporally ordered sequence of muscle activation patterns as observed in those analyses (Bobbert and Ingen Schenau, 1988).

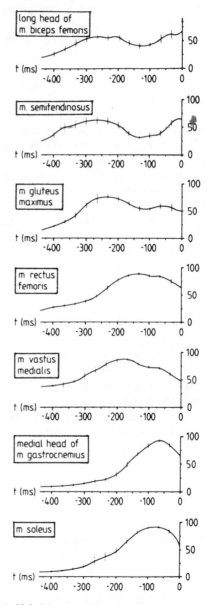

Figure 41.3: Mean muscle activation patterns of 10 experienced jumpers. Time is expressed in ms prior to the end of push off (toe off). Note the periods of co-activation of gluteus maximus and rectus femoris and of vasti and gastrocnemius.

41.5.2 Sequencing of Muscle Activations

Figure 41.3 presents as a function of time mean muscle activation levels of 10 experienced jumpers during the performance of counter movement jumps (see Bobbert and Ingen Schenau (1988) for details regarding this study). For the interpretation of muscle activation patterns one should keep in mind that there is a delay between a change in activation and a change in the mechanical response of the muscle of 80–100 ms (Thomas et al., 1988; Ingen Schenau, 1989b; Vos et al., 1990). These patterns can be shown to be highly functional with help of Figure 41.4, where the orientations of the jumpers' body segments are schematically depicted at four time intervals prior to the end of the pushoff. The thickness of the lines representing muscle actions is drawn in such a way that it gives an impression of the changes in mechanical responses of the muscles.

The vertical acceleration of the center of gravity is initiated by a rotation of the trunk following the increases in activity of hamstrings and gluteus maximus. Some 100 ms later the activity level of the quadriceps muscles is increased while the hamstring activity is decreased. In light of the discussed problems in the transfer of rotations into translation, this seems logical. Because of the large moment of inertia of the trunk, it takes a relatively long time to give this segment a large angular velocity. Activation of the hamstrings helps to increase the angular velocity of the trunk. At the same time, it prevents an early knee extension, which would hamper a fast trunk acceleration (an upward acceleration of the hip because of knee extension would cause an extra inertial force on the trunk). As soon as an increase in trunk rotation can no longer contribute to a vertical acceleration of the body center of gravity, rectus femoris activity is increased and hamstring activity is decreased. The hip flexing moment exerted by rectus femoris helps to reduce the angular acceleration of the trunk, and the power delivered by the gluteus maximus supports knee extension by tendinous action of the rectus femoris. During the last 50–100 ms the knee flexing moment of gastrocnemius helps to reduce the angular acceleration in the knee joint as explained above. In these last 50 ms all leg muscles can contribute to plantar flexion through tendinous actions of rectus femoris and gastrocnemius. The fact that hamstring activity is decreased but not terminated has

most likely to do with the fact that both the hamstrings and the rectus femoris shorten in this movement. This phenomenon was early on described by Lombard (1903), who showed that frogs can jump very efficiently by co-activation of these bi-articular "antagonists."

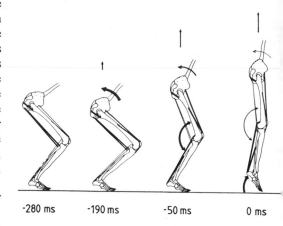

-280 ms   -190 ms   -50 ms   0 ms

Figure 41.4: A schematic outline of muscle actions as occurring in a sequential order during the pushoff. The thickness of the lines representing the muscles indicates the mechanical responses of the muscles on the changes in activity as presented in Figure 41.3. The curved arrows indicate the major angular accelerations.

A comparable proximo–distal sequence and co-activation of mono– and bi–articular muscles was recently found for the pushoff in the sprint start (Jacobs et al., in preparation). In speed skating, the proximo–distal sequence is not complete: the rectus femoris appears to play a comparable role in transporting energy from the mono–articular hip extensors to the knee, but coupling between knee and ankle is much less pronounced than in jumping and sprinting (de Koning et al., submitted for publication). This is of course due to the fact that speed skaters have learned to suppress a plantar flexion in order to prevent an increase in ice friction.

In an analysis of the overarm throw of female handball players, we also found a pronounced proximo–distal sequence in joint actions (Jöris et al., 1985). Though in that study no muscle activity patterns were obtained, that sequence too is likely to play a role in dealing with the constraints in the transfer of rotations into translations.

In this paragraph on timing it should be stressed that the observed sequence in changes in muscle activation and joint extensions could serve the purpose of transporting energy from joints where it can no longer be used effectively, to joints where it still can. Without these muscles, a proximo–distal sequence is not optimal, as recently shown in computer simulations by Alexander (1989). As indicated above, the actual transfer function between angular velocities in all joints and the translational velocity of the body center of gravity (which has no fixed anatomical location) is much more complicated than what is outlined in the basis of Figure 41.1. Moreover, other constraints may play a role as well (e.g. neurologically based constraints; intrinsic properties of muscles and tendons etc.). This makes it difficult to predict what would be an optimal timing. Moreover, one can imagine that rather large differences in timing exist between subjects, as indicated by Jensen et al. (1989). In our group of not specifically trained subjects, however, we found comparable patterns as for the trained subjects. The only significant difference was a shorter time interval between the onset of hip extension and the onset of knee extension than was found in the trained subjects. Despite the absence of reliable data on intrinsic muscle properties, muscle tendon lengths, tendon compliances and about possible "hard–wired" connections in the central nervous system, an understanding of why one activation pattern is better than another could be obtained by means of direct dynamics. In Chapter 42 (Pandy) such an application is presented for the vertical jump. Comparable results of our own group are presented below. Both Pandy's and our simulation seem to support the proximo–distal sequence.

41.6 Simulations of Vertical Jumping

The concepts presented in the previous paragraphs have been formulated on the basis of an inverse dynamical analysis of jumping. An assumption underlying the formulation of these concepts is that the coordination pattern found experimentally results in a (more or less) maximal jumping height. Experimentally, it is not possible to test this assumption. Also, it is impossible to ascertain from experiment exactly in what way the concepts described interact. Finally, it is not clear whether these concepts are sufficient (in contrast to necessary) to explain the coordination pattern found.

Simulation seems the most promising methodology to tackle these problems. Simulation and optimization of behavior of rather complex systems has become feasible in recent years thanks to a tremendous increase in computational power. In this section, basic ideas are introduced and some results are presented. Basically, direct dynamics simulation of "mechanical" systems can be described as follows [see also Chapter 8 (Zajac and Winters)]:

1) Given a position and velocity of the system and neural drive for the current time, the forces acting (usually a function of neural drive, position and/or velocity) are calculated.

2) Given the results of *(1)* plus external loading on the linkage system, and supposing that the equations of motion of the degrees of freedom have been derived in some way (which constitutes the major problem of direct dynamics), the accelerations of the degrees of freedom are calculated.

3) Given position, velocity and acceleration of the degrees of freedom, the former two are integrated numerically over a small time interval to yield new positions and velocities of the degrees of freedom. (If a dynamic muscle model is used, the muscle model state variables must also be integrated.)

4) Increment time. Given the new values for the state variables, as obtained results of *(3)*, calculate any other parameters that are static functions of the state variables (e.g. other kinematically-related positions and velocities within the assumed linkage system).

5) Return to *(1)*.

In our case, the (planar) model consists of four segments (foot, lower leg, upper leg, trunk–head–arms). Driving forces are generated by 6 groups of muscles: GLUtei, HAMstrings, VASti, RECtus femoris, SOLeus and GAStrocnemius. These groups represent the major mono–articular and bi–articular muscles contributing in vertical jumping. Input to these muscles is stimulation, which is transformed into active state using first order dynamics. The muscles are modelled using a three–component Hill–type muscle model. Parameter values for the muscle models are derived from known morphologic characteristics (e.g. number of sarcomeres in series) and experimental results (e.g. moment–angle relationships). Behavior of the muscle can be described by one first–order differential equation, in which contractile element velocity is calculated as a function of stimulation, contractile element length and total muscle length.

The behavior of the entire system thus defined is simulated using *SPACAR*, a finite element method with deformable links developed at the Technical University of Delft (van der Werff, 1977). All differential equations governing system behavior are integrated simultaneously using a variable order, variable stepsize predictor corrector integration algorithm (Shampine and Gordon, 1975).

For reasons of simplicity, simulations have been restricted to "squatting jumps", i.e. jumps started from a prescribed static squatted position; a countermovement is not allowed. At the start of the simulation, stimulation of the muscles is such that a static equilibrium is maintained.

The question addressed is: At what time should the stimulation of each muscle be switched from the starting value to the maximum value in order to jump as high as possible? This optimization problem is solved using a standard quasi–Newton optimization algorithm (*NAG*).

When comparing the jump performed with optimal timing with experimental data, a number of observations can be made (see Figure 41.5).

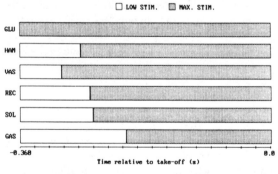

Figure 41.5: Optimal stimulation pattern for a mathematical model incorporating 6 muscle groups, performing a squatting jump from a prescribed static position as obtained from direct dynamics simulation. For all muscle groups, the time at which stimulation was switched to the maximal value was optimized. Time is expressed in seconds prior to the instant of takeoff.

* in the simulated optimal jump, a clear proximo–distal sequence in stimulation pattern is present; only the hamstrings group is a little out of line;

* during the major part of the simulated jump, co-activation of monarticular extensors with their bi-articular "antagonists" is present;

* kinematics of the simulated jump is very similar to the kinematics of experimentally observed squatting jumps (data not shown) except for the final part of the pushoff (see below).

What constraints play a major role in determining optimal timing? This question is currently being investigated in detail using simulation. Here, some observations will be made concerning the constraints discussed earlier in this chapter.

In the first place, the anatomical constraint, forcing a deceleration of joint extension just prior to takeoff, is absent in the model. In the behavior of the model, a continuous increase of the angular accelerations is observed. It can therefore be concluded that the anatomical constraint plays a significant role just before takeoff.

Secondly, the geometrical constraint clearly plays an important role in a fast ballistic movement such as jumping. From the previous paragraphs, it will be clear that a sequential action of the bi–articular rectus femoris and gastrocnemius muscles can contribute in dealing with this constraint (although full joint extension at takeoff probably cannot be reached). From the fact that both in experiment and in simulation such a sequential action is present, it is tempting to think that the stimulation pattern is aimed primarily at solving the problem posed by the geometrical constraint.

However, when performing these simulations it became evident that it is not at all easy to control the direction of the acceleration of the body center of mass. This "directional" constraint has not yet been subject to systematic research. In the near future, we hope to address the influence of this constraint, and the way it relates to the geometrical constraint.

41.7 Redistribution of Joint Moments and Powers

If the actions of bi–articular muscles discussed in this chapter are combined with those in Chapter 18 (Gielen et al.), it is possible to formulate a general role for bi–articular muscles. Generally speaking one might say that bi–articular muscles can redistribute net moments and net power over the joints that are crossed. Power delivered by mono–articular muscles which cross a particular joint can appear as joint power at an adjacent joint. This has been defined as a transport of energy (Ingen Schenau et al., 1987). Unfortunately, because the expression transport of energy has also been used to indicate a flow of energy between segments, our definition appears to cause confusion.   The following discussion is intended to prevent misunderstandings.

If the human body is modelled as a system consisting of rigid links connected by joints which predominantly allow rotations, an inverse dynamical analysis, as originally proposed by Elftman (1939a) and applied or further elaborated by many others (e.g. Capozzo et al., 1976; Robertson and Winter, 1980; Aleshinsky, 1986; Bobbert and Ingen Schenau, 1988), yields for each joint a net joint force and a net joint moment. Following Elftman (1939a) a number of authors have constructed an energy balance for each separate segment using the (external) joint forces and joint moments to calculate the flow of energy to the segment which, for rigid segments, should equal the rate of change of mechanical energy of the segment. By constructing these energy balances for all individual segments, these authors were able to calculate flows of energy between segments. It should be emphasized that the transport of energy as discussed in this chapter has nothing to do with these flows of energy between segments [which may be the reason why Pandy (Chapter 42), using the energy flow definition, did not find a unique action for bi–articular muscles in transporting energy].   What is meant here is based on an energy balance of the assembly of all segments together. As shown by Aleshinsky (1986) for running and for the more general case by Ingen Schenau and Cavanagh (1990), a summation of the energy balances for the separate links leads to one energy equation for the entire body. The time

derivative of this equation, called the instantaneous power equation of the system, is:

$$\sum P_{joints} = \sum (F_i * v_i) + dE_{segm}/dt \qquad (41.2)$$

The expression on the left–hand side equals the (instantaneous) sum of joint powers (sum of products of joint moment and joint angular velocity). The first expression on the right–hand side equals the power exchange with the environment, calculated as the product of forces external to the entire system (e.g. air friction, force on the pedal in cycling) and the velocity of their points of application. The last expression equals the rate of change of the sum of segmental energies. It should be noted that in this equation the joint forces are no longer present. In fact it shows the origin of power which can be used to fulfill a certain task (increase the mechanical energy of the entire system or do work against the environment). Though joint power is calculated for each joint separately it should be stressed that *Eq. 41.2* is only valid for the instantaneous sum of joint powers. This means that a negative power at one particular joint does not necessarily mean that power is degraded into heat or stored in elastic components of muscle-tendon complexes; if the summed power is positive, negative power at one particular joint may appear as positive power in an adjacent joint through action of bi-articular muscles. This is illustrated with the help of Figure 41.6.

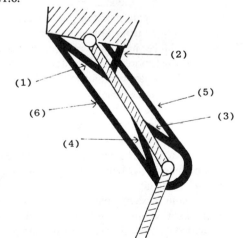

Figure 41.6: Schematic representation of a set of mono- and bi-articular muscles needed to meet the constraints discussed in this chapter and in Chapter 16 (Gielen et al.) for planar two-joint movements.

In the light of the constraints discussed in this chapter and in Chapter 16 (Gielen et al.), the indicated number of mono- and bi-articular muscles is sufficient to control a two bar system in one plane. Now imagine that this two bar system represents the upper and lower leg and that the leg is extended through actions of the mono-articular muscles (1) and (3). Let (1) deliver a moment $M1$ with respect to the hip joint and an amount of power $P1$. Let (3) deliver a moment $M3$ and power $P3$. If we now also activate the bi-articular muscle (5), the result will be that the net moment in the hip decreases and the net moment in the knee increases. If, for the sake of simplicity, we assume that the extension velocities remain the same and that (5) does not change in length (as the wire in the model of Figure 41.2b), it does not add power to the system while the other two muscles can still deliver the same amount of power. However, the power delivered by (1) does only in part appear as hip joint power since the net moment in the hip is decreased through action of muscle (5). The opposite is true at the knee. Here the power is larger than what is delivered by muscle (3) while muscle (5) in this example does not add power to the system. Note that strong activation of (5) can cause a negative moment (and power) at the hip while (1) is still producing the same amount of positive power. One can easily calculate that the decrease in power due to the decrease in net moment at the hip is exactly equal to the surplus power at the knee. When expressed in net joint powers we defined this phenomenon as a transport of power from hip to knee and basically this should be judged as a contemporary expression of what more than a century ago was defined as a "ligamentous action" (Cleland, 1867). This effect of transport of energy is of course also present if (5) changes in length. The total amount of joint power is then increased or decreased by the amount of power production or absorption by muscle (5). It will not be difficult to realize that activation of muscle (6) will cause a transport of power from knee to hip.

This versatile system makes it possible to redistribute joint moments and joint powers in such a way that the specific task demands are met. For jumping we have seen that this includes a proximo-distal shift in joint power: power is predominantly used in those joints where it can most effectively contribute to the increase of effective energy of the jumper.

For cycling we found that the shift in net moments was necessary to control the direction of the external force on the pedal in such a way that this force can do work on the pedal [see Ingen Schenau, 1989a, for details on cycling, and Chapter 18 (Gielen et al.) for a more general discussion of this topic].

Though the paper of Cleland came only recently to our notice we feel that he deserves the honor to be explicitly cited in this context. With respect to the effects of the actions of bi-articular muscles he wrote:

> ".... the total amount of muscular power is made available for overcoming the total amount of resistance, which ever be the joint on which that may to the greatest extent fall." (Cleland, 1867)

41.8 Conclusions and Future Directions

From the phenomena discussed above and those discussed in Chapter 18 (Gielen et al.), the conclusion can be drawn that a wide range of multi-*joint* movements require not only a control of joint displacements but also a particular distribution of net moments over the joints. These two requirements can to a large extent be judged as independent.

In preventing inefficient eccentric contractions and in interchanging net moments and energy between joints, bi-articular muscles appear to perform unique actions which could not be performed by a set of two mono-articular muscles. This means that the actions of these muscles cannot be qualified on the basis of a simple reference to the joint displacements which they are assumed to support.

The described actions require a co-activation of mono-articular and bi-articular muscles. Such co-activations are not inefficient. Instead, it is only by virtue of these co-activations that specific task demands can be satisfied efficiently. Thus, the co-activations may be qualified as synergistic rather than antagonistic. As such they are not in disagreement with Pettigrew's statement that "nature never works against herself".

On the basis of these conclusions, a number of plans, suggestions and speculations can be formulated to help direct future research.

41.8.1 Ongoing Research

A number of studies have been started to identify constraints in motor tasks other than jumping and cycling, and to determine whether the intermuscular coordination in these tasks can be understood from the mechanical aim. At present the research focusses on sprinting, running, walking and cycling at different loads and cycle frequencies. In a different set of studies it will also be attempted to quantitate the contributions of individual muscles to the net joint moments and powers, and to determine the amount of power transport via bi–articular muscles. Quantitation is tackled both with inverse and direct dynamics. Much attention is also paid to organizational principles using an experimental setup for leg tasks comparable to that described for arm work in Chapter 18 (Gielen et al.).

41.8.2 Movement Control and Learning

One of the ideas which needs further attention is that the control of mono- and bi-articular muscles is based on different organizational processes. Support for this idea comes from results of analyses of cycling (Ingen Schenau, 1989) and arm tasks [Chapter 9 (Hogan) and Chapter 18 (Gielen et al.)], as well as from other studies in the literature.

Especially from studies of walking, running or jumping cats, it can be deduced that the mono-articular hindlimb muscles seem to act simply as force or work generators while the bi–articular muscles show a considerably more complicated, and task dependent, behavior (Hoffer et al., 1987a,b; Perret and Cabelguen, 1980; Spector et al., 1980; Walmsley et al., 1978). An issue which also needs attention is the importance of sensory information. In many theories on motor control and especially in servo control theories, sensory information is necessary to modulate the α–motoneuron pools. For example, Suzuki et al (1982) suggested that in cycling, especially the bi-articular muscles show a pattern of reciprocal inhibition triggered on the basis of sensory information. Many more of such suggestions have been proposed, mainly based on reflexologic studies (see Smith, 1981 for references). Such reflexologic studies, however, and particularly those based on the characteristics of muscle spindles in lengthening muscles, may have overemphasized the importance of servo control

(Loeb, 1984). It has been shown that even in tasks with slow or moderate speed one can move well without any sensory information (Taub and Berman, 1968; Loeb, 1984).

In our analysis of experienced cyclists (Ingen Schenau, 1989a) the periods of activation of mono–articular muscles were highly cross–correlated with the periods of muscle shortening, the phase lag being about 90 ms. Since this phase lag agrees with the phase shift between *EMG* and mechanical response of these muscles, it seems that the central nervous system anticipates on the required mechanical responses. This, of course, can only be explained on the basis of an open loop control. In other words: experienced cyclists use a learned movement pattern.

On the basis of these observations and the referred literature our current position with respect to movement organization is that multi-joint movements require a process of learning both position control and control of the distribution of net moments and power over the joints as described in this and Chapter 18 (Gielen et al.). This learning process might occur along the lines described by Chapter 20 (Denier van der Gon et al.).

If it should be confirmed that mono–articular muscles mainly act as force and work generators while bi–articular muscles warrant the correct distribution of net moments and power over the joints, one can imagine that the learning process requires different sources of information for the learning to control these muscles. Learning to control mono–articular muscles might be possible with help of simple muscle length information provided by muscle spindles in lengthening as well as shortening muscles (Burke et al., 1978; Loeb, 1981); learning to control bi–articular muscles, however, might require a variety of multi–modal sources of information.

Acknowledgment

The authors gratefully acknowledge the permission of Prof. Dr. K. van der Werff (Delft Technical University) to use *SPACAR*, a software package developed by him for simulation of multi-link mechanical systems. The authors also acknowledge the help of M.G. Mullender in finding the papers published in the previous century and the beginning of this century, which were relevant to the topics addressed in this chapter.

References

Aleshinsky, S.Y. (1986) An energy "sources" and "fractions" approach to the mechanical energy expenditure problem I-V. *J. Biomechanics* **19**: 287-315.

Alexander, R.McN. (1989) Sequential joint extension in jumping. *Human Movement Science* **8**: 339-345.

Andrews, J.G. (1987) The functional roles of the hamstrings and quadriceps during cycling. Lombard's paradox revisited. *J. Biomechanics* **20**: 565-575.

Baeyer, H. von (1921) Zur Frage der mehrgelenkigen Muskeln. *Anat. Anz.* **54**: 289-301.

Bernstein, N.A. (1967) *The Coordination and Regulation of Movements*. London: Pergamon Press.

Bobbert, M.F., Hoek, E., Ingen Schenau, G.J. van, Sargeant, A.J. and Schreurs, A.W. (1987) A model to demonstrate the power transporting role of bi-articular muscles. *Journal of Physiology* **387**: 24P.

Bobbert, M.F. and Ingen Schenau, G.J. van (1988) Coordination in vertical jumping. *J. Biomechanics* **21**: 249-262.

Bogert, A.J. van den, Hartman, W., Schamhardt, H.C. and Sauren, A.A. (1989) In vivo relationship between force, EMG and length change in the deep digital flexor muscle in the horse. In: G. de Groot, A.P. Hollander, P.A. Huijing and G.J. van Ingen Schenau (eds.) *Biomechanics XI-A*, pp. 68-74. Free University Press, Amsterdam.

Borelli, J.A. (1685) De motu animalium. *Pars prima* (cited by Fick, 1879).

Burke, D., Hagbarth, K.E. and Lofstedt, L. (1978) Muscle spindle activity in man during shortening and lengthening contractions. *Journal of Physiology* **277**: 131-142.

Capozzo, A., Figura, F., Marchetti, M. and Pedotti, A. (1976) The interplay of muscular and external forces in human ambulation. *J. Biomechanics* **9**: 35-43.

Cleland, J. (1867) On the actions of muscles passing over more than one joint. *Journ. Anat. Physiol.* **1**: 85-93.

Descartes, R. (1662) *De Homine*, Leyden.

Duchenne, G.B. (1867) *Physiologie des Mouvements*, Paris.

Elftman, H. (1939a) Forces and energy changes in the leg during walking. *American Journal of Physiology* **125**: 339-356.

Elftman, H. (1939b) The function of muscles in locomotion. *American Journal of Physiology* **125**: 357-366.

Fenn, W.O. (1938) The mechanics of muscular contraction in man. *J. Appl. Physics* **9**: 165-177.

Fick, A.E. (1879) Uber zweigelenkige Muskeln. *Archiv. Anat. u. Entw. Gesch.* **3**: 201-239.

Fischer, O. (1902) Kritik der gebrauchlichen Methoden die Wirkung eines Muskels zu bestimmen. Abhandlungen der math-phys. Classe der Koenigl. *Sachs Gesellsch. d. Wissensch.* **22**: 483-590.

Fujiwara, M. and Basmajian, J.V. (1975) Electromyographic study of two-joint muscles. *American Journal of Physical Medicine* **54**: 234-241.

Gregoire, L., Veeger, H.E. Huijing,P.A. and Ingen Schenau, G.J. van (1984) The role of mono- and bi-articular muscles in explosive movements. *International Journal of Sports Medicine* **5**: 301-305.

Gregor, R.J., Cavanagh, P.R. and Lafortune, M. (1985) Knee flexor moments during propulsion in cycling - A creative solution to Lombard's Paradox. *J. Biomechanics* **18**: 307-316.

Hering, H.E. (1897) Über die Wirkung zweigelenkiger Muskeln auf drei Gelenke und über die pseudo-antagonistische Synergie. *Archiv für die gesammte Physiologie* **65**: 627-637.

Hoffer, J.A., Sugano, N., Loeb, G.E., Marks, W.B., O'Donovan, M.J. and Pratt, C.A. (1987) Cat hindlimb motoneurons during locomotion II. Normal activity patterns. *Journal of Neurophysiology* **57**: 530-553.

Hoffer, J.A., Loeb, G.E., Sugano, N., Marks, W.B.. O'Donovan, M.J. and Pratt, C.A. (1987) Cat hindlimb motoneurons during locomotion III. Functional segregation in sartorius. *Journal of Neurophysiology* **57**: 554-562.

Hogan, N. (1985) The mechanics of multi-joint posture and movement control. *Biological Cybernetics* **52**: 315-331.

Huter, C. (1863) Anatomische Studien an den Extremitätengelenken Neugeborenen und Erwachsener. *Arch. f. Path. Anat. und Physiol. und klinische Medizin* **28**: 253-281.

Ingen Schenau, G.J. van, Groot, G. de and Boer, R.W. de (1985) The control of speed in elite female skaters. *J. Biomechanics* **18**: 91-96.

Ingen Schenau, G.J. van, Bobbert, M.F. and Rozendal, R.H. (1987) The unique action of bi-articular muscles in complex movements. *Journal of Anatomy* **155**: 1-5.

Ingen Schenau. G.J. van (1989a) From rotation to translation: constraints on multi-joint movements and the unique action of bi-articular muscles. *Human Movement Science* **8**: 301-337.

Ingen Schenau, G.J. van (1989b) From rotation to translation: Implications for theories of motor control. *Human Movement Science* **8**: 423-442.

Ingen Schenau, G.J. van and Cavanagh, P.R. (1990) Power equations in endurance sports. *J. Biomechanics* (in press).

Jacobs, R. (in preparation) Intermuscular coordination in sprint running.

Jensen, J.L., Thelen, E. and Ulrich, B.D. (1989) Constraints on multi-joint movements: From the spontaneity of infancy to the skill of adults. *Human Movement Science* **8**: 393-402.

Jöris, H.J.J., Edwards van Muyen, A.J., Ingen Schenau, G.J. van and Kemper, H.C.G. (1985) Force, velocity and energy flow during the overarm throw in female handball players. *J. Biomechanics* **18**: 409-414.

Koning, J.J. de, Ingen Schenau, G.J. van and Groot, G. de (1989) The mechanics of the sprint start in Olympic speed skating. *Int. J. Sports Biomechanics* **5**:151-168.

Kumamoto, M. (1984) Antagonistic inhibition exerted between bi-articular leg muscles during simultaneous hip and knee extension movement. In: M. Kumamoto (ed.) *Neural and mechanical control of movement.* Kyoto: Yamaguchi Schoten, pp. 114-122.

Landsmeer, J.M.F. (1961) Studies in the anatomy of articulation. *Acta Morphologica Neerl. Scand.* **3**: 304-321.

Langer, C. (1879) Die Muskulatur der Extremitäten des Orang als Grundlage einer vergleichend-myologischen Untersuchung. *Sitzungsberichte der kaiserlichen Akademie der Wissenschaften Math-Naturwissens. Classe. Bd.* **79**: 177-219.

Loeb, G.E. (1981) Somatosensory unit input to the spinal cord during normal walking. *Can. J. Physiol. Pharmacol.* **59**: 627-635.

Loeb, G.E. (1984) The control and responses of mammalian muscles spindles during normally executed motor tasks. In: R.L. Terjung (ed.) *Exercise and Sport Sciences Reviews. Vol. 12*, 157-204.

Lombard, W.P. (1903) The action of two-joint muscles. *Am. Phys. Educ. Rev.* **9**: 141-145.

Markee, J.E., Logue, J.T., Williams, M. Stanton, W.B., Wrenn, R.N. and Walker, L.B. (1955) Two-joint muscles of the thigh. *J.of Bone and Joint Surgery* **37**: 125-142.

Molbech, S. (1966) On the paradoxical effect of some two-joint muscles. *Acta Morphol. Neerl. Scand.* **4**: 171-178.

Perret, C. and Cabelguen, J.M. (1980) Main characteristics of the hindlimb locomotor cycle in the decorticate cat with special reference to bifunctional muscles. *Brain Research* **187**: 333-352.

Robertson, D.E. and Winter, D.A. (1980) Mechanical energy generation, absorbtion and transfer amongst segments during walking. *J. Biomechanics* **13**: 845-854.

Shampine, L.F. and Gordon, M.K. (1975) *Computer Solution of Ordinary Differential Equations. The*

Initial Value Problem. W.H. Freeman & Co., San Francisco.

Sherrington, C.S. (1909) Reciprocal innervation of antagonistic muscles. Fourteenth note. On double reciprocal innervation. *Proc. R. Soc. London. Ser. B* **91**: 249-268.

Smith, A.M. (1981) The coactivation of antagonist muscles. *Can. J. Physiol. Pharmacol.* **59**: 733-747.

Soest, A.J. van, Schermerhorn, P., Huijing, P.A. and Ingen Schenau, G.J. van (1989) Influence of timing on vertical jumping performance: a simulation study. In: R.J. Gregor, R.F. Zernicke and W.C. Whiting (eds) *Proceedings Biomechanics XII.* University of California, Los Angeles, p. 244.

Spector, S.A., Gardiner, P.F., Zernicke, R.F., Rog, R.R. and Edgerton, V.R. (1980) Muscle architecture and force-velocity characteristics of cat soleus and medial gastrocnemius: implications for motor control. *Journal of Neurophysiology* **44**: 951-960.

Suzuki, S., Watanabe, S. and Saburo, H. (1982) EMG activity and kinematics of human cycling movements at different constant velocities. *Brain Research* **240**: 245-258.

Taub, E. and Berman, A.J. (1968) Movement and learning in the absence of sensory feedback. In: J.S. Feedman (ed.) *The Neurophysiology of Spatially Oriented Behavior.* Dorsey Press, Homewood, Ill., 173-192.

Thomas, D.O., Sagar, G., White, M.J. and Davies, C.T.M. (1988) Electrically evoked isometric and isokinetic properties of the triceps surae in young male subjects. *European Journal of Applied Physiology* **58**: 321-326.

Tilney, F. and Pike, F.H. (1925) Muscular coordination experimentally studied in its relation to the cerebellum. *Archives of Neurology and Psychiatry* **13**: 289-334.

Vos, E.J., Mullender,M.G. and Ingen Schenau, G.J. van (1990) Electro-mechanical delay in vastus lateralis muscle during dynamic isometric contractions. *Eur. J. Appl. Physiol.* (in press).

Walmsley, B., Hodgson, J.A. and Burke, R.E. (1978) Forces produced by medial gastrocnemius and soleus muscles during locomotion in freely moving cats. *Journal of Neurophysiology* **41**: 1203-1216.

Wells, R. and Evans, N. (1987) Functions and recruitment patterns of one- and two-joint muscles under isometric and walking conditions. *Human Movement Science* **6**: 349-372.

Wells, R.P. (1988) Mechanical energy costs of human movement: An approach to evaluating the transfer possibilities of two-joint muscles. *J. Biomechanics* **21**: 955-964.

Werff, K. van der (1977) Kinematic and dynamic analysis of mechanisms, a finite element approach. *Thesis*, Delft University, Delft.

Williams, P.L. and Warwick, R. (1980) *Gray's Anatomy* Churchill Livingstone, London.

Winter, D.A. (1984) Kinematic and kinetic patterns in human gait: variability and compensating effects. *Human Movement Science* 3: 51-76.

Yamashita, N. (1988) EMG-activities in mono- and bi-articular thigh muscles in combined hip and knee extension. *Eur. J. Appl. Physiol.* 58: 274-277.

CHAPTER 42

An Analytical Framework for Quantifying Muscular Action During Human Movement

Marcus G. Pandy

42.1 Introduction

A fundamental problem in human movement analyses is the quantification of individual, time-varying, muscle forces. Muscle forces not only play a major role in determining the stresses in bones and joints, but they also reflect the underlying neural control processes responsible for the observed movement patterns. Unfortunately, invasive techniques for measuring muscle forces are highly objectionable, whereas non-invasive techniques such as electromyography do not provide the quantitative accuracy needed to define muscle's action on the skeleton. In addition, the human musculoskeletal system is mechanically redundant (i.e., the number of muscles spanning a joint exceeds the number of degrees of freedom defining joint motion) so that a direct solution of the muscle force-joint torque equations is not possible.

To circumvent the mechanically redundant nature of the lower limb system, previous investigators have used both static (Crowninshield, 1978; Hardt, 1978) and dynamic (Chow and Jacobson, 1971; Hatze, 1976; Davy and Audu, 1987) optimization techniques to predict individual muscle forces during human movement [see also Chapter 8 (Zajac and Winters)]. Though the static optimization approach is computationally less intensive, it suffers from the assumption that muscle force is time independent (i.e., a static optimization is performed at discrete time intervals with muscles often modeled as ideal force actuators). Consequently, static optimization procedures can lead to less

meaningful muscle force predictions (e.g., Hardt, 1978; Davy and Audu, 1987). Dynamic optimization, on the other hand, allows both activation (excitation–contraction) and musculotendon dynamics to be taken into account, and is therefore a potentially powerful tool. However, previous attempts to use this technique have been noticeably scarce (Hatze, 1980). The reason is that a detailed dynamical model of the human musculoskeletal system is nonlinear, highly coupled, and highly dimensional (i.e., it has a large number of states). Therefore, dynamic optimization solutions for synthesizing body–segmental motions are computationally very demanding. In fact, previous dynamic optimization solutions for human movement are characterized by either over-simplified models of the overall musculoskeletal system (Chow and Jacobson, 1971; Ghosh and Boykin, 1976) or detailed models of isolated limbs (Hatze, 1976; Davy and Audu, 1987).

A major goal of our ongoing research is to understand how intermuscular control, inertial interactions among body segments, and musculotendon dynamics coordinate multi–articular human movement. With this in mind, we have been using dynamic optimization (or optimal control) theory to study maximum-height human jumping. Because this particular activity presents a relatively unambiguous performance criterion, it fits well into the framework of optimal control. Moreover, vertical jumping is an activity characterized by bilateral symmetry, which leads to a relatively simple representation of the body–segmental dynamical system. Most importantly, however, our rationale for using optimal

Multiple Muscle Systems: Biomechanics and Movement Organization
J.M. Winters and S.L-Y. Woo (eds.), © 1990 Springer-Verlag

control is based upon the belief that it is currently the most sophisticated methodology available for solving human movement synthesis problems. Optimal control requires not only that the system dynamics be formulated, but that the performance criterion be specified as well. Thus, differences between model and experiment indicate deficiencies in the modeling of either the system dynamics or the performance criterion [see also discussions in Chapter 8 (Zajac and Winters) and Chapter 19 (Seif-Naraghi and Winters)].

In this chapter, we describe the rudiments of an analytical framework which we have been using to quantify muscular action during complex motor tasks. The framework consists not only of a detailed musculoskeletal model driven by dynamic optimization theory (Sections 42.2 and 42.3), but also of mathematical equations which can be used to find the contribution of individual muscles to the acceleration and power of body segments during movement (Section 42.4). We then demonstrate the utility of this framework by applying it to understand how muscles coordinate the motions of all the body segments during a vertical jump (Sections 42.5 to 42.7).

42.2 Musculoskeletal Modeling

For vertical jumping, we modeled the human body as a four–segment, articulated, planar linkage, with adjacent links joined together by frictionless revolutes. A total of eight lower-extremity musculotendinous units provided the actuation in the model (Figure 42.1). The details of the musculotendinoskeletal model can be found in Pandy et al. (1990). Briefly, each musculoten-dinous unit was modeled as a three-element, lumped-parameter entity (muscle), in series with tendon. The mechanical behavior of muscle was described by a Hill–type contractile element which modeled its force–length–velocity–activation property, a series-elastic element which modeled its short-range stiffness, and a parallel-elastic element which modeled its passive response. Tendon was assumed to be elastic, and its properties were represented by a stress–strain curve (Pandy et al., 1990). The musculotendon model was driven by a first-order representation of activation (excitation–contraction) dynamics, which describes the time lapse between the incoming neural control signal (muscle excitation) and muscle force. In this model, we have assumed

that each musculotendinous unit has as its input only one control signal (i.e. we have not dis-sociated the net firing rate control of a muscle from the recruitment control; though others have (e.g., Hatze, 1978)). For a review of modeling musculotendon actuation, see Zajac (1989), Chapter 5 (Winters) and Chapter 8 (Zajac and Winters).

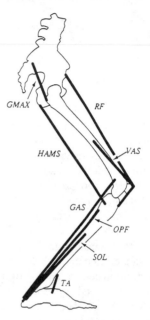

Figure 42.1: Schematic representation of the musculoskeletal model for the vertical jump. Symbols appearing in the diagram are: soleus (*SOL*), gastrocnemius (*GAS*), other plantarflexors (*OPF*), tibialis anterior (*TA*), vasti (*VAS*), rectus femoris (*RF*), hamstrings (*HAMS*), and gluteus maximus (*GMAX*).

42.3 Optimal Control Formulation

For maximum-height jumping, we chose the height reached by the center of mass of the body to be the measure of performance. Again, the mathematical details are provided in Pandy et al. (1990). Briefly, beginning from a prespecified, static, squat position, the optimal control problem was to maximize the height reached by the center of mass of the body (which depends explicitly upon the position and velocity of the mass centroid at the instant that the body leaves the ground) subject to: *a)* body-segmental, musculotendon, and activation dynamics; *b)* a set of inequality constraints which limit the magnitude of each neural control signal to lie between zero (no excitation) and one (full excitation); and *c*) a zero vertical ground reaction force at the instant that the body leaves the ground (lift-off). The optimal control problem for maximum-height jumping is bang–bang (i.e., the optimal controls can only take values of zero or one). The reason for this is that the first-order model used for activation dynamics is linearly related to the controls, and consequently the system Hamiltonian is also linear in the controls (Pandy et al., 1990). A computational solution to this problem was derived on the basis of a modified Polak-Mayne dynamic optimization algorithm (see Sim, 1988, for details).

42.4 Contribution of Muscles to Segmental Acceleration and Power

The dynamical equations of motion can be used to find the contribution of a muscle to the acceleration (Hatze, 1987; Zajac and Gordon, 1989; Pandy and Zajac, 1990) and instantaneous power (Aleshinsky, 1986; Pandy and Zajac, 1990) of individual body segments as well as the mass centroid of the body. From the equations of motion describing the dynamics of our four-segment, planar, skeletal model for jumping (Pandy et al., 1990):

$$\ddot{\theta} = A(\theta)^{-1}\{\, B(\theta)\, \dot{\theta}^2 + C(\theta) + D\, M(\theta)\, P^T + T(\theta, \dot{\theta})\}$$

$$(42.1)$$

where θ, $\dot{\theta}$, $\ddot{\theta}$ are vectors of limb angular displacement, velocity, and acceleration (all are *4 x 1* respectively; $T(\theta, \theta)$ is a *4 x 1* vector of externally applied joint torques (it contains only the moment applied to the foot segment from a damped torsional spring; see Pandy et al. (1990)); P^T is an *8 x 1* vector of musculotendon forces; $M(\theta)$ is a *3 x 8* moment-arm matrix formed by computing the perpendicular distance between each musculotendon actuator and the joint it spans; $A(\theta)$ is the *4 x 4* system mass-matrix; $C(\theta)$ is a *4 x 1* vector containing only gravitational terms; $B(\theta)\, \dot{\theta}^2$ is a *4 x 1* vector describing both Coriolis and centrifugal effects, where $\dot{\theta}^T$ represents $\dot{\theta}_i^T$ for *i = 1, 4*; and *D* is a *4 x 3* matrix which transforms joint torques into segmental torques. The details of *Eq. 42.1* are given in Pandy et al. (1990). Because each element of the vector $D\, M(\theta)\, P^T$ is the sum of the segmental torques developed by the muscles at-

taching to or crossing the corresponding body segment, the total contribution of muscles to the vector of segmental accelerations is thus:

$$\ddot{\theta}_m = A(\theta)^{-1} D\, M(\theta)\, P^T \qquad (42.2)$$

Since $A(\theta)^{-1}$ is, in general, non-diagonal, each muscle force accelerates all body segments. Thus, muscles not directly attached to or spanning a segment still contribute to its acceleration (see Zajac and Gordon, 1989, for review).

Similarly, knowing the total instantaneous power of a segment (i.e., the time rate of change of the total mechanical energy), the contribution of muscles, gravity, and inertia (i.e., Coriolis and centrifugal terms) to the instantaneous segmental power can also be found. For example, the net contribution of muscles to the instantaneous power of the trunk is composed of contributions from those intersegmental forces acting at the hip joint that arise from muscle forces, as well as from "direct" muscular forces that arise from the insertion of muscle on that segment. And, analogous to the contribution of muscles to segmental accelerations (*Eq. 42.2*), muscles not directly attached to a segment still contribute to its power because of the presence of the matrix $A(\theta)^{-1}$, which couples gravity, inertia forces, and muscle forces into *all* body-segmental motions (see Pandy and Zajac, 1990, for details).

42.5 Optimal Muscle-Force Sequencing

Quantitative comparisons between experimental results obtained from several subjects instructed to "jump as high as possible" and the model's predictions have indicated that the model successfully reproduces the major features of a maximum-height squat jump (i.e., the time histories of all muscle forces, body-segmental motions, and ground reaction forces, a proximal-to-distal sequence of muscular activity, overall jump height, and final lift-off time; see Figs. 1-3 in Pandy and Zajac, 1990). Subsequently, the optimal control solution was analyzed in detail to understand coordination. We have focused on quantifying the contribution of muscles to the vertical acceleration and the instantaneous power of the trunk during the ground contact phase because this segment represents approximately 70% of total body mass.

Muscles were found to dominate the angular acceleration of all segments for all but the final 10% of ground contact time. During the final 10% of ground contact time, inertial contributions to the acceleration of a segment outweighed those of muscles because segmental angular velocities increased rapidly near lift-off (Fig. 4 of Pandy and Zajac, 1990). Because muscles dominate the angular accelerations of the segments for almost the entire jump, we have analyzed individual muscular contributions (Figure 42.2) in order to better understand muscle-force sequencing during jumping. Notice that TA, VAS, and RF make negligible contributions to the joint angular accelerations (dashed lines, Figure 42.2). The muscles that dominate joint angular accelerations are GAS, VAS, GMAX, and the uniarticular ankle plantarflexors SOL and OPF (designated as UPF). Of all these muscles, only one (i.e., GAS) is biarticular.

HAMS and GMAX are activated first (approximately 20% and 30% of ground contact time, respectively) and these muscles accelerate the trunk towards an upright position to counteract the effects of gravity (Fig. 3 of Pandy and Zajac, 1990). VAS, on the other hand, induces an opposing angular acceleration of the trunk (Fig. 3 of Pandy and Zajac, 1990), and it is therefore excited later in the jumping cycle (about 40% of ground contact time). Even though VAS induces an undesirable forward (negative) angular acceleration of the trunk, this knee extensor actually accelerates the hip joint towards *extension* (Figure 42.2a; VAS) because of the large *positive* angular acceleration of the thigh it induces (Fig. 3 of Pandy and Zajac, 1990).

Because VAS is assumed to be the strongest muscle in the lower extremity (see Table 1 in Pandy et al. 1990), any action other than hip extension would clearly be undesirable. Thus, GMAX and VAS, the strongest hip and knee extensors in the model, work together to accelerate the knee and hip towards extension (Figure 42.2a–b), which in turn acts to accelerate the mass centroid of the body in the desired direction (upwards). Because VAS and GMAX simultaneously induce large ankle dorsiflexor accelerations (i.e., they accelerate the heel into the ground; Figure 42.2c), the uniarticular SOL and UPF are excited at 50–60% of ground contact time to counter these actions (Figure 42.2c). Finally, GAS is excited near lift-off to further accelerate the ankle towards plantarflexion (Figure 42.2c). In doing so, GAS

opposes the actions of VAS and GMAX at both the knee and hip [i.e. GAS accelerates the knee and hip towards flexion which opposes the efforts of VAS and GMAX at this time (Figure 42.2b)]. We conclude, therefore, that the increase in jump height due to an increase in foot angular velocity resulting from excitation of GAS outweighs its undesirable effects at the knee and hip. That is, the net effect of the biarticular GAS is to increase the vertical velocity of the center of mass of the body at lift-off.

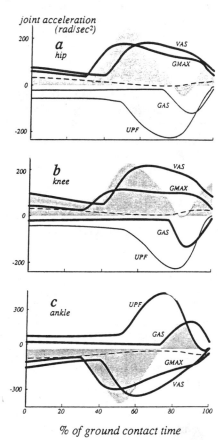

Figure 42.2: Contributions of muscles to the angular acceleration of the hip (*a*), knee (*b*), and ankle (*c*) derived from the model during the ground contact phase (0% to 100%) of a maximum-height squat jump. The shaded region is the summed effect of all the muscles. UPF is the combined contribution of the uniarticular ankle plantarflexors (*SOL* and *OPF*). The dashed lines represent the combined effect of *TA*, *HAMS*, and *RF*. Positive angular acceleration indicates joint extension. Prior to and at 0% ground contact time, muscle forces applied to the model are constant to maintain the body statically in the squat.

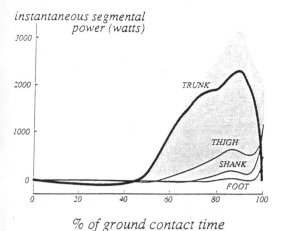

instantaneous segmental power (watts)

% of ground contact time

Figure 42.3: Instantaneous power of each segment in the model during the ground contact phase (0% to 100%) of a maximum-height squat jump. The shaded region is the summed power of all four body segments. The area under each curve is the energy of that segment at body lift-off. Notice that most of the energy (approximately 70%) resides in the trunk at lift-off.

42.6 Optimal Power Distribution

A large proportion of the total energy delivered to the segments (Figure 42.3, shaded region) resides in the trunk (Figure 42.3, compare the area under the heavy solid line with the shaded region). In fact, the combined energy of the thigh, shank, and foot amounts to only 30% of the total energy made available at body lift-off. Thus, almost 70% of the total input musculotendon energy is transferred to the trunk. This is not particularly surprising given that the trunk represents about 70% of total body mass.

Consistent with their contribution to body-segmental angular acceleration (Section 42.5), muscles also dominate the instantaneous power of individual body segments for all but the final 10% of ground contact time. Specifically, of all the muscles, *VAS* and *GMAX* contribute approximately 90% of the total energy of the trunk (Fig. 42.4a, summed area under *VAS* and *GMAX* compared to the shaded region). In contrast, the ankle plantarflexors (*SOL*, *OPF*, and *GAS*) contribute significantly only during the final 20% of ground contact time, during which time they account for about 30% of the total power delivered by the muscles to the trunk (Figure 42.4a, compare *PF* curve with the shaded region at body lift-off). Moreover, these ankle plantarflexors contribute as

much as 90% to the peak power delivered by the muscles to the thigh (Figure 42.4b, compare thigh power of *PF* curve with shaded region at 85% of ground contact time). Because of the combined effects of the ankle plantarflexors on the trunk and thigh, we conclude, in agreement with previous investigators (Gregoire et al., 1984; Bobbert and van Ingen Schenau, 1988) that the contribution of *SOL*, *OPF*, and *GAS* to overall jumping performance cannot be neglected.

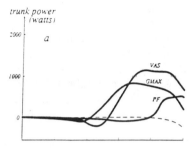

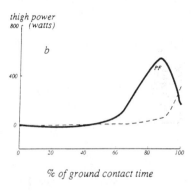

% of ground contact time

Figure 42.4: Individual muscular contributions to the instantaneous power of the trunk *(a)* and the thigh *(b)*. *(a)* Total contribution from all the muscles in the lower extremity is represented by the shaded region. *PF* represents the contribution of all ankle plantarflexors (*SOL*, *OPF*, and *GAS*). The dashed line is the combined contribution of *HAMS*, *RF*, and *TA*. The area under each curve is the energy contributed by each muscle to the trunk at body lift-off. Notice that *VAS* and *GMAX* are the major energy producers or prime movers of the lower extremity. *(b)* Total contribution from all the muscles in the lower extremity is the shaded region. *PF* is the contribution from all the ankle plantarflexors (*SOL*, *OPF*, and *GAS*). The dashed line represents the contribution from all other muscles. Notice that the ankle plantarflexors dominate the total energy delivered by all the muscles to the thigh at body lift-off.

Though our optimal control solution predicted a proximal-to-distal sequence of muscle activation [i.e., muscles spanning the hip (*GMAX* and *HAMS*) were activated first, followed by uniarticular muscles spanning the knee (*VAS*) and ankle (*SOL* and *OPF*)], it does not support the idea that energy "flows" proximally from the hip to the knee to the ankle [Gregoire et al., 1984; Bobbert and van Ingen Schenau, 1988; Chapter 41 (van Igen Schenau et al.)]. To the contrary, we found 'that energy production by the musculotendinous units is dominated by the uniarticular extensors *VAS* and *GMAX*, and that almost all of this energy (90%) is used to accelerate the center of mass of the trunk upwards (Figure 42.4 *VAS* and *GMAX*). Thus, uniarticular muscles, as well as biarticular muscles, transfer power proximally to the massive trunk segment in an effort to maximize the vertical velocity of the center of mass of the body at lift-off.

Of all the lower-extremity biarticular muscles, our analyses indicate, in agreement with others (Bobbert et al., 1986; Bobbert and van Ingen Schenau, 1988), that *GAS* is the most important. To assess the dependence of jumping performance on biarticular muscle function, we removed all biarticular muscles from the musculoskeletal model of Figure 42.1 and, with only uniarticular muscles available for propulsion, recomputed the optimal control solution. We found that even though the jump was still coordinated (Figure 42.5, stick figures), the net vertical displacement of the center of mass of the body from standing decreased by 20%. Because *RF* and *HAMS* contribute much less than *GAS* to the total energy of the trunk and thigh (not shown in Figure 42.4), we conclude that the absence of *GAS* is the major factor contributing to the decrease in jump height shown in Figure 42.5. However, we are opposed to the notion that jumping performance is increased by the unique **biarticular** action of *GAS* (Bobbert et al., 1986; Bobbert and van Ingen Schenau, 1988). We have found no evidence to support the claim that ankle power output is increased significantly as a result of power transferred by *GAS* from the knee to the ankle. To the contrary, by delivering power to the thigh and trunk, *GAS* behaves no differently from any uniarticular extensor muscle in the lower-extremity; it too transfers power in a proximal direction to the most massive body segments during upward propulsion.

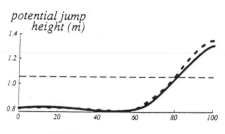

% of ground contact time

Figure 42.5: Simulated maximum-height squat jump using only uniarticular muscles. *HAMS*, *RF*, and *GAS* were removed from the model, and, given the remaining set of muscles, the optimal control solution was recomputed. The stick figures show how the body segments should be coordinated during the ground contact phase (0% to 100%) to achieve the maximum possible height with the new muscle set. The heavy solid and dashed lines show how high the center of mass of the body would go if the body could leave the ground at that instant, given that all the muscles, or just the uniarticular muscles, are used to coordinate the jump, respectively. The thin dashed line represents the standing height of the simulated jumper.

42.7 Influence of Musculotendon Properties on Performance and Coordination

To investigate the dependence of jumping performance and coordination on the mechanical properties of muscle and tendon, we have introduced large changes (typically 300–400%) to the parameters of our musculotendon model describing muscle strength, muscle-fiber contraction speed, and tendon compliance. Because each change introduced to the nominal musculotendon model imposes a new condition on the optimal control problem, changing the mechanical properties of the nominal musculotendon model necessitates a recomputation of the optimal controls producing a maximum-height jump. Changes to the parameters defining body strength-to-weight ratio, muscle-fiber contraction speed, and tendon compliance were all implemented over a physiologically justifiable range. Tendon compliance, for example, was increased only to the point where the strain in tendon became 10%, since this is the maximum strain defining tendon rupture (Zajac, 1989) (see Figure 42.8 and Pandy and Zajac 1989 for details).

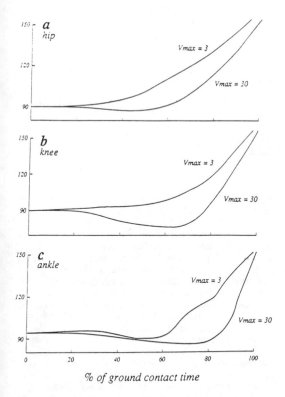

% of ground contact time

Figure 42.6: Angular displacements of the hip, knee, and ankle for a variety of simulated maximum-height squat jumps during ground contact (0% to 100%). The shaded region represents the range of joint trajectories produced as the value of muscle-fiber contraction speed was changed from $V_{max} = 3$ l_o/sec to $V_{max} = 30$ l_o/sec. In the model, muscle-fiber contraction speed was assumed to be invariant for all muscles in the lower extremity. Notice that the response of the model with fast muscles involves a considerable countermovement phase prior to upward propulsion, whereas this characteristic is noticeably absent during a simulated jump with slow muscles.

Increasing either muscle-fiber contraction speed or body strength-to-weight ratio delayed the onset of extension of all the joints (compare, for example, the outlines of the shaded regions in Figure 42.6). The reason is that as either muscle-fiber contraction speed or body strength-to-weight ratio was increased, all extensor muscles of the lower extremity were activated later in the jumping cycle. For example, at $V_{max} = 3$ l_o/sec, where l_o/sec is the contraction velocity expressed in terms of the number of muscle-fiber lengths per second, GMAX is activated at 10% of ground contact time,

compared with 40% of ground contact time when $V_{max} = 30$ l_o/sec (not shown). Therefore, characteristic of the response associated with "fast" muscles is the presence of a countermovement prior to upward propulsion (Figure 42.6, $V_{max} = 30$ l_o/sec; initial flexion at the hip, knee, and ankle prior to joint extension). On the other hand, an optimal performance generated by "slow" muscles is marked by immediate ankle, knee, and hip extension (Figure 42.6, $V_{max} = 3$ l_o/sec; no countermovement prior to upward propulsion).

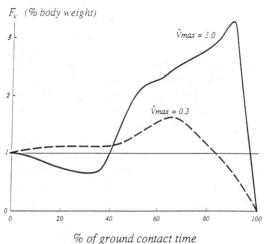

% of ground contact time

Figure 42.7: Vertical ground reaction forces generated by the model for two different values of muscle-fiber contraction speed ($\hat{V}_{max} = 0.3$ and $\hat{V}_{max} = 30$). $\hat{V}_{max}$ is the normalized muscle-fiber contraction speed obtained by dividing through by the value of muscle-fiber contraction speed assumed in the nominal musculotendon model (i.e., $V_{max} = 10 l_o$/sec). Notice that the duration of the preparatory countermovement phase increases and propulsion time decreases as muscle-fiber contraction speed increases (i.e., for $V_{max} = 0.3$ l_o/sec (dashed line) propulsion begins at 0% of ground contact time with no preparatory countermovement, whereas the countermovement phase for $V_{max} = 30$ l_o/sec (solid line) lasts until approximately 40% of ground contact time). Notice also that the area under the solid curve is much greater than that associated with the dashed curve, indicating that fast muscles produce higher jumps.

The appearance of a preparatory countermovement as either muscle-fiber contraction speed or body strength-to-weight ratio is increased is also evident in the vertical ground reaction force generated (e.g., Figure 42.7). In particular, as both muscle-fiber contraction speed and body strength-to-weight ratio are increased, the duration of the preparatory phase increases, which leads to a decrease in propulsion time (Figure 42.7, compare ground contact times at which minima in the vertical ground reactions occur). Moreover, peak magnitudes of the vertical ground reaction increase with an increase in either muscle-fiber contraction speed or body strength-to-weight ratio (e.g., Figure 42.7). Jump height (defined as the net vertical displacement of the center of mass of the body from standing) therefore increases since the area under the vertical ground reaction force is a measure of the total impulse delivered to the body, and greater impulses yield greater velocities of the center of mass at lift-off. For example, jump height increased by more than a factor of six when muscle-fiber contraction speed increased tenfold (i.e., from $V_{max} = 3 \, 1_o/\text{sec}$ to $V_{max} = 30 \, 1_o/\text{sec}$).

We found jump height to be **most** sensitive to changes in body strength-to-weight ratio. To quantify the dependence of jumping performance on muscle-fiber contraction speed, body strength-to-weight ratio, and tendon compliance, we plotted normalized jump height against a normalized version of each of the above musculotendon properties. Assuming that these relationships are linear, and by fitting straight lines to these data (Figure 42.8), we found that the slope of the line for body strength-to-weight ratio (1.2) was twice that for muscle-fiber contraction speed (0.6), and an order of magnitude greater than that for tendon compliance (0.08). Our model predicts, therefore, that jump height is relatively **insensitive** to tendon compliance. Because a squat jump involves considerably less countermovement than a countermovement jump per se, we should not expect tendon compliance to contribute significantly to overall jumping performance. In this respect, our results support a previous conclusion forwarded by Komi and Bosco (1978) that squat jumps yield lower values of performance because of their inability to utilize tendon's capacity for storing elastic energy. However, on the basis of the rationale that compliance can only be increased to the point where the strain in tendon becomes 10%, our preliminary calculations lead us to believe, in opposition to others [e.g., Bobbert and van Ingen Schenau, 1988; see also Chapter 41 (van Ingen Schenau)], that the contribution of tendon to countermovement jumping performance will also be negligible.

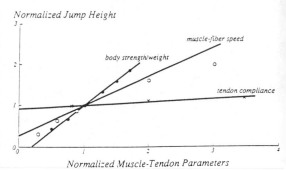

Normalized Jump Height

Normalized Muscle-Tendon Parameters

Figure 42.8: Dependence of jumping performance, as predicted by the model, on three musculotendon properties: muscle strength (actually body strength to weight ratio), muscle-fiber contraction speed (or maximum shortening velocity), and tendon compliance. Normalized jump height for the model was defined to be the ratio of jump height produced when one of the above musculotendon properties was varied independently, to jump height achieved when nominal musculotendon parameters were chosen (e.g. $V_{max} = 10 \, 1_i/\text{sec}$. Similarly, normalized musculotendon parameters were defined as follows: normalized muscle-fiber contraction speed for the model is the ratio of any given value of muscle-fiber contraction speed to the nominal value of muscle-fiber contraction speed assumed by the model; normalized body strength to weight ratio was the ratio of a given value of body strength-to-weight to the nominal value of body strength-to-weight; and normalized tendon compliance was the ratio of a given value of ankle plantarflexor tendon compliance in the model to the compliance of the ankle plantarflexor tendons computed under nominal conditions. In all cases, the influence of muscle-fiber contraction speed was studied by changing the maximum shortening velocity of all the muscles in the lower extremity simultaneously. Perturbations to body strength-to-weight ratio were introduced by either increasing the strength (i.e., the peak isometric force) of all the muscles in the lower extremity by as much as four times (relative to the nominal peak isometric force of each muscle), and adjusting the mass of all the body segments simultaneously (to account for the appropriate increase in muscle mass), or independently perturbing the mass of each body segment by as much as 20% from its nominal value, without altering muscle strength. Note that overall body strength was defined to be the sum of all peak isometric forces produced by all extensor muscles in the lower extremity, and that the strength of all extensor muscles was increased simultaneously. Only the compliance of the tendons in m. triceps surae (SOL and GAS) was changed because these tendons are much longer and therefore more compliant than those spanning either the knee or the hip (see Zajac 1989 for review). To estimate the overall compliance of m. triceps surae, the individual compliances of SOL and GAS were added together, assuming that these musculotendinous units act in parallel. Thus, the compliance in each of these tendons was increased to the point where the strain in each tendon became 10%, corresponding to a four-fold increase in the nominal value of compliance of these ankle plantarflexor tendons. Notice that the slope of the line for body strength to weight ratio (1.2) is steeper than those for muscle-fiber contraction speed (0.6) and tendon compliance (0.08). Thus, the model predicts jump height to be **most** sensitive to changes in body strength to weight ratio, and **least** sensitive to changes in tendon compliance.

As both body strength to weight ratio and muscle-fiber contraction speed increased, muscular coordination (defined here as the sequence and timing of muscle activation) became less important to the production of an optimal performance. In contrast, both slow and weak muscles demand early excitation. Under these circumstances, *GMAX* and *VAS* are especially important because they are responsible for accelerating the massive trunk segment towards an upright position. In fact, as muscle-fiber contraction speed and body strength to weight ratio are decreased, the trend for muscles to be activated in a proximal-to-distal sequence (Figure 3 in Pandy and Zajac, 1990) becomes more pronounced. In other words, weaker and slower muscles cause jumping to depend much *more* heavily upon the mechanism of muscle-force sequencing. The corollary is that faster and stronger muscles are not only activated later in the jumping cycle, but they are also all activated almost simultaneously.

42.8 Future Directions

To validate the model's predictions of how musculotendon properties alter jumping performance and coordination, we are currently conducting experiments on human subjects to independently determine the effects of muscle strength and muscle-fiber contraction speed. An estimate of overall body strength can be obtained by measuring the maximal, voluntary, isometric torque exerted at the knee as a function of knee joint angle using a Cybex dynamometer. Dividing the peak value of isometric knee torque by the subject's weight will then give the ratio of body strength to weight. Similarly, to study the effects of muscle-fiber contraction speed on performance and coordination, we have been conducting jumping experiments on both "fast" and "slow" athletes. Qualitatively, one would expect sprinters to possess a high percentage of fast-twitch muscle fibers, while elite distance runners should have a high concentration of slow-twitch fibers. Quantifying such differences, however, is somewhat problematic. While isokinetic measurements (i.e. maximal, voluntary, knee torque exerted at various (constant) speeds of knee-joint extension) are easily performed on a Cybex dynamometer, these data do not reveal an estimate of the maximum shortening velocity of muscles. Alternatively, muscle biopsies can supply such information, but these experiments are clearly invasive. Assuming that we are able to resolve this issue, the aim here will be to establish a set of curves much like those given in Figure 42.8 which define the relationships between jumping performance and muscle strength and speed. Thus, we will be able to assess the validity of our optimal control model's predictions.

Acknowledgments

I thank Felix Zajac for sharing with me his many valuable insights into muscular coordination of human movement in general, and vertical jumping in particular. I also acknowledge William Levine, Eunsup Sim, Eric Topp, and Melissa Hoy for their efforts in developing the optimal control model for jumping, and I thank Idd Delp for his help with figure preparation. This work was supported by NIH grant NS17662, the Alfred P. Sloan Foundation, and the Veterans Administration.

References

Aleshinsky, S.Y. (1986). An energy "sources" and "fractions" approach to the mechanical energy expenditure problem. II. Movement of the multi-link chain model. *J. Biomech.* **19**:295-300.

Bobbert, M.F., and van Ingen Schenau, G. (1988). Coordination in vertical jumping. *J. Biomech.* **21**:249-262.

Bobbert, M.F., Huijing, P.A., and van Ingen Schenau, G. (1986). An estimation of power output and work done by the human triceps surae muscle-tendon complex in jumping. *J. Biomech.* **19**:899-906.

Chow, C.K., and Jacobson, D.H. (1971). Studies of human locomotion via optimal programming. *Math. Biosci.* **10**:239-306.

Crowninshield, R.D. (1978). Use of optimization techniques to predict muscle forces. *J. Biomech. Engng.* **100**:88-92.

Davy, D.T., and Audu, M.L. (1987). A dynamic optimization technique for predicting muscle forces in the swing phase of gait. *J. Biomech.* **20**:187-201.

Ghosh, T.K., and Boykin, W.H. (1976). Analytic determination of an optimal human motion. *J. Opt. Theory Appl.* **19**:327-346.

Gregoire, L., Veeger, H.E., Huijing, P.A., and van Ingen Schenau, G. (1984). Role of mono- and biarticular muscles in explosive movements. *Int. J. Sports Med.* **5**:301-305.

Hardt, D.E. (1978). Determining muscle forces in the leg during human walking: An application and evaluation of optimization methods. *J. Biomech. Engng.* **100**:72-78.

Hatze, H. (1987). Gait analysis:Adequacy of current models and research strategies. *J. Motor Behavior*

19:280-287.

Hatze, H. (1980). Neuromusculoskeletal control systems modeling — A critical survey of recent developments. *IEEE Trans. Auto. Control* **AC-25**:375-385.

Hatze, H. (1978). A general myocybernetic control model of skeletal muscle. *Biol. Cybernetics* **28**:143-157.

Hatze, H. (1976). The complete optimization of a human motion. *Math. Biosci.* **28**:90-99.

Komi, P.V., and Bosco, C. (1978). Utilization of stored elastic energy in leg extensor muscles by men and women. *Med. Sci. Sports* **10**:261-265.

Pandy, M.G., Zajac, F.E., Sim, E., and Levine, W.S. (1990). An optimal control model for maximum-height human jumping. *J. Biomech.* (in press).

Pandy, M.G., and Zajac, F.E. (1990). Optimal muscular coordination strategies for jumping. *J. Biomech.* (in press).

Pandy, M.G., and Zajac, F.E. (1989). Dependence of jumping performance on muscle strength, muscle-fiber speed, and tendon compliance. In Stein, J.L. et al. (eds.):*Issues in the Modeling and Control of Biomechanical Systems*, 1989 ASME Winter Annual Meeting in San Francisco. New York:The American Society of Mechanical Engineers.

Sim, E. (1988). The application of optimal control theory for analysis of human jumping and pedaling. Ph.D. dissertation, Department of Electrical Engineering, University of Maryland, College Park.

Zajac, F.E., and Gordon, M.E. (1989). Determining muscle's force and action in multi-articular movement. *Exer. Sport Sci. Revs.* **17**:187-230.

Zajac, F.E. (1989). Muscle and tendon:Properties, models, scaling, and application to biomechanics and motor control. *CRC Critical Rev. Biomed. Engng.* **17**:359-411.

CHAPTER 43

Performing Whole-Body Simulations of Gait with 3-D, Dynamic Musculoskeletal Models

Gary T. Yamaguchi

43.1 Introduction

As walking is a basic activity, it is not surprising that the literature describing and analyzing human gait is prolific. Articles on walking appear regularly in newspapers, fitness magazines, medical research journals, technical publications, and the like. Most of these focus on the particular aspects of gait most interesting to the reader. Even sub-disciplines with relatively limited appeal have extensive bibliographies!

This is especially true regarding the subject of gait modeling, which typically attempts to reduce the complicated process of walking to one which can be analyzed mathematically. Doing so involves the formulation of musculoskeletal gait models, which in turn are constructed from mathematical models of various subcomponents, e.g. musculotendon actuator models (among others, Audu and Davy, 1985; Hatze, 1977; Hill, 1938; Zajac, 1989; see Figure 43.1), musculoskeletal geometry models (Brand et al., 1982; Delp et al., 1989; Friederich and Brand, 1990; Hardt and Mann, 1980; Hoy et al., 1990; Wickiewicz et al., 1983), and joint models (Crowninshield et al., 1976; Van Eijden et al., 1986; Wismans et al., 1980; Yamaguchi and Zajac, 1989a). Of course, the complexity of each subcomponent making up the aggregate gait model is dictated by the goals of the analysis (Davy and Audu, 1987) and limited by the computational facilities at hand.

Whether the gait models are formulated in two or three dimensions depends upon their intended use. Inverse-dynamic analyses (motion inputs, force/torque outputs), which seek to investigate or explain the mechanics of previously-observed (recorded) movements, often utilize the simplest

models possible in order to provide clear illustrations of mechanical principles. Because the data records for each body segment's movements are pre-recorded, and are uncoupled, the analyst can choose to ignore motions not having a direct bearing on the modeling study. Two-dimensional models are often sufficient for these purposes; however, extending the technique to three-dimensions is straightforward.

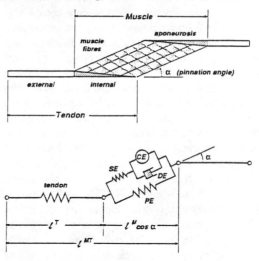

Figure 43.1: Simplified structure of the musculotendon actuator (top) and the musculotendon model used in the example of Section 43.4. The model is composed of lumped, idealized mechanical components such as springs (tendon, *SE*, *PE*), dashpots (*DE*), and force-generators (*CE*), that together mimic the macroscopic behavior of biological muscle-tendons. (Reprinted with permission from Yamaguchi & Zajac, *IEEE Trans. Biomed. Eng.* **37**. © 1990 - IEEE.)

Multiple Muscle Systems: Biomechanics and Movement Organization
J.M. Winters and S.L-Y. Woo (ed.), © 1990 Springer-Verlag

On the other hand, forward-dynamic analyses (control inputs, movement outputs) are performed in order to predict or simulate ambulatory movements in response to neural or muscular control inputs. Because resulting movements are not known *a priori*, and because body segmental motions are dynamically coupled (i.e., the movement of one segment affects the motions of every other body segment; see Zajac and Gordon, 1989), the quest for realism suggests that gait simulations be performed in three dimensions (*3-D*).

This requirement for accurate, *3-D* dynamic modeling leads to extreme computational demands. Not only must the calculational system compute the dynamic inner workings of the model (e.g. muscle activation dynamics, musculotendon contraction dynamics, and the reaction of the skeletal system to applied external and muscular forces), it must also determine some means of adequately controlling the muscles so that coordinated movements will result. Generally, this need for control becomes greater as modeling complexity increases. With relatively few degrees of freedom (*DOF*) to control, with few muscles to excite, and well-defined movements to execute, the control task is simpler, so that classical optimal control techniques may sometimes be utilized. When the numbers of *DOF* and/or muscles becomes large, constraints imposed by limitations in computational hardware forces compromises to be made in modeling accuracy, complexity, and/or control.

Recent dynamic, musculoskeletal gait models possess few *DOF* and muscles (Audu and Davy, 1987; Hardt and Mann, 1980; Ju and Mansour, 1988; Mena et al., 1981; Mochon and McMahon, 1980; Onyshko and Winter, 1980; Pandy and Berme, 1988a,b; Yen and Nagurka, 1987). Although these models were exemplary in light of available computing resources, these numbers of *DOF* and muscles are inadequate to describe normal and pathological gait in *3-D*, except in highly-constrained, rather artificial situations. To mimic the "important" movements during gait, approximately 20 *DOF* and many more musculotendon groups would be desirable (Yamaguchi, 1989).

To take but a small step toward this goal, this article will attempt to establish a framework for developing models with greater complexity and capability, and will address some of the tradeoffs involved in the formulation of *3-D* dynamic gait

models using the *8-DOF*, *10-muscle* model of Yamaguchi and Zajac (1990) as an example. It is hoped that as faster computers become available, and new, efficient control algorithms are developed, more accurate and comprehensive gait simulations will become possible, which in turn will lead to increases in understanding and the development of better medical treatment methods to correct gait abnormalities.

First, however, let us further discuss the difference between gait simulations (direct-dynamic analyses) and standard (inverse-dynamic) gait analysis techniques [see also Chapter 8 (Zajac and Winters)].

To begin our discussion, consider the natural "flow" of events by which movements are produced in a living organism. When an animal, say, desires to move its limbs, a sequence of motor commands emanates from its central nervous system and excites an appropriate set of muscles into developing precisely the right amount of contractile force. Since these muscles connect directly to bones (through tendons), the tensile forces acting on the bones exert moments about the joints of the skeletal system. It is the summation of the moments produced by all of these active moments, plus the kinematic constraints and inertial characteristics of the body, that govern how the body segments will move. Thus the natural flow of events proceeds outward from the nervous system to the muscular system, and the muscular system to the skeletal system (Zajac and Gordon, 1989).

Dynamic musculoskeletal models are useful in describing the interrelationship between the applied muscular forces and the resulting motions of the body segments. For instance, the mathematical gait model developed later in this chapter will be used as a geometric tool to define the position and velocity-dependent transformations between the inputs (here, muscle activations) and the outputs (limb trajectories) of the dynamic equations. These equations, which govern the motion of the body model, can be written in simplified form as

$$M\ddot{\theta} = T + V + G + E \qquad (43.1)$$

and can be restructured to emphasize the dependence of the segment accelerations vector ($\ddot{\theta}$) upon the mass characteristics and instantaneous configuration of the body (contained in *M*), and the segmental torques applied by the muscles (T), inertia (V), gravity (G), and external forces (E), respectively

$$\ddot{\theta} = M^{-1}[T + V + G + E]$$ (43.2)

In *Eq. 43.2*, M is the $n \times n$ "mass matrix" for the n-*DOF* model, and T, V, G, and E are $n \times 1$ vectors. As mentioned, M and G are functions of the segment orientations $\theta = (\theta_1,...,\theta_n)$, where the element θ_i ($i=1,...,n$), is the angular orientation of segment i relative to an inertial reference frame, and V is a function of θ and the segment angular velocity vector $\dot{\theta}$. The equations of motion written in this way clearly show motion outputs as functions of muscular inputs, and are therefore compatible with the natural flow of *neural-to-muscular-to-skeletal* events. For *dynamic simulations of movement*, or *motion syntheses*, *Eq. 43.2* is integrated forward in time to obtain motion trajectories in response to neuromuscular inputs,

$$\dot{\theta}(t_1) = \int_t^{t_1} \ddot{\theta}\ dt = \int_t^{t_1} M^{-1}[T + V + G + E]\ dt$$ (43.3)

where $t_1 = t + \Delta t$. The term "direct-dynamics" will be used to describe this type of analysis.

When such studies are applied to human walking, *pure* syntheses of gait can only be done without the imposition of trajectory constraints. Pure syntheses are complex because *all* of the interactions between the body segments must be taken into account when determining the controls. Adding constraints simplifies the control problem, but often introduces undesirable approximations. For example, a five-link human locomotion model by Yen and Nagurka (1987) was assumed to move at a constant forward speed. Moreover, the trunk was assumed to be rigidly fixed in a vertical orientation throughout their simulations. As a consequence, some undesirable motions of the trunk were not modeled, even though they should have been allowed to occur during realistic simulations (Onyshko and Winter, 1980). Of course, realism and accuracy improve as additional segments are included to model the actions of both legs and the trunk. Each additional segment and/or degree of freedom added, however, makes the gait synthesis problem more difficult.

"Inverse" dynamics, as opposed to direct-dynamic analyses, proceed in a direction opposite to the natural flow of events. In contrast, inverse-dynamic methods start from quantitative observations of a particular movement trajectory, and use dynamic musculoskeletal models to determine the joint moments or muscular forces that *must have been evident* to create the observed motion (see Zajac and Gordon, 1989). In inverse analyses, *Eq. 43.1* is rewritten as

$$T = (M\dot{\theta} - V - G - E)$$ (43.4)

which clearly expresses the output segmental torques as a function of the trajectory-dependent terms grouped on the right-hand side of the equation. To obtain the set of muscular forces once the torques T are known, typically some form of static optimization is used to distribute the tensile forces among the redundant muscle set at each instant of time.

Clearly, the inverse-dynamic approach applies itself readily to clinical and experimental settings where one must work **backwards** from observed motion trajectories. Since the trajectories of only the body segments of interest are needed, the experimenter can use inverse-dynamic analyses to focus upon few or many segments at will.

The drawbacks to inverse-dynamic methods have already been well-explained in the literature (Hardt, 1978). In particular, muscular controls are typically determined quasistatically, leading to control solutions that are unrelated to the controls at other instants of time, and hence the muscle forces can be discontinuous in time. Also, kinematic measurements often cannot resolve subtle movements that are crucial to the overall, whole-body gait pattern (Pandy and Berme, 1987). For instance, pelvic movements that vary by no more than a few degrees are often neglected in modeling studies, even though they can be important because large masses (e.g. the trunk) are displaced through these small movements, which potentially can involve significant amounts of mechanical work.

Direct-dynamic methods *do* determine continuous controls that can be optimized over the entire time-frame of the simulation, rather than just at discrete instants of time, and can produce subtleties of movement that are difficult to measure experimentally. Moreover, dynamic simulations have *predictive* value, in that the effects of musculoskeletal alterations (e.g. due to surgery, physical therapy, and orthotic interventions) potentially can be predicted *before* the actual alterations are performed. If a surgeon wished to perform a tendo-achilles lengthening, say, or transplant a particular ligament, the potential to predict the effects of these operations exists via direct-dynamic methods, but does not exist with inverse-dynamic analyses.

43.2 3-D Musculoskeletal Model Formulation

43.2.1 How Many Degrees of Freedom?

To really model gait well, many of the characteristics of normal gait should ultimately be included in a *3-D* gait model. In 1953, Saunders et al. classified these characteristics as the well-known "determinants" of gait: *(1)* pelvic rotation, *(2)* pelvic tilt (list), *(3)* knee flexion during stance, *(4,5)* knee and ankle interactions, and *(6)* lateral displacement of the pelvis. The "unifying principle" behind each of these determinants was the concept that "fundamentally locomotion is the translation of the center of gravity through space along a pathway requiring the least expenditure of energy...," i.e. the pathway in which the center of gravity had the least vertical fluctuation. Further, they found that the loss of one of these determinants could be compensated for by other mechanisms, but that compensating for the loss of two required a threefold increase in energy consumption [see also Chapter 44 (Winter et al.) for discussions regarding compensations in pathological gait].

A complete lower-extremity model with the capability of including all six determinants of gait would require joints possessing more than single *DOF*. It would be desirable, for instance, to have ankle joints with 3 *DOF*. Inversion/eversion would allow the pelvis to move laterally (determinant #6) while rotation of the foot about the tibial axis and subtalar joint axes would allow the ankle to have a full range of motion (Isman and Inman, 1969; Inman et al., 1981). Modeling the knee with more than 1 *DOF* would allow inclusion of the "screw-home" motion near full extension (Hallen and Lindahl, 1966), although there are indications that it may not occur during normal gait (Lafortune, 1984). And of course, the ball-and-socket hip joint would best be modeled with 3 *DOF*.

Some breakdown of the trunk into segments would also be desirable. Transverse rotations of the shoulder could be included for added realism, and the motions of the upper trunk could be separated from the motion of the pelvis. The analyst could then model pelvic list without compromising the position of the center of mass of the head, arms, and trunk (*HAT*). If the pelvis is included as part of a single-segment *HAT*, pelvic list equals "*HAT* list", which causes the moment exerted by gravity to increase in a highly unstable fashion (Figure 43.2). Furthermore, calculating musculotendon forces at the hip requires accurate relative orientations of the femur and *pelvis*. If pelvic list cannot be modeled for stability reasons, accuracy is compromised in the musculotendon force–length–velocity computations.

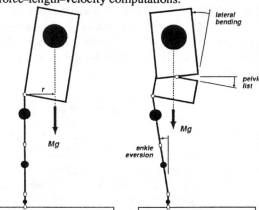

Figure 43.2: The model used here (left) and a proposed model (right) combining pelvic list, lateral trunk bending, and ankle eversion, as seen from the front. Note that the addition of lateral trunk bending allows pelvic list without destabilizing the system significantly. (Reprinted with permission from Yamaguchi & Zajac, *IEEE Trans. Biomed. Eng.* **37**. © 1990 - IEEE.)

A better model of the *HAT* would include a 2 *DOF* "sacral-pelvic" joint in the lumbar region allowing sagittal-plane flexion/extension and lateral bending in the frontal plane between the upper trunk and pelvis, plus an additional degree of freedom in the mid-thoracic region (near *T7*), where transverse axial rotation is maximal (Inman et al., 1981). These 3 *DOF*, plus 3 at each ankle and hip, and 2 at each knee, total 19 *DOF*. And since at least one more is required at the foot-ground interface to allow the heel to rise, a desirable whole-body, *3-D* gait model could easily possess 20 *DOF*.

43.2.2 Practical Limitations on the Number of DOF

Though one could argue that simple models are easier to understand, and thus serve to illustrate mechanical principles better, our current needs for accurately predicting the effects of musculoskeletal alterations requires a higher level of sophistication. Most dynamic musculoskeletal models described in the literature possess relatively few degrees of freedom (except for free-fall situations; see Dapena, 1981; Hatze, 1981; Ramey and Yang, 1981). Even the complex long-jump model of Hatze (1981), which had many segments, was presented in planar form, which simplified the analysis considerably. One might ask why there are no complex, *3-D* gait models at the present time? What imposes limitations on the degree of modeling complexity?

Until recently, probably the greatest roadblock to the creation of more complex gait models was the difficulty of deriving the dynamic equations of motion. Certainly Lagrange's classical method was simple enough in theory to understand, but very difficult and tedious to apply. The benefits of having the dynamic equations in algebraic form at the price of squaring and differentiating interminably long lists of trigonometric terms by hand could not usually be justified. Kane's vector-based method of analytically deriving the equations simplified this undesirable task immensely, but typical analyses could still take months to perform (Kane and Levinson, 1985). Numerical methods of simulating the dynamic motions, such as the iterative Newton-Euler method (Pandy and Berme, 1988b), were an alternative, but are computationally slow (however, see also Walker et al., 1981) and could not easily handle branches in the linked-segment models.

Relief to the tedium of deriving the dynamic equations by hand is now provided by computer programs proficient in algebraic manipulations. Combined with Kane's method, the latest of these programs can capably calculate off-line the entire set of equations in algebraic form for open-chain linked-segment models (*SYMBA*, by Nielan, 1986), or even perform Kane's step-by-step method on-line requiring only a personal computer (*AUTOLEV*, by Schaechter and Levinson, 1987). It is believed that programs such as these will enable researchers to perform dynamical simulations using musculoskeletal models of much greater complexity than ever before.

The advantage of performing the derivations step by step, as in *AUTOLEV*, is that one gains insight into the nature of the kinematical relationships inherent in the problem. This is much better than a "black box" approach to obtaining the equations of motion (as in *SYMBA*), where the output must often be accepted on faith. The more one understands the problem, the easier it becomes to analyze and interpret the results, not an easy task given the degree to which biomechanical segments are coupled together. Blind use of computer programs that generate complex dynamical equations from input specifications can easily extend oneself far beyond one's level of understanding, and can potentially lead to faulty analyses.

Since obtaining the equations of motion for reasonable biomechanical models is no longer an issue, the main source of difficulty is that of determining the controls needed to perform a coordinated task. Joint torque histories usually form the controls on a superficial level. In musculoskeletal models, these torques are really the summation of moments applied by the musculotendon actuators and other forces acting on the system (e.g., joint reaction, inertial, etc.). Therefore, the muscle activations, musculotendon lengths, contraction velocities, and moment arms about spanned joints all contribute to the difficulty of controlling a specified movement. And since muscles can only exert forces in tension, at least $n + 1$ muscle-tendon actuators are needed to control an n-*DOF* model (Nazarczuk, 1970). To add to the difficulty, "extra" or potentially redundant muscles are evident, some spanning multiple joints. One can see how much more difficult it is to control paralyzed extremities than a robot manipulator with well-characterized torque motors at each joint! Practical limits to the numbers of *DOF* incorporated in a musculoskeletal model are dictated somewhat by the difficulty of determining appropriate control information.

For example, linearization of the dynamic equations about an operating point or well-defined trajectory is often done to apply classic optimal control methods [e.g. Chapter 10 (Loeb and Levine)]. In order for the control to converge to an optimum, it is crucial that the linearized equations closely approximate the actual equations, and that the initial guess for the controls be close to ideal. The dynamics of intersegmental musculoskeletal motions are highly nonlinear and strongly coupled, making success unlikely. As the number of degrees of freedom gets large, the process of determining appropriate controls becomes very complex, and consequently, very time-consuming. In many cases it becomes doubtful that convergence can be obtained in a reasonable amount of time (see Section 43.3).

A means of systematically reducing the number of *DOF* to more manageable levels will now be explained.

43.2.3 Prioritizing the Modeled Degrees of Freedom

In this section, a hierarchy of the degrees of freedom, in order of their importance, is proposed for models of the normal gait cycle. First, however, a basis for prioritizing the degrees of freedom is given.

I. Support — trunk mass must be supported reliably

II. Functionality — the forward progression of the body's center of mass must be maintained

III. Efficiency — save energy

IV. Aesthetics — gait should look as natural as possible

By the above basis, the following motions are deemed to be important in any model of normal gait:

1. hip flexion/extension, stance and swing legs

2. knee flexion/extension, stance and swing legs

3. ankle dorsi/plantarflexion, stance and swing legs

4. foot rotation about the metatarsals, stance leg

5. hip abduction/adduction, stance leg

6. ankle inversion/eversion, stance leg

7. hip rotation about the axis of the stance-leg femur

8. sacral-pelvic lateral bending (frontal plane)

9. sacral-pelvic flexion/extension (sagittal plane)

10. hip abduction/adduction, swing leg

11. hip rotation about swing leg femoral axis

12. axial trunk rotation

Items *#1* to *#3* are important characteristics of gait without which normal walking would be impossible. If the legs could not move, the trunk would either stand still in the upright position or fall over. The degree of freedom about the metatarsal joints of the stance foot (*#4*) is important to allow the heel to rise, which includes the major power input to the gait cycle at push-off and is an important mechanism that enables the former stance hip and knee joints to flex during double-support without lowering the hip. Hip abduction/adduction (*#5*) is deemed to be important for stability in the frontal plane, and to allow a measure of pelvic list to occur.

Ankle inversion/eversion (*#6*) is considered to be an energy-conserving measure, which in conjunction with pelvic list allows the position of the body's center of mass to swing transversely over the stance foot. Item *#7*, transverse pelvic rotation, allows the step length to be increased, which is another method by which energy is conserved (Inman et al., 1981).

The inclusion of a 2 *DOF* sacral-pelvic joint (items *#8* and *#9*) separates the motions of the upper and lower trunk. In particular, lateral bending of the trunk would allow the pelvis to list without severely increasing the frontal-plane gravitational moment of the trunk about the supporting hip (Figure 43.2). Thus, vertical fluctuations of the body's center of mass and the stresses on the lateral hip muscles (abductors) could be minimized simultaneously.

Remaining items add realism and contribute somewhat to energy savings during normal gait. Adding swing-side hip abduction/adduction (*#10*) would allow the swing leg to extend in the direction of forward progression even if the pelvis were rotating transversely. The addition of this *DOF* would also decouple lateral leg motions from lateral motions of the trunk. Hip rotation (*#11*) on the swing side would allow the toes to be oriented correctly if the pelvis rotated transversely (*#7*). Axial trunk rotation (*#12*) would allow the effects of arm swing to be included.

43.3 Determining Musculotendon Controls in Dynamic Simulations

One of the classic problems in biomechanics is that of determining the forces or activation levels needed in each of the body's musculotendon actuators to create smooth, coordinated movements. In this section, muscle activation levels are considered to be the controls, while the phrase "coordinated movements" refers to *simultaneously* moving the joints through their movement trajectories. Thus, controlling coordinated movements is distinctly different from, and much more difficult than, sequentially moving multiple joints from their initial to final configurations as is commonly done with upper-extremity prostheses (Morecki, 1980). Furthermore, since the muscle set is redundant (there are many more muscles than joints), and they can only exert tensile forces, the problem becomes one of most efficiently distributing the forces among the muscle set in order to generate the joint moments required to produce a desired movement.

In the literature, most approaches to solving the problem of force distribution can be grouped into methods that either reduce the degree of muscle redundancy in the model, mathematically optimize the solution according to some criterion, or combine the two approaches. In the first method, constraints are introduced until the number of degrees of freedom just equals the number of controls, so that a determinate problem can be formulated. Since expressing such constraints often cannot be done without adding unphysiological artifacts to the analysis, optimization (the second method) is generally the favored, and most adaptable approach.

Optimization methods search for allowable combinations of the controls to find the "best" one according to some predetermined criterion. The presumption is that a mathematical expression called the cost or criterion function may be formulated to measure how good, say, the resulting movement of a biomechanical model would be in response to a particular control strategy. Typical cost functions for dynamic simulations penalize deviations from desired movement trajectories and/or excessive use of control efforts [see also Chapter 8 (Zajac and Winters)]. The main difficulty with expressing the cost mathematically is in formulating physiologically justifiable expressions that reflect energy expenditure, metabolic costs, muscle exertion, and fatigue. It is plausible, though, that the body naturally "optimizes" the energetics of movement control in a manner similar to mathematical optimization.

Human walking has been the subject of many such optimization studies. A commonly-used approach is linear programming, for example. Linear programming is one of the more popular optimization techniques because it requires no prior knowledge of the solution, and is easy to understand and apply to the biomechanical force distribution problem (Seireg and Arvikar, 1973, 1975; Chao and An, 1978; Crowninshield, 1978; Crowninshield et al., 1978; Hardt, 1978; Pedotti et al., 1978). As long as an optimization problem may be mathematically formulated with linear expressions for the cost and constraints, linear programming can be used.

Studies, however, suggest that the physiological costs associated with muscular exertions are in fact non-linear, as suggested by the inverse non-linear relationship between muscle stress and endurance (Crowninshield and Brand, 1981). Therefore, a linear cost function (as required by linear programming) would only be able to approximate the physiological cost, at best. Many of the constraints, especially if they contain geometrical or empirical relationships, also may not be readily cast into linear forms. Of course, non-linear expressions can be linearized about a nominal operating point or trajectory, but such approximations will be valid only if the deviations are small.

Linear programming imposes still other potentially undesirable limitations for use in solving movement and coordination problems. Since it is a quasistatic method, it optimizes the controls only at specific, unrelated instants of time. As a result, the muscle force solutions may not be optimal over the entire movement, and may be unrealistic as well (Hardt, 1978). More seriously, the linear programming method mathematically limits the number of muscles active at any given time to a number equal to the number of constraints. Thus linear programming would be hard-pressed to predict, for example, muscle synergies requiring the simultaneous control of many coactivated muscles.

Ultimately, dynamic optimization methods based on the calculus of variations should be used to solve the problem of finely controlling musculotendon forces to achieve normal gait. A physiologically based, time-integral cost function would have to be formulated, perhaps balancing energy expenditure with deviations from desired segment trajectories. If classical methods were employed, the dynamic system equations would require linearization about the desired trajectory [a difficult and tedious task; see also Chapter 10

(Loeb and Levine)], in order to predict the changes in cost due to infinitesimal changes in the controls. Gradient, or similar methods of efficiently searching the solution space for optimal solutions would probably be employed to determine the best patterns of control throughout the gait cycle.

Unfortunately, these dynamic optimal control algorithms are difficult to apply to complex systems. Unless one has prior knowledge of the character or form of the optimal control solution, or at least a good initial guess for it, the likelihood of obtaining meaningful results is very small. This is because the numerical gradient methods usually employed have difficulty converging to a minimum-cost solution unless they begin searching in the vicinity of the desired solution. Even if a solution is found, one cannot be sure without additional tests that the global, rather than a local, optimum has been reached. Additional difficulties arise because muscles, and particularly muscles with pathologic conditions, are limited in their ability to exert force. Thus the controls are bounded by nature, which undesirably applies restrictions to variational methods. Together with the other factors just mentioned, it is unlikely that good solutions can be found for movements utilizing complex models without preliminary study.

Though quite powerful dynamic optimization routines have been developed that are suitable for controlling biomechanical motions (for example, see Mayne and Polak, 1975), one might not need to determine optimal solutions as much as solutions that are *workable*. This depends to a large degree upon how the model is to be used. If one wishes to discover the way in which the neuromuscular system naturally operates, mathematical optimization is an appropriate tool through which various cost function formulations may be tested. If, on the other hand, one uses the model to design or predict a sequence of control signals that will enable a disabled individual to perform a specific movement, a workable solution may be all that is needed.

In the next section, dynamic programming is introduced as a potential means of obtaining workable, dynamic controls without requiring prior knowledge of the solution.

43.4 Example: The Use of Dynamic Programming to Control 10 Crudely Controlled Musculotendon Actuators in an 8-DOF Gait Model

Functional neuromuscular stimulation (*FNS*) is a technique whereby electrical currents are artificially applied to nerve and muscle tissues in order to elicit muscular contractions. In this example, the feasibility of utilizing *FNS* to enable paraplegics to walk with normal appearance and speed was evaluated through the use of dynamic, *3-D* movement simulations. What is illustrated here by this example are the interrelationships between model complexity, control determination, and hardware limitations. It is emphasized that this section, then, will focus upon these interrelationships, and not upon the results of the study itself, which have been adequately presented elsewhere (Yamaguchi, 1989; Yamaguchi and Zajac, 1989b, 1990).

In studying the use of *FNS* to restore normal gait to paraplegics, optimization was useful in order to determine: *a)* appropriate muscle sets, and *b)* their approximate patterns of stimulation to best utilize the limited capabilities of electrically-stimulated muscles. More specifically, *the goal of the study was to examine whether minimal sets of muscles could be used in order to generate approximately normal gait trajectories without requiring either high levels of force or unduly precise control of muscle activation.*

As this was a preliminary study, the question investigated here focused primarily upon feasibility, rather than on optimality of the controls. Dynamic programming was therefore suitable as a means of obtaining preliminary (e.g. *workable*) results. It was anticipated that this work would lay the necessary foundation for further, more refined work.

43.4.1 Dynamic Programming

Dynamic programming was attractive as an alternative to linear programming and variational methods for this preliminary work, in which the feasibility of restoring natural gait to paraplegics using *FNS* was studied. Compared to linear programming, dynamic programming does not require the criterion (cost) function or the constraints to be linear. Neither does it limit the optimization to quasistatic analyses, nor impose mathematical constraints on the number of active muscles as linear programming does. Many of the drawbacks of variational methods are avoided as well. For instance, without prior knowledge of the character of the control solution, obtaining meaningful results with variational optimal control methods was deemed to be a dubious proposition. Also, boundedness in the controls helps rather than hinders the process of numerical solution. Linearization of the dynamic equations of motion is not required, and a good initial guess of the solution is not needed, at least in principle. Above all, the optimal solution found by dynamic programming is guaranteed to be a good, global solution, though approximate.

A feature of dynamic programming is that the dynamic equations of motion are used *exactly* (i.e., without approximation) to compute and evaluate the movement trajectories resulting from particular patterns of control. Yet in searching for optimal control patterns, the control levels are considered to be available only in M discrete levels, rather than being infinitesimally variable. This is disadvantageous for some applications, since dynamic programming produces only approximate optimal control patterns. For exploring neuromuscular control issues in *FNS*, this drawback is not expected to be detrimental, as the activation levels of the muscles themselves are unlikely to be controlled with great precision.

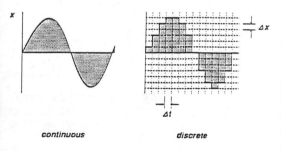

continuous   discrete

Figure 43.3: Discretization of a continuous state variable, x, and time.

The optimal controls, too, are not computed using the infinite array of possible configurations the model can attain. Rather, for purposes of computing the controls, the system configuration is itself approximated. The variables describing the state of the system, which are the angular orientations and velocities of the body segments, are each *discretized* in a manner similar to discretizing the controls. In other words, the element x_i of the

state vector X is considered for control purposes to be discretely variable in steps rather than infinitesimally variable over its range R_i (Figure 43.3); x_i is thus described by N discrete levels rather than by an infinity of possible values. Because all the elements are similarly discretized, the controls are determined only for a finite number of discrete state combinations, or admissible states. Considering that both the numbers of admissible controls and admissible states are finite suggests that the optimal controls might be found simply by trying each combination of the controls at each of the admissible states. Dynamic programming is simply an efficient method of doing just that.

For use in this study, dynamic programming has only one serious drawback, which has thus far prevented its widespread use. Normally, only very simple problems with relatively few variables or degrees of freedom can be solved due to the extreme requirements placed upon computational resources. For example, computation time increases exponentially with the degree of problem complexity. Given a system with n degrees of freedom ($2n$ state elements) and m control variables (muscle activations), and integers K, M, and N (all greater than 1), it would take

$$T_{CPU} = K \times N^{2n} \times M^m \times \Delta t_{CPU} \text{ sec} \qquad (43.5)$$

This expression for computation time reflects the process undertaken by dynamic programming, whereby every allowable combination (taking Δt_{CPU} seconds each) of the controls is tried and evaluated at every allowable state, for K "stages" (intervals of time within which the controls remain constant). Even if N and M were as small as practical ($N = M = 2$) for the 8 *DOF* model used here ($n = 8$), a theoretical minimum of $m = n + 1$ muscles (Nazarczuk, 1970), and a typical minicomputer were used ($\Delta t_{CPU} = 5$ msec), it would take approximately 513 hours of *CPU* time to analyze $K = 11$ such time-steps! Even if a computer could be monopolized for that long, the elapsed time required would stretch beyond the reasonable cycle or "up" times attained by current computers. Memory and storage requirements, too, would be severe for complex problems. Bellman (1957) coined this as the "curse of dimensionality." Yet by simplifying the model used to the utmost, and through efficient use of computer time and storage, biomechanical problems may be formulated, made tractable, and solved through the dynamic programming method.

A basic difference between dynamic programming and optimal control methods based on the calculus of variations should now be stated. In the variational approach, the system is linearized about an operating point (or trajectory) so that the effects of infinitesimal adjustments in the controls may be evaluated. In effect, the system is first approximated by one that is mathematically simpler to analyze, which is then used to finely tune the controls until the optimum is found. In dynamic programming, the system equations themselves are used exactly, but the states and controls are represented by discrete approximations. Since the numbers of discrete, admissible states and controls is finite, approximations to the optimal controls (determined from the set of admissible controls) can be found at every admissible state.

Once the approximate optimal controls are found for each of the admissible states, they are stored in computer memory for later usage. If applied in a movement simulation or in an actual control application, they can be used as a *control law*, whereby given any system configuration (e.g. state), the appropriate control over the next stage is known and can be retrieved from memory and applied. This phase is here referred to as the *forward* simulation phase, as it takes place in the forward-time direction (from the initial to the final time). In contrast, the optimal controls are determined by proceeding backward in time from the final time to the initial time.

Before going on, a subtle point should be emphasized concerning the dynamic nature of the controls obtained. Though the period from the initial to the final time has been divided into a series of discrete intervals of time (stages), the dynamic programming procedure optimizes the controls *dynamically* over many time intervals. That is to say, the optimal controls are determined from each admissible state of every stage to the desired *end* state (not to an intermediate state). When applied later during the forward simulation phase, the controls are periodically retrieved from storage and applied so as to guide the system state along an optimal pathway directed toward the final state. And, though the controls have been determined to optimize the *total* remaining pathway (over many stages) from the current time to

the final time, the optimal controls for only the current stage are applied. As each successive stage is reached, the state is re-evaluated to determine the control settings over the next stage. Therefore, the controls are updated in response to the state as each discrete interval of time is reached in any simulation or real-time application using a dynamic programming control law.

The reader is referred to Kirk (1970) and Larson and Casti (1978) for more complete descriptions of the dynamic programming method. As applied to the control of musculoskeletal movements, the reader is referred to Yamaguchi (1989).

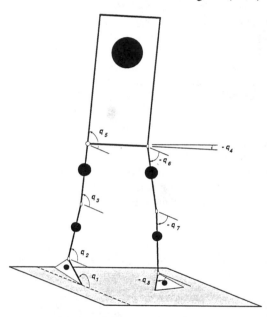

Figure 43.4: The 8 *DOF* model used in the example of Section 43.4. (Reprinted with permission from Yamaguchi & Zajac, *IEEE Trans. Biomed. Eng.* **37**. © 1990 - IEEE.)

43.4.2 Results

The dynamic programming controls were discretized in levels of 10% activation, which is commensurate with the coarse nature of muscle-fiber recruitment in typical *FNS* applications. Subject to the computational hardware limitations of a 1 MIP VAX 11/780 computer, only 8 *DOF* and 10 muscles were allowable even with suitable simplifications. Degrees of freedom were assigned according the hierarchy listed in Section 42.2.3, resulting in the 8 *DOF*, single-step model shown in Figure 43.4.

Highly damped, stiff linear springs were assumed to exert forces preventing the heel of the right leg, and the heel and forefoot of the left leg, from falling through the assumed ground-plane. The use of such "soft constraints" (Hemami et al., 1975) enabled the model to be used for the single-leg support and double-leg support phases of the step, as well as the transition in between. Assuming symmetry between left and right steps allowed the time interval of the analysis to be cut in half. Damping and passive constraints at the joints were defined using the method of Audu and Davy (1985), though parameter values were altered to more accurately fit the 8-*DOF* model.

Cost functions in this study were weighted sums of squared deviations from a nominal state trajectory (Winter, 1987; Inman et al., 1981) and a weighted sum of cubed muscle stresses, which has been shown to physiologically relate to muscle fatigue (Crowninshield and Brand, 1981).

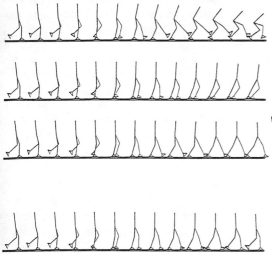

Figure 43.5: Stick-figure sequences of selected runs, showing (from top) stance leg instability, excessive swing-leg dorsiflexion leading to premature heel contact, too lengthy a step, and a near-normal step. (Reprinted with permission from Yamaguchi & Zajac, *IEEE Trans. Biomed. Eng.* **37**. © 1990 - IEEE.)

When implemented, typical runs took 14 hours of *CPU* time to both determine the controls and perform the dynamic simulations. Many months of effort (several hundred control determinations and dynamic simulations) were expended before meaningful results were obtained. The difficulty arose because a successful run required most of the unknown parameters to be simultaneously well-

adjusted before the 8-*DOF* model was able to even marginally follow the nominal trajectory under the derived control law. Even if one critical parameter was sufficiently off, the simulated walker would trip or begin falling catastrophically before one step had been accomplished (Figure 43.5)! This is likely due to the coarse nature of the control for such an unstable, dynamically coupled, multilink inverted pendulum. Fortunately, the analysis was able to utilize the published EMG record for guidance as to the selection of appropriate sets of muscles to try. Also, because of the stage-wise nature of the dynamic programming method, problems specific to the early-, mid-, late-stance, and double-leg support phases could be worked out separately.

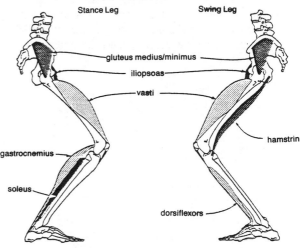

Active Muscle Groups

Figure 43.6: Muscles actually used to maintain the steady-state stepping motion. (Reprinted with permission from Yamaguchi & Zajac, *IEEE Trans. Biomed. Eng.* **37**. © 1990 - IEEE.)

The most difficult and critical aspects of the gait simulations proved to be obtaining: *(i)* stability of the knee and ankle during early to mid single-leg support; *(ii)* lateral stability of the hip; *(iii)* clearance of the toe and heel during swing; and *(iv)* appropriate step length and duration. A set of ten musculotendon actuators was determined to meet goals *(i)* to *(iv)*, as shown in Figure 43.6, to achieve a single step (14% to 62% of the gait cycle; Figure 43.7) under steady-state conditions. These were found to be adequate when coarsely-controlled except at the ankle, where plantarflexion strength was inadequate at push-off.

Dorsiflexion was also difficult to control when the foot was unloaded, necessitating the imposition of an ankle-foot orthosis (Figure 43.8).

The gait simulations were performed in two phases. In *Phase I*, dynamic programming was used to determine stepwise muscle activation patterns. Therefore the dynamics of muscle activation was neglected in this, more difficult, multiple-input/multiple-output (*MIMO*) portion of the study. Once these activation patterns were known, *Phase II* simulations were performed including the dynamics of muscle activation and musculotendon contraction, to confirm the dynamic programming results. These simulations were performed using trial-and-error to smooth the muscle activations and to more finely tune the simulated gait (e.g. *without* dynamic programming). This latter phase reduced to a series of simpler, single-input/single-output (*SISO*) problems, since muscle activations are not coupled and can therefore be determined on a muscle-by-muscle basis.

Figure 43.8: The ankle-foot orthosis (*AFO*), shown on the swing leg. Without resistance, even slight dorsiflexor activity caused the unloaded foot to dorsiflex excessively, resulting in abrupt, premature heel-strike during mid-swing (left drawing, 38% of the gait cycle). The *AFO* acts only when dorsiflexion exceeds 5° (right drawing).

In summary, dynamic programming was used to determine workable, discretized controls (Figure 43.9) for a complex, whole-body simulation using simplified musculature. Given these, a more complete musculotendon model was incorporated to more accurately model the gait simulations (Figure 43.10). Though the controls developed in this way were not optimal, it must be emphasized that at the least, workable results were obtained with a complex, highly unstable, dynamic musculoskeletal gait model.

43.5 Discussion

The example of the preceding section illustrates a process whereby workable controls were developed, via dynamic programming, for a musculoskeletal gait model. Practically speaking, dynamic programming will probably never be used to control normal walking movements with many DOF unless ultra-fast, portable controllers can be made. Computer memory chips with retrieval speeds and capacities many orders of magnitude beyond the current state of the art would need to be developed to overcome Bellman's "curse of dimensionality" (Bellman, 1957; see *Eq. 35.5*).

Certainly the method does lend itself to control problems having a limited number of bounded controls, highly coupled and nonlinear equations of motion, admissible states that are easily defined about a nominal set of segment trajectories, and a

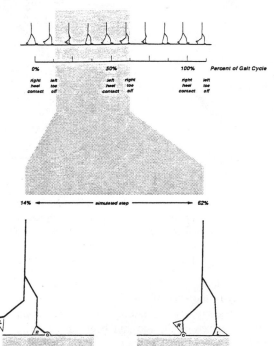

Figure 43.7: The portion of the complete (0-100%) gait cycle studied here begins just after left-toe-off (at 14%), and ends just before right-toe-off (62%). The right leg is considered to be the stance leg during single-leg support. (Reprinted with permission from Yamaguchi & Zajac, *IEEE Trans. Biomed. Eng.* **37.** © 1990 - IEEE.)

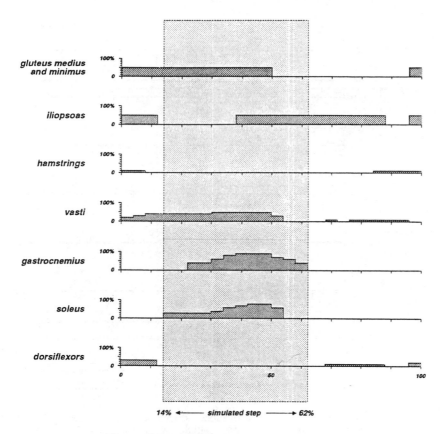

Figure 43.9: Muscle activations (shaded regions) over an entire stride, optimized with the dynamic programming method. Swing-side (left leg) muscle activations are used to extend the results from a single step to 0–12% and 64–100% of the gait cycle.

requirement for dynamic, rather than static, analyses. In particular, dynamic programming is attractive if the dynamic equations are difficult to linearize about the desired operating point or nominal trajectory. In such cases, much can be learned about the character of the control solutions using dynamic programming *before* steps are taken to linearize the dynamic equations. One can then use the approximate optimal control solutions as a good initial guess in more accurate optimization methods. For control problems involving relatively few degrees of freedom, the approximate controls delivered via dynamic programming might themselves be of sufficient accuracy.

What, then, constitutes sufficient accuracy in the controls for dynamic simulations of gait? It was surprising to the author that a set of muscle activations (in the *"closed-loop"* control process of *Phase I*) that were crudely modulated by levels of 10% and updated rather infrequently (at 4% gait cycle intervals, or 0.047 sec) were actually sufficient to generate near-normal gait patterns. The inherent instability of the model was obviously demonstrated by the sensitivity of ambulation to changes in the control parameters and cost-function weighting coefficients (Yamaguchi and Zajac, 1989b), which suggested a need for more finely adjusted muscular controls. When the complete muscle-tendon models were implemented,

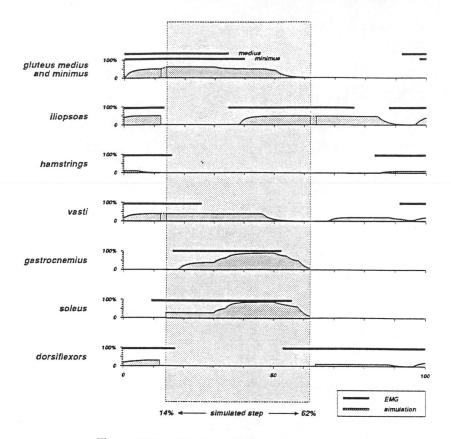

Figure 43.10: Muscle activations (shaded regions) compared with reported *EMG* activity (black bars; Shiavi, 1985). Swing-side activations are used to extend the simulated results to 0–12% and 64–100% of the gait cycle. (Reprinted with permission from Yamaguchi & Zajac, *IEEE Trans. Biomed. Eng.* **37**. © 1990 - IEEE.)

but controlled *"open-loop"* in *Phase II*, the muscular activations changed from stepped to continuously varying functions. Except for the time delay between the imposition of the "neural" controls and the buildup of muscle activation, no significant differences occurred between the simulations of Phases I and II.

Pandy and Berme (1987), too, reported surprise in that stable locomotion could be synthesized in a highly-coupled gait model, using *open-loop* control. Moreover, Pandy and Berme (1987) found simple, as opposed to complex, control functions to be sufficient. Why such a dynamically-unstable system can be controlled (at least in gait simulations) without requiring sophisticated controls is open to speculation.

One reason why some control inaccuracy can be tolerated may rest with the inherent properties of the musculotendon actuators. If the muscles are (i) appropriately sized and placed to control the inertial body segments, and (ii) self-regulating in terms of output force, then approximate controls may be sufficient. That is, if the musculotendon actuators deliver "appropriate" forces to the skeletal system no matter what the control inputs are, then proper accelerations/decelerations of the inertial segments of the lower extremity can occur in spite of the control inadequacies.

The author believes that the force–length-velocity properties of muscle provide such self-regulatory capabilities. For example, because a flexor musculotendon compartment is stretched when the joint is extended, large accelerating forces can be developed through both active and passive means to initiate joint closure. On the other hand, it is less highly accelerated into further closure if the joint is already flexed significantly. Damping in the muscles, too, allows the muscles to regulate tension by exerting maximal forces to change the accelerations of the segments, and decreases these forces automatically once the segment begins to move (Yamaguchi, 1989).

Furthermore, tendon elasticity increases the efficiency of the musculotendons during certain movements such as gait. If an active musculotendon is stretched, and then is allowed to shorten, energy can be stored in the elastic tendon much like energy storage in a spring. Such a "stretch-shorten cycle" allows the energy stored to be later released almost without loss since tendon is nearly elastic (Elftman, 1939; see also Chapters 5, 35–40).

In summary, the author believes that gait simulations can be performed without sophisticated control because musculotendon actuators are made to be efficient, self-regulating accelerators of the body segments. This conceptually opposes the view that muscles are simple generators of joint torque. A further point to be made is that determining *optimal* controls, at least in simulation studies, is not always worth the effort. Depending upon the goals of the particular analyses, *workable* controls may be all that are necessary to achieve desired function.

43.6 Future Directions

Existing computer models of human locomotion are constrained by: *i)* computational resource limitations; *ii)* difficulties in controlling the movements of the interconnected body segments; and *iii)* interpretion of the complex array of results. Hence, existing models of human locomotion are overly simple, and cannot adequately predict or explain the causal relationships of musculotendon contraction and human locomotor behavior.

What is desired are gait simulation models/procedures that have clinical relevance, and that can be applied to predict the effects of treatment procedures on pathological gait. Physicians would then be able to "optimize" the treatment of gait disorders prior to surgical, orthotic, and prosthetic interventions. *Predictive* usage of gait simulation models in this manner would be a step beyond the current, diagnostic usage of inverse-dynamic gait analyses (Sutherland, 1984; Gage et al., 1984; Gronley and Perry, 1984; Naumann et al., 1984; Simon et al., 1989).

To move toward this goal, future efforts should focus upon the development of human locomotion models that: *i)* are more complex, by introducing often-neglected but important *DOF*; *ii)* allow motions to occur in *3-D* rather than restricting them to a plane; and *iii)* more precisely represent the mechanical properties of muscles, tendons, ligaments, and joints. Improved methods of determining optimal/workable "neuromuscular" controls, and relating patient-specific parameters, capabilities, and disabilities to the gait model, should be developed.

New methods of coagulating, analyzing, and displaying the mountain of information that will be obtained with complex musculoskeletal models will also need to be developed. The massive array of information is likely to be unintelligible and of little use to anyone but the research engineer, and thus significant efforts should be expended to present the simulation data in formats that are as natural to interpret as possible.

Perhaps each of these goals will, in time, be achieved. And as computers continue to become faster and less expensive, perhaps biomechanical gait simulation will become a practical tool for the treatment of gait disorders in the not-too-distant future.

References

Audu, M.L., and Davy, D.T. (1985) The influence of muscle model complexity in musculoskeletal motion modeling. *J. Biomech. Engrg.*, **107**: 147-157.

Bellman, R.E. (1957) *Dynamic Programming.* Princeton University Press, Princeton, NJ.

Brand, R.A., Crowninshield, R.D., Wittstock, C.E., Pedersen, D.R., Clark, C.R., and Van Krieken, F.M. (1982) A model of lower extremity muscular anatomy. *J. Biomech. Engrg.*, **104**: 304-310.

Chao, E.Y., and An, K.N. (1978) Graphical interpretation of the solution to the redundant problem in biomechanics. *J. Biomech. Engrg.*, **100**: 159-167.

Crowninshield, R.D., Pope, M.H., and Johnson, R.J. (1976) An analytical model of the knee. *J. Biomechanics*, **9**: 397-405.

Crowninshield, R.D. (1978) Use of optimization techniques to predict muscle forces. *J. Biomech. Engrg.*, **100**: 88-92.

Crowninshield, R.D., and Brand, R.A. (1981) A physiologically based criterion of muscle force prediction in locomotion. *J. Biomechanics*, **14**: 793-801.

Crowninshield, R.D., Johnston, R.C., Andrews, J.G., and Brand, R.A. (1978) A biomechanical investigation of the human hip. *J. Biomechanics*, **11**: 75-85.

Dapena, J. (1981) Simulation of modified human airborne movements. *J. Biomechanics*, **14**: 81-89.

Davy, D.T., and Audu, M.L. (1987) A dynamic optimization technique for predicting muscle forces in the swing phase of gait. *J. Biomechanics*, **20**: 187-201.

Delp, S.L., Bleck, E.E., Zajac, F.E., and Bollini, G. (1989) Biomechanical analysis of the Chiari pelvic osteotomy: preserving hip abductor strength. *Clin. Orthop. Rel. Res.* **254**:189-198.

Delp, S.L., Loan, J.P., Hoy, M.G., Zajac, F.E., Topp, E.L., and Rosen, J.M. An interactive graphics-based model of the lower extremity to study orthopaedic surgical procedures. *IEEE Trans. Biomed. Engrg.*, Special Issue: Interaction with and Visualization of Biomedical Data, in press.

Elftman, H. (1939) The function of muscles in locomotion. *Am. J. Physiol.*, **125**: 357-366.

Friederich, J.A., and Brand, R.A. (1990) Muscle fiber architecture in the human lower limb. Technical Note, *J. Biomech.* **23**:91-95.

Gage, J.R., Fabian, D., Hicks, R., and Tashman, S. (1984) Pre- and post-operative gait analysis in patients with spastic diplegia: A preliminary report. *J. Pediatric Orthop.*, **4**: 715-725.

Gronley, J.K., and Perry, J. (1984) Gait analysis techniques. *J. American Physical Therapy Assn.*, **63**: 1831-1838.

Hallen, L.G., and Lindahl, O. (1966) The 'screw-home' movement in the knee joint. *Acta Orthop. Scand.*, **37**: 97-106.

Hardt, D.E. (1978) Determining muscle forces in the leg during normal human walking -- An application and evaluation of optimization methods. *J. Biomech. Engrg.*, **100**: 72-78.

Hardt, D.E., and Mann, R.W. (1980) A five body -- three dimensional dynamic analysis of walking. Technical Note, *J. Biomechanics*, **13**: 455-457.

Hatze, H. (1977) A myocybernetic control model of skeletal muscle. *Biol. Cybernetics*, **25**: 103-119.

Hatze, H. (1981) A comprehensive model for human motion simulation and its application to the take-off phase of the long jump. *J. Biomechanics*, **14**: 135-142.

Hemami, H., Jaswa, V.C., and McGhee, R.B. (1975) Some alternative formulations of manipulator dynamics for computer simulation studies. *Proc. 13th Allerton Conf. on Circuit Theory*, University of Illinois, October.

Hill, A.V. (1938) The heat of shortening and the dynamic constants of muscle. *Proc. Roy. Soc. B. (Lond.)*, **126**: 136-195.

Hoy, M.G., Zajac, F.E., and Gordon, M.E. (1990) A musculoskeletal model of the human lower extremity: The effect of muscle, tendon, and moment

arm on the moment-angle relationship of musculotendon actuators at the hip, knee, and ankle. *J. Biomechanics* **23**:157-169.

Inman, V.T., Ralston, H.J., and Todd, F. (1981) *Human Walking*. J. C. Lieberman (ed.), Williams & Wilkins, Baltimore, MD.

Isman, R.E. and Inman, V.T. (1969) Anthropometric studies of the human foot and ankle. *Bull. of Prosthetics Research*, Spring.

Ju, M.-S. and Mansour, J.M. (1988) Simulation of the double limb support phase of human gait. *J. Biomech. Engrg.*, **110**: 223-229.

Kane, T. R. and Levinson, D.A. (1985) *Dynamics: Theory and Applications*. A. Murphy and M. Eichberg (eds.), McGraw-Hill, New York.

Kirk, D.E. (1970) *Optimal Control Theory. An Introduction*. R. W. Newcomb (ed.), Prentice-Hall, Englewood Cliffs, NJ.

Lafortune, M.A. (1984) The use of intra-cortical pins to measure the motion of the knee joint during walking. *Ph.D. Dissertation*, College of Health, Physical Education and Recreation, The Pennsylvania State University, State College.

Larson, R.E., and Casti, J.L. (1978) *Principles of Dynamic Programming*. Part I. Basic Analytic and Computational Methods. J.M. Mendel (ed.), Marcel Dekker, New York.

Mayne, D.Q., and Polak, E. (1975) First-order strong variation algorithms for optimal control. *J. Optimization Theory and Applications*, **16**: 277-301.

Mena, D., Mansour, J.M., and Simon, S.R. (1981) Analysis and synthesis of human swing leg motion during gait and its clinical applications. *J. Biomechanics*, **14**: 823-832.

Mochon, S., and McMahon, T.A. (1980) Ballistic walking. *J. Biomechanics*, **13**: 49-57.

Morecki, A. (1980) Identification, modeling and rehabilitation problems in modern biomechanics In *Biomechanics of Motion*, A. Morecki (ed.) Springer-Verlag, New York, pp. 1-40.

Murphy, M.C., Zarins, B., Jasty, M., and Mann, R.W. (1985) In vivo measurement of the three-dimensional skeletal motion at the normal knee. *Proc. of the 31st Annual ORS*, Las Vegas, NV, Jan. 21-24, p. 142.

Naumann, S., Cairns, B., Mazliah, J., Silver, R., White, C., Rang, M., and Milner, M. (1984) Preoperative and postoperative gait assessments as a guide in planning tendon transfers about the ankle joint in children with cerebral palsy. *Proc. 2nd Internl. Conf. on Rehab. Engrg.*, Ottawa, Canada, pp. 629-630.

Nazarczuk, K. (1970) The theory of artificial muscle actuators and its application for synthesis and control of biomanipulators. *Ph.D. Dissertation*, Warsaw, Poland.

Nielan, P.E. (1986) Efficient computer simulation of motions of multibody systems. *Ph.D. Dissertation*, Department of Mechanical Engineering, Stanford University, Stanford, CA.

Onyshko, S., and Winter, D.A. (1980) A mathematical model for the dynamics of human locomotion. *J. Biomechanics*, 13:361-368.

Pandy, M.G., and Berme, N., (1987) Synthesis of human walking: A three-dimensional model for single support. Part 2: Pathological gait. *ASME Winter Annual Meeting*, Boston, December 13-18, pp. 9-15.

Pandy, M.G., and Berme, N., (1988) Synthesis of human walking: A planar model for single support. *J. Biomechanics*, 21: 1053-1060.

Pandy, M.G., and Berme, N., (1988) A numerical method for simulating the dynamics of human walking. *J. Biomechanics*, 21: 1043-1051.

Patriarco, A.G., Mann, R.W., Simon, S.R., and Mansour, J.M. (1981) An evaluation of the approaches of optimization models in the prediction of muscle forces during human gait. *J. Biomechanics*, 14: 513-525.

Pedotti, A., Krishnan, V.V., and Stark, L. (1978) Optimization of muscle-force sequencing in human locomotion. *Math. Biosci.*, 38: 57-76.

Ramey, M.R., and Yang, A.T. (1981) A simulation procedure for human motion studies. *J. Biomechanics*, 14: 203-213.

Saunders, J.B., deC. M., Inman, V.T., and Eberhart, H.D. (1953) The major determinants in normal and pathological gait. *J. Bone Jt. Surg.*, 35-A: 543-559.

Schaecter, D.B., and Levinson, D.A. (1987) *AUTOLEV User's Manual*. Available from: OnLine Dynamics, Inc., 1605 Honfleur Dr., Sunnyvale, CA 94087.

Seireg, A., and Arvikar, R.J. (1973) A mathematical model for evaluation of forces in lower extremities of the musculo-skeletal system. *J. Biomechanics*, 6: 313-326.

Seireg, A., and Arvikar, R.J. (1975) The prediction of muscular load sharing and joint forces in the lower extremities during walking. *J. Biomechanics*, 8: 89-102.

Shiavi, R. (1985) Electromyographic patterns in adult locomotion: A comprehensive review. *J. Rehab. R&D*, 22: 85-98.

Simon, S.R., Weintraub, M., Bylander, T., Hirsch, D., and Szolovits, P. (1989) Dr. Gait: An expert system for gait analysis. *RESNA 12th Annual Conf.*, New Orleans, pp. 93-94.

Sutherland, D.H. (1984) *Gait Disorders in Childhood and Adolescence*. Williams & Wilkins, Baltimore.

Van Eijden, T.M.G.J., Kouwenhoven, E., Verburg, J., and Weijs, W.A. (1986) A mathematical model of the patellofemoral joint. *J. Biomechanics*, 19: 219-229.

Wickiewicz, T.L., Roy, R.R., Powell, P.L., and Edgerton, V.R. (1983) Muscle architecture of the human lower limb. *Clin. Orthop. Rel. Res.*, 179:275-283.

Winter, D.A. (1987) *The Biomechanics and Motor Control of Human Gait* University of Waterloo Press, Waterloo, Ontario, Canada.

Wismans, J., Veldpaus, F., Janssen, J., Huson, A., and Struben, P. (1980) A three-dimensional mathematical model of the knee joint. *J. Biomechanics*, 13: 677-685.

Yamaguchi, G.T. (1989) Feasibility and conceptual design of functional neuromuscular stimulation systems for the restoration of natural gait to paraplegics based on dynamic musculoskeletal models. *Ph.D. Thesis*, Department of Mechanical Engineering, Stanford University, Stanford, CA Aug. 1989.

Yamaguchi, G.T., and Zajac, F.E. (1989) A planar model of the knee joint to characterize the knee extensor mechanism. *J. Biomechanics*, 22: 1-10.

Yamaguchi, G.T., and Zajac, F.E. (1989) Sensitivity of simulated human gait to neuromuscular control patterns. *XII International Congress of Biomechanics*, Los Angeles, June 26-30, paper #166.

Yamaguchi, G.T., and Zajac, F.E. (1990) Restoring unassisted natural gait to paraplegics via functional neuromuscular stimulation: A computer simulation study. *IEEE Trans. Biomed. Engrg.*, in press.

Yen, V., and Nagurka, M.L. (1987) Biomechanics of normal and prosthetic gait. *ASME Winter Annual Meeting*, Boston, December 13-18, pp. 17-22.

Zajac, F.E. (1989) Muscle and tendon: Properties, models, scaling, and application to biomechanics and motor control *In CRC Critical Reviews in Biomedical Engineering*. J.R. Bourne (ed.), CRC Press, Inc., Boca Raton, FL (in press).

Zajac, F.E. and Gordon, M.E. (1989) Determining muscle's force and action in multi-articular movement *In Exercise and Sport Science Reviews*. K. Pandolf (ed.), Williams & Wilkins, Baltimore, V. 17, pp. 187-230.

CHAPTER 44

Adaptability of Motor Patterns in Pathological Gait

David A. Winter, Sandra J. Olney, Jill Conrad
Scott C. White, Sylvia Ounpuu and James R. Gage

44.1 Introduction

Human walking is a complex motor control task requiring the integration of central and peripheral control of scores of muscles acting on a skeletal system with many degrees of freedom. Associated with the goal of forward progression is the overriding need for a safe transit: balance control to prevent falling over, a support control to prevent collapse against gravity, and a fine motor control of the foot during swing to ensure a safe toe clearance and a gentle heel contact.

When a pathology occurs, the *CNS* control is often drastically disrupted (stroke, cerebral palsy) or the musculoskeletal system is dramatically altered (amputation, joint surgery). With such chronic or traumatic alterations to the system it is important that the *CNS* have the capability of adapting to compensate for the loss of function. Most gait assessments have focused solely on the pathology with little regard for how the rest of the system was responding. Amputee gait assessments, for example, have zeroed in on the prosthesis, its alignment, and the forces the stump/socket interface. Almost totally neglected have been the drastically altered motor patterns of the residual muscle groups, on either the prosthetic or non-prosthetic side. The same is true for hemiplegic gait, and for joint surgery such as knee arthroplasty. The purpose of this chapter is to focus on the total motor system by first documenting the kinematic and kinetic changes at the site of the pathology and also the compensations by the residual intact system.

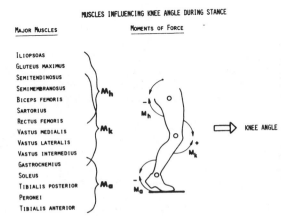

Figure 44.1: Illustrative example to demonstrate the three moments of force and major muscles that control the knee joint during stance. Redundancy of the system at the muscle level predicts considerable variability and adaptability at the motor level.

44.2 Inherent Redundancy and Variability of Motor System

Anyone who has taken a course in gross anatomy is faced with the task of learning the names of scores of muscles, many of which appear to have the same function. Considerable comment is made about the apparent excess redundancy of the system. Such redundancy even extends beyond the muscles crossing a given joint; in many situations the motor patterns at one joint can control the movement at one or two joints away. Figure 44.1 is presented to illustrate the motor patterns of the stance limb that can control the knee angle. It should be noted that the knee angle is under the control of moments of force at the hip and ankle as well as the knee. A stronger than normal ankle plantar-flexor moment early in stance will slow down or even reverse the forward rotation of the leg and thereby reduce the amount of knee flexion. In fact, in many cases of calf muscle spasticity the result is often a hyper-extended knee joint in mid-stance. Similarly, hyperactive hip extensor activity can cause a backward rotation of the thigh, thereby reducing the knee angle; above-knee amputees use their hip extensors on every stride to lock their knee joints. Thus there are an infinite number of combinations of moments at the three joints that can result in the same knee angle history. Thus we should anticipate highly variable moment patterns during stance and this has been documented (Winter, 1984). Furthermore, at the individual muscle level we see 15 of the more important muscles responsible for these three moments, and it is evident that even more variability manifests itself in the linear envelope *EMG* profiles which has been reported for intra- and inter-subject averages (Winter and Yack, 1987). Thus the inherent characteristic of the human motor system would predict tremendous variability, which, in turn, reflects the system redundancy and a great potential for adaptability.

44.3 Results and Discussion

A wide spectrum of case studies are now presented which will demonstrate the adaptability of the *CNS* to compensate for the loss of function. It is extremely important that atypical patterns which are adaptions be identified as such, because it is important to treat the primary problem rather than secondary adaptions. The examples are drawn from four gait laboratories in Canada and the United States and include: knee arthroplasty, above-knee amputee, candidates for unilateral total hip replacement (degenerative joint disease), hemiplegia, and cerebral palsy.

44.3.1 Case Study #1 - Knee Arthroplasty

This patient was a 53 year old female who had a total knee replacement. She was assessed at the University of Waterloo Gait Laboratory. Data were collected using a 16 mm cine camera at 50 Hz as the subject walked along a level walkway over an AMTI force platform. Raw coordinate data were digitized and digitally filtered using the

protocol described by Winter et al. (1974). Using an inverse link segment model (Bresler and Frankel, 1950), the moments of force at the ankle, knee, and hip on the surgical side were calculated. Normalization of these moment profiles by dividing the body mass permitted a comparison of these motor patterns with a data base of normals (Winter, 1987).

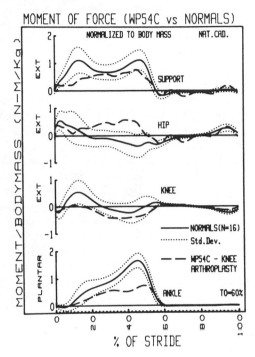

Figure 44.2: Moment-of-force profiles for a knee arthroplasty patient (dashed line) compared with ensemble average of normals (solid line ± 1 s.d. dotted line).

Figure 44.2 presents the comparative results of those moment curves. All extensor moments are shown as positive, and the support moment, M_s, was calculated (Winter, 1980). M_s reflected the algebraic sum of all three joint moments and represents the total extensor or flexor pattern of the lower limb. In normals (solid line) it can be seen to be extensor during most of stance, it becomes flexor at 55% of stride, as this limb unloads during double support. This flexor pattern continues until mid swing when it reverses to become a small extensor pattern during reach. The following changes (or adaptions) are noted. At the site of the surgery there was a flexor moment during all of stance. This can be interpreted as a primary change due to chronic pain at the knee joint due to the original osteoarthritis and any residual pain after the prosthetic knee was implanted. Thus this patient did not use her quadriceps because that would have increased the knee bone-on-bone compressive forces, which prior to surgery would have resulted in severe pain. After surgery the adaption appears to have remained. Because of the lack of quadriceps motor activity to control the knee joint from collapse, a major adaption was evident at the hip. Here we see an above-normal hip extensor pattern for all of stance, which means that her thigh collapse was being controlled from its proximal end; this in turn prevented knee collapse. The major muscles contributing to the hip extensor moment were gluteus maximus and hamstrings. These hamstring muscles then created a smaller flexor moment at the knee, but the hip motor pattern dominated in its control of the knee. A third compensation appears evident at the ankle, where the normally high plantar-flexor moment was drastically reduced. Such a reduction during normal push-off (40-60% stride) would dramatically reduce the normal "piston-like" drive of the ankle upwards and forwards during this dominant energy generating phase (Winter, 1983). It appears that such a reduction would also reduce any knee joint pain and instability during this critical period.

44.3.2 Case Study #2 - Above-Knee Amputee

This patient was a 30-year-old male who was fitted with an above-knee prosthesis. The prosthesis was fitted with a SACH (solid ankle cushioned heel) foot and had a knee mechanism which incorporated an extensor stop (to prevent hyper-extension of the knee during stance) and a spring/damper mechanism to assist during swing. He was analyzed at the University of Waterloo Gait Laboratory using identical data collection protocols as described for *Case Study #1*. In addition to the moment-of-force calculations, the mechanical power generation and absorption at each of the joints were calculated (Winter, 1983). Firstly, we were interested in determining the energy absorption/storage by the knee mechanism and also what adaption, if any, was evident by the hip muscles.

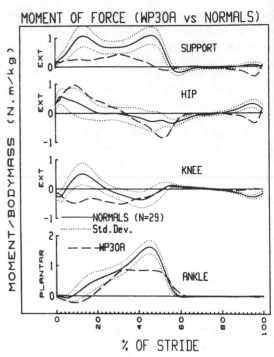

Figure 44.3: Moment-of-force profiles for an above-knee amputee (dashed line) compared with a data-base ensemble average for normals (solid line ± 1 s.d. dotted line). The moment patterns (Nm/kg) are normalized to remove the influence of body mass.

Figures 44.3 and 44.4 present the moments and powers respectively for this amputee compared with the data base of normals. The amputee's patterns are shown by the dashed line. The ankle moment (Figure 44.3) generated by the SACH foot and pylon was not insignificant; it was dorsiflexor for the first 18% of stride and then was plantar-flexor as the amputee moved forward over his prosthetic foot (which reached a flat foot position at 20% of stride). Then during the normal push-off period (40-60%) the plantar-flexor moment plateaued at slightly over 50% of normal. This moment profile is exactly what would be generated by a locked ankle (Winter and Sienko, 1988). The power profiles (Figure 44.4) associated with this moment pattern showed a minor absorption during mid stance which reflects the viscous nature of the foot. However, during the normal push-off period when there is rapid plantar-flexion and the major generation of power (A_2), there is, as expected, no generation.

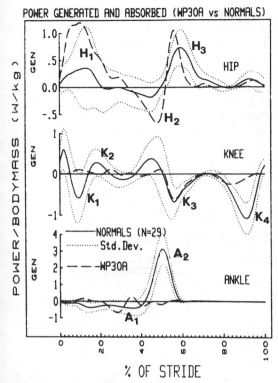

Figure 44.4: Power generation and absorption profile for the above-knee amputee reported in Figure 44.3. (The solid line is the mean and the dotted line ± s.d.)

At the knee and hip we see atypical motor patterns. The hip moment is above-normal extensor for the first half of stance while the knee has a strong flexor moment. This amputee prevents knee collapse by hyperactivity of his hip extensors immediately at heel contact. This locks his knee which is prevented from hyper-extending by an extensor stop (which results in the flexor knee moment). There is a second reason for this adaption, and this is apparent in the power profiles. Because of the total absence of the major push-off power (A_2), he had to adapt by generating at either the hip or the knee. Since there was no generation possible at the knee, his adaption was at the hip. He showed a very large generation by the hip extensors during the first half of stance (H_1); the hip extensors shorten concentrically after HC as the thigh rotates backwards from its initial flexed position. This power burst is referred to as "push-from-behind" and acts to propel H.A.T. and thigh forward. Then during the latter half of stance his hip flexors turn on to reverse the backward rotat-

ing thigh (H_2 during 30-50% of stride) then generate a pull-off burst (H_3) which had approximately the same energy generation as seen for normals (H_3 power peak was higher but its duration was less). Thus this amputee's main adaption was by the hip extensors. During late stance and early swing the knee moment pattern returns to normal; the knee mechanism in the prosthesis now controls the flexion of the knee and absorbs energy (K_3). During late swing a somewhat reduced energy absorption by the knee mechanism occurs. This reduction is due to a slower than normal swing velocity by the prosthetic limb and also the reduced mass of the prosthesis (compared to the intact lower leg and foot).

44.3.3 Case Study #3 - Candidate for Unilateral Total Hip Replacement

Gait assessments of subjects with hip pain have been extremely limited (Murray et al., 1971) and have reported only outcome measures (temporal, distance, and kinematic). No kinetics were reported, nor were possible compensations on the contralateral side. The purpose of a study conducted at the Biomechanics Laboratory in the Department of Physical Education, University of Wisconsin, Madison, was to document the full kinematic and kinetic changes from both ipsilateral and contralateral sides. Six candidates were assessed and one is reported in detail.

Bilateral 16 mm cine at 100 Hz and force platform records were collected for three repeat trials. Each stance period was normalized to 100%, and an ensemble average profile of all three trials was calculated for each of the variables. Joint moments and powers were normalized by dividing by body weight so that they could be compared with normative data (Winter, 1987) for significant differences.

The following significant differences ($p < .05$) were seen when comparing the unaffected side of the six subjects with their affected side: greater knee flexion at weight acceptance, increased hip flexion at heel contact, larger hip flexor moment at heel contact, greater peak ankle plantar-flexor moment during push-off, and increased peak ankle power at push-off.

The selected subject (OZ80) was an 80-year-old male who was observed on the affected side to have a hip hike, a shortened stance phase, a flexed but rigid knee during stance, and a weak push-off. The ankle angle history on the unaffected side fell within the normal range while the affected side has reduced ankle plantar-flexion during push off, probably an adaption to unload the painful leg more quickly. The knee angle history showed the normal side to have a normal profile, but the affected side flexed to 25° during weight acceptance and remained near 25° until push-off, when it flexed through a near-normal range into the swing period. The hip plots showed the unaffected hip to have a slightly larger than normal range of flexion and extension. However, the painful hip remained flexed for all of stance, partially owing to the forward inclination of the trunk.

The kinetic patterns at the ankle are presented in Figure 44.5a-b. The ankle moments (Figure 44.5a) showed a slight reduction in peak plantar-flexion moment. Although this reduction was small, it will be seen to result in a significantly reduced push-off power (Figure 44.5b). Small reductions of about 6% in the peak plantar flexor moment have been observed as resulting in a 50% decrease in push-off power (Winter, 1987) when similar motor patterns of natural and slow walkers were compared. The adaption by the ankle muscles to the painful hip appears to be to reduce the vigor of the push-off (A_2 burst) and thereby would reduce the joint reaction forces generated by the normally vigorous "piston-like" upwards drive by the plantar-flexors.

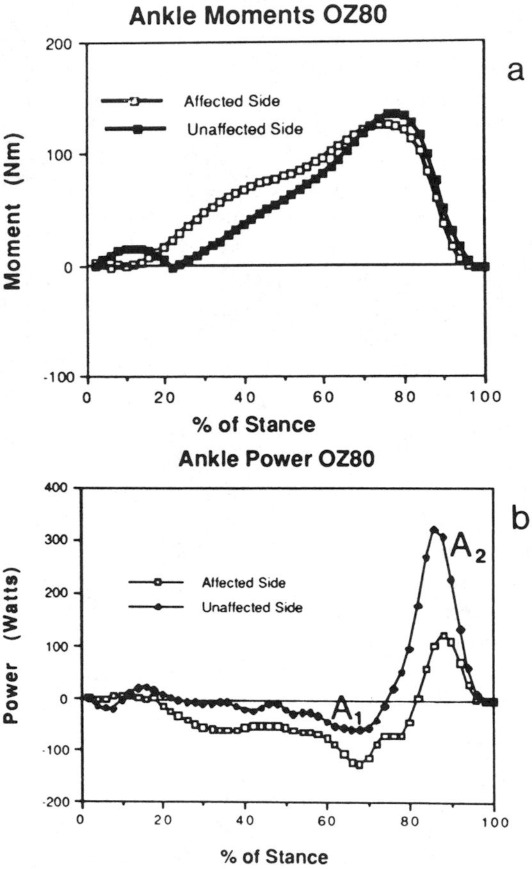

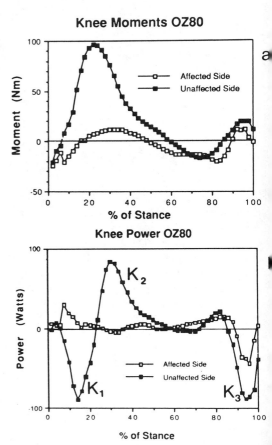

Figure 44.5: *a)* Ankle moments for affected and unaffected side for a patient with a degenerative hip prior to surgery. *b)* Ankle power generation and absorption for this same patient.

Figure 44.6: *a)* Knee moments for affected and unaffected side for the degenerative hip patient reported in Figure 44.5a. *b)* Knee power generation and absorption for this same patient.

The kinetic profiles for the knee are presented in Figure 44.6a–b. A larger than normal extensor moment is evident on the unaffected side as he controls an above-normal knee flexion during weight acceptance (0–25%), and which ultimately causes the knee to extend during most of mid-stance (25–65%). Then during push-off (65–100%) a second knee extensor pattern is seen to control the rapidly flexing knee. The knee power bursts (Figure 44.6b) associated with this knee moment profile are normal: K_1 absorption during weight acceptance as the knee flexed, K_2 generation during mid-stance as the knee extended, and K_3 absorption during push-off. The knee motor patterns on the affected side (Figure 44.6a) showed a moment profile that oscillated from slightly flexor to slightly extensor during the first half of stance, then slightly flexor and extensor again during the latter half of stance. It appears these small moment fluctuations were what was necessary to hold the knee stiff at 25° until late in stance when the knee was allowed to flex as it was unloaded prior to toe off. The power patterns associated with this stiff-legged knee support were almost non-existent. They remained essentially near-zero just until late stance when the small knee extensor moment absorbed a small amount of energy (K_3).

The motor patterns of the hip are seen in Figure 44.7a–b. The unaffected hip had a normal pattern, a hip extensor moment during early stance followed by a flexor pattern during the latter half of stance. The hip powers (Figure 44.7b) associated with this moment pattern are also well within normal: H_1 generation by hip extensors as the hip has an extensor velocity early in stance; H_2 absorption by hip flexors as they decelerate the backward rotated thigh; H_3 pull-off power as the hip flexors contract concentrically to accelerate thigh forward and upward. The painful hip showed a large and dominant extensor moment for virtually all of stance, and this was attributed to the forward lean of the trunk which required a biased extensor moment to maintain such a posture. This strong hip extensor pattern, because of hyperactive hamstrings, also explains why the knee moments on the affected side were not the normally strong extensor pattern. Rather, they had a drastically reduced extensor or even flexor pattern. The hip power associated with this extensor motor pattern (H_1) is a small but prolonged generation of energy as the thigh rotating slowly backwards during all of stance.

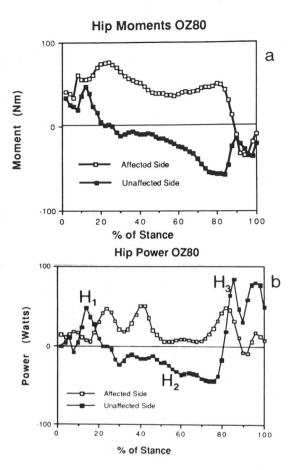

Figure 44.7: *a)* Hip moments for affected and unaffected side for the degenerative hip patient reported in Figure 44.5a. *b)* Hip power generation and absorption for this same patient.

We now summarize the adaptions evident by this subject with a painful hip. The primary adaption appears to be a flexed hip with very limited hip rotation during stance, combined with a forward lean of the trunk. This appears to be the least painful posture and articulation, and he has adjusted his hip motor patterns to accomplish those joint angle changes. A companion adaption of this hip pattern is a similarly flexed knee, also with a limited articulation during stance. The knee moments show an adaption that is compatable with the knee angle changes. Finally, the ankle motor patterns show a small reduction in peak plantarflex moment with a commensurate and large decrease in push-off power. Presumably such a reduction reduces the joint reaction forces at push-off, thus further reducing hip pain.

On the contralateral limb the above-normal knee extensor moment was necessary to adapt for the more rapid unloading of the painful hip. In fact, for this slow walking eighty year old man all of the moment and power profiles on the unaffected side approach or exceed those values seen for younger subjects walking a faster cadence. Thus the general adaption by the unaffected side was a general increase in the kinetic patterns to compensate for the drastic decrease in energy generation and absorption from the painful limb.

44.3.4 Case Study #4 - Stroke Subject with Hemiplegia (Q121)

Subjects who have suffered a stroke characteristically lose the ability to exert individual and graded control over muscle groups of their affected limbs (Knutsson and Richards, 1979). Instead, general extensor or flexor patterns tend to dominate their motor activity. In addition, the force generated by voluntary contractions tends to be less on the affected side, the loss varying with the extent of the damage. Sensory and perceptual problems are also common.

Ms. H. was a 41 year old subject who suffered a stroke seven years prior to the gait analysis in the Gait Laboratory at the School of Rehabilitation Therapy, Queen's University. She was in good health and held a job requiring standing and walking for several hours a day. A slight limp was apparent, and she walked with an average stride velocity of just under 0.80 m/s. Nevertheless, adaptations to compensate for gait deficiencies were apparent in the kinetic analysis.

A planar kinematic and kinetic analysis of Ms. H's usual walking was performed. Three strides from each side were analyzed. The kinematic profiles showed minor variations from normal profiles. The affected hip extended about 10° further than is usual for slow walking speeds, and further than the unaffected side. The affected knee flexed slightly more in early stance but otherwise both appeared normal. At initial contact and for the first 15% of the cycle the affected ankle showed more plantarflexion than is usual, but both the affected and unaffected sides reached normal levels of dorsiflexion at 50% of the cycle. However, the range of plantarflexion was about 5° less than normal at push-off on the affected side. Both hips demonstrated larger than normal hip flexor moments in late stance, at pull-off. The knee extensor moment occurring at the termina-

tion of stance was slightly higher than normal for both sides. Although the ankle profile was normal for the unaffected side, the moment of the affected ankle was reduced to about half the normal value.

The power profiles are shown in Figure 44.8. A comparison was made between the power bursts (W/kg) seen for the three trials for this patient (solid lines) and the normal values reported by Winter (1987). Larger than normal K_3 absorption phases are apparent for the knee on both sides. In contrast, the A_2 ankle power burst was increased by about 50% of its usual values for the unaffected side but reduced to less than a third of normal values for the affected side. At the hip, while H_3 values were normal on the unaffected side, they were increased to more than double their normal amplitudes on the affected side.

The stance phase of the unaffected limb is always increased in these subjects. This increase results in extended support at the end of stance, while the knee is flexing, with greater than normal K_3 absorptive power phases. Although it may assist balance, this represents a net loss of energy to the body compared to normal walking and should not be classified as a positive adaptation, but as a necessary accommodation to the pathological state.

Two positive adaptations are apparent in this pathological gait, one by muscles of the unaffected side and one by those of the affected side. First, the primary deficit appears to be diminished power production by the ankle plantar-flexors of the affected side at push-off. The ankle plantar-flexors of the unaffected side offered some compensation by generating considerably more energy than normal. This increase yielded a net increase of positive work, but, by itself, would result in quite asymmetric work inputs for successive left and right strides. The second adaptation produced positive work near the time of the deficient push-off. Almost concurrent with the weak push-off of the affected side is the event of pull-off by the hip flexors (H_3), which, as seen in Figure 44.8, is increased above normal values on the affected side. This extra power generation was caused by rapid flexion of the hip from an excessively extended position on the affected side, not from an above-average net moment. Thus the total input of work was augmented and, moreover, it was performed at the time that it was most compromised by deficient plantarflexor input.

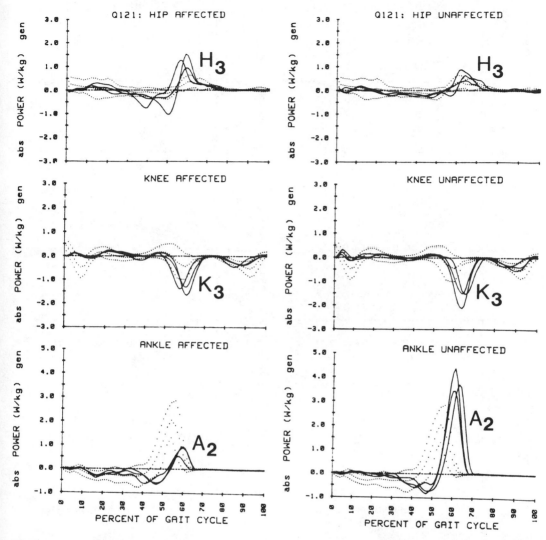

Figure 44.8: Power generation and absorption at the ankle, knee, and hip of both the affected and unaffected limbs of a hemiplegic patient. Solid lines are profiles of three repeat trials: dotted lines are ensemble averaged curves from a data base of normal subjects (average $\pm$ 1 s.d.).

44.3.5 Case Study #5 - Stroke Subject with Hemiplegia (Q070)

Mr. W. was a 57-year-old subject who suffered a stroke two years prior to the gait analysis conducted at the School of Rehabilitation Therapy, Queen's University. He walked with his affected knee held quite stiffly, and his average walking speed was 0.55 m/s. A two-dimensional kinematic and kinetic analysis of Mr. W's usual walking was undertaken. Three strides from each side were analyzed. Joint moment and power profiles appear in Figures 44.9 and 49.10.

The kinematic profiles showed several variations. The range of the affected hip was reduced to about 20° and neither hip extended beyond neutral. The affected knee showed a total range of only 20°. The unaffected knee was slightly flexed at initial contact and did not extend the usual amount prior to heel-off, but normal flexion occurred during swing phase. The affected ankle showed little dorsiflexion during mid-stance and reduced plantar-flexion during push-off. Dorsiflexion on the unaffected side was about 10° greater than normal and no planterflexion beyond neutral occurred.

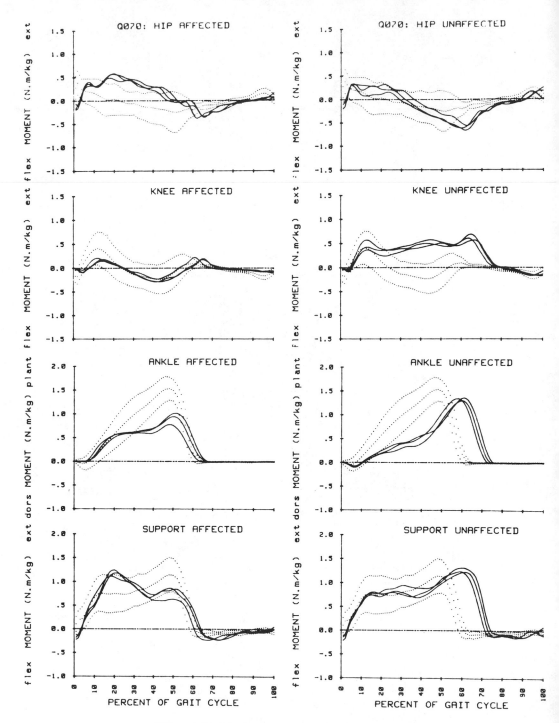

Figure 44.9: Moment-of-force profiles for 3 repeat trials from both limbs of a second hemiplegic patient compared with ensemble averaged curves (dotted lines) from normal subjects.

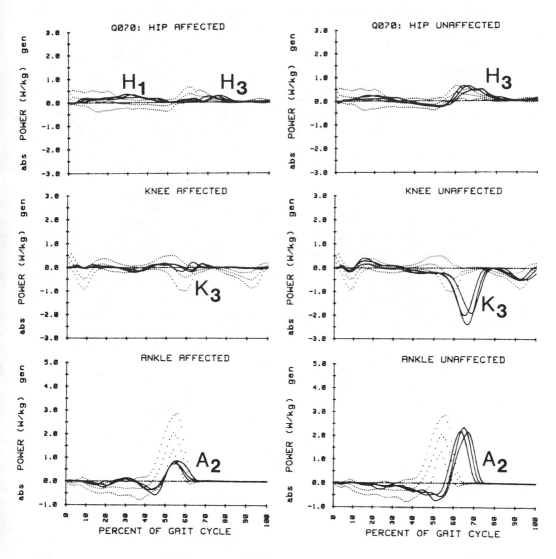

Figure 44.10: Power generation and absorption profiles for both limbs of the hemiplegic patient reported in Figure 44.9.

The extensor moment (Figure 44.9) of the affected hip was larger and longer than normal during stance. An above-normal flexor moment from mid-stance through pull-off was seen on the unaffected side. The knee extensor moment was lower than normal on the affected side, but considerably greater than expected on the unaffected side. The ankle plantarflexor profile was slightly lower than normal on the unaffected side, it was 75% of that of the unaffected side. Both sides showed reasonably normal support moments.

The power profiles (Figure 44.10) of the unaffected side were near normal except for a large K_3 absorption phase at the termination of stance. In contrast, the ankle power (A_2) of the affected side was reduced to less than half of normal values, and virtually no positive or negative work was performed at the knee. The positive burst of pull-off at the hip (H_3) was reduced on the affected side, but a long low-level period of hip extensor generation was evident in stance (H_1).

Despite relatively small variations from normal in the joint kinematics of the unaffected side, the net extensor moments at the hip and knee showed marked ability to "trade off" against each other to produce the normal support moment (Winter, 1980) shown in Figure 44.9. The unusually large knee extensor moment was matched by an unusually large flexor moment at the hip. While the rationale for the overall gait compensations remains unclear (and one suspects that it is far from optimal), the normality of the resulting power patterns attests to the remarkable adaptability of the individual components of the motor system.

The primary deficiency of the affected side appears to be diminished power generation of the ankle and hip, with associated reduction in absorption at the knee on the affected side. The increased stance phase resulted in extended support at the end of stance, while the knee was flexing, and an increased K_3 absorptive power phase for the unaffected side. In contrast, the affected knee did not flex much, and no absorption was evident. In contrast to subject Q121, little energy was lost by the body at this time. This strategy is probably dysfunctional to the gait, however, as the reduced knee flexion did not allow strong hip pull-off (H_3) to occur as an adaptation.

There is evidence of only one positive adaptation in the powers of this pathological gait - by muscles of the affected side. To produce more positive work in stance phase of the affected side, the hip extensors showed a greatly extended period of positive work (H_1). Although the amplitude of the power curve was not above normal, the net generation of energy was augmented by this strategy. In summary, this subject showed poorer compensations for his primary deficiencies than subject Q121.

44.3.6 Case Study # 6 - Cerebral Palsy Right Spastic Hemiplegia

This nine-year-old boy was assessed pre-surgically in the Kinesiology Laboratory of the Newington Children's Hospital, Newington, Conn. His clinical analysis revealed the following: right hip flexion contracture of 20°, tight hamstrings (hip flexed 90° and knee flexed 50°), neutral ankle position with knee fully extended, 10° dorsiflexion with knee flexed 90°, positive test for triceps and quadriceps spasticity. A leg length

discrepancy showed the right leg to be shorter than the left. Raw EMG recordings from four muscle groups on the affected side revealed the following: rectus femoris, hamstrings, tibialis anterior, and triceps surae all lacked normal phasic activity and were generally on all of the time. The same muscles on the non-involved side exhibited normal or near-normal phasic patterns.

The joint kinematics and kinetics for the ankle, knee, and hip are presented in Figure 44.11; the solid line is taken from normals, the long dashed line is left limb, the short dashed line is the affected limb. In addition the pelvic tilt (not plotted) averaged about 18° anteriorly vs the normal data base average of 10°; this coincided with hip angles which on both sides had the normal range but were biased towards greater flexion for all of the stride. Knee flexion at initial contact on both sides was 20° vs the normal 5°. On the affected side this was attributed to tight hamstrings preventing full extension at end of swing while on the unaffected side it was seen as a compensation for the leg length discrepancy. At the ankle, he had a "clonus" pattern on the right side due to the triceps spasticity. This was characterized by rapid dorsiflexion just after initial contact followed by a stretch-reflex induced plantar-flexion, then followed by a small dorsiflexion and another rapid plantar-flexion prior to toe-off. Finally, the triceps spasticity caused a drop foot during swing. The left side had a somewhat reduced form of these alternating patterns but without a drop foot in swing. Pelvic rotation in the coronal plane was within normal limits but the right hip showed above-normal hip adduction in stance as a result of hip abductor weakness. As compensation for this there was circumduction of the left limb to achieve foot clearance during swing. Pelvic rotation in the horizontal plane showed the right side to be posterior to the left during the gait cycle. Also the right foot was externally rotated 20° with respect to the thigh.

The joint kinetics associated with these kinematic and *EMG* profiles is presented in Figure 44.11. Both ankle moments were strongly plantarflexor after initial contact (on the ball of the feet) and exhibited the rapid oscillations characteristic of the "clonus" pattern. The corresponding ankle powers showed a sharp eccentric absorption of energy as a result of the first stretch reflex, followed by a high generation during the rapid

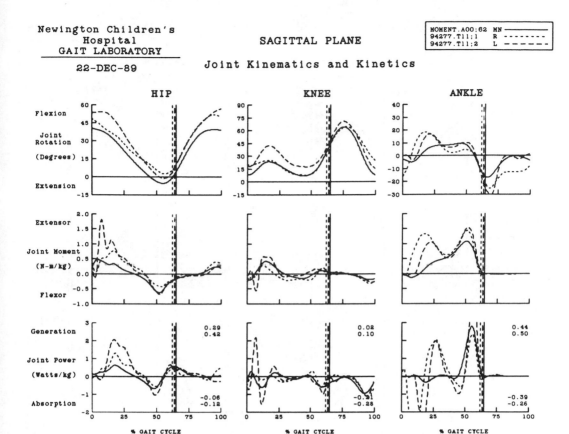

Figure 44.11: Joint kinematics and kinetics for both limbs of a cerebral palsy right hemiplegic patient prior to surgery. Solid line is average curve for normals, short dashed is for the right affected limb, long dashed line is for the left limb.

plantarflexion, then a small absorption and finally a short high generation during the normal push-off. At the knee there appears to be near-normal moments during both stance and swing. At the hip the crouch gait requires a much larger extensor moment during most of stance with an associated increase in hip extensor generation of energy, especially by the left side.

In summary, the primary problems were identified to be on the right side as follows:

(i) triceps surae spasticity and resulting clonus and drop-foot

(ii) tight hamstrings resulting in excessive knee flexion at initial contact

(iii) crouch gait due to hip flexor spasticity

(iv) a leg length discrepancy

(v) excessive external rotation of the foot

The compensations for these primary problems resulted in atypical left limb patterns which are classed as secondary problems.

(i) A crouch pattern on the left to compensate for the leg length discrepancy.

(ii) A circumduction to adapt for the right limb crouch gait.

(iii) The vaulting gait in mid-stance by the plantar-flexors was a compensation to assist in clearance by the right limb.

Surgery was planned for the right side only. The details of the surgery are beyond the scope of this chapter. The biomechanical details of the altered patterns not only on the surgical side but on the previously compensating limb are evident in Figure 44.12. Both right limb (solid line) and left limb (dashed line) kinematics are closer to normal as are the moment of force patterns at all three

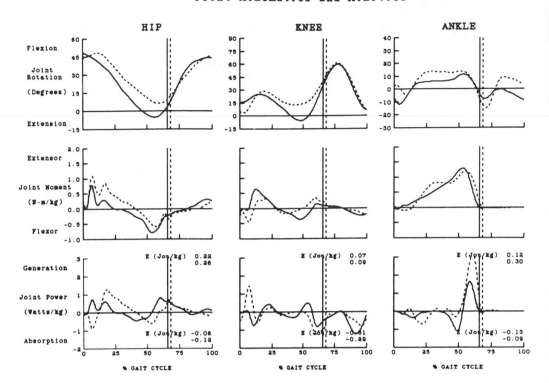

Figure 44.12: Post-surgical kinematics and kinetics
for the cerebral palsy patient reported in Figure 46.11.

limbs. Thus the resultant power profiles are also more normal. As well, not shown in these plots, were improvements in external rotation of the right foot (reduced from 30° to 15°) so that it progressed more normally.

44.4 Conclusions and Future Directions

It is extremely apparent from the biomechanical analyses of gait pathologies that well-defined adaptions by the intact neuromuscular system take place. It is important that these adaptions be recognized for what they are. They are atypical but they should not be considered as pathological. Some of those adaptions may be permanent, as in the case of amputees and joint replacement. However, in cases of major surgery (such as in cerebral palsy) or in long-term therapy it is important not to treat the adaptions (secondary problems) but to treat the primary problems. The plasticity of the *CNS* ensures that with successful post-surgery therapy the atypical adaptions will disappear or be drastically reduced.

The implications of the wide scope of the adaptability is far-reaching as far as data collection and analysis of gait is concerned. Because a wide range of adaptions are possible it is apparent that simultaneous data from both limbs and upper body must be recorded. Also, although *3–D* data would be preferred, a sagittal plus coronal plane analysis yields about 99% of the necessary information. Some researchers may be critical of such a "shotgun" approach in measuring everything. However, they would soon change their bias when they saw how adaptable the *CNS* really is, and when they realized the importance of pinpointing the motor adaptions versus the primary pathology.

References

Bresler, B., Frankel, J.P. (1950). The forces and moments in the leg during level walking. *Trans. ASME*, **72**: 27-36.

Knutsson, E., Richards, C. (1979). Different types of disturbed motor control in gait of hemiplegic patients. *Brain*, **102**: 405-430.

Murray, M.P., Gore, D.R., Clarkson, B.H. (1971). Walking patterns of patients with unilateral hip pain due to osteo-arthritis and avascular necrosis. *J. Bone Jt. Surgery* **53A**: 259-274.

Winter, D.A. (1980). Overall principle of lower limb support during stance phase of gait. *J. Biomech.*, **13**: 923-927.

Winter, D.A. (1983). Energy generation and absorption at the ankle and knee during fast, natural and slow cadences. *Clin. Orthop. & Rel. Res.*, **197**: 147-154.

Winter, D.A. (1984). Kinematic and kinetic patterns in human gait: Variability and compensating effects. *Human Movement Science*, **3**: 51-76, 1984.

Winter, D.A. (1987). Biomechanics and Motor Control of Human Gait. University of Waterloo Press, Waterloo, Ontario.

Winter, D.A., Sidwall, H.G., Hobson, D.A. (1974). Measurement and reduction of noise in kinematics of locomotion. *J. Biomech.* **7**: 157-159, 1974.

Winter, D.A., Sienko, S.E. (1988). Biomechanics of below-knee amputee gait. *J. Biomech.* **21**: 361-367.

Winter, D.A., Yack, H.J. (1987). EMG profiles during normal human walking: Stride-to-stride and inter-subject variability. *EEG and Clin. Neurophysiol.* **67**: 402-411.

CHAPTER 45

Whole Body Movement: Coordination of Arms and Legs in Walking and Running

Richard N. Hinrichs

45.1 Introduction

As Winter et al. so aptly pointed out in Chapter 33, the upper body has not received as much attention as the lower body in locomotion research. It is as if the arms and trunk were not important in the overall picture of gait. A few researchers, however, have wondered why the arms are swung as they are. Do the arms passively react to movements of the shoulders, or are the arms under muscular control? Do the arms serve any useful purpose other than helping to maintain one's balance? This chapter reviews the literature in the area of upper extremity function in locomotion and its relationship to that of the lower extremities. This review is divided into two parts: first, those studies dealing with walking, and second, those dealing with running. The chapter ends with a discussion of current research and future directions in this area of study.

45.2 Walking

45.2.1 Early Speculation

Nearly all references to upper extremity function in walking prior to 1965 were speculative in nature. The issue was the underlying mechanism of the arm swing. Do the arms passively react to the movements at the shoulders or are they under muscular control?

Gerdy (1829) believed that the arm swing was primarily a reaction to the rotation of the shoulders and trunk. He added, however, that sometimes it seemed that the movements of the arms are increased by the involuntary action of the biceps brachii, the pectoralis major, and one part of the deltoid. Whatever methods Gerdy used to arrive at this opinion were not disclosed. In 1867, Duchenne (English translation, 1949) provided the first real evidence of muscular control of the arm swing. He based his conclusions on clinical observations of patients in which the oscillatory motions of the upper extremities during walking were abolished following atrophy of the deltoid muscle. He believed the forward swing to be caused by action of the anterior deltoid and the backswing by the posterior deltoid. Gravity, he contended, played only a minor role. Du Bois-Reymond (1909) spoke of the arms as serving to protect the upper body from excessive rotation and lateral motion, and that the arm swing was partly passive and partly active in nature. He did not, however, speculate as to which muscles might be responsible for the arm swing.

Although Elftman (1939; see also Section 45.2.2) provided some further evidence in favor of the active role that the muscles play in the arm swing, Morton and Fuller (1952) evidently were not familiar with either Elftman's or Duchenne's work and continued to perpetuate in their textbook on locomotion the notion that the arm swing was primarily a passive movement. They argued that because it requires muscular effort to swing the arms while standing still and likewise to keep the arms from swinging while walking, it must follow that the arm swing arises without muscular contraction. Like Gerdy (1829), Morton and Fuller did recognize, however, the possible role of the muscles in increasing the magnitude of the arm swing as it is needed.

Multiple Muscle Systems: Biomechanics and Movement Organization
J.M. Winters and S.L-Y. Woo (ed.), © 1990 Springer-Verlag

It is clear from the early speculation on the nature of the arm swing in walking that there were considerable differences in opinion as to the nature of the arm swing in walking. Most acknowledged, however, that the muscles might play a role in the arm swing. It was clearly time for some scientific research in this area.

45.2.2 Kinematic Studies

By the mid-1960s scientific investigations into arm swing were finally underway. In a two-dimensional kinematic study, Murray et al. (1967) investigated the ranges of normal variability in upper limb displacement patterns of "normal" subjects. Thirty subjects, ranging in age from 21 to 66 years, walked overground at a "free" pace (averaging 114 steps/min and 1.54 m/s) and a "fast" pace (averaging 136 steps/min and 2.14 m/s). Successive positions of reflective targets placed on the arms were recorded using a camera with an open shutter and a stroboscope flashing at 20 Hz.

The investigators plotted the rotational patterns at the shoulder and elbow joints as functions of time. The mean patterns reported by Murray et al. of shoulder and elbow rotation are shown in Figure 45.1. The investigators found the arm swing to be the most variable (between subjects) of the 20 gait components they had measured. The main source of variability was the amplitude of the arm swing. The temporal patterns were quite consistent between subjects.

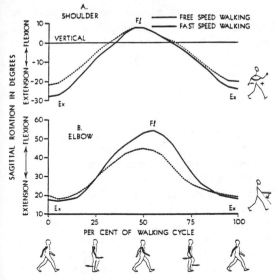

Figure 45.1: Mean patterns of shoulder and elbow rotation during walking as functions of time. (From Murray et al., 1967; reprinted with permission from Physical Therapy, copyright 1967, American Physical Therapy Association).)

Within subjects, the arm swing was highly reproducible, although some differences existed between right and left arms. These differences were not found to be related to hand dominance. Other comparisons included walking speed (greater arm swing amplitude at the fast speed than the free speed), stature (taller subjects were found to have a significantly greater total excursion at the elbow joint while walking at the free speed), and age (not a significant influence on arm swing).

Finally, Murray et al. calculated the path of the center of mass (CM) of the combined upper extremities and found that it oscillates vertically through two peaks and two valleys within each walking cycle. The peaks occurred when the arms were outstretched forward and backward, and the valleys occurred when the arms were in relatively vertical positions passing the trunk. They compared this to the results from a previous study by Fischer (1899) for the vertical oscillation of the entire body CM during walking and noted that the vertical oscillation of the arms would be 180° out of phase with that of the whole body. This suggests that the arm swing would tend to decrease the total vertical excursion of the whole body CM from what it would be if the arms did not swing. They suggested that this may lead to a decrease in energy expenditure in gait.

Hinrichs and Cavanagh (1981) provided some additional data on the two-dimensional kinematics of the arm swing in walking (their kinetic and EMG results are reviewed in Sections 45.2.3 and 45.2.4, respectively). These investigators used five subjects, each walking barefoot at two different speeds on a treadmill, a "slow" speed (0.9 m/s) and a "medium" speed (1.2 m/s). Films were taken from the left side of the body as the subjects walked on the treadmill. Only the left side of the body was analyzed from film. The movements of the right side were assumed to be 50% out of phase with the left.

Of particular relevance to this kinematics discussion are the investigators' findings on the vertical oscillations of the CM of the arms and the CM of the rest of the body (termed body-minus-arms, BMA). The results from a single subject are shown in Figure 45.2. Contrary to the findings of Murray et al. (1967) the oscillations of the CM of the arms were not found to be 180° out of phase with those of the BMA. In fact, at the slow speed, the two were rather close to being in phase with each other. At the faster speed, however, the two were further out of phase but not as much as 180° out of phase. These results could be explained as follows: The position of the *CM* of the arms is influenced by *(1)* the position of the shoulders and *(2)* the positions of the arm segments relative to the shoulders. The shoulders could be expected to oscillate up and down in a similar manner as the *CM* of the *BMA*. Added to this is the movement of the arms relative to the shoulders. It was this relative term that Hinrichs and Cavanagh found to be close to 180° out of phase with the *CM* of the *BMA*. The sum of these two terms produce the absolute movements of the *CM* of the arms. At the slow speed the arms *CM* seemed to follow the *CM* of the *BMA*, suggesting that the movements of the shoulders dominated the movements of the arms *CM* relative to the shoulders. At the faster speed, however, a more vigorous arm swing produced an oscillation of the arms *CM* which was substantially out of phase with the *CM* of the *BMA*. This would tend to reduce the magnitude of the vertical oscillations of the body *CM*.

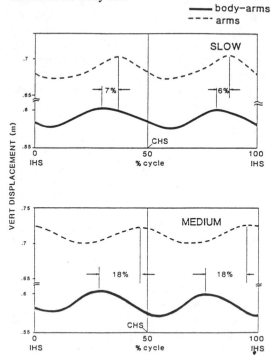

Figure 45.2: The results of Hinrichs and Cavanagh (1981) showing the vertical oscillations of the centers of mass of the arms and the body-minus-arms for a single subject walking at two different speeds, "slow" −0.9 m/s, and "medium" −1.2 m/s.

These results indicate that the picture may not be as clear-cut as Murray et al. (1967) suggested. However, Murray et al. do not clearly state how they calculated the *CM* of the arms, simply that it was calculated using the mean body weight and mean arm segment rotational patterns of their 30 subjects. They may have ignored the vertical motion of the shoulders themselves, giving rise to a calculation of the *CM* of the arms relative to the shoulders. If this were true, their results would be totally consistent with those of Hinrichs and Cavanagh.

At slow walking speeds the arms seem to do very little to reduce the vertical oscillation of the body *CM*. As the magnitude of the arm swing increases at faster walking speeds, however, the arms do appear to be capable of reducing the vertical oscillation of the body *CM*.

45.2.3 Kinetic Studies

The first study providing a kinetic analysis of the arm swing in walking was by Elftman (1939). This study was also the first major experimental study to be done on arm swing. His purpose was to investigate whether the arms react as pendulums to the movements of their points of attachment or whether the action is of muscular origin and plays some part in the integrated locomotion process.

Using the three-dimensional walking data of Braune and Fischer (1895) and Fischer (1899) (obtained from a single subject), Elftman calculated the linear and angular velocities and accelerations for all parts of the body during the interval in which the right foot was in contact with the ground. From this, he was able to compute the three components of each of *(1)* the angular momentum of each body segment relative to the body *CM*, *(2)* the net moments at the shoulder and elbow joints, and *(3)* the mechanical power at these joints.

The results of the angular momentum analysis are shown in Figure 45.3. Elftman retained Braune and Fischer's left-hand coordinate system which is defined as follows: *X*-forward, *Y*-to the right, and *Z*-up. Of particular interest is the vertical component of the angular momentum (H_z). With the exception of a short interval preceding heel strike, the H_z of the arms was approximately equal in magnitude and opposite in direction to the H_z of the rest of the body, and resulted in a net H_z of the whole body of approximately zero. He concluded that "It is therefore possible for the legs to go through the movements necessary for walking, without imparting marked rotation to the body as a whole" (p. 531). The arms also appear to make the changes which do occur in body rotation more gradual. About the X and Y axes (forward and to the right, respectively), the angular momentum of the arms was small compared to the rest of the body.

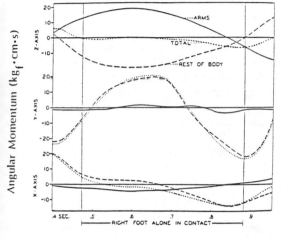

Figure 45.3: Three components of angular momentum about the body center of mass for the arms, the rest of the body, and the whole body in walking. (See text for axis definitions.) (From Elftman, 1939; reprinted by permission.)

The results of Elftman's analysis of the net moments at the shoulder are shown in Figure 45.4. About the Y axis a positive moment indicates a moment tending to flex the shoulder, a negative moment tends to extend. There is a net flexor moment as the right foot approaches heel strike. This is also true of the left arm during the period preceding left heel strike. Note that the shoulder is in maximum extension at ipsilateral heel strike (*IHS*). The flexor moment thus would tend to slow the backward swing and initiate the forward swing. In the period preceding contralateral heel strike (*CHS* − left heel strike for the right arm and right heel strike for the left arm), there is a net extensor moment. At this time, the shoulder is approaching maximum flexion. Thus, the moment would tend to slow the forward swing and initiate the backward swing.

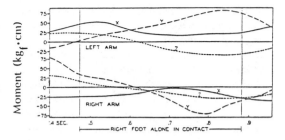

Figure 45.4: The three components of the net joint moment acting at the shoulder of each arm in walking. (See text for axis definitions.) (From E.F. Elftman, 1939.)

Through his analysis of the net moments at the shoulder, he concluded that because the muscles are exerting "a considerable torque" about the shoulder, "the swinging of the arms is not an example of pendulum action, but is brought about in large part by the action of the muscles" (p. 532). He neglected to consider, however, that net joint moments can result from sources other than muscular action, for example, from ligaments and from sources of friction in and around the joint. This author feels also that the words "considerable torque" may be a bit strong in describing the magnitudes of the net moments at the shoulder. Actually, they are relatively small, having peaks less than 8 N-m[1]. The relative contributions of passive and active sources to these net moments at the shoulder remain, of course, unknown.

Overall, however, Elftman's study was an excellent investigation, even by today's standards. A three-dimensional kinetic analysis requires a much greater degree of mathematical sophistication than is found in a strictly planar analysis. The nature of the arm swing clearly cannot be fully understood by studying the motion of the arms in two dimensions. Although there were many questions still left unanswered, and despite the use of only a single subject, Elftman's study was way ahead of its time judging from the complete lack of further scientific scrutiny in this area for the next 25 years.

---

[1] To put the magnitudes of these moments into perspective, one can estimate the moment required of the shoulder musculature to hold an out-stretched arm in a horizontal position. Using the mean anthropometric and *BSP* data of Clauser et al. (1969), the estimated value is 9.3 N-m. If a 2 kg mass is held in the hand, this would increase to 22.2 N-m.

Hinrichs and Cavanagh (1981) also presented results on the net moment in the sagittal plane at the shoulder and elbow joints during walking at two different speeds. Results for a typical subject walking at 1.2 m/s are presented in Figure 45.5. The moments about the shoulder joint were in close agreement with the Y component of the moments presented by Elftman. There is a net moment tending to flex the shoulder at *IHS* (when the shoulder is extended) changing to a net extension moment at *CHS* (when the shoulder is flexed). The elbow showed a similar pattern, but with moments generally of a lesser magnitude than at the shoulder. Also shown in Figure 45.5 are moment reference levels for the shoulder and elbow. These indicate approximately how much moment would be required to hold the entire arm (in the case of the shoulder) or just the forearm (in the case of the elbow) horizontally while standing. These were calculated using the mean anthropometric and *BSP* data of Clauser et al. (1969).

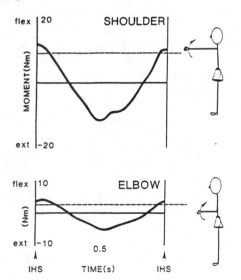

Figure 45.5: The results of Hinrichs and Cavanagh (1981) showing the net joint moments at the shoulder and elbow for a typical subject walking at 1.2 m/s. (See text for explanation of reference levels.)

45.2.4 EMG Studies

Four studies were found which have investigated upper extremity *EMG* during gait. Fernandez-Ballesteros et al. (1965) used telemetered *EMG* with needle electrodes in 23 normal subjects and two patients with upper extremity nerve damage. Subjects walked overground at their own preferred cadence averaging 80 steps/min. Hogue (1969) also used telemetered *EMG* with wire and needle electrodes in 15 subjects walking overground at various pre-set cadences and inclines. Jackson et al. (1978) collected surface *EMG* in eight subjects who walked at cadences ranging from 80 to 150 steps/min. Hinrichs and Cavanagh (1981) also used surface *EMG* in five subjects during barefoot treadmill walking at two fixed speeds. Table 45.1 summarizes the results of the four *EMG* studies on arm swing. The author has attempted to classify the magnitudes of the *EMG* activity using a five-point scheme: 0 = no activity, − = negligible activity, + = slight activity, ++ = moderate activity, and +++ = pronounced activity. A blank indicates that the muscle was not monitored by that particular investigator.

The Fernandez-Ballesteros et al. study was the most comprehensive of the four studies. They found that certain upper extremity muscles were indeed active during the walking cycle. In the normal subjects, the supraspinatus and upper trapezius muscles were found to be active throughout the entire walking cycle with the exception of brief "silent periods" around *IHS* and *CHS*. The posterior and middle deltoids were found to be active near the end of the forward swing and throughout the backward swing. Activity was also found in the teres major and upper latissimus dorsi during the first half of the backward swing coming on again near the end of the backward swing and persisting throughout the first half of the forward swing. The most striking result was that there was a complete absence of activity in the anterior deltoid and the clavicular head of the pectoralis major, both of which are shoulder flexors. Additionally, there was no activity in the triceps, biceps, and sternal head of the pectoralis major. (Jackson et al., 1978, did find triceps activity, but only at the faster cadences.) Of the muscles that were found to be active, this activity was small when compared to that obtained during a maximum voluntary contraction (*MVC*). The average activity was reported to range from 4-15% of that obtained during an *MVC*.

Fernandez-Ballesteros et al. also investigated some variations of arm swing. Four of the subjects performed "free swinging" pendular movements of the arms while standing and while seated. When the arm excursion was less than 30°, activity was confined to the posterior deltoid during the backward swing. No muscles were active during the forward swing, indicating a passive forward swing. When the excursion was greater than 30° (greater than that of normal walking), activity occurred in a number of other muscles as well. These included the anterior deltoid, the biceps, and the clavicular head of the pectoralis major, all active during the forward swing.

To see if muscle activity persisted when the arm swing was inhibited, the investigators had five subjects walk *(1)* with their arms loosely bound to their sides, and *(2)* unrestrained but voluntarily holding their arms at their sides. When the arms were restrained, some step-related activity similar to that in unrestrained walking was found in the posterior deltoid, the teres major, and the upper latissimus dorsi, providing evidence of an underlying motor program. Because the arms were only loosely bound to the sides in this condition, however, there may have been some movement of

the arms during walking. This, however, was not addressed by the authors. When the arms were voluntarily held at the sides while walking, they found that all muscle activity ceased. The authors did not discuss this very surprising result. This author would have expected to find some muscular activity present to keep the arms from swinging. This result implies that there is no passive pendular component to the arm swing; that the muscles are completely responsible for the arm action. Given the inertia of the system, this seems unlikely.

The posterior deltoid appears to arrest the forward swing and initiate the backward swing. This finding is highlighted by the results of one of their patients with recent nerve damage. In this subject, the three heads of the deltoid were paralyzed. There was still a forward swing, but in the backward swing, the arm did not extend beyond the trunk. Because the triceps were not recruited to replace the posterior deltoid, this provided further evidence of the existence of a motor program which probably includes inhibitory impulses to certain muscles. Activity in the middle deltoid and trapezius muscles suggests the need to slightly

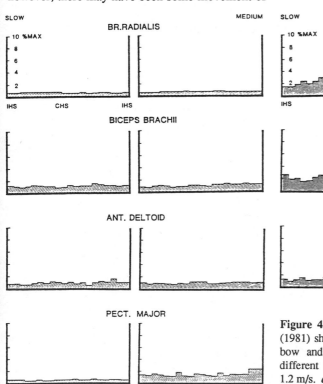

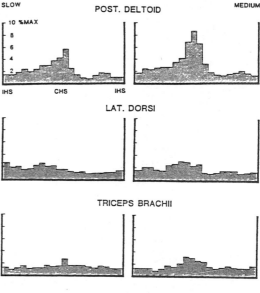

Figure 45.6: The results of Hinrichs and Cavanagh (1981) showing the mean *EMG* patters of selected elbow and shoulder muscles during walking at two different speeds, "slow" -- 0.9 m/s, and "medium" -- 1.2 m/s. *a.* Flexors, *b.* Extensors.

abduct the arms during the swing so that they clear the trunk.

Hinrichs and Cavanagh (1981) reported results similar to those found in Fernandez-Ballesteros et al. The Hinrichs and Cavanagh results are shown in Figure 45.6. Activity was negligible throughout the walking cycle for all the flexor muscles investigated at both speeds. The extensors, on the the other hand, showed activity occurring predominantly at *CHS*. The posterior deltoid showed the most activity with peaks of 6% and 9% of *MVC* at the slow and medium speeds, respectively. The latissimus dorsi and triceps showed slight activity (2-4% *MVC*). The triceps showed very little activity at the slow speed but seemed to come into play more at the medium speed. This result is consistent with the results of Jackson et al. (1978) who found the triceps active only at the faster cadences.

Overall, it appears that there are at least four muscles which show a moderate amount of activity during the arm swing in walking (see Table 45.1). The posterior and middle deltoid and the latissimus dorsi/teres major complex were shown to have at least moderate activity in every study. The triceps appears to be active primarily at

cadences greater than 120 steps/min. Although measured in only one study, the supraspinatus (an abductor of the shoulder) also had moderate activity.

The results of all four studies suggest that the backward swing is initiated by the posterior deltoid, the latissimus dorsi, and the teres major. The posterior fibers of the middle deltoid may assist in the extension (Rasch & Burke, 1973). There is some confusion, however, as to which muscles initiate the forward swing. The anterior deltoid does not appear to be involved except, perhaps, at high cadences. The forward swing may actually be primarily a passive movement at low cadences. Both flexion and extension at the elbow appear to be passive at low cadences. The triceps and brachioradialis may modify this motion at high cadences.

The concept of passive movements in the arm swing has not been adequately discussed in the literature. Besides gravity, which would always tend to seek the lowest position of the arm segments (as it would a simple pendulum), there are other sources of passive force affecting the arm swing. Among these are *(1)* resistance provided

Table 45.1: Summary of Results of *EMG* Studies on Arm Swing in Walking.

Muscle	Fernandez-Ballesteros 80 spm[1]	Hogue 70-120 spm	Jackson et al.[2] <120 spm	Jackson et al.[2] >120 spm	Hinrichs and Cavanagh[3] slow	Hinrichs and Cavanagh[3] medium
Posterior Deltoid	+++	+++	+++	+++	++	+++
Middle Deltoid	+++	+++	++	++		
Anterior Deltoid	0	−	+	+	−	−
Upper Lat. Dorsi	++		++	+++	+	+
Teres Major	++	++	++	+++		
Pect. Maj.--Sternal	0					
Pect. Maj.--Clavic.	0		0	−	0	−
Trapezius	+	−				
Supraspinatus	++					
Infraspinatus	0	−				
Rhomboids	−					
Biceps Brachii	0	0	0	0	−	−
Triceps Brachii	0	0	−	++	−	+
Brachialis			0	−		
Brachioradialis			−	+	0	0
Flex. Carpi Uln.	−					
Flex. Carpi Rad. Br.	−					

Key: " " = muscle not tested
"0" = no muscle activity
"−" = negligible activity
"+" = slight activity
"++" = moderate activity
"+++" = strong activity

Notes:
(1) Cadence: "spm" = steps/min
(2) Jackson et al. considered the lat. dorsi and teres major as one muscle.
(3) slow = 105 spm, 0.9 m/s, medium = 118 spm, 1.2 m/s

by the soft or bony structures in and around the joints and *(2)* movements of the shoulder joints themselves caused by motion of the trunk.

During the initial part of the forward swing, for example, the flexor moment at the elbow necessary to keep the elbow from extending could arise from the bony structure of the elbow as it approaches the fully extended position. A similar situation could occur at the shoulder. A net flexor moment could occur near the end of the backswing simply from a restriction on the range of motion of the shoulder in hyperextension.

There is a limitation that was not mentioned in any of the *EMG* studies on arm swing; that is, the unnatural situation created by having electrodes on the arms of the subjects. The electrodes may have drawn attention to the arms and caused the subjects to think about swinging their arms. This, in turn, may have altered both the natural pattern of muscular activity and the kinematics of the swing.

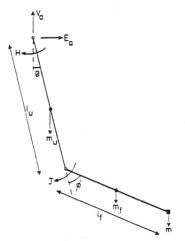

Figure 45.7: Mathematical model of the upper limb used to predict the kinematic features of the arm swing in walking. (From Jackson et al., 1978; reprinted with permission from Journal of Biomechanics, copyright 1978, Pergamon Press.)

45.2.5 Mathematical Modeling

The *EMG* data collected by Jackson et al. (1978), mentioned in the previous section, were part of a larger study intended to delineate a two-dimensional mathematical model of the arm swing in walking. The rotation of a two segment model, which is shown in Figure 45.7, was controlled by muscle moments (derived from *EMG*) at the shoulder and elbow joints (H, J), and the horizontal and vertical accelerations of the shoulder joint

(E_a, V_a). Also included in the model were passive resistive moments produced by elastic and viscous stiffness in the joints and surrounding structures and by the locking of the elbow at full extension. *EMG* was used to estimate the muscle moments driving the system.

The vertical and horizontal accelerations at the shoulder were measured on six subjects using two accelerometers placed at right angles to each other on an aluminum block. The block was held over the acromion with velcro straps attached around the trunk. The overall pattern was the same for both the vertical and horizontal accelerations and was consistent for all subjects. It consisted of two positive and two negative peaks per cycle. The positive peaks (representing upward and forward acceleration) occurred slightly after each heel strike. The negative peaks occurred between heel strikes. By mounting accelerometers on the left and right shoulders on one subject, the investigators were able to demonstrate that the horizontal acceleration of the shoulder was primarily due to a horizontal acceleration of the whole body, and that the angular acceleration of the body about its long axis contributed only a minor component of the horizontal acceleration at the shoulder.

The model was first used to predict what the arm swing would look like without input from the muscles. The results showed that the angular movements at both the shoulder and elbow were very small and were ragged. There was no recognizable arm swing. If the model is correct, this result might suggest that there is very little, if any, passive pendular action of the arms during walking.

It would have been convenient, however, if Jackson et al. had done some further investigation along this line. Although all the evidence points primarily to the posterior deltoid as being responsible for the backswing, there is no consensus as to which muscles, if any, produce the forward swing. Perhaps the oscillations are forced in only one direction and the forward swing is a passive pendular reaction to the backswing. The investigators could have simulated this by ignoring all *EMG* activity from muscles which might contribute to the forward swing (e.g., the anterior deltoid), and seeing if the forward swing still occurred.

Fernandez-Ballesteros et al. (1965) suggested that the latissimus dorsi, acting as an internal rotator of the upper arm, may contribute to the forward swing. By omitting the moment of the latissimus dorsi around the time of the forward swing and seeing its effect on the arm swing, the contribution of the latissimus dorsi to the forward swing might be revealed. It is unlikely, however, that this two-dimensional model could help in this case. Depending on how the three dimensional geometry of muscle insertions was reduced to two dimensions (something that was not reported by the authors), the latissimus dorsi would probably only be allowed to exert an extensor moment at the shoulder. By the same token, the primary role of the middle deltoid (i.e., abduction) could never be assessed in this two-dimensional model. Given all its limitations, though, the model of Jackson et al. was able to predict the main kinematic features of the arm swing as shown by Murray et al. (1967).

45.2.6 Summary of Walking Literature

From this review of the related literature, it can be seen that there is still much to be resolved concerning the function of the upper extremities in walking. Although it has been established that the arm swing in walking is under muscular control, the oscillations appear to be forced in only one direction, namely the backward swing. An analogy would be a child on a swing being pushed from just one direction.

There appears to be a motor program supplying the neural control of the arms swing. The phasic activity primarily in the posterior and middle deltoid, latissimus dorsi, and teres major is integrated into the overall pattern of locomotion. The arms swing in virtually exact opposition to the legs, thereby counteracting excessive rotation of the body produced by the striding of the legs. Elftman (1939) confirmed this by demonstrating the ability of the arms to balance the angular momentum of the rest of the body about the vertical axis. In addition, the arm swing may serve to reduce the vertical excursion of the body CM in walking, thereby possibly reducing energy cost.

45.3 Running

Although running differs from walking in many ways, the coordination of arms and legs is similar in both. The fact that the arms and legs act in opposition in both forms of locomotion suggests that angular momentum balance about the vertical axis is served in running as well as walking. The issue of the vertical oscillations of the CM, however, is not as clear, since this is where walking is different than running. In walking the body CM is highest in mid stance, whereas in running it is lowest. The effect of the arm swing on this vertical oscillation in running may be different than in walking. This section of the chapter briefly reviews the relevant literature on the coordination of arms and legs in running.

45.3.1. Speculation

There has been very little literature written on the function of the arms in running, although some of what has been written dates back more than 2300 years. The first reference to arm swing in the literature was written by Aristotle (English translation, 1961). In demonstrating a remarkable understanding of what would come to be known as Newton's laws of motion, Aristotle commented on the importance of swinging the arms in running:

> ... the animal that moves makes its change of position by pressing against that which is beneath it. . . Hence athletes jump farther if they have the weights in their hands than if they have not, and runners run faster if they swing their arms; for in the extension of the arms there is a kind of leaning upon the hands and wrists. (p. 489)

Exactly why do runners swing their arms? Do the arms some how propel the runner forward as Aristotle suggested? Can you actually press against your own hands and wrists? Or, as Mann (1981) and Mann and Herman (1985) have suggested, are the arms are only used for keeping one's balance? Recent research by the author has shed some light on this topic. The following represents a summary of a series of articles by the author. This research has focused on three major aspects: *1) EMG* and net joint moment considerations (Hinrichs, 1985); *2)* center of mass and propulsion considerations (Hinrichs, Cavanagh, & Williams, 1987); and *3)* angular momentum considerations (Hinrichs, Cavanagh, & Williams, 1983; Hinrichs, 1987). Because this work has recently been reviewed elsewhere (Hinrichs, 1990), this section will include only a short summary of the author's work (see Hinrichs (1990) for an expanded review).

45.3.2 The Author's Research

Hinrichs studied ten recreational distance runners and filmed them three-dimensionally from four cameras while each ran on a treadmill at three speeds (3.8 m/s, 4.5 m/s, and 5.4 m/s). Simultaneous *EMG*'s were recorded from eight upper extremity and trunk muscles with half of the subjects wearing the electrodes on the right side and half on the left side. The *EMG*'s were synchronized with the film data by means of an insole footswitch device. The raw *3D* kinematic data were smoothed and entered into a computer program which calculated the following variables:

1) The paths of the *CM* of the arms, the body-minus-arms, and the whole body over the running cycle.

2) The contributions of the arms, head and trunk, and legs to the changes in total-body linear momentum over each contact phase. These were called "lift" and "drive," respectively.

3) The angular momentum of each body segment and various systems of segments about the body *CM*.

4) The net moments at the shoulder and elbow joints.

The results revealed several interesting aspects about the coordination of arm and legs in running. First, the arms were found to contribute very little to angular momentum of the body about the mediolateral (*ML*) and anteroposterior (*AP*) axes. There was nothing found to indicate that the arms are able to "balance" the legs in any fashion about these two axes. About the vertical axis, however, angular momentum in the upper body (arms, head, and upper trunk) was found to offset the angular momentum of the lower body (legs and lower trunk), leaving only a small amount of total-body angular momentum about this axis (see Figure 45.8). This result, which was similar to that found by Elftman (1939) for walking, suggests that the upper and lower body "torque" off each other about the vertical axis. In running, the legs can rapidly change their vertical angular momentum during the airborne phase by receiving an angular impulse from the upper body. Without this impulse, a runner could only "leap" from one foot to another, waiting until the next foot contact to begin moving the swing leg through. The runner's legs would have to rely on the ground as the source of angular impulse.

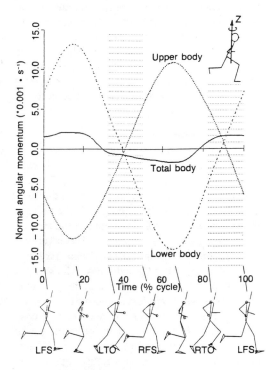

Figure 45.8: Mean curves for the vertical component of angular momentum (Hz) of the upper body, lower body, and whole body for running at 4.5 m/s. (From Hinrichs, 1987. Reprinted with permission from International Journal of Sport Biomechanics, copyright 1987, Human Kinetics.)

The arms were also found to make a small but meaningful contribution (5–10%) to the vertical linear impulse on the body during each contact phase and hence provided lift. The arms' contribution to lift increased as running speed increased. Somewhat unexpected was the finding that the arms contributed nothing to the forward propulsion of the body (drive). Instead, the arms were found to reduce the fluctuations in the velocity of the runner, both in the *ML* and *AP* directions. The arms were found to cross over slightly in front of the body, rather than swinging directly forward and back. While believed by some to be bad form in running, this crossover was actually found to be beneficial in that it tended to reduce the side-to-side motion of the body center of mass and hence promoted a more constant forward velocity.

Somewhat surprisingly, the *EMG* activity in the shoulder and elbow musculature showed moderate to strong activity in all the muscles sampled (three deltoids, latissimus dorsi, triceps, pectoralis major, biceps, and brachioradialis). This activity increased as running speed increased. As in walking, the posterior deltoid showed the strongest activity. The activity was substantially higher in running, however, reaching a peak level of 60% of maximum voluntary contraction at the fastest speed.

45.3.3 Other Research

Little else has been done on the mechanics of the arm action in running. Only two additional studies come to mind. In one, Williams (1980) reported a relationship between arm action and energy cost in distance running. Williams used wrist excursion as a global measure of the amount of arm swing used by a runner. Those runners who were more economical (i.e., who used less oxygen per minute per kg body mass) showed less excursion of the wrists over the running cycle than those runners who were less economical.

In another study which assessed energy cost, Claremont and Hall (1988) investigated the use of hand and ankle weights in running. The authors reported that carrying hand weights significantly reduced peak angular velocities and excursions of the upper arm and forearm compared to the no weight condition. This result is not surprising. The more massive arms would not have to swing as vigorously to generate an equal amount of angular momentum about the vertical axis.

45.3.4 Summary of Running Literature

It is interesting that what appear to be the two main functions of the arms in running both involve the vertical axis. Through their acceleration upward relative to the trunk, the arms help the legs in propelling the body upward (lift). Through their forward and backward swings, the arms (with the help of the upper trunk) provide the vast majority of the torque about the vertical axis needed to put the legs through their alternating patterns of stance and swing.

The role of the arms in the forward propulsion of the body has not been well defined. Based on the current definition of drive (that is, the forward acceleration of both arms combined relative to the trunk), the arms generally do not contribute to the forward propulsion of the body. Hopper (1964) had some early insight into this aspect of running. He basically said that at "constant speed" running, the main aim of the runner is not to speed up but rather to get off the ground. The arms, by providing lift, help the runner achieve this goal. Whether the arms contribute to the forward propulsion of the body in situations where the runner is speeding up is unknown. This is a topic for further research.

Secondary functions of the arm swing include the reduction of excursions of the body CM from side to side and, at least in the case of "constant speed" running studied here, in the direction of progression. Thus the arms help the runner achieve a more constant horizontal velocity which could lead to a reduction in energy cost.

45.4 Future Directions

There is much work to be done on the coordination of arms and legs in walking and running. Perhaps the most intriguing is the topic of energy cost. The link between arm action and energy cost in walking has not been established, but only speculated upon. Williams (1980) reported that runners who were more economical in terms of oxygen consumption tended to use less vigorous arm swings. Thus there may be a compromise between mechanical and physiological factors. A vigorous arm swing may provide more lift yet require more energy to execute. Further study is warranted in the area of uphill running where lift is of utmost importance.

Another topic for future research is in the area of sprinting. Some sprinters such as Carl Lewis demonstrate what is referred to as the "classic style" in which each arm swings nearly directly forward and back with little or no crossover. Perhaps this classic style has an advantage in sprinting that we don't understand yet. Most distance runners, however, do not adopt this form.

It may require more energy to swing the arms this way. The crossover of the arms in front of the chest, in fact, seems to be advantageous to the runner by reducing the side-to-side excursions of the body CM. From an angular momentum standpoint, just as much vertical angular momentum can be generated with a crossover as with the classic style. The classic style would seem to have an advantage when it comes to generating drive, but, at least in distance running, drive does not appear to be a factor.

Research is currently underway which investigates the arm action in sprint running and race walking. (Funding has been received from the

U.S. Olympic Committee and The Athletics Congress.) Contributions of the arms and legs to the propulsion out of the blocks in a sprint start are also being investigated. Questions which will be answered include: *1)* Do the arms propel the sprinter forward in the start and at top speed? *2)* What torques are left about the vertical axis for the walker or runner to get from the ground? *3)* Should the arms be swung directly forward and backward in sprinting, or is there an advantage, as in distance running, of letting them cross over in front slightly? It is assumed that the arm action in race walking serves similar functions as it does in walking and running. The influence of the arms on the motion of the body *CM* should be especially interesting, since it is here that running differs from walking.

Finally, a logical progression from the experimental work done so far is to create realistic mathematical models of gait which include the arms as well as the legs and be three-dimensional in nature. The complexity of such a model would undoubtedly greatly exceed that of the current two dimensional models of gait which lump head, arms, and trunk together as one point mass sitting on top of fully functional legs. We cannot truly understand gait without investigating the fine motor coordination of arms and legs together.

References

Aristotle. (English Translation by E.S. Forster, 1961) *Progression of Animals.* Harvard University Press, Cambridge.

Braune, C.W., and Fischer, C. (1895) Der gang des menschen:I. Teil. *Abhandl. d. Math. - Phys. Cl. d. k. Sachs Gesselsch. Wissench.* 21:153-322.

Claremont, A.D., and Hall, S.J. (1988) Effects of extremity loading upon energy expenditure and running mechanics. *Med. Sci. Sports Exerc.* 20:167-171.

Clauser, C.E., McConville, J.T., and Young, J.W. (1969) *Weight, Volume, and Center of Mass of Segments of the Human Body.* AMRL Technical Report 69-70, Wright-Patterson Air Force Base, Ohio. (NTIS# AD-710-622)

du Bois-Reymond, R. (1909) Specielle bewegungslehre. *Handbuch der Physiologie des Menschen* (Edited by Nagel, W.), F. Vieweg, Braunschweig, pp. 564-628.

Elftman, H. (1939) The function of the arms in walking. *Hum. Biol.* 11:529-535.

Fernandez-Ballesteros, M.L., Buchthal, F., and P. Rosenfalck. (1965) The pattern of muscular activity during the arm swing of natural walking. *Acta Physiol. Scand.* 63:296-310.

Fischer, O. (1899) Der gang des menschen:II. Teil. Die bewegung des gesammtschwerpunktes und die aussern krafte. *Abhandl. d. Math. - Phys. Cl. d. k. Sachs Gesellsch. Wissensch.* 25:1-163.

Gerdy, P.N. (1829) Memoires sur le mecanisme de la marche de l'homme. *J. Physiol. Exp. Path.* 9:1-28.

Hinrichs, R.N. (1985) A three-dimensional analysis of the net moments at the shoulder and elbow joints in running and their relationship to upper extremity EMG activity. *Biomechanics IX-B* (Edited by Winter, D.A., Norman, R.W., Wells, R.P., Hayes, and Patla, A.E.), Human Kinetics, Champaign, IL, pp. 337-342.

Hinrichs, R.N. (1987) Upper extremity function in running. II:Angular momentum considerations. *Int. J. of Sport Biomechanics* 3:242-263.

Hinrichs, R.N. (1990) Upper extremity function in distance running. *Biomechanics of Distance Running*, (Edited by Cavanagh, P.R.), Human Kinetics, Champaign, IL, pp. 107-133.

Hinrichs, R.N. and Cavanagh, P.R. (1981) Upper extremity function in treadmill walking. Paper presented at the 1981 Annual Meeting of the American College of Sports Medicine, Miami, FL.

Hinrichs, R.N., Cavanagh, P.R., and Williams, K.R. (1983) Upper extremity contributions to angular momentum in running. *Biomechanics VIII-B* (Edited by Matsui, H. and Kobayashi, K.), Human Kinetics, Champaign, IL, pp. 641-647.

Hinrichs, R.N., Cavanagh, P.R., and Williams, K.R. (1987) Upper extremity function in running. I:Center of mass and propulsion considerations. *Int. J. Sport Biomechanics* 3:222-241.

Hogue, R.E. (1969) Upper-extremity muscular activity at different cadences and inclines during normal gait. *Phys. The.* 49:963-972.

Hopper, B.J. (1964) The mechanics of arm action in running. *Track Technique* 17:520-522.

Jackson, K.M., Joseph, J., and Wyard, S.J. (1978) A mathematical model of arm swing during human locomotion. *J. Biomechanics* 11:277-289.

Mann, R.V. (1981) A kinetic analysis of sprinting. *Med. Sci. Sports Exerc.* 13:325-328.

Mann, R. and Herman, J. (1985) Kinematic analysis of Olympic sprint performance:Men's 200 meters. *Int. J. Sport Biomechanics* 1:151-162.

Murray, M.P., Sepic, S.P., and Bernard, E.J. (1967) Patterns of sagittal rotations of the upper limbs in walking, a study of normal men during free and fast walking. *Phys. Ther.* 47:272-284.

Rasch, P.J. and Burke, R.K. (1973) *Kinesiology and Applied Anatomy.* Lea and Febiger, Philadelphia.

Williams, K.R. (1980) A biomechanical and physiological evaluation of running efficiency. Unpublished doctoral dissertation, Pennsylvania State University, University Park.

CHAPTER 46

Brain Plans and Servo Loops in Determining Corrective Movements

John D. Brooke and William E. McIlroy

46.1 Introduction

There are research benefits to studying movements which correct a disturbance of the position or progression of a limb. For example, the prior state of the limb, the body or the brain can be manipulated. This can throw light on prior organization, or pre-planning, of the corrective movement. Further, the onset point for the need for the correction can be accurately specified. This serves as a temporal bench mark from which the corrective movement is initiated by the nervous system. Study of such movements can also highlight the role of peripheral feedback from sensory transducers in the organization of movement. (This then can be contrasted with pre-planning by the central nervous system.) The initial disturbance or 'perturbation' of the limb sets up additional discharge in the sensory receptors. When this discharge can be measured or estimated, correlation can be made between the input and the subsequent output of corrective muscular activity. This can test for the existence of servo relationships between the sensory discharge and the muscular activity. The possibility of investigating servomechanisms is improved when the concept is delimited, as presently, to 'simple' servos. 'Simple' means proceeding through brief pathways, with the effects of the input on the muscular discharge being those described as reflexes in basic texts of neuroscience.

Reflexes observed during corrective reactions have a long history, initiated by studies of perturbation at single joints. The first recorded use of load perturbation appears to have been by Hammond (1955). The paradigm involved loading of the human forearm to stretch the biceps brachii muscle. Two responses were observed, one excitation at a latency of approximately 45 ms and a second at 110 ms. The first was related to the monosynaptic stretch reflex and the second to a voluntary response to the external loading. A report in 1956 by Hammond reported the potential of an intermediate response (approximately 73 ms) to account for differences between responses observed when subjects were instructed to either 'let go' or 'resist' the perturbation. These studies represented the first evidence of multiple responses, distinguished by latency, to external loading.

In the early 1970s the technique was further developed (Newson et al., 1970; Melville-Jones and Watt, 1971). The emphasis was on the role of reflex function in normal behavior. Melville-Jones and Watt (1971) labelled a further response to those observed by Hammond (1956), the functional stretch reflex, which occurred at approximately 120 ms. Further categorization occurred when Tatton et al. (1975) reported four distinct responses to perturbation in the monkey, *M1*, *M2*, *M3* and *voluntary*. Hendrie and Lee (1978) revised this by combining *M2* and *M3* (*M2/M3*) in humans, where the two responses were not easily distinguishable. Many studies using different nomenclature were performed on different muscles and different species. It soon became difficult to compare the numerous, presumably distinct, responses which were distinguished primarily by latency.

Multiple Muscle Systems: Biomechanics and Movement Organization
J.M. Winters and S.L-Y. Woo (eds.), © 1990 Springer-Verlag, New York

In the mid 1970's individuals were also exploring using the load perturbation strategy to perturb more complex movements, such as postural stability. Nashner (1977) produced data suggesting that the responses to induced sway in standing subjects represented fixed synergies from a pattern generator in the central nervous system (*CNS*). There was a greater focus on the combination of responses in different muscles, as opposed to the segmented responses seen in individual muscles during single joint studies. Some of this new focus resulted from a lack of observation of simple responses such as the monosynaptic reflex. This lack could be due principally to the low rates of stretch around the joints. Allum (1983), studying posture, and Dietz et al. (1982), studying gait, also recorded responses which were not easily associated with those seen in single joint studies.

An immediate bi-product of focusing on more than one muscle was the view that the output arising from afferent input was much more complex than that seen in the single muscle case. Muscles which may not have been agonists to the perturbation were activated. Abbs and Cracco (1984) were critical of single joint studies for this limitation of not being able to observe 'open-loop control' processes because of the single muscle focus. They highlighted the importance of the non-autogenic muscle as evidence of the complexity of transformations.

In the last 5 years there has been a substantial rise in the number of studies evaluating responses to perturbation of complex movement, including study of posture (Horak et al., 1989; Diener et al., 1988; MacPherson et al., 1986; Horak and Nashner, 1986; Forssberg, 1985), gait (Dietz et al., 1987), pedalling (McIlroy & Brooke, 1987), and grip (Johansson and Westling, 1988; Cole and Abbs, 1987). These studies have concluded that a significant portion of the response to perturbation is the product of a centrally triggered (prestructured) program. The emphasis has clearly been on the complexity of the transformation rather than the afferents or the neuroanatomical characteristics.

In such studies, there are two potential contributors to the kinematic and electromyographic (*EMG*) responses due to the perturbation. The parameters of the afferent volleys (inferred from the characteristics of the perturbation) may determine the response in a scalar manner, through transmission over their briefest paths to the motoneuronal pools. We searched for such servo loops. We chose biomechanical measures to reflect particular sensory discharge. We then attempted to correlate these measures with the *EMG* responses.

In addition to the servo loops, the prior states of the *CNS* may partially or wholly determine the responses to perturbation. The identification of programmed responses, centrally derived, depends on two possible observations: *1)* independence of response characteristics from the peripheral inputs and *2)* anticipatory effects (Rosenbaum, 1985). The use of the first criterion is most common. In general, researchers have attempted to dissociate the response characteristics from peripheral input on the basis of inference of sensory input from changes in kinematics due to perturbation.

One possibility for such programmed responses would be that there are fixed patterns of neuronal discharge leading to *EMG* discharge, patterns which are released by a triggering stimulus; that is, 'hard-wired' neuronal circuits. A more flexible view would be that the *CNS* learns a number of patterns of *EMG* responses and that one of these is selected on the basis of prior sensory information about the task and the body condition that preceeded the onset of perturbation, and on the basis of the prior intentions of the subject. When we failed to find simple servo relationships between perturbation input and muscle output, we explored this possibility of preplanning. Specifically, could it be identified? Could it be shown to be flexible for different prior states?

46.2 Transduced Sensory Information from Limb Perturbation

There is a well documented range of biological transducers available to sense the position of the leg and on the types of perturbations that these sensory receptors receive [e.g., see table on physiological classification of nerve fibers and their receptors (p. 578) and section on the maintenance of equilibrium (p. 621) in Guyton (1986)]. Proprioception arises from kinesthetic and vestibular receptors. Kinesthetic receptors transduce the stretch of muscles, the muscle force and the position of joints. Vestibular receptors transduce the position of the head in space. In addition, information from limb contact may be transduced through haptic receptors (pressure-touch). The afferents serving these various transducers proceed in the spinal cord through a myriad of routes to the motoneurons which excite muscles (Baldissera et al., 1981). They also rapidly transmit information to brain sites, including the cerebral cortex (Guyton, 1986). The above list is not all-inclusive (e.g., special senses such as vision may contribute), but it probably contains the main sources of information transduced by subjects producing corrective muscular activity in the present tasks.

It must be noted that, in the past, study of response to perturbation has unduly emphasized transducers for muscle stretch and transformations of this information through conduction in *Ia* afferent fibres. A large number of types of sensory receptors may be activated when a limb is perturbed. The action potentials due to these receptors may be transmitted in the full range of afferents (certainly in *groups I, II* and *III*). This means that afferent conduction velocities from the initial disturbance of the limb can range in the human from approximately 60 m/s to 3 m/s. Accordingly, the initial latency of an evoked response in the *EMG* cannot easily be used to infer the time or modality of the onset of the transducer discharge. One cannot assume *group I* afferent velocities from muscle stretch without first showing how one rules out the other sensory possibilities.

Research on corrective movements has used some measure of the perturbing force to infer the volley evoked from a particular receptor. Two questions are pertinent. Can the parameters of the perturbing stimulus be used for such inference? If so, what parameter of the stimulus should be used?

We agree with earlier researchers that the characteristics of the stimulus of mechanical perturbation can be used to make reasonable inferences about the transducer discharge and subsequent afferent traffic, both for the muscle spindle stretch receptor (Gottlieb and Agarwal, 1979) and for haptic transducers. We note the low level of *EMG* response compared to that obtained in studies which used electrical stimulation of the afferents (McIlroy and Brooke, 1986). We are less secure about inferences for Golgi tendon organs transducing force through the musculotendinous junction, and do not feel secure about inferring discharge from joint transducers of posi-

tion or from the vestibular apparatus. Further, for muscle stretch and pressure-touch transducers, the parameter of the stimulus should be selected appropriately for the transducer tissue site. Figure 46.1 shows the appropriate delimitation. For example, if stretch discharge is anticipated from knee extensor muscles, measurement should be made of the evoked velocity of knee flexion, rather than forces or velocities at the ankle. [However, knee flexion velocity may in some cases be estimated via a mathematical model (e.g. Chapter 40 (Hull and Sanderson) for cycling).]

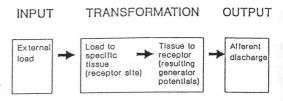

Figure 46.1: General perturbation load leads to specific transducer load, so as to infer the evoked sensory activity and afferent traffic.

It is concluded that a particular type of transducer needs to be separated and specifically activated when a role is to be assigned to it. The appropriate stimulus parameters of the perturbation must be obtained to represent that particular transducer discharge. Such specific activation and its adequate representation have not been well done in studies of human responses. This is partly because experimental control is difficult to obtain. It follows that confirmation of the *CNS* role, by showing that simple servo mechanisms are not a part of the *EMG* responses, is not an easy task. There are too many possible transducers and reflex loops.

46.3 Central Nervous System Plans

In the late 1900's, there was considerable research conducted on spinalized preparations which revealed that complex movement was possible in the absence of the brain. Chained reflexes were explored to account for behavior (Sherrington, 1906). These studies diverted attention away from the central nervous system as a control center for the patterning of complex behaviour. However, models for pattern generators were constructed by individuals such as Brown (1914) and later von Holst (1935). In recent years there has been substantial work, with the role of the central pattern generator well recognized in some preparations (e.g., Grillner, 1985). These observations, along with those from studies on deafferentated preparations (e.g., Deliagnia et al., 1975), contributed to the view of the motor physiologist that the *CNS* contains 'programs' for movement. These biological models of motor programs are complemented in the psychological literature by a more diffuse view of the brain's capacity to store and generate programs [e.g., see Schmidt (1982)].

The product of the outputs from motor programs has been observed as both modulation of ongoing reflex action and as distinct motor reactions. Hammond (1956) reported that the short latency reflex was attentuated when subjects were instructed not to correct the load disturbance. This has been repeatedly confirmed. More general evidence, outside the load perturbation paradigm, for the modulation of 'reflex' pathways by descending commands, reflecting some programmed discharge, has seen considerable growth in recent years (see Baldissera et al., 1981; Schieppati, 1987). Programmmed response can also mobilize muscle action directly. There is evidence that some of the corrective reaction to load disturbance represents a triggered 'pre-programmed' reaction (Crago et al., 1976).

There are a number of ways in which the *CNS* can deliver pre-planned patterns of response to limb perturbation. At the periphery, the *CNS* can modulate the information transmitted in the afferent line conducting from the transducer. One possibility is to alter the sensitivity of the transducer. For example, the primary stretch receptor of the muscle spindle receives gain control from the brain through the gamma motoneuronal innervation. Another possibility is to control afferent traffic by inhibiting the transmission. This occurs directly at spinal and brain stem levels through presynaptic inhibition. Indirectly, the effects of afferent traffic can be reduced in simple servo loops of the spinal cord by inhibiting the neuron which would be excited by the traffic. Thus pre-planning can, at the *CNS* periphery, control the flow of afferent information by modulations at the membranes of the receptor, the presynaptic and the postsynaptic neurons. Clearly, this pre-planning has the potential for considerable flexibility as the task alters.

There are additional ways of delivering pre-planned *EMG* responses, as the neural path from

transducer to muscle becomes more complex. This complexity might be through spinal or brain networks of neurons. Prior action potentials from environmental events result in memory and plans for future action. The neural circuits for the execution of these future actions will lead to descending modulations on the motoneuronal pools, resulting in a pattern of discharges from muscles. Species modification, through evolutionary adaptation from environmental events, may result in some such circuits becoming 'hard wired', fixed components of the *CNS* connections. An example is pattern generators (Grillner 1985). The brain and spinal cord circuits at complex levels, from phenotype or genotype memory, may need no more than trigger signals from the proprioceptive or haptic transducers. Alternatively, they may be flexible to changing task conditions.

No one knows what the quantified physical models are for these modes of *CNS* control that have the potential to be involved in pre-planning. There is little evidence for the linear scaling of input and output that is associated with simple servomechanisms. Pellionisz and Llinas (1980) have taken a novel approach to modeling motor coordination, viewing the brain as a geometric object, with its networks organized as vectors in reference to the three dimensional movement space. The sensory and motor vectors are felt to be treated in their natural coordinate systems, with the internal vectors being removed from these specific coordinates by a hypothesized neuronal network that acts as a metric tensor. This metric tensor determines the relationship between the input and output systems (Pellionisz and Llinas, 1985).

In conclusion, it is clear that at least some of the *EMG* responses to perturbation can arise from pre-planning in the *CNS*. Even the earliest latency responses can be influenced. Compared to the servo control view, there is much less direct experimental evidence for this pre-planning. This is partly due to the apparent ease of manipulating the peripheral afferent discharge and then studying spinal responses. It is also partly because valid numerical modeling of the neural aspects of multiple muscle movement systems has proved to be difficult, especially given out lack of knowledge regarding *CNS* pre-planning. One suspects that the earlier bias to servo control is also partly due to the ideological seduction of the Chained Reflex

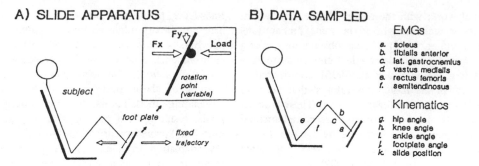

Figure 46.2: Outline of details of the methods used in a number of our experiments.

Theory, which for many decades directed the thoughts of experimenters.

46.4 Current Method of Investigating Perturbation of the Human Leg

Movements were performed from a seated position, on a custom designed slide, with ankle, knee, and hip angles initially at approximately 100°. Subjects tonically contracted leg extensor muscles to resist a flexor force of 120 N, and were instructed to return the slide to its starting position when perturbed. Perturbation flexor forces up to 350 N were applied at unpredictable intervals from 10 to 45 sec. The plantar surface of the foot contacted a footplate, which served as the interface between the limb and the external inertial loads imposed. Possible slide configurations defined specific trajectories or varied degrees of freedom for movement. The kinematic variables sampled with potentiometers included hip, knee and ankle angles, foot plate angle and slide displacement. Reaction forces of initial perturbation and of response were sampled. Surface electrodes sampled electromyograms from six leg muscles, as shown. The 13 data channels were analog-digital transformed and on-line sampled at a rate of 1 kHz.

Perturbations also were applied when subjects pedalled on a Monark friction brake ergometer at a rate of 50 rpm and power output of 160 W (McIlroy and Brooke, 1987). The ergometer was modified with a potentiometer to measure right pedal angle, strain gauge plates to measure pedal force, and a mechanical brake which rapidly increased the pedal resistance. The right pedal force signal was conditioned by a Vishay amplifier. Subjects were instructed to restore the cycling rate

as quickly as possible after the braking perturbation, which slowed the angular velocity of the pedals by 50% in approximately 30 ms. Ten to 30 seconds elapsed between perturbations, which were presented at three positions in the movement, near the top, middle and bottom of the first half of the leg cycle (when the leg is extending). At these points, the perturbation occurred at different knee angles and ankle angles and occurred at different levels of ongoing activity in the muscles.

46.5 Results on the Role of Inferred Afferent Volleys from Perturbation

46.5.1 Perturbation of Isometric Extension Activity of the Isolated Leg

In one study, loads of different and predictable magnitudes were applied. Figure 46.3A presents the kinematic changes resulting from such load changes. The slide displacement reveals the sharp change of limb position due to the onset of the external load and the eventual recovery of the limb position against the new load. As the magnitude of perturbing load increases, the slide is displaced further, then returned to the starting point. Initial and peak velocities were dependent on the magnitude of perturbation. It is interesting to note that the changes in knee and hip displacement paralleled, in shape, the changes in slide position but the ankle joint did not. In particular, note the early oscillation at the ankle. This information is not readily available from the slide kinematics. In this movement, part of the changes seen at the hip and knee also must arise from this oscillation at the ankle. The peak velocities of the early oscillation of the ankle are low, making it difficult to predict the afferent discharge resulting from muscle stretch transducers.

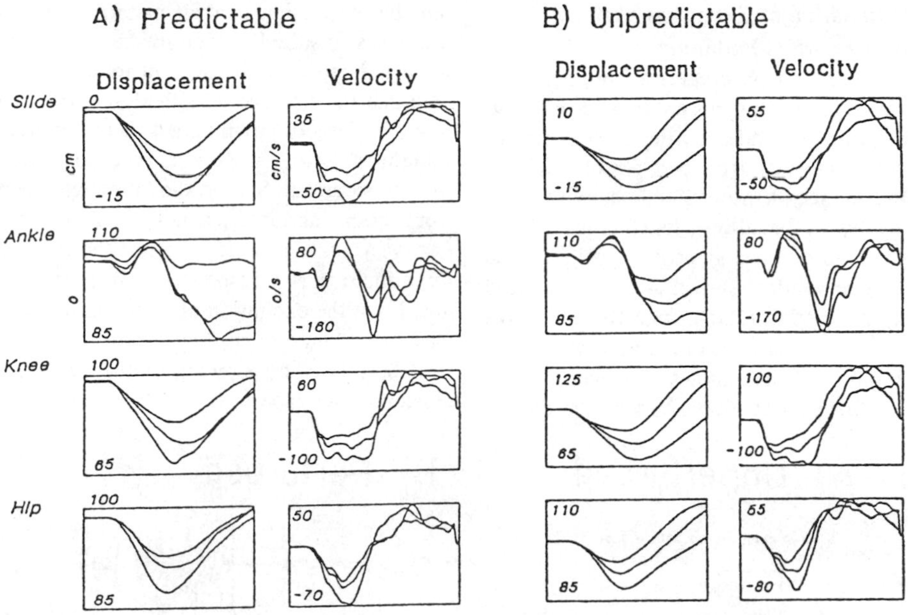

A) Predictable   ## B) Unpredictable

Figure 46.3: Displacements and velocities of slide, ankle, knee and hip, from 100 ms before to 500 ms after perturbation with three added loads of 120, 250 and 350 N, in one subject.

Figure 46.4A shows that on average, with predictable loading, *EMG*s rose as the velocity of perturbation of the appropriate joint rose. The peak velocities for dorsiflexion and plantar flexion were obtained from the initial period of oscillation reported in Figure 46.3. This correlation of *EMG* and joint velocity is not evidence for servo loops between the muscle stretch receptors and *EMG* response. These loads were presented in a predictable fashion, all of the samples of one load being presented together. (See 46.5.3 – Unpredictable Loading.) The main bursts of excitatory responses in ankle muscles were temporally separated. On average, the excitation in tibialis anterior started 50 ms before that in soleus. Also noteworthy is the one subject who corrected the perturbation, but showed no change in early soleus or tibialis anterior activity with increasing joint velocity.

Figure 46.4: Regression of *EMG* magnitude, 65-155 ms post perturbation, on joint velocity (deg/s). Lines are constructed through data points for individual subjects. *EMG*s were expressed as ratios of response (mV.s) evoked following the lowest load (*SOL*–soleus, *TA*–tibialis anterior, *VM*–vastus medialis).

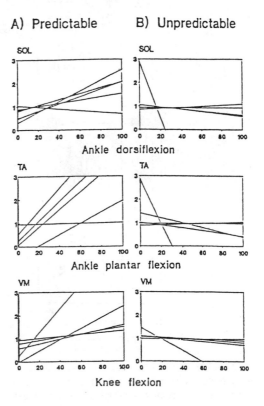

46.5.2 Perturbation of Extension Activity of the Leg when Pedalling

Some of the major observations from perturbation of isolated leg extensions on the slide can also be observed when natural activities such as pedalling are perturbed (McIlroy and Brooke, 1987). Figure 46.5A shows raw *EMG* activity over repeated cycles of pedalling, together with reaction force at the pedal and pedal angle. At the start of the time window shown, 48° past top dead centre for the pedal (*TDC*), the flywheel was rapidly braked and perturbed the movement of the limb. Figure 46.5B shows 15 samples of responses to this perturbation. As with the slide, it can be seen that stable patterns of muscular responses resulted. The pedal angle provides some indication of ankle joint rotation. It can be seen that the reactive dorsiflexion normally occurring is largely maintained with perturbation. Tibialis anterior is excited, yet no perturbation stretch occurred. Soleus and lateral gastrocnemius were both inhibited initially. Thus, the early *EMG*s in ankle muscles are not those anticipated for simple servo responses. It should also be noted that the excitation of the quadriceps muscles was not preceeded by perturbation-induced stretch, simply by slowing of the developing extension of the knee.

A) Unperturbed   B) Perturbed

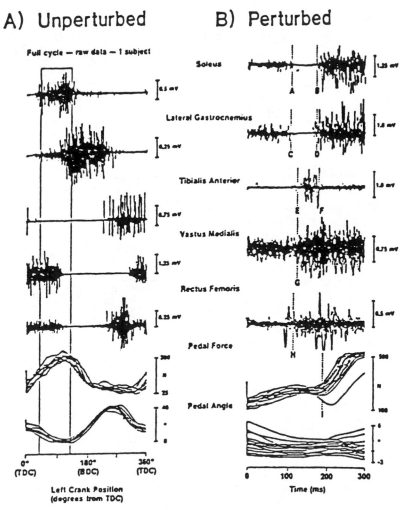

Figure 46.5: Raw *EMG* responses, pedal forces, and pedal angles, from a single subject, cycling at 0.8 *Hz*. The perturbed samples occurred in the time window shown on the full cycle trace. (Redrawn from McIlroy and Brooke Brain Res. (1987); printed with permission.)

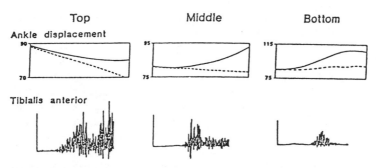

Figure 46.6: Displacement of the ankle joint and *EMG*s of tibialis anterior (mean full wave rectified, with ± 1 SD), for one subject, at the top, middle and end of the leg extensor phase following perturbation of pedalling.

We have perturbed pedalling at different positions of the leg during its extension. At the various points in the movement, there was a similar decrease in pedal crank velocity, which, when unperturbed, remained comparatively constant. However, as Figure 46.6 shows for the tibialis anterior at the ankle joint, even though the activity of muscles decreased over the extension phase, the initial *EMG* responses to perturbation could still be observed. No stretch occurred in muscles showing early excitatory responses. It is concluded that when pedalling is perturbed, the complex pattern of *EMG* responses is not primarily arising from servos from muscle stretch or force receptors.

46.5.3 Results on *CNS* Pre-Planning

CNS pre-planning appeared to be involved in the corrective movements of the leg. There was temporal separation in the activation of ankle muscles, and one subject showed *EMG* responses not correlated to the appropriate joint rotation. Therefore, this pre-planning was further explored in a different way by making the magnitude of the perturbing load unpredictable as subjects held leg extensor activity on the slide. Figure 46.3B shows the resulting effect on displacements and velocities of the slide and of ankle, knee and hip joints. Note that the early kinematics, which result from the perturbation as opposed to corrective *EMG* activity, were close to identical for unpredictable and predictable loads. That is, loading magnitude did not alter between predictable and unpredictable conditions. The main difference in kinematics between the two conditions occurred well into the corrective phase, when the *CNS* was presumably apprised of the magnitude of the load.

Figure 46.4B shows that, in the unpredictable condition, the subjects did not correlate the level of *EMG* activity with the magnitude of the load imposed. However, they should have done so if the peripheral sensory transduction of the load was in fact determining the magnitude of the *EMG* responses via simple servo loops. The evidence for such loops in controlling the corrective movement was not there. The failure to scale *EMG* activity to the magnitude of the perturbation when it was unpredictable contrasted with the linear regressions normally occurring with predictable loads, as shown in Figure 46.3A. We feel that the regressions in Figure 46.3A reflect the scaling of response due to the *CNS* pre-planning for a predictable load. We suggest that these results should not be taken as evidence for simple servo loops, however superficially attractive this idea may be when Figure 46.4A is viewed alone.

It can be seen from the displacements of the slide in unpredictable conditions (Figure 46.4B) that this subject was biased toward predicting one level of perturbation. In this case it was the high load, and consequently the return of low and medium perturbations overshot the return target. Different subjects showed biases to different levels of perturbation. Also, for a given subject there could be considerable variation over a number of trials.

Subjects quickly stabilized the pattern of muscular activation appropriate for a predictable load. Noting this rapid adaptation, and the changes that occurred in a pattern which is complex, we feel that the motor pre-planning was occurring in the brain rather than the spinal cord. However, we have no direct evidence for this opinion. There is evidence from primate studies that motor set involves the deep nuclei of the cerebellum.

Reversible cooling of these nuclei in Cebus monkeys abolishes the component of the perturbation-evoked *EMG* that is attributed to anticipatory set (Hore and Vilis, 1984). The prefrontal cortex[*] is also likely to be involved in the pre-planning (see, e.g., Sieb's hypotheses 1989).

46.6 Concluding Remarks, Future Directions

In the present work, sensory discharge, due to perturbation of the leg when isometrically exerting extensor force, was indirectly represented by the changes in displacement and angular velocity of the ankle, knee and hip joints, and by the loading force applied to the plantar surface of the foot. The *EMG*s of major extensor and flexor muscles of the leg were examined during the response to restore limb position, following the perturbation. A major conclusion is that such corrective muscular activity is not the output of simple servomechanisms through brief neural pathways from the inferred sensory input. (Inputs from muscle stretch and force, and from plantar surface pressure, were studied. There are not dynamic grounds for anticipating much Golgi force transducer discharge. Furthermore, joints did not move to the extremes at which the main discharge occurs from joint receptors.)

We have confidence in our conclusion, which suggests against the presence of servomechanisms, in part because it was not based simply on the relationship between the *EMG* responses and the global measure of perturbation, the loading force. There have been some recent research reports which arrive at our conclusion from a different set of observations (e.g., Diener et al. 1988). Clearly, the biomechanical measures need to approximate, as closely as possible, the perturbing phenomena which are appropriate for exciting the various types of sensory receptor that could be involved in servo loop responses. The present studies come close to this goal, and suggest that the peripheral servo loop is not the primary means of control in this task.

In one of the experiments, the loads of perturbation were delivered randomly in one trial and then in predictable series in another trial. After the trials, the *EMG* and kinematic data were grouped according to the load of perturbation and its predictability. It should be noted that the same stimuli were delivered to the sensory receptors, whether the magnitudes of perturbation were predictable or not. (This was confirmed by the same early kinematic changes due to perturbation in both predictable and unpredictable trials.) The identical stimuli should have resulted in the same *EMG* responses for the two types of trials if the signal was in fact processed through simple servo loops. This did not occur. The simple servo hypothesis again was discredited. However, the *EMG* responses were not random. Rather, subjects in unpredictable settings delivered an average *EMG* response which showed clear bias to one of the three loads. The brain anticipated a type of load. We conclude that motor plans were made before the perturbation occurs.

The rejection of the simple servo hypothesis lends strong support to the alternative, that the *CNS*, and probably the brain, pre-plans the responses. This conclusion has to be delimited presently to the characteristics of our experimental task. Particularly important is the fact that there was no penalty for inaccuracy and no reward for speed. Much more use of the servo control might have occurred if there had been a severe penalty for inaccuracy in returning the foot to the set position.

Our central conclusion, that a major role is played by the brain in corrective movements, signficantly directs whole limb modeling. The conclusion is supported by additional experiments we have conducted but not reported here. We have, under publication, further evidence on the role of the human brain in pre-planning. Our results suggest that: *1.* It is clear that truly novel perturbation events evoke *EMG* and kinematic responses which are quite different from those arising during known perturbations. *2.* We observe learning adaption, following novel disturbances (e.g., note the pedal force in the perturbed trial on Figure 46.5B). *3.* Diminished cutaneous drive from the foot results in response modulation at the comparatively long latencies associated with brain-processed responses and does not alter the initial *EMG* responses (McIlroy et al., 1989). *4.* Changes to the positions of the limb segments beforehand result in marked changes in the response pattern of the limbs (McIlroy et al., 1988).

The conclusion is that the brain determines much of the pattern of *EMG* and kinematic responses to a perturbation of the limb, when the cost of inaccuracy is low. The question arises:

what engineering model is appropriate to accomodate this brain processing? For these processes, in such regions as the cerebellum and the pre-frontal and motor-sensory cortices of the cerebrum, we do not know what sort of quantitative relationships exist between nerve cells. The system probably works in a parallel processing mode, altering the weights at synaptic junctions on the basis of previous outcomes, with this determining the current output pattern of responses to an appropriate initiating signal (in the present case, sensory signal). There may be a metric tensorial transformation intervening in the cerebellum (Pellionisz and Llinas, 1980; Hore and Vilis, 1984) between the sensory information transduced and the motor output observed in the *EMG*. Useful explorations may come through theoretically determined and experimentally derived computational neural networks (e.g., see the note by Churchland and Churchland, 1990). However, the biological understanding necessary for testing the validity of these models is not yet fully available. For example, we do not know the weights of the relevant central nervous synapses or how these weights are achieved.

The study of corrective movements overlaps the fields of biomechanics and neurophysiology. Statements on the neural control are constrained considerably by the combination of biomechanical and neurophysiological knowledge with measured information on the actual corrections made to the movement. Traditional concepts from one of the three tend to collapse when faced with the combined knowledge. Traditional words − e.g., reflex, synergy, feedback, motor programs, − do not meet the need in trying to express the reality emerging from the combination of the three areas. This difficulty in communication should ease when the new principles are clear and the taxonomy standardized.

The work we have reported clearly demonstrates the incisive effect of carefully planned biomechanical measurements. The measurements were selected to provide information on specific biological phenomena. Important hypotheses were formulated before the selection of the measures, hypotheses about the biological control of the movement. The data were then collected in order to test these hypotheses. We feel that such thrusts at deep etiology will be a major characteristic of biomechanical studies of human movement in the 1990s and beyond.

References

Abbs, J.H. and Gracco, V.L. (1984) Control of complex motor gestures: orofacial muscle responses to load perturbations of the lip during speech. *J. Neurophysiol.* **51**: 705-723.

Allum, J.H.J. (1983) Organization of stabilizing reflex responses in tibialis anterior muscles following ankle flexion perturbations of standing man. *Brain Res.* **264**: 297-301.

Baldissera, F, Hulborn, H. and Illert, M. (1981) Integration in spinal systems. *Handbook of Physiology, Sect. 1 (The nervous system), Vol. 2 (Motor control)* (Edited by Brooks, V.B.). American Physiological Society, Bethesda.

Brown, T.G. (1914) On the nature of fundamental activity of the nervous centres; together with an analysis of the conditioning of rhythmic activity in progression, and a theory of evolution of function in the nervous system. *J. Physiol.* **48**: 18-46.

Churchland, P.M. and Churchland, P.S. (1990) Can machines think? *Sci. Am.* **262**: 1: 32-37.

Cole, K.J. and Abbs, J.H. (1987). Kinematic and electromyographic responses to perturbation of a rapid grasp. *J. Neurophysiol.* **57(7)** 1498-1510.

Cole, K.J., Gracco, V.L. and Abbs, J.H. (1984) Autogenic and nonautogenic sensorimotor action in the control of multiarticulate hand movement. *Exp. Brain Res.* **56**: 582-585.

Crago, P.E., Houk, J.C. and Hasan, Z. (1976) Regulatory actions of human stretch reflex. *J. Neurophysiol.* **39**: 925-935.

Deliagnia, T.G., Fel'dman, A.G., Gelfand, I.M. and Orlovsky, G.N. (1975) On the role of central program and afferent inflow in the control of scratching movements in the cat. *Brain Res.* **100**: 297-313.

Diener, H.C., Horak, F.B. and Nashner, L.M. (1988) Influence of stimulus parameters on human postural responses. *J. Neurophysiol.* **59**: 1888-1905.

Dietz, V., Quintern, J. and Sillem, M. (1987) Stumbling reactions in man: significance of proprioceptive and pre-programmed mechanisms. *J. Physiol.* **386**: 149-163.

Forssberg, H. (1985) Phase dependent step adaptations during human locomotion. In *Feedback and Motor Control in Invertebrates and Vertebrates*. (Edited by Barnes, W.J.P, and Gladden, M.H.) Croomhelm, London, pp. 451-464.

Gottlieb, G.L. and Agarwal, G.C. (1979) Response to sudden torques about ankle in man: myotatic reflex. *J. Neurophysiol.* **42**: 91-106.

Grillner, S. (1985) Central pattern generators for locomotion, with special reference to vertebrates. *Ann. Rev. Neurosci.* **8**: 233-261.

Guyton, A.C. (1986) *Textbook of Medical Physiology* 7th Ed. Saunders, Philadelphia.

Hammond, P.H. (1955) Involuntary activity in biceps following sudden application of velocity to the abducted forearm. *J. Physiol.* **127**: 23-25P.

Hammond, P.H. (1956) The influence of prior instruction to the subject on an apparently involuntary neuromuscular response. *J. Physiol.* **128**: 17-18P.

Hendrie, A. and Lee, R.G. (1978) Selective effects of vibration on human spinal and long-loop reflexes. *Brain Res.* **157**: 369-375.

Horak, F.B. and Nashner, L.M. (1986) Central programming of postural movements: adaptation to altered support surface configurations. *J. Neurophysiol.* **55**: 1369-1381.

Horak, F.B., Diener, H.C. and Nashner, L.M. (1989) Influence of central set on human postural responses. *J. Neurophysiol.* **62**: 841-453.

Hore, J. and Vilis, T. (1984) Loss of set in muscle responses to limb perturbations during cerebellar dysfunction. *J. Neurophysiol.* **51**: 1137-1148.

Johansson, R.S. and Westling, G. (1988) Programmed and triggered reactions to rapid load changes during precision grip. *Exp. Brain Res.* **71**: 72-86.

Kuhn, T.S. (1970) *The Structure of Scientific Revolutions 2nd. Ed.* University of Chicago Press, Chicago.

MacPherson, J.M., Rushmer, D.S. and Dunbar, D.C. (1986) Postural responses in the cat to unexpected rotations of the supporting surface: evidence for a centrally generated synergic organization. *Exp. Brain Res.* **62**: 152-160.

McIlroy, W.E. and Brooke J.D. (1986) The magnitude of the H reflex of soleus substantially exceeds the myotatic response from rapid dorsi-flexion of the ankle joint. *Exp. Neurol.* **92**: 455-460.

McIlroy, W.E. and Brooke, J.D. (1987) Response synergies over a single leg when it is perturbed during the complex rhythmic movement of pedalling. *Brain Res.* **407**: 317-326.

McIlroy, W.E. and Brooke J.D. (1989) Contribution of prior expectation to corrective movements. *Neurosci. Abs.* **15**: 173.

McIlroy, W.E., Brooke, J.D. and Singh, R. (1989) Cutaneous contributions to corrective movements of the leg in humans. *Abs. Sth. Ont. Neurosci. Ass.* June 3: McMaster University, Hamilton, Ontario, Canada.

McIlroy, W.E., Yoon, P. and Brooke, J.D. (1988) The importance of foot position in the development of whole limb strategies for load bearing. *Neurosci. Abs.* **14**: 64.

Melville-Jones, G. and Watt, D.G.D. (1971) Observations on the control of stepping and hopping movements in man. *J. Physiol.* **219**: 709-727.

Nashner, L.M. (1977) Fixed patterns of rapid postural responses among leg muscles during stance. *Exp. Brain Res.* **30**: 13-24.

Newson Davis, J. and Sears, T.A. (1970) The proprioceptive reflex control of the intercostal muscles during their voluntary activation. *J. Physiol.* **209**: 711-738.

Pellionisz, A. and Llinas, R. (1980) Tensorial approach to the geometry of brain function: cerebellar coordination via a metric tensor. *Neurosci.* **5**: 1125-1136.

Pellionisz, A. and Llinas, R. (1985) Tensor network theory of the metaorganization of functional geometries in the central nervous system. *Neurosci.* **16**: 245-273.

Rosenbaum, D.A. Motor programming: a review of scheduling theory. In *Motor Behaviour.* (Edited by Heuer, H., Kleinbeck, U. and Schmidt, K.-H.)., Springer-Verlag, Berlin, pp. 1-33.

Schieppati, M. (1987) The Hoffmann reflex: a means of assessing spinal reflex excitability and its descending control in man. *Prog. Neurobiol.* **28**: 345-376.

Schmidt, R.A. (1982) *Motor Control and Learning.* Human Kinetics, Urbana.

Sherrington, C.S. (1906) *The Integrative Action of the Nervous System.* New Haven, Yale University Press, (2nd edition 1947).

Sieb, R.A. (1989) Proposed mechanism for cerebellar coordination, stabilization, and monitoring of movements and posture. *Medical Hypotheses.* **28**: 225-232.

Tatton, W.G., Forner, S.D., Gerstein, G.L., Chambers, W.W. and Liu, C.N. (1975) The effect of postcentral cortical lesions on motor responses to sudden upper limb displacement in monkeys. *Brain Res.* **96**: 108-113.

APPENDIX

A Survey of Human Musculotendon Actuator Parameters

G. T. Yamaguchi, A. G. U. Sawa, D. W. Moran, M. J. Fessler, and J. M. Winters

A.1 Introduction and Overview

Hopefully, this volume has convinced the reader that musculoskeletal modeling is inherently valuable. Good examples of how models are used to either interpret experimental data or to predict the movements resulting from muscular contractions have been included in the previous chapters. These models, incorporating a multiplicity of muscles, demand the availability of accurate anatomical data, especially with regard to the set of parameters required to describe the intricate system of physiological actuators. Hence, this appendix was created to contain, in one place, the musculotendon actuator parameters utilized by leading researchers in the field.

The information is catagorized into five tables as follows:

Table A.1: Lower Extremity Musculature

Table A.2: Trunk Musculature

Table A.3: Upper Extremity Musculature

Table A.4: Hand Musculature

Table A.5: Head and Neck Musculature

This appendix was intended to provide a benchmark which will allow us to measure both our current state of knowledge as well as to indicate where published information is lacking. Though much is known about the major muscles of the extremities, for example, less has been published concerning the small muscles of the trunk [although see Bogduk and Twomey (1987) for anatomical details that are difficult to catalog here]. Our hope is that by indicating the blank areas on the following tables, they will someday be filled. We have included published data from many sources in as close to original form as pos-sible (i.e. without scaling the data to a "normative" size or forcing the data to be consistent with regard to parameter definitions, reference frames, etc.). In doing so we hope that future studies will work to standardize the measurement techniques, parameter definitions, and reference frames used to define the musculoskeletal models, as well as to lay the groundwork for developing more accurate methods of scaling the data to specific subjects. *Users of these tables are strongly encouraged to check values tabulated here with the original sources for accuracy, for further descriptions of the methods employed to obtain the data, and for more precise definition of measurement reference frames.* The authors extend apologies in advance to those whose data has not yet been included in the following tables. Omissions of noteworthy data are purely unintentional, and will be corrected in future editions if such data and a description of the measurement technique is sent directly to the authors.

A.2 Parameter Definitions

PCSA muscle physiological cross-sectional area (in m^2); defined as the volume of the muscle divided by its "gross muscle length or its fiber length with or without accounting for pennation". Most authors use formulas presented by Alexander and Vernon (1975):

$$PCSA = \begin{cases} \dfrac{m}{\rho l}, & \text{(fusiform)} \\[2ex] \dfrac{m}{2\rho t} \sin(2\alpha) & \text{(unipennate)} \end{cases} \quad \text{(A.1)}$$

Multiple Muscle Systems: Biomechanics and Movement Organization
J.M. Winters and S.L-Y. Woo (eds.), © 1990 Springer-Verlag, New York

where m, ρ, and l are muscle mass, density (a typical value is 1.05 gm/cm^3), and length, t is the layer thickness of muscle pennation, and α is the pennation angle.

ML   muscle length (m)

LM   optimal muscle fiber length (m)

V   muscle volume (ml)

α average pennation angle (deg)

SO   denotes slow oxidative (*type I*, or "slow-twitch") muscle fibers; measured as a percentage of all muscle fibers (%)

FOG denotes fast oxidative glycolytic muscle fibers

FG   denotes fast glycolytic (*type II*, or "fast-twitch") muscle fibers

LT   resting tendon length

AT   tendon physiological cross sectional area

O,I   Origin and insertion of muscle-tendon system, usually the centroid of the areas of bone–tendon attachment.   Where several "origins" and "insertions" are given by one source for a single muscle, the intermediate origin/insertions represent "via" points, or points through which the musculotendon passes.

A.3 Principal Sources

[0] Alexander and Vernon (1975)

Muscles were dissected from the right leg of an embalmed male cadaver (166 cm, 64 kg, 48 yr. old) who "seemed superficially similar to the bodies of reasonably healthy middle-aged men". Muscles were weighed, then strategically cut for pennation angle and fiber length estimation.   Pennation angles were measured in two places for ext. digitorum longus and peroneus longus; both are given. *Eq. A.1* was used to estimate *PCSA*.

[1] Brand, Crowninshield, Wittstock, Pederson, Clark, and van Krieken (1982)

Three fresh cadavers (163 cm female, 172 cm male, 183 cm male) were used in this study. Muscles were dissected from their origins and insertions so that radiographically visible markers (single or multiple nails) could be attached in their place. Where muscle origin or insertion sites occurred over broad areas, the centroids of the attachment areas (estimated) were used to specify origin/insertion "points".   Some muscles (e.g. gluteus maximus) which exert forces upon excep-

tionally large regions of bone were considered to be composed of two or three components; thus, such muscles can be seen to have multiple numbered listings in the table. For muscles that do not act in a straight line, "effective" origin and/or insertion points were marked by a single point located where the estimated centroid of the cross section of the muscle or tendon crossed the joint and had the most realistic effect on moment arm predictions.

X-ray imaging was used to determine the Cartesian coordinates of the embedded metal markers. Locations of these points are specified with respect to tibial, femoral, and pelvic reference frames defined as follows: "The pelvic reference frame has its origin at the center of the acetabulum with axis directions defined as:

$$Y^P = P_3^P - P_2^P$$

$$X^P = Y^P \times (P_1^P - P_2^P) \qquad (A.2)$$

$$Z^P = X^P \times Y^P$$

where P_1^P is the right anterior superior iliac spine, P_2^P is the midpoint between the pubic tubercles, and P_3^P is the midpoint between the anterior superior iliac spines. The femoral reference frame has its origin at the midpoint between the medial and lateral epicondyle with axes directions defined as:

$$Y^F = P_3^F - P_2^F$$

$$X^F = Y^F \times (P_1^F - P_2^F) \qquad (A.3)$$

$$Z^F = X^F \times Y^F$$

where P_1^F is the lateral epicondyle, P_2^F is the midpoint of the medial and lateral epicondyle, and P_3^F is the femoral head center. The tibial reference frame has its origin at the midpoint between the medial and lateral malleolus with axis directions defined as:

$$Y^T = P_3^T - P_2^T$$

$$X^T = Y^T \times (P_1^T - P_2^T) \qquad (A.4)$$

$$Z^T = X^T \times Y^T$$

where P_1^T is the lateral maleolus, P_2^T is the midpoint between the medial and lateral malleolus, and P_3^T is the tibial tuberosity."

[2] Friedrich and Brand (1990)

The legs of two embalmed cadavers were dissected: *(1)* 37 year-old, 183 cm, 91 kg male, and *(2)* a 63 year-old, 168 cm, 59 kg female as denoted by S1 and S2, respectively, in Table A.1. Musculotendon (total) lengths and average pennation angles were measured with the bodies placed in a supine position ("with hips close to full extension and neutral abduction-adduction and rotation, knees extended, and feet in a neutral position neither dorsi- nor plantar-flexed"), prior to dissection. Excised muscle lengths were used instead of origin to insertion distances for muscles that wrapped around joints and hence followed curving, instead of straight, pathways.

Muscle volumes were obtained after excision via water displacement. Average fiber lengths were measured directly from single fibers, which were prepared from macerated fiber bundles obtained by sampling different parts of each muscle. Physiological cross-sectional areas were calculated by dividing the muscle volume by average muscle fiber length. Separate measurements of tendon were not included in these studies. The reader is cautioned that muscle volumes, and hence *PCSAs*, appear to shrink following death. Therefore *PCSAs* as measured in cadavers may be smaller than those in living subjects (Brand, 1990; Edgerton, 1990).

[3] Pierrynowski and Morrison (1985)

Muscle properties were obtained via a combination of dry-bone measurements (from a dried, disarticulated, male Caucasian skeleton), scaling, and literature surveys. A measurement device capable of 1 mm accuracy in *3-D* was built to collect data on the 38 muscles deemed to be important. Some muscles were partitioned into "two or more distinct structures due to functional considerations", so that data on 47 "muscles" were provided. Data includes useful measures such as tendon and fiber lengths as a percentage of total length, a shape factor *E*, the "maximum anatomical cross-sectional area divided by its mean anatomical cross-sectional area, tendon cross-sectional area, pennation, and muscle mass. To obtain muscle volumes, Eycleshymer and Shoemaker's cross-section anatomy text (1970) was used to define muscle cross-sections, from which areas could be measured via planimetry, and volumes estimated by summing the products of area and slice thickness. Area representations of three fiber types (percentages of *SO*, *FO*, *FG*) are also provided. However, many of these percentile values were estimated, for example, the subdivision of the fast twitch fibers into the oxidative and glycolytic group.

Dry-bone measurements of muscle–tendon origin and insertion locations, as well as up to four intermediate points describing a nonlinear pathway are reported to have been measured. Unfortunately, this data and the measured values of musculotendon lengths, were not available from this publication.

[4] Dostal and Andrews (1981)

Musculotendon origins and insertions were determined for the muscles of the pelvis. Points representing the centers of muscle attachment were marked on the bony pelvis and right femur of an adult, male, dry-bone specimen using anatomical texts as a guide (Gardner et al., 1969; Grant and Basmajian, 1965; Gray, 1959; Hollinshead, 1967; Morris, 1953). Each long, two-joint hip muscle that crosses the knee was given a fictitious distal attachment point on the femur. Likewise, two muscles that wrapped around the underlying bony pelvis prior to crossing the hip were also assigned fictitious pelvic attachments. These origins were taken as the last point of muscle contact with the bony pelvis.

Coordinates of bony landmark and muscle attachment points were measured in a convenient right-handed orthogonal laboratory reference system which was subsequently transformed to embedded pelvic and femoral coordinate systems. The origins of each of these coordinate systems are located at the center of the acetabulum/femoral head. The axes of each system were defined such that they would be coincident when the pelvis and femur were in the position referred to as the zero joint configuration, a position similiar to the anatomical position. In such a position, the plane containing the two anterior superior iliac spines and the more anterior of the two pubic tubercles would be parallel to the frontal (y,z)-plane. The straight line connecting the two anterior superior iliac spines was defined to be parallel to the z–axis . Hence, with the positive x–axis directed anteriorly, and with the positive z–axis directed laterally for this right hip specimen, the positive y–axis would therefore be defined superiorly from the origin.

[5] White, Yack, and Winter (1989)

Musculotendon origins and insertions measured from several dry-bone specimens (six disarticulated pelves, nine femurs and combined tibia-fibula, one reconstructed skeletal foot and one dissected cadaver foot) are provided in scaled form. All coordinates were scaled homogeneously and mapped to a subject of 66.5 kg mass and 1.77 m height. Measurements were made using graphical techniques (graph paper and drawing tools) to a precision of 2 mm. Four coordinate axes were used to measure bony landmarks and the origins and insertions of 40 muscle–tendons. The pelvis coordinate system used the right anterior superior iliac spine as the origin with left anterior superior iliac spine and right pubic tubercle as markers for defining axis orientation (x–axis points anteriorly, y–axis points superiorly, and the z–axis points laterally). The origin of the femoral coordinate system was located at the greater trochanter and orientation was defined using the medial lateral epicondyle. The three axes of the femoral system are all parallel to their respective axes in the the pelvic system. The tibial coordinate system used the tibial tuberosity as its origin with the same orientation as the other two systems. The origin of the foot segment is located at the heel with the same orientation as the other three systems.

[6] Spoor, Van Leeuwen, De Windt, and Huson (1989)

All anatomical measures were taken from one embalmed male specimen (a 48 cm baby) who had died during partus. Values of pennation angle, muscle length from origin to insertion, and physiological cross sectional area were determined for both the fetal and neonatal posture. The fetal posture, designated "fet" in the table, is described as hyperflexion of hip and knee, lateral rotation and very little abduction of the femur, the foot touches the trunk. The neonatal posture, designated "neo" in the table, is described as abduction, lateral rotation, and slight flexion of the hip.

[7] Seireg and Arvikar (1989)

Musculotendon parameter and coordinate data are based on *i)* direct measurements made on a skeleton of average size, *ii)* several approximately same-size (171 cm, 68 kg) cadavers (Seireg, 1990), and *iii)* scaled (2–D) diagrams based on Braus (1954). Data sets from this reference are used in all five appendices. The coordinate systems used to define the lower extremity muscles are defined as follows: pelvis, femur, tibia, calcaneus, retinaculum, talus, navicle, cuboid, cuneiforms one through three, metatarsals one through five, and finally the toes. The coordinate systems used to define the trunk muscles are thorax, pelvis, femur, and lumbar vertebrae ($L1$ through $L5$). The upper extremities musculature utilized five coordinate systems as follows: ulna, radius, humerus, scapula, and hand. The hand musculature used twenty different coordinate systems: four for each finger, three for the thumb and one for the ulna/radius. They are ulna/radius, five metacarpals ($MC1$ through $MC5$), five proximal phalanxes ($PP1$ through $PP5$), four middle phalanxes ($MP2$ through $MP5$), and five distal phalanxes ($DP1$ through $DP5$). Finally, the head musculature utilized ten reference frames. They include the first seven cervical vertebrae ($C1$ through $C7$), the skull, the thorax, and the clavicle.

In general, coordinate axes are located at the joint centers and aligned along bony axes. For vertebrae, reference frames are located at the centers of the inferior inter-vertebral disks, oriented such that Z points superiorly, and X points anteriorly. For the upper-extremity limbs, the reference frames appear to be located at the centers of the proximal joints, and oriented similarly to the vertebral axes in the anatomical position. Lower-extremity reference frames are located at distal joint centers. Finger-segment reference frames are located at the centers of the proximal joints, and oriented such that X points distally along the bone axes, and Y points laterally. The skull reference is located at the "level of the atlanto occipital joint, midway between the two occipital condyles." The clavicular frame is located at its junction with the sternum; orientations of the skull and clavicular frames are similar to that for the vertebral frames. The reader is referred to the original source for further definition.

[8] Wickiewicz, Roy, Powell, and Edgerton (1983)

Architectural features of the major knee flexors/extensors and ankle plantarflexors/dorsi-flexors were measured in three human cadavers (as denoted by $S1$, $S2$, and $S3$ in Table A.1). The hemipelvectomy sections were fixed in formalin with the hip and knee in maximum extension and the ankle in maximum dorsiflexion. Muscle lengths were measured as "the distance between the most proximal and the most distal" muscle fibers during dissection. Muscles were cleaned of fat and excessive connective tissue before weighing and preparation for architectural determinations (see Sacks and Roy, 1982). *PCSA* was computed as

$$PCSA = \frac{m}{\rho\, l} \cos \alpha \qquad (A.5)$$

where m was the wet weight of the fixed muscles and ρ was assumed to be 1.056 g/cm^3.

Their findings are characterized by a "marked uniformity of fiber length throughout a given muscle", similarity of fiber lengths within muscles of a synergistic group, "generally consistent" pennation angles throughout a given muscle, and "remarkably similar" ratios of muscle length to fiber length.

[9] Delp, Loan, Hoy, Zajac, Topp, and Rosen (1990); [10] Delp (1990)

These data sets are modifications of data presented

originally by Brand et al. (1982, 1986) and Friederich and Brand (1990). Musculotendon pathways have been defined relative to three-dimensional bone surface models. To acquire the bone surface data, bone surfaces were marked with a mesh of polygons, and digitized using a Polhemus electromagnetic digitizer to determine coordinates of the vertices. Bone surface coordinates [those measured by Delp et al. (1990) were used to define the pelvis and femur; other skeletal segments were derived by Stredney (1990)] were displayed on a computer graphics workstation (Silicon Graphics, IRIS 2400T) as Gouraud-shaded surfaces. Based on the anatomical landmarks of the bone surface models, the paths (i.e., the lines of action) of forty-three musculotendon actuators were defined. Each musculotendon path is represented as a series of line segments. Origins and insertions alone, in some cases, were sufficient for describing the muscle path (e.g., soleus). In other cases, where muscle wraps over bone

or is constrained by retinacula, intermediate "via points" were introduced to represent the muscle path more accurately. The number of via points was dependent on the body position. For example, the quadriceps tendon contacts the distal femur when the knee is flexed beyond about 80° (Yamaguchi and Zajac, 1989), but does not when the knee is extended. Thus, additional "wrapping points", were introduced for large knee flexion angles so that the quadriceps tendon would wrap over, rather than pass through, the bone. Wrapping points that are dependent on joint angle are designated by the symbol "**" in Table A.1.

Seven different coordinate systems were used to define the lower extremity muscles in this paper (Figure A.1). In the anatomical position, each of these reference frames are oriented such that the x–axes point anteriorly, the y–axes superiorly, and the z–axes point laterally.

[10] White (1989)

The average pennation angle was obtained from data from several sources. Fiber type compositions are provided as being of either *type I* (slow twitch, or *SO*), or *type II* (fast twitch, or *FG*), and were assigned using data from the literature. Fiber resting lengths were calculated from average numbers of sarcomeres per fiber (Wickiewicz et al., 1983), and an average sarcomere rest length of 2.6 μm. See also White et al. (1989).

[11] Hoy, Zajac, and Gordon (1990)

Grouped musculotendon actuators were constructed from the data sets of Brand et al. (1982), Friedrich and Brand (1990) according to sagittal-plane function. The coordinate systems used were consistent with those utilized by Brand et al. (1982), except that the longitudinal coordinate axis of the shank reference frame (the "Y^T" axis) was rotated 5° about a medial to lateral axis and displaced 1.4 cm distal to the tibial frame of Brand et al. (1982). By doing so, the local coordinate axes were more nearly placed at the rotational axes of the ankle joint, and the y-axis of the shank frame pointed toward the origin of the femoral coordinate frame instead of the tibial tuberosity. Curved musculotendon pathways were approximated by one to three straight-line segments defined in Cartesian space. Tendon slack lengths were determined by comparing the joint angles at which *(i)* passive moments initiate and *(ii)* active moments peak to *in-vivo* measurements. Origins and insertions were defined relative to the pelvic, femoral, tibial, and calcaneal reference frames.

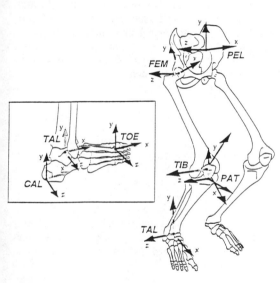

Figure A.1: Location of reference frames in the model of Delp et al. 1990. **PEL** (pelvis): fixed at the midpoint of the line connecting the two anterior superior iliac spines. **FEM** (femur): fixed at the center of the femoral head. **PAT** (patella): located at the most distal point of the patella. **TIB** (tibia): located at the midpoint of the line between the medial and lateral femoral epicondyles. **TAL** (talus): located at the midpoint of the line between the apeces of the medial and lateral maleoli. **CAL** (calcaneus): located at the most inferior, lateral point on the posterior surface of the calcaneus. **TOE**: located at the base of the second metatarsal. (From Delp (1990), reprinted with permission).

[12] Freivalds (1985)

Musculotendon origin/insertion coordinates were

measured from scaled photographs obtained from anatomical texts (among others, McMinn and Hutchings, 1977), and are given for the right side of the body. The reference frames relating to 10 of the 15 body segments are referred to by: pelvis, center torso, upper torso, neck, head, femur, tibia, foot, upper arm and lower arm. Origins of the segmental reference frames are defined relative to the center of mass of the segment identified, however, orientation was not specified. Joints are defined as: head junction, neck junction, waist, *L5/S1*, hips, knees, ankles, shoulders, and elbows. Muscle cross-sectional areas were determined by an independent investigator (C. S. Davis, not referenced) and are taken primarily from the Schumacher and Wolff (1966) reference.

[13] Reid and Costigan (1985)

Cross-sectional areas and average volumes for the rectus abdominis and erector spinae muscles were calculated based on computer tomography scans of 28 living subjects (16 normal males and 12 normal females). Transverse (coronal) scans were taken at 0.5 to 2.0 cm intervals between the xiphoid process and the symphysis pubis. The subjects ranged from 17-75 years old (mean 53.5 yrs), 147-185.4 cm in height (mean 166.1 cm), and 47-100.9 kg (mean 69.25). Absolute muscle moment arms about the spinal column were also calculated, though this data is not presented in the tables.

[14] Johnson, Polgar, Weightman and Appleton (1973)

The method of Jennekens et al. (1971) was used to measure the distribution of *Type I* and *II* muscle fibers in samples of 36 human muscles obtained from 6 normal, male adults at autopsy. The subjects ranged in age from 17 and 30 years (mean age 21.8 yrs), in height from 183-198 cm (mean 186.3), and in weight from 61-98 kg (mean 78.5). Muscle specimens were removed, placed in solution, and frozen within 24 hours after death in preparation for later measurement.

[15] Lehmkuhl and Smith (1983)

This source contained data on muscle cross-sectional area as adapted from Fick (1911); however, methods by which values were obtained were not stated in the current text. The cross-sectional areas of the upper extremity muscles were given for several positions of the arm (e.g. supination, midposition, pronation). The values of the arm in "midposition" were used in the table.

[16] Wood, Meek, and Jacobsen (1989)

The torso of an embalmed, male cadaver (from a person having good muscle definition and estimated to be 180 cm tall and 91 kg total weight) was used to define muscle parameters. The eviscerated torso was filled with expanding polyurethane foam and secured to a frame. Distributed sets of points defining musculotendinous attachment areas were collected using an electromechanical digitizer (accuracy estimated at better than 0.5 mm) and averaged to yield the single-point origin and insertion data presented in Table A.3. Computer-generated surface topographies (bicubic patches) of the muscles were also defined with the aid of the digitizer. Muscle lengths, surface areas, volumes, and cross-sectional areas were defined from these computerized representations. Reference frames ulna are shown graphically, and appear to be defined with respect to the eleven revolute axes of the arm linkage. These reference frames thus appear to be located along the elbow flexion/extension axis, at the humeral head (the intersection of the humeral flexion/extension, axial rotation, and ab/adduction axes), and at the approximate kinematic centers of the thorax/clavicle and clavicle/scapula joints. The reader is referred to Table 3 and Section VI.A. of the original reference for further definition.

[17] Högfors, Sigholm, and Herberts (1989)

Three dissected cadavers (range of subject ages: 55-71 yrs) and one human skeleton were used to determine the insertions of 21 muscles spanning the shoulder. Measurements were made using a ruler; accuracy was estimated to be within ± 4 mm in worst-case situations. Coordinates of the corresponding muscle insertions were defined using the following coordinate systems: thorax, scapula, clavicle, and humerus. Coordinates of the origin points, however, were not included.

The sternum system has its origin in the center of the sternoclavicular joint. It is such that the *x*-axis goes through the middle of both articular surfaces, i.e. it is normal to the sagittal plane and directed away from it. The entire system is orthonormal and the *x–y* plane contains the midpoint of the first thoracic vertebra. The *y*-axis and *z*-axis are directed forwards and upwards, respectively. The clavicular system has the same origin as the sternum system. Its *x*-axis goes through the center of the acromial articular surface. The *y*-axis is orthogonal to the *x*-axis and parallel to the upper planar surface on the lateral end of the clavicula. The *y*-axis is directed forwards and the *z*-axis upwards. The scapula system has its origin in the point common to the clavicula and the scapula.

The x–axis of the system is directed through the inferior angle. The x–y plane contains the superior angle in its first quadrant. The last coordinate system is that fixed in the humerus. The gliding surface of the humeral head has an essentially spherical shape. The center of this sphere is used as the origin of the humerus system. The x–axis is directed along the humerus through the end of the ridge between the coronoid fossa and the radial fossa. The y–axis lies in the plane of the x–axis and the angular mobility direction of the ulna, with the ulna having a positive y–coodinate.

[18] Daru (1989)

Muscle parameters (lengths, fiber compositions, tendon lengths, pennation angles, and PCSAs) of 38 muscles in the human head and neck are presented. Muscle lengths were measured from an anthropomorphic model representing a "small man". Origin/insertion points of the anthropomorphic model were based on data presented in Warfel (1973). Masses for the majority of the muscles were estimated from consideration of cross-sectional slices from Koritké and Sick (1983); other muscle masses were estimated based on their sizes relative to larger ones. Pennation angles were conservatively estimated at 10° for lack of better data. Thickness estimates for PCSAs were estimated based primarily on measurements made from Toldt and Hochstatter (1957). Percentages of slow muscle fibers were estimated (and scaled toward 50%) based on data giving the ranges of slow fiber distributions in cat muscle (Richmond, 1988). Tendon lengths were estimated from inspection of anatomy sources.

[19] Brand, Beach, and Thompson (1981)

Here, the ratios of muscle mass and PCSA were investigated, based on the observation that while muscle strength varies enormously with time, and from subject to subject, the ratios of muscle strengths within the same limb vary much less.

To develop a list of muscle qualities independent of time or the overall strength of an individual, the fleshy part of each of the muscles below the elbow (including portions of brachioradialis and extensor carpi radialis longus that extend above the elbow, but not including the anconeus) in 15 cadavers was weighed. Muscle "mass fraction" was defined as the percentage of individual muscle mass to total muscle mass. These enable comparisons to all arms as long as the total muscle weight is known or can be estimated. Mass fractions in atrophied and nonatrophied arms were found to be comparable.

Mean fiber lengths were measured in five of the

cadavers. These specimens were of normal build and were placed with the hands at a position of "physiologic rest" (wrist in 10° of extension, finger joints flexed at about 45°, thumb opposite the middle segment of the middle finger). Finally, the mass of each muscle was converted to volume and divided by the mean fiber length of that muscle to get the physiological cross-sectional area (PCSA) of the muscle. The PCSA's of the hand and forearm were then totaled. The percentage of each muscle in comparison to the total PCSA was termed "tension fraction". The average total PCSA of the muscles in each arm was 141 cm^2.

It must be clearly understood that these values for tension fraction are only proportional and involve no numerical statement of force, tension or PCSA. However, their figures do suggest that a muscle with a tension fraction of three is probably capable of 50% more tension than one with a value of two.

[20] Amis, Dowson, and Wright (1979)

Four limbs of stout muscular build were removed from embalmed cadavers, complete with shoulder girdle and pectoral muscles. Alexander's formula (1975) was used to compute each muscle's PCSA. To apply the formula, they measured the following upper-limb parameters: muscle mass, fiber length, and layer thickness of muscle pennation. For bipennate muscle fibers, each pennation angle was computed independently, as was each PCSA, then the PCSA's were added to obtain the total PCSA for the bipennate muscle. Muscle density was assumed to be 1050 (kg/m^3).

[21] Kleweno (1987)

Parameters for muscles crossing the elbow joint were estimated based on the data of An et al. (1981) and the previous estimations of Winters (1985), which in turn were based on anatomical sources combined with the data of Amis (1979) and An et al. (1981), scaled for an "average sized male." Additionally, origin-insertion values were checked and in a few cases modified slightly based on attachment of strings to an "average sized male" anthropomorphic replica with plastic bones (bones obtained from Carolina Biological Supply Co.). Not presented here are computer model results which estimate muscle moment arms and muscle length changes as a function of shoulder, elbow and wrist angles and the different assumptions utilized to curve muscle paths.

[22] An, Hui, Morrey, Linscheid, and Chao [1981]

The volumes, fiber lengths and physiological cross-sectional areas of muscles spanning the elbow joint were obtained from four fresh, unembalmed cadaver specimens "comparable in size". The specimens were dissected with the elbows in resting positions (70° flexion) following the technique developed by Brand (Brand et al. 1980). Volumes of all the muscles were measured by water displacement. The physiological cross-sectional area was calculated by dividing the muscle volume by its fiber length.

[23] An, Chao, Cooney, and Linscheid [1979]

Ten normal hand specimens were examined to develop a normative model of the hand. Small segments of each tendon and muscle near each of three finger joints (metacarpal-phalangeal (MP), proximal interphalangeal (PIP), and distal interphalangeal (DIP) joint in the finger; CMC, MCP, and IP joints in the thumb) were exposed via surgical incision. Surgical wire markers were inserted into the centers of each tendon and muscle at sites "immediately proximal and distal to the joint." Intermediate points between the tendon origins/insertions were defined "with reference to the pulley systems, retinacular ligaments and transverse bands which constrain the tendons and muscles at the joint" and "minimize bowstringing with flexion and extension", so as to best represent the direction of tendon force. Measurements were taken from anterior-posterior and lateral X-ray films.

Coordinates for each finger are normalized by the length of the middle phalanx. The length of the proximal phalanx normalized data for the thumb. The normative data presented was determined by averaging the "force" and "moment potentials" of each muscle-tendon as determined by the X-ray measures (see original source for definitions). Data is presented with respect to six Cartesian coordinate systems for each finger, numbered from distal (1) to proximal (6). Primary coordinate systems (numbers 2, 4, and 6) are "located at the approximate center of rotation of the phalangeal and metacarpal heads", and the secondary systems (numbers 1, 3, and 5) are located at "the centers of the concave articular surfaces". Therefore, reference frames are located immediately distal and proximal to each joint, i.e., odd-numbered frames are located immediately distal to each joint, while even-numbered frames are located immediately proximal. X-axes are directed proximally along the phalangeal or metacarpal shafts (from the origin of one frame to the origin of the next). Y-axes are projected dorsally from each segment, and the z-axes are projected radially for the right hand and ulnarly for the left hand.

Acknowledgements:

This project was supported in part by BRSG 2 S07 RR07112, Division of Research Resources, National Institutes of Health.

References

Alexander, R.McN. and Vernon, A. (1975). The dimensions of knee and ankle muscles and the forces they exert. *J. Human Movement Studies* **1**:115-123.

Amis, A.A., Dowson, D., and Wright, V. (1979). Muscle strengths and musculo-skeletal geometry of the upper limb. *Engineering in Medicine* **8**:41-48.

An, K.N., Chao, E.Y., Cooney, W.P. III, and Linsheid, R.L. (1979). Normative model of human hand for biomechanical analysis. *J. Biomech.* **12**:775-788.

An, K.N., Hui, F.C., Morrey, B.F., Linsscheid, R.L. and Chao, E.Y. (1981) Muscles across the elbow joint:a biomechanical analysis. *J. Biomech.* **14**:659-669.

Bogduk, N. and Twomey, L.T (1987) *Clinical Anatomy of the Lumbar Spine.* New York, Churchill Livingstone.

Brand, P.W., Beach, R.B., and Thompson, D.E. (1981). Relative tension and potential excursion of muscles in the forearm and hand. *J. Hand Surgery* **6**:209-219.

Brand, R.A. (1990). University of Iowa - personal communication.

Brand, R.A., Crowninshield, R.D., Wittstock, C.E., Pedersen, D.R., Clark, C.R., and Van Krieken, F.M. (1982). A model of lower extremity muscular anatomy. *J. Biomech. Engrg.* **104**:304-310.

Brand, R.A., Pedersen, D.R., and Friederich, J.A. (1986). The sensitivity of muscle force predictions to changes in physiologic cross-sectional area. *J. Biomech.* **19**:589-596.

Braus (1954). *Anatomic Der Menschen.* Springer-Verlag, Berlin.

Delp, S.L. (1990). A computer-graphics system to analyze and design musculoskeletal reconstructions of the lower limb. Ph.D. Thesis, Dept. of Mechanical Engineering, Stanford University.

Delp, S.L., Loan, J.P., Hoy, M.G., Zajac, F.E., Topp, E.L., and Rosen, J.M. (1990). An interactive graphics-based model of the lower extremity to study orthopaedic surgical procedures. *IEEE Trans. Biomed. Eng.* (in press).

Daru, K.R. (1989). Computer simulation and static analysis of the human head, neck, and upper torso. M.S. Thesis, Dept. of Chemical, Bio, and Materials Engineering, Arizona State University, Tempe, AZ.

Dostal, W.F. and Andrews J.G. (1981) A three dimensional biomechanical model of hip musculature. *J. Biomech.* **14**:803-812.

Edgerton, V.R. (1990). UCLA - personal communication.

Eychleshymer, A.C., and Schoemaker, D.M. (1911). *A Cross-Section Anatomy.* Appleton-Century- Crofts, New York.

Fick, R. (1911). Anatomie und Mechanik der Gelenke: Teil III, Spexielle Gelenk und Muskel Mechanik.

Fischer, Jena.

Freivalds, A. (1985). Incorporation of active elements into the articulated total body model. Paper #AAMRL-TR-85-061, Armstrong Aerospace Medical Research Laboratory, Wright-Patterson Air Force Base, OH.

Friederich, J.A. and Brand, R.A. (1990). Muscle fiber architecture in the human lower limb. Technical Note, *J. Biomech.* **23**:91-95.

Gardner et al. (1969). *Anatomy*, W.B. Saunders, Philadelphia.

Grant, J.C. and Basmajian, J.V. (1965). *Grants Method of Anatomy*, Williams and Wilkins, Baltimore.

Gray, H. (1959). *Anatomy of the Human Body* (Edited by Goss, C.), Lea and Febinger, Philadelphia.

Högfors, C., Sigholm, G., and Herberts, P. (1987). Biomechanical model of the human shoulder — I. Elements. *J. Biomech.* **20**:157-166.

Hollinshead, W.H. (1967). *Textbook of Anatomy*, Harper and Row, New York.

Hoy, M.G., Zajac, F.E., and Gordon, M.R. (1989). A musculoskeletal model of the human lower extremity: The effect of muscle, tendon, and moment arm on the knee and ankle. *J. Biomech.* **22**:157-169.

Jennekens et al. (1971). Data on the distribution of fiber types in five human limb muscles, *J. Neurol. Sci.* **14**:245-257 (as quoted in Johnson et al. (1973).

Johnson, M.A., Polgar J., Weightman, D., and Appleton, D. (1973). Data on the distribution of fiber types in thirty-six human muscles, *J. Neurol. Sci.* **18**:111-129.

Kleweno, D.G. (1987) Physiological and theoretical analysis of isometric strength curves of the upper limb. M.S. Thesis, Dept. of Chemical, Bio & Materials Engineering, Arizona State University.

Koritké, J.G., and Sick, H. (1983). *Atlas of Sectional Human Anatomy, Frontal, Sagittal, and Horizontal Planes, Vol.1, Head, Neck Thorax.* Urban & Scharzenburg, Baltimore.

Lehmkuhl, L.D. and Smith, L.K. (1983). *Brunnstrom's Clinical Kinesiology.* F.A. Davis Co., Philadelphia. pp. 391-401.

McMinn, R.M.H., and Hutchings, R.T. (1977). *Color Atlas of Human Anatomy.* Yearbook Medical Publishers, Chicago.

Pierrynowski, M.R. and Morrison, J.B. (1985). A physiological model for the evaluation of muscular forces in human locomotion: Theoretical aspects. *Math. Biosci.* **75**:69-101.

Reid, J.G. and Costigan P.A. (1985). Geometry of adult rectus abdominis and erector spinae muscles. *J. Orthop. and Sports Phys. Ther.* **5**:278-280.

Richmond, F.J. (1988) The motor system:joints and muscles of the neck. In *Control of Head Movements*, Chapter 1, pp. 1-22.

Sacks, R.D., and Roy, R.R. (1982). Architecture of the hind limb muscles of cats:Functional significance. *J. Morphol.* **173**:185.

Schaeffer, J. (1953). *Morris' Human Anatomy.* Blakiston, New York.

Schumacher, G.H. and Wolff, E. (1966). Trockengewicht and Physiologischer Querschnitt der Menschlichen Skelettmuskulatur, II, Physiologische Querschnitte, *Anat. Anz.* **119**:259-269.

Seireg, A. (1990). University of Wisconsin, Madison - personal communication.

Seireg, A. and Arvikar, R.J. (1989). *Biomechanical Analysis of the Musculoskeletal Structure for Medicine and Sports.* Hemisphere Publishing Corporation, New York.

Spoor, C.W., Van Leeuwen, J.L., De Windt, F.H.J., Huson, A. (1989). A model study of muscle forces and joint-force direction in normal and dysplastic neonatal hips. *J. Biomech.*, **8**:873-884.

Stredney, D. (1989). Ohio State University - personal communication.

Toldt, C., and Hochstetter, F. (1957). *Anatomischer Atlas für Studierende und Ärtze*, Vol. 1. Heinrich Hayek, ed., 23rd edit., Urban & Scharzenburg, Wien.

Warfel, J.H. (1973). *The Head, Neck, and Trunk Muscles and Motor Points*. 4th ed. Lea & Febiger, Philadelphia.

White, S.C. (1989) A deterministic model using EMG and muscle kinematics to predict individual muscle forces during normal human gait. Ph.D. Thesis, Dept. of Kinesiology, University of Waterloo.

White, S.C., Yack, H.J., and Winter, D.A. (1989). A three-dimensional musculoskeletal model for gait analysis. Anatomical variability estimates. *J. Biomech.* **22**:885-893.

Wickiewicz, T.L., Roy, R.R., Powell, P.L., and Edgerton, V.R. (1983). Muscle architecture of the human lower limb. *Clin. Orthop. Rel. Res.* **179**:275-283.

Winters, J.M. (1985) Generalized analysis and design of antagonistic muscle models:effect of nonlinear properties on the control of human movement. Ph.D. Dissertation, Bioengineering, Univ. of Calif., Berkeley and San Francisco.

Wood, J.E., Meek, S.G., and Jacobsen, S.C. (1989). Quantitation of human shoulder anatomy for prosthetic arm control — I. Surface modeling". *J. Biomech.* **22**:273-292.

Wood, J.E., Meek, S.G., and Jacobsen, S.C. (1989). Quantitation of human shoulder anatomy for prosthetic arm control — II. Anatomy matrices. *J. Biomech.* **22**:309-325.

Yamaguchi, G.T., Zajac, F.E. (1989). A planar model of the knee joint to characterize the knee extensor mechanism. *J. Biomech.* **22**:1-10.

TABLE A.1
LOWER EXTREMITY MUSCULATURE

MUSCLE		SOURCE	MUSCLE LENGTH ML (m)	FIBER LENGTH LM (m)	VOLUME V (ml)	MUSCLE PCSA =V/L (cm²)	PENNA-TION ANGLE (deg)	MUSCLE FIBER TYPE		
								%SO	%FOG	%FG
HIP MUSCLES										
adductor brevis	(S)	[1]	--	--	--	--	--	--	--	--
	(I)	[1]	--	--	--	--	--	--	--	--
	(1)	[2]-S1	0.110	0.0538	62	11.52	--	--	--	--
	(1)	[2]-S2	0.130	0.1200	23	1.92	0.0	--	--	--
	(2)	[2]-S1	0.145	0.1160	62	5.34	--	--	--	--
	(2)	[2]-S2	0.170	0.1170	23	1.97	0.0	--	--	--
		[3]	--	--	--	--	0.0	0.45	0.15	0.40
		[4]	--	--	--	--	--	--	--	--
		[5]	--	--	--	--	--	--	--	--
	(neo)	[6]	0.031	--	--	0.53	5.0	--	--	--
	(fet)	[6]	0.032	--	--	0.52	5.0	--	--	--
		[7]	--	--	--	--	--	--	--	--
		[8]-S1	0.178	0.115	--	3.9	0.0	--	--	--
		[8]-S2	0.152	0.093	--	6.6	0.0	--	--	--
		[8]-S3	0.138	0.101	--	3.5	0.0	--	--	--
		[9]	--	0.1330	--	--	0.0	--	--	--
		[10]	--	0.102	--	--	0.0	0.500	--	0.500
		[11]	--	0.132	--	--	3.6	--	--	--
		[12]	--	--	--	4.54	--	--	--	--
adductor longus		[1]	--	--	--	--	--	--	--	--
		[2]-S1	0.185	0.0827	188	22.73	--	--	--	--
		[2]-S2	0.165	0.1056	103	9.75	3.5	--	--	--
		[3]	--	--	--	--	0.0	0.65	0.15	0.20
		[4]	--	--	--	--	--	--	--	--
		[5]	--	--	--	--	--	--	--	--
	(neo)	[6]	0.028	--	--	0.53	4.0	--	--	--
	(fet)	[6]	0.024	--	--	0.62	4.0	--	--	--
		[7]	--	--	--	--	--	--	--	--
		[8]-S1	0.252	0.108	--	6.0	5.0	--	--	--
		[8]-S2	0.210	0.105	--	10.5	5.0	--	--	--
		[8]-S3	0.225	0.112	--	4.0	8.0	--	--	--
		[9]	--	0.1380	--	--	6.0	--	--	--
		[10]	--	0.102	--	--	6.0	0.500	--	0.500
		[11]	--	0.132	--	--	3.6	--	--	--
		[12]	--	--	--	5.03	--	--	--	--
adductor magnus	(1)	[1]	--	--	--	--	--	--	--	--
	(2)	[1]	--	--	--	--	--	--	--	--
	(3)	[1]	--	--	--	--	--	--	--	--
	(1)	[2]-S1	0.125	0.0870	222	25.52	--	--	--	--
	(1)	[2]-S2	0.125	0.0770	84	10.91	5.0	--	--	--
	(2)	[2]-S1	0.220	0.1210	222	18.35	--	--	--	--
	(2)	[2]-S2	0.220	0.1205	84	6.97	2.5	--	--	--
	(3)	[2]-S1	0.225	0.1311	222	16.95	--	--	--	--
	(3)	[2]-S2	0.235	0.1416	84	5.93	5.0	--	--	--
	(ant)	[3]	--	--	--	--	0.0	0.55	0.15	0.30
	(mid)	[3]	--	--	--	--	0.0	0.55	0.15	0.30
	(pos)	[3]	--	--	--	--	0.0	0.55	0.15	0.30
	(mid)	[4]	--	--	--	--	--	--	--	--
	(pos)	[4]	--	--	--	--	--	--	--	--
	(mid)	[5]	--	--	--	--	--	--	--	--
	(pos)	[5]	--	--	--	--	--	--	--	--
	(neo)	[6]	0.031	--	--	2.52	20.0	--	--	--
	(fet)	[6]	0.048	--	--	1.66	20.0	--	--	--
	(add)	[7]	--	--	--	--	--	--	--	--
	(ext)	[7]	--	--	--	--	--	--	--	--
		[8]-S1	0.327	0.131	--	20.1	0.0	--	--	--
		[8]-S2	0.299	0.106	--	20.8	0.0	--	--	--
		[8]-S3	0.288	0.109	--	13.6	0.0	--	--	--
	(1)	[9]	--	0.0870	--	--	5.0	--	--	--
	(2)	[9]	--	0.1210	--	--	3.0	--	--	--
	(3)	[9]	--	0.1310	--	--	5.0	--	--	--
		[10]	--	0.136	--	--	0.0	0.584	--	0.416
		[11]	--	0.144	--	--	0.0	--	--	--
	(1)	[12]	--	--	--	20.58	--	--	--	--
	(2)	[12]	--	--	--	--	--	--	--	--
	(surf)	[14]	--	--	--	--	--	0.535	--	0.465
	(deep)	[14]	--	--	--	--	--	0.633	--	0.367
adductor minimus		[4]	--	--	--	--	--	--	--	--
		[5]	--	--	--	--	--	--	--	--

TENDON LENGTH LT (m)	TENDON PCSA (mm²)	COORD. SYSTEM	ORIGIN			COORD. SYSTEM	INSERTION		
			X (m)	Y (m)	Z (m)		X (m)	Y (m)	Z (m)
--	--	pelvis	0.0312	-0.0373	-0.0611	femur	-0.0082	0.2828	0.0215
--	--	pelvis	0.0326	-0.0371	-0.0613	femur	-0.0112	0.2534	0.0211
--	--	--	--	--	--	--	--	--	--
--	--	--	--	--	--	--	--	--	--
--	--	--	--	--	--	--	--	--	--
--	0.0	--	--	--	--	--	--	--	--
--	--	pelvis	0.021	-0.045	-0.067	femur	-0.002	-0.131	0.038
--	--	pelvis	-0.047	-0.100	-0.102	femur	-0.012	-0.101	-0.054
--	--	--	--	--	--	--	--	--	--
--	--	pelvis	0.0400	0.0400	-0.0250	femur	0.0050	-0.0200	0.2800
--	--	--	--	--	--	--	--	--	--
--	--	--	--	--	--	--	--	--	--
0.0200	--	pelvis	-0.0587	-0.0915	0.0164	femur	0.0009	-0.1196	0.0294
0.040	--	pelvis	0.0416	-0.0340	-0.0611	femur	-0.0057	0.2269	0.0168
--	--	pelvis	0.0000	0.0394	0.0935	femur	-0.0300	0.1194	-0.0785
--	--	pelvis	0.0490	-0.0316	-0.0610	femur	-0.0031	0.1924	0.0134
--	--	--	--	--	--	--	--	--	--
--	0.0	--	--	--	--	--	--	--	--
--	--	pelvis	0.041	-0.031	-0.065	femur	0.005	-0.204	0.026
--	--	pelvis	-0.022	-0.087	-0.105	femur	-0.010	-0.198	-0.067
--	--	--	--	--	--	--	--	--	--
--	--	pelvis	0.0500	0.0470	-0.0150	femur	0.0150	-0.0100	0.1800
--	--	--	--	--	--	--	--	--	--
--	--	--	--	--	--	--	--	--	--
0.1100	--	pelvis	-0.0316	-0.0836	0.0169	femur	0.0050	-0.2111	0.0234
0.040	--	pelvis	0.0416	-0.0340	-0.0611	femur	-0.0057	0.2269	0.0168
--	--	pelvis	0.0000	0.0193	0.0833	femur	-0.0300	0.1194	-0.0335
--	--	pelvis	-0.0117	-0.0552	-0.0486	femur	-0.0122	0.2758	0.0290
--	--	pelvis	-0.0120	-0.0552	-0.0485	femur	-0.0036	0.1740	0.0163
--	--	pelvis	-0.0120	-0.0551	-0.0486	femur	-0.0064	0.0166	-0.0297
--	--	--	--	--	--	--	--	--	--
--	--	--	--	--	--	--	--	--	--
--	--	--	--	--	--	--	--	--	--
--	--	--	--	--	--	--	--	--	--
--	--	--	--	--	--	--	--	--	--
--	0.0	--	--	--	--	--	--	--	--
--	0.0	--	--	--	--	--	--	--	--
--	15.0	--	--	--	--	--	--	--	--
--	--	pelvis	-0.031	-0.061	-0.044	femur	0.005	-0.228	0.027
--	--	pelvis	-0.048	-0.059	-0.034	femur	0.001	-0.404	-0.031
--	--	pelvis	-0.092	-0.119	-0.073	femur	-0.010	-0.211	-0.067
--	--	pelvis	-0.108	-0.123	-0.068	femur	-0.028	-0.391	-0.117
--	--	--	--	--	--	--	--	--	--
--	--	pelvis	-0.0200	0.0100	-0.0550	femur	0.0000	0.0400	0.0200
--	--	pelvis	0.0000	0.0200	-0.0550	femur	0.0100	-0.0100	0.2202
--	--	--	--	--	--	--	--	--	--
--	--	--	--	--	--	--	--	--	--
0.0600	--	pelvis	-0.0732	-0.1174	0.0255	femur	-0.0045	-0.1211	0.0339
0.1300	--	pelvis	-0.0831	-0.1192	0.0308	femur	0.0054	-0.2285	0.0227
0.2600	--	pelvis	-0.0771	-0.1181	0.0276	femur	0.0070	-0.3837	-0.0266
--	--	--	--	--	--	--	--	--	--
0.070	--	pelvis	-0.0119	-0.0552	-0.0486	femur	-0.0080	0.1728	0.0088
--	--	pelvis	0.0000	0.0795	0.0935	femur	-0.0300	0.1194	-0.1636
--	--	pelvis	0.0000	0.0394	0.0935	femur	-0.0300	0.1194	-0.2235
--	--	--	--	--	--	--	--	--	--
--	--	pelvis	0.007	-0.049	-0.061	femur	-0.004	-0.124	0.040
--	--	pelvis	-0.072	-0.112	-0.088	femur	-0.015	-0.088	-0.049

MUSCLE	SOURCE	MUSCLE LENGTH ML (m)	FIBER LENGTH LM (m)	VOLUME V (ml)	MUSCLE PCSA =V/L (cm²)	PENNA-TION ANGLE (deg)	%SO	%FOG	%FG
gemelli	[9]	--	0.0240	--	--	0.0	--	--	--
gemellus inferior	[1]	--	--	--	--	--	--	--	--
	[2]-S1	0.060	0.0231	10	4.33	--	--	--	--
	[2]-S2	0.042	0.0265	4	1.51	0.0	--	--	--
	[3]	--	--	--	--	0.0	0.50	0.20	0.30
	[4]	--	--	--	--	--	--	--	--
(neo)	[6]	0.008	--	--	0.12	10.0	--	--	--
(fet)	[6]	0.011	--	--	0.10	10.0	--	--	--
gemellus superior	[1]	--	--	--	--	--	--	--	--
	[2]-S1	0.05(	0.0282	6	2.13	--	--	--	--
	[2]-S2	0.065	0.0275	4	1.45	0.0	--	--	--
	[3]	--	--	--	--	0.0	0.50	0.20	0.30
	[4]	--	--	--	--	--	--	--	--
(neo)	[6]	0.006	--	--	0.04	4.0	--	--	--
(fet)	[6]	0.016	--	--	0.02	4.0	--	--	--
gluteus maximus (1)	[1]	--	--	--	--	--	--	--	--
(2)	[1]	--	--	--	--	--	--	--	--
(3)	[1]	--	--	--	--	--	--	--	--
(1)	[2]-S1	0.155	0.1421	288	20.20	--	--	--	--
(1)	[2]-S2	0.160	0.0747	109	14.59	5.0	--	--	--
(2)	[2]-S1	0.165	0.1474	288	19.59	--	--	--	--
(2)	[2]-S2	0.160	0.1136	109	9.60	0.0	--	--	--
(3)	[2]-S1	0.185	0.1440	288	20.00	--	--	--	--
(3)	[2]-S2	0.155	0.1345	109	8.10	5.0	--	--	--
(deep)	[3]	--	--	--	--	0.0	0.50	0.20	0.30
(sup)	[3]	--	--	--	--	0.0	0.50	0.20	0.30
	[4]	--	--	--	--	--	--	--	--
	[5]	--	--	--	--	--	--	--	--
(neo-caud)	[6]	0.031	--	--	3.09	5.0	--	--	--
(fet-caud)	[6]	0.049	--	--	1.94	5.0	--	--	--
(neo-cran)	[6]	0.029	--	--	1.64	5.0	--	--	--
(fet-cran)	[6]	0.046	--	--	1.01	5.0	--	--	--
(1)	[7]	--	--	--	--	--	--	--	--
(2)	[7]	--	--	--	--	--	--	--	--
(3)	[7]	--	--	--	--	--	--	--	--
(1)	[9]	--	0.1420	--	--	5.0	--	--	--
(1)	[9]	--	--	--	--	--	--	--	--
(1)	[9]	--	--	--	--	--	--	--	--
(2)	[9]	--	0.1470	--	--	0.0	--	--	--
(2)	[9]	--	--	--	--	--	--	--	--
(2)	[9]	--	--	--	--	--	--	--	--
(3)	[9]	--	0.1440	--	--	5.0	--	--	--
(3)	[9]	--	--	--	--	--	--	--	--
(3)	[9]	--	--	--	--	--	--	--	--
	[10]	--	0.123	--	--	0.0	0.524	--	0.476
	[11]	--	0.180	--	--	3.4	--	--	--
(1)	[12]	--	--	--	29.42	--	--	--	--
(2)	[12]	--	--	--	--	--	--	--	--
	[14]	--	--	--	--	--	0.524	--	0.476
gluteus medius (1)	[1]	--	--	--	--	--	--	--	--
(2)	[1]	--	--	--	--	--	--	--	--
(3)	[1]	--	--	--	--	--	--	--	--
(1)	[2]-S1	0.135	0.0535	137	25.00	--	--	--	--
(1)	[2]-S2	0.100	0.0405	53	13.09	8.0	--	--	--
(2)	[2]-S1	0.125	0.0845	137	16.21	--	--	--	--
(2)	[2]-S2	0.110	0.0508	53	10.43	0.0	--	--	--
(3)	[2]-S1	0.130	0.0646	137	21.21	--	--	--	--
(3)	[2]-S2	0.130	0.0550	53	9.64	19.0	--	--	--
(ant)	[3]	--	--	--	--	0.0	0.50	0.20	0.30
(mid)	[3]	--	--	--	--	0.0	0.50	0.20	0.30
(pos)	[3]	--	--	--	--	0.0	0.50	0.20	0.30
(ant)	[4]	--	--	--	--	--	--	--	--
(mid)	[4]	--	--	--	--	--	--	--	--
(pos)	[4]	--	--	--	--	--	--	--	--
(ant)	[5]	--	--	--	--	--	--	--	--
(mid)	[5]	--	--	--	--	--	--	--	--
(pos)	[5]	--	--	--	--	--	--	--	--
(neo-dors)	[6]	0.014	--	--	1.65	15.0	--	--	--
(fet-dors)	[6]	0.037	--	--	0.63	15.0	--	--	--

TENDON LENGTH LT (m)	TENDON PCSA (mm²)	COORD. SYSTEM	ORIGIN			COORD. SYSTEM	INSERTION		
			X (m)	Y (m)	Z (m)		X (m)	Y (m)	Z (m)
0.0390	--	pelvis	-0.1133	-0.0820	0.0714	femur	-0.0142	-0.0033	0.0443
--	--	pelvis	-0.0426	-0.0165	-0.0095	femur	-0.0113	0.3949	0.0448
--	--	--	--	--	--	--	--	--	--
--	15.0	--	--	--	--	--	--	--	--
--	--	pelvis	-0.049	-0.012	-0.009	femur	-0.006	-0.005	0.047
--	--	--	--	--	--	--	--	--	--
--	--	pelvis	-0.0435	0.0009	-0.0201	femur	-0.0113	0.3947	0.0445
--	--	--	--	--	--	--	--	--	--
--	15.0	--	--	--	--	--	--	--	--
--	--	pelvis	-0.055	0.005	-0.028	femur	-0.006	-0.005	0.047
--	--	--	--	--	--	--	--	--	--
--	--	pelvis	-0.0338	0.1288	-0.0275	femur	-0.0158	0.4055	0.0350
--	--	pelvis	-0.0652	0.0842	-0.0429	femur	-0.0158	0.3609	0.0350
--	--	pelvis	-0.0747	0.0127	-0.0709	femur	-0.0158	0.2894	0.0350
--	--	--	--	--	--	--	--	--	--
--	--	--	--	--	--	--	--	--	--
--	--	--	--	--	--	--	--	--	--
--	--	--	--	--	--	--	--	--	--
--	--	--	--	--	--	--	--	--	--
--	69.0	--	--	--	--	--	--	--	--
--	23.0	--	--	--	--	--	--	--	--
--	--	pelvis	-0.087	0.068	-0.044	femur	-0.009	-0.103	0.047
--	--	pelvis	-0.151	-0.014	-0.068	femur	-0.015	-0.075	-0.042
--	--	--	--	--	--	--	--	--	--
--	--	--	--	--	--	--	--	--	--
--	--	--	--	--	--	--	--	--	--
--	--	pelvis	-0.0700	0.0200	0.0650	femur	-0.0150	-0.0400	0.3300
--	--	pelvis	-0.0750	0.1100	0.1000	femur	0.0000	-0.0800	0.3850
--	--	femur	0.0000	-0.0800	0.3850	tibia	0.0300	-0.0350	0.3800
0.1250	--	pelvis	-0.1195	0.0612	0.0700	pelvis	-0.1291	0.0012	0.0886
--	--	pelvis	-0.1291	0.0012	0.0886	femur	-0.0457	-0.0248	0.0392
--	--	femur	-0.0457	-0.0248	0.0392	femur	-0.0277	-0.0566	0.0470
0.1270	--	pelvis	-0.1349	0.0176	0.0563	pelvis	-0.1376	-0.0520	0.0914
--	--	pelvis	-0.1376	-0.0520	0.0914	femur	-0.0426	-0.0530	0.0293
--	--	femur	-0.0426	-0.0530	0.0293	femur	-0.0156	-0.1016	0.0419
0.1450	--	pelvis	-0.1556	-0.0314	0.0058	pelvis	-0.1529	-0.1052	0.0403
--	--	pelvis	-0.1529	-0.1052	0.0403	femur	-0.0299	-0.1041	0.0135
--	--	femur	-0.0299	-0.1041	0.0135	femur	-0.0060	-0.1419	0.0411
--	--	--	--	--	--	--	--	--	--
0.001	--	pelvis	-0.0578	0.0754	-0.0471	pelvis	-0.0158	0.3519	0.0350
--	--	pelvis	0.0000	-0.0800	-0.0165	femur	-0.0300	0.0351	-0.1135
--	--	pelvis	0.0000	0.0000	0.0000	femur	-0.0300	0.0351	-0.1135
--	--	--	--	--	--	--	--	--	--
--	--	pelvis	0.0168	0.0905	0.0356	femur	-0.0195	0.3899	0.0598
--	--	pelvis	-0.0239	0.1090	0.0054	femur	-0.0197	0.3902	0.0597
--	--	pelvis	-0.0546	0.0721	-0.0257	femur	-0.0195	0.3901	0.0596
--	--	--	--	--	--	--	--	--	--
--	--	--	--	--	--	--	--	--	--
--	--	--	--	--	--	--	--	--	--
--	--	--	--	--	--	--	--	--	--
--	13.0	--	--	--	--	--	--	--	--
--	13.0	--	--	--	--	--	--	--	--
--	13.0	--	--	--	--	--	--	--	--
--	--	pelvis	0.027	0.102	0.062	femur	-0.018	-0.026	0.073
--	--	pelvis	-0.002	0.132	0.018	femur	-0.018	-0.026	0.073
--	--	pelvis	-0.048	0.097	-0.015	femur	-0.018	-0.026	0.073
--	--	pelvis	-0.026	0.019	0.011	femur	-0.009	0.013	-0.020
--	--	pelvis	-0.077	0.042	-0.005	femur	-0.015	0.019	-0.023
--	--	pelvis	-0.112	0.029	-0.041	femur	-0.016	0.020	-0.024
--	--	--	--	--	--	--	--	--	--
--	--	--	--	--	--	--	--	--	--

MUSCLE		SOURCE	MUSCLE LENGTH ML (m)	FIBER LENGTH LM (m)	VOLUME V (ml)	MUSCLE PCSA =V/L (cm^2)	PENNA-TION ANGLE (deg)	MUSCLE FIBER TYPE %SO	%FOG	%FG
	(neo-vent)	[6]	0.010	--	--	3.84	13.0	--	--	--
	(fet-vent)	[6]	0.025	--	--	1.63	13.0	--	--	--
		[7]	--	--	--	--	--	--	--	--
	(1)	[9]	--	0.0535	--	--	8.0	--	--	--
	(2)	[9]	--	0.0845	--	--	0.0	--	--	--
	(3)	[9]	--	0.0646	--	--	19.0	--	--	--
		[10]	--	0.063	--	--	0.0	0.500	--	0.500
		[11]	--	0.081	--	--	9.7	--	--	--
		[12]	--	--	--	21.18	--	--	--	--
gluteus minimus	(1)	[1]	--	--	--	--	--	--	--	--
	(2)	[1]	--	--	--	--	--	--	--	--
	(3)	[1]	--	--	--	--	--	--	--	--
	(1)	[2]-S1	0.080	0.0680	46	6.76	--	--	--	--
	(1)	[2]-S2	0.080	0.0322	24	7.45	10.5	--	--	--
	(2)	[2]-S1	0.095	0.0561	46	8.20	--	--	--	--
	(2)	[2]-S2	0.085	0.0274	24	8.76	0.0	--	--	--
	(3)	[2]-S1	0.085	0.0384	46	11.98	--	--	--	--
	(3)	[2]-S2	0.080	0.0301	24	7.97	21.0	--	--	--
	(ant)	[3]	--	--	--	--	0.0	0.50	0.20	0.30
	(mid)	[3]	--	--	--	--	0.0	0.50	0.20	0.30
	(pos)	[3]	--	--	--	--	0.0	0.50	0.20	0.30
	(ant)	[4]	--	--	--	--	--	--	--	--
	(mid)	[4]	--	--	--	--	--	--	--	--
	(pos)	[4]	--	--	--	--	--	--	--	--
	(ant)	[5]	--	--	--	--	--	--	--	--
	(min)	[5]	--	--	--	--	--	--	--	--
	(pos)	[5]	--	--	--	--	--	--	--	--
	(neo-dors)	[6]	0.008	--	--	1.45	10.0	--	--	--
	(fet-dors)	[6]	0.031	--	--	0.37	10.0	--	--	--
	(neo-vent)	[6]	0.014	--	--	0.91	5.0	--	--	--
	(fet-vent)	[6]	0.022	--	--	0.58	5.0	--	--	--
		[7]	--	--	--	--	--	--	--	--
	(1)	[9]	--	0.0680	--	--	10.0	--	--	--
	(2)	[9]	--	0.0560	--	--	0.0	--	--	--
	(3)	[9]	--	0.0380	--	--	21.0	--	--	--
		[10]	--	0.050	--	--	0.0	0.500	--	0.500
		[11]	--	0.064	--	--	12.0	--	--	--
		[12]	--	--	--	9.6	--	--	--	--
iliacus		[1]	--	--	--	--	--	--	--	--
		[2]-S1	0.260	0.1003	234	23.33	--	--	--	--
		[2]-S2	0.180	0.0964	85	8.82	6.5	--	--	--
		[3]	--	--	--	--	0.0	0.50	0.20	0.30
	(neo)	[6]	0.030	--	--	1.39	10.0	-	--	--
	(fet)	[6]	0.020	--	--	2.08	10.0	--	--	--
		[9]	0.1000	--	--	--	7.0	--	--	--
		[9]	--	--	--	--	--	--	--	--
		[9]	--	--	--	--	--	--	--	--
ilio-psoas		[1]	--	--	--	--	--	--	--	--
		[4]	--	--	--	--	--	--	--	--
		[5]	--	--	--	--	--	--	--	--
		[7]	--	--	--	--	--	--	--	--
		[10]	--	0.121	--	--	0.0	0.492	--	0.508
		[11]	--	0.127	--	--	7.0	--	--	--
		[11]	--	--	--	--	--	--	--	--
	(1)	[12]	--	--	--	15.06	--	--	--	--
	(2)	[12]	--	--	--	--	--	--	--	--
		[14]	--	--	--	--	--	0.492	--	0.508
obturator externus		[1]	--	--	--	--	--	--	--	--
		[2]-S1	0.040	0.0295	8	2.71	--	--	--	--
		[2]-S2	0.100	0.0492	24	4.88	7.0	--	--	--
		[3]	--	--	--	--	0.0	0.50	0.20	0.30
		[4]	--	--	--	--	--	--	--	--
		[5]	--	--	--	--	--	--	--	--
	(neo)	[6]	0.013	--	--	0.74	5.0	--	--	--
	(fet)	[6]	0.009	--	--	1.07	5.0	--	--	--
		[10]	--	0.024	--	--	0.0	0.500	--	0.500
		[12]	--	--	--	4.95	--	--	--	--

TENDON LENGTH LT (m)	TENDON PCSA (mm²)	ORIGIN				INSERTION			
		COORD. SYSTEM	X (m)	Y (m)	Z (m)	COORD. SYSTEM	X (m)	Y (m)	Z (m)
--	--	--	--	--	--	--	--	--	--
--	--	--	--	--	--	--	--	--	--
--	--	pelvis	-0.0300	-0.0300	0.1100	femur	-0.0150	-0.0600	0.4000
0.0780	--	pelvis	-0.0408	0.0304	0.1209	femur	-0.0218	-0.0117	0.0555
0.0530	--	pelvis	-0.0855	0.0445	0.0766	femur	-0.0258	-0.0058	0.0527
0.0530	--	pelvis	-0.1223	0.0105	0.0648	femur	-0.0309	-0.0047	0.0518
--	--	--	--	--	--	--	--	--	--
0.035	--	pelvis	-0.0180	0.0891	0.0069	femur	-0.0196	0.3900	0.0597
--	--	pelvis	-0.0371	0.0000	-0.0566	femur	-0.0500	0.1494	-0.2085
--	--	pelvis	0.0236	0.0611	0.0305	femur	-0.0073	0.3810	0.0572
--	--	pelvis	-0.0084	0.0648	0.0130	femur	-0.0072	0.3810	0.0571
--	--	pelvis	-0.0293	0.0423	-0.0053	femur	-0.0073	0.3810	0.0572
--	--	--	--	--	--	--	--	--	--
--	--	--	--	--	--	--	--	--	--
--	--	--	--	--	--	--	--	--	--
--	--	--	--	--	--	--	--	--	--
--	--	--	--	--	--	--	--	--	--
--	19.0	--	--	--	--	--	--	--	--
--	19.0	--	--	--	--	--	--	--	--
--	19.0	--	--	--	--	--	--	--	--
--	--	pelvis	0.029	0.073	0.041	femur	0.004	-0.027	0.069
--	--	pelvis	-0.004	0.088	0.020	femur	0.004	-0.027	0.069
--	--	pelvis	-0.026	0.071	0.000	femur	0.004	-0.027	0.069
--	--	pelvis	-0.036	0.006	-0.003	femur	-0.006	0.021	-0.024
--	--	pelvis	-0.069	0.004	-0.011	femur	0.006	0.023	-0.026
--	--	pelvis	-0.092	-0.015	-0.030	femur	0.004	0.025	-0.025
--	--	--	--	--	--	--	--	--	--
--	--	--	--	--	--	--	--	--	--
--	--	--	--	--	--	--	--	--	--
--	--	pelvis	0.0200	-0.0200	0.0850	femur	0.0000	-0.0650	0.4000
0.0160	--	pelvis	-0.0467	-0.0080	0.1056	femur	-0.0072	-0.0104	0.0560
0.0260	--	pelvis	-0.0633	-0.0065	0.0991	femur	-0.0096	-0.0104	0.0560
0.0510	--	pelvis	-0.0834	-0.0063	0.0856	femur	-0.0135	-0.0083	0.0555
--	--	--	--	--	--	--	--	--	--
0.025	--	pelvis	-0.0097	0.0539	0.0093	femur	-0.0073	0.3810	0.0572
--	--	pelvis	0.0000	0.0000	-0.0366	femur	-0.0300	0.1494	-0.2085
--	--	pelvis	0.0199	0.0493	0.0025	femur	-0.0179	0.3350	0.0116
--	--	--	--	--	--	--	--	--	--
--	0.0	--	--	--	--	--	--	--	--
--	--	--	--	--	--	--	--	--	--
0.0900	--	pelvis	-0.0674	0.0365	0.0854	pelvis	-0.0230	-0.0550	0.0755
--	--	pelvis	-0.0230	-0.0550	0.0755	femur	0.0017	-0.0507	0.0081
--	--	femur	0.0017	-0.0507	0.0081	femur	-0.0193	-0.0621	0.0129
--	--	pelvis	0.0315	0.0111	-0.0102	femur	-0.0180	0.3352	0.0116
--	--	pelvis	0.028	0.024	0.005	femur	-0.002	-0.061	0.015
--	--	pelvis	-0.031	-0.045	-0.038	femur	-0.023	-0.034	-0.072
--	--	pelvis	0.0300	0.0000	0.0100	femur	-0.0100	-0.0270	0.3400
0.085	--	pelvis	0.0075	0.1350	-0.0400	pelvis	0.0260	0.0293	-0.0042
--	--	pelvis	0.0260	0.0293	-0.0042	femur	-0.0180	0.3351	0.0116
--	--	pelvis	0.0000	-0.0099	0.0866	femur	-0.0300	0.0201	-0.1636
--	--	Cen Torso	0.0000	0.0000	-0.0043	femur	-0.0300	0.0201	-0.1636
--	--	--	--	--	--	--	--	--	--
--	--	pelvis	0.0057	-0.0280	-0.0415	femur	-0.0242	0.3821	0.0415
--	--	--	--	--	--	--	--	--	--
--	15.0	--	--	--	--	--	--	--	--
--	--	pelvis	0.009	-0.035	-0.049	femur	-0.017	-0.016	0.050
--	--	pelvis	-0.069	-0.098	-0.078	femur	-0.014	0.018	-0.041
--	--	--	--	--	--	--	--	--	--
--	--	--	--	--	--	--	--	--	--
--	--	pelvis	0.0000	0.0495	0.0935	femur	-0.0300	-0.0500	-0.2235

MUSCLE	SOURCE	MUSCLE LENGTH ML (m)	FIBER LENGTH LM (m)	VOLUME V (ml)	MUSCLE PCSA =V/L (cm²)	PENNATION ANGLE (deg)	%SO	%FOG	%FG
obturator internus	[1]	--	--	--	--	--	--	--	--
	[2]-S1	0.090	0.0474	43	9.07	--	--	--	--
	[2]-S2	0.095	0.0344	32	9.30	25.0	--	--	--
	[3]	--	--	--	--	0.0	0.50	0.20	0.30
	[4]	--	--	--	--	--	--	--	--
	[5]	--	--	--	--	--	--	--	--
(neo)	[6]	0.008	--	--	1.62	25.0	--	--	--
(fet)	[6]	0.013	--	--	1.05	25.0	--	--	--
	[10]	--	0.039	--	--	0.0	0.500	--	0.500
	[12]	--	--	--	3.91	--	--	--	--
pectineus	[1]	--	--	--	--	--	--	--	--
	[2]-S1	0.120	0.0720	65	9.03	--	--	--	--
	[2]-S2	0.105	0.1047	13	1.24	0.0	--	--	--
	[3]	--	--	--	--	0.0	0.45	0.15	0.40
	[4]	--	--	--	--	--	--	--	--
	[5]	--	--	--	--	--	--	--	--
(neo)	[6]	0.027	--	--	0.34	5.0	--	--	--
(fet)	[6]	0.018	--	--	0.52	5.0	--	--	--
	[8]-S1	0.128	0.102	--	3.1	0.0	--	--	--
	[8]-S2	0.127	0.105	--	3.8	0.0	--	--	--
	[8]-S3	0.114	0.106	--	1.8	0.0	--	--	--
	[9]	--	0.1330	--	--	0.0	--	--	--
	[10]	--	0.095	--	--	0.0	0.500	--	0.500
	[11]	--	0.130	--	--	0.0	--	--	--
	[12]	--	--	--	2.47	--	--	--	--
piriformis	[1]	--	--	--	--	--	--	--	--
	[2]-S1	0.085	0.0258	53	20.54	--	--	--	--
	[2]-S2	0.085	0.0415	38	9.16	9.5	--	--	--
	[3]	--	--	--	--	0.0	0.50	0.20	0.30
	[4]	--	--	--	--	--	--	--	--
	[5]	--	--	--	--	--	--	--	--
(neo)	[6]	0.008	--	--	1.02	15.0	--	--	--
(fet)	[6]	0.023	--	--	0.34	15.0	--	--	--
	[9]	--	0.0260	--	--	10.0	--	--	--
	[9]	--	--	--	--	--	--	--	--
	[10]	--	0.028	--	--	0.0	0.500	--	0.500
psoas	[1]	--	--	--	--	--	--	--	--
	[2]-S1	0.255	0.1035	266	25.70	--	--	--	--
	[2]-S2	0.240	0.1217	45	3.70	7.5	--	--	--
	[3]	--	--	--	--	0.0	0.50	0.20	0.30
(neo)	[6]	0.039	--	--	1.08	5.0	--	--	--
(fet)	[6]	0.026	--	--	1.62	5.0	--	--	--
	[9]	--	0.1040	--	--	8.0	--	--	--
	[9]	--	--	--	--	--	--	--	--
	[9]	--	--	--	--	--	--	--	--
quadratus femoris	[1]	--	--	--	--	--	--	--	--
	[2]-S1	0.095	0.0538	113	21.00	--	--	--	--
	[2]-S2	--	--	--	--	--	--	--	--
	[3]	--	--	--	--	0.0	0.50	0.20	0.30
	[4]	--	--	--	--	--	--	--	--
	[5]	--	--	--	--	--	--	--	--
(neo)	[6]	0.008	--	--	0.29	5.0	--	--	--
(fet)	[6]	0.015	--	--	0.16	5.0	--	--	--
	[9]	--	0.0540	--	--	0.0	--	--	--
	[10]	--	0.020	--	--	0.0	0.500	--	0.500
	[12]	--	--	--	2.91	--	--	--	--
HIP AND KNEE MUSCLES									
biceps femoris long	[0]	--	--	--	21	17	--	--	--
	[1]	--	--	--	--	--	--	--	--
	[2]-S1	0.293	0.0794	217	27.34	--	--	--	--
	[2]-S2	0.255	0.0658	60	9.12	7.0	--	--	--
	[3]	--	--	--	--	15.0	0.65	0.10	0.25
	[4]	--	--	--	--	--	--	--	--
	[5]	--	--	--	--	--	--	--	--
(neo)	[6]	0.018	--	--	1.41	15.0	--	--	--
(fet)	[6]	0.038	--	--	0.68	15.0	--	--	--
	[7]	--	--	--	--	--	--	--	--
	[8]-S1	0.328	0.083	--	16.4	0.0	--	--	--
	[8]-S2	0.369	0.078	--	14.7	0.0	--	--	--

TENDON LENGTH LT (m)	TENDON PCSA (mm²)	ORIGIN				INSERTION			
		COORD. SYSTEM	X (m)	Y (m)	Z (m)	COORD. SYSTEM	X (m)	Y (m)	Z (m)
--	--	pelvis	-0.0488	-0.0091	-0.0135	femur	-0.0113	0.3947	0.0446
--	--	--	--	--	--	--	--	--	--
--	--	--	--	--	--	--	--	--	--
--	15.0	--	--	--	--	--	--	--	--
--	--	pelvis	-0.053	-0.011	-0.018	femur	-0.0006	-0.005	0.047
--	--	pelvis	-0.113	-0.074	-0.048	femur	-0.001	0.030	-0.037
--	--	--	--	--	--	--	--	--	--
--	--	--	--	--	--	--	--	--	--
--	--	--	--	--	--	--	--	--	--
--	--	pelvis	0.0000	0.0495	0.0935	femur	-0.0300	0.0886	-0.2256
--	--	pelvis	0.0318	-0.0096	-0.0299	femur	-0.0109	0.3146	0.0248
--	--	--	--	--	--	--	--	--	--
--	--	--	--	--	--	--	--	--	--
--	0.0	--	--	--	--	--	--	--	--
--	--	pelvis	0.044	-0.003	-0.038	femur	-0.004	-0.114	0.035
--	--	pelvis	-0.028	-0.073	-0.075	femur	-0.016	-0.064	-0.054
--	--	--	--	--	--	--	--	--	--
--	--	--	--	--	--	--	--	--	--
--	--	--	--	--	--	--	--	--	--
--	--	--	--	--	--	--	--	--	--
0.0010	--	pelvis	-0.0431	-0.0768	0.0451	femur	-0.0122	-0.0822	0.0253
--	--	--	--	--	--	--	--	--	--
0.001	--	pelvis	0.0318	-0.0096	-0.0299	femur	-0.0109	0.3146	0.0248
--	--	pelvis	0.0200	0.0394	0.0234	femur	-0.0300	0.0254	-0.1285
--	--	pelvis	-0.0559	0.0562	-0.0404	femur	-0.0132	0.3983	0.0484
--	--	--	--	--	--	--	--	--	--
--	15.0	--	--	--	--	--	--	--	--
--	--	pelvis	-0.078	0.055	-0.047	femur	-0.001	-0.001	0.055
--	--	pelvis	-0.133	-0.017	-0.068	femur	-0.008	0.034	-0.037
--	--	--	--	--	--	--	--	--	--
0.1150	--	pelvis	-0.1396	0.0003	0.0235	pelvis	-0.1193	-0.0276	0.0657
--	--	pelvis	-0.1193	-0.0276	0.0657	femur	-0.0148	-0.0036	0.0437
--	--	--	--	--	--	--	--	--	--
--	--	pelvis	0.0315	0.0111	-0.0102	femur	-0.0180	0.3352	0.0116
--	--	--	--	--	--	--	--	--	--
--	--	--	--	--	--	--	--	--	--
--	32.0	--	--	--	--	--	--	--	--
--	--	--	--	--	--	--	--	--	--
--	--	--	--	--	--	--	--	--	--
0.1300	--	pelvis	-0.0647	0.0887	0.0289	pelvis	-0.0250	-0.0570	0.0687
--	--	pelvis	-0.0250	-0.0570	0.0687	femur	0.0016	-0.0447	0.0062
--	--	femur	0.0016	-0.0447	0.0062	femur	-0.0188	-0.0597	0.0104
--	--	pelvis	-0.0319	-0.0479	-0.0231	femur	-0.0164	0.3446	0.0329
--	--	--	--	--	--	--	--	--	--
--	--	--	--	--	--	--	--	--	--
--	0.0	--	--	--	--	--	--	--	--
--	--	pelvis	-0.036	-0.046	-0.015	femur	-0.029	-0.040	0.047
--	--	pelvis	-0.100	-0.109	-0.051	femur	-0.028	-0.004	-0.047
--	--	--	--	--	--	--	--	--	--
--	--	--	--	--	--	--	--	--	--
0.0240	--	pelvis	-0.1143	-0.1151	0.0520	femur	-0.0381	-0.0359	0.0366
--	--	--	--	--	--	--	--	--	--
--	--	pelvis	-0.0201	0.0495	0.0935	femur	-0.0599	-0.0500	-0.1935
--	--	--	--	--	--	--	--	--	--
--	--	pelvis	-0.0414	-0.0474	-0.0146	tibia	-0.0383	0.3321	0.0431
--	--	--	--	--	--	--	--	--	--
--	--	--	--	--	--	--	--	--	--
--	20.0	--	--	--	--	--	--	--	--
--	--	pelvis	-0.053	-0.036	-0.013	femur	-0.023	-0.441	0.039
--	--	pelvis	-0.121	-0.098	-0.048	tibia	-0.039	0.012	0.031
--	--	--	--	--	--	--	--	--	--
--	--	--	--	--	--	--	--	--	--
--	--	pelvis	-0.0380	0.0000	-0.0250	tibia	-0.0250	-0.0350	0.3600
--	--	--	--	--	--	--	--	--	--
--	--	--	--	--	--	--	--	--	--

MUSCLE		SOURCE	MUSCLE LENGTH ML (m)	FIBER LENGTH LM (m)	VOLUME V (ml)	MUSCLE PCSA =V/L (cm²)	PENNATION ANGLE (deg)	MUSCLE FIBER TYPE %SO	%FOG	%FG
		[8]-S3	0.329	0.095	--	7.4	0.0	--	--	--
		[9]	--	0.1090	--	--	0.0	--	--	--
		[10]	--	0.091	--	--	0.0	0.669	--	0.331
		[12]	--	--	--	11.8	--	--	--	--
		[14]	--	--	--	--	--	0.669	--	0.331
gracilis		[0]	--	0.18	--	3.4	0.0	--	--	--
		[1]	--	--	--	--	--	--	--	--
		[2]-S1	0.270	0.2355	88	3.74	--	--	--	--
		[2]-S2	0.325	0.2543	20	0.79	0.0	--	--	--
		[3]	--	--	--	--	0.0	0.55	0.15	0.30
		[4]	--	--	--	--	--	--	--	--
		[5]	--	--	--	--	--	--	--	--
	(neo)	[6]	0.046	--	--	0.37	4.0	--	--	--
	(fet)	[6]	0.031	--	--	0.56	4.0	--	--	--
		[7]	--	--	--	--	--	--	--	--
		[8]-S1	0.375	0.300	--	1.7	0.0	--	--	--
		[8]-S2	0.316	0.271	--	2.3	5.0	--	--	--
		[8]-S3	0.315	0.260	--	1.3	5.0	--	--	--
		[9]	--	0.3520	--	--	3.0	--	--	--
		[9]	--	--	--	--	--	--	--	--
		[10]	--	0.215	--	--	0.0	0.500	--	0.500
		[11]	--	0.345	--	--	3.3	--	--	--
		[12]	--	--	--	1.63	--	--	--	--
hamstrings (lumped)		[11]	--	0.107	--	--	8.7	--	--	--
quadriceps	(1)	[12]	--	--	--	56	--	--	--	--
	(2)	[12]	--	--	--	--	--	--	--	--
rectus femoris		[0]	--	--	--	30	15	--	--	--
		[1]	--	--	--	--	--	--	--	--
		[2]-S1	0.332	0.0554	238	42.96	--	--	--	--
		[2]-S2	0.295	0.0652	60	9.20	14.0	--	--	--
		[3]	--	--	--	--	15.0	0.45	0.15	0.40
		[4]	--	--	--	--	--	--	--	--
		[5]	--	--	--	--	--	--	--	--
	(neo)	[6]	0.019	--	--	1.86	25.0	--	--	--
	(fet)	[6]	0.052	--	--	0.69	25.0	--	--	--
		[7]	--	--	--	--	--	--	--	--
		[8]-S1	0.324	0.067	--	15.2	5.0	--	--	--
		[8]-S2	0.319	0.068	--	14.1	5.0	--	--	--
		[8]-S3	0.305	0.063	--	8.9	5.0	--	--	--
	**	[9]	--	0.0840	--	--	5.0	--	--	--
	**	[9]	--	--	--	--	--	--	--	--
		[10]	--	0.071	--	--	10.0	0.381	--	0.619
		[11]	--	0.082	--	--	5.0	--	--	--
		[11]	--	--	--	--	--	--	--	--
	(lat surf)	[14]	--	--	--	--	--	0.295	--	0.705
	(lat deep)	[14]	--	--	--	--	--	0.420	--	0.580
	(med head)	[14]	--	--	--	--	--	0.428	--	0.572
sartorius		[0]	--	0.45	--	2.4	0.0	--	--	--
		[1]	--	--	--	--	--	--	--	--
		[2]-S1	0.519	0.4835	140	2.90	--	--	--	--
		[2]-S2	0.535	0.3911	105	2.68	0.0	--	--	--
		[3]	--	--	--	--	0.0	0.50	0.20	0.30
		[4]	--	--	--	--	--	--	--	--
		[5]	--	--	--	--	--	--	--	--
	(neo)	[6]	0.087	--	--	0.28	0.0	--	--	--
	(fet)	[6]	0.078	--	--	0.31	0.0	--	--	--
		[7]	--	--	--	--	--	--	--	--
		[8]-S1	0.552	0.482	--	2.1	0.0	--	--	--
		[8]-S2	0.499	0.464	--	1.9	0.0	--	--	--
		[8]-S3	0.458	0.419	--	1.1	0.0	--	--	--
		[9]	--	0.5790	--	--	0.0	--	--	--
		[9]	--	--	--	--	--	--	--	--
		[9]	--	--	--	--	--	--	--	--
		[9]	--	--	--	--	--	--	--	--
		[10]	--	0.385	--	--	0.0	0.496	--	0.504
		[11]	--	0.566	--	--	0.0	--	--	--
		[11]	--	--	--	--	--	--	--	--
		[11]	--	--	--	--	--	--	--	--
		[12]	--	--	--	1.55	--	--	--	--
		[14]	--	--	--	--	--	0.496	--	0.504

TENDON LENGTH LT (m)	TENDON PCSA (mm²)	ORIGIN COORD. SYSTEM	X (m)	Y (m)	Z (m)	INSERTION COORD. SYSTEM	X (m)	Y (m)	Z (m)
--	--	--	--	--	--	--	--	--	--
0.3410	--	pelvis	-0.1244	-0.1001	0.0666	tibia	-0.0081	-0.0729	0.0423
--	--	--	--	--	--	--	--	--	--
--	--	pelvis	-0.0450	0.0300	0.0632	tibia	0.0000	0.0399	-0.1814
--	--	--	--	--	--	--	--	--	--
--	--	pelvis	0.0303	-0.0441	-0.0691	tibia	-0.0586	0.3426	-0.0095
--	--	--	--	--	--	--	--	--	--
--	6.0	--	--	--	--	--	--	--	--
--	--	pelvis	0.010	-0.049	-0.068	femur	-0.014	-0.434	-0.041
--	--	pelvis	-0.053	-0.108	-0.104	tibia	-0.019	0.000	-0.014
--	--	--	--	--	--	--	--	--	--
--	--	pelvis	0.0350	0.0450	-0.0350	tibia	0.0100	0.0250	0.3300
--	--	--	--	--	--	--	--	--	--
--	--	--	--	--	--	--	--	--	--
0.1400	--	pelvis	-0.0563	-0.1038	0.0079	tibia	-0.0154	-0.0475	-0.0358
--	--	tibia	-0.0154	-0.0475	-0.0358	tibia	0.0060	-0.0836	-0.0228
--	--	--	--	--	--	--	--	--	--
0.080	--	pelvis	0.0303	-0.0441	0.0691	tibia	-0.0586	0.3426	-0.0095
--	--	pelvis	-0.0099	0.0193	0.0932	tibia	0.0099	-0.0201	-0.1313
0.385	--	pelvis	-0.0409	-0.0455	-0.0140	tibia	-0.0508	0.3321	0.0073
--	--	pelvis	0.0000	0.0000	-0.0058	tibia	0.0300	0.0000	-0.2014
--	--	femur	0.0000	0.0300	0.0165	tibia	0.0300	0.0000	-0.2014
--	--	--	--	--	--	--	--	--	--
--	--	pelvis	0.0326	0.0323	0.0174	tibia	0.0041	0.4084	-0.0006
--	--	--	--	--	--	--	--	--	--
--	24.0	--	--	--	--	--	--	--	--
--	--	pelvis	0.043	0.037	0.026	femur	0.043	-0.415	0.002
--	--	pelvis	-0.024	-0.040	-0.017	tibia	0.000	0.000	0.000
--	--	--	--	--	--	--	--	--	--
--	--	pelvis	0.0250	-0.0200	0.0250	femur	0.0500	-0.0200	0.0000
--	--	--	--	--	--	--	--	--	--
--	--	--	--	--	--	--	--	--	--
0.3460	--	pelvis	-0.0295	-0.0311	0.0968	femur	0.0334	-0.4030	0.0019
--	--	femur	0.0334	-0.4030	0.0019	patella	0.0121	0.0437	-0.0010
--	--	--	--	--	--	--	--	--	--
0.410	--	pelvis	0.0326	0.0323	0.0174	tibia	0.0041	0.4084	-0.0006
--	--	tibia	0.0041	0.4084	-0.0006	tibia	0.0000	0.3700	-0.0006
--	--	--	--	--	--	--	--	--	--
--	--	--	--	--	--	--	--	--	--
--	--	--	--	--	--	--	--	--	--
--	--	pelvis	0.0488	0.0649	0.0438	tibia	-0.0515	0.3478	-0.0205
--	--	--	--	--	--	--	--	--	--
--	24.0	--	--	--	--	--	--	--	--
--	--	pelvis	0.051	0.068	0.050	femur	-0.008	-0.435	-0.042
--	--	pelvis	-0.012	-0.011	-0.002	tibia	-0.016	-0.005	-0.011
--	--	--	--	--	--	--	--	--	--
--	--	pelvis	0.0450	-0.0300	-0.0850	tibia	0.0100	0.0250	0.3500
--	--	--	--	--	--	--	--	--	--
--	--	--	--	--	--	--	--	--	--
0.0400	--	pelvis	-0.0153	-0.0013	0.1242	femur	-0.0030	-0.3568	-0.0421
--	--	femur	-0.0030	-0.3568	-0.0421	tibia	-0.0056	-0.0419	-0.0399
--	--	tibia	-0.0056	-0.0419	-0.0399	tibia	0.0060	-0.0589	-0.0383
--	--	tibia	0.0060	-0.0589	-0.0383	tibia	0.0243	-0.0840	-0.0252
--	--	--	--	--	--	--	--	--	--
0.040	--	pelvis	0.0488	0.0649	0.0438	femur	0.0488	0.1300	-0.0440
--	--	femur	0.0488	0.1300	-0.0440	femur	-0.0515	0.0830	-0.0205
--	--	femur	-0.0515	0.0830	-0.0205	tibia	-0.0150	0.3478	-0.0200
--	--	pelvis	-0.0399	0.0000	-0.0267	tibia	0.0099	-0.0201	-0.1313
--	--	--	--	--	--	--	--	--	--

MUSCLE		SOURCE	MUSCLE LENGTH ML (m)	FIBER LENGTH LM (m)	VOLUME V (ml)	MUSCLE PCSA =V/L (cm²)	PENNA- TION ANGLE (deg)	%SO	%FOG	%FG
semimembranosus		[0]	--	--	--	30	16	--	--	--
		[1]	--	--	--	--	--	--	--	--
		[2]-S1	0.200	0.0749	347	46.33	--	--	--	--
		[2]-S2	0.215	0.0536	75	13.99	16.0	--	--	--
		[3]	--	--	--	--	15.0	0.50	0.15	0.35
		[4]	--	--	--	--	--	--	--	--
		[5]	--	--	--	--	--	--	--	--
	(neo)	[6]	0.015	--	--	1.69	15.0	--	--	--
	(fet)	[6]	0.034	--	--	0.73	15.0	--	--	--
		[7]	--	--	--	--	--	--	--	--
		[8]-S1	0.265	0.070	--	18.6	10.0	--	--	--
		[8]-S2	0.260	0.054	--	18.1	15.0	--	--	--
		[8]-S3	0.262	0.064	--	13.9	20.0	--	--	--
		[9]	--	0.0800	--	--	15.0	--	--	--
		[10]	--	0.066	--	--	0.0	0.500	--	0.500
		[12]	--	--	--	12.97	--	--	--	--
semitendinosus		[0]	--	0.15	--	8.5	0.0	--	--	--
		[1]	--	--	--	--	--	--	--	--
		[2]-S1	0.290	0.0911	212	23.27	--	--	--	--
		[2]-S2	0.275	0.0885	45	3.12	6.0	--	--	--
		[3]	--	--	--	--	0.0	0.50	0.15	0.35
		[4]	--	--	--	--	--	--	--	--
		[5]	--	--	--	--	--	--	--	--
	(neo)	[6]	0.037	--	--	0.88	10.0	--	--	--
	(fet)	[6]	0.050	--	--	0.64	10.0	--	--	--
		[7]	--	--	--	--	--	--	--	--
		[8]-S1	0.321	0.160	--	6.3	5.0	--	--	--
		[8]-S2	0.313	0.156	--	4.4	5.0	--	--	--
		[9]	--	0.2010	--	--	5.0	--	--	--
		[9]	--	--	--	--	--	--	--	--
		[9]	--	--	--	--	--	--	--	--
		[10]	--	0.155	--	--	15.0	0.500	--	0.500
		[12]	--	--	--	4.33	--	--	--	--
tensor fasciae latae		[1]	--	--	--	--	--	--	--	--
		[2]-S1	0.168	0.0950	76	8.0	--	--	--	--
		[2]-S2	0.145	0.1015	25	2.46	2.5	--	--	--
		[3]	--	--	--	--	0.0	0.70	0.10	0.20
		[4]	--	--	--	--	--	--	--	--
		[5]	--	--	--	--	--	--	--	--
	(neo)	[6]	0.025	--	--	0.36	5.0	--	--	--
	(fet)	[6]	0.017	--	--	0.54	5.0	--	--	--
	(1)	[7]	--	--	--	--	--	--	--	--
	(2)	[7]	--	--	--	--	--	--	--	--
		[9]	--	0.0950	--	--	3.0	--	--	--
		[9]	--	--	--	--	--	--	--	--
		[9]	--	--	--	--	--	--	--	--
		[10]	--	0.106	--	--	0.0	0.500	--	0.500
		[11]	--	0.118	--	--	2.5	--	--	--
		[12]	--	--	--	2.48	--	--	--	--
KNEE MUSCLES										
biceps femoris short		[0]	--	0.12	--	5.2	0.0	--	--	--
		[1]	--	--	--	--	--	--	--	--
		[2]-S1	0.205	0.1229	100	8.14	--	--	--	--
		[2]-S2	0.240	0.1108	52	4.69	15.0	--	--	--
		[3]	--	--	--	--	0.0	0.65	0.10	0.25
		[4]	--	--	--	--	--	--	--	--
		[5]	--	--	--	--	--	--	--	--
		[7]	--	--	--	--	--	--	--	--
		[8]-S1	0.258	0.145	--	--	23.0	--	--	--
		[8]-S2	0.292	0.133	--	--	22.0	--	--	--
		[8]-S3	0.262	0.140	--	--	25.0	--	--	--
		[9]	0.1730	--	--	--	23.0	--	--	--
		[10]	--	0.118	--	--	17.0	0.669	--	0.331
		[11]	--	0.173	--	--	23.3	--	--	--
		[12]	--	--	--	--	--	--	--	--
popliteus		[3]	--	--	--	--	0.0	0.50	0.15	0.35
		[8]-S1	0.115	0.036	--	6.5	0.0	--	--	--
		[8]-S2	0.101	0.022	--	9.3	0.0	--	--	--
		[12]	--	--	--	1.99	--	--	--	--

TENDON LENGTH LT (m)	TENDON PCSA (mm²)	ORIGIN COORD. SYSTEM	X (m)	Y (m)	Z (m)	INSERTION COORD. SYSTEM	X (m)	Y (m)	Z (m)
--	--	--	--	--	--	--	--	--	--
--	--	pelvis	-0.0382	-0.0448	-0.0143	tibia	-0.0564	0.3297	-0.0072
--	--	--	--	--	--	--	--	--	--
--	24.0	--	--	--	--	--	--	--	--
--	--	pelvis	-0.044	-0.031	-0.008	femur	-0.029	-0.428	-0.034
--	--	pelvis	-0.110	-0.099	-0.044	tibia	-0.046	0.014	-0.015
--	--	--	--	--	--	--	--	--	--
--	--	pelvis	-0.0380	0.0000	-0.0300	tibia	-0.0100	0.0300	0.3700
--	--	--	--	--	--	--	--	--	--
--	--	--	--	--	--	--	--	--	--
0.3590	--	pelvis	-0.1192	-0.1015	0.0695	tibia	-0.0243	-0.0536	-0.0194
--	--	--	--	--	--	--	--	--	--
--	--	pelvis	-0.0399	0.0000	0.0632	tibia	0.0000	-0.0300	-0.1814
--	--	--	--	--	--	--	--	--	--
--	--	pelvis	-0.0457	-0.0446	-0.0125	tibia	-0.0542	0.3369	-0.0058
--	--	--	--	--	--	--	--	--	--
--	12.0	--	--	--	--	--	--	--	--
--	--	pelvis	-0.053	-0.036	-0.013	femur	-0.022	-0.433	-0.040
--	--	pelvis	-0.122	-0.100	-0.049	tibia	-0.023	-0.005	-0.019
--	--	--	--	--	--	--	--	--	--
--	--	pelvis	-0.0400	0.0100	-0.0250	tibia	0.0000	0.0300	0.3300
--	--	--	--	--	--	--	--	--	--
0.2620	--	pelvis	-0.1237	-0.1043	0.0603	tibia	-0.0314	-0.0545	-0.0146
--	--	tibia	-0.0314	-0.0545	-0.0146	tibia	-0.0113	-0.0746	-0.0245
--	--	tibia	-0.0113	-0.0746	-0.0245	tibia	0.0027	-0.0956	-0.0193
--	--	--	--	--	--	--	--	--	--
--	--	pelvis	-0.0500	0.0000	0.0632	tibia	0.0099	-0.0201	-0.1115
--	--	pelvis	0.0327	0.0882	0.0547	tibia	-0.0099	0.3504	0.0292
--	--	--	--	--	--	--	--	--	--
--	5.0	--	--	--	--	--	--	--	--
--	--	pelvis	0.045	0.078	0.056	femur	0.022	-0.436	0.033
--	--	femur	-0.011	0.007	0.010	tibia	-0.019	0.031	0.023
--	--	--	--	--	--	--	--	--	--
--	--	pelvis	0.0400	-0.0300	0.1000	femur	0.0000	-0.0800	0.3850
--	--	femur	0.0000	-0.0800	0.3850	femur	0.0300	-0.0350	0.3800
0.4250	--	pelvis	-0.0311	0.0214	0.1241	femur	0.0294	-0.0995	0.0597
--	--	femur	0.0294	-0.0995	0.0597	femur	0.0054	-0.4049	0.0357
--	--	femur	0.0054	-0.4049	0.0357	tibia	0.0060	-0.0487	0.0297
--	--	--	--	--	--	--	--	--	--
0.430	--	pelvis	0.0327	0.0882	0.0547	tibia	-0.0099	0.3504	0.0292
--	--	pelvis	-0.0399	0.0000	-0.0566	femur	-0.0500	0.1394	-0.1336
--	--	--	--	--	--	--	--	--	--
--	--	femur	-0.0007	0.1784	0.0144	tibia	-0.0384	0.3323	0.0433
--	--	--	--	--	--	--	--	--	--
--	9.0	--	--	--	--	--	--	--	--
--	--	femur	0.005	-0.010	0.130	tibia	-0.025	-0.035	0.385
--	--	femur	-0.004	-0.186	-0.041	tibia	-0.039	0.012	0.031
--	--	pelvis	0.0500	-0.0100	0.1300	tibia	-0.0250	-0.0350	0.3600
--	--	--	--	--	--	--	--	--	--
--	--	--	--	--	--	--	--	--	--
0.1000	--	femur	0.0050	-0.2111	0.0234	tibia	-0.0101	-0.0725	0.0406
--	--	--	--	--	--	--	--	--	--
0.090	--	femur	-0.0007	0.1784	0.0144	tibia	-0.0384	0.3323	0.0433
--	--	femur	-0.0300	0.0300	-0.0135	tibia	0.0000	0.0399	-0.1814
--	24.0	--	--	--	--	--	--	--	--
--	--	--	--	--	--	--	--	--	--
--	--	femur	-0.0300	-0.0399	0.2464	tibia	-0.0201	0.0000	-0.1415

MUSCLE	SOURCE	MUSCLE LENGTH ML (m)	FIBER LENGTH LM (m)	VOLUME V (ml)	MUSCLE PCSA =V/L (cm²)	PENNA- TION ANGLE (deg)	MUSCLE FIBER TYPE		
							%SO	%FOG	%FG
quadriceps	[12]	--	--	--	--	--	--	--	--
vastus intermedius	[0]	--	--	--	28	18	--	--	--
	[1]	--	--	--	--	--	--	--	--
	[2]-S1	0.275	0.0739	606	82.0	--	--	--	--
	[2]-S2	0.330	0.0785	135	17.2	2.5	--	--	--
	[3]	--	--	--	--	10.0	0.50	0.15	0.35
	[5]	--	--	--	--	--	--	--	--
	[7]	--	--	--	--	--	--	--	--
	[8]-S1	0.328	0.064	--	39.5	5.0	--	--	--
	[8]-S2	0.355	0.078	--	11.0	5.0	--	--	--
	[8]-S3	0.303	0.063	--	16.4	0.0	--	--	--
	[9]	--	0.0870	--	--	3.0	--	--	--
**	[9]	--	--	--	--	--	--	--	--
**	[9]	--	--	--	--	--	--	--	--
	[10]	--	0.078	--	--	10.0	0.500	--	0.500
	[11]	--	0.084	--	--	4.5	--	--	--
	[11]	--	--	--	--	--	--	--	--
vastus lateralis	[0]	--	--	--	43	13	--	--	--
	[1]	--	--	--	--	--	--	--	--
	[2]-S1	0.305	0.0798	514	64.41	--	--	--	--
	[2]-S2	0.295	0.0807	133	16.48	13.0	--	--	--
	[3]	--	--	--	--	20.0	0.45	0.20	0.35
	[5]	--	--	--	--	--	--	--	--
	[7]	--	--	--	--	--	--	--	--
	[8]-S1	0.308	0.066	--	27.8	5.0	--	--	--
	[8]-S2	0.353	0.067	--	42.9	5.0	--	--	--
	[8]-S3	0.312	0.064	--	21.0	5.0	--	--	--
	[9]	--	0.0840	--	--	5.0	--	--	--
**	[9]	--	--	--	--	--	--	--	--
**	[9]	--	--	--	--	--	--	--	--
**	[9]	--	--	--	--	--	--	--	--
	[10]	--	0.075	--	--	10.0	0.480	--	0.520
	[11]	--	0.084	--	--	4.5	--	--	--
	[11]	--	--	--	--	--	--	--	--
(surf)	[14]	--	--	--	--	--	0.378	--	0.673
(deep)	[14]	--	--	--	--	--	0.469	--	0.531
vastus medialis	[0]	--	--	--	34	15	--	--	--
	[1]	--	--	--	--	--	--	--	--
	[2]-S1	0.280	0.0765	555	66.87	--	--	--	--
	[2]-S2	0.320	0.0790	123	15.60	7.0	--	--	--
	[3]	--	--	--	--	25.0	0.50	0.15	0.35
	[5]	--	--	--	--	--	--	--	--
	[7]	--	--	--	--	--	--	--	--
	[8]-S1	0.329	0.072	--	22.1	5.0	--	--	--
	[8]-S2	0.363	0.075	--	28.0	5.0	--	--	--
	[8]-S3	0.314	0.064	--	13.3	5.0	--	--	--
	[9]	--	0.0890	--	--	5.0	--	--	--
**	[9]	--	--	--	--	--	--	--	--
**	[9]	--	--	--	--	--	--	--	--
**	[9]	--	--	--	--	--	--	--	--
	[10]	--	0.088	--	--	10.0	0.470	--	0.530
	[11]	--	0.084	--	--	4.5	--	--	--
	[11]	--	--	--	--	--	--	--	--
(surf)	[14]	--	--	--	--	--	0.437	--	0.563
(deep)	[14]	--	--	--	--	--	0.615	--	0.385
patellar tendon	[7]	--	--	--	--	--	--	--	--
	[9]	--	0.0500	--	--	0.0	--	--	--
KNEE AND ANKLE MUSCLES									
gastrocnemius medialis	[0]	--	--	--	22	16	--	--	--
	[1]	--	--	--	--	--	--	--	--
	[2]-S1	0.235	0.0419	212	50.60	--	--	--	--
	[2]-S2	0.210	0.0358	61	17.04	6.5	--	--	--
	[3]	--	--	--	--	15.0	0.55	0.15	0.30
	[5]	--	--	--	--	--	--	--	--
	[7]	--	--	--	--	--	--	--	--
	[8]-S1	0.268	0.035	--	29.3	10.0	--	--	--
	[8]-S2	0.240	0.039	--	38.7	15.0	--	--	--
	[8]-S3	0.237	0.032	--	29.3	25.0	--	--	--
**	[9]	--	0.0450	--	--	17.0	--	--	--

TENDON LENGTH LT (m)	TENDON PCSA (mm²)	ORIGIN COORD. SYSTEM	ORIGIN X (m)	ORIGIN Y (m)	ORIGIN Z (m)	INSERTION COORD. SYSTEM	INSERTION X (m)	INSERTION Y (m)	INSERTION Z (m)
--	--	femur	0.0000	0.0300	0.0165	tibia	0.0300	0.0000	-0.2014
--	--	--	--	--	--	--	--	--	--
--	--	femur	0.0232	0.2067	0.0176	tibia	-0.0018	0.4110	0.0006
--	--	--	--	--	--	--	--	--	--
--	56.0	--	--	--	--	--	--	--	--
--	--	femur	0.017	-0.147	-0.040	tibia	0.000	0.000	0.000
--	--	femur	0.0300	-0.0200	0.2000	femur	0.0500	-0.0200	0.0000
--	--	--	--	--	--	--	--	--	--
--	--	--	--	--	--	--	--	--	--
0.1360	--	femur	0.0290	-0.1924	0.0310	femur	0.0335	-0.2084	0.0285
--	--	femur	0.0335	-0.2084	0.0285	femur	0.0343	-0.4030	0.0055
--	--	femur	0.0343	-0.4030	0.0055	patella	0.0058	0.0480	-0.0006
--	--	--	--	--	--	--	--	--	--
0.225	--	femur	0.0106	0.2026	0.0205	tibia	-0.0005	0.4056	0.0005
--	--	tibia	-0.0005	0.4056	0.0005	tibia	0.0000	0.3700	-0.0006
--	--	--	--	--	--	--	--	--	--
--	--	femur	0.0010	0.2127	0.0365	tibia	0.0089	0.4050	0.0151
--	--	--	--	--	--	--	--	--	--
--	89.0	--	--	--	--	--	--	--	--
--	--	femur	-0.003	-0.030	-0.022	tibia	0.000	0.000	0.000
--	--	femur	0.0250	-0.0300	0.1800	femur	0.0500	-0.0200	0.0000
--	--	--	--	--	--	--	--	--	--
--	--	--	--	--	--	--	--	--	--
--	--	--	--	--	--	--	--	--	--
0.1570	--	femur	0.0048	-0.1854	0.0349	femur	0.0269	-0.2591	0.0409
--	--	femur	0.0269	-0.2591	0.0409	femur	0.0361	-0.4030	0.0205
--	--	femur	0.0361	-0.4030	0.0205	femur	0.0253	-0.4243	0.0184
--	--	femur	0.0253	-0.4243	0.0184	patella	0.0103	0.0423	0.0141
--	--	--	--	--	--	--	--	--	--
0.225	--	femur	0.0106	0.2026	0.0205	tibia	-0.0005	0.4056	0.0005
--	--	tibia	-0.0005	0.4056	0.0005	tibia	0.0000	0.3700	-0.0006
--	--	--	--	--	--	--	--	--	--
--	--	--	--	--	--	--	--	--	--
--	--	femur	0.0043	0.1880	0.0088	tibia	-0.0079	0.3996	-0.0137
--	--	--	--	--	--	--	--	--	--
--	32.0	--	--	--	--	--	--	--	--
--	--	femur	-0.003	-0.090	-0.044	tibia	0.000	0.000	0.000
--	--	femur	0.0250	-0.0100	0.1800	femur	0.0500	-0.0200	0.0000
--	--	--	--	--	--	--	--	--	--
--	--	--	--	--	--	--	--	--	--
0.1260	--	femur	0.0140	-0.2099	0.0188	femur	0.0356	-0.2769	0.0009
--	--	femur	0.0356	-0.2769	0.0009	femur	0.0370	-0.4048	-0.0125
--	--	femur	0.0370	-0.4048	-0.0125	femur	0.0274	-0.4255	-0.0131
--	--	femur	0.0274	-0.4255	-0.0131	patella	0.0063	0.0445	-0.0170
--	--	--	--	--	--	--	--	--	--
0.225	--	femur	0.0106	0.2026	0.0205	tibia	-0.0005	0.4056	0.0005
--	--	tibia	-0.0005	0.4056	0.0005	tibia	0.0000	0.3700	-0.0006
--	--	--	--	--	--	--	--	--	--
--	--	femur	0.0400	-0.0100	0.3500	tibia	0.0500	-0.0200	0.0000
0.0050	--	patella	0.0021	0.0015	0.0001	tibia	0.0390	-0.0822	0.0000
--	--	--	--	--	--	--	--	--	--
--	--	femur	-0.0204	0.0077	-0.0157	calcaneus	-0.0368	-0.0429	0.0028
--	--	--	--	--	--	--	--	--	--
--	20.0	--	--	--	--	--	--	--	--
--	--	femur	-0.020	-0.380	-0.079	calcaneus	0.005	0.000	0.002
--	--	femur	-0.0200	0.0150	0.0000	calcaneus	-0.0450	-0.0100	-0.0370
--	--	--	--	--	--	--	--	--	--
--	--	--	--	--	--	--	--	--	--
0.4080	--	femur	-0.0127	-0.3929	-0.0235	femur	-0.0239	-0.4022	-0.0258

MUSCLE	SOURCE	MUSCLE LENGTH ML (m)	FIBER LENGTH LM (m)	VOLUME V (ml)	MUSCLE PCSA =V/L (cm²)	PENNA-TION ANGLE (deg)	MUSCLE FIBER TYPE %SO	%FOG	%FG
**	[9]	--	--	--	--	--	--	--	--
	[9]	--	--	--	--	--	--	--	--
	[10]	--	0.048	--	--	16.0	0.482	--	0.518
	[11]	--	0.048	--	--	14.8	--	--	--
	[12]	--	--	--	--	--	--	--	--
	[14]	--	--	--	--	--	0.508	--	0.492
gastrocnemius lateralis	[0]	--	--	--	13	8	--	--	--
	[1]	--	--	--	--	--	--	--	--
	[2]-S1	0.230	0.0769	110	14.30	--	--	--	--
	[2]-S2	0.205	0.0442	38	8.60	17.5	--	--	--
	[3]	--	--	--	--	10.0	0.55	0.15	0.30
	[5]	--	--	--	--	--	--	--	--
	[7]	--	--	--	--	--	--	--	--
	[8]-S1	0.226	0.059	--	--	5.0	--	--	--
	[8]-S2	0.229	0.053	--	--	10.0	--	--	--
	[8]-S3	0.195	0.040	--	--	10.0	--	--	--
**	[9]	--	0.0640	--	--	8.0	--	--	--
**	[9]	--	--	--	--	--	--	--	--
	[9]	--	--	--	--	--	--	--	--
	[10]	--	0.048	--	--	16.0	0.482	--	0.518
	[11]	--	0.048	--	--	14.8	--	--	--
(surf)	[14]	--	--	--	--	--	0.435	--	0.565
(deep)	[14]	--	--	--	--	--	0.503	--	0.497
	[12]	--	--	--	--	--	--	--	--
plantaris	[3]	--	--	--	--	5.0	0.45	0.15	0.40
	[8]-S1	0.101	0.041	--	1.4	0.0	--	--	--
	[8]-S2	0.099	0.050	--	1.7	5.0	--	--	--
	[8]-S3	0.055	0.027	--	0.5	5.0	--	--	--
ANKLE MUSCLES									
dorsiflexors (lumped)	[11]	--	0.101	--	--	6.9	--	--	--
	[11]	--	--	--	--	--	--	--	--
extensor digitorum longus	[0]	--	--	--	8	16,13	--	--	--
	[1]	--	--	--	--	--	--	--	--
	[2]-S1	0.270	0.0871	65	7.46	--	--	--	--
	[2]-S2	0.325	0.0743	30	4.04	13.0	--	--	--
	[3]	--	--	--	--	15.0	0.40	0.15	0.45
	[5]	--	--	--	--	--	--	--	--
(toeII)	[7]	--	--	--	--	--	--	--	--
(toeII)	[7]	--	--	--	--	--	--	--	--
(toeIII)	[7]	--	--	--	--	--	--	--	--
(toeIII)	[7]	--	--	--	--	--	--	--	--
(toeIV)	[7]	--	--	--	--	--	--	--	--
(toeIV)	[7]	--	--	--	--	--	--	--	--
(toeV)	[7]	--	--	--	--	--	--	--	--
(toeV)	[7]	--	--	--	--	--	--	--	--
	[8]-S1	0.346	0.094	--	4.4	5.0	--	--	--
	[8]-S2	0.381	0.082	--	6.4	10.0	--	--	--
	[8]-S3	0.338	0.065	--	5.9	10.0	--	--	--
	[9]	--	0.1020	--	--	8.0	--	--	--
	[9]	--	--	--	--	--	--	--	--
	[9]	--	--	--	--	--	--	--	--
	[9]	--	--	--	--	--	--	--	--
	[9]	--	--	--	--	--	--	--	--
	[10]	--	0.068	--	--	10.0	0.473	--	0.527
extensor hallucis longus	[1]	--	--	--	--	--	--	--	--
	[2]-S1	0.250	0.0462	30	6.49	--	--	--	--
	[2]-S2	0.175	0.0848	18	2.21	8.0	--	--	--
	[3]	--	--	--	--	10.0	0.50	0.20	0.30
	[5]	--	--	--	--	--	--	--	--
	[7]	--	--	--	--	--	--	--	--
	[7]	--	--	--	--	--	--	--	--
	[8]-S1	0.270	0.080	--	2.2	5.0	--	--	--
	[8]-S2	0.272	0.103	--	1.5	5.0	--	--	--
	[8]-S3	0.278	0.078	--	1.6	8.0	--	--	--
	[9]	--	0.1110	--	--	6.0	--	--	--
	[9]	--	--	--	--	--	--	--	--
	[9]	--	--	--	--	--	--	--	--
	[9]	--	--	--	--	--	--	--	--
	[9]	--	--	--	--	--	--	--	--
	[9]	--	--	--	--	--	--	--	--
	[10]	--	0.078	--	--	5.0	0.500	--	0.500

TENDON LENGTH LT (m)	TENDON PCSA (mm²)	ORIGIN				INSERTION			
		COORD. SYSTEM	X (m)	Y (m)	Z (m)	COORD. SYSTEM	X (m)	Y (m)	Z (m)
--	--	femur	-0.0239	-0.4022	-0.0258	tibia	-0.0217	-0.0487	-0.0295
--	--	tibia	-0.0217	-0.0487	-0.0295	calcaneus	0.0044	0.0310	-0.0053
0.425	--	--	--	--	--	--	--	--	--
--	--	femur	-0.0203	0.0071	-0.0073	calcaneus	-0.0368	-0.0289	0.0028
--	--	femur	0.0000	0.0000	0.2164	calcaneus	0.0000	0.0000	-0.0381
--	--	--	--	--	--	--	--	--	--
--	--	femur	-0.0198	0.0048	0.0226	calcaneus	-0.0369	-0.0430	0.0028
--	--	--	--	--	--	--	--	--	--
--	12.0	--	--	--	--	--	--	--	--
--	--	femur	-0.026	-0.393	-0.049	calcaneus	0.005	0.000	0.002
--	--	femur	-0.0200	-0.0250	0.0000	calcaneus	-0.0450	-0.0100	-0.0370
--	--	--	--	--	--	--	--	--	--
--	--	--	--	--	--	--	--	--	--
0.3850	--	femur	-0.0155	-0.3946	0.0272	femur	-0.0254	-0.4018	0.0274
--	--	femur	-0.0254	-0.4018	0.0274	tibia	-0.0242	-0.0481	0.0235
--	--	tibia	-0.0242	-0.0481	0.0235	calcaneus	0.0044	0.0310	-0.0053
0.425	--	--	--	--	--	--	--	--	--
--	--	femur	-0.0203	0.0071	-0.0073	calcaneus	-0.0368	-0.0289	0.0028
--	--	--	--	--	--	--	--	--	--
--	--	femur	0.0000	0.0000	0.2164	calcaneus	0.0000	0.0000	-0.0381
--	4.0	--	--	--	--	--	--	--	--
--	--	--	--	--	--	--	--	--	--
--	--	--	--	--	--	--	--	--	--
0.235	--	tibia	-0.0155	0.2175	0.0134	tibia	0.0259	0.0257	-0.0093
--	--	tibia	0.0259	0.0257	-0.0093	calcaneus	0.1850	-0.0510	-0.0330
--	--	--	--	--	--	--	--	--	--
--	--	tibia	-0.0228	0.2590	0.0280	calcaneus	0.0253	0.0116	-0.0021
--	--	--	--	--	--	--	--	--	--
--	13.0	--	--	--	--	--	--	--	--
--	--	tibia	-0.038	-0.090	0.011	calcaneus	0.131	-0.015	0.045
--	--	tibia	0.0000	-0.0250	0.1500	retinac	0.0200	-0.0100	-0.0100
--	--	retinac	0.0200	-0.0100	-0.0100	toes	0.0000	-0.0240	-0.0080
--	--	tibia	0.0000	-0.0250	0.1500	retinac	0.0200	-0.0100	-0.0100
--	--	retinac	0.0200	-0.0100	-0.0100	toes	-0.0070	-0.0340	0.0080
--	--	tibia	0.0000	-0.0250	0.1500	retinac	0.0200	-0.0100	-0.0100
--	--	retinac	0.0200	-0.0100	-0.0100	toes	-0.0130	-0.0460	0.0060
--	--	tibia	0.0000	-0.0250	0.1500	retinac	0.0200	-0.0100	-0.0100
--	--	retinac	0.0200	-0.0100	-0.0100	toes	-0.0240	-0.0560	0.0020
--	--	--	--	--	--	--	--	--	--
--	--	--	--	--	--	--	--	--	--
--	--	--	--	--	--	--	--	--	--
0.3450	--	tibia	0.0032	-0.1381	0.0276	tibia	0.0289	-0.4007	0.0072
--	--	tibia	0.0289	-0.4007	0.0072	calcaneus	0.0922	0.0388	-0.0001
--	--	calcaneus	0.0922	0.0388	-0.0001	calcaneus	0.1616	0.0055	0.0130
--	--	calcaneus	0.1616	0.0055	0.0130	toes	0.0003	0.0047	0.0153
--	--	toes	0.0003	0.0047	0.0153	toes	0.0443	-0.0004	0.0250
--	--	--	--	--	--	--	--	--	--
--	--	tibia	-0.0155	0.2175	0.0134	calcaneus	0.0259	0.0117	-0.0093
--	--	--	--	--	--	--	--	--	--
--	--	--	--	--	--	--	--	--	--
--	13.0	--	--	--	--	--	--	--	--
--	--	tibia	-0.042	-0.190	0.000	calcaneus	0.227	-0.026	0.030
--	--	tibia	0.0000	-0.0200	0.1600	retinac	0.0250	0.0040	0.0000
--	--	retinac	0.0250	0.0040	0.0000	toes	0.0000	0.0000	0.0100
--	--	--	--	--	--	--	--	--	--
--	--	--	--	--	--	--	--	--	--
--	--	--	--	--	--	--	--	--	--
0.3050	--	tibia	0.0012	-0.1767	0.0228	tibia	0.0326	-0.3985	-0.0085
--	--	tibia	0.0326	-0.3985	-0.0085	calcaneus	0.0970	0.0389	-0.0211
--	--	calcaneus	0.0970	0.0389	-0.0211	calcaneus	0.1293	0.0309	-0.0257
--	--	calcaneus	0.1293	0.0309	-0.0257	calcaneus	0.1734	0.0139	-0.0280
--	--	calcaneus	0.1734	0.0139	-0.0280	toes	0.0298	0.0041	-0.0245
--	--	toes	0.0298	0.0041	-0.0245	toes	0.0563	0.0034	-0.0186
--	--	--	--	--	--	--	--	--	--

MUSCLE		SOURCE	MUSCLE LENGTH ML (m)	FIBER LENGTH LM (m)	VOLUME V (ml)	MUSCLE PCSA =V/L (cm²)	PENNA- TION ANGLE (deg)	MUSCLE FIBER TYPE		
								%SO	%FOG	%FG
flexor digitorum longus		[1]	--	--	--	--	--	--	--	--
		[2]-S1	0.265	0.0469	30	6.40	--	--	--	--
		[2]-S2	0.205	0.0290	17	5.86	9.0	--	--	--
		[3]	--	--	--	--	20.0	0.40	0.20	0.40
		[5]	--	--	--	--	--	--	--	--
	(toeII)	[7]	--	--	--	--	--	--	--	--
	(toeII)	[7]	--	--	--	--	--	--	--	--
	(toeII)	[7]	--	--	--	--	--	--	--	--
	(toeIII)	[7]	--	--	--	--	--	--	--	--
	(toeIII)	[7]	--	--	--	--	--	--	--	--
	(toeIII)	[7]	--	--	--	--	--	--	--	--
	(toeIV)	[7]	--	--	--	--	--	--	--	--
	(toeIV)	[7]	--	--	--	--	--	--	--	--
	(toeIV)	[7]	--	--	--	--	--	--	--	--
	(toeV)	[7]	--	--	--	--	--	--	--	--
	(toeV)	[7]	--	--	--	--	--	--	--	--
	(toeV)	[7]	--	--	--	--	--	--	--	--
		[8]-S1	0.239	0.026	--	6.2	10.0	--	--	--
		[8]-S2	0.290	0.028	--	5.2	5.0	--	--	--
		[8]-S3	0.251	0.027	--	3.8	5.0	--	--	--
		[9]	--	0.0340	--	--	7.0	--	--	--
		[9]	--	--	--	--	--	--	--	--
		[9]	--	--	--	--	--	--	--	--
		[9]	--	--	--	--	--	--	--	--
		[9]	--	--	--	--	--	--	--	--
		[9]	--	--	--	--	--	--	--	--
		[10]	--	0.048	--	--	8.0	0.500	--	0.500
flexor hallucis longus		[1]	--	--	--	--	--	--	--	--
		[2]-S1	0.252	0.0502	93	18.52	--	--	--	--
		[2]-S2	0.195	0.0346	31	8.96	19.0	--	--	--
		[3]	--	--	--	--	20.0	0.50	0.15	0.35
		[5]	--	--	--	--	--	--	--	--
		[7]	--	--	--	--	--	--	--	--
		[7]	--	--	--	--	--	--	--	--
		[7]	--	--	--	--	--	--	--	--
		[7]	--	--	--	--	--	--	--	--
		[8]-S1	0.214	0.037	--	6.3	15.0	--	--	--
		[8]-S2	0.231	0.032	--	5.2	10.0	--	--	--
		[8]-S3	0.220	0.033	--	4.3	5.0	--	--	--
		[9]	--	0.0430	--	--	10.0	--	--	--
		[9]	--	--	--	--	--	--	--	--
		[9]	--	--	--	--	--	--	--	--
		[9]	--	--	--	--	--	--	--	--
		[9]	--	--	--	--	--	--	--	--
		[10]	--	0.034	--	--	11.0	0.500	--	0.500
peroneus brevis		[1]	--	--	--	--	--	--	--	--
		[2]-S1	0.250	0.0357	70	19.61	--	--	--	--
		[2]-S2	0.225	0.0434	23	5.29	12.0	--	--	--
		[3]	--	--	--	--	10.0	0.45	0.15	0.40
		[5]	--	--	--	--	--	--	--	--
		[7]	--	--	--	--	--	--	--	--
		[7]	--	--	--	--	--	--	--	--
		[8]-S1	0.252	0.046	--	4.9	5.0	--	--	--
		[8]-S2	0.230	0.038	--	7.7	5.0	--	--	--
		[8]-S3	0.208	0.034	--	4.5	5.0	--	--	--
		[9]	--	0.0500	--	--	5.0	--	--	--
		[9]	--	--	--	--	--	--	--	--
		[9]	--	--	--	--	--	--	--	--
		[9]	--	--	--	--	--	--	--	--
		[10]	--	0.035	--	--	8.0	0.625	--	0.375
peroneus longus		[0]	--	--	--	11	9,13	--	--	--
		[1]	--	--	--	--	--	--	--	--
		[2]-S1	0.265	0.0426	105	24.65	--	--	--	--
		[2]-S2	0.250	0.0460	35	7.61	5.5	--	--	--
		[3]	--	--	--	--	10.0	0.60	.010	0.30
		[5]	--	--	--	--	--	--	--	--
	(cun)	[7]	--	--	--	--	--	--	--	--

TENDON LENGTH LT (m)	TENDON PCSA (mm²)	ORIGIN				INSERTION			
		COORD. SYSTEM	X (m)	Y (m)	Z (m)	COORD. SYSTEM	X (m)	Y (m)	Z (m)
--	--	tibia	-0.0246	0.1996	0.0016	calcaneus	-0.0070	-0.0024	-0.0222
--	--	--	--	--	--	--	--	--	--
--	--	--	--	--	--	--	--	--	--
--	12.0	--	--	--	--	--	--	--	--
--	--	tibia	-0.040	-0.123	-0.016	calcaneus	0.133	-0.026	0.042
--	--	tibia	-0.0250	0.0900	0.0000	talus	0.0150	0.0150	-0.0340
--	--	talus	0.0150	0.0150	-0.0340	metatar2	0.0550	-0.0140	-0.0380
--	--	metatar2	0.0550	-0.0140	-0.0380	toes	0.0000	-0.0240	-0.0100
--	--	tibia	-0.0250	0.0090	0.0000	talus	0.0150	0.0150	-0.0340
--	--	talus	0.0150	0.0150	-0.0340	metatar3	0.0510	-0.0100	-0.0340
--	--	metatar3	0.0510	-0.0100	-0.0340	toes	-0.0070	-0.0340	-0.0100
--	--	tibia	-0.0250	0.0090	0.0000	talus	0.0150	0.0150	-0.0340
--	--	talus	0.0150	0.0150	-0.0340	metatar4	0.0520	-0.0120	-0.0260
--	--	metatar4	0.0520	-0.0120	-0.0260	toes	-0.0130	-0.0460	-0.0100
--	--	tibia	-0.0250	0.0090	0.0000	talus	0.0150	0.0150	-0.0340
--	--	talus	0.0150	0.0150	-0.0340	metatar5	0.0480	-0.0150	-0.0200
--	--	metatar5	0.0480	-0.0150	-0.0200	toes	-0.0240	-0.0560	-0.0100
--	--	--	--	--	--	--	--	--	--
--	--	--	--	--	--	--	--	--	--
--	--	--	--	--	--	--	--	--	--
0.4000	--	tibia	-0.0083	-0.2046	-0.0018	tibia	-0.0154	-0.4051	-0.0196
--	--	tibia	-0.0154	-0.4051	-0.0196	calcaneus	0.0436	0.0315	-0.0280
--	--	calcaneus	0.0436	0.0315	-0.0280	calcaneus	0.0708	0.0176	-0.0263
--	--	calcaneus	0.0708	0.0176	-0.0263	calcaneus	0.1658	-0.0081	0.0116
--	--	calcaneus	0.1658	-0.0081	0.0116	toes	-0.0019	-0.0078	0.0147
--	--	toes	-0.0019	-0.0078	0.0147	toes	0.0285	-0.0071	0.0215
--	--	toes	0.0285	-0.0071	0.0215	toes	0.0441	-0.0060	0.0242
--	--	--	--	--	--	--	--	--	--
--	--	tibia	-0.0266	0.1660	0.0204	calcaneus	-0.0092	-0.0065	-0.0159
--	--	--	--	--	--	--	--	--	--
--	15.0	--	--	--	--	--	--	--	--
--	--	tibia	-0.054	-0.197	-0.004	calcaneus	0.226	-0.038	0.028
--	--	tibia	-0.0180	-0.0050	0.0050	talus	-0.0200	0.0050	-0.0170
--	--	talus	-0.0200	0.0050	-0.0170	calcaneus	0.0000	0.0130	-0.0320
--	--	calcaneus	0.0000	0.0130	-0.0320	metatar1	0.0460	-0.0100	-0.0300
--	--	metatar1	0.0460	-0.0100	-0.0300	toes	0.0000	0.0000	-0.0100
--	--	--	--	--	--	--	--	--	--
--	--	--	--	--	--	--	--	--	--
0.3800	--	tibia	-0.0079	-0.2334	0.0244	tibia	-0.0186	-0.4079	-0.0174
--	--	tibia	-0.0186	-0.4079	-0.0174	calcaneus	0.0374	0.0276	-0.0241
--	--	calcaneus	0.0374	0.0276	-0.0241	calcaneus	0.1038	0.0068	-0.0256
--	--	calcaneus	0.1038	0.0068	-0.0256	calcaneus	0.1726	-0.0053	-0.0269
--	--	calcaneus	0.1726	-0.0053	-0.0269	toes	0.0155	-0.0064	-0.0265
--	--	toes	0.0155	-0.0064	-0.0265	toes	0.0562	-0.0102	-0.0181
--	--	--	--	--	--	--	--	--	--
--	--	tibia	-0.0226	0.1364	0.0253	calcaneus	-0.0081	-0.0058	0.0273
--	--	--	--	--	--	--	--	--	--
--	11.0	--	--	--	--	--	--	--	--
--	--	tibia	-0.049	-0.176	0.010	calcaneus	0.095	-0.024	0.058
--	--	tibia	-0.0150	-0.0280	-0.0180	calcaneus	0.0000	-0.0250	-0.0330
--	--	calcaneus	0.0000	-0.0250	-0.0330	metatar5	-0.0030	-0.0030	-0.0100
--	--	--	--	--	--	--	--	--	--
--	--	--	--	--	--	--	--	--	--
0.1610	--	tibia	-0.0070	-0.2646	0.0325	tibia	-0.0198	-0.4184	0.0283
--	--	tibia	-0.0198	-0.4184	0.0283	tibia	-0.0144	-0.4295	0.0289
--	--	tibia	-0.0144	-0.4295	0.0289	calcaneus	0.0471	0.0270	0.0233
--	--	calcaneus	0.0471	0.0270	0.0233	calcaneus	0.0677	0.0219	0.0343
--	--	--	--	--	--	--	--	--	--
--	--	tibia	-0.0268	0.2419	0.0356	calcaneus	-0.0094	-0.0076	0.0240
--	--	--	--	--	--	--	--	--	--
--	21.0	--	--	--	--	--	--	--	--
--	--	tibia	-0.046	-0.055	0.020	calcaneus	0.134	-0.008	0.024
--	--	tibia	-0.0150	-0.0200	-0.0180	calcaneus	-0.0100	-0.0250	-0.0400

MUSCLE		SOURCE	MUSCLE LENGTH ML (m)	FIBER LENGTH LM (m)	VOLUME V (ml)	MUSCLE PCSA =V/L (cm²)	PENNA-TION ANGLE (deg)	MUSCLE FIBER TYPE		
								%SO	%FOG	%FG
	(cun)	[7]	--	--	--	--	--	--	--	--
	(cun)	[7]	--	--	--	--	--	--	--	--
	(met)	[7]	--	--	--	--	--	--	--	--
	(met)	[7]	--	--	--	--	--	--	--	--
	(met)	[7]	--	--	--	--	--	--	--	--
		[8]-S1	0.282	0.044	--	12.3	10.0	--	--	--
		[8]-S2	0.318	0.039	--	17.3	10.0	--	--	--
		[8]-S3	0.258	0.033	--	7.4	10.0	--	--	--
		[9]	--	0.0490	--	--	10.0	--	--	--
		[9]	--	--	--	--	--	--	--	--
		[9]	--	--	--	--	--	--	--	--
		[9]	--	--	--	--	--	--	--	--
		[9]	--	--	--	--	--	--	--	--
		[9]	--	--	--	--	--	--	--	--
		[14]	--	--	--	--	--	0.625	--	0.375
peroneus tertius		[1]	--	--	--	--	--	--	--	--
		[2]-S1	0.190	0.0798	33	4.14	--	--	--	--
		[2]-S2	0.105	0.0652	8	1.23	13.0	--	--	--
		[3]	--	--	--	--	10.0	0.35	0.20	0.45
		[7]	--	--	--	--	--	--	--	--
		[7]	--	--	--	--	--	--	--	--
		[9]	--	0.0790	--	--	13.0	--	--	--
		[9]	--	--	--	--	--	--	--	--
plantarflexors (lumped)		[11]	--	0.038	--	--	12.1	--	--	--
		[11]	--	--	--	--	--	--	--	--
		[11]	--	--	--	--	--	--	--	--
soleus		[0]	--	--	--	67	20.0	--	--	--
		[1]	--	--	--	--	--	--	--	--
		[2]-S1	0.370	0.0308	575	186.69	--	--	--	--
		[2]-S2	0.305	0.0298	172	57.72	32.0	--	--	--
		[3]	--	--	--	--	15.0	0.75	0.15	0.10
		[5]	--	--	--	--	--	--	--	--
	(tib)	[7]	--	--	--	--	--	--	--	--
	(fib)	[7]	--	--	--	--	--	--	--	--
		[8]-S1	0.308	0.020	--	58.0	30.0	--	--	--
		[8]-S2	0.311	0.019	--	--	20.0	--	--	--
		[9]	--	0.0300	--	--	25.0	--	--	--
		[10]	--	0.035	--	--	23.0	0.750	--	0.250
		[11]	--	0.024	--	--	25.0	--	--	--
	(surf)	[14]	--	--	--	--	--	0.864	--	0.136
	(deep)	[14]	--	--	--	--	--	0.890	--	0.110
tibialis anterior		[0]	--	--	--	14	8	--	--	--
		[1]	--	--	--	--	--	--	--	--
		[2]-S1	0.290	0.0770	130	16.88	12.0	--	--	--
		[2]-S2	0.278	0.0684	58	8.48	--	--	--	--
		[3]	--	--	--	--	10.0	0.70	0.10	0.20
		[5]	--	--	--	--	--	--	--	--
	(cunl)	[7]	--	--	--	--	--	--	--	--
	(cunl)	[7]	--	--	--	--	--	--	--	--
	(metl)	[7]	--	--	--	--	--	--	--	--
	(metl)	[7]	--	--	--	--	--	--	--	--
		[8]-S1	0.288	0.070	--	12.7	5.0	--	--	--
		[8]-S2	0.322	0.093	--	7.9	5.0	--	--	--
		[8]-S3	0.284	0.069	--	9.0	5.0	--	--	--
		[9]	--	0.0980	--	--	5.0	--	--	--
		[9]	--	--	--	--	--	--	--	--
		[10]	--	0.073	--	--	6.0	0.730	--	0.270
	(surf)	[14]	--	--	--	--	--	0.734	--	0.266
	(deep)	[14]	--	--	--	--	--	0.727	--	0.273
tibialis posterior		[0]	--	--	--	17	20	--	--	--
		[1]	--	--	--	--	--	--	--	--
		[2]-S1	0.290	0.0354	93	26.27	--	--	--	--
		[2]-S2	0.250	0.0217	41	18.89	19.0	--	--	--
		[3]	--	--	--	--	20.0	0.55	0.25	0.20
		[5]	--	--	--	--	--	--	--	--
	(cunl)	[7]	--	--	--	--	--	--	--	--
	(cunll)	[7]	--	--	--	--	--	--	--	--
	(cunll)	[7]	--	--	--	--	--	--	--	--
	(cunlll)	[7]	--	--	--	--	--	--	--	--

TENDON LENGTH LT (m)	TENDON PCSA (mm²)	ORIGIN COORD. SYSTEM	X (m)	Y (m)	Z (m)	INSERTION COORD. SYSTEM	X (m)	Y (m)	Z (m)
--	--	calcaneus	-0.0100	-0.0250	-0.0400	cuboid	0.0350	-0.0300	-0.0460
--	--	cuboid	0.0350	-0.0300	-0.0460	cuneif1	0.0550	-0.0050	-0.0380
--	--	tibia	-0.0150	-0.0280	-0.0180	calcaneus	-0.0100	-0.0250	-0.0400
--	--	calcaneus	-0.0100	-0.0250	-0.0400	cuboid	0.0350	-0.0300	-0.0460
--	--	cuboid	0.0350	-0.0300	-0.0460	metatar1	-0.0020	-0.0100	-0.0080
--	--	--	--	--	--	--	--	--	--
--	--	--	--	--	--	--	--	--	--
0.3450	--	tibia	0.0005	-0.1568	0.0362	tibia	-0.0207	-0.4205	0.0286
--	--	tibia	-0.0207	-0.4205	0.0286	tibia	-0.0162	-0.4319	0.0289
--	--	tibia	-0.0162	-0.4319	0.0289	calcaneus	0.0438	0.0230	0.0221
--	--	calcaneus	0.0438	0.0230	0.0221	calcaneus	0.0681	0.0106	0.0284
--	--	calcaneus	0.0681	0.0106	0.0284	calcaneus	0.0852	0.0069	0.0118
--	--	calcaneus	0.0852	0.0069	0.0118	calcaneus	0.1203	0.0085	-0.0184
--	--	--	--	--	--	--	--	--	--
--	--	tibia	-0.0099	0.1202	0.0210	calcaneus	0.0205	0.0032	0.0097
--	--	--	--	--	--	--	--	--	--
--	4.0	--	--	--	--	--	--	--	--
--	--	tibia	-0.0100	-0.0200	0.1000	retinac	0.0200	-0.0100	-0.0100
--	--	retinac	0.0200	-0.0100	-0.0100	metatar5	0.0050	-0.0010	-0.0020
0.1000	--	tibia	0.0010	-0.2804	0.0231	tibia	0.0229	-0.4069	0.0159
--	--	tibia	0.0229	-0.4069	0.0159	calcaneus	0.0857	0.0228	0.0299
0.273	--	tibia	-0.0268	0.2419	0.0356	tibia	-0.0094	0.0064	0.0240
--	--	tibia	-0.0094	0.0064	0.0240	calcaneus	-0.0099	-0.0210	0.0240
--	--	calcaneus	-0.0099	-0.0210	0.0240	calcaneus	0.0715	-0.0420	-0.0260
--	--	--	--	--	--	--	--	--	--
--	--	tibia	-0.0292	0.2467	0.0006	calcaneus	-0.0365	-0.0428	0.0056
--	--	--	--	--	--	--	--	--	--
--	--	--	--	--	--	--	--	--	--
--	17.0	--	--	--	--	--	--	--	--
--	--	tibia	-0.042	-0.032	0.001	calcaneus	0.010	0.000	0.002
--	--	tibia	-0.0050	0.0100	0.2900	calcaneus	-0.0450	-0.0100	-0.0370
--	--	tibia	-0.0300	-0.0300	0.3100	calcaneus	-0.0450	-0.0100	-0.0370
--	--	--	--	--	--	--	--	--	--
0.2680	--	tibia	-0.0024	-0.1533	0.0071	calcaneus	0.0044	0.0310	-0.0053
--	--	--	--	--	--	--	--	--	--
0.270	--	tibia	-0.0292	0.2467	0.0006	calcaneus	-0.0365	-0.0288	0.0056
--	--	--	--	--	--	--	--	--	--
--	--	--	--	--	--	--	--	--	--
--	--	tibia	-0.0067	0.2397	0.0132	calcaneus	0.0221	0.0132	-0.0194
--	--	--	--	--	--	--	--	--	--
--	20.0	--	--	--	--	--	--	--	--
--	--	tibia	-0.016	-0.091	-0.003	calcaneus	0.141	0.005	0.003
--	--	tibia	0.0200	0.0000	0.2500	retinac	0.0250	0.0100	0.0070
--	--	retinac	0.0250	0.0100	0.0000	cuneif1	0.0550	0.0130	-0.0370
--	--	tibia	0.0200	0.0000	0.2500	retinac	0.0250	0.0100	0.0000
--	--	retinac	0.0250	0.0100	0.0000	Mf14	-0.0020	0.0050	-0.0110
--	--	--	--	--	--	--	--	--	--
--	--	--	--	--	--	--	--	--	--
0.2230	--	tibia	0.0179	-0.1624	0.0115	tibia	0.0329	-0.3951	-0.0177
--	--	tibia	0.0329	-0.3951	-0.0177	calcaneus	0.1166	0.0178	-0.0305
--	--	--	--	--	--	--	--	--	--
--	--	--	--	--	--	--	--	--	--
--	--	tibia	-0.0128	0.1786	0.0137	calcaneus	-0.0023	0.0023	-0.0276
--	--	--	--	--	--	--	--	--	--
--	17.0	--	--	--	--	--	--	--	--
--	--	tibia	-0.044	-0.136	-0.007	calcaneus	0.111	-0.002	0.002
--	--	tibia	-0.0250	0.0200	0.0000	cuneif1	0.0420	0.0120	-0.0430
--	--	tibia	-0.0250	0.0200	0.0000	navicle	0.0300	0.0200	-0.0360
--	--	tibia	-0.0250	0.0200	0.0000	talus	0.0180	0.0050	-0.0350
--	--	talus	0.0180	0.0050	-0.0350	cuneif2	0.0480	-0.0050	-0.0270

MUSCLE	SOURCE	MUSCLE LENGTH ML (m)	FIBER LENGTH LM (m)	VOLUME V (ml)	MUSCLE PCSA =V/L (cm²)	PENNA-TION ANGLE (deg)	MUSCLE FIBER TYPE %SO	%FOG	%FG
(cunlll)	[7]	--	--	--	--	--	--	--	--
(nav)	[7]	--	--	--	--	--	--	--	--
(metll)	[7]	--	--	--	--	--	--	--	--
(metll)	[7]	--	--	--	--	--	--	--	--
(metlll)	[7]	--	--	--	--	--	--	--	--
(metlll)	[7]	--	--	--	--	--	--	--	--
(metlV)	[7]	--	--	--	--	--	--	--	--
(metlV)	[7]	--	--	--	--	--	--	--	--
	[8]-S1	0.208	0.024	--	24.5	15.0	--	--	--
	[8]-S2	0.299	0.031	--	14.8	10.0	--	--	--
	[8]-S3	0.254	0.017	--	23.0	10.0	--	--	--
(1)	[9]	--	0.0310	--	--	12.0	--	--	--
(1)	[9]	--	--	--	--	--	--	--	--
(1)	[9]	--	--	--	--	--	--	--	--
(2)	[9]	--	0.0310	--	--	12.0	--	--	--
(2)	[9]	--	--	--	--	--	--	--	--
(2)	[9]	--	--	--	--	--	--	--	--
(2)	[9]	--	--	--	--	--	--	--	--
	[10]	--	0.028	--	--	14.0	0.500	--	0.500
FOOT MUSCLES									
abductor hallucis	[7]	--	--	--	--	--	--	--	--
	[7]	--	--	--	--	--	--	--	--
abductor digiti minimi	[7]	--	--	--	--	--	--	--	--
	[7]	--	--	--	--	--	--	--	--
adductor hallucis	[7]	--	--	--	--	--	--	--	--
(metll)	[7]	--	--	--	--	--	--	--	--
(metlll)	[7]	--	--	--	--	--	--	--	--
(metlll)	[7]	--	--	--	--	--	--	--	--
(metlV)	[7]	--	--	--	--	--	--	--	--
(metlV)	[7]	--	--	--	--	--	--	--	--
dorsal interossei	[7]	--	--	--	--	--	--	--	--
(toell)	[7]	--	--	--	--	--	--	--	--
(toell)	[7]	--	--	--	--	--	--	--	--
(toell)	[7]	--	--	--	--	--	--	--	--
(toelll)	[7]	--	--	--	--	--	--	--	--
(toelll)	[7]	--	--	--	--	--	--	--	--
(toelV)	[7]	--	--	--	--	--	--	--	--
(toelV)	[7]	--	--	--	--	--	--	--	--
extensor digitorum brevis	[7]	--	--	--	--	--	--	--	--
(toell)	[7]	--	--	--	--	--	--	--	--
(toelll)	[7]	--	--	--	--	--	--	--	--
(toelV)	[7]	--	--	--	--	--	--	--	--
	[14]	--	--	--	--	--	0.453	--	0.547
extensor hallucis brevis	[7]	--	--	--	--	--	--	--	--
	[7]	--	--	--	--	--	--	--	--
flexor digiti minimi	[7]	--	--	--	--	--	--	--	--
flexor digitorum brevis	[7]	--	--	--	--	--	--	--	--
(toell)	[7]	--	--	--	--	--	--	--	--
(toelll)	[7]	--	--	--	--	--	--	--	--
(toelll)	[7]	--	--	--	--	--	--	--	--
(toelV)	[7]	--	--	--	--	--	--	--	--
(toelV)	[7]	--	--	--	--	--	--	--	--
(toeV)	[7]	--	--	--	--	--	--	--	--
(toeV)	[7]	--	--	--	--	--	--	--	--
	[14]	--	--	--	--	--	0.445	--	0.555
flexor hallucis brevis	[7]	--	--	--	--	--	--	--	--
(cub)	[7]	--	--	--	--	--	--	--	--
(cunlll)	[7]	--	--	--	--	--	--	--	--
(cunlll)	[7]	--	--	--	--	--	--	--	--
(cub)	[7]	--	--	--	--	--	--	--	--
(cub)	[7]	--	--	--	--	--	--	--	--
(cunlll)	[7]	--	--	--	--	--	--	--	--
(cunlll)	[7]	--	--	--	--	--	--	--	--
plantar aponenrosis	[7]	--	--	--	--	--	--	--	--
(toel)	[7]	--	--	--	--	--	--	--	--
(toell)	[7]	--	--	--	--	--	--	--	--
(toell)	[7]	--	--	--	--	--	--	--	--
(toelll)	[7]	--	--	--	--	--	--	--	--
(toelll)	[7]	--	--	--	--	--	--	--	--

TENDON LENGTH LT (m)	TENDON PCSA (mm²)	ORIGIN COORD. SYSTEM	X (m)	Y (m)	Z (m)	INSERTION COORD. SYSTEM	X (m)	Y (m)	Z (m)
--	--	tibia	-0.0250	0.0200	0.0000	talus	0.0180	-0.0050	-0.0350
--	--	talus	0.0180	0.0050	-0.0350	cuneif3	0.0400	-0.0120	-0.0330
--	--	tibia	-0.0250	0.0200	0.0000	talus	0.0180	-0.0050	-0.0350
--	--	talus	0.0180	0.0050	-0.0350	metatar2	0.0030	-0.0020	-0.0080
--	--	tibia	-0.0250	0.0200	0.0000	talus	0.0180	0.0050	-0.0350
--	--	talus	0.0180	0.0050	-0.0350	metatar3	0.0050	0.0050	-0.0090
--	--	tibia	-0.0250	0.0200	0.0000	talus	0.0180	0.0050	-0.0350
--	--	talus	0.0180	0.0050	-0.0350	metatar4	0.0030	0.0050	-0.0070
--	--	--	--	--	--	--	--	--	--
--	--	--	--	--	--	--	--	--	--
0.3100	--	tibia	-0.0094	-0.1348	0.0019	tibia	-0.0144	-0.4051	-0.0229
--	--	tibia	-0.0144	-0.4051	-0.0229	calcaneus	0.0417	0.0334	-0.0286
--	--	calcaneus	0.0417	0.0334	-0.0286	calcaneus	0.0772	0.0159	-0.0281
0.3100	--	tibia	-0.0094	-0.1348	0.0019	tibia	0.0063	-0.3505	0.0146
--	--	tibia	0.0063	-0.3505	0.0146	tibia	0.0165	-0.3904	0.0176
--	--	tibia	0.0165	-0.3904	0.0176	calcaneus	0.0912	0.0463	0.0086
--	--	calcaneus	0.0912	0.0463	0.0086	calcaneus	0.1050	0.0325	0.0093
--	--	--	--	--	--	--	--	--	--
--	--	calcaneus	-0.0300	0.0050	0.0500	metatar1	0.0460	0.0055	-0.0300
--	--	metatar1	0.0460	0.0050	-0.0300	toes	0.0000	0.0060	-0.0100
--	--	calcaneus	-0.0380	-0.0100	-0.0580	metatar5	-0.0030	-0.0030	-0.0150
--	--	metatar5	-0.0030	-0.0030	-0.0150	toes	-0.0240	-0.0590	-0.0080
--	--	metatar2	0.0100	-0.0020	-0.0090	metatar1	0.0460	-0.0100	-0.0280
--	--	metatar1	0.0460	-0.0100	0.0280	toes	0.0000	-0.0090	-0.0080
--	--	metatar3	0.0090	0.0050	-0.0100	metatar1	0.0460	-0.0100	-0.0280
--	--	metatar1	0.0460	-0.0100	-0.0280	toes	0.0000	-0.0090	-0.0080
--	--	metatar4	0.0090	0.0050	0.0070	metatar1	0.0460	-0.0100	-0.0280
--	--	metatar1	0.0460	-0.0100	-0.0280	toes	0.0000	-0.0090	-0.0080
--	--	metatar1	0.0240	-0.0090	0.0000	toes	0.0000	-0.0200	-0.0070
--	--	metatar2	0.0300	-0.0020	-0.0080	toes	0.0000	-0.0200	-0.0070
--	--	metatar2	0.0300	-0.0100	-0.0080	toes	0.0000	-0.0280	-0.0070
--	--	metatar3	0.0270	0.0000	-0.0090	toes	0.0000	-0.0280	-0.0070
--	--	metatar3	0.0270	-0.0070	-0.0090	toes	-0.0070	-0.0370	-0.0070
--	--	metatar4	0.0290	-0.0010	-0.0060	toes	-0.0070	-0.0370	-0.0070
--	--	metatar4	-0.0290	-0.0080	-0.0060	toes	-0.0130	-0.0490	-0.0080
--	--	metatar5	-0.0300	-0.0060	-0.0050	toes	-0.0130	-0.0490	-0.0080
--	--	calcaneus	0.0170	-0.0220	-0.0250	cuneif3	0.0500	-0.0180	0.0210
--	--	cuneif3	0.0500	-0.0180	-0.0210	toes	0.0000	-0.0240	-0.0080
--	--	calcaneus	0.0170	-0.0220	-0.0250	toes	-0.0070	-0.0340	0.0080
--	--	calcaneus	0.0170	-0.0220	-0.0250	toes	-0.0130	-0.0460	0.0060
--	--	--	--	--	--	--	--	--	--
--	--	calcaneus	0.0170	-0.0220	-0.0250	metatar2	0.0940	-0.0070	0.0040
--	--	metatar2	0.0040	0.0070	0.0040	toes	0.0000	0.0000	0.0100
--	--	metatar5	0.0000	-0.0010	-0.0130	toes	-0.0240	-0.0590	-0.0080
--	--	calcaneus	-0.0250	0.0000	-0.0550	metatar2	0.0550	-0.0140	-0.0380
--	--	metatar2	0.0550	-0.0140	-0.0380	toes	0.0000	-0.0240	-0.0100
--	--	calcaneus	-0.0250	0.0000	-0.0550	metatar3	0.0510	-0.0100	-0.0340
--	--	metatar3	0.0510	-0.0100	-0.0340	toes	-0.0070	-0.0340	-0.0100
--	--	calcaneus	-0.0250	0.0000	0.0550	metatar4	0.0520	-0.0120	-0.0260
--	--	metatar4	0.0520	-0.0120	-0.0260	toes	-0.0130	-0.0460	-0.0100
--	--	calcaneus	-0.0250	0.0000	0.0550	metatar5	0.0480	-0.0150	-0.0200
--	--	metatar5	0.0480	-0.0150	-0.0200	toes	-0.0240	-0.0560	-0.0100
--	--	--	--	--	--	--	--	--	--
--	--	cuboid	0.0350	-0.0120	0.0370	metatar1	0.0460	0.0050	-0.0300
--	--	metatar1	0.0460	0.0050	-0.0300	toes	0.0000	0.0060	-0.0100
--	--	cuneif3	0.0400	-0.0100	0.0320	metatar1	0.0460	0.0050	-0.0300
--	--	metatar1	0.0460	0.0050	-0.0300	toes	0.0000	0.0060	-0.0100
--	--	cuboid	0.0350	-0.0120	0.0370	metatar1	0.0460	-0.0100	-0.0280
--	--	metatar1	0.0460	-0.0100	0.0280	toes	0.0000	-0.0090	-0.0080
--	--	cuneif3	0.0400	-0.0100	0.0320	metatar1	0.0460	-0.0100	-0.0280
--	--	metatar1	0.0460	-0.0100	-0.0280	toes	0.0000	-0.0090	-0.0080
--	--	calcaneus	-0.0300	0.0050	0.1500	retinac	0.0200	-0.0100	-0.0100
--	--	metatar1	0.0460	0.0050	-0.0100	toes	-0.0240	-0.0560	0.0020
--	--	calcaneus	-0.0300	0.0050	0.0000	toes	0.0000	0.0000	0.0100
--	--	metatar2	0.0550	-0.0140	0.0000	talus	0.0150	0.0150	-0.0340
--	--	calcaneus	-0.0300	0.0050	-0.0340	metatar2	0.0550	-0.0140	-0.0380
--	--	metatar3	0.0510	-0.0100	-0.0380	toes	0.0000	-0.0240	-0.0100

MUSCLE		SOURCE	MUSCLE LENGTH ML (m)	FIBER LENGTH LM (m)	VOLUME V (ml)	MUSCLE PCSA =V/L (cm²)	PENNA-TION ANGLE (deg)	MUSCLE FIBER TYPE		
								%SO	%FOG	%FG
	(toeIV)	[7]	--	--	--	--	--	--	--	--
	(toeIV)	[7]	--	--	--	--	--	--	--	--
	(toeV)	[7]	--	--	--	--	--	--	--	--
	(toeV)	[7]	--	--	--	--	--	--	--	--
	(toeI)	[7]	--	--	--	--	--	--	--	--
	(toeI)	[7]	--	--	--	--	--	--	--	--
	(toeII)	[7]	--	--	--	--	--	--	--	--
	(toeII)	[7]	--	--	--	--	--	--	--	--
	(toeIII)	[7]	--	--	--	--	--	--	--	--
	(toeIII)	[7]	--	--	--	--	--	--	--	--
	(toeIV)	[7]	--	--	--	--	--	--	--	--
	(toeIV)	[7]	--	--	--	--	--	--	--	--
	(toeV)	[7]	--	--	--	--	--	--	--	--
	(toeV)	[7]	--	--	--	--	--	--	--	--
	(toeI)	[7]	--	--	--	--	--	--	--	--
	(toeI)	[7]	--	--	--	--	--	--	--	--
	(toeII)	[7]	--	--	--	--	--	--	--	--
	(toeII)	[7]	--	--	--	--	--	--	--	--
	(toeIII)	[7]	--	--	--	--	--	--	--	--
	(toeIII)	[7]	--	--	--	--	--	--	--	--
	(toeIV)	[7]	--	--	--	--	--	--	--	--
	(toeIV)	[7]	--	--	--	--	--	--	--	--
	(toeV)	[7]	--	--	--	--	--	--	--	--
	(toeV)	[7]	--	--	--	--	--	--	--	--
quadratus plantae		[7]	--	--	--	--	--	--	--	--
	(toeII)	[7]	--	--	--	--	--	--	--	--
	(toeIII)	[7]	--	--	--	--	--	--	--	--
	(toeIII)	[7]	--	--	--	--	--	--	--	--
	(toeIV)	[7]	--	--	--	--	--	--	--	--
	(toeIV)	[7]	--	--	--	--	--	--	--	--
	(toeV)	[7]	--	--	--	--	--	--	--	--
	(toeV)	[7]	--	--	--	--	--	--	--	--
	(toeII)	[7]	--	--	--	--	--	--	--	--
	(toeII)	[7]	--	--	--	--	--	--	--	--
	(toeIII)	[7]	--	--	--	--	--	--	--	--
	(toeIII)	[7]	--	--	--	--	--	--	--	--
	(toeIV)	[7]	--	--	--	--	--	--	--	--
	(toeIV)	[7]	--	--	--	--	--	--	--	--
	(toeV)	[7]	--	--	--	--	--	--	--	--
	(toeV)	[7]	--	--	--	--	--	--	--	--
volar interossei		[7]	--	--	--	--	--	--	--	--
	(toeIII)	[7]	--	--	--	--	--	--	--	--
	(toeIV)	[7]	--	--	--	--	--	--	--	--

TENDON LENGTH LT (m)	TENDON PCSA (mm²)	ORIGIN COORD. SYSTEM	X (m)	Y (m)	Z (m)	INSERTION COORD. SYSTEM	X (m)	Y (m)	Z (m)
--	--	calcaneus	-0.0300	0.0050	0.0000	talus	0.0150	0.0150	-0.0340
--	--	metatar4	0.0520	-0.0120	-0.0340	metatar3	0.0510	-0.0100	-0.0340
--	--	calcaneus	-0.0300	0.0050	-0.0340	toes	-0.0070	-0.0340	-0.0100
--	--	metatar5	0.0480	-0.0150	0.0000	talus	0.0150	0.0150	-0.0340
--	--	calcaneus	-0.0300	-0.0120	-0.0340	metatar4	0.0520	-0.0120	-0.0260
--	--	metatar1	0.0460	0.0050	-0.0260	toes	-0.0130	-0.0460	-0.0100
--	--	calcaneus	-0.0300	-0.0120	0.0000	talus	0.0150	0.0150	-0.0340
--	--	metatar2	0.0550	-0.0140	-0.0340	metatar5	0.0480	-0.0150	-0.0200
--	--	calcaneus	-0.0300	-0.0120	-0.0200	toes	-0.0240	-0.0560	-0.0100
--	--	metatar3	0.0510	-0.0100	0.0050	talus	-0.0200	0.0050	-0.0170
--	--	calcaneus	-0.0300	-0.0120	-0.0170	calcaneus	0.0000	0.0130	-0.0320
--	--	metatar4	0.0520	-0.0120	-0.0320	metatar1	0.0460	-0.0010	-0.0300
--	--	calcaneus	-0.0300	-0.0120	-0.0300	toes	0.0000	0.0000	-0.0100
--	--	metatar5	0.0480	-0.0150	-0.0340	toes	-0.0070	-0.0340	-0.0100
--	--	calcaneus	-0.0300	-0.0220	0.0000	talus	0.0150	0.0150	-0.0340
--	--	metatar1	0.0460	0.0050	-0.0340	metatar4	0.0520	-0.0120	-0.0260
--	--	calcaneus	-0.0300	-0.0220	-0.0260	toes	-0.0130	-0.0460	-0.0100
--	--	metatar2	0.0550	-0.0140	-0.0380	toes	0.0000	-0.0240	-0.0100
--	--	calcaneus	-0.0300	0.0220	-0.0600	metatar3	0.0510	-0.0100	-0.0340
--	--	metatar3	0.0510	0.0150	-0.0340	metatar5	0.0480	-0.0150	-0.0200
--	--	calcaneus	-0.0300	-0.0220	-0.0600	metatar4	0.0520	-0.0120	-0.0260
--	--	metatar4	0.0520	-0.0120	0.0260	toes	-0.0130	0.0460	-0.0100
--	--	calcaneus	-0.0300	-0.0220	-0.0600	metatar5	0.0480	-0.0150	-0.0200
--	--	metatar5	0.0480	-0.0150	-0.0200	toes	-0.0240	-0.0560	-0.0100
--	--	calcaneus	-0.0120	0.0000	-0.0400	metatar2	0.0550	-0.0140	-0.0380
--	--	metatar2	0.0550	-0.0140	-0.0380	toes	0.0000	-0.0240	-0.0100
--	--	calcaneus	-0.0120	0.0000	-0.0400	metatar3	0.0510	-0.0100	-0.0340
--	--	metatar3	0.0510	-0.0100	-0.0340	toes	-0.0070	-0.0340	-0.0100
--	--	calcaneus	-0.0120	0.0000	-0.0400	metatar4	0.0520	-0.0120	-0.0260
--	--	metatar4	0.0520	-0.0120	-0.0260	toes	-0.0130	-0.0460	-0.0100
--	--	calcaneus	-0.0120	0.0000	0.0000	talus	0.0180	0.0050	-0.0350
--	--	metatar5	0.0480	-0.0150	-0.0350	metatar3	0.0050	0.0050	-0.0090
--	--	calcaneus	-0.0300	-0.0200	0.0000	talus	0.0180	0.0050	-0.0350
--	--	metatar2	0.0550	-0.0140	-0.0350	metatar4	0.0030	0.0050	-0.0070
--	--	calcaneus	-0.0300	-0.0200	0.1500	retinac	0.0200	-0.0100	-0.0100
--	--	metatar3	0.0510	-0.0100	-0.0100	toes	0.0000	-0.0240	0.0080
--	--	calcaneus	-0.0300	-0.0200	0.1500	retinac	0.0200	-0.0100	-0.0100
--	--	metatar4	0.0520	-0.0120	-0.0100	toes	-0.0070	-0.0340	0.0080
--	--	calcaneus	-0.0300	-0.0200	0.1500	retinac	0.0200	-0.0100	-0.0100
--	--	metatar5	0.0480	-0.0150	-0.0100	toes	-0.0130	-0.0460	0.0060
--	--	metatar3	0.0220	-0.0030	-0.0120	toes	-0.0070	-0.0310	-0.0070
--	--	metatar4	-0.0230	-0.0050	-0.0090	toes	-0.0130	-0.0430	-0.0080
--	--	metatar5	0.0250	-0.0050	-0.0080	toes	-0.0240	-0.0530	-0.0080

TABLE A.2
TRUNK MUSCULATURE

MUSCLE	SOURCE	MUSCLE LENGTH ML (m)	FIBER LENGTH LM (m)	VOLUME V (ml)	MUSCLE PCSA =V/L (cm²)	PENNA-TION ANGLE (deg)	MUSCLE FIBER TYPE %SO	%FO	%FG
1st dorsal interosseus	[14]	--	--	--	--	--	57.4	--	42.6
erector spinae	(1)	[7]	--	--	--	--	--	--	--
	(2)	[7]	--	--	--	--	--	--	--
	(3)	[7]	--	--	--	--	--	--	--
	(4)	[7]	--	--	--	--	--	--	--
	(5)	[7]	--	--	--	--	--	--	--
	(6)	[7]	--	--	--	--	--	--	--
	(7)	[7]	--	--	--	--	--	--	--
	(8)	[7]	--	--	--	--	--	--	--
	(9)	[7]	--	--	--	--	--	--	--
	(10)	[7]	--	--	--	--	--	--	--
	(11)	[7]	--	--	--	--	--	--	--
	(12)	[7]	--	--	--	--	--	--	--
	(13)	[7]	--	--	--	--	--	--	--
	(14)	[7]	--	--	--	--	--	--	--
	(15)	[7]	--	--	--	--	--	--	--
	(16)	[7]	--	--	--	--	--	--	--
	(L)	[13]	--	--	425.59	16.02	--	--	--
	(R)	[13]	--	--	417.87	15.73	--	--	--
	(surface)	[14]	--	--	--	--	--	58.4	--	41.6
	(deep)	[14]	--	--	--	--	--	54.9	--	45.1
external obliquus abdominis	[12]	--	--	--	6.85	--	--	--	--
	(1)	[7]	--	--	--	--	--	--	--
	(2)	[7]	--	--	--	--	--	--	--
iliocostalis dorsi	[12]	--	--	--	0.50	--	--	--	--
iliocostalis lumborum	[12]	--	--	--	1.00	--	--	--	--
iliopsoas	(1)	[7]	--	--	--	--	--	--	--
	(2)	[7]	--	--	--	--	--	--	--
	(3)	[7]	--	--	--	--	--	--	--
	(4)	[7]	--	--	--	--	--	--	--
	(5)	[7]	--	--	--	--	--	--	--
	(6)	[7]	--	--	--	--	--	--	--
	(7)	[7]	--	--	--	--	--	--	--
	(8)	[7]	--	--	--	--	--	--	--
	(9)	[7]	--	--	--	--	--	--	--
internal obliquus abdominis	[7]	--	--	--	--	--	--	--	--
	[12]	--	--	--	5.68	--	--	--	--
interspinalis	(1)	[12]	--	--	--	0.50	--	--	--
	(2)	[12]	--	--	--	--	--	--	--
intertransversarii	(1)	[12]	--	--	--	0.25	--	--	--
	(2)	[12]	--	--	--	--	--	--	--
latissimus dorsi	(1)	[7]	--	--	--	--	--	--	--
	(2)	[7]	--	--	--	--	--	--	--
	(3)	[7]	--	--	--	--	--	--	--
	(4)	[7]	--	--	--	--	--	--	--
	(5)	[7]	--	--	--	--	--	--	--
	[14]	--	--	--	--	--	50.5	--	49.5
longissimus dorsi	[12]	--	--	--	1.0	--	--	--	--
multifidus	(1)	[12]	--	--	--	1.25	--	--	--
	(2)	[12]	--	--	--	--	--	--	--
multifidus lumborum	(1)	[7]	--	--	--	--	--	--	--
	(2)	[7]	--	--	--	--	--	--	--
	(3)	[7]	--	--	--	--	--	--	--
	(4)	[7]	--	--	--	--	--	--	--
	(5)	[7]	--	--	--	--	--	--	--
	(6)	[7]	--	--	--	--	--	--	--
	(7)	[7]	--	--	--	--	--	--	--
	(8)	[7]	--	--	--	--	--	--	--
	(9)	[7]	--	--	--	--	--	--	--
	(10)	[7]	--	--	--	--	--	--	--
	(11)	[7]	--	--	--	--	--	--	--
quadratus lumborum	(1)	[7]	--	--	--	--	--	--	--
	(2)	[7]	--	--	--	--	--	--	--
	(3)	[7]	--	--	--	--	--	--	--
	(4)	[7]	--	--	--	--	--	--	--
	(5)	[7]	--	--	--	--	--	--	--
	[12]	--	--	--	2.80	--	--	--	--
rectus abdominis	[7]	--	--	--	--	--	--	--	--
	[12]	--	--	--	2.66	--	--	--	--
	[13]	--	--	356.16	10.50	--	--	--	--
	[14]	--	--	--	--	--	46.1	--	53.9
rotatores lumborum	(1)	[7]	--	--	--	--	--	--	--
	(2)	[7]	--	--	--	--	--	--	--
	(3)	[7]	--	--	--	--	--	--	--
	(4)	[7]	--	--	--	--	--	--	--
	(5)	[7]	--	--	--	--	--	--	--

TENDON LENGTH LT (m)	TENDON PCSA (mm²)	ORIGIN				INSERTION			
		COORD. SYSTEM	X (m)	Y (m)	Z (m)	COORD. SYSTEM	X (m)	Y (m)	Z (m)
--	--	--	--	--	--	--	--	--	--
--	--	Thorax	-0.0600	-0.0700	0.1000	Pelvis	-0.0650	-0.0600	0.1400
--	--	Thorax	-0.0600	-0.0700	0.1000	Pelvis	-0.0600	-0.0300	0.0800
--	--	L1	-0.0650	0.0000	0.0100	Thorax	-0.0400	-0.0200	0.1250
--	--	Thorax	0.0200	-0.0350	0.2200	Thorax	-0.0400	-0.0350	0.1250
--	--	L2	-0.0650	0.0000	0.0100	Thorax	-0.0400	-0.0230	0.1250
--	--	Thorax	0.0200	-0.0350	0.2200	Thorax	-0.0400	-0.0230	0.1250
--	--	L3	-0.0650	0.0000	0.0100	Thorax	-0.0400	-0.0250	0.1250
--	--	Thorax	0.0200	-0.0350	0.2200	Thorax	-0.0400	-0.0250	0.1250
--	--	L4	-0.0650	0.0000	0.0100	Thorax	-0.0400	-0.0270	0.1250
--	--	Thorax	0.0200	-0.0350	0.2200	Thorax	-0.0400	-0.0270	0.1250
--	--	L5	-0.0650	0.0000	0.0100	Thorax	-0.0400	-0.0290	0.1250
--	--	Thorax	0.0200	-0.0350	0.2200	Thorax	-0.0400	-0.0290	0.1250
--	--	Pelvis	-0.0660	-0.0300	0.0800	Thorax	-0.0400	-0.0350	0.1250
--	--	Thorax	0.0200	-0.0350	0.2200	Thorax	-0.0400	-0.0350	0.1250
--	--	Pelvis	-0.0650	-0.0600	0.1400	Thorax	-0.0400	-0.0420	0.1250
--	--	Thorax	0.0200	-0.0350	0.2200	Thorax	-0.0400	-0.0420	0.1250
--	--	--	--	--	--	--	--	--	--
--	--	--	--	--	--	--	--	--	--
--	--	--	--	--	--	--	--	--	--
--	--	--	--	--	--	--	--	--	--
--	--	Pelvis	0.0762	0.0894	-0.0782	CenTorso	0.0000	0.0894	0.0079
--	--	Thorax	-0.0400	-0.1200	-0.0450	Pelvis	0.0400	-0.1150	0.1150
--	--	Thorax	0.0700	-0.1300	0.0000	Pelvis	0.0600	0.0000	-0.0100
--	--	CenTorso	-0.0254	0.0762	0.0079	UpTorso	-0.0254	0.0762	0.0000
--	--	Pelvis	-0.0254	0.0381	0.0978	CenTorso	-0.0254	0.0762	0.0079
--	--	L1	-0.0050	-0.0200	0.0200	Pelvis	0.0250	-0.0750	0.0000
--	--	Femur	-0.0100	-0.0950	0.3450	Pelvis	0.0250	-0.0750	0.0000
--	--	L2	-0.0050	-0.0230	0.0200	Pelvis	0.0250	-0.0750	0.0000
--	--	L3	-0.0050	-0.0250	0.0200	Pelvis	0.0250	-0.0750	0.0000
--	--	Femur	-0.0100	-0.0950	0.3450	Pelvis	0.0250	-0.0750	0.0000
--	--	L4	-0.0050	-0.0270	0.0200	Pelvis	0.0250	-0.0750	0.0000
--	--	Femur	-0.0100	-0.0950	0.3450	Pelvis	0.0250	-0.0750	0.0000
--	--	L5	-0.0050	-0.0280	0.0200	Pelvis	0.0250	-0.0750	0.0000
--	--	Femur	-0.0100	-0.0950	0.3450	Pelvis	0.0250	-0.0750	0.0000
--	--	Thorax	0.1450	-0.0050	0.0200	Pelvis	0.0250	-0.1250	0.1250
--	--	Pelvis	0.0000	0.0894	-0.0782	CenTorso	0.0762	0.0894	0.0079
--	--	CenTorso	0.0254	0.0000	-0.0734	UpTorso	-0.0190	0.0000	0.1074
--	--	UpTorso	-0.0190	0.0000	-0.0518	Neck	-0.0190	0.0000	0.0716
--	--	CenTorso	-0.0254	0.0381	-0.0734	UpTorso	-0.0190	0.0254	0.1074
--	--	UpTorso	-0.0190	0.0254	-0.0518	Neck	-0.0190	0.0254	0.0716
--	--	L1	-0.0650	0.0000	0.0100	Thorax	-0.0400	-0.1150	0.0500
--	--	L2	-0.0650	0.0000	0.0100	Thorax	-0.0400	-0.1150	0.0400
--	--	L3	-0.0650	0.0000	0.0100	Thorax	-0.0400	-0.1150	0.0300
--	--	L4	-0.0650	0.0000	0.0100	Thorax	-0.0300	-0.1150	0.0200
--	--	L5	-0.0650	0.0000	0.0100	Thorax	-0.0300	-0.1150	0.0100
--	--	--	--	--	--	--	--	--	--
--	--	CenTorso	-0.0254	0.0381	-0.0048	UpTorso	-0.0190	0.0254	0.0180
--	--	CenTorso	-0.0254	0.0381	-0.0734	UpTorso	-0.0190	0.0000	0.1074
--	--	UpTorso	-0.0190	0.0254	-0.0518	Neck	-0.0190	0.0000	0.0716
--	--	Thorax	-0.0500	0.0000	0.0950	L1	-0.0450	-0.0200	0.0330
--	--	Thorax	-0.0550	0.0000	0.0650	L1	-0.0450	-0.0200	0.0330
--	--	Thorax	-0.0550	0.0000	0.0650	L2	-0.0450	-0.0200	0.0330
--	--	Thorax	-0.0550	0.0000	0.0300	L2	-0.0450	-0.0200	0.0330
--	--	Thorax	-0.0550	0.0000	0.0300	L3	-0.0450	-0.0200	0.0330
--	--	Thorax	-0.0550	0.0000	0.0000	L3	-0.0450	-0.0200	0.0330
--	--	Thorax	-0.0550	0.0000	0.0000	L4	-0.0450	-0.0200	0.0330
--	--	L1	-0.0650	0.0000	0.0100	L4	-0.0450	-0.0200	0.0330
--	--	L1	-0.0650	0.0000	0.0100	L5	-0.0450	-0.0200	0.0330
--	--	L2	-0.0650	0.0000	0.0100	L5	-0.0450	-0.0200	0.0330
--	--	L5	-0.0650	0.0000	0.0100	Pelvis	-0.0800	-0.0300	0.0650
--	--	Thorax	-0.0500	-0.0550	0.0100	Pelvis	-0.0400	-0.0750	0.1600
--	--	L1	-0.0350	-0.0350	0.0200	Pelvis	-0.0400	-0.0750	0.1600
--	--	L1	-0.0350	-0.0370	0.0200	Pelvis	-0.0400	-0.0750	0.1600
--	--	L3	-0.0350	-0.0420	0.0200	Pelvis	-0.0400	-0.0750	0.1600
--	--	L4	-0.0350	-0.0450	0.0200	Pelvis	-0.0400	-0.0750	0.1600
--	--	Pelvis	0.0000	0.0894	-0.0782	CenTorso	-0.0190	0.0254	-0.0043
--	--	Thorax	0.1500	-0.0400	0.0500	Pelvis	0.0550	-0.0080	-0.0100
--	--	Pelvis	0.0254	0.0127	0.0487	CenTorso	0.0762	0.0127	-0.1074
--	--	--	--	--	--	--	--	--	--
--	--	--	--	--	--	--	--	--	--
--	--	Thorax	-0.0550	0.0000	0.0300	L2	-0.0450	-0.0200	0.0330
--	--	Thorax	-0.0550	0.0000	0.0000	L2	-0.0450	-0.0200	0.0330
--	--	L1	-0.0650	0.0000	0.0100	L3	-0.0450	-0.0200	0.0330
--	--	L2	-0.0650	0.0000	0.0100	L4	-0.0450	-0.0200	0.0330
--	--	L3	-0.0650	0.0000	0.0100	L5	-0.0450	-0.0200	0.0330

TABLE A.3
UPPER EXTREMITY MUSCULATURE

MUSCLE		SOURCE	MUSCLE LENGTH ML (m)	FIBER LENGTH LM (m)	VOLUME V (ml)	MUSCLE PCSA =V/L (cm²)	PENNA- TION ANGLE (deg)	MUSCLE FIBER TYPE %SO	%FO	%FG
anconeus		[7]	--	--	--	--	--	--	--	--
		[12]	--	--	--	0.94	--	--	--	--
		[15]	--	--	--	3.18	--	--	--	--
		[22]	--	0.027	*6.7	2.5	--	--	--	--
biceps brachii	(LH)	[15]	--	--	--	3.33	--	--	--	--
	(LH)	[16]	0.3124	--	70.46	1.94	--	--	--	--
	(LH)	[21]	--	--	--	--	--	--	--	--
	(LH)	[22]	--	0.136	33.4	2.5	--	--	--	--
	(SH)	[15]	--	--	--	3.22	--	--	--	--
	(SH)	[16]	0.3327	--	37.69	1.29	--	--	--	--
	(SH)	[21]	--	--	--	--	--	--	--	--
	(SH)	[22]	--	0.150	30.8	2.1	--	--	--	--
	(1)	[17]	--	--	--	--	--	--	--	--
	(2)	[17]	--	--	--	--	--	--	--	--
	(surface)	[14]	--	--	--	--	--	42.3	--	57.7
	(deep)	[14]	--	--	--	--	--	50.5	--	49.5
	(LH)	[7]	--	--	--	--	--	--	--	--
	(LH)	[7]	--	--	--	--	--	--	--	--
	(LH)	[12]	--	--	--	3.55	--	--	--	--
	(SH)	[7]	--	--	--	--	--	--	--	--
	(SH)	[12]	--	--	--	--	--	--	--	--
brachialis		[7]	--	--	--	--	--	--	--	--
		[12]	--	--	--	4.63	--	--	--	--
		[15]	--	--	--	6.40	--	--	--	--
	(low head)	[16]	0.0762	--	34.41	4.52	--	--	--	--
	(upp head)	[16]	0.1499	--	65.55	4.52	--	--	--	--
		[21]	0.12	0.090	--	9	--	--	--	--
		[22]	--	0.090	59.3	7.0	--	--	--	--
brachioradialis		[7]	--	--	--	--	--	--	--	--
		[12]	--	--	--	1.37	--	--	--	--
		[15]	--	--	--	1.86	--	--	--	--
		[14]	--	--	--	--	--	39.8	--	60.2
		[16]	0.2870	--	40.97	1.29	--	--	--	--
		[21]	0.24	0.16	--	3	--	--	--	--
		[22]	--	0.164	21.9	1.5	--	--	--	--
coracobrachialis		[7]	--	--	--	--	--	--	--	--
		[12]	--	--	--	1.52	--	--	--	--
		[16]	0.1854	--	21.30	1.29	--	--	--	--
		[15]	--	--	--	5.8	--	--	--	--
	(1)	[17]	--	--	--	--	--	--	--	--
	(2)	[17]	--	--	--	--	--	--	--	--
deltoideus	(1)	[7]	--	--	--	--	--	--	--	--
	(2)	[7]	--	--	--	--	--	--	--	--
	(3)	[7]	--	--	--	--	--	--	--	--
	(1)	[12]	--	--	--	11.01	--	--	--	--
	(2)	[12]	--	--	--	--	--	--	--	--
	(1)	[17]	--	--	--	--	--	--	--	--
	(2)	[17]	--	--	--	--	--	--	--	--
	(3)	[17]	--	--	--	--	--	--	--	--
		[17]	--	--	--	--	--	--	--	--
	(superfic.)	[14]	--	--	--	--	--	53.3	--	46.7
	(deep)	[14]	--	--	--	--	--	61.0	--	39.0
	(acromial)	[16]	0.1651	--	222.86	13.55	--	--	--	--
	(clavicular)	[16]	0.1935	--	88.49	4.52	--	--	--	--
	(scapular)	[16]	0.1905	--	75.38	3.87	--	--	--	--
extensor carpi radialis brevis		[7]	--	--	--	--	--	--	--	--
		[15]	--	--	--	2.22	--	--	--	--
		[22]	--	0.053	15.8	2.9	--	--	--	--
extensor carpi radialis longus		[7]	--	--	--	--	--	--	--	--
		[15]	--	--	--	3.14	--	--	--	--
		[21]	0.22	0.080	--	7	--	--	--	--
		[22]	--	0.078	18.3	2.4	--	--	--	--
extensor carpi ulnaris		[7]	--	--	--	--	--	--	--	--
		[15]	--	--	--	5.30	--	--	--	--
		[22]	--	0.045	14.9	3.4	--	--	--	--
extensor digitorum communis		[7]	--	--	--	--	--	--	--	--
		[14]	--	--	--	--	--	47.3	--	52.7
		[15]	--	--	--	4.30	--	--	--	--

TENDON LENGTH LT (m)	TENDON PCSA (mm²)	ORIGIN				INSERTION			
		COORD. SYSTEM	X (m)	Y (m)	Z (m)	COORD. SYSTEM	X (m)	Y (m)	Z (m)
--	--	Humerus	-0.016	-0.025	-0.300	Ulna	-0.018	-0.006	-0.040
--	--	Humerus	0.000	0.030	0.132	Forearm	0.010	0.000	-0.176
--	--	--	--	--	--	--	--	--	--
--	--	--	--	--	--	--	--	--	--
--	--	Scapula	0.1829	0.1321	0.6985	Radius	0.0381	-0.1473	0.6045
--	--	Shoulder	0.025	0.000	-0.005	Shoulder	0.006	-0.318	-0.003
--	--	--	--	--	--	--	--	--	--
--	--	--	--	--	--	--	--	--	--
--	--	Scapula	0.1854	0.1270	0.7239	Radius	0.0406	-0.1448	0.6045
--	--	Shoulder	0.029	0.006	-0.015	Shoulder	0.006	-0.318	-0.003
--	--	--	--	--	--	Scapula	0.0180	0.0028	0.0025
--	--	--	--	--	--	Scapula	0.0090	0.0012	0.0402
--	--	--	--	--	--	--	--	--	--
--	--	Scapula	-0.005	0.020	-0.020	Humerus	0.005	0.005	0.025
--	--	Humerus	0.025	0.000	0.010	Radius	0.008	0.008	-0.055
--	--	UpTorso	0.0000	0.1595	-0.0724	Forearm	0.0099	0.0000	-0.1667
--	--	Scapula	0.010	0.036	-0.025	Radius	0.008	0.008	-0.055
--	--	UpTorso	0.0399	0.1595	-0.0823	Forearm	0.0099	0.0000	-0.1667
--	--	Humerus	0.010	-0.006	-0.200	Ulna	0.000	0.008	-0.030
--	--	Humerus	0.0099	0.0000	-0.0681	Forearm	0.0099	0.0000	-0.1816
--	--	--	--	--	--	--	--	--	--
--	--	Humerus	0.1016	-0.0762	0.5918	Ulna	0.0483	-0.1626	0.6020
--	--	Humerus	0.1270	-0.0178	0.6223	Ulna	0.0508	-0.1600	0.6020
--	--	Shoulder	0.001	-0.182	-0.003	Shoulder	-0.003	-0.304	-0.013
--	--	--	--	--	--	--	--	--	--
--	--	Humerus	0.000	-0.015	-0.270	Radius	0.000	-0.020	-0.240
--	--	Humerus	0.0099	0.0099	0.0917	Forearm	0.0000	0.0201	0.0084
--	--	--	--	--	--	--	--	--	--
--	--	Humerus	0.1041	-0.0737	0.5867	Radius	-0.0610	-0.1448	0.7595
--	--	Shoulder	-0.006	-0.195	0.005	Shoulder	0.010	-0.505	0.027
--	--	--	--	--	--	--	--	--	--
--	--	Scapula	0.010	0.036	-0.025	Humerus	0.010	0.008	-0.132
--	--	UpTorso	0.0399	0.1849	-0.0820	Humerus	0.0000	-0.0099	-0.0183
--	--	Scapula	0.1905	0.1219	0.7214	Humerus	0.1397	-0.0178	0.6274
--	--	--	--	--	--	Scapula	0.0209	-0.0025	0.0342
--	--	--	--	--	--	Humerus	0.1185	0.0042	-0.0054
--	--	Scapula	-0.065	0.060	-0.025	Humerus	-0.010	0.000	-0.120
--	--	Scapula	0.000	-0.025	-0.015	Humerus	0.010	-0.010	-0.120
--	--	Clavicle	0.008	-0.130	-0.005	Humerus	0.010	-0.008	-0.120
--	--	UpTorso	-0.0207	0.1648	-0.0968	Humerus	0.0099	0.0099	0.0018
--	--	UpTorso	0.0207	0.1648	-0.0968	Humerus	0.0099	0.0099	0.0018
--	--	--	--	--	--	Scapula	-0.0016	-0.0260	-0.0109
--	--	--	--	--	--	Scapula	0.0427	0.0352	-0.0308
--	--	--	--	--	--	Clavicle	0.1110	-0.0039	0.0021
--	--	--	--	--	--	Humerus	0.1185	-0.0012	0.0129
--	--	--	--	--	--	--	--	--	--
--	--	Scapula	0.1956	0.1422	0.6604	Humerus	0.1194	-0.0051	0.6223
--	--	Clavicle	0.1956	0.1473	0.7188	Humerus	0.1270	0.0203	0.6350
--	--	Scapula	0.2667	0.1270	0.6756	Humerus	0.1321	0.0127	0.6248
--	--	Humerus	-0.010	-0.030	-0.290	Hand	-0.006	-0.009	-0.036
--	--	--	--	--	--	--	--	--	--
--	--	Humerus	-0.010	-0.028	-0.280	Hand	-0.006	-0.013	-0.034
--	--	--	--	--	--	--	--	--	--
--	--	Shoulder	-0.008	-0.239	0.009	Shoulder	0.011	-0.490	0.024
--	--	--	--	--	--	--	--	--	--
--	--	Humerus	-0.014	-0.027	-0.300	Hand	-0.003	0.025	-0.030
--	--	--	--	--	--	--	--	--	--
--	--	Humerus	-0.012	-0.030	-0.295	Hand	-0.070	0.000	-0.015
--	--	--	--	--	--	--	--	--	--

MUSCLE		SOURCE	MUSCLE LENGTH ML (m)	FIBER LENGTH LM (m)	VOLUME V (ml)	MUSCLE PCSA =V/L (cm²)	PENNA-TION ANGLE (deg)	MUSCLE FIBER TYPE		
								%SO	%FO	%FG
flexor carpi radialis		[7]	--	--	--	--	--	--	--	--
		[15]	--	--	--	2.16	--	--	--	--
		[22]	--	0.058	12.4	2.0	--	--	--	--
flexor carpi ulnaris		[7]	--	--	--	--	--	--	--	--
		[14]	--	--	--	--	--	44.5	--	55.5
		[15]	--	--	--	5.0	--	--	--	--
		[22]	--	0.048	15.2	3.2	--	--	--	--
flexor digitorum profundus		[14]	--	--	--	--	--	47.3	--	52.7
		[15]	--	--	--	10.8	--	--	--	--
flexor digitorum sublimis		[7]	--	--	--	--	--	--	--	--
		[15]	--	--	--	10.7	--	--	--	--
flexor pollicis longus		[7]	--	--	--	--	--	--	--	--
		[15]	--	--	--	2.9	--	--	--	--
greater round	(1)	[17]	--	--	--	--	--	--	--	--
	(2)	[17]	--	--	--	--	--	--	--	--
infraspinatus		[14]	--	--	--	--	--	45.3	--	54.7
		[16]	0.1499	--	85.21	5.81	--	--	--	--
		[7]	--	--	--	--	--	--	--	--
		[12]	--	--	--	5.98	--	--	--	--
		[17]	--	--	--	--	--	--	--	--
	(1)	[17]	--	--	--	--	--	--	--	--
	(2)	[17]	--	--	--	--	--	--	--	--
infraspinatus and teres minor		[15]	--	--	--	16.5	--	--	--	--
latissimus dorsi	(1)	[7]	--	--	--	--	--	--	--	--
	(2)	[7]	--	--	--	--	--	--	--	--
	(3)	[7]	--	--	--	--	--	--	--	--
	(1)	[12]	--	--	--	5.37	--	--	--	--
	(2)	[12]	--	--	--	--	--	--	--	--
		[17]	--	--	--	--	--	--	--	--
		[16]	0.2692	--	339.21	12.90	--	--	--	--
levator scapulae	(1)	[7]	--	--	--	--	--	--	--	--
	(2)	[7]	--	--	--	--	--	--	--	--
	(3)	[7]	--	--	--	--	--	--	--	--
		[12]	--	--	--	17.75	--	--	--	--
		[17]	--	--	--	--	--	--	--	--
omohyoid		[17]	--	--	--	--	--	--	--	--
palmaris longus		[15]	--	--	--	0.93	--	--	--	--
		[22]	--	0.057	5.1	0.9	--	--	--	--
pectoralis major	(1)	[7]	--	--	--	--	--	--	--	--
	(2)	[7]	--	--	--	--	--	--	--	--
	(3)	[7]	--	--	--	--	--	--	--	--
	(1)	[12]	--	--	--	6.8	--	--	--	--
	(2)	[12]	--	--	--	--	--	--	--	--
	(clavicular)	[14]	--	--	--	--	--	42.3	--	57.7
	(sternal)	[14]	--	--	--	--	--	43.1	--	56.9
	(abdom.)	[16]	0.2184	--	78.66	3.87	--	--	--	--
	(clav.)	[16]	0.1854	--	93.40	5.16	--	--	--	--
	(stern.)	[16]	0.2159	--	96.68	4.52	--	--	--	--
	(1)	[17]	--	--	--	--	--	--	--	--
	(2)	[17]	--	--	--	--	--	--	--	--
pectoralis minor		[17]	--	--	--	--	--	--	--	--
		[16]	0.1346	--	52.44	3.87	--	--	--	--
pronator teres		[7]	--	--	--	--	--	--	--	--
		[12]	--	--	--	1.61	--	--	--	--
		[15]	--	--	--	3.24	--	--	--	--
		[21]	0.14	0.070	--	4	--	--	--	--
		[22]	--	0.056	18.7	3.4	--	--	--	--
pronator quadratus		[15]	--	--	--	2.22	--	--	--	--
		[21]	--	--	--	--	--	--	--	--
rhomboid	(1)	[7]	--	--	--	--	--	--	--	--
	(2)	[7]	--	--	--	--	--	--	--	--
	(3)	[7]	--	--	--	--	--	--	--	--
	(4)	[7]	--	--	--	--	--	--	--	--
	(5)	[7]	--	--	--	--	--	--	--	--
		[14]	--	--	--	--	--	44.6	--	55.4
	(major)	[16]	0.1143	--	40.97	3.87	--	--	--	--
	(minor)	[16]	0.0787	--	27.86	3.23	--	--	--	--
	(greater)	[17]	--	--	--	--	--	--	--	--
	(smaller)	[17]	--	--	--	--	--	--	--	--

TENDON LENGTH LT (m)	TENDON PCSA (mm²)	ORIGIN COORD. SYSTEM	X (m)	Y (m)	Z (m)	INSERTION COORD. SYSTEM	X (m)	Y (m)	Z (m)
--	--	Humerus	-0.003	0.036	-0.295	Hand	0.008	-0.018	-0.025
--	--	--	--	--	--	--	--	--	--
--	--	--	--	--	--	--	--	--	--
--	--	Humerus	-0.012	0.030	-0.295	Hand	0.007	0.018	-0.012
--	--	--	--	--	--	--	--	--	--
--	--	--	--	--	--	--	--	--	--
--	--	--	--	--	--	--	--	--	--
--	--	--	--	--	--	--	--	--	--
--	--	Humerus	-0.005	0.035	-0.295	Hand	0.007	0.000	-0.025
--	--	--	--	--	--	--	--	--	--
--	--	Humerus	-0.005	0.035	-0.295	Hand	0.007	-0.030	-0.030
--	--	--	--	--	--	--	--	--	--
--	--	--	--	--	--	Scapula	0.1494	-0.0103	-0.0011
--	--	--	--	--	--	Humerus	0.0428	0.0111	-0.0066
--	--	--	--	--	--	--	--	--	--
--	--	Scapula	0.2972	0.0787	0.6833	Humerus	0.1880	0.1346	0.6604
--	--	Scapula	-0.095	0.060	-0.080	Humerus	-0.008	-0.023	0.004
--	--	UpTorso	-0.0099	0.1148	-0.0071	Humerus	0.0000	0.0150	-0.1382
--	--	--	--	--	--	Humerus	-0.0054	-0.0075	0.0281
--	--	--	--	--	--	Scapula	0.1092	0.0425	-0.0188
--	--	--	--	--	--	Scapula	0.1345	0.0204	-0.0122
--	--	--	--	--	--	--	--	--	--
--	--	Thorax	-0.105	-0.090	-0.230	Humerus	0.010	0.000	-0.058
--	--	Thorax	-0.080	-0.115	-0.280	Humerus	0.015	0.006	-0.040
--	--	Pelvis	-0.110	-0.055	-0.380	Humerus	0.015	0.006	-0.040
--	--	CenTorso	-0.0508	0.0000	0.0978	Humerus	0.0000	0.0099	-0.0681
--	--	CenTorso	-0.0508	0.0000	0.0180	Humerus	0.0000	0.0099	-0.0681
--	--	--	--	--	--	Humerus	0.0509	0.0108	0.0051
--	--	Thorax	0.2845	-0.1219	0.6833	Humerus	0.1702	0.0660	0.6731
--	--	C1	0.0000	-0.0360	0.0070	Scapula	-0.0570	0.0900	-0.0040
--	--	C2	-0.0040	-0.0280	0.0050	Scapula	-0.0680	0.0900	-0.0120
--	--	C3	-0.0040	-0.0280	0.0050	Scapula	-0.0800	0.0870	-0.0280
--	--	UpTorso	-0.0508	0.1087	-0.0470	CenTorso	-0.0079	0.0114	-0.0470
--	--	--	--	--	--	Scapula	0.0392	0.0731	-0.0021
--	--	--	--	--	--	Scapula	0.0428	0.0358	0.0005
--	--	--	--	--	--	--	--	--	--
--	--	Thorax	0.040	0.010	0.070	Humerus	0.020	-0.009	-0.045
--	--	Thorax	0.040	-0.020	-0.190	Humerus	0.020	-0.009	-0.045
--	--	Clavicle	0.018	-0.050	0.007	Humerus	0.020	-0.009	-0.045
--	--	UpTorso	0.0508	0.0000	-0.0071	Humerus	0.0099	0.0099	-0.0483
--	--	UpTorso	0.0508	0.0848	-0.0670	Humerus	0.0099	0.0099	-0.0483
--	--	--	--	--	--	--	--	--	--
--	--	Thorax	0.1448	-0.0254	0.8763	Humerus	0.1473	0.0711	0.6629
--	--	Clavicle	0.2083	0.1194	0.8179	Humerus	0.1372	0.0432	0.6528
--	--	Thorax	0.2007	0.0483	0.8915	Humerus	0.1448	0.0660	0.6004
--	--	--	--	--	--	Humerus	0.0823	0.0119	0.0129
--	--	--	--	--	--	Clavicle	0.0522	0.0136	0.0038
--	--	--	--	--	--	Scapula	0.0289	0.0064	0.0347
--	--	Thorax	0.1549	0.0330	0.8001	Scapula	0.2007	0.1245	0.7290
--	--	Humerus	-0.005	0.035	-0.290	Radius	0.005	-0.010	-0.130
--	--	Humerus	0.0099	0.0000	0.1105	Forearm	0.0000	0.0180	-0.0968
--	--	Shoulder	-0.004	-0.247	-0.027	Shoulder	0.005	-0.390	0.025
--	--	--	--	--	--	--	--	--	--
--	--	Shoulder	0.010	-0.496	-0.010	Shoulder	0.014	-0.493	0.020
--	--	C3	-0.0350	-0.0040	-0.0070	Scapula	-0.1100	0.0920	-0.0400
--	--	C7	-0.0400	-0.0040	-0.0100	Scapula	-0.1150	0.0900	-0.0650
--	--	Thorax	0.0300	0.0000	0.2750	Scapula	-0.1180	0.0850	-0.0900
--	--	Thorax	0.0300	0.0000	0.2600	Scapula	-0.1200	0.0800	-0.1100
--	--	Thorax	0.0300	0.0000	0.2500	Scapula	-0.1200	0.0750	-0.1350
--	--	--	--	--	--	--	--	--	--
--	--	Thorax	0.3505	0.1016	0.7722	Scapula	0.3150	0.0533	0.6858
--	--	Thorax	0.3404	0.1321	0.7823	Scapula	0.3150	0.1168	0.7061
--	--	--	--	--	--	Scapula	0.1506	0.0428	-0.0246
--	--	--	--	--	--	Scapula	0.0646	0.0804	-0.0241

MUSCLE		SOURCE	MUSCLE LENGTH ML (m)	FIBER LENGTH LM (m)	VOLUME V (ml)	MUSCLE PCSA =V/L (cm²)	PENNA-TION ANGLE (deg)	MUSCLE FIBER TYPE		
								%SO	%FO	%FG
serratis anterior	(1)	[7]	--	--	--	--	--	--	--	--
	(2)	[7]	--	--	--	--	--	--	--	--
	(3)	[7]	--	--	--	--	--	--	--	--
	(4)	[7]	--	--	--	--	--	--	--	--
	(1)	[17]	--	--	--	--	--	--	--	--
	(2)	[17]	--	--	--	--	--	--	--	--
	(3)	[17]	--	--	--	--	--	--	--	--
	(lower)	[16]	0.1727	--	162.23	9.68	--	--	--	--
	(upper)	[16]	0.1041	--	31.14	3.23	--	--	--	--
smaller round	(1)	[17]	--	--	--	--	--	--	--	--
	(2)	[17]	--	--	--	--	--	--	--	--
sternocleido-mastoid		[17]	--	--	--	--	--	--	--	--
sternohyoid		[17]	--	--	--	--	--	--	--	--
subclavius		[7]	--	--	--	--	--	--	--	--
		[17]	--	--	--	--	--	--	--	--
		[16]	0.0762	--	--	--	--	--	--	--
subscapularis		[7]	--	--	--	--	--	--	--	--
		[12]	--	--	--	9.9	--	--	--	--
		[16]	0.1219	--	121.26	9.68	--	--	--	--
		[15]	--	--	--	25.24	--	--	--	--
		[17]	--	--	--	--	--	--	--	--
	(1)	[17]	--	--	--	--	--	--	--	--
	(2)	[17]	--	--	--	--	--	--	--	--
	(3)	[17]	--	--	--	--	--	--	--	--
supinator		[7]	--	--	--	--	--	--	--	--
		[7]	--	--	--	--	--	--	--	--
		[12]	--	--	--	1.77	--	--	--	--
		[15]	--	--	--	2.20	--	--	--	--
		[21]	--	0.033	--	4.5	--	--	--	--
		[22]	--	0.033	10.9	3.4	--	--	--	--
supraspinatus		[7]	--	--	--	--	--	--	--	--
		[14]	--	--	--	--	--	59.3	--	40.7
		[12]	--	--	--	3.3	--	--	--	--
		[16]	0.0864	--	39.33	4.52	--	--	--	--
		[15]	--	--	--	7.7	--	--	--	--
	(1)	[17]	--	--	--	--	--	--	--	--
	(2)	[17]	--	--	--	--	--	--	--	--
teres major		[7]	--	--	--	--	--	--	--	--
		[12]	--	--	--	4.97	--	--	--	--
		[16]	0.1219	--	70.46	5.81	--	--	--	--
		[15]	--	--	--	9.8	--	--	--	--
teres minor		[7]	--	--	--	--	--	--	--	--
		[12]	--	--	--	1.57	--	--	--	--
		[16]	0.1041	--	24.58	2.58	--	--	--	--
trapezius	(1)	[7]	--	--	--	--	--	--	--	--
	(2)	[7]	--	--	--	--	--	--	--	--
	(3)	[7]	--	--	--	--	--	--	--	--
	(4)	[7]	--	--	--	--	--	--	--	--
	(5)	[7]	--	--	--	--	--	--	--	--
	(6)	[7]	--	--	--	--	--	--	--	--
	(7)	[7]	--	--	--	--	--	--	--	--
	(8)	[7]	--	--	--	--	--	--	--	--
	(9)	[7]	--	--	--	--	--	--	--	--
	(10)	[7]	--	--	--	--	--	--	--	--
	(11)	[7]	--	--	--	--	--	--	--	--
	(12)	[7]	--	--	--	--	--	--	--	--
	(13)	[7]	--	--	--	--	--	--	--	--
	(14)	[7]	--	--	--	--	--	--	--	--
	(15)	[7]	--	--	--	--	--	--	--	--
	(16)	[7]	--	--	--	--	--	--	--	--
	(17)	[7]	--	--	--	--	--	--	--	--
	(18)	[7]	--	--	--	--	--	--	--	--
	(19)	[7]	--	--	--	--	--	--	--	--
	(20)	[7]	--	--	--	--	--	--	--	--
		[12]	--	--	--	10.60	--	--	--	--
	(lower)	[16]	0.2083	--	199.92	9.68	--	--	--	--
	(medial)	[16]	0.1245	--	90.13	7.10	--	--	--	--
	(upper)	[16]	0.1194	--	80.30	6.45	--	--	--	--

TENDON LENGTH LT (m)	TENDON PCSA (mm²)	ORIGIN				INSERTION			
		COORD. SYSTEM	X (m)	Y (m)	Z (m)	COORD. SYSTEM	X (m)	Y (m)	Z (m)
--	--	Thorax	0.0850	-0.0850	0.2250	Scapula	-0.0550	0.0900	0.0050
--	--	Thorax	0.0850	-0.1000	0.1900	Thorax	0.0300	-0.1100	0.1900
--	--	Scapula	-0.1250	0.0950	-0.0850	Thorax	0.0300	-0.1100	0.1900
--	--	Thorax	0.0800	-0.1400	0.0700	Scapula	-0.1300	0.0680	-0.1480
--	--	--	--	--	--	Scapula	0.0582	0.0804	-0.0191
--	--	--	--	--	--	Scapula	0.1242	0.0549	-0.0204
--	--	--	--	--	--	Scapula	0.1653	0.0147	0.0057
--	--	Thorax	0.1727	-0.0254	0.7264	Scapula	0.3124	0.0000	0.6680
--	--	Thorax	0.2261	0.1016	0.7518	Scapula	0.3073	0.0838	0.7188
--	--	--	--	--	--	Scapula	0.1062	-0.0087	-0.0042
--	--	--	--	--	--	Humerus	0.0117	-0.0135	-0.0111
--	--	--	--	--	--	Clavicle	0.0293	0.0047	0.0098
--	--	--	--	--	--	Clavicle	0.0092	-0.0094	0.0009
--	--	Thorax	0.1150	-0.0500	0.2130	Clavicle	0.0000	0.0950	0.0030
--	--	--	--	--	--	Clavicle	0.0768	-0.0050	-0.0003
--	--	Thorax	0.2811	0.1016	0.8280	Clavicle	0.2159	0.1872	0.7595
--	--	Scapula	-0.090	0.065	-0.085	Humerus	0.015	0.012	0.008
--	--	UpTorso	0.0099	0.1250	0.0130	Humerus	0.0150	0.0150	-0.1382
--	--	Scapula	0.3048	0.0711	0.6909	Humerus	0.2032	0.1067	0.6960
--	--	--	--	--	--	--	--	--	--
--	--	--	--	--	--	Humerus	0.0036	0.0254	0.0033
--	--	--	--	--	--	Scapula	0.0926	0.0543	-0.0242
--	--	--	--	--	--	Scapula	0.1182	0.0290	-0.0190
--	--	--	--	--	--	Scapula	0.1349	0.0122	-0.0113
--	--	Humerus	-0.013	-0.028	-0.300	Radius	0.006	-0.006	-0.065
--	--	Arm	-0.0130	-0.0280	0.1600	Radius	0.0060	-0.0060	0.0650
--	--	Humerus	0.0000	0.0300	0.1318	Forearm	0.0000	0.0150	-0.1509
--	--	--	--	--	--	--	--	--	--
--	--	Shoulder	0.015	-0.285	0.010	Shoulder	0.007	-0.340	-0.014
--	--	--	--	--	--	--	--	--	--
--	--	Scapula	-0.060	0.075	-0.006	Humerus	0.019	-0.020	0.025
--	--	--	--	--	--	--	--	--	--
--	--	UpTorso	0.0000	0.0950	-0.0470	Humerus	0.0000	0.0399	-0.1481
--	--	Scapula	0.2870	0.1397	0.7239	Humerus	0.1981	0.1397	0.6833
--	--	--	--	--	--	--	--	--	--
--	--	--	--	--	--	Scapula	0.0540	0.0616	-0.0198
--	--	--	--	--	--	Humerus	-0.0126	0.0093	0.0236
--	--	Scapula	-0.110	0.060	-0.140	Humerus	0.014	0.008	-0.064
--	--	UpTorso	0.0000	0.1250	0.0531	Humerus	0.0000	-0.0099	-0.0782
--	--	Scapula	0.2794	0.0305	0.6553	Humerus	0.1702	0.0610	0.6655
--	--	--	--	--	--	--	--	--	--
--	--	Scapula	-0.070	0.025	-0.095	Humerus	-0.006	-0.023	-0.005
--	--	UpTorso	0.0000	0.1450	0.0130	Humerus	0.0000	0.0099	-0.1382
--	--	Scapula	0.2667	0.0483	0.6604	Humerus	0.1905	0.1118	0.6604
--	--	Skull	-0.0660	-0.0100	0.0270	Clavicle	-0.0100	0.1300	0.0050
--	--	C1	-0.0270	0.0000	0.0060	Clavicle	-0.0100	0.1300	0.0050
--	--	C2	-0.0270	-0.0040	-0.0080	Clavicle	-0.0100	0.1300	0.0050
--	--	C3	-0.0350	-0.0040	-0.0070	Clavicle	-0.0100	0.1300	0.0050
--	--	C4	-0.0350	-0.0040	-0.0070	Clavicle	-0.0100	0.1300	0.0050
--	--	C5	-0.0350	-0.0040	-0.0070	Clavicle	-0.0100	0.1300	0.0050
--	--	C6	-0.0350	-0.0040	-0.0070	Clavicle	-0.0100	0.1300	0.0050
--	--	Skull	-0.0660	-0.0100	0.0270	Scapula	-0.0200	-0.0100	-0.0150
--	--	C1	-0.0270	0.0000	0.0060	Scapula	-0.0200	-0.0100	-0.0150
--	--	C2	-0.0270	-0.0040	-0.0080	Scapula	-0.0200	-0.0100	-0.0150
--	--	C3	-0.0350	-0.0040	-0.0070	Scapula	-0.0200	-0.0100	-0.0150
--	--	C4	-0.0350	-0.0040	-0.0070	Scapula	-0.0200	-0.0100	-0.0150
--	--	C5	-0.0350	-0.0040	-0.0070	Scapula	-0.0200	-0.0100	-0.0150
--	--	C6	-0.0350	-0.0040	-0.0070	Scapula	-0.0200	-0.0100	-0.0150
--	--	C7	-0.0400	-0.0040	-0.0100	Scapula	-0.0650	0.0500	-0.0200
--	--	Thorax	0.0300	0.0000	0.2550	Scapula	-0.0650	0.0500	-0.0200
--	--	Thorax	-0.0070	0.0000	0.1500	Scapula	-0.1150	0.0950	-0.0500
--	--	Scapula	-0.0900	0.0650	-0.0270	Scapula	-0.1150	0.0950	-0.0500
--	--	Thorax	-0.0450	0.0000	0.0800	Scapula	-0.1250	0.0900	-0.0800
--	--	Scapula	-0.0900	0.0650	-0.0270	Scapula	-0.1250	0.0900	-0.0800
--	--	UpTorso	0.0000	0.1849	-0.0597	Head	-0.0360	0.0099	0.0658
--	--	Thorax	0.3810	-0.0483	0.7544	Scapula	0.3073	0.1118	0.6833
--	--	Thorax	0.3480	0.1194	0.7772	Scapula	0.2616	0.1397	0.6731
--	--	Thorax	0.3048	0.1651	0.7772	Clavicle	0.2083	0.1575	0.7112

MUSCLE		SOURCE	MUSCLE LENGTH ML (m)	FIBER LENGTH LM (m)	VOLUME V (ml)	MUSCLE PCSA =V/L (cm²)	PENNA- TION ANGLE (deg)	MUSCLE FIBER TYPE		
								%SO	%FO	%FG
	(1)	[18]	0.1300	--	--	3.11	10	15	--	--
	(2)	[18]	0.1875	--	--	1.14	10	15	--	--
		[14]	--	--	--	--	--	53.7	--	46.2
	(1)	[17]	--	--	--	--	--	--	--	--
	(2)	[17]	--	--	--	--	--	--	--	--
	(3)	[17]	--	--	--	--	--	--	--	--
	(4)	[17]	--	--	--	--	--	--	--	--
triceps brachii		[17]	--	--	--	--	--	--	--	--
	(1)	[15]	--	--	--	6.78	--	--	--	--
	(2)	[15]	--	--	--	5.66	--	--	--	--
	(3)	[15]	--	--	--	4.75	--	--	--	--
	(surface)	[14]	--	--	--	--	--	32.5	--	67.5
	(deep)	[14]	--	--	--	--	--	32.7	--	19.6
	(lateral)	[16]	0.2464	--	108.15	4.52	--	--	--	--
	(lateral)	[22]	--	0.084	47.3	6.0	--	--	--	--
	(medial)	[16]	0.2134	--	62.27	3.23	--	--	--	--
	(medial)	[22]	--	0.063	38.7	6.1	--	--	--	--
	(long)	[16]	0.3124	--	127.82	3.87	--	--	--	--
	(long)	[22]	--	0.102	66.6	6.7	--	--	--	--
	(1)	[7]	--	--	--	--	--	--	--	--
	(2)	[7]	--	--	--	--	--	--	--	--
	(3)	[7]	--	--	--	--	--	--	--	--
	(1)	[12]	--	--	--	16.38	--	--	--	--
	(2)	[12]	--	--	--	--	--	--	--	--
	(3)	[12]	--	--	--	--	--	--	--	--

TENDON LENGTH LT (m)	TENDON PCSA (mm²)	ORIGIN COORD. SYSTEM	X (m)	Y (m)	Z (m)	INSERTION COORD. SYSTEM	X (m)	Y (m)	Z (m)
0.0325	--	--	--	--	--	--	--	--	--
0.01875	--	--	--	--	--	--	--	--	--
--	--	--	--	--	--	--	--	--	--
--	--	--	--	--	--	Scapula	0.0378	0.0354	-0.0290
--	--	--	--	--	--	Scapula	0.0526	0.0536	-0.0322
--	--	--	--	--	--	Scapula	0.0147	0.0156	-0.0178
--	--	--	--	--	--	Clavicle	0.1159	-0.0216	0.0046
--	--	--	--	--	--	Scapula	0.0648	-0.0193	0.0073
--	--	--	--	--	--	--	--	--	--
--	--	--	--	--	--	--	--	--	--
--	--	--	--	--	--	--	--	--	--
--	--	--	--	--	--	--	--	--	--
--	--	--	--	--	--	--	--	--	--
--	--	Humerus	0.1448	0.0051	0.6121	Ulna	0.0686	-0.1651	0.5537
--	--	--	--	--	--	--	--	--	--
--	--	Humerus	0.1346	-0.0660	0.6071	Ulna	0.0787	-0.1702	0.5588
--	--	--	--	--	--	--	--	--	--
--	--	Scapula	0.2210	0.0914	0.6706	Ulna	0.0838	-0.1651	0.5588
--	--	--	--	--	--	--	--	--	--
--	--	Scapula	-0.015	0.005	-0.060	Ulna	-0.015	0.000	0.025
--	--	Humerus	-0.010	0.012	-0.200	Ulna	-0.015	0.000	0.025
--	--	Humerus	-0.008	-0.010	-0.160	Ulna	-0.015	0.000	0.025
--	--	Humerus	-0.0099	0.0000	-0.0183	Forearm	-0.0150	0.000	-0.2167
--	--	Humerus	-0.0099	0.0000	-0.0381	Forearm	-0.0150	0.000	-0.2167
--	--	UpTorso	0.0000	0.1748	-0.0470	Forearm	-0.0150	0.000	-0.2167

TABLE A.4
HAND MUSCULATURE

MUSCLE		SOURCE	MUSCLE LENGTH ML	FIBER LENGTH LM	VOLUME V (ml)	MUSCLE PCSA =V/L (cm²)	PENNA-TION ANGLE (deg)	MASS FRACTION (%)	MASS (gm)
abductor digiti minimi		[19]	--	0.04	--	--	--	1.1	--
		[14]	--	--	--	--	--	51.8	--
		[7]	--	--	--	--	--	--	--
abductor pollicis brevis		[19]	--	0.037	--	--	--	0.8	--
		[14]	--	--	--	--	--	63.0	--
		[7]	--	--	--	--	--	--	--
		[7]	--	--	--	--	--	--	--
abductor pollicis longus		[19]	--	0.046	--	--	--	2.8	--
		[20]	--	0.033	--	2.62	6.09	--	13.0
		[15]	--	--	--	1.84	--	--	--
		[14]	--	--	--	--	--	80.4	--
		[7]	--	--	--	--	--	--	--
adductor pollicis longus		[19]	--	0.036	--	--	--	2.1	--
		[7]	--	--	--	--	--	--	--
		[7]	--	--	--	--	--	--	--
		[7]	--	--	--	--	--	--	--
		[7]	--	--	--	--	--	--	--
dorsal interosseous [I]	(met1)	[19]	--	0.031	--	--	--	0.8	--
	(met2)	[19]	--	0.016	--	--	--	0.6	--
		[7]	--	--	--	--	--	--	--
		[7]	--	--	--	--	--	--	--
		[7]	--	--	--	--	--	--	--
		[7]	--	--	--	--	--	--	--
		[7]	--	--	--	--	--	--	--
		[7]	--	--	--	--	--	--	--
	[II]	[19]	--	0.014	--	--	--	0.7	--
		[7]	--	--	--	--	--	--	--
		[7]	--	--	--	--	--	--	--
		[7]	--	--	--	--	--	--	--
		[7]	--	--	--	--	--	--	--
		[7]	--	--	--	--	--	--	--
		[7]	--	--	--	--	--	--	--
		[7]	--	--	--	--	--	--	--
	[III]	[19]	--	0.015	--	--	--	0.6	--
		[7]	--	--	--	--	--	--	--
		[7]	--	--	--	--	--	--	--
		[7]	--	--	--	--	--	--	--
		[7]	--	--	--	--	--	--	--
		[7]	--	--	--	--	--	--	--
		[7]	--	--	--	--	--	--	--
		[7]	--	--	--	--	--	--	--
		[7]	--	--	--	--	--	--	--
	[IV]	[19]	--	0.015	--	--	--	0.5	--
		[7]	--	--	--	--	--	--	--
		[7]	--	--	--	--	--	--	--
		[7]	--	--	--	--	--	--	--
		[7]	--	--	--	--	--	--	--
		[7]	--	--	--	--	--	--	--
		[7]	--	--	--	--	--	--	--
		[7]	--	--	--	--	--	--	--
extensor carpi radialis brevis		[19]	--	0.061	--	--	--	5.1	--
		[20]	--	0.049	--	3.47	9.40	--	24.5
		[7]	--	--	--	--	--	--	--
extensor carpi radialis longus		[19]	--	0.093	--	--	--	6.5	--
		[20]	--	0.108	--	2.73	6.38,3.72	--	37.5
		[7]	--	--	--	--	--	--	--
extensor carpi ulnaris		[19]	--	0.045	--	--	--	4.0	--
		[20]	--	0.034	--	3.25	8.46,6.76	--	17.0
		[7]	--	--	--	--	--	--	--
extensor digiti minimi		[19]	--	0.059	--	--	--	1.2	--
	(cen)	[7]	--	--	--	--	--	--	--
		[7]	--	--	--	--	--	--	--
		[7]	--	--	--	--	--	--	--
	(lat)	[7]	--	--	--	--	--	--	--
		[7]	--	--	--	--	--	--	--
		[7]	--	--	--	--	--	--	--
		[7]	--	--	--	--	--	--	--
	(med)	[7]	--	--	--	--	--	--	--
		[7]	--	--	--	--	--	--	--
		[7]	--	--	--	--	--	--	--
		[7]	--	--	--	--	--	--	--

TENSION FRAC. (%)	TENDON LENGTH LT (mm²)	TENDON PCSA	ORIGIN				INSERTION			
			COORD. SYSTEM	X	Y	Z	COORD. SYSTEM	X	Y	Z
1.4	--	--	--	--	--	--	--	--	--	--
48.2	--	--	--	--	--	--	--	--	--	--
--	--	--	CRP	-0.0170	0.0220	0.0160	PP5	0.0100	0.0070	0.0025
1.1	--	--	--	--	--	--	--	--	--	--
37.0	--	--	--	--	--	--	--	--	--	--
--	--	--	CRP	-0.0150	-0.0230	0.0200	MC1	0.0420	0.0060	0.0040
--	--	--	MC1	0.0420	0.0060	0.0040	PP1	0.0110	0.0050	0.0040
3.1	--	--	--	--	--	--	--	--	--	--
--	--	--	--	--	--	--	--	--	--	--
19.6	--	--	--	--	--	--	--	--	--	--
--	--	--	URD	-0.0100	-0.0200	0.0070	MC1	0.0080	0.0000	0.0060
3.0	--	--	--	--	--	--	--	--	--	--
--	--	--	CRP	-0.0040	-0.0030	0.0120	PP1	0.0080	0.0040	-0.0080
--	--	--	MC2	0.0100	0.0000	0.0070	PP1	0.0080	0.0040	-0.0080
--	--	--	MC3	0.0060	0.0000	0.0070	PP1	0.0080	0.0040	-0.0080
--	--	--	MC3	0.0300	0.0000	0.0030	PP1	0.0080	0.0040	-0.0080
1.8	--	--	--	--	--	--	--	--	--	--
1.9	--	--	--	--	--	--	--	--	--	--
--	--	--	MC1	0.0230	0.0000	-0.0060	PP2	0.0130	-0.0080	-0.0020
--	--	--	PP2	0.0330	-0.0060	0.0000	MP2	0.0060	-0.0060	-0.0020
--	--	--	MP2	0.0240	0.0000	-0.0025	DP2	0.0040	0.0000	-0.0030
--	--	--	MC2	0.0230	-0.0040	-0.0040	PP2	0.0130	-0.0080	-0.0020
--	--	--	PP2	0.0330	-0.0060	0.0000	MP2	0.0060	-0.0060	-0.0020
--	--	--	MP2	0.0240	0.0000	-0.0025	DP2	0.0040	0.0000	-0.0030
2.5	--	--	--	--	--	--	--	--	--	--
--	--	--	MC2	0.0270	0.0040	-0.0050	PP3	0.0130	-0.0090	-0.0030
--	--	--	PP3	0.0350	-0.0060	0.0000	MP3	0.0070	-0.0070	-0.0020
--	--	--	MP3	0.0310	0.0000	-0.0025	DP3	0.0040	0.0000	-0.0035
--	--	--	MC3	0.0300	-0.0040	-0.0040	PP3	0.0130	-0.0090	-0.0030
--	--	--	MC3	0.0300	-0.0040	-0.0040	PP3	0.0130	-0.0090	0.0000
--	--	--	PP3	0.0350	-0.0060	0.0000	MP3	0.0070	-0.0070	-0.0020
--	--	--	MP3	0.0310	0.0000	-0.0025	DP3	0.0040	0.0000	-0.0035
2.0	--	--	--	--	--	--	--	--	--	--
--	--	--	MC3	0.0300	0.0040	-0.0040	PP3	0.0130	0.0130	-0.0030
--	--	--	MC3	0.0300	0.0040	-0.0040	PP3	0.0130	0.0090	0.0000
--	--	--	PP3	0.0350	0.0060	0.0000	MP3	0.0070	0.0070	-0.0020
--	--	--	MP3	0.0310	0.0000	-0.0025	DP3	0.0040	0.0000	-0.0035
--	--	--	PP3	0.0130	0.0090	-0.0031	MC4	0.0280	-0.0040	-0.0025
--	--	--	MC4	0.0280	-0.0040	-0.0025	PP3	0.0130	0.0090	0.0000
--	--	--	PP3	0.0350	0.0060	0.0000	MP3	0.0070	0.0070	-0.0020
--	--	--	MP3	0.0310	0.0000	-0.0025	DP3	0.0040	0.0000	-0.0035
1.7	--	--	--	--	--	--	--	--	--	--
--	--	--	MC4	0.0280	0.0040	-0.0025	PP4	0.0120	0.0080	-0.0020
--	--	--	PP4	0.0300	0.0060	0.0000	MP4	0.0060	0.0060	-0.0020
--	--	--	MP4	0.0250	0.0000	-0.0020	DP4	0.0035	0.0000	-0.0030
--	--	--	MC5	0.0170	-0.0040	-0.0040	PP4	0.0120	0.0080	-0.0020
--	--	--	MC5	0.0170	-0.0040	-0.0040	PP4	0.0120	0.0080	0.0020
--	--	--	PP4	0.0300	0.0060	0.0000	MP4	0.0060	0.0060	-0.0020
--	--	--	MP4	0.0250	0.0000	-0.0020	DP4	0.0035	0.0000	-0.0030
4.2	--	--	--	--	--	--	--	--	--	--
--	--	--	URD	0.0000	-0.0100	-0.0100	MC3	0.0080	0.0050	-0.0080
3.5	--	--	--	--	--	--	--	--	--	--
--	--	--	--	--	--	--	--	--	--	--
--	--	--	URD	0.0000	-0.0150	-0.0080	MC2	0.0060	-0.0080	-0.0060
4.5	--	--	--	--	--	--	--	--	--	--
--	--	--	--	--	--	--	--	--	--	--
--	--	--	URD	-0.0150	0.0280	-0.0080	MC5	0.0060	0.0080	0.0000
1.0	--	--	--	--	--	--	--	--	--	--
--	--	--	URD	-0.0070	0.0150	-0.0110	MC5	0.0500	0.0000	-0.0060
--	--	--	MC5	0.0500	0.0000	-0.0060	PP5	0.0385	0.0000	-0.0025
--	--	--	PP5	0.0385	0.0000	-0.0025	MP5	0.0050	0.0000	-0.0035
--	--	--	URD	-0.0070	0.0150	-0.0110	MC5	0.0500	0.0000	-0.0060
--	--	--	MC5	0.0500	0.0000	-0.0060	PP5	0.0270	-0.0025	-0.0030
--	--	--	PP5	0.0270	-0.0025	-0.0030	MP5	0.0050	-0.0050	-0.0015
--	--	--	MP5	0.0215	0.0000	-0.0015	DP5	0.0030	0.0000	-0.0025
--	--	--	URD	-0.0070	0.0150	-0.0110	MC5	0.0500	0.0000	-0.0060
--	--	--	MC5	0.0500	0.0000	-0.0060	PP5	0.0270	0.0025	-0.0030
--	--	--	PP5	0.0270	0.0025	-0.0030	MP5	0.0050	0.0050	-0.0015
--	--	--	MP5	0.0215	0.0000	-0.0015	DP5	0.0030	0.0000	-0.0025

MUSCLE	SOURCE	MUSCLE LENGTH ML	FIBER LENGTH LM	VOLUME V (ml)	MUSCLE PCSA =V/L (cm²)	PENNA-TION ANGLE (deg)	MASS FRACTION (%)	MASS (gm)
extensor digitorum communis [II]	[19]	--	0.055	--	--	--	1.1	--
	[20]	--	0.055	--	0.77	7.84	--	5.5
	[22]	--	0.067	4.9	0.7	--	--	--
(cen)	[7]	--	--	--	--	--	--	--
	[7]	--	--	--	--	--	--	--
	[7]	--	--	--	--	--	--	--
(lat)	[7]	--	--	--	--	--	--	--
	[7]	--	--	--	--	--	--	--
	[7]	--	--	--	--	--	--	--
	[7]	--	--	--	--	--	--	--
(med)	[7]	--	--	--	--	--	--	--
	[7]	--	--	--	--	--	--	--
	[7]	--	--	--	--	--	--	--
	[7]	--	--	--	--	--	--	--
[III]	[19]	--	0.060	--	--	--	2.2	--
	[20]	--	0.057	--	1.38	4.02	--	12.5
	[22]	--	0.069	5.9	0.9	--	--	--
(cen)	[7]	--	--	--	--	--	--	--
	[7]	--	--	--	--	--	--	--
	[7]	--	--	--	--	--	--	--
(lat)	[7]	--	--	--	--	--	--	--
	[7]	--	--	--	--	--	--	--
	[7]	--	--	--	--	--	--	--
	[7]	--	--	--	--	--	--	--
(med)	[7]	--	--	--	--	--	--	--
	[7]	--	--	--	--	--	--	--
	[7]	--	--	--	--	--	--	--
	[7]	--	--	--	--	--	--	--
[IV]	[19]	--	0.058	--	--	--	2.0	--
	[20]	--	0.048	--	1.09	5.98,2.39	--	7.5
	[22]	--	0.061	8.0	1.4	--	--	--
(cen)	[7]	--	--	--	--	--	--	--
	[7]	--	--	--	--	--	--	--
	[7]	--	--	--	--	--	--	--
(lat)	[7]	--	--	--	--	--	--	--
	[7]	--	--	--	--	--	--	--
	[7]	--	--	--	--	--	--	--
	[7]	--	--	--	--	--	--	--
(med)	[7]	--	--	--	--	--	--	--
	[7]	--	--	--	--	--	--	--
	[7]	--	--	--	--	--	--	--
	[7]	--	--	--	--	--	--	--
[V]	[19]	--	0.059	--	--	--	1.0	--
	[20]	--	0.055	--	0.64	5.22	--	5.0
	[22]	--	0.059	4.7	0.8	--	--	--
(cen)	[7]	--	--	--	--	--	--	--
	[7]	--	--	--	--	--	--	--
	[7]	--	--	--	--	--	--	--
(lat)	[7]	--	--	--	--	--	--	--
	[7]	--	--	--	--	--	--	--
	[7]	--	--	--	--	--	--	--
	[7]	--	--	--	--	--	--	--
(med)	[7]	--	--	--	--	--	--	--
	[7]	--	--	--	--	--	--	--
	[7]	--	--	--	--	--	--	--
	[7]	--	--	--	--	--	--	--
extensor indicis proprius	[19]	--	0.055	--	--	--	1.1	--
	[15]	--	--	--	0.37	--	--	--
	[15]	--	--	--	1.20	--	--	--
	[20]	--	0.055	--	0.84	0.0	--	4.5
(cen)	[7]	--	--	--	--	--	--	--
	[7]	--	--	--	--	--	--	--
	[7]	--	--	--	--	--	--	--
(med)	[7]	--	--	--	--	--	--	--
	[7]	--	--	--	--	--	--	--
	[7]	--	--	--	--	--	--	--
(lat)	[7]	--	--	--	--	--	--	--
	[7]	--	--	--	--	--	--	--
	[7]	--	--	--	--	--	--	--
	[7]	--	--	--	--	--	--	--

TENSION FRAC. (%)	TENDON LENGTH LT (mm²)	TENDON PCSA	ORIGIN COORD. SYSTEM	X	Y	Z	INSERTION COORD. SYSTEM	X	Y	Z
1.0	--	--	--	--	--	--	--	--	--	--
--	--	--	--	--	--	--	--	--	--	--
--	--	--	--	--	--	--	--	--	--	--
--	--	--	URD	-0.0070	0.0030	-0.0130	MC2	0.0620	0.0000	-0.0080
--	--	--	MC2	0.0620	0.0000	-0.0080	PP2	0.0470	0.0000	-0.0030
--	--	--	PP2	0.0470	0.0000	-0.0030	MP2	0.0055	0.0000	-0.0040
--	--	--	URD	-0.0070	0.0030	-0.0130	MC2	0.0620	0.0000	-0.0080
--	--	--	MC2	0.0620	0.0000	-0.0080	PP2	0.0330	0.0030	-0.0030
--	--	--	PP2	0.0330	0.0030	-0.0030	MP2	0.0060	0.0060	-0.0020
--	--	--	MP2	0.0240	0.0000	-0.0025	DP2	0.0040	0.0000	-0.0030
--	--	--	URD	-0.0070	0.0030	-0.0130	MC2	0.0620	0.0000	-0.0080
--	--	--	MC2	0.0620	0.0000	-0.0080	PP2	0.0330	-0.0030	-0.0030
--	--	--	PP2	0.0330	-0.0030	-0.0030	MP2	0.0060	-0.0060	-0.0020
--	--	--	MP2	0.0240	0.0000	-0.0025	DP2	0.0040	0.0000	-0.0030
1.9	--	--	--	--	--	--	--	--	--	--
--	--	--	--	--	--	--	--	--	--	--
--	--	--	--	--	--	--	--	--	--	--
--	--	--	URD	-0.0070	0.0060	-0.0130	MC3	0.0620	0.0000	-0.0080
--	--	--	MC3	0.0620	0.0000	-0.0080	PP3	0.0520	0.0000	-0.0030
--	--	--	PP3	0.0520	0.0000	-0.0030	MP3	0.0060	0.0000	0.0040
--	--	--	URD	-0.0070	0.0060	-0.0130	MC3	0.0620	0.0000	-0.0080
--	--	--	MC3	0.0620	0.0000	-0.0080	PP3	0.0350	0.0040	-0.0040
--	--	--	PP3	0.0350	0.0040	-0.0040	MP3	0.0070	0.0070	-0.0020
--	--	--	MP3	0.0310	0.0000	-0.0025	DP3	0.0040	0.0000	-0.0035
--	--	--	URD	-0.0070	0.0060	-0.0130	MC3	0.0620	0.0000	-0.0080
--	--	--	MC3	0.0620	0.0000	-0.0080	PP3	0.0350	-0.0040	-0.0040
--	--	--	PP3	0.0350	-0.0040	-0.0040	MP3	0.0070	-0.0070	-0.0020
--	--	--	MP3	0.0310	0.0000	-0.0025	DP3	0.0040	0.0000	-0.0035
1.7	--	--	--	--	--	--	--	--	--	--
--	--	--	--	--	--	--	--	--	--	--
--	--	--	URD	-0.0070	0.0090	-0.0130	MC4	0.0540	0.0000	-0.0070
--	--	--	MC4	0.0540	0.0000	-0.0070	PP4	0.0470	0.0000	-0.0030
--	--	--	PP4	0.0470	0.0000	-0.0030	MP4	0.0050	0.0000	-0.0040
--	--	--	URD	-0.0070	0.0090	-0.0130	MC4	0.0540	0.0000	-0.0070
--	--	--	MC4	0.0540	0.0000	-0.0070	PP4	0.0300	-0.0030	-0.0030
--	--	--	PP4	0.0300	-0.0030	-0.0030	MP4	0.0060	0.0060	-0.0020
--	--	--	MP4	0.0250	0.0000	-0.0020	DP4	0.0035	0.0000	-0.0030
--	--	--	URD	-0.0070	0.0090	-0.0130	MC4	0.0540	0.0000	-0.0070
--	--	--	MC4	0.0540	0.0000	-0.0070	PP4	0.0300	-0.0030	-0.0030
--	--	--	PP4	0.0300	-0.0030	-0.0030	MP4	0.0060	-0.0060	-0.0020
--	--	--	MP4	0.0250	0.0000	-0.0020	DP4	0.0035	0.0000	-0.0030
0.9	--	--	--	--	--	--	--	--	--	--
--	--	--	--	--	--	--	--	--	--	--
--	--	--	--	--	--	--	--	--	--	--
--	--	--	URD	-0.0070	0.0090	-0.0130	MC5	0.0500	0.0000	-0.0060
--	--	--	MC5	0.0500	0.0000	-0.0060	PP5	0.0385	0.0000	-0.0025
--	--	--	PP5	0.0385	0.0000	-0.0025	MP5	0.0050	0.0000	-0.0035
--	--	--	URD	-0.0070	0.0090	-0.0130	MC5	0.0500	0.0000	-0.0060
--	--	--	MC5	0.0500	0.0000	-0.0060	PP5	0.0270	-0.0025	-0.0030
--	--	--	PP5	0.0270	-0.0025	-0.0030	MP5	0.0050	-0.0050	-0.0015
--	--	--	MP5	0.0215	0.0000	-0.0015	DP5	0.0030	0.0000	-0.0025
--	--	--	URD	-0.0070	0.0090	-0.0130	MC5	0.0500	0.0000	-0.0060
--	--	--	MC5	0.0500	0.0000	-0.0060	PP5	0.0270	-0.0025	-0.0030
--	--	--	PP5	0.0270	-0.0025	-0.0030	MP5	0.0050	-0.0050	-0.0015
--	--	--	MP5	0.0216	0.0000	-0.0015	DP5	0.0030	0.0000	-0.0025
1.0	--	--	--	--	--	--	--	--	--	--
--	--	--	--	--	--	--	--	--	--	--
--	--	--	--	--	--	--	--	--	--	--
--	--	--	URD	-0.0070	0.0030	-0.0130	MC2	0.0620	0.0000	-0.0080
--	--	--	MC2	0.0620	0.0000	-0.0080	PP2	0.0470	0.0000	-0.0030
--	--	--	PP2	0.0470	0.0000	-0.0030	MP2	0.0055	0.0000	-0.0040
--	--	--	MC2	0.0620	0.0000	-0.0080	PP2	0.0330	-0.0030	-0.0030
--	--	--	PP2	0.0330	-0.0030	-0.0030	MP2	0.0060	-0.0060	-0.0020
--	--	--	MP2	0.0240	0.0000	-0.0025	DP2	0.0040	0.0000	-0.0030
--	--	--	URD	-0.0070	0.0030	-0.0130	MC2	0.0620	0.0000	-0.0080
--	--	--	MC2	0.0620	0.0000	-0.0080	PP2	0.0330	0.0030	-0.0030
--	--	--	PP2	0.0330	0.0030	-0.0030	MP2	0.0060	0.0060	-0.0020
--	--	--	MP2	0.0240	0.0000	-0.0025	DP2	0.0040	0.0000	-0.0030

MUSCLE		SOURCE	MUSCLE LENGTH ML	FIBER LENGTH LM	VOLUME V (ml)	MUSCLE PCSA =V/L (cm²)	PENNA- TION ANGLE (deg)	MASS FRACTION (%)	MASS (gm)
extensor longus	[II]	[23]	--	--	--	--	--	--	--
	[III]	[23]	--	--	--	--	--	--	--
	[IV]	[23]	--	--	--	--	--	--	--
	[V]	[23]	--	--	--	--	--	--	--
extensor pollicis brevis		[19]	--	0.043	--	--	--	0.7	--
		[15]	--	--	--	1.84	--	--	--
		[20]	--	0.054	--	1.29	0.0	--	7.0
		[7]	--	--	--	--	--	--	--
		[7]	--	--	--	--	--	--	--
extensor pollicis longus		[19]	--	0.057	--	--	--	1.5	--
		[15]	--	--	--	0.56	--	--	--
		[20]	--	0.045	--	1.04	0.0	--	4.0
		[7]	--	--	--	--	--	--	--
		[7]	--	--	--	--	--	--	--
		[7]	--	--	--	--	--	--	--
extensor slip	[II]	[23]	--	--	--	--	--	--	--
	[III]	[23]	--	--	--	--	--	--	--
	[IV]	[23]	--	--	--	--	--	--	--
	[V]	[23]	--	--	--	--	--	--	--
flexor carpi radialis		[19]	--	0.052	--	--	--	4.2	--
		[20]	--	0.08	--	2.73	8.63	--	25.0
		[7]	--	--	--	--	--	--	--
		[7]	--	--	--	--	--	--	--
flexor carpi ulnaris		[19]	--	0.042	--	--	--	5.6	--
		[20]	--	0.048	--	5.39	9.59,1 79	--	30.5
		[7]	--	--	--	--	--	--	--
flexor digiti minimi		[19]	--	0.034	--	--	--	0.3	--
		[7]	--	--	--	--	--	--	--
flexor digitorum profundus [II]		[19]	--	0.066	--	--	--	3.5	--
		[20]	--	0.057	--	3.35	12.15	--	21.0
		[7]	--	--	--	--	--	--	--
		[7]	--	--	--	--	--	--	--
		[7]	--	--	--	--	--	--	--
		[7]	--	--	--	--	--	--	--
		[7]	--	--	--	--	--	--	--
	[III]	[19]	--	0.066	--	--	--	4.4	--
		[20]	--	0.088	--	2.68	11.14	--	25.0
		[7]	--	--	--	--	--	--	--
		[7]	--	--	--	--	--	--	--
		[7]	--	--	--	--	--	--	--
		[7]	--	--	--	--	--	--	--
		[7]	--	--	--	--	--	--	--
	[IV]	[19]	--	0.068	--	--	--	4.1	--
		[20]	--	0.074	--	2.9	13.45,6.68	--	43.0
		[7]	--	--	--	--	--	--	--
		[7]	--	--	--	--	--	--	--
		[7]	--	--	--	--	--	--	--
		[7]	--	--	--	--	--	--	--
	[V]	[19]	--	0.062	--	--	--	3.4	--
		[20]	--	--	--	2.61	--	--	--
		[7]	--	--	--	--	--	--	--
		[7]	--	--	--	--	--	--	--
		[7]	--	--	--	--	--	--	--
		[7]	--	--	--	--	--	--	--
		[7]	--	--	--	--	--	--	--
flexor digitorum sublimis [II]		[19]	--	0.072	--	--	--	2.9	--
		[20]	--	0.032	--	4.19	14.48	--	21.5
		[22]	--	0.045	11.1	2.5	--	--	--
		[7]	--	--	--	--	--	--	--
		[7]	--	--	--	--	--	--	--
		[7]	--	--	--	--	--	--	--
		[7]	--	--	--	--	--	--	--
	[III]	[19]	--	0.070	--	--	--	4.7	--
		[20]	--	0.058	--	3.61	7.93	--	25.5
		[22]	--	0.077	13.7	1.7	--	--	--
		[7]	--	--	--	--	--	--	--
		[7]	--	--	--	--	--	--	--
		[7]	--	--	--	--	--	--	--
		[7]	--	--	--	--	--	--	--

TENSION FRAC. (%)	TENDON LENGTH LT (mm²)	TENDON PCSA	ORIGIN COORD. SYSTEM	X	Y	Z	INSERTION COORD. SYSTEM	X	Y	Z
--	--	--	6	0.000	0.506	-0.024	5	0.000	0.455	-0.013
--	--	--	6	0.000	0.393	-0.039	5	-0.080	0.322	-0.018
--	--	--	6	0.000	0.388	0.010	5	-0.110	0.326	0.002
--	--	--	6	0.000	0.477	0.111	5	0.040	0.414	0.035
0.8	--	--	--	--	--	--	--	--	--	--
--	--	--	--	--	--	--	--	--	--	--
--	--	--	--	--	--	--	--	--	--	--
--	--	--	URD	-0.0100	-0.0200	0.0070	MC1	0.0420	-0.0060	0.0000
--	--	--	MC1	0.0420	-0.0060	0.0000	PP1	0.0100	-0.0050	0.0000
1.3	--	--	--	--	--	--	--	--	--	--
--	--	--	--	--	--	--	--	--	--	--
--	--	--	URD	-0.0070	-0.0050	-0.0130	MC1	0.0420	-0.0060	-0.0030
--	--	--	MC1	0.0420	-0.0060	-0.0030	PP1	0.0330	-0.0040	0.0000
--	--	--	PP1	0.0330	-0.0040	-0.0000	DP1	0.0070	-0.0040	0.0000
--	--	--	4	0.000	0.280	-0.040	3	-0.040	0.293	-0.036
--	--	--	4	0.000	0.236	-0.019	3	-0.030	0.247	-0.024
--	--	--	4	0.000	0.202	0.006	3	-0.040	0.208	-0.016
--	--	--	4	0.000	0.257	0.006	3	-0.040	0.272	-0.012
4.1	--	--	--	--	--	--	--	--	--	--
--	--	--	URD	0.0000	-0.0080	0.0120	CRP	-0.0130	-0.0150	0.0250
--	--	--	CRP	-0.0130	-0.0150	0.0250	MC2	0.0070	-0.0020	0.0060
6.7	--	--	--	--	--	--	--	--	--	--
--	--	--	--	--	--	--	--	--	--	--
--	--	--	URD	-0.0150	0.0300	0.0100	CRP	-0.0170	0.0200	0.0150
0.4	--	--	--	--	--	--	--	--	--	--
--	--	--	CRP	-0.0060	0.0160	0.0200	PP5	0.0100	0.0070	0.0025
2.7	--	--	--	--	--	--	--	--	--	--
--	--	--	--	--	--	--	--	--	--	--
--	--	--	URD	-0.0070	0.0100	0.0160	CRP	-0.0100	-0.0010	0.0130
--	--	--	CRP	-0.0100	-0.0010	0.0130	MC2	0.0600	0.0000	0.0090
--	--	--	MC2	0.0600	0.0000	0.0090	PP2	0.0420	0.0000	0.0025
--	--	--	PP2	0.0420	0.0000	0.0025	MP2	0.0210	0.0000	0.0020
--	--	--	MP2	0.0210	0.0000	0.0020	DP2	0.0060	0.0000	0.0020
3.4	--	--	--	--	--	--	--	--	--	--
--	--	--	--	--	--	--	--	--	--	--
--	--	--	URD	-0.0070	0.0140	0.0160	CRP	-0.0070	0.0020	0.0150
--	--	--	CRP	-0.0070	0.0020	0.0150	MC3	0.0580	0.0000	0.0080
--	--	--	MC3	0.0080	0.0000	0.0080	PP3	0.0470	0.0000	0.0030
--	--	--	PP3	0.0470	0.0000	0.0030	MP3	0.0280	0.0000	0.0025
--	--	--	MP3	0.0280	0.0000	0.0025	DP3	0.0070	0.0000	0.0025
3.0	--	--	--	--	--	--	--	--	--	--
--	--	--	--	--	--	--	--	--	--	--
--	--	--	URD	-0.0070	0.0160	0.0160	CRP	-0.0070	0.0060	0.0130
--	--	--	CRP	-0.0070	0.0060	0.0130	MC4	0.0500	0.0000	0.0070
--	--	--	MC4	0.0500	0.0000	0.0070	PP4	0.0430	0.0000	0.0030
--	--	--	PP4	0.0430	0.0000	0.0030	MP4	0.0220	0.0000	0.0020
--	--	--	MC4	0.0220	0.0000	0.0020	DP4	0.0060	0.0000	0.0025
2.8	--	--	--	--	--	--	--	--	--	--
--	--	--	--	--	--	--	--	--	--	--
--	--	--	URD	-0.0070	0.0180	0.0150	CRP	-0.0070	0.0100	0.0110
--	--	--	CRP	-0.0070	0.0100	0.0110	MC5	0.0460	0.0000	0.0070
--	--	--	MC5	0.0460	0.0000	0.0070	PP5	0.0350	0.0000	0.0020
--	--	--	PP5	0.0350	0.0000	0.0020	MP5	0.0190	0.0000	0.0017
--	--	--	MP5	0.0190	0.0000	0.0017	DP5	0.0060	0.0000	0.0020
2.0	--	--	--	--	--	--	--	--	--	--
--	--	--	--	--	--	--	--	--	--	--
--	--	--	URD	-0.0070	0.0100	0.0160	CRP	-0.0100	0.0000	0.0150
--	--	--	CRP	-0.0100	0.0000	0.0150	MC2	0.0600	0.0000	0.0090
--	--	--	MC2	0.0600	0.0000	0.0090	PP2	0.0420	0.0000	0.0025
--	--	--	PP2	0.0420	0.0000	0.0025	MP2	0.0130	0.0000	0.0020
3.4	--	--	--	--	--	--	--	--	--	--
--	--	--	--	--	--	--	--	--	--	--
--	--	--	URD	-0.0070	0.0100	0.0160	CRP	-0.0070	0.0050	0.0170
--	--	--	CRP	-0.0070	0.0050	0.0170	MC3	0.0580	0.0000	0.0080
--	--	--	MC3	0.0580	0.0000	0.0080	PP3	0.0470	0.0000	0.0030
--	--	--	PP3	0.0470	0.0000	0.0030	MP3	0.0160	0.0000	0.0025

MUSCLE		SOURCE	MUSCLE LENGTH ML	FIBER LENGTH LM	VOLUME V (ml)	MUSCLE PCSA =V/L (cm²)	PENNA-TION ANGLE (deg)	MASS FRACTION (%)	MASS (gm)
	[IV]	[19]	--	0.073	--	--	--	3.0	--
		[20]	--	0.050	--	2.48	0.0	--	16.5
		[22]	--	0.064	7.8	1.2	--	--	--
		[7]	--	--	--	--	--	--	--
		[7]	--	--	--	--	--	--	--
		[7]	--	--	--	--	--	--	--
		[7]	--	--	--	--	--	--	--
	[V]	[19]	--	0.070	--	--	--	1.3	--
		[20]	--	0.035	--	1.09	11.54	--	3.5
		[22]	--	0.045	3.3	0.7	--	--	--
		[7]	--	--	--	--	--	--	--
		[7]	--	--	--	--	--	--	--
		[7]	--	--	--	--	--	--	--
		[7]	--	--	--	--	--	--	--
flexor pollicis brevis		[19]	--	0.036	--	--	--	0.9	--
		[7]	--	--	--	--	--	--	--
		[7]	--	--	--	--	--	--	--
flexor pollicis longus		[19]	--	0.059	--	--	--	3.2	--
		[20]	--	0.038	--	3.08	11.38	--	18.0
		[7]	--	--	--	--	--	--	--
		[7]	--	--	--	--	--	--	--
		[7]	--	--	--	--	--	--	--
		[7]	--	--	--	--	--	--	--
flexor profundus	[II]	[23]	--	--	--	--	--	--	--
		[23]	--	--	--	--	--	--	--
		[23]	--	--	--	--	--	--	--
		[23]	--	--	--	--	--	--	--
		[23]	--	--	--	--	--	--	--
	[III]	[23]	--	--	--	--	--	--	--
		[23]	--	--	--	--	--	--	--
		[23]	--	--	--	--	--	--	--
		[23]	--	--	--	--	--	--	--
		[23]	--	--	--	--	--	--	--
	[IV]	[23]	--	--	--	--	--	--	--
		[23]	--	--	--	--	--	--	--
		[23]	--	--	--	--	--	--	--
		[23]	--	--	--	--	--	--	--
		[23]	--	--	--	--	--	--	--
	[V]	[23]	--	--	--	--	--	--	--
		[23]	--	--	--	--	--	--	--
		[23]	--	--	--	--	--	--	--
		[23]	--	--	--	--	--	--	--
		[23]	--	--	--	--	--	--	--
flexor sublimis	[II]	[23]	--	--	--	--	--	--	--
		[23]	--	--	--	--	--	--	--
		[23]	--	--	--	--	--	--	--
	[III]	[23]	--	--	--	--	--	--	--
		[23]	--	--	--	--	--	--	--
		[23]	--	--	--	--	--	--	--
	[IV]	[23]	--	--	--	--	--	--	--
		[23]	--	--	--	--	--	--	--
		[23]	--	--	--	--	--	--	--
	[V]	[23]	--	--	--	--	--	--	--
		[23]	--	--	--	--	--	--	--
		[23]	--	--	--	--	--	--	--
lumbrical	[II]	[19]	--	0.055	--	--	--	0.2	--
	[III]	[19]	--	0.066	--	--	--	0.2	--
	[IV]	[19]	--	0.060	--	--	--	0.1	--
	[V]	[19]	--	0.049	--	--	--	0.1	--
	[II]	[23]	--	--	--	--	--	--	--
	[III]	[23]	--	--	--	--	--	--	--
	[IV]	[23]	--	--	--	--	--	--	--
	[V]	[23]	--	--	--	--	--	--	--
opponens digiti minimi		[19]	--	0.015	--	--	--	0.6	--
		[7]	--	--	--	--	--	--	--
		[7]	--	--	--	--	--	--	--
		[7]	--	--	--	--	--	--	--
opponens pollicis		[19]	--	0.024	--	--	--	0.9	--
		[7]	--	--	--	--	--	--	--

TENSION FRAC. (%)	TENDON LENGTH LT (mm²)	TENDON PCSA	ORIGIN COORD. SYSTEM	X	Y	Z	INSERTION COORD. SYSTEM	X	Y	Z
2.0	--	--	--	--	--	--	--	--	--	--
--	--	--	--	--	--	--	--	--	--	--
--	--	--	--	--	--	--	--	--	--	--
--	--	--	URD	-0.0070	0.0150	0.0160	CRP	-0.0070	0.0100	0.0150
--	--	--	CRP	-0.0070	0.0100	0.0150	MC4	0.0500	0.0000	0.0070
--	--	--	MC4	0.0500	0.0000	0.0070	PP4	0.0430	0.0000	0.0030
--	--	--	PP4	0.0430	0.0000	0.0030	MP4	0.0130	0.0000	0.0025
0.9	--	--	--	--	--	--	--	--	--	--
--	--	--	--	--	--	--	--	--	--	--
--	--	--	--	--	--	--	--	--	--	--
--	--	--	URD	-0.0070	0.0150	0.0160	CRP	-0.0070	0.0120	0.0130
--	--	--	CRP	-0.0070	0.0120	0.0130	MC5	0.0460	0.0000	0.0070
--	--	--	MC5	0.0460	0.0000	0.0070	PP5	0.0350	0.0000	0.0020
--	--	--	PP5	0.0350	0.0000	0.0020	MP5	0.0100	0.0000	0.0020
1.3	--	--	--	--	--	--	--	--	--	--
--	--	--	CRP	-0.0120	-0.0200	0.0240	MC1	0.0420	0.0060	0.0040
--	--	--	MC1	0.0420	0.0060	0.0040	PP1	0.0090	0.0050	0.0040
2.7	--	--	--	--	--	--	--	--	--	--
--	--	--	--	--	--	--	--	--	--	--
--	--	--	URD	-0.0070	-0.0050	0.0140	CRP	-0.0130	-0.0150	0.0230
--	--	--	PP1	0.0330	0.0040	0.0000	DP1	0.0080	0.0040	0.0000
--	--	--	CRP	-0.0130	-0.0150	0.0230	MC1	0.0420	0.0060	0.0000
--	--	--	MC1	0.0420	0.0060	0.0000	PP1	0.0100	0.0050	0.0000
--	--	--	6	0.300	-0.605	-0.033	5	-0.100	-0.380	0.040
--	--	--	5	-0.100	-0.380	0.040	4	0.400	-0.375	0.039
--	--	--	4	0.400	-0.375	0.039	3	-0.210	-0.305	0.027
--	--	--	3	-0.210	-0.305	0.027	2	0.300	-0.270	0.056
--	--	--	2	0.300	-0.270	0.056	1	0.000	-0.159	0.028
--	--	--	6	0.300	-0.470	0.012	5	-0.180	-0.330	0.023
--	--	--	5	-0.180	-0.330	0.023	4	0.400	-0.266	-0.004
--	--	--	4	0.400	-0.266	-0.004	3	-0.280	-0.273	-0.004
--	--	--	3	-0.280	-0.273	-0.004	2	0.300	-0.254	0.033
--	--	--	2	0.300	-0.254	0.033	1	-0.050	-0.141	0.032
--	--	--	6	0.300	-0.463	0.012	5	-0.210	-0.278	0.006
--	--	--	5	-0.210	-0.278	0.006	4	0.400	-0.291	0.010
--	--	--	4	0.400	-0.291	0.010	3	-0.290	-0.278	-0.016
--	--	--	3	-0.290	-0.278	-0.016	2	0.300	-0.253	0.032
--	--	--	2	0.300	-0.253	0.032	1	-0.070	-0.145	0.001
--	--	--	6	0.300	-0.584	0.112	5	-0.060	-0.413	0.076
--	--	--	5	-0.060	-0.413	0.076	4	0.400	-0.359	0.058
--	--	--	4	0.400	-0.359	0.058	3	-0.210	-0.317	0.017
--	--	--	3	-0.210	-0.317	0.017	2	0.030	-0.236	-0.020
--	--	--	2	0.030	-0.236	-0.020	1	-0.010	-0.187	-0.075
--	--	--	6	0.300	-0.685	-0.082	5	-0.100	-0.483	-0.033
--	--	--	5	-0.100	-0.483	-0.033	4	0.400	-0.294	0.003
--	--	--	4	0.400	-0.294	0.003	3	-0.210	-0.240	0.005
--	--	--	6	0.300	-0.574	0.019	5	-0.180	-0.393	0.039
--	--	--	5	-0.180	-0.393	0.039	4	0.400	-0.217	-0.016
--	--	--	4	0.400	-0.217	-0.016	3	-0.280	-0.207	0.001
--	--	--	6	0.300	-0.517	0.036	5	-0.210	-0.338	0.038
--	--	--	5	-0.210	-0.338	0.038	4	0.400	-0.228	-0.004
--	--	--	4	0.400	-0.228	-0.004	3	-0.290	-0.234	-0.009
--	--	--	6	0.300	-0.653	0.129	5	-0.060	-0.489	0.117
--	--	--	5	-0.060	-0.489	0.117	4	0.400	-0.310	-0.008
--	--	--	4	0.400	-0.310	-0.008	3	-0.210	-0.271	0.004
0.2	--	--	--	--	--	--	--	--	--	--
0.2	--	--	--	--	--	--	--	--	--	--
0.1	--	--	--	--	--	--	--	--	--	--
0.1	--	--	--	--	--	--	--	--	--	--
--	--	--	6	0.400	-0.741	0.629	5	-0.300	-0.146	0.444
--	--	--	6	0.400	-0.623	0.422	5	-0.380	-0.140	0.328
--	--	--	6	0.400	-0.415	0.213	5	-0.410	-0.059	0.165
--	--	--	6	0.400	-0.363	0.469	5	-0.260	0.066	0.430
2.0	--	--	--	--	--	--	--	--	--	--
--	--	--	CRP	-0.0030	0.0160	0.0200	MC5	0.0160	0.0050	0.0000
--	--	--	CRP	-0.0030	0.0160	0.0200	MC5	0.0290	0.0040	0.0020
--	--	--	CRP	-0.0030	0.0160	0.0200	MC5	0.0390	0.0045	0.0000
1.9	--	--	--	--	--	--	--	--	--	--
--	--	--	CRP	-0.0120	-0.0220	0.0250	MC1	0.0250	0.0010	0.0020

MUSCLE		SOURCE	MUSCLE LENGTH ML	FIBER LENGTH LM	VOLUME V (ml)	MUSCLE PCSA =V/L (cm²)	PENNA- TION ANGLE (deg)	MASS FRACTION (%)	MASS (gm)
palmar interosseous	[II]	[19]	--	0.015	--	--	--	0.4	--
		[7]	--	--	--	--	--	--	--
		[7]	--	--	--	--	--	--	--
		[7]	--	--	--	--	--	--	--
		[7]	--	--	--	--	--	--	--
	[IV]	[19]	--	0.017	--	--	--	0.4	--
		[7]	--	--	--	--	--	--	--
		[7]	--	--	--	--	--	--	--
		[7]	--	--	--	--	--	--	--
		[7]	--	--	--	--	--	--	--
	[V]	[19]	--	0.015	--	--	--	0.3	--
		[7]	--	--	--	--	--	--	--
		[7]	--	--	--	--	--	--	--
		[7]	--	--	--	--	--	--	--
		[7]	--	--	--	--	--	--	--
radial band	[II]	[23]	--	--	--	--	--	--	--
	[III]	[23]	--	--	--	--	--	--	--
	[IV]	[23]	--	--	--	--	--	--	--
	[V]	[23]	--	--	--	--	--	--	--
radial interosseous	[II]	[23]	--	--	--	--	--	--	--
	[III]	[23]	--	--	--	--	--	--	--
	[IV]	[23]	--	--	--	--	--	--	--
	[V]	[23]	--	--	--	--	--	--	--
terminal extensor	[II]	[23]	--	--	--	--	--	--	--
	[III]	[23]	--	--	--	--	--	--	--
	[IV]	[23]	--	--	--	--	--	--	--
	[V]	[23]	--	--	--	--	--	--	--
ulnar band	[II]	[23]	--	--	--	--	--	--	--
	[III]	[23]	--	--	--	--	--	--	--
	[IV]	[23]	--	--	--	--	--	--	--
	[V]	[23]	--	--	--	--	--	--	--
ulnar interosseous	[II]	[23]	--	--	--	--	--	--	--
	[III]	[23]	--	--	--	--	--	--	--
	[IV]	[23]	--	--	--	--	--	--	--
	[V]	[23]	--	--	--	--	--	--	--

TENSION FRAC. (%)	TENDON LENGTH LT	TENDON PCSA (mm²)	ORIGIN				INSERTION			
			COORD. SYSTEM	X	Y	Z	COORD. SYSTEM	X	Y	Z
1.3	--	--	--	--	--	--	--	--	--	--
--	--	--	MC2	0.0400	0.0045	0.0000	PP2	0.0130	0.0080	0.0035
--	--	--	MC2	0.0400	0.0045	0.0000	PP2	0.0130	0.0080	0.0000
--	--	--	PP2	0.0330	0.0060	0.0000	MP2	0.0060	0.0060	-0.0020
--	--	--	MP2	0.0240	0.0000	-0.0025	DP2	0.0040	0.0000	-0.0030
1.2	--	--	--	--	--	--	--	--	--	--
--	--	--	MC4	0.0300	-0.0030	0.0010	PP4	0.0130	-0.0080	0.0025
--	--	--	MC4	0.0300	-0.0030	0.0010	PP4	0.0130	-0.0080	0.0000
--	--	--	PP4	0.0300	-0.0055	0.0000	MP4	0.0060	-0.0060	-0.0020
--	--	--	MP4	0.0250	0.0000	-0.0020	DP4	0.0035	0.0000	-0.0030
1.0	--	--	--	--	--	--	--	--	--	--
--	--	--	MC5	0.0280	-0.0040	0.0020	PP5	0.0100	-0.0080	0.0020
--	--	--	MC5	0.0280	-0.0040	0.0020	PP5	0.0100	-0.0080	0.0000
--	--	--	PP5	0.0270	-0.0050	0.0000	MP5	0.0050	-0.0050	-0.0015
--	--	--	MP5	0.0215	0.0000	-0.0015	DP5	0.0030	0.0000	-0.0025
--	--	--	4	0.100	0.202	0.261	3	-0.110	0.219	0.207
--	--	--	4	0.100	0.100	0.242	3	-0.180	0.172	0.227
--	--	--	4	0.100	0.110	0.237	3	-0.190	0.140	0.163
--	--	--	4	0.100	0.182	0.224	3	-0.110	0.203	0.184
--	--	--	6	0.400	-0.476	0.616	5	-0.300	0.020	0.478
--	--	--	6	0.400	-0.565	0.471	5	-0.380	-0.038	0.331
--	--	--	6	0.400	-0.211	0.283	5	-0.410	0.000	0.275
--	--	--	6	0.400	-0.145	0.473	5	-0.260	0.066	0.430
--	--	--	2	0.000	0.205	-0.020	1	0.000	0.217	-0.019
--	--	--	2	0.000	0.168	-0.015	1	-0.050	0.154	-0.022
--	--	--	2	0.000	0.141	-0.025	1	-0.070	0.144	-0.030
--	--	--	2	0.000	0.187	-0.119	1	-0.010	0.199	-0.132
--	--	--	4	0.100	0.160	-0.347	3	-0.110	0.149	-0.303
--	--	--	4	0.100	0.076	-0.279	3	-0.180	0.161	-0.247
--	--	--	4	0.100	0.072	-0.246	3	-0.190	0.141	-0.228
--	--	--	4	0.100	0.113	-0.293	3	-0.110	0.110	-0.321
--	--	--	6	0.400	-0.444	-0.477	5	-0.300	-0.067	-0.462
--	--	--	6	0.400	0.264	-0.358	5	-0.380	0.036	-0.357
--	--	--	6	0.400	-0.330	-0.207	5	-0.410	0.012	-0.199
--	--	--	6	0.400	-0.304	-0.590	5	-0.260	-0.100	-0.502

TABLE A.5
HEAD AND NECK MUSCULATURE

MUSCLE		SOURCE	MUSCLE LENGTH ML (m)	FIBER LENGTH LM (m)	VOLUME V (ml)	MUSCLE PCSA =V/L (cm²)	PENNA-TION ANGLE (deg)	MUSCLE FIBER TYPE %SO	%FO	%FG
frontalis		[14]	--	--	--	--	--	64.1	--	35.9
iliocostalis cervicis	(1)	[7]	--	--	--	--	--	--	--	--
	(2)	[7]	--	--	--	--	--	--	--	--
	(3)	[7]	--	--	--	--	--	--	--	--
longissimus	(1)	[18]	0.0700	--	--	1.88	10	30	--	--
	(2)	[18]	0.0800	--	--	1.88	10	40	--	--
	(3)	[18]	0.0850	--	--	1.88	10	50	--	--
longissimus capitis		[12]	--	--	--	0.50	--	--	--	--
	(1)	[7]	--	--	--	--	--	--	--	--
	(2)	[7]	--	--	--	--	--	--	--	--
	(3)	[7]	--	--	--	--	--	--	--	--
	(4)	[7]	--	--	--	--	--	--	--	--
	(5)	[7]	--	--	--	--	--	--	--	--
	(6)	[7]	--	--	--	--	--	--	--	--
		[18]	0.1200	--	--	1.22	10	30	--	--
longissimus cervicis		[12]	--	--	--	0.60	--	--	--	--
	(1)	[7]	--	--	--	--	--	--	--	--
	(2)	[7]	--	--	--	--	--	--	--	--
	(3)	[7]	--	--	--	--	--	--	--	--
	(4)	[7]	--	--	--	--	--	--	--	--
	(5)	[7]	--	--	--	--	--	--	--	--
	(6)	[7]	--	--	--	--	--	--	--	--
	(7)	[7]	--	--	--	--	--	--	--	--
	(8)	[7]	--	--	--	--	--	--	--	--
	(9)	[7]	--	--	--	--	--	--	--	--
	(10)	[7]	--	--	--	--	--	--	--	--
longus capitis		[12]	--	--	--	0.75	--	--	--	--
	(1)	[7]	--	--	--	--	--	--	--	--
	(2)	[7]	--	--	--	--	--	--	--	--
	(3)	[7]	--	--	--	--	--	--	--	--
	(4)	[7]	--	--	--	--	--	--	--	--
longus colli		[12]	--	--	--	0.75	--	--	--	--
		[18]	0.0850	--	--	0.82	10	50	--	--
longus colli vertical	(1)	[7]	--	--	--	--	--	--	--	--
	(2)	[7]	--	--	--	--	--	--	--	--
	(3)	[7]	--	--	--	--	--	--	--	--
	(4)	[7]	--	--	--	--	--	--	--	--
	(5)	[7]	--	--	--	--	--	--	--	--
	(6)	[7]	--	--	--	--	--	--	--	--
	(7)	[7]	--	--	--	--	--	--	--	--
	(8)	[7]	--	--	--	--	--	--	--	--
	(9)	[7]	--	--	--	--	--	--	--	--
	(10)	[7]	--	--	--	--	--	--	--	--
	(11)	[7]	--	--	--	--	--	--	--	--
	(12)	[7]	--	--	--	--	--	--	--	--
	(13)	[7]	--	--	--	--	--	--	--	--
	(14)	[7]	--	--	--	--	--	--	--	--
	(15)	[7]	--	--	--	--	--	--	--	--
	(16)	[7]	--	--	--	--	--	--	--	--
longus colli inferior oblique		[7]	--	--	--	--	--	--	--	--
	(2)	[7]	--	--	--	--	--	--	--	--
longus colli superior oblique		[7]	--	--	--	--	--	--	--	--
	(2)	[7]	--	--	--	--	--	--	--	--
	(3)	[7]	--	--	--	--	--	--	--	--
	(4)	[7]	--	--	--	--	--	--	--	--
	(5)	[7]	--	--	--	--	--	--	--	--
	(6)	[7]	--	--	--	--	--	--	--	--
multifidus cervicis	(1)	[7]	--	--	--	--	--	--	--	--
	(2)	[7]	--	--	--	--	--	--	--	--
	(3)	[7]	--	--	--	--	--	--	--	--
	(4)	[7]	--	--	--	--	--	--	--	--
	(5)	[7]	--	--	--	--	--	--	--	--
	(6)	[7]	--	--	--	--	--	--	--	--
	(7)	[7]	--	--	--	--	--	--	--	--
	(8)	[7]	--	--	--	--	--	--	--	--
	(9)	[7]	--	--	--	--	--	--	--	--
	(10)	[7]	--	--	--	--	--	--	--	--
	(11)	[7]	--	--	--	--	--	--	--	--
	(12)	[7]	--	--	--	--	--	--	--	--
oblique capitis superior		[7]	--	--	--	--	--	--	--	--
		[12]	--	--	--	1.00	--	--	--	--
		[18]	0.0600	--	--	1.75	10	50	--	--
oblique capitis inferior		[7]	--	--	--	--	--	--	--	--
		[18]	0.0550	--	--	0.89	10	30	--	--

TENDON LENGTH LT (m)	TENDON PCSA (mm²)	ORIGIN				INSERTION			
		COORD. SYSTEM	X (m)	Y (m)	Z (m)	COORD. SYSTEM	X (m)	Y (m)	Z (m)
--	--	--	--	--	--	--	--	--	--
--	--	C4	-0.0040	-0.0280	0.0050	Thorax	0.0150	-0.0400	0.2150
--	--	C5	-0.0040	-0.0280	0.0050	Thorax	0.0150	-0.0400	0.2150
--	--	C6	-0.0040	-0.0280	0.0050	Thorax	0.0150	-0.0400	0.2150
0.0175	--	--	--	--	--	--	--	--	--
0.0200	--	--	--	--	--	--	--	--	--
0.0212	--	--	--	--	--	--	--	--	--
--	--	Neck	-0.0190	0.0254	0.0549	Head	0.0020	0.0340	0.0678
--	--	Skull	-0.0180	-0.0430	0.0050	C3	0.0000	-0.0250	0.0100
--	--	Skull	-0.0180	-0.0430	0.0050	C4	0.0000	-0.0250	0.0100
--	--	Skull	-0.0180	-0.0430	0.0050	C5	0.0000	-0.0250	0.0100
--	--	Skull	-0.0180	-0.0430	0.0050	C6	0.0000	-0.0250	0.0100
--	--	Skull	-0.0180	-0.0430	0.0050	C7	0.0000	-0.0250	0.0100
--	--	Skull	-0.0180	-0.0430	0.0050	Thorax	0.0650	-0.0350	0.2600
0.0360	--	--	--	--	--	--	--	--	--
--	--	UpTorso	-0.0190	0.0254	-0.0320	Neck	-0.0190	0.0254	0.0137
--	--	C2	-0.0040	-0.0280	0.0050	Thorax	0.0450	-0.0350	0.2500
--	--	C3	-0.0040	-0.0280	0.0050	Thorax	0.0450	-0.0350	0.2500
--	--	C4	-0.0040	-0.0280	0.0050	Thorax	0.0450	-0.0350	0.2500
--	--	C5	-0.0040	-0.0280	0.0050	Thorax	0.0450	-0.0350	0.2500
--	--	C6	-0.0040	-0.0280	0.0050	Thorax	0.0450	-0.0350	0.2500
--	--	C2	-0.0040	-0.0280	0.0050	C7	-0.0050	-0.0280	0.0070
--	--	C3	-0.0040	-0.0280	0.0050	C7	-0.0050	-0.0280	0.0070
--	--	C4	-0.0040	-0.0280	0.0050	C7	-0.0050	-0.0280	0.0070
--	--	C5	-0.0040	-0.0280	0.0050	C7	-0.0050	-0.0280	0.0070
--	--	C6	-0.0040	-0.0280	0.0050	C7	-0.0050	-0.0280	0.0070
--	--	Neck	-0.0190	0.0254	-0.0137	Head	0.0391	0.0099	0.0808
--	--	Skull	0.0170	-0.0040	0.0080	C3	0.0050	-0.0230	0.0100
--	--	Skull	0.0170	-0.0040	0.0080	C4	0.0050	-0.0230	0.0100
--	--	Skull	0.0170	-0.0040	0.0080	C5	0.0050	-0.0230	0.0100
--	--	Skull	0.0170	-0.0040	0.0080	C6	0.0050	-0.0230	0.0100
--	--	UpTorso	0.0190	0.0000	-0.0508	Neck	0.0190	0.0000	-0.0457
0.0255	--	--	--	--	--	--	--	--	--
--	--	C1	0.0160	0.0000	0.0100	C5	0.0100	0.0000	0.0080
--	--	C2	0.0100	0.0000	0.0080	C5	0.0100	0.0000	0.0080
--	--	C3	0.0100	0.0000	0.0080	C5	0.0100	0.0000	0.0080
--	--	C4	0.0100	0.0000	0.0080	C5	0.0100	0.0000	0.0080
--	--	C1	0.0160	0.0000	0.0100	C6	0.0100	0.0000	0.0080
--	--	C2	0.0100	0.0000	0.0080	C6	0.0100	0.0000	0.0080
--	--	C3	0.0100	0.0000	0.0080	C6	0.0100	0.0000	0.0080
--	--	C4	0.0100	0.0000	0.0080	C6	0.0100	0.0000	0.0080
--	--	C1	0.0160	0.0000	0.0100	C7	0.0100	0.0000	0.0080
--	--	C2	0.0100	0.0000	0.0080	C7	0.0100	0.0000	0.0080
--	--	C3	0.0100	0.0000	0.0080	C7	0.0100	0.0000	0.0080
--	--	C4	0.0100	0.0000	0.0080	C7	0.0100	0.0000	0.0080
--	--	C1	0.0160	0.0000	0.0100	Thorax	0.1030	0.0000	0.2700
--	--	C2	0.0100	0.0000	0.0080	Thorax	0.1030	0.0000	0.2700
--	--	C3	0.0100	0.0000	0.0080	Thorax	0.1030	0.0000	0.2700
--	--	C4	0.0100	0.0000	0.0080	Thorax	0.1030	0.0000	0.2700
--	--	C5	0.0050	-0.0230	0.0100	Thorax	0.1030	0.0000	0.2700
--	--	C6	0.0050	-0.0230	0.0100	Thorax	0.1030	0.0000	0.2700
--	--	C1	0.0160	0.0000	0.0100	C3	0.0050	-0.0230	0.0100
--	--	C3	0.0100	0.0000	0.0080	C4	0.0050	-0.0230	0.0100
--	--	C1	0.0160	0.0000	0.0100	C5	0.0050	-0.0230	0.0100
--	--	C2	0.0100	0.0000	0.0080	C3	0.0050	-0.0230	0.0100
--	--	C1	0.0160	0.0000	0.0100	C4	0.0050	-0.0230	0.0100
--	--	C2	0.0100	0.0000	0.0080	C5	0.0050	-0.0230	0.0100
--	--	C2	-0.0270	-0.0040	-0.0080	C5	-0.0120	-0.0220	0.0200
--	--	C2	-0.0270	-0.0040	-0.0080	C6	-0.0120	-0.0220	0.0200
--	--	C3	-0.0350	-0.0040	-0.0070	C6	-0.0120	-0.0220	0.0200
--	--	C3	-0.0350	-0.0040	-0.0070	C7	-0.0120	-0.0220	0.0200
--	--	C4	-0.0350	-0.0040	-0.0070	C7	-0.0120	-0.0220	0.0200
--	--	C4	-0.0350	-0.0040	-0.0070	Thorax	0.0700	-0.0350	0.2800
--	--	C5	-0.0350	-0.0040	-0.0070	Thorax	0.0700	-0.0350	0.2800
--	--	C5	-0.0350	-0.0040	-0.0070	Thorax	0.0650	-0.0350	0.2600
--	--	C6	-0.0350	-0.0040	-0.0070	Thorax	0.0650	-0.0350	0.2600
--	--	C6	-0.0350	-0.0040	-0.0070	Thorax	0.0550	-0.0350	0.2500
--	--	C7	-0.0400	-0.0040	-0.0100	Thorax	0.0550	-0.0350	0.2500
--	--	C7	-0.0400	-0.0040	-0.0100	Thorax	0.0400	-0.0350	0.2400
--	--	Skull	-0.0370	-0.0320	0.0170	C1	0.0000	-0.0340	0.0070
--	--	Neck	-0.0190	0.0254	-0.0640	Head	-0.0140	0.0350	0.0678
0.0120	--	--	--	--	--	--	--	--	--
--	--	C1	0.0000	-0.0360	0.0070	C2	-0.0270	-0.0040	-0.0080
0.0138	--	--	--	--	--	--	--	--	--

MUSCLE		SOURCE	MUSCLE LENGTH ML (m)	FIBER LENGTH LM (m)	VOLUME V (ml)	MUSCLE PCSA =V/L (cm²)	PENNA- TION ANGLE (deg)	MUSCLE FIBER TYPE		
								%SO	%FO	%FG
orbicularis oculi		[14]	--	--	--	--	--	15.4	--	84.6
rectus capitis posterior major		[7]	--	--	--	--	--	--	--	--
		[12]	--	--	--	0.50	--	--	--	--
		[18]	0.0600	--	--	1.10	10	60	--	--
rectus capitis posterior minor		[7]	--	--	--	--	--	--	--	--
		[12]	--	--	--	0.385	--	--	--	--
		[18]	0.0300	--	--	1.03	10	60	--	--
rectus capitis anterior		[7]	--	--	--	--	--	--	--	--
		[12]	--	--	--	0.25	--	--	--	--
		[18]	0.0300	--	--	0.13	10	50	--	--
rectus capitis lateral		[7]	--	--	--	--	--	--	--	--
		[12]	--	--	--	0.25	--	--	--	--
		[18]	0.0250	--	--	0.21	10	50	--	--
rotatores cervicis	(1)	[7]	--	--	--	--	--	--	--	--
	(2)	[7]	--	--	--	--	--	--	--	--
	(3)	[7]	--	--	--	--	--	--	--	--
	(4)	[7]	--	--	--	--	--	--	--	--
	(5)	[7]	--	--	--	--	--	--	--	--
	(6)	[7]	--	--	--	--	--	--	--	--
	(7)	[7]	--	--	--	--	--	--	--	--
	(8)	[7]	--	--	--	--	--	--	--	--
	(9)	[7]	--	--	--	--	--	--	--	--
	(10)	[7]	--	--	--	--	--	--	--	--
	(11)	[7]	--	--	--	--	--	--	--	--
scalenus		[12]	--	--	--	1.75	--	--	--	--
		[18]	0.0922	--	--	2.39	10	30	--	--
scalenus anterior	(1)	[7]	--	--	--	--	--	--	--	--
	(2)	[7]	--	--	--	--	--	--	--	--
	(3)	[7]	--	--	--	--	--	--	--	--
	(4)	[7]	--	--	--	--	--	--	--	--
scalenus medial	(1)	[7]	--	--	--	--	--	--	--	--
	(2)	[7]	--	--	--	--	--	--	--	--
	(3)	[7]	--	--	--	--	--	--	--	--
	(4)	[7]	--	--	--	--	--	--	--	--
	(5)	[7]	--	--	--	--	--	--	--	--
	(6)	[7]	--	--	--	--	--	--	--	--
	(7)	[7]	--	--	--	--	--	--	--	--
scalenus posterior	(1)	[7]	--	--	--	--	--	--	--	--
	(2)	[7]	--	--	--	--	--	--	--	--
	(3)	[7]	--	--	--	--	--	--	--	--
semispinalis	(1)	[18]	0.1012	--	--	1.89	10	40	--	--
	(2)	[18]	0.1200	--	--	3.75	10	50	--	--
semispinalis capitis		[12]	--	--	--	2.38	--	--	--	--
	(1)	[7]	--	--	--	--	--	--	--	--
	(2)	[7]	--	--	--	--	--	--	--	--
	(3)	[7]	--	--	--	--	--	--	--	--
	(4)	[7]	--	--	--	--	--	--	--	--
	(5)	[7]	--	--	--	--	--	--	--	--
semispinalis cervicis		[12]	--	--	--	2.00	--	--	--	--
	(1)	[7]	--	--	--	--	--	--	--	--
	(2)	[7]	--	--	--	--	--	--	--	--
	(3)	[7]	--	--	--	--	--	--	--	--
	(4)	[7]	--	--	--	--	--	--	--	--
	(5)	[7]	--	--	--	--	--	--	--	--
	(6)	[7]	--	--	--	--	--	--	--	--
semispinalis dorsi		[12]	--	--	--	1.00	--	--	--	--
spinalis capitis		[12]	--	--	--	0.50	--	--	--	--
spinalis dorsi		[12]	--	--	--	1.00	--	--	--	--
splenius capitis		[12]	--	--	--	1.22	--	--	--	--
		[18]	0.1382	--	--	3.73	10	25	--	--
	(1)	[7]	--	--	--	--	--	--	--	--
	(2)	[7]	--	--	--	--	--	--	--	--
splenius cervicis		[18]	0.1653	--	--	3.73	10	50	--	--
	(1)	[12]	--	--	--	1.25	--	--	--	--
	(2)	[12]	--	--	--	0.70	--	--	--	--
	(1)	[7]	--	--	--	--	--	--	--	--
	(2)	[7]	--	--	--	--	--	--	--	--
	(3)	[7]	--	--	--	--	--	--	--	--
sternocleidomastoid		[12]	--	--	--	1.60	--	--	--	--
		[18]	0.2003	--	--	2.08	10	15	--	--
	(1)	[7]	--	--	--	--	--	--	--	--
	(2)	[7]	--	--	--	--	--	--	--	--
		[14]	--	--	--	--	--	35.2	--	64.8
temporalis		[14]	--	--	--	--	--	46.5	--	53.5

TENDON LENGTH LT (m)	TENDON PCSA (mm²)	ORIGIN COORD. SYSTEM	X (m)	Y (m)	Z (m)	INSERTION COORD. SYSTEM	X (m)	Y (m)	Z (m)
--	--	--	--	--	--	--	--	--	--
--	--	Skull	-0.0400	-0.0230	0.0120	C2	-0.0270	-0.0040	-0.0080
--	--	Neck	-0.0190	0.0000	-0.0472	Head	-0.0140	0.0239	0.0719
0.0090	--	--	--	--	--	--	--	--	--
--	--	Skull	-0.0370	-0.0100	0.0080	C1	-0.0270	0.0000	0.0060
--	--	Neck	-0.0190	0.0000	-0.0640	Head	-0.0140	0.0079	0.0757
0.0045	--	--	--	--	--	--	--	--	--
--	--	Skull	0.0100	-0.0060	0.0080	C1	0.0000	-0.0270	0.0080
--	--	Neck	-0.0190	0.0254	-0.0640	Head	0.0320	0.0099	0.0808
0.0045	--	--	--	--	--	--	--	--	--
--	--	Skull	-0.0090	-0.0300	0.0090	C1	0.0000	-0.0360	0.0070
--	--	Neck	-0.0190	0.0254	-0.0640	Head	0.0221	0.0300	0.0808
0.0038	--	--	--	--	--	--	--	--	--
--	--	C2	-0.0270	-0.0040	-0.0080	C4	0.0000	-0.0250	0.0100
--	--	C3	-0.0350	-0.0040	-0.0070	C4	0.0000	-0.0250	0.0100
--	--	C3	-0.0350	-0.0040	-0.0070	C5	0.0000	-0.0250	0.0100
--	--	C4	-0.0350	-0.0040	-0.0070	C5	0.0000	-0.0250	0.0100
--	--	C4	-0.0350	-0.0040	-0.0070	C6	0.0000	-0.0250	0.0100
--	--	C5	-0.0350	-0.0040	-0.0070	C6	0.0000	-0.0250	0.0100
--	--	C5	-0.0350	-0.0040	-0.0070	C7	0.0000	-0.0250	0.0100
--	--	C6	-0.0350	-0.0040	-0.0070	C7	0.0000	-0.0250	0.0100
--	--	C6	-0.0350	-0.0040	-0.0070	Thorax	0.0700	-0.0350	0.2800
--	--	C7	-0.0400	-0.0040	-0.0100	Thorax	0.0700	-0.0350	0.2800
--	--	C7	-0.0400	-0.0040	-0.0100	Thorax	0.0650	-0.0350	0.2700
--	--	UpTorso	0.0762	0.0508	-0.0470	Neck	-0.0190	0.0254	0.0051
0.0184	--	--	--	--	--	--	--	--	--
--	--	C3	0.0050	-0.0230	0.0100	Thorax	0.1050	-0.0530	0.2400
--	--	C4	0.0050	-0.0230	0.0100	Thorax	0.1050	-0.0530	0.2400
--	--	C5	0.0050	-0.0230	0.0100	Thorax	0.1050	-0.0530	0.2400
--	--	C6	0.0050	-0.0230	0.0100	Thorax	0.1050	-0.0530	0.2400
--	--	C1	0.0000	-0.0360	0.0070	Thorax	0.0750	-0.0530	0.2700
--	--	C2	-0.0040	-0.0280	0.0050	Thorax	0.0750	-0.0530	0.2700
--	--	C3	-0.0040	-0.0280	0.0050	Thorax	0.0750	-0.0530	0.2700
--	--	C4	-0.0040	-0.0280	0.0050	Thorax	0.0750	-0.0530	0.2700
--	--	C5	-0.0040	-0.0280	0.0050	Thorax	0.0750	-0.0530	0.2700
--	--	C6	-0.0040	-0.0280	0.0050	Thorax	0.0750	-0.0530	0.2700
--	--	C7	-0.0040	-0.0280	0.0050	Thorax	0.0750	-0.0530	0.2700
--	--	C5	-0.0040	-0.0280	0.0050	Thorax	0.0450	-0.0530	0.2550
--	--	C6	-0.0040	-0.0280	0.0050	Thorax	0.0450	-0.0530	0.2550
--	--	C7	-0.0040	-0.0280	0.0050	Thorax	0.0450	-0.0530	0.2550
0.0202	--	--	--	--	--	--	--	--	--
0.0240	--	--	--	--	--	--	--	--	--
--	--	Neck	-0.0190	0.0254	0.0549	Head	-0.0361	0.0201	0.0678
--	--	Skull	-0.0530	-0.0180	0.0250	C4	-0.0120	-0.0220	0.0200
--	--	Skull	-0.0530	-0.0180	0.0250	C5	-0.0120	-0.0220	0.0200
--	--	Skull	-0.0530	-0.0180	0.0250	C6	-0.0120	-0.0220	0.0200
--	--	Skull	-0.0530	-0.0180	0.0250	C7	0.0000	-0.0280	0.0100
--	--	Skull	-0.0530	-0.0180	0.0250	Thorax	-0.0400	-0.0300	0.2400
--	--	UpTorso	-0.0190	0.0254	-0.0660	Neck	-0.0190	0.0000	-0.0305
--	--	C2	-0.0270	-0.0040	-0.0080	Thorax	0.0150	-0.0300	0.2200
--	--	C3	-0.0350	-0.0040	-0.0070	Thorax	0.0150	-0.0300	0.2200
--	--	C4	-0.0350	-0.0040	-0.0070	Thorax	0.0150	-0.0300	0.2200
--	--	C5	-0.0350	-0.0040	-0.0070	Thorax	0.0150	-0.0300	0.2200
--	--	C6	-0.0350	-0.0040	-0.0070	Thorax	-0.0100	-0.0300	0.1600
--	--	C7	-0.0400	-0.0040	-0.0100	Thorax	-0.0100	-0.0300	0.1600
--	--	UpTorso	-0.0190	0.0254	0.0480	Neck	-0.0190	0.0254	0.0381
--	--	Neck	-0.0190	0.0254	0.0549	Head	-0.0221	0.0061	0.0719
--	--	CenTorso	-0.0254	0.0000	-0.1074	UpTorso	-0.0190	0.0254	0.0180
--	--	Neck	-0.0190	0.0000	0.0549	Head	-0.0099	0.0399	0.0678
0.021	--	--	--	--	--	--	--	--	--
--	--	Skull	-0.0300	-0.0440	0.0160	C7	-0.0400	-0.0040	-0.0100
--	--	Skull	-0.0300	-0.0440	0.0160	Thorax	0.0300	0.0000	0.2500
0.033	--	--	--	--	--	--	--	--	--
--	--	UpTorso	-0.0190	0.0000	-0.0660	Neck	-0.0190	0.0000	-0.0472
--	--	UpTorso	-0.0190	0.0000	-0.0320	Neck	-0.0190	0.0254	-0.0472
--	--	C1	0.0000	-0.0360	0.0070	Thorax	0.0150	0.0000	0.2250
--	--	C2	-0.0040	-0.0280	0.0050	Thorax	0.0150	0.0000	0.2250
--	--	C3	-0.0040	-0.0280	0.0050	Thorax	0.0150	0.0000	0.2250
--	--	UpTorso	0.0762	0.0000	-0.0508	Head	-0.0099	0.0399	0.0678
0.040	--	--	--	--	--	--	--	--	--
--	--	Skull	-0.0350	-0.0500	0.0230	Thorax	0.1300	-0.0100	0.2300
--	--	Skull	-0.0350	-0.0500	0.0230	Clavicle	0.0050	0.0350	0.0120
--	--	--	--	--	--	--	--	--	--
--	--	--	--	--	--	--	--	--	--

Index

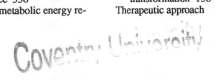